GW00390791

HANDBOOK OF BIOMEDICAL NONLINEAR OPTICAL MICROSCOPY

Photograph of (left to right): Victor Weiskopf, Maria Göppert-Mayer, and Max Born, all on bicycles on a street in Göttingen. Date is unknown. Credit: AIP Emilio Segre Visual Archives.

Handbook of Biomedical Nonlinear Optical Microscopy

Edited by

Barry R. Masters
Visiting Scientist
Department of Biological Engineering
Massachusetts Institute of Technology

Peter T. C. So
Professor
Department of Biological Engineering
Massachusetts Institute of Technology

UNIVERSITY PRESS
2008

Oxford University Press, Inc., publishes works that further
Oxford University's objective of excellence
in research, scholarship, and education.

Oxford New York

Auckland Cape Town Dar es Salaam Hong Kong Karachi
Kuala Lumpur Madrid Melbourne Mexico City Nairobi
New Delhi Shanghai Taipei Toronto

With offices in

Argentina Austria Brazil Chile Czech Republic France Greece
Guatemala Hungary Italy Japan Poland Portugal Singapore
South Korea Switzerland Thailand Turkey Ukraine Vietnam

Copyright © 2008 by Oxford University Press, Inc.

Published by Oxford University Press, Inc.
198 Madison Avenue, New York, New York 10016
www.oup.com

Oxford is a registered trademark of Oxford University Press

All rights reserved. No part of this publication may be reproduced,
stored in a retrieval system, or transmitted, in any form or by any means,
electronic, mechanical, photocopying, recording, or otherwise,
without the prior permission in writing of Oxford University Press.

Library of Congress Cataloging-in-Publication Data

Masters, Barry R.
    Handbook of biomedical nonlinear optical microscopy / Barry R. Masters and Peter So.
        p. cm.
    Includes bibliographical references and index.
    ISBN: 978-0-19-516260-8
    1. Microscopy. 2. Nonlinear optics. 3. Nonlinear optical spectroscopy.
4. Fluorescence microscopy. 5. Multiphoton excitation microscopy.
I. So, Peter T. C. II. Title.
    QH205.2.M372 2008
    570.28′2—dc22   2007004024

9 8 7 6 5 4 3 2 1

Printed in China
on acid-free paper

The editors of this book would like to dedicate this work to our mentors, Professor Britton Chance and Professor Enrico Gratton.

This book is further dedicated to those individuals who advance, promote, protect, and propagate science.

It is not possible to be a scientist unless you believe that the knowledge of the world, and the power which this gives, is a thing which is of intrinsic value to humanity, that you are using it to help in the spread of knowledge, and are willing to take the consequences.

J. Robert Oppenheimer
November 2, 1945

Foreword

Watt W. Webb

The origins of the fundamental physical concepts underlying the important, biophysically significant methods of the nonlinear optical microscopies have a long history. Multiphoton laser scanning fluorescence microscopy (MPM) utilizes "simultaneous" absorption of two or more photons to excite a molecule to an excited state vibrational band from which it can emit its energy as imageable fluorescence photons. MPM provides 3D imaging with tissue laser scanning. It had its original publication only in 1990 by Denk, Strickler, and Webb (1), but the idea that two photons can combine their energy to produce excitation to a level at the sum of the two was first advanced by Albert Einstein in his famous Nobel Prize paper of 1905 in *Ann Physik* (2). That's 85 years from the physical conception to the first practical application.

Along this discovery and development route, Maria Goeppert-Mayer reported the quantum mechanical formulation of two-photon molecular excitation in 1931 (3) after working it out in her doctorate thesis. But still another 30 years had to pass before bright enough lasers were produced to deliver the necessary $\sim 10^{16}$ photons/cm^2 to a molecule of $\sim 10^{-16}$ cross-sections for absorption of two photons during the Heisenberg uncertainty principle time of $\sim 10^{-16}$ sec. Thus, $\sim 10^{32}$ photons/cm^2 sec are required. Today the technology underlying the important biophysical tool of multiphoton microscopy (1) relies on the expensive femtosecond pulse trains of mode-locked lasers, which of course is motivating the development of more tractable laser sources, optical fibers to pass femtosecond pulses without distortion, and simpler, user-friendly MPM instruments.

A decade before the development of multiphoton fluorescence microscopy (1), the generation of harmonics of bright coherent wave laser illumination by nonlinear, non-centrosymmetric materials was developed as a macroscopy tool by Wilson and Sheppard in the 1970s (4). Its potential for useful applicability in biophysical microscopy was then pioneered by the elegant, understandable theoretical analysis and associated experiments of Mertz and coworkers early in this century (5–8). Early low-resolution biological applications to collagen go back to the 1970s (9–11), but recently Dr. Rebecca Williams demonstrated in our laboratory some elegant analytical imaging of collagen structures in tendon offering medically applicable images (12). Dombeck et al recently discovered that the elusive parallel polarization microtubule arrays in neural axons can be imaged usefully in brain by their second harmonic generation (13). Nonlinear, non-centrosymmetric dyes in cell membranes that are aligned or modified by the electric fields signaling in living cell membranes were demonstrated years ago by several groups, and their usefulness in imaging of action potential dynamics has recently been demonstrated (14,15).

Where do these nonlinear microscopy methods go from here? We think that they are ready for development of instrumentation for direct application in clinical medicine through in vivo real time multiphoton microscopy in medical endoscopy using intrinsic tissue fluorescence and second harmonic generation for application in a broad range of clinical specialties for diagnostics and guidance of surgery.

This volume, assembled by Barry R. Masters and Peter T. C. So, provides a generous summary of developments in these nonlinear optical microscopies.

REFERENCES

[1] Denk, W., Strickler, J.H. & Webb, W.W. Two-Photon Laser Scanning Fluorescence Microscopy. Science 248: 73–76 (1990).

[2] Einstein, A. Creation and Conversion of Light. Ann. Physik 17: 132–148 (1905).

[3] Goeppert-Mayer, M. Elementary processes with two quantum jumps. Ann. Physik 9: 273–294 (1931).

[4] Wilson, T. & Sheppard, C. Imaging and super-resolution in the harmonic microscope. Optica Acta 26 (1979).

[5] Moreaux, L., Sandre, O., Blanchard-Desce, M. & Mertz, J. Membrane imaging by simultaneous second-harmonic generation and two-photon microscopy (vol 25, pg 320, 2000). Optics Letters 25: 678–678 (2000).

[6] Moreaux, L., Sandre, O., Blanchard-Desce, M. & Mertz, J. Membrane imaging by simultaneous second-harmonic generation and two-photon microscopy. Optics Letters 25: 320–322 (2000).

[7] Moreaux, L., Sandre, O. & Mertz, J. Membrane imaging by second-harmonic generation microscopy. Journal of the Optical Society of America B-Optical Physics 17: 1685–1694 (2000).

[8] Mertz, J. & Moreaux, L. Second-harmonic generation by focused excitation of inhomogeneously distributed scatterers. Optics Communications 196: 325–330 (2001).

[9] Parry, D.A. & Craig, A.S. Quantitative electron microscope observations of the collagen fibrils in rat-tail tendon. Biopolymers 16: 1015–1031 (1977).

[10] Parry, D.A., Barnes, G.R. & Craig, A.S. A comparison of the size distribution of collagen fibrils in connective tissues as a function of age and a possible relation between fibril size distribution and mechanical properties. Proc. R. Soc. Lond. B Biol. Sci. 203: 305–321 (1978).

[11] Roth, S. & Freund, I. Second harmonic-generation in collagen. J. Chem. Phys 70: 1637–1643 (1979).

[12] Williams, R.M., Zipfel, W.R. & Webb, W.W. Interpreting Second Harmonic Generation Images of Collagen I Fibrils. Biophysical Journal 88: 1377–1386 (2005).

[13] Dombeck, D.A., Kasischke, K.A., Vishwasrao, H.D., Ingelsson, M., Hyman, B.T. & Webb, W.W. Uniform polarity microtubule assemblies imaged in native brain tissue by second-harmonic generation microscopy. PNAS 100: 7081–7086 (2003).

[14] Dombeck, D.A., Blanchard-Desce, M. & Webb, W.W. Optical Recording of Action Potentials with Second-Harmonic Generation Microscopy. Journal of Neuroscience 24: 999–1003 (2004).

[15] Sacconi, L., Dombeck, D.A. & Webb, W.W. Overcoming photodamage in second-harmonic generation microscopy: Real-time optical recording of neuronal action potentials. PNAS 103: 3124–3129 (2006).

Preface

In the past decade, multiphoton excitation microscopy and other forms of nonlinear optical microscopy have emerged as important techniques with ever-increasing numbers of important applications in the fields of biology, medicine, chemistry, and physics. The steadily increasing numbers of researchers that are involved with these nonlinear optical microscopic and spectroscopic techniques can be verified by several metrics. First, the annual numbers of publications in these fields is rapidly increasing. Second, the national and international conferences on these subjects are well attended with annual increases in the numbers of attendees. Third, the techniques of nonlinear optical microscopy and the development of associated technologies such as probe development are rapidly advancing. Forth, nonlinear optical microscopy and spectroscopy is providing new, innovative approaches for live cell imaging and endoscopic studies.

There are books that contain specific chapters on the special techniques of nonlinear optical microscopy; however, there is a need for a self-contained, single volume that would provide comprehensive, broad-based, up-to-date, and seminal coverage of the field.

The *Handbook of Biomedical Nonlinear Optical Microscopy* is designed to meet the eclectic needs of the advanced undergraduate student, the graduate student, the scientist, and the clinician-scientist who are involved in or wish to learn the principles and techniques of nonlinear optical microscopy and spectroscopy and wish to work with the myriad of associated fascinating studies and applications.

The handbook is developed from conception through final editing with a universal goal: to provide a physically accurate, clearly written volume that is readily understandable by both scientists and clinicians with different backgrounds. The handbook achieves this goal by deploying a two-level pedagogical approach. The basic sections provide explanation of the most important aspects of nonlinear microscopy described with minimal physics and mathematics formulism. These basic sections can be understood by researchers with no formal training in physics but nonetheless provide a comprehensive overview of the field and its relevancy to biology and medicine. This goal is achieved by presenting the historical experiments, providing simple physical models and analogies for the classical and quantum mechanical theories, and finally developing and discussing the mathematical equations that describe the physics. This allows the researchers to effectively use nonlinear microscopy as a tool in their biomedical research. The advanced sections contain mathematically and physically rigorous explanations of spectroscopic and imaging theories, instrumentation designs, and advanced applications. These sections are aimed for researchers with interests in further advancing nonlinear microscopy theory and instrumentation or in developing novel applications using these technologies. We use the symbol (*) to indicate an advanced topic in each chapter. In the first reading of the chapter, the reader can skip these advanced sections without a loss of continuity and still obtain a coherent discussion of the material. The handbook strives to be self-contained in that all of the appropriate physics is explained.

The handbook is divided into four parts; each part is prefaced with a section introduction providing an overview. Part I of the handbook is an analysis of the history of nonlinear optics and image formation. Since the theoretical work of Maria Göppert-Mayer is critical to the development of multiphoton excitation processes, this part includes chapters on her life and work. A brief chapter reviews the history of perturbation theory as it was developed for astronomy and later used for quantum mechanics. The two seminal papers of Maria Göppert-Mayer published in 1929 and 1931 are translated into English by Barry R. Masters. These papers are followed by a section containing extensive notes that serve as an introduction to the translations as well as a review of the prior theoretical work on atom–radiation field interactions. The seminal contributions of Paul Dirac to the theory of two-photon scattering via virtual intermediate states are discussed since they were so important to the subsequent theoretical work of Maria Göppert-Mayer two-photon processes. This aspect of the history of two-photon processes is neglected in the literature. It was only by the perusal of the original papers of both Paul Dirac and Maria Göppert-Mayer that this new information was discovered. Part II of the handbook describes the spectroscopic base for nonlinear microscopies. Three extensive and comprehensive self-contained tutorial chapters develop the theoretical basis for nonlinear fluorescence spectroscopy, second- and third-harmonic generation (SHG, THG) spectroscopy, and coherent anti-Stokes resonance spectroscopy (CARS). Part III of the handbook covers a variety of innovative instrumentation and techniques for nonlinear optical microscopy. The instrumentation section is replete with tutorial chapters that cover resolution, light sources, scanning and intermediate optics systems, detectors, molecular probes, and a wide variety of nonlinear optical microscopic techniques (FRET, STED, FCS, FRAP, and microprocessing). Many of these chapters include clear and concise tutorial introductions to the relevant theory that are necessary to understand the principles of these instrumentation and techniques. Part IV of the handbook covers the important biomedical applications of nonlinear optical microscopy. There are comprehensive chapters on neurobiology, intravital tumor biology, immunology, developmental biology, dermatology, and the optical monitoring of cellular metabolism. Each of these application chapters include a concise introduction to the biological or medical field together with a critical assessment of the nonlinear microscopic techniques as applied to biology or medicine.

The handbook is carefully balanced between the theoretical aspects of the field and the relevant instrumentation and applications. The emphasis of the handbook is to teach in a clear and comprehensive manner the principles and the techniques of nonlinear optical spectroscopy and microscopy. Numerous diagrams and figures, many of them in color, supplement the text and provide further understanding of the materials.

The role of the editors is to achieve a consistent book that is both comprehensive and balanced in coverage. In that role we have edited the entire contents of the volume. The editors, long-time researchers in the field of nonlinear spectroscopy and microscopy, have developed a handbook that is clear in exposition, comprehensive in analysis and coverage, and current in the state-of-the-art developments in the field.

In order to openly disclose the presence of any competing financial interests with the content of the contributor's chapters, the editors have asked each contributor to complete a declaration of financial interests. All of the authors replied; however, only those authors with competing financial interests have statements at the ends of their chapters. This requirement is consistent with the policies of many science journals and most medical journals. For an explanation of the rationale for this policy, the editors suggest an important editorial published in *Nature* (23 August 2001, volume 412, issue no. 6849). This article suggests that publication practices in biomedical research have been influenced by commercial interests of the authors. We hope that appropriate declaration of financial interests will prove to be helpful.

We thank the authors for their excellent contributions. We thank Kirk Jensen, Peter Prescott, Paul Hobson, and the editorial and production staff of Oxford University Press for their contribution to the success of the handbook.

Barry R. Masters and Peter T. C. So

Contents

Expanded Contents

Contributors

ABOUT THE EDITORS

BARRY R. MASTERS

Professor Barry R. Masters is a Visiting Scientist in the Department of Biological Engineering at the Massachusetts Institute of Technology, Cambridge, MA, and was formerly a Professor in Anatomy and Cell Biology at the Uniformed Services University of the Health Sciences, Bethesda, MD, and a Gast Professor in the Department of Ophthalmology, University of Bern, Bern, Switzerland. He is a Fellow of both the Optical Society of America (OSA) and The International Society for Optical Engineering (SPIE). He received a BS in chemistry from the Polytechnic Institute of Brooklyn, where he conducted his thesis research with Professor Murray Goodman; an MS in physical chemistry from the Florida State University (Institute of Molecular Biophysics), where he was a student of Professor Michael Kasha (chemistry) and Professor Dexter Easton (neurobiology); and a PhD in physical chemistry from the Weizmann Institute of Science in Israel, where he was a student of Professor Israel Miller. Professor Masters continued his research at the Max Planck Institute for Biophysical Chemistry in Göttingen with Professor Hans Kuhn, at Columbia University with Professor David Nachmansohn, at the Rockefeller University with Professor David Mauzerall, at the Johnson Research Foundation, University of Pennsylvania with Professor Britton Chance. He worked as an Assistant Professor at Emory University Eye Center and at Georgia Institute of Technology with Professor Bill Rhodes. Professor Masters has published 80 research papers in refereed journals, 110 book chapters, numerous papers in conference proceedings, and 103 scientific abstracts. He is the editor or author of several books: *Noninvasive Diagnostic Techniques in Ophthalmology* (1990); *Medical Optical Tomography: Functional Imaging and Monitoring* (1993); *Selected Papers on Confocal Microscopy* (1996); *Selected Papers on Optical Low-Coherence Reflectometry and Tomography* (2000); and *Confocal Microscopy and Multiphoton Excitation Microscopy: The Genesis of Live Cell Imaging* (2006), and (co-edited with Peter T. C. So) *Handbook of Biomedical Nonlinear*

Optical Microscopy (2008). Professor Masters and Professor Robert Alfano are the editors of a CD-ROM: *Biomedical Optical Biopsy, Vol.* 2. Classic Articles in Optics and Photonics on CD-ROM Series, Optical Society of America. He has been chair or co-chair of 44 international symposia and meetings on biomedical optics. He has taught many short courses on three-dimensional confocal microscopy and visualization in the United States and abroad. He has presented over 300 lectures on biomedical imaging. In 1999 Professor Masters and Professor Böhnke shared the Vogt Prize for Research (highest Swiss award for ophthalmology) for their research on the confocal microscopy of the cornea. Dr. Masters was a Program Director of Biological Instrumentation and Instrument Development at the National Science Foundation. He received an AAAS Congressional Science & Engineering Fellowship (OSA/SPIE) in 1999–2000, and worked on science policy as a Legislative Assistant in the U.S. Congress. He investigated the topics of genetically modified foods, medical ethics, and science education. Professor Masters has been a Visiting Professor at a number of universities: Netherlands Ophthalmic Research Institute, Amsterdam; Beijing Medical University, Beijing, PRC; Science University of Tokyo, Tokyo; University of Bern, Bern, Switzerland; Swinburne University of Technology, Victoria, Australia; National Taiwan University, Taipei; Fudan University, Shanghai, P.R. China; and a Research Fellow, Nuffield Laboratory of Ophthalmology, University of Oxford, Oxford, UK. He has been an Adjunct Professor in the Department of Engineering Science at the National Cheng Kung University, Tainan, Taiwan. Professor Masters received an OSA Fellow Travel Award to lecture in Russia on biomedical optics, spectroscopy and microscopy (2002). He received a Traveling Lecturer Award from the USAC, International Commission for Optics in 2003 to lecture on biomedical optics at India at several Universities and Indian Institutes of Technology. In 2004 he was the recipient of an OSA Fellow Travel Grant, from International Council Fellow Travel Grants to lecture on biomedical optics, nonlinear optical microscopy, and spectroscopy in biology and medicine, in Egypt at Cairo University, The National Center for Laser Enhanced Science, and at Ain Shams University. In 2006 he was the recipient of Traveling Lecturer Award from the International Commission for Optics (ICO) to lecture on biomedical optics and imaging at the Department of Optical Science and Engineering, Fudan University, Shanghai, and the Shanghai Institute for Optics and Fine Mechanics, Shanghai, P. R. China. Professor Masters is a member of the Editorial Board of three journals: *Computerized Medical Imaging and Graphics; Graefe's Archive for Clinical and Experimental Ophthalmology; and Ophthalmic Research.* He is a book reviewer for several physics, medical, and optics journals and a book reviewer for *Science Books & Films* (AAAS). He has published more than 80 book reviews. He was a member of the OSA *Applied Optics* Patent Review Panel for the last five years. He is actively working as a peer-reviewer for several medical and physics journals and is a special reviewer for medical and physics grants for the National Institutes of Health and the National Science Foundation (USA). His research interests include the development of in vivo confocal microscopy of the human eye and skin, cell biology of differentiation and proliferation in epithelial tissues, the application and development of multiphoton excitation microscopy to deep tissue imaging and spectroscopy, diagnostic and functional medical imaging, optical Fourier transform methods for cellular pattern recognition, and fractal analysis of biological branching vascular patterns. Professor Masters works to promote responsible conduct of research, international cooperation in science, biomedical ethics in research, science education from K–12 to college and university, and public understanding of the process of science.

PETER T.C. SO

Professor Peter T. C. So is a Professor of Mechanical Engineering and Biological Engineering at the Massachusetts Institute of Technology (MIT), Cambridge, MA. He received a BS degree in physics and mathematics in 1986 from the Harvey Mudd College and a PhD degree in physics in 1992 from Princeton University. As a postdoctoral associate and an electro-optical engineer, he worked with Professor Enrico Gratton in the Laboratory for Fluorescence Dynamics in the University of Illinois at Urbana-Champaign. Professor So joined MIT as a faculty member in 1996. Professor So has published over 75 research papers in refereed journals, over 30 refereed conference proceedings, many book chapters, and numerous scientific abstracts. He is also the co-inventor of five patents related to microscopy and instrumentation.

Starting in 2000, Professor So has been the conference co-chair for "Multiphoton Microscopy in the Biomedical Sciences" in the SPIE International Symposium on Biomedical Optics. He is also the co-editor for the SPIE proceeding series associated with this conference. Professor So is the conference co-chair elect for the 2008 Gordon Conference, "Lasers in Medicine and Biology." He has served regularly as session chair and program committee member in conferences and symposiums. He has presented over 100 invited lectures on biomedical optics. He has been guest instructor in over ten different short course series on spectroscopy and microscopy.

Professor So is a member on the editorial board for the *Journal of Fluorescence*. He has been a guest editor for a special issue in *Optics Express*. He is a member in many scientific societies such as the SPIE, the International Society for Optical Engineering, the Optical Society of America, and the Biophysical Society. He has served as reviewer for journals in different fields. He is a member in a number of scientific review panels for the National Institute of Health, the National Science Foundation, and the Department of Defense.

In MIT, Professor So teaches on both undergraduate and graduate levels in the fields of bioengineering and mechanical engineering. He has co-developed an undergraduate course on "Bioengineering Instrumentation and Measurement" and a graduate course on "Optical Microscopy and Spectroscopy for Biology and Medicine." From 2000 to 2003, he held the Esther and Harold E. Edgerton Chair. In 2003, he received the Ruth and Joel Spira Award for Distinguished Teaching from the Department of Mechanical Engineering. In 2004, he received the Frank E. Perkins Award for Excellence in Graduate Advising from MIT.

The laboratory of Professor So has the mission of creating novel optical imaging and manipulation technologies on the microscopic level realizing that many advances in biology and medicine are driven by the availability of new tools. The laboratory of Professor So has been operated with funding from the National Institute of Health, National Science Foundation, Department of Defense, private foundations, and industrial partners. In his career, he has participated in the development many new instruments and techniques such as multiphoton fluorescence lifetime resolved microscopy, multiphoton fluorescence correlation spectroscopy, photon counting histogram analysis, standing-wave total internal refraction microscopy, and image segmentation based global analysis. Professor So is also one of the first investigators who has successfully applied two-photon microscopy for the in vivo imaging of human tissues. Professor So has a broad scientific interest and his laboratory is engaged in many scientific and medical studies such as developing non-invasive optical biopsy technology for cancer diagnosis, understanding the mechanotransduction processes in cardiovascular diseases, and elucidating the effects of dendritic remodeling on memory plasticity.

CONTRIBUTORS

Karsten Bahlmann
Department of Biological Engineering
Massachusetts Institute of Technology
Cambridge, MA
E-mail: kbahlman@mit.edu

Wolfgang Becker
Becker & Hickl GmbH
Berlin, Germany
E-mail: becker@becker-hickl.de

Axel Bergmann
Becker & Hickl GmbH
Berlin, Germany
E-mail: bergmann@becker-hickl.de

Keith Berland
Department of Physics, Emory
    University
Atlanta, GA
E-mail: kberland@physics.emory.edu

Michael F. Booth
Edwin L. Steele Laboratory
Department of Radiation Oncology
Massachusetts General Hospital and
    Harvard Medical School
Boston, MA

Robert W. Boyd
The Institute of Optics and Department
    of Physics and Astronomy
University of Rochester
Rochester, NY
E-mail: boyd@optics.rochester.edu

Edward Brown
Department of Biomedical
    Engineering
University of Rochester
    Medical Center
Rochester, NY
E-mail:edward_brown@urmc.rochester.edu

Michael D. Cahalan
Physiology and Biophysics
University of California
Irvine, CA
E-mail: mcahalan@uci.edu

Paul J. Campagnola
University of Connecticut Health Center
Department of Physiology
Farmington, CT
E-mail: campagno@neuron.uchc.edu

Ye Chen
W. M. Keck Center for Cellular
    Imaging
Departments of Biology and Biomedical
    Engineering
University of Virginia
Charlottesville, VA

G. Omar Clay
Department of Physics
University of California San Diego
La Jolla, CA
E-mail: omar@physics.ucsd.edu

Mary E. Dickinson
Department of Molecular Physiology
   and Biophysics
Baylor College of Medicine
Houston, TX
E-mail: mdickins@bcm.tmc.edu

Chen-Yuan Dong
Microscopic Biophysics Laboratory
Department of Physics
National Taiwan University,
Taipei, Taiwan
E-mail: cydong@phys.ntu.edu.tw

Daniel L. Farkas
Minimally Invasive Surgical
   Technologies Institute and
   Department of Surgery
Cedars-Sinai Medical Center
Los Angeles, CA
E-mail: Daniel.Farkas@cshs.org

Dai Fukumura
Edwin L. Steele Laboratory
Department of Radiation Oncology
Massachusetts General Hospital and
   Harvard Medical School
Boston, MA
E-mail: dai@steele.mgh.harvard.edu

John Girkin
Institute of Photonics
University of Strathclyde
Wolfson Centre
Glasgow, Scotland
E-mail: j.m.girkin@strath.ac.uk

Min Gu
Centre for Micro-Photonics
Faculty of Engineering and
   Industrial Sciences
Swinburne University of Technology
Hawthorn, Victoria, Australia
E-mail: MGu@groupwise.swin.edu.au

Benjamin Harke
Department of NanoBiophotonics
Max Planck Institute for Biophysical
   Chemistry
Göttingen, Germany

Stefan W. Hell
Department of NanoBiophotonics
Max Planck Institute for Biophysical
   Chemistry
Göttingen, Germany
E-mail: shell@gwdg.de

Rakesh K. Jain
Edwin L. Steele Laboratory
Department of Radiation Oncology
Massachusetts General Hospital and
   Harvard Medical School
Boston, MA
E-mail: jain@steele.mgh.harvard.edu

Shiou-Hwa Jee
Department of Dermatology
National Taiwan University Hospital
Taipei, Taiwan

Jan Keller
Department of NanoBiophotonics
Max Planck Institute for Biophysical
   Chemistry
Göttingen, Germany

Daekeun Kim
Department of Mechanical Engineering
Massachusetts Institute of Technology
Cambridge, MA
E-mail: dkkim@mit.edu

Ki H. Kim
Wellman Center for Photomedicine
Massachusetts General Hospital
Boston, MA
E-mail: khkim@partners.org

David Kleinfeld
Department of Physics
University of California San Diego
La Jolla, CA
E-mail: dk@physics.ucsd.edu

Karsten König
Fraunhofer Institute of Biomedical
  Technology (IBMT)
St. Ingbert, Saarland, Germany
Faculty of Mechatronics and Physics,
  Saarland University
Saarbrücken, Germany
E-mail: karsten.koenig@ibmt.
  fraunhofer.de

Irina V. Larina
Department of Molecular Physiology
  and Biophysics
Baylor College of Medicine
Houston, TX
E-mail: larina@bcm.tmc.edu

Sung-Jan Lin
Department of Dermatology
National Taiwan University Hospital
Taipei, Taiwan

Wen Lo
Microscopic Biophysics Laboratory
Department of Physics
National Taiwan University,
Taipei, Taiwan

Ania Majewska
Department of Neurobiology and
  Anatomy
University of Rochester Medical Center
Rochester, NY
E-mail: ania_majewska@urmc.
  rochester.edu

Barry R. Masters
Department of Biological Engineering
Massachusetts Institute of Technology
Cambridge, MA
E-mail: bmasters@mit.edu

Jerome Mertz
Department of Biomedical Engineering
Boston University
Boston, MA
E-mail: jmertz@bu.edu

Lance L. Munn
Edwin L. Steele Laboratory
Department of Radiation Oncology
Massachusetts General Hospital and
  Harvard Medical School
Boston, MA

Quoc-Thang Nguyen
Department of Physics
University of California San Diego
La Jolla, CA
E-mail: qnguyen@physics.ucsd.edu

Nozomi Nishimura
Department of Physics
University of California San Diego
La Jolla, CA
E-mail: nnishimu@physics.ucsd.edu

Suzette Pabit
Department of Physics
Emory University
Atlanta, GA

Timothy P. Padera
Edwin L. Steele Laboratory
Department of Radiation Oncology
Massachusetts General Hospital and
  Harvard Medical School
Boston, MA

Ian Parker
Departments of Neurobiology & Behavior
  and Physiology & Biophysics
University of California, Irvine
Irvine, CA
E-mail: iparker@uci.edu

Ammasi Periasamy
W.M. Keck Center for Cellular Imaging
Departments of Biology and Biomedical
  Engineering
University of Virginia
Charlottesville, VA
E-mail: ap3t@virginia.edu

Eric O. Potma
University of California, Irvine
Department of Chemistry, Natural
    Sciences II
Irvine, CA
E-mail: epotma@uci.edu

Timothy Ragan
Department of Biological Engineering
Massachusetts Institute of Technology
Cambridge, MA
E-mail: tragan@mit.edu

Chris B. Schaffer
Department of Biomedical Engineering
Cornell University
Ithaca, NY
E-mail: cs385@cornell.edu

Andreas Schönle
Department of NanoBiophotonics
Max Planck Institute for Biophysical
    Chemistry
Göttingen, Germany
E-mail: aschoen@gwdg.de

Lee F. Schroeder
UCSD Neuroscience Graduate Program
University of California San Diego
La Jolla, CA
E-mail: lfschroe@ucsd.edu

Peter T. C. So
Departments of Mechanical and
    Biological Engineering
Massachusetts Institute of Technology
Cambridge, MA
E-mail: ptso@mit.edu

Jiunn Wen Su
Microscopic Biophysics Laboratory
Department of Physics
National Taiwan University
Taipei, Taiwan

Yen Sun
Microscopic Biophysics Laboratory
Department of Physics
National Taiwan University
Taipei, Taiwan

Bruce Tromberg
The Henry Samueli School of
    Engineering
University of California, Irvine
Irvine, CA
E-mail: bjtrombe@uci.edu

Philbert S. Tsai
Department of Physics
University of California, San Diego
La Jolla, CA
E-mail: ptsai@physics.ucsd.edu

Sebastian Wachsmann-Hogiu
Minimally Invasive Surgical
    Technologies Institute and
    Department of Surgery
Cedars-Sinai Medical Center
Los Angeles CA

Horst Wallrabe
W.M. Keck Center for Cellular Imaging
Departments of Biology and Biomedical
    Engineering
University of Virginia
Charlottesville, VA

Watt W. Webb
Applied and Engineering Physics
Cornell University
Ithaca, New York
E-mail: www2@cornell.edu

X. Sunney Xie
Harvard University
Department of Chemistry and Chemical
    Biology
Cambridge, MA
E-mail: xie@chemistry.harvard.edu

Chris Xu
School of Applied and Engineering
    Physics
Department of Biomedical Engineering
Cornell University
Ithaca, NY
E-mail: cx10@cornell.edu

Siavash Yazdanfar
Applied Optics Lab
General Electric Global Research
    Center
Niskayuna, NY
E-mail: yazdanfa@research.ge.com

Warren R. Zipfel
School of Applied and Engineering
    Physics
Department of Biomedical
    Engineering
Cornell University
Ithaca, NY
E-mail: wrz2@cornell.edu

Part I

HISTORICAL DEVELOPMENT OF NONLINEAR OPTICAL MICROSCOPY AND SPECTROSCOPY

Barry R. Masters

In the belief that the study of the history of science is both useful and beneficial to scientists, we include an extensive coverage on the historical development of nonlinear optical spectroscopy and microscopy. These themes are important for the understanding of the process of science, and they help one to gain insight into the education, influences, personality, and work of the scientists who made major contributions to the field.

Some questions that come to mind include the following: What are the early educational influences that affect the scientific output of an individual? What are the roles of the educational institutions, the teachers and mentors, the fellow students and colleagues? What is the role of the state and local governments? How does the political situation influence the development of a particular area of science (i.e., the development of quantum mechanics in Göttingen, Cambridge, Berlin, and Copenhagen)? More specifically concerning the life and early work of Maria Göppert-Mayer on the theoretical development of multiphoton processes, we may ask what role her mentor, Max Born, played in her scientific development. Furthermore, to what extent did her work build on Paul Dirac's quantum theory of radiation and his time-dependent perturbation theory?

The important contributions of Paul Dirac to the theory of two-photon transitions, theoretical developments that were subsequently incorporated in the thesis work of Maria Göppert-Mayer, are generally unknown to the general reader. Therefore, I explicitly list these points. The seminal contributions of Dirac in his 1927b paper include the following developments that were incorporated in the 1931 thesis paper of Maria Göppert-Mayer: (1) the use of second-order time-dependent perturbation theory, (2) the confinement of the radiation field into a cavity, and (3) the use of two-photon scattering via virtual intermediate states. Dirac states in his 1927b paper that the individual transitions (i.e., from the ground state to the virtual state, and a transition from the virtual state to the final state) each do not obey the law of conservation of energy; only the total transition, via two photons from the ground state to the virtual state to the final state, obeys the law of conservation of energy. These later points also appear in the 1931 paper of Maria Göppert-Mayer.

I hope that after reading Part I the reader will have some answers to these questions, and perhaps many additional questions will arise. Science is about questions and verifiable answers and predictive theories. It is also one of the great human adventures. That is why the framing of scientific advances in human terms is important: it stresses our unique personal thought processes, aspirations, influences, and interactions. Science is a universal, human enterprise and we should stress the human aspects.

Part I of the handbook consists of four chapters. Chapter 1, *The Genesis of Nonlinear Microscopies and Their Impact on Modern Developments*, is a historical account of nonlinear optics and their implementation in microscopies. We stress both the experimental and the theoretical developments in nonlinear optics, in particular two-photon excitation processes. The time scale is from the 1929 theoretical paper of Maria Göppert-Mayer to the experimental implementation of two-photon excitation microscopy by the group of Webb in 1990. A great emphasis is rightly placed on the work and life of Maria Göppert-Mayer. We also stress the seminal work on the scanning harmonic optical microscope developed at Oxford University by Sheppard, Kompfner, Gannaway, Choudhury, and Wilson. Furthermore we show that there was a real disconnect between the life scientists and the Oxford scientists.

Another important theme in this chapter is to point out the critical role of technology development in the advancement of science. In that theme we describe the role of lasers in the development of nonlinear microscopy. Finally, we discuss and analyze modern developments in nonlinear microscopy and their impact. New advances in detectors, new types of fluorescent probes, new techniques to modify the point-spread function, and new contrast mechanisms are rapidly advancing the field.

Chapter 2, *The Scientific Life of Maria Göppert-Mayer*, is devoted to an overview of the scientific life of Maria Göppert-Mayer. It contains several images of Maria Göppert-Mayer and Max Born at the time that she was a graduate student in Göttingen. The important message inherent in this chapter is that we must encourage women to study science and to become creative and productive scientists. That goal requires the cooperation of parents, teachers, peers, and cultural institutions. I hope that the life and work of Maria Göppert-Mayer will serve as a role model and encourage young women to pursue a life in science.

Chapter 3, *The History of Perturbation Theory from Astronomy to Quantum Mechanics*, is a brief introduction to the history of perturbation theory. Because the theoretical work of Maria Göppert-Mayer on two-photon processes relies heavily on Dirac's second- order, time-dependent perturbation theory, it is important to present a historical context for the development of perturbation theory used in Göppert-Mayer's work. The origins of perturbation theory come from the field of celestial mechanics; however, the development of the theory and its applications in the field of quantum mechanics are our main interest. After an introduction to various forms of perturbation theory that were devised for celestial mechanics, we describe several different approaches to perturbation theory in quantum mechanics. In Chapter 5 of the handbook we present a more detailed physical and mathematical description of second-order perturbation theory.

Chapter 4, *English Translations of and Translator's Notes on Maria Göppert-Mayer's Theory of Two-Quantum Processes*, contains the English translations of and the translator's notes on Maria Göppert-Mayer's theory of two-quantum processes. Section 1 is the translation of her 1929 paper, *Über die Wahrscheinlichkeit des Zusamenwirkens zweier Lichtquanten in einem Elementarakt* (*Die Naturwissenschaften*, 17, 932, 1929). Section 2 is the English translation of her 1931 thesis paper (written when she was in Baltimore, MD), *Über Elementarakte mit zwei Quantensprüngen* (*Ann. Phys Leipzig*, 9, 273–294, 1931). Section 3 is a comprehensive set of notes that serve as an introduction to the English translations of her two papers and also contain detailed ancillary materials on the theories of atom–radiation field interactions that were published during the time of her graduate work and have influenced her research. In the latter category are two papers published by Dirac in 1927, one on radiation theory and one on time-dependent perturbation theory as well as two-photon transitions via virtual intermediate states, a paper by Kramers and Heisenberg in 1925, and a detailed discussion of the experimental papers on collisional excitation of atoms. Finally, there is an overview of the theoretical approaches that are the basis of Maria Göppert-Mayer's theory of two-quantum processes. The key take-home lesson in this section is to understand the major role that Dirac's theories played in the 1931 thesis paper of Maria Göppert-Mayer.

—— **1**

The Genesis of Nonlinear Microscopies and Their Impact on Modern Developments

Barry R. Masters and Peter T. C. So

1.1 INTRODUCTION

In 1931, Maria Göppert-Mayer published her doctoral dissertation that developed the theory of two-photon quantum transitions (two-photon absorption and emission) in atoms. In this chapter we describe and analyze the theoretical and experimental work on nonlinear optics, in particular two-photon excitation processes, that occurred between 1931 and the experimental implementation of two-photon excitation microscopy by the group of Webb in 1990. More recent major developments in nonlinear microscopy and their impact are briefly described. In addition to Maria Göppert-Mayer's theoretical work, the invention of the laser has a key role in the development of two-photon microscopy. Nonlinear effects were previously observed in nonoptical frequency domains (low-frequency electric and magnetic fields and magnetization), but the high electric field strength afforded by lasers was necessary to demonstrate many nonlinear effects in the optical frequency range.

The physicist Maria Göppert-Mayer developed the theory of two-photon absorption and emission processes as part of her doctoral research in Göttingen. She published a brief summary paper in 1929 and her dissertation in 1931. Due to a lack of laser sources, the first high-resolution nonlinear microscope was not developed until 1978. The first nonlinear microscope with depth resolution was described by the Oxford (UK) group. Sheppard and Kompfner published a paper in *Applied Optics* that described microscopic imaging based on second-harmonic generation (Sheppard & Kompfner, 1978). In their paper they further proposed that other nonlinear optical effects, such as two-photon fluorescence, could also be applied. However, the developments in the field of nonlinear optical microscopy stalled due to a lack of a suitable laser source with high peak power and tunability. This obstacle was removed with the advent of femtosecond lasers in the 1980s. In 1990, the seminal paper of Denk, Strickler, and Webb on two-photon laser scanning fluorescence microscopy was published in *Science* (Denk et al., 1990). Their paper clearly demonstrated the capability of two-photon excitation microscopy for biology, and it served to convince a wide audience of biomedical scientists of the potential capability of this technique.

The purpose of this chapter is to examine the antecedents to two-photon scanning laser microscopy and other types of nonlinear microscopies up to 1990. We examine the impacts of the scientists and ancillary technologies, especially lasers, on this field. We conclude by discussing the most important advances in the biomedical optics field since the pioneering work of the Cornell group in 1990.

1.2 MARIA GÖPPERT-MAYER'S SCIENTIFIC LIFE

The scientific life of an individual provides insights into his or her creations and is therefore a valid topic in our historical discussion (Masters, 2000; Masters & So, 2004; McGrayne, 1996; Sachs, 1982). Maria Göppert-Mayer was proud of her heritage, which was linked to seven generations of professors on her father's side of the family. When she was 4, her family moved to Göttingen, where her father became a professor of pediatrics. Her father held and expressed high expectations for his daughter; therefore, he encouraged her to strive to achieve a university education. Early on, she decided to develop a professional career in science. Her father instilled in her a strong feeling of self-confidence and a free spirit. As she grew older there was never doubt in her mind, or that of her parents, that she would study at the University of Göttingen (Georg-August-Universität). The Göpperts' family friends in Göttingen included James Franck, who was from 1921 to 1933 the Director of the Second Physical Institute (II. Physikalischen Instituts der Georg-August-Universität); Max Born, who was from 1921 to 1933 Director of the Institute of Theoretical Physics (Instituts für Theoretische Physik) in the University of Göttingen; and David Hilbert of the Mathematics Institute. The Physics Institute in Göttingen consisted of three a divisions, and all three were located in the main physics building on Bunsenstrasse, where Göppert studied. Robert Pohl was the director of the other division, Erstes Physikalisches Institut. Robert Pohl was well known for his introductory textbooks on physics, *Einführung in die Physik* (Pohl, 1935).

Maria Göppert-Mayer entered the University in Göttingen to study mathematics. After she heard a lecture by Max Born on quantum mechanics, she changed her university major from mathematics to physics. At that time Göttingen was the world center of these fields, especially in quantum mechanics (Greenspan, 2005; Van der Waerden, 1967). The Department of Mathematics hosted Richard Courant, David Hilbert, Hermann Weyl, and Edmund Landau. Earlier Hermann Minkowski, Felix Klein, and David Hilbert worked to make Göttingen a world center of mathematics, and they followed in the steps of the great Göttingen 19th-century mathematicians: Carl Friedrich Gauss, Bernhard Riemann, and P. G. Dirichlet (Rupke, 2002).

Göppert-Mayer's thesis advisor, Max Born, came to Göttingen and created the Institute of Theoretical Physics and during a period of 12 years developed a world center of quantum mechanics until the Nazis forced him to emigrate in 1933 (Rupke, 2002). Born mentored several prominent doctoral students: J. Robert Oppenheimer, Max Delbruck, Maria Göppert-Mayer, and Victor Weisskopf. During that time he worked with many scientists, eight of whom eventually received the Nobel Prize: Werner Heisenberg, Wolfgang Pauli, Maria Göppert-Mayer, Max Delbrück, Enrico Fermi, Eugene Wigner, Linus Pauling, and Gerhard Herzberg. Other coworkers of Max Born included Pascual Jordan, Friedrich Hund, J. Robert Oppenheimer, Victor Weisskopf, Edward Teller, Fritz London, and Yakov Frenkel.

James Franck, who was the director of the Department of Experimental Physics in Göttingen during Göppert-Mayer's graduate studies, stands out in her life because of his teachings and his role as her lifetime mentor. In 1933 James Franck resigned from his university position in protest of the Nazi racial policies in the same year as Max Born was forced to leave.

In 1924 Max Born invited Maria Göppert to become his graduate student in physics at the Institute of Theoretical Physics. Her fellow students included Max Delbrück and Victor Weisskopf. For a deeper insight into the life of her thesis adviser and mentor, Max Born, as well as his role in the development of quantum mechanics, the reader is referred to a new biography of Max Born (Greenspan, 2005). During her Göttingen years, Maria interacted with the following eminent scientists: Arthur Compton, Paul Dirac, Enrico Fermi, Werner Heisenberg, John von Neumann, J. Robert Oppenheimer, Wolfgang Pauli, Linus Pauling, Leo Szilard, Edward Teller, and Victor Weisskopf.

Maria Göppert used Dirac's formulation of quantum mechanics in her 1931 dissertation. In fact, she acknowledged that Dirac's development of time-dependent perturbation theory was crucial for her theoretical work on two-photon processes. Paul Dirac, a developer of quantum theory, was among the visitors to Göttingen. Dirac was an undergraduate at Bristol, where he studied electrical engineering and applied mathematics. Dirac studied mathematics as a graduate student at Cambridge University, submitted his dissertation, "Quantum Mechanics," and was awarded his PhD degree in 1926. Dirac then spent 5 months in Göttingen in 1927, where he worked with Max Born, Oppenheimer, and Pascual Jordan, and interacted with James Franck, Johannes von Neumann, David Hilbert, and Andrew Weyl.

Maria Göppert-Mayer was a graduate student during this fatal transition in Germany, when many of the department heads were forced to leave their positions with the rise of National Socialism. Maria Göppert was a witness to this loss of Jewish professors. The effects on Göttingen University were devastating. Three heads of Institutes were removed: Richard Courant, director of the Mathematical Institute, as well Max Born and James Franck. Only 11 staff members of the four physics and mathematics institutes remained from 33 members (Medawar & Pyke, 2001).

About 20% of all German physicists and mathematicians were dismissed because they were Jews. The dismissal and forced migration of Jewish professors from German universities decimated the ranks in biology, medicine, mathematics, and physics (Medawar & Pyke, 2001; Weinreich, 1999). A recent paper on the pre-war and the post-war struggles of German scientists and the attempts of Otto Hahn to alter the historical record of the role of German scientists during the post-war period is recommended reading (Sime, 2006).

In 1930, Maria Göppert married Joseph Mayer, an American chemical physicist, and they moved to America, where she worked first at Johns Hopkins University and then at Columbia University. In 1931, Maria Göppert-Mayer submitted her doctoral dissertation in Göttingen on the theory of two-photon quantum transitions in atoms from the ground state to the excited state (Göppert-Mayer, 1929, 1931). The scientific content of her doctoral dissertation will be discussed more fully in the next section. She returned to Göttingen to work with Max Born during the summers of 1931, 1932, and 1933.

While at Columbia University she interacted with Harold Urey, Willard Libby, Enrico Fermi, I. I. Rabi, and Jerrold Zacharias. When Edward Teller joined the University of Chicago, Maria continued her work with him. It was during her work at the Argonne National Laboratory that she developed the nuclear shell model for which she was to earn the Nobel Prize in 1963. Several years earlier, in 1955, she and Hans Jensen published a book with the title *Elementary Theory of Nuclear Shell Structures*. Later, in 1963, she and Jensen would share the Nobel Prize in physics for their work on nuclear shell theory. The same physics prize was also shared with Eugene Wigner. She was the second woman to win the Nobel Prize in Physics; 60 years earlier it was won by Marie Curie. In addition to the Nobel Prize, Maria Göppert-Mayer was a member of the National Academy of Sciences and a Corresponding Member of the Akademie der Wissenschaften in Heidelberg and received honorary degrees from Russell Sage College, Mount Holyoke College, and Smith College.

1.3 MARIA GÖPPERT-MAYER'S 1931 DOCTORAL DISSERTATION

In 1931, Maria Göppert-Mayer submitted her doctoral dissertation in Göttingen on the theory of two-photon quantum transitions in atoms (Göppert-Mayer, 1931). She had published a preliminary paper on her theory 2 years earlier (Göppert-Mayer, 1929). She chose to publish her preliminary results in the prestigious weekly journal, *Die Natur-wissenschaften, Wochenschrift Für Die Fortschritte Der Reinen Und Angewandten Naturwissenschraften* ("The Natural Sciences, Weekly Journal for the Publication of Pure and Applied Natural Sciences"). It was the official journal of the German Society of Natural Scientists and Physicians (*Organ der Gesellschaft Deutscher Naturforscher und Ärzte*) and also the official journal of the Kaiser Wilhelm Society for the Promotion of Science (*Organ der Kaiser Wilhelm-Gesellschaft Zur Förderung der Wissenschaften*) and was first published in 1912. Arnold Berliner, a close friend of Max Born, was the editor for many years. This German journal was similar in form to the English journal *Nature*, which was first published in 1869.

In this one-page paper, which she signed "Maria Göppert," she formulated energy state diagrams for both two-photon emission and two-photon absorption processes. She indicated the presence of "virtual" intermediate states. She also concluded in second-order, time-dependent perturbation theory calculations that the probability for the two-photon absorption process is proportional to the square of the light intensity. Her calculations showed that the probability for a two-photon excitation process was very much less than that for a one-photon excitation process and therefore required a light source of very high intensity.

Her 1931 paper is an example of the theoretical approaches to quantum mechanics that were used by graduate students and faculty in Born's Institute. Maria Göppert-Mayer cited several references in her dissertation paper: the work of Dirac on the matrix formulation of quantum mechanics, his quantum theory of dispersion, and his quantum theory of emission and absorption of radiation based on time-dependent perturbation theory; the work of Kramers and Heisenberg on the quantum mechanical formulation of the problem of single quantum processes in absorption and emission of light; the work of Franck on the intensity of spectral lines from electronic transitions in an atom; and the work of Born on the theory of ionization processes. At the end of her paper she thanked Professor Born for friendly help and advice, and Victor Weisskopf, a fellow graduate student, for discussion and help. Maria Göppert-Mayer's complete 1931 paper is reprinted in the book *Selected Papers on Multiphoton Excitation Microscopy*

(Masters, 2003). The English translation of her 1929 and 1931 papers was undertaken by Barry R. Masters and is the subject of chapter 4 in this book.

Maria Göppert-Mayer followed the technique of Dirac for the use of second- order, time-perturbation theory to solve the quantum mechanical equations for the processes of absorption, emission, and dispersion of light in single-photon–atom interactions as well as two-photon processes. It is worthwhile to briefly look at the contributions to quantum theory by Paul Dirac between 1925 and 1928.

In 1925 Dirac independently constructed his quantum theory, after seeing a review copy of the first paper by Werner Karl Heisenberg on quantum theory, using a formalism based on operators. He first realized that the theory of Heisenberg involved noncommuting physical operators that corresponded to the Poisson brackets of the corresponding classical Hamiltonian mechanics theory. Later at Niels Bohr's Institute in Copenhagen Dirac developed his "transformation theory," which demonstrated that Schrödinger's wave mechanics and Heisenberg's matrix mechanics were both special cases of his operator theory. Heisenberg's matrix mechanics were developed in 1926, and a few months later Schrödinger's wave mechanics (the Schrödinger differential equation) was published. Heisenberg, Dirac, and also Schrödinger developed their forms of perturbation theory, which are approximation methods to solve physical problems without an exact solution. The Einstein transition probabilities for both emission and absorption were calculated by Heisenberg and also by Dirac. In 1927 Dirac developed quantum field theory, his method of "second quantization," and applied quantum theory to the electromagnetic field. With his new theory of the radiation field he derived the Einstein coefficients A and B, for the emission and absorption of photons by an atom, and also the Lorentzian line shape for the emission spectrum of the electronic transition. Dirac also devised a dispersion theory. In 1928 Dirac derived a relativistic equation for the electron—the Dirac equation. In 1930 he published his famous book *The Principles of Quantum Mechanics* (Dirac, 1958). Between 1926 and 1927 Dirac worked in Göttingen at the Institute of Theoretical Physics. There he wrote up his theory of dispersion. All of this occurred just prior to and during the doctoral research of Maria Göppert-Mayer at the Institute of Theoretical Physics.

Schrödinger developed a time-dependent perturbation theory in 1926. In his "semi-classical" theory, the atom was treated as a wave-mechanical system and the radiation field was treated classically (i.e., no field quantization). With his time-dependent perturbation theory he derived a dispersion formula that related the polarization of an atom by an oscillating electric field to the sum of the spectral transition probabilities for all the possible atomic spectral lines (electronic transitions).

Schrödinger showed that his quantum theory based on wave mechanics is equivalent to the quantum theory based on matrix mechanics that had been simultaneously developed by Werner Heisenberg, Max Born, and Pascual Jordan. Schrödinger showed that these two theories both predict the same physical behavior for simple harmonic oscillators. Afterwards, John von Neumann definitely demonstrated that both formulations of quantum mechanics were mathematically equivalent. Subsequently, Pauli applied the matrix mechanics theory to the hydrogen atom and obtained the correct result for the spectrum measured experimentally.

In her doctoral paper Maria Göppert-Mayer extended the quantum mechanical treatment of Dirac to electronic processes that involve two annihilation (or two creation) operators in order to describe both two-photon absorption and emission processes. The transition probability of a two-photon electronic process was derived by Maria Göppert-Mayer using second-order, time-dependent, perturbation theory (Peebles, 1992). Her derivation clearly states that the probability of a two-photon absorption process is quadratically related to the excitation light intensity. Her theory applied not only to two-photon absorption, two-photon emission, but also to Stokes and anti-Stokes Raman effects.

An important aspect of the Maria Göppert-Mayer's paper is that the process of two-photon absorption involves the interaction of two photons and an atom via an intermediate, "virtual" state. This interaction must occur within the lifetime of this "virtual" or intermediate state. What is a "virtual" state? It can be described as a superposition of states and is not a true *eigenstate* of the atom. The individual energy transitions from the ground state to the intermediate state, and also the transition from the intermediate state to the final state, are energy non-conserving transitions. However, in the transition from the initial state to the final state there is energy conservation. The probability of the two-photon transition has contributions from all of the "virtual" intermediate states. The first photon induces the transition from the ground state to the virtual state, and the second photon induces the transition from the virtual state to the excited state. Both photons interact to induce the transition from the ground state to the excited state. Since the probability of the two-photon absorption processes is very low, it is necessary to use a high-intensity light in order to achieve a measurable effect.

In respect for the work of Maria Göppert-Mayer, the units of two-photon absorption cross-section are measured in GM (Göppert-Mayer) units. One GM unit is equal to 10^{-50} cm^4 s/photon.

Since her 1931 Göttingen dissertation is written in German, it is helpful to follow the paper together with the derivation of Boyd, which closely follows the 1931 dissertation for those readers who cannot read German (Boyd, 2003). Boyd used time-dependent perturbation techniques to solve the quantum mechanical equations and derived the probabilities and cross-sections for single-photon absorption, two-photon absorption, and multiphoton absorption.

1.4 NONLINEAR OPTICAL PROCESSES AND MICROSCOPY

What are nonlinear optical processes? According to Boyd, they occur when "the response of a material system to an applied optical field depends in a nonlinear manner upon the strength of the optical field" (Boyd, 2003).

A bright light source is necessary to induce nonlinear optical effects. For example, in second-harmonic generation, the intensity of the light generated at the second-harmonic frequency increases as the square of the intensity of the incident laser light. Second-harmonic generation is the first nonlinear optical effect ever observed in which a coherent input generates a coherent output (Shen, 1984). For a two-photon absorption process, the intensity of the fluorescence generated in this process also increases as the square of the intensity of the incident laser light.

It is interesting to contrast the experimental discovery of spontaneous Raman scattering with the observation of two-photon absorption. The spontaneous Raman effect was discovered in 1927, but the two-photon absorption process was first demonstrated in 1961. Yet the two-photon absorption process was predicted by Maria Göppert-Mayer

30 years earlier. Why this long delay? This delay is caused by the required experimental conditions that could not be fulfilled earlier. In the Raman effect, the spontaneous emission of a Stokes shifted photon is dependent linearly on the intensity of the incident light. For the two-photon absorption process, the effect is dependent on the square of the intensity of the incident light and required the development of high-peak-power lasers in 1961.

The development of the laser that provides an intense source of monochromatic, coherent light ushered in the field of nonlinear optics (Maiman, 1960). When a medium has a quadratic relationship between the polarization density and the electric field, there are second-order nonlinear optical phenomena: second-harmonic generation with the frequency doubling of monochromatic light, frequency conversion with the sum or difference of two monochromatic waves giving the frequency of a third wave. Similarly, third-order relationship between the polarization density of the medium and the electric field gives rise to third-harmonic generation, self-focusing, optical amplification and optical phase conjugation (Bloembergen, 1965, 1996; Boyd, 2003; Shen, 1984). *The Quantum Theory of Radiation* is recommended as a source of general background to radiative processes (Heitler, 1954). A recommended graduate-level textbook is *The Quantum Theory of Light*, 3rd edition (Loudon, 2000).

Experimentally, stimulated Raman scattering (a two-photon process) was discovered in the early 1960s (Woodbury & Ng, 1962). Woodbury and Ng were investigating Q-switching in a ruby laser with a Kerr cell that contained nitrobenzene. They discovered that there was a component of the laser output that was at $1,345 \text{ cm}^{-1}$ and it was down-shifted from the laser frequency. Further analysis indicated that this frequency shift was identical to the frequency of vibration of the strongest Raman mode of the nitrobenzene molecule (Boyd, 2003). Raman scattering can be viewed as a two-photon process in which one photon is absorbed and one photon is emitted when the molecules undergo the transition from the initial state to the final state. The terms "Stokes" and "anti-Stokes" refer to a shift to lower frequencies, or a shift to higher frequencies of the scattered photon with respect to the frequency of the photons in the incident light. The Stokes components are usually orders of magnitude more intense than the anti-Stokes components.

Nonlinear optical spectroscopy preceded the development of nonlinear microscopy. The group of Franken made the first observation of second-harmonic generation in a quartz crystal irradiated with a ruby laser (Franken et al., 1961). The authors stated that the development of pulsed ruby optical masers that output monochromatic light (6,943 Å) can be focused and exhibit electric fields of 10^{+5} volts/cm. When this light is focused onto a quartz crystal the second-harmonic signal at 3,472 Å is produced. It is critical that the crystal lacks a center of inversion and is transparent to both the incident light at 6,943 Å and the second harmonic light at 3,472 Å. The authors were careful to present several controls: the light at 3,472 Å disappears when the quartz specimen is removed or is replaced by glass, and the light at 3,472 Å shows the expected dependence on polarization and crystal orientation. Furthermore, they predicted that isotropic materials such as glass could also show second-harmonic generation in the presence of a strong bias electric field.

Bloembergen, in his paper *Nonlinear Optics: Past, Present, and Future* (2000), relates an interesting historical aspect of scientific publication. Franken et al. (1961) passed a ruby laser pulse through a quartz crystal. They used a monochromator and a photographic plate to detect ultraviolet light from the experiment. However, since the spot of the second harmonic was so weak on the spectrographic plate, the editorial office of *Physical Review* removed the spots from the figure prior to publication. An erratum was never published!

Kaiser and Garrett, who worked at Bell Telephone Laboratories, Murray Hill, New Jersey, made the first experimental observation of two-photon excitation fluorescence in 1961. Their paper was published only a few weeks after the publication of Franken's paper, and they also used a ruby optical maser for their studies (Kaiser & Garrett, 1961). They generated blue fluorescent light around a wavelength of 4,250 Å by illuminating crystals of $CaF_2:Eu^{2+}$ with red light at 6,943 Å from a ruby optical maser. The authors state that the appearance of fluorescence indicated that the Eu^{+2} was in the excited state, and that the excited state was excited by a two-photon process. The authors show a log-log plot of fluorescent intensity versus incident intensity with a slope of two. They state the probability of an atom being excited by two photons, the sum of whose energies corresponds to the excited state of the atom. This optical process can be described by an imaginary third-order nonlinear susceptibility that was first treated in 1931 by Maria Göppert-Mayer. Maria Göppert-Mayer's theoretical paper was listed in the references. The authors calculated the efficiency for the two-photon excitation process under their experimental condition as 10^{-7}. The efficiency was calculated as the ratio of the number of fluorescence photons generated and the number of incident photons. The Kaiser and Garrett paper was published 30 years after two-photon absorption at optical frequencies was predicted by Maria Göppert-Mayer.

Another interesting nonlinear phenomenon observed at that time is optical mixing (Bloembergen, 1996; Boyd, 2003). In 1962 Bloembergen and coworkers published a classic paper on the theoretical basis of optical mixing (Armstrong et al., 1962). While Franken et al. (1961) showed that small quadratic terms in the optical polarizability of piezoelectric materials made possible the production of the optical second-harmonic when irradiated with intense red light from a ruby laser, these same terms would permit the mixing of light from two lasers at different frequencies. It was shown that two ruby laser beams of different frequencies that are coincident simultaneously on a crystal will produce the sum frequency in the near ultraviolet (Bass et al., 1962). The two frequencies were obtained from a ruby laser operating at liquid nitrogen temperature and a ruby laser operating at room temperature.

In 1962 Kleinman described the possibility of observing two-photon absorption with the intense light from a ruby optical maser (laser). Kleinman hypothesized that the transitions between initial and final states are connected with excited states. These transitions are electric dipole transitions with a total oscillator strength of approximately 1 (Kleinman, 1962a). His theory was an approximation of the general theory for these two-photon processes that was first developed by Marie Göppert-Mayer (1931) and cited as a reference in Kleinman's paper. Kleinman assumed that "charge transfer" bands in the crystal act as intermediate atomic states that are coupled through the electric moment to both the initial and the final states of the transition.

Higher-order multiphoton processes (for example, three-photon excitation fluorescence) were observed in the following years. Three-photon absorption processes were first reported in naphthalene crystals (Singh & Bradley, 1964). In the first report of a three-photon electronic transition, they used a ruby laser with a crystal of naphthalene and reported on the observed ultraviolet fluorescence. The intensity of the fluorescence signal increased as the third power of the laser intensity that is expected for this nonlinear process. Rentzepis and coworkers observed and photographed three-photon excited fluorescence in organic dye molecules (Rentzepis et al., 1970). Other spectroscopic studies using multiphoton processes followed (Birge, 1983, 1986; Friedrich, 1982; McClain, 1971; McClain & Harris, 1977).

The next step in the application of nonlinear optics was the development of a nonlinear optical microscope for the examination of the microscopic structure of polycrystalline ZnSe (Hellwarth & Christensen, 1974). They used a conventional harmonic microscope in which the entire specimen was wide-field illuminated with the laser beam, and the image of polycrystalline ZnSe was formed by imaging the emitted harmonic radiation. Their second-harmonic microscope used a repetitively Q-switched Nd:YAG laser giving 10^{+3} pulses per second at 1.06μ. Each pulse has 10^{-4} J energy and 2×10^{-7} second duration. The green second-harmonic signal was viewed by a microscope or on a Polaroid film. Their nonlinear second-harmonic generation microscope was used to observe single-crystal platelets that were not visible in an ordinary polarizing light microscope. Hellwarth and Christensen developed the theory for the second-harmonic generation in polycrystalline ZnSe. Their analysis indicated that phase-matched second-harmonic generation does not exist in a single-crystal ZnSe since it is isotropic; however, spatial variation of the second-harmonic nonlinear susceptibility (a third-rank tensor) in the polycrystalline ZnSe can create the necessary phase-matching that creates the second-harmonic light emission. The next major advance took place at Oxford University (UK).

1.5 THE SCANNING HARMONIC OPTICAL MICROSCOPE AT OXFORD UNIVERSITY

In 1977, the Oxford (UK) group of Sheppard, Kompfner, Gannaway, Choudhury, and Wilson first suggested and then demonstrated how a nonlinear optical phenomenon would be incorporated into their high-resolution, three-dimensionally resolved scanning laser microscope (Sheppard & Kompfner, 1978).

The principle of nonlinear scanning microscopy is simply explained in the *Theory and Practice of Scanning Optical Microscopy* (Wilson & Sheppard, 1984). If a high-intensity beam of light impinges on a specimen, the specimen would behave in a nonlinear manner and higher optical harmonics would be produced. This nonlinear harmonic generation would be a function of the molecular structure of the specimen. All materials possess third-order nonlinear susceptibility together with higher-order susceptibilities: $5, 7, 9, \ldots$ Even-order nonlinear susceptibility only exists in crystals that have non-centrosymmetric geometry; for example, crystals such as $LiNbO_3$ and some biological specimens.

The Oxford group suggested developing nonlinear scanning optical microscopes where the excitation light is focused to a small volume and an image is generated based on raster scanning of either the light or the specimen. Since the excitation light is focused to the diffraction limit, the electric field strength at the focal volume is significantly greater than wide-field illumination geometry of previous designs. The significantly higher field strength allows vastly more efficient generation of optical harmonics since the nth-order harmonic signal is generated in proportion to the nth-power of the fundamental intensity.

The Oxford group also pointed out an important consequence of the second-order dependence of the emission signal with the excitation light: super-resolution. If the fundamental radiation had a Gaussian distribution, the harmonic radiation will also have a Gaussian distribution, but the radius will be only $1/\sqrt{2}$ of that of the fundamental beam (Wilson & Sheppard, 1984). Axially, they further noted that this system will have depth discrimination because the intensity point-spread function of this microscope is a quadratic function of the intensity transfer function of the objective, similar to an incoherent confocal microscope.

Wilson and Sheppard further suggested that a laser microscope could be used to investigate the nonlinear optical properties of a specimen based on variations in the specimen's second-order susceptibility. In addition to harmonic generation, they further realized that other nonlinear effects, such as Raman scattering, sum frequency generation, or two-photon fluorescence, could be used to study the energy levels and hence the chemical structure of the specimen (Wilson & Sheppard, 1984).

Sheppard and his coworkers made the first conference presentation of a scanning second-harmonic generation microscopy in 1977 (Sheppard et al., 1977). They showed a schematic for a scanning harmonic microscope. They noted that the second harmonic is formed in the forward direction. The specimen was scanned relative to the focused laser beam and the focused beam produced optical second harmonics in the specimen itself. They proposed that a second-harmonic microscope could be used to image biological structures with a very high contrast. They stated the temperature rise must be small in biological samples. They calculated that 1 W of incident light may produce 10^{-10} W of second-harmonic light. To keep specimen temperature rise low, the laser beam and not the sample must be scanned. The authors built a specimen scanning second-harmonic microscope based on a 1-W CW neodymium YAG laser; while it was unsuitable for biological specimens, it was used to image crystals. They further illustrated the optical sectioning capability of their harmonic generation microscope by imaging various planes within a thin crystal.

Gannaway and Sheppard (1978) wrote the first full journal paper on scanning second-harmonic generation microscopy, with a discussion of the advantages of a pulsed beam and heating effects. This is an extended paper of their 1977 conference communication. They noted that second harmonics are generated only by molecules that are non-centrosymmetric, and the contrast in the microscope is a function of the molecular structure of the specimen and its orientation with respect to the laser beam. The authors stated their ultimate aim is to have a second harmonic microscope to examine biological specimens. They pointed out the advantage of scanning a focused spot in nonlinear microscopy, since a much lower laser power is necessary to achieve a given power density in the specimen. They also pointed out the advantage of using pulsed lasers to enhance the conversion of the fundamental to harmonic power in a scanning optical microscope. Their scanning laser microscope was a high-resolution system as compared to the previous microscope of Hellwarth and Christensen, which had a resolution in the range of 20 to 200 μm.

Later, the Oxford group proposed several microscope configurations for the enhancement of nonlinear interactions in the scanning optical microscope, such as placing the specimen in a resonant cavity and using beam pulsing (pulsed lasers) to improve the efficiency of conversion of fundamental to harmonic power (Sheppard & Kompfner, 1978). In the same paper, they proposed various types of nonlinear interactions for use in the scanning optical microscope: the generation of sum frequencies, Raman scattering, and two-photon fluorescence.

Thus, the development of the scanning laser harmonic microscope solved many of the problems of nonlinear microscopy: the advantages of diffraction-limited laser scanning, the analysis and mitigation of laser heating effects and subsequent thermal damage by laser scanning, the confinement of the second-harmonic generation to the focal volume of the microscope objective, and the origin of the optical sectioning effects based on the physics of the excitation process. The Oxford group also suggested that the efficiency of conversion of the fundamental to harmonic power would be increased by pulsing the laser (Masters, 1996).

1.6 THE ROLE OF LASERS IN THE DEVELOPMENT OF NONLINEAR MICROSCOPY

The invention of a new instrument is often an important antecedent in scientific discovery. See the appendix for a chronological list of major discoveries in laser science. This is obvious in the field of nonlinear optics. Following the development of the first laser in 1960 by Maiman, there were many discoveries, such as the second-harmonic generation of light (Franken et al., 1961) and the two-photon excited fluorescence (Kaiser & Garrett, 1961). The ruby laser was also used to demonstrate third-harmonic generation and anti-Stokes frequency mixing.

In 1961, the technique of Q-switching was used to obtain shorter laser pulses with higher peak intensities (Siegman, 1986). Laser pulses of picosecond duration were obtained by passive mode locking with a saturable dye cell. In fact, picosecond laser pulses have been used to induce two-photon absorption in organic molecules (Bradley et al., 1972).

In the 1970s it was possible to obtain femtosecond (1 fs is equal to 10^{-15} s) laser pulses based on a combination of saturable gain in a dye laser medium and a saturable dye absorber in a ring laser cavity mode together with compensation of the dispersion in group velocity (Schäfer & Müller, 1971). The early images from the Webb group using multiphoton excitation microscopy were obtained with a colliding-pulse 80-MHz mode-locked dye laser that generated 100-fs pulses at 630 nm (Valdemanis & Fork, 1986). The disadvantage of this laser is rather limited tunability and stability.

In the 1990s the development of the Ti-sapphire (Ti:Al$_2$O$_3$) femtosecond laser resulted in a laser source that was optimal for two-photon excitation microscopy (Spence et al., 1991). It is instructive to present a feel for the high intensities of ultrashort laser pulses and their extremely short durations (Dausinger & Stefan, 2004). A laser that outputs 100-fs pulses with 1 mm^2 area, with a pulse energy of 10 nJ, has an instantaneous intensity of 10^7 W/cm^2. To understand how short a 100-fs time interval is, it is necessary to realize that light will propagate only about 30 μm in this time interval; that is, less than the thickness of a human hair. In 1 second light propagates a distance equal to 7.5 times the Earth's circumference, which gives one an appreciation for the speed of light.

Initial designs of Ti:sapphire lasers are based on Kerr lens mode locking to generate output pulses with widths in the fs range. Kerr lens effect is due to the electric field generated by a strongly focused Gaussian laser beam causing an inhomogeneous change in the refractive index of the Ti:sapphire crystal. This change creates a weak lens in the crystal that results in a higher gain for mode-locked laser pulses than for continuous wave (cw) light. Ti:sapphire lasers offer a broad tunability that spans the range of 700 to 1,100 nm (Spense et al., 1991). The Ti:sapphire laser could be pumped with either a solid-state semiconductor or a high-power ion laser. The diode-pumped lasers form the basis of several commercial turnkey fs laser systems. Alternative laser sources providing broader spectral coverage can be obtained by coupling a Ti:sapphire laser with a frequency doubler/tripler, a regenerative amplifier (Salin et al., 1991), an optical parametric oscillator, or an optical parametric amplifier. The use of an optical parametric amplifier further provides a mean to obtain higher peak power at lower repetition rates.

A critical component of a multiphoton microscope is the laser light source (Masters et al., 1999). The fluorescence excitation rate is suboptimal for scanning microscopy applications unless fs pulsed lasers are used, although imaging using picosecond (ps) and continuous wave lasers has been demonstrated. For a pulsed laser the following parameters are important. The energy per pulse is the average power divided by the repetition rate. The peak power is approximately the energy per pulse divided by the pulse width. The duty cycle of the laser is given by the pulse width multiplied by the repetition rate.

Today, several ultrafast laser sources are available for multiphoton excitation microscopy. The important considerations in laser selection for multiphoton excitation microscopy include peak power, pulse width, repetition rate, range of wavelength tunability, cooling requirement, power requirement, and cost. The range of tunability should cover the multiphoton spectra of the fluorophores of interest.

Ti:sapphire laser systems are most commonly used for multiphoton excitation microscopy. These systems provide high average power (1–2 W), high repetition rate (80–100 MHz), and short pulse width (80–150 fs). The Ti:sapphire laser provides a wide tuning range from below 700 nm to above 1,000 nm. Ti:sapphire lasers require pump lasers. Pump lasers are typically solid state, accomplished by using an array of semiconductor lasers to excite optical materials such as neodymium:yttrium vanadate ($Nd:YVO_4$) crystals (Lamb et al., 1994; Spence et al., 1991).

Other single-wavelength solid-state systems are available for nonlinear microscopy, such as diode-pumped neodynamium-doped lithium yttrium fluoride (Nd:YLF) lasers, diode-pumped erbium-doped fiber laser systems, the chromium-doped lithium strontium aluminium fluoride (Cr:LiSAF) lasers, and the chromium-doped lithium calcium aluminium fluoride (Cr:LiCAF) lasers. The design and the application of some of these lasers have been reviewed by French et al. (1993).

The early work by Hellwarth and Christensen in 1974 on their second-harmonic microscope used a repetitively Q-switched neodymium yttrium aluminum garnate (Nd-YAG) laser. The early implementations of nonlinear optical microscopy at Oxford used a continuous-wave laser scanning microscope to image second-harmonic generation in crystals (Gannaway & Sheppard, 1978; Sheppard & Kompfner, 1978; Wilson & Sheppard, 1979). It was the development of the mode-locked laser that generated fs laser pulses at high repetition rate that provided sufficiently intense light sources for the next step in the development of the multiphoton excitation microscopy.

1.7 THE DENK, STRICKLER, AND WEBB *SCIENCE* REPORT (1990) AND
THEIR PATENT (1991)

It was the seminal work of Denk, Strickler, and Webb, which was published in *Science* in 1990, that launched a revolution in nonlinear optical microscopy (Denk et al., 1990). The patent application was filed on November 14, 1989. On July 23, 1991, a United States Patent on "two-photon laser microscopy" was assigned to the Cornell Research Foundation, Inc., Ithaca, New York (Denk et al., 1991). The patent lists Winfried Denk, James Strickler, and Watt W. Webb as the inventors. By integrating a laser scanning microscope (scanning mirrors, photomultiplier tube detection system) and a mode-locked laser that generates pulses of near-infrared light, they succeeded in demonstrating a new type of fluorescent microscope based on two-photon excitation of molecules. The pulses of red or near-infrared light (700 nm) were less than 100 fs in duration and the laser repetition rate was about 100 MHz. The patent states that "focused subpicosecond pulses of laser light" are used (Denk et al., 1991). These pulses have sufficiently high peak power to achieve two-photon excitation at reasonable rates at an average power less than 25 mW, which is tolerable to biological samples.

The high-intensity, short pulses of near-infrared light cause significant multiphoton excitations; however, the relative transparency of cell and tissue for infrared radiation and the lower average power minimizes photodamage. The benefits of two-photon excitation microscopy include improved background discrimination, reduced photobleaching of the fluorophores, and minimal photodamage to living specimens. The inventors proposed the application of two-photon excitation microscopy for optical sectioning in three dimensions and for uncaging of molecules inside of cells and tissues.

It is interesting to note the publications cited in the patent and their publication dates. Pages 8 and 9 of the Wilson and Sheppard book *Theory and Practice of Scanning Optical Microscopy* (1984) are cited. The two key points from these pages are (1) the discussion of the physical basis of the optical sectioning in a nonlinear second-harmonic microscope; the quadratic nature of the harmonic generation confines the effect to the focal region, and the concomitant sharpening of the beam in the lateral direction, and (2) the suggestion that other nonlinear optical effects, including Raman scattering or two-photon fluorescence, can be used for scanning laser microscopy. Many of the papers and patents that are relevant to multiphoton excitation microscopy have been reprinted in a book (Masters, 2003).

The patent can be summarized by the following sentence from the abstract, "A laser scanning microscope produces molecular excitation in a target material by [the] simultaneous adsorption of two photons to thereby provide intrinsic three-dimensional resolution." The patent also states, "the focused pulses also provide three-dimensional spatially resolved photochemistry which is particularly useful in photolytic release of caged effector molecules."

The patent gives the light source as "strongly focused subpicosecond pulses of laser light." The strong focusing occurs only in the focal region of the microscope objective and is similar to the origin of the optical sectioning in the second-harmonic microscope described by Wilson and Sheppard (Wilson and Sheppard, 1979). The laser scanning microscope described in the patent is similar to instruments described by others in prior articles (the patent cited 14 previous laser scanning instruments). The statement regarding subpicosecond opens a window for picosecond lasers to be used for two-photon laser microscopy without infringing on the Denk, Strickler, and Webb 1991 patent.

As early as 1972, picosecond lasers were used for two-photon absorption studies and measurement of absorption cross-sections (Bradley et al., 1972). In fact, in 1972 Bradley and coworkers measured the two-photon absorption cross-sections of Rhodamine 6G, Rhodamine B, Acridine Red, Disodium Fluorescein, and DODCI. For over eight orders of magnitude they demonstrated the intensity-squared dependence of the fluorescence using both Q-switched and mode-locked lasers.

In the section of the patent labeled "Background of the Invention," the authors review the various types of confocal microscopes, their light sources, and scanning mechanisms. The authors clearly state the limitations of confocal microscopy as applied to fluorescent molecules that are excited in the ultraviolet: (1) the lack of suitable microscope objectives for the ultraviolet that are chromatically corrected and transparent for both the absorption and emission wavelengths, (2) photodamage to living cells by the ultraviolet light, and (3) the problem of photobleaching of the fluorophores.

In the section labeled "Summary of the Invention," the authors propose that their invention overcomes these difficulties. The authors state that the two-photon excitation is made possible by (a) a very high, local, instantaneous intensity provided by the tight focusing of the laser scanning microscope in which the laser beam is focused to diffraction-limited waist of less than 1 micron, and (b) the temporal compression of the pulsed laser. This process yields improved background discrimination and reduced photobleaching, and minimized the photodamage to living specimens. The physics of the process is clearly described by the authors in the following sentence from their patent: "only in the region of the focal point on the object plane at the waist formed by the converging and diverging cones is the intensity sufficiently high to produce two-photon absorption in the specimen fluorophore, and this intensity dependence enables long wavelength light to provide the effect of short wavelength excitation only in the small local volume of the specimen surrounding the focal point."

The patent further provides a formula for the number of photons absorbed per molecules of fluorophore per pulse as a function of pulse duration, repetition rate, average power of the incident laser, the numerical aperture of the focusing lens, and the photon absorption cross-section. In a two-photon excitation process, the number of photons absorbed per molecule of fluorophore per pulse scales with the average incident laser power squared. This is the source of the experimental verification of two-photon excitation processes; on a log-log plot of detected intensity versus laser power, the slope of the plot is two. The authors also state that the two-photon excitation fluorescence can be increased by increasing the pulse repetition frequency until saturation of the excited state is achieved.

Another key feature of the patent is the description of a non-descanned detection of the fluorescence intensity derived from the two-photon absorption process. Since the fluorescence signal depends on the square of the excitation intensity, there is an optical sectioning effect through the specimen even in the absence of a pinhole being used as a spatial filter in front of the detector. Therefore, the detector can be a large-area detector such as a photomultiplier tube. This avoids many of the problems associated with conventional laser scanning confocal microscopes. With the publication of the

Science report and the patent from the Webb group in 1991, the reality of two-photon excitation microscopy began.

Webb and coworkers further stated that the microscope can be operated in sum or difference frequency mode. It is not necessary that the two photons that are absorbed in a two-photon excitation process be of the same wavelength. An excitation transition requiring energy, $\frac{hc}{\lambda_{ab}}$, can be achieved using lasers with wavelengths, λ_1 and λ_2, based on the following conservative equation:

$$\frac{1}{\lambda_{ab}} = \frac{1}{\lambda_1} + \frac{1}{\lambda_2}$$

where h and c are Planck's constant and the speed of light, λ_{ab} is the short wavelength of the absorber, and λ_1 and λ_2 are the laser incident beam wavelengths (Webb patent, 1991).

1.8 COMPARISON OF MULTIPHOTON EXCITATION MICROSCOPY AND CONFOCAL MICROSCOPY

The advantages of these nonlinear microscopes include improved spatial and temporal resolution without the use of pinholes or slits for spatial filtering. Since the optical sectioning capability of the two-photon excitation microscope derives from the physics of the excitation process, there is no requirement for a spatial filter in front of the detector as in confocal microscopy (Corle & Kino, 1996). Two-photon microscopy further allows deeper penetration into thick, highly scattering tissues and confines photobleaching and photodamage to the focal volume.

The main limitations of two-photon excitation microscopy are (1) multiphoton excitation microscopy is suitable only for fluorescent imaging; reflected light imaging is not possible, and (2) the technique is not suitable for imaging highly pigmented cells and tissues that absorb near-infrared light.

One can further compare these two techniques based on spatial resolution. Just after the Denk et al. *Science* report was published, the image formation properties in two-photon fluorescence microscopy were compared with confocal microscopy (Sheppard & Gu, 1990). These authors compared both transverse and axial resolution for both types of microscopy and concluded that confocal microscopy has a higher spatial resolution than multiphoton excitation microscopy. Further analysis and comparison of the transfer functions for a one-photon and a two-photon process are in agreement with the previous analysis (Gu, 1996; Gu & Sheppard, 1995).

Experimental comparisons of these two types of microscopies have been performed, including a study in which we imaged in vivo human skin with both multiphoton excitation microscopy and tandem scanning reflected light confocal microscopy (Masters & So, 1999; Masters et al., 1997). These studies demonstrate that multiphoton excitation microscopy has the capacity to image deeper within highly scattering tissues such as in vivo human skin. For observation of the dermis, the ability of the multiphoton excitation microscope to image the elastin fibers and the collagen fibers within the tissue is a significant advantage.

1.9 MODERN DEVELOPMENTS IN NONLINEAR MICROSCOPY AND
THEIR IMPACT

Major technological developments of instrumentation for nonlinear microscopy include the following:

1. Most of the commercial fs lasers are based on Ti:sapphire laser sources. A recent advance is the use of solid-state pump lasers such as the diode pumped neodymium:yttrium vanadate (Nd:YVO$_4$) laser (French, 1995; Kleinbauer et al., 2004a, 2004b; Silfvast, 2004). The commercial availability of these lasers provides the user with less expensive, more compact, and more reliable laser sources for nonlinear microscopy. The nonlinear microscope can be placed on a small vibration-isolation table and is becoming more portable because of these compact laser sources. Another advance is the development of fiber lasers. Wokosin and coworkers use an additive-pulse mode-locked neodymium-doped lithium yttrium fluoride (Nd:YLF) laser that generates ps pulses at 1.05 μm. These pulses are compressed to 150 fs outside of the laser (Wokosin et al., 1996). Alternatively, lasers based on Cr:fosterite crystal output fs pulses in the wavelength of 1,150 to 1,360 nm (Yanovsky et al., 1993). The fosterite laser is based on Cr^{+4} ions that are doped in a Mg$_2$SiO$_4$ crystal. With the use of second-harmonic generation these lasers output in the wavelength of 575 to 680 nm, which is an important range of absorption wavelengths for many fluorescent probes. The development of the extended cavity fs chromium-doped lithium strontium aluminium fluoride (Cr:LiSAF) laser pumped by inexpensive single spatial mode diodes further offers the performance of Ti:sapphire lasers at a much reduced cost (Prasankumar et al., 2003).

    Another major laser technology development that significantly affects the nonlinear microscopy field is the availability of fully automated, maintenance-free systems. These systems allow automated wavelength tuning, power and pulse width regulation, and good pointing stability. These fs laser systems can be operated by users with little or no laser tuning experience and allow nonlinear microscopy to be conducted in biological laboratories and clinical environments with no need of laser physicists.

2. The availability of new detectors with high sensitivity such as avalanche photodiodes and photomultiplier tubes has significantly reduced the inherent difficulties in performing nonlinear microscopy where the signal level is often low. New types of spatially resolved high-sensitivity photodetectors, such as multi-anode photomultiplier tubes, further facilitate the development of spectrally resolved nonlinear microscopy and higher-sensitivity multifocal multiphoton microscopes. There are now available CCD cameras with on-chip amplification that are useful two-dimensional detectors.

3. New types of fluorescent probes, including synthetic molecules with very high two-photon absorption cross-sections, numerous types of fluorescent molecules that can be genetically expressed in cells, and aggregates of inorganic compounds that show narrow band fluorescence over a wide range of wavelengths (quantum dots). One major advance is the application of

various green fluorescent proteins that are extremely useful reporters of gene expression. A new class of voltage-sensing dyes that utilizes second-harmonic excitation processes is becoming available, allowing high-speed sensing with good voltage sensitivity. These dyes will have utility in the field of cell signaling and neurobiology. A limitation of fluorescent probes is that they eventually fade due to photobleaching. A new class of fluorescent probes with photostability that is several orders of magnitude compared to conventional dyes has been developed and shows great potential for imaging live cells, for in vivo imaging, and for diagnostic techniques. These fluorescent probes are semiconductor nanocrystals and are called quantum dots or qdots (Michalet et al., 2005).

4. New techniques to modify the point-spread functions to increase spatial resolution have been developed (Hell, 1997). Stimulated emission depletion (STED) fluorescence microscopy provides a new technique to increase the spatial resolution of fluorescence microscopy (Klar et al., 2000). The idea is to nonlinearly quench the fluorescence of excited molecules based on stimulated emission in the periphery of the focal spot. The result is that the emission from the edge of the focal spot is quenched and the fluorescence of the sample is only detected from the central region of the focal spot with enhanced radial and axial spatial resolution.

5. New image contrast mechanisms based on novel fluorescence spectroscopic assays such as fluorescence correlation spectroscopy, fluorescence resonance energy transfer, fluorescence lifetime imaging, and fluorescence photobleaching recovery have resulted in new developments of cell and developmental biology (Xu & Webb, 1997). Other nonfluorescence-based nonlinear optical techniques that are also making a major impact include second-harmonic generation (SHG), third-harmonic generation (THG), and coherent anti-Stokes Raman scattering (CARS). Second-harmonic generation and third-harmonic generation microscopy is well suited to image structures in cells and tissues with a high degree of molecular order such as membrane interface, myosin crosslinks in muscles, and collagen fibers. Second-harmonic generation further provides specificity to non-centrosymmetric structures with chirality. CARS imaging has been implemented into microscopes and provides new molecular specificity that was previously unavailable (Cheng et al., 2001; Labarthet & Shen, 2003; Müller & Brakenhoff, 2003; Potma & Xie, 2004; Sun, 2003; Sun et al., 2004). Because CARS is resonantly enhanced at the vibrational levels of the molecules, this technique not only provides image contrast but also allows fingerprinting of specific molecules based on their chemical groups.

6. New high-speed nonlinear microscope systems that can acquire images at video rate and faster are becoming available (Brakenhoff et al., 1996; Buist et al., 1998; Fan et al., 1999; Kim et al., 1999; Nielsen et al., 2001; Oron et al., 2005), These high-speed imaging systems permit data acquisition at sufficient speed such that physiological motion of live specimens will not degrade the images. High-speed image acquisition is also important to study rapid transient events such as calcium signals and waves in excited cells and tissues. Finally, high-speed imaging further enables image cytometry measurements in tissues. Rapid data acquisition is limited by the number of photons available from the source at each pixel and by the saturation of the fluorescent

molecules. Photodamage and phototoxicity also place limits on the laser power delivered to the specimen. These limitations can be partially circumvented by parallelized data acquisition, as in multifocal multiphoton microscopes. Recently, Oron and coworkers (2005) developed a two-photon excitation microscope that depends on temporal focusing of the illumination pulse to achieve full-frame depth-resolved imaging without scanning. For fluorescent probes with very low quantum yields, this technique may not be useful. This technique could result in less expensive, more compact, and simpler nonlinear microscopes.

1.10 MAJOR BIOMEDICAL APPLICATIONS OF NONLINEAR MICROSCOPY

With the publication in 1990 of the *Science* report from the Webb group, the biological community was convinced of the capability of multiphoton excitation microscopy for the imaging of live cells, tissues, embryos, and organisms. The application of multiphoton excitation microscopy has resulted in many important experimental studies in biology and medicine. A few of these studies are described below by way of illustration.

A major biological and medical application of two-photon excitation microscopy is in the field of deep tissue imaging and spectroscopy. The enhanced confinement of the focal volume, the deeper penetration of the near-infrared radiation, and the reduced phototoxicity of the pulsed infrared radiation are all exploited to provide in vivo imaging of human skin (Masters et al., 1997; Masters et al., 1999; Ragan et al., 2003; So et al., 2000). The potential diagnostic impact of these techniques is in noninvasive optical biopsy of in vivo human skin as well as investigations of basal cell differentiation and proliferation; both of these are active research areas in basic and clinical dermatology. Multiphoton excitation microscopy and confocal microscopy of in vivo human skin have been compared and the deeper penetration of the nonlinear microscopy has been demonstrated (Masters & So, 1999).

Neurobiology is another field that has actively applied multiphoton excitation microscopy. An important area of research is that of neural plasticity, or how the neural system grows, changes, develops, and responds to sensory experience. An interesting application of multiphoton excitation microscopy is a study of long-term in vivo imaging of synaptic plasticity in adult cortex (Trachtenberg et al., 2002; Tsai et al., 2003). These investigators found that sensory experience drives the formation and the elimination of synapses and that these alterations may underlie adaptive remodeling of neural circuits.

The field of embryology and developmental biology is benefiting from studies that use multiphoton excitation microscopy to investigate the embryological development with less photodamage and more spatial resolution than that of previous confocal techniques. A good illustration of this application is a study on the orientation of the first cleavage in the sea urchin embryo (Summers et al., 1996). A major challenge to the study of developing embryos is to achieve long-term microscopic observation without photo-induced damage. A study that demonstrated this with mammalian embryos showed that multiphoton excitation microscopy, but not confocal microscopy, could be used to monitor embryos at frequent intervals over a 24-hour period (Squirrell et al., 1999).

A new and important application for multiphoton excitation microscopy is in the investigation of in vivo tumor biology. In vivo measurement of gene expression, angiogenesis, and physiological function in tumors using multiphoton laser scanning microscopy is an important emerging research area (Brown et al., 2001). The deeper penetration of the tumor with nonlinear microscopy as compared to confocal microscopy provides three-dimensional imaging of gene expression and function deep within the tumor, and these studies could result in new types of therapeutics for tumor biology.

In the area of cellular oxidative metabolism, two-photon excitation laser scanning microscopy was used to investigate alterations in the NAD(P)H levels in the basal cells of the in situ cornea (Piston, Masters, Webb, 1995). The two-photon excitation microscopy technique resulted in improved sensitivity and spatial resolution as compared to previous confocal techniques that used ultraviolet excitation (Masters & Chance, 1999).

All of these new techniques of nonlinear microscopy have the potential to damage live cells and tissues, and new techniques may be limited by the need to minimize cell damage from the laser sources. However, there are also opportunities, such as using nonlinear excitation for photodynamic therapy. In the technique of two-photon excitation imaging of in vivo skin, a new method has been developed based on a pulse picker that mitigates thermal mechanical damage during skin imaging (Masters et al., 2004).

1.11 CONCLUSION

The publication in 1931 of Maria Göppert-Mayer's doctoral dissertation on the theory of two-photon quantum transitions (two-photon absorption and emission) in atoms was the theoretical foundation of the chain of scientific and technology developments that led to multiphoton microscopy. The publications in the past decade in biology and medicine that are based on the multiphoton excitation microscope demonstrate well its capabilities, and they are a fitting tribute to the doctoral work of Maria Göppert-Mayer and to all the other scientists who have contributed to the field of nonlinear optical microscopy (Masters & So, 2004).

ACKNOWLEDGMENTS

Barry R. Masters is thankful for the support from the 1999 Alfred Vogt Prize for Ophthalmology (the highest award in Switzerland for scientific research in ophthalmology) from the Alfred Vogt-Stiftung zur Förderung der Augenheilkunde Zürich, which he shared with Professor M. Böhnke for their work "Confocal Microscopy of the Cornea."

NOTE

This chapter is based on an article by Barry R. Masters and Peter T. C. So in *Microscopy and Research Techniques*, 2004.

REFERENCES

Armstrong, J. A., Bloembergen, N., Ducuing, J., Pershan, P. S. 1962. Interactions between light waves in a nonlinear dielectric. *Phys. Rev.* 127:1918–1939.

Bass, M., Franken, P. A., Hill, A. E., Peters, C. W., Weinreich, G. 1962. Optical Mixing. *Phys. Rev. Lett.* 8:18.

Birge, R. R. 1983. One-photon and two-photon excitation spectroscopy. In: *Ultrasensitive laser spectroscopy*, (ed. D. S. Kliger) pp. 109–174, New York: Academic Press.

Birge, R. R. 1986. Two-photon spectroscopy of protein-bound fluorophores. *Acc. Chem. Res.* 19:138–146.

Bloembergen, N. 1965. *Nonlinear Optics*, New York: Benjamin.

Bloembergen, N. 1996. Nonlinear optics: a historical perspective. In: *Encounters in Nonlinear Optics, Selected Papers of Nicolaas Bloembergen*. Singapore: World Scientific, 122–133.

Bloembergen, N. 2000 Nonlinear optics: past, present, and future. *IEEE Journal on Selected Topics in Quantum Electronics* 6(6):876–880.

Born, M. 1969. *Atomic Physics*, New York: Dover Publications.

Boyd, R. W. 2003. *Nonlinear Optics*, second edition, New York: Academic Press.

Bradley, D. J., Hutchinson, M. H. R., Koetser, H. 1972. Interactions of picosecond laser pulses with organic molecules. Two-photon absorption cross-sections. *Proc. R. Soc. Lond. A.* 329:105–119.

Brakenhoff, G. J., Squier, J., Norris, T., Bilton, A.C., Wade, M. H., Athey, B. 1996. Real-time two-photon confocal microscopy using a femtosecond, amplified Ti:sapphire system. *J. Microscopy*, 181(3): 253–259.

Brown, E. B., Campbell, R. B., Tsuzuki, Y., Xu, L., Carmeliet, P., Fukumura, D., Jain, R. 2001. In vivo measurement of gene expression, angiogenesis and physiological function in tumors using multiphoton laser scanning microscopy. *Nature Medicine*, 7(7): 864–868.

Buist, A. H., Müller, M., Squier, J., Brakenhoff, G. J. 1998. Real time two-photon absorption microscopy using multipoint excitation. *J. Microscopy*, 192(2): 217–226.

Campagnola, P. J., Clark, H. A., Mohler, W. A., Lewis, A., Loew, L. M. 2001. Second-harmonic imaging microscopy of living cells, *J. Biomed. Opt.* 6: 277–286.

Campagnola, P. J., Wei, M.-de, Lewis, A., Loew, L. M. 1999. High resolution nonlinear optical imaging of live cells by second harmonic generation. *Biophysical Journal*, 77: 3341–3349.

Cheng, J-X, Volkmer, A., Book, L. D., Xie, X. S. 2001. An epi-detected coherent anti-Stokes Raman scattering (E-CARS) microscope with high spectral resolution and high sensitivity. *Journal of Physical Chemistry B*, 105(7): 1277–1280.

Corle, T. R. Kino, G. S.1996. *Confocal Scanning Optical Microscopy and Related Imaging Systems*, San Diego: Academic Press.

Dalitz, R. H. 1995. *The Collected Works of P.A. M. Dirac 1924–1948*, Cambridge, UK: Cambridge University Press.

Dausinger, F., Nolte, S. 2004. Introduction to femtosecond technology. In: *Femtosecond technology for technical and medical applications*. Eds. F. Dausinger, F. Lichtner, H. Lubatschowski, pp. 1–5, Berlin: Springer-Verlag.

Denk, W., Strickler, J. H., Webb, W. W. 1990. Two-photon laser scanning fluorescence microscopy, *Science*, 248:73–76.

Denk, W. J., Strickler, J. P., and Webb, W. W. 1991. Two-photon laser microscopy. United States Patent 5,034,613, July 23, 1991.

Diaspro, A. 2002. *Confocal and Two-Photon Microscopy: Foundations, Applications, and Advances*, New York: Wiley-Liss.

Dirac, P. A. M. 1958. *The Principles of Quantum Mechanics*, fourth edition, London: Oxford University Press.

Duck, I., Sudarshan, E. C. G. 2000. *100 years of Planck's Quantum*, Singapore: World Scientific.

Fan, G. Y., Fujisaki, H., Miyawaki, A., Tsay, R. -K., Tsien, R. Y., Ellisman, M. H. 1999. Video-rate scanning two-photon excitation fluorescence microscopy and ratio imaging with cameleons. *Biophysics J*. 76: 2412–2420.

Franken, P. A., Hill, A. E., Peters, C. W., Weinreich, G. 1961. Generation of optical harmonics, *Phys. Rev. Lett.* 7:118–119.

French, P. M. W. 1995. The generation of ultrashort laser pulses. *Rep. Prog. Phys.* 58: 169–267.

French, P. M. W., Mellish, R., Taylor, J. R., Delfyett, P. J., Florez, L. T. 1993. All-solid-state diode-pumped mode-locked Cr : LiSAF laser. *Electronics Letters*, 29:1263–1264.

Freund, I., Deutsch, M., Sprecher, A. 1986. Connective tissue polarity. *Biophysical Journal*, 50: 693–712.

Friedrich, D. M. 1982. Two-photon molecular spectroscopy. *J. Chem. Educ.* 59:472–483.

Friedrich, D. M., McClain, W. M. 198. Two-photon molecular electronic spectroscopy. *Annu. Rev. Phys. Chem.* 31:559–577.

Gannaway, J. N., Sheppard, C. J. R. 1978. Second harmonic imaging in the scanning optical microscope. *Optics and Quantum Electronics*, 10:435–439.

Göppert, M. 1929. Über die Wahrscheinlichkeit des Zusamenswirkens zweier Lichtquanten in einem Elementarakt. *Die Naturwissenschaften*, 17:932.

Göppert-Mayer, M. 1931. Über Elementarakte mit zwei Quantensprüngen, *Ann. Phys. (Leipzig)*, 9:273–294.

Greenspan, N. T. 2005. *The End of the Certain World: the life and science of Max Born: the Nobel physicist who ignited the quantum revolution.* Cambridge, MA: Basic Books.

Gu, M. 1996. *Principles of Three-Dimensional Imaging in Confocal Microscopes*, Singapore: World Scientific, 164–173.

Gu, M., and Sheppard, C. J. R. 1995. Comparison of three-dimensional imaging properties between 2-photon and single-photon fluorescence microscopy. *J. Micros.* 177:128–137.

Guo, Y., Ho, P. P., Tirksliunas, A., Liu, F., Alfano, R. R. 1996. Optical harmonic generation from animal tissues by the use of picosecond and femtosecond laser pulses. *Applied Optics*, 35:6810–6813.

Heisenberg, W. 1949. *The Physical Principles of the Quantum Theory*, New York: Dover Publications.

Heitler, W. 1954. *The quantum theory of radiation*. London: Oxford University Press. Reprinted in 1983 by Dover Publications, New York.

Hell, S. W. Increasing the resolution of far-field fluorescence light microscopy by point-spread-function engineering. In: *Topics in fluorescence Spectroscopy, volume 5, Nonlinear and Two-Photon-Induced Fluorescence*, Ed. J. Lakowicz, New York: Plenum Press, 361–426, 1997.

Hellwarth, R., Christensen, P. 1974. Nonlinear optical microscope examination of structures in polycrystalline ZnSe, *Opt. Commun.* 12:318–322.

Hunt, J. H. 2000. *Selected Papers on Nonlinear Optical Spectroscopy*, MS 160, Bellingham, WA., SPIE, The International Society for Optical Engineering.

Jammer, M. 1996. *The Conceptual Development of Quantum Mechanics*, New York: McGraw-Hill.

Kaiser, W., Garrett, C. G. B. 1961. Two-photon excitation in $CaF_2:Eu^{2+}$. *Phys. Rev. Lett.* 7:229–231.

Kim, K-H., Buehler, C., So, P. T., C. 1999. High-speed, two-photon scanning microscope. *Applied Optics*, 38(28): 6004–6009.

Klar, T. A., Jakobs, S., Dyba, M., Egner, A., Hell. S. W. 2000. Fluorescence microscopy with diffraction resolution barrier broken by stimulated emission. *Proc. Natl. Acad. Sci. USA*, 97: 8206–8210.

Kleinbauer, J., Knappe, R., Wallenstein, R. 2004a. Principles of ultrashort pulse generation. In: *Femtosecond technology for technical and medical applications.* Eds. F. Dausinger, F. Lichtner, H. Lubatschowski, pp. 9–16, Berlin: Springer-Verlag.

Kleinbauer, J., Knappe, R., Wallenstein, R. 2004b. Ultrafast $Nd:YVO_4$ and Yb:YAG bulk lasers and amplifiers. In: *Femtosecond technology for technical and medical applications.* Eds. F. Dausinger, F. Lichtner, H. Lubatschowski, pp. 17–33, Berlin, Springer-Verlag.

Kleinman, D. A. 1962a. Laser and Two-Photon Processes. *Physical Review*, 125(1): 87–88.

Kleinman, D. A. 1962b. Nonlinear dielectric polarization in optical media, *Phys. Rev.* 126: 1977–1979.

Kuhn, B., Fromhertz, P., Denk, W. 2004. High sensitivity of start-shift voltage-sensing dyes by one- or two-photon excitation near the red spectral edge. *Biophysical Journal*, 87(1): 631–639.

Lamb, K., Spence, D., Hong, J., Yelland, C., Sibbert, W. 1994. All-solid-state self-mode-locked Ti:sapphire laser. *Opt. Lett.* 19:1864–1866.

Labarthet, F. L., Shen, Y. R. 2003. Nonlinear optical microscopy. In: *Optical Imaging and Microscopy, Techniques and Advanced Systems*, Eds. P. Török, Fu-Jen Kao, pp. 169–196, Heidelberg, Springer-Verlag.

Loudon, R. 2000. *The Quantum theory of Light, third edition*, Oxford, Oxford University Press.

Maiman, T. H. 1960. Stimulated optical radiation in ruby. *Nature* 187:493–494.

Maiti, S., Shear, J. B., Williams, R. M., Zipfel, W., Webb, W. W. 1997. Measuring serotonin distribution in live cells with three-photon excitation. *Science*, 275:530–532.

Masters, B. R. 1996. *Selected Papers on Confocal Microscopy*, Milestone Series MS 131, Bellingham, WA: SPIE Optical Engineering Press.

Masters, B. R. 2000. The scientific life of Maria Göppert-Mayer. *Optics and Photonics News*, Optical Society of America, Washington, DC, 11:38–41.

Masters, B. R. 2003. *Selected Papers on Multiphoton Excitation Microscopy*, Milestone Series MS 175, Bellingham, WA: SPIE Optical Engineering Press.

Masters, B. R., Chance, B. 1999. Redox Confocal Imaging: Intrinsic Fluorescent Probes of Cellular Metabolism, in: *Fluorescent and Luminescent Probes for Biological Activity, Second Edition*, Ed. W. T. Mason, London, Academic Press.

Masters, B. R., So, P. T. C. 1999. Multi-Photon Excitation Microscopy and Confocal Microscopy Imaging of In Vivo Human Skin: A Comparison. *Microscopy and Microanalysis*, 5: 282–289.

Masters, B. R. and P. T. So. 2004. Antecedents of two-photon excitation laser scanning microscopy. *Microsc Res Tech*, 63(1):3–11.

Masters, B. R., and So, T. P. C. 2008. *Handbook of Biomedical Nonlinear Optical Microscopy*, New York: Oxford University Press.

Masters, B. R., So, P. T. C., Buehler, C., Barry, N., Sutin, J. D., Mantulin, W. W., Gratton, E. 2004. Mitigating thermal mechanical damage potential during two-photon dermal imagng. *Journal of Biomedical Optics*, 9(6): 1265–1270.

Masters, B. R., So, P. T. C., Gratton, E. 1997. Multiphoton excitation fluorescence microscopy and spectroscopy of in vivo human skin, *Biophysical J.*, 72:2405–2412.

Masters, B. R., So, P. T. C., Kim, K. H., Buehler, C., Gratton, E. 1999. Multiphoton excitation microscopy, confocal microscopy, and spectroscopy of living cells and tissues; functional metabolic imaging of human skin in vivo. In: *Methods in Enzymology, vol. 307, Confocal Microscopy* (Ed. P. M. Conn), pp. 513–536, San Diego: Academic Press.

McClain, W. M. 1971. Excited state symmetry assignment through polarized two-photon absorption studies in fluids. *J. Chem. Phys.* 55: 2789–2796.

McClain, W. M., Harris, R. A. (1977). Two-photon molecular spectroscopy in liquids and gases, In: *Excited States* (Ed. E. C. Lim), pp. 1–56, New York: Academic Press.

McGrayne, S. B. 1996. *Nobel Prize Women in Science: Their Lives, Struggles, and Momentous Discoveries*. New York: Birch Lane Press, pp. 175–200.

Medawar, J., Pyke, D. 2001. *Hitler's Gift: the true story of scientists expelled by the Nazi regime*. New York: Arcade Publishing, Inc.

Mehra, J., Rechenberg, H. 2001. *The Historical Development of Quantum Theory*. New York: Springer-Verlag.

Michalet, X., Pinaud, F. F., Bentolila, L. A., Tsay, J. M., Doose, S., Li, J. J., Sundaresan, G., Wu, A. M., Gambhir, S. S., Weiss, S. 2005. Quantum dots for live cells, in vivo imaging, and diagnostics. *Science*, 307: 538–544.

Millard, A. C., Wiseman, P. W., Fittinghoff, D. N., Wilson, K. R., Squier, J.A., Müller, M. 1999. Third-harmonic generation microscopy by use of a compact, femtosecond fiber laser source. *Applied Optics*, 38(36): 7393–7397.

Moreaux, L., Sandre, O., Mertz, J. 2000. Membrane imaging by second-harmonic generation microscopy, *J. Opt. Soc. Am. B*, 17(10): 1685–1694.

Müller, M., Brakenhoff, G. J. 2003. Parametric nonlinear optical techniques in microscopy, In: *Optical Imaging and Microscopy, Techniques and Advanced Systems*, Eds. P. Török, Fu-Jen Kao, pp. 197–217, Heidelburg: Springer-Verlag.

Nielsen, T., Fricke, M., Hellweg, D., Andressen, P. 2001. High efficiency beam splitter for multifocal multiphoton microscopy. *J. Microscopy*, 201(3): 368–376.

Oron, D., Tal, E., Silberberg, Y., 2005. Scanningless depth-resolved microscopy. *Optics Express*, 13(5): 1468–1476.

Oron, D., Yelin, D., Tal, E., Raz, S., Fachima, R., Silberberg, Y. 2004. Depth-resolved structural imaging by third-harmonic generation microscopy. *Journal of Structural Biology*, 147: 3–11.

Pauli, W. 2000. *Wave mechanics*, New York: Dover Publications, 2000.

Peebles, P. J. E. 1992. *Quantum Mechanics*, Princeton, New Jersey: Princeton University Press, pp. 259–322.

Piston, D. W., Masters, B. R., Webb, W. W. 1995. Three-dimensionally resolved NAD(P)H cellular metabolic redox imaging of the in situ cornea with two-photon excitation laser scanning microscopy. *J. Micros.* 178: 20–27.

Pohl, R. W. 1935. *Einführung in die Physik*, Berlin: Verlag von Julius Springer.

Potma, E. O., Xie, X. S. 2004. CARS microscopy for biology and medicine. *Optics and Photonics News*, pp. 40–45.

Prasankumar, R. P., Hirakawa, Y., Kowalevicz Jr., A. M., Kaertner, F. X., Fujimoto, J.G. 2003. An extended-cavity femtosecond Cr:LiSAF laser pumped by low-cost diode lasers. *Optics Express*, 11(11): 1265–1269.

Ragan, T. M., Huang, H., So, P. T. C. 2003. In Vivo and ex Vivo Tissue Applications of Two-Photon Microscopy. In: *Methods in Enzymology, Biophotonics*, Part B, Eds. G. Marriott, I. Parker, 361: 481–505. San Diego: Academic Press.

Rentzepis, P. M., Mitschele, C. J., Saxman, A. C. 1970. Measurement of ultrashort laser pulses by three-photon fluorescence. *Appl Phys. Lett.* 17:122–124.

Rupke, N. 2002. *Göttingen and the Development of the Natural Sciences*, Göttingen: Wallstein Verlag.

Sachs, R. G. 1982. Maria Goeppert Mayer—two-fold pioneer, *Physics Today*, 35(2): 46–51.

Salin, F., Squire, J., Mourou, G., Vaillantcourt, G. 1991. Multikilohertz Ti:Al$_2$O$_3$ amplifier for high-power femtosecond pulses. *Opt. Lett.* 16: 1964–1966.

Schäfer, F. P., Müller, H. 1971. Tunable dye ring-laser. *Opt. Commun.* 2: 407–409.

Shen, Y. R. 1984. *The Principles of Nonlinear Optics*, New York: John Wiley & Sons.

Sheppard, C. J. R., Gu, M. 1990. Image formation in two-photon fluorescence microscopy. *Optik* 86: 104–106.

Sheppard, C. J. R., Kompfner, R. 1978. Resonant scanning optical microscope. *Appl. Opt.* 17: 2879–2882.

Sheppard, C. J. R., Kompfner, R., Gannaway, J., Walsh, D. 1977. The scanning harmonic optical microscope. Presented at *IEEE/OSA Conf. Laser Eng. Appl.*; Washington, D.C., Vol. QE13, p. 100D.

Siegman, A. E.1986. *Lasers*, pp. 1004–1040, Mill Valley, California: University Science Books.

Silfvast, W. T. 2004. *Laser Fundamentals, Second Edition*, Cambridge, UK: Cambridge University Press.

Sime, R. L. 2006. The Politics of Memory: Otto Hahn and the Third Reich. *Physics in Perspective*, 8: 3–51.

Singh, S., Bradley, L. T. 1964. Three-photon absorption in naphthalene crystals by laser excitation. *Phys. Rev. Lett.*, 12: 612–614.

So, T. C., Dong, C. Y., Masters, B. R., Berland, K. M. 2000. Two-photon excitation fluorescence microscopy. *Annu. Rev. Biomed. Eng.* 2: 399–429.

Spence, D. E., Kean, P. N., Sibbert, W. 1991. 60-fsec pulse generation from a self-mode-locked Ti:sapphire laser. *Opt. Lett.*, 16: 42–44.

Squirrell, J. M., Wokosin, D. L., White, J. G., Bavister, B. D. 1999. Long-term two-photon fluorescence imaging of mammalian embryos without compromising viability. *Nature Biotechnology*, 17: 763–767.

Stoller, P., Celliers, P. M., Reiser, K. M., Rubenchik, A. M. 2003. Quantitative second-harmonic generation microscopy in collagen. *Applied Optics*, 42(25): 5209–5219.

Summers, R. G., Piston, D. W., Harris, K. M. and Morrill, J. B. 1996. The orientation of first cleavage in the sea urchin embryo, *Lytechinus variegatus*, does not specify the axes of bilateral symmetry. *Developmental Biology*, 175: 177–183.

Sun, C-K (2003). Second-harmonic generation microscopy versus third-harmonic generation microscopy in biological tissues. In: *Optical Imaging and Microscopy, Techniques and Advanced Systems*, Eds. P. Török, Fu-Jen Kao, pp. 219–232, Heidelberg, Springer-Verlag.

Sun, C-K, Chu, S-W, Chen, S-Y, Tsai, T-H, Liu, T-M, Lin, C-Y, Tsai, H-J. 2004. Higher harmonic generation microscopy for developmental biology. *Journal of Structural Biology*, 147: 19–30.

Trachtenberg, J. T., Chen, B. E., Knott, G. W., Feng, G., Sanes, J. R., Welker, E., Svoboda, K. 2002. Long-term *in vivo* imaging of experience-dependent synaptic plasticity in adult cortex. *Nature*, 420: 788–794.

Tsai, P. S., Nishimura, N., Yoder, E. J., Dolnick, E. M., White, G. A., Kleinfeld, D. 2002. Principles, Design, and Construction of a Two-Photon Laser Scanning Microscope for In Vitro and In Vivo Brain Imaging, In: *In Vivo Optical Imaging of Brain Function*, Ed. R. D. Frostig, Boca Raton, CRC Press.

Valdemanis, J. A., Fork, R. L. 1986. Design considerations for a femtosecond pulse laser: balancing self phase modulation, group velocity dispersion, saturable absorption, and saturable gain. *IEEE J. Quantum Electron QE-22*: 112–118.

Van der Waerden, B. L. 1967. *Sources of Quantum Mechanics*, Dover Publications, New York.

Weinreich, M. 1999. *Hitler's Professors: the part of scholarship in Germany's crimes against the Jewish people.* New Haven, Yale University Press.

Wilson, T. 1990. *Confocal Microscopy*, London: Academic Press.

Wilson, T., Sheppard, C. 1984. Nonlinear scanning microscopy. In: *Theory and Practice of Scanning Optical Microscopy*, pp. 196–209, London: Academic Press.

Wokosin, D. L., Centonze, V., White, J. G., Armstrong, D., Ropbertson, G., Ferguson, A. I., 1996. All solid-state ultrafast lasers facilitate multiphoton excitation fluorescence imaging. *IEEE Journal of Selected Topics in Quantum Electronics*, 2(4): 1051–1065.

Wokosin, D. L., Loughrey, C. M., Smith, G. L. 2004. Characterization of a range of Fura dyes with two-photon excitation. *Biophysical Journal*, 86: 1726–1738.

Woodbury, E. J., Ng, W. K. 1962. TITLE, *Proc. I.R.E.* 50: 2367.

Xu, C., Webb, W. W. 1997. Multiphoton excitation of molecular fluorophores and nonlinear laser microscopy. In: *Topics in fluorescence Spectroscopy*, volume 5, Nonlinear and Two-Photon-Induced fluorescence, Ed. J. Lakowicz, pp. 471–540. New York: Plenum Press.

Yanovsky, V., Pang, Y., Wise, F. 1993. Generation of 25-fs pulses from a Kerr-lens mode-locked Cr:fosterite laser with optimized group-delay dispersion. *Opt. Lett.* 18: 1541–1543.

Zoumi, A., Lu, X., Kassab, G. S., Tromberg, B. J. 2004. Imaging coronary artery microstructure using second-harmonic and two-photon fluorescence microscopy. *Biophysical Journal*, 87(4): 2778–2786.

Appendix: Key Developments of Lasers and Laser Technology

1958	C. Townes and A. L. Schawlow propose that masers can operate at optical frequencies.
1960	Theodore Maiman develops the first ruby laser.
1961	A. Javan, W. Bennett, and D. Hariott develop the first gas laser, the HeNe laser.
1962	R Hall develops the first semiconductor laser.
1962	F. J. McClung and R. W. Hellwarth develop the technique of Q-switching.
1963	C. K. N. Patel develops the infrared carbon dioxide laser.
1963	E. Bell develops the first ion laser using mercury vapor.
1964	W. Bridges develops the argon ion laser.
1964	L. E. Hargrove, R. L. Fork, and M. A. Pollack develop the technique of mode-locking.
1966	W. Silfvast, G. R. Fowles, and B. D. Hopkins develop the first blue HeCd metal vapor laser.
1966	P. P. Sorokin and J. R. Lankard develop a liquid organic dye laser.
1981	R. L. Fork, B. I. Greene, and C. V. Shank generate pulses shorter then 200 fs for the first time.
1986	P. Moulton develops the titanium-sapphire laser.
1991	M. Hasse develops the blue-green diode laser.
1991	D. E. Spence, P. N. Kean, and W. Sibbett develop the first Kerr lens mode-locking of a Ti:Sapphire laser.
2006	S. Nakamura develops the first blue diode laser.

2

The Scientific Life of Maria Göppert-Mayer

Barry R. Masters

Reprinted with permission from *Optics & Photonics News*, *11(9)*, 38–41, (2000).

Researcher and educator Maria Göppert-Mayer, winner of the 1963 Nobel Prize in physics for her work on the nuclear shell model, loved science and learning. That love sustained her through the difficulties she encountered in pursuing her profession, including relocation from Europe to America prior to the war years and never being granted a full-time salaried academic position until she was 53 years old. A study of Maria Göppert-Mayer's life reveals insights into the influence of her parents on her education and achievements, as well as into the role of her mentors, physicists James Franck and Edward Teller, on her professional development. [1] Göppert-Mayer's doctoral thesis, published in 1931, predicted multiphoton excitation processes and served as a precursor to the development of multiphoton excitation microscopy.

As we all know, the influence of parents on children is paramount, and this fundamental truth is nowhere better illustrated than by the life of Maria Göppert-Mayer. She was proud of her heritage which was linked, on her father's side, to seven generations of professors. Her father instilled a strong feeling of high self-esteem in the young Maria, and encouraged experimentation, discovery, and wonder at the natural world–all critical to the formation of a scientist.

When she was four the family moved to Göttingen where her father, who she considered her model in life, became a professor of pediatrics. The father held and openly expressed high expectations for his daughter, encouraging her to strive to be "more than a housewife." As a result, she decided at a young age to become a scientist. Her father instilled in her a strong feeling of self-confidence and a free spirit. As she grew older there was never a doubt in her mind, or that of her parents, that she would study at the university at which her father taught.

As Maria Göppert was growing up, and later during her studies at the University of Göttingen, she lived in a haven of mathematics and physics. In those scientifically fertile and exciting years before World War II, Göttingen was the world center of mathematics and physics, especially quantum mechanics. Close family friends included luminaries such as James Franck, Max Born and David Hilbert. The department of

Maria Göppert-Mayer in middle age (AIP Emilo Segré Visual Archives, Physics Today Collection). Victor Frederick Weisskopf (left), Maria Göppert-Mayer, and Max Born (American Institute of Physics [AIP], Emilo Segré Visual Archives).

mathematics at the University of Göttingen hosted Richard Courant, Hermann Weyl, and Edmund Landau. Maria's fellow students included Max Delbrück and Victor Weisskopf.

During her Göttingen years, Maria Göppert interacted with a number of eminent contributors to the theoretical development of quantum mechanics: Arthur Compton, Paul Dirac, Enrico Fermi, Werner Heisenberg, John von Neumann, J. Robert Oppenheimer, Wolfgang Pauli, Linus Pauling, Leo Szilard, and Edward Teller. At Göttingen, she studied theoretical physics under Max Born. In a scientific life studded with important intellectual influences, James Franck stands out because of his ongoing role as her lifetime mentor.

Göppert-Mayer with daughter Marianne, circa 1935 (American Institute of Physics [AIP], Emilo Segrè Visual Archives).

In January 1930, Maria Göppert married Joseph Mayer, an American chemical physicist. Two months later she completed her thesis and passed her final exam. Her examination committee consisted of three Nobel Prize winners. The couple relocated to America in the spring of 1930.

2.1 PROFESSIONAL LIFE

In Baltimore, Joseph Mayer obtained an appointment in the Department of Chemistry at Johns Hopkins University. Ostensibly because of the rules on nepotism, his wife was not considered for a faculty appointment; however, she was given a small stipend as an assistant, as well as access to the university facilities. She presented lecture courses in the graduate school. For three years while she was at Johns Hopkins, because she was technically not a professor, she was able to return to Germany and work on quantum mechanics with her former professor Max Born.

In 1938, after Johns Hopkins fired Joseph Mayer to cut expenses, the Mayers left Baltimore and moved to New York. Joseph Mayer was appointed an associate professor of chemistry at Columbia University; his wife did not receive an appointment but was given an office. While at Columbia, she interacted with the great scientists Harold Urey, Willard Libby, Enrico Fermi, I.I. Rabi and Jerrold Zacharias.

Joseph Mayer, an American chemical physicist, married Maria Göppert in 1930 (American Institute of Physics [AIP], Emilo Segrè Visual Archives).

In 1941, Sarah Lawrence College offered Göppert-Mayer a half-time teaching position. She accepted the position, electing to teach a unified science course. She concurrently accepted a half-time position at Columbia University, alongside Harold Urey, where the two worked on isotope separation. In the spring of 1945, she conducted research with Edward Teller at Los Alamos Laboratory.

After the war, the Mayers relocated to the University of Chicago where Joseph Mayer became a professor of chemistry and his wife accepted yet another voluntary position, this time as an associate professor of physics. When Edward Teller joined the University of Chicago, Maria Göppert-Mayer continued her work with him on the theory of nuclear structure. When Argonne National Laboratory was formed she worked there as a senior physicist with a half-time appointment. Incredibly, she continued to maintain her voluntary appointment at the University of Chicago: as an unpaid volunteer she lectured, served on university committees, directed students' theses, and was involved in other university work.

It was during her research at the Argonne National Laboratory that she developed the nuclear shell model for which she would earn the Nobel Prize in 1963. Her development of the nuclear shell model, based on her theory of spin-orbit coupling, is a fascinating story that merits examination in its own right. She discovered that there are nuclei with "magic numbers" of protons or neutrons that exhibit unusual stability. These "magic numbers" are 2, 8, 20, 28, 50, 82 and 126. Her theory predicted the nuclear stability of these elements.

In 1960, the Mayers moved to the University of California at San Diego, where Maria Göppert-Mayer accepted a full-time appointment as professor of physics, her first full-time university professorship with pay: she was 53 years old.

In addition to being awarded the Nobel Prize for physics in 1963, Maria Göppert-Mayer was elected a member of the National Academy of Sciences and a corresponding member of the Akademie der Wissenschaften in Heidelberg. She received honorary degrees from Russell Sage College, Mount Holyoke College, and Smith College.

2.2 EXPERIMENTAL VERIFICATION OF GÖPPERT-MAYER'S THEORY PREDICTING TWO-PHOTON ABSORPTION PROCESSES

What is the relevance of Maria Göppert-Mayer's doctoral thesis to the field of nonlinear optics? In 1930, she submitted her doctoral dissertation in Göttingen on the theory of two-photon quantum transitions in atoms. [2] She wrote her doctoral thesis on the decay of excited states by the simultaneous emission of two quanta. Her thesis work is often cited in modern publications on multiphoton excitation microscopy.

The field of nonlinear optics is the study of the interaction of intense laser light with matter. [3] Nonlinear optics may have begun with the experimental work of Franken and his group in 1961 on second harmonic generation of light. [4] They showed that if a ruby laser pulse at frequency v propagates through a quartz crystal, then light at the second harmonic frequency 2v is generated; light from the ruby laser at 694 nm which was incident on a quartz crystal generated light at 347 nm.

2.3 MICROSCOPIC IMPLEMENTATION OF NONLINEAR SPECTROSCOPY

The principle of nonlinear scanning microscopy is simply explained. If a high powered beam of light impinged on a specimen, the specimen would behave in a nonlinear manner and higher optical harmonics would be produced. [3] This nonlinear harmonic generation would be a function of the molecular structure of the specimen.

After Franken et al. first observed second harmonic generation, several groups demonstrated the practical applications of a second harmonic microscope. Kaiser and Garret in 1961 demonstrated the two-photon excitation of $CaF_2:Eu^{2+}$. In 1974, Hellwarth and Christensen developed a second harmonic microscope and observed microstructures in polycrystalline ZnSe materials. [6] Two-photon absorption processes were experimentally verified 32 years after they were predicted by Maria Göppert-Mayer.

The Oxford University group of Sheppard, Kompfner, Gannaway and Wilson, building on the earlier work of Hellwarth and Christensen, realized that nonlinear optical imaging could be combined with their development of the scanning optical microscope. [7] They pointed out that the nonlinear processes are confined to the focal plane of the objective and that the image intensity would depend quadratically on the illumination power. They also discussed improved resolution when combined with confocal detection, and the advantages of pulsed lasers as well as the potential for severe thermal damage due to the extremely high intensities used.

Early implementations of nonlinear optical microscopy used a laser scanning microscope to image second harmonic generation in crystals. The development of the mode-locked laser, which generated femtosecond laser pulses at a repetition rate of 100 MHz, provided the technical light source for the next step in the development of the multiphoton excitation microscope.

The seminal work of Denk, Strickler and Webb, published in Science in 1990, launched a new revolution in nonlinear optical microscopy. [8] By integrating a laser scanning microscope (scanning mirrors, photomultiplier tube detection system) and a mode-locked laser which generates pulses of near-infrared light, they succeeded in demonstrating a new type of microscope based on two-photon excitation of molecules. The pulses of red or near-infrared light (700 nm) were less than 100 fsec in duration and the laser repetition rate was about 80 MHz.

The benefits of two-photon excitation microscopy include: improved background discrimination, reduced photobleaching of the fluorophores and minimal photodamage to living specimens. The inventors proposed the application of two-photon excitation microscopy for optical sectioning three-dimensional microscopy and for uncaging of molecules inside cells and tissues. A recent application to the field of biomedical optics is the in vivo functional imaging of human skin with a multiphoton excitation microscope. [9]

NOTE

Barry R. Masters is a Gast Professor in the Department of Ophthalmology, University of Bern, Switzerland, and consults at MIT on nonlinear optics applied to dermatology. He is the 1999–2000 OSA/SPIE Congressional Science and Engineering Fellow (AAAS).

REFERENCES

[1] McGrayne, S.B., Maria Goeppert Mayer, chapter 8, pp. 175–200, in: Nobel Prize Women in Science, Birch Lane Press, New York, 1996.

[2] Göppert-Mayer, M., "Uber Elementarakte mit zwei Quantensprüngen," Ann. Phys. (Leipzig), 9, 273–294 (1931).

[3] Boyd, R.W., Nonlinear Optics (Academic Press, New York, 1992).

[4] Franken, P.A., et al., "Generation of optical harmonics," Phys. Rev. Lett. 7, 118 (1961).

[5] Kaiser, W. and Garrett, C.G.B., "Two-photon excitation in $CaF_2:Eu^{2+}$," Phys. Rev. Lett. 7, 229–231 (1961).

[6] Hellwarth, R. and Christensen, P., "Nonlinear optical microscope examination of structures in polycrystalline ZnSe," Opt. Commun. 12, 318–322 (1974).

[7] Gannaway, J.N. and Sheppard, C.J.R., "Second-harmonic imaging in the scanning optical microscope," Opt. Quant. Electron. 10, 435–439 (1978).

[8] Denk, W., et al., "Two-photon laser scanning fluorescence microscopy," Science, 248, 73–76 (1990).

[9] Masters, B.R., et al., "Multiphoton excitation fluorescence microscopy and spectroscopy of in vivo human skin," Biophysical J., 72, 2405–2412 (1997).

—— **3**

The History of Perturbation Theory from Astronomy to Quantum Mechanics

Barry R. Masters

3.1 INTRODUCTION

The goals of this chapter on perturbation theory are to introduce the concept of perturbation theory, to present the genesis of perturbation theory as a mathematical technique to solve complex problems in mechanics, mainly celestial mechanics, and to present a historical review of the application of the theory to quantum mechanics. An understanding of the historical development of perturbation theories allows the readers to put the modern quantum mechanical perturbation theory in the proper context. In a subsequent section we will develop the mathematical formulation of the various types of perturbation theory so that the reader can better understand the interaction of atoms and radiation, in particular the calculation of transition probabilities for the absorption, emission, and scattering of radiation by atoms.

First we present an overview of perturbation theory. Perturbation theory is a general set of mathematical approximation methods that are used to solve equations that have no exact analytical solutions. Perturbation theory yields approximate solutions to the perturbed problem that are based on the exact analytical solutions of the unperturbed problem. These approximation methods are used to obtain solutions that with each higher order of approximation approach closer to the exact solution.

I now explain the meaning of a perturbation. Perturbation theory assumes that the physical situation can be described by a known solution—for example, the energy levels of the hydrogen atom in which the eigenvalues (energy levels) are all known, and a very small perturbation on the system that results in a variation of the known eigenvalues of the unperturbed system. An example of the perturbed system is the Zeeman effect, in which the magnetic field results in small shifts in the eigenvalues (energy) of the hydrogen in the magnetic field as compared to the eigenvalues of the isolated hydrogen atom in the absence of the magnetic field.

The key question is how to define a very small perturbation. The next step is to define the quantum mechanical "energy operator" or Hamiltonian for the unperturbed system and a Hamiltonian for the perturbation. The Hamiltonian for the perturbation is much smaller in magnitude than the unperturbed system. Finally, there are algorithms that define the steps to be followed to use the approximation techniques for a number of different cases: time-independent perturbation theory or time-dependent perturbation theory. These approximation calculations are normally carried out for the first-order approximation or the second-order approximation. After perturbation theory is used for a calculation it is important to compare the calculated results with the experimental results: frequencies of spectroscopic lines, or transition probabilities for absorption, emission, or scattering.

3.2 THE GENESIS OF PERTURBATION THEORY

Next we describe the genesis of perturbation theory. While its origins come from the field of celestial mechanics, the applications in the field of quantum mechanics are our main interest. In particular, we are interested in the solutions of quantum mechanical problems that have no analytical solutions. Few physical systems can be described by quantum mechanics with analytical solutions for the eigenvalues (energies). They include the hydrogen atom and the harmonic oscillator. To solve the more complicated problems in quantum mechanics, such as many electron atoms, molecules, and non-linear interactions between atoms and radiation (multiphoton absorption and emission processes), it is necessary to use higher-order perturbation theory to calculate the tran-sition probabilities. In this section we describe the genesis of perturbation theory in physics, develop some applications as examples, and introduce the various types of perturbation theory that are used in quantum mechanical calculations of energies and transition probabilities.

In many cases the Hamiltonian for a soluble problem is only slightly different from the Hamiltonian for the more complex insoluble problem. The latter problem is a perturbation of the soluble problem.

A comprehensive discussion of the early development of perturbation theories in found in Goldstein (1981). The genesis of perturbation comes from early work on celestial mechanics. In mechanics there are problems such as the two-body Kepler (central force) problem, which has an exact solution. Newton's Law of gravitation is the only interaction between the two bodies. However, the three-body problem, that of three bodies interacting only by their mutual gravitational attraction, has not been solved analytically. For example, Newton worked out a simple form of perturb-ation theory for the oscillations of the motion of the moon as it revolves around the earth as due to perturbations of the attraction of the sun. In the early 19th century another perturbation method was developed. The "variation of coordinates" per-turbation technique was based on the perturbed motion of a two-body problem in which solutions of mechanical equations are written as expansions of a "perturbation parameter."

Newton's Laws describe the interaction between two bodies. Newton was able to derive Kepler's three empirical laws of planetary motion from Newton's law of uni-versal gravitation. The problem that occupied mathematicians for many years was the application to Newton's laws to many-body systems, such as real planetary systems in the solar system that are composed of more than two bodies (i.e., the earth, the moon, and the sun). The mathematical methods to solve these problems in an approximate manner are the subject of *Les Méthods nouvelles de la Mécanique céleste* (Poincaré, 1892–1899). Poincaré developed perturbation methods to solve the three-body problem and by extension the n-body problem.

The n-body problem is stated as follows. There exist n-point particles, each with arbitrary masses, initial position, and initial velocities. These particles are subject to Newton's law of mutual gravitational attraction. The problem is to develop a set of mathematical equations that describe the motion of the particles, and then to solve these equations. The solutions of these equations of motion should give the motion of the particles at any time. In volume I, Poincaré developed the problem and presented the theoretical techniques known at that time to describe the n-body problem.

Poincaré then stated that a complete and rigorous solution of the equations is not possible (i.e., integrating the equations of motion). Therefore, he suggested that the only possibility is the method of successive approximations. The entire second volume of *Les Méthodes nouvelles de la Mécanique céleste* is devoted to these perturbation methods. Again, the basis of all perturbation methods is that the solutions of the perturbed system are only slightly different from the solutions (integrated form of equations of motion) of the equations of motion of the unperturbed systems that are already integrated.

The main mathematical problem to overcome is that when series expansions were used as approximations of a function, they did not always converge or sum to a finite term; in many cases they diverged to infinity. Therefore, among the many contributions of Poincaré are his perturbation methods to solve n-body mechanical problems with an approximation technique. Much of the later work on perturbation theory that stems from Born, Schrödinger, and Dirac is built on the early work of Poincaré.

3.3 PERTURBATION THEORY IN QUANTUM MECHANICS

Time-dependent perturbation theory, also known as the method of "variation of constants," was first developed by Lagrange for celestial mechanics, and much later it was further incorporated into the time-dependent perturbation theory of quantum mechanics by Dirac (Tolman, 1938).

Many of the early formulations of perturbation theory were modified for applications in the old and the new quantum theories. For example, Max Born in his 1924 book *Atommechanik* (also published in 1960 in English as *The Mechanics of the Atom*) developed a time-independent perturbation theory (Born, 1989). The time-independent perturbation theory attempts to find the quantities that describe the mechanical system that are constant in the perturbed system. It was used to calculate the Zeeman effect, the effect of a perturbing magnetic field on the energy levels, and thus the spectral lines of the hydrogen atom.

The early applications of perturbation theory to the mechanics of atoms is found in the following references: N. Bohr, *Quantum Theory of Line Spectra*, Copenhagen, 1918, 1922; M. Born, E. Brody, Zeitschrift für Physik, vol. vi, p. 140, 1921; P. S. Epstein, Zeitschrift für Physik, vol. viii, pp.. 211, 305, 1922; vol. ix, p. 92, 1922.

Max Born, in his book *The Mechanics of the Atom*, explains that the methods of perturbation that he uses are essentially the same as those treated in detail by Poincaré in his *Les Méthods nouvelles de la Mécanique céleste*. Born explained that the three-body problem or the n-body problem has not been solved by the method of separation of the variables. Instead, he uses methods that give the motion of the particles as successive degrees of approximation.

The Hamiltonian function or operator (energy operator) is expanded in a converging power series of the variable λ, which is the expansion parameter that varies between zero and one. When the variable $\lambda = 0$ the Hamiltonian is exact for the problem that is soluble by the method of separation of variables. The additional terms in the power series are "perturbations" of the "unperturbed" motion that is described by the Hamiltonian function for which the solution exists.

Born and Jordan, in their 1930 book in German *Elementare Quantenmechanik*, devoted a long chapter to the subject of perturbation theory that attests to its importance in solving the equations of quantum mechanics that are nonsoluble by exact means (Born & Jordan, 1930). These authors derived the perturbation theory for several interesting cases that have been worked out by many different authors: nondegenerate systems (all the eigenvalues or energies of the atom are different), the anharmonic oscillator, coupling of vibration and rotation in diatomic molecules, perturbation theory for degenerate systems (those in which several eigenvalues are equal for a given quantum state), the Stark effect, the Zeeman effect, the Kerr effect, the Faraday effect, optical activity, the dispersion of light, and the intensity and the polarization of fluorescence.

In 1926 Erwin Schrödinger published five papers in the German journal *Annalen der Physik*: "Quantization as a problem of the eigenvalue problem (First Part)," "Quantization as a problem of the eigenvalue problem (Second Part)," "On the equivalence of the Heisenberg-Born-Jordan quantum mechanics and mine," "Quantization as a problem of the eigenvalue problem (Third Part) and perturbation theory with application to the Stark Effect of the Balmer lines," and "Quantization as a problem of the eigenvalue problem (Fourth Part)." The English translations of these papers are available in his book (Schrödinger, 1928).

Schrödinger developed his time-dependent wave equation with application to the Stark effect of the Balmer lines and thus was able to calculate the intensities and the polarization of the Stark effect on the Balmer series of transitions in the hydrogen atom. In the Stark effect, the external electric field results in shifts of the energy of the transitions of the hydrogen atom. He was able to derive an expression for the energy shifts that is equivalent to the formula derived by Epstein.

Epstein also independently developed his form of perturbation theory with applications to quantum mechanics (Epstein, 1922a, 1922b, 1922c). Here is another example of the utility of a broad theoretical knowledge of physics and mathematics. Epstein developed a method of perturbation theory in 1917 to treat the helium atom (Mehra & Rechenberg, 2001). Epstein based his version of perturbation theory on the similar work of the French astronomer Charles-Eugène Delaunay. In the 1860s Delaunay worked out a perturbation technique to solve the equations of motion for the moon.

Delaunay's perturbation technique is the basis of many of the later versions of the technique that were used in quantum mechanics. Here is a summary of his technique. First, he assumed that the Hamiltonian (energy operator) for the system can be expressed as the unperturbed term plus a term that describes the perturbation. The perturbation term is much smaller than the unperturbed term of the Hamiltonian. The solution of the unperturbed problem (the motion of the moon and the earth according to Newton's law of gravitation) was known as it could be solved exactly. Delaunay then suggested that to solve the perturbation problem (this perturbation was the effect of attraction of the other planets), he would use a method of successive approximations. Each solution at a given level of approximation would be used to solve the problem for the next higher level of approximation.

Bohr developed a perturbation method for the hydrogen atom. This theory was further developed by Kramers. Both Born and Pauli then generalized the Bohr-Kramers theory of perturbations and developed solutions for both nondegenerate and degenerate systems. It is of interest that Born and Pauli recognized that the perturbations in their theory were similar to the degenerate perturbations in celestial mechanics that were called "secular perturbations." This is another example of the unity of mathematical physics from the scale of planetary motions to atomic physics. The same equations and mathematical techniques for their solutions are developed.

I now discuss "secular perturbations." In the method to solve first-order perturbation theory for a degenerate system (several eigenfunction correspond to a given energy value), a series of homogeneous linear simultaneous equations arise. To solve these equations, the determinant of the coefficients in these equations is set equal to zero. This equation is called a secular equation, and the perturbation that requires a solution of this type is called a secular perturbation. The term "secular" comes from the Latin word *saeculum*, which means "generation ago." The word "secular" in the context of perturbations means something that occurs over a long period of time. It was first used in classical mechanics to describe a perturbation that causes a very slow and cumulative effect on the orbit of a planet.

How did perturbation theory engage Max Born and other physicists interested in theoretical mechanics? In 1921, Max Born worked with Brody on the subject of crystal dynamics. They applied perturbation theory to calculate the thermodynamic properties of solids. With that background, Born and his students applied the perturbation theory to the helium atom, the hydrogen molecule-ion, and the hydrogen molecule. In 1922 Born became interested in the various perturbation methods that were used by astronomers. Therefore, in the winter semester of 1922–1923, Born arranged for a course on perturbation theory at his Institute in Göttingen.

In 1926 Schrödinger published his papers on wave mechanics that included his development of perturbation theory. In the same year, Born, Heisenberg, and Jordan published a paper that included a matrix formulation of the perturbation theory. They developed a general perturbation theory that was applicable for systems with an arbitrary number of degrees of freedom, and was valid for both nondegenerate and degenerate systems (Born et al., 1926). The English translation of this very important paper is available (Van der Waerden, 1968).

The time-dependent perturbation theory was developed by Dirac in 1926 and 1927. This theory was also the basis of the thesis work of Maria Göppert-Mayer (1931) and will be described in a subsequent section on the quantum mechanical description of absorption and emission of photons (Dirac, 1926, 1927).

3.4 CONCLUSION

In summary, the perturbation techniques derive from the earlier work of astronomers who developed these methods to calculate problems in celestial mechanics (i.e., the n-body problem with Newton's law of gravitation). Much later these perturbation methods were used for quantum mechanical calculation of complex systems.

The various perturbation techniques are based on the following method. First, the equation of motion is known and its solutions are known analytical expressions. An example is the quantum mechanical calculation of the energies of the hydrogen atom; another example is the mechanical system of the earth and the moon. Second, for the case of a more complex system (i.e., the hydrogen atom in an electric or magnetic field, or the motion of the moon around the earth including the influence of the other planets), the equations are derived from the isolated system with a known analytical solution, and a perturbing term (electric field or n-body attractions). The Hamiltonian (energy operator) in the equations of motion is composed of the term in the absence of the perturbation (with known solutions) and another term that describes the perturbation. Third, the equations of motion or the equations of quantum mechanics are solved as a series of successive approximations. The zero-order approximation yields the known solution of the unperturbed system with terms that increase or decrease the energies; these solutions are input again into the equation of motion or of quantum mechanics, and the new solutions are the first-order solutions. The process can be continued to higher-order approximations. If the perturbation term in the total Hamiltonian is very much smaller than the Hamiltonian for the system in the absence of any perturbation, then the method yields solutions that successively approach the values for the perturbed system and are in agreement with experiment (i.e., the Stark effect and the Zeeman effect).

REFERENCES

Born, M. 1989. *Atomic Physics*, Eighth edition, New York: Dover Publications. First published in 1933 under the German title *Moderne Physik.*

Born, M., Heisenberg, W., Jordan, P. 1926. Zur Quantenmechanik II. *Zeitschrift für Physik*, 35: 557–615.

Born, M., Jordan, P. 1930. *Elementare Quantenmechanik, Zweiter Band der Vorlesungen Über Atommechanik*, Berlin: Julius Springer.

Dirac, P. A. M. 1926. On the Theory of Quantum Mechanics, *Proceedings of the Royal Society (London) A*, 112: 661–677.

Dirac, P. A. M. 1927. The Quantum Theory of Emission and Absorption of Radiation. *Proceedings of the Royal Society of London, Series A, Containing Papers of a Mathematical and Physical Character*, Vol. 114(767): 243–265.

Epstein, P. S. 1922a. Die Störungsrechnmung im Dienste der Quantentheorie. I. Eine Methode der Störungsrechnung. *Zeitschrift für Physik* 8: 211–228.

Epstein, P. S. 1922b. Die Störungsrechnmung im Dienste der Quantentheorie. II. Die numerische Durchführung der Methode. *Zeitschrift für Physik*, 8: 305–320.

Epstein, P. S 1922c. Problems of quantum theory in the light of the theory of perturbations. *Physical Review (2)*,19: 578–608.

Goldstein, H. 1981. *Classical Mechanics, Second Edition*, Reading, Massachusetts: Addison-Wesley Publishing Company.

Mehra, J., Rechenberg, H. 2001. *The Historical Development of Quantum Theory*. Volumes 1–6, New York: Springer-Verlag.

Poincaré, H. 1993. *New Methods of Celestial Mechanics*, three volumes. Daniel L. Goroff, Ed, Translation of: *Les Méthodes nouvelles de la Mécanique céleste*. It was originally published in 1892–1899. College Park, MD: American Institute of Physics.

Schrödinger, E. 1928. *Collected Papers on Wave Mechanics*, London: Blackie & Son Limited.

Tolman, R. C. 1938. *The Principles of Statistical Mechanics*, Oxford: Oxford University Press.

Van der Waerden, B. L. 1968. *Sources of Quantum Mechanics*, New York: Dover Publications.

——— 4

English Translations of and Translator's Notes on Maria Göppert-Mayer's Theory of Two-Quantum Processes*

Barry R. Masters

4.1 ENGLISH TRANSLATION OF: GÖPPERT, M. (1929). ÜBER DIE WAHRSCHEINLICHKEIT DES ZUSAMENWIRKENS ZWEIER LICHTQUANTEN IN EINEM ELEMENTARAKT. DIE NATURWISSENSCHAFTEN, 17, 932, (1929)*

Barry R. Masters

ON THE PROBABILITY OF TWO LIGHT QUANTA WORKING TOGETHER
IN AN ELEMENTARY ACT

2. H. A. Kramers, W. Heisenberg, Z. Physik 31, 681–708, (1925). [Über die Streuung von Strahlen durch Atome. On the dispersion of radiation by atoms]

Already from the systematic development of quantum mechanics have Kramers and Heisenberg[2] so completely deduced the quantum theoretical dispersion equation that today the famous Raman Effect ("Smekal jumps") can be quantitatively calculated. In addition, they mentioned another effect that until now has not yet been observed. In the light-quantum language it allows the following simple description:

A stimulated atom of energy $h\nu_{nm}$ becomes affected from a light quantum of small energy $h\nu < h\nu_{nm}$ and thereby jumps to the ground state with a forced [quantum] emission, in which the incident frequency ν is just the difference frequency $\nu' = \nu_{nm} - \nu$.

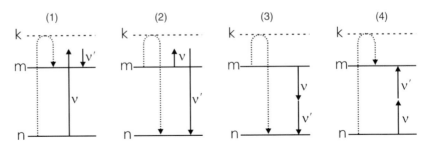

Figure 4.1.1 (1) Stokes case of Raman Effect. (2) Anti-Stokes case of Raman Effect. (3) Double Emission. (4) Double Absorption.

Occasionally the quantum mechanical representation of these things, through the aid of the soon to be published book of Born and Jordan in which I have contributed, I have noticed, that in the exact dispersion theory of Dirac there is also his delivery of the reverse effect. It occurs by the following process: An atom in the ground state, is acted upon by two light quantum $h\nu$ and $h\nu'$, the sum of which (within the linewidth) equals the excitation threshold [step] of the atoms, $v + v' = v_{nm}$, and through it becomes promoted to an excited state with the energy $h\nu_{nm}$. Consequently, one has here the working together of two light quanta in an elementary act and the probability of such processes can be calculated.

Here the two processes are schematically contrasted to the Raman effect. We will consider the transition of an atom between a lower state n and a higher state m. State k is some other atomic state. The dotted lines denote the behavior of an atom, the upwards pointing arrows are absorption, the downwards pointing [arrows] are emitted light-quanta.

In the Stokes' case of the Raman effect (Fig. 1), the emitted light of frequency $v' = v \pm v_{nm}$ is associated with a virtual oscillator with the moment.

$$P_{nm}^{R} = \sqrt{2\pi \int \rho(v)\, dv}\; p_{nm}^{R} \tag{1}$$

where,

$$p_{nm}^{R} = \sum_{k} \left(\frac{(P_{nk}\hat{e})\, P_{km}}{h\,(v_{kn} - v)} + \frac{P_{nk}\,(P_{km}\hat{e})}{h\,(v_{km} + v)} \right) \tag{2}$$

$\rho(v)$ is the radiation density
$\int \rho(v)dv$ extended over a spectral line
$\hat{e}$ is a unit vector in the direction of the incident [electric] field strength,
P_{nk} is a matrix element of the electrical moment of the unperturbed atom. In the anti-Stokes case (Fig. 2), the symbols n and m are exchanged in formula (2).

In case 3, the double emission, there stands in the place of the virtual moments P^{R} another P^{D} from the same formula (2), only in the place of (2) there is now

$$p_{nm}^{D} = \sum_{k} \left(\frac{(P_{mk}\hat{e})\, P_{kn}}{h\,(v_{km} + v)} + \frac{P_{mk}\,(P_{kn}\hat{e})}{h\,(v_{kn} - v)} \right) \tag{3}$$

One notices in that the great similarity of the two cases. The transition probability of the cases 1–3 is proportional to the light intensity:

$$w_{mn} = \frac{64\pi^4 v'^3}{3hc^3} \cdot 2\pi \int \rho(v)\, dv \; |p_{nm}|^2 \tag{4}$$

but in case 4 it is proportional to the square of the light intensity, namely

$$w_{mn} = \frac{8\pi^3}{h^2} \cdot 2\pi \int \rho\,(v_{mn} - v)\rho\,(v)\; dv \; \left| p_{nm}^{D}\, \hat{e}' \right|^2 \tag{5}$$

It is noticed that the expression in v and v', also $\hat{e}$ and $\hat{e}'$ are symmetrical. The probabilities (4) and (5) have the attribute that the radiation balance is not disturbed [conservation of energy].

The process 3 [Fig. 3] is probably difficult to observe, because the frequency is proportional to the number of atoms in the excited state; in case 4 on the contrary, it is proportional to the number of atoms in the ground state, whereby nevertheless the quadratic dependence of the light density works unfavorably.

Göttingen, Institute for Theoretical Physics, October 28, 1929,
Maria Göppert

Acknowledgment of the Translator

I thank Katie Lagoni-Masters for checking my translation. [Fig. 1], [Fig. 2], and [Fig. 3] refer to the German paper.

4.2 ENGLISH TRANSLATION OF: GÖPPERT-MAYER, M. (1931). ÜBER ELEMENTARAKTE MIT ZWEI QUANTENSPRÜNGEN. ANN. PHYS (LEIPZIG), 9, 273–294, (1931)

Barry R. Masters

German: Über Elementarakte mit zwei Quantensprüngen von Maria Göppert-Mayer (Göttingen Dissertation) **Elementary Acts With Two Quantum Jumps** Maria Göppert-Mayer (Göttingen Dissertation) Ann. Phys (Leipzig), **9**, 273–294, (1931)

INTRODUCTION

The first part of this work is concerned with two light quanta working together in one elementary act. With the help of the Dirac [1] dispersion theory [the variation of refractive index with wavelength], the probability of an analogous Raman effect process is calculated, namely, the simultaneous emission of two light quanta. It is shown that a probability exists for an excited atom to divide its excitation energy into two light quanta, the sum of their energy is equal to the excitation energy, but each can be an arbitrary value. If an atom is irradiated with light of a lower frequency than the frequency associated with an eigenfrequency in the atom, additionally there occurs a stimulated double emission, in which the atom divides the energy into a light quantum with the frequency of the irradiated light, and a quantum with the difference frequency. Kramers and Heisenberg [2] calculated the probability of this last process in a corresponding manner. [English translation: Van der Waerden, 1967. *Sources of Quantum Mechanics*, 223–252, New York: Dover Publications.]

The reverse of this process is also considered, namely the case that two light quanta, whose sum of frequencies is equal to the excitation frequency of the atom, work together to excite the atom.

It is further investigated how an atom responds to colliding particles, when at the same time it has the possibility of spontaneously emitting light. Oldenberg [3] experimentally found a broadening of the resonance lines of mercury, when he allowed the excited atoms to collide many times with slow particles.

[1] P. A. M. Dirac, Proc. of R. S. vol. **114**, S. 143 [read *S. 243, and S. 710, (1927)*]*
       *[Error in text, read *P. A. M. Dirac, Proc. of R. S. vol. **114**, p. 243, and p. 710, (1927).*]
[2] H. A. Kramers and W. Heisenberg, Ztschr. f. Phys. **31**, S. 681 (1925).
[3] O. Oldenberg, Ztschr. f. Phys. **51**, S. 605, (1928).

He interprets this with the assumption that a positive or a negative part of the excitation energy of the atom can be transferred as kinetic energy to the interacting particles, and the difference frequency is radiated. For this process, an equation is derived here that is analogous to the Raman effect or double emission.

Finally, in relation to a study by Franck [1], an attempt is made to explain the behavior of the intensity of excitation of spectral lines, induced by collision [of atoms] with fast electrons in such a double process.

Franck discusses the course of the excitation function of a spectrum line; that is the intensity [of the line] as a function of the velocity of the colliding electrons. This function is zero for low velocities, until the kinetic energy of the electrons becomes equal to the excitation energy of the particular spectral line of the original state. The excitation function sharply increases, has a maximum at a velocity that corresponds to a potential of a few volts above this critical potential, and then decreases towards zero again. This part of the curve is differentially calculated with the usual theory of collisions. A curve is obtained that closely represents the effect, especially the sharp increase of the function at the critical potential. For high velocities the theoretical curve shows a monotonic decrease to zero. However, studies with rapid electrons have shown that the intensity increases again, and that electrons with a velocity corresponding to about 100 volts, appears to reach a flat [intensity] maximum for all [spectral] lines. This maximum occurs at kinetic energies that cause strong ionization, and for which the theoretical value of the direct collision excitation as well as for the extrapolation of the experimental excitation function is already practically zero. The relative intensities of the individual [spectral] lines are completely displaced in relation to the behavior at low velocities; therefore, the excitation process appears to be based on a completely different process. Franck remarked that this effect is in many aspects similar to recombination luminescence.

[1] J. Franck, Ztschr. f. Phys. **47**, S. 509, (1928).

On the other hand Franck also shows that this process could not only be a matter of recombination, mainly because a great part of the experiments are performed in such high fields that secondary electrons released rapidly attain a velocity that is too high for them to be recaptured. Therefore, Franck explained this effect as the recombination of the atom with its own electron, a process which can be described as follows with the Bohr hypothesis: an atom's electron receives energy from the colliding particles and is thrown in a hyperbolic orbit, that is, a higher [energy] orbit than [that associated with] the ionization energy. Before the electron leaves the region of influence of the atom, it returns back to the elliptical orbit under the emission of light, so that the atom is now in an excited state.

Here the question should be discussed how such a process can be described according to the representation of quantum mechanics. Obviously we can no longer state that the atom recombines before the secondary electron has left the specific region of influence of the atom. However, it is reasonable to try to explain this effect by a process, in which simultaneously in an elementary act, the atom gains energy from the colliding electrons and emits light, so that it remains in a condition of discrete energy, and is now able in a second, independent process, to emit a spectral line of the discrete spectrum. Such an explanation contains a strong analogy to the Raman effect, which can also be described as the concurrence of two processes in a single elementary act. Since such a single act process occurs at the moment when the collision occurs with the atom, it would explain all the effects that cannot be explained by recombination luminescence.

The calculation shows a probability for such a process, the nature of which will be discussed.

1. THE COMBINED ACTION OF TWO LIGHT QUANTA IN ONE ELEMENTARY ACT

The following calculation is closely associated with the work of P. M. A. Dirac on emission, absorption, and dispersion [see reference 1 this paper, and R. H. Dalitz, 1995. *The Collected Works of P.A.M. Dirac, 1924–1948*, Cambridge, Cambridge University Press.]

Let us consider the interaction of an atom with a [electromagnetic] radiation field. To make the number of degrees of freedom countable, think of the radiation contained in a cubical box of volume V, which constrains the light waves to the condition of periodic repetition [standing-waves]. Later this box will be assumed to be infinitely large.

Such a radiation field is equivalent to a system of uncoupled harmonic oscillators. The radiation can be decomposed in plane, linear polarized waves; let A be the vector potential, then

$$A = \sum_{\alpha} Q^{\alpha} A^{\alpha} = \sum_{\alpha} Q^{\alpha} \hat{e}^{\alpha} e^{2\pi i \nu_{\alpha} \left(S_x^{\alpha} x + S_y^{\alpha} y + S_z^{\alpha} z \right)}$$

in which the components of the vector, $\nu_{\alpha} S^{\alpha}$ are determined in the usual manner by whole numbers which depend on the size of the box, and to each possible vector $\nu_{\alpha} S^{\alpha}$ there belong two unit vectors S^{α}, perpendicular to each other and to the unit vector $\hat{e}^{\alpha}$.

For large cavities, the asymptotic number of the standing waves, of which the frequency is between ν and $\nu + \Delta \nu$ is given by equation (1)

$$Z(\nu)\Delta \nu = \frac{8\pi \nu^2 V}{c^3} \Delta \nu \tag{1}$$

The Maxwell equations give for the values of Q^{α} the differential equation of the harmonic oscillators. Subjected to quantization the Hamilitonian function of the radiation is:

$$\sum H^{\alpha} = \sum_{\alpha} h\nu_{\alpha} \left(n_{\alpha} + \frac{1}{2} \right)$$

A state s of the radiation field is described by giving the state of all the oscillators $s = (n_1 \dots n_\alpha \dots)$, and the energy difference of two cavity states is:

$$hv_{ss'} = \sum_\alpha hv_\alpha (n_\alpha - n'_\alpha)$$

The matrix elements of Q^α are:

$$Q^\alpha_{ss'} = \sqrt{\frac{hc^2}{2\pi v_\alpha V}(n_\alpha + 1)} \qquad \text{for } n'_\alpha = n_\alpha + 1, \ n'_\beta = n_\beta \tag{2}$$

$$Q^\alpha_{ss'} = \sqrt{\frac{hc^2}{2\pi v_\alpha V} \cdot (n_\alpha)} \qquad \text{for } n'_\alpha = n_\alpha - 1, \ n'_\beta = n_\beta$$

$$Q^\alpha_{ss'} = 0 \qquad\qquad \text{otherwise.}$$

An atom with a stationary nucleus, a Hamiltonian function [Hamiltonian operator or energy operator] H_A, the stationary state n, and the eigenfrequency [resonance frequency] $v_{nn'}$ interacts with the radiation field. Then, the Hamilton function of the total system is:

$$H = \sum_\alpha H^\alpha + H_A + H'$$

The interaction energy H' is given from the function of the electron:

$$H_{el} = \frac{p^2}{2m} + V(q) + \frac{e}{mc}(pA) + \frac{e^2}{2mc^2}A^2$$

where p and q are respectively the momentum and coordinate vector of an atom's electrons, V is the potential energy of the electron in the atom, and A is the vector potential of the radiation field at the position of the electron. The two-terms of the interaction energy

$$H' = \frac{e}{mc}(pA) + \frac{e^2}{2mc^2}A^2$$

can be easily converted to a simpler form. Specifically, if we form the interaction [energy] function belonging to H_{el} as:

$$L = \dot{q}p - H_{el}$$

and express it with the help of $\dot{q} = \partial H_{el}/\partial p$ as a function of q and $\dot{q}$:

$$L(\dot{q}, q) = \frac{m}{2}\dot{q}^2 - V(q) - \frac{e}{c}(\dot{q}A)$$

then it is equivalent to another interaction [energy] function:

$$\bar{L}(\dot{q}, q) = \frac{m}{2}\dot{q}^2 - V(q) - \frac{e}{c}\left(q \frac{dA}{dt}\right)$$

which is derived from the first by adding the total time differentials $\frac{d}{dt}(qA)$. If we form the respective Hamilton function $\overline{H}_{el}$ from this, then we obtain:

$$\overline{H}_{el} = \frac{\overline{p}^2}{2m} + V(q) - \frac{e}{c}\left(q\frac{dA}{dt}\right)$$

If the wavelength of the light is large in relation to the atom's dimension, then the space variation of A within the atom can be neglected, so that

$$\frac{dA}{dt} = \frac{\partial A}{\partial t} = \dot{A}$$

After introducing the electric moments [(charge) × (displacement)]

$$P = \sum_r eq_r$$

(the sum ranges over all the atom's electrons), the term for the interaction energy is given as:

$$H' = -\frac{1}{c}\left(P\dot{A}\right)$$

For which $\dot{A}$ takes its value at the points x_0, y_0, z_0 from the atom's mid-point. Since $\frac{1}{c}\left(\dot{A}\right)$ is equal to the electric field strength E of the radiation, the energy of the interaction is simplified under this assumption to the potential energy of the electric moments P, against the light field.

 H' is the perturbation energy and the eigenfunction of the total system is developed as the eigenfunctions of the unperturbed system,

$$\Psi = \sum a_{ns}\Psi_{ns}$$

It is assumed that the unperturbed system at time, $t = 0$, is in the state n^0, s^0 then perturbation theory yields, for times that are short in relation to the average dwell time [lifetime], probability amplitude a_{ns}, in the first approximation is:

$$a_{ns}^{(1)} = H'_{ns,n^0s^0}\frac{1 - e^{2\pi i(v_{nn^0}+v_{ss^0})t}}{h(v_{nn^0} + v_{ss^0})}$$

and in a second approximation,

$$a_{ns}^{(2)} = \sum_{n's'}\frac{H'_{nn'ss'}H'_{n'n^0s's^0}}{h(v_{n'n^0} + v_{s's^0})}\left[\frac{1 - e^{2\pi i(v_{nn'}+v_{ss'})t}}{h(v_{nn'} + v_{ss'})} - \frac{1 - e^{2\pi i(v_{nn^0}+v_{ss^0})t}}{h(v_{nn^0} + v_{ss^0})}\right] \tag{3}$$

The first of these known equations gives absorption and emission, the second [of these known equations] gives the Raman effect and dispersion, and also the effects of simultaneous emission, and simultaneous absorption of two light quanta, in which here the details will be investigated.

First, in order to calculate the double emission let the atom at time, $t = 0$, be in the excited state n^0, and let there be only one standing wave in the cavity,

$$A^\alpha = \hat{e}\, e^{2\pi i v_\alpha \left(S_x^\alpha x + S_y^\alpha y + S_z^\alpha z \right)}$$

with a frequency v_α which differs from the eigenfrequency of the strongly excited atom, i.e.

$$n_\alpha^0 \gg 1; \quad n_\beta^0 = 0 \quad \text{for } \beta \neq \alpha; \quad v_\alpha \neq v_{nn'}$$

On account of the properties of the oscillator matrix Q^α (2) in

$$H'_{nn'ss'} = \sum_\alpha \dot{Q}^\alpha_{ss'} \left(P_{nn'} A^\alpha \right)$$

$H'_{nn'ss'} H'_{n'n^0s's^0}$ can only differ from zero, if [state] s proceeds from [state] s^0 either by the absorption of one light quantum v_α of the absorption frequency ["sent in"], and the emission of any other light quantum v_β, this gives the Raman effect and dispersion, or by the emission of two light quanta. It will then be shown that for the case with a fixed frequency v_α, while conserving the entire energy in any case only the emission of a specific frequency is possible. In order to calculate both cases simultaneously, in the following equation with doubled symbols the upper ones represent the Raman effect and the dispersion, and the lower ones represent the double emission.

$$a_{ns}^{(2)} = \frac{\dot{Q}^\alpha_{n_\alpha^0 \pm 1,\, n_\alpha^0}\, \dot{Q}^\beta_{1,0}}{c^2} \sum_{n'} \left| \frac{(P_{nn'}A^\alpha)(P_{n'n^0}A^\beta)}{h(v_{n'n^0} + v_\beta)} \right.$$

$$\times \left\{ \frac{1 - e^{2\pi i (v_{nn'} \mp v_\alpha)t}}{h(v_{nn'} \mp v_\alpha)} - \frac{1 - e^{2\pi i (v_{nn^0} + v_\beta \mp v_\alpha)t}}{h(v_{nn^0} + v_\beta \mp v_\alpha)} \right\}$$

$$+ \frac{(P_{nn'}A^\beta)(P_{n'n^0}A^\alpha)}{h(v_{n'n^0} \mp v_\alpha)}$$

$$\left. \times \left\{ \frac{1 - e^{2\pi i (v_{nn'} + v_\beta)t}}{h(v_{nn'} + v_\beta)} - \frac{1 - e^{2\pi i (v_{nn^0} + v_\beta \mp v_\alpha)t}}{h(v_{nn^0} + v_\beta \mp v_\alpha)} \right\} \right| \quad (4)$$

$a_{ns}^{(2)}$ is only then significantly different from zero when v_β is equal to or close to zero in one of the three denominators.

$$v_{n'n^0} + v_\beta, \quad v_{nn'} + v_\beta, \quad v_{nn^0} + v_\beta \mp v_\alpha$$

The regions of the zero value of the first two denominators yield the probabilities for processes in which the energy rule is not followed, namely for the transitions of the atom from n^0 to n, with respectively absorption or emission of a quantum with the ("sent in") frequency v_α, and emission of a quantum at the eigenfrequency of the atom.

These transitions do not correspond to real processes; they are based on a particularity of the method of variation of constants used here. Specifically, it is assumed that the perturbation energy H' begins to act at time $t = 0$, when it is actually continuous. This "turning on" [of the purturbation energy] is the cause of the occurrence of the abnormal transitions.

However, the location of the zero point in the third denominator represents the change of the total energy in the process. For such frequencies v_β in the region of $v' = v_{n^0 n} \pm v_\alpha$, the additive terms can be ignored,

$$\frac{1 - e^{2\pi i (v_{nn'} \mp v_\alpha)t}}{h(v_{nn'} \mp v_\alpha)}, \qquad \frac{1 - e^{2\pi i (v_{nn'} + v_\beta)t}}{h(v_{nn'} + v_\beta)}$$

and we obtain for the matrix elements Q their value from equation (2).

$$\left| a_{ns}^{(2)} \right|^2 = \frac{4\pi^2 h^2}{V^2} \left(n_\alpha^0 + \begin{pmatrix} 0 \\ 1 \end{pmatrix} \right)$$

$$\times \left| \sum_{n'} \frac{(P_{nn'} n^\alpha)(P_{n'n^0} \hat{e}^\beta)}{h(v_{n'n^0} + v_\beta)} + \frac{(P_{nn'} n^\beta)(P_{n'n^0} \hat{e}^\alpha)}{h(v_{n'n^0} + v_\alpha)} \right|^2$$

$$\times \frac{4 \sin^2 \pi (v_{nn^0} + v_\beta \mp v_\alpha)t}{h^2 (v_{nn^0} + v_\beta \mp v_\alpha)^2} \tag{5}$$

(The symbol $\begin{pmatrix} 0 \\ 1 \end{pmatrix}$ means that for dispersion and the Raman effect the value is 0, and for double emission the value is 1.) The probability of the process is given through the sum of $|a_{ns}^{(2)}|^2$ over all β for frequency v_β that is close to the frequency v'. Proceeding in the usual way for a large cavity, from $\sum_\beta$ the integral summed over the number of standing waves,

$$A^\beta, \quad \int Z(v_\beta) dv_\beta \tag{1}$$

one obtains the following equation due to the sharp maximum of the integrand at the positions of resonance.

$$\sum_\beta \left| a_{ns}^{(2)} \right|^2 = \frac{8\pi^2 h^2}{V^2} \cdot \frac{1}{3} \left(n_\alpha^0 + \begin{pmatrix} 0 \\ 1 \end{pmatrix} \right)$$

$$\times \left| \sum_{n'} \frac{(P_{nn'} \hat{e}^\alpha) P_{n'n^0}}{h(v_{n'n^0} + v_\beta)} + \frac{P_{nn'} (P_{n'n^0} \hat{e}^\alpha)}{h(v_{n'n^0} \mp v_\alpha)} \right|^2$$

$$\times \int Z(v_\beta) \frac{2 \sin^2 \pi (v_{nn^0} + v_\beta \mp v_\alpha)t}{h^2 (v_{nn^0} + v_\beta \mp v_\alpha)^2} dv_\beta$$

$$\sum_\beta \left| a_{ns}^{(2)} \right|^2 = \frac{64\pi^4 v'^3}{3hc^3} \cdot \frac{2\pi h v_\alpha \left(n_\alpha^0 + \begin{pmatrix} 0 \\ 1 \end{pmatrix} \right)}{V}$$

$$\times \left| \sum_{n'} \frac{(P_{nn'} \hat{e}^\alpha) P_{n'n^0}}{h(v_{n'n^0} \pm v_\alpha)} + \frac{P_{nn'} (P_{n'n^0} \hat{e}^\alpha)}{h(v_{n'n^0} \mp v_\alpha)} \right|^2 \cdot t$$

This equation is only meaningful if $v' > 0$. Double emission is only possible if $v_{n^0 n} > 0$ and the absorbed ["sent in"] frequency is $v_\alpha < v_{n^0 n}$.

Assume that not only a single eigenfunction [standing wave] of the cavity is strongly excited at the beginning, but also that a spectrum line of infinite width is sent in [absorbed], and consider for the case of double emission that the probability of the emission of v_α is not exactly in one eigenfunction A^α, but in the narrow frequency range Δv, then the above equation is to be summed over all the eigenfunctions A^α with frequencies between v and $v + \Delta v$. Using the function $\rho(v)$ which defines the average monochromatic radiation density defined as:

$$V\rho(v)\Delta v = \sum_{v < v_\alpha < v+\Delta v} h v_\alpha n_\alpha^0 \tag{6}$$

one obtains the probability of the double emission per unit time as:

$$w_{nn^0}^{de} = \frac{64\pi^4 v'^3}{3hc^3} \cdot 2\pi \left(\rho(v) + \frac{8\pi v^3 h}{c^3} \right) \Delta v$$

$$\times \frac{1}{3} \left| \sum_{n'} \left(\frac{P_{nn'} P_{n'n^0}}{h(v_{n'n} - v)} + \frac{P_{nn'} P_{n'n^0}}{h(v_{n'n^0} + v)} \right) \right|^2. \tag{7}$$

Thus it is demonstrated that in unoccupied [no radiation field] cavity, a probability for a spontaneous, simultaneous emission of two light quanta exist, and all frequency divisions are possible. If light of frequency v is sent in [absorbed], a portion of the stimulated emission of frequency v' behaves with regard to its intensity, as if it is a virtual oscillator with the moment

$$P_{nn^0}^\alpha = \sqrt{2\pi\rho(v)} p_{nn^0}^\alpha$$

$$p_{nn^0}^\alpha = \sum_{n'} \left(\frac{(P_{nn'}\hat{e})P_{n'n^0}}{h(v_{n'n} - v)} + \frac{P_{nn'}(P_{n'n^0}\hat{e})}{h(v_{n'n^0} + v)} \right) \tag{8}$$

with the frequency v' that is spontaneously emitted. For normal radiation densities, the strength of the stimulated double emission is much less than the corresponding spontaneous double emission. (The same relations exist between ordinary stimulated and spontaneous emission.) This equation (8) is fully similar to that for the Raman effect in which the standard moment is:

$$p_{nn^0}^r = \sum \left(\frac{(P_{nn'}\hat{e})P_{n'n^0}}{h(v_{n'n} + v)} + \frac{P_{nn'}(P_{n'n^0}\hat{e})}{h(v_{n'n^0} - v)} \right)$$

Since the observed intensities also depend on the number of atoms in the original state, which in this case is the excited state, the effect can hardly be observed due to a strong spontaneous emission. Maybe we could find it under metastable states, if the spontaneous emission probability $|P_{nn^0}|^2$ is negligible.

The inverse process of double absorption [two-photon absorption], in contrast, is proportional to the number of atoms in the ground state. The calculations in this case proceed in a similar manner. Only in this case the initial state n^0 of the atom is in the ground state, and before the start of the perturbation in the cavity only the light in two narrow spectral ranges of average frequencies v and v' are present, the average frequency sum is equal to the resonance frequency of the atom, $v + v' = v_{nn^0}$. The eigenfunction for each spectral range can have the same propagation direction s, s' and the same polarization, $\hat{e}, \hat{e}'$ respectively.

From similar considerations as before, the amplitude of the transition probability from n^0 to n, (3) is only different from zero when the state of the cavity evolves from s to s^0 through the absorption of a light quantum v_α in the spectrum range v and another [light quantum] v_β from the spectrum range v'. For such transitions, neglecting the corresponding terms as in (5):

$$a_{ns}^{(2)} = \frac{\dot{Q}_{n_\alpha^0-1,n_\alpha^0}^\alpha \dot{Q}_{n_\beta^0-1,n_\beta^0}^\beta}{c^2}$$

$$\times \sum \left[\frac{(P_{nn'}\hat{e}^\beta)(P_{n'n^0}\hat{e}^\alpha)}{h(v_{n'n^0} - v_\alpha)} + \frac{(P_{nn'}\hat{e}^\alpha)(P_{n'n^0}\hat{e}^\beta)}{h(v_{n'n^0} - v_\beta)} \right]$$

$$\times \left\{ \frac{1 - e^{2\pi i \left(v_{nn^0} - v_\alpha - v_\beta \right)t}}{h(v_{nn^0} - v_\alpha - v_\beta)} \right\}$$

$$= \frac{2\pi h}{V} \sqrt{n_\alpha^0 v_\alpha n_\beta^0 v_\beta}$$

$$\times \sum_{n'} \left[\frac{(P_{nn'}\hat{e}')(P_{n'n^0}\hat{e})}{h(v_{n'n^0} - v_\alpha)} + \frac{(P_{nn'}\hat{e})(P_{n'n^0}\hat{e}')}{h(v_{n'n^0} - v_\beta)} \right]$$

$$\times \left\{ \frac{1 - e^{2\pi i \left(v_{nn^0} - v_\alpha - v_\beta \right)t}}{h(v_{nn^0} - v_\alpha - v_\beta)} \right\}$$

The probability of the process is obtained by the sum of $\left| a_{ns}^{(2)} \right|^2$ over s that is for large cavities through the integration over v_α and v_β. By using in (6) the defined function $\rho(v)$ for the monochromatic radiation intensity per unit volume we obtain in the usual way the sharp resonance at the point $v_\alpha + v_\beta = v_{nn^0}$,

$$\sum_s \left| a_{ns}^{(2)} \right|^2 = 4\pi^2 \cdot \left| \sum_{n'} \left[\frac{(P_{nn'}\hat{e}')(P_{n'n^0}\hat{e})}{h(v_{n'n^0} - v)} + \frac{(P_{nn'}\hat{e})(P_{n'n^0}\hat{e}')}{h(v_{n'n^0} - v')} \right] \right|^2$$

$$\times \int \rho(v_\alpha)\rho(v_\beta) \frac{4\sin^2 \pi \left(v_{nn^0} - v_\alpha - v_\beta \right)t}{h^2 \left(v_{nn^0} - v_\alpha - v_\beta \right)^2} dv_\alpha dv_\beta$$

$$w_{nn_0}^{da} = \frac{16\pi^4}{h^2} \int \rho\left(v_{nn^0} - v \right) \rho(v)dv \left| p^\alpha \hat{e}' \right|^2 \qquad (9)$$

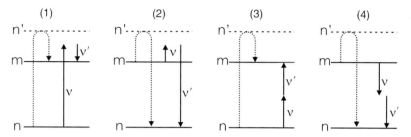

Figure 4.2.1. The broken lines show the behavior of the atom, upward arrows show light quantum absorption, and downward arrows show light quantum emission.

where as p^α signifies the vector defined in (8). The integral ranges over the width of one of the in-going spectral lines [absorption]. It is shown for ordinary light intensities that the probability of the simultaneous absorption is less than that for simultaneous emission. The relation of the probabilities of double emission to double absorption is the same as for ordinary emission and absorption. The processes will therefore not affect the radiation equilibrium [conservation of energy].

The frequency of this process of simultaneous absorption is increased because it is proportional to the number of atoms in the ground state. Conversely, the quadratic dependence on light intensity is unfavorable, so high light intensities are required for observation [of the effect].

It should be noted that both of the processes discussed, like the Raman effect, act as if the two processes, neither of which satisfies the energy rule [energy conservation], take place in one act: the atom with the emission or the absorption of frequency v goes from the state n^0 to an intermediate state n', and from there, under emission or absorption of frequency v', to the final state n.

All of these processes are compared schematically below, for simplicity, all the transitions between two states n and m, for which $v_{nm} > 0$ are illustrated. n' is any intermediate [virtual] state. Figure 1 and 2 show the Stokes' and the anti-Stokes' cases of the Raman effect, Figure 3 the double absorption [two-photon absorption], and Figure 4 the double emission.

2. THE WORKING TOGETHER OF LIGHT AND COLLISIONS [ATOM-ELECTRONS] IN ONE ELEMENTARY ACT

In this part of the work the working together of light and colliding particles, such as electrons, with one atom will be considered.

The electron waves are enclosed in a cubic box of volume V, with the same condition as for the radiation; namely periodic repeated [standing] waves.

In the absence of interaction, the energy of the electrons is only kinetic energy T, and the eigenfunctions are plane waves

$$\Psi_\chi = \frac{1}{\sqrt{V}} e^{\frac{2\pi i}{h}(p_x^\chi x + p_y^\chi y + p_z^\chi z)}$$

in which the components of p^χ are determined by integral numbers related to the definite size of the box and the number of standing waves, for which the energy

E_χ between E and $E + \Delta E$, the moment vector p_χ in the frequency range $\Delta\omega_p$ is given by:

$$N(E)\Delta E \Delta\omega_p = \frac{mV}{h^3}\sqrt{2mE}\,\Delta E \Delta\omega_p \tag{10}$$

The interaction of such a field of free electrons and a radiation field with one atom will be investigated. The Hamiltonian function of the total system is:

$$H = H_A + \sum_\alpha H^\alpha + T + H'$$

in which the interaction energy H' is separated into two parts:

$$H' = V + U$$

the first term V is the interaction between the atom and the radiation,

$$V = -\frac{1}{c}(P\dot{A})$$

and the second term U is the interaction between the atom [nucleus] and the electron which is approximately represented by the Coulomb field. The interaction between the radiation and the electron is neglected.

As in the first part [of the paper] the probability amplitudes $a_{n\chi s}$ are investigated, for the case in which initially there is no radiation in the cavity, thus there is only pure emission processes. Let there be only *one* electron with energy $E_0 = p_{\chi^0}^2/2m$ in the cavity, and initially the atom is in the excited or unexcited state n^0. Since V is independent of the coordinates of the electron, and U is independent of the radiation, the matrix element of H'

$$H'_{n\chi s; n'\chi's'} = V_{ns; n's'}\delta_{\chi\chi'} + U_{n\chi; n'\chi'}\delta_{ss'}$$

thus to a first approximation, the transition probability is split into two additive terms [no cross terms]; that of the light alone, and that of the collision alone. The working together of both can only be obtained in the second approximation [perturbation theory].

The probability amplitude in the second approximation, due to the properties of the perturbation potential is:

$$
\begin{aligned}
a_{n\chi s}^{(2)} = {}& a_{n\chi}^{(2)}\delta_{ss^0} + a_{ns}^{(2)}\delta_{\chi\chi^0}\\
& + \sum_{n'}\left[\frac{U_{nn'\chi\chi^0}V_{n'n^0ss^0}}{h(v_{n'n^0}+v_{ss^0})}\left(\frac{1-e^{\frac{2\pi i}{h}(hv_{nn'}+E_\chi-E_0)t}}{hv_{nn'}+E_\chi-E_0} - \frac{1-e^{\frac{2\pi i}{h}(hv_{nn^0}+hv_{ss^0}+E_\chi-E_0)t}}{h(v_{nn^0}+v\alpha)+E_\chi-E_0} \right) \right.\\
& \left. + \frac{V_{nn'ss^0}U_{n'n^0\chi\chi^0}}{hv_{n'n^0}+E_\chi-E_0}\left(\frac{1-e^{\frac{2\pi i}{h}(v_{nn'}+v_{ss^0})t}}{hv_{nn'}+v_{ss^0}} - \frac{1-e^{\frac{2\pi i}{h}(hv_{nn^0}+hv_{ss^0}+E_\chi-E_0)t}}{h(v_{nn^0}+v_{ss^0})+E_\chi-E_0} \right) \right]
\end{aligned}
\tag{11}
$$

In the second approximation, the terms $a_{ns}^{(2)}\delta_{\chi\chi^0}$ and $a_{n\chi}^{(2)}\delta_{ss^0}$ are the contributions only due to collisions and those only due to radiation. For processes in which both light and collisions contribute, and for which both χ [letter k corrected by translator] and s vary, they are zero.

The emission probability of light is investigated for the case in which the emission frequency ν is not equal to the eigenfrequency of the atom. From the same considerations as in the previous paragraphs, $a_{n\chi s}^{(2)}$ is only different from zero for the state χ in which the energy rule is approximately followed, that is for which $h\nu_{nn^0}+h\nu_\alpha+E_\chi-E_0$ is small, and for such states in which the additive terms

$$\frac{1-e^{\frac{2\pi i}{h}\left(h\nu_{nn'}+E_\chi-E_0\right)t}}{h\nu_{nn'}+E_\chi-E_0}, \qquad \frac{1-e^{\frac{2\pi i}{h}\left(\nu_{nn'}+\nu_\alpha\right)t}}{h\left(\nu_{nn'}+\nu_\alpha\right)}$$

can be neglected. Therefore,

$$a_{n\chi s}^{(2)} = \sqrt{\frac{2\pi h\nu_\alpha}{V}}\sum_{n'}\left(\frac{U_{nn'\chi\chi^0}(P_{n'n^0}A^\alpha)}{h(\nu_{n'n^0}+\nu_\alpha)}+\frac{(P_{nn'}A^\alpha)U_{n'n^0\chi\chi^0}}{h\nu_{n'n^0}+E_\chi-E_0}\right)$$

$$\times \frac{1-e^{\frac{2\pi i}{h}\left(h\nu_{nn^0}+h\nu_\alpha+E_\chi-E_0\right)t}}{h\nu_{nn^0}+h\nu_\alpha+E_\chi-E_0} \tag{12}$$

To obtain the probability of the emission of light in a narrow frequency range, $\Delta\nu$, with an average frequency ν, in the transition of an atom from the state n^0 to the state n, the term $\left|a_{n\chi s}^{(2)}\right|$ must be multiplied with the number of eigenfunctions in the range $\Delta\nu$, that is $Z(\nu)\Delta\nu$, and summed over all χ, which in a large cavity is an integral over $N(E)dE$ [see equation (10)].

Due to the sharp maximum in the integrand we obtain in the usual manner:

$$\sum_{s\Delta\nu}\sum_\chi\left|a_{n\chi s}^{(2)}\right|^2 = \frac{2\pi h\nu}{V}Z(\nu).$$

$$\frac{1}{3}\int\left|\sum_{n'}\frac{U_{nn'\chi\chi^0}P_{nn^0}}{h(\nu_{n'n^0}+\nu)}+\frac{P_{nn'}U_{n'n^0\chi\chi^0}}{h\nu_{n'n^0}+E_\chi-E_0)}\right|^2 d\omega_{p\chi}$$

$$\times \int N(E)\frac{2-2\cos 2\pi\left(\nu_{nn^0}+\nu_\alpha+\frac{E-E_0}{h}\right)t}{(h\nu_{nn^0}+h\nu_\alpha+E-E_0)^2}$$

$$= \frac{8\pi^3}{3V}\nu Z(\nu)N(E_0-h(\nu_{nn^0}+\nu_\alpha))\cdot\Delta\nu\cdot t$$

$$\times \int\left|\frac{P_{n'n^0}U_{nn'\chi\chi^0}}{h(\nu_{n'n^0}+\nu)}+\frac{P_{nn'}U_{n'n^0\chi\chi^0}}{h(\nu_{n'n}-\nu)}\right|^2\cdot d\omega_{p\chi}$$

This probability is normalized on a density $1/V$ for the incident electrons. If we consider a stream of incident electrons, of current density one, then the result must be multiplied by:

$$\frac{V}{v_0} = \frac{V}{\sqrt{\frac{2}{m}E_0}}$$

Finally, by substituting the values for z and N, the probability for the combined effect is obtained.

$$w_{nn^0} = \frac{64\pi^4}{3} \cdot \frac{v^3 m^2}{h^3 c^3} \sqrt{1 - \frac{h(v_{nn^0} + v_\alpha)}{E_0}} V^2$$

$$\times \int \sum_{n'} \left| \frac{U_{nn'\chi\chi^0} P_{n'n^0}}{h(v_{n'n^0} + v)} + \frac{P_{nn'} U_{n'n^0\chi\chi^0}}{h(v_{n'n} - v)} \right|^2 d\omega_{p\chi} \tag{13}$$

The result does not depend on the size of the box V, because $U_{n'n^0} \chi' \chi^0 V$ is inversely proportional.

Let n^0 be an excited state and n be a lower state of the atom, therefore $v_{n^0 n} > 0$. It can be shown that a transition in an atom from n^0 to n can emit light of frequencies v that are higher or lower than the exciting frequency $v_{n^0 n}$. The probability is greater when only small amounts of the energy is transferred to the colliding particles, and when the frequency v is close to $v_{n^0 n}$ There occurs in the above equation in the summation over n' other terms in which $n' = n^0$ or $n' = n$. They have the corresponding values:

$$\frac{P_{nn^0} U_{n^0 n^0 \chi\chi^0}}{h(v_{n^0 n} - v)} \quad \text{for } n' = n^0$$

$$\frac{P_{nn^0} U_{nn\chi\chi^0}}{h(v_{nn^0} + v)} \quad \text{for } n' = n$$

These two terms give a resonance for $v = v_{n^0 n}$. When v is close to $v_{n^0 n}$ they will dominate all other terms. In the corresponding equations for the Raman effect, and the double emission effect, respectively, this resonance does not occur. This result agrees with the experimental results of Oldenberg.

For the excitation luminescence effect discussed in the introduction, the initial conditions are as before: the atom at the start of the process is in an initial state n^0, which can be the ground state, but the kinetic energy of the colliding electron $E_0 = p_{\chi 0}^2/2m$ is now much greater than the ionization energy of the atom.

The transition probability of the atoms from a state n of discrete spectrums is analyzed, and it will be shown that there is a possibility that the atom arrives at this state by way of a detour through a continuous spectrum.

In order to calculate the probability for such processes, it is necessary to return to equation (11). The summation over n' implies the integral dE' is over a continuous spectrum. The assumptions made that lead from equation 11 to equation 12 are here not justified since the denominator of some of the dropped terms can be zero, even when the energy rule is followed. But due to the initial conditions a few further simplifications can be made.

Since n^0 represents the ground state of the atoms and s^0 is the unoccupied cavity, $v_{nn^0} > 0$ and $v_{ss^0} = v_\alpha > 0$. The denominator $v_{n'n^0} + v_\alpha$ of the first summation

in (11) is always large. Conversely, for the second summation, all three denominators can at the same time be zero. When we only look at those processes, the last
term dominates over the first term by far, and reflects by itself the main aspects of
the effect.

Therefore, mainly

$$a_{n\chi s}^{(2)} = \sqrt{\frac{2\pi h\nu_\alpha}{V}} \sum_{n'} \frac{(P_{nn'}A^\alpha)U_{n'n^0\chi\chi^0}}{h\nu_{n'n^0} + E_\chi - E_0}$$

$$\times \left(\frac{1 - e^{2\pi i(\nu_{nn'} + \nu_\alpha)t}}{h(\nu_{nn'} + \nu_\alpha)} - \frac{1 - e^{\frac{2\pi i}{h}(h\nu_{nn^0} + h\nu_\alpha + E_\chi - E_0)t}}{h(\nu_{nn^0} + \nu_\alpha) + E_\chi - E_0} \right)$$

Through rearrangement of terms the equation yields:

$$a_{n\chi s}^{(2)} = \sqrt{\frac{2\pi h\nu_\alpha}{V}} \cdot \frac{1}{h(\nu_{nn^0} + \nu_\alpha) + E_\chi - E_0} \sum_{n'} (P_{nn'}A^\alpha)U_{n'n^0\chi\chi^0}$$

$$\times \left[\frac{1 - e^{2\pi i\left(\nu_{n'n^0} + \frac{E_\chi - E_0}{h}\right)t}}{h\nu_{n'n^0} + E_\chi - E_0} \cdot e^{2\pi i\left(\nu_{nn^0} + \nu_\alpha + \frac{(E_\chi - E_0)}{h}\right)t} \right.$$

$$\left. - \frac{1 - e^{2\pi i(\nu_{nn'} + \nu_\alpha)t}}{h(\nu_{nn'} + \nu_\alpha)} \right]$$

Here the summation is over n' respectively, and integration over dE', those terms are
mainly considered for which the denominator becomes small.

If the resonances are situated in the continuum, then the integrand for $\nu_{E'n^0} = E_0 - E_\chi$, $\nu_{E'n} = \nu_\alpha$ shows a strong maximum and becomes small at larger distances
[from maximum]. The product $P_{nE'}U_{E'n^0\chi\chi^0}$ is slowly varying so it can be taken as
a constant, within the range, equal in value to that at the resonance point, and the
integral can be approximately evaluated. For the discrete eigenfunctions an error is
caused when discrete values of $P_{nn'}$, $U_{n'n^0\chi\chi^0}$ in n' are interpolated from the continuous
function E' in which the summation is replaced by an integration. It will be shown later
that for the special case of high speed colliding particles that this error is small, since
the energy changes $h\nu_{En^0}$, $h\nu_{E'n}$ which occur have resonance points which are in the
continuum.

Thus the approximation is made:

$$a_{n\chi s}^{(2)} = \sqrt{\frac{2\pi h\nu_\alpha}{V}} \cdot \frac{1}{h(\nu_{nn^0} + \nu_\alpha) + E_\chi - E_0} \left[(P_{nE^1}A^\alpha)U_{E^1n^0\chi\chi^0} \right.$$

$$\times e^{2\pi i\left(\nu_{nn^0} + \nu_\alpha + \frac{E_\chi - E_0}{h}\right)t} \cdot \int \frac{1 - e^{-2\pi i\left(\frac{E_\chi - E_0}{h} + \nu_{E'n^0}\right)t}}{\nu_{E'n^0} + \frac{E_\chi - E_0}{h}} d\nu_{E'n^0}$$

$$+ (P_{nE^2}A^\alpha)U_{E^2 n^0 \chi \chi^0} \int \frac{1 - e^{2\pi i(\nu_{nE'} + \nu_\alpha)t}}{\nu_{nE'} + \nu_\alpha} d\nu_{nE'} \Bigg]$$

$$a_{n\chi s}^{(2)} = \sqrt{\frac{2\pi h \nu_\alpha}{V}} \cdot \frac{i\pi}{h(\nu_{nn^0} + \nu_\alpha) + E_\chi - E_0} \Bigg[(P_{nE'}A^\alpha)U_{E'n^0 \chi \chi^0}$$

$$\times e^{2\pi i \left(\nu_{nn^0} + \nu_\alpha + \frac{E_\chi - E_0}{h} \right)t} + \left(P_{nE^2}A^\alpha \right) U_{E'n^0 \chi \chi^0} \Bigg]$$

[The last term in the equation above was corrected by the translator.]
The atomic states E^1, E^2 are the points of resonance, and depend on χ and s so:

$$h\nu_{E^1 n^0} + E_\chi - E_0 = 0, \quad \nu_{nE^2} + \nu_\alpha = 0 \tag{14}$$

The transition probability of atom from state n^0 to state n is $\left| a_{n\chi s}^{(2)} \right|^2$ summed over χ and s. First, the sum is made over χ, which for a large cavity is an integration over $N(E)dE_\chi d\omega_\chi$ (see equation 10). Again only the domain is considered in which the denominator, which represents the conservation of the total energy of the process is small, since for this point the atomic states E^1 and E^2 are equal, $E^1 = E^2 = E$, we obtain:

$$\sum_\chi \left| a_{n\chi s}^{(2)} \right|^2 = \frac{2\pi^3 h \nu_\alpha}{V} N(E_0 - h\nu_{En^0}) \left| P_{nE}A^\alpha \right|^2$$

$$\times \int \left| U_{En^0 \chi \chi^0} \right|^2 d\omega_{p\chi} \cdot \frac{4\pi^2}{h} \cdot t \tag{15}$$

Here the state E and the energy given up by the colliding electrons, $E_\chi - E_0$, shown in (14) depends on s, ν_α thus:

$$h\nu_{nn^0} + h\nu_\alpha + E_\chi - E_0 = 0, \quad E_0 - E_\chi = h\nu_{En^0}, \quad \nu_\alpha = \nu_{En} \tag{16}$$

Equation (15) gives the probability for an atomic transition to the state n with emission of a light quantum of frequency $\nu_\alpha = \nu_{En}$ to the standing wave A^α within the cavity. The equation (16) shows that this process acts as if it is two separate processes, each one satisfying the energy rule, takes place in a single action, specifically, the ionization of an atom through collision, in which the electron give up the energy $h\nu_{En^0}$ and then there is recombination with emitted radiation [outgoing radiation] of frequency $\nu_\alpha = \nu_{En}$.

To obtain the total transition probability of the atom in the state, equation (15) must be summed over all s or over all α, which is as in the first paragraph, an integration over $Z(\nu_\alpha)d\nu_\alpha$. Referring to this probability, instead of to a density $1/V$ of the incident electrons, to a current density of one electron per second across a cross section perpendicular to the direction of propagation of the electrons, we must multiply by

$Vm/\sqrt{2mE_0}$, then for the transition probability of atoms from state n^0 to n:

$$w_{nn^0} = \frac{8\pi^5}{3}\sqrt{\frac{m}{2E_0}}\int \nu_{En}Z(\nu_{En})\,|P_{nE}|^2 \cdot N(E_0 - h\nu_{En^0})$$

$$\times \int |U_{En^0\chi\chi^0}|^2 d\omega_\chi d\nu_{En}$$

$$= \frac{64\pi^6}{3}\frac{m^2}{h^3c^3}V^2\int \nu_{En}^3\,|P_{nE}|^2\sqrt{1 - \frac{h\nu_{En^0}}{E^0}}\times \int |U_{En^0\chi\chi^0}|^2 d\omega_\chi d\nu_{En}$$

The matrix elements in the equation are expressed by the emission coefficients:

$$A_{nE} = \frac{64\pi^4\nu_{En}^3}{3c^3h}\,|P_{nE}|^2$$

and through the theoretical value of the ionization probability through collision [compare with Born[1]] which in this relation is:

$$S_{En^0} = \frac{4\pi^2m^2}{h^4}\sqrt{1 - \frac{h\nu_{En^0}}{E_0}}\,V^2\int |U_{En^0\chi\chi^0}|^2 d\omega_{p\chi}$$

we obtain the final equation:

$$w_{nn^0} = \frac{h}{4}\int A_{nE}S_{En^0}dE \tag{17}$$

Subsequently, the previous method of interpolation of discrete matrix elements through a continuous function of energy is justified. Since the collision excitation probability is low for high velocities of colliding electrons, and is only essential for transitions in the continuum which follows from consideration of the Born equations. Therefore, in the final equation it is sufficient that the integral is made only over the continuum.

This process of collision ionization and simultaneous recombination is an analogue to the process described by Franck which was discussed in the introduction [to this paper]. This process [and the calculation relates to this case] is really a unified act, and not two independent processes, which follows from the proportionality of the frequency with t. Since the frequency of succession for two independent processes— such as excitation by collision, and later light emission after a delay of 10^{-8}sec—must always be proportional to t^2 for time intervals less that 10^{-8}sec, and the calculation is valid for these cases.

This equation yields the transition probability from state n through collision with high velocity electrons. The atom can then in a second, independent process, emit lines from the discrete spectrum.

---

[1] M. Born, Ztschr. f. Phys. 38, S. 803. 1926; Nachrichten der Göttingen Ges. der Wissenschaften 1926, S. 146

Equation (17) makes it understood that the maximum intensities of these lines is in the region where the total ionization $\int S_{En^0}dE$ is large. Also explained is the complete displacement of the intensities as compared with direct excitation by collision, for which the excitation from the state n is proportional to the matrix element $\left|U_{nn^0 \chi \chi^0}\right|^2$, and thus gives qualitatively the effects that were discussed in the introduction.

Finally, an estimate of the magnitude of the probability for such excitation in relation to the probability of ionization should be given.

The total probability is:

$$w_I = \int S_{En^0} dE$$

and therefore:

$$\frac{w_{nn^0}}{w_I} = \frac{h}{4} \frac{\int A_{nE} S_{En^0} dE}{\int S_{En^0} dE}$$

$\overline{A}$ is the mean value of A_{nE}, thus:

$$w_{nn^0} \approx \frac{h}{4} \overline{A} \int S_{En^0} dE$$

and

$$\frac{w_{nn^0}}{w_I} \approx \frac{h}{4} \overline{A}$$

Therefore:

$$\int A_{nE} dE \approx \overline{A} \Delta E$$

the emission coefficient for the entire continuum is the same order of magnitude as the emission coefficient for a line in the discrete spectra. This shows the inverse of the average time interval, and is about $[10^{-8} \ \text{sec}^{-1}]$.

Therefore,

$$\overline{A} \approx \frac{10^8}{\Delta E}$$

and

$$\frac{w_{nn^0}}{w_I} \approx \frac{h \cdot 10^8}{\Delta E} = \frac{10^8}{\Delta \nu}$$

The width of the continuous line spectra is about

$$\Delta \nu = 10^{15} \ \text{sec}^{-1}$$

then:

$$\frac{w_{nn^0}}{w_I} \approx \frac{10^8}{10^{15}} \approx 10^{-7}$$

This is a very rough estimate. A more exact consideration of the behavior of this transition probability could only be obtained though numerical evaluation of the matrix elements P_{nE}, U_{En^0}, perhaps for hydrogen.

The fraction w_{nn^0}/w_I is the number of atoms, that in a single act, double process are brought to the state n, compared to the number of the atoms that are ionized. This is measured experimentally by the intensity of a line, for example, the resonance line $v_{n^1n^0}$. The atoms induced to a higher state n usually radiate the resonance line $v_{n^1n^0}$, the observed intensity is not only proportional to w_{nn^0}, but also depends on the transition probability w_{nn^0} of all the higher states.

I greatly thank Professor Born for friendly help and stimulation, without which this work would not come to this state. I am also indebted and give great thanks to V. Weisskopf for discussions and help.

John [read *Johns*] Hopkins University
(Received 7 December 1930).

ACKNOWLEDGMENT

I thank Katie Lagoni-Masters for checking my translation.

NOTES ON THE TRANSLATION

1. The numbering of the equations corresponds to the equation numbering in the original paper.
2. When a word was added in the English translation it is placed within square brackets.
3. When the translator made a correction to either the text or to the equations a note to that effect was placed within square brackets and italics.
4. The references in the original German paper are not translated into English and appear in their original form and position.
5. When the translator added an important reference it was placed within square brackets and italicized.

4.3 NOTES ON MARIA GÖPPERT-MAYER'S THEORY OF TWO-QUANTUM PROCESSES*

Barry R. Masters

4.3.1 Introduction

What experimental studies influenced Maria Göppert-Mayer's selection of her thesis topic? Did Einstein's paper published in 1905 stimulate her thinking about two-photon processes? What was the influence of Dirac's quantum theory? How does Dirac's 1927b paper on the quantum theory of dispersion differ from Maria Göppert-Mayer's 1931 thesis? In this section I present background material that the reader may find helpful in order to answer these questions.

In the late 1920s Maria Göppert-Mayer worked out the theory of two-photon excitation and emission processes. Under the supervision of her mentor and doctoral thesis advisor Professor of Theoretical Physics Max Born, Maria Göppert-Mayer developed her doctoral research on the theory of "two quantum processes." Specifically she worked on the theory of two light quanta working together in an elementary act. Her research resulted in two publications written in German: a one-page paper in 1929 in the journal *Die Naturwissenschaften*, and an extensive and detailed dissertation paper published in 1931 in the journal *Annalen der Physik*.

At the start of her theoretical work, she focused on a theoretical explanation, based on quantum theory, to explain the experimental findings of Franck and Oldenberg (Franck, 1928; Oldenberg, 1928). These two experimental physicists worked in the Zweites Physikalisches Institut, which was located in the same building as Born's Institut für Theoretische Physik. Their experiments involved the collision of electrons with atoms and the induced electronic excitation of the atoms. Franck and Oldenberg asked the following question: Is the energy transfer quantized when the energy comes from the kinetic energy of the colliding particles? Maria Göppert-Mayer worked out a theory that not only explained the experimental data of Franck and Oldenberg, but also calculated the transition probabilities for two-photon absorption, two-photon emission, and two-photon Raman processes for the Stokes and the anti-Stokes cases.

The theoretical work of Maria Göppert-Mayer provided the theory of multiphoton excitation processes and provided the groundwork for the development of multiphoton excitation microscopy half a century later. Her complete doctoral publication of 1931 written in German is reprinted in *Selected Papers on Multiphoton Excitation Microscopy* (Masters, 2003).

The origin of her thesis work was the work of Oldenberg (1928) and Franck (1928) on electronic excitation of atoms due to collisions with electrons and the subsequent luminescence. The key finding was that the kinetic energy of the electrons in the inelastic collision was equal to the energy of the emitted light. The significance of these important

Figure 4.3.1. Maria Göppert-Mayer in middle age (AIP Emilo Segré Visual Archives, Physics Today Collection).

experiments is that they demonstrated the discrete energy levels of atoms. The inelastic collisions of electrons and atoms can result in the transfer of energy to the atoms and can excite the atoms without ionizing them. These studies demonstrated the discrete energy levels of atoms independently of the optical spectroscopic studies.

These works were conducted at the Institute of Experimental Physics, or the Zweites Physikalisches Institut, headed by Franck at the University of Göttingen and located in the same building where Maria Göppert worked under Born.

Earlier, in 1914, James Franck and Gustav Hertz investigated the elastic scattering of electrons from mercury atoms. When the electrons were subjected to a potential difference of 4.9 volts there was an abrupt change in their scattering with mercury atoms. At kinetic energies up to 4.9 volts the electrons were elastically scattered from the mercury atoms; however, at that critical kinetic energy the electrons lose all their kinetic energy (inelastic collision with mercury atoms) and the mercury atoms emit monochromatic light.

These experiments demonstrate a reversal of the photoelectric effect that was earlier described and interpreted by Einstein. In that case incident light of the proper frequency caused the release of electrons from a metal surface. To summarize, when the kinetic energy of the colliding electrons was below $h\nu$, where ν is the frequency of the resonance line, the electrons are scattered elastically. When the kinetic energy of the electrons attains the value of $h\nu$, the kinetic energy is transferred to the mercury atoms in the inelastic collision and light emission with a frequency ν is emitted from the atom.

The proposal that light acts as if it consists of discrete units of energy or quanta of energy was made by Einstein in 1905. Einstein concluded from his statistical thermodynamic analysis that monochromatic radiation behaves as if it consisted of independent energy quanta (Einstein, 1905). That was five years after Max Planck derived his law for the frequency distribution of black body radiation. Planck's central assumption in his derivation was that the energy is distributed into discrete quanta on a large number of charged harmonic oscillators that are in thermal equilibrium with the thermal radiation contained in a black body cavity.

Exactly what words did Einstein write in 1905, and how do they relate to the current conception of multiphoton excitation processes that derive from the 1929 publication of Maria Göppert? Einstein published his paper in 1905 in the German journal *Annalen der Physik*. The paper had the title "On a heuristic point of view concerning the production and transformation of light." On page 145 of his paper, in the section on Stokes rule, Einstein wrote

Abweichungen von der Stokesschen Regel sind . . .
1. wenn die Anzahl der gleichzeitig in Umwandlung begriffenen Energiequanten pro Volumeneinheit so gross ist, dass ein Energiequant des erzeugten Lichtes seine Energie von mehreren erzeugenden Energiequanten erhalten kann.

I translate the German into English as follows:

... deviations from Stokes rule are conceivable in the following cases:
1. when the number of simultaneously transforming energy quanta per unit volume is so large that an energy quantum of the light produced could obtain its energy from several producing energy quanta.

That statement of Einstein appears to indicate that several energy quanta could contribute to the energy quanta contained in the emitted light. To conclude: *hoc non est demonstratum, sed suppositio quaedam.* One can loosely interpret Einstein's 1905 statement to be related to the modern theory of multiphoton excitation processes as developed in the 1931 dissertation of Maria Göppert.

Einstein wrote that deviations from Stokes rule can occur in several cases. By way of review, Stokes rule states that there is a Stokes shift or gap between the maximum of the first absorption band and the maximum of the fluorescence spectrum (expressed in wavenumbers). Therefore, the energy of the emitted light is less than the energy of the excitation light. There is also the case where Stokes rule is not followed, and that is called anti-Stokes emission. When Stokes rule is not observed, the energy of the emitted photon is greater than that of the excitation photon.

One possibility to account for the case in which there are deviations from Stokes rule is that several incident quanta contribute to the energy of the emitted quanta. That is the case that Einstein suggested in his 1905 paper.

I now return to Göttingen in the years 1926–1931. There are many multiphoton processes in which the processes of sequential excitation are paramount. For example, the first photon results in excitation to the first excited state, and the second photon results in excitation to a higher excited state. These processes were well known to Franck and coworkers in Göttingen during the time Maria Göppert-Mayer worked on her theory of two-photon excitation processes in atoms. As shown by Maria Göppert-Mayer in her 1931 theory, two-photon excitation processes occur when a near-simultaneous interaction of two photons, each with insufficient energy to excite the upper state of an atom, interact via a "virtual" intermediate state. The "virtual" intermediate states are not true eigenstates of the atom. To conclude, the ambiguous statement contained in Einstein's 1905 paper regarding the conception that several light quanta can contribute to the energy quanta contained in the emitted light appears to be related to multiphoton excitation processes; however, perusal of both the 1905 *Annalen* paper by Einstein and the 1931 *Annalen* paper by Maria Göppert-Mayer demonstrate that the linkage between these two phenomena is at best a weak link.

In order to obtain a sense of the physical theories and techniques that were in use at the time of Maria Göppert's graduate work I recommend that the reader study the book *Elementare Quantenmechanik* (in German), which was written by Born and Jordan in 1930. Maria Göppert contributed a section to this book on Dirac's theory of emission, absorption, and dispersion that is similar to her thesis work, and Born and Jordan acknowledged her contribution in their foreword. Note that in her 1929 paper she stated that Dirac's (1926) theory described not only the Raman effect but also the reverse process in which two photons act together in a single elementary act to promote an atom from the ground state to an excited state. In her subsequent more complete calculation published in 1931 on two-photon processes, she used the Dirac formulation of the time-dependent perturbation theory. In addition, it is useful to read the Göttingen dissertation on resonance fluorescence by her fellow graduate student, Victor Weisskopf (Weisskopf, 1931).

In the spring of 1930 she married Joseph Mayer, completed her thesis and exam, and moved to Baltimore. She worked in the Physics Department at Johns Hopkins University, where her husband Joseph Mayer obtained a position in the Department of Chemistry. She also worked during the next three summers in Göttingen together with Max Born. Her thesis publication was received by *Annalen der Physik* on December 7, 1930. This is the first publication in which she used her married name Maria

Göppert-Mayer, and she ended the paper with her new home: John [read *Johns*] Hopkins University (see error notes below).

Maria Göppert-Mayer worked on the theory of atom–photon interactions. Building on the dispersion theory of Kramers and Heisenberg, and the time-dependent perturbation theory of Dirac, she developed analytical expressions of the transition probability for multiphoton absorption and stimulated emission as well as Raman scattering processes.

An alternative modern approach to these problems that is based on nonperturbative calculations of transition amplitudes is contained in the book *Atom–Photon Interactions, Basic Processes and Applications* (Cohen-Tannoudji et al., 1992). The following general definitions taken from their book are useful:

Emission processes: a new photon appears.

Absorption processes: a photon disappears.

Scattering processes (i.e., Raman scattering): a photon disappears and another photon appears.

Multiphoton processes: several photons appear or disappear.

Because of the historical importance of this scientific publication as well as her 1929 paper, I translated both papers into English. Perusal of both publications in the course of the translations uncovered several errors made by the author; however, they were corrected by the translator and the changes are indicated in the text. In order to understand the quantum physics as it was known in the period of her doctoral work I studied the papers cited in her two publications as well as other books and publications on quantum theory that were published in the period from 1925 to 1931. In addition, the six-volume work *The Historical Development of Quantum Theory* provides a wealth of historical details (Mehra & Rechenberg, 2001). These readings provide the correct framework for the state of quantum theory at the time of her graduate work.

Maria Göppert-Mayer was very familiar with the latest advances of quantum theory through the published works of her mentor Max Born and related papers of Kramers and Heisenberg (Kramers & Heisenberg, 1925). She gained insight and advice from Victor Weisskopf, who was a fellow doctoral student under Max Born (Weisskopf, 1931). In her doctoral thesis she freely acknowledged her use of Paul Dirac's second-order, time-dependent perturbation theory that was published in 1927. Because of the importance of the Dirac theory in her work I include a section that presents a qualitative discussion of his seminal papers (Dirac, 1927a, 1927b).

A direct reading of the English translation of Maria Göppert-Mayer's publications of 1929 and 1931 may not be clear to those readers who are not familiar with quantum theory; therefore, I provide these introductory notes, which present the central concepts, her assumptions, and her theoretical approaches. Similar notes are provided as a qualitative introduction to the papers cited in her papers. I hope that these qualitative descriptions and summaries of the theory will make the English translations of her two papers more comprehensible. Many of the assumptions and approximations that she used in her work are also developed and discussed in Chapter 5, which is devoted to the theory of nonlinear fluorescence spectroscopy.

To facilitate the organization of these notes I list the complete citations of papers that are analyzed and described in the following sections:

1. Kramers, H. A. and Heisenberg, W. (1925). Über die Streuung von Strahlen durch Atome [On the Dispersion of Radiation by Atoms]. *Zeitschrift für Physik*, 31: 681–708.
2. Oldenberg, O. 1928. Über ein Zusammenwirken von Stoss und Strahlung. Zeitschrift für Physik, 51: 605–612.
3. Franck, J. 1928. Beitrag zum Problem der Wiedervereinigung von Ionen und Elektronen. Zeitschrift für Physik, 47: 509–516.
4. Fermi, E. 1932. Quantum Theory of Radiation, *Rev. Mod. Phys.* 4: 87–132.
5. Dirac, P. A. M. 1926. On the theory of Quantum Mechanics, *Proceedings of the Royal Society (London), Series A, Containing Papers of a Mathematical and Physical Character*, Vol. 112, No. 767 (October 1, 1926), 661–677. [This paper is not discussed but it is highly recommended for the reader.]
6. Dirac, P. A. M. 1927a. The Quantum Theory of Emission and Absorption of Radiation. *Proceedings of the Royal Society of London, Series A, Containing Papers of a Mathematical and Physical Character*, Vol. 114, No. 767 (March 1, 1927), 243–265.
7. Dirac P. A. M., 1927b. The Quantum Theory of Dispersion, *Proceedings of the Royal Society (London), Series A, Containing Papers of a Mathematical and Physical Character*, Vol. 114, No. 767 (May 2, 1927), 710–728. (Received April 4, 1927). This paper was communicated to the Royal Society by R. H. Fowler, F. R. S.).
8. Göppert, M. 1929. Über die Wahrscheinlichkeit des Zusamenswirkens zweier Lichtquanten in einem Elementarakt. *Die Naturwissenschaften*, 17: 932.
9. Göppert-Mayer, M. 1931. Über Elementarakte mit zwei Quantensprüngen, *Ann. Phys.* (Leipzig), 9: 273–294.

Prior to a more detailed discussion of the theoretical work on two-photon processes that was the focus of Maria Göppert's thesis work, I present a more general overview of the contemporary developments in quantum theory and how she built her theoretical work on the prior theory.

The following qualitative introduction to Dirac's radiation theory is based on Fermi's excellent review article (Fermi, 1932). Dirac had a simple idea for his theory: he treated the atom and the radiation field with which it interacts as a single system. The energy of the single system is composed of three terms: the energy of the atom is the first term, the electromagnetic energy of the radiation field is the second term, and the very small energy of the interaction between the atom and the radiation field is the third term. Without the third interaction term the atom could not absorb nor emit any radiation.

Fermi presents a simple analogy, that of a pendulum of mass M and an oscillating string located near the pendulum. The atom is represented by the pendulum, and the radiation field is represented by the oscillating string. When the string and the pendulum are not connected, both oscillate independently and the total energy is the sum of the energy of the pendulum and the string; there is no interaction term. Fermi represents the coupling term by a very thin elastic thread that connects the oscillating pendulum and the oscillating string.

First, he makes the analogy for the absorption of energy by the atom from the radiation field. At time $t=0$, the string is vibrating and the pendulum is stationary. When the thin elastic thread is connected to the oscillating pendulum, the vibrating string energy can be transferred from the vibrating string to the pendulum. If the period of oscillation of the pendulum and the period of vibration of the string are very different, then the amplitude of oscillation of the pendulum remains very small. However, if the period of the vibration of the string is equal to the period of oscillation of the pendulum, then the amplitude of the oscillation of the pendulum will increase by a large amount. This represents absorption of light by the atom.

Second, Fermi describes the case that corresponds to the emission of radiation by an atom. At time $t=0$, the pendulum is oscillating and the string is stationary. The forces are transmitted through the string from the oscillating pendulum to the string. The string will begin to vibrate. However, only the harmonics of the string whose frequencies are close to the frequency of the pendulum results in a vibration with a large amplitude. That is the analogy of the emission of radiation of the atom.

Maria Göppert-Mayer skillfully applied and developed the previous theoretical approaches of Paul Dirac to her thesis problem of "double transitions" or two-photon transitions.

In her 1931 paper she mentioned the "method of variation of constants." Dirac developed this method in his papers of 1926 and 1927. The method of variation of constants is the formulation of a time-dependent perturbation theory that was developed by Dirac (1926, 1927a, 1927b). The state of an atom is represented by an expansion in terms of the unperturbed energy eigenfunctions. The Hamiltonian operator is different from the true Hamiltonian by a very small term that is the perturbation. It is called the method of variation of constants because the constant coefficients used in the expansion of the wave function in terms of the true energy eigenfunction vary with time. Further background information on the historical development of perturbation theory is given in Chapter 3.

In a double transition or a two-photon transition via intermediate states or virtual state, each part of the transition does not obey the conservation of energy law; however, the total transition from the initial state to the final state follows the law of conservation of energy. A two-photon transition has a probability that is proportional to the square of the intensity of the light, or to the fourth power of the electric field of the radiation. This is consistent with the probability of each of the steps being proportional to the intensity of the light.

Since the double or two-photon transitions are related to the square of the intensity of light they were very improbable with the light sources available prior to the development of high-power pulsed lasers. However, prior to that development many examples were observed of the two-photon transition that occurs under the combined action of radiation and another perturbation that does not depend on time—for example, the experimental studies of Oldenberg and Franck that are cited in the 1931 paper of Maria Göppert-Mayer. In the second part of her 1931 thesis, she calculated the probabilities of the combined action of light and electron collisions in the transition of atoms.

Another example of double transitions or the working together of two elementary processes in a single act is the scattering of light. This can be viewed as the absorption of a photon and the spontaneous emission of another photon. These processes were also calculated in the Maria Göppert-Mayer's 1931 paper.

Maria Göppert-Mayer built on Dirac's time-dependent perturbation theory to calculate transitions between discrete energy levels as well as in physical systems with continuous energy levels. For example, in particle collisions, the eigenfunctions of the free particles, i.e., colliding electrons with atoms, are described as plane waves, and the energy of the particles is not quantized, but can take different positive values. If the particles are now confined to a box, of side length L, the eigenvalues or the energy of the particle is now quantized. As the side of the box increases to infinity, the free particle eigenfunctions and energy eigenvalues approach those of the free particle. For a free particle in a box, the quantized energy eigenvalues can be calculated by perturbation theory for discrete energy levels. Then the size of the box is increased to infinity, and the result obtained is valid for continuous energy levels.

Raman scattering is another example of a two-photon process; a photon is absorbed and another photon is emitted, and the atom makes a transition from the initial to the final state. The energy difference between the initial and the final states is equal to the energy difference of the two photons. Second-order perturbation can be used to calculate the Raman transition probabilities. Transition probabilities are the square of the transition amplitudes for the process. Time-dependent perturbation is required to calculate the rates of transitions.

With that general introduction to the experimental background and the concurrent theoretical developments in quantum theory of atom–radiation interactions, Maria Göppert-Mayer began her thesis work on two-quantum transitions. Within a few years she was able to publish a brief one-page summary of her theoretical results (Göppert, 1929). She signed her paper with the name "Maria Göppert" as she did not marry Joseph Mayer until 1930.

4.3.2 Summaries of Experimental Papers Cited in Maria Göppert-Mayer's 1931 Paper

4.3.2.1 Introduction to the 1928 Paper of Oldenberg

Oldenberg, O. 1928. Über ein Zusammenwirken von Stoss und Strahlung. Zeitschrift für Physik, 51: 605–612.

In the this section I proceed to briefly review two experimental papers which initiated the theoretical research of Maria Göppert-Mayer on two-quantum processes.

Oldenberg performed these electron–atom collision experiments in the "Zweites Physikalisches Institut," in Göttingen. The basis of his experiments was the question: Could an atom become excited [its electrons in higher energy states than the ground electronic state] through the working together, in a single act, of collisions with electrons and an incident light field. He also discussed the concept that two light quanta can work together in one elementary act to excite an atom or molecule (i.e., the Smekal-Raman effect; Smekal, 1923).

Oldenberg produced experimental evidence on the broadening of the resonance lines of mercury atoms when the excited atoms collided many times with slow particles. He showed that the excitation energy of the mercury atoms can be transferred as kinetic energy to the particles, and the difference frequency is radiated as light. The publication contains an equation that show how two light quanta, with two different frequencies can work together in a single elementary act to excite an atom [double absorption or two-photon absorption].

4.3.2.2 Introduction to the 1928 Paper of Franck

Franck, J. 1928. Beitrag zum Problem der Wiedervereinigung von Ionen und Elektronen. Zeitschrift für Physik, 47: 509–516.

As Director of the Zweites Physikalisches Institut, Göttingen, Franck experimented with collisions of fast electrons and atoms. He explored the effect of the velocity of the colliding electrons on the spectral lines of the atoms. Franck explained the experimental results on the basis of the Bohr hypothesis. He studied the ionization of atoms due to collisions with slow and fast electrons and the subsequent luminescence that was observed. He interpreted this process as due to the recombination of ions and electrons. This publication is remarkable since it contains neither figures nor illustrations that show the experimental data, which is only described in the text.

4.3.3 Summaries of Theoretical Papers Cited in Maria Göppert-Mayer's 1931 Thesis

4.3.3.1 Introduction to the 1925 Paper of Kramers and Heisenberg

H. A. Kramers and W. Heisenberg, On the dispersion of radiation by atoms. Zeitschrift für Physik, 31, 681–708, (1925).

An English translation of this paper may be found in: Van der Waerden, B. L. (1968). *Sources of Quantum Mechanics*. This book contains introductions and English translations of full papers (1916–1926) from: Einstein, Ehrenfest, Bohr, Ladenburg, Kramers, Born, Van Vleck, Kuhn, Heisenberg, Jordan, Pauli, and Dirac. Also, a discussion of this paper, together with an English translation of excerpts, appears in: Duck, I, Sudarshan, E. C. G. (2000). *100 years of Planck's Quantum. Singapore*. In addition, it contains introductions and translated excerpts of papers from: Planck, Wien, Einstein, Compton, Bohr, de Broglie, Kramers, Heisenberg, Born, Jordan, Dirac, and Schrödinger.

The Kramers and Heisenberg (1925) calculation pertains to an excited atom that is irradiated with light of a lower frequency than the resonance frequency of the atom. This results in a stimulated double or two-photon emission, in which the atom divides the energy into a light quantum with the frequency of the irradiated light, and a quantum with the difference frequency.

According to Van der Waerden, the methods used in the 1925 paper by Kramers and Heisenberg are similar to those expressed by Max Born in his 1924 paper (Born, 1924). In this paper Born acknowledges his use of the Kramers' dispersion equation and also that his work is in close relation to Heisenberg's development of the rules for the anomalous Zeeman effect. In fact, Born cites the discussion with Niels Bohr, and also Born thanks Heisenberg for advice and help with the calculations. Thus, many papers published in the early 1920s contain modifications and amplifications based of the previous works of others.

Now I discuss the Kramers and Heisenberg paper "On the Scattering of Radiation by Atoms" of 1925 in more detail. I begin with a quote from the abstract of their paper:

When an atom is exposed to external monochromatic radiation of frequency ν, it not only emits secondary monochromatic spherical waves of frequency ν which are coherent with the incident radiation; but the correspondence principle [Bohr] also demands, in general, that spherical waves of other frequencies are emitted as well. These frequencies are all of the form $|\nu \pm \nu^*|$, where $h\nu^*$ denotes the energy difference of the atom in the state under consideration and some other state. The incoherent scattered radiation corresponds in part to certain processes which have recently been discussed by Smekal [Raman scattering], in connection with investigations which are linked to the concept of light quanta.

What is the Bohr correspondence principle? It is an approach that requires that in the region of high quantum numbers the actual properties of the atom can be described asymptotically with the help of the classical electrodynamic laws. The authors address the following problem: how to find the electric moment of the atom as a function of time when the atom is radiated with plane monochromatic light whose wavelength is much larger than the atomic dimensions. Their solution to the problem was as follows: in the classical equations for the probability of electronic transitions, they replaced derivatives in their equation by differences divided by h, frequencies by quantum energy differences, and dipole moments by amplitudes that are related to quantum transitions. Note that the words "transition amplitude" are from the Kramers and Heisenberg paper.

To repeat, the authors' theory describes the following process. An atom exposed to incident light of angular frequency ω will emit light whose intensity is proportional to the intensity of the incident light. The emitted light splits into harmonic components that contain the frequency of the incident light, and frequencies that are sums and differences of the incident light frequency and a frequency that is characteristic of the atom.

Optical phenomena such as dispersion and absorption arise from the passage of monochromatic light through a gas. This was interpreted by assuming that each irradiated atom emits a secondary spherical wave whose frequency is the same as that of the incident light and which are coherent with it. Classical electrodynamics can provide a detailed description of the scattering effect of the atoms.

The paper assumes, following classical electrodynamics, that the electrons of an atom are quasielastically bound in the atom and execute harmonic oscillations. In the interaction of monochromatic light with the atom, the plane monochromatic polarized waves, of frequency ν, interact with the atom, and the oscillating dipole due to the oscillating electron causes the emission of spherical waves with the frequency of the oscillating dipole. This approach accounts classically for the phenomena of dispersion, absorption, and scattering of light. The Kramers and Heisenberg paper could not explain spontaneous emission (i.e., the emission of light from an atom in an excited state in the absence of incident radiation).

What is new in the paper is expressed in the following quote:

> Under the influence of irradiation with monochromatic light, an atom not only emits coherent spherical waves of the same frequency as that of the incident light: it also emits systems of incoherent spherical waves, whose frequencies can be represented as combinations of the incident frequency with other frequencies that correspond to possible transitions to other stationary states.

The paper now proceeds to solve the following problem, first classically and then quantum mechanically: the effect of external radiation on a periodic system—more specifically, to calculate the electrical moment of the system as a function of time when the system is exposed to plane, monochromatic light in which the wavelength is large compared to the size of the atom. They calculate a quantum mechanical expression for the absorption and the emission transitions. Their work, as in many similar works of the period, shows the use of the Bohr correspondence principle to bridge classical physics and quantum mechanics in the limiting case of very large quantum numbers. In summary, their work may be stated as follows: with monochromatic radiation, an atom emits not only coherent spherical waves of the same frequency as that of the incident light; it also emits systems of incoherent spherical waves, whose frequencies can be represented as a combination of the incident frequency with other frequencies that correspond to all possible transitions to other stationary states. The bulk of the paper works out the mathematical description of this phenomenon, the scattering of incoherent radiation.

4.3.3.2 Introduction to Dirac's First Paper on Radiation Theory (Dirac, 1927a)

Dirac, P. A. M. 1927a. The Quantum Theory of Emission and Absorption of Radiation. *Proceedings of the Royal Society of London, Series A, Containing Papers of a Mathematical and Physical Character*, Vol. 114, No. 767 (March 1, 1927), 243–265.

The papers are reproduced in the following book: R. H. Dalitz, *The Collected Works of P.A.M. Dirac, 1924–1948*, Cambridge University Press, Cambridge, (1995). The biographical notes for the year 1927 which provide the context for the science were also derived from this work [chronology for the life of P. A. M. Dirac].

First I present some historical context about Paul Dirac. Dirac studied electrical engineering at the University of Bristol, where in his later years of university studies he gained first-class honors in mathematics. In October 1923 he began his postgraduate studies in theoretical physics. His research mentor was Dr. R. H. Fowler of the Cavendish Laboratory, Cambridge.

In 1925 Werner Heisenberg lectured at Cambridge University. Shortly after this visit Professor Fowler received a copy of Heisenberg's paper on his semiclassical formulation of quantum theory. Professor Fowler showed the paper to his student Dirac, who later stated that this paper stimulated his own formulation of a new quantum mechanics that is based on noncommuting dynamical operators (i.e., $\hat{x}\hat{y}$ is not equal to $\hat{y}\hat{x}$).

Starting in February 1927, Dirac worked for five months at the University of Göttingen. During that period he worked with J. Robert Oppenheimer, Pascual Jordan, and Max Born. Dirac also met with James Franck, Johannes von Neumann, David Hilbert, and Andrew Weyl. During the previous year, 1926, he submitted his Dissertation of the Degree of Ph.D. on Quantum Mechanics, as well as several papers on his theory of quantum mechanics. His new quantum theory was based on the assumption that the dynamical variables do not obey the commutative law of multiplication. Provided that a dynamical system can be described by a Hamiltonian function, Dirac could in principle interpret the mathematics physically by a general method. Extracting the analytical solution (closed form) was another matter. This thesis work in its expanded form was published in the book *Quantum Mechanics* (Dirac, 1935).

I now present some additional background material on the theoretical work of Paul Dirac. Some time after Heisenberg published his quantum theory, Dirac attempted to adapt it to the formal theory of Hamiltonian functions. In 1925 Dirac was able to modify the original quantum theory of Heisenberg by introducing a novel algebraic algorithm that allowed him to derive the key equations of quantum theory without the use of Heisenberg's matrices. Dirac in his 1925 paper used the term "Heisenberg product" since he did not know that they were matrices. Max Born is credited with that realization. Dirac formulated his quantum theory in terms of Poisson brackets instead of derivatives, as were used in the previous theory of Heisenberg. Note that Poisson brackets were first formulated for mechanics in 1809 by Simeon Denis Poisson in his work on the perturbation theory of planetary orbits. In summary, Dirac showed that the differential operators of classical dynamics could be reformulated into algebraic operations in his new quantum theory.

Dirac showed how to calculate the probability for two-photons scattering processes that can occur via a virtual state. This process is analogous to the Raman process in which two photons are involved; one photon goes in and one photon comes out.

At the time Dirac wrote this paper he was at St. John's College, Cambridge, and he also spent some time at the Institute for Theoretical Physics in Copenhagen. Niels Bohr, the Director of the Institute for Theoretical Physics in Copenhagen and a Foreign Member of the Royal Society, communicated Dirac's paper to the Royal Society of London. Now I summarize the key points of Dirac's (1927a) paper:

1. Dirac states that the field of quantum electrodynamics is wide open and there-fore ripe for new developments. Some of the questions that Dirac pondered were as follows: how to treat systems in which forces are propagated with the speed of light, not instantaneously; how the moving electron produces an electromagnetic field; and how the electromagnetic field interacts with an electron. His new theory of the emission of radiation and the reaction of the radiation field on the emitting system was not strictly relativistic; however, the variation of the mass with velocity can be easily incorporated into his theory. Dirac stated that "the underlying ideas of the theory are very simple." This may be typical British understatement.

2. The basis of Dirac's theory for the interaction of an atom with a field of radiation is based on the following assumptions. First, the radiation field is confined in a box or cavity in order that the number of degrees of freedom can be discrete and not become infinite. Second, the radiation can be resolved into its Fourier com-ponents, with the energy and the phase of each component being the dynamical variables to describe the radiation field. In the mathematical description of a plane, linear-polarized, monochromatic wave that is resolved into its Fourier components, there appears the frequency of the wave, an amplitude which is a complex vector, and two complex components of the wave amplitude; they are each multiplied by a unit polarization vector, e_1 and e_2, which represent the two independent states of linear polarization. In Dirac's theory the radiation in a cavity can be described by giving the amplitude of each standing wave at a particular time; therefore, the amplitude can be considered as a coordinate that follows the laws of quantum mechanics. In the interaction of atoms and radia-tion he calculated the probabilities of both induced emission and spontaneous emission [no radiation present]. This description was used by Maria Göppert-Mayer in her 1931 thesis paper. Third, in the absence of interaction between the atom and the radiation, the Hamiltonian (energy operator) of the entire system is composed of two terms, the energy of the field and the Hamiltonian of the atom alone. Fourth, when there is interaction between the radiation and the atom, an interaction term would then be added to the total Hamiltonian.

This term gives the effect of radiation on the atom and the effect of the atom to the radiation field. Fifth, to put these assumptions into a quantum theory, the commutation relation of the dynamical variables is set equal to $i\hbar/2\pi$. These assumptions result in an equation that when solved according to the rules of quantum mechanics gives the correct laws for the emission and the absorption of radiation and the correct values for Einstein's 'A-' and 'B coefficients'. The basic idea of Dirac's theory of radiation is to describe the total system of the radiation and the atom as the sum of three terms: one represents the energy of the atom, one represents the electromagnetic energy of the radiation field, and the third term represents the interaction energy of the atom and the radiation field. The last interaction term is the coupling term for the atom and the radiation field. In the absence of the third interaction, the atom could not absorb nor emit radiation. Dirac considered the perturbation of an assembly that satisfying the Bose–Einstein statistics.

3. Dirac stated that these assumptions have been used by Born and Jordan [Z. f. Physik, vol. 34, p. 886 (1925)] in their adoption of the classical theory of the radiation emission of a dipole into quantum theory, and also by Born, Heisenberg, and Jordan [Zeitschrift für Physik, vol. 35, p. 606 (1925)] in their calculation of the energy fluctuation of a field of black body radiation.

4. Dirac previously developed his formulation of time-dependent perturbation theory (Dirac, 1926). This theory can be used to solve problems with quantum mechanics in which the exact solution cannot be obtained, but solutions can be approximated based on the exact solution of simpler cases (the hydrogen atom, the harmonic oscillator). See P. A. M. Dirac, *On the Theory of Quantum Mechanics*, Proceedings of the Royal Society (London) A, vol. 112, pp. 661–677, October 1, 1926.

5. With the above assumptions and the use of Dirac's time-dependent perturbation theory, the probabilities for the induced emission [the emission of a photon from an excited state atom in the presence of a radiation field], the spontaneous emission [emission of a photon from an atom in an excited stated in the absence of a radiation field], and the absorption of a photon by an atom in the presence of a radiation field could be calculated. In this paper Dirac explains how to calculate the probability of electronic transitions, and also how to calculate the probability of transitions that are induced by colliding electrons.

6. In order to calculate the Einstein coefficients for emission and absorption Dirac must establish the relation between the number of light quanta per stationary state and the intensity of the radiation. To do this Dirac proceeds by confining the radiation to a fixed cavity of finite volume V.

7. Here are some comments on the Dirac's perturbation theory. Two cases are discussed, time-dependent and time-independent perturbations. An example of the time-dependent case is as follows: in the calculation of absorption of light or the induced emission of light by an atom in a radiation field, the energy-perturbing operator is given by the incident radiation field. An example of the time-independent case used by Dirac is as follows: since the integrals in the Dirac radiation theory are integrated from 0 to t, it is assumed that a constant

perturbation is "turned on" at time $t=0$, for example in the transition of a system to an excited state. This same argument is used by Maria Göppert-Mayer in her 1931 dissertation paper.

8. The Dirac theory on the interaction of radiation and an atom as developed in his 1927 paper on the quantum theory of emission and absorption of radiation is as follows (Dirac, 1927a). The interaction Hamiltonian gives in the first-order perturbation theory the transitions of the radiation oscillators that correspond to the absorption or the emission of photons. The energy of the total system, atoms and photons, follows the conservation of energy law. The transition probabilities, which are the squares of the matrix elements, are for absorption proportional to the initial photon occupation numbers, (n), and for emission they are proportional to (n + 1).

    First, Dirac decided not to consider the radiation in infinite space, but to represent the radiation as confined to a cavity of finite volume, V, and with perfectly reflecting walls. Later, the cavity will be expanded to become infinite, and that will represent the radiation in free space.

    Then the oscillations of the confined electromagnetic field are represented as the superposition of a finite number of fundamental vibrations, each one corresponding to a system of standing waves. This representation is only valid for the radiation field [the part of the electromagnetic wave that causes radiation], not a general electromagnetic wave. The Coulomb forces are not part of the radiation field. The number of standing waves with a frequency between ν and $\nu + d\nu$ is the known equation (1) in the Maria Göppert-Mayer 1931 paper. The electromagnetic field of a monochromatic, plane standing wave in the cavity can be described as a vector potential A. The electric field can be mathematically derived from the vector potential. A fuller discussion of vector potentials is contained in Chapter 5.

    Next the Hamiltonian of the atom and the radiation field are described. The electromagnetic energy of the radiation field can be shown to have the same Hamiltonian as a system of uncoupled [independent] harmonic oscillators. Then Dirac develops his time-dependent perturbation theory to calculate the probabilities of transitions of energy for the atom and for the radiation field. This is studied for a variety of cases—absorption, emission, induced emission. For example, an atom in the cavity that is in an excited state will over time be found with a lower energy state, and the radiation field will gain the energy difference lost by the atom.

9. Dirac has shown that the electric moment calculated for two different wave functions that each represents the initial and the final state of the atom described the transition moment for the process or changing from the initial to the final state of the atom.

    In summary, Dirac shows that the Hamiltonian that describes the interactions of an atom and the electromagnetic waves is identical to the Hamiltonian for the problem of the interaction of an atom with an assembly of particles moving with the velocity of light and satisfying the Bose–Einstein statistics [Dirac writes "the Einstein–Bose statistics"]. Dirac treated an assembly of similar systems satisfying the Bose–Einstein statistical mechanics as they interact with another different system. Dirac used a Hamiltonian function to describe the motion. He applied his theory to the interaction of an assembly of light quanta with an atom and derived the Einstein laws (the correct expressions for Einstein's 'A-' and 'B coefficients') for the emission and the absorption of radiation. Dirac's theory could account for spontaneous emission, which the classical theories developed prior to his 1927 paper were unable to do.

4.3.3.3 Introduction to Dirac's Second on Radiation
        Theory (Dirac, 1927b)

Dirac, P. A. M., 1927b. The Quantum Theory of Dispersion, *Proceedings of the Royal Society (London), Series A, Containing Papers of a Mathematical and Physical Character*, Vol. 114, No. 767 (May 2, 1927), 710-728. (Received April 4, 1927). This paper was communicated to the Royal Society by R. H. Fowler, F. R. S.

In his second paper of 1927, on the quantum theory of dispersion, Dirac used his second-order time-dependent perturbation theory to the problem of the atom–radiation field interaction and was able to rederive the Kramers–Heisenberg equation for the scattering of light by atoms (Dirac, 1927b). Dirac used the Fourier amplitudes in the classical theory and reformulated them into the "transition amplitudes" in his new quantum theory.

The seminal contributions of Dirac in his 1927b paper include the following developments that were incorporated in the 1931 thesis paper of Maria Göppert-Mayer: (1) the use of second-order time-dependent perturbation theory, (2) the confinement of the radiation field into a cavity, and (3) the use of two-photon transitions (scattering processes) via virtual intermediate states. Dirac states in his 1927b paper that the individual transitions (i.e., from the ground state to the virtual state, and a transition from the virtual state to the final state) each do not obey the law of conservation of energy; only the total transition via two photons from the ground state to the virtual state to the final state obeys the law of conservation of energy. These later points also appear in the 1931 paper of Maria Göppert-Mayer.

Dirac was able to quantize not the entire electromagnetic field, but only the part that was composed of plane traverse light waves (the "radiative part"). He included the electrostatic fields by incorporating the Coulomb interaction energies [nucleus and electrons] into the interaction Hamiltonian.

In this paper Dirac applies his theory to determine the radiation scattered by atoms. He also presents the case in which the incident radiation coincides with the spectral line of the atom.

1. Dirac begins the paper (1927b) by writing that the new quantum mechanics could only deal with atom–radiation interactions through analogies with the classical theory. For example, in Heisenberg's matrix theory, it is assumed that the matrix elements of the polarization of an atom determine the emission and absorption properties of light that is completely analogous the Fourier components in the classical theory

   Dirac then stated that the recent papers of Heisenberg and those of Schrödinger cannot be used to calculate the breadths of spectra lines and for the phenomenon of resonance radiation. Dirac states that the perturbation method must be used in cases for which one cannot solve the Schrödinger equation exactly. That is valid only when the interaction energy [radiation field and the atom] is much smaller than the energy of the atom. This assumption corresponds to the physical situation where the mean lifetime of the atom is much larger than its periods of vibration.

2. Dirac assumes the radiation field is resolved into components that are plane waves, plane-polarized, and of a definite frequency, direction of motion, and state of polarization. Any arbitrary type of polarization can be expressed as the combination of two states of polarization. He assumes that the radiation is confined in a cavity, which implies stationary waves. That assumption results in a discrete number of degrees of freedom of the radiation field.

3. Dirac postulates that a field of radiation can be treated as a dynamical system in which the interaction of the field with an atom may be described by a Hamiltonian function. In the next step, the Hamiltonian function is described for the interaction of the radiation field and the atom. Dirac assumes that the dynamical variables that describe the field can be specified in terms of energies and phases of its various harmonic components, each of which is a harmonic oscillator.

   Dirac gives the Hamiltonian function in terms of the unperturbed system of an electron in the absence of the radiation field and the perturbed system due to the presence of the radiation field. Dirac states that some of the terms in the Hamiltonian relate to two light quanta being emitted or absorbed simultaneously; however, he states that these terms are negligible.

4. In the next section, which deals with the theory of dispersion, Dirac uses his time-dependent second-order perturbation theory to calculate the probabilities of transition for absorption and emission. He considers the case not of a single final state, but of a set of final states lying close together. The final result of the probability of transitions is in agreement with the work of Kramers and Heisenberg.

5. For the case of resonance, in which the frequency of the incident radiation is equal to that of an absorption or an emission line of the atom, a modification must be made in the previous equation since the denominator vanishes. Dirac suggests that instead of a single frequency of the incident radiation, a distribution of light quanta over a range of frequencies that includes the resonance frequency should replace the single frequency of the incident light. He then calculates the probabilities for the emission of a light quantum with the absorption of another light quantum. With the above approximations and second-order time-dependent perturbation theory, Dirac then concludes that resonance-scattered radiation is due to absorptions and emissions that follow the Einstein absorption and emission laws. When the time t is large as compared to the periods of the atom, Dirac found that the terms in the probability of transitions are negligible compared to those terms that are proportional to t^2; therefore, Dirac concluded that the resonance-scattered radiation is due to absorptions and emissions of radiation in agreement with the laws of Einstein and the Einstein coefficients.

6. Dirac states that the exact interaction energy of the field and the atom is too complicated; therefore, he uses the dipole energy. That assumption results in a

divergent series that appears in the calculation. In the calculation of dispersion and resonance radiation there is no divergent series, but when he attempts to calculate the breadth of a spectral line a divergent series appears.

We now continue to a more complete discussion of the key points, assumptions, and techniques that are contained in the 1929 brief paper and the 1931 thesis paper of Maria Göppert-Mayer.

4.3.4 Notes on Maria Göppert's October 28, 1929, Paper Published in *Die Naturwissenschaften*

Below is a summary of the ideas and concepts contained in her paper (Göppert, 1929).

1. Maria Göppert first acknowledges the work of Kramers and Heisenberg (1925) that allows the Raman effect ("Smekal jumps") to be quantitatively calculated. One references is given in her 1929 paper, but the reference to the publication of Kramers and Heisenberg is given in her 1931 Göttingen dissertation [see below].

   Adolf Smekal was a professor and the director of the Institute of Theoretical Physics at the University of Halle, Germany. He worked on the theory of the two-photon process that later became know as the Raman effect. In 1923 Smekal published his quantum theory of dispersion (Smekal, 1923). He showed that when light is scattered by transparent materials, there are frequencies in the scattered light that are not present in the incident light. He interpreted these new frequencies as being due to a positive or a negative net exchange of energy between the molecules and the incident light.

2. Maria Göppert then states that she contributed a review of the Dirac dispersion theory to the book *Elementare Quantenmechanik* (Born & Jordan, 1930). She noticed that Dirac's (1926) theory described not only the Raman effect, but also the reverse process in which two photons act together in a single elementary act to promote an atom from the ground state to an excited state. Her statement thus directly links the theoretical work of Dirac with her subsequent calculation of two-photon processes. She used the Dirac formulation of second-order time-dependent perturbation theory in her thesis work.

   An experimental dispersion curve is a graph of the refractive index of a material as a function of wavelength. In normal dispersion, the refractive index of a material increases from longer to shorter wavelengths. The refractive index is the ratio of the velocity of light in a vacuum to the velocity of light in the material. In regions of light absorption (anomalous dispersion) there is an absorption maximum and also a reflection maximum. In the first quarter of the 20th century there were many theoretical models that attempted to define a dispersion equation based on the properties of atoms that fit the experimental data.

3. The author states that for cases 1–3 (Stokes case, Raman effect; Anti-Stokes case, Raman effect; and double emission), the transition probability is proportional to the light intensity, but in case 4 (double absorption), the transition probability is proportional to the square of the light intensity.

4. The author states case 3 (double emission) is difficult to observe because the frequency is proportional to the number of atoms in the excited state.

5. The author states for case 4 (double absorption, two-photon absorption), even though the probability is proportional to the number of atoms in the ground state, the quadratic dependence of the light intensity makes the observation of the two-photon absorption an event of very low probability.

6. Maria Göppert submitted this summary paper on October 28, 1928. Since she was not yet married to Dr. Joseph Mayer she did not use his last name.

7. In her four figures (energy diagrams) illustrating the Stokes case of the Raman effect, the anti-Stokes case of the Raman effect, the case of double emission, and the case of double-absorption, Maria Göppert placed the dotted line that represents the virtual state above the upper state of the atom. In modern illustrations of the two-photon processes, the dotted line that represents the virtual state would be placed between the lower ground state and the upper excited state energy of the atom.

4.3.5 Notes on Maria Göppert-Mayer's 1931 Doctoral Paper Published in *Annalen der Physik*

There are two types of assumptions and approximations made in Maria Göppert-Mayer's 1931 Göttingen dissertation (Göppert-Mayer, 1931). The first type consists of physical assumptions and the second type is mathematical assumptions. Both types of assumptions and approximations were widely employed in the years 1926–1931, and many of them still appear in modern textbooks on quantum mechanics. See, for example, *Elementare Quantenmechanik* (Born & Jordan, 1930) for the former, and *Nonlinear Optics*, Second Edition (Boyd, 2003) for the latter.

Maria Göppert-Mayer relied strongly on the work of her thesis advisor Max Born, as well as the 1927 publications of Dirac and the work of Weisskopf, who was a fellow graduate student of Born. She also used the theoretical results of Kramers and Heisenberg (1925). Maria Göppert-Mayer employed all of the above assumptions as well as the time-dependent perturbation theory of Dirac in her thesis work on double emission and double absorption and the Raman effect. In the treatment of an atom and its interaction with a radiation field by Maria Göppert-Mayer, the process of the absorption of a photon by an atom involves the increase in the energy of the atom by a quantum of energy $h\nu$ and the decrease of the harmonic oscillators making up the radiation field by a quantum of energy $h\nu$. The combined energies of the electron and the radiation oscillators follow the law of conservation of energy.

The following papers were cited, and I copy the citations as they appear in her paper:

(1) P. A. M. Dirac, Proc. of R. S. vol. **114**, p. 143 [should read 243], and p. 710, (1927)* *Error in text, [read *P. A. M. Dirac, Proc. of R. S. vol. **114**, p. 243, and p. 710, (1927)*].

(2) H. A. Kramers and W. Heisenberg, Zeitschrift für Physik, **31**, S. 681. 1925.

(3) O. Oldenberg, Ztschr. f. Phys. 51. S. 605. 1928.

(4) J. Franck, Ztschr. f. Phys. 47. S. 509. 1928

(5) M. Born, Ztschr. f. Phys. 38. S. 803, 1926; Nachrichten der Göttingen Ges. der Wissenschaften 1926, s. 146.

Assumptions and Techniques Used by Maria Göppert-Mayer in Her 1931 Göttingen Dissertation

Göppert-Mayer, M. 1931. Über Elementarakte mit zwei Quantensprüngen, *Ann. Phys.* (Leipzig), 9: 273–294.

   Maria Göppert-Mayer's 1931 dissertation paper was divided into two parts, section 1 and section 2. Our discussion follows the same order.

Section 1. The Combined Action of Two Light Quanta in One Elementary Act

The theoretical basis follows the experimental studies of Oldenberg, Franck, and others on collisions of atoms with electrons and their subsequent luminescence. The theoretical calculation strongly follows the previous work of Dirac (time-dependent perturbation theory) and the theory of Kramers and Heisenberg. All of these key papers were previously discussed.

THE INTERACTION OF ATOMS WITH AN
ELECTROMAGNETIC FIELD

The general idea is to replace the electromagnetic field with a set of uncoupled, harmonic oscillators within a large cavity. An atom with one electron is placed in the cavity and the probability of several physical processes are calculated: Raman scattering, double or two-photon emission, and double or two-photon absorption. These processes are mediated by the emission or absorption of photons during interaction of the electron in the atom and the radiation field (standing waves within the cavity). The law of conservation of energy holds for the total system (atom and electromagnetic field) and not for partial transitions such as the single-photon transition in a two-photon process.

   The idea of confining the radiation to a cavity as with the subsequent semiclassical quantization of the radiation field is developed as follows. The electromagnetic radiation is assumed to be monochromatic transverse waves. When restricted to a confined cavity these waves form standing waves with nodes at the walls of the cavity. The standing waves oscillate with a frequency v and they are equivalent to a system of uncoupled harmonic oscillators. The allowed energies of these harmonic oscillators are given by quantum mechanics and this procedure quantizes the radiation field. The detailed steps of this quantization of the radiation field are given below. Once the radiation field is quantized, the atoms can exchange quanta of energy with the radiation field.

   Following this introduction I follow the logical progression of Maria Göppert-Mayer's thesis paper and I extract and discuss her assumptions and the techniques.

1. Assume the radiation is contained in a cubic box, of volume V, and side L, which causes the waves to become standing waves with nodes at the surface or walls of the box [classical physics]. This assumption is required to make the number of degrees of freedom of the radiation countable or finite. Later the box will be made to be infinitely large. The well-known Equation (1) in her paper gives the limiting number of standing waves in a large cavity for a given range of frequencies of the harmonic oscillators.

2. Assume the radiation field is equivalent to a system of uncoupled harmonic oscillators. Quantum mechanics provides an exact solution for the energies,

including the zero-point energy of a harmonic oscillator. It is known that the quantum number of a harmonic oscillator can only change by unity during a radiation process.

3. Assume the radiation field can be decomposed into plane, linear polarized waves that can be described by a vector potential A. Note: in many problems in electromagnetics, the vector potential is used to simplify the mathematics. The electric and the magnetic vectors are easily derived from the vector potential by mathematical operations [classical physics].

4. The next equations are all previously known from quantum mechanics. They give the Hamiltonian function or the energy function H for the radiation and describe the radiation field by giving the state of all the harmonic oscillators. The energy difference of two cavity states, s and s', is given in terms of the energy of all the harmonic oscillators. Equation (2) gives the quantum mechanical matrix elements for the different states of the cavity.

5. Assume that the Hamiltonian function for an atom with a stationary nucleus [i.e., no kinetic energy of the nucleus] and an electron interacts with the radiation field, and the total energy of the system can be separated into three additive terms: the Hamiltonian of the radiation field [the uncoupled harmonic oscillators], the Hamiltonian of the atom, and the Hamiltonian of the interaction between the atom and the radiation field. In several steps the interaction term of the electron is derived.

6. Assume that the wavelength of the light is much larger than the atom's dimension; then it is possible to assume that the electromagnetic field is constant over the size of the atom. Here she invokes the electric dipole approximation but does not call it by name.

7. The Hamiltonian function for the electron is given in terms of the momentum of the electron, the coordinate vector of the electrons, the vector potential of the radiation field at the position of the electrons, and the potential energy of the electron in the atom (Coulomb field).

8. Assume classically that the electron interacts with the electromagnetic field through an electrical moment, p, given by the product of the electronic charge and the displacement due to the electromagnetic field. This interaction is summed over all of the atom's electrons. This is further simplified to yield the interaction energy of the electron, and the radiation field is given by the potential energy of the electric moments and the radiation field.

9. The following equations [3–9] in her thesis incorporate all of the previous assumptions together with first-order and second-order time-dependent perturbation theory to solve the quantum mechanical problem interaction of the atom and the radiation field (uncoupled harmonic oscillators in a cavity). The Dirac formulation of the time-dependent perturbation theory is closely followed and the probability amplitudes for the various processes (Raman scattering, double emission, and double absorption) are calculated by using various assumption about the ground and excited states together with statement of conservation of energy and the quantum mechanical selection rules that state which transitions are allowed and which transitions are forbidden.

10. Assume that at time, $t = 0$, the atom and the cavity are in their unperturbed states [normal unexcited states]. First- and second-order, time-dependent perturbation theory together with this assumption yields Equation (3). These known equations give the probability amplitude of absorption and emission (the first approximation), and the second approximation yields the Raman effect and dispersion and the processes of simultaneous emission and simultaneous absorption of two light quanta.

11. To calculate the probability amplitude for double emission, assume at time, $t = 0$, the atom is in the excited state and only one standing wave (the radiation field) exists in the cavity. This standing wave has a frequency that differs from the eigenfrequency of the excited atom in the cavity. This assumption yields Equation (4) for the Raman effect and the dispersion and also for double or two-photon emission as well as double or two-photon absorption.

12. These transitions are a result of the assumption that the perturbation energy begins to act on the atom at time, $t=0$, when it is actually continuous. The "turning on" of the perturbation energy is the cause of these abnormal transitions. Note that the square of the absolute magnitude of the probability amplitudes is the probability of a given transition. The probability amplitudes are functions of time; therefore, the probability of different quantum states varies with time. This implies that the effect of the perturbation on the system is to induce transition probabilities among the quantum states of the unperturbed system.

13. Assume that not only a single standing wave of the cavity is strongly excited, but also a spectrum line of a finite width is sent in. Then, using Equation (6), the final equation of Equation (7) is obtained. This equation gives the probability of double emission per unit of time.

14. This is an important result because it demonstrates that in a cavity without a radiation field, there is the probability for a spontaneous, simultaneous emission of two light quanta (two-photon emission), and all frequency divisions are possible for the two quanta. For normal radiation densities, the strength of the stimulated double emission is much less than the spontaneous double emission.

15. From similar considerations as for double emission, the probability of two-photon absorption or double absorption is derived. This is given in Equation (9).

16. For the double processes described, such as the Raman effect, they act as if the two processes take place in one act; each process does not satisfy the conservation of energy, but the total of the two process does satisfy the conservation of energy. Since the double absorption or two-photon absorption has a quadratic dependence on light intensity, very high light intensities are required to observe the effect.

17. In her four figures (energy diagrams) illustrating the Stokes case of the Raman effect, the anti-Stokes case of the Raman effect, the case of double absorption, and the case of double emission, Maria Göppert placed the solid line, n', which

represents the virtual state above the upper state of the atom. In modern illustrations of the two-photon processes, the upper solid line that represents the virtual state, n', would be placed between the lower ground state and the upper excited state energy of the atom, and following the figures in her 1929 paper the virtual state would be represented as a dotted line.

Section 2. The Working Together of Light and Collisions
in One Elementary Act

In the second section of her 1931 Göttingen dissertation, Maria Göppert-Mayer develops the calculations, using methods similar to those in the first section, for the working together of light and colliding electrons with atoms. Again, her work aimed to theoretically explain the experimental studies of both Oldenberg and Franck. As in Section 1 of her thesis, she developed the theory as the concurrence of two processes in a single elementary act, in strong analogy to the Raman effect.

1. Assume that the electrons are contained in a cubic box, of volume V. The electron waves are assumed to be standing waves, similar to the standing waves of the radiation in Section 1.

2. Assume that in the absence of interaction the electrons have only kinetic energy T. Assume the eigenfunctions are plane waves, related to the size of the box and the number of standing waves. Assume their energy in the small energy range is given by Equation (10).

3. First the interaction between a field of free electrons and a radiation field with one atom is investigated. Assume the total Hamiltonian of the system is composed of the following additive terms: energy of the atom, energy of the radiation field, kinetic energy of the electrons, energy of the interaction.

4. Assume that the last term, the energy of interaction, is composed of two terms: the interaction between the atom and the radiation, and the interaction between the atom [nucleus] and the electron. This latter term is approximated by the Coulomb field. The interaction between the radiation and the electron is neglected here.

5. In the first part of Section 2, the probability amplitudes are calculated for the case in which there is no radiation in the cavity; there is only pure emission. Assume there is only one electron in the cavity, and the atom is in the excited, or in the unexcited [ground] state.

6. Time-dependent perturbation theory is used for the calculations of the probability amplitudes in the first approximation and then in the second approximation. In the first approximation, the perturbation theory gives two terms [no cross-terms]: that of light alone, and that of collision alone. Only the second approximation of the perturbation theory yields the probability amplitude of the working together of both processes in a single act.

7. Next the emission probability of light is investigated for the case in which the emission frequency is not equal to the eigenfrequency of the atoms. This calculation is made for a narrow range of frequencies of light. The probability is normalized on a density $1/V$ for the incident electrons. This results in the

probability for the combined effect of light and collisions, Equation (13), which is independent of the size of the box. These theoretical results are in agreement with the experimental results of Oldenberg.

8. To explain the excitation luminescence effect, it is assumed that the atom at the start of the process is in the ground state, and it is assumed that the kinetic energy of the colliding electrons is much greater than the ionization energy of the atoms.

9. Equation (15) is the calculated probability for an atomic transition to the state *n*, with the emission of a light quantum from the atom to the standing wave (of the radiation field) within the cavity. Equation (16) indicated that there are two separate processes, each one of which satisfies the conservation of energy rule; the two processes take place in a single act—that is, the ionization of atoms through collisions, in which the electrons give up kinetic energy to the atoms, and then recombination of the electrons and the atoms with emitted radiation. These results are an analog to the process described by Franck that was discussed in the first part of the paper. The calculations show that this process is really a unified act, and not two sequential independent processes.

Some Errors in Maria Göppert-Mayer's 1931 Paper Noted by the Translator

Page 271:

P. A. M. Dirac, Proc. of R. S. vol. **114**, p. 143 [read 243], and p. 710, (1927)*

*Error in text, [read *P. A. M. Dirac, Proc. of R. S. vol. **114**, p. 243, and p. 710, (1927)*].

Page 286:

The two additive terms have incorrect denominators; the correct terms should be:

$$\left[\frac{1 - e^{\frac{2\pi i}{h}(h\nu_{nn'} + E_\chi - E_0)t}}{h\nu_{nn'} + E_\chi - E_0} \right], \quad \left[\frac{1 - e^{\frac{2\pi i}{h}(\nu_{nn'} + \nu_{ss0})t}}{h(\nu_{nn'} + \nu_{ss0})} \right]$$

Page 294:

John [read *Johns*] Hopkins University.

In summary, the preceding introductory notes and biographical materials introduce the qualitative concepts that are contained in the cited papers of Maria Göppert-Mayer's 1929 and 1931 papers. They also stress the connections between her theoretical work and the previous theoretical work of Paul Dirac.

REFERENCES

Born, M. 1924. Über Quantenmechanik, *Zeitschrift für Physik*, 26: 379–395.

Born, M. 1960. *The Mechanics of the Atom*, translated from the German by J. W. Fisher, and revised by D. R. Hartree. New York: Frederick Ungar Publishing Company. First published in 1924.

Born, M. 1989. *Atomic Physics*, Eighth Edition, New York: Dover Publications. This work was first published in 1935.

Born, M., Jordan, P. 1930. *Elementare Quantenmechanik, Zweiter Band der Vorlesungen Über Atommechanik*, Berlin: Julius Springer. (Dr. Max Born was a Professor at the University of Göttingen. Dr. Pascual Jordan was a Professor at the University of Rostock.)

Boyd, R. W. 2003. *Nonlinear Optics, Second Edition*, San Diego: Academic Press. (Professor Boyd is working on a new edition.)

Cohen-Tannoudji, C., Dupont-Roc, J., Grynberg, G. 1992. *Atom-Photon Interactions. Basic Processes and Applications*. New York: John Wiley & Sons, Inc.

Dirac, P. A. M. 1926. On the theory of quantum mechanics, *Proceedings of the Royal Society (London), Series A, Containing Papers of a Mathematical and Physical Character*, Vol. 112, No. 767 (October 1, 1926), 661–677.

Dirac, P. A. M. 1927a. The quantum theory of emission and absorption of radiation. *Proceedings of the Royal Society of London, Series A, Containing Papers of a Mathematical and Physical Character*, Vol. 114, No. 767 (March 1, 1927), 243–265.

Dirac, P. A. M. 1927b. The quantum theory of dispersion. *Proceedings of the Royal Society of London, Series A, Containing Papers of a Mathematical and Physical Character*, Vol. 114, No. 767 (March 1, 1927), 710–728.

Dirac, P. A. M. 1935. *Quantum Mechanics*, New York: Oxford University Press.

Dirac, P. A. M. 1995. *The Collected Works of P. A. M. Dirac 1924–1948*, Ed. R. H. Dalitz, Cambridge, UK: Cambridge University Press.

Einstein, A. 1905. On a heuristic point of view concerning the production and transformation of light. Ann. der. Physik, 17: 132–148.

Fermi, E. 1932. Quantum Theory of Radiation, *Rev. Mod. Phys.* 4: 87–132. (Enrico Fermi delivered a series of lecture at the Symposium for Theoretical Physics in the summer of 1930 at the University of Michigan. This paper is based on his lectures and includes a discussion of radiation theory from 1926–1931. The paper includes a clear discussion of Dirac's theory of radiation. The double transition theory of Maria Göppert-Mayer is cited.)

Franck, J. 1928. Beitrag zum Problem der Wiedervereinigung von Ionen und Elektronen. Zeitschrift für Physik, 47: 509–516.

Frenkel, J. 1950. *Wave Mechanics, Advanced General theory*. New York: Dover Publications. First published by Oxford University Press in 1934. (Frenkel was a professor at the Physico-Technical Institute, Leningrad. This book presents a clear overview of the convergence of classical mechanics and quantum mechanics and the role of the variation principle in the formulation of the Schrödinger equation. The theory of double transitions (two-photon transitions) is clearly described as well as the correspondence between wave mechanics and matrix mechanics.)

Göppert, M. 1929. Über die Wahrscheinlichkeit des Zusamenswirkens zweier Lichtquanten in einem Elementarakt. *Die Naturwissenschaften*, 17: 932.

Göppert-Mayer, M. 1931. Über Elementarakte mit zwei Quantensprüngen, *Ann. Phys.* (Leipzig), 9:273–294.

Heisenberg, W. 1930. *The Physical Principles of Quantum Theory*, Chicago: The University of Chicago Press.

Kramers, H. A. and Heisenberg, W. (1925). Über die Streuung von Strahlen durch Atome [On the Dispersion of Radiation by Atoms]. *Zeitschrift für Physik*, 31: 681–708.

Masters, B. R. 2003. *Selected Papers on Multiphoton Excitation Microscopy*. Milestone Series, MS175, Bellingham, WA: SPIE Optical Engineering Press.

Mehra, J., Rechenberg, H. 2001. *The Historical Development of Quantum Theory*. New York: Springer Verlag.

Oldenberg, O. 1928. Über ein Zusammenwirken von Stoss und Strahlung. Zeitschrift für Physik, 51: 605–612.

Smekal, A., 1923. Zur Quantentheorie der Dispersion, *Die Naturwissenschaften*. 11: 873–875.

Weisskopf, V. 1931. Zur Theorie der Resonanzfluoreszenz, Göttingen Dissertation. Annalen der Physik, Fünfte Folge, Band 9, pp. 23–66.

Part II

NONLINEAR OPTICAL SPECTROSCOPY

Robert W. Boyd and Barry R. Masters

This part of the handbook presents a detailed account of the various physical processes that lie at the heart of biological microscopy and are based on nonlinear optics. It consists of three chapters that cover the basic nonlinear optical processes that are used in nonlinear microscopy. These processes are saturated fluorescence, harmonic generation, and stimulated light scattering. It is the intent of the authors of these chapters to present a clear and comprehensive exposition of the physical processes that are crucial to the success of the implementation of these nonlinear optical techniques.

The beginning of the field of nonlinear optics is often, and somewhat arbitrarily, taken to be the discovery of optical second-harmonic generation by Franken and coworkers in 1961 (Franken et al., 1961) but the origins of this field can be traced back much earlier. The earliest report of a nonlinear optical effect known to the present authors is that of G. N. Lewis, who in 1941 reported saturation with increasing excitation strength of the fluorescence intensity of the organic dye fluorescein in a boric acid glass host (Lewis et al., 1941). In broad terms, the interest in using nonlinear optical techniques for microscopy is to extend the capabilities of traditional microscopy. For example, traditional, linear-optics microscopy is limited in the transverse and longitudinal spatial resolution that can be achieved to a value of approximately the wavelength of the light being used. In many circumstances, it is not possible to obtain increased resolution by using a shorter wavelength, because many biological materials do not even transmit light at these wavelengths. Nonlinear techniques hold the promise of achieving high spatial resolution while making use of light of visible wavelengths. A unifying theme in the description of nonlinear optical interactions is the use of the nonlinear susceptibility to quantify the strength of a nonlinear optical interaction. According to this procedure, the nonlinear polarization $\tilde{P}(t)$ is related to the electric field strength $\tilde{E}(t)$ of the applied optical field according to the expansion:

$$\tilde{P}(t) = \epsilon_0 \left[\chi^{(1)}\tilde{E}(t) + \chi^{(2)}\tilde{E}^2(t) + \chi^{(3)}\tilde{E}^3(t) + \cdots \right], \tag{II.1}$$

Here $\chi^{(1)}$ is the linear susceptibility, $\chi^{(2)}$ is the second-order susceptibility, $\chi^{(3)}$ is the third-order susceptibility, etc.

The procedure of describing nonlinear interactions in terms of the susceptibility is developed in some detail within this section; see especially the chapter by Boyd.

II.1 TWO-PHOTON EXCITATION FLUORESCENCE SPECTROSCOPY

In microscopic implementations of multiphoton excitation (MPE), in which two or more photons induce the electronic transition from the ground state to the excited state via virtual states, the signal generation is proportional to the intensity of the excitation light squared. Therefore, only in the focal volume of the microscope objective is there significant MPE of fluorescent. Thus, the physics of MPEM defines the optical sectioning capability of the technique. The advantages of two-photon-excitation microscopy as compared to the single photon case typically include minimal volume of signal generation, increased depth penetration, and reduced photodamage of live cells and tissues. These same advantages apply for thick-tissue microscopy using two-photon excitation as compared to confocal microscopy with ultraviolet excitation. The major impediment to using multiphoton spectroscopy is the cost of the femtosecond laser that is the basis of the spectroscopic and imaging instrumentation. Two-photon-excitation fluorescence (TPEM) techniques are currently extremely popular, in part due to the wide variety of fluorescent probes that are available and the rapid development of new probes: organic probes, genetically expressed probes, and quantum dots. All three classes of fluorescent probes are in a state of very active development, and we can expect continuing development of new probes with important utility in neurobiology, cell biology, developmental biology, and clinical diagnostics. In addition, there are a variety of intrinsic probes for TPEM: NAD(P)H, flavoproteins, elastin, and neurotransmitters that are useful for live cell imaging. Genetically expressed fluorescent probes are becoming more common in the laboratory, and ease of use of this technique is one factor that is driving their rapidly expanding usage. We should note that all genetically expressed fluorescent probes are overexpressed, and the appropriate controls must be performed to demonstrate the validity of all experiments. For example, it is critical that the experimenter validate the experiment by demonstrating that the fusion protein is correctly localized and is functional. The large size of green fluorescent protein is a major limitation. In general, it is necessary to demonstrate that the use of extrinsic fluorescent probes is not causing alterations in the cells and tissue.

II.2 SECOND-HARMONIC AND THIRD-HARMONIC GENERATION

Second-harmonic generation (SHG) and third-harmonic generation (THG) both provide the ability to image collagen and elastin in living biological cells and tissues. While SHG is allowed only in acentric systems or noncentrosymmetric media (those without a center of inversion symmetry), THG is allowed in both acentric and centric systems (centrosymmetric media) (i.e., in those systems both with and without a center of inversion symmetry). The depth of focus is due to the physics of the harmonic generation. The generation of the harmonic is proportional to the intensity squared; therefore, the region of harmonic generation occurs in the region of the focus where the intensity is very large. The major advantage of these techniques is that the SHG and the THG signal is generated within live cells and tissues without the need for extrinsic probes, and there is no damage. While intrinsic molecules that provide contrast for SHG and THG are useful for live-tissue imaging, the types of biological molecules that generate signals are very limited. In addition, the efficiency of SHG and THG is very low. A major thrust in this field is to develop new types of contrast probes for use with SHG and THG in live-cell imaging. SHG microscopy can be used to image collagen, myosin heads in muscle, and microtubules. THG microscopy gives signals when there is a change in the refractive index within the sample and will yield a signal in the presence or the absence of centers of symmetry. These imaging techniques can be used to investigate patterns of collagen in normal biological specimens as well as alterations in collagen structure and organization that can occur with development, aging, and injury and pathology. These noninvasive optical techniques may provide new knowledge on the extremely important process of wound healing and scar formation. We can expect to see the future development of extrinsic probes with specific biological specificity to aid in the further application of these optical imaging techniques. The further development of potential probes that are excited with SHG will be another tool to investigate membrane potentials with applications in neuroscience. For example, these probes may find utility in optical imaging of the brain in vivo. With appropriate SHG probes, studies of membrane potentials can be measured with four times the sensitivity compared to two-photon-excitation fluorescence methods. SHG methods do not involve the formation of excited states; however, resonance-enhanced SHG occurs with two-photon excitation fluorescence, and therefore photodamage and photobleaching may still occur. The combination of two or more of these nonlinear optical microscopic techniques will provide new insights into both the basic sciences and clinical and diagnostic medicine.

II.3 COHERENT ANTI-STOKES RAMAN SCATTERING

Coherent anti-Stokes Raman scattering (CARS) spectroscopy and microscopy are emerging techniques for the visualization of live biological specimens with chemical specificity and do not require extrinsic labeling. The technique of CARS requires stable picosecond lasers and/or optical parametric oscillators (OPOs) in the wavelength range appropriate to excite biomolecules in live cells. Two laser beams are required, a pump beam and a Stokes beam. If the beat frequency (difference of these two beam frequencies) is identical to the Raman vibrational frequency of a molecule, then the vibrations are driven coherently and a strong anti-Stokes signal is obtained at two times the pump frequency minus the Stokes frequency. For tissue imaging and live-cell imaging the signal is detected in the epi-direction.

CARS shows promise based on three important factors: it is a noninvasive optical technique, it has high sensitivity, and it is specific to the chemical structure of the molecules that are involved in the CARS processes. The vibronic levels that are excited in CARS are unique chemical makers of individual molecules; therefore, CARS provides a high measure of chemical selectivity. CARS microscopy offers three-dimensional optical sectioning ability due to the fact that the signal generation occurs only at the focal volume of the microscope objective.

With present sensitivity, CARS is an important microscopic technique to image biomolecules such as lipids and water in cells and in cell membranes. A recent paper on chemical imaging of tissue in vivo with video-rate coherent anti-Stokes Raman scattering microscopy points out how this technique can be used to image the lipid-rich structures in the skin of a live mouse (Evans, Potma, Puoris'haag, Cote, Lin, Xie, 2005).

We can expect to see future developments in the application of CARS microscopy to the basic studies of lipid movements in cells and tissues, and there will be applications to the study of normal and diseased skin and the study of skin grafts, scar formation, wound healing, and the aging process in skin. CARS may also provide a diagnostic tool for skin pathology. The chemical selectivity with CARS is very high; specific molecules in a cell can be imaged. The nonlinear technique of CARS generates only a signal within the focal volume of the laser. This is the physical origin of the optical sectioning effect and is similar to that of MPEM.

II.4 COMPARISON OF THE THREE NONLINEAR OPTICAL TECHNIQUES

We now present an overview and comparison of the three different techniques in terms of sensitivity, specificity, selectivity, types of probes required, unique information available from each technique, and the experimental complexity of each method.

II.4.1 Sensitivity

It is possible to use MPEM to detect a single molecule that is spatially restrained. SHG, THG, and CARS microscopic techniques have a comparable sensitivity. Recently, CARS imaging and spectroscopy have been demonstrated at video rates in in vivo tissue. The sensitivity of SHG and THG is considerably less, as it requires 10^4 to 10^6 molecules for detection.

II.4.2 Selectivity

CARS techniques have high chemical specificity and work with the intrinsic molecules within cells and tissues. For example, CARS can be tuned into the CH_2 vibrational band of lipids. Therefore, the microscopic implementation yields images of the location of specific chemical structures such as lipids. SHG is specific for biological molecules without centers of symmetry. SHG and THG spectroscopic techniques are sensitive to molecular orientation and they show specificity for membranes. CARS spectroscopic techniques are sensitive to chemical structure and therefore can differentiate different cellular and tissue structures based on their chemical composition. Both MPEM and CARS can be used to track the diffusion of chemicals in skin.

II.4.3 Endogenous Versus Exogenous Probes

MPEM has available a wide variety of highly specific molecular probes. Many studies with MPEM used genetically expressed probes that fluorescen. These genetically expressed probes can be used to image the location of specific proteins in live cells and tissues. Some new and promising probes for MPEM are a variety of three-dimensionally confined quantum dots. The use of highly specific fluorescent probes also introduces the possibility of altering the structure and the function of the cells and tissues that are being studied. SHG, THG, and CARS spectroscopic techniques applied to optical microscopes can be used to image many endogenous molecules. There is a great need to synthesize contrast agents that respond to SHG and THG spectroscopic techniques.

SHG and THG microscopy is useful for the imaging of organized and orientated protein assemblies; however, these techniques are not useful to image other biological components due to the low specificity and sensitivity. This limitation points out the need for the development of contrast agents for SHG and THG microscopy. The great advantage of SHG and THG is that they image intrinsic molecules in tissues. CARS microscopic techniques, a nonlinear Raman technique, can provide high-resolution optical imaging together with a high degree of chemical specificity and do not depend on exogenous probes to generate contrast.

II.4.4 Experimental Complexity (Equipment, Data Interpretation)

The main limitation in the microscopic implementation of these nonlinear spectroscopic techniques is the cost and complexity of the femtosecond or picosecond lasers and the technique of their synchronization. While the size and the maintenance of these lasers are constantly being improved, the high cost is still an impediment to many investigators.

II.5 CONCLUSIONS

Photodamage of live cells and tissues is always an important limitation for live-cell imaging. The use of near-infrared excitation minimized both the linear absorption and the photodamage to live cells and tissues. In summary, results obtained over the past several years demonstrate the good promise that nonlinear optical techniques hold for microscopy. For example, Gustafsson (2000, 2005) has shown that the transverse resolution of optical microscopy can be increased through use of spatially structured illumination. The resolution is increased by only a factor of two for the case of linear optics, but is in principle unlimited for a nonlinear response. Also, the use of multiphoton effects for applications in endoscopy has been described by Jung and Schnitzer (2003). Reviews of various nonlinear techniques for biological microscopy have been presented by Masters, 2003; Masters and So, 2004; Zipfel et al., 2003; and Hell et al., 2004. The merging of several of these nonlinear techniques, for example by combining SHG and MPEM, may provide unique information on cellular structure, and function and new insights for diagnostic biomedical imaging.

REFERENCES

Alfano, R. R., Masters, B. R. 2003. Biomedical Optical Biopsy, Classic Reprints on CD-ROM, Volume 2, Washington, D.C., Optical Society of America.

Evans, C. L., Potma, E. O., Puoris'haag, M, Cote, D., Lin, C. P., Xie, X. S. 2005. Chemical imaging of tissue in vivo with video-rate coherent anti-stokes Raman scattering microscopy. PNAS, 102(36), 16807–16812.

Franken, P. A., A. E. Hill, C. W. Peters, and G. Weinreich 1961. Generation of optical harmonics, Phys. Rev. Lett 7, 118–121.

Gustafsson, M. G. L., 2000. Surpassing the lateral resolution limit by a factor of two using structured illumination microscopy, Journal of Microscopy, 198, 82–87.

Gustafsson, M. G. L., 2005. Nonlinear structured illumination microscopy: Wide-field fluorescence imaging with theoretically unlimited resolution, Proceedings of the National Academy of Sciences, 102, 13081–13086.

Hell, S. W., M. Dyba and S. Jakobs, 2004. Concepts for nanoscale resolution in fluorescence microscopy, Current Opinion in Neurobiology, 14, 599–609.

Jung, J. C. and M. J. Schnitzler, 2003. Multiphoton microscopy, Optics Letters, 28, 902–904.

Lewis, G. N. D. Lipkin, and T. T. Magel, 1941. Reversible photochemical processes in rigid media. A study of the phosphorescent state, J. Am. Chem. Soc. 63, 3005.

Masters, B. R. Selected Papers on Multiphoton Excitation Microscopy, Bellingham, SPIE Press, 2003.

Masters, B. R. Selected Papers on Optical Low Coherence Reflectometry & Tomography, Bellingham, SPIE Press, 2000.

Masters, B R., So, P.T.C. Antecedents of two-photon excitation laser scanning microscopy, Microscopy Research and Technique, 63, 3–11, 2004.

Zipfel, W. R., R. M. Williams, and W. W. Webb, 2003. Nonlinear magic: multiphoton microscopy in the biosciences, Nature Biotechnology, 21, 1369–1377.

5

Classical and Quantum Theory of One-Photon and Multiphoton Fluorescence Spectroscopy

Barry R. Masters and Peter T. C. So

5.1 INTRODUCTION

In this chapter we introduce the basics of fluorescence spectroscopy. The fundamentals of experimental spectroscopy are presented. Then we develop the classical and the quantum mechanical theory of molecular spectroscopy. These sections provide an introduction to the subsequent chapters on the theory and applications of nonlinear spectroscopic techniques that are used in nonlinear microscopies to study biological processes.

Quantum mechanics is used to calculate the electronic energy levels of an atom, and also to calculate the transition probabilities, or the probability that an atom will change from one electronic energy state to another with the absorption or emission of a photon. Our aim is to present a physical interpretation to the mathematics and thus present a coherent picture of the physics that is correct and understandable to readers without formal training in quantum theory and its mathematical formulations (matrix algebra, infinite dimensional Hilbert space, in particular the Heisenberg, Schrödinger, and Dirac interaction, representations or pictures). An extensive list of annotated references is provided to help the reader to understand the historical development and to achieve a deeper appreciation of the modern theory of quantum mechanics.

The classical theory is described and contrasted with the semiclassical quantum theory in which the atom is quantized and the light field is not quantized. Then we describe one-photon interactions with atoms. That is followed by a description of multiphoton processes in terms of quantum mechanics. Finally, we discuss and compare single-photon processes with multiphoton excitation processes: transition probabilities, parity of states, and absorption cross-sections.

In general our pedagogic approach is to present the physical and mathematical assumptions together with the key equations. We stress the physical insights, define the principal terms and concepts, and where appropriate present quantitative values in order that the reader can obtain a feeling for the magnitude of the physics.

5.1.1 History of Photoluminescence

During the 20th century the following pioneers established both the experimental and the theoretical basis for molecular luminescence and spectroscopy. A partial list of these pioneers includes the following: Enrique Gaviola, Jean Perrin, Francis Perrin, Peter Pringsheim (1928, 1963), Sergei Vavilov, F. Weigert, F. Dushinsku, Alexander Jablonski, Theodor Förster, and Gregorio Weber (Jameson et al., 2003).

The etymology of the words *fluorescence*, *phosphorescence*, and *luminescence* is the subject of a very interesting historical article (Valeur, 2002). Valeur points out that the word *phosphorescence* comes from the Greek and means "that substance that glows in the dark after exposure to light." He then explains that the word *luminescence* comes from the Latin word *lumen* meaning light, and relates to light given off by any means except heating. The term *fluorescence* (from fluorospar, a mineral) was introduced by Sir George Gabriel Stokes at Cambridge University. The term is coined by analogy to opalescence (from the mineral opal). The word *spectrum* comes from the Greek word for appearance. The term *photon* for the light quantum was introduced by G. N. Lewis in 1926 (Kasha, 1990). The word *radiation* derives from Ra, the name of the sun god in ancient Egypt.

We now present some historical experiments. Spectroscopy became the subject of scientific investigation with Newton's 1670 studies on the dispersion of sunlight by prisms as well as the use of prisms to recombine the dispersed sunlight.

By 1814 Josef Fraunhofer had measured and catalogued 574 dark lines in the solar spectrum. David Brewster in 1833 reported that chlorophyll shows a red light when excited with white light and observed from the side. In 1842 Edmond Becquerel first showed that the emitted light is of a longer wavelength than the exciting light. In 1852 Stokes showed that the light emitted following absorption by quinine was always of a longer wavelength than the excitation light.

In 1860 the physicist Gustav Kirchhoff and the chemist Robert Bunsen demonstrated that chemical elements excited by the high temperature of flames exhibited unique line spectra when the emitted light was observed with a prism spectroscope. Future investigations with improved spectroscopes resulted in the technique of characterizing the elements by their emission spectra, and this technique was applied to light from terrestrial elements and also to light emitted from the sun and other stars.

It was the attempts to model the energy distribution (energy as a function of wavelength or frequency) of light from heated cavities, the so-called black bodies, that led to the emergence of Max Planck's quantum theory of energy. While Max Planck is credited with the first use of the quantum of energy $h\nu$, it was Albert Einstein who equated $h\nu$ with the "light quantum." The symbol h is Planck's constant, $h = 6.626 \times 10^{-34}$ Js, and ν is the frequency of the light.

At the beginning of the 20th century, there were two conflicting approaches to the physical understanding of light. On one hand, the second-order partial differential equations of Maxwell predicted the propagation of electromagnetic radiation in a vacuum in terms of continuous transverse electromagnetic waves with a propagation speed equal to that measured for light waves propagating in a vacuum. This result demonstrated that light was a propagating transverse electromagnetic wave. On the other hand, the discrete nature of the spectral lines from excited atoms, as well as Einstein's theory of the quantized photoelectric effect, did not fit with the concept of continuous waves, and supported the concept that light consisted of particles with discrete energy, $h\nu$. The theory of the Compton effect, in which x-rays are scattered from crystals of atoms with low atomic number that resulted in an increase in the wavelength, further confounded the understanding of the nature of radiation–matter interactions.

The birth of quantum theory in its various formulations is in part the result of extremely precise experimental studies of the spectroscopic properties of atoms and molecules (Kuhn, 1987; Jammer, 1989; Mehra & Rechenberg, 2001). The attempts to understand the interaction of matter and radiation (absorption, spontaneous emission, induced emission) resulted in further understanding of the structure of atoms and molecules through the modern developments of quantum theory in which not only is the energy of matter quantized into discrete units, but also the energy of radiation is quantized into discrete units.

The calculation of the energy states of atoms and molecules together with the prediction of the intensities of the emission lines was a major goal of the emerging quantum theory. Verification of the theoretical calculations was made by comparing the theoretical values with the precise experimental values obtained from spectroscopy. Spectroscopy was therefore intimately connected with the development and the verification of quantum theory and its applications to light–matter interactions. It provided the experimental data that stimulated Bohr's early development of the atomic theory, as well as verification of the quantum theory that was subsequently developed by Bohr, Born, Heisenberg, Jordan, Schrödinger, and Dirac.

The historical development of quantum mechanics and the genesis of relativity theory represent one of the outstanding intellectual achievements of humankind. For those with sufficient time and energy, the original papers are highly recommended for further reading. Many of the key papers are available in English translation in the sources cited in the references. In addition to study of the original papers, we recommend several books that cover the genesis of quantum mechanics and describe the transition from classical mechanics to quantum mechanics. Thomas Kuhn's *Black-Body Theory and the Quantum Discontinuity, 1894–1912* (1987) describes the emergence of discontinuous physics in the beginning of the 20th century. Max Jammer's *The Conceptual Development of Quantum Mechanics* is an outstanding one-volume classic. And the six volumes by Jagdish Mehra and Helmut Rechenberg, *The Historical Development of Quantum Theory*, provide the reader with a definitive historical study of the people and the events related to the development of quantum theory.

We attempt to describe only those aspects of physics that are important for our subsequent development of the quantum theory of the absorption and the emission of radiation. While we cite the appropriate books and papers for a comprehensive study of the subject, we discuss selected topics in order to build the foundation for the quantum mechanical description of atom–radiation interaction, in particular the nonlinear interactions that are at the foundation of the topics described in this book on nonlinear spectroscopy and microscopy.

The application of spectroscopy to biology and medicine also has a long and innovative history. In the late 1800s Charles Alexander MacMunn investigated the spectra of heme proteins in different states of oxygenation. He summarized his spectroscopic findings in two important books: *The Spectroscope in Medicine* (1880) and *Spectrum Analysis Applied to Biology and Medicine* (1914). Many of these early investigations were extended by David Keilin from 1925 to 1960, during which he used optical spectroscopy to study the respiratory chain and cytochromes common to plants, yeasts, and higher animals. In 1950, Torbjoern O. Caspersson of the Karolinska Institute, Sweden, published a book *Cell Growth and Cell Function, a Cytochemical Study*, which summarized his 20 years of research on microspectrophotometry of cell organelles, nucleotides, and proteins during the cell cycle and during growth and differentiation. Thus, spectroscopy and microscopy were combined to investigate the structure and the function of live cells, tissues, and organs.

Today, fluorescence imaging and spectroscopy is one of the most useful microscopic techniques in biology and clinical medicine. This is due to the high sensitivity and specificity of fluorescence. The single-photon excited fluorescence occurs at a longer wavelength than the wavelength of the light used to excite the fluorescence; this property permits the separation of the weak intensity of fluorescence from the much higher intensity of the exciting light. The appropriate use of fluorescent probes can image the protein expression profiles in a cell, a tissue, or an organism.

In the next section we introduce the principles of fluorescence spectroscopy.

5.2 PRINCIPLES OF FLUORESCENCE AND PHOSPHORESCENCE

In this section we define several terms used in molecular spectroscopy and augment this tutorial with figures to illustrate the experimental aspects of fluorescence. For further details on the photophysical processes involved with molecular fluorescence, we suggest the excellent book by Bernard Valeur (2002).

Before any type of luminescence occurs there must be the absorption of a photon to populate an excited state. A word often used in spectroscopy is *chromophores* (Greek word for "color bringers"); they are portions of a molecule in which optical transitions can occur. The fluorescence signal from single chromophores can be detected. To obtain an absorption spectrum of a chromophore (usually a dilute solution of the molecules in a solvent) we use a grating or a prism to disperse white light into a spectrum of wavelengths. Each narrow group of wavelengths of a specific energy or color is sequentially incident on the sample, and the transmitted intensity is detected as a function of the incident wavelengths. The resulting curve is known as the absorption spectrum. In absorption spectrophotometry the detector measures small changes in intensity on a high background of incident light.

The Beer-Lambert law relates the absorption of molecules in solution in a cuvette to the concentration. The law states that the fraction of light absorbed by a layer of solution is proportional to the number of absorbing molecules. The law is given as:

$$A(\lambda) = 2.3 \ln \frac{I_0(\lambda)}{I(\lambda)} = N_a \sigma(\lambda)cl = \varepsilon(\lambda)cl = -\log T(\lambda), \qquad (5.1)$$

where $A(\lambda)$ is the absorbance, $T(\lambda)$ is the transmittance, and $\varepsilon(\lambda) = N_a \sigma(\lambda)$ is the molar extinction coefficient. I_0 is the intensity of the light entering the solution. I is the intensity leaving the solution. N_a is Avogadro's number, $\sigma(\lambda)$ is the molecular absorption cross-section, which is a function of the wavelength, c is the concentration of molecules in solution, and l is the path length in the cuvette.

Note that the molar extinction coefficient is a function of wavelength or frequency with units of M^{-1} cm^{-1}. Alternative units are cm^2 mol^{-1}, where c is concentration [mol L^{-1}], and l the pathlength [cm].

The molar absorption coefficient, $\varepsilon(\lambda)$, can be related to the oscillator strength, f, of a classical oscillating dipole:

$$f = 2.303 \frac{mc_0^2}{N_a \pi e^2 n} \int \varepsilon(\bar{\nu}) d\bar{\nu} = \frac{4.3 \times 10^{-9}}{n} \int \varepsilon(\bar{\nu}) d\bar{\nu} \qquad (5.2)$$

where m is the mass of the electron, e is the charge of the electron, c_0 is the speed of light, n is the index of refraction, and $\bar{\nu}$ is the wavenumber in cm^{-1}.

Wavelength is given in nanometers, or alternatively and historically as wavenumber, $\bar{\nu}$, or waves per unit length, with units in reciprocal centimeters (cm^{-1}). These units are interconverted by taking the reciprocal of each value. While the wavenumber scale is linear in energy, most commercial spectrometers use the units of wavelength. Higher values of the oscillator strength for a transition correspond to the higher probability for the transition to occur and a larger intensity is associated with the transition. In quantum mechanics the oscillator strength of a transition between two states corresponds to the transition moment between energetic states of polyatomic molecules.

Polyatomic molecules have discrete energetic states (see Fig. 5.3). In this diagram, the ordinate represents the energy of the molecule and the abscissa represents the configuration of the molecule. The configuration of a molecule is a vector that represents the positions and velocities of all the electrons and nuclei making up the molecule. This is in reality a multiple-dimensional vector but is presented in this diagram as a single axis for simplicity. In a polyatomic molecule, the total energy depends on the configurations of the electrons (electronic states), the vibrational motions of the nuclei (vibronic states), and the rotational motion of the molecule (rotational states). In this diagram, the electronic and vibronic states are represented. The electronic state separations are much larger than the vibrational level that lies within each electronic state. The electronic state energy difference is on the order of 10,000 cm^{-1}, corresponding to the energy of ultraviolet and visible photons; the vibronic level energy difference is on the order of 100 cm^{-1}, which corresponds to the energy of infrared photons. In fact, there are rotational levels that lie within each vibronic level, but their energies are an order of magnitude less, corresponding to photons in the microwave region, and are not shown in this diagram.

The electronic states in the diagram are characterized as either singlet or triplet. Typically singlet states are denoted with the symbol S_n and triplet states with the symbol T_n, where the letter $n = 0, 1, 2, \ldots$ represents the first, second, or higher electronic states. An understanding of singlet and triplet states requires first the introduction of a quantum property of subatomic particles called "spin" (Tomonaga, 1997). Stern and Gerlach observed in 1922 that silver atoms are deflected by an inhomogeneous magnetic field into two discrete groups. In 1925, George Uhlenbeck and Samuel Goldsmit concluded that this observation was due to the "intrinsic rotation" of subatomic particles and generating magnetic moments either parallel or antiparallel to the magnetic field. This property was later called "spin" of the particle, and spin can have two possible states with magnitudes of $\pm \frac{h}{2\pi}$. As a shorthand, these states are referred to as spin $\frac{1}{2}$ or spin-up state and spin $-\frac{1}{2}$ or spin-down state. Since the term $\frac{h}{2\pi}$ often occurs in quantum theory it is given a special symbol, $\hbar$, where $\hbar = \frac{h}{2\pi} = 1.055 \times 10^{-34} J\, s$. The symbol $\hbar$ had been used by Kramers in his doctoral thesis during the time he was a graduate student of Bohr in Copenhagen. One should note that these subatomic particles are not literally spinning spatially, but the "spin" should be considered as an intrinsic property of subatomic particles.

Singlet and triplet states of an atom refer to the spin configurations of its electrons. In a molecule, the electrons populate the orbitals (a word introduced by R. S. Millikan) of the ground, singlet state in pairs; the two electrons in each orbital have opposite spins and thus the total electron spin is zero, which defines the state as a singlet state. The excited singlet state can have a single electron in each higher-energy orbital, but the two electrons still have opposite spins. In the excited triplet state, the lower orbitals have two paired electrons each, but not the higher-energy orbitals, which contain one electron each, and spins are parallel so the triplet state has a net spin of 1. The term "triplet state" refers to the three possible energy levels that correspond to that state.

These polyatomic molecules are capable of many complex transitions between electronic, vibronic, and rotational energy states. Not all the possible transitions actually occur; quantum mechanical selection rules state which transitions are allowed (intense absorption or emission) and which are forbidden (very weak absorption or emission).

For absorption processes there are two types of selection rules. The first class consists of symmetry-forbidden transitions. The application of group theory to the quantum mechanical transition probabilities results in some transitions being forbidden (low probability) due to symmetry of the initial and final states. Even for symmetry-forbidden transitions the vibronic coupling of the two states may allow these "forbidden" transitions to be observed.

The second class consists of spin-forbidden transitions. In a classical view the electron has orbital angular momentum due to its motion around the nucleus. Since moving charges generate magnetic fields, the electron also has a magnetic moment associated with this orbital motion. There is also another magnetic moment due to the spin of the electron. Transitions between singlet–singlet and triplet–triplet are allowed, but transitions between singlet–triplet multiplicities and triplet–singlet are forbidden. Even these weak transitions between states of different multiplicities may be observed due to spin–orbit coupling, the interaction of spin and orbital magnetic moments of an electron.

While a molecule can be excited to higher electronic states, the emission usually occurs from the lowest singlet state, S_1. This is a statement of Kasha's Rule. As stated by Kasha, the emitting electronic level of a given multiplicity is the lowest excited level of that multiplicity (Kasha, 1950). Fluorescence is typically defined as the emission of photons accompanied by a $S_1 \rightarrow S_0$ relaxation, after optical excitation. There are rare cases of fluorescence occurring from higher excited singlet states. The fluorescence is composed of photons with an energy difference that corresponds to the lowest vibrational level of the excited singlet electronic state to the vibrational level of the ground state. Therefore, the spectral properties of fluorescence do not depend on the excitation wavelength. A transition from the excited triplet state to the singlet ground state is "forbidden" and the rates are very slow; they are on the order of milliseconds to seconds. This process of electronic transitions from the first triplet electronic state to the various vibronic energy levels of the singlet ground state is called phosphorescence. The triplet states have energy levels that are lower than the corresponding set of singlet state energy levels. This follows from Hund's Rule, which states that states of electrons with parallel spins have lower energies than states with antiparallel spins. A consequence is that phosphorescence occurs at longer wavelengths than does fluorescence. Typically, we do not observe phosphorescence from solution at room temperature due to competing nonradiative processes and quenching. The experimental evidence for phosphorescence from the triplet state as well as the paramagnetic susceptibility of an organic molecule in its phosphorescent state is contained in these important papers in molecular spectroscopy (Lewis & Kasha, 1944; Lewis et al., 1949).

This scheme discussed previously in which electronic transitions occur from the lowest vibronic level is valid for absorption and emission of light in condensed media (solids or liquids). For very dilute gases, emission may be observed from higher vibronic levels of an excited electronic state due to a lack of thermal de-excitation mechanisms.

In summary, fluorescence is the emission of a photon and the electronic transition from the first excited singlet state to the singlet ground state after optical excitation. The energy of the emitted photon corresponds to the energy difference between the lowest vibrational level of the excited state and the vibrational level of the singlet ground state. Phosphorescence is the emission of a photon and the electronic transition from the first excited triplet state to the singlet ground state after optical excitation. These radiative transitions are called electric dipole transitions.

Why are electronic transitions shown in energy-level diagrams as vertical transitions? In polyatomic molecules the time for an electronic transition (about 10^{-15} s) is much shorter than the time for nuclear rearrangement (about 10^{-13} s); therefore, during an electronic transition from the ground state to the excited state, the molecules do not have time to realign and hence remain in approximately the same configuration.

This phenomenon is exploited in the Born-Oppenheimer approximation that is used to solve the quantum mechanical equations for the energy states of the molecule. This approximation is used to solve the Schrödinger equation for diatomic molecules. An approximate wave function is used, which is the product of two functions: the first is the electronic part, and the second is the vibronic-rotation part. It is based on the approximation in which all interactions between the nuclear and the electronic motions are ignored. The key point is that the time for the electronic transitions is much faster than the time for the nuclei to change their position. The Franck-Condon principle is based on the Born-Oppenheimer approximation and states that transitions between electronic states correspond to vertical transitions, as shown in diagrams of energy versus molecular configuration. The most intense transition is from the ground vibration state to the vibrational state lying vertically above it.

In fluorescence spectroscopy, the incident light is restricted to a narrow range of wavelengths centered on the wavelength of the absorption peak of the molecule. A fluorescence filter is used to separate the incident light from the fluorescence light (which occurs at longer wavelengths). A prism or a grating can also be used to disperse and isolate the emitted light. This shift in wavelength between the excitation light and the emitted light is called the Stokes' shift. It permits the extremely weak fluorescence to be detected with a high signal-to-noise ratio since the high-intensity excitation light was excluded. This is the source of the high sensitivity of fluorescence techniques. A plot of the intensity of the fluorescence versus the wavelength is called the emission spectrum (Fig. 5.1). Note from the figure that the fluorescence intensity is maximum with optimal excitation wavelengths, and the intensity decreases as the excitation wavelength is

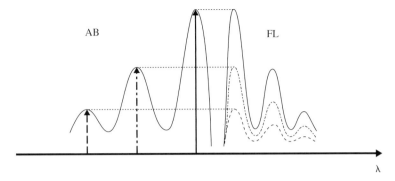

Figure 5.1. Effect of the excitation wavelengths within an absorption band (AB) on the profile and the intensity of the fluorescence spectrum (FL). The position of three absorption bands is shown on the left portion of the figure with the corresponding changes of the intensity (solid and dashed lines) of the emission spectra (FL) on the right portion of the figure. Excitation at different absorption bands of a molecule does not affect the emission profile, but it does affect the emission intensity.

displaced from the absorption maximum. The excitation-induced fluorescence spectra are typically independent of the excitation wavelength (Kasha's rule); however, the intensity of the emission is dependent on the excitation wavelength.

When the excitation and the emission spectra of a molecule in a dilute solution are compared, it is observed that the absorption and emission spectra have mirrored features; however, there are some exceptions to this rule due to excited state reactions. This effect is a called the mirror rule (Fig. 5.2). The approximate mirror symmetry of the intensity pattern of the absorption and the fluorescence spectra is described by Levschin's (1931) "law of mirror correspondence." The absorption spectrum is a plot of the extinction coefficient versus wavenumber (cm^{-1}) or wavelength (nm) and the emission spectra are plotted as the fluorescence intensity (arbitrary units) versus wavenumber or wavelength. Much of the early experimental studies and their theoretical foundations of light–matter interaction were worked out for the simplest case of light interacting with a hydrogen atom composed of a proton and an electron. For example, under electric arc discharge, discrete spectral emission lines corresponding to the Lyman, Balmer (visible region), and higher-order series are observed. This corresponds to electronic transitions between various stable (stationary) electronic states of the atom in which photons are absorbed or emitted.

A very important parameter is the lifetime of the luminescence (fluorescence or phosphorescence). In order to understand the term "fluorescence lifetime" it is helpful to show the generalized energy diagram for molecular radiative and nonradiative processes that can occur and their kinetics (Fig. 5.3). The transition from the excited singlet state to the singlet ground state with the emission of a photon (fluorescence) is spin-allowed and readily occurs in about 10 ns. In 1926, Gaviola (1927) made the direct measurement of nanosecond lifetimes using a phase fluorometer that was based on a Kerr cell. Gaviola constructed the lifetime instrument in Pringsheim's laboratory in the Physikalisches Institut der Universität, Berlin. The experimental aspects of time-domain lifetime measurements and frequency-domain lifetime measurement are well described (Lakowicz, 2006).

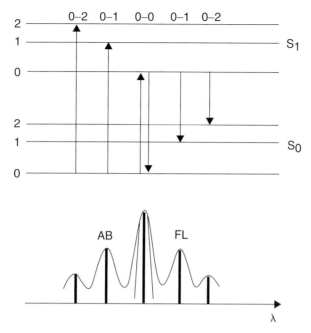

Figure 5.2. The mirror rule for absorption (AB) and fluorescence (FL) of a molecule. The upper portion of the figure shows the various electronic-vibronic transitions from the ground state S_0 to the first excited singlet state S_1. The vertical transitions follow from the Franck-Condon principle. The 0–2 transition is an absorption from the 0 vibronic level of the S_0 electronic state to the second (2) vibronic level of the first excited singlet state . The 0–1 fluorescence emission is a transition from the 0 vibronic state of the first excited single state S_1 to the first vibronic state (1) of the ground singlet state S_0. The lower portion of the figure shows the lines and bands of the absorption spectra (AB) and the corresponding fluorescence (FL) spectra.

The fluorescence lifetime is one of the most important parameters of fluorescence because it determines the interval for the various dynamic processes that can affect the emission. The time ranges for the various processes shown in Figure 5.3 determine the observed spectroscopic properties of a given molecule in a given environment (concentration, other molecules present, temperature, etc.). Absorption occurs in about 10^{-15} s. Intersystem crossing occurs in the range of 10^{-10} to 10^{-8} s. Internal conversion occurs in the range of 10^{-11} to 10^{-9} s. Vibrational relaxation occurs in the range of 10^{-12} to 10^{-10} s. The lifetime of the first excited singlet state is in the range of 10^{-10} to 10^{-7} s and occurs with the emission of fluorescence. The lifetime of the first excited triplet state T_1 occurs in the range of 10^{-6} to 1 s with the emission of phosphorescence.

The emission lifetime is defined as the time for the light emission intensity of the first electronic state of an ensemble of fluorophores to decrease by a factor of $\frac{1}{e}$ immediately following a short pulse of excitation light. The measured lifetime is not always identical with the intrinsic emission lifetime of fluorophores. They are only identical when there are no nonradiative processes that are occurring and compete with the photon emission. When the quantum efficiency or quantum yield (a measure of the ratio of the number of fluorescence photons to the number of absorbed photons) is equal to one, then the intrinsic and the observed fluorescence lifetimes are equal.

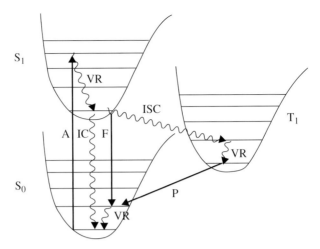

Figure 5.3. Modified Franck-Condon energy state diagram shows the electronic-vibronic energy levels and energy transitions of a molecule. The ground singlet state is S_0, the first excited singlet state is S_1, and the first excited triplet state is T_1. Transitions from absorption (A), fluorescence (F), and phosphorescence (P) are indicated by boldfaced lines with arrowheads. The wavy lines indicate various types of non-radiative transitions: internal conversion (thermal relaxation) (IC) and intersystem crossing (S_1 to T_1 transitions) (ISC). The horizontal lines within each electronic energy level represent the vibronic levels.

The fluorescence lifetime of the excited singlet state S_1 for the process:

$$A^* \xrightarrow{k_r^S} A + \text{photon} \tag{5.3}$$

is defined by the reciprocal of the rate constant, k_r^S, for the radiative decay of the excited singlet state. A^*, the excited state of the fluorophore, is de-excited to the ground state with the emission of a photon.

When de-excitation only occurs due to radiative processes, the fluorescence lifetime, τ_s:

$$\tau_S = \frac{1}{k_r^S}. \tag{5.4}$$

However, if there are radiative and nonradiative de-excitation processes occurring, then the lifetime is given by:

$$\tau_s = \frac{1}{k_r^S + k_{nr}^S}. \tag{5.5}$$

The time constant or the lifetime of this process is given by the reciprocal of the sum of the rate constants for all the processes that result in de-excitation of the excited state—the rate constants for the radiative processes and the rate constants for the nonradiative processes, k_{nr}^S.

The fluorescence intensity is defined as the number of photons (mol or einsteins, where 1 mole of photons is one einstein unit) emitted per unit time (s), per unit volume of solution (L).

$$[A^*](t) = [A^*]_0 \exp\left(-\frac{t}{\tau_S}\right). \qquad (5.6)$$

This equation gives the single-exponential decay of the concentration of the excited state, $[A^*](t)$, as a function of time following a very short pulse of excitation light. $[A^*]_0$ is the concentration of the excited state immediately following the excitation, since the fluorescence intensity is proportional to $[A^*]$,

$$i_F(t) = k_r^S [A^*](t) = k_r^S [A^*]_0 \exp\left(-\frac{t}{\tau_s}\right), \qquad (5.7)$$

$$i_F = i_0 \exp\left(-\frac{t}{\tau_S}\right). \qquad (5.8)$$

where i_0 is the intensity immediately following the excitation.

Note that the amplitude of the detected fluorescence decay will depend on the instrument collection efficiency but the decay constant can be recovered independently.

We have described the decay kinetic of a fluorophore as a single-exponential process. However, fluorescence decay represented by multi-exponential processes or continuous distribution may also occur due to the presence of multiple decay pathways.

Photosaturation is experimentally observed when a plot of excitation intensity versus fluorescence reaches a plateau. Further increases in excitation intensity will not result in increased fluorescence. Photosaturation occurs due to the finite lifetime of the molecule. Increasing excitation intensity depletes the number of the fluorescent molecules in the ground state, transiently trapping them in the singlet or triplet excited state, and results in a lower absorption and fluorescence.

5.2.1 Radiationless (Nonradiative) Transitions in Complex Molecules

Internal conversion (IC) is the decay from a higher vibrational state to the ground state of the same multiplicity. Energy can be nonradiatively dissipated by internal conversion from the first excited singlet state, S_1, to vibronic levels of the singlet ground state, S_0 (thermal relaxation of the excited molecule without the emission of photons). Internal conversion is a rapid, nonradiative process between two electronic states of the same spin-multiplicity (i.e., $S_1 \rightarrow S_0$). In solution it is usually followed by vibrational relaxation to the lowest vibrational level. The internal conversion process occurs at least 10^4 times faster than the spontaneous fluorescence emission.

Vibronic relaxation (VR) is the transfer of a fluorophore's vibrational energy to solvent molecules, which results in a lowering of the vibrational energy to that of the lowest vibronic level of a given electronic state. Thermodynamically, a molecule resides in the singlet ground state. The absorption of a photon can excite the molecule to the first excited singlet state or to higher excited singlet states. The absorption spectrum corresponds to various vibronic levels in the excited state. Vibrational relaxation occurs in the time range of 10^{-12} to 10^{-10} s. These electronic transitions can occur to various vibronic levels of the singlet ground state. Since the time for VR processes to occur is much faster than the time for electronic transitions, the emission spectrum corresponds to relaxation from the lowest vibration level of the first excited electronic state to various ground-state vibronic levels.

Intersystem crossing (ISC) is a process in which the energy is transferred nonradiatively from the singlet state to a lower triplet electronic state, $S_1 \rightarrow T_1$, following vibronic relaxation in the first excited singlet electronic state (see Fig. 5.3). Intersystem crossing is forbidden by the Born-Oppenheimer approximation but occurs at an extremely low rate due to spin–orbit coupling. Intersystem crossing is the spin–orbit-coupling–dependent internal conversion (Kasha, 1950). In most cases, ISC is the radiationless transition from the lowest excited singlet level to the lowest triplet level of the molecule.

5.2.2 Characterization of Fluorescence and Phosphorescence

Fluorescence and phosphorescence can be characterized by excitation spectra, emission spectra, lifetimes, quantum yields, and polarization. We have discussed some of these quantities except quantum yield and polarization.

The emission quantum efficiency or quantum yield is a measure of the ratio of the number of fluorescence photons to the number of absorbed photons. Vavilov is credited with the definition of quantum yield and its first measurement (Jameson et al., 2003). Warburg developed experimental techniques to measure the quantum yield in photochemical processes (Neckers, 1993). A quantum efficiency that is approximately one indicates that most of the absorbed photons are emitted as fluorescence. Fluorescein has a quantum yield approximately equal to one in water. When the quantum efficiency is low, then other nonradiative processes such as thermal relaxation compete with the radiative fluorescence. The quantum yield for fluorescence, Φ_F, is given by the relative rates of the radiative processes and the rates of all the nonradiative processes.

We define the fluorescence quantum yield, Φ_F, as:

$$\Phi_F = \frac{k_r^S}{k_r^S + k_{nr}^S} = k_r^S \, \tau_S. \tag{5.9}$$

The excitation efficiency of a fluorophore depends on the relative orientation of the molecule and the excitation electric field. The excitation dipole of a fluorophore is a function of its molecular structure and the associated symmetry of its ground electronic state. The excitation dipole quantifies the directional polarizability of the molecule in the presence of an electric field. The excitation efficiency of this molecule by polarized light is a cosine square function of the angle between the excitation dipole of the molecule and the polarization direction of the light. The directional polarizability of the molecule in the excited state, characterized by the emission dipole, may not be the same as that of the ground state. The emission dipole determines the polarization of the emitted light by the same cosine square factor.

Depolarization of the fluorescence can be caused by rotational diffusion of the fluorescent molecules. The measured polarization of luminescence is related to the intrinsic polarization of the molecule by the Perrin equation, which was derived and experimentally confirmed by Francis Perrin (Perrin, 1926). The Perrin equation is correct for spheres and was later modified for ellipsoids. The measured polarization of luminescence can be modified either by the rotation of the molecule during the emission lifetime or by nonradiative energy transfer. Experimentally, the polarization of the fluorescence is based on exciting the molecules with polarized light using a polarizer. This process is called photoselection. The polarized light photoselects those molecules with the correct molecular orientation, with respect to the polarization of the excitation, and results in a partially polarized fluorescence. The emission is measured with a polarizer (called an analyzer in this context) with orientations either parallel or perpendicular to the excitation polarization. In solutions, the polarization is near zero for molecules with a rotational diffusion coefficient shorter than their fluorescence lifetime. The rotational diffusion coefficient characterizes the time for a molecule to randomize its orientation in a solvent due to rotation. The polarization is finite for larger molecules when its rotational diffusion is slow compared with the fluorescence lifetime. In viscous and constrained environments (i.e., cell membranes), the measured fluorescence polarization of an embedded fluorophore can be larger and the Perrin equation can be used to investigate the rigidity of domains in membranes and cellular compartments. Energy transfer between fluorophores can also reduce the polarization. The polarization is a dimensionless quantity that is independent of the total fluorescence intensity of the molecules.

Experimentally, the polarization ratio, P, is given by:

$$P = \frac{I_{//} - I_{\perp}}{I_{//} + I_{\perp}}. \tag{5.10}$$

A related quantity, the emission anisotropy ratio, r, is given by:

$$r = \frac{I_{//} - I_{\perp}}{I_{//} + 2I_{\perp}}, \tag{5.11}$$

where $I_{//}$ is the measured fluorescence intensity when the orientations of the polarizer and the analyzer are parallel, and $I_{\perp}$ is the measured fluorescence intensity when the orientations of the polarizer and the analyzer are perpendicular to each other.

The relation between the polarization ratio and the emission anisotropy is given by:

$$r = \frac{2P}{3 - P}. \tag{5.12}$$

Measurements of fluorescence polarization can be used to determine the rotational diffusion constant of a molecule and perhaps to infer information about its shape and apparent volume.

The older literature contains the equation derived by Francis Perrin:

$$\left(\frac{1}{P} - \frac{1}{3}\right) = \left(\frac{1}{P_0} - \frac{1}{3}\right)(1 + 6D_r\tau) \tag{5.13}$$

where P is the measured polarization ratio, P_0 is the intrinsic polarization ratio, D_r is the rotational diffusion coefficient, and τ is the fluorescence lifetime.

The intrinsic polarization is related to the orientation difference between excitation and emission dipoles and is high when they are coincident.

The Stokes-Einstein relation allows the rotational diffusional coefficient, D_r, to be related to the local viscosity and the molecular volume assuming spherical geometry.

$$D_r = \frac{RT}{6V\eta} \qquad (5.14)$$

where D_r is the rotational diffusion coefficient, R is the molar gas constant, T is the temperature, V is the apparent molecular volume, and η is the viscosity of the medium in which the fluorescent molecule is situated.

It is very important to note that the Stokes-Einstein equation may not be valid on molecular dimensions.

A more modern version of the Perrin equation in terms of emission anisotropy is given below:

$$\frac{1}{r} = \frac{1}{r_0} + \frac{\tau RT}{r_0 \eta V} \qquad (5.15)$$

where r is the measured anisotropy, r_0 is the fundamental anisotropy and is the analog of the intrinsic polarization ratio, τ is the fluorescent lifetime, η is the viscosity of the solution, R is the molar gas constant, T is the temperature, and V is the apparent volume of the rotating unit.

5.2.3 Resonance Energy Transfer (RET)

There is another important process that occurs in excited states: resonance energy transfer (RET) (Förster, 1946, 1951; Clegg, 1996). RET is the process in which energy is transferred nonradiatively from an excited donor molecule to the excited state of another molecule called the acceptor through an intermolecular long-range dipole–dipole coupling.

The process occurs when the emission spectrum of the fluorophore (called the donor) partially overlaps with the absorption spectrum of another molecule (called the acceptor). When the spectra overlap occurs and the donor and acceptor molecules are sufficiently close to each other (about 10–100 Å), and the transition dipoles of the donor and the acceptor are properly oriented with respect to each other, then the donor can transfer its energy to the acceptor. There is a dipole–dipole coupling between the donor molecule and the acceptor molecule. The emission from the donor is quenched and the emission from the acceptor is increased (sensitized emission).

The energy transfer between the donor and the acceptor is not mediated by the emission and the subsequent absorption of a photon: it is mediated by nonradiative energy transfer without the emission and absorption of a photon. The efficiency of the resonance energy transfer is dependent on the spectral overlap of the donor and the acceptor as well as the inverse sixth power of their separation distance. Therefore, this Förster resonance energy transfer can be used to measure intra- or intermolecular distances.

The Förster distance, R_0, is defined as the distance at which resonance energy transfer is 50% efficient. This critical Förster distance is the donor–acceptor distance for which the energy transfer rate equals the donor de-excitation rate in the absence of the acceptor molecule. The rate of energy transfer from donor to acceptor is given by:

$$k_T = \frac{1}{\tau_D^0} \left(\frac{R_0}{r} \right)^6 \tag{5.16}$$

where τ_D^0 is the decay constant of the donor in the absence of acceptor and r is the donor–acceptor distance.

$$\text{If } r = R_0, \text{ then } k_T = \frac{1}{\tau_D^0}. \tag{5.17}$$

Because the Förster distance is the dimensions of biological molecules, RET provides a sensitive method to investigate configuration changes in a wide variety of biological interactions (i.e., cell signaling). The efficiency, E, of the energy transfer is given by the number of quanta transferred from donor to acceptor, divided by the number of quanta absorbed by the donor. The transfer efficiency and the distance between the donor molecule and the acceptor molecule can be calculated from measurements of the emission lifetime of the donor in the presence and in the absence of the acceptor. Fluorescence resonance energy transfer is a widely used technique to measure molecular distances as a "spectroscopic ruler" (Latt et al., 1965; Stryer, 1978; Stryer & Haugland, 1967).

$$E = \frac{1}{1 + \left(\frac{r}{R_0} \right)^6} = 1 - \frac{\tau_D}{\tau_D^0} \tag{5.18}$$

where τ_D is the donor excited lifetime in the presence of acceptor, and τ_D^0 is the donor excited lifetime in the absence of acceptor.

A detailed discussion of the classical and the quantum mechanical theories of Förster RET is presented by Masters and So in Chapter 22 of this book.

5.2.4 Environmental Effects on Fluorescence

The fluorescence of a molecule is very sensitive to environmental effects such as temperature, quenching, photobleaching, solvent polarity, pH, and viscosity.

Temperature is a very important parameter. As the temperature increases, the intensity of the interaction between the molecules (fluorophore and solvent) increases. There are more frequent collisions between molecules and they are more energetic. That results in a decrease in the fluorescence quantum yield and the fluorescence lifetime since the nonradiative processes related to collisions are more efficient at higher temperatures.

Quenching is the sum of the processes that cause nonradiative relaxation from the excited state to the ground state. In collisional or dynamic quenching or Stern-Volmer quenching, the molecule in the excited state collides with another molecule (quencher) and transfers its energy to the other molecule. The excited state lifetime and the quantum yield are reduced as a result of collisional quenching.

The decrease in fluorescence intensity due to collisional (diffusive process) quenching can be described by the Stern-Volmer equation, in which Φ_0 is the quantum yield of the fluorescence in the absence of quencher, Φ is the quantum yield in the presence of quencher, τ_0 is the fluorescence lifetime in the absence of quencher, and τ is the lifetime in the presence of quencher.

The Stern-Volmer quenching constant is K_{sv}, the bimolecular quenching constant is k_q, the unquenched lifetime is τ_0, and the concentration of the quencher is $[Q]$.

The Stern-Volmer relation is:

$$\frac{\Phi_0}{\Phi} = \frac{I_0}{I} = 1 + k_q \tau_0 [Q] = 1 + K_{sv}[Q]. \tag{5.19}$$

In another quenching process, termed static or complex quenching, the fluorophore molecule combines with another molecule, reducing the number of molecules that can become excited. The fluorophore can form a nonfluorescent complex with quenchers. This process occurs in the ground state. Then the complex of the fluorophore and quencher absorbs light, and it immediately returns to the ground state without the emission of a photon. It is not dependent on diffusion or collisional processes. In this case the fluorescence lifetime is unchanged, but the fluorescence intensity is reduced.

Figure 5.4 shows plots of the ratio $\frac{I_0}{I}$ and the ratio $\frac{\tau_0}{\tau}$ versus the concentration of the quencher $[Q]$ for the cases of static and dynamic quenching respectively. It is important to note that only lifetime measurements in the presence and absence of quencher can distinguish between the processes of static and dynamic quenching.

Photobleaching is a very important process that destroys the fluorescence in an irreversible manner during illumination. Many fluorescent molecules undergo a photochemical reaction in the excited state that converts them into nonfluorescent compounds. In many cases the excited state of the fluorescent molecule reacts with oxygen and forms

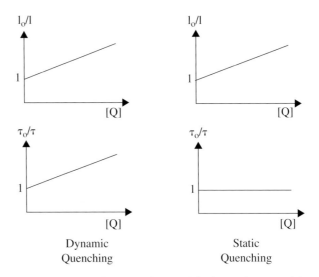

Figure 5.4. Experimental procedures to distinguish dynamic quenching and static quenching. The ratios of either intensities or lifetimes are plotted against the concentration of quencher $[Q]$. The fluorescence intensity in the presence and the absence of quencher is $[I]$ and $[I_0]$ respectively. The fluorescence lifetime in the presence and the absence of the quencher is τ and τ_0 respectively.

a nonfluorescent compound. Therefore, under constant illumination, the fluorescence is observed to decrease with time.

This overview of molecular spectroscopy is by necessity brief, and the reader is advised to explore the references for more comprehensive treatments (Herman, 1998; Lakowicz, 2006; Preamps, 1995; Valeur, 2002).

5.3 ATOM–RADIATION INTERACTIONS: TRANSITION FROM CLASSICAL TO QUANTUM APPROACHES

5.3.1 Atom–Radiation Interactions: the Damped Harmonic Oscillator

Fluorescence is a form of light–matter interaction. Physicists at the beginning of the 20th century were trying to understand light–matter interactions. One observation that they could not explain is how the refractive index and the absorption of a material vary as a function of the wavelength of the incident light; these relationships are called the dispersion curves (Fig. 5.5). Dispersion is a phenomenon in which the refractive index, $n(\nu)$, depends on the frequency, ν, of the electromagnetic radiation. Historically, the term "dispersion" derives from the dispersion or separation of light by a prism or diffraction grating into its spectrum.

The refractive index is a complex number:

$$n_c = n_r + i\kappa \tag{5.20}$$

where n_c is the complex refractive index.

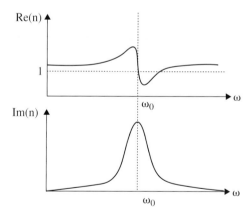

Figure 5.5. Normal and anomalous dispersion. Normal dispersion occurs away from resonance absorption. Anomalous dispersion occurs at the resonance absorption frequency. The figure shows typical curves of the real part of the refractive, Re(n), and the imaginary part of the refractive index, Im(n), versus the frequency of the radiation, ω. The resonance frequency of the absorption is at the frequency ω_0. Note that the real part of the refractive index changes rapidly near the absorption frequency. The real part of the refractive index takes positive and negative values. The imaginary part of the refractive index peaks at the resonance absorption frequency ω_0, and it only takes positive values from zero.

The imaginary part of the refractive index, κ, gives the absorption of the electromagnetic wave, and the real part of the refractive index, n_r, gives the dispersion. This complex refractive index gives rise to a complex permittivity, ε_c. For transparent materials, such as water or air, the absorption coefficient for visible light is very small, and the refractive index is given by the real part of the index.

For the dispersion curve (see Fig. 5.5), in regions where n increases with ω (the angular frequency, $\omega = 2\pi\nu$), or when $\frac{dn}{d\omega} > 0$, the dispersion is called normal dispersion. For transparent materials, that is the case for visible, ultraviolet, or infrared radiation. Where there is a resonance (absorption), then $\frac{dn}{d\omega} < 0$, and the dispersion is called anomalous dispersion.

The absorption of radiation by a material is given as:

$$I = I_0 \, e^{-a\Delta z}, \tag{5.21}$$

where I_0 is the incident intensity of the radiation, and after passage through a thin section of thickness, Δz, the intensity is I.

The absorption coefficient, a, is proportional to the imaginary part of the complex refractive index:

$$a = \frac{4\pi\kappa}{\lambda_0}, \tag{5.22}$$

where λ_0 is the wavelength in vacuum.

In this section we discuss simple physical models, in particular the damped harmonic oscillator model, that predict these experimental dispersion curves. Classical theoretical physics was taken to its limits in this task. All of this work occurred prior to the development of quantum theory.

At the beginning of the 20th century Maxwell's equation, together with relativistic corrections, summarized all the experimental observations of electromagnetic waves. Light was shown to be a transverse electromagnetic wave because the calculated velocity from Maxwell's equations agrees with the measured velocity. Maxwell's equations yielded the laws of refraction and reflection and explained how light propagates.

The wave equation in a vacuum can be derived from the Maxwell equations, and it governs the free propagation of light.

$$\nabla^2 \vec{E} = \frac{\varepsilon\mu}{c^2} \frac{\partial^2 \vec{E}}{\partial t^2}, \tag{5.23}$$

where c is the speed of light, ε is the dielectric constant, and μ is the magnetic permeability. The Laplacian operator, ∇^2, is defined as:

$$\nabla^2 = \frac{\partial^2}{\partial x^2} + \frac{\partial^2}{\partial y^2} + \frac{\partial^2}{\partial z^2}. \tag{5.24}$$

Maxwell showed that the speed of an electromagnetic wave in free space is:

$$c = 3 \times 10^8 \, ms^{-1}$$

This theoretical value was in excellent agreement with the precisely measured speed of light. Maxwell therefore assumed that light is an electromagnetic wave.

The index of refraction of a material, n, is the ratio of the speed of an electromagnetic wave in vacuum, c, to the speed of the electromagnetic wave in matter, $v = c/\sqrt{\varepsilon\mu}$. In Gaussian units,

$$n = \frac{c}{v} \tag{5.25}$$

A solution to the wave equation is the plane wave:

$$\vec{E} = \vec{E}_0\, e^{-i(\vec{k}\cdot\vec{r}-\omega t)} \tag{5.26}$$

where $\vec{E}$ is the electric field in Gaussian units of statvolt/cm and $\vec{k}$ is the propagation wave vector whose magnitude is the propagation constant or wavenumber equal to $2\pi/\lambda$. The $\vec{k}$ vector is defined in terms of its components:

$$\vec{k} = k_x\,\mathbf{i} + k_y\,\mathbf{j} + k_z\,\mathbf{k}, \tag{5.27}$$

where $\mathbf{i}, \mathbf{j}, \mathbf{k}$ are unit vectors.

The angular frequency is defined as:

$$\omega = 2\pi\nu \tag{5.28}$$

where ν is the frequency of light.

Furthermore,

$$\frac{\omega}{k} = c \tag{5.29}$$

The spatial vector $\vec{r}$ is defined in terms of its components:

$$\vec{r} = x\,\mathbf{i} + y\,\mathbf{j} + z\,\mathbf{k}. \tag{5.30}$$

Light is a transverse electromagnetic wave in which the electric and magnetic fields are perpendicular, and both fields are perpendicular to the direction of propagation. For a linear polarized wave there are two orthogonal components of polarization for a wave propagating in the z-direction.

$$\vec{E} = E_{0x}\,\mathbf{i}\,e^{-i(k_z z - \omega t)} + E_{0y}\,\mathbf{j}\,e^{-i(k_z z - \omega t + \theta)}, \tag{5.31}$$

where θ is a phase function.

The next step is the development of microscopic models to describe the interaction of light with matter, in particular how to explain the phenomenon of dispersion. This section follows the work of Drude, Lorentz, and later Kramers on their theories for the interaction of light (an oscillating electromagnetic field) and matter (consisting of oscillating electric charges). Their microscopic models of the interaction resulted in dispersion relations (i.e., the refractive index as a function of frequency), which until the development of quantum theory were a good description of the experimental observations.

The central concept of these microscopic models is as follows: an oscillator consisting of an electric charge is placed in an electromagnetic field, and energy can be gained or lost between the charged oscillator and the electromagnetic field. An electromagnetic wave that propagates through a material induces electric dipole moments in the material; that is, the field causes oscillating movements of electric charges from their equilibrium positions in the absence of the field. H. A. Lorentz (1880) and L. Lorenz (1881) both developed what is now known as the Lorentz-Lorenz formula. They assumed that an electric dipole has dipole moment, $\vec{p}$, defined by:

$$\vec{p} = e\vec{r}, \tag{5.32}$$

where a positive and a negative charge e is separated by a distance $\vec{r}$.

For these dipoles, the dipole moment is proportional to the electric field, $\vec{E}$, in the linear response regime. We can define polarizability, α as:

$$\vec{p} = \alpha\vec{E}. \tag{5.33}$$

For a collection of these dipoles, it is observed that the induced dipole moment per unit volume, $\vec{P}$, is also proportional to the electric field, $\vec{E}$:

$$\vec{P} = \chi_e \vec{E} \tag{5.34}$$

where χ_e, the proportionality constant, is called the electric susceptibility of the material. The electric susceptibility is a function of the frequency of the electromagnetic field.

The polarization $\vec{P}$, and the macroscopic field, $\vec{E}$, are related by:

$$\vec{P} = N\alpha\left(\vec{E} + \frac{4\pi}{3}\vec{P}\right) \tag{5.35}$$

where N is the number of atoms per unit volume. The first term accounts for the contribution of all the dipoles in the volume and the second term, $\frac{4\pi}{3}\vec{P}$, accounts for the shielding of the external electric fields by these dipoles in the material.

Further, the presence of $\vec{P}$ is equivalent to creating additional charges in the volume. The induced volume charge density, ρ', can be calculated as:

$$\rho' = -\vec{\nabla} \cdot \vec{P} \tag{5.36}$$

where $\vec{\nabla}\cdot$ is the divergence.

Gauss Law in a dielectric can be generalized as:

$$\vec{\nabla} \cdot \vec{E} = 4\pi(\rho + \rho') = 4\pi\rho - 4\pi\vec{\nabla} \cdot \vec{P} \tag{5.37}$$

where ρ is the free charge density in the material (Jackson, 1999).

Defining the displacement vector $\vec{D}$:

$$\vec{D} \equiv \vec{E} + 4\pi \vec{P} = \vec{E} + 4\pi \chi_e \vec{E}, \tag{5.38}$$

using equation (5.34).

We further define the proportionality constant between $\vec{D}$ and $\vec{E}$ as the dielectric constant, ε:

$$\vec{D} = \varepsilon\vec{E}. \tag{5.39}$$

Therefore, the dielectric constant is related to susceptibility as:

$$\varepsilon = 1 + 4\pi \chi_e \tag{5.40}$$

The wave equations derived from Maxwell's equations in the dielectric medium dictate that:

$$n^2 = \varepsilon\mu \approx \varepsilon \tag{5.41}$$

where μ is the magnetic permeability in materials, which is very close to one except for ferromagnetic materials.

The Lorentz-Lorenz formula can be derived from solving equations (5.34, 5.35, 5.40, 5.41) algebraically:

$$\alpha = \frac{3}{4\pi N} \cdot \frac{n^2 - 1}{n^2 + 2}. \tag{5.42}$$

The Lorentz-Lorenz formula is very important because it relates the macroscopic index of refraction, n, to the microscopic polarizability, α, of each dipole.

H. A. Lorentz then further developed a theory to account for the frequency dependence of refractive index, the dispersion curve, by treating the dipole as a classical damped harmonic oscillator as follows. In this treatment, the effect of the magnetic field is neglected and we first give a physical explanation of why this approximation is valid, except for ferromagnetic materials.

The Lorentz force, $\vec{F}$, is the force on an electron that is moving with a velocity, $\vec{v}$, in an electromagnetic field, $\vec{E}$.

$$\vec{F} = e\left(\vec{E} + \frac{\vec{v}}{c} \times \vec{B}\right). \tag{5.43}$$

$\times$ is the vector cross product, and $\vec{B}$ is the magnetic field. If the velocity of the electron is much less than the speed of light, $|\vec{v}| \ll c$, we can neglect the portion of the Lorentz force due to the magnetic field. Then the Lorentz force is given by:

$$\vec{F} = e\vec{E} \tag{5.44}$$

where e is the charge on the electron.

To calculate the polarizability of a dipole, Lorentz assumed that the negative charge has mass m, and magnitude, e. This charge is bound to its equilibrium position by an elastic restoring force $\vec{F}$, with a force constant k:

$$\vec{F} = -k\vec{r}. \tag{5.45}$$

The quantity, $\vec{r}$, is the distance that an electron is displaced from its equilibrium position. The driving force of the electrons in the material is the oscillating electromagnetic field:

$$\vec{E} = \vec{E}_0 e^{-i\omega t}. \tag{5.46}$$

The angular frequency of the electromagnetic wave, ω, is given by:

$$\omega = 2\pi \nu. \tag{5.47}$$

Then the equation of motion for the displacement of the electron from its equilibrium position due to the driving force of the electromagnetic field is:

$$m\frac{d^2\vec{r}}{dt^2} + k\vec{r} = e\vec{E}. \tag{5.48}$$

A solution to this equation is:

$$\vec{r} = \frac{e\vec{E}}{m(\omega_0^2 - \omega^2)}. \tag{5.49}$$

The driving frequency is ω, and the resonance frequency ω_0 is given by:

$$\omega_0 = \sqrt{k/m}. \tag{5.50}$$

Thus, the electron will oscillate at the frequency of the incident electromagnetic field. Since each electron will contribute an electric moment, $\vec{p}$, where $\vec{p} = e\vec{r}$, to the polarization:

$$\vec{p} = e\vec{r} = \frac{e^2\vec{E}}{m(\omega_0^2 - \omega^2)}. \tag{5.51}$$

From the Lorentz-Lorenz formula (5.42), then:

$$\frac{n^2 - 1}{n^2 + 2} = \frac{4\pi}{3}\frac{Ne^2}{m(\omega_0^2 - \omega^2)}. \tag{5.52}$$

For a gas, $n \cong 1$, so:

$$n^2 - 1 = \frac{4\pi Ne^2}{m(\omega_0^2 - \omega^2)}. \tag{5.53}$$

This theory produces an infinite sharp resonance line of the natural resonance frequency of the harmonic oscillator, which is inconsistent with the observed dispersion curve, which has a finite width and amplitude (see Fig. 5.5). Realizing that an oscillating electron emits radiation, which removes energy, it is necessary to include a damping term that is proportional to the speed that the electron moves, $\gamma \frac{d\vec{r}}{dt}$. The new equation of motion is:

$$m\frac{d^2\vec{r}}{dt^2} + k\vec{r} + \gamma\frac{d\vec{r}}{dt} = e\vec{E}. \tag{5.54}$$

A solution is:

$$\vec{r} = \frac{e\vec{E}}{m(\omega_0^2 - \omega^2) - i\omega\gamma}. \tag{5.55}$$

and

$$\vec{p} = e\vec{r} = \frac{e^2\vec{E}}{m(\omega_0^2 - \omega^2) - i\omega\gamma}. \tag{5.56}$$

Now the dipole moment, $\vec{p}$, is a complex quantity, since $i = \sqrt{-1}$. From the Lorentz-Lorenz formula, we get:

$$\frac{n^2 - 1}{n^2 + 1} = \frac{4\pi}{3}\frac{Ne^2}{m\left(\omega_0^2 - \omega^2\right) - i\omega\gamma} \tag{5.57}$$

In general, there are many resonance frequencies, and in the absence of damping:

$$\frac{n^2 - 1}{n^2 + 2} = \frac{4\pi}{3}\frac{Ne^2}{m}\sum_k\frac{f_k}{(\omega_{0k}^2 - \omega^2)}. \tag{5.58}$$

$N f_k$ is the number of electrons per cm^3 at the resonance frequency ω_{0k}, where $k = 1, 2, 3\dots$. The Thomas-Reiche-Kuhn-sum rule states that:

$$\sum_k f_k = 1, \tag{5.59}$$

and

$$0 \leq f_k \leq 1. \tag{5.60}$$

Drude called the term f_k the fraction of "dispersion" electrons. It was Pauli who labeled the term f_k as the "strength" of the oscillator, and thus coined the term oscillator strength to indicate the relative intensity of electronic transitions in an atom that are induced by the radiation field.

It was known that these electronic transitions can occur either as absorption of light from a state of lower energy to a state of higher energy, or as emission of light from a state of higher energy to a state of lower energy. This was expressed as the Kuhn sum rule:

$$\sum_{absorption} f_a - \sum_{emission} f_e = 1. \tag{5.61}$$

Kramers developed his dispersion theory, which accounted for oscillators that interacted with the radiation field by both absorption (upward energy transitions), denoted with the label a under the summation sign, and by emission of light (downward energy transitions), denoted with the label e under the summation sign. Kramers' dispersion formula for a gas, assuming an index of refraction that is close to unity, is:

$$n = 1 + \frac{2\pi e^2 N}{m} \left[\sum_a \frac{f_a}{(\omega_a^2 - \omega^2)} - \sum_e \frac{f_e}{(\omega_e^2 - \omega^2)} \right]. \tag{5.62}$$

Remarkably, the Kramers' dispersion equation is analogous to that derived from rigorous quantum theory.

What was achieved with the classical approach to atom–radiation interactions with the model of the damped harmonic oscillator? The model explained the experimentally observed dispersion curves and allowed the resonance frequency to be calculated (see Fig. 5.5). The correct form of the dispersion curve, a plot of the refractive index versus the frequency of the light, was obtained for normal dispersion that is far from resonance and for anomalous dispersion that occurs at resonance when there is an absorption or emission of radiation.

What was not included in the theory is the following: the calculation of the probability of a given electronic transition (related to the intensity of absorption or emission), and the theory of nonlinear optical phenomena (i.e., two-photon excitation processes and second-harmonic generation). These theoretical developments required the development of a new theory: quantum mechanics. In the subsequent sections we develop the background of quantum mechanics and then explain one-photon and two-photon electronic transitions. We show how quantum mechanics can be used to calculate the probabilities of these electronic transitions.

5.3.2 History of Theories and Experiments Leading to Quantum Mechanics

We divide this historical section into two subsections. First, we include a subsection that summarizes the quantized nature of atomic structure and the quantized exchange of energy in matter–radiation interactions. Second, we summarize the emergence of quantum and wave mechanics in the period of 1925 to 1926. It was in the latter period that techniques for calculating the probability of electronic transitions became available.

5.3.3 Descriptions of Atomic Structure and Matter–Radiation Interactions

A detailed history of the development of atomic theory and quantum theory is provided in the sources listed in the bibliography (Jammer, 1989; Mehra & Rechenberg, 2001).

There is no pedagogical equivalent to reading the original papers, but below is a brief overview of some of the key developments. This section will focus on the historical development of quantum mechanics, leaving formal quantitative treatment of the subject to a subsequent section.

While classical physics provides an adequate description for macroscopic phenomena, it totally fails to explain and predict many of the phenomena that constitute this handbook. Classically, matter can have energy values that are continuous and all values are possible. Quantum theory is based on the idea that both matter and radiation can possess and exchange energy only in discrete amounts (quanta). Quantum theory, together with Einstein's Special Theory of Relativity in 1905 and his General Theory of Relativity in 1915, represents a revolution in physics.

Before the advent of quantum theory all of classical physics was expressed in the following laws: Maxwell's four equations describing static and dynamic electromagnetic fields, the conservation of charge, the force law for a moving charge, and Newton's law relating the force to the time change of the momentum of a moving body together with the relativistic correction for mass, and the law of gravitation.

Max Planck is credited with the hypothesis of quantization of energy or the quantum of energy, which allows matter to possess and exchange energy only in discrete units. Max Planck lived 90 years. His life spanned two world wars and the development of modern physics, in which he played a major role. It is worthwhile to read his book *Scientific Autobiography*, which contains several of his writings and a memorial address delivered by Max von Laue in the Albani Church in Göttingen on October 7, 1947 (Planck, 1949). Planck wrote a series of articles in the last years of his life that addressed these topics: his scientific autobiography, phantom problems in science, the meaning and limits of exact science, the concept of causality in physics, and religion and natural science. Planck was a professor in Kiel before he moved to Berlin. Planck lost his oldest son Karl at Verdun in 1916. His second son, Erwin, was killed in January 1945 after he participated in an abortive plot against Hitler. With that biographical background we continue with the seminal contribution of Max Planck to the development of quantum theory.

Planck derived his energy distribution law for radiation in a black body (1905). The term "black body" describes a heated body containing a cavity with perfectly reflecting walls that is filled with radiation in thermal equilibrium with the walls. A small hole permitted the radiation to be analyzed externally. Classical physics predict that the spectrum of the black body radiation is an increasing function of frequency. This theory is in stark contrast with the experimental observation that the black body radiation has a maximum that increases with increasing temperature.

Planck's seminal contribution lies in proposing a radically different theory that derives the correct energy distribution law for black body radiation. Planck made the following hypothesis to derive his black body energy distribution law: the emission and absorption of radiation by matter does not occur continuously, but only in finite quanta of energy.

Planck assumed that the walls of the cavity were composed of harmonic oscillators, and the energy of each oscillator $E_{osc} = nh\nu$ where $n = 1, 2, 3 \ldots$ This assumption converted a physical problem with continuous, uncountable degrees of freedom, to one based on a large number, but a countable number, of identical harmonic oscillators, each of which has one degree of freedom. The mathematical description of a harmonic oscillator has a closed analytical form. Planck announced his quantum hypothesis and his frequency law before the Berlin Physical Society on October 19, 1900. That date is the birth of quantum theory.

Therefore, the radiation at a frequency v can only be exchanged with the cavity wall's oscillators in units of energy,

$$E = hv. \tag{5.63}$$

Later the constant h was called Planck's constant, which is in units of action, or (energy) $\times$ (frequency)$^{-1}$, or (energy) $\times$ (time). The energy density in the range of v and $v + dv$ for a black body is given by:

$$\rho(v) = \frac{8\pi v^2}{c^3} \frac{hv}{e^{(hv/kT)} - 1}. \tag{5.64}$$

Albert Einstein made the hypothesis that light also consists of quanta. Each quantum has energy that is proportional to the light's frequency (i.e., $E = hv$). Therefore, radiation consists of packets of energy that depend on the frequency of the radiation.

Einstein's light-quantum hypothesis was able to explain the photoelectric effect (1905), which was first observed by Heinrich Hertz in 1887. The photoelectric effect describes the phenomenon that light incident on a metal surface (the cathode) could be used to expel electrons. While the number of electrons expelled was proportional to the light intensity, the kinetic energy of the electrons was related to the frequency, v, of the light. Earlier, in 1902, Philipp Lenard published a paper (Über die lichtektrische Wirkung, Ann. der Phys., 8:149-198, 1902) in which he demonstrated that the maximum velocity of cathode particles did not depend on the incident light intensity on the cathode, but it did vary with the type of illumination.

Einstein's equation for the photoelectric effect based on his light-quantum hypothesis is:

$$E_{kin} = \frac{mv^2}{2} = hv - \varphi \tag{5.65}$$

where the kinetic energy of the electrons emitted from the metal surface is E_{kin}, the mass of the electron is m, the velocity of the electron is v, Planck's constant is h, the frequency of the incident light is v, and the work function of the metal or the work to remove an electron from the metal surface is φ.

Further evidence that atoms can absorb and emit radiation in quantized amounts (quanta) came from the experiments of James Franck and Gustav Hertz (1914). They studied the collisions between electrons and atoms in the gas phase. In their experiments, they ask the following question: Were transfers of energy from the electrons to the atoms quantized when the kinetic energy of the electron is continuous? In a collision between an electron and an atom, the energy loss of the electron was equal to the energy difference between the ground state of the atom and the excited state according to a theory by Bohr developed almost concurrently. When the excited atom emits radiation and returns to the ground state, it was found that this radiation corresponds to the Bohr frequency condition,

$$v = \frac{E_2 - E_1}{h}. \tag{5.66}$$

In 1913 Niels Bohr presented his atomic theory, which was based on a combination of classical physics and a number of assumptions. Bohr assumed that atoms exist in stationary states that are time-independent; their energy remains constant. This contradicted the expected result based on Maxwell's theory that an accelerating electron in an atom would continuously radiate energy. He assumed the existence of orbits with a discrete orbital angular momentum, L.

$$L = \frac{nh}{2\pi}, \quad n = 1, 2, 3, \ldots \tag{5.67}$$

The values of n are the quantum numbers. Furthermore, Bohr postulated that these transitions or "jumps" occur from one atomic energy level to another. Bohr assumed that atoms in stationary states can only absorb and emit energy,

$$\Delta E = E_2 - E_1 \tag{5.68}$$

where $E = nhv$.

This equation is the Bohr frequency rule. His theory was able to predict the size of the hydrogen atom, the Balmer series of the spectral lines emitted from excited hydrogen atoms, and the Rydberg constant for hydrogen that was used to calculate the lines in atomic spectra. The excellent agreement between theory and experiment gave strong support to Bohr's atomic theory. Another important part of the Bohr atomic theory was his correspondence principle. That principle formed a bridge between classical and quantum mechanics: it stated that for the case of very large quantum numbers, n (i.e., very high energy), the quantum theory-derived equations must become equivalent to the equations of classical theory.

One of the triumphs of the Bohr model is the derivation of the hydrogen Balmer series. The Balmer formula yields the energy of the transitions as differences of terms containing integers. In an electrical discharge in hydrogen gas at low pressure, more than 40 lines are observed in the emission spectrum. The atomic lines in the visible and the near-ultraviolet portion of the emission spectrum of the hydrogen atom are given by the Balmer formula.

$$\frac{1}{\lambda} = \frac{v}{c} = R_H \left(\frac{1}{2^2} - \frac{1}{n^2} \right). \quad n = 3, 4, 5, 6 \ldots \tag{5.69}$$

The Rydberg constant for hydrogen, R_H, is 109,677 cm^{-1}. In the Bohr theory the energy of a stationary state is given by:

$$E_n = -\frac{2\pi^2 m e^4 Z^2}{n^2 h^2} \tag{5.70}$$

where the mass of the electron is m, the charge of the electron is e, the principle quantum number is n, Planck's constant is h, and the atomic number of the atom is Z.

The energy difference between two states with quantum numbers n_2 and n_1 is given by:

$$E_2 - E_1 = hv = \frac{2\pi^2 m e^4 Z^2}{h^2} \left(\frac{1}{n_1^2} - \frac{1}{n_2^2} \right). \tag{5.71}$$

The emitted frequency for this transition is:

$$\nu = \frac{E_2 - E_1}{h} = \frac{2\pi^2 m e^4 Z^2}{h^3}\left(\frac{1}{n_1^2} - \frac{1}{n_2^2}\right). \qquad (5.72)$$

The constant $\frac{2\pi^2 m e^4 Z^2}{h^3}$ yielded the correct value for the Rydberg constant for hydrogen. The entire equation yielded the correct frequency for the electronic transition in hydrogen from $n = 1$ to $n = 2$ quantum numbers.

Arnold Sommerfeld extended the Bohr theory to elliptical orbits and calculated the effects of magnetic and electric fields on the energy levels of atoms. He applied relativistic corrections to the motion of the electron in elliptical orbits where the rest mass of the electron varies with its velocity. He added the azimuthal quantum number, which is related to angular momentum, to extend the Bohr principle quantum number. In addition, he introduced the reduced mass of the atom, which accounted for the mass of the nucleus and the electron. With these additions Sommerfeld calculated the expected fine structure of the spectral lines from ionized helium, and these results were experimentally verified by Paschen.

The particle characterization of light was further supported by a series of scattering experiments. In 1923, Arthur Compton discovered that when x-rays (photons) are scattered from a graphite target, there is a slight increase in their wavelength. This experiment was first proposed by Stark in 1909. Compton derived the following equation that was later called the Compton effect:

$$\lambda_\theta - \lambda_0 = \left(\frac{2h}{mc}\right)\sin^2\left(\frac{\theta}{2}\right) \qquad (5.73)$$

where the left side of the equation is the change in wavelength of the scattered x-ray, the mass of the electron is m, the speed of light in vacuum is c, and the scattering angle is θ, which is defined as the angle between the incident and the scattered x-ray.

Since the frequency of the scattered x-rays is independent of the element in the target, he assumed the scattering is from a photon and an individual electron in the target. To derive this equation Compton (and independently Peter Debye) assumed that both the law of conservation of energy and the law conservation of momentum are valid for an elastic scattering process between an a light quantum (x-ray) and a quasifree electron. Compton further assumed that the x-ray beam was composed of photons, each with an energy $E = h\nu$. For a photon, the velocity is c (the speed of light in a vacuum) and its rest mass is zero. The magnitude of the momentum of a photon is

$$p = \frac{h\nu}{c}. \qquad (5.74)$$

In addition, Compton assumed that photons can be associated with a wavelength and therefore they can behave as a wave. The significant result was that a photon, a massless particle, can be associated with a momentum similar to a material particle, and that in collisions between photons and electrons (in condensed matter) the conservation laws for energy and momentum are valid. Compton's investigation gives material characteristics to a photon, but it was de Broglie's work that gives wave characteristics to a material particle.

While he was a graduate student de Broglie made the hypothesis that the wave particle duality of light also occurs in matter. In 1924 de Broglie presented a remarkable doctoral thesis to the Faculty of Science at the University of Paris: *Recherches sur la théorie des quanta* (*Investigations on quantum theory*). It contained his hypothesis that a material particle will have a corresponding matter wave associated with it. De Broglie was influenced by both Fermat's principle in optics and the principle of least action in mechanics.

Historically, it is helpful to describe the de Broglie's wave theory (1924) for a material particle before we describe the Schrödinger equation (1926) in the next section. The reason for this digression is that Schrödinger began his development of wave mechanics with de Broglie's wave theory.

To understand the de Broglie wave theory we proceed as follows. For a particle of mass m, moving in the x-direction, with a velocity $\vec{v}$, there exists a corresponding de Broglie matter wave:

$$\Psi(x,t) = Ae^{i(kx-\omega t)} = Ae^{\left(\frac{i}{\hbar}\right)(\vec{p}\cdot\vec{x}-Et)}. \tag{5.75}$$

In this equation we have the following terms: $\hbar = \frac{h}{2\pi}$, $\vec{p} = \hbar\vec{k}$, and $k = \frac{2\pi}{\lambda}$, where h is the Planck constant, $\vec{p}$ is the momentum of the particle, $\vec{k}$ is the propagation wave vector, $k = |\vec{k}| = \frac{2\pi}{\lambda}$ is the wavenumber that is a scalar quantity, and λ is the wavelength of the material particle. Then the classical de Broglie wave formula is given as:

$$\lambda = \frac{h}{|\vec{p}|} = \frac{h}{\sqrt{2mE}}, \tag{5.76}$$

$$\text{and } \lambda = \frac{h}{m|\vec{v}|}, \tag{5.77}$$

which expresses the wavelength of a material particle of mass m that is moving with a velocity $\vec{v}$.

In summary, the de Broglie hypothesis ascribes wave properties to particles. It is important to note that the classical de Broglie theory matter of matter waves is not identical with the Schrödinger equation.

De Broglie described a particle of matter, having momentum $\vec{p}$, and energy E, as an infinite wave. The energy is given by $E = h\upsilon$, and the momentum is given by $p = \frac{h}{\lambda}$, where λ is the wavelength of the associated matter wave. Every mass particle, m, moving with a velocity, $\vec{v}$, would have in the nonrelativistic limit the wavelength:

$$\lambda = \frac{h}{m|\vec{v}|}. \tag{5.78}$$

In 1927 Clinton Davisson and Lester Germer demonstrated the diffraction of electrons from single crystals of metal. Prior to this experiment, diffraction effects were observed only in wave systems such as light and water waves. Diffraction theory describes how a periodic intensity pattern arises when multiple waves interact and summing constructively and destructively at different spatial locations. These experiments provided further evidence that matter (electrons) can be associated with wave properties.

Furthermore, it is interesting to note that it is confinement that quantizes the energy levels of a particle. The energy of a particle in free space ($\hat{E}_{pot} = 0$) is not quantized; it can take on a continuous range of values. On the other hand, the energy of a particle in a confined space is quantized and can occur only in discrete values. An example of a confined particle is the electron in a quantum dot: the electron is confined in a three-dimensional potential well, and thus the quantum dot particles exhibit transitions between discrete electronic states.

5.3.4 The Emergence of Quantum and Wave Mechanics in the Period from 1925 to 1926

Bohr's theory provided for the calculation of the transition energy for hydrogen atoms but does not provide a way to calculate the transition between electronic states. This was resolved by the development of quantum mechanical theory in 1925 and 1926. During this time there were three independent formulations of quantum theory, which were later shown to be mathematically identical. The expression "quantum mechanics" is attributed to Max Born and was first used by him in his 1924 paper (Zeits f. Phys. 26, 379, 1924).

In 1925 Heisenberg developed a new quantum theory that used only observable quantities and applied it to the anharmonic oscillator, which yielded half-integral quantum numbers. His technique was an algebraic approach that is based on the idea of discontinuity.

Heisenberg showed that the lowest energy of the harmonic oscillator was not zero, but had the value $\frac{h\nu}{2}$. Max Born realized that the multiplication rule in Heisenberg's quantum theory was identical for the multiplication rule of matrices, and he showed the commutation relation (the rule for multiplication) for momentum $\vec{p}$ and position q as:

$$\hat{p}\hat{q} - \hat{q}\hat{p} = \frac{h}{2\pi}\hat{1}. \tag{5.79}$$

Note that $\hat{p}$ and $\hat{q}$ are operator matrices, which are denoted with the "hat ^ sign", and Born uses $\hat{1}$ to denote the unit matrix on the right side of the equation. These operators are based on infinite Hermitean matrices, which represented observable quantum mechanical variables. This is the origin of Born's matrix mechanics.

In 1925, Paul Dirac developed his own formulation of quantum theory based on a new form of algebra that involves noncommuting ($\hat{A}\hat{B} \neq \hat{B}\hat{A}$) "q-numbers." Dirac's method reformulated the classical Poisson brackets of Hamilton-Jacobi dynamics into a new quantum theory. He derived the frequencies of the hydrogen atom spectral lines with his "quantum algebra."

In 1925 Born, Heisenberg, and Jordan published their theory of quantum mechanics that provided a matrix-mechanical technique for solving periodic systems with many degrees of freedom and also included their perturbation theory.

In the winter of 1925 Max Born worked with Norbert Wiener at the Massachusetts Institute of Technology in Boston. They developed the mathematical technique of operators in their quantum mechanics and applied these methods to the quantization of periodic and aperiodic phenomena. In their theory they represented physical quantities such as position coordinates or momenta by mathematical operators.

In 1926 Pauli published his theory for the energy of the hydrogen atom. He gave the correct derivation of the Stark effect and calculated the effect of crossed electric and magnetic fields on the hydrogen atom spectrum; all of these calculations were based on matrix mechanics.

In 1926 Schrödinger published his theory of wave mechanics in four papers, each of which has four parts. He was stimulated by his reading of the doctoral thesis of de Broglie, which Einstein had suggested to him. He was also influenced by Hamilton's variational principle and Fermat's principle of least time, which is the foundation of geometrical optics.

Erwin Schrödinger was well educated in both the physics of continuous media and the mathematics of eigenvalue problems. He studied Rayleigh's *Theory of Sound* and was also well versed in the theory of vibrations, differential equations, partial differential equations, the role of group velocity in wave phenomenon, as well as the theory of elasticity.

In 1926 Schrödinger, inspired by Louis de Broglie's doctoral thesis, redefined the eigenfunction in the classical Hamilton-Jacobi equations and developed a variational principle that resulted in the Schrödinger wave equation. He was able to demonstrate that its eigenvalues (energy values) for the case of a Coulomb central force problem (the hydrogen atom) correspond to the Bohr spectrum.

His wave equations were formulated as linear, homogeneous differential equations, similar to those for fluid mechanics. They were formulated as an analytical technique that was based on a generalization of the laws of motion in classical mechanics; they stressed the idea of continuity and the concept of waves. All of these factors made Schrödinger's wave equations readily understandable to contemporary physicists. This was in contrast to the quantum theory formulation of Heisenberg, Born, and Jordan based on matrices, and to the algebra of q-numbers that was the basis of Dirac's quantum theory. In another 1926 paper Schrödinger showed the equivalence between the quantum mechanics of Heisenberg, Born, and Jordan and his theory of wave mechanics. Eventually, all of these different mathematical theories were shown to be mathematically equivalent by von Newmann.

Schrödinger's first paper, "Quantization as a problem of eigenvalues (Part I)," presented his wave equation. In the nonrelativistic approximation it yielded the correct Balmer spectrum and the continuous positive energy eigenvalues for the ionized hydrogen atom. In Schrödinger's second paper, "Quantization as a problem of eigenvalues (Part II)," he derived a wave equation to describe atomic systems having several electrons and more than three degrees of freedom. He presented his time-independent wave equation for atomic systems of arbitrary many degrees of freedom. He also solved the one-dimensional Planck oscillator and calculated the correct energy values. In Schrödinger's third paper, "Quantization as a problem of eigenvalues (Part III)," he developed the wave-mechanical perturbation theory for time-independent perturbations and calculated the Stark effect of the Balmer lines. In Schrödinger's fourth paper, "Quantization as a problem of eigenvalues (Part IV)," he developed the theory of time-dependent perturbations and his time-dependent Schrödinger wave equation. He extended his perturbation theory to perturbations that contained time explicitly.

Schrödinger showed how to calculate the transition probabilities between two states (lower and upper states for an absorption).

Schrödinger's technique was an analytical method that was based on differential equations. It was built on the concept of continuity and the notion of a wave. Therefore, physicists found the method of Schrödinger easier to work with and easier to understand as compared to matrix mechanics or Dirac's quantum algebra.

In 1926, Max Born proposed a probability interpretation of Schrödinger's wave function and calculated the scattering of atomic particles by atoms. Born proposed that the square of the modulus of the wave function $\Psi(x,t)$ is proportional to the probability, P, of finding a particle in the interval x, and $x + dx$, when measurements were made on a very large number of identical systems.

$$P(x,t)\,dx = |\Psi(x,t)|^2 dx \tag{5.80}$$

$|\Psi(x,t)|^2$ is proportional to the probability. Therefore, Ψ can be a complex quantity but the square of its modulus is real.

To summarize the historical developments of quantum theory we note the following. Born, Heisenberg, and Jordan of Göttingen's Theoretical Physics Institute first developed "matrix mechanics." Pauli immediately applied the new "matrix mechanics" to the hydrogen atom and obtained the correct result. It took Schrödinger more than a year to prove that although mathematically quite different, matrix and wave mechanics were physically completely equivalent. Eventually, all these different theories of quantum mechanics were shown to be mathematically equivalent.

5.4 QUANTUM MECHANICS PRIMER

This brief primer introduces the basic definitions and postulates of quantum mechanics, allowing further discussion of perturbation theory. This background will permit the reader to follow the discussion of how one may calculate the Einstein transition probabilities with quantum theory. The discussion of the quantum theory of one- and two-photon electronic transitions will follow, as our emphasis is on the calculation of transition in linear and nonlinear spectroscopy. For ease of understanding we limit our discussion of quantum theory to Schrödinger's wave mechanics, in which we neglect the effects of spin and relativity theory.

We now present the Schrödinger equations and define the key terms. First we define the terms operator, eigenfunction, and eigenvalue since they were used in mathematics (the eigenvalue problem) prior to Schrödinger's equations for wave mechanics.

The eigenvalue problem is given as $\hat{A}\Psi = a\Psi$, where $\hat{A}$ is a mathematical operator, Ψ is a function, and a is a number. Solutions that yield unique values of the number a are called eigenvalues. For example, $\hat{A}\Psi_n = a_n\Psi_n$, where the number a_n is an eigenvalue, and the complex function Ψ_n is an eigenfunction. If several eigenvalues are equal $a_i = a_j = a_k$, then the eigenvalues and their eigenfunctions are called degenerate. The degeneracy of an energy level is defined as being equal to the number of eigenstates or eigenfunctions that have the same eigenvalue; in this example, they are triply degenerate. If the eigenvalue a_n has only a single value that corresponds to a given eigenfunction it is called nondegenerate.

An operator $\hat{A}$ transforms a mathematical function, f, into another function, g. For example, $\hat{A}f = g$. In quantum mechanics these functions could be multidimensional and are described by matrices. As an example in the case of eigenvalue problems, the operator, $\hat{A}$, operates on function, f, to give a product of a scalar and the original function f: $\hat{A}f = af$.

There are quantum mechanical operators for all observable physical quantities: position, potential energy, kinetic energy, total energy, etc. The energy operator is particularly important and is given a special name: the Hamiltonian operator $\hat{H}$. The eigenfunctions of the Hamiltonian operator are called the wavefunctions Ψ of the system. The eigenvalues, E, correspond to the allowed energy levels of the system:

$$\hat{H}\Psi = \left(\hat{E}_{kin} + \hat{E}_{pot}\right)\Psi = E\Psi, \tag{5.81}$$

where $\hat{E}_{kin}$ and $\hat{E}_{pot}$ are the kinetic and potential energy operators.

The physical interpretation of wave function and its eigenvalues is that the wave function, Ψ, completely describes the system at a stationary energetic state with energy equal to its corresponding eigenvalue, E. As in any eigenfunction in quantum mechanics, the wave functions are continuous, have a continuous slope, are single-valued, and are bound.

The kinetic energy operator is related to the momentum operator. The momentum operator is given by:

$$\hat{p} = \frac{\hbar}{i}\left(\frac{\partial}{\partial x}\mathbf{i} + \frac{\partial}{\partial y}\mathbf{j} + \frac{\partial}{\partial z}\mathbf{k}\right) = -i\hbar\nabla. \tag{5.82}$$

This is for Cartesian coordinates, in which $\mathbf{i}$, $\mathbf{j}$, $\mathbf{k}$ are unit vectors.

The kinetic energy operator, , is given as $\frac{\vec{p}^2}{2m}$ similar to classical definition:

$$\hat{E}_{kin} = \frac{\vec{p}^2}{2m} = -\frac{\hbar^2}{2m}\left(\frac{\partial^2}{\partial x^2} + \frac{\partial^2}{\partial y^2} + \frac{\partial^2}{\partial z^2}\right) = -\frac{\hbar^2}{2m}\nabla^2. \tag{5.83}$$

∇^2 is the Laplacian operator. Note that in different coordinate systems (e.g., Cartesian, spherical, cylindrical), the forms of the Laplacian operator are different.

5.4.1 Postulates of Quantum Mechanics

Before presenting the Schrödinger equations we state some of the postulates of quantum mechanics:

1. The state of a dynamic system at a given time t (i.e., its energy), or its momentum, or any other observable quantity, is completely described by a complex valued state function, $\Psi(X,t)$, where X is a complete set of variables including spatial coordinates of a particle. Focusing on the spatial coordinates, the state function can be written as $\Psi(\vec{r},t)$. These state functions exist in infinite Hilbert space. Ψ is a complex function, $|\Psi|e^{i\theta}$ with an amplitude and a phase θ.

The state functions are the solutions that satisfy the Hamiltonian eigenvalue problem.

2. Consider a system consisting of one particle with state function: $\Psi_1(\vec{r}, t)$. From the work of Max Born, the probability $P(\vec{r}, t) \, d^3\vec{r}$ of finding a particle in the volume element, $d^3\vec{r}$, about the point $\vec{r}$, at time t, when a large number of measurements are taken on independent particles, is described by a one-particle state function,

$$P(\vec{r}, t) \, d^3\vec{r} = \Psi_1^*(\vec{r}, t) \, \Psi_1(\vec{r}, t) \, d^3\vec{r} = |\Psi_1(\vec{r}, t)|^2 \, d^3\vec{r}. \qquad (5.84)$$

Note that Ψ_1 is called the probability amplitude, and $\Psi_1^* \Psi_1$ is the probability density. The wave functions are normalized so that the total probability is equal to 1.

$$\int |\Psi_1(\vec{r}, t)|^2 d^3\vec{r} = 1. \qquad (5.85)$$

While the eigenfunction can be a complex quantity, the square of its modulus is real and proportional to the probability of finding the value in a series of measurements on many identical systems.

3. A linear operator can be uniquely defined for each observable variable (i.e., there is an operator for energy, one for momentum, one for position, etc.). When an operator acts on a state function, the state function is altered. In general, the order of operators on a wave function cannot be interchanged; they do not "commute." The order of measurements matters! $\hat{A}\hat{B} \neq \hat{B}\hat{A}$. The "commutator" is defined as the difference of two operators applied in different order: $\hat{A}\hat{B} - \hat{B}\hat{A} = [\hat{A}, \hat{B}]$. When the "commutator" of two operators is zero, the two operators commute. The Heisenberg Uncertainty Principle is a result that some operators do not commute. The Heisenberg Uncertainty Principle states that $\Delta x \Delta p_x \geq \frac{\hbar}{2}$ and $\Delta t \Delta E \geq \frac{\hbar}{2}$. Note the uncertainty is the standard deviation of a large number of repeated measurements of identically prepared systems.

4. The eigenfunctions, Ψ_i^q, of any operator, q, that correspond to a physical observable are said to form a complete, orthonormal set. Orthonormal is defined as $\int \Psi_i^{q*} \Psi_j^q d^3\vec{r} = 1$ if $i = j$ and is equal to 0 if $i \neq j$, where $d^3\vec{r} = dx\, dy\, dz$. The idea of a complete set means that we can take an arbitrary state function Ψ and express it as a sum of a complete set of orthonormal eigenfunctions of any operator: $\Psi = \sum_n c_n \Psi_n^q$. These functions are sometimes called the basis states of the system.

5. The average value or the expectation value of an observable quantity corresponding to operator, $\hat{A}$, of a physical system is $\langle A \rangle$. This value is the average of many repeated measurements on identically prepared systems.

$$\langle A \rangle = \frac{\int_{-\infty}^{+\infty} \Psi^*(\vec{x}, t) \, \hat{A}\Psi(\vec{x}, t) \, d^3\vec{x}}{\int_{-\infty}^{+\infty} \Psi^*(\vec{x}, t) \, \Psi(\vec{x}, t) \, d^3\vec{x}} = \int_{-\infty}^{+\infty} \Psi^*(\vec{x}, t) \, \hat{A}\Psi(\vec{x}, t) \, d^3\vec{x}. \qquad (5.86)$$

Note that for normalized wave functions the denominator is equal to one in the expression given above.

6. Schrödinger presented the time-independent Schrödinger equation in 1926. For a physical system in a stationary state, its total energy is conserved and is constant with respect to time. The Hamiltonian of a stationary state is time independent. The eigenstate of the time-independent Hamiltonian describes the stationary state of the system. The time-independent Schrödinger equation is:

$$\hat{H}(\vec{r})\,\varphi_n(\vec{r}) = -\frac{\hbar^2}{2m}\nabla^2\varphi_n + E_{pot}(\vec{r})\varphi_n = E_n\varphi_n(\vec{r}). \qquad (5.87)$$

$\varphi_n(\vec{r})$ is the state function of a conservative system with constant energy E_n.

7. The time evolution of a quantum mechanical system or the time development of the state functions, $\Psi(\vec{r},t)$, is given by the time-dependent Schrödinger equation. This equation can be used to calculate transition probabilities between two energy states (e.g., in the process of absorption of a photon in which the electron's energy is changed from the lower ground state to the upper excited state). The time-dependent Schrödinger equation is:

$$i\hbar\frac{\partial\Psi(\vec{r},t)}{\partial t} = \hat{H}(\vec{r},t)\Psi(\vec{r},t) \qquad (5.88)$$

where $\hat{H}$ is the Hamiltonian operator.

Note that the Hamiltonian is time-dependent and corresponds to the total energy of the system (i.e., the kinetic and the potential energy). The potential energy is written as $E_{pot}(\vec{r},t)$ or alternatively as $V(\vec{r},t)$.

$$\hat{H} = -\frac{\hbar^2}{2m}\nabla^2 + E_{pot}(\vec{r},t). \qquad (5.89)$$

The time-dependent Schrödinger equation is valid for all state functions, Ψ, but only the eigenfunctions of the Hamiltonian satisfy the time-independent Schrödinger equation.

8. The Schrödinger equations are linear, homogeneous differential equations. Therefore the superposition principle is valid: if Ψ_1 and Ψ_2 are solutions to the Schrödinger equation, then the sum of these functions, $\Psi_3 = a\Psi_1 + b\Psi_2$, is also a solution. Note that a and b are arbitrary complex numbers. Since the probability distribution is proportional to Ψ^2, there are cross-terms or interference terms, and the relative phases of the two probability functions are critical.

9. The general solution of the time-dependent Schrödinger equation is a superposition of the eigenstates of the Hamiltonian with time-dependent phase factor $e^{-iE_nt/\hbar}$.

$$\Psi(\vec{r},t) = \sum_n C_n e^{-iE_nt/\hbar}\varphi_n(\vec{r}) \qquad (5.90)$$

where $\hat{H}_0\varphi_n = E_n\varphi_n$ and $i\hbar\frac{\partial}{\partial t}\Psi = \hat{H}_0\Psi$.

In this superposition, the coefficients, C_n, give the probabilities of the respective eigenstates.

5.4.2 Dirac Bra-Ket Notation

Dirac introduced a very elegant bra-ket notation, sometimes called the Dirac notation, which is often used in quantum mechanics (Dirac, 1958; Peebles, 1992; Tang, 2005). The bracket notation is extremely useful in quantum mechanics because it permits us to write equations that are independent of a particular choice of variables.

A state function or a state vector that completely describes a quantum mechanical system is represented by a ket vector $|\alpha\rangle$ or in short form a ket, and its Hermitian adjoint by a bra vector $\langle\alpha|$ or in short form a bra. The inner product of these two vectors $\Psi_\alpha^* \Psi_\beta = \langle\alpha|\beta\rangle$ is called a bra-ket expression, which is a number. For example,

$$\int \Psi^*(\vec{r})\Phi(\vec{r})\,d^3\vec{r} \equiv \langle\Psi|\Phi\rangle, \tag{5.91}$$

which is valid for all coordinate systems (Cartesian, cylindrical, or spherical systems). Two further examples are:

$$\hat{O}|\beta\rangle = |\beta'\rangle \quad \text{and} \quad \langle\alpha|\hat{O} = \langle\alpha'|. \tag{5.92}$$

The matrix elements connecting two states (i.e., the transition from state $\Psi_\alpha \rightarrow \Psi_\beta$) is:

$$\int \Psi_\alpha^* \hat{O} \Psi_\beta\,d^3\vec{r} = \langle\alpha|\hat{O}|\beta\rangle. \tag{5.93}$$

For example, the time-independent Schrödinger equation $\hat{H}\Psi = E\Psi$ can be expressed in the Dirac notation as:

$$i\hbar\frac{\partial|\Psi\rangle}{\partial t} = \hat{H}|\Psi\rangle. \tag{5.94}$$

5.4.3 Perturbation Theory Applied to Quantum Mechanics

The chapter on the history of perturbation theory introduced the historical aspects and presented a qualitative picture of how this approximation theory could be used to solve mechanical problems in terms of the known solutions to the unperturbed cases. In this section the application of perturbation theory to quantum mechanics is discussed. The basic concept is that the Schrödinger equation can be solved exactly for only a very few cases (i.e., the hydrogen atom, and the harmonic oscillator). Perturbation theory is a mathematical approximation technique to solve the more complex problems as small perturbations to the Schrödinger equations with known solutions. For example, the Stark effect, the hydrogen atom in a uniform static electric field, can be solved by assuming that the static uniform electric field only slightly changes (perturbs) the known eigenfunctions and the eigenvalues of the hydrogen atom in the absence of the field.

Time-independent perturbation theory describes the situation when the perturbation of the system is static (i.e., it does not depend on time). An example is the Zeeman effect in which a static, homogeneous magnetic field affects the energy levels of the hydrogen atom. Only the nondegenerate case will be discussed. Recall that Ψ_1 and Ψ_2 are eigenfunctions of a system with the corresponding eigenvalues E_1 and E_2. If $E_1 = E_2$, then the eigenfunctions are said to be degenerate.

The Schrödinger equation for the unperturbed system is:

$$\hat{H}_0 \varphi_n^{(0)} = E_n^{(0)} \varphi_n^{(0)}. \tag{5.95}$$

This equation can be solved exactly in analytical form for both the eigenfunctions and the eigenvalues. For the perturbed system, assume that the Hamiltonian $\hat{H}'$ is given by the sum of two terms: the unperturbed Hamiltonian $\hat{H}_0$ and the extremely small perturbation term $\lambda \hat{H}_1$:

$$\hat{H}' = \hat{H}_0 + \lambda \hat{H}_1. \tag{5.96}$$

We define λ as a number between 0 and 1. When $\lambda = 0$, the total Hamiltonian $\hat{H}'$ is equal to the Hamiltonian $\hat{H}_0$ in the absence of the perturbation. For a nondegenerate system:

$$\hat{H}' = \hat{H}_0 + \lambda \hat{H}_1 \tag{5.97}$$

$$E_n = E_n^{(0)} + \lambda E_n^{(1)} + \lambda^2 E_n^{(2)} + \cdots \tag{5.97a}$$

$$\varphi_n = \varphi_n^{(0)} + \lambda \varphi_n^{(1)} + \lambda^2 \varphi_n^{(2)} + \cdots \tag{5.97b}$$

We require the solutions to the Schrödinger equation for the perturbed system in terms of the solution for the unperturbed system:

$$\left(\hat{H}_0 + \lambda \hat{H}_1 \right) \varphi_n = E_n \varphi_n. \tag{5.98}$$

Substituting (5.97a) and (5.97b) into (5.98), collecting the terms with order λ provides the first-order approximation. The first-order correction to the energy of the system is:

$$E_n^{(1)} = \int \varphi_n^{(0)*} \hat{H}_1 \varphi_n^{(0)} \, d^3 \vec{r}, \text{ or in Dirac notation:} \tag{5.99}$$

$$E_n^{(1)} = \left\langle \varphi_n^{(0)} | H_1 | \varphi_n^{(0)} \right\rangle \tag{5.100}$$

Therefore, in the first-order approximation the first-order energy shift due to the perturbation is the expectation value of the perturbing potential in that state.

The second-order correction to the energy is:

$$E^{(2)} = \sum_{k \neq n} \frac{\left\langle \varphi_n^{(0)} | \hat{H}_1 | \varphi_k^{(0)} \right\rangle \left\langle \varphi_k^{(0)} | \hat{H}_1 | \varphi_n^{(0)} \right\rangle}{E_n^{(0)} - E_k^{(0)}} = \sum_{k \neq n} \frac{\left| \left\langle \varphi_n^{(0)} | \hat{H}_1 | \varphi_k^{(0)} \right\rangle \right|^2}{E_n^{(0)} - E_k^{(0)}}. \tag{5.101}$$

The second-order shift in energy due to the perturbation is given by the square of the matrix element that connects a given state $\varphi_n^{(0)}$ to the other states $\varphi_k^{(0)}$, weighted by the reciprocal of the energy difference between the states.

In the presence of a perturbation, the energy of the system is given by the following approximation:

$$E = E_n + \lambda \left\langle \varphi_n^{(0)} \left| \hat{H}_1 \right| \varphi_n^{(0)} \right\rangle + \lambda^2 \sum_{k \neq n} \frac{\left\langle \varphi_n^{(0)} | \hat{H}_1 | \varphi_k^{(0)} \right\rangle \left\langle \varphi_k^{(0)} | \hat{H}_1 | \varphi_n^{(0)} \right\rangle}{E_n^{(0)} - E_k^{(0)}} + higher\ orders$$

$$(5.102)$$

Recall that there are two cases of time-independent perturbation theory: the nondegenerate case and the degenerate case. In the nondegenerate case, which is the case described above, the first-order effect of the time-independent perturbation theory is to shift the unperturbed energy levels. In the degenerate case of time-independent perturbation theory, the unperturbed degenerate levels are split into several new energy levels.

The previous discussion was adequate to calculate the stationary states of an isolated atom in the presence of a static perturbation. In order to calculate transition probabilities it is necessary to use time-dependent perturbation theory. Dirac's method of "variation of constants" was published in 1927 and was used by Maria Göppert-Mayer in her 1931 dissertation. In a subsequent section this method is illustrated in a quantum mechanical calculation of transition probabilities and the Einstein coefficients, so in this section we present only the highlights of the method (Tolman, 1938).

The total Hamiltonian is the sum of two terms: the first term is the time-independent energy operator of the isolated atom $\hat{H}_0$, and the second term is the time-dependent perturbation $\hat{H}_1(t)$ (for example, a time-varying electromagnetic field).

$$\hat{H}' = \hat{H}_0 + \lambda \hat{H}_1(t) \qquad (5.103)$$

where λ is the perturbation parameter between zero and one.

It is assumed that the exact solution for the unperturbed time-independent system is known in terms of the eigenfunctions (or eigenstates) and the eigenvalues (or energies). $\Psi^{(0)}(\vec{r}, t) = \sum_n c_n^{(0)} e^{-iE_n^{(0)} t/\hbar} \varphi_n^{(0)}(\vec{r})$ is a general solution of the time-dependent Schrödinger equation: $i\hbar \frac{\partial}{\partial t} \Psi^{(0)} = \hat{H}_0 \Psi^{(0)}$. The unperturbed wave function consists of a superposition of the eigenstates, $\varphi_n^{(0)}$, of the unperturbed Hamiltonian with time-dependent phase factor $e^{-iE_n^{(0)} t/\hbar}$ where $\hat{H}_0 \varphi_n^{(0)} = E_n^{(0)} \varphi_n^{(0)}$. In this superposition, the time-independent coefficients, $c_n^{(0)}$, give the probabilities, $|c_n^{(0)}|^2$, of the respective eigenstate.

The time-dependent Schrödinger equation for the total Hamiltonian is:

$$i\hbar \frac{\partial \Psi(\vec{r}, t)}{\partial t} = \left(\hat{H}_0 + \lambda \hat{H}_1(t) \right) \Psi(\vec{r}, t). \qquad (5.104)$$

An approximation method similar to time-independent perturbation theory is applied. The perturbed wave function is assumed to remain a linear superposition of unperturbed eigenstates, $\varphi_n^{(0)}$, with the same temporal evolutions governed by $e^{-iE_n^{(0)} t/\hbar}$ containing unperturbed eigenvalues, $E_n^{(0)}$. The effect of the time-dependent perturbation is to "shift" the probabilities of these eigenstates in time. The coefficient c_n is made time-dependent and is expressed as a power series of expansion coefficient λ.

$$\Psi(\vec{r},t) = \sum_n \sum_l \lambda^l c_{nl}(t) e^{-iE_n^0 t/\hbar} \varphi_n^{(0)}(\vec{r}) \tag{5.105}$$

Substituting (5.105) into (5.104) and multiplying by $<\varphi_m^{(0)}|$, we obtain:

$$i\hbar \sum_l \left[\dot{c}_{ml}(t)\lambda^l + c_{ml}(t)\lambda^l \left(-\frac{i}{\hbar}E_m^{(0)}\right) \right] e^{-\frac{i}{\hbar}E_m^{(0)}t} =$$

$$\sum_n \sum_l c_{nl}(t)\lambda^l (\langle\varphi_m^{(0)}|\varphi_n^{(0)}\rangle E_n^{(0)} + \lambda\langle\varphi_m^{(0)}|\hat{H}_1(t)|\varphi_n^{(0)}\rangle) e^{-\frac{i}{\hbar}E_n^{(0)}t} \tag{5.106}$$

Note that a variable with a dot over it means the time derivative of that variable. We can evaluate $c_{ml}(t)$ by collecting terms at successively higher powers of λ.

From λ^0 terms, we obtain $\dot{c}_{m0}(t) = 0$ and $c_{m0}(t) = c_{m0}(0) = c_m^{(0)}$. The 0th-order perturbation term gives just the initial state in the absence of perturbation as expected.

From λ^1 terms, we obtain:

$$\dot{c}_{m1}(t) = -\frac{i}{\hbar}\sum_n c_n^{(0)} \left\langle \varphi_m^{(0)} \left| e^{\frac{i}{\hbar}E_m^{(0)}t}\hat{H}_1 e^{-\frac{i}{\hbar}E_n^{(0)}t} \right| \varphi_n^{(0)} \right\rangle$$

$$= -\frac{i}{\hbar}\sum_n c_n^{(0)} \left\langle \varphi_m^{(0)} \left| e^{\frac{i}{\hbar}\hat{H}_0 t}\hat{H}_1 e^{-\frac{i}{\hbar}\hat{H}_0 t} \right| \varphi_n^{(0)} \right\rangle$$

$$c_{m1}(t) = -\frac{i}{\hbar}\int_{-\infty}^t dt' \sum_n c_n^{(0)} \left\langle \varphi_m^{(0)} \left| e^{\frac{i}{\hbar}\hat{H}_0 t'}\hat{H}_1 e^{-\frac{i}{\hbar}\hat{H}_0 t'} \right| \varphi_n^{(0)} \right\rangle$$

$$= -\frac{i}{\hbar}\int_{-\infty}^t dt' \left\langle \varphi_m^{(0)} \left| e^{\frac{i}{\hbar}\hat{H}_0 t'}\hat{H}_1 e^{-\frac{i}{\hbar}\hat{H}_0 t'} \right| \Psi^{(0)}(0) \right\rangle \tag{5.107}$$

From λ^2 terms, we similarly obtain:

$$\dot{c}_{m2}(t) = -\frac{i}{\hbar}\sum_n c_{n1}(t) \left\langle \varphi_m^{(0)} \left| e^{\frac{i}{\hbar}\hat{H}_0 t}\hat{H}_1 e^{-\frac{i}{\hbar}\hat{H}_0 t} \right| \varphi_n^{(0)} \right\rangle$$

$$c_{m2}(t) = -\frac{i}{\hbar}\int_{-\infty}^t dt' \sum_n c_{n1}(t') \left\langle \varphi_m^{(0)} \left| e^{\frac{i}{\hbar}\hat{H}_0 t'}\hat{H}_1 e^{-\frac{i}{\hbar}\hat{H}_0 t'} \right| \varphi_n^{(0)} \right\rangle$$

$$= \left(-\frac{i}{\hbar}\right)^2 \sum_n \int_{-\infty}^t dt' \int_{-\infty}^{t'} dt'' \left\langle \varphi_m^{(0)} \left| e^{\frac{i}{\hbar}\hat{H}_0 t'}\hat{H}_1(t') e^{-\frac{i}{\hbar}\hat{H}_0 t'} \right| \varphi_n^{(0)} \right\rangle$$

$$\times \left\langle \varphi_n^{(0)} \left| e^{\frac{i}{\hbar}\hat{H}_0 t''}\hat{H}_1(t'') e^{-\frac{i}{\hbar}\hat{H}_0 t''} \right| \Psi^{(0)}(0) \right\rangle \tag{5.108}$$

Consider an example particularly relevant for one- and two-photon excitation processes: a two-level atom with the unperturbed, time-independent Hamiltonian, $\hat{H}_0$. Assume the eigenstates and eigenvalues for these two levels, a and b, are known. Specifically,

$$\hat{H}_0\,\varphi_a = E_a\,\varphi_a \quad \text{and} \quad \hat{H}_0\,\varphi_b = E_b\,\varphi_b \tag{5.109}$$

Any arbitrary state can be written as a linear combination of these two eigenfunctions:

$$\Psi = c_a \, \varphi_a + c_b \, \varphi_b. \tag{5.110}$$

The time dependence of this linear combination of states is:

$$\Psi(t) = c_a \, \varphi_a \, e^{-iE_a t/\hbar} + c_b \, \varphi_b \, e^{-iE_b t/\hbar}. \tag{5.111}$$

In the presence of a time-dependent perturbation, the total Hamiltonian depends on time $\hat{H}_0 + \hat{H}_1(t)$. For this case the coefficients now depend on time.

$$\Psi(t) = c_a(t) \, \varphi_a \, e^{-iE_a t/\hbar} + c_b(t) \, \varphi_b \, e^{-iE_b t/\hbar}. \tag{5.112}$$

Recall that the probability to find the system at a given state at a given time given by t, is given by $c_n^*(t)c_n(t)$, where the $*$ is the complex conjugate and where n is either a or b.

We now illustrate how this works. For the two levels, a and b, assume the atom in initially in the lower energy state a. If there are no perturbations, then the atom will remain in that lower energy state for all time; it is a stationary state.

$$c_a(0) = 1 \quad \text{and} \quad c_b(0) = 0. \tag{5.113}$$

In the presence of a time-dependent perturbation (i.e., an electromagnetic field), after a period of time the atom is in the higher energy state b.

$$c_a(t) = 0 \quad \text{and} \quad c_b(t) = 1. \tag{5.114}$$

This transition occurs with the absorption of a photon with the energy that is the energy difference between the two states:

$$\Delta E = h\nu = E_b - E_a. \tag{5.115}$$

First-order time-dependent perturbation theory gives the time evolution of the coefficients for one-photon absorption and therefore the transition probabilities. Using (5.107), the general expression for the probability of finding the system in a final state b is $c_b(t)^* c_b(t)$ where

$$c_b(t) = -\frac{i}{\hbar} \int_0^t dt' \, \left\langle b | \hat{H}_1(t') | a \right\rangle e^{i(E_b - E_a)t'/\hbar}. \tag{5.116}$$

The quantity, $|c_b(t)|^2$, yields the probability that the perturbation, $\hat{H}_1(t)$, has induced a transition from the initial state to the final state at time t. Similarly, the expression for finding the system in the final state b via a nonlinear, second-order process can also be obtained by second-order time-dependent perturbation theory.

5.5 QUANTUM MECHANICAL TRANSITION PROBABILITIES FOR ONE-PHOTON AND TWO-PHOTON PROCESSES

5.5.1 Einstein's Contributions on Emission and Absorption of Radiation

In this section we describe Einstein's approach for the interaction of radiation and a gas composed of atoms of one electron and a nucleus (Einstein, 1916). The significant aspect of the Einstein coefficients is that they depend on the two states involved in the transition. This is very different from the classical oscillator strengths, which contain a single subscript that denotes a given electronic transition.

The atom is assumed to have two electronic states: the ground state and the excited state. We now describe three different processes and the relations among them: stimulated (induced) absorption, stimulated (induced) emission, and spontaneous emission. Within linear response regime, the stimulated absorption or emission processes are proportional to the radiation density of the external radiation field. In the spontaneous emission process the excited atom emits radiation independently of the radiation field near the atom; therefore, spontaneous emission can occur in the absence of external radiation.

These three processes are schematically shown in Figure 5.6. A vertical line with an arrowhead pointing from the ground state to the excited state represents the absorption of a photon. A vertical line with the arrowhead pointing from the excited state to the ground state represents the emission of a photon. Horizontal wavy lines represent radiation. The lower energy level is the ground state. The upper energy level is the excited state. The coefficients for the transition were derived by Einstein and are called the Einstein coefficients. They are not universal physical constants (Einstein, 1916). In this classic paper, in addition to proving a new derivation of the Planck radiation law, the relationships of spontaneous emission to induced emission and absorption were first found by Einstein. Einstein assumed that in the process of emission or absorption only "directed" radiation bundles are emitted or absorbed, and that the laws of conservation of energy and momentum hold for these elementary processes.

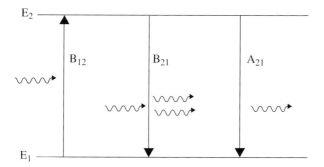

Figure 5.6. The Einstein coefficients related to the probability of a transition for a molecule with two electronic states; the lower ground state is E_1 and the upper excited state is E_2. The induced absorption coefficient is B_{12}. The induced emission coefficient is B_{21}. The spontaneous emission coefficient is A_{21}.

In Figure 5.6, the first process, characterized by the Einstein B_{12} coefficient, represents the process of light absorption. Prior to the radiation being turned on, the electron is in the ground energy state. When the light is turned on, the atom absorbs a photon and the atom jumps to an excited state with the electron in the upper excited state. There is a second process shown in Figure 5.6, characterized by the Einstein coefficient B_{21}, in which the electron is in the excited state prior to the light being turned on. The subsequent radiation field induced the electronic transition from the excited state to the ground state with the emission of a photon. This process is called stimulated emission and was introduced by Einstein. Stimulated emission is coherent; all emitted photons have the same direction, phase, and polarization. Stimulated emission is a critical process in the operation of lasers.

Finally, there is a third process shown in Figure 5.6, characterized by the Einstein coefficient A_{21}, in which there is no radiation field that interacts with the atom. Initially, the electron is in the excited state, and it spontaneously jumps to the ground state with the emission of a photon. Spontaneous emission is incoherent radiation. Typically in fluorescence microscopy we use the processes of absorption and spontaneous emission for cellular imaging.

In an assembly of atoms, the total intensity of the emission is proportional to the number of atoms in the excited state multiplied by the probability of transition (jumps) from the higher-energy excited state to the lower-energy ground state. In this chapter we will develop the classical and quantum mechanical approaches to calculate the so-called transition probability for absorption, stimulated emission, and spontaneous emission.

We now relate the emission and the absorption coefficients for transitions between two states of an atom (Einstein, 1916, 1917). Einstein stated that in thermal equilibrium the number of absorptions per unit time, from the lower-energy state to the upper-energy state, must be equal to the number of emissions per unit of time from the upper-energy state to the lower-energy state.

Einstein then defined the rates, or the probabilities per unit time, for the various transitions. The rate of absorption (induced absorption) is proportional to the Einstein coefficient B_{12}. For the transition, $E_1 \rightarrow E_2$, the rate is further proportional to the number of atoms n_1 in the lower state E_1 and the intensity of the radiation at frequency v, $\rho(v)$, $n_1 B_{12} \rho(v)$.

The rate of stimulated or induced emission is proportional to the Einstein coefficient B_{21}. For the transition $E_2 \rightarrow E_1$ in the presence of radiation, is further proportional to the number of atoms, n_2, in the higher energy state E_2, and the intensity of the radiation at frequency v, $\rho(v)$, $n_2 B_{21} \rho(v)$.

The rate constant for spontaneous emission is proportional to the Einstein coefficient A_{21}. For the process, $E_2 \rightarrow E_1$, the rate is further proportional to the number of atoms and n_2 in the excited state E_2, $A_{21} n_2$.

Einstein knew from the Boltzmann Law in thermodynamics that the relative numbers of molecules in the lower (n_1) and upper states (n_2) is given as:

$$\frac{n_2}{n_1} = e^{-E/kT} = e^{-(E_2-E_1)/kT} = e^{-hv/kT} \qquad (5.117)$$

where E is the energy difference between the two states, h is Planck's constant, and v is the frequency of the light.

Since the induced emission process $E_2 \rightarrow E_1$ occurs at the same rate as the absorption process due to thermal equilibrium, it follows that

$$B_{12}\,\rho(v)\,n_1 = [A_{21} + B_{21}\rho(v)]n_2. \tag{5.118}$$

Combining equations (5.117) and (5.118),

$$\rho(v) = \frac{\frac{A_{21}}{B_{21}}}{\frac{B_{12}}{B_{21}}e^{hv/kT} - 1} \tag{5.119}$$

Einstein further noted that as the radiation intensity approaches infinity, the temperature of the system must also approach infinity. This fact allowed Einstein to deduce the important result that the transition rate for stimulated emission is the same as for absorption:

$$B_{12} = B_{21} = B. \tag{5.120}$$

Einstein compared this result with the Planck black body radiation distribution law, which relates the energy density as a function of frequency:

$$\rho(v) = \frac{8\pi v^2}{c^3}\frac{hv}{e^{hv/kT} - 1}. \tag{5.121}$$

Einstein then can relate the spontaneous emission rate to the stimulated emission rate as:

$$A_{21} = \frac{8\pi hv^3}{c^3}B_{21}, \tag{5.122}$$

where h is Planck's constant and c is the speed of light in vacuum.

In the visible wavelength region, spontaneous emission predominates because:

$$\frac{A_{21}}{B_{21}} = \frac{8\pi hv^3}{c^3} \gg 1. \tag{5.123}$$

The Einstein B-coefficient can be calculated from the Einstein A-coefficient, which can be measured from the radiative lifetime of atoms in level 2.

The lifetime of an excited state can be expressed in terms of the Einstein coefficients. For example, the Einstein coefficient A_{21} is also called the transition probability for de-excitation from the upper energy state to the lower energy state. If this is the only possible transition with the emission of radiation, then the lifetime, τ, is given as:

$$\frac{1}{\tau} = A_{21}. \tag{5.124}$$

If an atom is in an excited state that can emit several different lines, then the mean lifetime, τ_m, is given as:

$$\frac{1}{\tau_m} = \sum_m A_{nm} \tag{5.125}$$

where n is the ground state.

Therefore, in this case the lifetime depends on the sum of the transition probabilities of all of the possible transitions.

In summary, what Einstein showed in 1916 is the following. First, the coefficients for stimulated emission and stimulated absorption are equal. Second, the ratio of the spontaneous emission coefficient to the stimulated emission varies with the cube of the frequency. Both stimulated absorption and stimulated emission processes are enhanced by the presence of photons of the correct frequency, v,

$$v = \frac{E_2 - E_1}{h}. \tag{5.126}$$

Therefore, the larger the energy difference between the two states (ground and excited state), the more likely is the process of spontaneous emission as compared to stimulated emission. That is correct for visible radiation; however, for microwave radiation (much longer wavelengths), the reverse condition will prevail and stimulated emission will predominate.

5.5.2 Quantum Mechanical Calculation of Transition Probabilities and the Einstein Coefficients

Einstein calculated the ratios between the spontaneous emission rate, the stimulated emission rate, and the stimulated absorption rate based on thermodynamics and a semiclassical approach, but he could not calculate the absolute magnitudes of these quantities. Quantum mechanics is required to calculate their absolute magnitudes.

Now we show the use of the Schrödinger equation to calculate the probability of the light-induced atomic transition from the ground state energy of an atom to the excited state. An atom in the ground state absorbs a photon from the radiation field and makes the electronic transition to an excited state (higher energy state). That is the process of absorption; the opposite transition, emission, from the excited state to the ground state (lower energy) occurs with the emission of a photon.

Time-dependent perturbation theory derived in the previous section will be used to calculate the Einstein coefficients for the stimulated absorption and stimulated emission. This will demonstrate the general use of the technique to calculate the effects of a radiation field on an atom. This treatment follows the technique of solving time-dependent perturbations by the method of variation of constants that was introduced by Dirac in 1926. This technique was used by Maria Göppert-Mayer in her 1931 dissertation.

In order to apply time-dependent perturbation theory to atom–radiation interaction, it is necessary to have an explicit form for the time-dependent perturbation Hamiltonian, $\hat{H}_1(t)$, corresponding to the incident radiation field.

5.5.3 The Electric Dipole Approximation

Electric dipole approximation is used in the derivation of the perturbation Hamiltonian corresponding to the incident radiation. We have seen that an electromagnetic field can induce electric dipoles in materials, which can explain the refractive index of materials. For many cases the interaction term between the electric field and the atom is given by the scalar product of the induced electric dipole, $\vec{\mu}$, and the electric field, $\vec{E}$:

$$\vec{E} \cdot \vec{\mu} = e\vec{E} \cdot \vec{r} \tag{5.127}$$

This is called the electric dipole approximation. Marie Göppert-Mayer used this approximation in her calculations of two-photon transitions.

Our treatment follows an important recent paper on the electric dipole approximation that clarifies the correct form for the Hamiltonian for atom–radiation interactions (Rzazewski & Boyd, 2004). An atomic system in the absence of an electric field can be described by the Schrödinger equation:

$$\left(\frac{\vec{p}^2}{2m} + V(\vec{r}) \right) \Psi(\vec{r}, t) = i\hbar \frac{\partial \Psi(\vec{r}, t)}{\partial t} \tag{5.128}$$

In the presence of a radiation field that consists of monochromatic waves, correcting the mistakes in many former treatments, Rzazewski and Boyd showed that the correct form of the Schrödinger equation is:

$$\left\{ \frac{1}{2m} \left[\vec{p} - \frac{e}{c}\vec{A}(\vec{r}, t) \right]^2 + V(\vec{r}) - e\vec{E}(\vec{r}, t) \cdot \vec{r} \right\} \Psi(\vec{r}, t) = i\hbar \frac{\partial \Psi(\vec{r}, t)}{\partial t} \tag{5.129}$$

where $\vec{A}(\vec{r}, t)$ is the vector potential and and $-e\vec{E}(\vec{r}, t) \cdot \vec{r}$ is the electric dipole interaction term.

The concept of a vector potential, $\vec{A}(\vec{r}, t)$, and scalar potential, $\phi(\vec{r}, t)$ has long been used in electrodynamics to simplify mathematical treatments. The electric and the magnetic vectors are easily derived from these potentials. The vector potential that describes a given vector field (i.e., the electric field) is not unique. It can be changed by adding the gradient of any scalar quantity, which is represented by a scalar potential, ϕ. The electric field can be given in terms of both the vector potential and the scalar potential:

$$\vec{E} = -\nabla\phi - \frac{1}{c}\frac{\partial \vec{A}}{\partial t}. \tag{5.130}$$

In the absence of a source, this is commonly simplified in the Coulomb gauge (choosing scale potential, ϕ, such that $\nabla\phi = 0$) as:

$$\vec{E} = -\frac{1}{c}\frac{\partial \vec{A}}{\partial t}. \tag{5.131}$$

Furthermore, because the size of the atom is very small compared to the wavelength of the incident electromagnetic field, the spatial variation of the vector potential, $\vec{A}(\vec{r}, t)$, can be considered constant. Therefore, the spatial dependence of the vector potential is removed and the vector potential is now only a function of time, $\vec{A}(t)$. The form of the Hamiltonian that describes the interaction between the atom and the incident electromagnetic field can be expressed as:

$$\hat{H} = \left\{ \frac{1}{2m} \left[\vec{p} - \frac{e}{c}\vec{A}(t) \right]^2 + V(\vec{r}) - e\vec{E}(t) \cdot \vec{r} \right\}, \tag{5.132}$$

While Rzazewski and Boyd showed that the vector potential term should be included in the Hamiltonian to correctly describe an atomic system in the electromagnetic field, this term is often neglected in the classical treatment. Since it often represents a relatively small correction, we will also neglect this term in our future treatment.

5.5.4 Quantum Mechanical Calculation of Transitions

The incident electric field of the radiation is polarized along the x-direction and can be expressed by:

$$\vec{E}(v,t) = \mathbf{i}\, E_x^0(v)(e^{2\pi ivt} + e^{-2\pi ivt}) \tag{5.133}$$

where $\mathbf{i}$ is the unit vector along the x-direction.

We now use perturbation theory to calculate the probability of the transition of an atom from the state n to a higher energy state m. Extending the perturbation term to include a collection of charged particles exposed to an electric field polarized along the x-axis is:

$$\hat{H}'(t) = \sum_j e_j \vec{E}(v,t) \cdot \vec{r}_j = E_x^0(v,t)\left(e^{2\pi ivt} + e^{-2\pi ivt}\right)\sum_j e_j x_j \tag{5.134}$$

where e_j is the electronic charge and x_j is the x-coordinate of the jth particle of the system.

Note that the term $\sum_j e_j x_j$ is the x-component of the electric dipole moment, $\vec{\mu}$, of the system:

$$\mu_x \equiv \sum_j e_j x_j. \tag{5.135}$$

The complete time-dependent Schrödinger's equation for this system becomes:

$$\left\{\frac{1}{2m}\vec{p}^2 + V(\vec{r}) - \sum_j e_j \vec{E}(v,t) \cdot \vec{r}_j\right\}\Psi(\vec{r},t) = i\hbar\frac{\partial \Psi(\vec{r},t)}{\partial t}$$

$$\left\{\frac{1}{2m}\vec{p}^2 + V(\vec{r}) - E_x^0(v,t)\left(e^{2\pi ivt} + e^{-2\pi ivt}\right)\mu_x\right\}\Psi(\vec{r},t) = i\hbar\frac{\partial \Psi(\vec{r},t)}{\partial t} \tag{5.136}$$

In the absence of the field, assume the atom can exist in two stationary states m and n, with corresponding eigenvalues $E_m^{(0)}$ and $E_n^{(0)}$, and that $E_m^{(0)} > E_n^{(0)}$. The eigenfunctions that correspond to these states (in the absence of any radiation) are

$$\varphi_m^{(0)}(\vec{r}) \quad \text{and} \quad \varphi_n^{(0)}(\vec{r}). \tag{5.137}$$

Assume that at time $t = 0$, the system is in the state n (the ground state of lowest energy), giving initial conditions:

$$c_m(0) = 0 \quad \text{and} \quad c_n(0) = 1 \tag{5.138}$$

The atom is exposed to radiation field with a single frequency v. For single-photon–induced absorption and emission processes, first-order time-dependent perturbation theory can be applied. The probability that the atom will be in the excited state can be calculated by evaluating $c_m(t)$ using (5.107):

$$\frac{dc_m(t)}{dt} = -\frac{i}{\hbar} \int \varphi *_m^{(0)}(\vec{r}) \hat{H}' \varphi_n^{(0)}(\vec{r}) \, d^3\vec{r}. \tag{5.139}$$

$$\frac{dc_m(t)}{dt} = -\frac{i}{\hbar} \int \varphi *_m^{(0)}(\vec{r}) e^{\frac{i}{\hbar} E_m^{(0)} t} E_x^0(v)(e^{2\pi i v t} + e^{-2\pi i v t}) \mu_x$$

$$\times \varphi_n^{(0)}(\vec{r}) e^{-\frac{i}{\hbar} E_n^{(0)} t} d^3\vec{r} \tag{5.140}$$

Let

$$\mu_{x_{mn}} = \int \varphi_m^{*(0)} \mu_x \varphi_n^{(0)} d^3\vec{r}. \tag{5.141}$$

Then,

$$\frac{dc_m(t)}{dt} = -\frac{i}{\hbar} \mu_{x_{mn}} E_x^0(v) \left[e^{\frac{i}{\hbar}(E_m^{(0)} - E_n^{(0)} + hv)t} + e^{\frac{i}{\hbar}(E_m^{(0)} - E_n^{(0)} - hv)t} \right] \tag{5.142}$$

When this equation is integrated:

$$c_m(t) = \int_0^t \frac{dc_m(t')}{dt'} dt' \tag{5.143}$$

$$c_m(t) = \mu_{x_{mn}} E_x^0(v) \left[\frac{1 - e^{\frac{i}{\hbar}\left(E_m^{(0)} - E_n^{(0)} + hv\right)t}}{E_m^{(0)} - E_n^{(0)} + hv} + \frac{1 - e^{\frac{i}{\hbar}\left(E_m^{(0)} - E_n^{(0)} - hv\right)t}}{E_m^{(0)} - E_n^{(0)} - hv} \right]. \tag{5.144}$$

$$\text{(induced emission)} \quad \text{(induced absorption)}$$

The first term inside the brackets corresponds to induced emission, and the second term corresponds to induced absorption. Only if v is extremely close to $\frac{\left(E_m^{(0)} - E_n^{(0)}\right)}{h}$ is the probability of a transition appreciable.

For the case of absorption we neglect the first term.

$$c_m(t) = \mu_{x_{mn}} E_x^0(v) \left[\frac{1 - e^{\frac{i}{\hbar}\left(E_m^{(0)} - E_n^{(0)} - hv\right)t}}{E_m^{(0)} - E_n^{(0)} - hv} \right]. \tag{5.145}$$

The probability of finding the system in the excited state at time t is:

$$c_m^*(t) c_m(t) = \frac{4}{\hbar^2} (\mu_{x_{mn}})^2 E_x^{0^2}(v) \frac{\sin^2 \left(\left[\frac{E_m^{(0)} - E_n^{(0)} - hv}{\hbar} \right] \frac{t}{2} \right)}{\left(\frac{E_m^{(0)} - E_n^{(0)} - hv}{\hbar} \right)^2}. \tag{5.146}$$

This equation can be rewritten in terms of the Bohr frequency, v_{mn},

$$\omega_{mn} = 2\pi v_{mn} = \frac{E_m^{(0)} - E_n^{(0)}}{\hbar}. \tag{5.147}$$

$$c_m^*(t)c_m(t) = \frac{4}{\hbar^2}(\mu_{x_{mn}})^2 E_x^{0^2}(v) \frac{\sin^2\left((\omega_{mn} - \omega)\frac{t}{2}\right)}{(\omega_{mn} - \omega)^2} \tag{5.148}$$

Instead of monochromatic radiation, consider that the excitation field has a distributed spectrum; this will allow coupling to other states of the atom within energy conservation constraints. Let $\rho(\omega)$ be the density of states of the atom normalized as:

$$\int_0^\infty \rho(E)dE = \int_0^\infty \hbar\rho(\omega)d\omega = 1. \tag{5.149}$$

The probability of the system in the excited state at time t due to one-photon absorption becomes:

$$P_m^{(1)}(t) = \int dE\rho(\omega)c_m^*(t)c_m(t) = \int d\omega \frac{4}{\hbar}\rho(\omega)(\mu_{x_{mn}})^2 E_x^{0^2}(v) \frac{\sin^2\left((\omega_{mn} - \omega)\frac{t}{2}\right)}{(\omega_{mn} - \omega)^2} \tag{5.150}$$

Since the sinc function, $\frac{\sin^2((\omega_{mn} - \omega)\frac{t}{2})}{(\omega_{mn} - \omega)^2}$, is sharply peaked at the Bohr frequency, a consequence of energy conservation, a transition occurs only close to Bohr frequency range. Within this small frequency range, both the density of states, $\rho(\omega)$, and the radiation field strength, $E(v)$, can be approximated as constants for the integration.

$$P_m^{(1)}(t) = \frac{4}{\hbar}\rho(\omega = \omega_{mn})(\mu_{x_{mn}})^2 E_x^{0^2}(v = v_{mn}) \int d\omega \frac{\sin^2\left((\omega_{mn} - \omega)\frac{t}{2}\right)}{(\omega_{mn} - \omega)^2}$$

$$= \frac{2\pi t}{\hbar}\rho(\omega = \omega_{mn})(\mu_{x_{mn}})^2 E_x^{0^2}(v = v_{mn}) \tag{5.151}$$

$P_m^{(1)}(t)$ can be considered as the transition probability at time t. The probability of the transition increases with time. The transition rate is the time derivative of the transition probability, and the result is a special case of Fermi's Golden Rule:

$$R_{mn}^{(1)} = \frac{dP_m^{(1)}(t)}{dt} = \frac{2\pi}{\hbar}\rho(\omega = \omega_{mn})(\mu_{x_{mn}})^2 E_x^{0^2}(\omega = \omega_{mn}) \tag{5.152}$$

Recall that the probability of an absorption transition for a two-state system per unit time for an atom is $B_{12}\rho(v)$. From classical electromagnetic theory (Jackson, 2000), the density of radiation as a function of the frequency, v, is:

$$\rho(v) = \frac{1}{8\pi}E_x^{0^2}(v) \tag{5.153}$$

The Einstein coefficient for induced absorption, $B_{n \to m}$ (B_{12}), is:

$$B_{n \to m} = \frac{16\pi^2}{\hbar}\rho(\omega_{mn})(\mu_{x_{mn}})^2 \tag{5.154}$$

It follows by an equivalent treatment for induced emission, where $c_n(0) = 0$, and $c_m(0) = 1$. The Einstein coefficient for the induced emission (B_{21})

$$B_{m \to n} = \frac{16\pi^2}{\hbar} \rho(\omega_{mn})(\mu_{x_{mn}})^2. \tag{5.155}$$

Using equation (5.122) the Einstein coefficient for spontaneous emission is (A_{21}):

$$A_{m \to n} = \frac{256\pi^4 v_{mn}^3}{c^3} \rho(\omega_{mn})(\mu_{x_{mn}})^2 \tag{5.156}$$

Thus we have demonstrated how to calculate the Einstein transition probabilities with perturbation theory.

5.5.5 Single-Photon and Multiphoton Transitions

In this section we describe and compare one- and two-photon electronic transitions. In the previous section time-dependent perturbation theory was shown to be a useful technique to calculate the transition probabilities for single-photon transitions.

We alert the reader to the distinction between two-photon excitation processes that occur via virtual intermediate states, and sequential or stepwise excitation that occurs via the absorption of one photon followed by the absorption of a second photon. The process of stepwise excitation or stepwise radiation was first investigated by Füchtbauer in 1920. Füchtbauer showed that an atom could absorb one photon in the first step, which induced the transition from the ground state to the first excited state, and then the excited state would absorb a second photon, which induced the transition from the first excited state to a higher excited state. Finally, emission from the higher excited state occurred and the atom returned to the ground state. In this mechanism at least two photons are successively absorbed before the light emission occurs. In 1928, Wood and Gaviola experimentally verified that the sequential two-photon excitation of mercury had the following characteristic: that intensity of the emission at 3,654 Å from the second excited state was proportional to the second power of the exciting light intensity.

The 1931 dissertation of Maria Göppert-Mayer developed the Dirac formulation of second-order time-dependent perturbation theory for the case of two-photon absorption and emission. The problem was to make a theory for the interaction of an atom and the radiation field. The radiation field was confined to a cavity. This resulted in standing waves or modes in which the electromagnetic field vibrates at a definite frequency. This classical formulation is transformed in quantum theory by assuming that each of these modes of the electromagnetic field is given by a quantized harmonic oscillator. Now we describe the general description of multiphoton transitions together with their absorption cross-sections and parity. First we compare the one-photon and the two-photon transitions and their cross-sections (Boyd, 2003).

We have solved for the case of single-photon absorption explicitly for linearly polarized radiation earlier. This treatment can be easily generalized for arbitrary polarized radiation. Defining:

$$\vec{E}(v,t) = \vec{E}^0(v)\left(e^{2\pi ivt} + e^{-2\pi ivt}\right) \tag{5.157}$$

and

$$\vec{\mu}_{mn} = \sum_j \int \varphi_m^{*(0)} \, \vec{\mu}_j \varphi_n^{(0)} \, d^3\vec{r} = \sum_j e_j \int \varphi_m^{*(0)} \, \vec{r}_j \varphi_n^{(0)} \, d^3\vec{r} \tag{5.158}$$

The probability of transition to the excited state for large t is:

$$P_m^{(1)}(t) = c_m(t)c_m^*(t) = \frac{2\pi |\vec{\mu}_{mn} \cdot \vec{E}|^2 t}{\hbar} \rho(\omega = \omega_{mn}), \tag{5.159}$$

where ρ is the density of final states.

The transition rate is the time derivative of the transition probability:

$$R_{mn}^{(1)} = \frac{2\pi |\vec{\mu}_{mn} \cdot \vec{E}|^2}{\hbar} \rho(\omega = \omega_{mn}). \tag{5.160}$$

The one-photon absorption cross-section is $\sigma_{mn}^{(1)}(v)$, which is proportional to the light intensity I, in units of $\frac{photon}{cm^2 \cdot s}$,

$$R_{mn}^{(1)} = \sigma_{mn}^{(1)}(\omega)I. \tag{5.161}$$

The cross-section unit is cm^2.

Since the intensity of the light I is related to the electric field $\vec{E}$ by the following relation:

$$I = \left(\frac{nc}{2\pi \hbar \omega}\right) |\vec{E}|^2, \tag{5.162}$$

then the one-photon-cross section is given by:

$$\sigma_{mg}^{(1)}(\omega) = \frac{4\pi^2 \omega}{nc} |\vec{\mu}_{mn}|^2 \rho(\omega = \omega_{mn}). \tag{5.163}$$

Similarly, the use of second-order perturbation time-dependent theory (5.108) yields the following results for two-photon absorption (Boyd, 2003). The transition probability for the absorption of two photons from the ground state, n, to an excited state, n, via an intermediate or virtual set of states m is:

$$P_m^{(2)}(t) = \frac{2\pi t}{\hbar^3} \left| \sum_k \frac{\vec{\mu}_{mk} \, \vec{\mu}_{kn} \, \vec{E}^2}{\omega_{kn} - \omega} \right|^2 \rho(2\omega = \omega_{mn}). \tag{5.164}$$

The sum is over all of the "virtual" intermediate states k, as was first described in the 1931 dissertation of Maria Göppert-Mayer.

These two-photon transitions are described as follows: the squared sum of single-photon transition amplitudes occurs through intermediate states k, and all of these intermediate transitions are weighed (the term in the denominator) by the difference in energy between the photon $E = \hbar\omega$ and the transition energy $E_{kn} = \hbar\omega_{kn}$.

It is possible for a transition to occur from the states $n \rightarrow m$, even when there is no possibility of a direct transition. The transition can occur through a virtual state. For example, the transition can occur from states $n \rightarrow k$, and from $k \rightarrow m$. In the language of quantum mechanics, there exists a probability amplitude for the transition $n \rightarrow m$ to occur via the different virtual or intermediate states k, and all of the contributions to the probability amplitude interfere (Feynman & Hibbs, 1965). Note that the intermediate or virtual states differ in energy from the initial and the final states. The law of conservation of energy applies to the total transition from $n \rightarrow m$, and not for the transitions involving the intermediate states. The probability for the two-photon transition via the intermediate or virtual states is inversely related to the difference in energy between the initial and the final state. These intermediate or virtual states occur from the formulation of the time-dependent perturbation theory. This is clearly illustrated in the second-order time-dependent perturbation theory of Dirac and was used by Maria Göppert-Mayer in her two-photon theory of light-induced electronic transitions. When transitions can occur only by two or more intermediate states (virtual states), then the calculation requires third-order or higher perturbation theory.

The two-photon process described above can be visualized as a two-step process. First there is a transition from the state n to the state k. In this transition energy is not conserved. Next there is another transition from the state k to the final state m. This transition is also characterized by the nonconservation of energy. The important point is that the total transition from state n to the final state m obeys the law of energy conservation. The transitions to and from the intermediate states are called virtual transitions. Note the summation must be taken over all of the intermediate states. In general we perform experiments with monochromatic light and the final states form a continuum of final states of density ρ. For these conditions the probability of a two-photon transition is given by equation (5.146).

The transition rate of the two-photon process is:

$$R_{mn}^{(2)} = \frac{dP_m^{(2)}(t)}{dt} = \sigma_{mn}^{(2)}(\omega)I^2 \quad \text{where} \quad I = \frac{nc}{2\pi\hbar\omega}|\vec{E}|^2. \tag{5.165}$$

The transition rate for the two-photon absorption is proportional to the light intensity squared and independent of time. Note that the constant transition rate is a consequence of the fact that the transition is driven by an electric field that varies in time as a harmonic wave of frequency ω. The two-photon absorption rate is proportional to the imaginary part of the third-order nonlinear optical susceptibility.

The cross-section for the two-photon absorption process is:

$$\sigma_{mn}^{(2)}(\omega) = \frac{8\pi^3\omega^2}{n^2c^2\hbar} \left| \sum_k \frac{\vec{\mu}_{mk}\,\vec{\mu}_{kn}}{(\omega_{kn} - \omega)} \right|^2 \rho(2\omega = \omega_{mn}). \tag{5.166}$$

The units of $\sigma_{mn}^{(2)}(\omega)$ are given in $\frac{cm^4 s}{photons^2}$.

5.5.6 Generalization of the Theory for N-Photon Processes

Now that we have presented the transition probabilities, the transition rates, and the absorption cross-sections for both one-photon and two-photon processes, it is appropriate to present a generalization of the theory for N-photon processes.

A good review of multiphoton absorption and emission in atoms in terms of general equations written in terms of N, where $N = 1, 2, 3, \ldots$, for one-photon, two- photon, and higher-order photon processes is recommended (Lambropoulos, 1976). An alternative approach that is based on modern theories of multiphoton processes is also highly suggested (Cohen-Tannoudji, 2004). For the classical approach based on the second-order, time-dependent perturbation theory of Dirac, see the English translation of the 1931 doctoral dissertation of Maria Göppert-Mayer in the previous chapters.

Following Lambropoulos (1976) we present the cross-section for an N-photon transition between initial state i and final state f:

$$\sigma_{fi}^{(N)} \propto \left| \sum_{k_{N-1}, \cdots k_2, k_1} \frac{\langle f | \vec{E} \cdot \vec{\mu} | k_{N-1} \rangle \langle k_{N-1} | \vec{E} \cdot \vec{\mu} | k_{N-2} \rangle \cdots \langle k_2 | \vec{E} \cdot \vec{\mu} | k_1 \rangle \langle k_1 | \vec{E} \cdot \vec{\mu} | i \rangle}{[E_{k_{N-1}i} - (N-1)E] \cdots [E_{k_2 i} - 2E][E_{k_1 i} - E]} \right|^2$$

$$(5.167)$$

In this equation the photon energy is E, the transition energy between the any intermediate states k_l and ground state i is $E_{k_l i}$, the electric dipole moment operator is $\vec{\mu}$, and the electric field amplitude vector is $\vec{E}$.

Lambropoulos (1976) also gives an expression for the transition rates of N-photon processes $R_{fi}^{(N)}(\nu)[s^{-1}]$ and the selection rules for the transitions. The transition rate is expressed as the number of atoms excited per second at the frequency ν:

$$R_{fi}^{(N)}(\nu) = I^{(N-1)}I(\nu)\sigma_{fi}^{(N)}(\nu). \qquad (5.168)$$

Note that the choice of units for the cross-section $\sigma_{fi}^{(N)}$ depends on the intensity units. For the case in which the intensity is the number of photons per cm^2 per second $(cm^2 s^{-1} Hz^{-1})$, $I(\nu)$ is the number of photons per cm^2 per second per unit of bandwidth (Hz) $(cm^{-2}s^{-1}Hz^{-1})$. Therefore, the units of the cross-section $\sigma^{(N)}$ are $cm^{2N}s^{N-2}cm^{2N}s^{N-1}$.

5.6 SELECTION RULES AND PARITY

Recall we have discussed selection rules and parity in a previous section. Our better understanding of quantum mechanics allows us to better elucidate this concept.

For hydrogen atoms it is observed that there are many pairs of energy levels that are possible, but they are not associated with lines in the observed hydrogen spectrum. The observed lines correspond to allowed transitions, and the possible lines that do not occur in the hydrogen spectrum correspond to the forbidden transitions.

Parity is that quantity that indicates if a wave function will change its sign (positive to negative or negative to positive) when the coordinate axes are inverted. For example, the wave function $\Psi(-x)$ could equal $\Psi(x)$ or $-\Psi(x)$. In the first case the wave function has even parity (g), and in the second case it has odd parity (u). States with odd orbital angular momentum quantum numbers have odd parity or symmetry. Those states with even orbital angular momentum quantum numbers have even parity or symmetry.

Selection rules define which electronic transitions are allowed. Photon excitation affects parity because excitation introduces a dipole term , $e\vec{r}$, which has odd parity. The dipole transition probability, P_{fi}, is given by:

$$P_{fi} = |\langle f|e\vec{r}|i\rangle|^2 \tag{5.169}$$

$P_{fi} \neq 0$ only if $\langle f|$ and $|i\rangle$ have opposite parity. That is, if one state is of even parity, then the other state must be of odd parity. One-photon "allowed" electronic transitions can occur as $g \rightarrow u$ or $u \rightarrow g$. One-photon electronic transition of the same parity such as $g \rightarrow g$ or $u \rightarrow u$ are "forbidden". These selection rules are for cases in the absence of spin–orbit perturbations, which can produce very weak "forbidden" transitions of low intensity and longer lifetime as compared to the "allowed" transitions. Transitions that are forbidden in one-photon process are allowed by two-photon process since two-photon process contains two dipole terms.

The probability for a two-photon transition is:

$$P_{fi} = \sum_m \left| \langle f|e\vec{r}|m\rangle \langle m|e\vec{r}|i\rangle \right|^2. \tag{5.170}$$

Thus, two-photon transitions connect electronic states of the same parity. Two-photon electronic excitation can reach states that are not accessible by one-photon transitions, and that is an important advantage for the experimental study of these states.

5.7 SUMMARY OF IMPORTANT CONCEPTS AND DEFINITIONS RELATING TO ELECTRONIC TRANSITIONS

1. Excitation and emission transitions are described by the electric dipole approximation. $\vec{\mu} = e \sum_k \vec{r}_k$, $\vec{\mu}_{mn} = \langle m|\vec{\mu}|n\rangle$ is the transition moment between the levels m, n.

   The transition moment regulates the transition probability between two levels. The selection rules for electronic transitions follow from the parity of the transition moment for electric dipole transitions; these selection rules differ for one-photon and two-photon transitions.

2. The application of first-order time-dependent perturbation theory to the problem of radiation-induced transition in atoms and molecules results in the equation that is called Fermi's golden rule:

$$P_m^{(1)}(t) = \frac{2\pi t}{\hbar} \rho(\omega = \omega_{mn})(\mu_{x_{mn}})^2 E_x^{0^2}(\omega = \omega_{mn}) \tag{5.171}$$

This equation assumes that that transition is being induced by a plane-polarized, monochromatic incident light wave. This result provides the transition probability for a transition from an initial state to a continuum of states as a function of time. Due to the linear time dependence, the transition rate is constant. Typically the final state of the transition is modeled as a continuum due to processes such as vibrational relaxation or ionization physically. This is observed in real physical situations where the radiation consists of a spectral distribution of finite width. The case of "homogeneous line broadening" occurs if the finite lifetimes of the states in the transition result in the line broadening. For this case the line shape is given by a Lorentzian distribution. The second type of line broadening is due to variations in the local environment of the atoms involved in the transition. For this case the broadening of the transitions is defined as "inhomogeneous broadening" and the line shape of the transitions is given by a Gaussian distribution.

3. The transition probability for the absorption of two photons from the ground state n, to an excited state n, via an intermediate or virtual set of states, m, is:

$$P_m^{(2)}(t) = \frac{2\pi t}{\hbar^3} \left| \sum_k \frac{\vec{\mu}_{mk} \, \vec{\mu}_{kn} \, \vec{E}^2}{\omega_{kn} - \omega} \right|^2 \rho(2\omega = \omega_{mn}). \qquad (5.172)$$

The sum is over all of the "virtual" intermediate states k, as was first described in the 1931 dissertation of Maria Göppert-Mayer. The transition rate for the two-photon excitation process is:

$$R_{mn}^{(2)} = \frac{dP_m^{(2)}(t)}{dt} = \sigma_{mn}^{(2)}(\omega)I^2 \quad \text{where} \quad I = \frac{nc}{2\pi\hbar\omega} |\vec{E}|^2. \qquad (5.173)$$

4. The radiative lifetime, τ_i, is the property of a single level, and τ_i^{-1} is the rate at which the population of level i decays. That is usually given by a single exponential decay.

NOTE

This chapter is dedicated to Michael Kasha, my teacher [BRM].

REFERENCES

Books on Fluorescence

Herman, B. 1998. *Fluorescence Microscopy, Second Edition*, New York: Springer-Verlag.

Lakowicz, J. R. 2006. *Principles of Fluorescence Spectroscopy, Third Edition*, New York: Springer-Verlag.

Lefebvre-Brion, H., Field, R. W. 2004. *The spectra and dynamics of diatomic molecules*. Amsterdam: Elsevier.

Preamps, H.-H. 1995. *Encyclopedia of Spectroscopy*, Weinheim, Germany: VCH.

Valeur, B. 2002. *Molecular Fluorescence: Principles and Applic*ations, Weinheim, Germany: Wiley-VCH. This is highly recommended for understanding the physical principles of fluorescence and its applications.

Books and Articles on the History of Fluorescence

Berberan-Santos, M. N. (2001). Pioneering contributions of Jean and Francis Perrin to molecular luminescence. In: *New Trends in Fluorescence Spectroscopy, applications to chemical and life sciences*, Eds. B. Valeur, J-C. Brochon, Berlin, Springer-Verlag, pp. 7–34, (2001).

Clark, G., Kasten, F. H. (1983). *History of Staining, Third Edition*. Baltimore, Williams & Wilkins.

Einstein, A. (1916). Strahlungs-Emission und -Absorption nach der Quantentheorie. "Emission and Absorption of Radiation in Quantum Theory," Deutsche Physikalische Gesellschaft. Verhandlungen 18: 318–323.

Einstein, A. (1917). Zur Quantentheorie der Strahlung. "On the Quantum Theory of Radiation." Physikalische Zeitschrift, 18, 121–128, (1917). This paper was also published in 1916 in Physikalische Gesellschaft Zürich. Mitteilungen 18: 47–62, 1916.

Einstein, A. (19). *The Collected Papers of Albert Einstein*. Available in both German and in English translation in a series of volumes through Princeton University Press, Princeton, New Jersey.

Franck, J., Jordan, P. 1926. *Anregung von Quantensprügen durch Stösse*, Berlin: Verlag von Julius Springer.

Gaviola, E. 1927. Ein fluorometer. Apparat zur Messung von Fluoreszenzabklingungzeiten. Z. für Physik, 42: 853–861.

Grotrian, W. 1928. *Graphische Darstellung der Spektren von Atomen und Ionen mit ein, zwei und drei Valenzelektronen*, [Graphrical representation of spectra from atoms and ions with one, two, and three valence electrons]. Berlin: Springer.

Jackson, M. 2000. *Spectrum of belief: Joseph von Fraunhofer and the craft of precision optics*. Cambridge, MA: The MIT Press.

Kasha, M. 1950. Characterization of electronic transitions in complex molecules. *Disc. Faraday Soc.* 9: 14–19. This important papers defines internal conversion, intersystem crossing, the intersystem crossing ratio, $\pi \to \pi^*$ and $n \to \pi^*$ transitions.

Kasha, M. 1984. The triplet state. An example of G. N. Lewis' research style. *Journal of Chemical Education*, 61(3): 204–215. This paper illustrates the Jablonski diagram for the lowest metastable state of dyes together with other energy level diagrams by Franck and Livingston, Lewis, Lipkin, and Magel, Lewis and Kasha, and Kasha and McRae. It also teaches the important lesson to be wary of instrument performance-calibrate, calibrate, calibrate.

Kasha, M. 1990. Four great personalities of science: G. N. Lewis, J. Franck, R. S. Mulliken and A. Szent-Györgyi, *Pure and Applied Chemistry*, 62(8): 1615–1630.

Kasha, M. 1999. From Jablonski to femtoseconds. Evolution of molecular photophysics. *Acta Physica Polonica A*, 95(1): 15–35.

Kasten, F. H. 1989. The Origins of Modern Fluorescence Microscopy and Fluorescent Probes. In: *Cell Structure and Function by Microspectrofluorometry*. Eds. E. Kohen, J. G. Hirschberg, 3–50. San Diego: Academic Press.

Lewis, G. N., Calvin, M., Kasha, M. 1949. Photomagnetism. Determination of the paramagnetic susceptibility of a dye in its phosphorescent state. *The Journal of Chemical Physics*, 17(9): 804–812.

Lewis, G. N., Kasha, M. 1944. Phosphorescence and the triplet state. *Journal of the American Chemical Society*, 66: 2100–2116.

Lewis, G. N., Kasha, M. 1945. Phosphorescence in fluid media and the reverse process of singlet-triplet absorption. *Journal of the American Chemical Society*, 67: 994–1003.

McGucken, W. 1969. *Nineteenth-Century Spectroscopy: Development of the Understanding of Spectra 1802–1897*. Baltimore: The Johns Hopkins Press.

Mitchell, A. C. G., Zemansky, M. W. 1933. *Resonance Radiation and Excited Atoms*, London: Cambridge University Press.

Newton, H. E. 1957. *A history of luminescence, from the earliest times until 1900*.

Philadelphia: The American Philosophical Society. This book is perhaps the most comprehensive work in English on the historical aspects of all types of luminescence. It includes an extensive biography of works in English, German, and French.

Perrin, M. F. 1926. Polarisation de la lumière de fluorecence. Vie moyenne des molécules dans l'etat excité. *J. Phys. Radium* 7: 390–401.

Pringsheim, E. G. 1928. *Fluorescenz und Phosphorescenz*, [1921, 1923, 1928], Berlin. Germany: Verlag von Julius Springer.

Pringsheim, E. G. 1963. *Fluorescence und Phosphorescence*, (English). New York, Interscience.

Renn, J. 2005. Supplementary Issue, Einstein's Annalen Papers, J. Renn, Guest Editor, *Annalen der Physik*, 14, SI, 1–586. Following a series of introductory essays on the contributions of Einstein to several areas of physics, the original papers in German that were published in *Annalen der Physik* by Einstein are reproduced.

Valeur, B. 2002. Introduction: on the origin of the terms fluorescence, phosphorescence, and luminescence, In: *New Trends in Fluorescence Spectroscopy: applications to chemical and life sciences*, Eds. B. Valeur, J-C. Brochon, 3–6. Berlin: Springer-Verlag.

Books and Articles on Theory and Instrumentation

Berland, K. 2001. Basics of Fluorescence, In: *Methods in Cellular Imaging*, Ed. A. Periasamy, 5–19. New York: Oxford University Press.

Clegg, R. M. 1996. Fluorescence Resonance Energy Transfer, In: *Fluorescence Imaging Spectroscopy and Microscopy*, Eds. X. F Wang, B Herman, 179–252. New York: John Wiley & Sons. This chapter is highly recommended as a review of Förster's energy transfer theory.

Demtröder, W. 2002. *Laser spektroskopie: Grundlagen und Techniken, 4. Auflage*, Berlin: Springer-Verlag.

Einstein, A. 1916. Strahlungs-Emission and -Absorption nach der Quantentheorie [Emission and Absorption of Radiation in Quantum Theory]. *Deutsche Physikalische Gesellschaft. Verhandlungen* 18: 318–323.

Feynman, R. P., Leighton, R. B., Sands, M. 2006. *The Feynman Lectures on Physics, The Definitive and Extended Edition*, San Francisco: Benjamin Cummings. Volume I has a good introduction the origin of the refractive index, the harmonic oscillator, and light scattering and polarization. Volume III is a very clear development of the theory and applications of quantum mechanics with the emphasis on understanding of the basic physics.

Förster, T. 1946. Energiewanderung und Fluoreszenz. *Naturwissenschaften*, 6: 166–175.

Förster, T. 1951. *Floreszenz Organischer Verbindungen*. Göttingen: Vandenhoeck & Ruprecht.

Jackson, J. D. 1999. *Classical Electrodynamics, third edition*. New York: John Wiley & Sons, Inc.

Jameson, D. M., Croney, J. C., Moens, P. D. J. 2003. Fluorescence: Basic Concepts, Practical Aspects, and Some Anecdotes. In: *Methods in Enzymology, Volume 360, Biophotonics, Part A*, Eds. G. Marriott, I. Parker, 1–43. San Diego: Academic Press.

Kasha, M. 1944. Phosphorescence and the Triplet State. *Journal of the American Chemical Society*, 66: 2100–2116.

Kasha, M. 1950. Characterization of electronic transitions in complex molecules. *Disc. Faraday Society*, 9: 14–19.

Kasha, M. 1960. Paths of molecular excitation. *Radiation Research*, 2: 243–275.

Latt, S. A., Cheung, H. T., Blout, E. R. 1965. Energy Transfer. A System with Relatively Fixed Donor-Acceptor Separation. *J. Am. Chem. Soc.* 87(5): 995–10003.

Lewis, G. N., Clavin, M., Kasha, M. 1949. Photomagnetism. Determination of the Paramagnetic Susceptibility of a Dye in Its Phosphorescent State. *The Journal of Chemical Physics*, 17(9): 804–812. This paper describes a seminal experiment on paramagnetic susceptibility of molecules in the triplet state. The experimental work was done after the death of Professor Lewis; however, the theoretical treatment and the experimental design are largely his.

Rzazewski, K., Boyd, R. W. 2004. Equivalence of interaction Hamiltonians in the electric dipole approximation. *Journal of Modern Optics* 51(8): 1137–1147.

Stryer, L. 1978. Fluorescence energy transfer as a spectroscopic ruler. *Annu. Rev. Biochemistry*, 47: 819–846.

Stryer, L., Haugland, R. P. 1967. Energy transfer: a spectroscopic ruler. *Proc. Natl. Acad. Sci. U.S.A.* 58: 719–726.

Vo-Dinh, T., Cullum, B. M. 2003. Fluorescence spectroscopy for biomedical diagnostics. In: *Biomedical Photonics Handbook*, Ed. T. Vo-Dinh, 28–1–28–50, Boca Raton: CRC Press.

Books on Quantum Mechanics

Books on the History and Interpretation of Quantum Mechanics

Born, M. 2005. *The Born-Einstein Letters 1916–1955, Friendship, Politics and Physics in Uncertain Times*, New York: Macmillan.

Cassidy, D. C. 1992. *Uncertainty: The Life and Science of Werner Heisenberg*. New York: W. H. Freeman.

Eckert, M. 1993. *Die Atomphysiker: eine Geschichte der theoretischen Physik am Beispiel der Sommerfeld*schule, Braunschweig, Wiesbaden: Vieweg. A very readable book that covers the development of atomic physics in Germany from its beginnings to the end of World War II. The emphasis is on those individuals who led the revolution in physics.

Ghirardi, G. 1997. *Sneaking a look at God's cards: Unraveling the mysteries of quantum mechanics*. Princeton: Princeton University Press. This book gives the reader a deep understanding of the problems related to the interpretation of quantum mechanics.

Goldstein, H. 1981. *Classical Mechanics*, Second Edition, Reading, Massachusetts: Addison-Wesley Publishing Company.

Greenspan, N T. 2005. *The End of the Certain World, the life and science of Max Born: the Nobel physicist who ignited the quantum revolution*. New York: Basic Books.

Jammer, M. 1989. *The Conceptual Development of Quantum Mechanics, second edition*. Philadelphia: The American Institute of Physics, 1989. This well-documented and -referenced book discusses all aspects of the historical development and the interpretation of quantum mechanics.

ter Haar, D. 1967. *The Old Quantum Theory*. Oxford, UK: Pergamon Press.

ter Haar, D. 1998. *Master of Modern Physics. The Scientific Contributions of H. A. Kramers*. Princeton, NJ: Princeton University Press.

Hentschel, A. 2000. *Quantum Theory Centenary*, Berlin: Deutsche Physikalische Gesellschaft.

Hughes, R. I. G. 1989. *The Structure and Interpretation of Quantum Mechanics*. Cambridge, MA: Harvard University Press.

Kuhn, T. S., Heilbron, J. L., Forman, P., Allen, L. 1967. *Sources for History of Quantum Physics*, Philadelphia: The American Philosophical Society.

Kuhn, T. S. 1987. *Black-Body Theory and the Quantum Discontinuity, 1894–1912*. Chicago: The University of Chicago Press.

Massimi, M. 2005. *Pauli's Exclusion Principle, The Origin and Validation of a Scientific Principle*. Cambridge: Cambridge University Press. This important book presents the historical background and the physics behind Pauli's Exclusion Principle. The principle evolved from Pauli's attempt to understand the spectroscopic anomalies in the 1920s. The problem was solved when he introduced a fourth degree of freedom to the electron; that also destroyed the Bohn theory of the atomic core model. What Pauli called the "twofoldness" [Zweideutigkeit] of the electron's angular momentum was shortly afterwards understood to be the electron's spin. The Pauli Exclusion Principle is extremely important since it stimulated the transition from the old quantum theory to the new quantum theory in 1925.

Mehra, J., Rechenberg, H. 2001. *The Historical Development of Quantum Theory*. Volumes 1–6. New York: Springer-Verlag. These volumes provide a comprehensive and well-researched history of quantum mechanics.

Miller, A. I. 1994. *Early Quantum Electrodynamics, A Source Book*. Cambridge: Cambridge University Press. The first part of the book contains frame-setting essays. The second part of the book contains translations of selected papers of Heisenberg, Dirac, Weisskopf, Pauli, Fierz, and Kramers.

Moore, W. 1989. *Schrödinger: Life and Thought*. New York: Cambridge University Press.

Pais, A. 1991. *Niels Bohr's Times, In Physics, Philosophy, and Polity*, Oxford: Oxford University Press.

Planck, M. 1949. *Scientific Autobiography and Other Papers*. New York: Philosophical Library.

Poincaré, H. 1993. *New Methods of Celestial Mechanics*, three volumes. Daniel L. Goroff, Editor, Translation of: Les méthodes nouvelles de la mécanique céleste. Originally published in 1892–1899. American Institute of Physics.

Tomonaga, S. 1997. *The Story of Spin*. Chicago, The University of Chicago Press. A clear explanation of electron spin and modern quantum theory.

Yourgrau, W., Mandelstam, S. 1968. *Variational Principles in Dynamics and Quantum Theory*. New York: Dover Publications. This book places the classical developments in mechanics in proper context with the development of quantum mechanics. The first part of this book presents a clear discussion of the following topics: Fermat's Principle of Least Time, Principle of Least Action of Maupertuis, the Equations of Lagrange, Hamilton's Principle, and the Hamilton-Jacobi Equation.

Books with Translations of Key Papers about Quantum Mechanics

Duck, I, Sudarshan, E. C. G. 2000. *100 years of Planck's Quantum*. Singapore: World Scientific. Contains introductions and translated excerpts of papers from Planck, Wien, Einstein, Compton, Bohr, de Broglie, Kramers, Heisenberg, Born, Jordan, Dirac, and Schrödinger.

Hettema, H. 2000, *Quantum Chemistry, classic scientific papers*. Singapore: World Scentific. This book contains translations of many of the papers on the theory of the chemical bond, spectroscopy, and other molecular interaction and approximation methods. The paper of M. Born and R. Oppenheimer, "On the quantum theory of molecules," first published in Ann. Phys. (Leipzig) 84, 457 (1927) is a very complicated paper on the separation of the electronic and the nuclear coordinates in an approximate solution of a quantum mechanical problem of a molecule that uses second- and third-order perturbation theory.

Van der Waerden, B. L. 1968. *Sources of Quantum Mechanics*, New York: Dover Publications. Contains introductions and English translations of full papers (1916–1926) from Einstein, Ehrenfest, Bohr, Ladenburg, Kramers, Born, Van Vleck, Kuhn, Heisenberg, Jordan, Pauli, and Dirac.

Wheeler, J. A., Zurek, W. H., 1983. *Quantum Theory and Measurement*, Princeton, NJ: Princeton University Press.

Books by Inventors and Developers of Quantum Mechanics

Bohr, N. 1981. *Niels Bohr Collected Works*, Volume 2, Work on Atomic Physics (1912–1917). Ed. Hoyer, U. Amsterdam: North-Holland Publishing Company.

Born, M. 1989. *Atomic Physics, Eighth Edition*, New York: Dover Publications. First published in 1933 under the title "Moderne Physik."

Born, M. 1960. *The Mechanics of the Atom*, translated from the German by J. W. Fisher, and revised by D. R. Hartree. New York: Frederick Ungar Publishing Company. First published in German in 1924 under the title "Atommechanik."

Born, M., Jordan, P. 1930. *Elementare Quantenmechanik, Zweiter Band der Vorlesungen Über Atommechanik*, Berlin: Julius Springer. Dr. Max Born was a Professor at the University of Göttingen. Dr. Pascual Jordan was a Professor at the University of Rostock.

Broglie, L. De 1924. *Recherche sur la théorie des quanta* (doctoral thesis in French), (Investigations on quantum theory), Paris, Masson et Cie. Reprinted: Annales de la Fondations Louis De Broglie, 17(1): 1–109 (1992).

Condon, E. U., Morse, P. M. 1929. *Quantum Mechanics*. New York: McGraw-Hill Book Company.

Dirac, P. A. M. 1935. *Quantum Mechanics*, New York: Oxford University Press.

Dirac, P. A. M. 1958. *The Principles of Quantum Mechanics, fourth edition*, London: Oxford University Press. The first edition was published in 1930.

Dirac, P. A. M. 1995. *The Collected Works of P. A. M. Dirac 1924–1948*, Ed. R. H. Dalitz, Cambridge, UK: Cambridge University Press.

Feynman, R. P., Hibbs, A. R. 1965. *Quantum Mechanics and Path Integrals*. New York: McGraw-Hill, Inc. This book is a wonderful mixture of the physical theory and the mathematics that make up quantum theory. The chapter on perturbation theory includes a very clear description of electronic transitions, virtual states, and vector potentials.

Frenkel, J. 1950. *Wave Mechanics, Advanced General theory*. New York: Dover Publications. First published by Oxford University Press in 1934. Frenkel was a professor at the Physico-Technical Institute, Leningrad. This book presents a clear overview of the convergence of classical mechanics and quantum mechanics and the role of the variation principle in the formulation of the Schrödinger equation. The theory of double transitions (two-photon transitions) is clearly described as well as the correspondence between wave mechanics and matrix mechanics.

Heisenberg, W. 1930. *The Physical Principles of Quantum Theory*, Chicago: The University of Chicago Press.

Heisenberg, W. 1949. *The Physical Principles of the Quantum Theory*, New York: Dover Publications. First published in 1930 by The University of Chicago Press, Chicago.

Heisenberg, W. 1985. *Werner Heisenberg, Gesammelte Werke* (Collected Works), *Series A / Part I, Wissenschaltliche Originalarbeiten*, Eds. Blum, W., Dürr, H.-P., Rechenberg, H., Berlin: Springer-Verlag.

Pauli, W. 2000. *Wave mechanics*, New York: Dover Publications. First published in 1973 by The MIT Press, Cambridge MA.

Schrödinger, E. (1927) *Die Wellenmechanik, Dokumente Der Naturwissenschaft, Abteilung Physik*, Band 3, Ed. A. Hermann, Stuttgard: Ernst Battenberg Verlag.

Schrödinger, E. 1929. *Collected Papers on Wave Mechanics*, Translated from the Second German Edition, London: Blackie & Son Limited. This book contains the English translations of Schrödinger's four key papers on wave mechanics that he published in 1926.

Schrödinger, E. 1984. *Gesammelte Abhandlungen*, Ed. W. Thirring, 4 vols., Braunschweig, Vieweg.

Sommerfeld, A. 1919. *Atombau und Spektrallinien*. Braunschweig: Fr.. Vieweg und Sohn. For decades this work was the classic for theoretical physics of atoms.

Sommerfeld, A. 1930. *Wave Mechanics*, London: Methuen & Company, Ltd.

Sommerfeld, A. 1978. *Atombau und Spektrallinien*, Nachdruck der 8. Auflage, Band I, II. Frankfurt/M.: Verlag Harri Deutsch.

Tomonaga, S. 1997. *The Story of Spin*, Chicago: The University of Chicago Press. This book is an English translation of the original title *Spin wa merguru* in Japanese. Sin-itiro Tomonaga shared the Nobel Prize in physics in 1965 for his work on the development of quantum electrodynamics. This book contains the best historical description of how Dirac, building on the work of Pauli and the relativistic equation, derived the theory of electron spin and its magnetic moment.

Von Neumann, J. 1932. *Mathematische Grundlagen der Quantenmechanik*, Berlin: Julius Springer.

Von Neumann, J. 1955. *Mathematical Foundations of Quantum Mechanics*, Princeton: Princeton University Press.

Weyl, H. 1950. *The Theory of Groups and Quantum Mechanics*, New York: Dover Publications. The English translation of the 1931 German edition of Gruppentheorie und Quantenmechanik was published by Methuen and Company, London.

Books on the Theory of Quantum Mechanics

Atkins, P. W. (1990). *Physical Chemistry, Fourth Edition*, New York, W. H. Freeman and Company, pp. 228–534. A clear introduction to the quantum theory of atomic and molecular structure and spectroscopy.

Bethe, H. A., Jackiw, R. 1997. *Intermediate Quantum Mechanics, Third Edition*, Reading, MA: Addison-Wesley.

Bohm, D. 1989. *Quantum Theory*. First published in 1951 by Prentice-Hall, Inc. Mineola: Dover Publications.

Cohen-Tannoudji, C., Diu, B., Laloë, F. 1977. *Quantum Mechanics*, English translation of "Mécanique Quantique," New York: John Wiley & Sons.

Condon, E. U., Morse, P. M. 1929. *Quantum Mechanics*. New York: McGraw-Hill Book Company, Inc. A good description of Dirac's 1927 treatment of spontaneous and induced (stimulated) emission of an atom interacting with radiation contained in a cavity with reflecting walls.

Eisberg, R. Resnick, R. 1985. *Quantum Physics of Atoms, Molecules, Nuclei, and Solids, second edition*. New York: John Wiley and Sons.

Feynman, R. P., Leighton, R B., Sands, M. 1965. *The Feynman Lectures on Physics, Volume III, Quantum Mechanics*, Reading, MA: Addison-Wesley Publishing Company.

Fong, P. 1962. *Elementary Quantum Mechanics*, Reading, MA: Addison-Wesley Publishing Company.

Friedrich, H. 2006. *Theoretical Atomic Physics, Third Edition*, Berlin: Springer-Verlag. This is an advanced book with an emphasis on theory.

Gasiorowicz, S. 2003. *Quantum Physics, Third Edition*, New York: John Wiley & Sons, Inc.

Griffiths, D J. 2005. *Introduction to Quantum Mechanics, Second Edition*. Upper Saddle River, NJ: Prentice Hall.

Haken, H., Wolf, H. C. 2000. *The Physics of Atoms and Quanta, Introduction to Experiments and Theory, Sixth Edition*. Berlin: Springer-Verlag.

Jackson, J. D. 1999. *Classical Electrodynamics, Third Edition*, New York: John Wiley & Sons.

Kramers, H. A. 2003. *Quantum Mechanics*, translated by D. ter Haar from 1937 edition, Mineola, NY: Dover Publications.

Liboff, R. L. 2003. *Introductory Quantum Mechanics, fourth edition*. San Francisco: Addison Wesley.

Merzbacher, E. 1999. *Quantum Mechanics, Third Edition*, New York: John Wiley & Sons, Inc.

Messiah, A. 1999. *Quantum Mechanics*, English translation of "Mécanique Quantique," first published in English in 1958 by John Wiley & Sons, Mineola: Dover Publications.

Pauling, L., Wilson, Jr., E. B. 1985. *Introduction to Quantum Mechanics with Applications to Chemistry*, Mineola: Dover Publications (1985). First published by McGraw-Hill Book Co., New York in 1935.

Peebles, P. J. E. 1992. *Quantum Mechanics*, New Jersey: Princeton University Press, Princeton.

Sakurai, J. J. 1994. *Modern Quantum Mechanics, Revised Edition*, Reading, MA: Addison-Wesley.

Schiff, L. I. 1968. *Quantum Mechanics, Third Edition*, New York: McGraw-Hill Book Company.

Shankar, R. 1994. *Principles of Quantum Mechanics, Second Edition*, New York: Springer-Verlag.

Tang, C. L. 2005. *Fundamentals of Quantum Mechanics, for solid state electronics and optics.* Cambridge: Cambridge University Press. I strongly recommend this book for those readers who have not taken a course on quantum mechanics. It is concise and easy to read. There is a perfect balance between theory and applications.

Tolman, R. C. 1938. *The Principles of Statistical Mechanics*, Oxford: Oxford University Press.

Tomonaga, S-I. 1962. *Quantum Mechanics, Volume I. Old Quantum Theory*, Amsterdam: North-Holland Publishing Company. This book contains a very clear introduction to the Heisenberg formulation of quantum theory and a clear explanation of the transition from classical mechanics to quantum theory based on Bohr's correspondence principle.

Tomonaga, S-I. 1966. *Quantum Mechanics, Volume II. New Quantum Theory*, Amsterdam: North-Holland Publishing Company.

Townsend, J. S. 2000. *A Modern Approach to Quantum Mechanics*, Sausalito, CA, University Science Books.

Books on the Quantum Theory of Light

Cohen-Tannoudji, C. 2004. *Atoms in Electromagnetic Fields, Second Edition*, Singapore: World Scientific Publishing Company.

Cohen-Tannoudji, C., Dupont-Roc, J., Grynberg, G. 1992. *Atom–Photon Interactions. Basic Processes and Applications.* New York: John Wiley & Sons, Inc.

Heitler, W. 1954. *The Quantum Theory of Radiation.* London: Oxford University Press. Reprinted in 1983 by Dover Publications, New York.

Loudon, R. 2000. *The Quantum Theory of Light, Third Edition*, Oxford: Oxford University Press.

Books on Nonlinear Optics

Bloembergen, N. 1965. *Nonlinear Optics*, New York: Benjamin.

Bloembergen, N. 1996. *Encounters in Nonlinear Optics, Selected Papers of Nicolaas Bloembergen*. Singapore: World Scientific.

Boyd, R. W. 2003. *Nonlinear Optics, Second Edition*, San Diego, Academic Press. An excellent choice, as it covers all aspects of modern nonlinear optics in an extremely clear manner. Multiphoton absorption processes are derived.

Shen, Y. R. 1984. *The Principles of Nonlinear Optics*, New York: John Wiley & Sons. Text contains derivation of two-photon absorption and Raman scattering.

Books with Reprints of Papers on Multiphoton Excitation Microscopy and Other Nonlinear Microscopies

Hunt, J. H., ed. 2000. *Selected Papers on Nonlinear Optical Spectroscopy*, Bellingham: SPIE/The International Society for Optical Engineering.

Masters, B. R., ed. 2003. *Selected Papers on Multiphoton Excitation Microscopy*∗, Bellingham: SPIE/The International Society for Optical Engineering. ∗Contains a reprint of thesis paper of Göppert-Mayer, M. (1931). Über Elementarakte mit zwei Quantensprüngen, Ann. Phys. (Leipzig), 9:273–294.

Neckers, D. C. ed. 1993. *Selected Papers on Photochemistry*. Bellingham: SPIE/The International Society for Optical Engineering. This volume contains reprints of the important papers related to light–matter interactions from the following authors: Draper, Planck, Einstein, Stark, Warburg, Lewis, Cario and Franck, and Kasha.

Original Papers

Göppert-Mayer, M. 1929. Über die Wahrscheinlichkeit des Zusamenswirkens zweier Lichtquanten in einem Elementarakt. *Die Naturwissenschaften*, 17:932.

Göppert-Mayer, M. 1931. Über Elementarakte mit zwei Quantensprüngen, *Annalen der Physik*. (Leipzig), 9:273–294.

Lambropoulos, O. 1976. Topics on Multiphoton Processes in Atoms. In: *Advances in Atomic and Molecular Physics*, Eds. D. R. Bates, B. Bederson, Volume 12, 87–164, New York: Academic Press.

Weisskopf, V. 1931. Zur Theorie der Resonanzfluoreszenz, *Annalen der Physik*, Fünfte Folge, Band 9, pp. 23–66.

6

Second- and Higher-Order Harmonic Generation

Robert W. Boyd

6.1 INTRODUCTION

Second-harmonic generation is the prototypical nonlinear optical process. Its discovery by Franken et al. (1961) is often taken as the birth of the field of nonlinear optics. Even today, this process is extremely important for various applications, including shifting the output frequency of lasers, as a diagnostic tool to determine the surface properties of various materials, and for use in nonlinear optical microscopy.

The process of second-harmonic generation is illustrated in Figure 6.1. A laser beam at frequency ω illuminates a nonlinear optical material and a beam of light at frequency 2ω is created. Under proper circumstances, the efficiency of this process can exceed 50% (Seka et al., 1980). The transfer of energy from the input field to the output field can be visualized in terms of the energy-level diagram shown on the right-hand side of the figure. One visualizes the process as one in which two photons from the input beam are lost and one photon in the output beam is created. The process of second-harmonic generation generalizes straightforwardly to direct third- and higher-order harmonic generation. The process of N-th harmonic generation, for arbitrary order N, is illustrated in Figure 6.2. Here N photons are lost to the input beam and one photon at frequency $N\omega$ is created. These intuitive descriptions of the nature of harmonic generation can be justified more formally by means of the Manley-Rowe relations (Manley & Rowe, 1959).

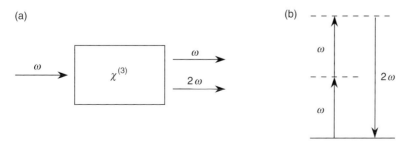

Figure 6.1. The process of second-harmonic generation.

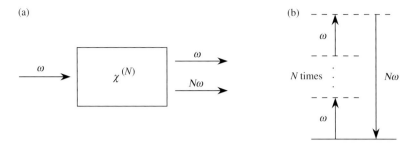

Figure 6.2. The process of direct *N*-th harmonic generation.

6.2 THEORY OF THE NONLINEAR OPTICAL SUSCEPTIBILITY

Theoretical treatments of harmonic generation can be found both in the early research papers (Armstrong et al. 1962) and in various textbook accounts (Boyd, 2003; Butcher & Cotter, 1990; Shen, 1984; Zernike & Midwinter, 1973). These treatments usually begin by considering how the response of the material medium, specified by means of the polarization P (dipole moment per unit volume), depends on the amplitude E of the electric field of the applied optical wave. Under the simplest circumstances, this relationship can be expressed in the time domain as

$$\tilde{P}(t) = \epsilon_0 \left[\chi^{(1)} \tilde{E}(t) + \chi^{(2)} \tilde{E}^2(t) + \chi^{(3)} \tilde{E}^3(t) + \cdots \right], \tag{6.1}$$

where the presence of a tilde over a quantity indicates that that quantity is a rapidly varying function of time. Here $\chi^{(1)}$ is the linear susceptibility, $\chi^{(2)}$ is the second-order susceptibility, $\chi^{(3)}$ is the third-order susceptibility, etc.

Second-harmonic generation occurs as a result of the second-order response described by $\chi^{(2)}$. A slight variant of this process is sum-frequency generation, which is illustrated in Figure 6.3. The situation is a bit more complicated for third-harmonic generation, which can occur either directly as a consequence of the third-order response $\chi^{(3)}$, or indirectly as a two-step process involving two second-order processes. In this latter case, the first step involves second-harmonic generation involving $\chi^{(2)}$ and the second step involves sum-frequency mixing, also involving $\chi^{(2)}$, of frequencies ω and 2ω to produce 3ω. This circumstance is illustrated in Figure 6.4. In well-designed optical systems, the sequential process can be far more efficient than the direct process, although in situations involving biological materials the direct process usually dominates. Similar considerations regarding direct and indirect processes apply to higher-order harmonic generation.

A more complete description of the nonlinear response requires that the vector nature of the electric field and polarization be taken into account. Also, if the various orders of the nonlinear susceptibility are frequency dependent, the relationship between the electric field and polarization is best expressed in the frequency domain. For example, the second-order polarization can be expressed more generally as:

$$P_i(\omega_n + \omega_m) = \epsilon_0 \sum_{jk} \sum_{(nm)} \chi_{ijk}^{(2)}(\omega_n + \omega_m, \omega_n, \omega_m) E_j(\omega_n) E_k(\omega_m). \tag{6.2}$$

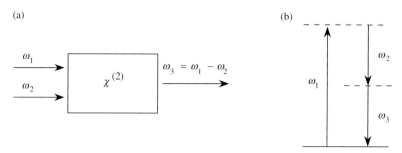

Figure 6.3. The process of sum-frequency generation.

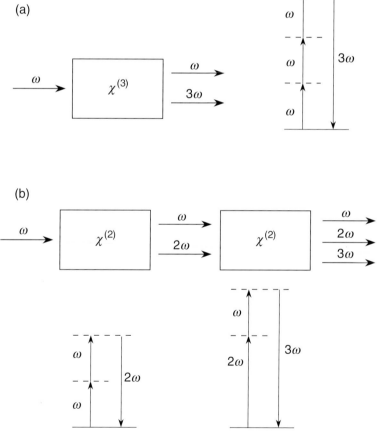

Figure 6.4. (a) Direct third-harmonic generation. (b) The process of sequential third-harmonic generation.

Here i, j, and k represent various cartesian components of the field vectors and the notation (n, m) implies that the expression is to be summed over ω_n and ω_m but only those contributions that lead to the particular frequency $\omega_n + \omega_m$ given in the argument on the left-hand side of the equation are to be retained. Likewise, the third-order polarization can be expressed as:

$$P_i(\omega_o + \omega_n + \omega_m) = \epsilon_0 \sum_{jkl} \sum_{(mno)} \chi_{ijkl}^{(3)}(\omega_0 + \omega_n + \omega_m, \omega_o, \omega_n, \omega_m)$$

$$\times E_j(\omega_o)E_k(\omega_n)E_l(\omega_m). \tag{6.3}$$

These relations generalize in obvious ways to higher-order contributions to the nonlinear polarization.

The values of the nonlinear susceptibility elements can be obtained either by measurement or, in principle, by calculation. Extensive tables of values of the nonlinear susceptibility can be found in the scientific literature (Cleveland Crystals, 2005; Smith, 2005; Sutherland, 1996). Even when explicit calculation of the nonlinear susceptibility is hopelessly difficult, as it is for many biological materials, theoretical models of the nonlinear response are still very useful as they provide insight into the nature of the nonlinear response and show how the magnitude of the response depends on physical properties of the material system.

6.3 SIMPLE MODEL OF THE NONLINEAR SUSCEPTIBILITY

Considerable insight can be obtained by considering the form of the potential energy function that binds the electron to the atomic core. Two different possibilities are shown in Figure 6.5. According to linear reponse theory, the potential well would have the form of a perfect parabola, and the restoring force would be described by Hooke's law. In a real material, the force law need not obey Hooke's law and the potential well need not be a parabola. For a material that possesses a center of inversion symmetry, illustrated on the right-hand side of the figure, the potential well must be a symmetric function of the displacement x. Such a nonlinear response can produce only even

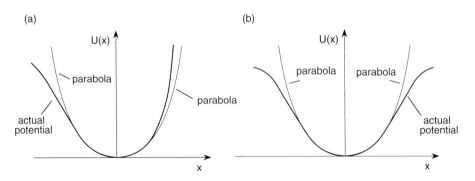

Figure 6.5. (a) The potential well for a material that lacks a center of inversion symmetry. Both even- and odd-order nonlinearities can occur for such a material. (b) The potential well for a material that possesses a center of inversion symmetry. Only even-order nonlinearities can occur for such a material.

harmonics of the applied optical frequency. However, a material that lacks a center of inversion symmetry will have a potential function of the sort shown on the left-hand side of the figure. A potential function of this form can produce both even and odd harmonics of an applied field frequency. This argument leads to the important conclusion that second-harmonic generation can occur only in materials that lack a center of inversion symmetry. Most biological materials lack a center of inversion and thus in principle can produce second-harmonic generation. But it should be noted that a material must be non-centrosymmetric over macroscopic distances in order to produce appreciable radiation at the second-harmonic frequency. Thus, for example, a liquid sample of biological molecules, each of which lacks a center of inversion symmetry, cannot produce second-harmonic generation.

6.4 QUANTUM MECHANICAL TREATMENT OF THE NONLINEAR SUSCEPTIBILITY*

More exacting models of the nonlinear optical response are provided by quantum mechanical calculation (Armstrong et al., 1962; Boyd, 2003; Butcher & Cotter, 1990; Hanna et al., 1979; Shen, 1984; Zernike & Midwinter, 1973). For example, for the case of usual interest in which the applied and generated fields are detuned by at least several line widths from the closest material resonance, the second-order susceptibility can be expressed as:

$$\chi_{ijk}^{(2)}(\omega_\sigma, \omega_q, \omega_p) = \frac{N}{\epsilon_0 \hbar^2} \mathcal{P}_F \sum_{mn} \frac{\mu_{gn}^i \mu_{nm}^j \mu_{mg}^k}{(\omega_{ng} - \omega_\sigma)(\omega_{mg} - \omega_p)}, \tag{6.4}$$

where $\omega_\sigma = \omega_p + \omega_q$, N is the number density of molecules, μ_{nm}^j represents the j-th cartesian component of the electric-dipole moment matrix element connecting levels n and m, and ω_{mg} is the energy separation of levels m and g divided by $\hbar$. The symbol $\mathcal{P}_F$ is the full permutation operator, defined such that the expression that follows it is to be summed over all permutations of the frequencies ω_p, ω_q, and $-\omega_\sigma$.

The cartesian indices are to be permuted along with the related frequencies, and the final result is to be divided by the number of distinct permutations of the input frequencies ω_p and ω_q. In the general case in which ω_p and ω_q are distinct, this equation thus expands to six separate terms. Three of these six terms are illustrated in Figure 6.6; the other six terms result from simply interchanging ω_p and ω_q in these figures.

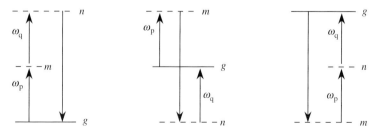

Figure 6.6. Various quantum-mechanical contributions to the second-order nonlinear optical response.

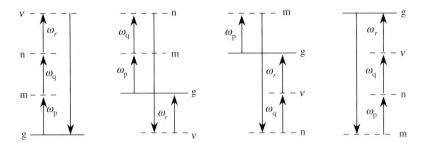

Figure 6.7. Various quantum-mechanical contributions to the third-order nonlinear optical response.

Similarly, the expression for the third-order susceptibility can be expressed in the limit of nonresonant interaction as:

$$\chi^{(3)}_{kjih}(\omega_\sigma, \omega_r, \omega_q, \omega_p) = \frac{N}{\epsilon_0 \hbar^3} \mathcal{P}_F \sum_{mnv} \frac{\mu^k_{gv}\mu^j_{vn}\mu^i_{nm}\mu^h_{mg}}{(\omega_{vg}-\omega_\sigma)(\omega_{ng}-\omega_q-\omega_p)(\omega_{mg}-\omega_p)},$$

(6.5)

where $\omega_\sigma = \omega_p + \omega_q + \omega_r$.

When expanded, this expression represents 24 separate terms. Four of these terms are represented in the diagrams shown in Figure 6.7. The other terms can be found from the six interchanges of the frequencies ω_p, ω_q, and ω_r. The quantum mechanical expressions given here describe the nonlinear response of bound electrons. Even free electrons can produce a nonlinear response, for instance, by means of relativistic effects (Park et al., 2002). Also, it has been found that extremely large harmonic orders can be generated under conditions such that the laser field is sufficiently large to nearly remove the electron from the atomic core. For example, harmonic orders as large as 221 have been observed by Chang et al. (1997).

6.5 WAVE EQUATION DESCRIPTION OF HARMONIC GENERATION*

The intensity of the radiation emitted in the harmonic generation process can be predicted by means of a propagation calculation. We begin with the wave equation in the form:

$$\nabla^2 \tilde{E} - \frac{n^2}{c^2}\frac{\partial^2 \tilde{E}}{\partial \tilde{p}^{NL}t^2} = \mu_0 \frac{\partial^2 \tilde{p}^{NL}}{\partial t^2},$$

(6.6)

where μ_0 denotes the magnetic permeability of free space.

For the present, we take the fundamental- and second-harmonic waves to be plane waves of the form:

$$\tilde{E}_j(z, t) = E_j(z)e^{-i\omega_j t} + \text{c.c.} = A_j(z)e^{i(k_j z - \omega_j t)} + \text{c.c.}$$

(6.7)

where $j = 1, 2$, with $\omega_1 = \omega$ and $\omega_2 = 2\omega$, and where c.c. stands for complex conjugate. Here $E_j(z)$ represents the complex amplitude of field j, and $A_j(z)$ represents the spatially slowly varying field amplitude of this wave.

The nonlinear polarization is then given by:

$$\tilde{P}_2(z,t) = P_2(z)e^{i(2k_1 z - 2\omega t)} + \text{c.c.}, \qquad (6.8)$$

where $P_2 = P(\omega_2) = \epsilon_0 \chi^{(2)} E_1^2$ represents the complex amplitude of the nonlinear polarization.

If these expressions are introduced into the wave equation (6.6), we find that:

$$\frac{d^2 A_2}{dz^2} + 2ik_2 \frac{dA_2}{dz} - k_2^2 A_2 - \frac{(2\omega)^2 n_2^2}{c^2} A_2 = -(2\omega)^2 \epsilon_0 \mu_0 \chi^{(2)} A_1^2 \, e^{i(2k_1 - k_2)z} \qquad (6.9)$$

This expression can be simplified by omitting the first term on the basis of the argument that it is much smaller than the second term. This simplification is known as the slowly varying amplitude approximation. We also see that the third and fourth terms cancel exactly. We are thus left with the equation:

$$\frac{dA_2}{dz} = \frac{2i\omega}{n_2 c} \chi^{(2)} A_1^2 \, e^{i(2k_1 - k_2)z}. \qquad (6.10)$$

Under the assumption that A_1 is not appreciably modified by the nonlinear interaction, we can take A_1 to be a constant and solve this equation by direct integration. We find that after propagation through a distance L the amplitude of the second-harmonic field is given by:

$$A_2(L) = \frac{2i\omega \chi^{(2)} A_1^2}{n_2 c} \frac{e^{i\Delta k} - 1}{i \Delta k L}, \qquad (6.11)$$

where $\Delta k = 2k_1 - k_2$.

Since the intensity is related to the field strength according to $I = 2n\sqrt{\epsilon_0/\mu_0}\,|A|^2$, we find that the intensity of the generated radiation is given by:

$$I_2(L) = \sqrt{\frac{\mu_0}{\epsilon_0}} \frac{2\omega^2}{n_1^2 n_2 c^2} [\chi^{(2)}]^2 I_1^2 L^2 \, [\text{sinc}^2(\Delta k \, L/2)]. \qquad (6.12)$$

The condition $\Delta k = 0$ is known as the condition of perfect phase matching and is a requirement for efficient gereration of second-harmonic radiation. When this condition is fulfilled, the last factor in this equation is equal to unity, and one sees that the intensity of the second-harmonic radiation scales with the square of the length L of the interaction region. For technological applications of nonlinear optics, the phase-matching condition is often fulfilled by making use of a birefringent material. When using such a material, the fundamental and harmonic fields are chosen to have orthogonal polarizations, and the difference in refractive index for the two polarization directions can compensate for the intrinsic wavelength dependence of the refractive index. The phase-matching condition can also be mimicked by using quasi–phase-matching in structured materials, in which a periodic variation of the sign of $\chi^{(2)}$ is used to compensate for wave vector mismatch (Lim et al., 1989). Furthermore, structured materials in the form of photonic crystals and photonic bandgap materials can lead to phase matching by means of the large contribution to the refractive index associated with the periodic arrangement of constituent particles (Markowicz et al., 2004). However, these methods are generally not suitable for utilization with most biological materials, and harmonic generation must therefore occur in the presence of a large wave vector mismatch, leading to greatly reduced efficiency of the harmonic generation process. Nonetheless, the conversion efficiency is typically sufficiently large to produce a measurable signal at the harmonic frequency, and thus harmonic generation is useful as a method to study certain biological structures such as collagen fibrils.

6.6 N-TH HARMONIC GENERATION*

For definiteness, the discussion of the previous paragraph was restricted to the case of second-harmonic generation, but this discussion is readily extended to direct harmonic generation of arbitrary order N. In this case, the amplitude of the nonlinear polarization is given by $P_N = P(N\omega) = \epsilon_0 \chi^{(N)} E_1^N$ and equation (6.11) is replaced by:

$$A_N(L) = \frac{2i\omega\chi^{(N)}A_1^N}{n_N c} \frac{e^{i\Delta k} - 1}{i\Delta k\, L}. \tag{6.13}$$

Also, equation (6.14) for the intensity of the emitted radiated intensity is replaced by:

$$I_N(L) = (\mu_0/\epsilon_0)^{(N-1)/2} \frac{N^4\omega^2}{2^{N+1}n_1^N n_N c^2} [\chi^{(N)}]^2 I_1^N L^2 [\text{sinc}^2(\Delta k\, L/2)]. \tag{6.14}$$

In this case, the wave vector mismatch is given by $\Delta k = Nk_1 - k_N$.

6.7 HARMONIC GENERATION WITH FOCUSED LASER BEAMS*

Harmonic generation can display quite different behavior when excited by a focused laser beam. In such a situation, the peak amplitude $\mathcal{A}_N$ (that is, the amplitude on axis at the beam waist) of the generated N-th harmonic wave is related to the peak amplitude $\mathcal{A}_1$ of the fundamental wave by:

$$\mathcal{A}_N(z) = \frac{iN\omega}{2nc} \chi^{(N)} \mathcal{A}_1^N J_N(\Delta k, z_0, z), \tag{6.15}$$

where $J_N(\Delta k, z_0, z)$ represents the integral

$$J_N(\Delta k, z_0, z) = \int_{z_0}^{z} \frac{e^{i\Delta kz'} dz'}{(1 + 2iz'/b)^{N-1}}. \tag{6.16}$$

Here z_0 represents the value of z at the entrance to the nonlinear medium, $b = 2\pi w_0^2/\lambda_1$ is the confocal parameter of the fundamental laser beam, and w_0 is its beam radius measured to the $1/e$ amplitude point.

For the case of a beam that is tightly focused into the bulk of the interaction region, the limits of integration can be replaced by $\pm\infty$, and the integral can be evaluated to obtain:

$$J_N(\Delta k, z_0, z) = \begin{cases} 0, & \Delta k \leq 0, \\ \dfrac{b}{2} \dfrac{2\pi}{(N-2)!} \left(\dfrac{b\Delta k}{2}\right)^{N-2} e^{-b\Delta k/2}, & \Delta k > 0. \end{cases} \quad (6.17)$$

This result shows that harmonic generation vanishes whenever Δk is negative, which it is for the usual case of normal dispersion. The vanishing of the harmonic generation is a result of destructive interference between the contributions to the harmonic field from both sides of the beam waist.

6.8 SURFACE NONLINEAR OPTICS

One important application of second-harmonic generation is as an exacting diagnostic of the surface properties of optical materials. As noted above, second-harmonic generation is a forbidden process for a material that possesses a center of inversion symmetry. The surface of a material clearly lacks inversion symmetry, and thus second-harmonic generation can occur at the surface of a material of any symmetry. For the same reason, the intensity and angular distribution of surface second-harmonic generation depends critically on the morphology of a surface and on the presence of impurities on the surface of the material. Good reviews of the early work in this area are given by Shen (1985, 1989), and procedures for calculating the intensity of the second-harmonic light are given by Mizrahi and Sipe (1988).

6.9 NONLINEAR OPTICAL MICROSCOPY

Another application of harmonic generation is its use in nonlinear microscopy. Since this topic is developed in considerable detail in succeeding chapters, only a few comments of a general nature will be made here. One motivation for using nonlinear effects and in particular harmonic generation in microscopy is to provide enhanced transverse and longitudinal resolution. The resolution is enhanced because nonlinear effects are excited most efficiently only in the region of maximum intensity of a focused laser beam. Microscopy based on harmonic generation also offers the advantage that the signal is far removed in frequency from the unwanted background light that results from linear scattering of the incident laser beam. Also, light at a wavelength sufficiently long that it will not damage biological materials can be used to achieve a resolution that would normally require a much shorter wavelength. Harmonic-generation microscopy can either make use of the intrinsic nonlinear response of biological materials or can be used with materials that are labeled with nonlinear optical chromophores. Also, harmonic-generation microscopy can be used to form images of transparent (phase) objects, because the phase-matching condition of nonlinear optics depends sensitively on the refractive index variation within the sample being imaged (Muller et al., 1998).

Gao et al. (1997) have used tomography based on second-harmonic generation to characterize biological materials. Gauderon et al. (1998) have demonstrated three-dimensional imaging based on second-harmonic generation with fs laser pulses.

They used this method to characterize the microcrystal structure of lithium triborate. Campagnola et al. (1999) have used second-harmonic generation to produce images of live cells. Moreaux et al. (2000) have used styrl dyes as labels to image membranes using second-harmonic generation microscopy.

Third-harmonic generation has also been used for imaging applications. Muller et al. (1998) have demonstrated imaging of transparent objects using microscopy based on third-harmonic generation. Yelin and Silbenberg (1999) have constructed a scanning microscope based on third-harmonic generation and have used it for the imaging of biological materials.

6.10 HARMONIC GENERATION AS A PROBE OF MEMBRANE POTENTIALS

Second-harmonic generation can also be used to probe potential differences across biological membranes. Bouevitch et al. (1993) observed that the second-harmonic signal shows a pronounced dependence on this potential difference. Styrl dyes imbedded into membranes also show a strong dependence on the potential difference (Millard et al., 2004). Similar effects have been observed in mammalian brain tissue (Dombeck et al., 2005).

In summary, harmonic generation is a well-established physical process that holds considerable promise for applications in nonlinear microscopy.

REFERENCES

Armstrong, J. A., N. Bloembergen, J. Ducuing, and P. S. Pershan, 1962. Interactions between Light Waves in a Nonlinear Dielectric, *Phys. Rev.* 127, 1918–1939.

Bouevitch, O., A. Lewis, I. Pinevsky, J. P. Wuskell, and L. M. Loew 1993. Probing membrane potential with nonlinear optics, Biophys. J. 1993 65: 672–679.

Boyd, R.W. 2003. Nonlinear Optics, Second Edition, Academic Press, Amsterdam, 2003.

Butcher, P. N. and D. Cotter, 1990. *The Elements of Nonlinear Optics*, Cambridge University Press.

Campagnola, P. J., M. Wei, A. Lewis, and L. M. Loew, 1999. High-resolution nonlinear optical imaging of live cells by second harmonic generation, Biophys. J. 77, 3341–3349.

Chang, Z., A. Rundquist, H. Wang, M. M. Murnane, and H. C. Kapteyn 1997. Generation of coherent soft X rays at 2.7 nm using high harmonics, Phys. Rev. Lett. 79, 2967–2970.

Cleveland Crystals, Inc, 2005. 19306 Redwood Road, Cleveland, Ohio 44110 USA, provides a large number of useful data sheets which may also be obtained at http://www.clevelandcrystals.com

Dombeck, D. A., L. Sacconi, M. Blanchard-Desce, and W. W. Webb, 2005. Optical recording of fast neuronal membrane potential transients in acute mammalian brain slices by second-harmonic generation, Microscopy J Neurophysiol, 94 3628–3636.

Franken, P. A., A. E. Hill, C. W. Peters, and G. Weinreich, 1961. Generation of Optical Harmonics, *Phys. Rev. Lett.* 7, 118–119.

Gauderon, R., P. B. Lukins, and C. J. R. Sheppard, 1998. Three-dimensional second-harmonic generation imaging with femtosecond laser pulses, Opt. Lett. 23, 1209–1211.

Guo, Y., P. P. Ho, H. Savage, D. Harris, P. Sacks, S. Schantz, F. Liu, N. Zhadin, and R. R. Alfano, 1997. Second-harmonic tomography of tissues, Opt. Lett. 22, 1323–1325.

Hanna, D. C., M. A. Yuratich, and D. Cotter, 1979. *Nonlinear Optics of Free Atoms and Molecules*, Springer-Verlag, Berlin.

Lim, E. J., M. M. Fejer, R. L. Byer, and W. J. Kozlovsky, 1989. Blue light generation by freuency doubling in a periodically poled lithium niobate channel waveguide, Electron. Lett. 25 731–732.

Manley, J. M. and H. E. Rowe, 1959. General energy relations in nonlinear reactance, *Proc. IRE* 47, 2115 (1959).

Markowicz, P. P., et al., 2004. Dramatic enhancement of third-harmonic generation in three-dimensional photonic crystals, Phys. Rev. Lett. 92, 089903.

Mizrahi, V. and J. E. Sipe 1988. Phenomenological treatment of surface second-harmonic generation, J. Opt. Soc. Am. C 5 660–667.

Midwinter, J. E. 1965 and J. Warner, The effects of phase matching method and of uniaxial crystal symmetry on the polar distribution of second-order non-linear optical polarization, Br. J. Appl. Phys. 16 1135–1142.

Moreaux, L., O. Sandre, and J. Mertz, 2000. Membrane imaging by second harmonic generation microscopy, J. Opt. Soc. Am. B17, 1685–1694.

Millard, A. C., L. Jin, M.-d. Wei, J. P. Wuskell, A. Lewis, and L. M. Loew, 2004. Sensitivity of second harmonic generation from styryl dyes to transmembrane potential, Biophys. J. 86, 1169–1176.

Muller, M., J. Squier, K. R. Wilson, and G. J. Brakenhoff, 1998. 3D-microscopy of transparent objects using third-harmonic generation, J. Microsc. 191, 266–269.

Q.-H. Park, R. W. Boyd, J. E. Sipe and A. L Gaeta 2002. Theory of relativistic optical harmonic generation, IEEE Journal of Selected Topics in Quantum Electronics, 8, 408–413.

Peleg, G., A. Lewis, M. Linial, and L. M. Loew, 1999. Non-linear optical measurement of membrane potential around single molecules at selected cellular sites, Proc. Natl. Acad. Sci. USA 96, 6700–6704.

Seka, W., S. D. Jacobs, J. E. Rizzo, R. Boni and R. S. Craxton, 1980. Demonstration of high efficiency third harmonic conversion of high power Nd-glass laser radiation, Optics Communications 34, 474–478.

Shen, Y. R. 1984. *The Principles of Nonlinear Optics*, Wiley, New York.

Shen, Y. R. 1985. Surface studies by optical second harmonic generation: An overview, J. Vac. Sci. Technol. B 3, 1464–1466.

Shen, Y. R. 1989. Surface properties probed by second-harmonic and sum-frequency generation, Nature 337 519–525.

Smith, A. V., 2005. SNLO, a public-domain nonlinear optics data base which can be obtained at http://www.sandia.gov/imrl/XWEB1128/xxtal.htm

Sutherland, R. L. 1996. *Handbook of Nonlinear Optics*, Marcel Dekker, Inc., New York.

Yelin, D and Y. Silberberg, 1999. Laser scanning third-harmonic-generation microscopy in biology, Opt. Express 5, 169–175.

Zernike, F. and J. E. Midwinter, 1973. *Applied Nonlinear Optics*, Wiley, New York.

—————— **7**

Theory of Spontaneous and Coherent Raman Scattering

Eric O. Potma and X. Sunney Xie

> The universality of the phenomenon, the convenience of the
> experimental technique and the simplicity of the spectra
> obtained enable the effect to be used as an experimental aid to
> the solution of a wide range of problems in physics and
> chemistry.
>
> —C.V. Raman on the Raman effect,
> Nobel Lecture (1930)

7.1 INTRODUCTION

In 1928, Raman observed that when monochromatic light of frequency ω_p is incident on molecules, the scattered light contains an array of different colors (Raman & Krishnan, 1928). The strongest component has the same frequency as the incident light, but weaker contributions with shifted frequencies $\omega_p \pm \omega_R$ are also observed. The first contribution is referred to as the Rayleigh component, whereas the shifted contributions are now known as the Raman scattered components, with the red-shifted frequencies called the Stokes components and the blue-shifted frequencies called the anti-Stokes components. Raman quickly realized that the frequencies ω_R correspond to characteristic frequencies of molecular motions, and that the spectrum obtained can be used to identify and characterize molecules.

The molecular motions identified by Raman correspond to molecular vibrations, most commonly vibrations of individual chemical bonds or groups of neighboring chemical bonds. A nonresonant technique, Raman scattering permits access to these low-frequency vibrational bands with the use of a high-frequency visible or near-infrared light source. Because of this unique feature, Raman scattering has become a widely applied technique for molecular analysis. Moreover, when applied to microscopy, the Raman technique allows visualization of unstained samples with vibrational contrast at high spatial resolution (Otto et al., 1997; Uzunbajakava et al., 2003).

A disadvantage of the Raman technique is the intrinsic weakness of the scattering process. Only 1 part out of 10^6 of the incident radiation will be scattered into the Stokes frequency when propagating through 1 cm of a typical Raman active medium. This feebleness severely complicates several applications, including vibrational imaging of biological samples. Here nonlinear methods come to the rescue.

In 1965, Maker and Terhune showed that very strong signals could be obtained when two coherent light beams of frequency ω_1 and ω_2 were used to drive a vibrational Raman mode at frequency $\omega_R = \omega_1 - \omega_2$ (Maker & Terhune, 1965). This nonlinear technique, which was later dubbed anti-Stokes Raman scattering (CARS) spectroscopy, generates signals that are about 10^5 stronger than in spontaneous Raman spectroscopy (Begley et al., 1974). Not surprisingly, the CARS technique has found many applications in physics and in chemistry. In addition, the CARS approach proved to be very successful for microscopy of cells and tissues (Cheng et al., 2002a; Zumbusch et al., 1999), adding cellular biology to the list of applications of Raman-based techniques.

In this chapter we briefly review the theoretical basics of CARS. A short introduction to Raman scattering is provided to underscore the connection between the linear and nonlinear forms of vibrational spectroscopy. Special emphasis is paid to the unique features of CARS in the tight focusing limit, which are essential for understanding the imaging properties of CARS microscopy. Our discussion will be qualitative in nature, highlighting the physical interpretation rather than giving elaborate mathematical descriptions.

7.2 THE ORIGIN OF THE CARS SIGNAL

The classical description of Raman scattering and CARS takes the vibrational motion explicitly into account, in the form of a harmonic oscillator. Although the classical model provides an intuitive picture of the process, a quantum mechanical description is required to predict the intensities of the Raman and CARS vibrational signals. In this section we discuss both the classical and quantum mechanical approaches and point out key differences between spontaneous Raman and nonlinear CARS spectroscopy.

7.2.1 Spontaneous Raman Scattering

Figure 7.1 displays the energy diagrams of spontaneous Raman scattering. Typically, ω_p lies in the visible or near-infrared range of the spectrum ($1.5 - 3.7 \times 10^3 THz$, or $500 - 1250 nm$), while the nuclear vibrational motions ω_R are in the mid- to far-infrared range ($75 - 470 THz$, or $4 - 25 \mu m$). Because the frequency of the applied light exceeds by far the nuclear vibrational frequencies, the nuclei are unable to respond instantaneously to the electromagnetic field. On the contrary, the electrons follow the oscillating field adiabatically. The Raman effect is therefore understood as a field-induced modulation of the electron cloud, which exhibits shifted frequency components due to the presence of nuclear normal modes.

7.2.1.1 A Classical Description: Harmonic Oscillator

Light scattering phenomena can generally be interpreted as arising from spatial variations of the dielectric properties of the medium. In the case of scattering by molecules in transparent media, these variations are based on fluctuations of the molecules under the influence of the applied light. The electromagnetic field induces an electric dipole moment, which in turn radiates at its oscillation frequencies. With the applied field of

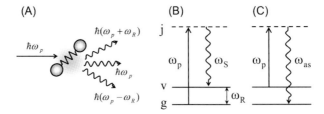

Figure 7.1. Schematic of spontaneous Raman scattering. (A) Incident light is scattered into Rayleigh (ω_p), Stokes ($\omega_S = \omega_p - \omega_R$), and anti-Stokes components ($\omega_{as} = \omega_p + \omega_R$). (C) Energy diagram for Stokes Raman scattering and (B) anti-Stokes Raman scattering. The ground state level is labeled g, the vibrational level v, and the intermediate level j. Far from electronic resonance, the intermediate states j have infinitely short lifetimes and are thus considered virtual states. Solid arrows correspond to applied light; wiggled arrows indicate the emitted radiation.

frequency ω_p expressed as $E(t) = E(r)e^{-i\omega_p t}$, the induced dipole moment is:

$$\mu(t) = \alpha(t)E(t) \tag{7.1}$$

where $\alpha(t)$ is the polarizibility of the material.

The Raman response follows from the presence of nuclear modes that affect the polarizability of the electron cloud. To incorporate this effect, the polarizibility can be expressed as a function of the nuclear coordinate Q, and expanded to the first order in a Taylor series (Placzek, 1934):

$$\alpha(t) = \alpha_0 + \frac{\partial \alpha}{\partial Q}Q(t) \tag{7.2}$$

If a simple harmonic oscillator $Q(t) = 2Q_0 \cos(\omega_R t)$ is assumed for the nuclear mode, we find for the electric dipole moment:

$$\mu(t) = \alpha_0 E(r)e^{-i\omega_p t} + \frac{\partial \alpha}{\partial Q}E(r)Q_0\left\{e^{i(\omega_p + \omega_R)t} + e^{-i(\omega_p - \omega_R)t}\right\} \tag{7.3}$$

The dipole moment is seen to oscillate at different frequencies. The first term on the right-hand side of equation (7.3) corresponds to Rayleigh scattering of light at the incident frequency. The second term describes the Raman shifted scattered frequencies at $\omega_p \pm \omega_R$. Note that the Raman term depends on $\partial \alpha / \partial Q$, the change of the polarizibility along Q. Raman scattering therefore occurs only in molecules in which the applied field brings about a polarizibility change along the nuclear mode. The condition of polarizibility change forms the basis of the selection rules for Raman spectroscopy, which depend on the symmetry of the nuclear mode in the molecule.

The strength of the Raman signal is commonly expressed in terms of the differential scattering cross-section, defined as the amount of scattered intensity within the elementary solid angle $d\Omega$, divided by the incident intensity:

$$\sigma_{diff} \equiv \frac{\partial \sigma}{\partial \Omega} = \frac{\omega \bar{I}_R r^2}{\omega_R \bar{I}} \tag{7.4}$$

where $\bar{I}$ ($\bar{I}_R$) is defined as the cycle-averaged Poynting vector of the incoming (scattered) light.

The differential cross-section can be related to square modules of the polarizibility change (i.e., the first-order term in the Taylor expansion of the polarizibility):

$$\sigma_{diff} \propto \left| \frac{\partial \alpha}{\partial Q} \right|^2 \tag{7.5}$$

The total intensity of the Raman scattered light is then found by integrating over the solid angle and by summing the scattering contributions of all N molecules incoherently:

$$I_{tot} = NI \int \sigma_{diff} \, d\Omega \tag{7.6}$$

We see that the spontaneous Raman signal scales linearly with the incident light intensity I and linearly with the number of scattering molecules. The classical model provides an empirical connection between the Raman scattering cross-section and the observed intensities of the Raman lines, but it is unable to predict the Raman transition strengths. To take the scattering interactions explicitly into consideration, a quantum mechanical description is necessary.

7.2.1.2 The Quantum Mechanical Approach: Second-Order Interaction*

In the quantum mechanical description of light scattering, the energy of the molecule is quantized. From energy diagram in Figure 7.1B we see that the incoming field changes the initial state of the molecule, after which the state is changed for a second time under the emission of a photon. The Raman effect, and light scattering in general, is therefore a second-order process as it involves two light–matter interactions. The most common quantum mechanical treatment of the Raman effect is based on the calculation of the transition rate between the states of the molecule. Here we give a short summary of the essentials of the quantum mechanical description of Raman as a second-order process. Although this approach is very insightful, it fails to make a direct connection to the theoretical description of CARS in terms of the material's nonlinear susceptibility. In Section 7.3.1 we will attempt to bridge the linear and nonlinear Raman techniques from the material's response perspective.

In quantum mechanics, the transition rate between states is described by Fermi's golden rule. It states that the transition rate between an initial state $|g\rangle$ and a final state $|v\rangle$ is proportional to the square modulus of the transition dipole μ_{vg} between the two states. Fermi's golden rule assumes one light–matter interaction to establish this transition and is a good description for first-order radiative transitions such as absorption and emission events. To describe the Raman process, however, we need to calculate the second-order radiative transition rate to account for two light–matter interactions. Fermi's golden rule can be expanded to the second order to incorporate Raman interactions. The second-order contribution to the transition rate τ^{-1} can be expressed as (Kramers & Heisenberg, 1925; Loudon, 1978):

$$\frac{1}{\tau} = \sum_v \sum_R \frac{\pi e^4 \omega_p \omega_R n_R}{2\varepsilon_0^2 \hbar^2 V^2} \left| \sum_j \left\{ \frac{\mu_{vj}\mu_{jg}}{\omega_j - \omega_p} + \frac{\mu_{vj}\mu_{jg}}{\omega_j + \omega_R} \right\} \right|^2 \delta(\omega_v + \omega_R - \omega_p) \quad (7.7)$$

where e is the electron charge, ε_0 is the vacuum permittivity, $\hbar$ is Planck's constant, n_R is the refractive index at the Raman emission frequency, and δ is the Dirac delta function.

In equation (7.7), ω_p is the incident pump frequency and ω_R is the frequency of the Raman scattered light. The frequencies ω_v and ω_j are the transition frequencies from the ground state to the final state $|v\rangle$ and intermediate states $|j\rangle$, respectively. The total transition rate includes the summation over all final and intermediate levels. Note also that the summation runs over all Raman scattered photon modes, labeled by R. To evaluate this expression further, we need to have a closer look at the nature of these "photon modes."

In the first light–matter interaction, an incoming field converts the system from the ground state to an intermediate level $|j\rangle$. In the second light–matter interaction, the transition from the intermediate level to the final state $|v\rangle$ is made. This second interaction is a spontaneous emission process, which cannot be properly described if the electric field is treated classically. A full description of Raman scattering thus incorporates a quantized radiation field (Scully & Zubairy, 1996). This is where the photon modes R come in. The second interaction can be thought of as the interaction of the molecule with the fluctuating vacuum field, which perturbs the molecule and induces the emission of a photon R.

From quantized field theory we can find the number of photon modes at frequency ω_R in the volume V (Scully & Zubairy, 1996), which enables us to perform the summation in equation (7.7) over R. From the argument of the Dirac delta function it is seen immediately that the only nonzero contributions are the ones for which the emission frequency ω_R equals $\omega_p - \omega_v$, the red-shifted Stokes frequencies. Using $\sigma_{diff} = d/d\Omega(V/cn_R\tau)$, the expression for the transition rate can now be directly related to the scattering cross section (Loudon, 1978). The result is:

$$\sigma_{diff} = \sum_v \frac{e^4 \omega_p \omega_R^3}{16\pi^2 \varepsilon_0^2 \hbar^2 c^4} \left| \sum_j \left\{ \frac{\mu_{vj}\mu_{jg}}{\omega_j - \omega_p} + \frac{\mu_{vj}\mu_{jg}}{\omega_j + \omega_R} \right\} \right|^2 \quad (7.8)$$

This is the Kramers-Heisenberg formula, which plays a central role in quantum mechanical light scattering (Kramers & Heisenberg, 1925). For our Raman process, the differential cross-section for a particular vibrational level v can be further simplified to:

$$\sigma_{diff} = \frac{\omega_p \omega_R^3}{16\pi \hbar^2 \varepsilon_0^2 c^4} |\alpha_R|^2 \quad (7.9)$$

where we have written the Raman transition polarizibility as:

$$\alpha_R = \sum_j \left\{ \frac{\mu_{vj}\mu_{jg}}{\omega_j - \omega_p} + \frac{\mu_{vj}\mu_{jg}}{\omega_j + \omega_R} \right\} \tag{7.10}$$

From equation (7.9) we see that the Raman transition polarizibility plays a role similar to the classical polarizibility change $|d\alpha/d\Omega|$ of equation (7.5). From equation (7.9) and the discussion above we can learn the following about the spontaneous Raman process:

1. The scattering strength scales as ω^4. The Raman process is linear in the incoming frequency field ω_p and linear in the scattered frequency field ω_R. The remaining ω_R^2-dependence results from the density of photon modes at the emission frequency. Because of this frequency dependence, Raman scattering grows increasingly stronger at higher frequencies or, similarly, shorter wavelengths.
2. From the denominator of the first term in equation (7.10) it is seen that the scattering is very strong when the incident frequency approaches the frequency of a real electronic state $|j\rangle$ of the molecule. This phenomenon is known as resonance Raman and is used to enhance the efficiency of the otherwise weak Raman effect.
3. Although Raman scattering is a second-order radiative process, the total Raman process depends linearly on the intensity of the incoming light. The second light–matter interaction occurs with the vacuum field, which is independent of the incident light.
4. The second interaction with the vacuum field has an arbitrary phase. As a result, the emission field is incoherent. Note that this is true even when we address the molecules with a coherent laser beam, as the final phase of the emitted field is scrambled by the second light–matter interaction.
5. Because there is no phase relation between the vibrationally addressed molecules, the established polarization in the sample dephases instantaneously. Hence, after the applied light field is switched off, there is no delayed emission in the spontaneous Raman process.

Spontaneous Raman scattering is a weak effect because the spontaneous interaction through the vacuum field occurs only rarely. This weakness can be overcome when the spontaneous nature of the $j \rightarrow v$ transition is eliminated by *applying* a second field of frequency ω_R. Nonlinear coherent Raman scattering techniques, of which coherent anti-Stokes Raman scattering is an example, are based on this principle.

7.2.2 Coherent Anti-Stokes Raman Scattering

The CARS energy level diagram is given in Figure 7.2A. We see that the first two light–matter interactions, one with pump frequency ω_p and another with Stokes frequency ω_S, bear much resemblance to the spontaneous Raman scattering process. The important difference is that in the CARS process, the Stokes frequency stems from an applied laser field. In a classical (non-quantum mechanical) picture we can think of the joint

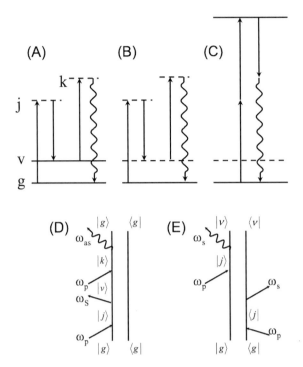

Figure 7.2. Schematic of coherent anti-Stokes Raman scattering. (A) Energy level diagram of vibrationally resonant CARS. Solid lines indicate real states (g and v); dashed lines denote virtual states (j and k). (B) Nonresonant electronic contribution and (C) electronically enhanced nonresonant contribution. (D) Double-sided Feynman diagram that contributes to the vibrationally resonant CARS signal. (E) Double-sided Feynman diagram describing the spontaneous Raman process as a third-order perturbation to the density matrix. See Mukamel, 1995, for the interpretation of the Feynman diagrams.

action of the pump and Stokes fields as a source for driving the Raman mode actively at the difference frequency $\omega_p - \omega_S$. We will first describe the classical model, after which the quantum mechanical model will be used for finding the correct expression for the nonlinear susceptibility.

7.2.2.1 Classical Model: Driving the Raman Mode

A model based on a damped harmonic oscillator conveniently describes the essentials of active driving of the vibrational mode. The equation of motion for the molecular vibration along Q is expressed as (Garmire et al., 1963):

$$\frac{d^2 Q(t)}{dt^2} + 2\gamma \frac{dQ(t)}{dt} + \omega_v^2 Q(t) = \frac{F(t)}{m} \qquad (7.11)$$

with γ the damping constant and m the reduced nuclear mass. The damping is included to account for any mechanism, such as collisions between molecules, that reduces the extent of the oscillation over time. As we will see, the damping will manifest itself through the line width of the vibrational transition. The driving force $F(t)$ of the oscillation is provided by the pump and Stokes fields:

$$F(t) = \left(\frac{\partial \alpha}{\partial Q}\right)_0 E_p E_S^* e^{-i(\omega_p - \omega_S)t} \tag{7.12}$$

where the asterisk denotes the complex conjugate of the field.

The time-varying force oscillates at the beat frequency of the applied fields. A solution to equation (7.11) can be found by adopting the trial solution of the form

$$Q = Q(\omega_p - \omega_S)\left\{ e^{-i(\omega_p - \omega_S)t} + e^{i(\omega_p - \omega_S)t} \right\}$$

We can then solve for the amplitude of the vibration the following form:

$$Q(\omega_p - \omega_S) = \frac{(1/m)\,[\partial \alpha / \partial Q]_0 \, E_p E_S^*}{\omega_v^2 - (\omega_p - \omega_S)^2 - 2i(\omega_p - \omega_S)\gamma} \tag{7.13}$$

We thus notice that the nuclear mode is efficiently driven when the difference frequency between the pump and the Stokes fields approaches the resonance frequency ω_v of the vibration. The amplitude of the nuclear mode further scales with the product of the amplitudes of the driving fields and the polarizability change. Because the nuclear modes of all N molecules oscillate in phase, the refractive index of the medium experiences a periodic modulation of frequency ω_v. A probe field of frequency ω_{pr} propagating through the medium will now be modulated by the vibrational frequency, resulting in components at the anti-Stokes frequency $\omega_{pr} + \omega_p - \omega_S$. The total nonlinear polarization is the sum of all N dipoles—that is,

$$P(t) = N \cdot \mu(t) = N\left[\left(\frac{\partial \alpha}{\partial Q}\right)_0 Q\right] E_{pr}(t) \tag{7.14}$$

Assuming that the pump beam provides the probing field (i.e., $\omega_{pr} = \omega_p$), the nonlinear polarization at the anti-Stokes frequency $\omega_{as} = 2\omega_p - \omega_S$ can be written in terms of an amplitude and a phase:

$$P(t) = P(\omega_{as})e^{-i(\omega_{as})t} \tag{7.15}$$

Combining (7.13), (7.14), and (7.15), we can find for the polarization amplitude:

$$P(\omega_{as}) = \frac{(N/m)\,[\partial \alpha / \partial Q]_0^2}{\omega_v^2 - (\omega_p - \omega_S)^2 - 2i(\omega_p - \omega_S)\gamma} E_p^2 E_S^* = 3\chi_r^{(3)}(\omega_{as}) E_p^2 E_S^* \tag{7.16}$$

It is seen that the amplitude of the polarization depends on three incoming fields. The material response is described by the vibrational resonant third-order susceptibility $\chi_r^{(3)}(\omega)$, which in this classical model is given as:

$$\chi_r^{(3)}(\omega_{as}) = \frac{(N/3m)\,[\partial \alpha / \partial Q]_0^2}{\omega_v^2 - (\omega_p - \omega_S)^2 - 2i(\omega_p - \omega_S)\gamma} \tag{7.17}$$

The nonlinear susceptibility is a material property that describes the medium's response to the applied laser fields. It is a function that peaks when $\omega_p - \omega_S = \omega_v$

(i.e., when the difference frequency matches a vibrational resonance). The CARS signal intensity is proportional to the square modulus of the total polarization:

$$I_{CARS} \propto \left| P^{(3)}(\omega_{as}) \right|^2 = 9 \left| \chi_r^{(3)}(\omega_{as}) \right|^2 I_p^2 I_S \tag{7.18}$$

We see that the CARS response scales quadratically with the pump intensity, linearly with the Stokes intensity, and quadratically with the nonlinear susceptibility of the medium. Note also that because $\chi_r^{(3)}(\omega)$ is proportional to the number of scatterers N, the intensity shows a square dependence on the number of Raman active molecules. Although not apparent from equation (7.18), the signal intensity also strongly depends on the phase-matching of the CARS waves, which will be discussed in more detail in Section 7.4.

The classical model provides limited insight into the internal symmetries of the nonlinear susceptibility and is unable to account for the interaction of the fields with the quantized states of the molecule. As a consequence, the classical approach fails to deliver numerical estimates of $\chi_r^{(3)}(\omega)$. More accurate expressions of the nonlinear susceptibility are derived from a quantum mechanical treatment of the CARS process.

7.2.2.2 Quantum Mechanical Model: Third-Order Perturbation Theory*

The quantum mechanical CARS theory describes the light–matter interactions as perturbations to the molecular eigenstates. The three interactions with the applied fields are effectively described by time-dependent third-order perturbation theory. To fully account for the dynamics induced by the field–matter interactions, the system is usually expressed in terms of the density operator:

$$\rho(t) \equiv |\psi(t)\rangle \langle \psi(t)| = \sum_{nm} \rho_{nm}(t) |n\rangle \langle m| \tag{7.19}$$

where the wave functions are expanded in a basis set $\{|n\rangle\}$ with time-dependent coefficients $c_n(t)$, and $\rho_{nm}(t) \equiv c_n(t) c_m^*(t)$.

The expectation value for the electric dipole moment is then given by:

$$\langle \mu(t) \rangle = \sum_{n.m} \mu_{mn} \rho_{nm}(t) \tag{7.20}$$

The third-order nonlinear susceptibility for the CARS process is found by calculating the third-order correction to the density operator through time-dependent perturbation theory and using the relation $\chi_r^{(3)} = N \langle \mu(t) \rangle / 3 E_p^2 E_S^*$:

$$\chi_r^{(3)} = \frac{N}{V} \sum_v \frac{A_v}{\omega_v - (\omega_p - \omega_S) - i\Gamma_v} \tag{7.21}$$

where Γ_v is the vibrational decay rate, which is associated with the line width of the Raman mode R.

The amplitude A_v can be related to the differential Raman cross-section:

$$A_v \propto \frac{(\pi \varepsilon_0)^2 c^4 n_p}{\hbar \omega_p \omega_S^3 n_S}(\rho_{gg} - \rho_{vv})\sigma_{diff,R} \qquad (7.22)$$

where n_p (n_S) is the refractive index at the pump (Stokes) frequency, and ρ_{gg} (ρ_{gg}) is the element of the density matrix of the ground state (vibrationally excited state). The CARS signal intensity is again found by substituting equation (7.21) into (7.8).

We can interpret the quantum mechanical CARS model qualitatively by considering the time-ordered action of each laser field on the density matrix $\rho_{nm}(t)$. The evolution of the density matrix is conveniently visualized with the double-sided Feynman diagrams depicted in Figure 7.2D. Each electric field interaction establishes a coupling between two quantum mechanical states of the molecule, changing the state of the system as described by the density matrix. Before interaction with the laser fields, the system resides in the ground state ρ_{gg}. In the first diagram of Figure 7.2D, an interaction with the pump field changes the system to ρ_{jg}, followed by a Stokes field interaction that converts the system into ρ_{vg}. The density matrix now oscillates at frequency $\omega_{vg} = \omega_{jg} - \omega_{vj}$, which we call the (vibrational) coherence. The coherence can be converted into a radiating polarization through the interaction with the third field, creating ρ_{kg}, which propagates at $\omega_{kg} = \omega_{jg} + \omega_{vg}$. After emission of the radiation field, the system is brought back to the ground state. With three incident–field interactions and the four levels involved, many time-ordered diagrams can be drawn by permutation. However, the diagram contributes to the vibrational CARS process only if the density matrix after two interactions is resonant with the vibrational frequency ($\omega_{vg} = \omega_R$ for diagram 7.2a), and if the third "propagator" oscillates at the anti-Stokes frequency ($\omega_{kg} = 2\omega_p - \omega_S$ in 7.2a). Four distinct resonant diagrams can be found, one of which is shown in Figure 7.2, giving a full description of the allowed quantum paths that give rise to the vibrational CARS signal.

7.2.3 Differences between Spontaneous Raman and CARS

Although CARS spectroscopy is based on the Raman response of molecules, the nonlinear technique exhibits a number of distinct features. With signal intensities of typically more than five orders of magnitude greater than linear Raman scattering, CARS spectroscopy overcomes many of the problems associated with the inefficient linear Raman response. The stronger CARS signals are a direct result of:

1. The active driving of the nuclear mode through the beat frequency of the pump and Stokes fields as opposed to spontaneous scattering
2. The coherent summation of the radiating polarization, leading to a CARS intensity quadratic in the number of Raman scatterers. In contrast, the spontaneous Raman response is linearly dependent on the number of scatterers.
3. The directionality of the CARS signal due to constructive interference of the radiating polarization in certain directions only. This CARS property stems from the phase-matching criterion, discussed in Section 7.4, and allows a

much more efficient signal collection in CARS compared to capturing only a portion of the isotropically scattered, incoherent Raman emission.

Besides stronger signal intensities, CARS has several other advantages over spontaneous Raman spectroscopy. A serious limitation in linear Raman scattering is that the red-shifted Stokes signals are often collected against a one-photon excited fluorescence background. This is especially relevant in heterogeneous samples such as biological specimens. In CARS, the signal is blue-shifted from the incident beams, which allows detection free of any one-photon excited fluorescence contributions. Furthermore, the coherent nature of the CARS emission opens up various ways to control the signal, such as selective excitation of certain modes through phase control of the CARS process (Oron et al., 2003) and amplification through heterodyne detection (Evans et al., 2004; Yacoby & Fitzgibbon, 1980). Finally, using ultrashort laser pulses, the CARS response can be time-resolved when a time delay between the pump–Stokes pulse pair and the probe pulse is applied (Laubereau & Kaiser, 1978). The CARS signal recorded as a function of this time delay reflects the evolution of the time-dependent vibrational coherences ρ_{vg}. Spectral information related to the Raman spectrum can then be obtained by a proper Fourier transformation of the time-resolved profiles. The advantage of time-resolving the CARS signal is that an intuitive time-ordered picture is obtained for the vibrational response. In the case that intermediate electronic levels contribute to the nonlinear response, time-resolved CARS can reveal more information than accessible when measuring the spontaneous Raman spectrum directly.

Though related, the information probed by the CARS technique differs from Raman scattering. As a result, the shape of the CARS spectrum is substantially different from the ordinary Raman spectrum, which will be discussed in greater detail in the next section.

7.3 CARS SPECTRAL PROPERTIES*

7.3.1 Resonant and Nonresonant Contributions

Strong CARS signals are obtained when the frequency difference between the pump and Stokes fields equals the vibrational frequency of the Raman mode. This vibrationally resonant signal is, however, not the only contribution to the total anti-Stokes emission. In the absence of Raman active nuclear modes, the electron cloud still has oscillating components at the anti-Stokes frequency $\omega_{as} = 2\omega_p - \omega_S$, which can couple with the radiation field. We call this purely electronic part the nonresonant contribution. Hence, the total nonlinear susceptibility is composed of a resonant ($\chi_r^{(3)}$) and a nonresonant ($\chi_{nr}^{(3)}$) part:

$$\chi^{(3)} = \chi_r^{(3)} + \chi_{nr}^{(3)} \tag{7.23}$$

Two nonresonant energy diagrams are depicted in Figure 7.2B,C for the case when all three incident fields overlap in time. A radiating polarization at $2\omega_p - \omega_S$ is established via field interactions with virtual levels. Diagram 7.2C may be two-photon enhanced when $2\omega_p$ is close to the frequency of a real electronic state. By considering all the distinct quantum paths that give rise to a polarization at the anti-Stokes frequency, we

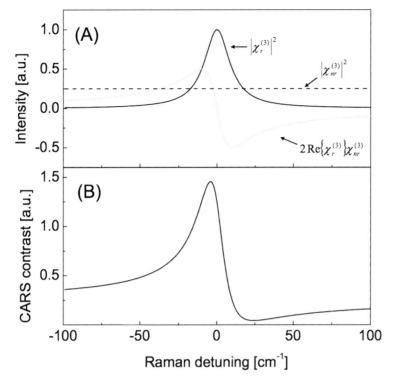

Figure 7.3. Contributions to the CARS spectrum. (A) The spectral variation of the real and imaginary part of the resonant component, modeled with a single Lorentzian line with $\Gamma_v = 10\,\text{cm}^{-1}$, and the nonresonant component. (B) Total CARS signal found by summing the contributions given in (A).

find that $\chi^{(3)}(\omega_{as})$ consists of 24 different terms (Maeda et al., 1988). As we have seen, four of these terms contain coherences at vibrational frequencies and contribute to the vibrationally resonant CARS signal. The remaining 20 terms are electronic nonresonant contributions.

Far from electronic resonance, the nonresonant contributions show negligible variation upon tuning the frequency difference between the pump and Stokes fields. The nonresonant susceptibility is therefore frequency-independent and real. The resonant contribution is small far from a vibrational resonance but grows larger when approaching a vibrational resonance. Importantly, the resonant contribution is complex, with real and imaginary components. The reason for this is that near the vibrational resonance, the phase of the oscillator starts to change, which is typical for resonances in general. At resonance, for instance, the oscillator is $\frac{1}{2}\pi$ out of phase with the driving field. We can decompose the frequency-dependent amplitude and phase of the nonlinear susceptibility into real and imaginary parts. Figure 7.3A shows the profiles of the real and imaginary components of the susceptibility near a single vibrational resonance.

The total CARS signal is proportional to the square modulus of the nonlinear susceptibility (Shen, 1984):

$$I(\omega_{as}) \propto \left|\chi^{(3)}(\omega_{as})\right|^2 = \left|\chi_r^{(3)}(\omega_{as})\right|^2 + \left|\chi_{nr}^{(3)}\right|^2 + 2\chi_{nr}^{(3)} Re\left\{\chi_r^{(3)}(\omega_{as})\right\} \qquad (7.24)$$

The CARS spectrum is composed of a resonant contribution $|\chi_r^{(3)}(\omega_{as})|^2$, a nonresonant background $|\chi_{nr}^{(3)}|^2$, and a mixing term $2\chi_{nr}^{(3)}Re\{\chi_r^{(3)}(\omega_{as})\}$. This last term gives rise to the dispersive line shape of the vibrational resonances in the CARS spectrum, as indicated by Figure 7.3B.

What is the relation of the CARS spectrum to the spontaneous Raman spectrum? To make this connection, we will have to relate the Raman response to the third-order nonlinear susceptibility. In Section 7.1.2, we found the Raman signal by describing it as a second-order light–matter interaction. The third-order susceptibility, however, describes the material's third-order response to the incoming fields. To reconcile the CARS and the Raman process, we need an alternative approach for calculating the spontaneous Raman response. In describing the Raman process as a second-order scattering event, we have previously considered only the transitions between states $|n\rangle$. We now need to consider the Raman process in the framework of the density matrix, so that we can account for coherences $|n\rangle \langle m|$. Both interactions from the bra and ket side need to be considered in order to convert the system from the ground state ρ_{gg} to the final state ρ_{vv} through intermediate levels j. The lowest-order expansion of the density operator that accounts for these transitions is a third-order expansion. A Feynman diagram that describes the Raman process using the density matrix formalism is given in Figure 7.2E.

The Raman intensity can be calculated by considering the expectation value of the photon-mode emission rate operator within the density matrix framework. An expression of the following form can then be obtained (Hellwarth, 1977; Mukamel, 1995):

$$I_{Raman} \propto Im\left\{\chi_r^{(3)}\right\} \tag{7.25}$$

Although this relation should be used with caution, as the nonlinear susceptibility contains more quantum paths than are probed by the spontaneous Raman scattering process (Mukamel, 1995), equation (7.25) points out a fundamental relation between Raman and CARS spectra. Spontaneous Raman probes the imaginary component of the nonlinear susceptibility and is sensitive only to the resonant part of the material response.

The presence of a nonresonant background in CARS can sometimes be problematic. For weak vibrational resonances, the background may overwhelm the resonant information. In biological samples, the concentration of molecular species of interest is usually low, while the nonresonant background from the aqueous surrounding is generally ubiquitous. In addition to introducing an offset, the mixing of the nonresonant field with the resonant field gives rise to broadened and distorted line shapes relative to the Raman spectrum. Suppression of the nonresonant contribution is therefore essential for certain applications.

7.3.2 Suppression of the Nonresonant Background

Several schemes for suppression of the nonresonant background have been developed. Key in reducing the background contribution is to minimize detection of $\chi_{nr}^{(3)}$ while retaining the detection sensitivity for $\chi_r^{(3)}$. Making use of the differences in the polarization orientation properties of $\chi_r^{(3)}$ and $\chi_{nr}^{(3)}$ is a popular approach to realize this goal.

In this subsection we will discuss briefly the basics of polarization-sensitive detection in CARS spectroscopy (Ahkmanov et al., 1977; Brakel & Schneider, 1988; Oudar et al., 1979). Other background suppression methods based on time-resolved methods (T-CARS), spatial properties of the sample (epi-CARS), and phase-sensitive techniques (heterodyne CARS) will be discussed in Section 4 of Chapter 17.

The nonlinear susceptibility is a tensor of rank four, which can be decomposed into elements based on the polarization orientation of the interacting fields. For isotropic media, the polarization-dependent $\chi_{ijkl}^{(3)}$ consists of 21 elements, where i, j, k, l indicate the Cartesian polarization directions of the anti-Stokes, pump, probe, and Stokes fields respectively. Only three elements are independent, and their sum is given by (Butcher, 1965):

$$\chi_{1111}^{(3)} = \chi_{1122}^{(3)} + \chi_{1212}^{(3)} + \chi_{1221}^{(3)} \tag{7.26}$$

Here $(1,2)$ are a combination of two of the orthogonal axes (x, y, z). The tensor elements can be related to the polarization-dependent Raman polarizibility α_{ij}, with i, j the polarization orientation of the scattered and incident fields, respectively (Dick, 1987; Woodward, 1967). A useful relation is $\chi_{1111}^{(3)} \propto (\alpha_{11})^2$, and $\chi_{1212}^{(3)} \propto (\alpha_{12})^2$. The CARS depolarization ratio can be defined as follows:

$$\rho_{\text{CARS}} = \frac{\chi_{1221}^{(3)}}{\chi_{1111}^{(3)}} \tag{7.27}$$

For purely symmetric modes, ρ_{CARS} is equal to the Raman depolarization ratio (Woodward, 1967). In transparent media, the nonresonant susceptibility follows Kleinman's symmetry (Kleinman, 1962), and the depolarization ratio of the nonresonant background is $\rho_{\text{CARS}}^{NR} = 1/3$. The depolarization ratio ρ_{CARS}^{R} of the resonant susceptibility depends, however, on the symmetry of the molecule and may vary from 1/3. Polarization-sensitive CARS is based on the difference between ρ_{CARS}^{R} and ρ_{CARS}^{NR}, which gives rise to a net difference in the polarization orientation of the resonant and nonresonant fields. The nonlinear polarization, polarized at an angle θ, can be written as a function of the angle between the pump and Stokes fields (Fig. 7.4A):

$$P_{\text{CARS}}^i(\theta) = \tfrac{3}{4}\chi_{1111}^i \cos\phi \left\{ \bar{e}_x + \rho_{\text{CARS}}^i \tan\phi \bar{e}_x \right\} E_p^2 E_S^* \tag{7.28}$$

where i denotes either the resonant or nonresonant component.

Using $\rho_{\text{CARS}}^{NR} = 1/3$, it is found that the nonresonant field is linearly polarized along an angle $\theta_{NR} = \tan^{-1}(\tan(\phi)/3)$. When detecting the signal at an angle orthogonal to the linearly polarized nonresonant background (i.e., $\theta_\perp = \theta_{NR} - \tfrac{1}{2}\pi$), the resonant signal is:

$$P_{\text{CARS}}^R(\theta_\perp) = \tfrac{3}{4}\chi_{1111}^R \cos\phi \sin\theta_{NR} \left(1 - 3\rho_{\text{CARS}}^R\right) E_p^2 E_S^* \tag{7.29}$$

The maximum resonant-to-nonresonant signal ratio is obtained for $\phi = 71.6°$ and $\theta_{NR} = 45°$. For these parameters, the nonresonant background is negligible so that the measured signal is now proportional to $I_{\text{CARS}}(\theta_\perp) \propto |\chi_r^{(3)}|^2$. Figure 7.4B shows

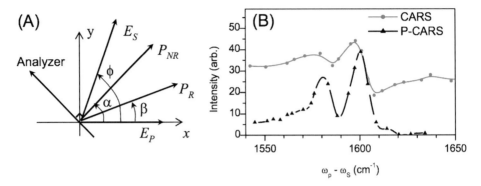

Figure 7.4. Polarization CARS spectroscopy. (A) Schematic of the polarization vectors of the participating fields. (B) Regular and polarization-detected CARS spectrum of a 1-μm polystyrene bead.

the CARS spectrum of polystyrene for regular detection and polarization-sensitive detection. The polarization-CARS (P-CARS) spectrum is free of nonresonant background and shows much more resemblance with the spontaneous Raman spectrum than the ordinary CARS spectrum. P-CARS has been successfully applied for background suppression in spectroscopy and microscopy (Cheng et al., 2001).

7.4 CARS UNDER TIGHT FOCUSING*

In the previous sections we have seen how the CARS signal depends on the frequency-dependent nonlinear response of the material. Furthermore, we learned that the CARS signal is coherent—that is, the CARS waves emitted from different positions in the sample have a well-defined phase relation with one another. In this section, we will see how the spatial addition of the coherent waves determines the total detected CARS signal. The spatial properties of the CARS signal are particularly relevant for microscopy studies, where the interference between the generated fields introduces interesting effects.

7.4.1 CARS Signal Generation

The third-order polarization $P^{(3)}(r, t)$ acts as a source for the coherently emitted field. The correct relation between the polarization and the CARS field is given by Maxwell's wave equation (Boyd, 2003; Shen, 1984):

$$\nabla \times \nabla \times \mathrm{E}(\mathrm{r}, t) + \frac{n^2}{c^2} \frac{\partial^2 \mathrm{E}(\mathrm{r}, t)}{\partial t^2} = -\frac{4\pi}{c^2} \frac{\partial^2 \mathrm{P}^{(3)}(\mathrm{r}, t)}{\partial t^2} \tag{7.30}$$

Solving the three-dimensional wave equation is difficult for an arbitrary form of $\mathrm{P}^{(3)}(\mathrm{r}, t)$. However, a simple analytical solution of the wave equation can be obtained when it is assumed that the polarization is spatially invariant. In a one-dimensional scalar approximation, a trail function for the signal field can be expressed as:

$$E(z, t) = E_{as}(z)e^{-i(\omega_{as}t - k_{as}z)} + c.c. \tag{7.31}$$

Here *c.c.* denotes the complex conjugate. The field is written as a wave that oscillates at the anti-Stokes frequency ω_{as} and travels with wave vector k_{as} along the propagation axis $+z$ of the incident beams. It can furthermore be assumed that the growth of the field amplitude will not significantly change within interaction lengths on the order of an optical wavelength. This approximation is known as the slowly varying envelope approximation (SVEA), which states that $\left|\partial^2 E(z)/\partial z^2\right| << \left|k\partial E(z)/\partial z\right|$. Under these assumptions, the following expression for the spatially dependent field amplitude is found when equation (7.31) is substituted into the wave equation:

$$E_{as}(L) = -\frac{4\pi i}{n_{as}}\frac{\omega_{as}}{c}P_{as}^{(3)} \cdot L \frac{\sin\left[\Delta k L/2\right]}{\Delta k L/2}e^{-i\Delta k L/2} \qquad (7.32)$$

with L as the interaction length and $\Delta k = k_{as} - 2k_p + k_S$ as the wave vector mismatch between the CARS field and the incident fields. The wave vectors of the incident beams are indicated by k_p and k_S for the pump and Stokes fields, respectively. The total CARS intensity is then given by taking the square modulus of the field:

$$I_{as}(L) = \frac{n_{as}c}{8\pi}\left|E_{as}(L)\right|^2 \propto \frac{\omega_{as}^2}{n_{as}c^2}\left|\chi^{(3)}(\omega_{as})\right|^2 I_p^2 I_S \cdot L^2\left(\frac{\sin\left[\Delta k L/2\right]}{\Delta k L/2}\right)^2 \qquad (7.33)$$

From equation (7.33) we see that the CARS intensity grows quadratically with inter-action length L. The signal also depends on the phase mismatch $\Delta\phi = \Delta k \cdot L$. Maximum signals are obtained if all the emitted waves oscillate in phase throughout the sample, which is realized when $\Delta\phi \approx 0$. A small phase mismatch is attained if $\Delta k \approx 0$, which is satisfied when the medium is dispersionless. Condensed phase materials, however, are generally dispersive and the condition of a small vector wave mismatch is not automat-ically fulfilled. To compensate for material dispersion in CARS spectroscopy, special directions of the interacting wave vectors are required for guaranteeing a minimum wave vector mismatch. A popular phase-matched geometry is the folded BoxCARS beam configuration (Shirley et al., 1980).

Another way to keep the phase mismatch small is to reduce the interaction length L. For interaction lengths on the order of the optical wavelength in a typically dispersive material, the signal waves add up in phase almost completely. This condition applies to CARS microscopy, which will be further explained in Section 7.4.3.

7.4.2 The CARS Field Distribution under Tight Focusing

The one-dimensional solution to the wave equation takes into consideration wave vec-tors in the direction of the propagation axis only. Near the focal point of highly focused laser beams, however, wave vectors of different angles are important in the signal generation process as well. A three-dimensional solution of the wave equation is now required for understanding the spatial distribution of the CARS field.

When the laser field is tightly focused, the nonlinear polarization can no longer be assumed to be spatially invariant. First, the power density of the product of the driving fields is highest near focus, which implies that $P^{(3)}(r)$ peaks in the vicinity of the focal volume. Second, the nonlinear susceptibility may be position-dependent; such is the

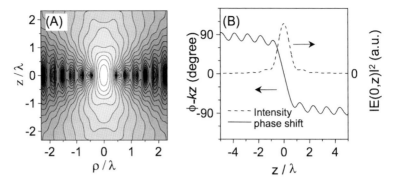

Figure 7.5. (A) Intensity distribution on a log scale of a Gaussian beam focused by a NA = 1.4 objective lens. (B) Intensity distribution (dashed line) and the phase shift (solid) along the optical axis under the same conditions. The π-phase shift is also known as the Gouy phase shift.

case for the spatial inhomogeneous media studied in microscopy. The third-order source term to the wave equation can thus be expressed as:

$$P^{(3)}(\mathbf{r}) = 3\chi^{(3)}(\mathbf{r})E_p^2(\mathbf{r})E_S^*(\mathbf{r}) \tag{7.34}$$

The amplitude and phase of the pump and Stokes driving fields near focus can be found from diffraction theory. A rigorous expression for the focal field is given by Richards and Wolf (1959), while the simpler paraxial approximation (Fraunhofer diffraction) is more frequently used for describing the imaging properties of optical microscopes (Born & Wolf, 1999; Wilson & Sheppard, 1984). Figure 7.5 shows the focal field produced by an objective lens of NA = 1.4.

A general solution of the wave equation for arbitrary $P^{(3)}(\mathbf{r})$ can be found with Green's function methods (Cheng et al., 2002b; Novotny, 1997). At the point R in the far field, the transverse CARS signal field for parallel linearly polarized incident fields is:

$$E_{as}(\mathbf{R}) = -\frac{\omega_{as}^2}{c^2}\frac{e^{ik_{as}|\mathbf{R}|}}{|\mathbf{R}|}\int_V e^{-ik_{as}\mathbf{n}\cdot\mathbf{r}} \times \mathbf{M}(\theta,\phi) \times P^{(3)}(\mathbf{r})dV \tag{7.35}$$

where θ is the angle between R and the optical axis, and ϕ is the polar angle (Figure 7.6).

Equation (7.35) shows that the emitted CARS field looks like a spherical wave. The exact amplitude and phase of this wave depend on the integrand of the nonlinear polarization in the vicinity of the focal volume. The phase factor $\exp(-ik_{as}\mathbf{n}\cdot\mathbf{r})$, with unit vector n in the direction of R, accounts for the phase difference between the different points in V. The projection matrix $\mathbf{M}(\theta,\phi)$ transforms the Cartesian components of $P^{(3)}(\mathbf{r})$ onto a spherical surface defined by θ and ϕ (Cheng et al., 2002b; Moreaux et al., 2000). The total CARS signal is now obtained by integrating the square modulus of the field over the cone angle of the intercepting lens:

$$I_{as} = \frac{nc}{8\pi}\int_{\theta_1}^{\theta_2}d\theta\int_0^{2\pi}d\phi\,|E_{as}(\mathbf{R})|^2R^2\sin\theta \tag{7.36}$$

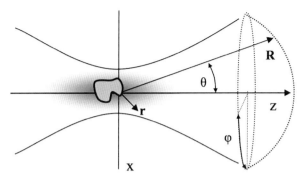

Figure 7.6. Schematic of CARS signal detection in the far field with tight focusing.

It is clear that by adding the emitted waves with different phases along each vector R, the total CARS signal has a rather complicated dependence on the spatial details of $\chi^{(3)}(\mathbf{r})$. In Chapter 17 we will see that because of the phase-coherent nature of CARS, simple models commonly used for explaining the imaging properties of microscopes cannot be straightforwardly applied to CARS microscopy.

7.4.3 Phase-Matching and Tight Focusing

Similar to the one-dimensional model, the total CARS signal in three dimensions is strongly dependent on the phase mismatch between the signal waves. The larger the phase mismatch between the CARS field and the incident waves, the more destructive interference will occur. The signal is called phase-matched if $\Delta\phi$ is smaller than π in which case destructive interference has depleted less than half of the maximum signal. There are two important differences between phase matching in the tight focusing limit and extended focus limit ($L \gg \lambda_{as}$). The first one is the much shorter interaction length, which reduces the signal depletion due to wave–vector mismatch. In the forward propagation direction, the wave–vector mismatch is defined by $\Delta k = k_{as} - 2k_p + k_S$ and is generally nonzero due to the wavelength dependence of the wave vectors. However, the total phase shift due to the wave–vector mismatch is $\Delta\phi = \Delta k \cdot L$, and with L on the order of a wavelength, the phase shift amounts to only $\sim 10^{-2}$ radian for typical conditions in aqueous media. Clearly, the wave–vector mismatch introduces no phase mismatch and the signal is hardly affected at all (Cheng et al., 2002b; Potma et al., 2000). The second difference is the role of the π-phase shift associated with focused laser beams, the so-called Gouy phase shift (Gouy, 1890; Siegman, 1986). For high numerical aperture lenses (typically NA > 1), the full π-phase shift occurs within a distance that is on the order of the optical wavelength λ_{as} (see Fig. 7.5B). It is known that the Gouy phase shift introduces a phase jump ($\phi_g > \pi$) that gives rise to destructive interference in second- and third-harmonic generation (SHG, THG) spectroscopy under tight focusing (Boyd, 2003; Cheng & Xie, 2002). In CARS, the effective Gouy phase shift, induced by the phase shifts of the pump and Stokes fields, is $\phi_g = 2\phi_p - \phi_S \approx \pi$ (Cheng et al., 2002b). This phase shift corresponds to the phase profile of a tightly focused field (Boyd, 2003). Therefore, unlike in SHG or THG, the Gouy phase shift in CARS introduces only nominal destructive inference and the signal is virtually unaffected.

In the epi-direction, the situation is quite different. Due to the reverse direction of the wave vector of the signal field, Δk is now given as $(-k_{as} - 2k_p + k_S)$. Under the same conditions as considered above, the wave–vector mismatch is now much larger than π. Consequently, destructive interference dominates and the signal from objects that span the full interaction length is generally negligible in the epi-direction. The epi-CARS signal from a homogeneous medium is therefore insignificant in microscopy, similar to the situation in regular CARS spectroscopy. For objects much smaller than the interaction length, however, emission in the epi-direction is allowed in CARS under tight focusing. This will be discussed in the next subsection.

7.4.4 Forward and Backward CARS

From the previous subsections it is evident that the strength and direction of the CARS signal strongly depend on the interaction length (i.e., size of the object) and the wave–vector matching. In Figure 7.7, CARS radiation patterns are shown for objects of different size and geometry. For a single Raman active scatterer, L is negligibly small and phase matching is not important. The CARS radiation goes forward and backward symmetrically, similar to radiation from an induced Hertzian dipole in the plane normal to the light propagation. For larger objects, the waves add up in phase in the forward direction. The phase matching is optimal along the optical axis, leading to strong and highly directional signals in the forward direction when the object size is increased.

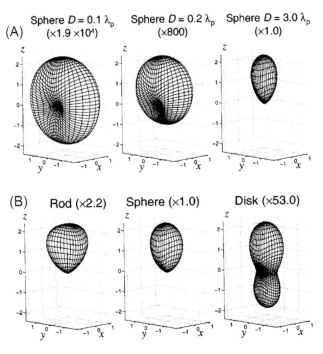

Figure 7.7. Far-field CARS emission patterns from various objects. (B) CARS emission from spheres of different sizes, centered at the focus. (A) Emission patterns from a rod, sphere, and disk of identical volume in focus. Note the difference in the emission patterns from the different objects. Signals were calculated assuming parallel polarized incident beams tightly focused with a 1.4 NA lens. (From Cheng et al., 2002b).

In the backward direction, wave addition is unfavorable and destructive interference will reduce the epi-signal for larger object sizes. For objects smaller than $\lambda_{as}/4$, destructive interference is still incomplete and radiation in the epi-direction remains. In other words, although the backward-going wave has a large $|\Delta k|$, the phase mismatch $|\Delta k| \cdot L$ is compensated by the short interaction length $L < \lambda_{as}/4$ of the object, and the phase mismatch condition is still satisfied. Note that this reasoning holds only for a single object in focus: if the focal volume is filled with multiple small objects, destructive interference of the epi-CARS emission may occur if the objects are found at different positions along the optical axis.

The contrast in epi-CARS microscopy of particles is thus size-selective: smaller objects are visualized efficiently, while the signal from larger objects is suppressed through destructive interference. The object's size is, however, not the sole criterion for observing a particle in epi-CARS. A signal is detected only if the $\chi^{(3)}$ of the particle is different from the nonlinear susceptibility of its surroundings. In CARS microscopy of biological specimens, the $\chi^{(3)}$ of the surrounding is commonly the nonresonant susceptibility of the aqueous medium. Hence, signals from small objects can be expected when $\left| \chi_{object}^{(3)} - \chi_{medium}^{(3)} \right|$ is large. Such is the case, for instance, when $\chi_{object}^{(3)}$ exhibits a strong vibrational resonance. Note that epi-CARS depends on modulus of the difference between the nonlinear susceptibility of the object and its surrounding. This implies that contrast in epi-CARS is also observed when $\chi_{object}^{(3)} - \chi_{medium}^{(3)}$ is negative—that is, "holes" in the medium generate a backward-going signal as well. In this case, a nonzero polarization in the backward direction is derived from the incomplete interference of the signal generated by the material surrounding the hole.

More generally, epi-CARS contrast arises because of spatial differences of $\chi^{(3)}$ in the focal volume that give rise to incomplete destructive interference of the backward propagating waves. This condition is satisfied for (1) small particles, as discussed above, and (2) abrupt changes in $\chi^{(3)}$ along the optical axis, as in the case of an interface. The latter mechanism can be thought of as the nonlinear equivalent of linear reflection from an interface. Similarly, $\chi^{(3)}$ reflection is stronger when the materials on each side of the interface have very different third-order susceptibilities, with the strongest signals observed when the $\chi^{(3)}$ of one the materials is vibrationally resonant. A third mechanism of contrast in epi-CARS is linear reflection or scattering of the forward propagating signal. The latter mechanism is particularly important in CARS microscopy studies of highly scattering media such as tissues.

7.5 CONCLUSION

In this chapter we highlighted the similarities and differences between spontaneous Raman and CARS spectroscopy. The much stronger signals in CARS are the result of the active pumping of the Raman mode in a coherent fashion. Although the CARS signal is based on the Raman activity of molecular oscillators, the CARS spectrum is very different from the spontaneous Raman spectrum. The main reason for this is the existence of an electronic nonresonant background in CARS, which interferes with the vibrationally resonant CARS field. Because of the much shorter signal integration times, CARS is very well suited for vibrational microscopy of live biological samples. The reduced dimensions of signal generation in tight foci give rise to new effects not seen in conventional CARS spectroscopy. In particular, the epi-phase matching conditions under tight focusing allow observation of structures smaller than the wavelength in the backward direction. Applications of CARS microscopy techniques to cellular biology are discussed in Chapter 17.

REFERENCES

Ahkmanov, S.A., A.F. Bunkin, S.G. Ivanov, and N.I. Koroteev. 1977. Coherent ellipsometry of Raman scattered light. *JETP Lett.* 25:416–420.

Begley, R.F., A.B. Harvey, and R.L. Byer. 1974. Coherent anti-Stokes Raman spectroscopy. *Appl. Phys. Lett.* 25:387–390.

Born, M., and E. Wolf. 1999. Principles of Optics. Cambridge University Press, Cambridge.

Boyd, R.W. 2003. Nonlinear Optics. Academic Press, San Diego.

Brakel, R., and F.W. Schneider. 1988. Polarization CARS spectroscopy. *In* Advances in Nonlinear Spectroscopy. Clark RJH, Hester RE, editors. John Wiley & Sons Ltd., New York. 149.

Butcher, P.N. 1965. Nonlinear Optical Phenomena. Engineering Experiment Station, Ohio State University, Columbus.

Cheng, J.X., L.D. Book, and X.S. Xie. 2001. Polarization coherent anti-Stokes Raman scattering microscopy. *Opt. Lett.* 26:1341–1343.

Cheng, J.X., E.O. Potma, and X.S. Xie. 2002a. Coherent anti-Stokes Raman scattering correlation spectroscopy: Probing dynamical processes with chemical selectivity. *J. Phys. Chem. A* 106:8561–8568.

Cheng, J.X., A. Volkmer, and X.S. Xie. 2002b. Theoretical and experimental characterization of coherent anti-Stokes Raman scattering microscopy. *J. Opt. Soc. Am. B* 19:1363–1375.

Cheng, J.X., and X.S. Xie. 2002. Green's function formulation for third-harmonic generation microscopy. *J. Opt. Soc. Am. B* 19:1604–1610.

Dick, B. 1987. Response function theory of time-resolved CARS and CSRS of rotating molecules in liquids under general polarization conditions. *Chem. Phys.* 113:131.

Evans, C.L., E.O. Potma, and X.S. Xie. 2004. Coherent anti-Stokes Raman scattering spectral interferometry: determination of the real and imaginary components of nonlinear susceptibility for vibrational microscopy. *Opt. Lett.* 29:2923–2925.

Garmire, E., F. Pandarese, and C.T. Townes. 1963. *Phys. Rev. Lett.* 11:160.

Gouy, G. 1890. Sur une propriete nouvelle des ondes lumineuses. *Compt. Rendu Acad. Sci. Paris* 110:1251–1253.

Hellwarth, R.W. 1977. Third-order optical susceptibilities of liquids and solids. *Prog. Quant. Electr.* 5:1–68.

Kleinman, D.A. 1962. Nonlinear dielectric polarization in optical media. *Phys. Rev.* 126:1977.

Kramers, H.A., and W. Heisenberg. 1925. Uber die streuung von strahlen durch atome. *Z. Phys.* 31:681.

Laubereau, A., and W. Kaiser. 1978. Vibrational dynamics of liquids and solids investigated by picosecond light pulses. *Rev. Mod. Phys.* 50:607–665.

Loudon, R. 1978. The Quantum Theory of Light. Oxford University Press, Oxford.

Maeda, S., T. Kamisuki, and Y. Adachi. 1988. Condensed phase CARS. *In* Advances in Nonlinear Spectroscopy. Hester RJHCaRE, editor. John Wiley and Sons Ltd., New York.

Maker, P.D., and R.W. Terhune. 1965. Study of optical effects due to an induced polarization third order in the electric field strength. *Phys. Rev.* 137:A801–818.

Moreaux, L., O. Sandre, and J. Mertz. 2000. Membrane imaging by second harmonic generation microscopy. *J. Opt. Soc. Am. B* 17:1685–1694.

Mukamel, S. 1995. Principles of Nonlinear Optical Spectroscopy. Oxford University Press, New York.

Novotny, L. 1997. Allowed and forbidden light in near-field optics. II. Interacting dipolar particles. *J. Opt. Soc. Am. A* 14:105–113.

Oron, D., N. Dudovich, and Y. Silberberg. 2003. Femtosecond Phase-and-polarization control for background-free coherent anti-Stokes Raman spectroscopy. *Phys. Rev. Lett.* 90:213902, 213901–213904.

Otto, C., C.J.d. Grauw, J.J. Duindam, N.M. Sijtsema, and J. Greve. 1997. Applications of micro-Raman imaging in biomedical research. *J. Raman Spectrosc.* 28:143–150.

Oudar, J.L., R.W. Smith, and Y.R. Shen. 1979. Polarization-sensitive coherent anti-Stokes Raman spectroscopy. *Appl. Phys. Lett.* 34:758–760.

Placzek, G. 1934. Rayleigh-Streuung und Raman-Effekt. *In* Handbuch der Radiologie. Marx E, editor. Akademische Verlagsgesellschaft, Leipzig. 205.

Potma, E.O., W.P.d. Boeij, and D.A. Wiersma. 2000. Nonlinear coherent four-wave mixing in optical microscopy. *J. Opt. Soc. Am. B* 25:1678–1684.

Raman, C.V., and K.S. Krishnan. 1928. A new type of secondary radiation. *Nature (London)* 121:501.

Richards, B., and E. Wolf. 1959. Electromagnetic diffraction in optical systems. II. Structure of the image field in an aplanatic system. *Proc. R. Soc. London, Ser A* 235:358–379.

Scully, M.O., and M.S. Zubairy. 1996. Quantum Optics. Cambridge University Press, Cambridge.

Shen, Y.R. 1984. The Principles of Nonlinear Optics. John Wiley and Sons Inc., New York.

Shirley, J.A., R.J. Hall, and A.C. Eckbreth. 1980. Folded BOXCARS for rotational Raman studies. *Opt. Lett.* 5:380–382.

Siegman, A.E. 1986. Lasers. University Science, Mill Valley, CA.

Uzunbajakava, N., A. Lenferink, Y. Kraan, E. Volokhina, G. Vrensen, and J. Greve. 2003. Nonresonant confocal Raman imaging of DNA and protein distribution in apoptotic cells. *Biophys. J.* 84:3968–3981.

Wilson, T., and C. Sheppard. 1984. Theory and Practice of Scanning Optical Microscopy. Academic Press, London.

Woodward, L.A. 1967. *In* Raman Spectroscopy. Theory and Practice. Szymanski HA, editor. Plenum Press, New York.

Yacoby, Y., and R. Fitzgibbon. 1980. Coherent cancellation of background in four-wave mixing spectroscopy. *J. Appl. Phys* 51:3072–3077.

Zumbusch, A., G.R. Holtom, and X.S. Xie. 1999. Three-dimensional vibrational imaging by coherent anti-Stokes Raman scattering. *Phys. Rev. Lett.* 82:4142–4145

────── # Part III

NONLINEAR OPTICAL INSTRUMENTS FOR
MICROSCOPIC IMAGING AND ANALYSIS

Review and Forecast

Peter T. C. So and Daekeun Kim

While the root of nonlinear optical microscopy lies in the distant past with Maria Göppert-Mayer in the 1930s (Göppert-Mayer, 1931), the application of nonlinear optical microscopy and microanalysis in biology and medicine remained essentially dormant until the paper by Denk, Webb, and coworkers in 1990 that unequivocally demonstrated the potential of this exciting new tool (Denk et al., 1990). Although the theoretical and practical foundations of nonlinear microscopy were previously recognized by microscopists such as Sheppard and coworkers (Sheppard et al., 1977), a number of factors made the 1990s the "right time" for the emergence of this field. The most decisive factor was clearly the invention of robust femtosecond lasers at that time. The innovations in laser sources continues to drive this field today (Chapter 8). Another less considered technical factor that makes nonlinear optical imaging and spectroscopy feasible for biology and medicine is clearly the availability of high-sensitivity and low-noise photodetectors (Chapter 12). Nonlinear processes are inherently "weak" and many biomedical applications of nonlinear optics could not become feasible until measurements with approximately single-photon sensitivity have become routine. An understanding of the image formation theories for nonlinear optics (Chapter 11) and the development of microscopes optimized for nonlinear optical processes (Chapter 10) are also clearly important. The availability of novel biomarkers for nonlinear imaging (Chapter 13) and the understanding of nonlinear photodamage mechanisms (Chapter 14) are two other factors that play crucial roles in the increasing adoption of nonlinear optical methods by the biomedical community.

The development of all successful technologies for biology and medicine follows a similar trajectory. After the inception, an incubation period follows where most work focuses on refining and extending the instrument and the technique. For nonlinear optical microscopy, this period lasted approximately a decade and was pursued by a relatively small number of research groups with strong engineering and physics orientations. In the past several years, the technology has become mature and there has been a broad adoption of this technique by biomedical researchers. The research emphasis of this field is now shifting towards applications and application-driven instrumentation adaptation.

During the incubation period of this field, a number of technical developments in nonlinear optical microscopy were particularly important. First, while fluorescence is arguably the most important and clearly the most widely used modality among non-linear optical methods, other contrast mechanisms are emerging. The dominance of fluorescence in this field is not expected to subside due to fluorescence being one of the strongest nonlinear optical processes, given its resonance nature. Fluorescence further has the advantage of having available many genetically expressible or biochemically sensitive probes. However, the prospect of performing bioimaging with molecular specificity and without adding a label is very attractive. Two of the most promising directions are imaging based on second-harmonic and higher harmonic generation (Chapters 15 and 16) and coherent anti-Stokes' Raman scattering (CARS) methods (Chapter 17).

Second, it is well recognized today that bioimaging that focuses only on morphology without biochemical or genetic specific information is ultimately limited in this post-genomic era. One of the most fruitful areas is in the development of nonlinear optical analysis methods for biological specimens. Many spectroscopic assays, traditionally performed in cuvettes using ex vivo biofluids, are now being adapted for studies inside living cells and tissues. The investigation volume is confined to a subfemtoliter region by the inherent localization property of the nonlinear excitation processes. Nonlinear optical microscopy has been combined with spectroscopic assays such as fluorescent emission spectroscopy (Chapter 19), fluorescent lifetime spectroscopy (Chapter 21), and fluorescent resonance energy transfer (Chapters 22 and 23). There are also a number of new techniques where the subfemtoliter interaction volume defined by nonlinear optics is critical in providing sufficient sensitivity for the assay, such as fluorescence correlation spectroscopy (Chapter 25), or in providing unique 3D-resolved information, such as nonlinear photobleaching recovery (Chapter 26), 3D-resolved photoactivation of biochemicals (Denk, 1994; Denk et al., 1990), and 3D-resolved nanoprocessing (Chapter 27).

Third, as the biology is driven towards high-content imaging to unravel the complex regulatory network in cells and tissues, as pharmaceutical and biotechnology industry are driven towards high-throughput drug discovery (Buehler et al., 2005) and high-content assays to monitor drug–biosystem interactions, high-throughput, high-content nonlinear optical microscopic imaging methods have been developed (Chapter 18).

Fourth, nonlinear optical processes can also be applied to manipulate molecular systems on the quantum level that were not possible previously. One of the most important works in this area is the application of nonlinear processes to break the diffraction barrier and to perform optical imaging in the far field with tens of nanometer resolution (Chapter 24).

As the technique for nonlinear optical microscopy reaches maturity, where are the areas of opportunity for future technology development? It is always difficult to predict the future, but there are three major areas that should see major growth over the next decade. The first major area is the quantum-level manipulation of molecular states. In addition to the elegant work by the Hell group in high-resolution imaging (Hell & Wichmann, 1994; Kittel et al., 2006; Willig et al., 2006), the selective access of specific molecular states based on coherent control is also extremely promising (Dela Cruz et al., 2004; Dudovich et al., 2002).

A second growth area lies in the continuing development of nonfluorescence-based nonlinear optics modalities. Second-harmonic imaging has found its applications in the imaging of collagen and muscle structures. There are now promising clinical applications in studying cancer metastasis based on extracellular matrix degradation and in studying muscular dystrophy. CARS imaging may have just found its major breakthrough in the study of demyelination diseases (Kennedy et al., 2005; Wang et al., 2005a). It is clear that new applications of these techniques are just emerging and will required more optimized imaging systems that fully utilize the unique features of these new modalities. Other nonlinear contrast mechanisms are also possible, and promising ones including nonlinear luminescence imaging based on the excitation of surface plasmon resonance in metal nanoparticles (Wang et al., 2005b).

Finally, another challenging but important area for new instrumentation is the drive to develop clinical diagnosis techniques based on nonlinear optical microscopy. Clearly, the need for more flexible and versatile delivery methods for ultrashort light pulses is critical, and the development of photonic bandgap fibers provides new opportunities (Chapter 9). However, this is clearly not sufficient, as many clinical applications require robust and miniaturized devices that can reach different internal organs in noninvasive procedures. Therefore, the nonlinear optical microscopy community must utilize the rapid advances of the micro-electro-mechanical systems (MEMS) field to create millimeter-size-scale nonlinear microscopes. Equally importantly, many clinical applications will demand nonlinear imaging to examine tissue depths that are beyond the current reach today.

What are the new ideas to extend the depth of imaging while maintaining submicron-level resolution? Although harmonic generation imaging and CARS imaging are very powerful methods, they are severely limited by the type of molecules that can provide sufficiently strong signal and contrast. What are other medically important endogenous molecules that can be used with these modalities, and what are the prospects of developing exogenous molecular probes specifically for these methods?

The development of instruments for nonlinear optical microscopic imaging and analysis has provided exciting research opportunities. Through the ingenuity and collaborative effort of many research groups, a rich set of tools are now available for the use of biomedical researchers and are solving fundamental problems in many diverse areas ranging from neurobiology to cancer biology. There is no doubt that novel devices and techniques will continue to fuel new biological and medical applications.

REFERENCES

Buehler, C., J. Dreessen, K. Mueller, P. T. So, A. Schilb, U. Hassiepen, K. A. Stoeckli and M. Auer (2005). "Multi-photon excitation of intrinsic protein fluorescence and its application to pharmaceutical drug screening", Assay Drug Dev Technol 3(2): 155–67.

Dela Cruz, J. M., I. Pastirk, M. Comstock, V. V. Lozovoy and M. Dantus (2004). "Use of coherent control methods through scattering biological tissue to achieve functional imaging", Proc Natl Acad Sci U S A 101(49): 16996–7001.

Denk, W. (1994). "Two-photon scanning photochemical microscopy: mapping ligand-gated ion channel distributions", Proc. Natl. Acad. Sci. U S A 91(14): 6629–33.

Denk, W., J. H. Strickler and W. W. Webb (1990). "Two-photon laser scanning fluorescence microscopy" Science 248(4951): 73–6.

Dudovich, N., D. Oron and Y. Silberberg (2002). "Single-pulse coherently controlled nonlinear Raman spectroscopy and microscopy", Nature 418(6897): 512–4.

Göppert-Mayer, M. (1931). "Über Elementarakte mit zwei Quantensprüngen", Ann Phys (Leipzig) 5: 273–94.

Hell, S. W. and J. Wichmann (1994). "Breaking the diffraction resolution limit by stimulated emission: stimulated-emission-depletion fluorescence microscopy", Optics Letters 19(11): 780–2.

Kennedy, A. P., J. Sutcliffe and J. X. Cheng (2005). "Molecular composition and orientation in myelin figures characterized by coherent anti-stokes Raman scattering microscopy", Langmuir 21(14): 6478–86.

Kittel, R. J., C. Wichmann, T. M. Rasse, W. Fouquet, M. Schmidt, A. Schmid, D. A. Wagh, C. Pawlu, R. R. Kellner, K. I. Willig, S. W. Hell, E. Buchner, M. Heckmann and S. J. Sigrist (2006). "Bruchpilot promotes active zone assembly, Ca2+ channel clustering, and vesicle release", Science 312(5776): 1051–4.

Sheppard, C. J. R., R. Kompfner, J. Gannaway and D. Walsh (1977). "The scanning harmonic optical microscope". IEEE/OSA Conference on Laser Engineering and Applications, Washington.

Wang, H., Y. Fu, P. Zickmund, R. Shi and J. X. Cheng (2005a). "Coherent anti-stokes Raman scattering imaging of axonal myelin in live spinal tissues", Biophys J 89(1): 581–91.

Wang, H., T. B. Huff, D. A. Zweifel, W. He, P. S. Low, A. Wei and J. X. Cheng (2005b). "In vitro and in vivo two-photon luminescence imaging of single gold nanorods", Proc Natl Acad Sci U S A 102(44): 15752–6.

Willig, K. I., S. O. Rizzoli, V. Westphal, R. Jahn and S. W. Hell (2006). "STED microscopy reveals that synaptotagmin remains clustered after synaptic vesicle exocytosis", Nature 440(7086): 935–9.

8

Laser Sources for Nonlinear Microscopy

John Girkin

8.1 OVERVIEW

Since the first images were recorded by multiphoton excitation, probably the single biggest improvement has been in the laser sources. The initial experiments using colliding pulse dye lasers (1) seem a long way removed from the compact, reliable femtosecond (fs) sources now routinely used for nonlinear imaging. The current range of sources available vary in wavelength, tunability, and pulse duration and in more mundane but critically important matters such as cost, reliability, and ease of use. Before examining the operation of mode-locked lasers and the wide range of sources now available, the actual requirements on the laser source for multiphoton imaging should be considered. Throughout this chapter "multiphoton" will be used as a term for all forms of nonlinear excitation microscopy.

In the simplest case two-photon excitation is proportional to the square of the photon density; thus, if one can halve the pulse duration of the excitation source, one has the potential to increase the signal by four times. In practice, though, the actual pulse is present for only half the time, and hence the overall effect is that the two-photon excitation rate is approximately inversely proportional to the pulse width. The net result is that in general it is advantageous to use as short a pulse as possible in order to maximize the peak power for a given average power at the sample. However, as is so often the case, there is a limit imposed by another physical phenomenon. Due to the large spectral width of ultrashort pulses (as will be discussed later), group velocity dispersion in the optical train in the microscope and scan head has the effect of stretching the pulses and hence lowering the peak power. As this pulse-stretching effect is related to the original pulse length, one tends to find that in practical situations the optimal pulse length for most imaging is around 70 to 200 fs.

Imaging has been undertaken with significantly longer pulses, and picosecond multiphoton imaging has been accomplished successfully (2); taking the pulse length to the limit, continuous-wave multiphoton imaging has been demonstrated (3). In general, with the longer pulse lengths one is more likely to suffer from thermal problems within the sample, as the average power has to be increased to obtain a sufficient number of detectable excitation events.

One also needs to consider the repetition rate of the laser, which for mode-locked lasers is determined by the laser cavity geometry. The interval between laser pulses should be longer than the fluorescence lifetime of the fluorophore being imaged;

otherwise, excited molecules are being "wasted." Typical lifetimes for the commonly used fluorophores are around 5 ns, which would indicate a repetition frequency of 200 MHz to ensure a maximum fluorescence yield before unwanted saturation effects start to take place. However, the higher the repetition rate of the laser, the lower the pulse peak power for a given average power on the sample, assuming a constant pulse width. If one were operating well away from the saturation point of the fluorophore, this would mean that an 80-MHz laser would produce 2.5 times as many excitation events than a 200-MHz laser (due to the higher peak power). However, as one approaches saturation (every available fluorescent molecule being excited), the 200-MHz system would eventually produce 2.5 times more events than the 80-MHz source. It appears as though the 80-MHz repetition rate of most commercial lasers (originally due to the use of a 4-ft argon tube for pumping dye lasers) is fortuitously close to the optimal compromise. The added complication, and in the end the overriding consideration, is to ensure that the excitation source does not compromise the biological events being observed. Eventually the excitation source will cause damage to the sample either through heating (higher repetition rates) or phototoxic events (higher peak laser powers).

The remaining consideration for the laser source is the wavelength. The most commonly used laser sources for nonlinear microcopy at present, Ti:sapphire lasers, are broadly tunable from 700 to 1000 nm, giving them the ability to excite a wide range of fluorophores. At present the other commercial wavelength sources are around 1,050 nm, which can be used for both three-photon excitation of UV fluorophores (4) or two-photon excitation of the longer wavelength dyes. A more detailed consideration of practical excitation wavelengths is presented elsewhere in the book.

8.2 MODE-LOCKING

Lasers are normally considered to emit a continuous output of monochromatic light, but for a short pulse laser the output contains a wide range of wavelengths (frequencies) in a repetitive series of bursts. In a conventional laser cavity only certain "modes" are allowed, as the waves circulating within the cavity must fulfill a mathematical formula to ensure stable and constant feedback (Fig. 8.1a). In such a laser a number of these modes oscillate, but there is no relationship between the various frequencies. This is similar to a number of church bells all ringing together, each one operated independently of the next, with the result being a steady cacophony. If, however, the allowed modes within the laser cavity can be forced to oscillate with a constant relationship between themselves, they will add up at one point to produce a very intense pulse (see Fig. 8.1b), which can circulate around the cavity. The more modes (wavelengths) that can be made to oscillate together, the shorter the pulse and hence the broader the optical spectrum. For a typical 100-fs pulse one has a spectral output of around 15 nm, which is large compared to the 0.01-nm output from a typical argon ion laser. In the church bell analogy, when they are all forced to swing together, one hears a series of intense chords with silence in between. The process of forcing the various cavity modes to operate together is known as mode-locking.

In order to achieve this in a laser system, one has to fulfill two requirements. Firstly, the laser gain must have a broad spectral output, to support as many oscillating modes as possible, and secondly there must be some mechanism within the laser so that the

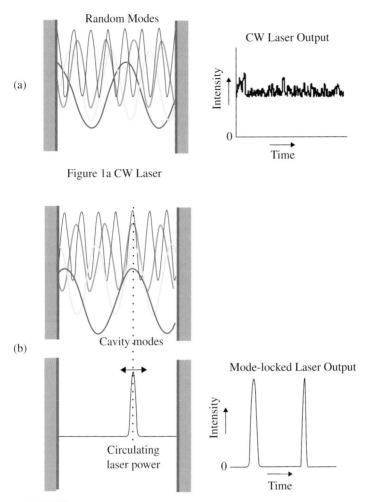

Figure 8.1. Laser modes and output power in a continuous wave laser illustrates the effect of locking the phase between the permitted cavity modes. In (a) the modes oscillate at random, leading to a continuous output, with minor intensity fluctuations. (b) Laser modes and output power in a mode-locked laser. A constant phase relation between the modes is established, giving rise to a single high-intensity pulse circulating within the cavity.

preferred operation is with the laser modes locked together—in other words, so there is a selection process preferring an intense pulse ahead of a continuous output. The first laser materials used to produce ultrashort pulses were organic dyes. Such systems could routinely produce pulses of a few picoseconds, but it was hard to operate them below a picosecond, and they were most efficient in the visible portion of the spectrum. In addition to these technical limitations, they were difficult to maintain, and most laboratories using such sources had large stains on the wall where the high-pressure dye jets had developed rapid and spectacular leaks! In the mid-1980s a new range of broadly tunable crystals became available, in particular titanium-doped sapphire crystals (Ti:Al$_2$O$_3$). These systems operate from 700 to nearly 1,100 nm in the near-infrared (5) and ushered in a new generation of lasers suitable for mode-locked operation.

8.3 METHODS OF MODE-LOCKING

Having outlined the broad principles behind the operation of mode-locked laser sources, we now need to consider how such a physical effect can actually be introduced into a laser cavity.

8.3.1 Active Mode-Locking

Active mode-locking takes place when a component placed within the laser cavity introduces a periodic loss, as illustrated in Figure 8.2. The modulator is driven at the cavity mode spacing frequency (c/2L) such that when the laser light reaches the modulator there is a large loss for most of the time. Light that experiences a low loss will pass once round the cavity and the pulse will again experience a low loss as the modulator will again be open when the pulse returns. After many successive round trips, all of the light will eventually be compressed to pass through the modulator at low loss. Active mode-locking provides a universal technique for the generation of short pulses, but the pulse compression mechanism is not strong; in fact, as the pulses become shorter and shorter in time, the mode-locking mechanism becomes progressively weaker, typically limiting the pulse length to around 10 ps. It should also be noted that when the active device is switched off, the laser will return to continuous-wave operation. An intercavity modulator can, however, be used to instigate the now more commonly used technique of Kerr lens mode-locking, which although not strictly an active method requires a disturbance to the cavity optics to initiate the effect.

8.3.2 Kerr Lens Mode-Locking

Kerr lens mode-locking is based upon the optical Kerr effect first reported by John Kerr in Glasgow in 1875. He observed that when an electric field is applied to a transparent material, the optical properties of the material could be altered. In the case of a DC electric field, the polarization of the light can be altered as it passes through the material,

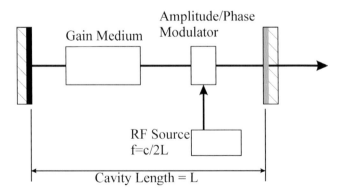

Figure 8.2. Block diagram description of active mode-locking. The loss within the cavity is modulated such that it is high except for a short time at a frequency equal to the time taken for light to travel around the cavity, permitting the formation of a single pulse.

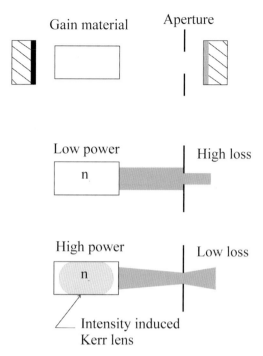

Figure 8.3. Principle of Kerr lens mode-locking. At low power there is insufficient intensity in the light beam to create a significant Kerr effect within the gain material. At high powers the AC electric field from the light causes a higher refractive in the more intense central portion of the beam, effectively producing a positive lens. Thus, by using an aperture in the cavity, only high-intensity pulses experience low loss.

while with an AC electric field the refractive index is altered. At sufficiently high intensities the AC electric field associated with the light itself can change the refractive index. The Kerr effect depends on the square of the electric field and thus, in the situation of interest for laser designers, the intensity profile of the laser beam. At the center of a laser beam the intensity is at its highest, and hence the effective refractive index is larger (in most materials), decreasing as one moves towards the edge of the beam. The result is that the light at the center of the beam travels a greater optical distance, and hence a positive lens has appeared in the cavity (6). Figure 8.3 illustrates how this lens is then used to ensure that the operation with the highest lensing effect (most intense and therefore shortest pulse) can be made to oscillate in preference to a continuous output. The practical requirements for such a system will be considered below as Kerr lens mode-locking is the most common method employed for commercial Ti:sapphire lasers.

8.3.3 Additive Pulse Mode-Locking

An alternative use of the Kerr effect to mode-lock lasers is in the use of additive pulse mode-locking using an optical fiber as the nonlinear Kerr material (7). This technique is also regarded as active mode-locking as the mode-locked output will be lost if the electronic locking mechanism is switched off. A schematic diagram is shown

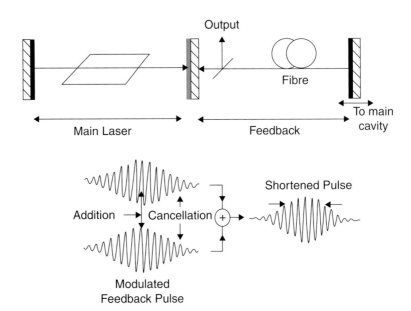

Figure 8.4. Principles of additive pulse mode-locking. When the laser cavity and the feedback cavity are the same length, then one has constructive interference for a short period of time as the pulses overlap. This short pulse then preferentially circulates around the cavity, leading to high-power short pulses.

in Figure 8.4. Due to the natural amplitude fluctuations in a laser cavity, at some time a higher-intensity pulse will leave the laser cavity and enter the nonlinear (fiber) cavity. Here the pulse will experience the nonlinear Kerr effect, causing what is known as "self-phase modulation." If one considers the pulse intensity in time, the leading edge of the pulse is shifted down in frequency and the trailing edge is shifted to a higher frequency. The pulse has therefore been "chirped," and if the nonlinear cavity is the same length as the laser cavity, the chirped pulse will meet the remains of the original pulse that has passed once round the laser cavity. These two pulses can then be made to interfere constructively where the two pulses have the same frequency.

At the ends of the pulse the frequencies will be different (as one has been chirped by the fiber) and one will not have constructive interference. One thus has an enhancement of the pulse in terms of intensity and a shortening (outside the region of constructive interference) and on returning back through the fiber again the chirp will be greater (due to the higher power). This will mean that the constructive interference will take place over a shorter period and hence an ever-decreasing pulse length. In the end the pulse length is limited either by the gain of the laser medium or the dispersion in the cavity. The main advantage of this technique is the very powerful compression process, but the disadvantage is the need to align and maintain two cavities with interferometric accuracy.

One commercial system used this technique where Nd:YLF was the laser gain material. In this laser system the output from the laser directly is around 4 ps at 1,047 nm, but by passing the output through a second single mode fiber, further self-phase modulation takes place, increasing the frequencies present within the pulse. The pulse then has an increased spectral bandwidth still spread over the few-ps time period. One can, however, use pulse compression techniques to delay the leading edge of the pulse in time such that the back end "catches up," resulting in a shorter pulse. The optical delay of the leading pulse edge is normally achieved using a pair of diffraction gratings (8).

8.3.4 Passive Mode-Locking

Passive mode-locking was one of the first mode-locking techniques to be explored and is now re-emerging as a technique of choice for the new sources currently under development in laser laboratories. In the simplest form a material is placed in the cavity that absorbs the laser light at low intensities, but at higher intensities the absorption saturates and the transmission rises (Fig. 8.5). If the laser gain and absorption are well balanced the laser will oscillate, but due to the fluctuations in the laser power, a higher-power pulse will start to oscillate around the cavity. This pulse will notice less loss through the saturable absorber and hence will start to dominate the laser, stripping the gain as it passes through the laser gain region at the expense of the lower-power modes, and eventually resulting in a single pulse containing all of the available power.

In the original passively mode-locked laser the saturable absorber was a dye, but recently this has been replaced with a semiconductor quantum well structure (9). This has the advantage of being "clean" but has to be designed for a specific wavelength, reducing the "tunability" of the source but making mode-locked operation easier, as there are now few adjustments to be made and the system is self-starting. Such passively mode-locked lasers have been used successfully for multiphoton operation (10), and although now available on a commercial basis they have not become well established. It is likely that should multiphoton imaging become a diagnostic tool for clinicians, then such sources are likely to be the only practical and commercial route to such applications (11).

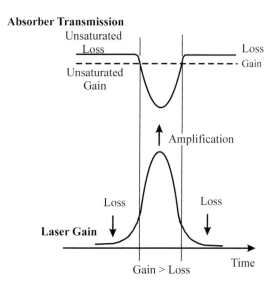

Figure 8.5. Transmission and gain in a laser cavity containing a saturable absorber. This figure shows a plot of the round-trip gain of a laser with a saturable absorber. With low power one has high loss, but for high powers the absorption saturates, giving low loss and high gain within the cavity and preferentially allowing high-power short pulses of continuous operation.

8.4 PRACTICAL MODE-LOCKED LASERS

8.4.1 Kerr Lens Mode-Locked Ti:Sapphire Lasers

Having now considered, in outline, the principles behind mode-locked laser operation, we will consider how such technical requirements may be implemented in a practical system.

For laser operation, Ti:sapphire crystals require optical excitation to establish the gain in the crystal, and this is their one main drawback, as this excitation is required from a blue, or preferable green, laser source, originally a large-frame, water-cooled, argon laser operating at 514 nm. Such systems require around 30 kW of three-phase electrical power from the wall, 99% of which is then used to heat up the water and the working environment. All laboratories that have used such a source have numerous stories of water leaks, electrical supply problems, and frequent calls to service engineers. In the mid-1990s the large-frame ion laser largely became obsolete as a source of green laser light as diode-pumped frequency-doubled Neodymium (Nd)-based lasers became routinely available.

These all-solid-state sources utilize an infrared laser diode (at 810 nm) to optically excite a Nd-doped crystal, with the output from this laser then being frequency-doubled in a nonlinear crystal to produce light at around 532 nm. With an operation efficiency of around 10% to 20%, these lasers operate directly from a standard electrical supply and can be cooled using closed-loop portable chiller units with, in principle, lifetimes of 10,000 hours operation, and have made nonlinear imaging a practical biological tool.

A typical configuration for a Ti:sapphire laser is illustrated in Figure 8.6. The output from the pump laser is focused into the Ti:sapphire crystal and the laser cavity is then formed using external optics around the crystal. Generally the Ti:sapphire crystal is water-cooled to reduce the thermal-induced lensing and thus ensure stable operation. To induce mode-locking within the cavity via the Kerr lens route, a disturbance is needed so that the laser output develops a slightly higher peak power. One commercial company (Coherent) achieves this using a pair of glass plates that wobble when the system is not mode-locked, inducing a perturbation into the cavity and ensuring the development of a Kerr lens in the system. As shown in Figures 8.3 and 8.6 a slit is then used such that only the laser mode with the induced Kerr lens can circulate through the cavity (in a similar way to a confocal pinhole allowing only the detectors to see light from the focus of an objective lens). Once mode-locked operation has been established the glass plates are held stationary.

In an alternative formulation (Spectra Physics) the glass plates are replaced by an acousto-optic modulator that gives a constant "kick" to the system, ensuring that one pulse has a preference over the random modes of the laser. This "seed" is then fed into the main cavity, and the induced optical effect in the main cavity ensures continual mode-locked operation. This technique is known as regenerative mode-locking, but the actual short-pulse generation is due to the nonlinear effects induced in the Ti:sapphire crystal due to the high peak powers, not the acousto-optic modulator.

With mode-locked operation established within the laser, there are still two other critical issues to consider, both related to the wavelength. A Ti:sapphire laser is capable of routine operation from around 700 to 1,000 nm, and selection of the center operating wavelength is normally accomplished using a birefringent or Lyot filter (12). This

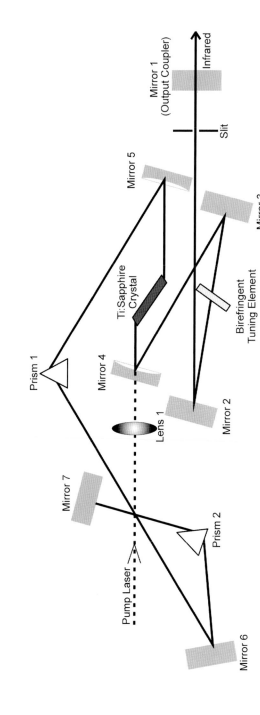

Figure 8.6. Typical layout for a commercial Ti:Sapphire laser. (Figure courtesy of Coherent Inc.)

moves the central wavelength of the cavity to the required position. However, as this happens the position of the slit has to be altered as the different wavelengths propagate at slightly different positions around the cavity due to the dispersive elements in the optical design that are required for short-pulse operation. As has been stated above, fs pulses possess spectral widths of around 10 to 20 nm. All optical elements tend to be dispersive, and hence the optical spectrum of the pulse is spread out in time (as some wavelengths will travel further through the cavity and hence take longer to return to the same point). This means that although the optical pulse has the required spectral width, the blue end of the pulse and the red end of the pulse may not emerge from the cavity together, resulting in a pulse that is temporally longer than it should be.

This effect is overcome by adding prisms into the cavity that are adjusted such that by the time the light pulse has traveled round the cavity once, the various spectral components of the pulse will all have traveled the same optical distance, and hence are grouped together to produce a short pulse. The disadvantage of this is that as the center wavelength of the laser is tuned, the prisms need to be altered to ensure that there is the correct compensation within the cavity for that "group of wavelengths" to all experience the same optical distance. This is known as dispersion compensation, and the effect is discussed later in relation to compensating for the dispersive effects of other optical elements (notably the objective lens) once the fs pulses have left the laser cavity.

When all of these considerations are designed into a practical laser, one typically has a system with around 10 control knobs, which can be daunting to many non-laser experts. For most of the time only two adjustments are required, and as long as each knob is adjusted separately one can avoid constant calls to the service engineers. More recently the demand from the life science market has led to the development of "user-friendly," no-adjustment systems by both major laser companies. All the cavity adjustments are made under computer control, resulting in a "hands-free" system.

8.5 PULSE SHAPES FROM MODE-LOCKED LASERS

8.5.1 Laser Pulse Shape

Considerable thought and time has been spent by laser physicists in producing fs laser systems with the "perfect" laser pulse, and several excellent general review articles are available on the subject (13). It is not the aim of this text to explore the minutiae of laser physics involved in such sources but to provide practical guides to biological researchers.

In a "perfect" mode-locked laser system, the laser pulse is said to be "transform limited." This means that the product of the pulse width and pulse duration is a number that is constant. In principle this seems an easy undertaking, but actually measuring and defining the pulse duration is a nontrivial task, as certain assumptions must be made over its shape, but several guiding principles are worth understanding. For a "good" mode-locked pulse, the time evolution should be symmetrical around the peak intensity. Section 8.8 in this chapter briefly mentions instrumentation to measure the duration of these very short pulses, with all of the methods inferring the pulse length rather than measuring the duration with an ultrafast stopwatch.

In order to characterize fully the laser pulse, one observes both the spectral profile and time evolution of the pulse and then relates these two using the mathematics mentioned above. A typical fs pulse spectral profile is shown in Figure 8.7a. The symmetrical shape demonstrates that the correct quantity of dispersion compensation has been included in the laser cavity and that the system is well adjusted. If too much or insufficient dispersion is present, then the asymmetrical profile shown in Figure 8.7b is the result. By adding or removing glass in the optical train through the use of the dispersion-compensating prisms discussed above, the correct profile can be restored. The spectrum can appear to become very "spiky," as illustrated in Figure 8.7c. The system is now "Q-switching," as there is insufficient loss in the cavity to ensure that mode-locked operation is taking place. In this regime the output is a series of pulses, each of which contains a number of

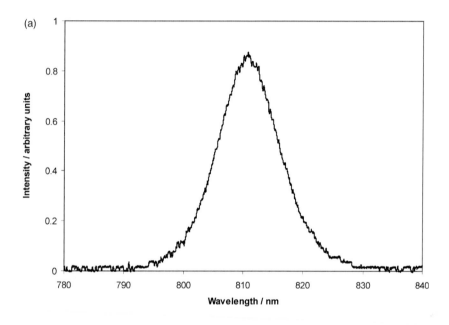

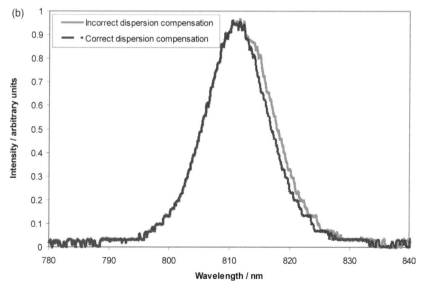

Figure 8.7. Continued

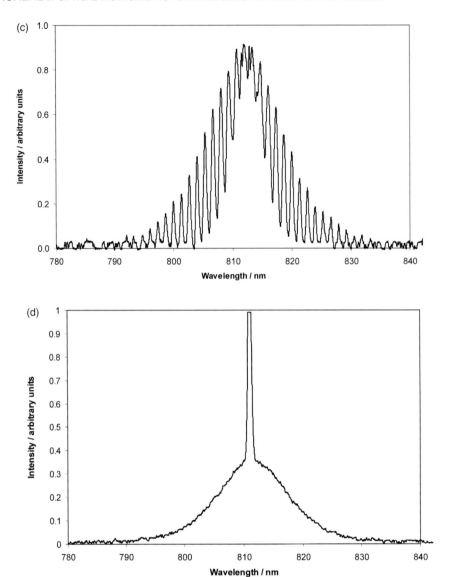

Figure 8.7. Spectral profiles obtained from a Ti:Sapphire laser running under various conditions. (a) Well mode-locked. (b) Asymmetric profile with incorrect dispersion compensation. (c) Q-switched/mode-locked operation. (d) CW and mode-locked operation.

fs pulses inside the overall envelope. This mode of operation can normally be corrected by adjustment of the slit controlling the laser mode allowed to oscillate around the cavity. The other common occurrence is that the laser output contains a continuous-wave component as well as the fs pulses. This is illustrated in Figure 8.7d and is seen as the spike superimposed onto the correct spectral profile. This can be removed either by decreasing the pump power or altering the slit width.

   The question, however, should be asked whether these less-than-perfect pulses affect the multiphoton imaging process. Generally the answer is yes, as the spectrally incorrect pulses indicate that the pulse length is longer than it might be and hence the peak power at the sample will have fallen, resulting in a reduction in the detected signal. This tempts the operator to increase the average power on the sample to correct for the loss of signal, going against the general rule of minimizing power on the sample. If the system is running in the Q-switched mode-locked mode, then the energy in the packet of mode-locked pulses can be very high, causing damage to the sample through ablation or localized heating. The effect of a continuous signal on the mode-locked pulse train again tends to increase local thermal load within the sample.

8.5.2 Dispersion in the Optical Train

The foregoing discussion was primarily concerned with the performance of the laser system, but many of the effects mentioned above can also caused by the optical elements present before the pulses reach the sample. The most significant consideration in the transport of fs laser pulses is the effect of dispersion caused by the broad spectral width of the light pulses. Technically this effect is known as "group velocity dispersion" (GVD). In nonmathematical terms all optical materials, from different glasses to optical coatings, have refractive indices that vary with the wavelength of light. This effect in more conventional optical microscopy causes chromatic aberration and through the use of different glasses can be overcome. A rigorous explanation of the effect can be found in standard optical textbooks (14,15).

In the case of fs pulses the solution is not quite so straightforward and the effects can be significant, especially if the light is sent through a long optical distance through glass (for example, in an optical fiber). The spectral width of fs pulses is dependent on both the center wavelength and the pulse length. Around 800 nm, a 60-fs pulse has a spectral width of around 30 nm. If such a pulse enters a glass with a refractive index that changes significantly with wavelength, this pulse can easily be stretched to around 200 fs in a thickness equivalent to a lens (16). This is not, however, the end of the story. As this now-lengthened pulse progresses through the optical system, it may now have a longer pulse length but it still has the same spectral width, and it will continue to suffer from GVD as it moves through the optical train. This can mean that pulses that start at 60 fs may be 500 fs at the sample.

If, however, one started with a slightly longer pulse, with therefore a smaller original spectral bandwidth, then the lengthening effect will not be so significant. For a 150-fs pulse traveling through the same block of glass, one would expect the pulse to lengthen to around 200 fs but perhaps only be 250 fs at the sample. This idea of pulse lengthening not being related to the pulse length but the spectral width is slightly counterintuitive but can be a significant effect in certain multiphoton applications. The effect can be overcome through the use of pulse compensation techniques. The principle is straightforward in that the original fs pulse is stretched in time in one direction so that after it has passed through the complete optical train, the trailing edge of the pulse has caught up with the leading edge. The net effect of this is to once again have the original pulse length on the sample.

Dispersion compensation schemes are not, however, easy to align and maintain. The dispersion has to be readjusted for each center wavelength, and some power is lost passing through the extra optical elements. As a general rule it is more practical not to consider such schemes unless one has to work with the shortest possible pulses at the sample, either to keep the average power low or simply to maximize the peak power and hence the signal for very deep imaging. This is also the reason why fs pulses are best "moved around" in air rather than through optical fibers, and this is before consideration is given to the nonlinear effects that can take place described above in the section on mode-locking methods.

Several schemes have been developed (17) in which optical fibers can be used as the delivery system. Typically these systems "chirp" the pulse before it is made to enter the optical fiber, perhaps with a pulse length of around 2 ps. This has the effect of reducing the peak power and hence unwanted nonlinear effects within the fiber (self-phase modulation, as discussed earlier). The dispersion within the fiber then starts to "shrink" the pulse in time as it progresses along the fiber. The output pulse length is normal around 500 fs and a final block of glass applies the remaining compensation for the originally applied chirp. All of the compensation cannot be undertaken in the fiber as the peak power will rise along the fiber and hence self-phase modulation will take place towards the output end of the fiber. The output from the fiber can be collimated and a large spot can be used through the final glass block to reduce unwanted nonlinear effects. Even using such a complex scheme, the maximum reliable power that can be delivered is around 200 mW for 100-fs pulses, and the system has to be altered for each center wavelength.

8.6 CURRENTLY AVAILABLE COMMERCIAL LASER SYSTEMS

As was stated at the outset of the chapter, the most significant recent advances for the nonlinear microscopist have been the improvements that have been made in fs laser systems (18). The two major laser companies, at the time of writing, Coherent and Spectra Physics (19), both supply excellent mode-locked Ti:sapphire lasers suitable for multiphoton imaging applications. The two systems operate in slightly different modes and have different methods of initiating the mode-locking as described above, but both work! The systems are currently supplied with mirror sets that allow the laser to operate between around 720 and 980 nm, which has been the most recent innovation for conventional Ti:sapphire lasers. Beyond these regions it is advisable to change the mirrors for ones enhanced to operate at the extremes of the gain curve of the laser system.

As important as the Ti:sapphire laser itself is the pump laser. Most systems being purchased currently use all-solid-state sources with output powers from 5 to 10 W, and it is really up to personal preference, and budget, as to the source selected. All of these pump sources are self-contained and require only a conventional electrical mains connection to operate.

In addition, Nd-based lasers are available, with the original system being designed specifically as a simple-to-use source for multiphoton microscopy (7). Operation is straightforward, as there are in essence no external adjustments to be made except to set an electrical level using an oscilloscope. The sources are again self-contained and in more recent versions can be operated in temperatures up to 40°C.

In the past 2 years simpler-to-operate Ti:sapphire lasers have become available. The aim with these sources is to make the system operate with minimal user intervention. Currently Spectra Physics and Coherent have systems that can be controlled using a computer to tune from around 710 to 1,000 nm, with no user-adjustable controls except the output power and the wavelength through the computer interface.

With both of these systems there is one potential drawback: they are supplied as hermetically sealed units with no user-controllable components. In fact, there are very few adjustments that can be made by the service engineer, and so if the system does fail it has to be returned to the factory for repair. The compensation for this is systems that can be operated with no laser expertise.

8.7 FUTURE SOURCES

The ultimate goal for the multiphoton microscopist is to have a laser that is easy to operate and low in cost and has a long lifetime. As has been discussed above, the various laser companies have made significant advances, but the ideal system is still not here, though new options are emerging.

8.7.1 Passive Mode-Locking Using Saturated Bragg Reflectors

The principle of this mode-locking technique has been described above. A passive mode-locking element is inserted into the cavity based upon an optical semiconductor (9) with a loss of tunability but a significant reduction in complexity of the laser system. Where these simple devices are useful is in their application to lasers with limited tunability and where the system can be directly pumped by a laser diode. Two systems have been explored in detail to this end.

One crystal that can be made to operate from 800 to 900 nm is chromium-doped into lithium, strontium, aluminum fluoride (Cr:LiSAF). This material has absorption around 670 nm and so is suitable for direct optical excitation using high-powered laser diodes. A typical configuration is shown in Figure 8.8. The output from the laser diode has to be shaped using a series of lenses in order to create the correct spot size in the laser crystal. The saturable Bragg reflector (SBR or SESAM) is mounted at a focus at one end of the laser cavity. In a typical SBR design around 32 pairs of high- and low-refractive index material are used as the mirror, with one or more quantum wells placed on top of the mirror. For higher-gain laser material, several wells are required in order to produce sufficient loss in the laser cavity. The cavity then contains the standard pair of dispersion-compensating prisms and a slit to encourage the laser to operate at the wavelength preferred by the SBR mirror. Depending on the SBR design, the laser can be made to operate at a center wavelength from around 820 to 870 nm. It should, however, be noted that each laser operates over only a limited (perhaps 20 nm) center wavelength range.

The technique is self-starting once a certain threshold power is reached on the SBR. For a single 500-mW pump diode, the output is typically 50 mW, with pulses around 100 fs (11). Higher output powers are possible through the use of two or more pump diodes, but there is a practical limit in Cr:LiSAF lasers as the crystals have poor thermal conductivity, and hence at higher powers crystal damage can ensue.

Direct diode pumping of Nd glass lasers also produces pulses of around 100 fs with up to 500 mW of output power using an SBR as the mode-locking element. Again, the power limitation is due to thermal constraints in the glass material, and a range of ingenious designs are being explored in order to increase the removal of heat from

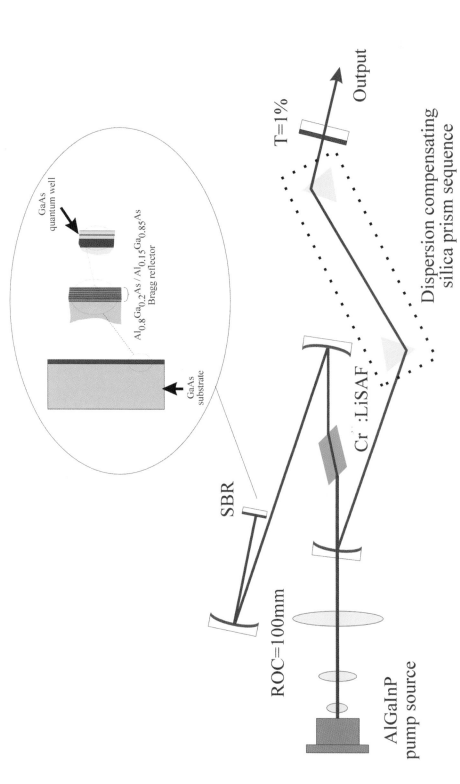

Figure 8.8. A diode-pumped Cr:LiSAF laser operating around 850 nm using a semiconductor saturable absorber as the mode-locking element. Inset shows the structure of the absorber.

the system. These systems all operate around 1,050 nm, though the exact host glass determines the precise operating wavelength.

The advantage of these systems is that once aligned, the laser can be used without adjustment for long periods of time. If a material can be found to operate at the correct wavelength for a particular application, the SBR mode-locking technique does open up the possibility of applying multiphoton techniques in a clinical setting (20). The other advantage of the technique is that the component cost of the system can be as low as $5,000, which is significantly lower than that required for the ubiquitous Ti:sapphire laser, in which the pump laser alone is in excess of $38,000.

8.7.2 Optically Pumped Semiconductor Lasers

In the past few years there has been significant interest in a new range of laser systems. In so-called VECSELs (Vertical External Cavity Surface Emitting Lasers), the gain of the laser is provided by an optically pumped semiconductor arranged in an external cavity (21,22). In a structure similar to the SBR devices discussed above, a Bragg mirror is grown onto a semiconductor substrate. A series of quantum wells are then grown onto this mirror and the entire structure is "capped" with a final semiconductor layer. Through careful design of the Bragg mirror and gain region, the laser can be made to operate at a selected wavelength region with typically around 20 to 30 nm of tunability. As the wafer is grown using conventional semiconductor techniques, these sources offer the possibility of "mass production."

VECSELs should not, however, be seen as a panacea for all laser problems. In order to obtain good mode-locking at suitable powers, the excess pump power in the semiconductor device has to be removed. Due to the structure of the device, heating causes the center wavelength of the mirror to move while the peak gain from the quantum wells moves in the other direction. This means that as the pump power to the laser is increased, the output power starts to fall beyond a certain power. In principle this problem is surmountable and has been demonstrated for CW systems through the use of modern engineering and thermal management techniques (23). In addition, each laser will operate at a limited wavelength range only, and so the technique should be thought of as being capable of supplying wavelength-selectable rather than wavelength-tunable sources.

Although systems have been shown to operate in a continuous manner, mode-locked pulses have yet to fall below a few hundred fs (24,25). The present limitation is mainly due to the thermal problems mentioned above and the gain region acting as a wavelength-selective element, reducing the spectral gain of the laser, along with dispersion problems within the gain element. There are also issues of energy storage within the gain region, and currently a number of routes are being investigated around these limitations. The concept of a low-cost, wavelength-selectable source costing under $10,000 is not just a dream for the multiphoton microscopist and may be realized in the near future.

The laser systems described above are, in general, commercially available and used routinely in multiphoton imaging systems. There are, however, several techniques that have a potential for use in multiphoton imaging systems. The techniques have been demonstrated but are a long way from practical and commercial exploitation.

8.7.3 Optical Parametric Oscillators

The use of nonlinear materials in conjunction with mode-locked lasers has enabled the generation of pulsed radiation, suitable for multiphoton imaging, at wavelengths not easily achieved by conventional sources. Some nonlinear techniques have been used routinely in multiphoton systems where the second harmonic of a Neodymium laser is used to pump the Ti:sapphire laser as described earlier. Another nonlinear technique (which provides tunable radiation) is the use of optical parametric oscillators (OPO) synchronously pumped by a mode-locked source, which are commercially available as additions to standard Ti:sapphire lasers.

A generic OPO is shown in Figure 8.9. An intense optical field (provided by a pump laser source) with a frequency ω_3 is applied to a nonlinear crystal. The birefringent nature of the crystal gives rise to the creation of two new optical fields, namely the signal (ω_1) and idler frequencies (ω_2), and by convention $\omega_1 > \omega_2$. By the principle of energy conservation $\omega_3 = \omega_1 + \omega_2$. The signal and idler frequencies generated through the application of the pump light are dependent on the properties of the nonlinear medium. In theory there exist an infinite number of signal and idler frequencies that can satisfy the equation. In practice, however, phase-matching conditions limit the frequencies that can actually be generated. A full detailed description of this process is beyond the scope of this book but can be found in standard texts on nonlinear laser optics (for example, 26). The exact wavelengths that can be generated depend on the pump wavelength and the crystal material and orientation.

By the use of an optical cavity around the nonlinear crystal, it is possible to resonate one, or more, of these frequencies and hence increase the conversion efficiency. In the simplest system the OPO is described as being singly resonant, as the cavity is matched to just one of the pump, signal, or idler frequencies. When two of these are made to resonate together, the system is known as a doubly resonant OPO. By resonating two or more of the beams, the threshold for the system falls, but the alignment, and more crucially maintenance of the alignment, becomes significantly more complex. The efficiency of this generation process depends crucially on the incident pump laser power, and ultrashort pulses offer high peak powers suitable for efficient conversion.

In order to maintain the higher power in the crystal, the length of the cavity around the nonlinear crystal is selected such that the time taken for the pulse to travel once round the cavity is the same as the separation of the pump laser pulses. This means that the power in the nonlinear crystal is always as high as possible, as well as maintaining the ultrashort nature of the initial pump pulse. Through the careful selection of the crystal (e.g., KTP,

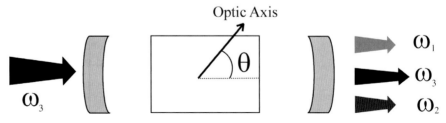

Figure 8.9. Principles behind an optical parametric oscillator. A pump source with an optical frequency ω_3 is converted into two new frequencies ω_1 and ω_2

LBO, BBO, or more recently periodicals polled lithium niobate [PPLN] (27, 28)) and the correct pump laser wavelength, it is possible to produce light from around 400 nm to 3 μm. Recently multiphoton images at 620 nm using a mode-locked Nd:YLF laser and PPLN have been recorded for several short-wavelength UV-excited fluorophores, including tryptophan, NADH and indo, with the longer-term aim of generating light suitable for efficient multiphoton uncaging of cage compounds (29).

8.7.4 Photonic Bandgap Optical Fibers

At the time of writing, photonic bandgap (PBG) optical fibers have just been demonstrated as sources for multiphoton imaging (30). In a conventional optical fiber the light is confined to the central core through differences in the refractive index of the glasses or polymer making up the fiber (i.e., by total internal reflection), but a different principle is at work in a PBG-based fiber.

In its simplest form, if one introduces an array of tiny air holes into a dielectric material, photons within a particular wavelength range will not propagate through the otherwise transparent material. This effect is known as the photonic bandgap and is best understood by analogy with the physics of semiconductors, where the interaction of electrons with a crystal lattice produces allowed energy states and a prohibited bandgap region. The full explanation can be derived using complex quantum mechanical calculations and can be found in several review articles (31–33).

To produce a photonic bandgap, the air holes need to be separated by a distance roughly equal to the photon wavelength divided by the refractive index of the dielectric, which implies a spacing of around 250 nm for light at 750 nm in silica. The width and properties of the bandgap then depend on the difference in the refractive index of the two materials, with a larger contrast giving rise to a wider bandgap. As with the electrical analogy, when light is injected into these devices close to the bandgap cutoff points, a number of nonlinear effects start to take place within the fiber. It is thus possible for a fs pulse with an initial spectral width of around 20 nm to be spread to produce a continuum of white light from around 400 to 650 nm (34). Such a broad spectrum can then be compressed using conventional grating pairs (as described earlier) to produce very short pulses that have had their initial wavelength shifted. As a concept this could mean that a single-wavelength fs laser source could have an array of such fibers at the output (the lengths of fibers need only be several mm in length) and by switching between the fibers a tunable source could be produced. In addition, such fibers could be used for dispersion compensation (35) with the longer-term potential for fiberoptically coupled multiphoton microscopes. This class of material offers excellent opportunities for the multiphoton microscopist in the future.

8.8 PRACTICAL ADVICE FOR USE OF FEMTOSECOND LASERS

8.8.1 Operating Environment

In order to maximize the use from a fs laser source (or alternatively to prevent unnecessary calls to the local service engineer or friendly laser physicist), a few simple rules can be followed that in the long term can save a great deal of time, frustration, and potentially cost. Like most pieces of complex scientific equipment, fs lasers are best operated in a stable environment.

Ideally the complete multiphoton system should be mounted on a large optical table (with vibration isolation legs). The laser and microscope should both be on the same surface; provided there is space available in the laboratory, select as large a table as possible; the space will soon be used. For a typical manual-tuning Ti:sapphire laser (including the pump laser) and scan head, one ideally requires an 2.4×1.2 m (8×4 feet) surface, with the table having a thickness of around 200 mm (8 inches). Although vibration isolation may not be required in some environments, the small additional cost will reduce the source of potential problems.

The laboratory itself should ideally be temperature controlled to between 20 to 25°C, with a stability of around 1°C. It should be remembered that this should be a true air conditioning system and not one that just cools! The relative humidity should be kept below around 50%, though this is not critical. The heat load in the laboratory will vary considerably. Typically air-cooled argon lasers (for use with the system in a conventional confocal mode) will produce 5 to 10 kW of heat, and a modern diode-pumped frequency-doubled Nd laser (the Ti:sapphire pump source) can produce in excess of 3 kW of additional heat.

8.8.2 Diagnostic Equipment

The following is not meant to provide an extensive description of all of the equipment needed to run and maintain a fs laser system for multiphoton microscopy, but rather to provide outline description of the operating principles. Detailed descriptions of various forms of these items are available in a number of references or from potential suppliers. It is also assumed that the user will purchase a laser power meter suitable for measuring the average power from the source, and the laser supplier often provides these.

8.8.2.1 Auto-Correlators for Pulse Length Measurement

One of the most crucial figures to the laser physicist using fs lasers is to know their exact pulse length. Although this is of less direct interest to the biological user, there are times when access to such an item is useful. However, measuring the exact pulse length of a burst of light only lasting 100 fs is a nontrivial exercise. You cannot just direct the light into a fast photodiode and display the results on an oscilloscope, as no electronics or detectors react fast enough.

All auto-correlators work on the same basic principle of splitting the incoming pulse, delaying one of the beams, and then recombining the two beams and detecting how they are overlapped in time (Fig. 8.10). The incoming laser light (only a few mW at most is required) is directed onto a thin beam splitter (in order to ensure that the beam splitter does not lengthen the pulse). One of the resulting beams is then directed along an optical path to the detector. The second passes through an optical delay line in which the pulse of light is made to travel a variable extra distance compared to the first beam. The second beam is then overlapped with the direct beam at the detector. The detector is nonlinear in that it does not see any light unless both beams are present at the same time. Typical methods here involve the use of a frequency-doubling crystal in which no second harmonic light is present from one beam as the light intensity is too low, but if

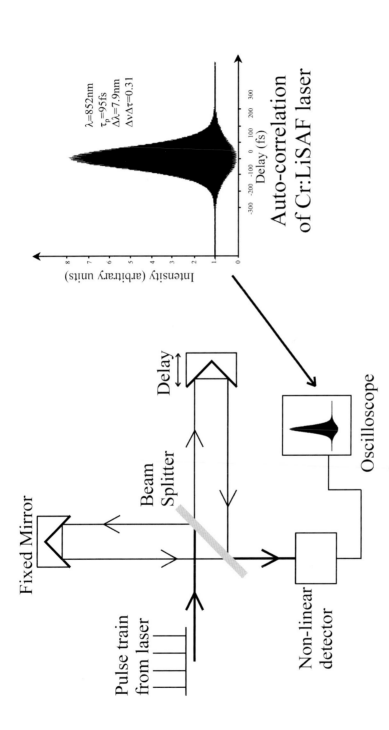

Figure 8.10. Basic building blocks of an autocorrelator. The incoming beam is split into two and then recombined on a nonlinear detector. When the two beams arrive at the detector together, there is sufficient power to produce a signal; when the extra distance traveled by one means that it arrives after the other, then no signal is detected and the optical delay can be converted into the temporal profile of the beam.

both beams are present then the second harmonic is generated and detected either on a photomultiplier tube or an avalanche photodiode. An alternative detector is a nonlinear photodiode as described below.

The optical delay in one beam is then altered by moving a mirror, or retro-reflector cube, and the detected signal is plotted on an oscilloscope. A physical movement of 1 micron is equivalent to a time delay of 3 fs; thus, by moving the cube over 100 microns, one can observe the time profile of a 300-fs pulse (though in practice typical movements are several mm).

A full detailed description of the information that can be gleaned from examining the pulse profile is beyond the scope of this chapter but can be found in most texts on femtosecond lasers (e.g., 26). More advanced systems, but still generally based around the principle outlined above, are also possible but are of more interest to the fs laser physicists than a routine user of such sources.

8.8.2.2 Spectrometers for Laser Diagnosis

As has been described above, fs pulses have a broad spectral width, and it is significantly easier to measure the spectrum of the pulse rather than its actual pulse length. Therefore, the most useful instrument to a user of such laser sources is a simple optical spectrometer. Most laser suppliers, or the multiphoton microscope supplier, suggest that such a small spectrometer is built into a system. A small fraction of the incoming beam is split off (under 1 mW) and directed into a spectrometer with a resolution of around 1 nm. Typically these spectrometers either use a revolving grating (Reece Instruments Ltd.) or a fixed grating and the optical spectrum is then viewed using a linear photodiode array of CCD camera (Ocean Optics Inc., or Avantes Inc.).

The shape of the spectrum provides a great deal of practical information on the performance of the laser, and these have been illustrated in Figure 8.7. All of the fs lasers in the author's laboratories have spectrometers of one form or another attached to the laser systems, and these provide the day-to-day performance of the laser. If an image disappears from the screen, a quick glance at the spectral output provides information on whether the problem is with the laser or elsewhere with the experiment. In addition, the spectrometer provides an accurate method of tuning the laser to the required wavelength, as the calibrated micrometer supplied with the laser should not be relied upon.

8.8.2.3 Photodiodes as Two-Photon Detectors

When undertaking any form of quantitative biological or other scientific investigation, it is important to ensure that the conditions when experiments are repeated are the same. When this requirement is placed on multiphoton imaging, there are a very large number of variables present that need to be measured. In respect to the imaging, the optical system can be assumed to be the same, but the laser parameters can change both from day to day and even during a single day. The critical requirements on the laser source are:

1. Wavelength
2. Average power
3. Pulse length

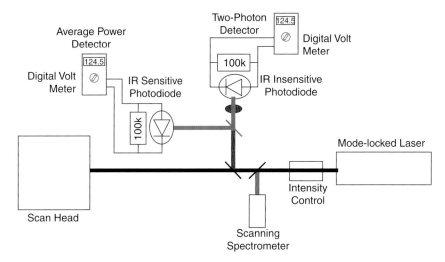

Figure 8.11. Circuit details for a combined average-power and peak-power detection system using photodiodes. Infrared sensitive gives average power and visible sensitive, via two-photon excitation, provides peak power response.

The instruments described above provide a method of measuring the wavelength (spectrometer), average power (power meter), and pulse length (auto-correlator), but the latter is not a trivial measurement, and they are not suitable for measuring the pulse length at the sample (after the objective). A simple system is used by the author to monitor the two-photon dose on the sample based upon nonlinear absorption in a semiconductor device (36).

As illustrated in Figure 8.11, a GaAsP infrared blind photodiode (e.g., Hamamatsu G1116) is mounted in a holder behind a focusing lens (12-mm diameter, focal length between 10 and 30 mm). A small portion of the incoming laser beam is sent into the detector before the beam enters the scan head but after any intensity-controlling optics (neutral density filters, AOTFs or Pockels' cell). A 100-k ohm resistor is mounted across the diode and a 200-mV digital voltmeter is connected across this output. When the Ti:sapphire pulses are incident on the detector, the fundamental gives no signal on the meter, but when the focal volume is matched to the depletion region of the diode, then two-photon excitation of electrons into the conduction band takes place and a signal can be detected. This voltage is directly related to the peak power in the same way as the two-photon excitation fluorescent signal in the sample. As the pulse lengthens, so the voltage falls. Through the correct choice of resistor or focusing lens, up to four decades of useful range can be found. This gives a quantified measure of the output from the laser related to the pulse length, and with a measurement of the average one-photon power, the laser output is characterized.

A second diode can then have its outer "can" removed and be mounted, with a similar resistor and meter, after the objective, and this will then give a measurement of the two-photon "dose" on the sample. By undertaking a simple calibration process changing the power and wavelength, the dose on the sample stage can be linked to that measured by the first detector. The stage detector can then be removed and the user can now make a measurement of the incoming two-photon dose and relate this to the actual sample dose. In the author's multiphoton system, such a device is mounted for every laser source, along with a small infrared-sensitive photodiode, which has been calibrated to measure the average power.

8.8.3 Deciding on a Commercial System

To most multiphoton users, the selection of the microscope and laser provides a daunting challenge, and probably the selection of the laser provides the most worry, as the potential user is unlikely to be an experienced researcher of fs laser sources. The following section aims to provide a few guiding principles to aid the first-time buyer through the potential pitfalls of the process.

Determine which fluorophores are going to be used both in the short term and long term. If the system is for a single user who uses only one fluorophore, then a single-wavelength source may be the most suitable. However, care should be exercised for longer-term projects. If you are not going to use dyes that need light below 750 nm or above 930 or 950 nm, do not place these on the wish list. At these extreme wavelengths for the Ti:sapphire laser, higher-power pump sources are required and potentially optics changes in the laser cavity. In addition, above 920 nm, due to water absorption, the laser cavity has to be purged with clean, dry nitrogen or air.

Consider the advantages of a computer-tuned versus manual system. In principle they are easier to operate but are more expensive and may have a more limited tuning range. If something goes wrong, the system probably needs to be returned to the factory for repair. In this case a purchasing deal including a loan laser from the supplier may be worth considering.

Discuss with local purchasers which company provides the best after-sales service. In most situations once the salesman has made the sale it will be the service engineers who are potentially the most frequent visitors. They will also install the system and explain the operation to the users and are thus crucial to your understanding of the system. In different areas within the same country the local engineer may be different, and this should be considered.

Seriously consider a service contract purchased at the time of sale of the main system. This can save problems later on.

The most likely fault (apart from the user misaligning the system) is for the pump laser to fail. Although all Ti:sapphire lasers are now sold with diode-pumped frequency-doubled Nd lasers as the pump source, the cost of replacement diodes for these systems varies (typically lifetimes are in excess of 5,000 hours, and so this is likely to happen after the service contract expires). In some systems the diode replacement can be undertaken simply by the user, and this means that no realignment of the laser and microscope system is required.

Request user training as part of the purchase deal, preferably in your own lab. This should consist of two parts, firstly shortly after installation (or at installation) and then about a month later, when people have used the system and have developed a second list of questions, or wish to move on to more advanced techniques.

- In summary, all of the commercially systems currently available work for multiphoton imaging.
- Be sure in your own mind what is required, and do not allow yourself to be "talked up" into accepting a higher-specified system than you really need.

- Discuss the deal with the company or university purchasing department, who should be able to help with obtaining a good price for a complete package.
- Find out about the service engineers in the area from other users (which may mean going to see colleagues in a different department within a university, for example).
- Discuss systems with other users (website discussion boards and past correspondence play an important role here).
- Obtain quotations from more than one company, but examine what is being offered in full and do not automatically take the cheapest deal.

When all of these points have been considered, the user should be in a position to make an informed decision on the selection of the laser.

ACKNOWLEDGMENTS

The author wishes to acknowledge funding support from the Wellcome Trust, SHEFC, JIF, EPSRC, and University of Strathclyde and very helpful discussions with David Wokosin, Allister Ferguson, Gail McConnell, and Amanda Wright.

REFERENCES

[1] W Denk, J H Strickler, W W Webb, "Two-photon laser scanning fluorescence microscopy", Science 248:73–76 (1990).

[2] P E Hannienen, M Schrader, E Soni, S W Hell, "Two-photon excitation fluorescence microscopy using a semiconductor laser", Bioimaging 3:70–75 (1995).

[3] P E Hannienen, E Soni, S W Hell, "Continuous wave excitation two-photon fluorescence microscopy," J. Micros. 176:222–225 (1994).

[4] D L Wokosin, V E Contonze, S Crittenden, J G White, "Three photon excitation fluorescence imaging of biological specimens using an all-solid-state laser", Bioimaging 4 1–7 (1997).

[5] P F Moulton, "Spectroscopic and laser characteristics of Ti_2O_3", JOSA B 3 125–133 (1986).

[6] D E Spence, P N Kean, W Sibbett, "60-fsec pulse generation from a self-mode-locked Ti:Sapphire laser", Opt Lett 16 42–44 (1991).

[7] G P A Malcolm, A I Ferguson, "Mode-locking of diode laser-pumped solid-state lasers", IEEE Opt and Quantum Elec 24 705–717 (1992).

[8] A Robertson, N Langford, A I Ferguson, "Pulse compression of the output from a diode pumped additive-pulse mode-locked Nd:YLF laser", Opt. Comm. 115 516–522 (1995).

[9] S Tsuda, W H Knox, S T Cundiff, W Y Jan, J E Cunningham, "Mode-locked Ultrafast Solid-State Lasers with Saturable Bragg Reflectors", IEEE Journal of Selected Topics in Quantum Electronics, Vol 2 No 3 465–472 (1996).

[10] K Svoboda, W Denk, W H Knox, S Tsuda, "Two-photon-excitation scanning microscopy of living neurons with a saturable Bragg reflector mode-locked diode-pumped Cr:LiSAlFl laser", Optic. Lett. 21 1411–1413 (1996).

[11] J M Girkin, D Burns, M D Dawson, "Macroscopic multiphoton biomedical imaging using semiconductor saturable Bragg reflector mode-locked lasers", *Proc. SPIE*, 3616, 92–97 (1999).

[12] W Demtroder, "Laser Spectroscopy" Springer-Verlag 3rd Edition 2002 (for example).

[13] H Kapteyn, M Murnane, "Ultrashort Light Pulses", Physics World 12 31–35 (1999).

[14] R S Longhurst, "Geometrical and Physical Optics", Longman 3rd Edition (1974).

[15] M Born, E Wolf, "Principles of Optics", Cambridge University Press, 7th Edition 1999.

[16] J B Guild, C Xu, W W Webb, "Measurement of group delay dispersion of high numerical aperture objective lenses using two-photon excited fluorescence", App. Opt. 36:397–401 (1997).

[17] B W Atherton, M K Reed, "Pre-chirped fiber transport of 800 nm 100 fs pulses" Proc SPIE 3269 22–25 (1998).

[18] J M Girkin, "Optical Physics Enables Advances in Multiphoton Imaging", J Phys D 36 R250–258 (2003).

[19] Coherent Inc, Santa Clara, CA, USA, Spectra Physics Inc, Mountain View, CA, USA.

[20] J M Girkin, A F Hall, S Creanor, "Two-photon Imaging of Intact Dental Tissue", Dental Caries 2 317–325 (2000).

[21] M Kuznetsov, F Makimi, R Sprague, A Mooradian, "High-power (>0.5-W CW) diode-pumped vertical-external-cavity surface-emitting semiconductor lasers with circular TEM_{00} beams" Photon, Tech Lett. 9 1063–1069 (1997).

[22] M A Holm, D Burns, P Cusumano, A I Ferguson, M D Dawson, "High-power diode-pumped AlGaAs surface-emitting laser", Appl. Opt 38 5781–5784 (1999).

[23] J E Hastie, J M Hopkins, S Calvez, C W Jeon, H W Choi, D Burns, M D Dawson, R Abram, E Riis, A I Ferguson, "Characteristics of highpower VECSELs with Silicon Carbide heatspreaders", CLEO-Europe June 2003.

[24] A Garnache, S Hoogland, A C Tropper, I Sagnes, G Saint-Girons, J S Roberts, "Sub-500-fs soliton-like pulse in a passively mode-locked broadband surface-emitting laser with 100 mW average power" Appl. Phys. Lett. 80 3892–3894 (2002).

[25] J M Girkin, "Novel compact sources for multiphoton microscopy", Progress in Biomedical Optics and Imaging SPIE, 2 186–191 (2001).

[26] A E Siegman "Lasers", University Science Books, 1986.

[27] C T Chen, Y C Wu, A D Jiang, "New non-linear optical crystal- LiB_3O_5", Journal of the Optical Society of America B, 6 616–621 (1989).

[28] J G Hush, M J Johnson, B J Orr, "Spectroscopic and non-linear optical applications of a tunable beta-barium borate optical parametric oscillator", Journal of the Optical Society of America B, 10 1765–1777 (1993).

[29] G McConnell, G L Smith, J M Girkin, A M Gurney, A I Ferguson, "Two-photon microscopy of rabbit ventricular cardiomyocytes using an all-solid-state tunable and visible femtosecond laser source", Optic Letters 28, 1742–1744 (2003).

[30] G McConnell, Private Communication.

[31] R D Meade, A M Rappe, K D Brommer, J D Joanopoulos, O L Alerhand, "Accurate theoretical analysis of Photonics bandgap materials", Phys Rev B 55 8434–8437 (1993).

[32] J Broeng, T Sondergaard, S E Barkou, P M Barbeito, A Bjarklev, "Waveguidance by the Photonic bandgap effect in optical fibers", J Opt. A, 1 477–482 (1999) and references therein.

[33] T A Birks, J C Knight, B J Mangan, P StJ Russell, "Photonics Crystal Fibers: An Endless Variety", IEICE Trans Electron, E84-C 585–592 (2001) and references therein.

[34] J K Ranka, R S Windeler, A J Stentz, "Efficient visible continuum generation in air-silica microstrucutre optical fibers with anomalous dispersion at 800nm" CLEO'99 post deadline paper CPD8, Baltimore, (1999).

[35] T A Birks, D Mogilevtsev, J C Knight P StJ Russell, "Dispersion compensation using single material fibers" IEEE Phot. Tech Lett 11 674–676 (1999).

[36] J K Ranka, A L Gaeta, A Baltuska, M S Pshenichnikov and D A Wiseman, "Autocorrelation measurement of 6fs pulses based on the two-photon-induced photocurrent in a GaAsP photodiode", Opt. Lett. 22 1344–1346 (1997).

Ultrashort Optical Pulse Delivery for Nonlinear Optical Microscopy

Daekeun Kim

9.1 INTRODUCTION

After John Tyndall's demonstration of guiding light with a jet of water in the 19th century, it was shown that light can travel along arbitrary paths through high-refractive index materials (Goff & Hansen, 2002). An optical fiber is generally defined as a cylindrical waveguide where wave can propagate along its axis based on total internal reflection. A fiber has a high index material forming its core at the center and is surrounded by a cladding material, which has a slightly lower refractive index. This refractive index difference enables total internal reflection to occur at the boundary between the core and the cladding when the light is incident within at a given angle, the critical angle. Since the introduction of the fiberscope in 1950s, flexible optical fibers have been widely used in the imaging and telecommunication fields, given its versatility to conduct light into tight spaces where free space delivery is difficult.

There are several drawbacks in using optical fibers, mainly optical loss and dispersion. Due to the commercial significance of fiber technology, considerable effort has been devoted to overcome these limitations for pulse propagation through optical fibers. Optical loss has been dramatically reduced by improving the fiber manufacturing process over the years. Dispersion management has also been done extensively with the use of single-mode step-index fibers or multimode-graded index fibers to eliminate or alleviate intermodal dispersion, which results from optical path differences between propagating modes. Chromatic dispersion is minimized by properly selecting fiber material with the optimal optical window. Since the refractive index is frequency-dependent, choosing the appropriate combination of waveguide dispersion and material dispersion can minimize total dispersion over a broad frequency range.

In 1990, Denk and associates introduced two-photon excitation microscopy (Denk et al., 1990), and it has gained popularity for cellular and tissue imaging over the past decade. It enables intrinsic three-dimensional imaging based on two-photon nonlinear excitation, providing deep penetration into the tissues, subcellular resolution, low photodamage and photobleaching, and high cell viability (So et al., 2000). The term "multiphoton" is now commonly used to describe two- and n-photon fluorescence excitation processes. However, multiphoton imaging using a standard laboratory microscopy with free space light delivery can be limited its applications, especially for clinical or animal studies. The efficiency of multiphoton excitation microscopy significantly improves when ultrashort optical pulses are used, with typical temporal duration on the femtosecond and picosecond range. Ultrashort optical pulses provide high instantaneous optical power essential to efficient nonlinear excitation. However, due to time-bandwidth limit, a large spectral bandwidth is needed to produce a narrow temporal pulse. Since the chromatic dispersion is strongly dependent on the wavelength bandwidth or spectral bandwidth, chromatic dispersion management is critical for the ultrashort pulse delivery.

The majority of the work in delivering ultrashort pulses through fibers is in the telecommunication field. Even though telecommunication and multiphoton excitation have the same objective (to maintain optical pulse shape through the fiber), they have major differences in terms of optical power and operating wavelength. In fiberoptic telecommunication, optical pulse shape is only considered as a major factor of pulse delivery, and dispersion elimination is a key to increase communication bandwidth. However, optical power and dispersion are both crucial for multiphoton excitation microscopy applications. As optical power through a fiber increases, nonlinear effects become more important. These nonlinear effects include self-phase modulation to enlarge spectral bandwidth and broaden temporal pulse width (Agrawal, 2001). Since the optical window, where chromatic dispersion in the fiber is minimized, is far from the near-infrared wavelength needed for multiphoton fluorescence excitation, chromatic dispersion is a concern and can result in lower performance in fiberoptic-based multiphoton excitation microscopy. Although it can be compensated with prechirping techniques (Helmchen et al., 2002), temporal pulses can still broaden due to nonlinear effects.

Recently developed photonic crystal materials have been used in manufacturing optical fibers called photonic crystal fibers. It was shown that the dispersion of ultrashort optical pulses can be significantly reduced by using photonic crystal fibers (Gobel et al., 2004; Kim et al., 2005; Tai et al., 2004). Importantly, photonic crystal fibers also offer better performance than traditional fibers in conducting high-power ultrashort pulses. Despite its promises, the effects of ultrashort pulse delivery using photonic crystal fibers on the performance of nonlinear microscopes still need to be quantified.

In this chapter, we briefly discuss basic physics of ultrashort optical pulse delivery using optical fibers and the experimental design considerations. We further describe experimental characterization of the performance of multiphoton excitation microscopy using fiber delivery. A variety of commonly available optical fibers are characterized and compared based on parameters including spectral bandwidth and temporal pulse width, multiphoton efficiency, and microscope optical resolution. A theoretical discussion regarding the differences in the performance of these fibers is provided.

9.2 EXPERIMENTAL DESIGN CONSIDERATIONS FOR ULTRASHORT OPTICAL PULSE DELIVERY

Figure 9.1 shows the experimental apparatus designed to characterize the performance of a multiphoton excitation microscope when ultrashort pulses are conducted through

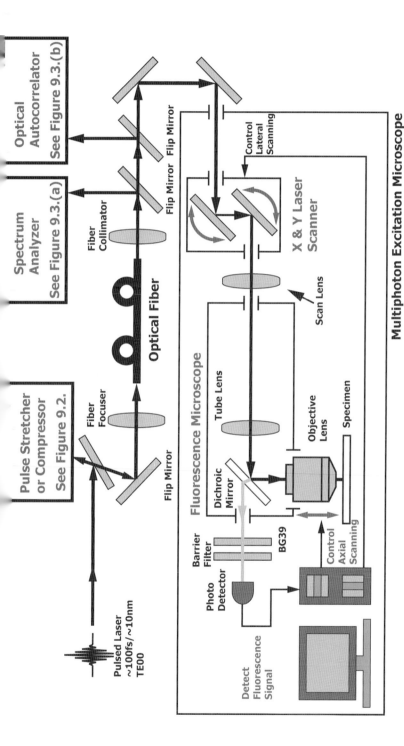

Figure 9.1. An experimental design for measuring ultrashort optical pulse characteristics through optical fibers and their multiphoton microscopy performance. The experimental apparatus consists of a pulse stretcher or compressor (see details in Fig. 9.2), an optical fiber coupling system including a fiber focuser and a fiber collimator, and multiphoton scanning microscopy. The pulse stretcher provides the prechirping in order to compensate for the dispersion associated with fiber delivery. Light rays are focused into the fiber core and are collimated at the fiber output by a fiber focuser and a fiber collimator, respectively. The optical pulse shape can be measured with either a spectrum analyzer or an optical autocorrelator (see details in Fig. 9.3). The delivered optical pulses also go into a multiphoton excitation microscope, where the multiphoton efficiency and the optical resolution are measured.

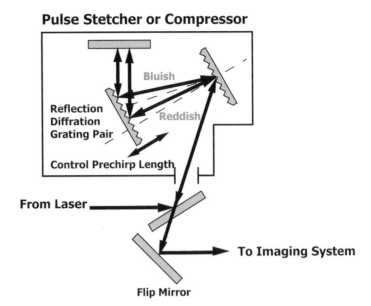

Figure 9.2. A pulse compressor gives dispersion in the opposite direction of normal dispersion. The length between diffractive grating pairs is adjusted to provide enough prechirping for ultrashort optical pulses.

a variety of optical fibers. First, we need an ultrashort optical pulse laser. The pulses should be delivered via a delivery system consisting of a pulse compressor (Fig. 9.2), a fiber focuser and a fiber coupler, and an optical fiber. After the delivery, the optical pulse shape needs to be examined both spectrally and temporally with a spectrum analyzer (Fig. 9.3a) and an optical autocorrelator (Fig. 9.3b). Delivered pulses are used as the light source of multiphoton excitation microscopy. We measure the microscope's two-photon excitation efficiency and its resolution (i.e., its point spread function [PSF]). The following subsections briefly describe the operational principles of these components and how to integrate them for an ultrashort pulse characterization experiment.

9.2.1 Optical Fibers

An optical fiber is a dielectric cylindrical waveguide that conducts light along its axis. There are two different types of optical fibers: total internal reflection (TIR)-based optical fiber and photonic crystal fiber (PCF). TIR-based optical fiber has been routinely used for years, but PCF has been introduced only recently. TIR is an optical phenomenon that occurs when light is incident from a high-refractive-index medium toward a low refractive medium. When the incident angle is sufficiently large (larger than the critical angle), the refraction components of light will vanish at the boundary and all the lights are reflected at the interface. TIR-based optical fiber basically consists of a high refractive index core and a low refractive index cladding. Based on the TIR process, the light travels only inside the core of the fiber if its incident angle is larger than the critical angle at the boundary between core and cladding. Otherwise, it disappears through the cladding. In general, refractive index distribution in TIR-based optical fiber is given by

Spectrum Analyzer

Optical Autocorrelator

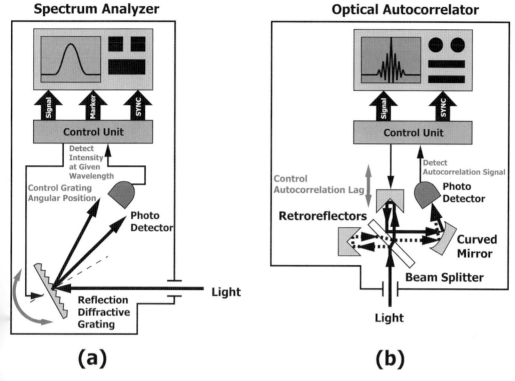

(a)

(b)

Figure 9.3. (a) A spectrum analyzer and (b) an optical autocorrelator. In the spectrum analyzer, the power spectrum over the wavelength range can be obtained by measuring the intensity in the photodetector at a given wavelength selected by a rotating grating. In the optical autocorrelator, interferometric optical autocorrelation is measured by moving one of the retroreflector arms back and forth to delay one of two identical optical pulses. A photodetector measures the two-photon absorption when two optical pulses interact interferometrically.

(Ghatak & Thyagarajan, 1998):

$$n^2(r) = n_1^2 \left(1 - 2\Delta \left(\frac{r}{a}\right)^q\right) \quad for \quad 0 < r < a$$

$$= n_2^2 = n_1^2(1 - 2\Delta) \quad for \quad r > a \qquad (9.1)$$

$$where \quad \Delta = \frac{1}{2}\left(1 - \left(\frac{n_2}{n_1}\right)^2\right)$$

It is usually referred to as the power law profile. r represents the cylindrical radial coordinate, and a corresponds to the radius of the core in the fiber. n_1 is the refractive index on the axis ($n(0) = n_1$), and n_2 is the refractive index in the cladding of fiber, which is homogeneous or constant through the cladding. q is the factor for determining the refractive index profile in the core. At the boundary between core and cladding ($r = a$), TIR happens, since the refractive index of the outer face in the core is a little greater than that of the inner face in the cladding ($n_1 > n_2$). As seen in equation (9.1),

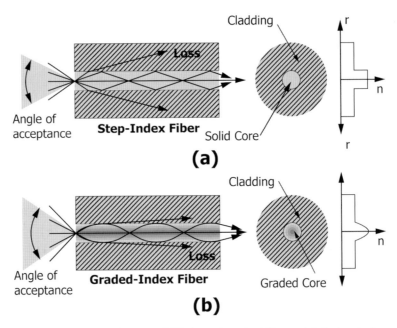

Figure 9.4. Total internal reflection (TIR)-based optical fibers. (a) A step index optical fiber and (b) a graded index optical fiber. The light delivery mechanisms are briefly described (left). The fiber face profiles normal to the optical axis are shown (middle). The refractive index distributions are plotted along the radial direction (right).

depending on the refractive index distribution along the radial direction in the fiber core, TIR-based fibers are categorized into step-index ($q = \infty$) and graded-index fibers ($0 < q < \infty$). Figure 9.4 shows how light travels inside a step-index and a graded-index fiber, and their radial index distribution. As seen in the figure, the refractive index at the core–cladding boundary changes abruptly for step-index fibers, whereas the refractive index in the core changes gradually, and it reaches the refractive index of the cladding for graded-index fibers. Step-index fibers deliver more optical modes; the maximum number of modes to be delivered through the fiber is determined by V number, which is given by

$$V = \frac{2\pi a}{\lambda}\sqrt{n_1^2 - n_2^2} \tag{9.2}$$

where λ is the wavelength.

Single-mode fiber is defined as the optical fiber that transmits only one fundamental mode through the fiber. The condition to be a single-mode step-index fiber is $0 < V < 2.4048$, based on modal analysis. All other step-index fibers are multimode fibers, which deliver at least two modes. For highly multimoded fibers ($V \geq 10$), the number of modes can be approximated to

$$N \approx \frac{q}{2(2 + q)}V^2 \tag{9.3}$$

Therefore, graded-index fibers have a smaller number of modes than step-index fibers, since multimode step-index fiber ($q = \infty$) has $N \approx V^2/2$ modes. When fiber is coupled with rays, they propagate with TIR process inside the core of the fiber only within a certain cone along the fiber-optical axis. This cone is called the acceptance cone, and its half-angle is the acceptance angle. The numerical aperture (NA) in the fiber optics is set to the full angle of the acceptance cone. Fiber coupling has to satisfy both refraction law and TIR condition, and NA in a step-index optical fiber is given by

$$NA = \sqrt{n_1^2 - n_2^2} \qquad (9.4)$$

It is also important to choose the appropriate NA of the rays in order to minimize coupling loss.

Optical fibers work very well as waveguides, but their performance needs to be enhanced in terms of optical efficiency and optical pulse delivery. Three different mechanisms limit the performance of fibers in delivering ultrashort pulses. First is optical loss or attenuation, which results from Rayleigh scattering; absorption induced by the impurities of silica; and coupling loss caused by NA mismatch between fiber and incoming rays. This attenuation has been dramatically reduced for years by better manufacturing processes. Second, dispersion effects consist of intermodal dispersion and intramodal dispersion, including material and waveguide dispersions. Intermodal dispersion originates from optical path differences between modes. Material dispersion results from group velocity dispersion (GVD) induced by the second-order wavelength dependence of refractive index. Waveguide dispersion is due to waveguide effects where the phase and the group velocity of the pulses are functions of fiber geometry and wave propagation constant, but independent of material dispersion. It is insignificant in multimode fiber. Finally, nonlinear effect associated with high power, such as self-phase modulation (SPM), can also limit performance. It is due to the intensity dependence of the refractive index in nonlinear optical medium, and the nonlinear effect is stronger in smaller-core fibers (Agrawal, 2001). Multimode step-index fiber has been used to deliver the optical pulse for telecommunication, since it has both low nonlinear effect and high coupling efficiency. However, intermodal dispersion is one of the most dominant factors to broaden optical pulse due to the many modes in multimode step-index fiber, and it limits optical bandwidth in delivering ultrashort optical pulses.

There are two solutions to alleviate intermodal dispersion: multimode graded-index fiber and single-mode step-index fiber. Using multimode graded-index fiber reduces the number of modes from (9.3) and results in intermodal dispersion reduction. For example, a parabola graded-index fiber ($q = 2$) has only $N \approx V^2/4$ modes, which is half the number of modes in a step-index fiber with the same V number. A single-mode step-index fiber has low coupling efficiency; however, since it has only a single mode, it does not suffer from intermodal dispersion. On the other hand, material dispersion becomes important, and waveguide dispersion is not negligible in single-mode step-index fiber.

PCF has been introduced recently (Birks et al., 1995; Cregan et al., 1999; Knight et al., 1998b), and it has various potential applications in nonlinear microscopy imaging and telecommunications, since it has better optical characteristics than conventional optical fibers in terms of guiding, dispersion, and refraction (Lourtioz, 2005). PCF is a waveguide constructed with photonic crystal, which is dielectric or metallic material with the periodic structure of the refractive index. This material has photonic band gap (PBG), the range of optical frequencies where the light cannot propagate through.

The bandgap location depends on fiber microstructure and can be made independent of the light's propagation direction and polarization. One of the common examples for 1D photonic crystal is Bragg mirror (Zolla, 2005), which contains a periodic pile of alternating dielectric films along the optical axis. Light travels along the optical axis with consecutive reflection and transmission at the face of each film, and wave is fully reflected at a given wavelength range for appropriate film thickness. Extension to two or three dimensions of PBG effect has been partially successful recently. If a defect is inserted in an infinitely periodic microstructure, the defect supports some modes inside the bandgap, which is localized in the outskirts of the defect (Johnson & Joannopoulos, 2002; Lourtioz, 2005). Line defects create a waveguide, whereas point defects generate a resonant cavity. Therefore, the two-dimensional PBG effect with a line defect makes PCF useful as a waveguide. There are many kinds of PCFs, depending on both the core and cladding structures. Figure 9.5 represents the description of light delivery with several types of PCFs. One of the popular PCFs is the solid-core PCF (Knight et al., 1996, 1997), which has a solid core with a microstructured cladding (see Fig. 9.5a). It is sometimes called index-guiding or high-index core fiber. Its cladding consists of silica-air photonic crystal material. Even though it is made of photonic crystal, the PBG

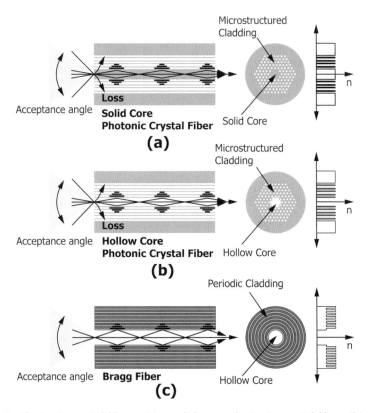

Figure 9.5. Photonic crystal fibers: (a) a solid-core photonic crystal fiber, (b) a hollow-core photonic crystal fiber, and (c) a Bragg fiber. The light delivery mechanisms are briefly described (left). Fiber face profiles normal to the optical axis are shown (middle). The refractive index distributions are plotted along the radial direction (right).

effect does not actually occur during the light propagation. Depending on its structure, effective refractive index in the cladding can be controlled, and it should always be smaller than the refractive index of the core, although the silica refractive indexes for both core and cladding are same. Therefore, the light is confined in the fiber core by a mechanism analogous to TIR called modified total internal reflection (MTIR). These PCFs provide single-mode delivery over the broadband wavelength regardless of core size (Birks et al., 1995; Knight et al., 1996, 1997), and even for an unusually large mode area (Knight et al., 1998a). Other applications of these PCFs are optical soliton formation (Wadsworth et al., 2000), supercontinuum generation (Jusakou & Herrmann, 2001), and dispersion management (Renversez et al., 2003).

Another form of PCF is hollow-core PCF, or low-index core fiber (Cregan et al., 1999). It has a hollow core instead of a solid core, and the cladding is microstructured photonic crystal (see Fig. 9.5b). The light is confined to the hollow core by the PBG effect in the microstructured cladding, even if the refractive index in the core is lower than that of the cladding. Its applications are divided into different classes: removing constraints due to the core material's optical properties, and reinforcing interactions between light and gas filling the core of the fiber. Some of its applications are high optical power delivery with lower nonlinearity, megawatt optical soliton generation with the core filled with xenon gas (Ouzounov et al., 2003), and stimulated Raman scattering in the core filled with hydrogen (Benabid et al., 2002).

Bragg fiber, the third type of PCF, uses Bragg mirror as cladding in the fiber (see Fig. 9.5c). The cladding is made with a concentric periodic distribution of alternating materials and surrounds a hollow or a solid core. It is feasible to deliver the light since it has been discovered that Bragg mirror has omnidirectional reflection that happens based on the PBG effect (Fink et al., 1998; Hart et al., 2002). One of its applications is to improve light transmission over certain frequency ranges (Temelkuran et al., 2002).

One of challenges in PCF research is to invent the PCF fabrication process. Microstructured PCFs can be manufactured by drawing a silica-air hole preform acquired by stacking silica capillaries or by drilling holes, or by extrusion (Knight, 2003; Russell, 2003). Bragg fiber is made with a cigar-rolling technique. Multilayer mirror is rolled up to make a preform with a hollow core, and it is drawn (Temelkuran et al., 2002). PCFs are mostly made of silica and air to form silica-air holes, but they can be also formed with polymer instead of silica (Argyros et al., 2006).

In the fiber characterization experiments described in this chapter, five different optical fibers are evaluated. For step-index fibers, three fibers are selected: a single-mode fiber with 5 μm/125 μm dimensions (core/cladding dimensions), denoted as SMF5um; a slightly larger-core fiber with 10 μm/125 μm dimensions, denoted as QMMF10um; and a multimode fiber with 50 μm/125 μm dimensions, denoted as QMMF50um. This wide selection allows the effect of different core size and different number of propagating modes to be evaluated. Additionally, QMMF10um is reported to enhance optical characteristics of fiber delivery with prechirping (Helmchen et al., 2002). Experiments with this fiber using prechirping are denoted as QMMF10um.PC. Graded-index fiber with 50 μm/125 μm dimension, denoted as MMF50um, is used to investigate nonlinear effects, compared with step-index fiber. Hollow-core photonic crystal fiber with 6 μm/125 μm dimensions, denoted as HCPCF6um, is also used in our experiments. We choose a fiber length of 2 meters for these characterization experiments since it is long enough for nonlinear effects to be important. This length is also reasonable for microscopic applications instead of telecommunications, where a length of hundreds of kilometers of fiber is needed to transmit signal. Every fiber except for hollow-core photonic crystal fiber has FC/APC connectors on both ends to avoid back-reflection to the light source. We use just bare fiber without connectors in hollow-core photonic crystal fiber. This is because the air-silica hole structure, which has a major role in the PBG effect, may be contaminated during the polishing process needed in connecterizing the fiber.

9.2.2 Ultrashort Pulsed Laser

To generate ultrashort optical pulses, the Ti:sapphire (Ti:Sa) tunable pulsed laser is the most common light source and is used in these characterization experiments (Tsunami, Spectra-Physics). The Ti:Sa laser was introduced in 1986 (Moulton, 1986). It is pumped with continuous wave (CW) diode-pumped solid state (DPSS) laser (Millennia V, Spectra-Physics). Titanium-doped sapphire (Ti^{3+}:Sa), a crystal of sapphire doped with titanium ions, is used as transition-metal–doped gain medium. It is tunable over a broad range in the near-infrared (NIR) wavelength range. It delivers continuously tunable output with wavelength range from 690 to 1,080 nm and also produces the optical pulse width as short as 50 fs with a pulse repetition rate of $\sim$80 MHz. It uses active mode-locking (Rullière, 2005), instead of passive, with an acousto-optical modulator (AOM) to produce chirp-free ultrashort optical pulses. It supplies several Watts of average power, depending on the capability of the pumped laser. In our experimental design, we choose $\sim$100-fs ultrashort optical pulses with the center wavelength of 780 nm as a light source. The pulse width is sufficiently short to provide efficient two-photon excitation. The wavelength is selected since it is suitable for the excitation of many endogenous and exogenous fluorophores in biological specimens.

9.2.3 Pulse Compressor

GVD is the most dominant factor responsible for the broadening of optical pulses as they travel through highly dispersive medium. GVD is more severe when shorter pulses are used (Agrawal, 2001). It characterizes the wavelength dependence of the group velocity in a medium. GVD is expressed as the dispersion parameter, $D(\lambda)$, time delay per unit length of material along the optical axis per unit width of wavelength at a given center wavelength ($ps/km/nm$). It is also calculated as follows:

$$D(\lambda) = -\frac{\lambda}{c}\left(\frac{d^2n}{d\lambda^2}\right) \qquad (9.5)$$

where λ is the wavelength and c is the speed of light.

There is no GVD if D is zero. If D is negative, the medium has positive or normal GVD; otherwise, it has negative or anomalous GVD. In addition, waveguide dispersion dominance in a small mode area such as a single-mode fiber should also be considered, and total chromatic dispersion is a combination of both dispersions. To alleviate the chromatic dispersion problem, prechirping, which counterbalances linear chromatic dispersion, can be applied. A negative chromatic dispersion can be imposed before fiber delivery when the medium has a positive chromatic dispersion, and vice versa.

A pulse compressor can be used in some cases. Since optical pulse broadens as it is transmitted through the dielectric optical components, such as the objective, in the microscope system, the broadening effect can be reduced by compensating dispersion with a diffractive grating pair (Agostinelli et al., 1979; Tai & Tomita, 1986) or a prism pair (Fork et al., 1984; Kafka & Baer, 1987; Muller et al., 1998). Alternatively, double-chirped mirror (DCM) is also popular in dispersion compensation of the mode-locked pulsed laser (Szipocs et al., 1994). Using a prism pair provides higher efficiency than a grating pair, but it needs a relatively long compression length since glass or quartz is less dispersive than the grating, and a narrower spectral bandwidth of the pulse due to the geometrical limitation of prisms. In case of DCM, it accurately controls dispersion, but group delay difference is fixed since it depends on the thickness of the coating. Moreover, DCM doesn't provide large group delay of the light. The purpose of the pulse compressor in this experimental design is to provide enough negative dispersion to compensate for the effects of all the dispersions induced by both chromatic dispersion and SPM (Agrawal, 2001), which happens when the light travels through the small fiber core. One can assume that dispersion induced before the fiber is trivial in most cases. Since the GVD due to the fibers can be very large, a grating pair pulse compressor (see Fig. 9.2) is used to produce sufficient negative dispersion compensation.

9.2.4 Fiber Focuser and Collimator

A fiber coupler is generally described as the device to connect fiber-to-fiber, or fiber-to-lens in terms of the light delivery. In this chapter, the fiber coupler is limited to the pair of fiber and lens. It includes a fiber focuser, which couples light from free space to fiber, and a fiber collimator, which collimates the emission from fiber in free space. The fiber focuser consists of optomechanical components to hold the fiber. Fine adjustment of fiber location and tilt allows a lens to direct light efficiently into the fiber core. The fiber collimator includes a fiber holder and a (collimating) lens to collimate the diverging rays from fiber output. Depending on the NA and the core diameters of the fiber, the focal length of the focusing lens and the collimating lens should be carefully selected. Otherwise, coupling loss may be significant, since most of the optical loss originates from fiber coupling due to NA mismatch between the fiber and the focused beam, or the difference between the core diameter of the fiber and the focused spot size of the beam. The collimated beam diameter is also controlled by the NA of the collimating lens. Therefore, different focusing lens and collimating lens pairs should be used to match a specific fiber used. In our fiber characterization experiments, measuring the maximum achieved coupling efficiency is also included since most of optical power loss originates from this coupling loss.

9.2.5 Spectrum Analyzer

It is important to measure the optical power or intensity spectrum to provide complementary information to direct temporal measurement of the optical pulse profile. In the estimation of the optical pulse temporal profile, the highly chirped optical pulse shape sometimes cannot be identified well with optical autocorrelation technique alone. Spectral measurement for the optical pulse is performed with a laser spectrum analyzer (E201LSA03A, Imaging System Group). As described in Figure

9.3a, it works based on a monochromator and uses a continuously rotating grating in a typical monochromator configuration to swipe the given wavelength range. All the data taken from it are digitized with a digital oscilloscope and stored in a personal computer for future analysis.

9.2.6 Optical Autocorrelator

Direct temporal electromagnetic wave or intensity measurement for ultrashort optical pulse is not possible, since the order of time scale ($\sim$100 fs) is beyond the limit of high-speed electronics ($\sim$15 GHz bandwidth). Although ultrashort optical pulse cannot be measured directly, two alternatives have been proposed with optical methods: optical autocorrelation (Rullière, 2005) and frequency-resolved optical gating (FROG) (Trebino, 2000).

Optical autocorrelation is based on a modified Michelson interferometer. It consists of a 50:50 beam splitter and two retroreflector mirror arms that prevent the pulse from reflecting back to the laser source, and enable easy alignment. Incoming pulses at 45° of the beam splitter are split into two identical pulses: a reflected one and a transmitted one. They reflect on the mirror arms, come back to the beam splitter again, and then are recombined. Eventually, two identical pulses travel collinearly or in parallel to the detector after the beam splitter. Without any motions of the mirror arms, the optical paths for two identical pulses are designed to be exactly the same (without any lag or delay of the autocorrelation). As one of the mirror arms moves back and forth, the delay τ of the autocorrelation, the temporal version of optical path difference between the two split pulses, is determined by the position of the movable arm. If the output of the interferometer is collinear, and the detector measures its intensity, the measurement I_1 is given by

$$
\begin{aligned}
I_1(\tau) &= \int_{-\infty}^{+\infty} |E(t) + E(t - \tau)|^2 \, dt \\
&= \int_{-\infty}^{+\infty} (E(t) + E(t - \tau))(E(t) + E(t - \tau))^* \, dt \\
&= \int_{-\infty}^{+\infty} |E(t)|^2 \, dt + \int_{-\infty}^{+\infty} 2E(t)E^*(t - \tau) \, dt + \int_{-\infty}^{+\infty} |E(t - \tau)|^2 \, dt \\
&= 2 \int_{-\infty}^{+\infty} |E(t)|^2 \, dt + 2R_1(\tau)
\end{aligned}
\tag{9.6}
$$

where $R_1(\tau) = \int_{-\infty}^{+\infty} E(t)E^*(t - \tau) \, dt$

$E(t)$ is the complex electromagnetic field of the light, and $R_1(\tau)$ is autocorrelation of the complex electromagnetic field of the light called field autocorrelation. Field autocorrelation can be obtained easily without major modification of the collinear Michelson interferometer, but it provides a contrast of only 2:1, since the intensity measurement includes field autocorrelation (second term of I_1 in equation (9.6)) as well as the baseline signal (first term of I_1 in equation (9.6)).

The signal-to-noise ratio (SNR) can be improved by adding a focusing lens after the interferometer output to take advantage of nonlinear effects at its focus. Two different nonlinear processes may be employed. In one implementation, a nonlinear crystal may be placed at the focal plane, resulting in second-harmonic generation. The nonlinearly induced signal can be collection via a short-pass filter and can be detected using either a photodiode or a photomultiplier. When using a nonlinear crystal, an additional tuning process such as phase-matching (Boyd, 2003) is required to generate frequency doubling effectively. A second implementation uses a GaAsP photodiode at the focal plane. The photocurrent produced at the diode junction is found to respond quadratically to the excitation light (Reid et al., 1997). Unlike a nonlinear crystal, the diode response is not dependent on phase-matching and facilitates alignment. The diode method also has the advantage that it is inexpensive. With collinear geometry, interferometric autocorrelation I_2 is obtained as follows:

$$I_2(\tau) = \int_{-\infty}^{+\infty} |(E(t) + E(t - \tau))^2|^2 dt \qquad (9.7)$$

It supplies very high SNR, which has a contrast of 8:1, compared with the field autocorrelation, but it needs higher power to take advantage of nonlinear processes. This interferometric signal needs to be filtered to calculate temporal pulse width. The interferometer output from its two arms can also be combined in noncollinear (or in-parallel) geometry. This geometry spatially separates the SHG generated by the interference signal from the background signals from the two individual arms of the interferometer. Collecting only the interference signal, the signal at the detector I_3 is given by

$$I_3(\tau) = 4 \int_{-\infty}^{+\infty} I(t)I(t - \tau)dt \qquad (9.8)$$

I_3 represents intensity autocorrelation, which is the autocorrelation of light intensity. It has no background, unlike previous measurements. The pulse width and profile can be directly measured without further filtering.

The spectral-temporal information of ultrashort optical pulses can also be obtained using a FROG. A FROG instantaneously gates the pulse with a variable delayed replica of itself in a nonlinear optical medium. Spectrally resolving the autocorrelation signal, a FROG produces a spectrogram corresponds to the Fourier transform of optical autocorrelation function I_{FROG} as below:

$$I_{FROG}(\tau, \omega) = \left| \int_{-\infty}^{+\infty} E_{sig}(t, \tau) \exp(-i\omega t)dt \right|^2 \qquad (9.9)$$

$E_{sig}(t, \tau)$ is the modified optical autocorrelation. A FROG is capable of measuring both spectral and temporal information for ultrashort optical pulses simultaneously. However, a FROG often needs more alignment degree of freedom than a conventional optical autocorrelator.

For most fiber characterization experiments described in this chapter, an autocorrelator is used instead of FROG since commercially available FROGs are designed to measure pulse widths in the range of 50 to 200 fs, but the pulses broadened by some fibers can have widths of several picoseconds. An interferometric optical autocorrelator (Autocorrelator MINI, APE GmbH) is used. As described in Figure 9.3b, it generates interferometric autocorrelation with the photodiode. It is capable of measuring autocorrelation up to 15 ps.

9.2.7 Multiphoton Excitation Laser Scanning Microscopy

The fiber delivers the light into a multiphoton excitation laser scanning microscope. Multiphoton excitation microscopy is a generalized term including nonlinear fluorescence excitation based on the simultaneous absorption of two or more photons. The two-photon process was predicted by Göppert-Mayer in 1931, and this process was observed with the introduction of the pulsed laser in the 1960s. Two-photon microscopy was implemented by Denk and colleagues (1990). Details about two-photon excitation microscopy design are covered in other chapters. In this section, we provide a brief summary of the microscopy setup used to characterize the two-photon efficiency and the optical resolution of fiber-delivered light.

This multiphoton excitation microscopy system consists of an inverted fluorescence microscope (Axiovert S100TV, Zeiss), intermediate optics including filters, laser scanners, a piezoelectric objective actuator, an objective lens, and a detector. We use a galvanometric system (6350, Cambridge Technology) for lateral raster scanning and install a piezoelectric actuator (P-721.00, PI) on the objective lens for axial scanning. We also use a 40× oil-immersion objective lens (Fluar, 40×/1.30 Oil, Zeiss) for high-resolution imaging. A photomultiplier tube (PMT) is used as a photodetector (R7400P, Hamamatsu) to collect fluorescent emission lights through multiphoton excitation. For each optical fiber tested, the excitation power after the objective is measured. The fluorescence signal produced in a fluorescein solution (fluoroescein-5-isothiocyanate, Molecular Probes) as a function of excitation power can be used to calculate two-photon efficiency. For the resolution measurement, 0.1-μm carboxylate-modified yellow-green (505/515) fluorescent beads (F8803, Molecular Probes) are imaged in 3D, allowing a direct evaluation of the microscope PSF. All the fiber delivery results can be compared with those of free space delivery (delivery without fiber).

9.3 CHARACTERISTICS OF FIBER CONDUCTION OF ULTRASHORT OPTICAL PULSES

9.3.1 Maximum Coupling Efficiency of Ultrashort Optical Pulse through the Fibers

The coupling efficiency is defined as the ratio of the output power delivered through the fiber after the collimator to the input power before the fiber focuser. It's quite different from optical fiber attenuation, or light loss that happens when the light travels in the fiber core due to material absorption. Coupling loss is caused by a mismatch between the size of the focused rays and the fiber core, or between the NA of focused rays and the acceptance angle of fiber when the rays are coupled into fiber. Optical fiber attenuation mostly results from the absorbance of light by the fiber material. In general, optical loss includes both coupling loss and optical fiber attenuation. Thanks to fiber manufacturing technology, optical fiber attenuation is negligible compared with coupling loss in the

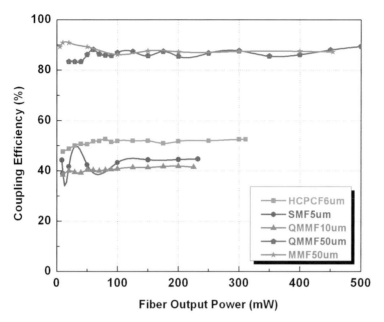

Figure 9.6. The maximum coupling efficiency for different fibers. The x-axis represents the fiber output power on a logarithmic scale. The fibers with a small number of modes have 40% to 50% coupling efficiency, whereas the fibers with a large number of modes have 80% to 90% coupling efficiency. The coupling efficiency is plotted against fiber output power. It is observed to be power-independent.

short length of a few meters used in microscopy. Therefore, the ratio of the output power after the fiber collimator to the input power before the fiber focuser can be considered to be the fiber coupling efficiency.

Figure 9.6 presents the maximum achieved coupling efficiency for each fiber with respect to fiber output power. All the fibers with small core have 40% to 50% coupling efficiency, and the others have 80% to 90%. Coupling efficiency is also mostly independent of fiber output power. Most coupling loss occurs at the fiber input end. Spot diameter of focused rays on the face of the fiber core can be calculated based on diffraction theory. If we assume the lens aperture is circular and the intensity distribution of collimated rays before the lens along the radial direction is uniform, normalized intensity of spot I_{norm} is given by:

$$I_{norm}(r) = \left[2\frac{J_1(\pi x)}{\pi x} \right]^2 \quad where \quad x = \frac{NA}{\lambda}r \qquad (9.10)$$

r is the radial position from the center of spot intensity. Assuming the NA of the lens is less than the NA of the fiber, the theoretical coupling efficiency $\varepsilon_{coupling}$ can be estimated as follows:

$$\varepsilon_{coupling} = \frac{\int_0^{\frac{d_{core}}{2}} I_{norm}(r)r\,dr}{\int_0^{+\infty} I_{norm}(r)r\,dr} \qquad (9.11)$$

where d_{core} is the core diameter of fiber.

A quick estimate of coupling efficiency can be obtained by assuming that intensity distribution of spot I_{norm} is uniform:

$$I_{norm}(r) = \begin{cases} 1 & 0 \leq r \leq \dfrac{d_{FWHM}}{2} \\ 0 & Otherwise \end{cases} \quad where\ d_{FWHM} = 0.5145\dfrac{\lambda}{NA} \quad (9.12)$$

d_{FWHM} is the FWHM of spot intensity for the circular aperture. Not all the focused rays may be transmitted when the core diameter is smaller than d_{FWHM}.

$$\varepsilon_{coupling} \approx \begin{cases} \left(\dfrac{d_{core}}{d_{FWHM}}\right)^2 = \left(\dfrac{d_{core}NA}{0.5145\lambda}\right)^2 & if\ d_{FWHM} > d_{core} \\ 1 & Otherwise \end{cases} \quad (9.13)$$

The coupling efficiency of the fiber increases as the core size of fiber increases and is also function of the NA of the coupling lens. Increase of core size at a given wavelength provides a large number of modes to be delivered with fiber. Therefore, coupling efficiency depends on the number of supported modes and the NA of the fiber but not the fiber output power or the interaction between light and material forming the fiber core.

9.3.2 Temporal Broadening of Ultrashort Optical Pulse through the Fibers

As described above, temporal pulse width cannot be measured directly. We quantify the pulse width instead, based on the full-width-at-half-maximum (FWHM) of optical autocorrelation. If the signal is bandwidth limited, or transform limited, we can easily convert it to the actual temporal pulse width.

Figure 9.7 illustrates the temporal pulse width broadening of an ultrashort optical pulse as a function of fiber output power for different fibers. Some TIR-based fibers (SMF5um and QMMF10um) have strong power-dependent temporal pulse broadening. For large-core fiber, temporal pulse width is still power dependent even with prechirping (QMMF10um.PC), although the prechirping significantly reduces pulse width and its power dependence. TIR-based multimode fibers (QMMF50m and MMF50um) are power independent until the fiber output power is very high and temporal pulse width is relatively short. Hollow-core photonic crystal fiber does not show any temporal pulse width change with dispersion and nonlinear effect (no temporal pulse broadening or power dependence).

In nonlinear fiber optics, SPM is a major spectral distortion source in optical pulse delivery. The pulse propagation equation for ultrashort optical pulse through the single-mode fiber is given by:

$$i\frac{\partial A}{\partial z} = -\frac{i\alpha}{2}A + \frac{\beta_2}{2}\frac{\partial^2 A}{\partial T^2} + \frac{i\beta_3}{6}\frac{\partial^3 A}{\partial T^3} - \gamma|A|^2 A \quad (9.14)$$

$$where \quad T = t - \frac{z}{v_g}$$

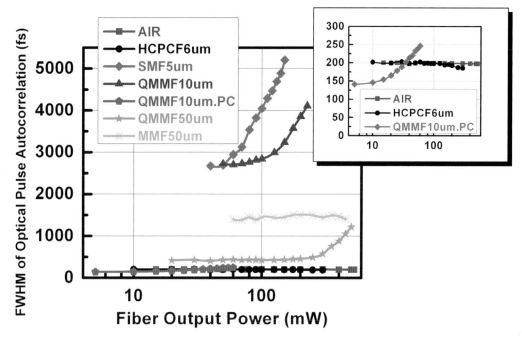

Figure 9.7. The full width at half maximum (FWHM) of the optical autocorrelation function of the transmitted light for different fiber delivery methods. Free space delivery is included for comparison. The x-axis represents fiber output power on a logarithmic scale. The TIR-based optical fibers have dispersion mostly induced by GVD. The temporal pulse width increases as power increases due to nonlinear effects. The prechirped 10-μm core step index optical fiber has minimum dispersion at low power, but its temporal pulse width is high at high power. Hollow-core photonic crystal fiber has a constant temporal pulse width without dispersion (see insert).

In this equation, third-order dispersion (TOD) is considered since it takes an import-ant role for pulse width <5 ps (Agrawal, 2001). $A(z, t)$ is the pulse envelope, v_g is group velocity, α is the absorption coefficient of fiber core material, β_2 is GVD , and β_3 is TOD. γ is a nonlinear parameter. On the right side of equation (9.14), the first term represents material absorption, the second term and the third term represent dispersion, and the last term is responsible for SPM. We assume that the other nonlinear effects are negligible. The length scale for dispersion and nonlinearity can be introduced to the above equation:

$$i\frac{\partial U}{\partial z} = \frac{sgn\,(\beta_2)}{2L_D}\frac{\partial^2 U}{\partial \tau^2} + i\frac{sgn\,(\beta_3)}{6L'_D}\frac{\partial^3 U}{\partial \tau^3} - \frac{\exp\,(-\alpha z)}{L_{NL}}\,|U|^2\,U$$

$$where \quad \tau = \frac{T}{T_0} = \frac{t - z/v_g}{T_0}, \quad A\,(z, \tau) = \sqrt{P_0}\exp\,(-\alpha z/2)\,U\,(z, \tau) \qquad (9.15)$$

$$L_D = \frac{T_0^2}{|\beta_2|}, \quad L'_D = \frac{T_0^3}{|\beta_3|}, \quad L_{NL} = \frac{1}{\gamma P_0}$$

L_D, L'_D, and L_{NL} represents dispersion length with GVD, dispersion length with TOD, and nonlinear length respectively. T_0 is initial pulse width, and P_0 is peak power of initial optical pulse. The nonlinear term is dependent on the power but not the dispersions. Extrapolating to low fiber output power where the nonlinear effect is negligible, since all the fibers have similar pulse widths and are comparable to free space delivery, the effect of dispersion on pulse broadening is small compared with the nonlinear effect at higher power. If the nonlinear effect is dominant, $L_D/L_{NL} \gg 1$ and $L'_D/L_{NL} \gg 1$. Both dimensionless lengths can be rewritten as follows:

$$\frac{L_D}{L_{NL}} = \frac{\gamma P_0 T_0^2}{|\beta_2|} = \frac{n_2 \omega_0}{c A_{eff}} \frac{P_0 T_0^2}{|\beta_2|} = \left(\frac{n_2 \omega_0 T_0^2}{c |\beta_2|} \right) \frac{P_0}{A_{eff}}$$

$$\frac{L'_D}{L_{NL}} = \frac{\gamma P_0 T_0^3}{|\beta_3|} = \frac{n_2 \omega_0}{c A_{eff}} \frac{P_0 T_0^3}{|\beta_3|} = \left(\frac{n_2 \omega_0 T_0^3}{c |\beta_3|} \right) \frac{P_0}{A_{eff}}$$

$$(9.16)$$

n_2 is a nonlinear index coefficient, ω_0 is the center frequency of the initial optical pulse, c is the speed of light, and A_{eff} is the effective core area related to core size and core–cladding index difference. As shown in equation (9.16), the nonlinear effect becomes stronger in the fiber as the peak power of the initial pulse increases and the core size of fiber decreases.

Based on pulse propagation equations, a nonlinear effect such as SPM has a significant contribution to temporal pulse broadening, since the pulse widths of most fibers show strong dependence of power. Extrapolating temporal pulse width to low power, the temporal pulse broadening is not negligible for some fibers when compared with free space delivery, indicating that dispersion effects are significant in some cases, especially for small-core TIR-based fibers (SMF5um and QMMF10um). The two fibers have similar levels of dispersion, demonstrating that material dispersion is mainly responsible for the observed dispersion, since this broadening effect is not dependent on the fiber core size or the transmitted power. In addition, waveguide dispersion (Ghatak & Thyagarajan, 1998), sometimes important in single-mode fiber, is much smaller than material dispersion at this wavelength and can be neglected. With the prechirping, the dispersion of small-core fiber (QMMF10um.PC) can be eliminated, but the nonlinear effect cannot. The experiment result for QMMF10um.PC implies that chromatic dispersion is compensated by the prechirping at low power, but temporal pulse width still increases due to nonlinear effects as power increases. It is shown in the Figure 9.6 insert and is also consistent with Helmchen and coworkers (2002). For large-core fibers (QMMF50um and MMF50um), temporal pulse width starts increasing only when the peak power is very high, since nonlinear effects are weak in fibers with a large effective core area. Temporal pulse width in graded-index fiber (MMF50um) is less dependent on power than in step-index fiber (QMMF50um). Temporal broadening due to dispersion in multimode fiber is less well understood since it is not well described by the pulse propagation equation. Temporal broadening is the result of both dispersion and nonlinear effect of refractive index for TIR-based fibers. Even though GVD is compensated with the prechirping, a nonlinear effect still exists for high-power delivery. Hollow-core photonic crystal fiber has no material dispersion since the light is delivered through the air.

9.3.3 Spectral Broadening of Ultrashort Optical Pulse through the Fibers

The intensity spectrum complements the pulse temporal profile in the characterization of ultrashort optical pulses. The FWHM of the intensity spectrum can serve as a first-order estimate of the spectral bandwidth. Figure 9.8 presents measurements of the intensity spectral bandwidth through various fibers as a function of output power. At low fiber output power, the spectral bandwidth remains narrow for all the fibers. However, the spectra broaden as the delivered power increases in TIR-based fibers (SMF5um, QMMF10um, QMMF50um, and MMF50um). The spectral broadening is greater for smaller-core fibers. Graded-index fiber (MMF50um) has less spectral broadening than step-index fiber (QMMF50um) of the same core size. For a little larger-core TIR fiber with prechirping (QMMF10um.PC), spectral bandwidth decreases as power increases, which has been reported previously by Helmchen et al. (2002). Unlike those fibers described above, hollow-core photonic crystal fiber (HCPCF6um) transmits ultrashort optical pulse without any spectral distortion. Moreover, spectral bandwidth and shape have no relation with fiber output power.

The results for small-core TIR-based fibers (SMF5um and QMMF10um) are consistent with the pulse propagation equation, which describes spectral broadening caused by SPM. Spectral bandwidth increases as the delivered power increases, as expected. For the TIR-based multimode fibers (QMMF50um and MMF50um), their spectral

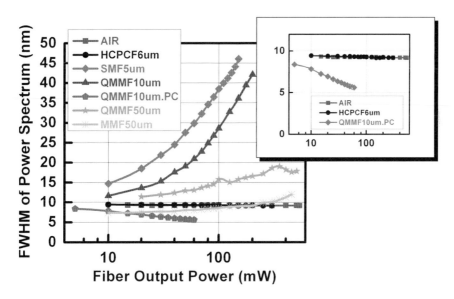

Figure 9.8. The full width at half maximum (FWHM) of the spectral bandwidth of the transmitted light for different fiber delivery methods. Free space delivery is included for comparison. The x-axis represents the fiber output power on a logarithmic scale. For the TIR-based optical fibers, their spectral bandwidths widen due to nonlinear effects. The nonlinear effects are stronger in smaller-core fibers. Using the prechirping, the 10 μm core step index optical fiber's spectral bandwidth slightly decreases at high power. In the case of hollow-core photonic crystal fiber, it has constant spectral bandwidth over the fiber output power range (see insert).

bandwidths are quite different because the shape of the optical pulse depends on the propagating modes in the fiber. It's difficult to measure spectral bandwidth for multimode fiber due to the sensitivity of propagating modes to fiber bending. Nonetheless, we observe that the nonlinear effect is less in large-core fibers since their spectral bandwidths have only weak dependence on optical power. Although the pulse propagation equation is limited to single-mode fiber, it can predict that nonlinear effects in multimode fibers are not dominant since their effective areas are large. Further, spectral bandwidth in graded-index fiber (MMF50um) is less affected by the nonlinear effect than in step-index fiber (QMMF5um). This difference comes from the difference in the refractive index structures of these fibers. For hollow-core photonic crystal fiber (HCPCF6um), both nonlinearities and dispersion are minimal since the light is confined in free space within the core of the fiber.

9.3.4 Multiphoton Efficiency with Fiber-Delivered Optical Pulses

Two-photon excitation has quadratic dependence on excitation power. Two-photon efficiency measurement provides how well the light delivered by the fiber excites fluorophores. Figure 9.9 shows two-photon efficiency for various fibers as a function of power. In a log-log plot, a quadratic dependence on power implies that the power

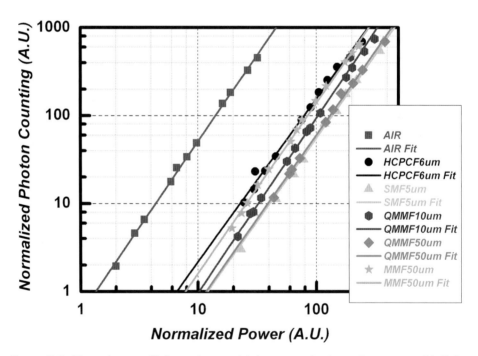

Figure 9.9. Two-photon efficiency in a multiphoton excitation microscopy with light delivered by different fibers. Both the x- and y-axes are in logarithmic scales. Quadratic dependence of sample fluorescence as a function of power indicates two-photon process. Fiber delivery is much less efficient for two-photon excitation than for free space delivery. Among all the fibers tested, the hollow-core photonic crystal fiber is the most efficient pulse delivery method.

dependence should be linear with a slope of 2. Therefore, all fibers show quadratic dependence as expected for a two-photon process. In our experiments, hollow-core photonic crystal fiber (HCPCF6um) delivery has the best two-photon efficiency after fiber delivery. Due to large temporal pulse width, single-mode fiber (SMF5um) delivery has the worst two-photon efficiency.

To observe how efficient two-photon excitation is, the coefficient of two-photon efficiency should be calculated. In two-photon excitation theory, the two-photon efficiency is given by (Denk et al., 1990):

$$n_a \approx \frac{p_0^2 \delta}{\tau_p f_p^2} \left(\frac{(NA)^2}{2\hbar c \lambda} \right)^2 = \left[\frac{\delta}{\tau_p f_p^2} \left(\frac{(NA)^2}{2\hbar c \lambda} \right)^2 \right] p_0^2 \tag{9.17}$$

where n_a is the number of photons absorbed per fluorophore per pulse, τ_p is the pulse duration, δ is the fluorophore's two-photon absorption at a given wavelength λ, P_o is averaged laser intensity, f_p is laser repetition rate, and $\hbar$ is Planck's constant. c and NA are the same as defined earlier.

From equation (9.17), fiber delivery does not change any parameters except for pulse width, τ_p, and effective NA. If the effective NA remains constant, the coefficient of two-photon efficiency is inversely proportional to temporal pulse width. Since temporal pulse width often broadens with fiber delivery, the two-photon efficiency is expected to decrease when light is conducted through fibers.

It is interesting to note that the two-photon efficiency for hollow-core photonic crystal fiber (HCPCF6um) is not identical with the free space delivery one, even though both spectral bandwidths and temporal pulse widths are the same for both delivery methods. The discrepancy between theory and measurement may originate from the fact that the fiber output light is not single mode, and hence there is an effectively lower NA. Another possibility is that this theory assumes that optical pulse delivered through fibers is bandwidth-limited or chirp-free optical pulse. It has been reported that TOD can be significant in photonic crystal fibers. In some cases, the balance between normal GVD and TOD results in soliton-like behavior resisting dispersion of ultrashort optical pulses (Ouzounov et al., 2003). However, the phase shift induced by these high-order dispersions can also generate chirping of the pulses. The width of the temporal autocorrelation profile cannot determine if the pulse is chirped or chirp-free. Therefore, chirping effect can be another major source to deteriorate two-photon efficiency. If inter-modal dispersion is induced, two-photon efficiency can decrease significantly, since the two-photon process is sensitive to the wavefront mismatch caused by intermodal dispersion.

9.3.5 Optical Resolution with Fiber-Delivered Optical Pulse

Optical resolution is usually defined as the minimum distance that two point sources can be discriminated. It is not easy to quantify optical resolution because it depends on SNR. Therefore, the PSF FWHM is used as a measure of optical resolution.

Figure 9.10 shows measurements of both lateral and axial PSF for multiphoton excitation microscopy after light is delivered by different fibers. For each fiber delivery method, the top figures are intensity profiles of a point source in the solution, and the

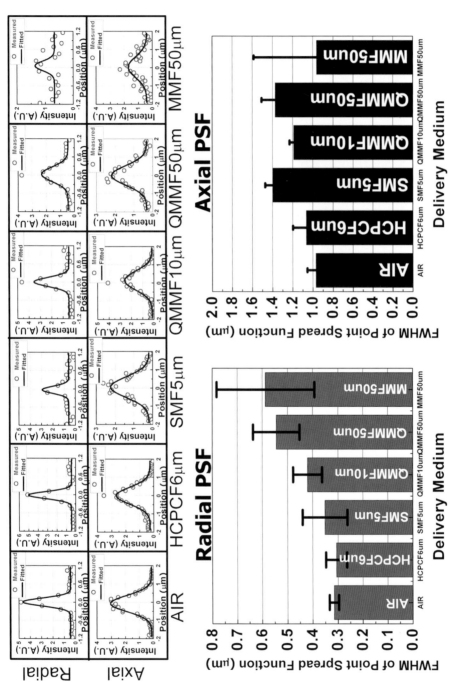

Figure 9.10. The full width at half maximum (FWHM) of the point spread function (PSF) of a multiphoton excitation microscope, using different fiber delivery methods. The top figures show both lateral and axial PSF measurements with light delivered through various fibers. Free space delivery is included for comparison. Bottom figures represent statistical analysis of the lateral and axial PSFs and their FWHMs for each delivery method. As shown, the lateral resolution is dependent on the number of modes that the fiber supports, whereas the axial resolution is almost independent of the delivery method.

bottom figures show the FWHM measurements. As seen in the figures, the FWHM of lateral PSF increases with the number of modes that each fiber delivers, whereas the FWHM of axial PSF is not sensitive to the number of delivered modes. In theory, the excitation PSF determines the optical resolution in laser scanning microscopy. The excitation PSF intensity profile is calculated as (So et al., 2000):

$$I(u,v) = \left| 2 \int_0^1 J_0(v\rho) e^{-\frac{i}{2}u\rho^2} \rho \, d\rho \right|^2$$

$$where \quad \alpha = \sin^{-1}(NA), \quad u = 4k\sin^2\left(\frac{\alpha}{2}\right)z, \quad and \quad v = k\sin(\alpha)r \qquad (9.18)$$

J_0 is the zero-th-order Bessel function, and $k = 2\pi/\lambda$ is wave number. z and r are radial and axial coordinates of intensity profile, respectively. (u and v are their dimensionless version, respectively.) In the two-photon process, the fluorescence signal proportional to intensity squared, $I^2\left(\frac{u}{2}, \frac{v}{2}\right)$, is produced by this illumination, and $I^2\left(\frac{u}{2}, \frac{v}{2}\right)$ is defined as a PSF in two-photon excitation microscopy. It can be fitted to Gaussian function. One definition of the optical resolution is FWHM of PSF, and the lateral and the axial optical resolution can be expediently estimated using a Gaussian fit of the PSF (Zipfel et al., 2003).

$$FWHM_{xy} = \begin{cases} \dfrac{0.533\lambda}{\sqrt{2}NA} & NA \leq 0.7 \\[3mm] \dfrac{0.541\lambda}{\sqrt{2}NA^{0.91}} & NA > 0.7 \end{cases} \qquad (9.19)$$

$$FWHM_z = \frac{0.886\lambda}{\sqrt{2}}\left[\frac{1}{n - \sqrt{n^2 - NA^2}}\right] \qquad (9.20)$$

NA is the numerical aperture of the objective lens, n is the effective refractive index of the specimen, $FWHM_{xy}$ is the FWHM of lateral PSF, and $FWHM_z$ is the FWHM of axial PSF function. As expressed in equations (9.18), (9.19), and (9.20), optical resolution is determined by the NA of the objective lens and wavelength. However, the basic assumption for the above equations is that the illumination light has a distortion-free wavefront and its intensity is uniform in the lateral direction. The lateral PSF profile depends on how many modes are delivered. When the number of delivered modes is close to unity, the experimentally measured PSF is close to the theoretical situation; otherwise, the lateral optical resolution is worse than the theoretical one. The axial resolution is found to be less dependent on the delivery modes and may be a result of the lateral nature of these spatial modes. In any case, the effect of spatial modes on axial resolution requires further investigation. Axial resolution is found to be more sensitive to the objective NA and the sample's refractive index.

Based on PSF measurement and theoretical evaluation, the delivered modes affect mostly lateral optical resolution, and axial optical resolution is more sensitive to the NA of the objective lens than any other factors. Therefore, for all the fiber deliveries as well as free space delivery, lateral PSFs are different, depending on the modes of the light, but axial PSFs remain the same in multiphoton excitation microscopy regardless of the pulse shape.

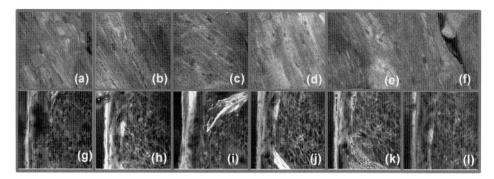

Figure 9.11. Multiphoton microscopy image for tissue specimens using different fiber delivery methods. (a–f) Autofluorescence images of myocardial cells in a mouse heart. (g–l) Fluorescence images of H&E-stained ex vivo human skin (provided by Penn State Medical Center). The fiber delivery methods are (a & g) AIR, (b & h) HCPCF6um, (c & i) SMF5um, (d & j) QMMF10um, (e & k) QMMF50um, and (f & l) MMF50um.

9.3.6 Images Taken from Multiphoton Excitation Microscopy with Optical Pulse Delivery through the Fibers

Multiphoton microscopy can effectively image biological specimens, using optical fiber for light delivery. Two tissue specimens, mouse heart and human skin, are imaged. Mouse heart tissue is fixed without staining and imaged based on autofluorescence contrast. Human tissue is fixed and H&E stained, which produces fluorescence signal through the multiphoton process. We use a $40\times$ oil immersion objective lens, and the image field of view is 120 μm $\times$ 120 μm. Figure 9.11 shows both autofluorescence and fluorescence images taken in multiphoton excitation microscopy with different light delivery methods. The top figure shows autofluorescence images obtained from mice heart tissue, and the bottom figure represents the fluorescence images acquired from human skin tissue. In terms of image resolution and details, the differences between fiber delivery modes are not large. In order to better quantify these differences, we introduce power spectral image analysis (Lim, 1990; Oppenheim et al., 1999). Since human skin tissues have more details than mouse heart tissues, power spectra for human skin tissue are measured for different fiber delivery methods (Fig. 9.12). As expected, the results of the image power spectrum conform to the results of lateral PSF measurements. Single-mode fiber images have more power than multimode fiber images within the spatial frequency range where tissue details exist. At the high spatial frequency, there is no power spectral density difference for different delivery methods since these high-frequency features are associated with noise in the image.

9.4 SUMMARY

In this chapter, we describe the methods to transmit ultrashort optical pulses through fiber optics so that they can be guided along an arbitrary path. We observe the ultrashort optical pulse profile both spectrally and temporally. Spectral bandwidth is sensitive to

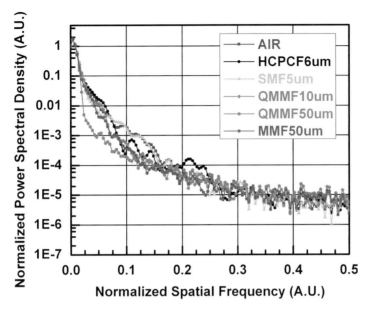

Figure 9.12. The power spectrum of images obtained with different fiber delivery methods. Power spectral image analysis is performed on a series of images depicting H&E-stained ex vivo human skin. The images with higher resolution have broader power spectra containing more detailed information. Fibers with small number of modes provide better optical resolution. The associated images have higher power spectral densities than images acquired using fibers with larger number of modes. Above a normalized frequency of 0.2, power spectral density levels for all the fiber delivery methods, including free space delivery, are same since noise dominates at this regime.

nonlinear effects such as SPM, whereas temporal pulse width is affected by both SPM and dispersion. These effects originate from the nonlinearity of refractive index in the fiber core materials. We further measure multiphoton excitation microscopy's performance in terms of two-photon efficiency and optical resolution. A two-photon process is observed for all the fiber delivery methods, but the signal level of the two-photon process depends on spectral bandwidth, temporal pulse width, spatial pulse shape, and phase mismatch induced by mode shape, chirping, and intermodal dispersion. Optical resolution is evaluated with PSF FWHM. We observe that optical resolution is not sensitive to the laser pulse temporal profile. However, the lateral resolution is sensitive to the delivered modes, while the axial resolution is sensitive to the NA of the objective lens.

In conclusion, hollow-core photonic crystal fiber has the potential to deliver ultrashort optical pulses with the least distortion of both spectral and temporal pulse shapes compared to any other fiber deliveries. It also provides lateral and axial optical resolution the same as those with free space delivery. However, hollow-core photonic crystal fiber has lower two-photon efficiency than that of free space delivery, even though they have same spectral bandwidth and temporal pulse width. Further investigations, based on modal analysis and spatiotemporal analysis using FROG, are needed to better understand this efficiency loss. Nevertheless, hollow-core photonic crystal fiber provides better overall performance than conventional optical fiber.

REFERENCES

Agostinelli, J., G. Harvey, T. Stone, and C. Gabel. 1979. Optical Pulse Shaping with a Grating Pair. Applied Optics 18(14):2500–2504.

Agrawal, G. P. 2001. Nonlinear fiber optics. Academic Press, San Diego.

Argyros, A., M. A. van Eijkelenborg, M. C. J. Large, and I. M. Bassett. 2006. Hollow-core microstructured polymer optical fiber. Optics Letters 31(2):172–174.

Benabid, F., J. C. Knight, G. Antonopoulos, and P. S. J. Russell. 2002. Stimulated Raman scattering in hydrogen-filled hollow-core photonic crystal fiber. Science 298(5592):399–402.

Birks, T. A., J. C. Knight, and P. S. Russell. 1997. Endlessly single-mode photonic crystal fiber. Optics Letters 22(13):961–963.

Birks, T. A., P. J. Roberts, P. S. J. Russel, D. M. Atkin, and T. J. Shepherd. 1995. Full 2-D Photonic Bandgaps in Silica/Air Structures. Electronics Letters 31(22):1941–1943.

Boyd, R. W. 2003. Nonlinear optics. Academic Press, San Diego, CA.

Cregan, R. F., B. J. Mangan, J. C. Knight, T. A. Birks, P. S. Russell, P. J. Roberts, and D. C. Allan. 1999. Single-mode photonic band gap guidance of light in air. Science 285(5433):1537–1539.

Denk, W., J. H. Strickler, and W. W. Webb. 1990. 2-Photon Laser Scanning Fluorescence Microscopy. Science 248(4951):73–76.

Fink, Y., J. N. Winn, S. H. Fan, C. P. Chen, J. Michel, J. D. Joannopoulos, and E. L. Thomas. 1998. A dielectric omnidirectional reflector. Science 282(5394):1679–1682.

Fork, R. L., O. E. Martinez, and J. P. Gordon. 1984. Negative Dispersion Using Pairs of Prisms. Optics Letters 9(5):150–152.

Ghatak, A. K., and K. Thyagarajan. 1998. An introduction to fiber optics. Cambridge University Press, Cambridge; New York.

Gobel, W., A. Nimmerjahn, and F. Helmchen. 2004. Distortion-free delivery of nanojoule femtosecond pulses from a Ti:sapphire laser through a hollow-core photonic crystal fiber. Optics Letters 29(11):1285–1287.

Goff, D. R., and K. S. Hansen. 2002. Fiber optic reference guide: a practical guide to communications technology. Focal Press, Amsterdam; Boston.

Hart, S. D., G. R. Maskaly, B. Temelkuran, P. H. Prideaux, J. D. Joannopoulos, and Y. Fink. 2002. External reflection from omnidirectional dielectric mirror fibers. Science 296(5567):510–513.

Helmchen, F., D. W. Tank, and W. Denk. 2002. Enhanced two-photon excitation through optical fiber by single-mode propagation in a large core. Applied Optics 41(15):2930–2934.

Husakou, A. V., and J. Herrmann. 2001. Supercontinuum generation of higher-order solitons by fission in photonic crystal fibers. Physical Review Letters 8720(20).

Johnson, S. G., and J. D. Joannopoulos. 2002. Photonic crystals : the road from theory to practice. Kluwer Academic Publishers, Boston.

Kafka, J. D., and T. Baer. 1987. Prism-Pair Dispersive Delay-Lines in Optical Pulse-Compression. Optics Letters 12(6):401–403.

Kim, D., K. H. Kim, S. Yazdanfar, and P. T. C. So. 2005. Optical biopsy in high-speed handheld miniaturized multifocal multiphoton microscopy. In Multiphoton Microscopy in the Biomedical Sciences V. SPIE, San Jose, CA, USA. 14–22.

Knight, J. C. 2003. Photonic crystal fibres. Nature 424(6950):847–851.

Knight, J. C., T. A. Birks, R. F. Cregan, P. S. Russell, and J. P. de Sandro. 1998a. Large mode area photonic crystal fibre. Electronics Letters 34(13):1347–1348.

Knight, J. C., T. A. Birks, P. S. Russell, and D. M. Atkin. 1996. All-silica single-mode optical fiber with photonic crystal cladding. Optics Letters 21(19):1547–1549.

Knight, J. C., T. A. Birks, P. S. J. Russell, and D. M. Atkin. 1997. All-silica single-mode optical fiber with photonic crystal cladding: Errata. Optics Letters 22(7):484–485.

Knight, J. C., J. Broeng, T. A. Birks, and P. S. J. Russel. 1998b. Photonic band cap guidance in optical fibers. Science 282(5393):1476–1478.

Lim, J. S. 1990. Two-dimensional signal and image processing. Prentice Hall, Englewood Cliffs, N.J.

Lourtioz, J. M. 2005. Photonic crystals: towards nanoscale photonic devices. Springer, Berlin.

Moulton, P. F. 1986. Spectroscopic and Laser Characteristics of Ti-Al2o3. Journal of the Optical Society of America B-Optical Physics 3(1):125–133.

Muller, M., J. Squier, R. Wolleschensky, U. Simon, and G. J. Brakenhoff. 1998. Dispersion pre-compensation of 15 femtosecond optical pulses for high-numerical-aperture objectives. Journal of Microscopy-Oxford 191141–191150.

Oppenheim, A. V., R. W. Schafer, and J. R. Buck. 1999. Discrete-time signal processing. Prentice Hall, Upper Saddle River, N.J.

Ouzounov, D. G., F. R. Ahmad, D. Muller, N. Venkataraman, M. T. Gallagher, M. G. Thomas, J. Silcox, K. W. Koch, and A. L. Gaeta. 2003. Generation of megawatt optical solitons in hollow-core photonic band-gap fibers. Science 301(5640):1702–1704.

Reid, D. T., M. Padgett, C. McGowan, W. E. Sleat, and W. Sibbett. 1997. Light-emitting diodes as measurement devices for femtosecond laser pulses. Optics Letters 22(4):233–235.

Renversez, G., B. Kuhlmey, and R. McPhedran. 2003. Dispersion management with microstructured optical fibers: ultraflattened chromatic dispersion with low losses. Optics Letters 28(12):989–991.

Rullière, C. 2005. Femtosecond laser pulses: principles and experiments. Springer Science + Business Media, New York.

Russell, P. 2003. Photonic crystal fibers. Science 299(5605):358–362.

So, P. T. C., C. Y. Dong, B. R. Masters, and K. M. Berland. 2000. Two-photon excitation fluorescence microscopy. Annual Review of Biomedical Engineering 2399–2429.

Szipocs, R., K. Ferencz, C. Spielmann, and F. Krausz. 1994. Chirped Multilayer Coatings for Broad-Band Dispersion Control in Femtosecond Lasers. Optics Letters 19(3):201–203.

Tai, K., and A. Tomita. 1986. 1100 X Optical Fiber Pulse-Compression Using Grating Pair and Soliton Effect at 1.319-Mu-M. Applied Physics Letters 48(16):1033–1035.

Tai, S. P., M. C. Chan, T. H. Tsai, S. H. Guol, L. J. Chen, and C. K. Sun. 2004. Two-photon fluorescence microscope with a hollow-core photonic crystal fiber. Optics Express 12(25):6122–6128.

Temelkuran, B., S. D. Hart, G. Benoit, J. D. Joannopoulos, and Y. Fink. 2002. Wavelength-scalable hollow optical fibres with large photonic bandgaps for CO2 laser transmission. Nature 420(6916):650–653.

Trebino, R. 2000. Frequency-resolved optical gating: the measurement of ultrashort laser pulses. Kluwer Academic, Boston.

Wadsworth, W. J., J. C. Knight, A. Ortigosa-Blanch, J. Arriaga, E. Silvestre, and P. S. J. Russell. 2000. Soliton effects in photonic crystal fibres at 850 nm. Electronics Letters 36(1):53–55.

Zipfel, W. R., R. M. Williams, and W. W. Webb. 2003. Nonlinear magic: multiphoton microscopy in the biosciences. Nature Biotechnology 21(11):1368–1376.

Zolla, F. 2005. Foundations of photonic crystal fibres. World Scientific, Hackensack, N.J.

——— 10

An Optical Design Primer for Nonlinear Optical Microscopes

Peter T. C. So and Daekeun Kim

10.1 INTRODUCTION

Nonlinear microscopes can operate based on contrast mechanisms such as fluorescence, harmonic generation, or coherent anti-Stokes Raman scattering. The microscope optical design should be adapted for each mode of contrast. Furthermore, the optical configuration of the microscope should be optimized for a given biological application. This chapter describes the basic design considerations of nonlinear optical microscopes. Specifically, this chapter will predominately focus on fluorescence-based systems, since they are most widely used. The instrument considerations specific to the other contrast mechanisms will be addressed in later chapters.

The design of optical microscopes and their components has been discussed in many books (Diaspro, 2002; Inoué & Spring, 1997; Masters, 2006; Pawley, 1995; Wilson, 1990; Wilson & Sheppard, 1984), book chapters (Stelzer, 1995), and articles (Diaspro et al., 1999a, 1999b). In this chapter, we do not seek to comprehensively restate materials that have been thoroughly discussed. Only a brief overview will be provided for the design considerations that are common to all forms of optical microscopes. We will instead focus on design issues that are specific to optimizing the generation and detection of nonlinear optical signals.

In this chapter, we will first consider the "conditioning" of excitation light from femtosecond (fs) laser sources before the microscope. A short review of the basic microscope optical design will be provided. Key design considerations will be examined for the three main parts of a nonlinear optical microscope: the excitation light path, the microscope objective, and the detection light path.

10.2 EXCITATION LIGHT CONDITIONING

Efficient excitation of nonlinear optical effects requires high electric field strength. Femtosecond or picosecond pulse laser sources are needed without depositing excess energy onto biological specimens. However, the use of ultrafast light sources entails unique challenges in the design of nonlinear optical microscopes. Most nonlinear microscopes use free space optics to direct light from the laser source into the microscope with minimal dispersion. However, the availability of low-dispersion photonic crystal fibers may allow more versatile and compact light delivery schemes. The issues related to fiber delivery have been considered more thoroughly in Chapter 9.

10.2.1 General Considerations

In fluorescence mode, the efficiency of the n-photon excitation process as a function of pulse laser parameters can be expressed as (Denk et al., 1990):

$$F(t) \propto \delta^{(n)} \eta \, p^n(t) \tag{10.1}$$

where n is the order of the nonlinear optical process, $F(t)$ is the instantaneous fluorescent photon emission rate per molecule $\left(\text{in units of } \frac{photons}{sec \cdot molecule}\right)$, $\delta^{(n)}$ is the n-photon absorption cross-section $\left(\text{in units of } \frac{m^{2n} \cdot s^{n-1}}{molecule}\right)$, η is the quantum yield of the fluorophore (unitless), and $p(t)$ is the instantaneous flux of the laser $\left(\text{in unitsof } \frac{photons}{m^2 \cdot sec}\right)$.

For pulse laser, the instantaneous power, $p(t)$, is higher than the average power, P $\left(\text{also in units of } \frac{photons}{m^2 \cdot sec}\right)$. We can express the number of fluorescent photons produced per second, F_{pulse} (in units of $\frac{photons}{sec \cdot molecule}$), for pulsed laser as:

$$F_{pulse} \propto \frac{\delta^{(n)} \eta \, P^n}{\tau^{n-1} f^{n-1}} \tag{10.2}$$

where τ is the laser pulse width (in units of sec), and f (in units of sec^{-1}) is the laser pulse repetition rate.

For a continuous wave laser, the instantaneous and average powers are equal,

$$F_{cont} \propto \delta^{(n)} \eta \, P^n \tag{10.3}$$

Therefore, with the same average power, a pulse laser is more efficient in producing nonlinear optical effect than a continuous wave laser by a factor:

$$\frac{F_{pulse}}{F_{cont}} = \tau^{1-n} f^{1-n} \tag{10.4}$$

From this equation, it is clear that there is no advantage of using pulsed laser light for one-photon linear excitation. However, the use of ultrafast pulse laser becomes more important with increasing order of nonlinear process, n.

From equation (10.2), the excitation efficiency can be optimized by varying the adjustable parameters of the laser: the average power, P, the pulse width, τ, and the pulse repetition rate, f. All three parameters can be modified by beam conditioning components prior to the entrance of the laser pulses into the microscope.

10.2.1.1 Excitation Power

It is important to regulate the laser power incident upon the specimen. It is clear that multiphoton excitation efficiency increases nonlinearly with laser average power.

However, the maximum power is limited by two factors. First, the maximum usable power is limited by photophysics. We can calculate the probability of excitation per fluorophore per laser pulse Pr $\left(\text{in units of } \frac{n-photons\ absorbed}{pulse\cdot molecule}\right)$:

$$\Pr \propto \frac{\delta^{(n)} P^n}{\tau^{n-1} f^n} \tag{10.5}$$

Since the fluorophores have a lifetime on the order of nanoseconds, each fluorophore can be excited only once per laser pulse. When this excitation probability approaches unity, the molecules in the ground state are depleted and the fluorescence excitation becomes less efficient. The fluorescence signal scales with excitation intensity in a less than n^{th} power dependence. Since the localization of the excitation volume in a nonlinear optical microscope depends on the presence of this nonlinearity across the spatial profile of the excitation point spread function (PSF), this decrease of the nonlinearity results in a broadening of the excitation PSF and resolution degradation of the microscope. This is the process of excitation saturation (Cianci et al., 2004; Nagy et al., 2005a, 2005b). The power level for the saturation of typical fluorophores during two-photon excitation ($\delta^{(2)} \approx 10\ \text{GM} = 10 \times 10^{-50} cm^2 \cdot \sec$) is about tens of milli-Watts.

Second, the maximum excitation average power is further limited by specimen photodamage. Depending on the specimen, photodamage can be caused by the linear absorption of the infrared light inducing thermal damage or by higher-order absorption resulting in photodamage through an oxidative process. Typically, the damage threshold on individual cells is a few milli-Watts. For a more in-depth discussion of photodamage, please see Chapter 14.

The most common light source for multiphoton microscope is titanium-sapphire (Ti:Sa) laser. The output power level of Ti:Sa laser is typically a few Watts, which must be attenuated prior to its introduction into the microscope. Furthermore, some of the most important applications of multiphoton microscope are in the imaging of thick and highly scattering specimens. In a multiple scattering medium, the power delivered to a specific depth of the specimen, $I(z)$, is an exponentially decaying function of the reduced scattering coefficient of the specimen, μ_s, and the imaging depth, z:

$$I(z) = I(0)e^{-\mu_s z} \tag{10.6}$$

In order to achieve equivalent excitation efficiency at different depths, it is often desirable to deliver higher excitation power at greater imaging depth. Beam conditioning components are needed to regulate the average laser power entering the microscope.

There are several methods to regulate the average of the incident laser light. Neutral-density filters are often used for power control. However, absorptive neutral-density filters should be used with caution in nonlinear microscopes since the high laser peak power can cause microscopic damage to the filter coating, degrading the transmitted beam profile. Another common method that is often used for beam power control is the combination of a half-wave plate and a polarizer (Fig. 10.1a). Ti:Sa laser output is linearly polarized. A polarizing element with high power tolerance such as a polarizing

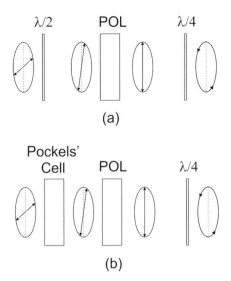

Figure 10.1. Regulation of laser average power using polarization elements. (a) The laser output is horizontal polarized (represented by the direction of the arrow in the ellipse). After the half-wave plate ($\lambda/2$), linearly polarized light of arbitrary polarization can be produced. A polarizer is set with polarization direction perpendicular to the original laser polarization. The transmitted light is vertically polarized. Final polarization can be set by another polarization element. In this case, a quarter-wave plate ($\lambda/4$) produces circularly polarized light. (b) The polarization-based power regulation scheme can be automated by replacing the half-wave plate with a Pockels cell.

beam splitter cube or a Glan-laser polarizer is placed at cross polarization to the laser output, resulting in almost complete attenuation of the laser light. A half-wave plate is inserted between the laser and the polarizer. The polarization axis of laser light can be rotated with minimal power loss and minimal beam pointing deviation using a half-wave plate. The final transmitted power scales as the square of the cosine of the angle between the polarization axis of the wave plate and the polarizer. The transmitted light is linearly polarized perpendicular to the laser output. For the imaging of oriented structures, such as the membrane of a unilamellar vesicle, the final polarization may need to be adjusted. Linear-polarized light with arbitrary orientation may be produced by introducing a second half-wave plate. When linear polarization is not desirable, circularly polarized light may be produced by a quarter-wave plate. This polarization-based power control system has the virtue of having a very high damage threshold, it does not affect beam pointing when power is varied, and it allows convenient control of final polarization of light upon the specimen.

It is sometimes important to rapidly and automatically vary the transmitted power level. First, different power levels are needed to produce images with comparable signal-to-noise levels at different specimen depths due to the attenuation of excitation and emission light. Second, the laser power used in nonlinear microscopy is high and can cause significant fluorophore bleaching or specimen photodamage if the laser beam is parked at a single position for any period of time. Therefore, it is desirable to attenuate, to "blank," the laser beam after completing the scan a line or the scan a frame. Since the "blanking" time required between scans can be as little as tens of milliseconds, the power control component should have kHz bandwidth for this application. Rapid and automated power control can be accomplished by replacing the half-wave plate in the polarization-based power control scheme with an electro-optical polarization control device such as a Pockel's cell (see Fig. 10.1b). A Pockel's cell is an electro-optical element consisting of a birefringent optical crystal, such as lithium niobate. The polarization axis of the crystal is sensitive to an imposed electric field. By varying the electric field, the polarization of the Pockel's cell can be rapidly varied with bandwidth over the MHz level.

10.2.1.2 Pulse Width

Group velocity dispersion (GVD) has been discussed in Chapter 8. A number of optical elements, such as the microscope objective and the Pockel's cell, in a nonlinear optical microscope introduce significant GVD. GVD results in the effective broadening of the excitation pulse width at the specimen, reducing the nonlinear excitation efficiency. The output pulse width, τ_{out}, depends on the input pulse width, τ_{in}, and the GVD of the optical system as (Wokosin, 2005):

$$\tau_{out} = \tau_{in} \sqrt{1 + 7.68 \left(\frac{GVD}{\tau_{in}^2} \right)^2} \tag{10.7}$$

To compensate for GVD, a pulse compressor or stretcher can be introduced to optimize the pulse profile at the specimen. Prism- or grating-based pulse compressors can be introduced. The design and operation of these devices have been discussed at length in Chapter 8 and will not be repeated here. Is the optimization of pulse width critical, given the added beam alignment complexity of introducing the pulse compressor unit? As discussed previously, most laser sources provide significantly more power than is needed for imaging. A nonoptimal pulse profile can often be compensated by using slightly higher input power; this is in fact the common practice. Two situations merit the use of a pulse compressor. First, when there is significant linear absorption in the specimen, the optimization of the pulse profile allows more efficient nonlinear excitation while minimizing total power deposition. Second, even the high-power Ti:Sa lasers often do not provide sufficient power for high-speed imaging using the multiple foci method as discussed in Chapter 18. In this case, pulse compression is recommended to optimize excitation efficiency.

10.2.1.3 Pulse Repetition Rate

The optimization of the laser pulse repetition rate should also be considered in some experiments. For fluorescence excitation, the optimal pulse repetition rate should be commensurate with the inverse of the fluorescence lifetime, allowing the molecules to return to the ground state just before the arrival of the next excitation pulse. This timing scheme optimizes the fluorescence signal without causing excitation saturation. The typical 100-MHz repetition rate of the Ti:Sa laser happens to match well with the nanosecond lifetimes of many common fluorophores. While the upper laser repetition rate for nonlinear microscope imaging is limited by fluorophore lifetime, this upper limit does not apply to scattered light imaging modes, such as second-harmonic generation, given the instantaneous nature of these nonlinear processes. More efficient second-harmonic imaging using a 2-GHz repetition rate ultrafast laser has been demonstrated (Liu et al., 2001; Sun, 2005). There are also situations where the repetition rate of the Ti:Sa lasers should be reduced. Specifically, thermal diffusion time in water is on the order of 100 ns, which is approximately an order of magnitude longer than Ti:Sa laser pulse separation. In a specimen with a high infrared absorption coefficient, such as melanin in some tissues, it may be desirable to decrease the pulse repetition rate to match the thermal relaxation time; this approach limits temperature buildup in the specimen and the resultant thermal damage (Masters et al., 2004).

The laser pulse repetition rate can be reduced using a laser pulse picker. A laser pulse picker selectively deflects a fraction of laser pulses into the excitation light path, decreasing the repetition rate. The laser pulse picker is based on an acousto-optical deflector. A periodic index of refraction variation in a crystal, such as tellurium oxide, functions as an optical grating. This variation in the index of refraction is produced transiently by an acoustic wave using a piezoelectric actuator. Light incident upon this transient grating is deflected by the Bragg effect. In the ideal situation, the pulse picker reduces only the pulse train repetition rate but not the instantaneous power of the pulses (Fig. 10.2a). It should be noted that the excitation efficiency is often decreased due to pulse broadening since the tellurium oxide crystal has a fairly high GVD coefficient of 10,000 fs^2. This dispersive effect can be compensated by the pulse-shaping techniques described earlier. The acousto-optical modulator may also introduce spherical aberration, which affects beam focusing. Investigations into compensating for aberration effects based on adaptive optics and other methods are underway (Booth et al., 2001, 2002a, 2002b).

Another reason to reduce the laser pulse repetition is to improve image penetration depth (Potma et al., 1998). Since nonlinear excitation becomes more efficient

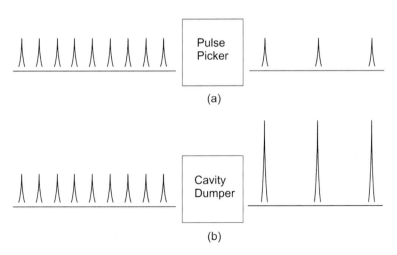

Figure 10.2. (a) Pulse train before and after a pulse picker. Picking one out of three pulses is shown. Pulse height is not changed for an ideal device. (b) Pulse train before and after a cavity dumper. Picking one out of three pulses is shown. Pulse height increases after the device.

with higher laser peak power, a method to improve imaging depth in highly scattering specimen is to use higher laser peak power. From equation (10.2), this is equivalent to reducing the pulse repetition rate while keeping the average power and the pulse width constant. Experimentally, this can be accomplished using a laser cavity dumper where multiple pulses emergent from the laser are trapped in a second laser cavity (see Fig. 10.2b). The second laser cavity length is adjusted such that each laser pulse entering the cavity overlaps the previous pulses in space and in time, summing to a combined laser pulse with higher peak power. An acousto-optical deflector in the cavity periodically deflects the combined pulse to the output port, providing a light source with lower repetition rate but higher peak power. Due to the finite loss in the cavity dumper optics, pulse peak power typically does not improve beyond a factor of five to ten.

Another method to increase pulse peak power is the use of an optical parameter amplifier, but at the cost of significantly increasing the price of the microscope system and severely reducing the repetition rate to tens of kHz.

10.2.2 Coherent Control of Nonlinear Optical Excitation*

Nonlinear microscopy uses fs pulse laser sources that have broad spectral bandwidths. The broad spectral bandwidth offers a unique opportunity to manipulate the excitation process based on coherent control (Dela Cruz et al., 2004; Lozovoy & Dantus, 2005; Meshulach & Silberberg, 1998; Pastirk et al., 2005). The idea of using coherent control in multiphoton microscopy, multiphoton intrapulse interference (MII), is based on manipulating the phase and magnitude of each frequency component available in the fs laser pulse (Walowicz et al., 2002). The interference of these spectral components can selectively enhance or suppress a specific molecular transition.

Theoretically, we will consider the simple case of two-photon excitation via a virtual state, but the same approach can be extended to a high-order fluorescence process or other nonlinear excitation processes (Christov et al., 2001; Lee et al., 2001; Lozovoy & Dantus, 2005). Two-photon excitation probability, S_2, is proportional quadratically to the electric field, $|E(t)|^2$. A Fourier transform of the electric field, $\tilde{E}(\omega)$, gives the spectral contents of the light:

$$\tilde{E}(\omega) = |\tilde{E}(\omega)| \exp(i\varphi(\omega)) \tag{10.8}$$

The phase, $\varphi(\omega)$, and magnitude, $|\tilde{E}(\omega)|$, of the light at each spectral frequency can be manipulated and controlled experimentally. The two-photon excitation probability can be expressed spectrally as:

$$S_2 = \int_{-\infty}^{\infty} g^{(2)}(2\omega) \left| \int_{-\infty}^{\infty} |\tilde{E}(\omega + \Omega)| |\tilde{E}(\omega - \Omega)| \exp[i\varphi(\omega + \Omega) + i\varphi(\omega - \Omega)] d\Omega \right|^2 d\omega \tag{10.9}$$

where $g^{(2)}(2\omega)$ is the two-photon absorption spectra of the molecule and Ω is the intermediate spectral detuning.

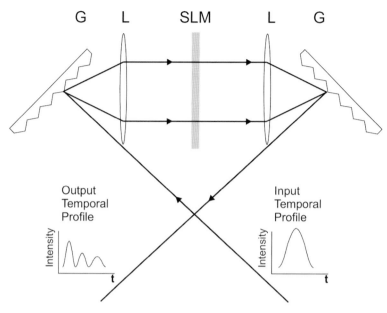

Figure 10.3. Schematics for coherent control (G = grating, L = lens, SLM = spatial light modulator). Transform-limited light pulse from laser enters from bottom right (the temporal profile of the light is illustrated by the left insert). The light pulse is dispersed spectrally by the grating on the left and is then collimated by a lens. The phases of the different spectral components are modified by a spatial light modulator. The lens on the right focuses the transmitted beam from the spatial light modulator onto the grating on the right. The grating recombines the different spectral components, generating a pulse with arbitrary temporal profile (illustrated by the right insert).

By controlling the phase and the magnitude of the different spectral bands, we can control the power delivered into each band. This method allows matching of the excitation spectrum with the absorption spectrum of the molecule, thus optimizing excitation efficiency. One can further selectively excite a specific allowed transition level within the absorption band of a molecule. Furthermore, in the presence of multiple fluorophores, this approach allows selective imaging of these fluorophores, provided the laser has sufficiently broad spectral bandwidth.

Experimentally, a number of methods have been developed to control the magnitude and the phase of the spectral components of the laser pulses. The most common approach uses a spatial light modulator (Fig. 10.3). The incident ultrafast fs light is spectrally dispersed with a grating. The dispersed light is transmitted through a spatial light modulator, allowing selective control of magnitude and phase of the spatial components. After pulse shaping, the spectral component is recombined and transmitted to the nonlinear microscope.

This technique has been used successfully to control the excitation processes in a multiphoton fluorescence microscope. This technique was first demonstrated in controlling two- and three-photon excitation efficiencies of organic dye molecules (Verluise et al., 2002; Walowicz et al., 2002). Selective excitation and imaging of a fluorophore with pH-dependent spectral shift were demonstrated (Fig. 10.4a) (Pastirk et al., 2003). Recently, this technique has been applied in the imaging of green fluorescent

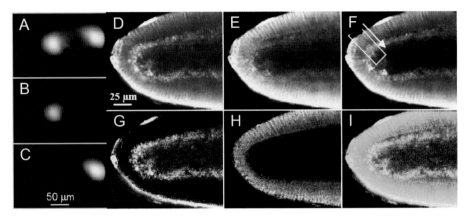

Figure 10.4. Selective multiphoton excitation using coherent control. (a–c) Selective imaging of fluorescent probe 8-hydroxypyrene-1,3,6-trisulfonic acid. This probe is pH-sensitive and its spectrum red shift is over 50 nm as pH is varied from 6 to 10. (The images are adapted from Pastirk et al., 2003.). (a) Excitation of both pH 6 and 10 region with transform-limited pulses. (b) Selective excitation of pH 6 region (left) with coherent control. (c) Selective excitation of pH 10 region (right) with coherent control. (d–i) Selective imaging of eGFP and autofluorescence in Drosophila embryo. (The images are adapted from Ogilvie et al., 2005.) (d, e) Coherent control to preferentially excite eGFP and autofluorescence respectively. (f) Excitation of the same region using transform-limited pulses. (g, h) Using linear unmixing to isolate regions of eGFP and autofluorescence respectively. (i) False color merged image of eGFP (green) and autofluorescence (blue) after unmixing.

protein-labeled structures and the autofluorescence in Drosophila embryos (see Fig. 10.4b) (Ogilvie et al., 2005). The use of coherent control in harmonic generation and coherent anti-Stokes' Raman microscopes have also been demonstrated (Dudovich et al., 2002; Oron et al., 2002).

10.3 ELEMENTS OF MICROSCOPE DESIGN

The optical design of a nonlinear microscope is often very simple. The main complexity lies in the microscope objective. The objective is required to focus infrared light with minimal aberration and pulse dispersion and to optimize the transmission of the emitted signal. Excellent microscope objectives are currently available commercially and can be readily incorporated into nonlinear microscope design. The rest of the microscope design is often satisfactorily modeled by paraxial approximation and can be understood using simple ray tracing.

Very basic ray tracing can be codified in four simple rules (Fig. 10.5): (1) A light ray emitted from the front focal point propagates parallel to the optical axis after the lens. (2) A light ray incident upon the lens parallel to the optical axis goes through the back focal point after the lens. (3) A divergent fan of light rays emitted from a point at the front focal plane becomes collimated after the lens. Coupled with rule #2, one can determine the propagation angle of the collimated light after the lens. (4) Collimated light rays incident upon a lens are focused at the back focal plane. The focus position at the back focal plane can be determined with the aid of rule #1. All the microscope

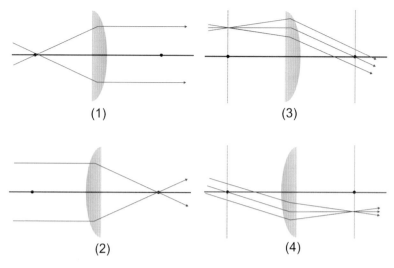

Figure 10.5. Four simple rules of ray tracing. The solid horizontal lines denote the optical axis. The front and back focal points are denoted by black solid circles on the optical axis. The arrow-headed lines are the light rays.

designs in the following sections contain lens pairs with their separation equal to the sum of their focal lengths. In this case, their design can be understood based on just these four ray-tracing rules.

10.3.1 Basic Configuration of Microscopes

The most basic microscope design consists of three elements: a microscope objective, a tube lens, and a spatially resolved detector such as a charge coupled device (CCD) (Fig. 10.6). A more extensive discussion of microscope objectives will be presented in a following section. In this section, the microscope objective can be considered as an ideal singlet lens. An extensive discussion of detectors will be presented in Chapter 12. Simple ray tracing of two objects at the specimen plane shows that they are imaged at the detector surface with a separation increased by a factor equal to the ratio of the focal length of the tube lens to that of the objective. This factor is called the magnification of the objective.

For practical imaging, most microscopes further require an illumination source, which can be arranged in either epi- or trans-illumination geometry (Fig. 10.7). Most microscopes have wide-field white light illumination, which is extremely useful in the rapid inspection of the specimen. The white light illuminator is often arranged in a trans-illumination geometry based on scattering, absorption, phase change, or polarization change as contrast mechanism. Microscope illumination uses the Koehler design. The light source intensity distribution is typically inhomogeneous, such as the filament of a xenon arc lamp. It is undesirable to directly project an image of the light source onto the specimen plane, resulting in an uneven illumination. Instead, the light emitted from each point of the light source is collimated and broadly illuminates the whole specimen plane, resulting in more uniform exposure. Koehler illumination is achieved by focusing the light from each point of the illuminator at the back aperture of the microscope objective.

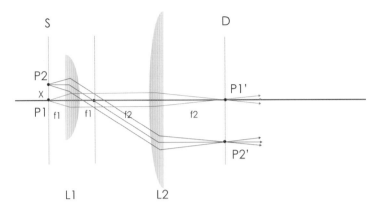

Figure 10.6. Basic microscope design S = specimen plane, D = detector, L1 = lens (objective) with focal length f1, L2 = lens (tube lens) with focal length f2. Point P1 and P2 at specimen plane are mapped to P1′ and P2′ on the detector. The separation x between P1 and P2 and the separation x′ between P1′ and P2′ are related by the magnification factor M where $M = \frac{f2}{f1}$.

Fluorescence wide-field illumination is also very useful in nonlinear microscopes for the initial inspection of the specimen. A common fluorescence wide-field light source is a mercury arc lamp. Similar to white light illumination, the mercury arc lamp uses Koehler illumination geometry to ensure uniform illumination. However, for fluorescence imaging mode, epi-illumination is typically employed instead of trans-illumination (see Fig. 10.7). The importance of epi-illumination can be understood by noting that the illumination power at one pixel of the image can be easily 10^{10} higher than the fluorescence signal originating from the same pixel. Therefore, for high-sensitivity imaging, the direct illumination of the detector by the excitation light results in lower sensitivity even in the presence of a barrier filter that blocks the excitation light. On the other hand, in the epi-illumination geometry, the excitation propagates away from the specimen and the detector, allowing higher sensitivity imaging.

Another difference between fluorescence and white light imaging is the need for highly spectrally selective interference filters. An excitation filter is needed to isolate a specific emission light in the mercury lamp emission spectrum. For epi-illumination, a dichroic filter is needed to reflect the shorter-wavelength excitation light towards the objective while allowing the longer-wavelength emission light to pass toward the detector. Although the majority of the excitation light propagates away from the detector and most optical surfaces in the microscope are anti-reflection coated, typically several percents of excitation light intensity are reflected back towards the detector. As discussed previously, fluorescent signal is weak and is low, even compared with this small fraction of the excitation light. Therefore, a barrier filter is still needed for further excitation light rejection. Barrier filters used in fluorescence microscope can have over 4 to 6 O.D. blocking power.

10.3.2 Design of Nonlinear Optical Microscopes

Unlike white light or fluorescent wide-field imaging, nonlinear optical microscopes often require point excitation at the focal plane to maximize intensity and nonlinear

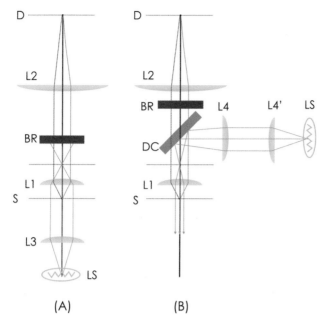

Figure 10.7. Wide-field microscope configuration. (a) Trans-illumination and (b) epi-illumination. LS = light source, L1, L2, L3, L4, L4′ = lenses, S = specimen plane, D = detector, DC = dichroic mirror, BR = barrier filter.

excitation efficiency. In fact, the nonlinear excitation light path is often simpler than the wide-field illumination light path. Since the laser source is a collimated Gaussian beam, there are no issues related to spatial inhomogeneity as in a lamp source. Further, the laser used for nonlinear excitation is nearly monochromatic, and chromatic aberration in the beam path is not a serious concern. The typical design of a nonlinear microscope consists of three major compartments: the excitation light path, the microscope objective, and the detection light path (Fig. 10.8).

10.3.2.1 Microscope Objective Considerations

The objective is the heart of the microscope optical system. A microscope objective is a complex optical component that typically contains over ten individual optical elements, aiming to correct for many of the monochromatic and chromatic aberrations. The design considerations for microscope objectives are beyond the scope of this chapter.

The most important parameter of the objective is its numerical aperture (NA). This specifies its ability (solid angle) to gather light and determines the microscope's optical resolution. The NA of the objective is defined as (Fig. 10.9):

$$NA = n \sin \theta \tag{10.10}$$

where n is the index of refraction of the medium between the objective and its focal point, and θ is the half-angle subtended by the objective lens aperture and its focal point.

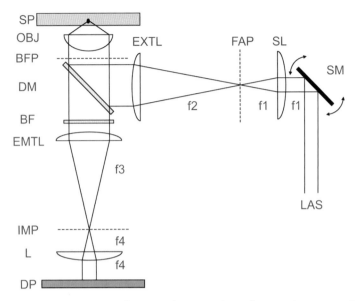

Figure 10.8. Schematic design of a typical inverted nonlinear microscope. LAS = laser source, SP = specimen plane, DP = detector plane, BFP = back focal plane of the objective, FAP = field aperture plane, OBJ = objective, EXTL = excitation tube lens, EMTL = emission tube lens, SL = scanning lens, L = collimating lens for detector, SM = scanning mirrors, DM = dichroic mirror, BF = barrier filter. The focal length of SL is f1. The distances between SM-SL and SM-FAP are both equal to f1. The focal length of EXTL is f2. The distances between EXTL-FAP and ETL-BFP are both equal to f2. The focal length of EMTL is f3. The distances between EMTL-BFP and EMTL-IMP are both equal to f3. The focal length of L is f4. The distances between L-IMP and L-DP are both equal to f4. The addition of lens L ensures that the projection of the signal rays on the detector is stationary, independent of scanning mirror motion. This arrangement gives intensity uniformity across the image even if the detector surface sensitivity is non-uniform.

Objectives are often classified for air, water, or oil immersion, depending on the coupling fluid between the objective and the specimen. Clearly, a microscope objective designed for using a higher index coupling medium can achieve a NA. Today, the highest NA objective has an NA of about 1.65 using a special immersion oil.

A higher NA objective will maximize light collection. Light collection efficiency of an objective scales as NA^2. For a fixed lens diameter, a higher NA also implies that the focal distance shortens proportionally. The working distance of the objective is related to this focal distance. For a typical objective that is designed to image through a piece of cover glass, the working distance is defined as the difference of the focal distance and the cover glass thickness. On the other hand, for objectives designed for use without a cover glass, the working distance is just the focal distance.

Most importantly, the microscope objective determines the resolution of the microscope. The microscope objective is effectively a circular aperture. Since light behaves as a wave, how finely that light can be focused through the aperture is limited by the diffraction of light waves as formulated by Abbe. In the Fraunhofer limit, the diffraction theory predicts the light distribution at the focal point. Mathematically, the intensity distribution resulting from the imaging of a point object is called the point spread

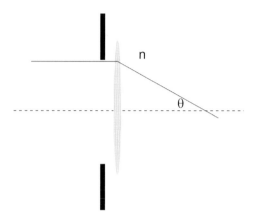

Figure 10.9. Numerical aperture (NA) of a lens. The NA of a lens is equal to the product of the index of refraction, n, of the medium in the focal region and the sine of the angle, θ, subtended by the lens aperture radius and its focal length.

function (PSF). The PSF can be quantified for an n-photon process as:

$$PSF(u, 0) \propto \left[\frac{\sin(u/4)}{(u/4)} \right]^{2n} \tag{10.11}$$

$$PSF(0, v) \propto \left[\frac{2J_1(v)}{v} \right]^{2n} \tag{10.12}$$

where J_1 is the first-order Bessel function.

The dimensionless optical coordinates along axial (z) and radial (r) directions are defined: $u = 4k\,NA^2z$ and $v = kNAr$ where $k = \frac{2\pi}{\lambda}$ is the wavenumber, λ is the wavelength, and NA is the numerical aperture of the objective. A complete and in-depth discussion of nonlinear microscope resolution is presented in Chapter 11.

Note that the magnification of the microscope objective is not an important parameter for nonlinear scanning optical microscopes. Magnification is not an inherent property of the objective itself. Recall that it is the combination of the objective and the tube lens that defines the magnification (angular deviation of the light rays). Therefore, the magnification of the objective lens is only correct as long as it is used with a specific matching tube lens (i.e., objectives should be used with the tube lens provided by the same manufacturer).

10.3.2.1.1 TRANSMISSION AND PULSE DISPERSION IN
            OBJECTIVES

The construction of microscope objectives is beyond the optical design capability of most laboratories. Excellent microscope objectives are made by major microscope manufacturers for linear optical applications in the visible range. These objectives have low intrinsic fluorescence and have high transmission efficiency throughout the near-ultraviolet and visible spectrum. Unfortunately, many commercial microscope objectives are not optimized for nonlinear optical microscopy. First, many objectives do not have high transmission throughout the spectrum spanning the near-ultraviolet to the near-infrared wavelengths. The issue is particularly severe for applications such as third-harmonic generation using excitation light at 1.2 to 1.3 μm. Second, nonlinear microscopes require the transmission of laser light with minimal pulse dispersion. However, due to the presence of multiple optical components in these microscope objectives, many microscope objectives have a GVD on the order of 3,000 fs^2 (Wokosin, 2005). There are microscope objectives with a significantly highly GVD and these are less suitable for nonlinear microscopy. This level of GVD causes a factor of two or three broadening of the 100-fs pulses.

10.3.2.1.2 INDEX MATCHING BETWEEN MICROSCOPE OBJECTIVE AND SPECIMEN

While microscope objectives are aberration-corrected for specimen right on top of the cover slip, their PSF can degrade significantly as imaging depth approaches 50 to 100 μm. This effect is due to index mismatch between the immersion medium and the specimen, and resolution degradation is observed when oil immersion or air objectives are used for aqueous biological specimens (Visser et al., 1992). This index of refraction-induced spherical aberration can often be minimized by using water immersion objectives for thick specimens.

10.3.2.2 Excitation Light Path Considerations

The excitation beam path of nonlinear microscopes typically consists of two components: a beam-scanning actuator and beam-shaping optics to couple excitation light to the microscope objective. In addition to these components, dichroic filters and barrier filters are needed to separate and direct the excitation and emission light. In nonlinear microscopy, the excitation light spectrum is far from the emitted light spectrum and is relatively monochromatic. Typically no excitation filter is needed to refine the excitation spectrum, as is needed in wide-field one-photon fluorescence microscopy using a mercury lamp. However, sometimes a small fraction of the light from the pump laser of the Ti:Sa laser can transmit through the Ti:Sa laser and enter the microscope. Since the pump laser light often falls within the signal spectrum, it is not rejected by the dichroic filter or the barrier filter and may interfere with the detection of very weakly emitting object. This effect can be eliminated by using an excitation filter to refine the Ti:Sa laser input.

10.3.2.2.1 LASER BEAM SCANNER SYSTEM

The depth discrimination of most nonlinear microscopes results from the focusing of the excitation light to a diffraction limited point. Therefore, the scanning of this excitation volume in three dimensions is required to produce an image stack. For lateral scanning, two methods have been implemented: stage scanning and beam scanning. The stage scanning system is straightforward. The optics can be kept perfectly paraxial, resulting in minimal aberration. Further, the movement of a piezoelectric-driven stage can be controlled within nanometer precision. However, while stage movement is precise, it is also slow. Stage scanning microscopes can rarely exceed a scan rate of one frame per minute. Further, the mechanical vibration of the scanning stage may also compromise the registration of soft biological specimens and may cause potential mechanical damage. Today, stage-scanning microscopes are rarely used in biomedical research.

Instead of stage scanning, most nonlinear microscopes use a beam-scanning method. The most common beam-scanning system utilizes a galvanometric scanner. The galvanometric scanning has a significantly higher bandwidth than stage scanning because the weight of the scanning mirror is orders of magnitude less than that of the specimen stage. The typical galvanometric scanner consists of two synchronized mirrors that move in tandem to produce an x-y raster scanning pattern. The scanner typically runs in the servo mode with a bandwidth up to about 1 kHz. For higher-speed scanning, the scanner can also run in resonance mode. The bandwidth of resonance scanners can reach about tens of kHz, but they are more difficult to synchronize with the other subsystems of the microscope because their motion is no longer under feedback control. Another important high-speed mechanical scanner is a polygonal scanner. A rotating cylinder with multiple mirror facets at its perimeter is rotated at high speed. The swipe of each facet across the beam results in a line scan. This system has similar bandwidth as a resonance scanner. For a nonmechanical scanner, acousto-optical deflectors have also been used for beam scanning. An acousto-optical system has the advantage that one can generate an arbitrary scanning pattern with submillisecond point-to-point settling time. However, acousto-optical scanners produce significant spatial aberration and temporal dispersion. A more detailed discussion of integrating these fast scanners in nonlinear microscope for high-speed imaging will be given in Chapter 18.

An additional actuator is required for axial scanning. The most common approach is to translate the microscope objective using a piezo-positioner. Due to the weight of the microscope objective, the bandwidth of this system is at best a few 100 Hz. Due to its lower bandwidth compared to the lateral scanners, most nonlinear microscopes produce 2D en-face images first and then scan in depth. One limitation of a piezoelectric positioner is the limited translation range, on the order of 100 to 300 microns. This limitation can be overcome by including a slower servo motor system that controls the height of the microscope specimen stage, working in conjunction with the faster piezoelectric positioner.

10.3.2.2.2 BEAM EXPANSION, OBJECTIVE OVERFILLING, AND RESOLUTION

The excitation light path is configured such that the field aperture plane is "telecentric" to the specimen plane (see Fig. 10.8). Two planes are telecentric if the light distribution on one plane is a scaled version of the other plane. Since the specimen plane is telecentric to the field aperture plane, the laser scan pattern at the specimen is a scaled version of the scan pattern of the laser focus at the field aperture plane. Telecentricity is accomplished by placing a tube lens between the objective and the field aperture plane; the tube lens is positioned such that its front focal point coincides with the back focal point of the microscope objective and its back focal point lies on the field aperture plane. The scale factor of the scan pattern between the two planes is equal to the magnification factor for the objective–tube lens pair.

Because of telecentricity, the task of generating the appropriate scan pattern at the specimen plane is equivalent to producing a scan pattern at the field aperture plane.

A scan lens focuses the deflected beam from the scanner at the field aperture plane, forming the scan pattern. The scan lens is positioned such the scanners are at its front focal point (eyepoint) while the field aperture plane is at its back focal point. The scan lens is a "$\theta - \varepsilon$" lens, a compound lens with the property that the displacement of its focal point from the optical axis is proportional to the incident angle of the incident ray. This relationship is approximately true for most lenses at a small angle, but this relationship holds even at a large angle for a "$\theta - \varepsilon$" lens. This arrangement ensures that the translation of the focal point at the specimen plane is proportional to the angular deflection of the scan lens. Note that this optical design assumes that the pivot points of both the x- and y- scan mirrors are placed at the eyepoint of the scan lens. Of course, it is impossible to place the pair of galvanometric scanning mirrors both at the same eyepoint. This problem can be solved approximately by positioning the two mirrors very close to each other and placing the eyepoint equidistant between the two pivot points of the mirrors. This arrangement results in slightly nonplanar scanning of the focus in the specimen and also loss of excitation efficiency at a large scan angle. This problem can be solved by using either a pair of relay lenses or a pair of spherical mirrors that conjugates the pivot point of one scan mirror to the pivot point of the other scan mirror.

The resolution (and depth discrimination) of the nonlinear optical microscope depends on the use of a high NA objective. The nominal NA of an objective given by its manufacturer assumes that a plane wave incidents upon and fills the microscope back aperture. Therefore, since the laser beam has a Gaussian profile, to use the full NA of the microscope objective, it needs the beam-expanded laser such that only the central, relatively flat, portion of the Gaussian beam profile illuminates the back aperture of the objective. This is accomplished by expanding the laser beam such that it overfills the back aperture of the objective. The scan lens and the tube lens function together as a beam expander. The focal length ratio of the tube lenses and the scan lens is the magnification factor of the beam expander and should be chosen such that the incident laser beam is expanded to overfill the back aperture of the microscope objective to achieve the best spatial resolution. With the development of field-flattening optical components, it may be possible to produce a plane incident wave at the back aperture of the objective without overfilling, resulting in more efficient use of the power of the excitation laser.

10.3.2.3 Emission Light Path Considerations

Since most laser light sources provide far more power than needed for nonlinear excitation, the efficiency of the excitation light path is typically not critical. On the other hand, nonlinear effects are typically weak, and it is vital to optimize the sensitivity of the microscope system. Four factors in the optical design of emission light path should be considered: descanning, optical component solid angle, confocal detection, and light detection in transconfiguration.

10.3.2.3.1 ADVANTAGE AND DISADVANTAGE OF CONFOCAL DETECTION

Confocal microscopes obtain depth discrimination by placing a pinhole aperture at a position conjugated to the excitation focus in the detection light path. While a pinhole aperture in the emission light path is not necessary for nonlinear optical microscopes, the inclusion of a confocal pinhole can improve the microscope's spatial resolution with a resulting loss of signal. Optical theory has shown that the inclusion of a confocal pinhole can allow a multiphoton microscope to achieve resolution superior to that of a confocal microscope and significantly better than a multiphoton microscope without pinhole (Gu, 1996; Gu & Sheppard, 1995). Experimentally, this has been verified in optically thin specimens. However, this resolution advantage in thin specimens is quickly overcome by the reduction in image signal-to-noise ratio in optically thick samples due to the scattering of the emission photons (Centonze & White, 1998).

In the case that confocal detection is desirable, it is also important to consider the size of the pinhole aperture. Theoretically, the ideal PSF can be achieved only using an infinitely small pinhole aperture. It is intuitively clear that if we have an infinitely small pinhole, we will also have an infinitely small signal. Although we will achieve ideal resolution in this limit, we will measure no signal. In practice, the pinhole aperture clearly should not be set infinitely small. One would expect that the pinhole should have dimensions comparable to the lateral dimension of the PSF to maximize the detection signal-to-noise ratio. Sandison and Webb have confirmed this idea theoretically and experimentally (Sandison et al., 1995; Sandison & Webb, 1994).

10.3.2.3.2 DETECTION OPTICS SOLID ANGLE AND DESCANNING
          GEOMETRY

Ideally, the light produced at the focal region of the specimen should be collimated after the objective; the design of relay optics to take a collimated ray to the detector should be relatively straightforward. However, when imaging is deep within an optically thick specimen, the emission photons must first escape the sample, a significant fraction of the emission photons may be scattered, and their trajectories are deviated from their theoretical paths. In this case, the emission light distribution from the specimen more resembles an extended object instead of a point source. To improve light collection efficiency from this extended source, it is recommended that the solid angles of optical components in the detection light path should be maximized and the detection light path should be minimized. Furthermore, detectors with large active surface areas should be implemented. In the future, the detection light path of nonlinear optical microscopes may be further optimized based either on Monte Carlo modeling (Blanca & Saloma, 1998; Gan & Gu, 2000) or experimental measurement of the distribution of these scattered emission photons (Beaurepaire & Mertz, 2002; Dong et al., 2003; Dunn et al., 2000; Oheim et al., 2001; Ying et al., 1999).

For confocal microscopy, a major hurdle in the design of a beam-scanning system is the need to descan the emission light. Since the emission light originates from different locations in the specimen during a raster scan, the emission light travels back through the microscope along different paths. It is clearly difficult to dynamically position a pinhole aperture in the optical path during beam scanning. Considering the emission light path of the microscope, the beam is nonstationary in the light path between the scanning mirrors and the specimen. However, since the mirror movement is slow compared with the speed of light, the scan mirrors remain approximately at the same angles for the passage of the emission light as for that of the excitation light. By symmetry, it is clear that the emission light ray will become stationary after it passes through the scan mirrors again. By placing a dichroic mirror in between the scan mirrors and the light source, one can separate the excitation light from the emission light. The emission light can then be focused through a stationary confocal pinhole aperture. This is the process of descanning, and it is implemented in most commercial confocal microscopes.

While descanning is critical for the design of a confocal microscope, it is often not necessary in the design of a nonlinear optical microscope. Since the effects of resolution enhancement and depth discrimination originate from the nonlinear excitation process itself, the use of a confocal pinhole aperture is not required. Removing the constraint of descanning has a number of advantages. First, the number of optical surfaces in the emission light path can be reduced to increase signal transmission. Second, since the emission light path optical design is simpler in non-descanning geometry, the emission light optical path can be shortened, resulting in higher collection efficiency for scattered photons.

10.3.2.3.3 TRANSILLUMINATION GEOMETRY FOR THE DETECTION OF SECOND-HARMONIC SIGNAL

In a previous section, we have explained why transillumination geometry is often not suitable for the detection of weak nonlinear optical signal in the presence of a large excitation background. However, one exception to this rule is second- and higher-harmonic generation microscopy. Since the harmonic generation signal is coherent, the signal emitted from the focal point is not isotropic, as in the fluorescence case. Constructive interference occurs only within a narrow forward-going cone for the second- and higher-harmonic generation. There is almost no backward-going harmonic generation signal due to destructive interference. Therefore, transillumination geometry is needed in this case to maximize signal collection. In addition to a barrier filter, the signal-to-noise ratio can be improved by taking advantage of the narrow solid angle of the higher harmonic emission when placing an annular aperture to block the excitation light without blocking the transmission of the signal. The physical origin of the directed emission of higher harmonic signal and the instrument designs for its detection will be discussed extensively in Chapters 15 and 16.

10.4 CONCLUSION

The optical design of nonlinear optical microscopes is in general very simple. This simplicity originates from three facts. First, since only a near-monochromatic excitation source is used, the excitation path is relatively immune to chromatic aberration, requiring only that the excitation light is focused at the diffraction limit. Second, most nonlinear microscopes operate in non-descanning detection mode, significantly relaxing the optical requirement in the detection path. Since no imaging is required during detection, the detection optics only needs to maximize photon collection efficiency, but becomes completely insensitive to chromatic and nonchromatic aberrations. Third, the large separation between excitation and emission spectra allows the signal to be readily isolated with minimal loss, using highly efficient interference filters.

With the many advantages of nonlinear optical microscopes, there are also two challenging areas in their design. First, nonlinear optical microscopy requires ultrafast laser pulses for efficient excitation. The dispersion properties of all the optical components must be evaluated, and sometimes a dispersion compensation technique must be applied. Second, one of the major advantages of nonlinear microscope is its ability to image a highly scattering, thick specimen. The effects of tissue scattering and the index of refraction variation on imaging quality still need to be better quantified. The minimization of these effects should allow the next-generation nonlinear microscopes to image into thicker specimens.

REFERENCES

Beaurepaire, E. and J. Mertz (2002). "Epifluorescence collection in two-photon microscopy." Appl. Opt. 41(25): 5376–5382.

Blanca, C. M. and C. Saloma (1998). "Monte Carlo analysis of two-photon fluorescence imaging through a scattering medium." App. Opt. 37(34): 8092–8102.

Booth, M. J., M. A. Neil, R. Juskaitis and T. Wilson (2002a). "Adaptive aberration correction in a confocal microscope." Proc Natl Acad Sci U S A 99(9): 5788–5792.

Booth, M. J., M. A. Neil and T. Wilson (2002b). "New modal wave-front sensor: application to adaptive confocal fluorescence microscopy and two-photon excitation fluorescence microscopy." J Opt Soc Am A Opt Image Sci Vis 19(10): 2112–2120.

Booth, M. J. and T. Wilson (2001). "Refractive-index-mismatch induced aberrations in single-photon and two-photon microscopy and the use of aberration correction." J Biomed Opt 6(3): 266–272.

Centonze, V. E. and J. G. White (1998). "Multiphoton excitation provides optical sections from deeper within scattering specimens than confocal imaging." Biophys. J. 75(4): 2015–2024.

Christov, I. P., R. Bartels, H. C. Kapteyn and M. M. Murnane (2001). "Attosecond time-scale intra-atomic phase matching of high harmonic generation." Phys Rev Lett 86(24): 5458–5461.

Cianci, G. C., J. Wu and K. M. Berland (2004). "Saturation modified point spread functions in two-photon microscopy." Microsc Res Tech 64(2): 135–141.

Dela Cruz, J. M., I. Pastirk, M. Comstock, V. V. Lozovoy and M. Dantus (2004). "Use of coherent control methods through scattering biological tissue to achieve functional imaging." Proc Natl Acad Sci U S A 101(49): 16996–17001.

Denk, W., J. H. Strickler and W. W. Webb (1990). "Two-photon laser scanning fluorescence microscopy." Science 248(4951): 73–76.

Diaspro, A., Ed. (2002). Confocal and Two-Photon Microscopy: Foundations, Applications and Advances. New York, Wily-Liss.

Diaspro, A., M. Corosu, P. Ramoino and M. Robello (1999a). "Adapting a compact confocal microscope system to a two-photon excitation fluorescence imaging architecture." Micros. Res. Tech. 47: 196–205.

Diaspro, A., M. Corosu, P. Ramoino and M. Robello (1999b). "Two-photon excitation imaging based on a compact scanning head." IEEE Engineering in Medicine & Biology Magazine 18(5): 18–30.

Dong, C. Y., K. Koenig and P. So (2003). "Characterizing point spread functions of two-photon fluorescence microscopy in turbid medium." J Biomed Opt 8(3): 450–459.

Dudovich, N., D. Oron and Y. Silberberg (2002). "Single-pulse coherently controlled nonlinear Raman spectroscopy and microscopy." Nature 418(6897): 512–514.

Dunn, A. K., V. P. Wallace, M. Coleno, M. W. Berns and B. J. Tromberg (2000). "Influence of optical properties on two-photon fluorescence imaging in turbid samples." Applied Optics 39: 1194–1201.

Gan, X. and M. Gu (2000). "Spatial distribution of single-photon and two-photon fluorescence light in scattering media: Monte Carlo simulation." App. Opt. 39(10): 1575–1579.

Gu, M. (1996). Principles of Three-Dimensional Imaging in Confocal Microscopy. Singapore, World Scientific.

Gu, M. and C. J. R. Sheppard (1995). "Comparison of Three-Dimensional Imaging Properties between Two-Photon and Single-Photon Fluorescence Microscopy." J. Microsc. 177: 128–137.

Inoué, S. and K. R. Spring (1997). Video microscopy: the fundamentals. New York, Plenum Press.

Lee, D. G., J. H. Kim, K. H. Hong and C. H. Nam (2001). "Coherent control of high-order harmonics with chirped femtosecond laser pulses." Phys Rev Lett 87(24): 243902.

Liu, T. M., S. W. Chu, C. K. Sun, B. L. Lin, P. C. Cheng and I. Johnson (2001). "Multiphoton confocal microscopy using a femtosecond Cr:forsterite laser." Scanning 23(4):249–254.

Lozovoy, V. V. and M. Dantus (2005). "Systematic control of nonlinear optical processes using optimally shaped femtosecond pulses." Chemphyschem 6(10): 1970–2000.

Masters, B. R. (2006). Confocal Microscopy and Multiphoton Excitation Microscopy: the Genesis of Live Cell Imaging. Bellingham, SPIE Press.

Masters, B. R., P. T. So, C. Buehler, N. Barry, J. D. Sutin, W. W. Mantulin and E. Gratton (2004). "Mitigating thermal mechanical damage potential during two-photon dermal imaging." J Biomed Opt 9(6): 1265–1270.

Meshulach, D. and Y. Silberberg (1998). "Coherent quantum control of two-photon transitions by a femtosecond laser pulse." Nature 396: 239–242.

Nagy, A., J. Wu and K. M. Berland (2005a). "Characterizing observation volumes and the role of excitation saturation in one-photon fluorescence fluctuation spectroscopy." J Biomed Opt 10(4): 44015.

Nagy, A., J. Wu and K. M. Berland (2005b). "Observation volumes and {gamma}-factors in two-photon fluorescence fluctuation spectroscopy." Biophys J 89(3): 2077–2090.

Ogilvie, J. P., D. Débarre, X. Solinas, J. L. Martin, E. Beaurepaire and M. Joffre (2005). "Use of coherent control for selective two-photon fluorescence microscopy in live organisms." Opt. Exp. 14(2): 759–766.

Oheim, M., E. Beaurepaire, E. Chaigneau, J. Mertz and S. Charpak (2001). "Two-photon microscopy in brain tissue: parameters influencing the imaging depth." J Neurosci Methods 111(1): 29–37.

Oron, D., N. Dudovich and Y. Silberberg (2002). "Single-pulse phase-contrast nonlinear Raman spectroscopy." Phys Rev Lett 89(27): 273001.

Pastirk, I., J. M. Dela Cruz, K. A. Walowicz, V. V. Lozovoy and M. Dantus (2003). "Selective two-photon microscopy with shaped femtosecond pulses." Opt. Exp. 11(14): 1695–1701.

Pastirk, I., M. Kangas and M. Dantus (2005). "Multidimensional analytical method based on binary phase shaping of femtosecond pulses." J Phys Chem A Mol Spectrosc Kinet Environ Gen Theory 109(11): 2413–2416.

Pawley, J. B., Ed. (1995). Handbook of Confocal Microscopy. New York, Plenum.

Potma, E. O., W. P. de Boeij, M. S. Pshenichnikov and D. A. Wiersma (1998). "30-fs, cavity-dumpedoptical parametric oscillator." Opt Lett 23(22): 1763–1765.

Sandison, D. R., D. W. Piston, R. M. Williams and W. W. Webb (1995). "Quantitative comparison of background rejection, signal-to-noise ratio, and resolution in confocal and full-field laser scanning microscopes." Applied Optics 34(19): 3576–3588.

Sandison, D. R. and W. W. Webb (1994). "Background rejection and signal-to-noise optimization in confocal and alternative fluorescence microscopes." Applied Optics 33(4): 603–615.

Stelzer, E. H. (1995). The Intermediate Optical System of Laser-Scanning Confocal Microscopes. Handbook of Biological Confocal Microscopy. J. Pawley. New York, Plenum Press: 139–154.

Sun, C. K. (2005). "Higher harmonic generation microscopy." Adv Biochem Eng Biotechnol 95: 17–56.

Verluise, F., V. Laude, Z. Cheng, C. Spielmann and P. Tournois (2002). "Amplitude and phase control of ultrashort pulses by use of an acousto-optic programmable dispersive filter: pulse compression and shaping." Opt Lett 25(8): 575–577.

Visser, T. D., J. L. Oud and G. J. Brakenhoff (1992). "Refractive index and axial distance measurements in 3-D microscopy." Optik 90: 17–19.

Walowicz, K. A., I. Pastirk, V. V. Lozovoy and M. Dantus (2002). "Multiphoton Intrapulse Interference. 1. Control of Multiphoton Processes in Condensed Phases." J. Phy. Chem. A 106: 9369–9373.

Wilson, T. (1990). Confocal Microscopy. London, Academic Press.

Wilson, T. and C. J. R. Sheppard (1984). Theory and Practice of Scanning Optical Microscopy. New York, Academic Press.

Wokosin, D. L. (2005). Pulse duration spectra and measurements for laser scanning microscope systems. Multiphoton Microscopy in the Biomedical Sciences V, SPIE Proc., SPIE.

Ying, J., F. Liu and R. R. Alfano (1999). "Spatial distribution of two-photon-excited fluorescence in scattering media." App. Opt. 38(1): 224–229.

11

Image Formation in Multiphoton Fluorescence Microscopy*

Min Gu

11.1 INTRODUCTION

This chapter is dedicated to image formation in multiphoton fluorescence microscopy. In particular, the comparison of image formation is based on the three-dimensional intensity point spread function (IPSF), the three-dimensional optical transfer function (OTF), and axial and transverse image resolution for thin and thick objects.

Since its inception, confocal scanning microscopy has become a widely used and important tool in many fields, including biology, biochemistry, chemistry, physics, and industrial inspection (1–6). One of the main advantages in confocal microscopy is its ability of three-dimensional (3-D) imaging of a thick object. Because of the 3-D imaging property in confocal scanning microscopy, confocal fluorescence microscopy was achieved in the same period as confocal bright-field microscopy. Under the illumination of intermediate power, one incident photon can be absorbed in the sample under inspection to excite the electron transition from the ground state to an excited state. The excited electron returns to the ground state by radiating fluorescence light. The energy of the radiated fluorescence photon is slightly less than the incident one due to the nonradiation relaxation during the downward transition, and the corresponding microscopy in which the fluorescence light is imaged is termed single-photon (1-p) fluorescence microscopy.

The ultrashort pulsed laser technology has recently been combined with confocal microscopy for novel imaging modes. Such a combination also allows many novel applications to be possible. One of the emerging areas is nonlinear optical microscopy, which uses the nonlinear radiation generated from the sample by the high peak power of an ultrashort pulse. A multiphoton fluorescence process is one of the nonlinear processes caused by the simultaneous absorption of two or more incident photons under the illumination of an ultrashort pulsed beam. The energy of the resulting fluorescence photon is slightly less than the sum of the energy of the absorbed incident photons.

The original idea of two-photon (2-p) fluorescence scanning microscopy was proposed by Sheppard and colleagues along with other nonlinear scanning microscope modes (6,7). 2-p fluorescence microscopy was first demonstrated by Denk in 1990 (8). It has been reported that strong three-photon (3-p) fluorescence can also be generated in some organic solutions (9), where three incident photons are absorbed simultaneously and the radiating photon has energy approximately three times as large as the incident one. It was Hell and his colleagues who measured the first 3-p fluorescence microscopic image using BBO (10). With multiphoton absorption, one can have access to ultraviolet (UV) photon excitation without using a UV source. Due to the cooperative multiphoton excitation, the photobleaching associated with 1-p fluorescence is confined only to the vicinity of the focal region, and 3-D imaging becomes possible without the necessity for a confocal pinhole mask (8,10). However, axial resolution in multiphoton fluorescence microscopy can be improved if a confocal mask is used (11–17).

These imaging properties can be well understood in terms of the method based on the concept of the 3-D transfer function (5). With this method, the relationship between different optical arrangements can be gained and the image quality can be improved in image processing. This chapter is organized as follows: Section 11.2 presents a general description of image formation in multiphoton fluorescence microscopy with a finite-sized detector. In Sections 11.3 and 11.4, 3-D IPSF and the 3-D OTF are used to understand the performance of the multiphoton fluorescence imaging systems. Resolution in the transverse and axial directions is discussed in Section 11.5 in terms of the calculated images of layer and edge objects. The effect of the fluorescence wavelength on image resolution is presented in Section 11.6.

11.2 THREE-DIMENSIONAL IMAGE FORMATION

It should be pointed out that the method presented in this chapter is based on continuous-wave illumination. However, practical multiphoton excitation is usually performed under the ultrashort pulsed-beam illumination (8–10). The analysis in this paper can be generalized to the case under the pulsed illumination (14). Conclusions presented in this chapter may remain the same if the pulse width is longer than 10 femtoseconds (fs).

A schematic diagram of a confocal microscope consisting of an objective and a collector can be found elsewhere (5). Consider a point source and assume that the light from the source has a wavelength λ. Therefore, the intensity distribution at a point r_1 in object space may be expressed, under the paraxial approximation, as (5,16,18):

$$I_o(r_1) = |h_1(M_1 r_1)|^2 . \tag{11.1}$$

Here the parameter M_1 is a matrix including demagnification factors for the objective (5,16) and r_1 is a position vector in object space. $h_1(r)$ is the 3-D amplitude point spread function for the objective lens at incident wavelength λ (5,16,18).

For a multiphoton fluorescent object, the fluorescence light from a given point at r_1 is proportional to the square (2-p) or cube (3-p) of the light intensity impinging on the point and has a wavelength λ_f, which is slightly larger than $\lambda/2$ (2-p) or $\lambda/3$ (3-p). Assume that the strength of the fluorescence light from the object is $o_f(r)$. The total fluorescence strength emitted from the point r_1 in the object under the illumination $I_o(r_1)$ can be expressed as:

$$I_1(r_1) = |h_1(M_1 r_1)|^{2n} o_f(r_s - r_1), \tag{11.2}$$

where a scan position r_s has been introduced to include the scanning movement of the object, so that equation (11.2) is also a function of r_s. Here $n = 1, 2$, and 3 correspond to 1-p, 2-p, and 3-p processes, respectively.

For a single fluorescent point object at r_1, the intensity at r_2 in detection space is given, according to reference (18), by:

$$I_2(r_1, r_2) = |h_2(r_1 + M_2 r_2)|^2,\qquad(11.3)$$

where $h_2(r)$ is the 3-D amplitude point spread function for the collector lens at wavelength λ_f and M_2 is a matrix including magnification factors for the collector lens (16,18). Since a multiphoton fluorescence process is incoherent, the total intensity at r_2 from the 3-D fluorescent object with the fluorescence strength $I_1(r_1)$ is the superposition of the intensity contributed from all points in the object (i.e., the integration of the product of $I_1(r_1)$ and $I_2(r_1, r_2)$ with respect to r_1):

$$I(r_2, r_s) = \int_{-\infty}^{\infty} |h_1(M_1 r_1)|^{2n} o_f(r_s - r_1) |h_2(r_1 + M_2 r_2)|^2 \, dr_1,\qquad(11.4)$$

which has been explicitly expressed as a function of the scan position. If a finite-sized detector of an intensity sensitivity $D(r)$ is employed to collect enough signal, the final measured intensity from the sample is the summation of equation (11.4) over the detector aperture:

$$I(r_s) = \int_{-\infty}^{\infty} \left[\int_{-\infty}^{\infty} |h_1(M_1 r_1)|^{2n} o_f(r_s - r_1) |h_2(r_1 + M_2 r_2)|^2 \, dr_1 \right] D(r_2) dr_2,$$
$$(11.5)$$

which can be rewritten as a form of the 3-D convolution operation:

$$I(r_s) = h_i(r_s) \otimes_3 o_f(r_s),\qquad(11.6)$$

where $\otimes_3$ denotes the 3-D convolution operation and $h_i(r)$ is given by

$$h_i(r) = |h_1(M_1 r)|^{2n} \left[|h_2(r)|^2 \otimes_3 D\left(\frac{r}{M_2}\right) \right].\qquad(11.7)$$

Clearly, equation (11.7) represents the image intensity of a single fluorescent point and is thus the 3-D effective IPSF for multiphoton fluorescence microscopy.

11.3 THREE-DIMENSIONAL INTENSITY POINT SPREAD FUNCTION

To understand further the property of the 3-D IPSF for multiphoton fluorescence imaging, let us consider that the confocal system has two identical circular lenses. In this case, the following radial and axial optical coordinates v and u can be introduced:

$$v = \frac{2\pi}{\lambda_f} r \sin \alpha$$

$$u = \frac{8\pi}{\lambda_f} z \sin^2(\alpha/2), \quad (11.8)$$

where r and z are real radial and axial coordinates and $\sin \alpha$ is the numerical aperture of the objective in the object space.

The 3-D effective IPSF can be therefore expressed as:

$$h_i(v, u) = \left| h\left(\frac{v}{\beta}, \frac{u}{\beta}\right) \right|^{2n} \left[|h(v, u)|^2 \otimes_3 D(v) \right], \quad (11.9)$$

where $\beta = \lambda/\lambda_f$ and

$$|h(v, u)|^2 = \left| \int_0^1 \exp\left(\frac{iu\rho^2}{2}\right) J_0(v\rho)\rho d\rho \right|^2. \quad (11.10)$$

Here J_0 is a Bessel function of the first kind of order zero and

$$D(v) = \begin{cases} 1, & v < v_d \\ 0, & otherwise, \end{cases} \quad (11.11)$$

is the detector sensitivity for a circular detector of radius r_d. The normalized radius v_d is thus defined as

$$v_d = \frac{2\pi}{\lambda_f} r_d \sin \alpha_d, \quad (11.12)$$

where $\sin\alpha_d$ is the numerical aperture of the objective in detection space.

For confocal ($v_d = 0$) and conventional ($v_d \to \infty$) systems, the 3-D IPSFs are reduced to:

$$h_i(v, u) = \left| h\left(\frac{v}{\beta}, \frac{u}{\beta}\right) \right|^{2n} |h(v, u)|^2, \quad (11.13)$$

and

$$h_i(v, u) = \left| h\left(\frac{v}{\beta}, \frac{u}{\beta}\right) \right|^{2n}, \quad (11.14)$$

respectively.

Recalling that v and u are normalized by λ_f (see equation (11.8)), we find that the imaging performance in conventional multiphoton fluorescence microscopy is independent of the fluorescence wavelength. This is understood because the role of the collector lens is completely degraded when $v_d \to \infty$. The effect of a finite-sized detector on the 3-D IPSF for multiphoton fluorescence microscopy is demonstrated elsewhere (12,16,17).

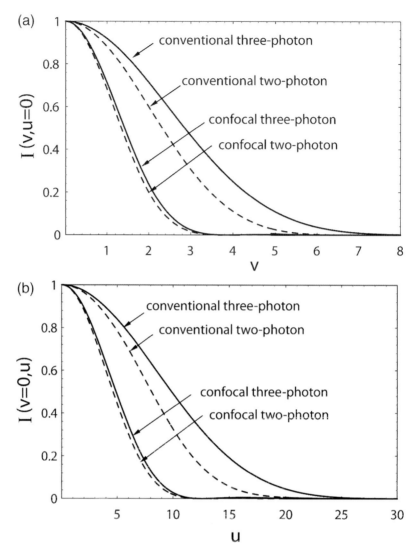

Figure 11.1. Transverse (a) and axial (b) cross-sections of the 3-D intensity point spread function. The dashed curves correspond to two-photon fluorescence microscopy, the solid curves to three-photon fluorescence microscopy. The fluorescence wavelength in 3-p imaging is assumed to be identical to that in 2-p imaging and to be one third of the incident wavelength ($\beta = 3$).

For a comparison, Figure 11.1 shows the cross-sections of the 3-D IPSF for confocal ($v_d = 0$) and conventional ($v_d \to \infty$) 2-p and 3-p fluorescence microscopy. This comparison is based on the assumption of the equal fluorescence wavelength in 2-p and 3-p fluorescence imaging. The transverse and axial half-widths at half-maximum of the 3-D IPSF for 2-p and 3-p fluorescence microscopy, $v_{1/2}$ and $u_{1/2}$, are given in Table 11.1, including a comparison with those for 1-p fluorescence microscopy. The values of $v_{1/2}$ and $u_{1/2}$ in 3-p fluorescence microscopy between the conventional case are increased approximately by a factor of 2, compared with those in the confocal case. In the case of confocal microscopy, they are close to those for 2-p microscopy. However, $v_{1/2}$ and $u_{1/2}$ are degraded approximately by 22% for conventional imaging.

Table 11.1. Comparison of resolution between two-photon and three-photon fluorescence microscopy under the condition of the equal fluorescence wavelength, v and u are normalized by the fluorescence wavelength. $v_{1/2}$ and $u_{1/2}$ are the transverse and axial half widths at half maximum of the 3-D IPSF. $\Delta u_{1/2}$ is the half width at half maximum for an axial response to a thin layer, γ the gradient of the image intensity at the surface of a thick layer, γ the gradient of the image intensity at the edge of a thin edge, and γ'' the gradient of the image intensity at the edge of a thick edge

		Conventional (1-photon)	Confocal (1-photon)	Conventional (2-photon)	Confocal (2-photon)	Conventional (3-photon)	Confocal (3-photon)
Half-width of 3-D IPSF	$v_{1/2}$	1.62	1.17	2.34	1.34	2.86	1.41
	$u_{1/2}$	5.56	4.01	8.02	4.62	9.87	4.89
Axial resolution	$\Delta u_{1/2}$	∞	4.3	8.6	5.1	9.98	5.67
	γ	0	0.09	0.045	0.093	0.047	0.082
Transverse resolution	γ'	0.27	0.417	0.208	0.361	0.171	0.349
	γ''	0	0.333	0.167	0.340	0.166	0.305

11.4 THREE-DIMENSIONAL OPTICAL TRANSFER FUNCTION

The point spread function analysis cannot provide a complete description of the imaging performance since only one point object is considered. For a complete description, an OTF can be introduced, which describes the strength with which each periodic component in a 3-D object is imaged. Mathematically, a 3-D OTF for an incoherent imaging system is given by the Fourier transform of the 3-D IPSF (5). For the system described by equation (11.9), the corresponding 3-D OTF can be expressed as:

$$C(l,s) = F_3\left[\left|h\left(\frac{v}{\beta},\frac{u}{\beta}\right)\right|^{2n}\right] \otimes_3 \left\{F_3\left[|h(v,u)|^2\right]F_2\left[D(v)\right]\right\}. \qquad (11.15)$$

F_2 and F_3 denote the two-dimensional (2-D) and the 3-D Fourier transforms, respectively. The variables l and s in $C(l,s)$ are the radial and axial spatial frequencies and have been normalized by $\sin\alpha/\lambda_f$ and $4\sin^2(\alpha/2)/\lambda_f$, respectively. Equation (11.15) can be rewritten as:

$$C(l,s) = F_3\left[\left|h\left(\frac{v}{\beta},\frac{u}{\beta}\right)\right|^{2n}\right] \otimes_3 \left\{F_3\left[|h(v,u)|^2\right]F_2\left[D(v)\right]\right\}, \qquad (11.16)$$

where

$$F_3\left[|h(v,u)|^2\right] = \frac{2}{l}\left\{\mathrm{Re}\left[\sqrt{1-\left(\frac{|s|}{l}+\frac{l}{2}\right)^2}\right]\right\} \qquad (11.17)$$

$$F_2[D(v)] = v_d\left[\frac{J_1(lv_d)}{l}\right], \qquad (11.18)$$

For confocal and conventional cases, equation (11.18) reduces to a constant and a delta function, respectively. For a finite-sized detector, equation (11.18) together with equations (11.16) and (11.17) should be used to calculate the 3-D OTF for a multiphoton fluorescence (5,16,11,12,14,17).

The 3-D OTFs for 1-p, 2-p, and 3-p confocal and conventional fluorescence microscopy are shown in Figure 11.2. Figures 11.2b through 11.2f have been normalized to unity at the origin, whereas Figure 11.1a has not because there exists a singularity at the origin. It is seen that Figure 11.2c has the same shape as Figure 11.2b but its scale is shrunk by a factor of two, which implies that the cutoff spatial frequencies for 2-p confocal fluorescence microscopy are twice as large as those in the conventional case, and thus more information can be imaged in the 2-p confocal case. In both cases, no missing cone of spatial frequencies exists (see Figs. 11.2c and 11.2d). This property promises good 3-D image formation in a 2-p conventional fluorescence microscope. In this respect, it contrasts with the 1-p conventional fluorescence microscope, which has a missing cone of spatial frequencies (see Fig. 11.2a). It should be noted that the cutoff spatial frequency in 2-p confocal imaging is the same as that in 1-p confocal imaging (see Fig. 11.2b).

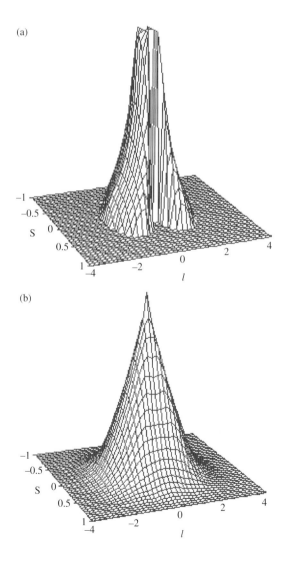

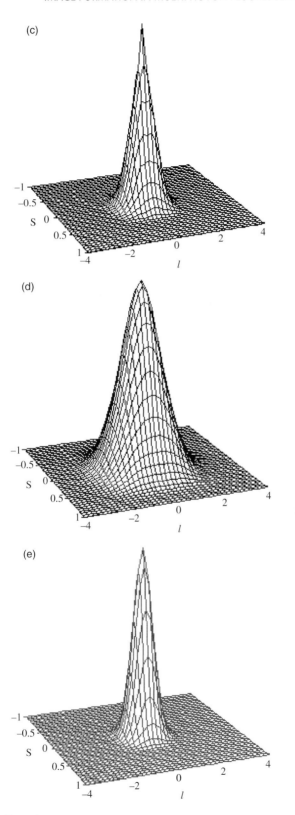

Figure 11.2. Continued

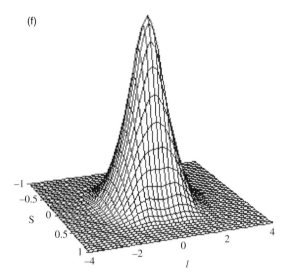

Figure 11.2. 3-D optical transfer functions, normalized to unity at the origin (except for the plot in (a)), for multiphoton confocal ($v_d = 0$) and conventional ($v_d \longrightarrow \infty$) fluorescence microscopy: (a) 1-p conventional; (b) 1-p confocal; (c) 2-p conventional; (d) 2-p confocal; (e) 3-p conventional; (f) 3-p confocal.

The 3-D OTF for a 3-p confocal fluorescence system has the same passband as that for 1-p and 2-p fluorescence microscopy for the assumption of equal same fluorescence wavelength (5,12,14,17). The effective size of the passband becomes smaller when the detector size is enlarged, as expected. However, even for a large detector, the 3-D OTF for 3-p imaging does not show a missing cone of spatial frequencies along the axial direction. This property is similar to that in 2-p fluorescence imaging but different from that in 1-p fluorescence imaging. It implies that a 3-D image of a thick 3-p fluorescent object can be formed without necessarily using a small detector. In other words, a conventional 3-p fluorescence microscope exhibits an optical sectioning property. In fact, due to the nonlinear dependence of the fluorescence intensity on the incident intensity, any multiphoton fluorescence microscopy does not necessarily need a small detector if an optical sectioning effect is required.

Figures 11.3a and 11.3b represent the transverse and axial cross-sections through the 3-D OTF at s = 0 and l = 0, respectively. The former can be used to describe imaging for a thick object structure with no variations in the axial direction, whereas the latter is for a planar structure such that there is no variation in the transverse directions. For example, the image of a thin planar object, giving the optical sectioning strength of a fluorescence microscope system, can be derived by the inverse Fourier transform of the axial cross-section of the 3-D OTF. For 2-p imaging, as a result of the enhanced OTF at low spatial frequencies, the slope of the 3-D OTF for 2-p confocal fluorescence is zero at the origin, while that for the conventional 2-p case is not.

Although the passbands in 3-p and 2-p confocal microscopy are the same, the strength of the 3-D OTF in 3-p imaging is weaker in comparison with that in 2-p imaging. This property can be clearly seen from Figure 11.3, where the transverse and axial

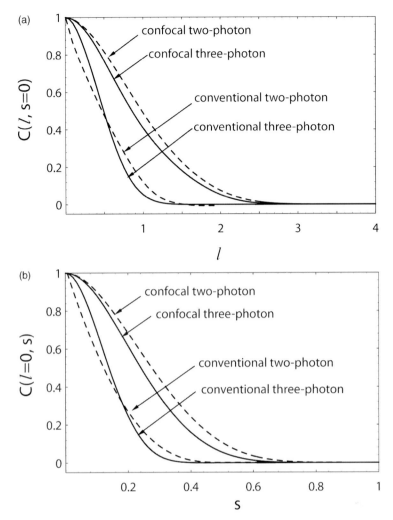

Figure 11.3. Transverse (a) and axial (b) cross-sections of the 3-D optical transfer function. The other conditions are the same as those in Figure 11.1.

cross-sections of the 3-D OTF are depicted. These narrower responses accordingly result in poor resolution, which will be described in the next section.

The description of imaging using the 3-D OTF is a generalized method that describes completely imaging of any object (5,16). For example, imaging of thin objects can be described by the 2-D in-focus OTF given by the projection of the 3-D OTF in the focal plane. Another special case is the imaging of a line object placed along the axis and varying in strength along the length. In this case, the image can be determined by the 1-D on-axis OTF produced by the projection of the 3-D OTF on the axis (5,16). Figures 11.4a and 11.4b give the 2-D in-focus OTF and the 1-D on-axis OTF. Note that the 3-p imaging performance is almost the same as 2-p fluorescence in the confocal case but poorer than that in 2-p imaging in the conventional case.

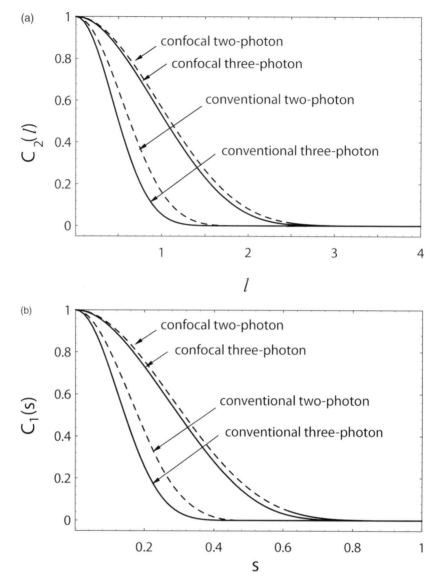

Figure 11.4. 2-D in-focus (a) and 1-D on-axis (b) optical transfer functions. The other conditions are the same as those in Figure 11.1.

11.5 IMAGE RESOLUTION

One of the advantages of the 3-D OTF description is that it enables the calculation of the image of a 3-D object (5,16). To characterize a confocal imaging system, one can measure images of various simple objects. For example, a uniform fluorescent layer scanned along the axial direction is usually used for the characterization of the axial resolution, and an edge object scanned in the transverse direction can be imaged to characterize the transverse resolution (5,15,16). These images can be modeled by performing the inverse Fourier transform of the object spectrum multiplied by the 3-D

OTF. Methods for calculating these images in terms of the 3-D OTF are presented in this section (5,15,16).

11.5.1 Axial Image of a Planar Fluorescent Layer

In multiphoton fluorescence microscopy, axial resolution can be evaluated by considering the axial response of a thin planar fluorescence layer scanned in the axial direction. This result is the so-called optical sectioning property. The narrower the axial response, the higher the axial resolution and therefore the stronger the optical sectioning effect.

As the planar layer does not include transverse structures, the transverse spatial frequency is zero (i.e., $l = 0$). Therefore, the axial image of this layer can be calculated by the inverse Fourier transform of the axial cross-section of the 3-D OTF shown in Figure 11.2b:

$$I(u) = F^{-1}[C(l = 0, s)], \qquad (11.19)$$

where F^{-1} denotes the operation of the inverse Fourier transform with respect to s.

The calculated 2-p and 3-p axial responses normalized to unity at the focal plane are shown in Figure 11.5a. When the thickness of the fluorescent layer becomes much larger than the focal depth of the objective, the object can be considered to be a thick object, the image of which can be calculated by (5,16):

$$I(u) = \frac{1}{2} + \frac{1}{\pi} \int_0^{S_c} C(l = 0, s) \frac{\sin(us)}{s} ds, \qquad (11.20)$$

where s_c is the cutoff axial spatial frequency of the 3-D OTF.

The axial responses to 2-p and 3-p thick fluorescent layers are shown in Figure 11.5b. The gradient, $\gamma = I'(u = 0)$, of the image at the surface of the layer is simply determined by the area under the axial cross-section of the 3-D OTF divided by π. It actually defines the sensitivity with which one can locate the surface of the thick layer.

It is seen from Figure 11.5 and Table 11.1 that the axial response to a thin fluorescent layer for 2-p conventional fluorescence is broader than that for 2-p confocal fluorescence. It is confirmed that 2-p conventional fluorescence gives an optical sectioning effect, although the half-width of the axial response for 2-p confocal fluorescence is decreased approximately by 41%, compared with that for the 2-p conventional case. This demonstrates an advantage over the conventional 1-p fluorescence imaging method, which does not give rise to the optical sectioning property. The improvement in sharpness between 2-p confocal and 1-p confocal microscopy is approximately 3.4%. The sharpness of the image of the thick fluorescent layer in 2-p confocal imaging is approximately twice that for 2-p conventional imaging.

Compared with those in 2-p confocal fluorescence microscopy, $\Delta u_{1/2}$ and γ demonstrate a degradation of the axial resolution by approximately 11%. For a comparison in conventional imaging, γ is almost unchanged, while $\Delta u_{1/2}$ is degraded about by 16%. It is shown that using a confocal geometry in 3-p fluorescence leads to an improvement in axial resolution by approximately 43% or 74% for thin or thick layer objects, respectively.

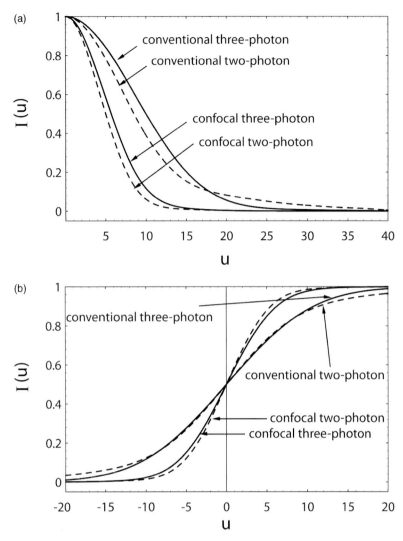

Figure 11.5. Image intensity of thin (a) and thick (b) fluorescent layers. The other conditions are the same as those in Figure 11.1.

11.5.2 Image of a Sharp Fluorescent Edge

For a sharp fluorescent edge scanned in the focal plane, the image can be expressed as:

$$I(v) = \frac{1}{2} + \frac{1}{\pi} \int_0^{l_c} C_2(l) \frac{\sin(vl)}{l} dl \qquad (11.21)$$

for a thin edge, and

$$I(v) = \frac{1}{2} + \frac{1}{\pi} \int_0^{l_c} C(l, s = 0) \frac{\sin(vl)}{l} dl \qquad (11.22)$$

for a thick edge. Here $C_2(l)$ and $C(l, s = 0)$ are the 2-D in-focus OTF and the transverse cross-section of the 3-D OTF, respectively. l_c is the transverse cutoff spatial frequency.

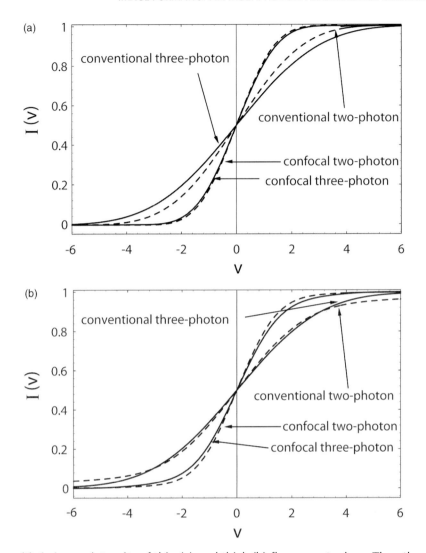

Figure 11.6. Image intensity of thin (a) and thick (b) fluorescent edges. The other conditions are the same as those in Figure 11.1.

The gradient of the image at the edge is determined by the area under $C_2(l)$ for a thin edge and under $C(l, s = 0)$ for a thick edge and gives the transverse resolution.

Figure 11.6a displays the image of the thin edge for 2-p confocal, 2-p conventional, 3-p confocal, and 3-p conventional fluorescence imaging, and the corresponding gradient γ' at the edge is shown in Table 11.1. In both confocal and conventional imaging methods, the 1-p fluorescence method results in sharper images compared with those using 2-p excitation (see Table 11.1).

As can be expected from Figure 11.4a, the transverse resolution for a thin 3-p fluorescent edge in the confocal case remains almost the same as that for 2-p fluorescence, but it is slightly degraded in the conventional case. γ' in 3-p imaging is improved approximately by 100% when a confocal pinhole is used.

The images of the thick edge for 2-p confocal, 2-p conventional, 3-p confocal, and 3-p conventional fluorescence microscopy are presented in Figure 11.6b. Table 11.1 also gives the gradient γ'' of such images for 1-p confocal, 1-p conventional, 2-p confocal, 2-p conventional, 3-p conventional, and 3-p confocal fluorescence imaging. As a result of the enhanced response of $C(l, 0)$ at low spatial frequency, 2-p confocal imaging (see Table 11.1) gives the sharpest image of the thick edge among the six cases. The confocal 1-p image is sharper than the 2-p conventional image. The conventional 1-p imaging method gives no image of a thick edge as $C(l, 0)$ has a singularity at $l = 0$ (5,12,16). 2-p and 3-p conventional microscopy exhibits almost the same transverse resolution, while confocal 3-p microscopy gives a slightly reduced resolution. The improvement of γ'' in 3-p imaging is approximately 84% when a confocal pinhole is used.

11.6 EFFECT OF FLUORESCENCE WAVELENGTH

Table 11.1 summarizes various comparisons of resolution between 1-p, 2-p, and 3-p fluorescence microscopy in terms of the 3-D IPSF, the 3-D OTF, and the images of layer and edge objects. These comparisons are based on the assumption of the equal fluorescence wavelength, therefore implying that the illumination wavelength in 3-p imaging is 1.5 times longer than that in 2-p imaging. This assumption is not necessarily true in practice. One of the advantages of the 3-p fluorescence imaging is to access the UV region of a shorter wavelength than 2-p fluorescence imaging for a given illumination wavelength. If the incident wavelength is given, v and u should be normalized by the incident wavelength:

$$v = \frac{2\pi}{\lambda} r \sin \alpha$$

$$u = \frac{8\pi}{\lambda} z \sin^2 \left(\frac{\alpha}{2} \right). \tag{11.23}$$

The passband and the cutoff spatial frequencies of the 3-D OTF for 3-p fluorescence are therefore increased by a factor of 1.5 in comparison with those in 2-p fluorescence imaging. Thus, more information in 3-p microscopy is imaged, which can lead to a better resolution. Under the condition of equation (11.23), the values of $\Delta u_{1/2}$, γ, γ', and γ'' for 2-p and 3-p fluorescence microscopy are summarized in Table 11.2, which shows that resolution in both conventional and confocal 3-p fluorescence imaging is approximately up to 40% to 50% better than that in 2-p imaging. For example, the improvement in axial resolution (γ) in 3-p conventional imaging is approximately 56%, which agrees with the experimentally measured value of 50% (10). It should be mentioned that the improvement of resolution in 3-p confocal fluorescence imaging originates from two effects, the nonlinear absorption and the confocal detection of the fluorescence of a shorter wavelength. However, only the first effect plays a role in conventional 3-p microscopy.

Another effect associated with the fluorescence wavelength is that the 3-p fluorescence wavelength is slightly longer than one third of the incident wavelength due to the nonradiation relaxation from the upper energy level. To discuss this effect further, let us consider confocal and conventional cases only. If v and u are normalized by the incident wavelength λ (see equation (11.23)), the corresponding 3-D IPSFs are

Table 11.2. Comparison of resolution between two-photon and three-photon fluorescence microscopy under the condition of the equal illumination wavelength, v and u are normalized by the incident wavelength. The others conditions are the same as those in Table 11.1

		Conventional (1-photon)	Confocal (1-photon)	Conventional (2-photon)	Confocal (2-photon)	Conventional (3-photon)	Confocal (3-photon)
Half-width of 3-D IPSF	$v_{1/2}$	0.81	0.58	1.17	0.67	0.953	0.47
	$u_{1/2}$	2.78	2.0	4.01	2.31	3.29	1.63
Axial resolution	$\Delta u_{1/2}$	∞	2.15	4.3	2.50	3.33	1.89
	γ	0	0.18	0.09	0.186	0.141	0.246
Transverse resolution	γ'	0.54	0.84	0.416	0.722	0.513	1.047
	γ''	0	0.67	0.334	0.68	0.498	0.915

then expressed as:

$$h_i(v, u) = |h(v, u)|^{2n} |h(\beta v, \beta u)|^2 \qquad (11.24)$$

and

$$h_i(v, u) = |h(v, u)|^{2n} \qquad (11.25)$$

respectively. Here $\beta = \lambda/\lambda_f$. Clearly, the imaging property in the conventional case does not depend on the fluorescence wavelength because the collector lens does not contribute to the imaging process; it simply plays a role of collecting the signal. However, the imaging performance in the confocal case is degraded as the fluorescence wavelength is increased (i.e., as β is decreased) (17).

REFERENCES

[1] Wilson T and Sheppard C J R 1984 *Theory and Practice of Scanning Optical Microscopy* (Academic, London).

[2] Sheppard C J R 1987 Scanning optical microscopy *Advances in Optical and Electron Microscopy* Vol. 10 ed R Barer and V E Cosslett (Academic, London) 1–98.

[3] Wilson T 1990 *Confocal Microscopy* (Academic, London).

[4] Pawley J B 2006 *Handbook of Biological Confocal Microscopy, 3rd edition* (Springer, New York).

[5] Gu M 1996 *Principles of Three-Dimensional Imaging in Confocal Microscopes* (World Scientific, New Jersey).

[6] C. J. R. Sheppard, R. Kompfner, J. Gannaway, and D. Walsh, *IEEE J. Quantum Electron.* QE-13 (1977) 100D.

[7] C. J. R. Sheppard and R. Kompfner, *Applied Optics*, 17 (1978) 2879–2883.

[8] Denk W, Strickler J H and Webb W W 1990 Two-photon laser scanning fluorescence microscopy *Science* 248 73–76.

[9] He G S, Bhawalker J B, Prasad P N and Reinhardt B A 1995 Three-photon-absorption-induced fluorescence and optical limiting effects in an organic compound *Opt. Lett.* 20 1524–1526.

[10] Hell S, Bahlmann K, Schrader M, Soini A, Malak H, Gryczynski I and Lakowicz J R 1996 Three-photon excitation in fluorescence microscopy *J. Biomedical Optics* 1 71–74.

[11] Sheppard C J R and Gu M 1990 Image formation in two-photon fluorescence microscopy *Optik* 86 104–106.

[12] Gu M and Sheppard C J R 1993 Effects of a finite-sized pinhole on 3D image formation in confocal two-photon fluorescence microscopy *J. Mod. Opt.* 40 2009–2024.

[13] Nakamura O 1993 Three-dimensional imaging characteristics of laser scan fluorescence microscopy: two-photon excitation vs. single-photon excitation *Optik* 93 39–42.

[14] Gu M and Sheppard C J R 1995 Three-dimensional confocal fluorescence imaging under ultrashort pulse illumination *Opt. Commun* 117 406–412.

[15] Gu M and Sheppard C J R 1995 Comparison of three-dimensional imaging properties between two-photon and single-photon fluorescence microscopy *J. Microscopy* 177 128–137.

[16] Gu M 1996 Resolution in three-photon fluorescence scanning microscopy *Opt. Lett.* 21 988–990.

[17] Gu M and Gan X 1996 Effect of the detector size and the fluorescence wavelength on the resolution of three- and two-photon confocal microscopy *Bioimaging* 4 129–137.

[18] Gu M 1995 Three-dimensional space-invariant point-spread function for a single lens *J. Opt. Soc. Am. A* 12 1602–1604.

12

Signal Detection and Processing in Nonlinear Optical Microscopes

Siavash Yazdanfar and
Peter T. C. So

12.1 INTRODUCTION

The inherent three-dimensional (3D) resolution is an important strength of nonlinear microscopy. However, 3D resolution comes at the price that the optical signal originating from a subfemtoliter size volume is inherently weak. The need to rapidly acquire data, for example to observe sample dynamics or to image volumetric data, restricts pixel dwell time. These factors jointly limit the image signal level and demand highly sensitive optical detectors, since the signal-to-noise ratio (SNR) directly determines image quality. The advent of optoelectronic detector technology has been partly responsible for the practical applications of nonlinear microscopy in biology and medicine.

The requirement for high-sensitivity detectors can be better appreciated by estimating the signal strength of a typical two-photon fluorescence microscope. Using a high numerical aperture (NA >1) objective, the two-photon excitation volume is on the order of 0.1 femtoliter. The fluorophore concentration in a biological specimen is limited. For a reasonable concentration of 10 μM, a mere 60 fluorophores reside inside the excitation volume. These fluorophores are excited by femtosecond pulse lasers with a repetition rate of about 100 MHz. The maximum excitation efficiency is about 10% without appreciably depleting the fluorophores in the ground state and broadening the excitation point spread function. Therefore, the maximum excitation rate for each fluorophore is 10 MHz, assuming that the fluorophore has perfect quantum efficiency and its lifetime is significantly shorter than the pulse repetition rate. The total optical signal (i.e., photon flux) generated from the excitation volume is 600 MHz. While the microscopes are highly optimized, the combination of the limited detection solid angle and the inevitable loss at various optical surfaces often limits the transmission efficiency of the microscope to at best 20%. This factors result in an incident photon rate at the detector of 120 MHz. For a 512×512 pixel image acquired at a one-second frame rate, the pixel dwell time is 3 μs. Within this pixel dwell time, only 420 photons are present at the detector, and the accurate measurement of this weak signal is clearly a challenge. Although the fundamental physical and biochemical limitations are different for other modalities of nonlinear microscopy, such as second-harmonic generation (SHG) or coherent anti-Stokes Raman (CARS) microscopy, their typical signal levels are often comparable or weaker than the fluorescence case. Therefore, optoelectronic detectors with close to single photon detection sensitivity are required in most biological applications of nonlinear optical microscopy.

This chapter will focus on optical detector technologies that are suitable for low-light-level measurements. In Section 12.2, the basic physical principles of the photoelectric and photovoltaic effects for high-sensitivity optical detectors are described. In Section 12.3, the key characteristics of optical detectors will be defined and their relevancy for microscopy imaging will be discussed. In Section 12.4, we will specifically consider the noise sources of optical detectors and how signal processing affects SNR. In Section 12.5, a survey of commonly used high-sensitivity optical detectors in nonlinear microscopy will be provided.

12.2 PHYSICAL PRINCIPLES FOR HIGH-SENSITIVITY OPTICAL DETECTION

Common optical detectors are sensitive to the intensity of light but not its amplitude or phase. Optical detectors can be based on many different physical properties of light. For example, a thermopile detector measures light intensity by measuring temperature changes due to light absorption. Since we are interested in nonlinear optical microscopy applications, we will consider only highly sensitive and highly quantitative detectors. Today, the photodetectors with this level of sensitivity are based on either photoelectric or photovoltaic principles. A brief overview of these phenomena is helpful to understand the operation of most common detectors.

12.2.1 Photoelectric Effect

The photoelectric effect describes the emission of electrons from a metallic surface upon the absorption of light. Historically, the photoelectric effect was first observed by Becquerel in 1839 as he observed current generation in a conductive solution as an electrode is exposed to light (Becquerel, 1892; Harvey, 1957). The photoelectric effect was later observed by Hertz as he observed ultraviolet light assist the discharge through a spark gap in 1887. The dependence of the photoelectric effect on the frequency of light was observed by von Lenard in 1902. Theoretical explanation of the photoelectric effect was accomplished by Einstein in 1905, firmly establishing the quantum nature of light.

The work function ϕ characterizes the electrostatic interaction of the electron with the bulk material. Upon illumination with light at frequency, v, electrons are emitted if the photon energy, hv, is greater than the work function of the material, where h is the Planck's constant. If the photon energy is less than the work function, no electron will be emitted. When photon energy is greater than the work function, the number of electrons emitted is proportional to the intensity of light. The emitted electrons will have kinetic energies, E_k, characterized by:

$$hv = \phi + E_k \tag{12.1}$$

12.2.2 Photovoltaic Effect

The photovoltaic effect describes the formation of free charge carriers in semiconductor materials upon the exposure of light. The observation of the photovoltaic effect, similar to the photoelectric effect, can be traced to the experiments of Becquerel (Ashcroft & Mermin, 1976; Komp, 1995; Shive, 1959; Sze, 2002). Quantum mechanical consideration of electrons in a periodic atomic potential of a solid demands that they are arranged in energy bands separated by gaps that the electrons cannot occupy (Fig. 12.1a). Electrons in a completely filled band, the valence band, are electrically inert and do not contribute to electrical conduction. Conduction is carried by only electrons in partially filled bands, known as conduction bands. Therefore, materials with completely filled bands are insulators, whereas materials with partially filled bands are conductors. The photovoltaic effect occurs in a class of material called semiconductors. Semiconductors are insulators at absolute zero temperature but possess energy gaps that are relatively small relative to thermal energy such that a number of electrons occupy the conduction band at room temperature. Semiconductors that are optically active have bandgap separation comparable to the incident photon energy. Upon illumination, an electron can be promoted to the conduction band; a hole, a lack of an electron, is left in the valence band. The motions of the electron and the hole can be driven by an external electric field until recombined. This generation of charge carriers resulting in electric field conduction through the material constitutes the photovoltaic effect. Semiconductors can be modified by "doping," or mixing in during the fabrication process, impurity atoms that are either electron donors or acceptors. Doping provides excess electrons, or holes, with energies close to the conduction, or valence, bands respectively (see Fig. 12.1b). Semiconductors with electron donor doping are called n-type semiconductors, whereas those with electron acceptor doping are called p-type. The principle of constructing practical photovoltaic devices, such as photodiodes, will be covered in Section 12.5.

12.3 CRITERIA FOR DETECTOR SELECTION IN NONLINEAR MICROSCOPY

The selection of a detector for imaging applications requires a careful examination of its performance characteristics. While specific applications will dictate different choices, a small set of parameters determines detector performance. These parameters, defined below, include quantum efficiency, internal amplification (gain), dynamic range, response speed, geometric format, and noise.

Quantum efficiency (QE) may be the most important parameter of an optical detector, since it characterizes the probability of the photoelectron production in a detector. The maximum QE of detectors used in nonlinear microscopy is about 0.8 but can be significantly lower, as it is a function of the material property of the sensor. For photoelectric devices, the QE depends on the work function of the cathode material. For photovoltaic devices, the QE is a function of the semiconductor material's bandgap magnitude and the electron–hole recombination probability. The QE of the photocathode also depends on the geometric design of the sensor, such as its thickness. Furthermore, QE is a function of the energy of the incident photon and hence its wavelength.

Many photodetectors, such as photomultiplier tubes and avalanche photodiodes, feature internal amplification. A single electron is generated upon the absorption of one incident photon by either photoelectric or photovoltaic processes. Direct detection of this single electron is virtually impossible in the presence of other electronic noise.

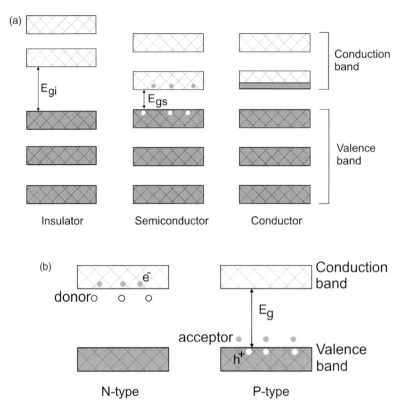

Figure 12.1. (a) Bandgap structures of insulator, semiconductor, and conductors. (b) n-type and p-type doped semiconductors.

To overcome this difficulty, some photodetectors feature internal amplification, where the photoelectron generated is amplified electronically, resulting in a large cascade of secondary electrons. Operationally, the internal amplification of a detector can be defined as the ratio of the photocurrents at the input (e.g., cathode) and the output (e.g., anode) of the detector. Internal amplification of a detector can be very high; for photomultiplier tubes, the internal amplification reaches a level of 10^6 to 10^7. Detectors with sufficient internal amplification allow the detection of even a single photon.

Dynamic range measures the input intensity range where the detector responds before saturation. The dynamic range of detectors is important in biological microscopy as the signal amplitude can vary over several orders of magnitude both spatially and temporally. Spatial signal variations may result from factors such as local variation in probe concentration (with linear dependence in fluorescence case and quadratic dependence in SHG and CARS) or local biochemical processes such as quenching. Temporally, significant intensity variations may occur on picosecond and nanosecond regimes due to excitation decay processes, on microsecond and millisecond regimes due to diffusion and chemical reactions, and on the longer time scale dominated by active biological processes such as transport. An example where fluorescence signal strength may vary by a couple orders of magnitude temporally is in the imaging of calcium signaling. In this case, the intensity of calcium sensors, such as Fluo3 or

CalciumGreen, can vary greatly between a cell in the resting and the activated states. The ability to quantify signal over multiple decades in amplitude is important to study many of these processes. Photodetector circuitry with linearity over six to eight orders of magnitude can be readily fabricated.

Photodetector selection also depends on the temporal resolution required to study specific biological processes. Since optoelectronic devices are inherently fast, processes that are on the microsecond time scale and slower do not pose significant challenges except in terms of acquiring sufficient photons to achieve an acceptable SNR. However, for processes that require picosecond and nanosecond time resolution, such as measuring the lifetime of fluorescence decay and the anisotropy decay of rotational diffusion, high-speed detectors are required. High temporal resolution requires the transition time of the photoelectron, and its secondary electrons, to be reproducible. In the case of common high-speed detectors such as photomultiplier tubes or microchannels plates, the temporal characteristics can be optimized by minimizing the cathode–anode distance and maximizing their relative voltage.

The geometric factors of a detector can influence the design of a nonlinear optical microscope. Two important geometric factors will be considered here, spatial resolution and detector active area. First, photodetectors can be divided into two classes: (1) single pixel detectors without spatial resolution and (2) array detectors that record the spatial coordinates of the arrival photons. Most nonlinear microscopes produce 3D images by raster scanning a single excitation focus, so single pixel detectors without spatial resolution are sufficient since the scan position is encoded temporally. However, the usage of array detectors is increasing due to their applications in spectrally resolved and high-speed imaging systems. Second, the size of the detector active area is critical for nonlinear microscopy and should be sufficiently maximized. The 3D spatial resolution is a result of the nonlinear dependence of the emission signal strength upon the excitation light intensity, and thus spatial filtering in the detection path is not necessary as in confocal microscopy. Since descanning is not required, the detection optical path of nonlinear microscopes can be greatly simplified to achieve higher signal throughput. However, without descanning, the emission light does not remain motionless at the detector plane while the excitation beam is scanned across the sample. Therefore, the detector active area must be sufficiently large and uniform such that the emission light remains within its boundary throughout the scanning process. Furthermore, using a detector with large excitation area is also critical in maximizing the collection of scattered emission photons in the imaging of thick tissue specimens, which is discussed extensively in Chapter 10.

An understanding of the origins of device noise sources and their characteristics is critically important for nonlinear microscopy design and will be discussed extensively in the following section.

12.4 CHARACTERIZATION OF DETECTOR NOISE CHARACTERISTICS AND SIGNAL PROCESSING*

The dynamic range of a nonlinear microscopy system is limited by the aggregate noise of the detector and signal processing electronics. Careful consideration of the noise sources allows for optimization of system performance.

12.4.1 Characterization of Noise in the Temporal and Spectral Domain

Under constant illumination, the optical power incident on the detector, S_O, can be quantified in terms of the mean number of photons, $\langle m \rangle$, detected within a time interval Δt:

$$S_O = \langle m \rangle \frac{E_\gamma}{\Delta t} \qquad (12.2)$$

where E_γ is energy per photon.

Using optoelectronic transducers, optical signal is also conveniently measured as electrical power. The data collection time, Δt, is typically chosen to match the bandwidth, Δf, of the instrument such that $\Delta f = \frac{1}{2\Delta t}$ where the factor of one-half is due to the sampling theorem. The photocurrent produced by optical signal can be described as:

$$I = \alpha \langle n \rangle \frac{q}{\Delta t} \qquad (12.3)$$

where q is the charge of an electron, $\langle n \rangle$ is the mean number of photoelectrons generated, and α is the amplification factor of the detector.

Note that:

$$\langle n \rangle = \beta \langle m \rangle \qquad (12.4)$$

where β is the QE of the detector.

The signal power detected is:

$$S = I^2 R \qquad (12.5)$$

where R is the impedance of the detection circuit.

In practice, the measured current is limited by the measurement bandwidth, determined by the transit time of charge carriers through the detector, as well as the output impedance of the detection circuit. The detection bandwidth is considered in more detail in Section 4.3.

The quality of nonlinear microscopy images is limited by the presence of noise in the measurement. Although noise sources have different physical origins, they can be similarly described as random processes. The magnitude of noise can be quantified by its power variance:

$$N = \langle \Delta I_N^2(t) \rangle R \qquad (12.6)$$

where $\Delta I_N(t)$ is the noise current and $\langle \Delta I_N^2(t) \rangle$ is its variance.

In the case where the total noise has a number of constituents,

$$\langle \Delta I_N^2(t) \rangle = \langle \Delta I_{N1}^2(t) \rangle + \langle \Delta I_{N2}^2(t) \rangle + \langle \Delta I_{N3}^2(t) \rangle + \cdots \qquad (12.7)$$

where $\langle \Delta I_{Ni}^2(t) \rangle$ is the ith component contributing to the noise.

The performance of the detector can be characterized by its SNR:

$$SNR = \frac{S}{N} \tag{12.8}$$

Another figure of merit that is commonly used to characterize the performance of a detector is noise equivalent power (NEP), defined as the optical power level S where SNR is unity.

Note that ΔI_N is random and represents fluctuation of the photocurrent from its mean. Specifically,

$$\Delta I_n(t) = \frac{\alpha q(n(t) - \langle n \rangle)}{\Delta t} \tag{12.9}$$

where $n(t)$ is the number of photoelectrons at a given time.

Note that the temporal variations in the number of photoelectrons may have several physical origins, such as variations in the number of incident photons and the stochastic nature inherent in photon–electron conversion.

As in many random processes, the noise in photocurrent can be described by its power spectrum. The noise power spectrum density, $\tilde{P}(f)$, is defined as:

$$\tilde{P}(f) = R\Delta f \left| \Delta \tilde{I}_n(f) \Delta \tilde{I}_n(f)^* \right| \tag{12.10}$$

where Δf is the bandwidth of the measurement and $\Delta \tilde{I}_n(f)$ is the Fourier transform of the photocurrent noise:

$$\Delta \tilde{I}_n(f) = \int_{-\infty}^{\infty} \Delta I_n(t) e^{-i2\pi ft} dt \tag{12.11}$$

For clarity, functions in the frequency domain are distinguished from those in the time domain by the tilde. The noise power contained within spectral band, Δf, is:

$$N(f, \Delta f) = \tilde{P}(f) \Delta f \tag{12.12}$$

The autocorrelation function of the photocurrent is defined as:

$$\Re_{II}(\tau) = R\Delta f \int_{-\infty}^{\infty} \Delta I(t + \tau) \Delta I(t)^* dt \tag{12.13}$$

The noise power spectrum density and the autocorrelation function are related through the Wiener-Khintchine theorem via a Fourier transform:

$$R\Delta f \left| \Delta \tilde{I}_n(f) \Delta \tilde{I}_n(f)^* \right| = \int_{-\infty}^{\infty} \Re_{II}(\tau) e^{-i2\pi f\tau} d\tau \tag{12.14}$$

Therefore, the noise spectrum can also be expressed in terms of the autocorrelation function as:

$$\tilde{P}(f) = \int_{-\infty}^{\infty} \Re_{II}(\tau) e^{-i2\pi f\tau} d\tau \tag{12.15}$$

Consider the autocorrelation function at $\tau = 0$:

$$\Re_{II}(0) = R\Delta f \int_{-\infty}^{\infty} \Delta I(t)\Delta I(t)^* dt = R\sigma_I^2 \qquad (12.16)$$

where σ_I^2 is the variance of the photocurrent.

12.4.2 Photodetector Noise Sources

Noise sources that are important in photodetectors include photon shot noise, detector dark noise, Johnson noise, and multiplicative noise.

12.4.2.1 Photon Shot Noise

Photon shot noise has its origin in the quantized nature of light. The incidence of photons at a detector follows Poisson statistics (Fig. 12.2). For Poisson statistics, the probability of detecting m photons during one measurement when the mean number of photons

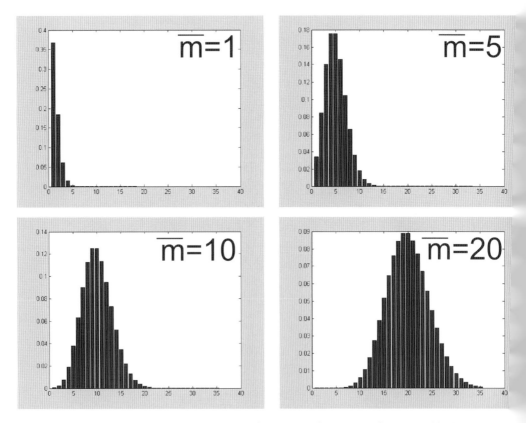

Figure 12.2. Poisson distribution for mean photon number $\bar{m}$ equal to 1, 5, 10, 20. Note that as the mean photon number increases, the Poisson distribution approaches the Gaussian distribution.

being $\langle m \rangle$ is given by:

$$P(m|\langle m \rangle) = e^{-m}\frac{(\langle m \rangle)^m}{m!} \tag{12.17}$$

Poisson statistics describes processes where the outcome of the present measurement is independent of when the measurement is made and all past history. Poisson statistics belong to the class of independent and identically distributed (iid) statistical processes. A key property of the Poisson statistics is that the variance is always equal to the mean. Specifically,

$$\langle (m - \langle m \rangle)^2 \rangle = \langle m \rangle \tag{12.18}$$

Since the photocurrent is proportional to the photoelectron n, and the photoelectron is proportional to the incident photon number m, we expect the variance of the photocurrent to be equal to its mean:

$$\sigma_I^2 = 2\alpha q \Delta f \langle I \rangle \tag{12.19}$$

where the factor of 2 represents positive and negative frequencies.

Therefore, the autocorrelation function of the photocurrent at $\tau = 0$ can be expressed as:

$$\Re_{II}(0) = R\sigma_I^2 = 2R\alpha q \Delta f \langle I \rangle \tag{12.20}$$

Further, since the Poisson process is iid, the autocorrelation function must be a delta function, indicating a lack of temporal correlation except at the present time (Fig. 12.3a):

$$\Re_{II}(\tau) = 2R\alpha q \langle I \rangle \delta(\tau) \tag{12.21}$$

Therefore, the power spectrum of the photon shot noise is white (frequency independent) and is presented in Fig. 12.3b:

$$\tilde{P}(f) = 2R\alpha q \langle I \rangle \tag{12.22}$$

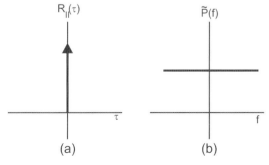

(a)   (b)

Figure 12.3. (a) The autocorrelation function $\Re_{II}(\tau)$ and (b) the power spectrum $\tilde{P}(f)$ of photon shot noise.

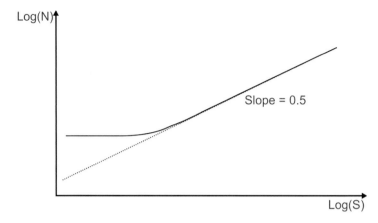

Figure 12.4. Noise versus signal of a typical detector on log scale. At sufficiently high signal level, the device is limited by only photon shot noise (ideal detector) and the slope is 0.5.

The photon shot noise power at frequency f within a frequency band of Δf is:

$$N_s(f, \Delta f) = 2R\alpha q \Delta f \langle I \rangle \tag{12.23}$$

It is important to emphasize that photon shot noise originates from the quantized nature of light and is present even in the absence of any device-dependent noise from the optoelectronic detector. A photodetector is considered "ideal" when the noise of the measurement is dominated by only photon shot noise (Fig. 12.4). In the case that photon shot noise is dominant, the SNR of the detector is:

$$SNR = \frac{S}{N_s} = \frac{\langle I \rangle}{2\alpha q \Delta f} = \frac{\alpha \langle n \rangle q / \Delta t}{2\alpha q \Delta f} = \langle n \rangle \tag{12.24}$$

using equations (12.2) and (12.23), and assuming that the measurement bandwidth is set to match the collection time as $\Delta f = 1/2\Delta t$. Thus, in a system dominated by photon shot noise, SNR of the detection increases linearly with the mean number of photons.

12.4.2.2 Detector Dark Noise

Detector dark noise originates from the thermally induced, spontaneously generated electrons in either photoelectric or photovoltaic devices. Given the thermal origin of dark noise, the probability for the generation of these dark electrons is governed by the Boltzmann factor; the magnitude of the dark noise is an exponential function of the absolute temperature of the device. Let $\langle n_d \rangle$ be the mean number of dark electrons. These electrons produce a dark current of magnitude:

$$\langle I_d \rangle = \alpha \langle n_d \rangle q \Delta f \tag{12.25}$$

The average dark current produces only a constant offset. However, since the thermally induced electrons are a discrete event, the dark current fluctuates according to

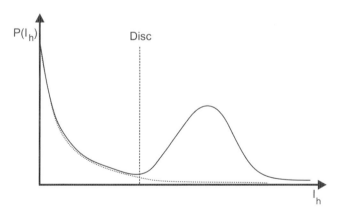

Poisson statistics similar to the photon shot noise. Specifically, dark current exhibits a white spectrum similar to photon shot noise:

$$N_d(f, \Delta f) = 2R\alpha q \Delta f \langle I_d \rangle \qquad (12.26)$$

12.4.2.3 Multiplicative Noise

Some detectors have internal gain; the electron generated by an incident photon is electrically amplified to produce many electrons. This amplification process increases the sensitivity of the detector but introduces another source of noise. Similar to the probabilistic generation of a photoelectron at a detector surface, amplification processes within the detector are stochastic and can result in significant uncertainty in the amplification magnitude. In the photomultiplier literature, this effect is often referred to as multiplicative noise, and in the semiconductor literature, this is referred to as the excess noise factor, F, which is greater than unity (McIntyre, 1966). To account for this effect, both the shot and dark noise terms should be modified as:

$$N_s(f, \Delta f) = 2RF\alpha q \Delta f \langle I \rangle \quad N_d(f, \Delta f) = 2RF\alpha q \Delta f \langle I_d \rangle \qquad (12.27)$$

The excess noise factor is device dependent and can be quantified by the pulse height distribution. Pulse height distribution represents the probability distribution of the photocurrent amplitude and is typically measured at low-light levels with single photon sensitivity. The current amplitude distribution of conventional photomultipliers has a single peak at zero amplitude corresponding to the dark current of the device (Fig. 12.5). In the presence of light, an additional peak at high amplitude is recorded

Figure 12.5. Photon current burst distribution plot for single photon counting devices with internal gain, such as a photomultiplier. The x-axis, I_h, is the current burst amplitude of a single photon and the y-axis, $P(I_h)$, is the probability of observing certain burst amplitude I_h. The typical distribution in the single photon counting regime has two peaks. One peak maximizes at zero amplitude and corresponds to the dark noise. The second peak maximizes at some higher value dependent upon the current gain of the detector. The correct level to set the discrimination level is at the minima between the two peaks, as indicated by the dotted vertical line.

corresponding to the current pulses produced by single photons. The nonzero variance of the single photon peak is the origin of the multiplicative noise. Pulse height distribution provides a direct measurement of this uncertainty.

12.4.3 Signal Conditioning Circuitry Noises

In addition to the noises originating from the sensor itself, the downstream signal conditioning electronics can also contribute to measurement noise.

12.4.3.1 Typical Design of Signal Conditioning Circuit

Since a typical photodetector used in high-sensitivity microscopy produces a current output, the measurement of this signal requires converting the current to a voltage signal and appropriately amplifying it, a process called transimpedance amplification. Two typical simple transimpedance circuits are shown below. The simplest transimpedance circuit is shown in Fig. 12.6a where the current generated by the photodetector produces a voltage drop across a parallel resistor. The transimpedance gain of this circuit is equal to the resistance of the resistor, measured in ohms: $\Omega = V/A$. For weak current signals, a high resistance value must be used to produce a readily measurable voltage. This passive transimpedance circuit has large output impedance and can often cause loading in the subsequent detection circuitry. A better transimpedance circuit utilizes active elements that do not cause loading downstream (see Fig. 12.6b).

Ultimately, the voltage signal produced by the transimpedance circuitry of the detector must be quantified and recorded. For light intensity measurement, two main classes of digitization circuitry are employed: single photon counting and analog current digitization. More sophisticated signal conditioning circuitries are needed in advanced spectroscopic measurements, such as lifetime-resolved imaging, which is discussed in Chapter 21.

Typical single photon counting circuitry in nonlinear microscopy uses a discriminator with a single threshold level. Recall that the amplitude of the voltage bursts from the detector can be characterized by the pulse amplitude distribution plot (see

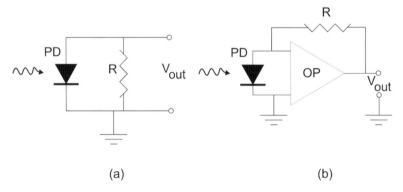

(a)   (b)

Figure 12.6. (a) Passive and (b) active transimpedance circuit for a detector that is a current source. In these examples, the detector is a photodiode (PD). R is a resistor, OP is an operational amplifier, and V_{out} is the output node.

Fig. 12.5). Single photon counting circuitry seeks to distinguish the signal from the photons from the dark noise. A discriminator is a device used to threshold voltage bursts with amplitude higher than the dark noise, passing the signal resulting from the incidence of single photons and discarding voltage bursts with lower amplitude (see Fig. 12.5). A high-speed comparator circuit can be designed such that voltage pulses with amplitude higher than the threshold are amplified to become TTL or ECL digital pulses and are counted by a digital counter, discarding pulses below threshold. The main advantage of single photon counting circuitry is its ability to reject dark noise, allowing a higher SNR measurement than analog current detection. The main disadvantage of single photon counting circuitry is that it is limited to low photon flux. The temporal spread of the detector electron pulses and the bandwidth of the transimpedence amplifier/discriminator circuitry limit the maximum counting rate possible in single photon counting systems. Today, the typical limitation lies in the amplifier/discriminator circuitry, which typically has bandwidth on the order of 100 to 200 MHz. Since photon arrival follows Poisson statistics, the actual maximum single photon counting rate without pulse "pile-up" is on the order of 1 to 5 MHz. For detectors such as hybrid photomultiplier tubes, multiple-level discrimination circuitry can in principle be used, allowing detection at higher photon rates.

Instead of single photon counting, voltage signal produced by the transimpedance amplifier can also be digitized using an analog-to-digital converter circuitry (ADC). The voltage output of the transimpedance amplifier is either directly compared with the voltage ladder in a digitizer or compared following additional voltage amplification. Typical 8-bit to 16-bit ADCs are used in nonlinear microscopy imaging, corresponding to a voltage resolution of 256 to 65,536 levels, respectively. However, 24-bit and higher ADCs are also available. The main advantage of analog detection circuitry is the applicability of these circuitries at high light levels. The photon "pile-up" that plagues single photon counting circuitry is not a problem for analog detection since the voltage signal is proportional to the current level and the resolution of the electron burst is not necessary. Therefore, analog detection circuits can handle high dynamic range measurement at high light levels, limited only by the digitization level available in the ADC and current saturation of the detector. The main disadvantage of analog detection is its inability to reject detector dark noise and digitization noise, which will be discussed below.

12.4.3.2 Johnson Noise

An element common to all transimpedance amplification circuitry is the presence of a shunt resistor through which the photocurrent is converted into a voltage. Johnson noise, also known as thermal noise, is the voltage fluctuations in the shunt resistor resulting from thermally driven electron charge motions in the conductor (Johnson, 1928). Nyquist provided the first theoretical description of this phenomenon based on elegant thermodynamics considerations (Nyquist, 1928). The quantification of the energy residing in the thermally induced electromagnetic field around a conductor of an arbitrary shape is a difficult problem. Alternatively, Nyguist suggests considering an isolated transmission line of length L impedance matched to two resistors R shorting both ends (Fig. 12.7). For a transmission line shorted on both ends, the electromagnetic

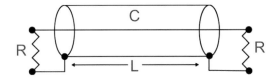

Figure 12.7. A transmission line with length L and capacitance C, terminated with matching resistors R.

field is characterized by standing wave modes with wavelengths:

$$f = \frac{2nL}{c} \quad n = 1, 2, 3 \ldots \tag{12.28}$$

The number of modes in a frequency band, Δf, is $2L\Delta f/c$. From equipartition theorem, each standing wave mode must have energy equal to kT (composed of an electric and a magnetic mode). Therefore, the energy in the transmission line within a given frequency band is:

$$E(f)\Delta f = \frac{2LkT}{c}\Delta f \tag{12.29}$$

Since the system is isolated and the two resistors are at the same temperature, half this energy corresponds to the energy transmission from one resistor to the other and vice versa. The time constant for energy transmission across the line is simply $\Delta t = L/c$. Therefore, the power transmitted in a given spectral band is $kT\Delta f$ along one direction.

Let V be the voltage fluctuation produced in the resistor. The current produced in the transmission line is:

$$I = \frac{V}{2R} \tag{12.30}$$

The Joule heating power within a given spectral band is:

$$\frac{V^2}{4R^2}R\Delta f = \frac{V^2}{4R}\Delta f \tag{12.31}$$

Since the system is isolated, the power transmitted in the electromagnetic field must be equal to Joule heating:

$$\frac{V^2}{4R}\Delta f = kT\,\Delta f \tag{12.32}$$

We can calculate the magnitude of thermally driven voltage fluctuation:

$$V_J^2 = 4RkT \tag{12.33}$$

Therefore, Johnson current can be calculated:

$$I_J = \sqrt{\frac{kT}{R}} \tag{12.34}$$

Johnson noise within a spectral band is:

$$N_J(f, \Delta f) = kT\Delta f \qquad (12.35)$$

12.4.3.3 Other Electronic Considerations

In addition to the Johnson noise contribution in the shunt resistor, the use of active elements in some transimpedance circuitry introduces additional systematic offsets and stochastic noises. In terms of systematic offsets, the main contributors are the input bias and offset currents of the operational amplifiers. Ideal operational amplifiers draw no current in their inputs, but practical devices do. The bias and offset currents can be as high as a few hundred nA, which can be comparable to the current input from the photodetector. The use of low input current amplifiers, such as field effect transistor-based devices and balanced input impedance circuitry, can have an impact on input bias currents. Other nonideal amplifier behavior such as output offset voltage may further need to be eliminated in transimpedance circuit design. Some operation amplifier characteristics such as gain are temperature dependent and may vary with device ambient conditions. Another important stochastic noise is the ADC in analog detection circuitry. Typical ADCs are designed such that their output accuracy is stable only to the last bit of its conversion range. Therefore, if the total conversion range is 10 V, an 8-bit ADC has a noise level of 40 mV, whereas a 16-bit ADC has a noise level of 0.15 mV.

12.5 A SURVEY OF DETECTOR TYPES AND THEIR CHARACTERISTICS

Detectors with higher sensitivity, lower noise, broader spectral range, and higher speed are becoming available at increasingly lower cost. A complete survey of all detectors is beyond the scope of this chapter, but we will focus on some of the most important detectors that are either routinely used or may find potential applications in nonlinear optical microscopy.

12.5.1 Photomultiplier Tube

Among all optical detectors, photomultiplier tubes (PMTs) are probably the most broadly used in nonlinear microscopy. Most importantly, a PMT is a photoelectric effect device with substantial internal gain (Hamamatsu, 2006; Kaurman, 2005). The cathode of the device is typically held at high negative voltage relative to the anode, which is set at ground. A series of electrodes, known as dynodes, are placed between the cathode and the anode at successively lower voltages set by a voltage divider chain (Fig. 12.8). Photoelectrons are generated at the cathode due to the photoelectric effect. Photoelectrons are accelerated from the cathode to the first dynode due to the voltage differential and the electrons in the process gain substantial kinetic energy. The bombardment of the first dynode by the high-kinetic-energy electrons results in the production of multiple secondary electrons from each incident electron. The amplified electron burst is then accelerated again toward the next dynode. As a result of this multiplicative process, a photoelectron is amplified exponentially as a function of the number of the dynode

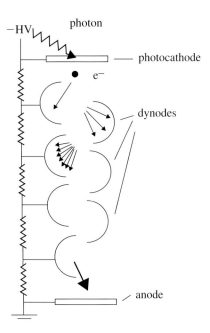

Figure 12.8. The design of a photomultiplier. A photon incident on the photocathode is converted into an electron. The electron is accelerated by an electric field toward successive dynodes, resulting in electron multiplication. The amplified photoelectron current is read out at the anode. High voltage (HV) differences are held between the cathode, the dynodes, and the anode using a voltage divider chain. (This figure is adapted from Kaufmann et al., 2005.)

stages. The amplification efficiency is a function of dynode material, geometry, and voltage differential. Typical PMTs have eight to ten dynode stages, providing an electron gain factor of 10^6 to 10^8. The electron burst is subsequently incident upon the anode and is directed to the preamplifier electronics of the readout circuit, where it is converted to a current or voltage pulse. Due to the large gain possible, single photons can be readily detected using a PMT.

The choice of photocathode materials is critical in PMT selection (Fig. 12.9). Photocathodes composed of traditional bi-alkali materials have a QE of 20% to 25% in the 300- to 500-nm wavelength region. Multi-alkali cathodes extend the usable range to about 700 nm but have a QE of only a few percent in the red. An important recent development is the introduction of GaAsP cathode materials that have extended maximum QE to 40% in the blue-green and maintain sensitivity above 10% up to almost 700 nm. Another new class of PMTs features GaAs cathodes that have over 10% QE from 500 nm up to almost 850 nm.

Other relevant parameters of the PMT include dark noise, pulse height distribution, and electron transition time and its spread. The PMT dark noise is a result of thermally driven emission of electrons at the cathode and the dynodes. Due to the amplification nature of the PMT, electrons emitted at the cathode and the initial stages of the dynode chain will contribute to higher dark current at the anode. The dark current of a PMT is dependent on its cathode material and the design of its dynode chain. The dark current of typical PMTs at room temperature can vary from tens to thousands of electrons

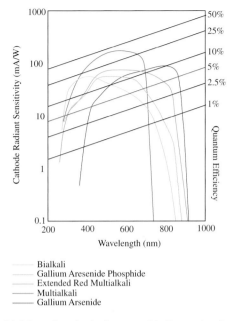

Bialkali
Gallium Aresenide Phosphide
Extended Red Multialkali
Multialkali
Gallium Arsenide

Figure 12.9. The sensitivities of typical photomultiplier cathode materials as a function of incident light wavelength. The left vertical axis is the cathode radiant sensitivity and the right vertical axis is the quantum efficiency of these materials. The horizontal axis is incident light wavelength. The black slanted lines are lines of constant quantum efficiency. (This figure is adapted from the literature of Hamamatsu Ltd., 2006.)

per second. As in any thermally driven process, reducing its absolute temperature can exponentially reduce the dark current rate of PMT. Since electron generation at each dynode is a stochastic process, the current burst produced from each photon at the anode can vary, resulting in the multiplicative noise discussed earlier. The typical pulse height distributions of PMTs have been shown earlier (see Fig. 12.5) and the width of its single photon peak is typically comparable to its median amplitude. Finally, the time required for the detection of a current burst after photon–electron conversion at the cathode is dependent on the geometry of the dynode chain design. The typical electron transit time is on the order of nanoseconds. Again, due to the stochastic nature of electron generation at each dynode, there is a substantial variation in the arrival time of the electrons generated from a single photon. This is characterized by the transit time spread, which is on the order of hundreds of picoseconds to nanoseconds. The transit time spread sets an upper limit on the measurable photon flux of the detector.

12.5.2 Multi-anode Photomultiplier Tubes

Multi-anode PMTs belong in the PMT family, but they are sufficiently important in nonlinear microscopy applications to warrant discussion. While most PMTs are point detectors without spatial resolution, the multi-anode PMT is a position-sensitive detector that can resolve the incident position of the incident photon on its cathode surface. Unlike a charge-coupled device camera (discussed in a forthcoming section), which has on the order of one million pixels, a multi-anode PMT typically has fewer than one hundred pixels. The cathode of a multi-anode PMT is similar to a conventional PMT, but its dynodes are designed such that the electron cascade resulting from the incident at a photon at a specific location at the cathode is directed toward a corresponding location at the anode, minimizing the spatial spread of the electron burst. The anode of these PMTs is segmented into pixels and individual electrode contacts are provided for each pixel element. This allows the photocurrent from each pixel to be read out independently, providing both spatial and intensity information of the photons incident upon the photocathode. While most nonlinear microscopes raster scan a single point and do not require detectors with spatial resolution, 1-D and 2-D multi-anode PMTs are now used extensively in spectrally resolved imaging (Dickinson et al., 2003) (see Chapter 19) and in high-speed multiphoton multifocal devices (Kim et al., 2004) (see Chapter 18). The overall performance of the multi-anode PMTs is similar to conventional PMTs. They typically are faster devices than most PMTs with lower electron transition time and spread due to their compact design. However, it should be noted that the confinement of the electron cascade within individual pixels is not perfect, resulting in typical cross-talk between adjacent pixels on the order of a few percent.

12.5.3 Photodiodes

Photodiodes do not have sufficient sensitivity for most nonlinear microscopy applications and typically have not been used. However, the design of photodiodes forms the basis of other advanced detectors such as avalanche photodiodes and CCD cameras, which are used frequently. A photodiode is a photoconductive solid state device composed of p-type and n-type semiconductor layers in contact (Fig. 12.10). At the junction of the p-type and n-type semiconductors, electrons from the n-type region diffuse into the p-type region and holes from the p-type region diffuse into the n-type region. This diffusion is stopped as an electrical potential develops due to carrier separation. At steady state, a central region, the depletion zone, is formed where there are no free charge carriers. A positive potential is present on the n-type side of the depletion zone,

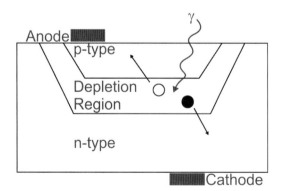

Figure 12.10. The design of a photodiode. A depletion region is formed between a p-type and an n-type semiconductor. An incident photon (γ) creates an electron (solid circle) and hole (empty circle) at the depletion region. The migration of these electron and holes can be detected as a current by completing a circuit via the cathode and the anode terminals.

while a negative potential is present on the p-type side. When a photon is absorbed in the depletion zone, resulting in the promotion of an electron to the conduction band, a pair of free charge carriers is produced. The potential across the depletion zone causes the electron to drift toward the n-type region, the hole toward the p-type. If an electrical circuit is connected between the p- and n-type terminals, a current will flow, resulting in the recombination of the electron and holes. The current produced in the circuit is proportional to the intensity of light impinged upon the device. It should be noted that photons absorbed outside the depletion zone have minimal effect as they result in charge carriers that are often quickly recombined locally. Therefore, increasing the size of the depletion zone by placing a lightly doped region between the p–n junction increases the effective responsive area of the detector.

Photodiodes are configured in either of two modes: nonbiased and biased (see Fig. 12.6a,b). In either case, due to the small current produced by the photodiode, high input impedance operational amplifiers are used downstream of the sensor. The nonbiased configuration has the advantage that impedance matching essentially eliminates the effect of bias current in the amplifier stage and results in a low noise operation. However, in the nonbiased configuration, the speed of the photodiode is relatively low due to the low drift velocity of the charge carriers. A biased photodiode circuit improves electron drift velocity and the response speed of the system but suffers from greater electronic noise.

As photodiodes are fabricated using silicon semiconductors, their bandgaps match well with light in the visible and infrared spectrum and typically have excellent QE, on the order of 70% to 80%. The difference between silicon devices and PMTs is their near-infrared sensitivity; at 800 nm, the best PMTs have QE of less than 5%, while silicon devices can still have QE above 50%. They are also highly linear, with a dynamic range of about six to eight orders of magnitude. The only disadvantage is that photodiodes have no internal gain; each photon converted will generate a single electron–hole pair. Due to their susceptibility to dark, Johnson, and electronic noises, typical photodiodes do not have single photon sensitivity.

12.5.4 Avalanche Photodiodes

Avalanche photodiodes (APDs) are diodes with internal gain (Cova et al., 1996; Dautet, 1993) . APDs are designed with high reverse-bias voltage across the depletion region (Fig. 12.11). The drift velocity of the carriers increases with increasing bias voltage. At the field strength at about 10^4 V/cm, drift velocity plateaus due to charge carrier collision with the crystal lattice. With further increase in field strength beyond 10^5 V/cm, secondary carriers are created upon collision. This process is called impact ionization. These secondary carriers further gain kinetic energy and again create more carriers in an avalanche fashion, providing internal gain for APDs.

In the consideration of APD noise characteristics, it should be noted that its dark current contains two terms. One term, I_{db}, corresponds to thermally induced charge carriers in the depletion zone inside the bulk of the device, which is amplified. Another term, I_{ds}, corresponds to the carrier generation close to the device surface and is not amplified. The total dark current is:

$$I_d = I_{ds} + \alpha I_{db} \qquad (12.36)$$

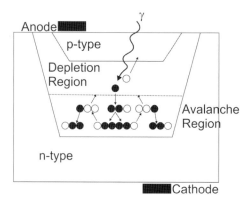

Figure 12.11. The design of an avalanche photodiode. A depletion region is formed between a p-type and an n-type semiconductor. An incident photon (γ) creates an electron (solid circle) and hole (empty circle) at the depletion region. The presence of high voltage creates an avalanche region where the electrons and holes created are accelerated, and their collision creates more electron–hole pairs. The migration of these amplified electron and holes can be detected as a current by completing a circuit via the cathode and the anode terminals.

Since the amplitude of the bulk dark current depends on the volume of the semi-conductor substrate, the bulk dark current can be reduced by minimizing the size of the detector. Today, the most sensitive, single photon counting APDs have active areas slightly less than 1 mm. The small active area limits the use of these small APDs in nonlinear optical microscopes, where non-descan geometry is critical to max-imize signal level. On the other hand, these high-sensitivity APDs find applications in situations where little or no scanning is required, such as fluorescence correlation spectroscopy.

The response time of the APD depends on a number of factors. In addition to a time constant resulting from the capacitance of the detector and the shunt resistance in the readout circuit, two factors limit the response speed of these devices. First, the avalanche processes is a result of repeated collision of the avalanching electrons with the material lattice. The collision processes introduce significant response delay. Second, in the presence of high light intensity, a large population of electrons and holes causes local shielding of the bias voltage across the depletion layer, resulting in slower drift speed of the carrier. This process is called the space charge effect.

Two common electronic circuits are used with APDs: linear mode and Geiger mode. In the linear mode, the configuration of the APDs is similar to normal photodiodes, with typical current gain on the order of hundreds. Geiger mode configuration of APDs allows their use for single photon counting. In this case, the bias on the APD is higher than the reverse breakdown voltage of the photodiode, achieving gain levels on the order of 10^5 to 10^6. In the absence of current limiting circuitry, the large current flow in this configuration will destroy the APD. An active current limiting and quenching circuit is employed that quenches the avalanche current after its onset. The reverse breakdown process is initiated by the spontaneous generation of dark carriers in the depletion region as characterized the APD's bulk dark current. Reducing the device size and its temperature lowers the rate of dark carrier formation and results in fairly infrequent dark current bursts. The APD is in its off-state most of the time until the absorption of an incident photon triggers a large current burst, permitting single photon counting. Since the termination of the avalanche photocurrent after trigger is accomplished using an operational amplifier quenching circuit, the dead time between successive photons is restricted by the bandwidth of the quenching circuit. Therefore, the maximum single photon counting rate of APDs is significantly less than that of PMTs.

12.5.5 Charge-Coupled Device Cameras

Charge-coupled devices (CCDs) are commercially the most important optical sensors and are widely used in consumer products. Similar to multi-anode PMTs, linear and two-dimensional CCD imagers have found applications in spectrally resolved nonlinear microscopes and high-speed systems. CCDs can be thought of as an array of photo-diodes (Fig. 12.12) (Inoué & Spring, 1997). A matrix of control electrodes fabricated out of thin polysilicon materials is placed on the front of the silicon semiconductor. These series of electrodes define a periodic potential array, alternating regions of reverse biasing and no voltage, forming the pixel structure of the sensor. In the sensor, an electron–hole pair is created as in a photodiode upon the absorption of a photon. The electron migrates to the n-layer and the hole migrates to the p-layer. The electron is then trapped at the potential well created by the nearest reverse biasing electrode. The CCD is exposed to light during the data acquisition cycle; the electrons trapped in each pixel are proportional to the number of incident photons absorbed.

   The front surface of the CCD is defined as the surface where the polysilicon gate electrodes are placed. Most CCDs are front illuminated and light transmits through these thin electrodes before being absorbed in the silicon layer (see Fig. 12.12a). Although the polysilicon gates are fairly transparent, they limit this type of CCD to a quantum efficiency of about 40%. In contrast, the silicon back layer can be etched thin, allowing photons to penetrate from the backside of the chip into the depletion region

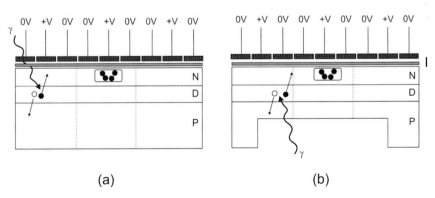

Figure 12.12. The design of a charge couple device (CCD). (a) A front-illuminated CCD. (b) A back-illuminated CCD. An incident photon (γ) creates an electron (solid circle) and hole (empty circle) pair at the depletion region formed between a p-type and an n-type semiconductor. The electrons are migrated to the positive electrodes, which define the pixel structure of the CCD. Thinning of the p-type semiconductor region is necessary in the back-illuminated CCD to maximize the throughput of light to the depletion region.

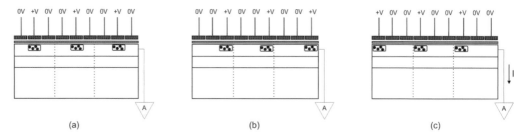

Figure 12.13. The read out sequence (a), (b), (c) of electron packets is achieved by cycling the positive voltage through a triplet of electrodes. In (c), when the electron packet reaches the output node, it is read out as a current *I* and amplified by on-chip amplifier A.

(see Fig. 12.12b). These are called back-illuminated CCDs, and they have quantum efficiency over 80% from the visible through the near-infrared spectrum.

The charges stored in each pixel of the CCD are then read out in a sequential manner. Most common CCD chips have a single readout node consisting of a low noise amplifier circuitry. The readout of the CCD chip is based on periodically varying the potentials between adjacent electrodes to "shift" charges toward the readout node. After the pixels from the whole row are read, the charges from the next row are shifted down in an analogous fashion. This process is best understood graphically (Fig. 12.13).

The electronic noise, called the readout noise, in reading out a single CCD pixel is dominated by the noise of the readout amplifier and increases with the readout bandwidth. For commercial-grade CCDs, read noise amplitude can easily correspond to thousands of electron charges; for scientific-grade CCDs with highly optimized readout electronics, read noise amplitude corresponding to a few electron changes has been achieved. Since conventional CCDs have no gain, these devices do not have single photon sensitivity.

Unlike all other sensors discussed previously, the readout of a whole CCD chip can be a fairly slow process, especially when read noise must be minimized. Therefore, it is important that the image is not distorted by light exposure during the readout process. A number of CCD chip designs have been developed to accomplish this task. For example, an interline CCD has alternative lines that are blocked from light and the charges are shift into the blocked region during readout. A frame transfer CCD is similar, where half the chip is shielded and the rows are rapidly moved into the shielded region before readout.

12.5.6 Intensified CCD Cameras

While back-illuminated CCDs have excellent QE, they nonetheless do not have single photon sensitivity due to the lack of gain and finite read noise. CCDs are not sufficient for applications characterized by extremely low illumination intensity or applications requiring extremely high speed, where only a few photons will be stored in each pixel. For full-frame imaging in these applications, a method is to use intensified CCD cameras. An intensified CCD consists of a standard CCD camera with a microchannel plate

(MCP) coupled to its front end. MCPs are composed of a highly conductive glass substrate with a dense array of micron-size holes coated with secondary electron emitters. The cathode and anode surfaces of the MCP are held at high voltage similar to a PMT. The front end of the MCP has a photocathode composed of similar material as a PMT. Electrons are produced at the cathode due to the photoelectric effect. The high voltage accelerates the photoelectron into a nearby hole in the array. An electron avalanche is produced into the hole similar to a PMT, resulting in an amplified electron burst exiting the glass substrate. The anode of the MCP is typically a phosphor screen that emits light due to the incident electron burst. Since the electrons are confined by the hole spatially, the spatial location of the incident photon is preserved. MCPs have quantum efficiency similar to a PMT but provide spatial resolution. Furthermore, since MCPs have a more compact design than the bulkier PMT, the transition time spread of the electron burst can be very short, allowing tens of picosecond time resolution measurement using these devices. The optical output of the MCP can be transferred to the CCD by either direct or lens coupling. The readout noise is no longer a limiting factor in the sensitivity of intensified CCDs due to the gain of the MCP. Instead, the dark current of MCP influences the SNR, and high-sensitivity devices require cooling of both the MCP and CCD camera. Further, while the MCP glass channels are in a closely packed array, some spatial resolution degradation is often observed in intensified CCDs as compared with systems without MCP.

12.5.7 Electron Bombardment CCD Cameras

Electron bombardment CCDs (EBCCDs), also known as electron multiplying CCDs, have been recently introduced to improve the sensitivity limitation of standard CCDs (Fig. 12.14) (Coates et al., 2004; Hynecek & Nishiwaki, 2003; Oshiro & Moomaw, 2003; Robbins & Hadwen, 2003). Unlike standard CCDs, the output node of the CCD

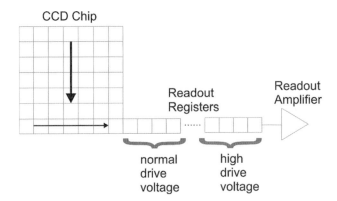

Figure 12.14. The design of an electron bombardment charge couple device (EBCCD). The electron packets are read out on the imaging region of the CCD as normal. However, unlike normal CCD, EBCCD has many nodes in the readout stage. Some of the readout nodes have normal-level readout voltages, but some nodes have higher readout voltage levels, resulting in electron generation in the high-voltage nodes. The shifting of the electron packet through these high-voltage nodes results in an amplification.

chip is fed directly to a readout amplifier. The EBCCD has a multiple shift register stage. The charge from a pixel is shifted from one register to the next through a few hundred stages. As the charges are accelerated into the next register, impact ionization occurs, similar to the process in an APD although at a significantly lower gain. The drive voltage is controlled such that a gain of approximately 1% is achieved at each stage. After a few hundred stages, a charge amplification factor up to almost 1,000 can be achieved. This amplification process, of course, results in multiplicative noise and has a typical excess noise factor of approximately 1.5. Nonetheless, the presence of internal gain allows single photon detection using a CCD in a completely solid-state platform. The sensitivity of EBCCD is actually quite similar to intensified CCDs. Further, these two types of devices are similar in permitting faster readout compared with standard CCDs without degrading image SNR. EBCCD may have advantages over intensified systems in spatial resolution and resistance to damage from exposure to high light level.

12.5.8 Streak Cameras

Streak cameras are another class of photoemissive device that are extremely fast and can resolve photon arrival time difference by fractions of a picosecond; they are used in high temporal resolution lifetime-resolved microscopes (Hamamatsu, 2002; Krishnan et al., 2003) (Fig. 12.15). After electron emission from the photocathode, the electrons are accelerated towards an MCP. In between the photocathode and the MCP is a pair of

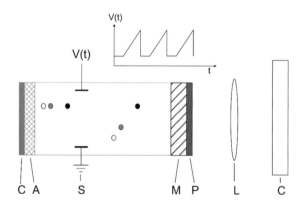

Figure 12.15. The design of a streak camera. Photons arriving at the photocathode (C) are converted into electrons. The electrons are accelerated through a high-voltage region (A). Three electrons are depicted that are generated by photons arriving at successive times. The black circle denotes the electron generated by the earliest photon, the gray circle denotes the electron generated by the next photon, and the empty circle denotes the electron generated by the last photon. These electrons go through a region in the camera where there is a scanning transverse voltage (S), which is a repetitive ramp in time. As a result of the linearly increasing transverse voltage, the electrons are deflected laterally depending on arrival time. This process encodes the electron's timing information along the horizontal spatial dimension. The electrons then enter a microchannel plate (M), where they are amplified with spatial resolution. The amplified electron bursts are then incident upon a phosphor screen (P) and converted back to light. A CCD camera (C) images the phosphor screen via a lens L.

electrodes positioned transversely (denote that as x-axis). The voltage across these pair of electrodes can be ramped up rapidly, called a swipe. The transverse displacement along the x-axis of the electron on the phosphor screen is then a function of the ramp voltage. Electrons that arrive at the start of the swipe are minimally deflected laterally, while electrons that arrive later have greater deflection. Therefore, the arrival time of the electron at the photocathode is then encoded by the electron lateral position at the MCP. The electrons are amplified and are converted by a photon burst at the output phosphor screen of the MCP. The output of the MCP can be imaged by a high-sensitivity CCD detector, providing both intensity and timing information of the photons. Since the MCP and CCD are 2D devices, multiple optical channels can be positioned along the y-axis, allowing simultaneous measurement. An example is to spectrally resolve the emission light along the y-axis, allowing simultaneous spectral and lifetime measurements. Streak cameras are mostly used in ultrafast lifetime imaging microscopes. While the streak camera cathode has similar sensitivity as PMTs, the complex multiple-stage design introduces noise for intensity measurement and degrades spatial resolution. The duty cycle of a streak camera suffers from long setup time between swipes, typically over tens of nanoseconds.

12.5.9 CMOS Cameras

Complementary metal oxide semiconductor (CMOS) cameras are competing favorably against CCDs for applications where sensitivity is less critical, such as white light wide-field microscopy (Janesick, 2002). Due to their lower sensitivity, they have hardly been used in nonlinear optical microscopy. However, with the continuing improvement in CMOS fabrication technology and some of its unique advantages, this type of detector may one day become more prevalent. CMOS cameras are in many ways similar to CCDs. Both devices are semiconductor devices but they are produced using different fabrication protocols. CMOS fabrication is similar to the production of many high-volume electronic components such as microprocessors, allowing very low production cost. In contrast, CCDs are fabricated by a custom technology developed and optimized over the past 20 years specifically for these imaging sensors. The smaller market dictates higher cost. Ignoring cost concerns, CMOS cameras also have the significant advantage of allowing the integration of readout amplifiers and logics into each pixel since the chip is produced using standard electronic fabrication technology. This opens the possibility of ultrafast readout and on-chip signal processing. However, today, in terms of optical performance, CCDs remain superior to CMOSs. Since CMOS has integrated nonoptical electronic components at each pixel and requires more interconnections, the optical throughput from the front of the chip is low, resulting in significantly lower QE due to a lower fill factor. This problem may potentially be overcome by the introduction of back-thinned CMOS chips. Other issues include charge collection and readout electronic quality. CCD chips are fabricated with higher-resistance silicon substrate; coupled with higher bias voltage, this creates more well-defined potential wells, allowing better charge localization and broader spectral response. Since the readout electronics of the CMOS are present in each pixel, the complexity and sophistication of the readout amplifier circuitry must be reduced to improve chip fill factor, size, and cost. This limitation results in more noise readout circuitry for CMOS systems.

12.5.10 Hybrid Photomultipliers

Hybrid photomultiplier (HPD) tubes are also novel detector devices that are in the development pipeline of a number of sensor manufacturers and have not been used in the nonlinear microscopy field (Cushman et al., 1997; Cushman & Heering, 2002; Hayashidaa et al., 2005). HPD merges PMT and APD technology (Fig. 12.16a). Photoelectrons are produced in the HPD cathode. High voltage accelerates these photoelectrons through towards an APD. As the electron is deposited into the silicon substrate of the APD, the electron bombardment effect causes the generation of many secondary electrons, with a gain on the order of one thousand. These electrons subsequently are further amplified in the APD by an impact ionization effect in the depletion zone, producing another gain of 50 to 100, resulting in a total amplification factor on the order of 10^4 to 10^5. The distinguishing properties of these devices are their excellent pulse height distribution (see Fig. 12.16b) (Hayashidaa et al., 2005). Since the current amplitude produced by one, two, and higher number of photons can be readily distinguished, HPD opens the opportunity of using these devices for higher-rate single photon counting applications.

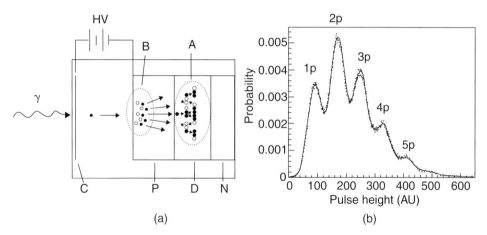

(a)   (b)

Figure 12.16. (a) The design of a hybrid photomultiplier tube. The incident photon (γ) is converted into an electron (solid circle) at the photocathode (C). The electron is accelerated by high voltage (HV) towards essentially an avalanche photodiode consisting of a pair of p-type (P) and n-type (N) semiconductors. Due to the high voltage, instead of a single electron (solid circle) and hole (empty circle) pairs being generated at the depletion region (D), many electron–hole pairs are generated by the high-kinetic-energy electron in a process called bombardment gain (B). These electron–hole pairs undergo further amplification by the standard avalanche gain processes (A). The electron burst is then read out after this two-step amplification process. (b) The current burst amplitude distribution of a hybrid photomultiplier tube. The unique feature of this device is that multiple photon processes can be cleanly separated. In the device described by Hayashidaa et al. (Hayashidaa et al., 2005), the current bursts due to the simultaneous incidence of one (1p), two (2p), three (3p), four (4p), and five (5p) photons can be distinguished. (This portion of the figure is adapted from Hayashidaa et al., 2005).

12.6 CONCLUSION

With today's high-sensitivity detectors such as PMTs, APDs, and CCDs, the signals from single molecules can be readily imaged. High-speed detectors such as PMTs and MCPs allow lifetime imaging to distinguish fluorescence lifetime with only tens of picosecond difference. High-speed CMOS detectors allow high frame rates, beyond tens of thousands, at relatively low cost. The continuing development of these exciting detector technologies and their incorporation into nonlinear optical microscopes will provide new opportunities for biomedical investigations.

REFERENCES

Ashcroft, N. W. and N. D. Mermin (1976). Solid State Physics. New York, Holt, Rinehart and Winston.

Becquerel, H. (1892). La chaire de physique du Muséum, Revue Scientifique 49: 674–678.

Coates, C. G., D. J. Denvir, N. G. McHale, K. D. Thornbury and M. A. Hollywood (2004). Optimizing low-light microscopy with back-illuminated electron multiplying charge-coupled device: enhanced sensitivity, speed, and resolution, J Biomed Opt 9(6): 1244–1252.

Cova, S., M. Ghioni, A. Lacaita, C. Samori and F. Zappa (1996). Avalanche photodiodes and quenching circuits for single photon detection, Appl. Opt. 35: 1956–1976.

Cushman, P., A. Heering, J. Nelson, C. Timmermans, S. R. Dugad, S. Katta and S. Tonwar (1997). Multi-pixel hybrid photodiode tubes for the CMS hadron calorimeter, Nucl. Instrum. Methods A387: 107–112.

Cushman, P. B. and A. H. Heering (2002). Problems and solutions in high-rate multichannel hybrid photodiode design: The CMS experience, IEEE Transactions on Nuclear Science 49(3): 963–970.

Dautet, H. l. (1993). Photon-counting Techniques with Silicon Avalanche Photodiode, Applied Optics 32(21): 3894–3900.

Dickinson, M. E., E. Simbuerger, B. Zimmermann, C. W. Waters and S. E. Fraser (2003). Multiphoton excitation spectra in biological samples, J Biomed Opt 8(3): 329–338.

Hamamatsu (2002). Guide to Streak Cameras. Hamamatsu, Japan, Hamamatsu Inc.

Hamamatsu (2006). Photomultiplier Tubes: Basics and Applications. Hamamatsu, Japan, Hamamatsu, Inc. 3.

Harvey, E. N. (1957). A History of Luminescence from the Earliest times Until 1900. Philadelphia, American Philosophical Society.

Hayashidaa, M., J. Hosea, M. Laatiaouia, E. Lorenza, R. Mirzoyana, M. Teshimaa, A. Fukasawab, Y. Hottab, M. Errandoc and M. Martinez (2005). Development of HPDs with an 18-mm-diameter GaAsP Photo cathode for the MAGIC-II project. 29th International Cosmic Ray Conference, Pune.

Hynecek, J. and T. Nishiwaki (2003). Excess noise and other important characteristics of low light level imaging using charge multiplying CCDs, IEEE Trans. Electron Devices 50(1): 239–245.

Inoué, S. and K. R. Spring (1997). Video microscopy: the fundamentals. New York, Plenum Press.

Janesick, J. (2002). Dueling Detectors. OE Magazine. 2: 30–33.

Johnson, J. B. (1928). Thermal Agitation of Electricity in Conductors, Physical Review 32: 97–109.

Kaufman, K. (2005). Choosing Your Detector. OE Magazine. 5: 25–29.

Kim, D., K. H. Kim, S. Yazdanfar and P. T. C. So (2004). High-Speed Handheld Multiphoton Multifoci Microscopy, SPIE Press.

Komp, R. J. (1995). Practical photovoltaics: electricity from solar cells. Ann Arbor, Aatec Publications.

Krishnan, R. V., A. Masuda, V. E. Centonze and B. Herman (2003). Quantitative imaging of protein-protein interactions by multiphoton fluorescence lifetime imaging microscopy using a streak camera, J Biomed Opt 8(3): 362–367.

McIntyre, R. J. (1966). Multiplication Noise in Uniform Avalanche Diodes, IEEE Trans. Electron Devices 13: 164–168.

Nyquist, H. (1928). Thermal Agitation of Electricity in Conductors, Physical Review 32: 110–113.

Oshiro, M. and B. Moomaw (2003). Cooled vs. intensified vs. electron bombardment CCD cameras–applications and relative advantages, Methods Cell Biol 72: 133–56.

Robbins, M. S. and B. J. Hadwen (2003). The noise performance of electron multiplying charge coupled devices, IEEE Transactions on Electron Devices 50: 1227–1232.

Shive, J. N. (1959). The Properties, Physics, and Design of Semi-Conductor Devices. Princeton, D. Van Nostrand Company Inc.

Sze, S. M. (2002). Semiconductor devices, physics and technology. New York, Wiley & Sons.

13

Multiphoton Excitation of Fluorescent Probes

Chris Xu and Warren R. Zipfel

13.1 INTRODUCTION AND OVERVIEW

Molecular two-photon excitation was first theoretically predicted by Göppert-Mayer in her Ph.D. thesis in 1931 (Göppert-Mayer, 1931). It is arguable, however, that Einstein has speculated on the possibility of multiphoton absorption in his Nobel Prize paper on the quantum nature of light in 1905 (Einstein, 1905). Confirmation in the laboratory (even though not conclusively), first by Hughes and Grabner in 1950 (Hughes & Grabner, 1950), came out of atomic and molecular beam spectroscopy in the radio and microwave frequency regions. Like most other nonlinear processes, two- or three-photon excitation requires high instantaneous photon intensities, typically, for the observation of two-photon absorption. Months after its invention in 1960, the pulsed ruby laser was used by Kaiser and Garrett to excite two-photon absorption in $CaF_2:Eu^{3+}$, manifested by fluorescent de-excitation in the blue spectral region (Kaiser & Garrett, 1961). This was the second demonstration of nonlinear optics, closely following the demonstration of second-harmonic generation. Three-photon excited fluorescence was observed in naphthalene crystals by Singh and Bradley in 1964 (Singh & Bradley, 1964) and the three-photon absorption cross-sections were estimated.

One of the basic parameters in quantitative fluorescence microscopy is the fluorescence excitation cross-sections. Although the one-photon absorption spectra of common biological fluorophores are well documented, it is often difficult to predict the multiphoton excitation spectra, especially the two-photon spectra, from the known one-photon absorption data because of the different selection rules involved and the effects of vibronic coupling. Before the 1990s, two-photon excitation cross-section measurements were almost always carried out at 694 nm (ruby laser) and 1,064 nm (Nd:glass laser) on laser dyes (Smith, 1986). Less effort was devoted to accurate quantitative studies of common fluorophores widely used in two-photon laser scanning microscopy (TPLSM). In addition, substantial disagreement (sometimes over one order of magnitude) between published values of two-photon cross-sections often exists. The lack of knowledge of two-photon excitation cross-sections and spectra for common fluorophores used in biological studies has been a significant obstacle in the use of TPLSM.

The emergence of the mode-locked solid-state femtosecond lasers, most commonly the Ti:Sapphire (Ti:S) lasers (Curley et al., 1992; Spence et al., 1991), has greatly facilitated the measurement of multiphoton excitation cross-section. When compared to earlier ultrafast lasers (e.g., ultrafast dye lasers), the Ti:S lasers are highly robust and widely tunable, making femtosecond pulses from ~690 nm to 1,050 nm easily accessible. A large number of fluorescent indicators, intrinsic fluorescent molecules, genetically engineered probes, and some nanoparticles (quantum dots, for example) have since been measured. New measurement techniques have also been developed to further improve the absolute accuracy.

This review on multiphoton excitation cross-sections is motivated by the application of multiphoton microscopy as a powerful tool for three-dimensionally resolved fluorescence imaging of biological samples (Denk et al., 1990; Helmchen & Denk, 2005; Masters, 2003; Masters & So, 2004; So et al., 2000; Xu et al., 1996b; Zipfel et al., 2003a, 2003b). We will review the known two-photon excitation (2PE) cross-sections of biological indicators and discuss several related issues, such as how to experimentally gauge the two-photon cross-section of an indicator. An effort is made to compare cross-section values obtained from various research groups in the past 10 years or so. While we will concentrate on 2PE, three-photon excitation will also be briefly discussed. This chapter is divided into four sections. Section 13.2 provides simple estimates of multiphoton excitations of dyes and quantum dots from theory. In the method section, we will review the multiphoton excitation (MPE) cross-section measurement methods and provide practical guides for experimentally estimating the excitation cross-section. The last section of this chapter is a compilation of multiphoton excitation cross-sections of extrinsic and intrinsic fluorophores, genetically engineered probes, and nanoparticles.

13.2 ESTIMATION OF MULTIPHOTON EXCITATION

The essence of the theory of multiphoton processes can be represented in perturbation theory of one form or another. The time-dependent semiclassical interaction can be used to derive the multiphoton transition amplitude because a more accurate description using quantum field theory generates completely equivalent results for virtually all processes in laser field. Details of the rigorous derivation can be found elsewhere (Faisal, 1987). Here, for the purpose of an order of magnitude estimation, a much simplified approach will be presented to describe the MPE processes, and only the lowest-order dipole transition will be considered.

Intuitively, multiphoton processes require two or more photons to interact *simultaneously* with the molecule. For example, in a two-photon process, the "first" photon excites the molecule to an intermediate state and the "second" photon further excites the molecule to its final state. Because the intermediate state can be a superposition of molecular states instead of an eigenstate of the molecule, it is usually referred to as the virtual intermediate state. Although the two-photon transition strength consists of contributions from all eigenstates as possible intermediate states, the single intermediate state (SIS) approximation (Birge, 1983) can be used to give an order of magnitude estimation. Figure 13.1 shows a schematic illustration of the SIS approximation. Therefore, by simple examination of the absorption rates, the two-photon absorption cross-section (σ_2) can be obtained:

$$\sigma_2 = \sigma_{ij}\,\sigma_{jf}\,\tau_j, \tag{13.1}$$

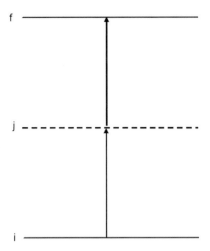

Figure 13.1. The single intermediate state (SIS) approximation. i, j, and f represent the initial state, the intermediate state, and the final state, respectively.

where σ_{ij} and σ_{jf} represent the appropriate one-photon absorption cross-sections and τ_j is the intermediate state lifetime that determines the time scale for photon coincidence (simultaneous arrival).

τ_j can be estimated from the uncertainty principle (i.e., τ_j must be short enough in order not to violate energy conservation). Thus,

$$\tau_j \approx \frac{1}{\Delta\omega} = \frac{1}{|\omega_{ij} - \omega|}, \tag{13.2}$$

where ω_{ij} and ω are the transition frequency and the incident photon frequency, respectively, for an electronic transition (ω_{if}) in the visible frequency range and further assuming that the intermediate state and the final state are close in energy. We then have $\tau_j \approx 10^{-15}$ to $10^{-16}s$. The one-photon absorption cross-section of a molecule can be estimated by its dipole transition length (typically $\sigma_1 \approx 10^{-16}$ to 10^{-17} cm^2 for a transition length of 10^{-8} cm). Hence, the estimated two-photon absorption cross-sectionsghould be approximately 10^{-49} cm^4 s/photon (equation (13.1)), or 10 Göppert-Mayer (GM, 1 GM$= 10^{-50}$ cm^{4}s/photon).

It is important to note that the above estimates are valid only for simultaneous MPE (i.e., virtual states with lifetimes on the order of 10^{-15} s serve as the intermediate states). However, in a sequential MPE process, real states with lifetimes approximately 10^{-9} to $10^{-12}s$ serve as the intermediate states. Consequently, the cross-section for a sequential MPE process can be orders of magnitude larger than the above estimate. Although the enhanced MPE cross-sections may seem desirable for some applications, there are many limitations in such multistep MPE, which we choose not to discuss in the chapter.

The above description and estimation of multiphoton excitation for dye molecules need to be modified in the case of nanocrystals such as quantum dot (QD) (Alivasatos, 1996; Efros & Efros, 1982). QDs are made out of semiconductor materials. If one can synthesize a piece of the semiconductor so small that the electron feels confined, the nearly continuous spectrum of the bulk material will become discrete and the energy gap increases. Such a strong quantum confinement occurs when the size of the dot is

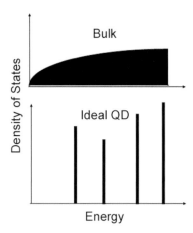

Figure 13.2. Density of electron states in bulk- and size-quantized semiconductor. The optical absorption spectrum is roughly proportional to the density of states.

much smaller than the exciton Bohr radius of the bulk material (i.e., the average physical separation between the electron and hole in a bulk material). Even though such a electron–hole pair (called an exciton) is analogous to a hydrogen atom, it should be emphasized that the charge carriers in a QD are bound by the confining potential of the boundary rather than the Coulomb potential as in the bulk (Wise, 2000). A zero-th order model for the description of QD is a single particle confined in an infinite potential well. One direct consequence of this quantum confinement is the bandgap engineering. As the size of the bandgap in a QD is controlled simply by adjusting the size of the dot, it is possible to control the output wavelength (determined by the bandgap) of a dot with extreme precision. In fact, it is possible to tune the bandgap and therefore specify its "color" output depending on the needs of the user. Another desirable effect of the reduced dimension in a QD is the concentration of the density of state into discrete bands (Fig. 13.2). Because light always interacts with one electron–hole pair regardless of the size of the material, the total or integrated absorption does not change. Thus, the concentration of the density of states significantly increases absorption at certain photon energies, at the expense of reduced absorption at other energies. A rough order of magnitude estimation of the absorption enhancement can be obtained by examining the density of states in a QD. There are approximately 10^4 atoms in a QD of 5-nm radius, resulting in a total integrated number of states of $\sim 10^4$. Assuming all these available states are concentrated in a few discrete bands, a QD will have a peak absorption strength of $\sim 10^4$ of that of a single atom. Using the estimation obtained above for a dye molecule, the two-photon absorption cross-section of a QD will be approximately 10^5 GM.

13.3 MEASUREMENT METHODS OF MULTIPHOTON EXCITATION CROSS-SECTION

For much spectroscopic research, the primary use of two- and three-photon absorption data is to elucidate the energy states of matter. Hence, transition energies and *relative* strengths are more informative than *absolute* values. For some technological applications, however, such as nonlinear laser scanning microscopy, the absolute strengths of multiphoton transition are required as well as the transition energies and relative strengths. In general, methods for measuring the absolute multiphoton excitation cross-section can be divided into two categories: those using fluorescence excitation, and those employing transmission measurements. We will mainly focus on the fluorescence method, which is directly applicable to the field of multiphoton microscopy. A brief review of the transmission method will also be provided in this section.

13.3.1 Fluorescence Excitation Method

In the case of fluorescent materials, which are most interesting in biological imaging, two- and three-photon generated fluorescence can be measured to determine the action cross-sections or absorption cross-sections (if the fluorescence emission quantum efficiencies are known) (Hermann & Ducuing, 1972). The fluorescence technique provides very high detection sensitivity and two-photon cross-sections can be measured in dilute fluorophore solution (usually less than 0.1 mM) or with low-power CW lasers (Catalano & Cingolani, 1981).

A generic setup for measuring the multiphoton fluorescence cross-section is shown in Figure 13.3. A 90-degree setup, where the detection path is perpendicular to the excitation path, can also be used.

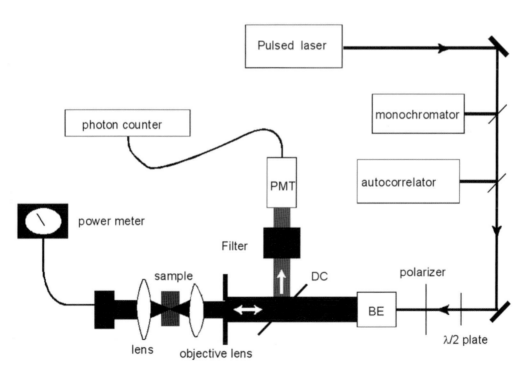

Figure 13.3. Generic setup for multiphoton cross-section measurements. BE=beam expander (optional). DC=dichroic mirror. The half waveplate and the polarizer are used to adjust the excitation intensity.

The crucial problem in MPE cross-section measurements is that the absorption rates strongly depend on the high-order (≥ 2) spatial and temporal coherence of the excitation light. Knowing the relationship between the experimentally measured fluorescence power and the excitation power is essential in MPE cross-section measurements. In this section, we first examine the relationship in general and discuss the relevant parameters in any MPE cross-section measurements. Detailed analysis of several special cases will follow.

Because n-photon excitation is an nth-order process, the number of photons absorbed per molecule per unit time via n−photon excitation is proportional to the n−photon absorption cross-section σ_n and to the nth power of the incident intensity I (McClain & Harris, 1977). In a particular experiment, the total number of photons absorbed per unit time N_{abs} is also a function of dye concentration C and the illuminated sample volume V.

$$N_{abs}(t) = \int_V d\mathbf{r}\, \sigma_n\, C(\mathbf{r}, t)\, I^n(\mathbf{r}, t). \tag{13.3}$$

In the absence of ground state depletion and photobleaching, $C(\mathbf{r}, t)$ can be assumed a constant in a solution. Moreover, in the case where one can separate time and space dependence of the excitation intensity, we have:

$$N_{abs}(t) = C\, \sigma_n\, I_0^n(t) \int_V d\mathbf{r}\, S^n(\mathbf{r}), \tag{13.4}$$

where $S(\mathbf{r})$ and $I_0(t)$ describe the spatial and temporal distribution of the incident light (i.e., $I(\mathbf{r},t) = S(\mathbf{r}) I_0(t)$).

We chose $S(\mathbf{r})$ to be a unitless spatial distribution function and $I_0(t)$ to be the intensity at the geometric focal point.

Assuming no stimulated emission and no self-quenching, the number of fluorescence photons collected per unit time (F) is given by:

$$F(t) = \frac{1}{n}\, \phi\, \eta\, N_{abs}, \tag{13.5}$$

where η and ϕ are the fluorescence quantum efficiency of the dye and fluorescence collection efficiency of the measurement system, respectively.

Here the factor $1/n$ simply reflects the fact that n photons are needed for each excitation event. In practice, only the time-averaged fluorescence photon flux $\langle F(t) \rangle$ is measured:

$$\langle F(t) \rangle = \frac{1}{n}\, \phi\, \eta\, C\sigma_n\, <I_0^n(t)> \int_V d\mathbf{r} S^n(\mathbf{r}). \tag{13.6}$$

We note that $\langle F(t) \rangle$ is proportional to $\langle I_0^n(t) \rangle$, not $\langle I_0(t) \rangle^n$. Because most detectors give only a signal that is proportional to $\langle I_0(t) \rangle$, we rewrite equation (13.6) in terms of the average intensity:

$$\langle F(t) \rangle = \frac{1}{n} g^{(n)} \phi \eta C \sigma_n <\,<I_0(t)>^n \int_V d\mathbf{r} S^n(\mathbf{r}), \qquad (13.7a)$$

$$g^{(n)} = \frac{<I_0^n(t)>}{<I_0(t)>^n}, \qquad (13.7b)$$

where $g^{(n)}$ is a measure of the *nth*-order temporal coherence of the excitation source.

All of the relevant quantities in a MPE cross-section measurement are presented in equation (13.7). Experimental determination of σ_n usually involves the characterization of three parameters: the spatial distribution of the incident light ($\int_V S^n(\mathbf{r})d\mathbf{r}$), the *nth*-order temporal coherence ($g^{(n)}$), and the fluorescence collection efficiency of the system (ϕ).

13.3.1.1 Spatial Considerations

The spatial distributions of two limiting cases—the Gaussian-Lorentzian (GL) focus and the diffraction limited focus—will be discussed in detail. In practice, only these two profiles have been used in quantitative measurements in order to avoid the complex nature of beam profiles that results from the partial filling of the back aperture of the focusing lens. We will only consider two- and three-photon cross-section measurements in the following sections. The general formula for measuring the *n*-photon cross-section for higher *n* can be obtained by parallel analysis.

13.3.1.1.1 EXCITATION WITH A GAUSSIAN-LORENTZIAN BEAM

A laser that operates in TEM$_{00}$ mode has a GL spatial profile. However, diffraction by a finite aperture distorts the wavefront. Detailed theory of a GL beam propagating through such finite apertures has been developed in the past (Dickson, 1970). The resulting spatial profile after the aperture is complex and depends on the ratio of the GL beam waist (w_0) and the aperture size. Practically, such difficulties can be circumvented by keeping the incident beam diameter much smaller than that of the back aperture of the focusing lens. As a practical guide, a ratio of the back aperture radius to the $1/e^2$ radius of the incident beam of 3 or greater is generally sufficient to neglect the consequences of diffraction, and therefore a nearly perfect GL profile can be obtained after the lens.

The intensity distribution of a GL beam is in the following form:

$$I(\rho, z, t) = \frac{2P(t)}{\pi w^2(z)} \exp\left[-\frac{2\rho^2}{w^2(z)}\right], \qquad (13.8a)$$

$$w(z) = w_0 \left[1 + \left(\frac{z}{z_R}\right)^2\right]^{1/2}, \qquad (13.8b)$$

$$z_R = \frac{n_0 \pi w_0^2}{\lambda}, \qquad (13.8c)$$

where z is the distance along the optical axis, ρ is the distance away from the optical axis, λ is the wavelength of excitation light in vacuum, n_0 is the refractive index of the sample media, and $P(t)$ is the incident power. z_R, frequently referred to as the Rayleigh length of the beam, characterizes the focal depth of the Gaussian beam.

We can easily obtain the expressions for $S(\mathbf{r})$ and $I_0(t)$ from equation (13.8):

$$I_0(t) = \frac{2P(t)}{\pi w_0^2}, \tag{13.9a}$$

$$S(\mathbf{r}) = \frac{w_0^2 \, Exp\left[-\frac{2\rho^2}{w^2(z)}\right]}{w^2(z)}. \tag{13.9b}$$

Assuming the sample thickness is much greater than the Rayleigh length, an analytical expression can be readily obtained for the integral of $S^n(\mathbf{r})$ (for $n \geq 2$):

$$\int_{V \to \infty} S^n(\mathbf{r})d\mathbf{r} = \frac{(2n-5)!!}{2n(2n-4)!!} \frac{n_0 \, \pi^3 w_0^4}{\lambda}, \tag{13.10}$$

where !! denotes double factorial (i.e., $2n!! = 2 \cdot 4 \cdot 6 \cdots 2n$):

$$(2n-1)!! = 1 \cdot 3 \cdot 5 \cdots (2n-1), (-1)!! = 1, 0!! = 1.$$

13.3.1.1.2 A PLANE WAVE FOCUSED BY DIFFRACTION-LIMITED OPTICS

We first consider the three-dimensional intensity distribution at the focal spot of a diffraction-limited objective lens with uniform illumination. In practice this is achieved by expanding the incident laser beam so that the beam diameter is much larger than the back aperture of the focusing lens. Properties of the lens are then determined by its point spread function (PSF). The dimensionless distance from the optic axis v and the distance from the in-focus plane u are given by:

$$v = \frac{2\pi (NA)\rho}{\lambda}, \tag{13.11a}$$

$$u = \frac{2\pi (NA)^2 z}{n_0 \, \lambda}, \tag{13.11b}$$

where $NA = n \sin\theta$ and θ as the half-angle of collection for the lens.

We use the paraxial form of the normalized intensity PSF ($h^2[u, v]$) for a diffraction-limited lens (Wilson & Sheppard, 1984):

$$I_0(t) = \frac{\pi (NA)^2}{\lambda^2} P(t), \tag{13.12a}$$

$$S(\mathbf{r}) = h^2(u, v) = \left| 2 \int_0^1 J_0(v \, \theta) \exp[-(1/2) \, i \, u \, \theta^2] \theta d\theta \right|^2. \tag{13.12b}$$

The paraxial approximation is adequate for $\sin\theta < 0.7$, or $NA < 1.0$ with an oil immersion objective (Sheppard & Matthews, 1987). After changing ρ and z to their

Table 13.1. Numerical integrations of a_n under paraxial approximation

n	2	3	4	5	6	7	8	10	12
a_n^*	70	28.1	18.3	13.2	10.1	8.06	6.62	4.76	3.64

* Because numerical integrations can be done over a limited space only, the estimated uncertainty of the integral is < 4% for n = 2 and negligible for n > 2. Absolute values of a_n are expected to be different in the high NA limit where paraxial approximation is no longer valid. However, numerical calculations for NA = 1.2 showed that the *relative* values of a_n (n = 2, 3, 4, 5, 6) are approximately the same as those listed in Table 13.2.

optical units v and u by using equation (13.11), the integration of S(**r**) becomes:

$$\int_{V \to \infty} S^n(\mathbf{r})d\mathbf{r} = a_n \left[\frac{n_0 \lambda^3}{8\pi^3 (NA)^4} \right], \tag{13.13}$$

where $a_n = \int_0^\infty 2\pi\, v\, dv \int_{-\infty}^\infty du\, [h(u,v)]^{2n}$.

Here we have again assumed that the sample thickness is much greater than the focal depth of the lens. It is interesting to note that, in the paraxial approximation, the volume integrals are proportional to $(NA)^{-4}$ regardless of the order of photon processes. Because analytical expressions of a_n cannot be obtained, numerical integrations of $[h(u,v)]^{2n}$ over the entire space have been performed for selected n values. Results are listed in Table 13.1

13.3.1.1.3 TWO-PHOTON EXCITATION

In the thick sample limit, the expression for experimentally detected two-photon excited fluorescence in a GL focus can be obtained using equations (13.7) and (13.10):

$$\langle F(t) \rangle = \frac{1}{2} g^{(2)} \phi \eta C \sigma_2 n_0 \frac{\pi <P(t)>^2}{\lambda}. \tag{13.14}$$

A similar equation can also be obtained for diffraction-limited focus using equations (13.7) and (13.13):

$$<F(t)> \approx \frac{1}{2} g^{(2)} \phi \eta C \sigma_2 n_0 \frac{8.8<P(t)>^2}{\pi \lambda}. \tag{13.15}$$

Although the collection efficiency ϕ is dependent on the NA of the collecting lens, we note that the total fluorescence generation is independent of the NA of the focusing lens (or the beam waist size of the GL beam) in thick samples. Intuitively, one sees that the effect of the increased intensity by tighter focusing is just compensated by the shrinking excitation volume. Thus, the amount of fluorescence summed over the entire space remains unchanged for 2PE. Working with thick samples in 2PE measurements takes advantage of this interesting property: the generated fluorescence power is insensitive to the size of the focal spot, assuming no aberration. Hence, a small variation of the laser beam size (with excitation power monitored after the objective lens) or sample thickness would not affect the results of measurements.

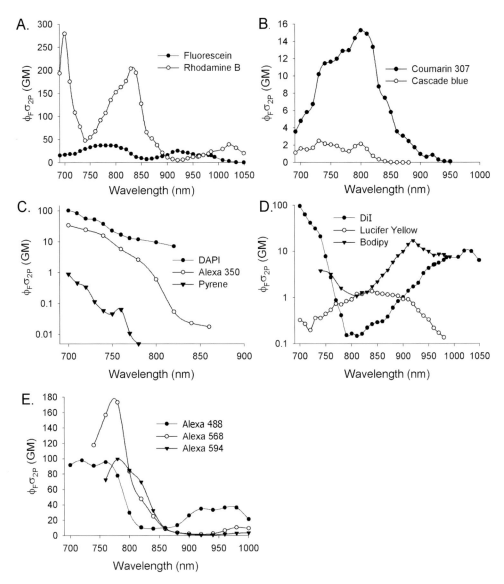

Figure 13.4. Two-photon action cross-sections of conventional fluorophores used in MPM. (A) Fluorescein (water, pH 11) and Rhodamine B (MeOH). (Numbers for fluorescein and Rhodamine B are absorption cross-sections calibrated assuming the one- and two-photon quantum efficiencies are the same.) (B) Cascade Blue (water) and coumarin 307 (MeOH). (C) DAPI (measured in water without DNA; values shown are multiplied by 20 to reflect the known QE enhancement from binding), Alexa 350 hydrazide (water) and pyrene (MeOH). (D) DiI C-18 (MeOH), Lucifer yellow (water) and Bodipy (water). Note logarithmic y-axis for C and D. (E) Alexa 488, 568 and 594 hydrazide (water). Units are Goppert-Mayer (GM); 1 GM= 10^{-50} cm^4 s/photon.

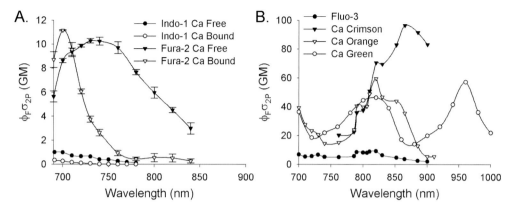

Figure 13.5. Two=photon action cross-sections of calcium-indicating dyes. (A). Indo-1 (±Ca in water) and Fura-2 (±Ca in water). Error bars on Furs-2 represent the standard error of the mean of three independent measurements of the action cross-section spectra. (B) Calcium-bound forms of Fluo-3, Calcium Crimson, Calcium Orange, and Calcium Green-1N. Units are Goppert-Mayer (GM); 1 GM= $10^{-}50 \, cm^4$ s/photon.

One possible concern of using thick samples is lens aberration. Although microscope objectives are highly corrected, index mismatch between the immersion medium and the sample introduces aberration that generally reduces the amount of two-photon fluorescence (Hell et al., 1993). However, such aberration effects are negligible when low NA lenses are used.

13.3.1.1.4 THREE-PHOTON EXCITATION

In the thick sample limit, the expression for experimentally detected three-photon excited fluorescence in a GL focus can be obtained using equations (13.7) and (13.10):

$$<F(t)> = \frac{1}{3} g^{(3)} \phi \eta C \sigma_3 n_0 \frac{2<P(t)>^3}{3\lambda w_0^2}. \tag{13.16}$$

A similar equation can also be obtained for diffraction-limited focus using equations (13.7) and (13.13):

$$<F(t)> \approx \frac{1}{3} g^{(3)} \phi \eta C \sigma_3 n_0 \frac{3.5(NA)^2 <P(t)>^3}{\lambda^3}. \tag{13.17}$$

It is clear that the amount of three-photon excited fluorescence depends on the size of the focal spot (i.e., NA of the lens or the beam waist size). Thus, unlike 2PE experiments with thick samples, quantitative measurements of three-photon excited fluorescence require the characterization of the focal spot size. However, working with thick samples still provides the advantage of less sensitivity to the beam waist measurements because $<F(t)> \propto w_0^{-6}$ or NA^6 in the thin sample limit, in which case a small deviation of the focal spot size can cause a much larger error in the measured cross-section.

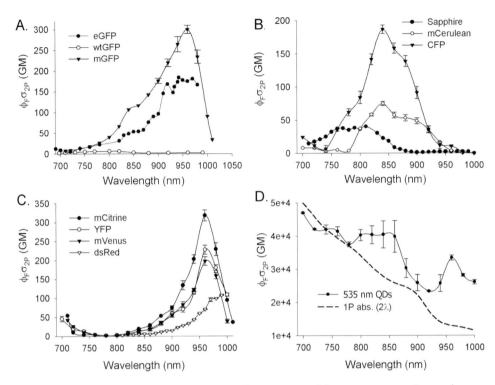

Figure 13.6. Two-photon action cross-section spectra of fluorescent proteins and quantum dots. (A) Three variants of green fluorescent protein: eGFP, monomeric eGFP (mGFP), and wild-type. Error bars for mGFP represent the standard error of the mean for two independent measurements. Concentrations for wtGFP and eGFP were based on measurements of protein concentration; mGFP concentration was measured by FCS using G(0) of the autocorrelation and a focal volume calibration based on a known concentration of Rhodamine Green. (B) Three blue and cyan fluorescent proteins: Sapphire, CFP, and a monomeric form of the cerulean protein. Error bars for CFP and mCerulean represent the standard error of the mean for two independent measurements. Concentrations of CFP and mCerulean were measured by FCS; Sapphire concentration was based on measurement of protein concentration. (C) Four yellow and red fluorescent proteins: monomeric Citrine, YFP, monomeric Venus, and dsRed. Error bars for mCIT, YFP, and mVenus represent the standard error of the mean for two independent measurements. Concentrations of mCIT, YFP, and mVenus were measured by FCS; dsRed concentration was based on measurement of protein concentration. (D) Two-photon action cross-sections of water-soluble, high-quantum-yield (> 90%) batch of 535 nm emitting cadmium selenide quantum dots. Dotted line is the single photon absorption line shape plotted at double the wavelength for comparison. Error bars represent the standard error of the mean from two independent cross-section determinations from the same batch of quantum dots. The concentration of nanoparticles was determined using FCS. Units are Goppert-Mayer (GM); 1 GM= 10^{-50} cm^4 s/photon.

13.3.1.2 Temporal Considerations

In addition to the characterization of the spatial profiles of the excitation beam, experimental determination of MPE cross-sections is difficult mainly because the multiphoton absorption rates depend crucially on the temporal profiles of the excitation beam (i.e., $g^{(n)}$ as defined by equation (13.7b)) (Swofford & McClain, 1975). An effective and

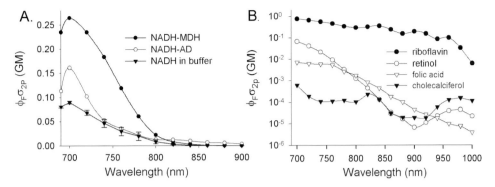

Figure 13.7. Two-photon action cross-sections for five intrinsic biological compounds. (A) NADH at pH 7.0 in free and enzyme-bound form. Protein-NADH complexes were prepared by adding excess purified enzyme to free NADH. Measurements of both enzymes free of NADH showed no two-photon-generated fluorescence in the 450- to 530-nm range using excitation in the wavelength range shown. Error bars on free NADH spectrum represent the standard deviation from four independent measurements. MDH=malate dehydrogenase, AD or ADH=alcohol dehydrogenase. (B) Four intrinsic fluorophores commonly found in cells and tissue measured in PBS solution (pH 7.0) in free form. As with bound and free NADH, the brightness (QY) of these compounds can vary greatly depending on the environment and binding. Note: y-axis is logarithmically scaled; units are Goppert-Mayer (GM); 1 GM= $10^{-}50 \, cm^4$ s/photon.

simple experimental approach to solve this problem is to compare the generated fluorescence of the specimen to some known two- or three-photon references, provided that reliable two- or three-photon standards in the wavelength range of interest exist (Jones & Callis, 1988; Kennedy & Lytle, 1986). Such reference technique also takes into account the spatial distribution if the same excitation geometry is used for both the standard and the specimen to be measured. However, several measurement methods can be used in the absence of such multiphoton standards. They will be briefly reviewed in the following sections.

13.3.1.2.1 MEASUREMENTS WITH A SINGLE-MODE CW LASER

An ideal single-mode CW laser has $g^{(n)} = 1$. Measuring MPE cross-sections using such lasers requires only knowledge of the value of the average excitation intensity. However, single-mode CW excitation usually requires many orders of magnitude more average power than pulsed excitation to obtain the same rate of excitation, because of the much smaller value of $g^{(n)}$. Using the same average power, the excitation rate achieved with a single-mode CW laser is only approximately 10^{-5} and 10^{-10} for two- and three-photon excitation, respectively, of that using a mode-locked femtosecond Ti:Sa laser with 80-MHz repetition rate and 100-fs pulse width. Nevertheless, CW excitation is feasible for dyes with relatively large cross-section in 2PE. The prospect of CW three-photon excitement is not good except for molecules with extraordinarily large three-photon cross-sections (such as those reported by He et al., 1995). For example, with an average power of 100 mW at the specimen of 1 mM concentration and using the three-photon fluorescence action cross-section ($\eta\sigma_3$) of

$10^{-82} cm^6 (s/photon)^2$, equation (13.17) predicts that the detected fluorescence rate is only 0.2 photon/s, assuming $NA = 1.3, \lambda = 1,000$ nm, and a high detection efficiency ϕ of 1%.

It is important to stress the "true CW" nature of a single (longitudinal)-mode CW laser. A multimode CW laser may retain enough RMS noise (such as caused by mode beating) to prevent its use as a stable CW source.

13.3.1.2.2 MEASUREMENTS WITH A PULSED LASER

Measurements performed with a pulsed laser require careful evaluation of the temporal coherence factor $[g^{(n)}]$. The focused intensity obtained from a mode-locked laser is the periodic function in time:

$$I_0(t) = I_0 \left(t + \frac{m}{f} \right), \quad m = 1, 2, 3 \dots, \tag{13.18}$$

where f is the pulse repetition rate.

Let $t = 0$ be at the peak of one excitation pulse. Because of the periodical nature of the pulse train, one needs to calculate $g^{(n)}$ for only one cycle. Defining τ as the excitation pulse width (FWHM) and $f \bullet \tau$ as the duty cycle, $g^{(n)}$ can be expressed in terms of the ratio of the dimensionless quantity $g_p^{(n)}$, which depends only on the shape of the laser pulse and the duty cycle:

$$g^{(n)} = g_p^{(n)} / (f \tau)^{n-1}, \tag{13.19a}$$

$$g_p^{(n)} = \frac{\tau^{n-1} \times \int_{-1/(2f)}^{1/(2f)} I_0^n(t) dt}{\left[\int_{-1/(2f)}^{1/(2f)} I_0(t) dt \right]^n}. \tag{13.19b}$$

For pulses with a Gaussian temporal profile, one finds $g_p^{(2)} = 0.66$ and $g_p^{(3)} = 0.51$, whereas for a hyperbolic-secant square pulse, one finds $g_p^{(2)} = 0.59$ and $g_p^{(3)} = 0.41$. Intuitively, $g^{(n)}$ represents the n-photon-excitation efficiency of the excitation light at a fixed time-averaged intensity. We note that $g^{(n)} \langle I_0(t) \rangle$ is generally not the commonly referred *peak* intensity or *pulse* intensity (I_p), which should be given as $I_p = <I_0(t)> / (f \tau)$.

Mode-locked lasers with their high peak intensities and broad tunability offer tremendous advantages in multiphoton fluorescence excitation. However, determination of MPE cross-sections depends crucially on the estimation of the temporal coherence at zero time delay (Loudon, 1983) $g^{(n)} = <I_0^n> / <I_0>^n$ of the excitation light *inside the sample*. Although the pulse width and the laser repetition rate can be routinely measured in any laser laboratory, the determination of $g_p^{(n)}$ in the femtosecond time region is difficult. In addition, external measurements of the pulse shapes do not provide reliable estimates of $g^{(n)}$ *inside the sample*, because ultrashort pulses are distorted by the optical nonlinearities and group delay dispersion (GDD) (e.g., a GDD of $-2,000$ fs^2 at 730 nm for a Zeiss Neofluo 10X 0.3NA lens) (Guild et al., 1997). Thus, special caution must be taken to assign the value of $g^{(n)}$, and there is one reported technique that can provide the absolute value of $g^{(2)}$ (Xu et al., 1995).

13.3.1.3 System Collection Efficiency (ϕ) and Fluorescence Quantum Efficiency (η)

System collection efficiency (ϕ) is a function of the collection efficiency of the collection lens, transmission of the optics, and detector quantum efficiency. The most convenient approach to determine ϕ is to use one-photon excited fluorescence. Direct comparison with one-photon excited fluorescence also gives the absolute values of 2PE cross-section, assuming equal emission quantum efficiencies. However, care must be taken to ensure that the collection efficiencies for one- and two-photon excited fluorescence are the same. For example, the use of short path-length sample cuvettes ensures uniform fluorescence collection efficiency throughout the sample, which is necessary because one-photon excitation (1PE) occurs throughout the entire sample thickness (i.e., it is not localized to the focal region, as in 2PE). In general, it is assumed that one-, two-, and three-photon excited fluorescence emission spectra are the same. As discussed later, we observe no exceptions to this assumption for the fluorophores discussed in this chapter.

One can also determine ϕ directly based on the fluorescence emission spectrum, the measured NA of the collection lens, and the transmission of the optics. The detector response can be obtained from the manufacturer's catalog or can be independently calibrated using a known intensity source (e.g., with a laser and calibrated neutral density filters). In our experiments, the collection efficiencies determined by the above two methods were quite consistent.

A measurement of the n-photon-excited fluorescence quantum efficiency η is very difficult because direct measurement of n-photon absorption (usually very small) is required. It is usually assumed that fluorescence quantum efficiencies are the same for one-, two-, and three-photon excitations and are constant over the entire spectral range. One method of measuring η by investigating the fluorescence behavior under saturation was proposed (Xu & Webb, 1997). However, we are not aware of any direct measurements of fluorescence quantum efficiency under multiphoton excitation.

13.3.2 Transmission Method

Direct measurements of multiphoton absorption cross-sections by recording the attenuation of the incident beam can be achieved only in highly concentrated (usually $>10\,mM$) fluorophore solutions unless high-power laser systems are employed. Nonetheless, with the increasing availability of amplified femtosecond laser systems and the development of more sensitive techniques, direct measurement of multiphoton absorption in fluorophore solutions can be performed. Several experimental techniques are available to directly measure the nonlinear absorption: nonlinear transmission (Boggess et al., 1986), Z-scan techniques (Sheik-Bahae et al., 1990), and modulation techniques (Kang et al., 1997; Tian & Warren, 2002).

Nonlinear transmission methods measure the light transmission as a function of the incident intensity. To improve the measurement sensitivity and reduce the single photon background, Z-scan techniques and variations based on the original Z-scan techniques were developed. In a typical Z-scan setup, the transmitted signal is recorded as a sample is translated through the focal region of a focused Gaussian beam. The Z-scan technique is simple and has proven to be particularly useful in measuring the nonlinearities (both nonlinear absorption and nonlinear refraction) of solid-state materials such as glasses and semiconductors. There are also several reports of measuring 2PE cross-sections using the modulation techniques, where spectral regions (or frequency regions) that are specifically sensitive to 2PE were measured to further improve the measurement sensitivity.

Because of the weak nonlinear absorption process, however, a high incident intensity and/or a high sample concentration was typically required in the transmission methods. The use of high incident intensity is a concern when the transmission methods are applied to measure organic fluorophore solutions. Processes such as heating, saturation, photobleaching, excited state absorption, stimulated emission, etc., all compete with 2PE and can easily lead to erroneous interpretation of the measurements. Furthermore, a high fluorophore concentration ($> 10\,\mathrm{mM}$) was typically used in transmission methods in order to obtain a clean signal. Depending on the dye and solvent, such a high concentration can possibly cause significant dye aggregation and dimerization, changing the spectroscopic properties of the dye (Selwyn & Steifeld, 1972).

When compared to the fluorescence methods, the main advantages of the transmission method are measuring the absolute multiphoton absorption cross-section without assumption of the fluorescent quantum efficiency and measuring nonfluorescent species. However, because the fluorescence measurement can be performed at a much lower excitation intensity and orders of magnitude lower dye concentration ($< 0.1\,\mathrm{mM}$), the fluorescence excitation method is a preferred approach for measuring fluorescent molecules. Furthermore, the multiphoton action cross-sections (i.e., the product of the quantum efficiency and absorption cross-section) are values of importance in multiphoton microscopy. Thus, knowledge of the absolute absorption cross-section is typically not necessary in practical imaging applications. In Section 13.4 of this chapter, we have included cross-sections obtained by fluorescence and transmission methods when published values are available for both methods.

13.3.3 Estimation and Practical Laboratory Technique for Measuring the Excitation Cross-Section

13.3.3.1 Estimations of Multiphoton Excited Fluorescence

For historical reasons, the units for two- and three-photon cross-sections are a little confusing. While the one-photon absorption cross-section has an intuitive unit of area (i.e., cm^2), two- and three-photon cross-sections are in units of $\mathrm{cm}^4\mathrm{s/photon}$ (or GM) and $\mathrm{cm}^6(\mathrm{s/photon})^2$, respectively. A simple method of understanding the units is to note that two-photon cross-section multiplied by the photon flux density (i.e., intensity in unit of photons per cm^2 per second) gives the unit of one-photon cross-section. The same logic applies for three-photon cross-section (multiplied by the photon flux density squared). We hope that the following discussion will provide the readers a practical guide of using the cross-section numbers provided in Section 13.4. Equations 13.15 provided the detected two-photon excited fluorescence in the thick sample limit with diffraction-limited focus. In this section, we will express equation 13.15 in practical physical parameters so that a quantitative calculation can be quickly performed. Assuming that the pulse has a Gaussian temporal profile (i.e., $g_p^{(2)} = 0.66$) and using the two-photon cross-section (σ_2) in units of GM, the expression for the detected fluorescence (F) can be obtained as:

$$F \approx 1.41 \phi C \eta \sigma_2 n_0 \frac{\lambda^2 P^2}{f \tau}. \tag{13.20}$$

where η is the fluorescence quantum efficiency of the dye, ϕ is the fluorescence collection efficiency of the measurement system, C (in μM) is the indicator concentration, λ (in μm) is the wavelength of excitation light in vacuum, n_0 is the refractive index of the sample media, P (in mW) is the average incident power, f is the pulse repetition rate, and τ is the excitation pulse width (FWHM).

For example, if $\eta * \sigma_2 = 1$ GM, C = 100 μM, $\tau = 100$ fs, $\lambda = 1 \mu$m, $n_0 = 1.3$, and a laser power of $P = 10$ mW at a repetition rate of 100 MHz, we will have $F \approx \phi \times 1.83 \times 10^9$. Equation 13.20 can also be used as a quick measurement of the 2PE cross-section of an indicator if the collection efficiency can be estimated.

13.3.3.2 Measuring Multiphoton Cross-Section by Referencing to a Known Standard

Accurate determination of TPE action cross-sections requires knowledge of the temporal and spatial coherence of the optical field inside the sample, which is a nontrivial task. An effective and simple experimental approach to solve this problem is to compare the generated fluorescence of the specimen to some known two- or three-photon references, provided that reliable two- or three-photon standards in the wavelength range of interest exist (Kennedy & Lytle, 1986; Jones & Callis, 1988). (We note that there was an error in the published Bis-MSB cross-section value by Kennedy and Lytle.) We note that the collection efficiencies must be taken into account when comparing the standard with the new indicators. Thus, it is more convenient to compare the indicator with a standard of similar fluorescence emission spectra.

For a given excitation wavelength, $g^{(2)}/(f\tau)$ is the same for both the calibration sample and new fluorophore. The ratio of the experimentally measured fluorescence signals becomes (Albota et al., 1998b):

$$\frac{\langle F(t) \rangle_{cal}}{\langle F(t) \rangle_{new}} = \frac{\phi_{cal} \eta_{cal} \sigma_{2cal} C_{cal} \langle P_{cal}(t) \rangle^2 n_{cal}}{\phi_{new} \eta_{new} \sigma_{2new} C_{new} \langle P_{new}(t) \rangle^2 n_{new}}, \tag{13.21}$$

where the subscripts *cal* and *new* indicate the parameters for the calibration and new fluorophore, respectively.

The TPE action cross-section of a new molecular fluorophore is then related to known experimental wavelength-dependent parameters, including the TPE action

Table 13.2. Two-photon absorption cross-section of fluorescein in water (pH = 13)

Wavelength (nm)	2PE Cross-Section (GM)	Wavelength (nm)	2PE Cross-Section (GM)
691	16	870	9.0
700	19	880	11
710	17	890	14
720	19	900	16
730	25	910	23
740	30	920	26
750	34	930	23
760	36	940	21
770	37	950	18
780	37	960	16
790	37	970	17
800	36	980	16
810	32	992	13
820	29	1008	5.3
830	19	1020	3.2
840	13	1034	1.0
850	10	1049	0.23
860	8.0		

cross-section of the calibration standard, as described by the equation:

$$\sigma_{2new}(\lambda)\eta_{new} = \frac{\phi_{cal}\eta_{cal}\sigma_{2cal}(\lambda)C_{cal}}{\phi_{new}C_{new}}\frac{\langle P_{cal}(t)\rangle^2}{\langle P_{new}(t)\rangle^2}\frac{\langle F(t)\rangle_{new}}{\langle F(t)\rangle_{cal}}\frac{n_{cal}}{n_{new}}. \tag{13.22}$$

We list in Table 13.2 our measured two-photon cross-section values for fluorescein in the wavelength range of 690 to 1050 nm (Xu & Webb, 1996). These values can be used to calibrate the two-photon cross-sections of new indicators.

13.4 CROSS-SECTION DATA

In this section, we summarize our measurements of a variety of dyes, intrinsic molecules, fluorescent proteins, and QDs. We realize that various other groups have also reported cross-section measurements on many of the same molecules shown here. Thus, the first part of this section will compare the published cross-sections and/or two-photon excitation spectra after 1996. We collected the literature by doing a forward citation search of the papers that we published in cross-section measurements. Thus, the literature covered in this section may not include all relevant work.

Xanthene dyes (i.e., rhodamine B, rhodamine 6G, and fluorescein) are inexpensive and widely available and have good photostability and reasonably low toxicity. These dyes are natural candidates for use as calibration standards. The fact that the most reported cross-sections are those of xanthene dyes ensures careful examinations by many independent research groups and further enhances the confidence of using

Table 13.3. General comments on the probes

Probe	Advantages with MPE	Disadvantages with MPE
Conventional organic dyes	Commonly and readily available. Some have reasonably high 2P cross-sections (>100 GM). Loading and staining protocols are well developed and available. There are many good UV dyes that work extremely well under MPE (e.g., DAPI).	Some key organic fluorophores have low 2P cross-sections (e.g., Fura, Indo). Certain probes that are photostable under 1PE photobleach more rapidly under 2PE (e.g., Cy3).
Genetically encodable proteins	Readily available and have revolutionized fluorescence imaging. Are relatively good 2P fluorophores with cross-sections often >100 GM.	As redder versions are introduced, it becomes more difficult to excite them within the wavelength range of a Ti:S laser.
Nanoparticles	Quantum dots have the largest reported 2P cross-sections and can be as much as 5,000 times brighter than good conventional fluorophores. Other nanoparticles are being developed that may have advantages (for both 1PE and 2PE). For example, silica-encased particles can stabilize many dyes that are normally photo-unstable.	Nanoparticles are relatively large; for example, water-soluble quantum dots are on the order of 15 nm in hydrodynamic radius. Nearly all types of nanoparticles are multivalent-that is, it is not possible to label one antibody with one nanoparticle. Toxicity of certain nanoparticles is still an open question if being used in chronic imaging experiments.
Intrinsic fluorophores	Enables imaging without any added labels. It is sometimes possible to obtain physiological information using intrinsic fluorophores (e.g., NADH). Most intrinsic fluorophores excite in the UV so are well suited for MPE imaging. MPE restricts excitation to the focal plane, producing much less overall photodamage than direct UV excitation. Other nonlinear signals are simultaneously available (e.g., SHG).	Most are extremely poor fluorophores (true for both 1PE as well as 2PE). Imaging intrinsic molecules may be directly causing photodamage; very often the fluorescence quantum yield is low for these molecules, but they still absorb and the energy is dissipated in more deleterious ways. (Also true for both 1PE as well as 2PE.)

them as standards. In addition to our own measurements (Albota et al., 1998b; Xu & Webb, 1996), these dyes have also been measured using both fluorescence and nonlinear transmission methods. Kaatz and Shelton measured the two-photon cross-section of rhodamine B, rhodamine 6G, and fluorescein at 1,064 nm by calibrating two-photon fluorescence to hyper-Rayleigh scattering (Kaatz & Shelton, 1999). The results for rhodamine B are in reasonable agreement with previous data, taking into account various factors such as wavelength difference. Measurements of fluorescein cross-sections were repeated by Song et al. (Song et al., 1999). The measured value at 800 nm is ~1.5 times of what we reported in 1996. Two-photon cross-section of rhodamine 6G has been measured by three other groups at 800 nm, using 2PE fluorescence calibrated against a luminance meter (Kapoor et al., 2003), and nonlinear transmission methods (Sengupta et al., 2000; Tian & Warren, 2002). While the 2PE method obtained results nearly identical to what we published in 1998, values obtained using the nonlinear transmission methods are consistently lower by about a factor of 2. This discrepancy between the fluorescence method and nonlinear transmission methods has been discussed in the past (Oulianov et al., 2001). By comparing the excited-state methods (i.e., fluorescence and transient spectroscopy following two-photon excitation) and nonlinear transmission methods, it is found that although the values obtained by these two excited-state methods are comparable and agree with values obtained by the fluorescence method in the past, they differ considerably from the value obtained using the nonlinear transmission method. As discussed in the methods section, caution is warranted when using the transmission method.

Intrinsic fluorescent molecule flavine mononucleotide (FMN) has also been measured by Blab and colleagues (Blab et al., 2001). The reported value and spectral shape of FMN are very close to what we reported in 1996. The authors further reported two-photon action cross-sections of fluorescent proteins, including a value of 41 GM for eGFP (Cormack et al., 1996). We have measured eGFP using the fluorescence technique and *estimated* an action cross-section value of ~100 GM at 960 nm (Xu & Webb, 1997). An alternative measurement, using the FCS method that essentially measures fluorescence per single molecule, gave a two-photon absorption cross-section of eGFP at 180 GM (Schwille et al., 1999), which agreed very well with our measurement, assuming a quantum efficiency of 0.6 for eGFP (Patterson et al., 1973). Two-photon absorption of eGFP has also been measured using a combination of nonlinear transmission method and absorption saturation. However, the reported value is ~600,000 GM at 800 nm (Kirkpatrick et al., 2001). Not only is this value four orders of magnitude larger than the values obtained by the fluorescence method, it is also two orders of magnitude larger than any reported two-photon absorption cross-section. We are currently not certain about the origin of this large discrepancy.

The importance of fluorescent proteins cannot be overstated. The discovery of green fluorescent protein in the early 1960s catalyzed a new era in biology by enabling investigators to apply molecular cloning methods to fuse a fluorophore moiety to a wide variety of protein and enzyme targets that could be monitored in vivo. There are now multiple mutated forms of the original jellyfish protein with improved functionality available, as well as many new fluorescent proteins from other organisms such as coral (Tsien, 2005; Zacharias & Tsien, 2006). Thus, for experiments involving intact tissue or live animals, where multiphoton microscopy has real advantages, measurements of the cross-sections of the available fluorescent protein are important.

Below we provide a compilation of the majority of the two-photon action cross-sections (two-photon absorption cross sections for Fluorescein and Rhodamine B) that we have measured at Cornell over the past decade, including some general comments on the properties of the probes (Table 13.3). The data is presented in four figures starting with a set of conventional dyes and calcium ion indicators (figures 13.4 and 13.5), action cross-sections of 10 commonly used fluorescent proteins (Figure 13.6) (for a detailed review of fluorescent proteins, please see Shaner, Steinbach et al., 2005), quantum dots (Figure 13.6D), and finally action cross-sections of several intrinsic biological molecules found in cells or in the extracellular matrix (Figure 13.7). About 70% of the data has been previously presented (Xu and Webb, 1996; Xu, Williams et al., 1996; Xu, Zipfel et al., 1996; Larson, Zipfel et al., 2003; Zipfel, Williams et al., 2003), with the exception of the Alexa dyes (Figure 13.4, C and E) and several of the fluorescent proteins. The measured cross sections for QDs are ensemble-averaged values, given the uncertainties caused by blinking and/or nonradiant dark fractions (Yao, Larson et al., 2005). We have not included measurements of some of the newly synthesized large cross-sections molecules (Albota, Beljonne et al., 1998) since these dyes have not yet found many actual uses in biological imaging due to their highly lipophilic nature and toxicity. These compounds, however, have cross-sections an order of magnitude higher than the conventional dyes and indicators in use today and suggest that two-photon imaging could be further improved by rational design of fluorophores specifically for nonlinear excitation.

REFERENCES

Albota, M., D. Beljonne, et al. (1998a). Design of organic molecules with large two-photon absorption cross sections. Science 281: 1653–1656.

Albota, M. A., C. Xu, et al. (1998b). Two-photon excitation cross sections of biomolecular probes from 690 to 980 nm. Appl. Opt. 37: 7352–7356.

Alivasatos (1996). Science 271: 933–937.

Birge, R. R. (1983). One-photon and two-photon excitation spectroscopy. Ultrasensitive Laser Spectroscopy. D. S. Kliger. New York, Academics: 109–174.

Blab, G. A., P. H. M. Lommerse, et al. (2001). Two-photon excitation action cross-sections of the autofluorescent proteins. Chem. Phys. Lett. 350: 71–77.

Boggess, T. F., K. M. Bohnert, et al. (1986). IEEE J. Quantum Electron. 22: 360.

Catalano, I. M. and A. Cingolani (1981). Absolute two-photo fluorescence with low-power cw lasers. Appl. Phys. Lett. 38: 745–747.

Cormack, B. P., R. H. Valdivia, et al. (1996). FACS-optimized mutants of the green fluorescent protein (GFP). Gene (Amsterdam) 173(1): 33–38.

Curley, P. F., A. I. Ferguson, et al. (1992). Application of a femtosecond self-sustaining mode-locked Ti:sapphire laser to the field of laser scanning confocal microscopy. Optical and Quantum Electronics 24: 851–859.

Denk, W., J. H. Strickler, et al. (1990). Two-photon laser scanning fluorescence microscopy. Science 248: 73–76.

Dickson, L. D. (1970). Characteristics of a propagating Gaussian beam. Appl. Opt. 9: 1854–1861.

Efros, A. L. and A. L. Efros (1982). Interband absorption of light in a semiconductor sphere. Sov. Phys. Semicond. 16: 772–774.

Einstein, A. (1905). On a heuristic point of view concerning the production and transformation of light. Ann. Physik. 17: 132.

Faisal, F. H. M. (1987). Theory of multiphoton processes. New York and London, Plenum Press.

Göppert-Mayer, M. (1931). Über Elementarakte mit zwei Quantensprüngen. Ann. Physik. 9: 273–295.

Guild, J., C. Xu, et al. (1997). Measurement of group delay dispersion of high numerical aperture objectives using two-photon excited fluorescence. Appl. Opt. 36: 397–401.

He, G. S., J. D. Bhawalkar, et al. (1995). Three-photon-absorption-induced fluorescence and optical limiting effects in an organic compound. Opt. Lett. 20: 1524–1526.

Hell, S., G. Reiner, et al. (1993). Aberrations in confocal microscopy induced by mismatches in refractive index. J. Microsc. 169: 391–405.

Helmchen, F. and W. Denk (2005). Deep tissue two-photon microscopy. Nat. Methods 2: 932–940.

Hermann, J. P. and J. Ducuing (1972). Absolute measurement of two-photon cross-sections. Phys. Rev. A 5: 2557–2568.

Hughes, V. and L. Grabner (1950). The radiofrequency spectrum of Rb85F and Rb87F by the electric resonance method. Phys. Rev. 79: 314.

Jones, R. D. and P. R. Callis (1988). A power-square sensor for two-photon spectroscopy and dispersion of second-order coherence. J. Appl. Phys. 64: 4301–4305.

Kaatz, P. and D. P. Shelton (1999). Two-photon fluorescence cross-section measurements calibrated with hyper-Rayleigh scattering. J. Opt. Soc. Am. B. 16: 998–1006.

Kaiser, W. and C. G. B. Garrett (1961). Two-photon excitation in $CaF_2:Eu^{2+}$. Phys. Rev. Lett. 7: 229.

Kang, I., T. Krauss, et al. (1997). Sensitive measurement of nonlinear refraction and two-photon absorption by spectrally resolved two-beam coupling. Opt. Lett. 22: 1077–1079.

Kapoor, R., C. S. Friend, et al. (2003). Two-photon-excited absolute emission cross section measurements calibrated with a luminance meter. J. Opt. Soc. Am. B 20: 1550–1554.

Kennedy, S. M. and F. E. Lytle (1986). p-Bis(o-methylstyryl)benzene as a power-square sensor for two-photon absorption measurements between 537 and 694 nm. Anal. Chem. 58: 2643–2647.

Kirkpatrick, S. M., R. R. Naik, et al. (2001). Nonlinear saturation and determination of the two-photon absorption cross section of green fluorescent protein. J. Phys. Chem. 105: 2867–2873.

Larson, D. R., W. R. Zipfel, et al. (2003). Water soluble quantum dots for multiphoton fluorescence imaging in vivo. Science 300: 1434–1436.

Loudon, R. (1983). The Quantum Theory of Light. Oxford, Clarendon Press.

Masters, B. R. (2003). Selected papers on multiphoton excitation microscopy. Bellingham, SPIE Press.

Masters, B. R. and P. T. C. So (2004). Antecedents of two-photon excitation laser scanning microscopy. Microscopy Research and Technique 63: 3–11.

McClain, W. M. and R. A. Harris (1977). Two-photon molecular spectroscopy in liquids and gases. Excited States. E. C. Lim. New York, Academic: 1–56.

Oulianov, D. A., I. V. Tomov, et al. (2001). Observations on the measurement of two-photon absorption cross-section. Opt. Commun. 191: 235–243.

Patterson, G. H., S. M. Knobel, et al. (1973). Use of the green fluorescent protein and its mutants in quantitative fluorescence microscopy. Biophys J. 73: 2782–2790.

Schwille, P., U. Haupts, et al. (1999). Molecular dynamics in living cells observed by fluorescence correlation spectroscopy with one- and two-photon excitation. Biophys J. 77: 2251–2265.

Selwyn, J. E. and J. I. Steifeld (1972). Aggregation equilibria of Xanthene dyes. J. Phys. Chem. 76: 762–774.

Sengupta P., B. J., Banerjee S, Philip R, Kumar GR, Maiti S (2000). Sensitive measurement of absolute two-photon absorption cross sections. J. Chem. Phys. 112: 9201–9205.

Shaner, N. C., P. A. Steinbach, et al. (2005). A guide to choosing fluorescent proteins. Nat. Methods 2(12): 905–909.

Sheik-Bahae, M., A. A. Said, et al. (1990). Sensitive measurement of optical nonlinearities using a single beam. IEEE J. Quantum Electron. 26: 760–769.

Sheppard, C. J. R. and H. J. Matthews (1987). Imaging in high-aperture optical systems. J. Opt. Soc. Am. A 4: 1354–1360.

Singh, S. and L. T. Bradley (1964). Three-photon absorption in naphthalene crystals by laser excitation. Phys. Rev. Lett. 12: 612–614.

Smith, W. L. (1986). Two-photon absorption in condensed media. Handbook of Laser Science and Technology. J. Weber. Boca Raton, CRC: 229–258.

So, P. T. C., C. Y. Dong, et al. (2000). Two-photon excitation fluorescence microscopy. Annu. Rev. Biomed. Eng. 2: 399–429.

Song, J. M., T. Inoue, et al. (1999). Determination of two-photon absorption cross section of fluorescein using a mode-locked Titanium Sapphire laser. Analytical Sciences 15: 601–603.

Spence, D. E., P. N. Kean, et al. (1991). 60-fsec pulse generation from a self-mode-locked Ti:sapphire laser. Opt. Lett. 16: 42.

Swofford, R. and W. M. McClain (1975). The effect of spatial and temporal laser beam characteristics on two-photon absorption. Chem. Phys. Lett. 34: 455–460.

Tian, P. and W. S. Warren (2002). Ultrafast measurement of two-photon absorption by loss modulation. Opt. Lett. 27: 1634–1636.

Tsien, R. Y. (2005). Building and breeding molecules to spy on cells and tumors. FEBS Lett. 579: 927–932.

Wise, F. W. (2000). Lead salt quantum dots: the limit of strong quantum confinement. Acc. Chem. Res. 33: 773–780.

Wilson T and Sheppard C J R 1984 *Theory and Practice of Scanning Optical Microscopy* (Academic, London).

Xu, C., J. Guild, et al. (1995). Determination of absolute two-photon excitation cross-sections by in situ second-order autocorrelation. Opt. Lett. 20: 2372–2374.

Xu, C. and W. W. Webb (1996). Measurement of two-photon excitation cross-sections of molecular fluorophores with data from 690 nm to 1050 nm. J. Opt. Soc. Am. B 13: 481–491.

Xu, C. and W. W. Webb (1997). Multiphoton excitation of molecular fluorophores and nonlinear laser microscopy. Topics in fluorescence spectroscopy. J. Lakowicz. New York, Plenum Press. 5: 471–540.

Xu, C., R. M. Williams, et al. (1996a). Multiphoton excitation cross sections of molecular fluorophores. Bioimaging 4: 198–207.

Xu, C., W. Zipfel, et al. (1996b). Multiphoton fluorescence excitation: new spectral windows for biological nonlinear microscopy. Proc. Nat. Acad. Sci. USA 93: 10763–10768.

Yao, J., D. R. Larson, et al. (2005). Blinking and nonradiant dark fraction of water-soluble quantum dots in aqueous solution. Proc Natl Acad Sci U S A 102: 14284–14289.

Zacharias, D. A. and R. Y. Tsien (2006). Molecular biology and mutation of green fluorescent protein. Methods Biochem Anal. 47: 83–120.

Zipfel, W. R., R. M. Williams, et al. (2003a). Live tissue intrinsic emission microscopy using multiphoton-excited native fluorescence and second harmonic generation. Proc Natl Acad Sci U S A 100(12): 7075–80.

Zipfel, W. R., R. M. Williams, et al. (2003b). Nonlinear magic: multiphoton microscopy in the biosciences. Nat Biotechnol 21(11): 1369–77.

14

Multiphoton-Induced Cell Damage

Karsten König

14.1 INTRODUCTION

Conventional one-photon fluorescence microscopes employ ultraviolet (UV) and visible (VIS) light sources, such as high-pressure mercury/xenon lamps, the argon ion laser at 364/488/515-nm emission, the frequency-converted Nd:YAG laser at 355/532 nm, and the helium neon laser at 543/633 nm. Typically, the radiation power on the target is some microwatt that corresponds to light intensities in the range of kW/cm^2 when focused to diffraction-limited spots by objectives with high numerical aperture (NA > 1). Multiphoton absorption requires higher light intensities.

In spite of the low μW radiation power, photodestructive effects occur during one-photon microscopy. UVA (315–400 nm) radiation sources, such as the mercury lamp at 365 nm, the Ar^+ laser at 364 nm, and the nitrogen laser at 337 nm, have a relatively high damage potential compared to VIS radiation (e.g., König et al., 1996b). Multiphoton fluorescence microscopy has revolutionized high-resolution live cell imaging. Multiphoton microscopes including two-photon fluorescence microscopes are based on the application of low-energy photons in the near-infrared (NIR) range between 700 nm and 1,200 nm. This spectral range is also referred as "optical window of cells and tissues" where the one-photon absorption and scattering coefficients of unstained cells and tissues are relatively low. Most cells appear transparent. The light penetration depth in tissue is high in this spectral window (Fig. 14.1).

Nonresonant multiphoton absorption requires high light intensities in the range of MW/cm^2 up to TW/cm^2. When using continuous wave (CW) NIR radiation in combination with NA > 1 objectives, laser powers of at least 100 mW have to be applied to obtain sufficient intensity values (Hänninen et al., 1994; König et al., 1995,1996a). These powers are three orders higher than in one-photon microscopes. A side effect of 100 mW microbeams is the generation of mechanical picoNewton forces, which results in optical trapping. In order to avoid trapping effects and to reduce the mean power, most multiphoton microscopes employ femtosecond laser pulses at MHz repetition rate with high kilowatt peak power, P, and low mean power in the μW and mW range. The multiphoton efficiency of an n-photon process depends on an P^n relation. In the case of two-photon microscopy, the fluorescence intensity depends on P^2/τ. The shorter the pulse width τ and the higher the laser power, the more fluorescence photons. This enables a fast scanning rate with microsecond beam dwell time per pixel. However, cell damage may occur during multiphoton microscopy in spite of this low beam dwell time, the low photon energy (1 eV), and the low picojoule laser pulse energy. The reason for the photodamage is the enormous transient light intensity. Obviously, NIR

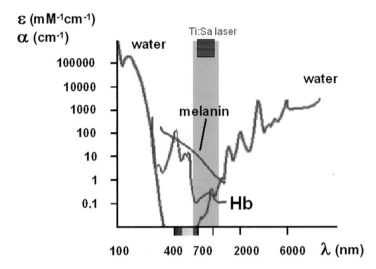

Figure 14.1. Spectral window of cells and tissues.

radiation may induce UVA effects, including photodamage, in particular when using wavelengths below 800 nm due to nonresonant two-photon absorption. Three-photon effects could result in direct absorption by DNA.

When four photons or more are involved in the nonlinear process, ionization of the target combined with the formation of free or quasi-free electrons may occur, which can lead to optical breakdown and plasma (Fig. 14.2). Plasma-filled bubbles expand and generate shock waves.

In addition, high intense laser beams are able to change the refractive index during the laser–target interaction and to change the beam direction and the light intensity by self-focusing effects. However, this phenomenon does not apply when using high-NA objectives. A significant advantage of multiphoton high-NA microscopy compared to conventional one-photon microscopy is the tiny subfemtoliter focal volume where the nonlinear absorption occurs. Absorption in out-of-focus regions can be avoided. In a first approximation, the probability of two-photon absorption decreases with the distance d from the focal point according to a d^{-4} relation.

In particular when studying 3D objects including large cells, multicell aggregates, embryos, and tissues by optical sectioning, multiphoton microscopy with its tiny excitation volume is the superior method compared to one-photon confocal scanning microscopy with its large excitation cones and the subsequent problem of out-of-focus damage.

It has been demonstrated in long-term studies that multiphoton microscopy on living specimens can be performed without photodamage under certain conditions. In particular, single Chinese hamster ovarian (CHO) cells have been laser-exposed for 4 hours with a high 200 GW/cm^2 peak intensity without any impact on cellular reproduction and vitality (König, 2000). In another long-term study, living hamster embryos were exposed in vitro for 24 hours with NIR femtosecond laser pulses of a multiphoton fluorescence microscope and implanted in the mother animal without impact on embryo development in contrast to control studies performed with a conventional one-photon VIS laser scanning microscope (Squirrel et al., 1999). The first commercial multiphoton

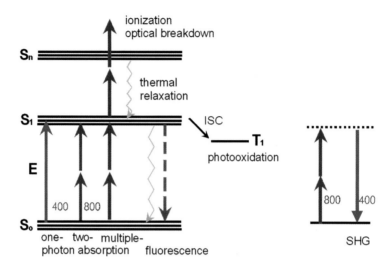

Figure 14.2. Principle of two-photon absorption and multiphoton ionization.

imaging device *DermaInspect* (JenLab GmbH) is in clinical use for melanoma diagnosis and in situ intradermal drug targeting (König & Riemann, 2003; König et al., 2006).

When ignoring certain laser exposure parameters, NIR multiphoton microscopes will induce undesirable destructive multiphoton effects (Diaspro & Robello, 1999; König et al., 2000). Indeed, a further "optical window" exists for the safe, nondestructive observation of the specimens of interest, which is mainly defined by light intensity thresholds with a dependence on NIR wavelength. Outside this window, cells may undergo (1) photothermal damage in the case of pigmented cells, (2) a slow destructive effect by multiphoton excitation of cellular absorbers and subsequent cytotoxic photochemical reactions, or (3) an instantaneous destructive optomechanical effect by optical breakdown phenomena and plasma formation due to multiphoton-mediated ionization of molecules (Table 14.1).

Table 14.1. Sources of multiphoton damage

Damage type	Effects
Photochemical	Formation of reactive oxygen species Apoptosis
Photothermal	Protein denaturation (coagulation) Enzyme inactivation, boiling, carbonization
Optomechanical	Photodisruption (shock waves, bubble formation, jet formation)

14.2 MULTIPHOTON MICROSCOPES AND IMAGING DEVICES

We used a variety of laser scanning Zeiss microscopes for our studies on multiphoton-induced cell damage, such as a modified LSM410 with baseport detector and a Meta-LSM510-NLO confocal microscope. It is more convenient for cell damage studies using the TauMap pinhole-free ultracompact laser scanning microscope (Fig. 14.3), where the galvoscanners and the photon detectors were attached to the side ports of an inverted microscope (JenLab GmbH, Jena, Germany) (Ulrich et al., 2004). The beam was focused to a diffraction-limited spot by conventional high-NA objectives onto the sample. The typical NIR transmission of the objective was found to be 50%, whereas the overall transmission was typically in the range of 30%.

In most experiments compact solid-state 80/90 MHz titanium:sapphire lasers (Vitesse and Chameleon from Coherent, USA, and MaiTai from Spectra Physics, USA) as turnkey laser sources have been used. The laser was coupled to an adapter module (JenLab GmbH, Jena, Germany), which consisted of a beam expander, a step-motor-driven beam attenuator, a fast photodiode for power measurements, a trigger pulse supply, a control unit of laser operation, and a shutter. The TauMap system enables time-resolved single photon counting and fluorescence lifetime imaging (FLIM).

The high sensitivity of the photomultiplier enabled the detection of exogenous fluorescent probes with high fluorescence quantum yield (e.g., DAPI and Hoechst) with a mean power of 25 μW to 1 mW at a frame rate of 1 Hz (512 × 512 pixels). The endogenous intracellular fluorophore NAD(P)H could be imaged with less than 2 mW mean power at appropriate NIR wavelengths.

Optical dispersion results in pulse broadening during transmission through the microscope. Typically, the pulse width at the sample was determined with a special

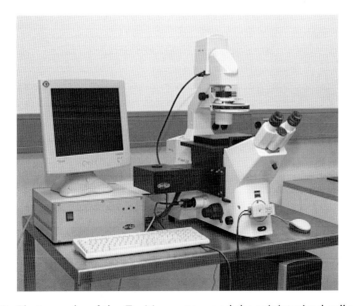

Figure 14.3. Photographs of the *TauMap* system and the miniaturized cell chamber *MiniCem-grid*.

autocorrelator with a nonlinear diode detector to be about 150 to 200 fs (König, 2000). At 10 mW mean power, the peak power and the peak intensities reach values of 0.8 kW and 1.2×10^{12} W/cm^2 (1.2 TW/cm^2), respectively, when assuming a full-width half-maximum (FWHM) beam size of $\lambda/2NA \approx 310$ nm. A typical beam dwell time (time of exposure per pixel) during one scan was about 30 μs, which resulted in a frame rate of 8 s/frame. Typical images consist of 512×512 pixels covering an area of 160 μm $\times$ 160 μm. For cell damage studies, typically single cells of interest were scanned 10 times in the same focal plane at 30 μs beam dwell time.

When performing long-term studies on living cells, closed microchambers with 170-μm-thick glass windows with an etched grid and silicon gaskets have been used (MiniCeM-grid, JenLab GmbH, Jena, Germany). After laser exposure, the cell chamber was placed in an incubator. The medium change and the incubation with probes occurred through the gasket using standard injection needles.

Damage studies on tissue biopsies have been performed with the novel clinical 80/90 MHz femtosecond laser system *DermaInspect* (JenLab GmbH, Jena, Germany). The CE-certified system is employed for cancer detection and in vivo intradermal drug monitoring (Fig. 14.4). The letters CE are the abbreviation of the French phrase "Conformité Européene." CE marking certificates are obtained from Notified Bodies under control of European authorities and mean that the product complies with the requirements of the relevant European health, safety, and environmental regulations, such as the European Union Directive 93/68/EEC and Medical Device Directives.

The *DermaInspect* system has a power feedback system that tunes the laser power at the target depending on the intratissue focal depth. The maximum output power is limited to 50 mW mean power for intratissue imaging. Typically, the mean laser power at the skin surface (incident power) is about 2 mW when imaging the stratum corneum, and up to 40 mW when imaging skin structures at 100 μm tissue depth. Note that the

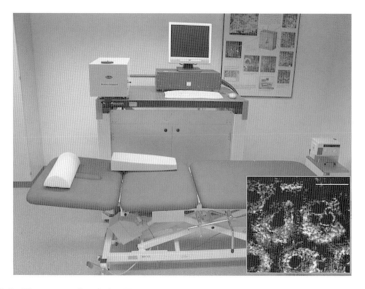

Figure 14.4. Photograph of the DermaInspect system.

laser power at the focal spot and the laser intensity, respectively, at an intraskin target is significantly reduced due to multiple scattering.

The working distance of the focusing optics is 200 μm. Due to this limitation, the system is eye safe and is classified as a class 1M device according to the European laser safety regulations EN 60825 and EN 60601 (König & Riemann, 2002; König et al., 2006).

14.3 PHOTODAMAGE PHENOMENA

Immediate lethal photodamage occurs during scanning microscopy of living cells in aqueous medium at high TW/cm^2 light intensity. In particular, complete cell fragmentation accompanied by the formation of transient bubbles can be monitored. When reducing the power and light intensity, respectively, morphological changes such as the formation of blebs, membrane disruption, organelle destruction, etc. can still be detected. Typically, the laser exposed cells lost vitality. Lethal effects can be confirmed by adding the blue stain Trypan Blue or the red fluorescent Propidium Iodide, which do not enter intact cells.

Membrane damage to erythrocytes results in leakage of hemoglobin and the transformation from a discocytic shape into an echinocytic shape with irregular spicula formation. The empty cell becomes transparent.

When reducing the laser intensity further, localized damage to some intracellular organelles such as the mitochondria may be monitored. Still, the cell can survive. However, cell damage may occur even without any morphological changes. The cell remains alive. A sensitive parameter of cell damage is the analysis of the reproduction behavior. Many laboratories are using CHO cells. These cells are relatively easy to handle, they can grow as a single cell without the need of contact with neighboring cells, and they divide twice a day. We have studied the influence of intense NIR exposure on the reproduction behavior of more than 3,500 single CHO cells. For that purpose, adherent CHO cells were maintained in special miniaturized cell chambers and exposed through the 170-μm-thick glass window. Following laser scanning, the cell division of the exposed cell as well as of its daughter cells was monitored for up to 3 days. If the cell was not affected by the NIR light, a clone of up to 64 cells was produced within 3 days.

We found that the monitoring of just one cell division is not an indicator for biosafe microscopy. Many laser-exposed cells completed one division. However, the daughter cells were not able to reproduce. Sometimes the laser-exposed cell and the daughter cells turned into unlimited cell growth. So-called giant cells have been generated with sizes that are more than five times the normal size. These giant cells are not able to divide.

Further, changes in the intracellular ion concentration, the formation of DNA strand breaks, and modifications of the cellular autofluorescence can be used as indicators of photoinduced cell damage. The two-photon autofluorescence can be detected simultaneously during laser scanning microscopy. The major two-photon fluorophore is the reduced coenzyme NAD(P)H. The oxidized form NAD(P) does not fluoresce. The coenzyme acts, therefore, as a sensitive bioindicator of intracellular redox state. In a first phase, photodamage will result in a decrease of the NAD(P)H signal, which in part is based on auto-oxidation processes. In a second phase, when the mitochondrial membrane is ruptured, the autofluorescence increases and extramitochondrial NAD(P)H will contribute to the signal.

Of course, the findings obtained with CHO cells cannot be transferred automatically to other biological systems. However, when comparing the multiphoton behavior of CHO cells with PTK2 cells, animal and human fibroblasts, as well as human pulpa stem cells, the intensity thresholds for the onset of significant cell damage differed by less than a factor of 2. Nevertheless, the user of multiphoton microscopes should determine this for his or her particular optical system and target the biosafe window for live cell microscopy.

According to our studies on thousands of cells, CHO cells can be scanned with an 800-nm femtosecond and picosecond laser beams (0.07–5 ps) at 2 mW mean power for hours without damage to the exposed cells and their derivatives. This implies that below certain power (intensity) thresholds, "safe" multiphoton microscopy is possible.

However, at higher mean powers and intensities, a significant decrease in cloning efficiency occurred. Interestingly, the photodamage scales with the power squared. When the P_{50} values versus the square root of the pulse broadening are depicted in a graph, a nearly linear function is obtained (König et al., 1999). This indicates that the photodamage process is likely based on a two-photon rather than a one-photon or three-photon absorption process. This was confirmed by Koester and colleagues (1999) based on calcium measurements in neocortical neurons in rat brain slices. Therefore, the origin of this slow photodamage is a nonlinear two-photon process.

We found less damage at wavelengths in the 800 to 900 nm range compared to the 700 to 800 range. Photodamage was found to be more pronounced in the case of femtosecond laser pulses compared to picosecond laser pulses (1–5 ps). For example, using a wavelength of 780 nm, impaired cell division was found in half of the cells at 11 mW mean power (P_{50}) in the case of 2-ps pulses and at 3 mW for 240-fs pulses. CW scanning beams did not impair the cellular reproduction up to 35 mW power (Fig. 14.5).

In order to compare the photodamage in NIR multiphoton microscopes versus UV one-photon microscopes, the same cloning assay was performed on CHO cells using a 364-nm scanning beam of a conventional Ar^+ laser scanning microscope. The parameter 50% cloning efficiency was obtained at the low UV power of 4 μW. Therefore,

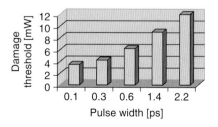

Figure 14.5. Photodamage dependent on pulse width.

UV-induced damage occurs at about 3 orders lower powers (4 µW) compared to the mean values of NIR femtosecond laser pulses (2–8 mW, 750–800 nm).

Ultrastructural electron microscopy studies on CHO cells reveal that mitochondria are the major targets of NIR irradiation. Swelling accompanied by loss of christae and the formation of electron dense bodies in the mitochondrial matrix have been observed (Oehring et al., 2000). For a mean laser power of more than 12 mW (800 nm), the complete destruction of mitochondria was detected. At higher power levels, the outer membranes also ruptured.

14.4 PHOTOTHERMAL DAMAGE

In the case of nonpigmented cells, water, with a typical absorption coefficient of less than 0.1 cm^{-1}, is considered the most important absorber in the "optical window of cells and tissues." Interestingly, the cell damage at 920 nm is less pronounced compared to exposure with 780 nm radiation, where water absorbs less. It is therefore conceivable that photodamage by high intense femtosecond laser radiation is thus not exclusively based on linear water heating.

A typical low heating of 1 K/100 mW was determined for CHO cells with the thermosensitive fluorescent membrane probe Laurdan during exposure with CW 1064-nm microbeams at MW/cm^2 intensity and GJ/cm^2 fluence (Liu et al., 1995). Optically trapped spermatozoa showed no photothermal damage when exposed for 15 minutes with 100 mW microbeams (MW/cm^2 intensity). Further, temperature calculations performed for cells exposed to femtosecond and picosecond pulses revealed only a slight increase in temperature (0.1 K/100 mW at 850 nm) (Schönle & Hell, 1998).

However, pigmented cells with melanin or hemoglobin may undergo photothermal destruction because the one-photon absorption coefficients of these pigments are higher than that of water. The absorption of melanin decreases with increasing wavelength. Melanin NIR absorption coefficients in skin of 78 to 130 cm^{-1} have been calculated (Jacques & McAuliffe, 1991; Kollias & Bager, 1985). Reduced hemoglobin has a 760-nm absorption band ($1 \text{ mM}^{-1}\text{cm}^{-1}$). We found that human erythrocytes ($\alpha \approx 30$ cm^{-1}) at room temperature exposed to CW 60 mW microirradiation at 780 nm undergo hemolysis within 2 minutes. This corresponds to a heating rate of about 60 K/100 mW that can be estimated from these experimental data. The heating rate decreases with wavelength > 760 nm.

Interestingly, photoinduced hemolysis occurred at a much lower mean power (factor of 6) of pulsed radiation at high repetition frequency compared to power values of CW radiation. Using 130-fs pulses at 780 nm, a mean power of 10 mW was found to be sufficient to induce hemolysis (König et al., 1996c). It is assumed that the damage mechanism is based on linear heating and nonlinear effects.

Pigmented human skin such as non-Caucasian skin, nevi, and melanoma can be imaged with 80/90-MHz NIR femtosecond laser pulses without thermal destructive effects (König & Riemann, 2002; König et al., 2006; Masters et al., 1997). However, low pulse energies and low beam dwell time have to be taken. Interestingly, the luminescence of melanin is brighter than that of NAD(P)H. Therefore, melanin can easily be imaged with lower mean powers than typically required to detect intracellular coenzymes. Typically, we perform noninvasive optical sectioning of deep-tissue basal cells, which contain melanin, nevi, and melanoma, with (incident) laser powers of less than 20 mW (0.25 nJ). With a scan rate of 1 s per frame (512×512 pixels), the beam dwell time per pixel is 3.8 µs. For high-contrast images, the beam dwell time can be increased up to 27 µs.

When increasing the mean power and/or the beam dwell time significantly higher, thermal damage due to one-photon absorption by melanin has to be considered. Vaporization, the formation of gas-filled bubbles, and destructive mechanical microexplosions occur when reaching temperatures of about 100°C to 110°C (Jacques & McAuliffe, 1991). Although the laser beam–tissue interaction time of <300 fs is shorter than the thermal relaxation time of melanosomes, cumulative thermal effects have to be taken into account. For example, 308 laser pulses are applied at the same "pixel" during the 3.8-µs beam dwell time of an 80-MHz system. In order to minimize these cumulative effects in cells with melanin, Masters and colleagues (2004) suggested reducing the repetition frequency with the use of pulse pickers. They performed experiments on ex vivo human skin and measured morphological changes after 120 s of beam laser illumination depending on pulse energy and repetition frequency. According to their ex vivo results, the laser pulse energy can be increased up to 0.35 nJ when using 10-kHz instead of 1-GHz systems for safe skin imaging, even with long beam dwell times of 10 µs to 10 ms. High repetition rates of up to 1 GHz can be employed when keeping the pulse energies below 150 pJ.

Photothermal effects have to be considered in the case of tissue processing with NIR femtosecond laser pulses, in particular during refractive eye surgery. NIR photons can easily be transmitted through the eye and are absorbed by the retina (LeHarzic et al., 2005).

14.5 PHOTOCHEMICAL DAMAGE

Endogenous as well as exogenous absorbers can induce photochemical damage. Phototoxic reactions often include the formation of reactive oxygen species (ROS), which result in oxidative stress. A variety of endogenous cellular absorbers are known to generate ROS.

Amino acids with a one-photon absorption maximum below 300 nm can be excited via a three-photon process. The reduced coenzymes NAD(P)H absorb in the 700- to 800-nm range by two-photon excitation. It is known from one-photon studies that excitation of amino acids as well as of NAD(P)H can induce irreversible DNA damage (Bertling et al., 1996; Cunningham et al., 1985; Tyrell & Keyse, 1990). Excitation of flavins, which can also be excited by a two-photon process, may also cause cell damage by the UV/blue-light induced production of H_2O_2 and its metabolite, which are hydroxyl radicals (Hockberger et al., 1999).

In contrast, metal-free porphyrins produce singlet oxygen by a type II photo-oxidation process (photodynamic reaction). The efficient ROS producers coproporphyrin and protoporphyrin are used as photosensitizers in photodynamic tumor therapy. Also, destructive photodynamic type I photo-oxidation has to be considered (Fisher et al., 1997). As known from photodynamic porphyrin therapy, ROS formation in mitochondria leads often to apoptosis, whereas ROS formation in other organelles may also induce necrosis (Doiron & Gomer, 1984; Moor, 2000).

The major application of multiphoton microscopes in life sciences involves the excitation of routinely used intracellular fluorescent probes without one-photon absorption bands in the NIR by the simultaneous absorption of two or three NIR photons (Denk et al., 1990). Typical two-photon and three-photon absorption cross-sections of these fluorophores are about 10^{-48} to 10^{-50} cm^{4}s/photon and 10^{-75} to 10^{-84}cm^6(s/photon)^2 (Maiti et al., 1997; Xu et al., 1995).

It has to be considered that the excitation of the exogenous fluorescent probes results in fluorescence but may also induce a phototoxic response, in particular via photodynamic reactions. In particular we studied the effect of 780-nm excited protoporphyrin IX, which was synthesized in the mitochondria of CHO cells after external administration of the porphyrin precursor aminolevulinic acid (ALA). Four hours after incubation, cells were exposed to 170-fs pulses at a mean power of 2 mW and a frame rate of 0.0625 s^{-1} (16 s/frame). This corresponds to a mean beam dwell time of $\approx$400 ms per scan on a cell. Lethal effects occurred in most of cells after 100 scans in the same focal plane. In contrast, control cells without ALA could be scanned for more than 3 hours without impact on the vitality.

Using 170-fs laser pulses at 800 nm, we found that a mean power of $\geq$7 mW was sufficient to evoke ROS in laser-exposed PTK2 cells exactly within a scanned region of interest (Tirlapur et al., 2001). As an ROS indicator the fluorescent probe *Jenchrom px blue* (JenLab GmbH, Jena, Germany) and the membrane permeable dye dihydrofluorescein (DHF) (Hockberger et al., 1999) were used. Strong Jenchrom/DHF fluorescence was detected in cells exposed to mean laser power of 15 mW.

ROS formation can induce DNA breaks. Shafirovich and colleagues (1999) have shown during in vitro studies that excitation of supercoiled plasmid DNA with intense 810-nm NIR femtosecond laser pulses resulted in single-strand breaks. We studied the effect of 800-nm NIR laser pulses on the DNA of PTK2 cells using the TdT-mediated dUTP-nick end labeling (TUNEL) assay (e.g., Gavrieli et al., 1992; Negri et al., 1997). When exposed to 10 mW of power, many cells had TUNEL-positive nuclei. At 20 mW, green fluorescence was seen in all nuclei, indicative of DNA strand breaks. Because the structural changes and DNA breaks in the nuclei of NIR-exposed cells are quite similar to those occurring during apoptosis in different cell types (Bedner et al., 2000; Negri et al., 1997), it is conceivable that multiphoton microscopy may under certain exposure conditions lead to apoptosis-like cell death.

In order to evaluate the parameters for safe biomonitoring of human skin with the *DermaInspect* system, fresh Caucasian skin biopsies were laser scanned with mean powers up to 60 mW and analyzed for DNA photodamage by detecting photoinduced cyclobutane pyrimidine dimers. The results were compared with those obtained from UV-exposed biopsies using standard sunlight-mimicking lamps for dermatological use. The results show clearly that skin imaging with the *DermaInspect* system induces less DNA photodamage than 15 minutes of sunlight. The system can therefore be considered as a safe biomonitoring tool. When imaging intraocular structures with NIR femtosecond laser pulses, second-harmonic generation (SHG) by collagen in the cornea, the lens, etc., has to be considered. SHG leads to the formation of visible radiation at half the incident laser wavelength which could cause photodamage of the photosensitive cells ("blue light effect").

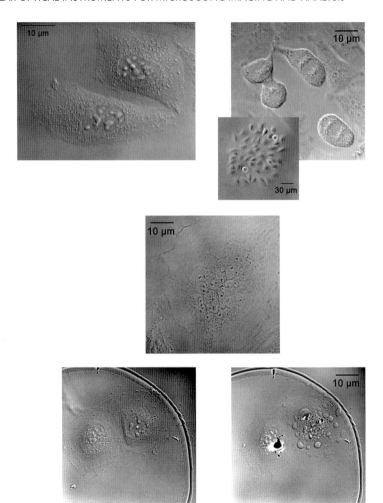

Figure 14.6. Cell cloning assay. Laser-exposed cells may develop (i) cell clones (upper part) or (ii) giant cells (middle), or undergo an immediate lethal process (lower part).

14.6 DAMAGE BY MULTIPHOTON IONIZATION

Light intensities in the TW/cm^2 range can cause multiphoton ionization phenomena resulting in optical breakdown, plasma formation, and destructive thermomechanical effects (König et al., 2005). The threshold for optical breakdown decreases with the pulse width. In water, the threshold for the onset of optical breakdown for 100-fs pulses was found to be 2 orders lower than for 6-ns pulses (Vogel et al., 1999).

Optical breakdown and plasma formation are often accompanied by the occurrence of a nearly white luminescence and complex subnanosecond emission decay (König et al., 1996d). The optical breakdown occurred preferentially in the mitochondrial region and did not start in the extracellular medium, nor in the nucleus. Evaluation with the fluorescent live/dead fluorophore kit (calcein and ethidium homodimer) confirmed immediate death in laser-exposed cells where extended plasma formation was monitored. Cavitation microbubbles are generated during high-intensity exposure to cells in aqueous environment (medium). Jets arise when the bubble is expanded and also when it collapsed. In addition, shock waves are formed. These mechanical effects lead to photodisruption. Material ejection by explosive tissue ablation occurs when the pressure exceeds the tensile strength of the extracellular matrix (Vogel & Venugopalan, 2003). A detailed explanation of multiphoton ionization effects and their possible use for precise nanoprocessing (König, 2005; Tirlapur & König, 2002) can be found in the chapter on nanoprocessing by König in this book.

Ocular tissues exposed to intense femtosecond laser pulses have been found to be damaged in out-of-focus regions by self-focusing effects. In that case the intense laser beam changes the refractive index of the exposed material by nonlinear processes and remains focused in a forward direction. This leads to uncontrolled destructive side effects. Self-focusing can be avoided when using high-NA focusing optics.

14.7 CONCLUSION

Multiphoton microscopes and imaging tools can be used as safe diagnostic tools within certain spectral and laser intensity windows. Above certain laser exposure parameters, destructive effects in cells and tissues are produced that can induce impaired/failed reproduction, formation of giant cells, modifications of ion exchange and autofluorescence, DNA breaks, and even cell death.

Photothermal effects have to be considered when imaging cells and tissues with hemoglobin and melanin at high mean powers, long beam dwell times, and high repetition frequency. Photochemical photodamage is based on a slow two-photon destructive process when using MHz systems and high-NA objectives at GW/cm^2 light intensities and pJ laser pulse energies. A fast (immediate) destructive process is obtained at higher TW/cm^2 intensities and pulse energies of more than 0.5 nJ based on multiphoton ionization phenomena. Photodisruptive side effects occur when working far above the threshold for optical breakdown. The fast, highly localized destructive effects for optical breakdown can be used to realize nanoprocessing in cells and tissues without collateral side effects when working near the threshold. Safe in vivo biomonitoring of human tissue ("optical biopsy") is possible with low-energy NIR femtosecond laser pulses. Possible effects due to SHG and self-focusing have to be considered.

Declaration of Financial Interests

Karsten König is the cofounder of the spin-off company JenLab GmbH.

REFERENCES

Bedner, E, Smolewski P, Amstad P, Darzynkiewicz Z (2000): Activation of caspases measured *in situ* by binding of fluorochrome-laneled inhibitors of caspases (FLICA): Correlation with DNA fragmentation. Exp Cell Res 259:308–313.

Bertling CJ, Lin F, Girotti AW (1996): Role of hydrogen peroxide in the cytotoxic effects of UVA/B radiation on mammalian cells. Photochem Photbiol 64:137–142.

Cunningham ML, Johnson JS, Giovanazzi SM. Peak MJ (1985): Photosensitised production of superoxide anion by monochromatic (290–405 nm) ultraviolet irradiation of NADH and NADPH coenzymes. Photochem Photobiol 42(2):125–128.

Denk W, Strickler JH, Webb WW (1990): Two-photon laser scanning microscope. Science 248:73–76.

Doiron DR, Gomer CJ (1984): *Porphyrin localisation and treatment of tumours.* New York: Alan R. Liss.

Fisher WG, Partridge WP, Dees C, Wachter EA (1997): Simultaneous two-photon activation of type I photodynamic therapy agents. Photochem Photobiol 66:141–155.

Gavrieli Y, Y. Sherman Y, Ben-Sasson SA (1992): Identification of programmed cell death *in situ* via specific labelling of nuclear DNA fragmentation. J Cell Biol 119:493–501.

Hänninen PE, Soini E, Hell SW (1994): Continuous wave excitation two-photon fluorescence microscopy. J Microsc 176:222–225.

Hockberger PE, Skimina TA, Centonze VE, Lavin C, Chu S, Dadras S, Reddy JK, White JG (1999): Activation of flavin-containing oxidases underlies light-induced production of H_2O_2 in mammalian cells. Proc Natl Acad Sci USA 96:6255–6260.

Jacques SL and McAuliffe DJ (1991): The melanosome: threshold temperature for explosive vaporization and internal absorption coefficient during pulsed laser irradiation. Photochem Photobiol 53:769–775.

Koester HJ, Baur D, Uhl R, Hell SW (1999): Ca2+ fluorescence imaging with pico- and femtosecond two-photon excitation: signal and photodamage. Biophys J. 77:2226–2236.

Kollias N, Bager A. Spectroscopic characteristics of human melanin in vivo. J Invest Derm 85: 38–42 (1985).

König K. Invited review: Multiphoton microscopy in life sciences. J Microsc 200:83–104 (2000).

König K. Cellular response to laser radiation in fluorescence microscopes. In: A. Perisami (editor): *Cellular Imaging.* pp. 236–251 (2001), Oxford University Press.

König K, Becker TW, Fischer P, Riemann I, Halbhuber KJ (1999): Pulse-length dependence of cellular response to intense near-infrared laser pulses in multiphoton microscopes. Opt Lett 24:113–115.

König K, Ehlers A, Stracke F, Riemann I. In vivo drug screening in human skin using femtosecond laser multiphoton tomography. Skin Pharmacol Physiol. (2006).

König K, Liang H, Berns MW, Tromberg BJ (1995): Cell damage by near-IR microbeams. Nature 377:20–21.

König K, Liang H, Berns MW, Tromberg BJ (1996a): Cell damage in near infrared multimode optical traps as a result of multiphoton absorption. Opt Lett 21:1090–1092.

König K, Krasieva T, Bauer E, Fiedler U, Berns MW, Tromberg BJ, Greulich KO (1996b): Cell damage by UVA radiation of a mercury microscopy lamp probed by autofluorescence modifications, cloning assay, and comet assay. J Biomedical Optics 1(2):217–222.

König K, Simon U, Halbhuber KJ (1996c): 3D resolved two-photon fluorescence microscopy of living cells using a modified confocal laser scanning microscope. Cell Mol Biol 42:1181–1194.

König K, So PTC, Mantulin WW, Tromberg BJ, Gratton E (1996d): Two-photon excited life-time imaging of autofluorescence in cells during UVA and NIR photostress. J Microsc 183: 197–204.

König K, So P, Mantulin WW, Tromberg BJ, Gratton E. (1997): Cellular response to near infrared femtosecond laser pulses in two-photon microscopes. Opt Lett 22:135.

König K, Riemann I. High-resolution multiphoton tomography of human skin with subcellular spatial resolution and picosecond time resolution. J Biomed Opt 8: 432–439 (2003).

König K, Riemann I, Stracke F, LeHarzic R. Nanoprocessing with nanojoule near-infrared femtosecond laser pulses. Medical Laser Application 20: 169–184 (2005).

LeHarzic R, Bückle R, Wüllner C, Donitzky C, König K. Laser safety aspects for refractive eye surgery with femtosecond laser pulses. Medical Laser Application 20: 233–240 (2005).

Liu Y, Cheng D, Sonek GJ, Berns MW, Chapman CF, Tromberg BJ (1995): Evidence for localised cell heating induced by near infrared optical tweezers. Biophys J 68:2137–2144.

Maiti S, Shear JB, Williams RM, Zipfel WR, Webb WW (1997): Measuring serotonin distribution in live cells with three-photon excitation. Science 275:530–532.

Masters BR, So PT, Gratton E (1997): Multiphoton excitation fluorescence microscopy and spectroscopy of in vivo human skin. Biophys J 72:2405–2412.

Masters BR, So PTC, Buehler C, Barry N, Sutin JD, Mantulin WW, Gratton E (2004): Mitigating thermal mechanical damage potential during two-photon dermal imaging. JBO 9:1265–1270.

Moor ACE (2000): Signaling pathways in cell death and survival after photodynamic therapy. J Photochem Photobiol 57:1–13.

Negri C, Donzelli, M, Bernardi, R, Rossi L, Bürkle A, Scovassi IA (1997): Multiparametric staining to identify apoptotic human cells. Exp Cell Res 234:174–177.

Oehring H, Riemann I, Fischer P, Halbhuber KJ, König K. (2000): Ultrastructure and reproduction behaviour of single CHO-K1 cells exposed to near infrared femtosecond laser pulses. Scanning 22:263–270.

Schönle A, Hell S (1998): Heating by absorption in the focus of an objective lens. Opt Lett 23:325–327.

Shafirovich V, Dourandin A, Luneva NP, Singh C, Kirigin F, Geacintov NE (1999): Multiphoton near-infrared femtosecond laser pulse-induce DNA damage with and without the photosensitizer proflavine. Photochem Photobiol 68:265–274.

Squirrel JM, Wokosin DL, White JG, Bavister BD (1999) Long-term two-photon fluorescence imaging of mammalian embryos without compromising viability. Nature Biotechnology 17:763–767.

Tirlapur UK, König K, Peuckert C, Krieg R, Halbhuber KJ (2001): Femtosecond near-infrared laser pulses elicit generation of reactive oxygen species in mammalian cells leading to apoptosis-like death. Exp Cell Res 263:88–97.

Tirlapur UK, König K (2002): Targeted transfection by femtosecond laser. Nature 418: 290–291.

Tyrell RM, Keyse SM (1990): The interaction of UVA radiation with cultured cells. J Photochem Photobiol 4:349–361.

Ulrich V, Fischer P, Riemann I, König K (2004): Compact multiphoton/single photon laser scanning microscope for spectral imaging and fluorescence lifetime imaging. Scanning 26:217–225.

Vogel A, Nahen K, Theisen D, Birngruber R, Thomas RJ, Rockwell BA (1999): Influence of optical aberrations on laser-induced plasma formation in water and their consequences for intraocular photodisruption. Appl. Opt. 38, 3636–43.

Vogel A, Venugopalan V. Mechanisms of pulsed ablation of biological tissues. Chem Rev 103: 577–644 (2003).

Xu C, Guild J, Webb WW, Denk W (1995): Determination of absolute two-photon excitation cross sections by in situ second-order autocorrelation. Opt Lett 20(23):2372–2374.

Applications of Second-Harmonic Generation Microscopy

Jerome Mertz

15.1 INTRODUCTION

By far the most best-known form of nonlinear microscopy is based on two-photon excited fluorescence (TPEF) (Denk et al., 1990), which is described in considerable detail in Chapters 4 and 5. However, a lesser-known form of this microscopy was used several years prior to the invention of TPEF microscopy, based on the generation of second-harmonic light either from surfaces (Hellwarth & Christensen, 1964; Sheppard et al., 1977) or from endogenous tissue structures such as rat-tail tendons (Freund et al., 1986). Because of difficulties in signal interpretation and perhaps because of its seemingly arcane utility, at least in biological imaging, second-harmonic generation (SHG) microscopy has gone by relatively unnoticed until only very recently (Campagnola et al., 1999; Guo et al., 1997; Moreaux et al., 2001). The discovery that exogenous markers (Chemla & Zyss, 1984; Prasad, 1991) can lead to exceptionally high signal levels has been a leading cause for the revival of SHG microscopy. In particular, SHG markers, when properly designed and collectively organized, can produce signal levels comparable to those encountered in standard TPEF microscopy (Moreux et al., 2000a, 2000b). Moreover, the spatial resolutions provided by SHG and TPEF microscopies are commensurate, meaning that the two contrasts can be conveniently derived from the same instrument.

Despite their similarities, SHG and TPEF are based on fundamentally different phenomena (Bloembergen, 1965; Butcher & Cotter, 1990). The first relies on nonlinear scattering, whereas the second relies on nonlinear absorption followed by fluorescence emission. In other words, the first is coherent, whereas the second is not. The consequences of this basic difference will be described below. The organization of this chapter is as follows. First, the basic principles of SHG will be qualitatively described at the molecular level. Second, typical experimental configurations for combined SHG and TPEF imaging will be briefly described, along with basic protocols for sample preparation and labeling. Finally, various applications of SHG microscopy will be addressed, with an emphasis on high-sensitivity membrane potential imaging and on endogenous imaging in thick tissue.

15.2 BASIC PRINCIPLES

15.2.1 Second-Harmonic Generation Versus Hyper-Rayleigh Scattering

To begin, we consider radiative scattering from a single isolated molecule. To first order, the molecule may be regarded as a simple electrical dipole. The molecular dipole moment is defined as $\vec{\mu} = q\vec{r}$, where q is the difference between the net nuclear and electronic charges and $\vec{r}$ is their relative displacement. Changes in $\vec{\mu}$ are occasioned by forces applied to the electrons. We consider here only electric dipole forces—that is, forces whose interaction energies are given by $W = \vec{\mu} \cdot \vec{E}$, where $\vec{E}$ is the applied electric field. In its simplest form, the molecule may be regarded as a one-dimensional rod along which its electron cloud can shift up or down.

We begin by examining the case when the molecule is perfectly centrosymmetric, meaning that the molecular dipole moment vanishes at rest. Upon illumination, the electron cloud, depicted in Figure 15.1 as a point electron, is subject to a sinusoidal

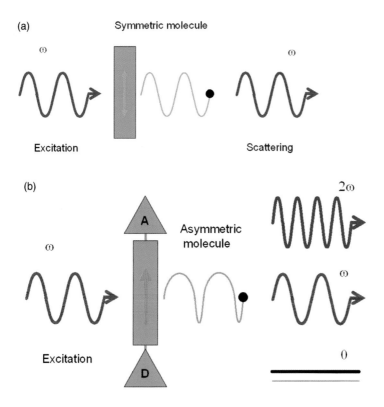

Figure 15.1. (a) When driven by light of frequency , the induced electron motion in a centrosymmetric uni-axial molecule produces Rayleigh scattering at the same frequency ω. (b) The induced electron motion in a non-centrosymmetric molecule produces scattered light at frequencies ω and 2ω. The latter is called hyper-Rayleigh scattering (HRS).

force along the molecular axis, whose frequency is ω. Because the molecule is centrosymmetric, the resultant oscillation of the electron cloud is also centrosymmetric and generates radiation at the same frequency ω. Such radiation is called Rayleigh scattering (Born & Wolf, 1993). It is linear in the sense that the scattered frequency is exactly the same as that of the driving illumination beam. In this example of a centrosymmetric molecule, no second-harmonic light (i.e., light of frequency 2ω) can be generated.

Charge non-centrosymmetry may be imparted on a molecule by grafting electron donor (D) and acceptor (A) moieties to its opposite ends, producing what is known as a "push–pull" chromophore (Liptay, 1974) since both moieties act in tandem to favor electronic motion in the direction D→A. When a push–pull chromophore is subjected to the same sinusoidal driving field as above, then the resultant electron motion, though still periodic, is no longer symmetric, as shown in Figure 15.1b. The electron motion (or more precisely its acceleration) contains additional frequency components. In addition to the usual linear component at frequency ω, the scattered radiation also contains nonlinear components at frequencies 0 and 2ω. The 0 frequency component is called optical-rectification (Butcher & Cotter, 1990) and technically should not be referred to as radiation since it represents a static electric field. The 2ω frequency component is called hyper-Rayleigh scattering (HRS) (Clays et al., 1994). This is the radiation that will interest us throughout the remainder of this chapter. As a general rule, *the generation of second-harmonic light requires a non-centrosymmetric source*.

We consider now some important features of hyper-Rayleigh scattering (HRS) that distinguish it from Rayleigh scattering or, for that matter, from fluorescence. As an illustration, we consider two non-centrosymmetric molecules that are located in close proximity, separated by a distance much smaller than an optical wavelength, and oriented in parallel directions (we neglect any possibility of chemical interactions). When these molecules are illuminated by a driving field, their respective electron motions will be identical, and the resultant HRS from both molecules will be in phase (Fig. 15.2a).

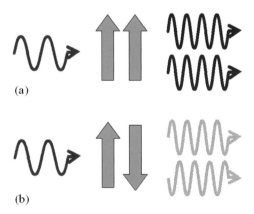

(a)

(b)

Figure 15.2. The HRS from two molecules located close together. (a) If the molecules are parallel, their HRS is in phase and interferes constructively. (b) If the molecules are anti-parallel, their HRS is out of phase and cancels.

In other words, the net HRS amplitude will be doubled, meaning that the net HRS power will be quadrupled relative to the HRS power obtained from a single molecule. If, on the other hand, the two molecules are oriented anti-parallel, then the result is very different. The asymmetric components in their electronic motions are now inverted relative to one another, meaning that their respective HRS amplitudes are out of phase and exactly cancel (see Fig. 15.2b). In other words, the net HRS power produced by the anti-parallel molecules is zero! Of course, we could have inferred this from the start since two non-centrosymmetric molecules, when located at the same position with opposite orientation, effectively become centrosymmetric. We note that the dependence of radiative phase on molecular orientation is a characteristic property of HRS that occurs neither in Rayleigh scattering nor in fluorescence. In particular, the Rayleigh scattered components from both molecules in Figure 15.2 remain in phase regardless of whether the molecules are oriented parallel or anti-parallel. Similarly, if the molecules were fluorescent, then the phases of their fluorescence emissions would essentially be random, also regardless of molecular orientation.

In practice, for imaging applications, one is interested in looking at the HRS from not one molecule, or even two molecules, but from an entire population of molecules. We first consider a configuration where N molecules are in solution (Fig. 15.3a). Because the orientations of these molecules are random, so too are the phases of their individual HRS contributions. The molecules then produce HRS *incoherently*, and the total generated HRS power simply scales as the number of radiating molecules: $P_{HRS} \propto N$. This same scaling law applies to fluorescence. Alternatively, if the molecules are globally aligned along a same direction, in the manner of a liquid crystal (see Fig. 15.3b), their individual HRS contributions are then no longer random but instead are prescribed by the phase of the driving field, presumably well defined. The HRS is then produced

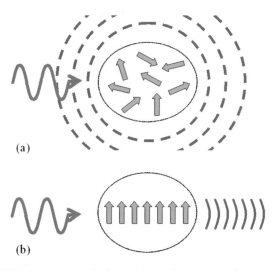

(a)

(b)

Figure 15.3. The HRS from a population of N molecules. (a) Randomly oriented molecules scatter incoherently and the total HRS power scales as N. (b) Aligned molecules scatter coherently, producing second-harmonic generation (SHG) whose power is well collimated and scales as N^2.

coherently, meaning that interference effects play a major role in determining its global properties. When HRS is produced coherently by an organized collection of molecules, it is called second-harmonic generation (SHG) (Boyd, 1992), whereas the term HRS is reserved for the case when molecules are unorganized. The distinction between SHG and HRS (as well as fluorescence) is noteworthy. Because SHG is coherent, its total power scales as the square of the number of radiating molecules: $P_{SHG} \propto N^2$ (as in the example of Fig. 15.2b). Moreover, also because it is coherent, its angular distribution is highly structured. In general, SHG is constrained to propagate along the same direction as the driving field. Much more will be said about this below. For imaging applications, it is therefore highly desirable that the radiating molecules be organized rather than disorganized because (1) much more signal is generated for the same number of molecules, and (2) the signal can be collected more efficiently because of its high directionality. We turn to molecular design strategies for obtaining these conditions.

15.2.2 Molecular Design Strategies

Molecules produce efficient SHG when two conditions are met. First, the molecules must individually be capable of generating significant HRS. Second, the molecules must globally be organized to favor the coherent summation of their individual HRS contributions. We address these conditions separately.

15.2.2.1 Chromophore Design

As described above, a molecule can produce HRS when it is push–pull in design—that is, it comprises donor and acceptor moieties spanned by a uni-axial charge transfer path, known as a conjugation path. All the molecules we will consider in this chapter will be of this geometry, as illustrated by the specific example of a stilbazolium-type dye. Here, aniline serves as an electron donor and pyridinium as an electron acceptor, both being spanned by a carbon bridge. The length of this bridge can be variable. For stilbene, the prototypical push–pull molecule, the bridge comprises one bond. For ASP, the example considered here, it comprises three bonds, etc. These molecules fall into the family of polyenic stilbene derivatives (Oudar, 1977). In addition to being able to generate HRS, these molecules can also produce fluorescence. The effects of bridge length on both HRS and fluorescence will be addressed below.

In most cases, push–pull molecules can be adequately described by a quantum two-state model (Blanchard-Desce & Barzoukas, 1998; Loew & Simpson, 1981; Zyss, 1979). The "aromatic" configuration of ASP, which possesses only a small dipole moment, is shown at the top of Figure 15.4. In contrast, the "quinoid" configuration of ASP, which possesses a much larger dipole moment, is shown at the bottom of Figure 15.4. The change in dipole moment between these configurations is due to the global shift of all the π-electron bonds (shown as double bonds) from the donor to the acceptor ends of the molecule. In the absence of illumination the molecule resides in its ground state $|g\rangle$, which typically resembles the aromatic configuration. Upon illumination, however, the ground state becomes coupled to the excited state, which more closely resembles the quinoid configuration. As a result, the molecule incurs a net change in dipole moment. We emphasize that this change in dipole moment is not the same as the oscillating dipole moment induced in the molecule at the driving optical

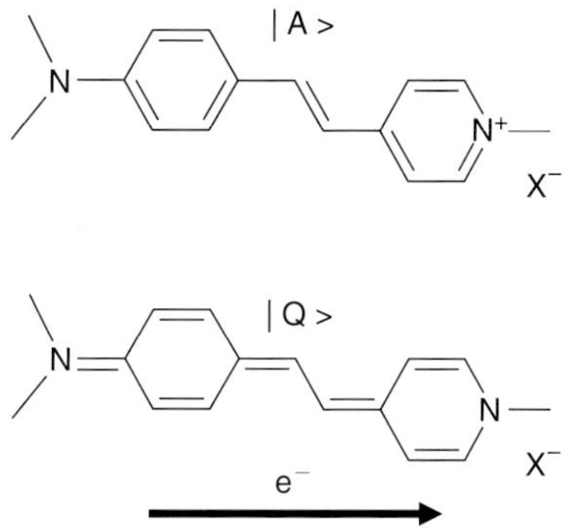

Figure 15.4. A polyenic stilbene derivative in its aromatic (top) and quinoid (bottom) configurations. There is a net change in dipole moment between both configurations.

frequency. Instead, it corresponds to the zero-frequency optical rectification alluded to in Figure 15.1b. Its magnitude is proportional both to the illumination intensity and to the net difference in dipole moments between the excited and ground states, $\Delta\mu = \mu_e - \mu_g$. Furthermore, we note that because neither $|g\rangle$ is completely aromatic nor $|e\rangle$ completely quinoid, $\Delta\mu$ is in general not as large as the one illustrated in Figure 15.4 which only represents its maximum limiting value. Finally, we state without proof that for a molecule to be capable of generating significant HRS, it *must* exhibit a non-zero $\Delta\mu$ as large as possible. $\Delta\mu$ may therefore be regarded as an approximate measure of a molecule's asymmetry, which is indispensable for HRS. For polyenic stilbene derivatives, an increase in conjugation path length leads to an increase in $\Delta\mu$ (Blanchard-Desce et al., 1995, 1997). On the other hand, it also leads to a decrease in the energy gap between the excited and ground states, and in the efficiency of fluorescence emission, both of which may be undesirable. For Di-6-ASPBS, $\Delta\mu$ is roughly 10 to 20 Debye (1 Debye = 3.3×10^{-30}C-m) corresponding to an electron charge displacement of about 0.2 to 0.4 nm. In the literature, $\Delta\mu$ is also referred to as "charge transfer", and for this reason, push–pull molecules are often referred to as charge transfer molecules.

15.2.2.2 Molecular Organization

Even the best push–pull chromophores produce HRS with only modest efficiencies, typically orders of magnitude less than they produce fluorescence. To obtain adequate HRS signals comparable to those found in fluorescence microscopy, further steps must be taken. The chromophores must be designed such that, in addition to efficiently producing HRS, they may be organized with relatively high number densities and high degrees of spatial alignment, as depicted for example in Figure 15.3b. In this example HRS becomes SHG, and the signal enhancement resulting from coherence can be fully exploited. Possible strategies for favoring molecular alignment include poling with an electric field (Chemla & Zyss, 1984; Oudar, 1977). We consider here a different strategy that involves modifying the chromophores to make them behave like surfactants. In particular, as shown in Figure 15.5, hydrocarbon sidechains may be grafted onto one end of the chromophore and a polar head-group onto the other (alternatively, the polar head-group may consist of an unattached counter-ion, as shown in Fig. 15.4). The

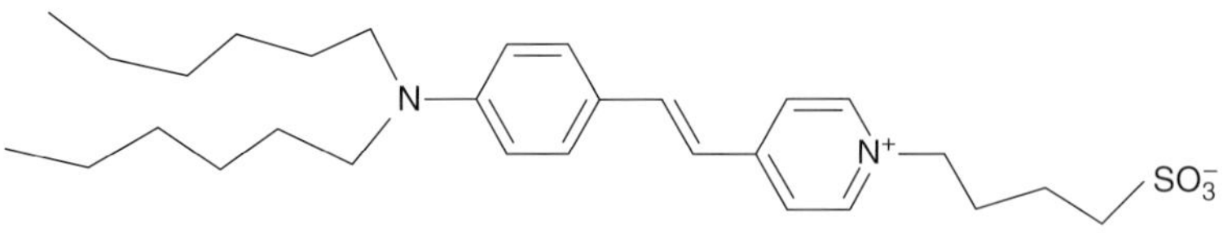

Figure 15.5. An amphiphilic stilbene derivative with an attached SO_3^- counterion (Di-6-ASPBS).

hydrocarbon sidechains are fatty in nature, and therefore hydrophobic, whereas the polar head-group is hydrophilic. As a result, the chromophore has been "functionalized" to be a molecular membrane marker. When perfused onto a lipid bilayer, say from above as shown in Figure 15.6, the modified molecules invariably insert themselves into the membrane with the same orientation, with the hydrophobic sidechains as far as possible from the aqueous surroundings (that is, directed into the membrane interior) and the polar head-group pointing outward, into the aqueous surroundings. Significant energy is required for the flipping of a molecule from one lipid bilayer leaflet to the other, since such a passage entails the traversing of a polar head-group through a nonpolar membrane interior. Though such a flipping event is rare (Devaux, 1993), it is measurable, as will be shown below in the applications section.

Figure 15.6, depicting Di-6-ASPBS markers in a phospholipid membrane, is drawn approximately to scale (except for the phosphatidyl head-groups, depicted here only as ovals). Exactly how deeply Di-6-ASPBS resides in the membrane under normal conditions is unknown, though the chromophore core of the molecule is known to be well isolated from the aqueous surroundings (Radda, 1971). The transverse area covered by the molecule is approximately that of the phospholipids themselves (Israelachvili, 1991) (that is, <1 nm^2). Labeling densities used in experiments are typically on the order of 1 dye molecule for every 100 phospholipids (more will be said about this).

In short, we have achieved our desired goal of producing reasonably high labeling densities with high degrees of chromophore alignment. As will be shown below, such labeling configurations can easily produce enough SHG signal to be visualized in imaging applications.

15.2.3 Second-Harmonic Generation Radiation Patterns and Powers

To simplify matters, both conceptually and experimentally, we first consider model membranes consisting of phospholipid bilayers, and nothing else. Such membranes can readily be synthesized in the form of giant unilamellar vesicles (GUVs) (Angelova et al., 1992), which are single bilayer spheres roughly 10 to 50 microns in diameter, comparable in size to typical biological cells. When GUVs are synthesized in aqueous media and subsequently labeled from the exterior with push–pull membrane dyes, such as Di-6-ASPBS, the dyes reside in the external leaflets of the GUV membranes and are directed outward. Illuminating the GUVs with a laser beam then produces SHG. The reason the illumination must be from the side, as shown in Figure 15.7, is because the radiation–molecule interaction is dominantly dipolar in nature (see above), and is most efficient when the principal axes of the dye molecules are oriented parallel to the electric-field polarization axis of the illumination beam.

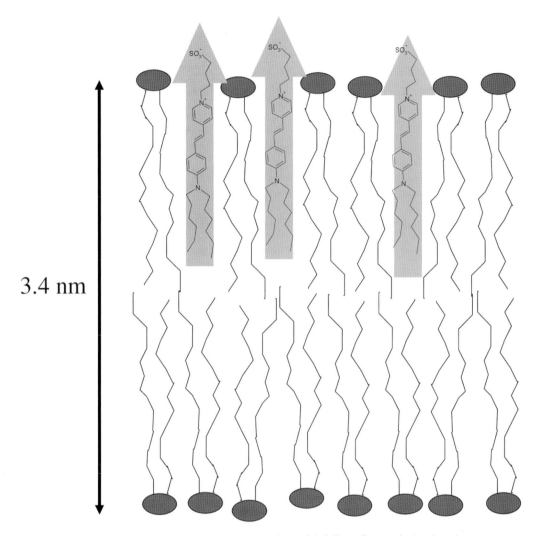

Figure 15.6. Presumptive insertion geometry of amphiphilic stilbene derivatives in a lipid bilayer, when perfused from the top.

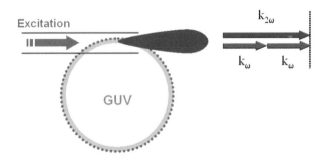

Figure 15.7. SHG from a labeled GUV when illuminated by a collimated laser beam. The SHG propagates on-axis because of momentum conservation. The excitation polarization is perpendicular to the membrane plane.

We consider the interaction area in more detail, approximating it here to be planar and the dye molecules to be roughly uniformly distributed throughout the outer leaflet. We recall that the phase of the HRS produced by each molecule is governed by the phase of the driving field. However, in our case the driving field is propagating, meaning that its phase is propagating. The HRS produced by the molecules must propagate as well: molecules that are located upstream (leftmost) generate HRS slightly earlier than those that are located downstream (rightmost). The resulting SHG radiation pattern can be evaluated by treating each molecule in the interaction area as a local emitter of second-harmonic radiation. The phase and amplitude of this radiation is precisely determined by the phase and amplitude of the driving field at the same location. This method of calculation is borrowed directly from the theory of phased-array antennas (Krauss, 1950). When the second-harmonic contributions from each emitter (molecule) are added, taking into account their respective phases and amplitudes, then the resulting total SHG is found to propagate in the same direction as the driving field. This result follows directly from momentum conservation (or, in nonlinear optics parlance, phase-matching [Boyd, 1992]). In effect, the phenomenon of SHG locally "converts" two incoming driving photons into a single outgoing photon at the second-harmonic frequency. To ensure momentum conservation in the axial direction, the second-harmonic photon, whose momentum is twice that of a driving photon, must propagate in the same direction as the driving photons, as illustrated vectorially in Figure 15.7 (we have neglected here the possibility of a frequency dependence in the refractive index of the surrounding medium, which is tantamount to assuming that the interaction area is small).

The illumination geometry depicted in Figure 15.7 is overly simplistic and does not correspond to geometries used in laser scanning microscopy. In particular, 3-dimensional microscopic resolution can be obtained only when the illumination beam is focused into the sample to a microscopic spot size. This is the case in confocal and TPEF microscopies, and is also the case in SHG microscopy. A focused beam differs from a collimated beam (see Fig. 15.7) in two respects. First, the intensity of a focused beam is significantly enhanced near its focal center. Second, its phase fronts are no longer evenly distributed along its propagation axis. We examine the effects of these differences separately.

It is well known in TPEF microscopy that when a laser beam is focused into a sample, essentially all the fluorescence excitation occurs within a well-defined volume around the focal center—the tighter the focus, the smaller the volume. The rigorous confinement of TPEF to a small volume stems from its quadratic dependence on excitation intensity (as is explained in detail in Chapter 5). Inasmuch as SHG also depends quadratically on excitation intensity, we may expect the active SHG volume, and hence the effective resolution afforded by SHG microscopy, to be the same as that in TPEF microscopy. This turns out to be true in almost all practical cases of interest, and in general the SHG and TPEF volumes may be defined identically (Moreaux et al., 2000a, 2000b). However, one should remember that the samples in TPEF microscopy are usually 3-dimensionally distributed, whereas in SHG membrane microscopy they are 2-dimensionally distributed. In the latter case, it is more accurate to speak of an active SHG "area."

Specifically, we consider a geometry where the excitation beam is focused with an objective of moderate to high numerical aperture (NA > 0.8). Though the intensity profile of such an excitation beam about the focal center cannot be expressed analytically, it can be approximated as being Gaussian along both the radial and axial coordinates (we emphasize that this approximation is valid only when considering nonlinear inter-actions). The active SHG area can then be defined as $(\pi/2)\, w_r w_z$, where w_r and w_z are the $1/e$-halfwidths of the radial and axial beam amplitude profiles respectively (called beam "waists"). For example, when focusing with NA = 0.9 in water, then $w_r \approx 0.5\,\mu m$ and $w_z \approx 1.9\,\mu m$, and the SHG active area is approximately $1.5\,\mu m^2$. This active area is centered at the beam focus, and any SHG generated outside this area may be safely neglected. We note that this is very different from the configuration shown in Figure 15.7 where the intensity of the illumination beam is uniform along the entire propagation axis, meaning that the SHG active area is not intrinsically confined axially.

In TPEF microscopy, the fluorescence signal is determined by the intensity profile of the excitation beam alone. In SHG, however, phase plays a crucial role and cannot be overlooked. A fundamental difference between a collimated beam and a focused beam is that the phase of a focused beam is not evenly distributed along the propagation axis. Specifically, near the focal center, the phase fronts do not appear to travel as quickly in a collimated beam. Intuitively, this arises from the fact that a focused beam comprises a cone of illumination directions, most of which are off axis and hence slow the overall axial phase. As a result of this phase retardation, a focused beam necessarily incurs a net axial phase lag of π as it travels through its focal center, relative to the corresponding phase of an unfocused (collimated) beam. This phase lag is called a Gouy shift (Born & Wolf, 1993). It is localized near the focal center, but it is not abrupt. On the scale of the SHG active area it may be approximated as varying linearly about its $\pi/2$ midpoint. The consequences of this Gouy shift on SHG are dramatic, as shown in Figure 15.8. The SHG from a focused beam, instead of propagating on-axis as in Figure 15.7, now propagates off-axis in two well-defined symmetric lobes (Moreaux et al., 2000a, 2000b). This may be explained again by momentum conservation (phase-matching). Because the phase of the excitation beam has effectively been retarded near the focal center, then its effective axial momentum has accordingly been decreased and may be written as $\xi\, k_\omega$, where k_ω is the momentum of the beam if it were unfocused and ξ a factor less than 1. As before, the propagation momentum of the SHG is $k_{2\omega}$. Momentum conservation

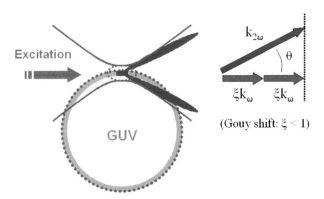

Figure 15.8. SHG from a labeled GUV when illuminated by a focused laser beam. The active SHG area is confined near the focal center. The SHG propagates off-axis because of momentum conservation, taking into account the Gouy phase slippage.

along the axial direction then forces the SHG to propagate at an off-axis angle given by $\theta \approx \pm \cos^{-1}(\xi)$, as may be readily verified experimentally by placing an ordinary camera downstream from the emission and observing two well-defined SHG lobes.

We emphasize that the SHG angular pattern shown in Figure 15.8 is specific to the case when the radiating molecules in a membrane are uniformly distributed, at least on the scale of the SHG wavelength. This is the case usually encountered in practice, particularly when imaging GUVs; however, it should be kept in mind that more complicated molecular distributions can easily occur. For example, if the molecular distribution possesses an axial periodicity in the vicinity of $\lambda/4$ (where λ is the wavelength in the sample medium), then up to 25% of the SHG power is radiated in backward-directed lobes (Mertz & Moreaux, 2001). Alternatively, if the molecules are not distributed throughout the SHG active area but instead are tightly clustered around the focal center, then the SHG radiation is dipolar in nature, meaning that it is equally distributed in the forward and backward directions (we neglect chemical interactions that might occur between the molecules at such small separations). Momentum conservation arguments can again be invoked to explain these patterns: When the molecular distribution possesses axial inhomogeneities, then in effect these impart their own pseudo-momenta to the SHG emission. In the examples above, the pseudo-momenta impart backward-directed "kicks" to the SHG of sufficient magnitude to provoke partially backward-directed emission.

The total power contained in the SHG radiation, whether the excitation is focused or unfocused, or whether the molecular distribution is homogeneous or inhomogeneous, can be calculated by integrating the SHG intensity over all emission angles (Moreaux et al., 2000a, 2000b). We return to the case of HRS emission from a single *isolated* molecule located at the excitation focal center. The total HRS power produced by the molecule, which is dipolar in nature and hence distributed equally in the forward and backward directions, may be written as

$$P_{2\omega}^{(1)} = \tfrac{1}{2}\sigma_{HRS}\, I_\omega^2,$$

where I_ω is the intensity of the excitation beam at the focal center, and σ_{HRS} is the HRS cross-section of the molecule when it is oriented parallel to the excitation polarization.

The factor $\tfrac{1}{2}$ is conventionally used when all the radiated powers are defined in units of photons/s, as opposed to Watts. σ_{HRS} can then be defined in Göppert-Mayer units (see Chapter 5), allowing it to be directly compared to a TPEF cross-section, as will be discussed below. Instead of considering a single molecule, we turn now to a cluster of N molecules identically oriented and located exactly at the excitation focal center (again, we neglect the possibility of chemical interactions). In this case, the molecules are all driven at exactly the same phase, and their individual HRS contributions interfere exactly constructively. The resultant SHG power is therefore $P_{2\omega}^{(N)} = N^2 P_{2\omega}^{(1)}$ (again, distributed equally in the forward and backward directions). Now let us take this same number N of molecules and distribute them within the SHG active area more or less uniformly. Because the molecules are no longer exactly at the same locations, they are no longer driven at the same phases, for the two reasons as described above: (1) there is a time delay between the excitation of molecules at different locations because the excitation beam is propagating, and (2) the location-dependent Gouy phase shift must be taken into account. These effects, in addition to drastically modifying the SHG radiation pattern (see Figs. 15.7 and 15.8), also tend to drastically reduce the SHG power, and we write in general:

$$P_{2\omega}^{(N)} = \Theta \, N^2 \, P_{2\omega}^{(1)},$$

where $\Theta \leq 1$ is the "antenna" factor resulting from the molecular distribution profile.

In the case of a homogeneous molecular distribution and moderate excitation beam focusing, Θ is very roughly given by $0.01 \, \lambda^2/w_0^2$. As an example, for an excitation wavelength of 800 nm and a focal spot radius of $w_0 = 0.5 \, \mu m$ in water, then $\Theta \approx 0.01$, meaning that the arrangement of the N molecules into a uniform distribution has indeed had a very deleterious effect on their capacity to efficiently produce SHG. Nevertheless, a sufficient labeling density (i.e., N) can often more than compensate for such a reduced Θ, as will be amply illustrated in the sections below.

15.3 INSTRUMENTATION

15.3.1 Microscope Design

The efficiency with which a molecule can produce HRS depends on the wavelength of the illuminating excitation beam. The closer the excitation is to resonance (that is, the closer the energy of two excitation photons corresponds to the energy difference between the molecular ground and excited states), the greater the probability of HRS. At the same time, however, near-resonance excitation also enhances the probability of TPEF, assuming the molecule is fluorescent. In general, though not always, efficient SHG and TPEF go hand in hand, and a microscope can in principle benefit from both contrast mechanisms simultaneously. To ensure a proper discrimination of SHG and TPEF, two techniques can be employed. A first technique makes use of the fact that nonlinear scattering of light is essentially an instantaneous phenomenon, occurring on time scales of order the inverse resonance frequency (i.e., femtoseconds). In contrast, molecular fluorescence lifetimes are typically on the order of nanoseconds (i.e., a million times longer). A time-gating technique can therefore effectively distinguish SHG from TPEF. A second technique, however, is far simpler. The wavelength of SHG is rigorously equal to half that of the excitation beam, and its linewidth is equal to $1/\sqrt{2}$ that of the excitation beam. This differs fundamentally from TPEF, whose wavelength and linewidth depend not on the excitation laser but on the dye molecules. For example, Figure 15.9 shows an emission spectrum produced by a GUV uniformly labeled with Di-6-ASPBS. Two identifiable signals are apparent. The leftmost peak is the SHG signal, centered at half the excitation wavelength (here about 880 nm and out of range). The rightmost lobe is the simultaneously generated TPEF signal, centered at the Di-6-ASPBS fluorescence emission wavelength (about 560 nm when in a lipid environment). If the excitation laser beam were tuned to a new wavelength, the SHG peak would track this new wavelength, whereas the TPEF lobe would not. In Figure 15.9, the excitation beam has been tuned so that the SHG and TPEF signals are widely disparate in wavelength and do not overlap, meaning that they can easily be separated with dichroics and/or interference filters, as will be shown below.

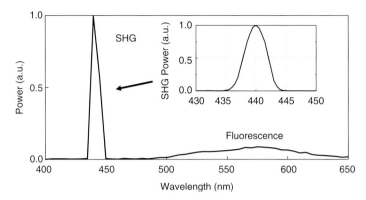

Figure 15.9. The emission spectrum obtained from a Di-6-ASPBS-labeled GUV illuminated by a laser of wavelength 880 nm. The leftmost peak is SHG. The rightmost lobe is TPEF. The signals are comparable in power.

On further examining Figure 15.9, we observe that the areas under the SHG and TPEF signals are roughly equal, indicating that their measured powers are approximately the same. The reader is reminded that such parity in powers is possible only because SHG is a coherent phenomenon, whereas TPEF is not—that is, even though $\sigma_{HRS} \ll \sigma_{TPEF}$ for a single molecule, and Θ is small for a uniform labeling distribution (see Section 15.2.3), SHG scales with N^2 whereas TPEF scales only with N, meaning that the two powers can be made comparable by choosing a sufficiently large labeling density. For the particular example of Figure 15.9, N is roughly 10^4.

In essence, there is little difference between a SHG and a TPEF microscope, and the two are ideally suited to be combined within the same instrument. Both are based on a scanning geometry, meaning that a laser beam is focused into a sample and either the sample is scanned, or alternatively, when high speed is required or the sample must remain fixed, the laser focal spot is scanned in three dimensions. In the latter case, lateral X-Y scanning is usually performed with a pair of galvanometer-mounted mirrors, while the axial Z scanning is obtained by adjusting the objective height. Because both SHG and TPEF depend quadratically on excitation intensity, their signals arise from inherently confined focal volumes that have the same dimensions. This is particularly convenient because it means that SHG and TPEF microscopies intrinsically provide the same spatial resolutions.

The fact that both SHG and TPEF depend quadratically on excitation intensity means that they also benefit in the same way from pulsed excitation (see Chapters 5 and 8). In practice, the most commonly employed excitation source for these microscopies is a mode-locked Ti:Sapphire laser, which typically delivers 100-fs light pulses up to 10 nJ in energy, at a repetition rate of about 80 MHz. Because SHG and TPEF usually benefit from chromophore resonance enhancements at about the same excitation wavelengths, they are typically maximized for the same laser tuning.

Figure 15.10 illustrates the main components of a combined SHG–TPEF microscope. Basically, the SHG portion of this microscope consists of the addition to a standard TPEF microscope of a detector in the transmitted light direction. Such an addition is relatively straightforward provided a few precautions are taken into account. First, because the excitation beam is focused, the SHG is dominantly emitted off-axis,

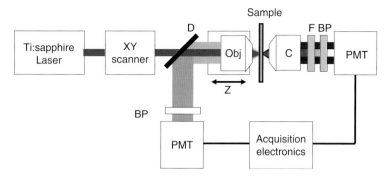

Figure 15.10. A combined scanning TPEF-SHG microscope. A pulsed laser beam is focused into a thin sample. The TPEF is collected in the backward direction using a dichroic beamsplitter. The SHG is collected in the forward direction after blocking both transmitted laser light and fluorescence. The TPEF and SHG signals are acquired simultaneously.

meaning that the collection optics must have a wide enough NA to effectively accept the entire SHG signal (Moreaux et al., 2000a, 2000b). In the case of a 2-dimensional sample geometry (e.g., a membrane) with uniform labeling, the angular spread between the two outgoing SHG lobes is roughly $1/\sqrt{2}$ that of the incoming excitation beam. In the case of a 3-dimensional sample geometry (not discussed so far), the SHG is emitted even closer to axis. A simple rule of thumb may generally be applied: to collect all the SHG, the collection optics should have an NA no smaller than that of the excitation optics. Moreover, precautions must be taken to completely block both the transmitted laser light and any simultaneously generated TPEF from impinging on the detector. The laser light can very effectively be blocked with color glass filters. In turn, the fluorescence can be blocked with appropriate interference filters, assuming that the SHG and TPEF emission spectra are well separated (see above). Finally, we consider the choice of detectors. When using Ti:Sapphire excitation, which can be tuned from 700 nm to 1000 nm in wavelength, we can expect SHG wavelengths in the range 350 nm to 500 nm. Currently the most appropriate detectors in this range are photomultipliers with bi-alkali cathodes. These have the added benefit of being red-insensitive and hence relatively blind to laser light leakage. They also happen to be the most commonly used detectors in TPEF microscopy. We remind the reader that, as in TPEF microscopy, the SHG signal is produced from an intrinsically confined volume, which entirely defines its 3-dimensional resolution. As such, no particular efforts need be made to precisely image the SHG signal onto the detector. Even if the signal is aberrated, blurred, or nonstationary at the detector surface, resolution is maintained as long as the detector surface is wide enough to collect the entire signal. Moreover, because SHG signal powers can be made comparable to TPEF powers, the output from the SHG detector can often be plugged directly into the same acquisition electronics normally meant for a fluorescence channel, without modifying sampling rates, integration times, or even amplification levels. In other words, the added operation of SHG imaging in no way impairs the normal operation of TPEF imaging. Both contrast mechanisms provide the same resolution and comparable signal levels and hence operate with compatible image pixel sizes and acquisition times.

15.3.2 Labeling

Illustrations of combined SHG and TPEF imaging are shown below. Figure 15.11 depicts an isolated neuron from a primary cell culture labeled with an amphiphilic stilbene derivative. Though the dye is meant to label the cytoplasmic membrane only, it has also been incorporated into the cell body, as is apparent from the TPEF image. The mechanisms responsible for this incorporation have not been fully identified, though they probably involve some active form of endocytosis. We note that dye molecules inside the cell cytoplasm are randomly oriented in general and hence cannot produce coherent SHG. SHG imaging is therefore insensitive to intracellular dye incorporation and reveals only outer cell membranes with exceptionally high contrast. The labeling protocol for Figure 15.11 consisted of the extracellular perfusion of the dye markers with the aid of cyclodextrin "cages" to enhance the solubility of the amphiphilic markers in aqueous media (Redenti et al., 2000). Other labeling protocols include dissolving the markers with the aid of ethanol (0.1% concentration). The estimated labeling density in Figure 15.11 is about 10^4 molecules per square micron.

Figure 15.12 illustrates a simultaneous SHG-TPEF image of a 50-μm-thick hippocampal slice culture. The labeling procedure in this case involved depositing a small crystal of RH237 below the glial cell surface and incubating overnight at 36°. Again, in this illustration, the cell membranes are revealed with much higher contrast with SHG than with TPEF. We point out that although no SHG is produced in the

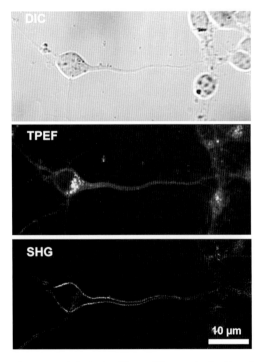

Figure 15.11. Isolated neuron labeled with a stilbene derivative, in primary cell culture from the mouse neocortex. (Top) Differential interference contrast (DIC) image. (Middle) TPEF image. (Bottom) SHG image. (Cells provided courtesy of C. Métin, ENS Paris).

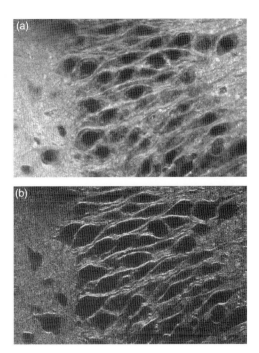

Figure 15.12. Simultaneous (top) TPEF and (bottom) SHG images of a hippocampal slice culture labeled with a stilbene derivative. (Culture provided courtesy of B. Gähwiler, University of Zürich.)

intracellular regions, a fair amount of SHG is still produced in the extracellular neuropile, which consists mainly of lipid membrane and apparently imparts some degree of spatial organization.

Attempts to label cell membranes by intracellular injection have largely been marred by the fact that amphiphilic dyes, by nature, tend to be somewhat oily and clog the injection pipette. Recently, however, it has been discovered that a derivative of the highly popular membrane dye FM1-43, known as FM4-64, can successfully label individual neurons deep inside tissue by intracellular injection (Dombeck et al., 2005; Yuste et al., 2005). This SHG dye has the added advantage that it is sensitive to electric fields, making it a highly attractive candidate for membrane potential imaging (see next section).

15.4 APPLICATIONS

The following sections present examples of applications of SHG microscopy.

15.4.1 Molecular "Flip-Flop" Dynamics in Membranes

As emphasized throughout this chapter, SHG is sensitive to molecular centrosymmetry. This is illustrated in Figure 15.13, which depicts simultaneously acquired SHG and TPEF images of two adhering GUVs. The GUVs were labeled by external perfusion of Di-6-ASPBS after they were created but presumably before they adhered. The dye,

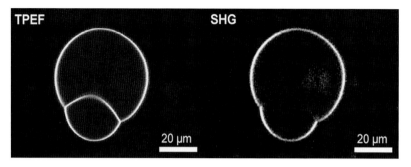

Figure 15.13. Two labeled GUVs in adhesion. Dye molecules in adhesion zone produce TPEF but no SHG because they are centrosymmetrically distributed.

which is both fluorescent in membrane and push–pull in nature, initially labels only the external leaflet of the GUVs and is therefore non-centrosymmetrically distributed and preferentially oriented. This non-centrosymmetry is canceled, however, in the GUV adhesion zone, where on average an equal number of molecules point in opposite directions. SHG is therefore forbidden, whereas TPEF, which is insensitive to orientation, remains unaltered. Though Figure 15.13 illustrates the extreme case of a complete absence of non-centrosymmetry, it is clear that SHG is highly sensitive to the degree of centrosymmetry, and in particular can monitor the dynamics of this centrosymmetry as it evolves in time.

This occurs, for example, when one images a labeled GUV over a long period of time. At an initial perfusion time ($t = 0$), the dye molecules label only the external leaflet of a GUV. With time, however, these molecules undergo random "flips" to the internal membrane because of thermal redistribution (Moreaux et al., 2001; Srivastava & Eisenthal, 1998). As pointed out above, the energy barrier opposing such a flip is high, and so the probability of flipping is small. Nevertheless, all the molecules eventually redistribute themselves evenly between both leaflets, and the "flopping" rate (passage from the internal to the external leaflet) becomes equal to the flipping rate. We define the density of molecules in the external and internal leaflets as $N_{ext}(t)$ and $N_{int}(t)$, both of which explicitly depend on time. The net SHG signal is then proportional to $(N_{ext} - N_{int})^2$, whereas the net TPEF signal is proportional to $(N_{ext} + N_{int})$. With time the SHG signal decays, whereas the TPEF signal remains roughly constant. In practice, the TPEF signal may vary somewhat as the result of photobleaching, long-term perfusion dynamics, etc., and may be used to monitor of the total number of molecules actually present in the excitation volume at any given instant. Together, the SHG and TPEF signals provide a measure of a normalized asymmetry parameter, defined by $R(t) = (N_{ext} - N_{int})/(N_{ext} + N_{int})$. As shown in Figure 15.14, this parameter exponentially decays with a time constant of about 2 hours (specific to Di-6-ASPBS molecules in dioleoylphosphatidylcholine [DOPC] membranes at room temperature). An alternative technique for observing flip-flop dynamics in membranes makes use of nuclear magnetic resonance (NMR) spectroscopy of spin-labeled markers (Devaux, 1993). The comparative ease of use of the SHG microscopy currently makes it a tool of choice for such measurements.

The flip-flop rates described above are "passive" in the sense that they are thermally driven. In certain cases, these rates can be enhanced several orders of magnitude by what

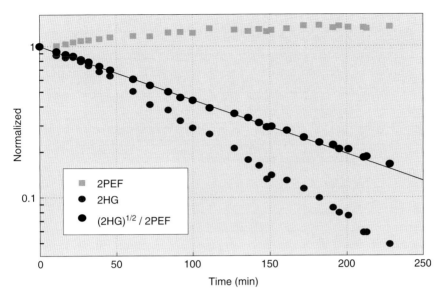

Figure 15.14. Dye molecules perfused onto a GUV initially label only the outer leaflet. Because of thermally induced flip-flop, the SHG signal decays with time, but not the TPEF signal. At large times, the dyes are equally distributed in the outer and inner leaflets.

is known as "photoinduced" flip-flop (Pons et al., 2002). This applies to amphiphilic molecules that are photoisomerizable, as depicted in Figure 15.15. For example, Di-6-ASPBS is about one thousand times more likely to undergo flip-flop when in a bent, or *cis*, configuration, than when in a straight, or *trans*, configuration. As a result, its 2-hour passive flip-flop time constant can be reduced to only a few seconds by a simple increase in laser illumination power.

15.4.2 Intermembrane Separation Measurements

In section 15.2.3 we considered geometries where the active SHG volume encompassed a single membrane only. More complicated geometries can certainly be envisaged. For example, we consider what happens when two membranes parallel to one another are in such close proximity that they fall within the same SHG volume. This situation can be realized experimentally when two labeled GUVs almost touch one another (Moreaux et al., 2001). We assume here that the SHG markers label only the external membrane leaflets, meaning they are oriented in opposing directions. If the membranes touch, then no SHG is possible—this is the same adhesion zone geometry as was encountered in Figure 15.13. On the other hand, if the membranes do not touch, the detailed formation of SHG must be examined more carefully. Figure 15.15 depicts the SHG lobes produced by each membrane individually (only the upward-directed lobes are shown for simplicity). Both lobes propagate with the same off-axis angle and hence, when observed from a distance, essentially overlap. Because the SHG lobes are driven by a common excitation beam, they are coherent, and one must take into account their respective phases. In particular, we already know that there is a 180-degree phase shift between the lobes that arises from the opposition in the molecular orientations. But

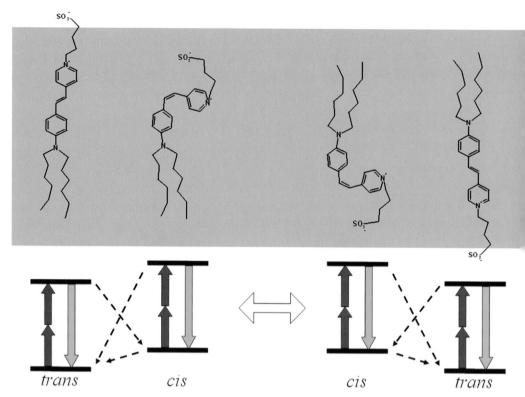

Figure 15.15. Photoinduced flip-flop of photoisomerizable amphiphilic molecules in lipid bilayer membrane.

there is also an extra phase shift that arises from the slight optical pathlength difference between each membrane and the observer. This pathlength difference, depicted in Figure 15.16>, is given by $\Delta l = d \sin \theta_0$, where θ_0 is the off-axis propagation angle (see above), and d is the intermembrane separation. At an optimal d, the extra phase shift occasioned by the membrane separation exactly counteracts the baseline 180-degree phase shift and provokes a completely *constructive* interference between the SHG lobes. This optimal distance is given by $d_{opt} \approx 0.6\, w_r$, where w_r is the excitation beam waist at the focal center. The net SHG intensity produced by the two membranes therefore varies as a function of intermembrane separation, as shown in Figure 15.17. When the separation is zero, there is no extra phase shift and the SHG interferes destructively. When the separation is d_{opt}, the SHG interferes constructively and is maximum. When the separation is much larger than w_r the two membranes cannot be simultaneously illuminated and they become effectively independent of one another. We observe that the SGH signal increases approximately linearly as the separation increases from 0 to d_{opt}. As defined here, d_{opt} roughly corresponds to the microscope diffraction-limited resolution (typically a few hundred nanometers), meaning that SHG interference provides a measure of membrane separation with a resolution window that is inaccessible to standard microscopy techniques. The three possibilities depicted graphically in Figure 15.17 are all occasioned in Figure 15.18, which is an SHG image of two GUVs in close proximity. Presumably, the GUVs are not in complete contact here because the SHG nowhere vanishes. This may be the result of steric hydration forces that maintain

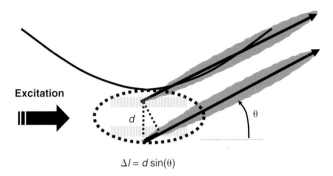

Figure 15.16. SHG radiated from two parallel, labeled membranes in close proximity (only upward-directed lobes shown). The optical path length difference Δl is proportional to the intermembrane separation d.

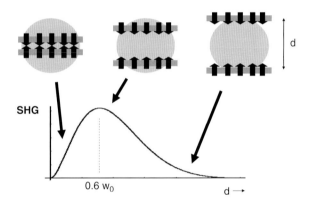

Figure 15.17. The SHG produced by two membranes interferes destructively or constructively, depending on the membrane separation d. The net SHG power is maximum when $d \approx 0.6w_0$, where w_0 is the excitation beam waist at the focus.

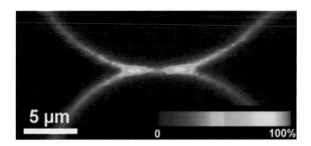

Figure 15.18. SHG "hot spots," corresponding to the constructive interference of the SHG from two labeled membranes in close proximity.

the GUVs at a minimum separation (Israelachvili, 1991). At larger separation distances, in particular those close to d_{opt}, the SHG constructively interferes and exhibits "hot spots" in the membrane signals. At still larger separations, the GUV membranes are imaged independently.

The occurrence of hot spots stems directly from the fact that the SHG propagates off-axis, as is clear from the definition of the intermembrane optical pathlength difference Δl. If the SHG were generated on-axis, or equivalently if the excitation beam were unfocused, these hot spots would not exist.

15.4.3 Membrane Potential Imaging

The resting potential of living cells relative to their environment is typically about -60 mV. Recalling that cell membranes are about 4 nm thick, this corresponds to a transmembrane field on the order of 10^5 V/cm. The optical imaging of such fields with microscopic resolution has been the goal of considerable research.

Currently, several techniques for membrane potential imaging are available. These are traditionally cataloged as being fast or slow, depending on their response times (Cohen & Salzberg, 1978). Slow dyes, with response times longer than a millisecond, typically operate by a mechanism of voltage-induced dye redistribution. This redistribution can modify global dye populations, as in the case of Nernstian markers, or local dye arrangements within membranes by way of electric field-induced molecular reorientation or displacement. The literature describing such markers is exhaustive, and the reader is referred to review papers for more information (Chen, 1988; Loew et al., 1990).

Though slow markers can be highly sensitive, they are generally inadequate for imaging the fast dynamics of excitable cells, which occur on time scales of a millisecond or less. At present, the most robust technique for fast imaging is based on a phenomenon called molecular electrochromism. The principle behind this technique is that an externally applied electric field provokes shifts in the electronic energy levels of a molecule, known as Stark shifts. These shifts, in turn, lead to changes in the molecular absorption or fluorescence emission wavelengths, which can be monitored by techniques of ratiometric excitation or fluorescence detection. The use of molecular electrochromism to monitor membrane potential has been established for several decades, and literally hundreds of candidate molecules have been screened for their voltage sensitivities both in model membranes and in live cells (Fluhler et al., 1985; Grinvald et al., 1988; Loew, 1988). Recent advances in the synthesis of Stark shift dyes have led to membrane potential sensitivities greater than 50% relative (one-photon excited) fluorescence power changes for 100-mV transmembrane voltages (Kuhn et al., 2004).

The use of SHG, as opposed to fluorescence, to monitor electric fields is well known (Ben-Oren et al., 1996; Bouevitch et al., 1993). For example, it has been demonstrated that SHG can report membrane potential by a mechanism of internal molecular charge redistribution. Intuitively, one can think of an applied field as provoking a change in the molecular charge non-centrosymmetry. The resulting change in SHG efficiency can be regarded as stemming from an electric field-induced modification in the molecular hyperpolarizability (Boyd, 1992). Alternatively, it has been recently recognized that, to first order, this change in hyperpolarizability is a simple manifestation of the Stark effect (Pons & Mertz, 2006)—that is, molecules that exhibit strong electrochromism are also expected to exhibit strong SHG electric field response. In both cases, the field response critically depends on the molecule's being charge transfer in nature, and high field sensitivity in general requires large $\Delta\mu$'s. In both cases, also, the maximum electric field sensitivity (or, more properly, the minimum resolvable field) occurs at off-resonant excitation wavelengths (Kuhn et al., 2004; Moreaux et al., 2002; Pons & Mertz, 2006). However, electrochromic and SHG responses differ when performing actual measurements: electric field variations are identified with electrochromism by shifts in the fluorescence excitation or emission wavelengths, whereas with SHG they are identified by direct changes in signal power (the SHG wavelength cannot change since it is fixed by that of the excitation laser). Despite this difference, electrochromic and SHG response sensitivities are expected to be about the same because they stem from the same physical phenomenon (again, to first order). Both electro-optic and electrochromic responses in molecules are fast since they rely only on internal charge redistributions. Typical response times are likely to be on the order of picoseconds or less, which are essentially instantaneous as far as any biological dynamics are concerned.

An electric field can alter molecular centrosymmetry by other mechanisms than the Stark effect. For example, if dipolar molecules are initially randomly aligned (centrosymmetric), then an applied DC field provokes a dipolar force that can lead to global molecular alignment, in turn provoking SHG. This technique is known as EFISH (electric field-induced second harmonic). Because EFISH establishes alignment, or "poling," of molecules from a nonaligned configuration, it tends to be slow. It has been shown, however, that electric fields can provoke small modulations in molecular alignment with response times easily in the milliseconds (Gross et al., 1986; Pons & Moreaux et al., 2003).

A compilation of different membrane potential-sensing techniques is shown in Figure 15.19. This compilation is by no means exhaustive and provides only a general survey of the variety of techniques available. These include techniques based on

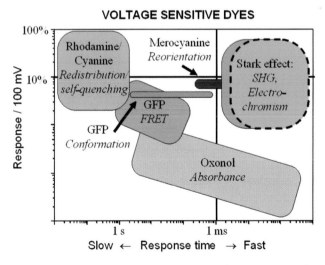

Figure 15.19. Optical techniques for monitoring membrane potential are traditionally classified as fast or slow. The most common technique for fast monitoring is based on fluorescence electrochromism. SHG is expected to exhibit similar response as electrochromism (dashed box; not yet demonstrated for best electrochromic dyes).

exogenous (Gross et al., 1986; Wagonner, 1979) or endogenous (Gonzalez & Tsien, 1995; Siegel & Isacoff, 1997) fluorescence, absorbance (Ross et al., 1977), and, of interest here, SHG. We note again that both fluorescence electrochromism and SHG are in fact much faster than suggested by this diagram, which is meant to cover only biologically relevant response times.

The combination of high speed and reasonably high sensitivity makes both fluorescence electrochromism and SHG attractive techniques for membrane potential imaging. However, it should be emphasized that applications to live-cell imaging have often been frustrated by ongoing problems of phototoxicity. Though the labeling of cells with charge-transfer chromophores is relatively straightforward, the actual illumination of these chromophores with laser excitation is usually deleterious and seems to provoke an increase in membrane permeability. The cause of this phototoxicity is not well understood and has been the subject of much speculation.

While fluorescence electrochromism and SHG are expected to yield similar sensitivities, SHG nevertheless presents distinct advantages that tend to favor its use, particularly when imaging in thick tissue. In particular, SHG exhibits significantly higher spatial contrast than fluorescence when imaging membranes (e.g., see Fig. 15.12). Moreover, it exhibits higher spectral contrast, in the sense that the SHG emission wavelength is well defined to be exactly half that of the laser excitation wavelength. Fluorescence emission wavelengths, on the other hand, are molecule dependent and tend to be spread over several tens of nanometers. As a result, wide spectral filters are required to collect fluorescence signals, whereas only narrow filters are required for SHG, significantly restricting the collection of any autofluorescence background. Finally, SHG is essentially an instantaneous phenomenon compared to fluorescence in the sense that molecules do not reside in the excited state, as they do in fluorescence (fluorescence lifetimes are typically on the order of nanoseconds). SHG is therefore immune to excited-state decay mechanisms that are nonradiative, such as quenching or photobleaching, and to environmentally induced variations in fluorescence quantum yield that can provoke nonlinear or generally unreliable electric field responses. The advantages of SHG over fluorescence for membrane potential imaging are further detailed in (Pons & Mertz, 2005).

In its current state of the art, SHG membrane potential sensitivity has been used to reveal action potentials in Aplysia neurons (Dombeck et al., 2004) (Fig. 15.20) and even in mammalian neurons (Dombeck et al., 2005; Yuste et al., 2005), though only after substantial image averaging. The latter demonstrations made use of the intracellular marker FM4-64, which exhibits a sensitivity of about 7.5% per 100 mV. The recent development of highly efficient Stark effect dyes (Kuhn et al., 2004) is likely to considerably boost this sensitivity in the near future.

15.4.4 Endogenous Second-Harmonic Generation

Most of this chapter has been devoted to the study of SHG from exogenous chromophores. Historically, however, biological SHG microscopy was first used to image endogenous tissue, and there has been a revival in interest in this application owing to the wide variety of structures that can naturally generate SHG (Campagnola et al., 2002; Chu et al., 2003). The physical principle behind SHG, of course, remains the same, meaning that local non-centrosymmetry is required along the direction of

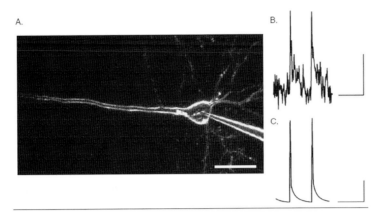

Figure 15.20. High-resolution SHG recording of action potentials deep in hippocampal brain slices. (A) SHG image of a neuron patch clamped and filled with FM4-64 in slice. (B) SHG line-scan recording of elicited action potentials (55 line scans were averaged). (C) Electrical patch pipette recording of action potentials seen in B. Calibrations: (A) 20 μm; (B) 2.5% ΔSHG/SHG, 50 ms; (C) 20 mV, 50 ms. (Courtesy of the W.W. Webb lab, Cornell University).

excitation polarization. While this chapter has concentrated on non-centrosymmetries involving uniaxial polar molecules, other non-centrosymmetries may readily be found in nature that also produce SHG. These include helical geometries exhibiting "chirality," as are found in sugars or many amino acids, or on a larger scale, filaments or fibrous bundles. For example, it is well known that collagen, an extracellular protein that is the main fibrous component of skin, bone, tendon, cartilage and teeth, can readily produce SHG microscopy (Fig. 15.21) (Roth & Freund, 1979). This has led to the use of SHG microscopy to monitor cell motility in diseased or abnormal tissue structure (Brown et al., 2003; Wang et al., 2002). Though the physical origin of collagen's remarkably high SHG efficiency has not been fully established, its structure consisting of three intertwined polypeptide helices very likely plays a significant role. From a polarization analysis, the main component of collagen hyperpolarizability is known to be roughly parallel to collagen axis (Freund et al., 1986; Stoller et al., 2002), and the SHG is seemingly generated from the fibril outer shell (Williams et al., 2005). Other endogenous structures known to produce SHG are acto-myosin complexes, mitotic spindles, microtubules (Campagnola et al., 2002), and cellulose (Malcolm Brown et al., 2003).

Recently, endogenous SHG microscopy has been applied to native brain tissue imaging, where it can reveal the presence of microtubule assemblies, such as those found in high density in the CA1 region of the hippocampus (Dombeck et al., 2003). The requirement of non-centrosymmetry has led here to a remarkable application of SHG microscopy involving the observation of neuron maturation. At an early stage of development, hippocampal neurons produce protoprocesses, most of which are destined to become dendrites and one of which is destined to become the axon. The microtubules in these immature protoprocesses exhibit uniform polarity, presumably defined by the direction of the process growth. After maturation, however, only the axon maintains a uniform microtubule polarity, whereas in the dendrites the polarity becomes mixed.

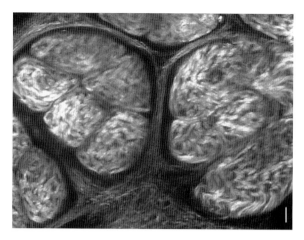

Figure 15.21. SHG produced by endogenous collagen in a rat tail. The shrouding matrices are dominantly revealed in reflection mode, whereas the enclosed fibrous collagen is dominantly revealed in transmission mode. Scale bar = 20 μm. (Image provided courtesy of the W.W. Webb lab, Cornell University).

This transition from uniform to mixed polarity can readily be observed with SHG microscopy (Fig. 15.22).

SHG, whether exogenous or endogenous, is coherent, meaning that it is subject to the rules of phase-matching (see Section 15.2.3). Initial demonstrations of endogenous SHG imaging all involved signal collection in the transmitted light direction (Kim et al., 1999; Roth & Freund, 1979), where most of the SHG power is expected to be confined. Such collection geometries preclude the possibility of SHG imaging in thick tissue. However, more recently endogenous SHG microscopy using backward signal detection has become common, despite the reduced signal power (Campagnola & Loew, 2003). The fact that backward-directed SHG occurs at all is, in itself, intriguing. At least two phenomena can account for this. First, the SHG can be initially forward-directed, and subsequently be rerouted into the backward direction either by reflection from internal sample structures or by multiple scattering if the sample is diffusive. Though in principle such SHG backscattering would allow imaging in thick tissues, certain constraints must be taken into account. In particular, to ensure SHG coherence, the average distance that a SHG photon travels in a sample before it is scattered (or "scattering length") must be large compared to the SHG active volume dimensions. Moreover, the average total pathlength that a SHG photon travels before it escapes the sample by multiple backscattering must be short compared to the average pathlength it travels before it is absorbed. This latter condition can severely restrict SHG imaging depth because biological samples tend to be highly absorptive for short wavelengths.

Alternatively, SHG can be intrinsically backward-directed. This occurs if the radiation sources producing SHG, endogenous or exogenous, contain specific local inhomogeneities in their axial concentration distribution. As described in Section 15.2.3, these inhomogeneities can arise, for example, from molecular clusters or interface geometries, and can provide the necessary pseudo-momentum "kick" to allow SHG propagation in the backward direction (Mertz & Moreaux, 2001). A comparison between forward and backward signal powers can provide structural information on

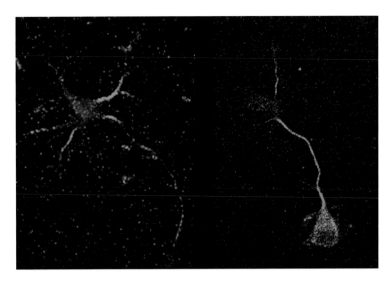

Figure 15.22. Rat hippocampal neurons are imaged by SHG (green pseudocolor) after 5 days (left) and 7 days (right) in culture. At the later development stage, protoprocesses have presumably matured into dendrites whose microtubules are of mixed polarity and cannot produce SHG. Only the axon microtubules, which maintain a uniform polarity, are revealed by SHG (red pseudocolor represents autofluorescence). (Images provided courtesy of W.W. Webb lab, Cornell university)

sample geometries that is unavailable with fluorescence techniques, and in particular can provide an indication of molecular aggregation (Williams et al., 2005).

15.5 CONCLUSION

The widespread establishment of nonlinear microscopy began with the development of TPEF microscopy. This not only prompted a re-examination of initial results in SHG imaging, but also launched several alternative forms of nonlinear microscopy based on higher-order contrast mechanisms. These include, for example, third-harmonic generation (THG) microscopy (Barad et al., 1997; Muller et al., 1998) and coherent anti-stokes Raman scattering (CARS) microscopy (Zumbusch et al., 1999), both of which involve various combinations of four-wave mixing and have been successfully applied to endogenous biological imaging. In the years to come, certainly more varieties of nonlinear microscopies will be developed, each providing different specificities and advantages.

As for SHG microscopy, its panoply of applications, particularly in membrane potential and structural protein imaging, as well as its ease of implementation in combination with TPEF microscopy, assure it an important role in the future of biological imaging.

ACKNOWLEDGMENTS

Much of this chapter was written when the author was at the Ecole Supèrieure de Physique et Chimie Industrielles in Paris, France. I particularly thank my colleagues L. Moreaux, T. Pons, M. Blanchard-Desce, and S. Charpak for their considerable help.

REFERENCES

Angelova, M. I., S. Soleau, et al. (1992). *Prog. Colloid. Polym. Sci.* 89: 127–131.

Barad, Y., H. Eisenberg, et al. (1997). Nonlinear scanning laser microscopy by third harmonic generation. *Appl. Phys. Lett.* 70: 922–924.

Ben-Oren, I., G. Peleg, et al. (1996). Infrared nonlinear optical measurements of membrane potential in photoreceptor cells. *Biophys. J.* 71: 1616–1620.

Blanchard-Desce, M., V. Alain, et al. (1997). Intramolecular charge transfer and enhanced quadratic optical nonlinearities. *J. Photchem. Photobiol. A* 105: 115–121.

Blanchard-Desce, M. and M. Barzoukas (1998). Two-form two-state analysis of polarizabilities of push-pull molecules. *J. Opt. Soc. Am. B* 154: 302–307.

Blanchard-Desce, M., R. Wortmann, et al. (1995). Intramolecular charge transfer in elongated donor-acceptor conjugated polyenes. *Chem. Phys. Lett.* 243: 526–532.

Bloembergen, N. (1965). *Nonlinear Optics*, World Scientific.

Born, M. and E. Wolf (1993). *Principles of Optics*. Oxford, Pergamon Press.

Bouevitch, O., A. Lewis, et al. (1993). Probing membrane potential with nonlinear optics. *Biophys. J.* 65: 672–679.

Boyd, R., W. (1992). *Nonlinear Optics*. London, Academic Press.

Brown, E., T. McKee, et al. (2003). Dynamic imaging of collagen and its modulation in tumors in vivo using second-harmonic generation. *Nat. Med.* 9: 796.

Butcher, P. N. and D. Cotter (1990). *The elements of nonlinear optics*. Cambridge, Cambridge University Press.

Campagnola, P. J. and et al. (2002). Three-dimensional high-resolution second harmonic imaging of endogenous structural proteins in biological tissues. *Biophys. J.* 82: 493–508.

Campagnola, P. J. and L. M. Loew (2003). Second-harmonic imaging microscopy for visualizing biomolecular arrays in cells, tissues and organisms. *Nat. Biotech.* 21: 1356–1360.

Campagnola, P. J., M. Wei, et al. (1999). High-Resolution Nonlinear Optical Imaging of Live Cells by Second Harmonic Generation. *Biophys. J.* 77: 3341–3349.

Chemla, D. S. and J. Zyss (1984). *Nonlinear Optical Properties of Organic Molecules and Crystals*. New York, Academic Press.

Chen, L. B. (1988). *Ann. Rev. Cell Biol.* 4: 155–181.

Chu, S.-W., S.-Y. Chen, et al. (2003). In-vivo developmental biology study using noninvasive multi-harmonic generation microscopy. *Opt. Express* 11: 3093–3098.

Clays, K., A. Persoons, et al. (1994). Hyper-Rayleigh scattering in solution. *Advances in Chemical Physics*. M. Evans and S. Kiclich. New York, John Wiley & Sons. 85.

Cohen, L. B. and B. M. Salzberg (1978). *Rev. Physiol. Biochem. Pharmacol.* 83: 35–88.

Denk, W., J. H. Strickler, et al. (1990). Two-photon laser scanning fluorescence microscopy. *Science* 248: 73–76.

Devaux, P. F. (1993). Lipid transmembrane asymmetry and flip-flop in biological membranes and in lipid bilayers. *Curr. Opin. Struct. Biol.* 3: 489–494.

Dombeck, D. A., M. Blanchard-Desce, et al. (2004). Optical recording of action potentials with second-harmonic generation microscopy. *J. Neurosci.* 24: 999–1003.

Dombeck, D. A., K. A. Kasischke, et al. (2003). Uniform polarity microtubule assemblies imaged in native brain tissue by second harmonic generation microscopy. *Proc. Nat. Acad. Sci. USA* 100: 7081–7086.

Dombeck, D. A., L. Sacconi, et al. (2005). Optical recording of fast neuronal membrane potential transients in acute mammalian brain slices by second-harmonic generation microscopy. *J. Neurophys.* doi:10.1152/jn.00416.2005.

Fluhler, E., V. G. Burnham, et al. (1985). Spectra, membrane binding, and potentiometric responses of new charge shift probes. *Biochemistry* 24: 5749–5755.

Freund, I., M. Deutsch, et al. (1986). Connective tissue polarity. Optical second-harmonic microscopy, crossed-beam summation and small angle scattering in rat-tail tendon. *Biophys. J.* 50: 693–712.

Gonzalez, J. E. and R. Y. Tsien (1995). Voltage sensing by fluorescence resonance energy transfer in single cells. *Biophys. J.* 69: 1272–1280.

Grinvald, A., R. Frostig, et al. (1988). *Physiol. Rev.* 68: 1285–1366.

Gross, D., L. M. Loew, et al. (1986). Optical imaging of cell membrane potential changes induced by applied electric fields. *Biophys. J.* 50: 339–348.

Guo, Y., P. P. Ho, et al. (1997). Second-harmonic tomography of tissues. *Opt. Lett.* 22: 1323.

Hellwarth, R. and P. Christensen (1974). Nonlinear optical microscopic examination of structure in polycrystaliline ZnSe. *Opt. Commun.* 12: 318–322.

Israelachvili, J. (1991). *Intermolecular and surface forces.* San Diego, Academic Press.

Kim, B.-M., J. Eichler, et al. (1999). Frequency doubling of ultrashort laser pulses in biological tissues. *Appl. Opt.* 38: 7145–7150.

Krauss, J. D. (1950). *Antennas.* New York, McGraw Hill.

Kuhn, B., P. Fromherz, et al. (2004). High sensitivity of Stark-shift voltage-sensing dyes by one- or two-photon excitation near the red spectral range. *Biophys. J* 87: 631–639.

Liptay, W. (1974). *Excited states.* New York, Academic Press.

Loew, L. M. (1988). *Spectroscopic membrane probes.* Boca Raton, CRC Press.

Loew, L. M., D. L. Farkas, et al. (1990). *Optical Microscopy for Biology.* B. Herman and K. Kacobson. New York, Wiley-Liss: 131–142.

Loew, L. M. and L. L. Simpson (1981). Charge shift probes of membrane potential. A probable electrochromic mechanism for ASP probes on a hemispherical lipid bilayer. *Biophys. J.* 34(34): 353.

Malcolm Brown Jr., R., A. C. Millard, et al. (2003). Macromolecular structure of cellulose studie by second-harmonic generation imaging microscopy. *Opt. Lett* 28: 2207–2209.

Mertz, J. and L. Moreaux (2001). Second harmonic generation by focused excitation of inhomogeneously distributed scatterers. *Opt. Commun.* 196: 325–330.

Moreaux, L., T. Pons, et al. (2003). Electro-optic response of second harmonic generation membrane potential sensors. *Opt. Lett.* 28: 625–627.

Moreaux, L., O. Sandre, et al. (2000a). Membrane imaging by simultaneous second-harmonic generation and two photo-photon microscopy. *Opt. Lett.* 25(5): 3220–3222.

Moreaux, L., O. Sandre, et al. (2001). Coherent Scattering in Multi-Harmonic Light Microscopy. *Biophys. J.* 80(3): 1568–1574.

Moreaux, L., O. Sandre, et al. (2000b). Membrane imaging by second-harmonic generation microscopy. *J. Opt. Soc. Am. B* 17(10): 1685–1694.

Muller, M., J. Squier, et al. (1998). 3D-microscopy of transparent objects using third-harmonic generation. *J. Microscopy* 191: 266.

Oudar, J. L. (1977). Optical nonlinearities of conjugated molecules. Stilbene derivatives and highly polar aromatic compounds. *J. Chem. Phys.* 67(2): 446–457.

Pons, T. and J. Mertz (2006). Membrane potential detection with second-harmonic generation and two-photon excited fluorescence: A theoretical comparison. *Opt. Commun.* 258: 203–209.

Pons, T., L. Moreaux, et al. (2002). Photoinduced flip-flop of amphiphilic molecules in lipid bilayer membranes. *Phys. Rev. Lett.* 89: 288104.

Pons, T., L. Moreaux, et al. (2003). Mechanisms of membrane potential sensing with second-harmonic generation microscopy. *J. Biomed. Opt.* 8(3): 428–431.

Prasad, P. N. (1991). *Introduction to nonlinear optical effects in molecules and polymers.* New York, Wiley.

Radda, G. K. (1971). *Biochem. J.* 122: 385–396.

Redenti, E., L. Szente, et al. (2000). Drug/Cyclodextrin/Hydroxy acid multicomponent systems. Properties of pharmaceutical applications. *J. Pharmac. Sci.* 89: 1–8.

Ross, W. N., B. M. Salzberg, et al. (1977). *J. Membr. Biol.* 33.

Roth, S. and I. Freund (1979). Second harmonic generation in collagen. *J. Chem. Phys.* 70: 1637–1643.

Sheppard, C. J. R., R. Kompfner, et al. (1977). Scanning harmonic optical microscope. *IEEE J. Quant. Electron.* 13E: 100D.

Siegel, M. S. and E. Y. Isacoff (1997). A genetically encoded optical probe of membrane voltage. *Neuron* 19: 735–741.

Srivastava, A. and K. B. Eisenthal (1998). Kinetics of molecular transport across a liposome bilayer. *Chem. Phys. Lett.* 292: 345–351.

Stoller, P., B.-M. Kim, et al. (2002). Polarization dependent optical second-harmonic imaging of a rat-tail tendon. *J. Biomed. Opt.* 7: 205–214.

Wagonner, A. S. (1979). *Ann. Rev. Biophys. Bioeng.* 8: 847–868.

Wang, W., J. B. Wyckoff, et al. (2002). Single cell behavior in metastatic primary mammary tumors correlated with gene expression patterns revealed by molecular profiling. *Cancer Res.* 62: 6278–6288.

Williams, R. M., W. R. Zipfel, et al. (2005). Interpreting second-harmonic generation images of collagen I fibrils. *Biophys. J.* 88: 1377–1386.

Yuste, R., B. Nemet, et al. (2005). *Second harmonic generation imaging of membrane potential.* Imaging neurons and neural activity: New methods, new results, Cold Spring Harbor, NY.

Zumbusch, A., G. R. Holtom, et al. (1999). Vibrational microscopy using coherent anti-Stokes Raman scattering. *Phys. Rev. Lett.* 82: 4142–4146.

Zyss, J. (1979). Hyperpolarizabilities of substituted conjugated molecules. III. Study of a family of donor-acceptor disubstituted phenyl-polyenes. *J. Chem. Phys.* 71: 909–916.

16

Second-Harmonic Generation Imaging Microscopy of Structural Protein Arrays in Tissue

Paul J. Campagnola

16.1 INTRODUCTION AND HISTORY OF SECOND-HARMONIC GENERATION IMAGING

In the past several years, second-harmonic generation (SHG) has emerged as a powerful nonlinear optical contrast mechanism for biological imaging applications. SHG is a coherent process where two lower-energy photons are up-converted to exactly twice the incident frequency (or half the wavelength). This effect was first demonstrated by Kleinman (Kleinman, 1962) in crystalline quartz in 1962, where this advance was made possible with the invention of the ruby laser. Since that discovery, SHG in uni-axial birefringent crystals has been exploited to frequency double pulsed lasers to obtain shorter wavelengths, thereby producing multiple colors from a single source. SHG from interfaces was later discovered by Bloembergen in 1968 and rapidly became a versatile spectroscopic tool to study chemical and physical processes at air–solid, air–liquid, and liquid–liquid interfaces (for reviews, see Eisenthal, 1996; Shen, 1989). The first integration of SHG and optical microscopy was achieved in 1974 by Hellwarth and coworkers, who used SHG as an imaging tool to visualize the microscopic crystal structure in polycrystalline ZnSe (Hellwarth & Christensen, 1974). Sheppard then implemented the method on a scanning microscope in 1977 (Sheppard et al., 1977).

To the best of our knowledge, the first spectroscopic demonstration of SHG from a purely biological specimen was by Fine and Hansen in 1973 (Fine & Hansen, 1971); they showed its existence in type I collagen. Biological SHG imaging was first used in1986, when Freund and coworkers (Freund et al., 1986) used this contrast mechanism to investigate the polarity of collagen fibers in rat-tail tendon. While this work was performed at low resolution (ca. 50 microns), it showed initial proof of concept that, in part, laid the foundation for modern versions of SHG imaging. For example, in 1993 Loew and Lewis (Bouevitch et al., 1993) demonstrated that second-harmonic signals could be produced from model and cell membranes labeled with voltage-sensitive styryl dyes. Subsequently, Campagnola and Loew (Campagnola et al., 1999) implemented high-resolution SHG imaging on a laser scanning microscope, which afforded data acquisition rates comparable to those of confocal fluorescence imaging.

Since this development, two general of forms of SHG imaging have emerged in parallel. The first has further exploited the interfacial aspect of SHG to study membrane biophysics through the use of voltage-sensitive dyes (Millard et al., 2003, 2004; More-aux et al., 2000a, 2000b, 2001). The second approach has used this contrast mechanism to investigate structural protein arrays (e.g., collagen) in tissues at higher resolution and with more detailed analysis than previously possible. In this chapter, we focus on the latter implementation. The reader is directed to Chapter 15 in this volume by Mertz for details on membrane imaging by SHG microscopy.

SHG has several advantageous features that make it an ideal approach for imaging tissues in situ. A particularly strong attribute of the SHG method lies in the property that in tissues the contrast is produced purely from endogenous species. This, coupled with the physical basis that the SHG signals arise from an induced polarization rather from absorption, leads to substantially reduced photobleaching and phototoxicity relative to fluorescence methods (including multiphoton). Additionally, because the excitation typically uses the same near-infrared wavelengths (800–1,000 nm) produced by titanium-sapphire lasers that are also used for two-photon excited fluorescence (TPEF), this method is well suited for studying intact tissue samples, since excellent depths of penetration can be obtained. For example, we have acquired optical sections into 550 microns of mouse muscle tissue with high contrast throughout the entire axial profile (Campagnola et al., 2002). Furthermore, detailed information about the organization of protein matrices at the molecular level can be extracted from SHG imaging data. This is because the SHG signals have well-defined polarizations with respect to the laser polarization and specimen orientation that can be used to determine the absolute orientation of the protein molecules in the array, as well as the degree of organization of proteins in tissues.

These attributes have recently been exploited to examine a wide range of structural proteins arrays in situ (Boulesteix et al., 2004; E. Brown et al., 2003; R.M.J. Brown et al., 2003; Campagnola et al., 2002; Chu et al., 2004; Cox et al., 2003; Dombeck et al., 2003; Mohler et al., 2003; Sun et al., 2004; Williams et al., 2005; Yeh et al., 2002; Zoumi et al., 2002, 2004). The plurality of the recent reports has focused on visualizing collagen fibers in natural tissues, including skin, tendon, blood vessels, and cornea (Han et al., 2004; Stoller et al., 2001, 2002; Yeh et al., 2002, 2003; Zoumi et al., 2004). Other work has examined the fibrillar structure of self-assembled or reconstituted collagen structures (Ramanujan et al., 2002; Zoumi et al., 2002). A smaller body of publications have described SHG imaging of acto-myosin complexes in muscle (Campagnola et al., 2002) as well as microtubule-based structures in live cells (Campagnola et al., 2002; Dombeck et al., 2003). Some of these efforts have focused on elucidating the physical principles that give rise to SHG in tissues as well as taking steps to use this contrast mechanism in the application of diagnosing disease states or other pathological conditions.

In this chapter we first describe both the general physical principles of SHG as well as the specific issues with respect to tissue imaging. We next provide details regarding the experimental considerations of SHG imaging, and describe the steps required to create an optimized SHG microscope for tissue applications. We then present an overview of recent results from our lab and those of other investigators that serves as a biophysical framework by which to understand the underlying contrast mechanism, as well as provide a context for the types of questions that can be addressed through this methodology. We further provide an explicit comparison of the strengths and weaknesses of the method with those of linear (fluorescence and polarization microscopy) and other nonlinear optical imaging modalities (TPEF, third-harmonic generation [THG], and coherent anti-Stokes Raman scattering [CARS]), and then provide some outlook for the future.

16.2 THEORETICAL AND PHYSICAL CONSIDERATIONS

16.2.1 Physical Description of Nonlinear Polarization

In general, the total polarization for a material can be expressed as:

$$P = \chi^{(1)}E^1 + \chi^{(2)}E^2 + \chi^{(3)}E^3 + \cdots \tag{16.1}$$

where P is the induced polarization, $\chi^{(n)}$ is the nth-order nonlinear susceptibility tensor, and E is the electric field vector.

The first term describes linear absorption, scattering, and reflection of light; the second term, SHG, sum and difference frequency generation; and the third term, two- and three-photon absorption, third-harmonic generation, and stimulated Raman processes, such as CARS. Typically, each term in this expansion is many orders of magnitude smaller than that which preceded it, and in general, most materials possess vanishingly small nonlinear coefficients. It is thus challenging to find biological materials that can imaged by nonlinear schemes.

Since SHG is a second-order nonlinear optical process, it can arise only from media lacking a center of symmetry. This is because the SHG wave is a vector quantity and the induced polarization (second term, equation (16.1)) in a centrosymmetric sample (such as dye molecules in solution) would be equal and opposite and vector sum to zero. It is well known that this criterion can be satisfied at interfaces and by anisotropic crystals (Shen, 1984). This constraint can also be met by structural protein arrays because the dipole moments of the molecules align in a highly ordered fashion and additionally because the chirality in the helices increases the overall asymmetry of the assembly.

The second-harmonic signal intensities per laser pulse from structural proteins will scale as follows (Shen, 1989):

$$SHG_{sig} \propto \left[\frac{p}{a^z}\right]^2 \tau(\chi^2)^2 a \tag{16.2}$$

where the term in brackets represents the peak power, p is the laser pulse energy, a is the area of the focused spot, and τ is the laser pulse width; χ^2 is the second-order nonlinear susceptibility of the protein and off resonance is essentially wavelength-independent.

The net inverse dependence on the pulse width arises because while the signal scales as the square of the peak power, due to the coherence of the process, second-harmonic signal will be produced only within the duration of the laser pulse. Although differing in physical origins, it should be noted that SHG and two-photon excitation probabilities have this same inverse pulse width dependence. Similarly, as the signal is proportional to the square of the flux (photons/area), second harmonic will be generated only within the focal area. The squared dependence of χ^2 on the signal magnitude arises because SHG is a cooperative process among aligned molecules and is manifested in a squared dependence of the molecular concentration. This is a fundamental differentiating aspect with respect to fluorescence, where in the limit of no quenching, the overall intensity is the simple linear sum of the fluorescence from individual molecules.

The nonlinear susceptibility tensor, χ^2, is a bulk property and is the quantity measured in an experiment. However, the molecular-level property of the nonlinearity (i.e., the first hyperpolarizability, β) needs to be considered as it forms the basis of the contrast. The hyperpolarizability is defined in terms of the permanent dipole moment:

$$d^{(2)} = \beta EE \tag{16.3}$$

The molecular and bulk properties are then related by:

$$\chi^{(2)} = N_s \langle \beta \rangle \tag{16.4}$$

where N_s is the density of protein molecules and the brackets denote an orientational average of those molecules, which further underscores the need for an environment lacking a center of symmetry.

Thus we see that to produce second-harmonic signals the molecule must have a permanent dipole moment to satisfy the molecular-level requirements (equation (16.3)). Additionally, these dipole moments must then be aligned in a nonrandom fashion to satisfy the bulk property constraints, such that χ^2 is nonvanishing (equation (16.4)).

We can consider the general molecular properties that give rise to SHG within the two-level system model. The first hyperpolarizability, β, and thus the SHG probability at laser frequency ω is given by (Heinz et al., 1982):

$$\beta \approx \frac{3e^2}{2\hbar^3} \frac{\omega_{ge} f_{ge} \, \Delta\mu_{ge}}{[\omega_{ge}^2 - \omega^2][\omega_{ge}^2 - 4\omega^2]} \tag{16.5}$$

where e is electron charge and ω_{ge}, f_{ge}, and $\Delta\mu_{ge}$, are the energy difference, oscillator strength (i.e., integral of the absorption spectrum), and change in dipole moment between the ground (g) and excited (e) states, respectively. This change in polarity requires molecules that are highly polarizable (i.e., have strongly deformable electron densities). Since the polarizability generally increases with size, large structural proteins such as collagen are expected to satisfy this requirement.

A second consequence of equation (16.5) is that the magnitude of the SHG intensity can be strongly enhanced when the energy of the SHG signal ($2\hbar\omega$) is in resonance with an electronic absorption band ($\hbar\omega_{ge}$). Because of the denominator in this equation, resonance-enhanced SHG has a dependence on the wavelength of the incident light similar to the two-photon excitation spectrum. For example, the SHG from type I collagen could be enhanced when using wavelengths ($\sim$800 nm) that excite the autofluorescence band via two-photon excitation. This enhancement was observed by Tromberg and coworkers in investigating the SHG wavelength dependence in a RAFT organotissue model (Zoumi et al., 2002). However, this enhancement comes at the expense of increased photobleaching as SHG, like TPEF, becomes a resonant process. Due to low signal strengths, this mode, however, is required for imaging dye-labeled membranes with SHG (Campagnola et al., 1999).

16.2.2 SHG Signal Parameters Relative to Fluorescence

Like the more widespread TPEF modality, SHG is also a nonlinear optical process. However, the underlying physical natures differ substantially, and here we present a comparison of the respective photophysics. A Jablonski diagram of the two processes is shown in Figure 16.1. Two-photon excitation proceeds from a ground state S_0 through a virtual state to a final bound excited state S_1. In a time-independent quantum mechanical picture, the final state, after internal conversion, is identical to that accessed in one-photon excitation; thus, the emission properties of one-photon and multiphoton excitation are identical. By contrast, SHG, in general, does not arise from an absorptive process. Instead, an intense laser field induces a second-order polarization (second term, equation (16.1)) in the assembly of molecules, resulting in the production of a coherent wave at exactly twice the incident frequency (or half the wavelength). In this case, there is no bound state (unless the laser wavelength overlaps with a long-lived singlet state; see equation (16.5)) and both the intermediate levels are virtual, being created by the laser field. This coherence of the process has the effect that the spectral and temporal profiles of TPEF and SHG are very different. In the incoherent process of TPEF, the wavelength and width of the emission spectrum are determined by the relative energies and geometries of the ground and first excited electronic state of the fluorophore (i.e., its Frank-Condon factors). The emission lifetime is related to the oscillator strength and is typically 1 to 4 nanoseconds for strongly absorbing, highly fluorescent molecules. Thus, the emission properties are dependent on the molecular properties only and are independent of the excitation laser characteristics (i.e., pulse width and bandwidth). In contrast, both the temporal and spectral characteristics of SHG are derived exclusively from the laser source. Due to the coherence, the SHG pulse is exactly temporally synchronous with the excitation pulse. A Gaussian laser profile will result in a SHG

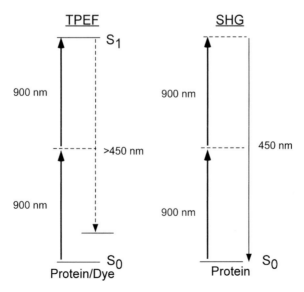

Figure 16.1. Jablonski diagram showing the relevant states for two-photon excited fluorescence (left) and second-harmonic generation (right). Both intermediate levels are virtual for second-harmonic generation.

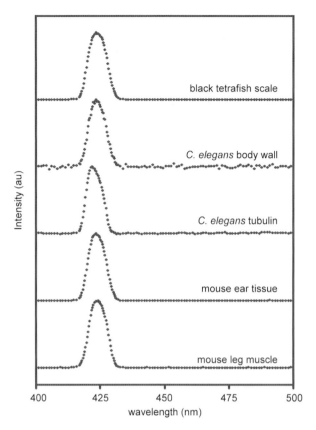

Figure 16.2. Second-harmonic generation spectra for several tissue types acquired on a fiberoptic spectrometer. The laser excitation wavelength was 850 nm. The spectra are free of any contaminating autofluorescence and have the appropriate bandwidth. (Reproduced with permission from *Biophysical Journal*, volume 82, 493–508, 2002).

frequency bandwidth that scales as $1/\sqrt{2}$ of that of the excitation laser. For example, a 100-femtosecond laser pulse (as produced by commercially available titanium-sapphire lasers) has a Fourier transform limited full-width half-maximum (FWHM) bandwidth of about 10 nm, and thus results in a SHG spectrum of about 7 nm FWHM. Representative SHG "spectra" for several tissues are shown in Figure 16.2, where the laser fundamental wavelength was 850 nm. These were acquired through a fiber-coupled spectrometer, where the SHG was imaged and integrated on a linear CCD array. It is observed that the spectra all have the same (and appropriate) bandwidth centered around the expected 425-nm wavelength. Further, they do not contain any autofluorescence components. It should be noted, however, that at shorter wavelengths (< 800 nm), the collagen autofluorescence band ($\sim$450–500 nm) will be excited (Mohler et al., 2003; Zoumi et al., 2002).

16.2.3 Comparison with Other Forms of SHG

The two well-known forms of SHG are frequency doubling in uniaxial birefringent crystals (e.g., potassium dihydrogen phosphate [KDP]) (Yariv, 1989) and that arising at interfaces (Guyot-Sionnest et al., 1986), where both these environments lack inversion symmetry. By contrast, SHG from structural protein arrays is distinct from both of these situations. While several groups have now shown that SHG can be observed from dye-labeled cell and model membranes (Campagnola et al., 1999; Millard et al., 2004; Moreaux et al., 2000a, 2000b), this is an interfacial process as the membrane delineates the extracellular and intracellular regions and is thus only probing the membrane surface. By contrast, SHG can readily section through tissues where it functions as a volumetric imaging probe. For example, Figure 16.3 shows four slices through a tetrafish scale of approximately 35 microns thickness, where the orientation of the crosshatched collagen fibers and their evolution through the axial profile is readily visualized. Collagen matrices have been described as resembling nematic liquid crystals (Prockop & Fertala, 1998), and in this context one might consider SHG in collagen to be classified in the same rubric as the well-known birefringent doubling crystals. In fact, many tissues are known to be highly birefringent, having long been studied by polarization microscopy (Inoue, 1986). However, a fundamental difference exists that does not allow us to interpret the SHG of protein arrays in this context. In negatively birefringent bulk uniaxial crystals, it is necessary to satisfy a phase-matching condition between the extraordinary index of refraction of the SHG and the ordinary index of the fundamental wavelength (Yariv, 1989). Proteins in tissues, however, do not possess type I phase-matching conditions (Bolin et al., 1989). Additionally, in a uniaxial crystal, the polarization of the SHG wave is rigorously orthogonal to that of the laser fundamental, as this is the only nonvanishing matrix element of the χ^2 tensor (Yariv, 1989). However, we have measured the polarization anisotropy in the tetrafish scale and found the SHG to be predominantly parallel to the fundamental (Campagnola et al., 2002). Furthermore, the coherence length of these tissues is approximately 20 microns (Bolin et al., 1989), which is longer than the axial point spread function of lenses used in high-resolution imaging. Thus, SHG produced from protein arrays in a microscope is a different process than in the more well-known bulk forms in that it is volumetric rather than interfacial in nature and it is not phase-matched. We can consider this SHG to arise from alignment of the dipoles of the protein molecules in the array within the focal volume of the excitation lens.

16.2.4 Directionality of the SHG

Here we consider the effects of coherence on the directionality of the SHG signal. For coherent processes governed by the electric–dipole interaction (i.e., SHG, THG, and CARS), the directionality is dependent upon the size and packing of the molecules (Cheng et al., 2001a, 2001b). Specifically, a single dipole will emit equally forward and backwards, in a "figure 8" pattern. But in the limit where molecules align in the plane perpendicular to the direction of the laser propagation, the emission becomes more forward-directed. As this alignment extends in the plane parallel to the laser direction, and the extent of the axial alignment approaches the size of the wavelength of the excitation, essentially the entire signal becomes forward-directed. In strong contrast, fluorescence emission is in general emitted isotropically over 4 pi steradians; thus, in a conventional epifluorescence microscope, even at high numerical aperture (NA), less than half of the fluorescence is collected.

a)

b)

c)

50 μm

Figure 16.3. Three second-harmonic generation optical sections through a tetrafish scale of 35-micron overall thickness. The fibrillar contrast is from type I collagen.

For highly turbid specimens, such as tissues, a new factor plays a role in the observed directionality. The forward-directed signal undergoes multiple scattering events, and a fraction is scattered and emitted in the backward direction. The SHG emission then becomes analogous to ordinary scattering in tissues, where such multiple scattering has been previously well described in the literature (Tromberg et al., 2000). This backscattered component will be better suited for in vivo imaging applications where the specimen is too thick for transmission mode detection. In the experimental section we will discuss the details of adapting a two-photon microscope for optimal SHG imaging for both forward and backward collection geometries.

16.2.5 Molecular and Structural Aspects of Intrinsic Harmonophores

We here define SHG-producing proteins as "harmonophores" to explicitly distinguish them from fluorescent dyes or fluorescent proteins (e.g., green fluorescent protein [GFP]). We have already shown that second-order harmonophores must possess a permanent dipole moment (equation (16.3)) and also must be highly polarizable (equation (16.5)). Since most of the current literature has focused on collagen-based tissues, and to a lesser extent acto-myosin complexes in muscle, we can initially conclude that a common aspect of these species is that their quarternary structure is fibrillar as opposed to globular. While a fibrillar structure does appear to be necessary, it is not entirely sufficient, and we must consider also secondary and tertiary structures. For example, elastin fibers do not produce SHG (Zoumi et al., 2004). This can be understood by considering that this protein consists of a mixture of α helices, β sheets, and random coils. In tissue the elastin exists as a random ensemble that corresponds to an environment with inversion symmetry and thus lacks the regular organization required by equations (16.1) and (16.4).

A more complex situation arises from the observation that keratin fibers seemingly do not produce SHG contrast (Campagnola et al., 2002). The SHG and simultaneously acquired TPEF from an intact region of a mouse ear containing hair follicles (keratin-containing) are shown in Figures 16.4a and 16.4b, respectively. A two-color overlay of the two channels is shown in Figure 16.4c, where the SHG and TPEF channels are violet and green, respectively. The SHG (a) arises from the dermal collagen matrix, while the TPE autofluorescence (b) is detected both in the hair shaft and in cells of the epidermis (regular polygons). While keratin is well known to be strongly autofluorescent, collagen produces essentially no TPEF signal at this fundamental wavelength (850 nm). By contrast, collagen is an efficient SHG source, while the keratin in the hair shafts and epidermis is invisible by SHG. Since keratin fibers consist almost entirely of α helices, it is not clear why it is not an effective second-order harmonophore. It is apparent that further experimental and theoretical effort is required to more fully understand the requisite molecular and structural properties that give rise to efficient SHG from proteins in tissue.

A specific important aspect that needs to be understood is the role of chirality of the proteins in the second-order response in tissue. All proteins are intrinsically chiral, and it has long been known that chirality can enhance $\chi^{(2)}$ at interfaces, thin films, and membranes (Beyers et al., 1994; Campagnola et al., 1999; Hicks & Petralli-Mallow, 1999;

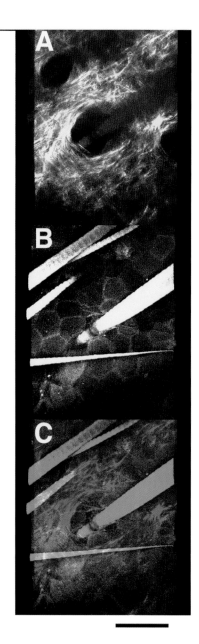

Figure 16.4. Three-dimensional imaging of endogenous second-harmonic generation and fluorescence in tissue layers of a mouse ear. (a) Second-harmonic generation signal from deep collagen-rich layer within dermis. Note the complete absence of signal from the central hair follicle, in contrast with the dense arrangement of fibrils in the bulk of the deep tissue. Extremely high absorbance by the overreaching hairshafts (see b) creates a shadowing of second-harmonic generation signal in the upper right and left corners of the frame. (b) TPEF autofluorescence signal from keratin-rich hair shafts and epidermis. (c) Color overlay of both second-harmonic generation (violet) and TPEF (green) channels, allowing spatial correlation of the complementary signals. Scale bar = 50μm. (Reproduced with permission from *Biophysical Journal*, volume 82, 493–508, 2002).

Verbiest et al., 1994, 1998). The chirality is essentially randomized in the structure of elastin, and this may be the underlying cause of its lack of contrast. It should be noted that chirality alone is not sufficient, as SHG from a neat chiral solution is symmetry-forbidden (Belkin et al., 2001). We further speculate that collagen is much brighter in SHG contrast than myosin (~10–50 fold) because it comprises a triple versus double helical structure. This notion is consistent with the concept of "supramolecular chirality" put forth by Verbiest and colleagues in their work on synthetic chiral organic thin films (Verbiest et al., 1998). In this work, they showed that greatly enhanced SHG was observed from stacked structures acting cooperatively, and further that single enantiomers produced much more SHG than racemic mixtures (~50/1). We showed in Section 16.2.1 above that the alignment of the dipoles in the array is necessary for SHG, and now we conclude that the efficiency is enhanced through the chirality arising from the helical structure.

16.3 EXPERIMENTAL CONSIDERATIONS

Given the significant overlap of the required equipment, in this section we discuss the relevant issues in adapting a SHG microscope from a two-photon fluorescence microscope. A schematic of our combined SHG/TPEF microscope is shown in Figure 16.5 (Campagnola et al., 1999). The main components are a femtosecond laser, a laser scanning confocal system, and either an upright or inverted microscope.

16.3.1 Laser and Microscope

Laser system requirements for SHG imaging are comparable to those of two-photon fluorescence imaging and typically consist of a solid-state Nd:YAG laser (5–10 watts) pumped tunable (700 and 1,000 nm) titanium sapphire (ti:sap) oscillator. Since SHG is not a resonant process, the choice of laser fundamental wavelength is not as important as that of fluorescence excitation. We typically use wavelengths in the range of 850 to 900 nm, as this represents a good compromise between using relatively noninvasive wavelengths and efficient operation of the ti:sap laser. Commercially available lasers produce approximately 100-femtosecond pulses, and while very short pulses (<50 femtoseconds) can broaden through optics and reflecting from optical surfaces, there is essentially no dispersion at wavelengths longer than 850 nm for these pulses. Thus, no external precompensation is necessary to compensate for the minimal group delay in the optical path, scan head, or objective. Note that although shorter pulses will increase peak power and subsequent SHG efficiency, pulses of duration less than about 50 femtoseconds broaden nonlinearly, and it is difficult to maintain these widths at the focal plane. Required average powers at the sample for good signal-to-noise are usually in the range of 1 to 50 mW, depending upon the structural protein and tissue type under study. For example, collagen is at least 50-fold brighter than tubulin and requires only a few mW.

Several modifications are made to optimize the optical path, detection geometry, and detection electronics to optimize a laser scanning microscope for SHG imaging. First, since the output of the ti:sap laser is typically a couple meters from the input of the microscope (to allow adequate space for coupling, especially on an upright

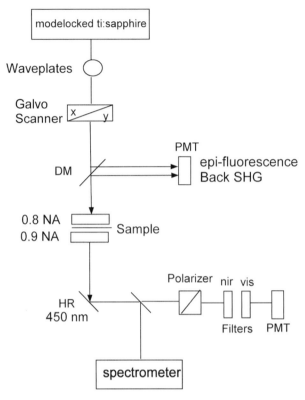

Figure 16.5. Second-harmonic generation microscope. Second-harmonic generation can be detected in either the forward or backward geometries. Both channels use single photon counting electronics. $\lambda/2$ and $\lambda/4$ are half and quarter waveplates, respectively, and provide precise control of the laser polarization. DM = dichroic mirror, HR = hard reflector.

platform), we need to consider the divergence of the laser and resulting spot size at this point. Ti:sap lasers have divergences of 1 mrad, and assuming a 2-meter path-length, the initial 1-mm spot becomes approximately 3 mm at the entrance of the microscope. Microscopes are typically designed for a 1-mm input spot size to properly fill the back aperture of the objective, and the larger spot will result greatly overfill the lens and result in a loss of usable photons. We compensate for the divergence with a 1-meter focal length lens in the coupling space to achieve the proper spot size and thus maximize the peak power at the specimen. Both half-wave ($\lambda/2$) and quarter-wave ($\lambda/4$) plates are used to precisely control the polarization of the laser at the focal plane of the microscope. These are necessary in order to compensate for the ellipticity introduced by the scanning mirrors and other non-45-degree reflections. This level of control is important to perform accurate polarization methods that allow the determination of the structural properties of the protein array (see Section 16.3.2).

The use of an upright platform has some advantages for SHG imaging over an inverted microscope. This is particularly true for the forward-detection geometry, as the optical path is simpler, and it is also more convenient for the use of long working distance water-immersion lenses (1–3 mm) for thick tissue imaging. The upright geometry is also more straightforward to implement the additional external optics required for polarization anisotropy measurements.

As discussed in Section 16.2.4, SHG is a coherent process, and consequently the majority of the signal wave co-propagates with the laser. This requires the construction of an efficient transmitted detection path. We typically use a long working distance 40×0.8 NA (3 mm) water-immersion lens and a 0.9 NA condenser for excitation and signal collection, respectively. The use of a higher-NA condenser yields more efficient collection for highly scattering tissue specimens. Since unlike the epigeometry that can be used for backscattered SHG, the transmission path does not contain an infinity space, an additional lens or telescope is required to recollimate the SHG signal and provide a loose focus on the detector. Two additional elements are added to the transmission path. A Glan–Laser polarizer with high extinction is used for analysis of the SHG signal polarization that can be used to provide a measure of the organization of the protein array (Campagnola et al., 2002). Second, a fiberoptic-based spectrometer (as used for the data shown in Fig. 16.2) is inserted into the collection path with a kinematic mount to verify the SHG wavelengths.

An important consideration in SHG microscopy is the efficient isolation of the desired wavelength from the fundamental laser. This is because the latter is least six orders of magnitude more intense than the desired second-harmonic signal. In our microscope, the SHG signal is first reflected with a reflector centered around the signal ($\sim$20-nm bandwidth), which provides some discrimination from the laser as well as autofluorescence. The SHG is further isolated from the laser and any remaining fluorescence by BG-39 color glass filters and a bandpass filter ($\sim$10-nm bandwidth) respectively. This filtering arrangement has an optical density (OD) for the laser of about 8 and a SHG throughput of about 40%, where the majority of the loss is due to the bandpass filter. While this throughput is not ideal, other options for near-ultraviolet detection do not perform any better. For example, tunable acousto-optic filters are somewhat more transmissive ($\sim$70–90%) but typically have extinction coefficients of only 10^3 and are not adequate for the required spectra purity. Short-wave pass filters have comparable transmission to bandpass filters but may transmit contaminating components of lower wavelengths (e.g., tryptophan autofluorescence) than the SHG signal. Prisms or gratings can be more efficient but require too much space to separate out components in a scanning beam.

While the signal levels are smaller, it is often more convenient to detect the backscattered SHG signal. This is because it can then be detected using the standard epifluorescence pathway on commercial instruments with a simple change in the dichroic mirror and filters. This geometry may be in fact necessary for true in vivo applications. In our instrument this is implemented in non-descanned geometry to optimize the collection efficiency.

16.3.2 Detection of Second-Harmonic Generation

While absolute values of $\chi^{(2)}$ for structural proteins are not known, SHG signal levels are typically smaller than those of TPEF. Thus, we use single photon counting electronics to maximize detection efficiency. This approach provides higher sensitivity and reduced noise over the analog detection and integration that is used in commercial microscopes. Photomultipliers (PMTs) with GaAsP (gallium arsenide phosphine) photocathodes are ideal for SHG detection as they have a quantum efficiency of almost 40% at 400 nm. Thermoelectrically cooled single photon modules are commercially available, or alternatively the electronics can be assembled from individual components. We have found that single TTL pulses can be integrated by the acquisition electronics of several commercial confocal microscopes (Campagnola et al., 1999). We have further found that responses of the integrator are linear (until saturation occurs), affording the ability to make accurate, quantitative measurements with high sensitivity. It should be noted that while single photon counting provides increased sensitivity and signal-to-noise relative to analog detection, the dynamic range of the PMT is smaller. However, single photon counting can be required for detection of weak signals and further allows the use of lower laser power. Using this detection scheme, we find that typical data acquisition times for SHG imaging are between 1 and 10 seconds per 512×512 frame, comparable to two-photon imaging.

16.3.3 Depth Limits

The depth of SHG imaging through thick tissue is generally limited by the Rayleigh scattering of both the fundamental and SHG waves as well as absorption and scattering of the latter. The intensity, I(z), as a function of depth, z, is expected to scale as:

$$I(z) = I(0)e^{-\mu_{\text{eff}}^z} \tag{16.6}$$

where I(0) is the intensity at the top of the stack and μ_{eff} is the effective attenuation coefficient and is the reciprocal of the mean penetration depth (Tromberg et al., 2000).

For the near-infrared wavelengths where SHG microscopy is performed ($\sim$800–900 nm) most tissues have mean penetration depths of approximately 100 to 200 microns ($1/e$ point). Lasers operating further into the infrared, such as Cr:Fosterite ($\sim$1,230 nm), would further increase the limit of penetration. This could be especially valuable for imaging more turbid tissue such as bone. Damage considerations are also generally less at lower energy wavelengths (Konig et al., 1997). For example, Sun and coworkers used a Cr:Fosterite laser for combined SHG and THG imaging for developmental biology applications and found excellent viability of developing fish embryos (Sun et al., 2004). It should be noted that 1,300 nm is approximately the longest desirable wavelength for biological SHG imaging, as longer wavelengths fall outside of the well-known "transparency window" of 700 to 1,300 nm (Tromberg et al., 2000). Longer wavelengths excite water absorption bands (vibrational overtones) and will result in adverse effects from local heating of the tissue.

16.4 SECOND-HARMONIC GENERATION OF STRUCTURAL PROTEIN ARRAYS

In this section we present a survey of current work using SHG to image tissues. Through these examples, we will show how SHG can be used to examine the structure of the protein matrices and also provide some insight into the origin of the contrast. We break this section down into subsections explicitly describing collagen, acto-myosin, tubulin, and cellulose harmonophores. While cellulose comprises polysaccharides, its SHG properties provides some interesting parallels to the protein harmonophores, shedding some light on the contrast mechanism.

16.4.1 Collagen

Type I collagen is the most prevalent protein in the human body, being a primary component in such diverse tissues as skin, bone, tendon, and cornea. These tissues are all birefringent and consist of highly ordered structures and thus are expected to be amenable to study by SHG imaging microscopy. Indeed, the first report of SHG imaging of tissue was by Freund in 1986 (Freund et al., 1986): they examined the polarity in type I collagen fibers from rat-tail tendon (RTT). Several recent efforts have revisited the RTT at higher resolution and in greater depth (Stoller et al., 2001, 2002; Williams et al., 2005) as well as investigated the applicability of using SHG to visualize other collagen-based tissues (Han et al., 2004; Zoumi et al., 2004). Here we provide demonstrative examples of current work in this area.

16.4.1.1 Connective Tissue

We begin with looking at the morphology of collagen in connective tissue. Mammalian tail tendon has the greatest regularity of the tissues we have examined and consequently produces very intense SHG signals. Figure 16.6a shows the x-y maximum point projection from mouse tail tendon, where the collagen fibers are seen to be very highly aligned. This morphology is similar to that reported previously for RTT (Stoller et al., 2001). The collagen connective tissue adjacent to muscle can also be readily visualized. Figure 16.6b shows an optical section from a mouse leg muscle at a depth 5 μm into sample, revealing intense SHG from a matrix of collagen fibrils in the epimysium layer surrounding muscle fibers (Campagnola et al., 2002).

16.4.1.2 Bone

Bone is highly turbid, consisting largely of collagen and minerals. Thus, the same level of penetration is not expected to be achievable as that in less highly scattering tissues such as tendon or muscle, where the mean penetration depths are ~100 microns. However, here we show that high-contrast SHG images can still be obtained in the forward collection geometry. Figure 16.7 shows two optical sections from mouse leg bone. While these do not possess the striking regularity of the tendon in Figure 16.6a, the collagen fibers are clearly identifiable and possess good image contrast. An important question is the depth of penetration that can be achieved. To determine this, an x-z projection was acquired by digital reslicing through the image volume (from the same specimen as in Fig. 16.7a,b), and the resulting section is shown in Figure 16.7. We find that SHG could be collected from depths through approximately 100 microns. Thicker regions were too opaque to collect sufficient signal in transmission mode. However, demineralization of the bone with a chelating agent would have provided increased

a)

b)

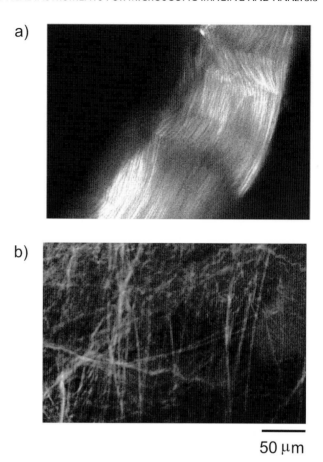

50 μm

Figure 16.6. Second-harmonic generation images of collagen connective tissue. (a) Mouse tail tendon, showing highly organized and regular fibers. (b) Extracellular matrix (epimysium) adjacent to muscle fibers in mouse leg muscle.

imaging depth. This work opens the door to the examination of bone pathological conditions such as osteogenesis imperfecta, where differences in morphology and collagen organization may be apparent by SHG contrast.

16.4.1.3 Corneal Collagen

The cornea comprises predominantly epithelial cells and keratocytes on a type I collagen extracellular matrix well suited for SHG imaging. SHG may indeed become an important clinical imaging tool for this tissue. For example, in 2002, Tromberg and coworkers (Yeh et al., 2002) used combined TPEF and SHG (backscattered geometry) imaging to examine ex vivo rabbit cornea and showed that the cells and stromal layers could be selectively imaged with these two techniques, suggesting the possibility of using SHG to diagnose corneal pathologies and also to monitor the effects of surgical procedures. Indeed, as a step in this direction Han used SHG imaging (transmission mode) to examine cornea ex vivo following mini-invasive laser cornea surgery, and there were differentiated cells from the laser-dissected stroma (Han et al., 2004).

a)

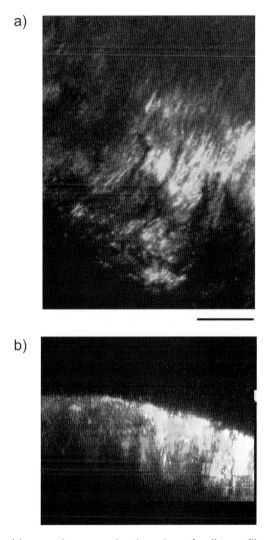

b)

Figure 16.7. Second-harmonic generation imaging of collagen fibers in bone. (a) x-y maximum point projection of mouse leg bone. (b) x-z section through the same specimen showing the achievable depth of penetration of ∼100 microns. Scale bar = 50 microns.

16.5 SECOND-HARMONIC GENERATION DIRECTIONALITY OF COLLAGEN FIBERS: STRUCTURAL CONSIDERATIONS

For future potential in vivo clinical applications (e.g., skin and corneal SHG imaging), the backscattered collection geometry will be required. It is thus imperative that we understand the mechanisms that govern the forward and backward signals. Of particular importance are the signal strengths, the polarization characteristics, and the structural features that give rise to the contrast. To this end, Williams and associates isolated type I collagen fibrils and studied their SHG emission in detail using polarization analysis and also measuring the relative forward-to-backward signal strength ratios (Williams et al., 2005). These measurements, in conjunction with the known ultrastructural aspects of fibrils, allowed them to conclude that the SHG is produced from the shells of fibrils rather than from the bulk. In addition, they determined the thickness of the shells and overall fibril diameter that govern the SHG emission directionality. These aspects are just beginning to be considered, and they need to be thoroughly investigated to advance SHG from the lab to clinical applications.

16.5.1 Three-Dimensional Second-Harmonic Generation Imaging of Muscle

16.5.1.1 Acto-Myosin Complexes

Muscle structure has long been studied by ultrastructural methods, including electron microscopy and x-ray crystallography. Given the parallels of ultrastructural studies on collagen and acto-myosin (Beck & Brodsky, 1998), as well as the existence of strong SHG in the former, we questioned whether the latter would yield bright SHG images. A previous report suggesting that a source of SHG lies within striated muscle further led us to examine muscle tissue on a high-resolution laser scanning microscope (Guo et al., 1997). Indeed, we found that detailed, high-contrast features could be resolved in SHG optical sections throughout the full ~550–μm thickness of a freshly dissected, unfixed sample of mouse lower leg muscle (Campagnola et al., 2002). Figure 16.8a shows an optical section at a depth 45 μm into the sample, where the SHG contrast clearly reveals the repetitive sarcomere pattern in the myofilament lattice. A second optical section at a depth of 200 μm (see Fig. 16.8b) also shows the pattern extending continuously across the full width of muscle fibers. Figure 16.8c shows an x-z section through the same volume of tissue, where the top edges on the images at left are positioned at their corresponding depth in the image on the right (Mohler et al., 2003). The SHG contrast shows the whole muscle cells, and the image is comparable to that observed in standard histological transverse sections, but here it is acquired within unfixed, unsectioned tissue and unstained tissue.

Two important points can be gleaned by examination of the slices shown in Figure 16.8a and 16.8b. First, at the qualitative level the resolution remains similar (and sufficient to resolve the sarcomeres at 0.8 NA) through this range of depth (and in fact the entire 550-micron profile). There are disparate reports in the literature on the effects of turbid tissue on fluorescence image resolution, but in this regime the SHG resolution appears unaffected (Dunn et al., 2000; Konig, 2000). Second, good SHG signal intensity is obtainable throughout this relatively thick muscle tissue sample. Even at 550 microns, only a factor of 5-fold reduction in signal from the top of the stack was observed, which is consistent with predicted depth decay with an effective scattering length of 100 microns (equation (16.6)).

16.5.1.2 Molecular Origin of Second-Harmonic Generation in Acto-Myosin Complexes

SHG has great potential as a diagnostic imaging tool for pathological conditions of muscle (e.g., aging and muscle dystrophy), but before this goal is realized we must fully understand the underlying physical and biochemical origins of the contrast mechanism. To this end, we imaged the myofilament lattice in the optically compliant and

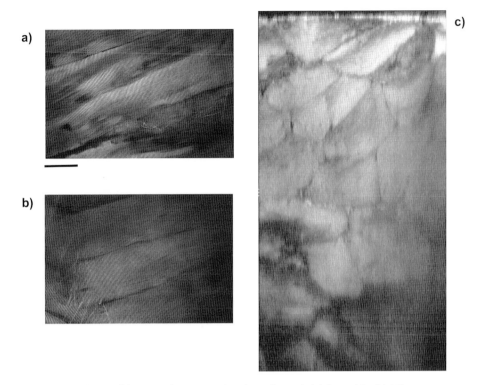

Figure 16.8. Second-harmonic generation imaging of thick and turbid tissues. (a) striated muscle within native mouse muscle tissue. A freshly dissected sample of leg muscle was optically sectioned through its full thickness of ~550μm. Individual slices 45 and 200 microns into the stack are shown in (a) and (b), respectively. The sarcomere repeat pattern is observed across the whole field of view in both slices. (c) An x-z slice through this same stack, where the slices in (a) and (b) are top aligned with the corresponding origin in the x-z slice. Scale bar = 50 microns. (Reproduced with permission from *Methods*, Vol 29, 2003, pp. 97–109).

genetically tractable model organism *Caenorhabditis elegans*. Our approach was to combine SHG/TPEF imaging to observe muscle in a *C. elegans* strain expressing green fluorescent protein-tagged myosin heavy chain A (GFP::MHC A), the minor MHC isoform known to be restricted to the longitudinal midzone of thick filaments (Miller et al., 1983; Waterston, 1988). Figures 16.9a and 16.9b show simultaneously acquired high-magnification SHG and TPEF images, respectively. These reveal that the signals are spatially distinct and complementary, as seen in the two-color overlay in Figure 16.9c, where the color scheme is the same as in Figure 16.4. Fluorescence from GFP::MHC A is observed in the dim middle portion of the SHG-bright band, while SHG is strongest in regions not containing GFP::MHC A. This result suggests that the SHG arises from the thick filaments. Next we question the role of actin in SHG contrast formation. In contrast to myosin, actin has little birefringence (and is thus denoted the "I" or isotropic band in the context of sarcomeres), and based on these grounds, this protein would not be expected to be an efficient second-order harmonophore. Indeed, we have performed imaging and biochemical assays that confirm that actin does not directly yield SHG contrast; this may also explain the dim central stripe.

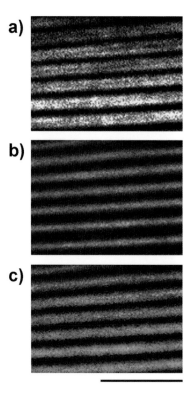

Figure 16.9. Correlative second-harmonic generation and TPEF imaging to determine the SHG source in nematode myofilament lattice. Endogenous second-harmonic generation (a) and TPEF fluorescence from GFP::MHC-A (b) in *C. elegans* body wall muscle are shown separately and in color overlay (c; Second-harmonic generation = violet, TPEF = green). A central green/white stripe in the center of each violet band (c) indicates that GFP::MHC-A is localized at the center of the second-harmonic generation bright band. Scale bar = 10μm. This correlative imaging shows the second-harmonic generation arises from thick filaments. (Reproduced with permission from *Biophysical Journal*, volume 82, 493–508, 2002).

Other factors may also explain the lack of contrast in the middle of the sarcomeres. For example, a symmetry-induced cancellation effect (as will also be seen for tubulin structures, Section 16.3.2) may arise at the center of the thick filament, where proximal portions of oppositely oriented filament arms are close enough to break down the conditions of local asymmetry required for SHG. Due to its coherence, SHG emission will arise only from objects separated at distances on the order of or larger than the optical coherence length, L_c

$$\Delta k \bullet L_c \approx \pi \tag{16.7}$$

where Δk is the difference in wave vectors between the fundamental and second-harmonic waves.

For SHG arising from the electric dipole interaction, the lower limit for producing SHG empirically appears to be about $\lambda/10$. At smaller distances, the SHG signals undergo complete destructive interference (Dadap et al., 1999). Thus, a partial or near-complete cancellation of the SHG may be operative in the midregion of the sarcomeres.

Recent work by other researchers has further examined the molecular and structural aspects that give rise to SHG in muscle tissue. Sun and coworkers (Chu et al., 2004) used detailed polarization analysis and extracted the relative values of the d_{nn} coefficients of the hyperpolarizability tensor (equation (16.3)). One principal finding was that the angular dependence of the SHG signal on the laser polarization was complex, displaying maxima and minima every 45 degrees and never decreasing to zero intensity. Interestingly, Stoller and colleagues performed an analogous analysis on RTT and found a highly similarly shaped angular dependence (Stoller et al., 2001). Although they are highly differently shaped molecules, myosin and collagen both contain coiled-coil domains. The similarity in the polarization dependence perhaps suggests that this form of tertiary structure is important in determining the second-order properties of the harmonophores.

An open question regards whether SHG in myosin arises from the coiled-coil tails or from the double-head region of the molecule. Boulesteix (Boulesteix et al., 2004) inferred that the latter was most likely the "active" part of the molecule. However, our work using polarization and biochemical analysis suggests that the coiled-coil tail region is mainly operative (Plotnikov et al., 2005). Given the complex nature of myosin in terms of the actin interactions as well as the double head and tail morphology of the molecule, more work clearly needs to be performed in this area to obtain definitive answers.

16.5.2 Microtubules

The previous two sections examined large structural protein arrays comprising collagen and acto-myosin complexes, both of which at the molecular level consist of coiled-coil structures. While tubulin does not have this structure, it does assemble into linear, polar microtubules that possess a characteristic birefringence (Cassimeris & Salmon, 1988). In order to test whether the second-harmonic signals could also be generated from mitotic spindles, asters, and centrosomes, we attempted to image these structures in *C. elegans* embryos (Campagnola et al., 2002). This investigation is of twofold interest. First, it may provide a new method of imaging cellular developmental processes such as mitosis without the use of exogenous labels, and it may also provide more detail and depth discrimination than possible by polarization microscopy. Second, the simplicity of the structures (e.g., relative to acto-myosin complexes) allows a straightforward examination and determination of the specific contexts giving rise to SHG signals in microtubules, namely if any lateral and radial symmetry constraints exist, or if these can be then measured. Further, this may allow us to infer the role of coiled-coils in protein harmonophores.

To this end we again use the simultaneous SHG/TPEF approach (described in Section 16.4.2) where a strain of *C. elegans* expressing a GFP-tagged beta tubulin (β-tubulin::GFP) was created. Figures 16.10a and 16.10b show the respective SHG and TPEF images of a mitotic spindle, and striking differences are observed between the two contrast mechanisms (Campagnola et al., 2002). Namely, the region in the spindle midzone, where the chromosomes are located, appears dark by SHG. However, the bright TPEF image proves that it contains abundant microtubules. The two-color overlay in Figure 16.10c, where the color scheme is the same as in Figure 16.4, further delineates the differences in the contrast mechanism. We can understand this effect

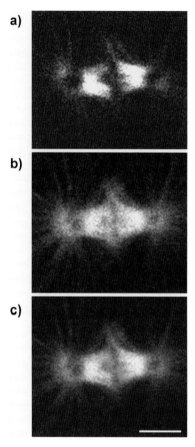

Figure 16.10. Combined second-harmonic generation (a) and TPEF (b) of an embryo shown at higher magnification during the first mitosis. In (b), fluorescent microtubules compose the dense spindle array as well as sparse astral arrays surrounding both spindle poles. In (a), however, second-harmonic generation arises from only the subset of the microtubules seen that are overlapped with the laser polarization. (b) Bright second-harmonic generation in the spindle is interrupted by a discrete dark space at the spindle midzone, and the spindle poles appear distinctly more hollow by second-harmonic generation than by TPEF. (c) Two-color overlay further delineating the difference in second-harmonic generation and TPEF contrast. Scale bar $= 10\mu$m. (Reproduced with permission from *Biophysical Journal*, volume 82, 493–508, 2002).

by considering that the polar microtubules interdigitate anti-parallel to each other in the midzone (Haimo, 1985). Thus, this region is dark by SHG because of a symmetry cancellation caused by this anti-parallel alignment, which in effect creates a local environment with inversion symmetry, violating the conditions defined by equations (16.1) and (16.4). In contrast, the fluorescence is an incoherent process and is not subject to these symmetry constraints.

We further use SHG imaging to probe the radial organization of microtubule-based structures. In Figure 16.11 the SHG (left) and TPEF (right) images of interphase centrosomes in early embryonic cells are shown. As in the case of the mitotic spindles, the two contrast mechanisms again yield highly different images. Specifically, the TPEF signal of the centrosomes appears as a uniform disk, whereas double-crescent profiles

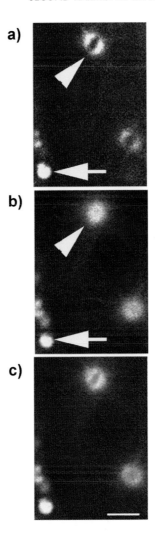

Figure 16.11. Combined second-harmonic generation (a) and TPEF (b) and color overlay (c) of three interphase centrosomes in early embryonic cells shown at high magnification (arrowheads in (a) and (b). In (b), centrosomes are uniformly labeled with fluorescent tubulin. In (a), centrosomes yield a double arc, hollow second-harmonic generation pattern. By contrast, the TPEF contrast is a filled-in disk. Scale bar = 10μm. (Reproduced with permission from *Biophysical Journal*, volume 82, 493–508, 2002).

are seen in the SHG channel. Since the microtubules of the centrosome are radially arrayed around the organizing center, contrast is observed only for the overlap of the laser polarization with the orientation of the microtubule direction. Furthermore, rotation of the laser polarization results in concomitantly the same degree of rotation of the double arc pattern (not shown) (Campagnola et al., 2002). This finding is consistent with the dipole moment of the protein lying parallel to the long, symmetry axis of the microtubule. This situation is in strong contrast to the case of the coiled-coil myosin and collagen, where the complex polarization dependence may suggest that in addition to the long axis of the protein, the helical structure is important in the SHG contrast. Moreover, tubulin is a much weaker harmonophore than collagen or myosin.

We can understand the "filled-in" pattern in the fluorescence images of the centrosomes by considering that the GFP label is not part of the tubulin domain and can freely rotate. While molecules must still be oriented with respect to the laser polarization for the two-photon absorption, a random ensemble of absorption dipoles exists due to this rotation, and light is emitted isotropically. These two examples of the mitotic spindles and centrosomes show the ability of SHG to directly visualize the symmetry of the protein array, whereas this information is lost in the fluorescence imaging of the GFP label.

16.5.3 Cellulose

Cellulose is the most abundant biomacromolecule, being the primary component in plant cell walls (Brown, 1996). Although it consists of polysaccharides rather than amino acids, cellulose has many parallels to the structural proteins discussed above. Thus we have extended the use of endogenous SHG imaging microscopy to studying the micron-scale structure of cellulose (R.M.J. Brown et al., 2003) to provide further context to understand the origin of the contrast mechanism in the protein harmonophores. Like the structural proteins arrays, cellulose forms structures that are highly birefringent. Further, because it comprises polysaccharide chains, it is also strongly chiral (also like the coiled-coil protein structures). Additionally, cellulose has a supramolecular structure that parallels that of collagen: collagen assembles into progressively higher-ordered structures, beginning with the nanometer level (microfibril), which then become organized into fibrils (~100 nm), which assemble into fibers (~ microns), and finally fascicles (~10–100 microns) (Beck & Brodsky, 1998; Ottani et al., 2001). Cellulose similarly evolves through a progression of well-ordered molecular chains that crystallize into microfibrils (Itoh & Brown, 1984). Thus, it is an interesting proposition to compare the SHG from cellulose with that of collagen and other protein harmonophores.

Here we show representative SHG images from celluloses derived from two different organisms, the gram-negative bacterium *Acetobacter* and the giant algal cell *Valonia*. The fibrous nature of *Acetobacter* cellulose on the micron-size scale is demonstrated in Figure 16.12a, where the field size is approximately 300×300 microns. The morphology seen in these SHG images strongly resembles those of collagen from various animal sources described above. We also observe similar morphology in the microfibrils of *Valonia* (see Fig. 16.12b). The fibrous structure is more obvious for *Valonia* because the individual microfibrils consist of more glucan chains and are very much larger and more perfectly ordered than in *Acetobacter*.

To more closely examine the alignment and orientation of *Valonia* microfibrils and to provide a comparison with the protein harmonophores, we investigated the polarization dependence of the SHG signal. The top two panels in Figure 16.13 show the resulting images for one axial slice of a *Valonia* cell wall with orthogonal laser excitations (Figs. 16.13a and 16.13b, respectively) (R.M.J. Brown et al., 2003). In Figure 16.13a we see that the fibrils are strongly aligned in one direction; with the orthogonal polarization, these fibrils essentially disappear and only a vestige of fibrils oriented in the orthogonal direction are observed (see Fig. 16.13b). This finding is consistent with

a)

b)

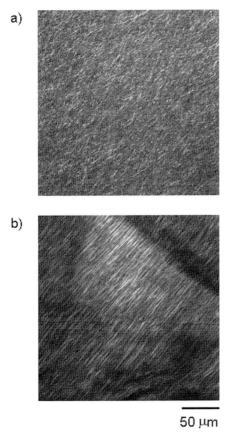

50 μm

Figure 16.12. Single second-harmonic generation optical sections showing the microfibril structure of two celluloses: (a) the gram-negative bacteria *Acetobacter* and (b) giant algal cell *Valonia*.

the SHG in cellulose arising from the electric–dipole interaction, in analogy to the structural proteins. This analysis is extended by measuring the magnitude of the SHG signal through rotation of 180 degrees of laser fundamental polarization (data points every 4 degrees) and measuring the integrated SHG signal as a function of this angle. In the limit of having the dipole moment oriented completely parallel to the symmetry axis of the fiber, the angular dependence will yield a sinusoidal pattern with a maximum and minimum with complete extinction 90 degrees out of phase and should follow a $\cos^2\theta$ distribution. As shown in Figure 16.13c, this is indeed borne out. There is some deviation from the theoretically expected distribution because the thickness of the optical section includes some fibrils from adjacent layers that have orthogonal absolute orientations. We do observe that this polarization dependence is strikingly similar to the simple case of interphase centrosomes discussed in the last section, where the polarization data suggest that the dipole lies along the physical symmetry axis of the microtubules. This result is most interesting since cellulose is a polysaccharide consisting of sugar monomer units rather than a protein consisting of amino acids.

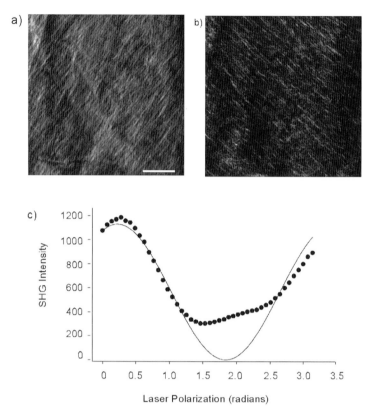

Figure 16.13. Second-harmonic generation polarization dependence of microfibrils in *Valonia*. Orthogonal excitations are shown in (a) and (b). (c) Integrated second-harmonic generation intensity as a function of laser polarization, where the data were taken every 4 degrees, and the result was fit to a $\cos^2\theta$ distribution. Scale bar = 40 micron. The deviation from the fit occurs because some adjacent, orthogonally oriented fibrils were also present in the optical section. (Reproduced with permission from *Optics Letters*, volume 28, 2207–2209, 2003).

16.6 COMPARISON WITH OTHER OPTICAL IMAGING METHODS

In this section we compare the salient and advantageous features of SHG imaging to those of linear (polarization and fluorescence microscopy) and other nonlinear optical modalities (TPEF, THG, and CARS). Table 16.1 provides a comparison of the relative strengths of several attributes of the linear and nonlinear methods, where we qualitatively rate the performance and information content as "E", "G", and "P" for excellent, good, and poor, respectively. This table is meant to provide a general comparison of the methods and is not inclusive of every possible scenario.

16.6.1 Linear Methods

16.6.1.1 Polarization Microscopy

Polarization microscopy has often been used to observe the birefringence of ordered structural proteins, including those of contractile muscle (Waterston, 1988) and

Table 16.1. Comparison of attributes of existing imaging modalities

Modality	3D Section Sectioning	Penetration Depth	Molecular Structural Information	Chemical Specificity	Low Photo-bleaching	Tissue Imaging
1-photon	P	P	P	G	P	P
2-photon	E	E	P	G	E	E
Polarization microscopy	P	P	G	P	G	P
CARS	E	E	G	E	E	E
Second-harmonic generation	E	E	E	G	E	E
Third-harmonic generation	E	G	G	G	E	G

Note: P, G, and E are poor, good, and excellent, respectively.

microtubule assemblies (Inoue, 1986). Since both polarization contrast and SHG arise from birefringence, we compared the respective images of *C. elegans* sarcomeres in a polarization and SHG microscope in Figure 16.14a and 16.14b, respectively. It should be noted that these images were not from the same specimen, but all *C. elegans* of the same type (here N2) are genetically identical. Although somewhat comparable, the polarization and SHG microscopes do not produce exactly the same patterns of contrast (Campagnola et al., 2002). Both modes reveal bright bands with dim central stripes, yet the SHG-bright bands appear relatively broader than the anisotropic A-bands seen by polarization microscopy. The dark region between bright SHG bands (arrow) is devoid of bright nodular signals corresponding to dense bodies in the isotropic band of the polarization image. Thus, there is no simple one-to-one correspondence between SHG and polarization images, although the same thick filaments generate both effects. These differences may exist because the contrast in a polarization scope image arises from linear birefringence, whereas SHG is a quadratic process with respect to molecular density and laser power. Additionally, contrast in a polarization microscope can arise from either form (structural) or intrinsic birefringence, whereas SHG requires the former to satisfy the symmetry constraints described by equations (16.1) and (16.4).

SHG may be a more powerful imaging modality than polarization microscopy. The latter only probes residual birefringence between orthogonal polarizers. While the use of a compensator allows the extraction of the retardance or thickness of the specimen, it is not generally possible to extract quantitative *molecular-level* data. By contrast, through polarization analysis, SHG imaging can be used to directly extract the coefficients of the hyperpolarizability (Chu et al., 2004) and nonlinear susceptibility tensors and thereby determine the absolute molecular orientation of the dipole moments of the proteins in the array. Finally, because of the nonlinear nature, SHG provides intrinsic optical sectioning; while achievable, this is difficult by polarization microscopy (Shribak & Oldenbourg, 2003).

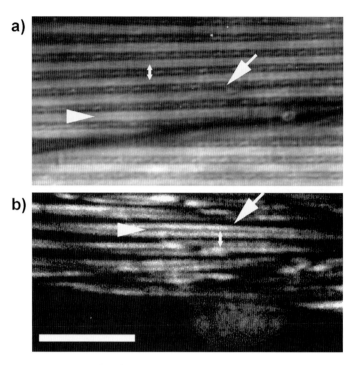

Figure 16.14. Comparison of polarization and second-harmonic generation microscopy in nematode muscle. Body wall muscle of live *C. elegans* was imaged by both methods and is shown at identical magnification in (a) and (b); double arrows indicate spacing between centers of adjacent sarcomere A-bands in both panels. (a) Polarization image, where the bright A-bands are characterized by a dim central stripe (arrowhead). The dark I-band region is punctuated by birefringent dense bodies (arrow). (b) Second-harmonic generation image, where the bright bands also show a pronounced dark central stripe (arrowhead). The dark region between bright bands (arrow) is devoid of nodular signal corresponding to dense bodies. Scale bar = 10 microns. (Reproduced with permission from *Biophysical Journal*, volume 82, 493–508, 2002).

16.6.1.2 Fluorescence Microscopy

SHG imaging of tissues has distinct advantages over imaging tissues labeled with fluorescent dyes. The most obvious advantage lies in the lack of any staining preparation. In addition to photobleaching considerations, excitation of fluorescent dyes also results in the creation of toxic reactive oxygen species. While multiphoton excitation of fluorescence greatly reduces these problems away from the focal plane, they still do occur in plane. By contrast, there is no resonant component of the SHG arising from these structural proteins at the wavelengths we use ($\lambda > 850$ nm). Indeed, we have previously shown that SHG is essentially bleach-free relative to GFP labels, where repetitive high-intensity exposure did not lead to a decrease in the SHG intensity, whereas the GFP intensity decreased dramatically (Campagnola et al., 2002). However, it should still be noted that no imaging method is completely free from adverse phototoxic effects. For example, in the present case, endogenous species such as cytochromes can absorb the SHG emission.

The strongest rationale for SHG imaging of structural proteins relative to fluorescence lies in the ability to directly extract information about the molecular orientation and organization in the tissue. This is due to the ability to directly determine the matrix elements of the hyperpolarizability, β, and nonlinear susceptibility tensor, $\chi^{(2)}$ that describe these parameters (Shen, 1984). While fluorescence anisotropy measurements can be performed on dye- or GFP-labeled proteins, these measurements probe the dynamics of the tag, and information about the protein can at best only be inferred. While in principle the fluorescence anisotropy of tryptophan could be measured, this amino acid lacks sufficient fluorescence contrast in practice, and furthermore the result would be an average over all the occurrences of this residue in the protein. Furthermore, we showed in Figures 16.10 and 16.11 that GFP labeling of tubulin does not allow the visualization of the anisotropy of the centrosomes or spindle. In contrast, SHG has exquisite sensitivity to these properties.

Additional comparisons can be drawn between SHG and GFP imaging of protein structures. While the use of GFP has added new dimensions to the capabilities of imaging live cells, the approach can still be limiting for the case of "non-model" organisms. On the other hand, virtually any protein in a cell can be expressed with a GFP tag. As SHG relies exclusively on signals generated by endogenous structures, it should be universally applicable to all animal species and may be well suited for clinical applications. However, as discussed earlier, not all structural proteins give rise to SHG contrast. Furthermore, we have demonstrated that when used together, they can provide complementary information not possible by either method alone (Campagnola et al., 2002).

16.6.2 Nonlinear Optical Imaging Methods

16.6.2.1 Third-Harmonic Generation

The most closely related nonlinear contrast mechanism to SHG is third-harmonic generation (THG). Like SHG, THG is also a nonresonant, coherent process and has been used as a tool by which to up-convert lasers to shorter wavelengths. However, unlike SHG, THG does not require a non-centrosymmetric environment to produce signal (Shen, 1984). This can be understood because THG is described by the third-order nonlinear susceptibility tensor, $\chi^{(3)}$, and has different symmetry conditions than the second-order susceptibility, $\chi^{(2)}$. While both SHG and THG have been used to image membranes, they consequently do so through different mechanisms. For example, THG probes the volume around a membrane, whereas interfacial SHG probes the two-dimensional membrane itself. This principle was demonstrated by Squier and coworkers, who obtained high-resolution, high-contrast THG images of interfacial regions of specimens, including plant leafs and red blood cells (Muller et al., 1998).

While proving a new contrast mechanism, THG imaging has some limitations. A fundamental issue is that the sensitivity is highest in regions where there is a large change in refractive index, such as at the interface at the top of a stack. While sectioning can be performed into bulk tissues, the signals are much smaller than at the surface of the specimen. A second limiting factor is that this scheme is not readily compatible with the use of ti:sap excitation lasers. For example, even with 1,000-nm excitation, the THG signal is too far in the UV to be efficiently collected by conventional glass optics, where they become highly absorptive at $\sim$350 nm. Sun has circumvented this problem through the use of a home-built Cr:Fosterite laser, with a wavelength of 1,230 nm; they showed that SHG and THG can be produced concurrently from several plant and mammalian specimens (Sun et al., 2004). Unfortunately, these lasers are not currently commercially available. Perhaps if these or other infrared lasers become more readily accessible, THG will find greater utility.

16.6.2.2 Coherent Anti-Stokes Raman Scattering (CARS)

Like SHG, CARS has long been used by physical chemists as a spectroscopic tool (Shen, 1984) and similarly is now emerging as a tool for biological imaging applications. In this scheme two lasers are tuned such that their energy difference corresponds to a vibrational frequency, such as a specific stretching or bending mode of a protein or nuclei acid. The CARS signal is then observed at $2\omega_p - \omega_s$, where ω_p and ω_s are the pump and Stokes frequencies, respectively. For example, the Xie group has used CARS to image live cells during both interphase and metaphase by tuning to the C-H aliphatic stretch of the lipid membrane and the PO_2^- symmetric stretch of the DNA backbone, respectively (Cheng et al., 2002). They were also able to monitor apoptosis, where they visualized shrinking and loss of integrity of the cell membrane, as well as vesicle formation in the cytoplasm.

The strongest aspect of CARS lies in its chemical specificity. For example, Müller used CARS to image different components in lipid multilamellar vesicles (Wurpel et al., 2002). The fact that the lasers can be tuned to probe specific vibrational modes positions CARS as a powerful tool to image a wide range of dynamic processes in cells. This is a unique aspect of CARS relative to SHG and the other methods described above. On the other hand, while some polarization analysis can be used in CARS imaging (Cheng et al., 2001a, 2001b), the signal does not possess the information about the organization of the molecules in a structural protein array that is intrinsic to the SHG contrast. For example, CARS would not be as sensitive as SHG to pathologies arising from a disorder in the protein array. On the other hand, CARS would be more sensitive to any accompanying changes in chemical composition. In this light, SHG and CARS are best viewed as highly complementary, non-overlapping techniques.

Currently CARS is a very expensive and technologically demanding imaging scheme, as two tunable laser colors are required. Furthermore, these colors must be precisely overlapped both temporally and spatially. This scenario will likely improve with advances in ultrafast laser technology.

16.7 LIMITATIONS OF SECOND-HARMONIC GENERATION IMAGING

Here we discuss some limitations of SHG imaging in terms of both experimental aspects as well as its applicability to biological imaging. A primary issue that needs to be considered is the SHG signal strength relative to TPEF of dye-labeled cells or tissues. Since the absolute values of the $\chi^{(2)}$ susceptibility tensors of structural proteins are at present unknown, we can only provide a qualitative assessment based on our experience. The attainable contrast from collagen-based tissues is very high and certainly comparable to that of typical TPEF imaging with dyes. However, acto-myosin complexes produce signal levels smaller by at least an order of magnitude, and those from tubulin-based structures are at least an order of magnitude yet weaker (Campagnola et al., 2002). Efficient detection of these weak signals requires careful alignment of the optical path, noise reduction, high extinction filtering, and single photon counting schemes.

A related issue is the primarily forward-directed nature of the SHG signal (in the absence of multiple scattering). As discussed previously, it is more convenient to perform SHG in an epigeometry on commercial microscopes, since an external path to collect a scanning beam is not required. However, the available backscattered SHG contrast for muscle- and microtubule-based structures is small. Additionally, it is likely that significant polarization anisotropy will be lost due to the multiple scattering that gives rise to the backscattered signal. While it will not increase the absolute signal levels, the use of time-gated or interferometric detection would improve the signal-to-noise in this geometry. The predominantly forward-directed nature also may limit the ultimate application of SHG to in vivo imaging in the clinic. On the other hand, ex vivo biopsies can be imaged in either the forward or backward geometries. Additionally, we have begun to examine the use of hyperosmotic agents such as glycerol to achieve greater SHG imaging depth into tissue. We have shown that an increase of threefold in imaging depth is readily achievable into striated muscle while not adversely affecting the structure of the sarcomeres (Plotnikov et al., 2006).

A second limitation of SHG imaging lies in its lack of general applicability. As described earlier, a high concentration of harmonophores is required to produce sufficient contrast. Furthermore, not all proteins assemble properly or have the appropriate secondary, tertiary, or quarternary structure to satisfy the symmetry constraints. For example, while type I and II collagen produces bright SHG signals, type IV appears to be transparent in this scheme (E. Brown et al., 2003;). SHG from elastin has also not been observed (Zoumi et al., 2004). A second aspect that limits the flexibility is that, unlike fluorescence, SHG does not result from a process at the level of a single molecule, but rather from the cooperativity between adjacent molecules in an array. This has the consequence that it is not possible to track single particles such as peptides or proteins in cytoplasm and their effect on cellular structures with the SHG contrast mechanism.

16.8 PROSPECTS

SHG is already rapidly expanding as a biophysical imaging tool. It may ultimately find great applicability as an imaging tool for probing disease states. Many diseases are characterized by defects or changes in the assembly of the proteins in tissue. For example, it is well known that the collagen content in tumors differs from normal tissue, and indeed Jain has demonstrated that SHG has sufficient sensitivity to follow the progression of a tumor in mouse by imaging the collagen content (E. Brown et al., 2003). Additionally, diseases such as osteogenesis imperfecta, scleroderma, and osteoporosis arise from defects in the collagen matrix that should be observable by SHG. These structural differences might be transparent by other optical modalities. For example, we have showed that there are clear and striking morphological differences in the collagen fibers from osteogenesis imperfecta and normal tissue. Based on these early applications, and with the improvements and miniaturization of lasers and microscopes, SHG may find a niche as a powerful diagnostic imaging tool.

ACKNOWLEDGMENTS

I thank my coworkers Prof. William Mohler, Dr. Andrew Millard, Dr. Oleg Nadiarnykh, and Dr. Sergey Plotnikov on this project. I also thank Prof. Leslie Loew for continuing helpful discussions. Support from NIH NCRR R21 RR13472 and NIBIB R01 EB1842 is gratefully acknowledged.

REFERENCES

Beck, K., and B. Brodsky. 1998. Supercoiled Protein Motifs: The collagen triple-helix and the alpha-helical coiled coil. *J. Struct. Biol.* 122:17–29.

Belkin, M.A., S.H. Han, X. Wei, and Y.R. Shen. 2001. Sum-frequency generation in chiral liquids near electronic resonance. *Phys Rev Lett* 87(11):113001.

Beyers, J.D., H.I. Lee, T. Petralli-Mallow, and J.M. Hicks. 1994. Second-harmonic generation circular-dichroism spectroscopy from chiral monolayers. *Phys. Rev. B* 49:1464–14647.

Bolin, F.P., L.E. Preuss, R.C. Taylor, and R.J. Ference. 1989. Refractive index of some mammalian tissue using a fiber optic cladding method. *Applied Optics* 28:2297–2303.

Bouevitch, O., A. Lewis, I. Pinevsky, J.P. Wuskel, and L.M. Loew. 1993. Probing membrane potential with non-linear optics. *Biophys. J.* 65:672–679.

Boulesteix, T., E. Beaurepaire, M.P. Sauviat, and M.C. Schanne-Klein. 2004. Second-harmonic microscopy of unstained living cardiac myocytes: measurements of sarcomere length with 20-nm accuracy. *Opt Lett* 29(17):2031–2033.

Brown, E., T. McKee, E. diTomaso, A. Pluen, B. Seed, Y. Boucher, and R.K. Jain. 2003. Dynamic imaging of collagen and its modulation in tumors in vivo using second-harmonic generation. *Nat Med* 9(6):796–800.

Brown, J., R. M. 1996. The biosynthesis of cellulose. *Pure Appl. Chem.* 10:1345–1373.

Brown, R.M.J., A.C. Millard, and P.J. Campagnola. 2003. Macromolecular structure of cellulose studied by second-harmonic generation imaging microscopy. *Opt. Lett.* 28:2207–2209.

Campagnola, P.J., A.C. Millard, M. Terasaki, P.E. Hoppe, C.J. Malone, and W.A. Mohler. 2002. 3-Dimesional High-Resolution Second Harmonic Generation Imaging of Endogenous Structural Proteins in Biological Tissues. *Biophys. J.* 82:493–508.

Campagnola, P.J., M.D. Wei, A. Lewis, and L.M. Loew. 1999. High resolution non-linear optical microscopy of living cells by second harmonic generation. *Biophys. J.* 77:3341–3349.

Cassimeris L, S. Inoue, and E.D. Salmon. 1988. Microtubule dynamics in the chromosomal spindle fiber: analysis by fluorescence and high-resolution polarization microscopy. *Cell Motil. Cytoskel.* 10:185–196.

Cheng, J.-X., L.D. Book, and X.S. Xie. 2001a. Polarization coherent anti-Stokes Raman scattering microscopy. *Opt Lett* 26:1341–1343.

Cheng, J.-X., A. Volkmer, L.D. Book, and X.S. Xie. 2001b. An Epi-Detected Coherent Anti-Stokes Raman Scattering (E-CARS) Microscope with High Spectral Resolution and High Sensitivity. *J. Phys. Chem. B* 105:1277–1280.

Cheng, J.X., Y.K. Jia, G. Zheng, and X.S. Xie. 2002. Laser-scanning coherent anti-Stokes Raman scattering microscopy and applications to cell biology. *Biophys J* 83(1): 502–509.

Chu, S.-W., S.-Y. Chen, G.-W. Chern, T.-H. Tsai, Y.-C. Chen, B.-L. Lin, and C.-K. Sun. 2004. Studies of (2)/(3) Tensors in Submicron-Scaled Bio-Tissues by Polarization Harmonics Optical Microscopy. *Biophys J* 86:3914–3922.

Cox, G., E. Kable, A. Jones, I. Fraser, F. Manconi, and M.D. Gorrell. 2003. 3-Dimensional imaging of collagen using second harmonic generation. *J Struct Biol* 141(1):53–62.

Dadap, J.I., J. Shan, K.B. Eisenthal, and T.F. Heinz. 1999. Second-Harmonic Rayleigh Scattering from a sphere of centrosymmetric material. *Phys. Rev. Lett.* 83:4045–4048.

Dombeck, D.A., K.A. Kasischke, H.D. Vishwasrao, M. Ingelsson, B.T. Hyman, and W.W. Webb. 2003. Uniform polarity microtubule assemblies imaged in native brain tissue by second-harmonic generation microscopy. *Proc Natl Acad Sci U S A* 100(12):7081–7086.

Dunn, A.K., V.P. Wallace, M. Coleno, M.W. Berns, and B.J. Tromberg. 2000. Influence of optical properties on two-photon fluorescence imaging in turbid samples. *Appl. Opt.* 39:1194–1201.

Eisenthal, K.B. 1996. Liquid interfaces probed by second-harmonic and sum-frequency spectroscopy. *Chem. Rev.* 96:1343–1360.

Fine, S., and W.P. Hansen. 1971. Optical second harmonic generation in biological systems. *Appl Opt* 10:2350–2353.

Freund, I., M. Deutsch, and A. Sprecher. 1986. Connective Tissue Polarity. *Biophys. J.* 50: 693–712.

Guo, Y., P.P. Ho, H. Savage, D. Harris, P. Sacks, S. Schantz, F. Liu, N. Zhadin, and R.R. Alfano. 1997. Second-harmonic tomography of tissues. *Opt. Lett.* 22:1323–1325.

Guyot-Sionnest, P., W. Chen, and Y.R. Shen. 1986. General considerations on optical second-harmonic generation from surfaces and interfaces. *Phys. Rev. B.* 33:8254–8263.

Haimo, L. 1985. Microtubule polarity in taxol-treated isolated spindles. *Can. J. Bochem. Cell Biol.* 63:519–532.

Han, M., L. Zickler, G. Giese, M. Walter, F.H. Loesel, and J.F. Bille. 2004. Second-harmonic imaging of cornea after intrastromal femtosecond laser ablation. *J Biomed Opt* 9(4): 760–766.

Heinz, T.F., C.K. Chen, D. Ricard, and Y.R. Shen. 1982. Spectroscopy of Molecular Monolayers by Resonant Second-Harmonic Generation. *Phys. Rev. Lett.* 48:478–481.

Hellwarth, R., and P. Christensen. 1974. Nonlinear optical microscopic examination of structure in polycrystalline ZnSe. *Optics Comm.* 12:318–322.

Hicks, J.M., and T. Petralli-Mallow. 1999. Nonlinear optics of chiral surface systems. *Appl. Phys. B* 68:589–593.

Inoue, S. 1986. Video Microscopy. Plenum Press, New York, NY.

Itoh, T., and J. R.M. Brown. 1984. The assembly of cellulose microfibrils in Valonia macrophysa. *Planta* 160:372–381.

Kleinman, D.A. 1962. Nonlinear dielectric polarization in optical media. *Phys. Rev.* 126: 1977–1979.

Konig, K. 2000. Multiphoton microscopy in life sciences. *J Microsc* 200 (Pt 2):83–104.

Konig, K., P.T.C. So, W.W. Mantulin, and E. Gratton. 1997. Cellular response to near-infrared femtosecond laser pulses in two-photon microscopes. *Opt. Lett.* 22:135–136.

Millard, A.C., L. Jin, A. Lewis, and L.M. Loew. 2003. Direct measurement of the voltage sensitivity of second-harmonic generation from a membrane dye in patch-clamped cells. *Opt Lett* 28(14):1221–1223.

Millard, A.C., L. Jin, M.D. Wei, J.P. Wuskell, A. Lewis, and L.M. Loew. 2004. Sensitivity of second harmonic generation from styryl dyes to transmembrane potential. *Biophys J* 86(2):1169–1176.

Miller D.M, I. Ortiz, G.C. Berliner, and H.F. Epstein. 1983. Differential localization of two myosins within nematode thick filaments. *Cell* 34:477–490.

Mohler, W., A.C. Millard, and P.J. Campagnola. 2003. Second harmonic generation imaging of endogenous structural proteins. *Methods* 29(1):97–109.

Moreaux, L., O. Sandre, M. Blanchard-desce, and J. Mertz. 2000a. Membrane imaging by simultaneous second-harmonic generation and two-photo microscopy. *Opt. Lett.* 25: 320–322.

Moreaux, L., O. Sandre, S. Charpak, M. Blanchard-Desce, and J. Mertz. 2001. Coherent scattering in multi-harmonic light microscopy. *Biophys. J.* 80:1568–1574.

Moreaux, L., O. Sandre, and J. Mertz. 2000b. Membrane imaging by second-harmonic generation microscopy. *J. Opt. Soc. Am. B* 17:1685–1694.

Muller, M., J.A. Squier, T. Wilson, and G. Brakenhoff. 1998. 3D microscopy of transparent objects using third-harmonic generation. *J. Microscopy* 191:266–272.

Ottani, V., M. Raspanti, and A. Ruggeri. 2001. Collagen structure and functional implications. *Micron* 32:251–260.

Plotnikov, S., V. Juneja, A.B. Isaacson, W.A. Mohler, and P.J. Campagnola. 2006. Optical clearing for improved contrast in second harmonic generation imaging of skeletal muscle. *Biophys J* 90(1):328–339.

Plotnikov, S.V., A.C. Millard, P.J. Campagnola, and W. Mohler. 2005. Characterization of the myosin-based source for second-harmonic generation from muscle sarcomeres. *Biophys J.*

Prockop, D.J., and A. Fertala. 1998. The collagen fibril: the almost crystalline structure. *Journal of Structural Biology* 122:111–118.

Ramanujan, S., A.Pluen, T.D. McKee, E.B. Brown, Y.Boucher, and R.K. Jain. 2002. Diffusion and Convection in Collagen Gels: Implications for Transport in the Tumor Interstitium. *Biophysical Journal* 83:1650–1660.

Shen, Y.R. 1984. The principles of nonlinear optics. John Wiley and Sons, New York, NY.

Shen, Y.R. 1989. Surface properties probed by second-harmonic and sum-frequency generation. *Nature* 337:519–525.

Sheppard, C.J.R., R. Kompfner, J. Gannaway, and D. Walsh. 1977. Scanning harmonic optical microscope. *IEEE J. Quantum Electronics* 13E:100D.

Shribak, M., and R. Oldenbourg. 2003. Techniques for fast and sensitive measurements of two-dimensional birefringence distributions. *Appl Opt* 42(16):3009–3017.

Stoller, P., B.-M. Kim, A.M. Rubinchik, K.M. Reiser, and L.B. Da Silva. 2001. Polarization-dependent optical second-harmonic imaging of a rat-tail tendon. *J.Biomed. Opt.* 7: 205–214.

Stoller, P., K.M. Reiser, P.M. Celliers, and A.M. Rubinchik. 2002. Polarization-Modulated Second Harmonic Generation in Collagen. *Biophys. J.* 82:3330–3342.

Sun, C.-K., S.-W. Chu, S.-Y. Chen, T.-H. Tsai, T.-M. Liu, C.-Y. Lin, and H.-J. Tsai. 2004. Higher harmonic generation microscopy for developmental biology. *J. Struct. Biol* 147:19–30.

Tromberg, B.J., N. Shah, R. Lanning, A. Cerussi, J. Espinoza, T. Pham, L. Svaasand, and J. Butler. 2000. Non-invasive in vivo characterization of breast tumors using photon migration spectroscopy. *Neoplasia* 2(1–2):26–40.

Verbiest, T., S.V. Elshocht, M. Kauranen, L. Hellemans, J. Snauwaert, C. Nuckolls, T.J. Katz, and A. Persoons. 1998. Strong enhancement of nonlinear optical properties through supramolecular chirality. *Science* 282:913–915.

Verbiest, T., M. Kauranen, A. Persoons, M. Ikonen, J. Kurkela, and H. Lemmetyinen. 1994. Nonlinear optical activity and biomolecular chirality. *J. Am. Chem. Soc.* 116:9203–9205.

Waterston, R.H. 1988. The Nematode Caenorhabaditis elegans. Wood WB, editor. Cold Spring Harbor Laboratory, Cold Spring Harbor.

Williams, R.M., W.R. Zipfel, and W.W. Webb. 2005. Interpreting second-harmonic generation images of collagen I fibrils. *Biophys J* 88(2):1377–1386.

Wurpel, G.W.H., J.M. Schins, and M. Mueller. 2002. Chemical specificity in three-dimensional imaging with multiplex coherent anti-Stokes Raman scattering microscopy. *Opt. Lett.* 27:1093–1095.

Yariv, A. 1989. Quantum electronics. Wiley, New York.

Yeh, A.T., B. Choi, J.S. Nelson, and B.J. Tromberg. 2003. Optical clearing affects collagen structure reversibly. *Lasers in Surgery and Medicine*:3–3.

Yeh, A.T., N. Nassif, A. Zoumi, and B.J. Tromberg. 2002. Selective corneal imaging using combined second-harmonic generation and two-photon excited fluorescence. *Opt. Lett.* 27:2082–2084.

Zoumi, A., X. Lu, G.S. Kassab, and B.J. Tromberg. 2004. Imaging coronary artery microstructure using second-harmonic and two-photon fluorescence microscopy. *Biophys J* 87(4):2778–2786.

Zoumi, A., A. Yeh, and B.J. Tromberg. 2002. Imaging cells and extracellular matrix in vivo by using second-harmonic generation and two-photon excited fluorescence. *Proc Natl Acad Sci U S A* 99(17):11014–11019.

Coherent Anti-Stokes Raman Scattering (CARS) Microscopy

Instrumentation and Applications

Eric O. Potma and X. Sunney Xie

Key to the success of optical microscopy in cellular biology is the ability to detect structures with a clear chemical contrast. Identifying the distribution and dynamics of targets in the cell is accomplished only if these compounds can be visualized selectively. Fluorescence microscopy achieves this chemical selectivity through associating structures with a fluorescent tag. Thanks to the vast number of specific fluorescent probes, this recipe has unveiled many of the secrets of the living cell.

Despite the triumph of fluorescence microscopy, some cellular imaging studies require complementary methods to achieve the desired chemical selectivity. Not every cellular compound can be easily labeled; examples include the water molecule and other small molecules in the cell. Furthermore, at low concentrations of the fluorescent probe, photobleaching can severely restrict the observation time. Finally, putative interference of the probe with cellular functioning and possible toxicity limit the applicability of the fluorescence approach.

In the previous chapter it was shown that harmonic generation microscopy allows imaging of certain cellular structures without the addition of labels. In particular, second-harmonic generation (SHG) microscopy has proved very sensitive to fibrillar collagen (Campagnola et al., 2002; Freund & Deutsch, 1986; Williams et al., 2005). The contrast arises because of the unique non-centrosymmetric structure of the collagen fibrils and other molecular assemblies such as microtubules and myosin, which dresses the protein bundles with a nonvanishing second-order nonlinearity. Selectivity in SHG imaging is thus a result of structural arrangement at the supramolecular level.

On the other hand, the contrast in coherent anti-Stokes Raman scattering (CARS) microscopy is spectroscopic in nature. Molecules are addressed through the Raman active vibrational signatures of their chemical bonds, which permits selective visualization of a wide variety of molecular species. Other than in spontaneous Raman, however, the nonlinear CARS signals are very strong and allow for rapid scanning of biological samples. Avoiding the use of extrinsic probes, vibrational microscopy shows great promise as an invaluable tool for cellular imaging.

CARS microscopy was first demonstrated in 1982 by Duncan and associates (Duncan et al., 1982). This initial design was based on a noncollinear beam geometry that is often used in CARS spectroscopy measurements to achieve phase matching. In 1999, Zumbusch and colleagues showed that the use of high numerical objective lenses significantly relaxed the phase-matching condition and that a simple collinear beam geometry was the method of choice (Zumbusch et al., 1999). The intrinsic three-dimensional spatial resolution obtained in this fashion paved the way for the first CARS images of living cells. Subsequent improvements have culminated in a mature nonlinear microscopy technique with imaging properties matching those of two-photon fluorescence microscopy in terms of resolution and imaging speeds. The unique imaging capabilities of CARS microscopy have been used to study dynamic processes in living cells that are difficult to study with fluorescent techniques. Examples include the trafficking and growth of lipid droplets (Nan et al., 2003a), intracellular water diffusion (Potma et al., 2001), and biomedical imaging of tissues in vivo (Evans et al., 2005).

In this chapter we summarize the experimental fundamentals of CARS microscopy. Several CARS imaging modalities will be discussed, including epi-CARS, polarization CARS, and CARS microspectroscopy. In addition, we review recent applications in cellular biophysics and point out prospective applications in biomedical imaging. An account of the underlying physics of CARS microscopy can be found in Chapter 7. For a concise overview of CARS microscopy the reader is also referred to excellent review papers in literature (Cheng & Xie, 2004; Volkmer, 2005).

17.1 IMAGING PROPERTIES OF CARS MICROSCOPY*

CARS microscopy is a solution to the low signal yield in spontaneous Raman microscopy. Because CARS is a nonlinear coherent technique, the imaging properties of the CARS microscope differ substantially from those of a linear Raman imaging system. Here we briefly outline the key differences.

The CARS signal scales quadratically with the pump beam intensity and linearly with the Stokes beam power. This nonlinear power dependence ensures that CARS signals are generated almost exclusively in the vicinity of the focal volume where the excitation density is the highest. Because the signal originates only from the focal spot, a high three-dimensional resolution is attained, similar to other forms of nonlinear optical microscopy. Hence, unlike spontaneous Raman imaging, no confocal pinhole is required for depth-resolved imaging in CARS microscopy.

In incoherent microscopy, such as confocal fluorescence and spontaneous Raman microscopy, the signal measured at a given position of the sample depends on the overlap of the object function $|O(r)|^2$ with an effective intensity point spread function $|h(r)|^2$. The point spread function is defined as the position-dependent signal obtained from a point object, and is proportional to the focal intensity distribution. Every time the position r' of the object is changed, the overlap of $|O(r')|^2$ and $|h(r)|^2$ will vary accordingly. The position-dependent signal is thus a convolution of the object function and the intensity point spread function:

$$I(r) = |h(r)|^2 \otimes |O(r)|^2 \qquad (17.1)$$

where $\otimes$ indicates the convolution sign.

When the object image is a point, we may replace the object function with a Dirac delta function. The microscope response is then:

$$I(r) = |h(r)|^2 \otimes \delta(r) = |h(r)|^2 \qquad (17.2)$$

The point spread function is a convenient measure of the resolution of an incoherent optical microscope.

In CARS, the focal excitation intensity distribution is a well-defined function of the incident pump (E_p) and Stokes (E_S) fields, as discussed in Section 7.2 of Chapter 7. The effective excitation field is given by a product of the focal fields of the incident beams:

$$h_{eff}(r) = E_p^2(r)E_S^*(r) \qquad (17.3)$$

It is therefore tempting to relate the resolution of the microscope to the focal excitation intensity distribution $|h_{eff}|^2$. There is, however, no direct linear relationship between the excitation intensity and the total CARS signal generated. Rather, as explained in Chapter 7, the signal is found from solving the wave equation, which requires a full three-dimensional integration over all the coherent waves emanating from the vicinity of focus.

We can find an expression for the position-dependent CARS signal comparable to equation (17.1). The object at point r is now best characterized in terms of the nonlinear susceptibility $\chi^{(3)}(r)$. If it is assumed that the excitation field is linearly polarized and that only the nonlinear polarization in the x-direction is important, we find from the equations in Chapter 7 that the CARS signal at the far field point R can be approximated as:

$$|E_{as}(\mathrm{R})|^2 \propto \left|\frac{M(\theta,\phi)}{R}\right|^2 \left|\int_V e^{-ik_{as}\mathrm{n\cdot r}}h_{eff}(\mathrm{r'})\chi^{(3)}(r'-r)dV\right|^2 \qquad (17.4)$$

where $M(\theta,\phi)$ is the function that projects the CARS field onto a far field surface Ω defined by the polar angle ϕ and the azimuth angle θ. As before, $\chi^{(3)}$ is the nonlinear susceptibility, k_{as} is the wave vector, and r runs over the volume in the vicinity of the focal spot.

The total signal is found by integrating over Ω, from which we obtain:

$$I(r) \propto \int_\Omega M^2(\theta,\phi)\left|\left(e^{-ik_{as}\mathrm{n\cdot r}}h_{eff}(\mathrm{r})\right)\otimes\chi^{(3)}(r)\right|^2 d\Omega \qquad (17.5)$$

It is clear that there are some striking differences between equations (17.1) and (17.5). Unlike equation (17.1), the convolution is performed before the square modulus is taken, reflecting the coherent imaging properties of CARS. Furthermore, the effective excitation field distribution is multiplied with the phase function $e^{-ik_{as}\mathrm{n\cdot r}}$, which accounts for the coherent addition of the emitted CARS waves in the far field. The phase function depends on the far field angles (θ,ϕ) and cannot be taken out of the integration. Whenever the object is moved, the angular distribution of the signal may change, and the full far field integration in equation (17.5) has to be performed in order to calculate the detected CARS signal. As a consequence, standard deconvolution algorithms based on the simple convolution relation of equation (17.1) cannot be applied straightforwardly to CARS microscopy (Cheng et al., 2002d; Potma et al., 2000).

Because the direct proportionality between the focal intensity distribution function and the spatial-dependent CARS signal is lost, the use of an effective point spread function as a general measure of resolution is compromised in CARS microscopy. However, the point spread function can still give us a rough idea of the resolution of the CARS microscope. When the object is a true point, we may write $\chi^{(3)}(r) = \delta(r)$ and the signal measured is proportional to $\left|h_{eff}(r)\right|^2$. This simple relation results from the fact that when the signal emerges from a single point only, there are no other waves to interfere with. Consequently, its phase becomes irrelevant. For isolated objects that are small enough, much smaller than the optical wavelength, the effective point spread function gives an approximate estimate of the resolving power of the CARS microscope. For larger and multiple objects, however, the concept of the point spread function breaks down and a full evaluation of equation (17.5) is required for an accurate interpretation of the CARS image.

17.2 THE CARS LIGHT SOURCE

CARS is a nonlinear process that is proportional to the square modulus of the third-order susceptibility of the material. Typical values of $\chi^{(3)}$ are on the order of 10^{-14} esu (cm^2/V^2), so strong electric fields are required for inducing a detectable third-order signal (Kajzar & Messier, 1985). The third-order signal becomes comparable to a linear optical signal when the beam intensity is on the order of 10^8 W/cm^2 (or ~ 1 W/μm^2). Such intensities are readily reached with pulsed laser sources. For instance, the power of an individual 100-fs pulse, taken from an 80-MHz pulse train with an average power of 1 mW, is 125 W. Consequently, when focused down to a spot size on the order of μm^2, the light intensity is strong enough to induce a detectable nonlinear optical response. In this section we touch on several considerations important for choosing the optimal ultrafast laser source for CARS microscopy.

17.2.1 Femtosecond or Picosecond?

The integrated CARS signal intensity per pixel scales as:

$$S_{CARS} \propto \frac{t_{dwell}}{(f\tau)^2} \langle P(t) \rangle^3 \tag{17.6}$$

where t_{dwell} is the pixel dwell time, f is the repetition rate, τ is the temporal width of the pump and Stokes pulses, and $\langle P(t) \rangle$ is the total time-integrated average power of the incident radiation.

We see that the total signal scales inversely proportional to the square of the temporal pulse width. Keeping the average power of the pulse train constant, highest CARS signals are thus attained for the shortest pulses.

Next to the signal strength, another important parameter in CARS imaging is the resonant-to-nonresonant background (S_R/S_{NR}) ratio. Generally, this ratio grows smaller for pulses with broader spectra. The width of a typical Raman line in the condensed phase is about 10 cm^{-1}, while the spectral width of a bandwidth-limited 100-fs (Gaussian) pulse measures 333 cm^{-1}. Due to the poor spectral overlap, not all the spectral

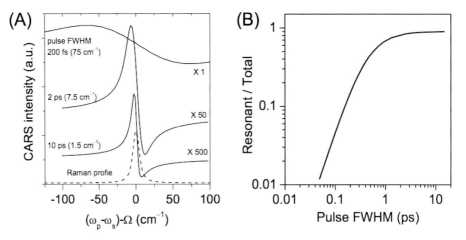

Figure 17.1. Influence of the temporal width of bandwidth limited pulses on the CARS signal. (A) Higher signals are obtained for shorter pulses, at the expense of loss of spectral resolution and stronger nonresonant background contributions. The frequency difference of the pump and the Stokes is indicated by $(\omega_p - \omega_s)$, while the frequency of the Raman vibration is indicated with Ω. (B) Signal-to-nonresonant background ratio as a function of pulse duration. The contrast saturates around 1 to 2 ps, providing an excellent balance between signal strength and signal contrast. For these calculations, a Raman line of width $2\Gamma = 10\,\text{cm}^{-1}$ was used.

components participate in efficiently driving the Raman mode, whereas the entire pulse spectrum contributes to the generation of the nonresonant background. Better overlap is obtained with (bandwidth-limited) ps pulses, leading to higher resonant-to-nonresonant signal ratios. Figure 17.1 shows that for typical Raman lines a balance between generating strong signals, on the one hand, and maintaining high S_R/S_{NR} ratios on the other, is obtained for 1- to 2-ps pulses (Cheng et al., 2001b).

Matching the pulse spectrum to the spectral width of the Raman bands thus yields better contrast in CARS imaging. Hashimoto and colleagues used a 2-ps light source to achieve a spectral resolution high enough for resolving the narrow features in the fingerprint region (Hashimoto et al., 2000), and Cheng and associates showed that two synchronized picosecond Ti:sapphire lasers constituted improved S_R/S_{NR} ratios in CARS imaging relative to femtosecond pulse trains (Cheng et al., 2001b). Similarly, to match the spectral width of the much broader water spectrum ($>300\ \text{cm}^{-1}$), Potma and coworkers used 100-fs pulses from an optical parametric oscillator (Potma et al., 2001).

High spectral resolution can also be attained with femtosecond pulses when the spectral phase is shaped accordingly. Overlapping an fs pump pulse with a stretched, linearly chirped Stokes pulse allows selection of certain frequency components, narrowing down the spectral resolution of the microscope (Knutsen et al., 2004). By stretching both pulses linearly, a similar resolution enhancement can be achieved (Hellerer et al., 2004). In this scheme, spectral tuning is accomplished by varying the time delay between the pulses. More advanced pulse shaping is realized with a spatial light modulator, which has been applied to suppress the nonresonant background generated by broad-bandwidth fs pulses (see also Section 17.3.4) (Dudovich et al., 2002; Oron et al., 2002).

17.2.2 Repetition Rate and Pulse Energy

Picosecond pulses from a Ti:sapphire oscillator typically have an energy of $\sim$6 nJ at an average power level of about 500 mW and a repetition rate of around 80 MHz. Such laser pulses have been successfully applied for high-speed CARS imaging (Cheng et al., 2002a; Nan et al., 2003a). The CARS signal can be enhanced by a factor of m^2 ($= m^3/m$) if the pump and Stokes pulse energies are increased by a factor of m and the repetition frequency is lowered by the same factor (see equation (17.6)). Keeping the pulse energy high while reducing the repetition rate is readily achieved by pulse picking the laser beams, allowing generation of strong CARS signals at low average powers (Cheng et al., 2001a). Higher pulse energies can be obtained by cavity dumping the pulsed laser source, yielding pulse energies in the 100-nJ range (Killi et al., 2005). For pulses even more intense, pulse amplification schemes can be used. For instance, the ps-Ti:sapphire pulses can be passively amplified with an high finesse external cavity, yielding pulse amplification factors of 30 while scaling down the repetition rate to the 100-kHz range (Potma et al., 2003).

In practice, there is a limit to the maximum tolerable pulse power as set by the photodamaging of the sample. Studies by Hopt and Neher on cellular photodamaging induced by ultrafast near-infrared laser sources under typical in vitro imaging conditions revealed that the number of scans before photodamaging occurs is empirically given by (Hopt & Neher, 2001):

$$\#scans \propto \frac{(f\tau)^{1.5}}{t_{dwell}\,\langle P(t)\rangle^{2.5}} \tag{17.7}$$

Equation (17.2) indicates that photodamaging is a nonlinear function of the input beams, underlining that higher-order light–matter interactions contribute to the damaging process. Indeed, it was shown that below illumination doses of 100 mW, linear heating of the sample is relatively insignificant for near-infrared radiation (Booth & Hell, 1998; Schonle & Hell, 1998), and photodamaging is predominantly nonlinear in nature (Hopt & Neher, 2001; Konig et al., 1999). At 800 nm, it was found that 2.5 mW from a 150-fs, 82-MHz pulse constitutes a safe illumination condition for cellular imaging in two-photon fluorescence microscopy (Hopt & Neher, 2001). For 2-ps pulses used in CARS, this would translate into a maximum allowable peak energy of 2.1 nJ (1.1 kW) at 100 kHz and 13 nJ (6.7 kW) at 1 kHz. From these values and equation (17.6) it can be learned that because of the limitation of photodamage, there is no need to lower the repetition rate from 100 kHz to 1 kHz. In general, for CARS microscopy, repetition rates in the range of 0.1 to 100 MHz give good imaging results for picosecond pulses of up to a few nJ.

17.2.3 Wavelength Range

To limit linear heating of the aqueous sample and nonlinear absorption by endogenous chromophores, near-infrared light sources in the range of 700 to 1,300 nm are well matched for CARS microscopy. Triggered by the development of two-photon microscopes, microscope optics optimized for the near-infrared range have recently become available from manufacturers, guaranteeing excellent transmission and focusing properties. In addition, far removed from electronic resonances, less two-photon enhanced nonresonant background is induced if near-infrared radiation is employed relative to the use of visible excitation.

17.2.4 Synchronization

CARS requires a pump and a Stokes beam of different color. When two independent laser sources, such as Ti:sapphire lasers, are used, active synchronization of the pulse trains is required. Temporal fluctuations between the pulse trains, called pulse jitter, introduce intensity variations of the CARS signal. Tight synchronization can be achieved with higher harmonic locking techniques, which reduce pulse jitter to less than 100 fs (Jones et al., 2002; Potma et al., 2002). Electronic modules for synchronizing two independent laser sources have recently become commercially available from laser manufacturers.

The timing jitter problem can be avoided if the second beam is generated through frequency conversion of part of the fundamental laser beam. This is readily realized if a portion of an ultrafast laser beam is spectrally broadened and shifted in an optical fiber. Using a femtosecond Ti:sapphire laser as the fundamental, frequency-shifted beams have been generated in photonic crystal fibers (Kano & Hamaguchi, 2005; Paulsen et al., 2003) or a tapered fiber (Kee & Cicerone, 2004), producing intrinsically synchronized pump and Stokes pulse trains. With the aid of spectral phase shaping, Dudovich and colleagues generated CARS signals with a single broadband pulse (Dudovich et al., 2002; Oron et al., 2002). The use of a single fs Ti:sapphire laser represents a cost-efficient solution for CARS. While less efficient for single-band CARS imaging, broadband light sources are well suited for CARS microspectroscopy, which will be further discussed in Section 17.4.

Intense, spectrally narrow pulses can be attained with synchronously pumped optical parametric oscillators (OPOs). This approach offers the advantage of broadly tuneable, synchronized pulses. Next to Ti:sapphire lasers, mode-locked ps-solid-state lasers equipped with semiconductor saturable absorber mirrors (SESAM) technology constitute an attractive and compact pump source for OPOs. In combination with commercially available OPOs based on periodically poled crystals, synchronously pumped parametric devices are a very robust and affordable light source for CARS microscopy.

17.3 THE CARS LASER SCANNING MICROSCOPE

The heart and soul of the CARS imaging instrument is the optical microscope. To ensure the relatively high power density required for engendering the CARS signal, the CARS microscope is based on point illuminating the sample with a tight focal spot. The optical microscope comprises the raster scanning devices for generating the image, filters and detectors, and, most importantly, the objective lens and relay optics. With minor modifications, many commercially available (confocal) scanning microscopes can be converted into a CARS imaging system.

A schematic of the CARS microscope setup is shown in Figure 17.2. Before coupling into the microscope, the properties of the synchronized pump and Stokes laser beams are tailored for optimal CARS generation in focus. To control the divergence of the beams, a telescope is placed in each of the laser beam paths. Optionally, the quality of the beam

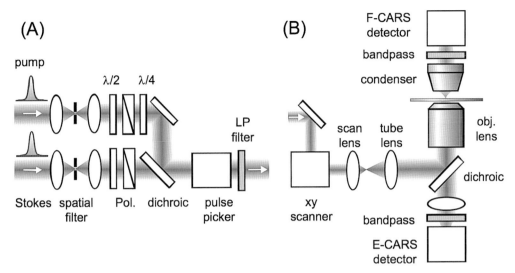

Figure 17.2. CARS microscope setup. (A) The pump and Stokes beam parameters are optimized in terms of spatial mode and polarization orientation. The beams are spatially overlapped before coupling into the microscope. (B) The CARS microscope shows much similarity with the (confocal) scanning microscope. For rapid scanning a beam scanner is employed. Strongest signals are collected at the non-descanned forward and backward detectors. A proper bandpass filter is used to reject all incident radiation while passing the anti-Stokes radiation.

mode can be improved with a spatial filter. A half waveplate and a polarizer are inserted to control the polarization orientation of the beams for polarization-sensitive CARS. To compensate for birefringent downstream optical elements, a quarter waveplate is often used. Reduced repetition rates can be achieved by pulse picking the beams with a Pockel's or a Bragg cell. The beams are temporally overlapped, collinearly combined on a dichroic mirror, and interfaced with the optical microscope. In some cases a long pass filter may need to be placed in the beam path to prevent leakage of the lasers' broadband superfluorescence into the microscope.

A beam scanner is the method of choice for rapid scanning. Many commercial laser scanning microscopes are outfitted with a pair of galvanometric mirrors that can be utilized for CARS imaging. The plane between the scanning mirrors is imaged onto the back aperture of the objective lens with a telescope formed by a scan lens and a tube lens. The beam's inclination angle applied with the galvanometric mirror pair is translated into a spatial shift of the focus in the focal plane. Subsecond frame rates can be conveniently accomplished this way. Even faster point scanning can be achieved by replacing one of the galvos with a fast-spinning polygon, bringing the frame scanning time down to video rates. Alternatively, the sample can be scanned with a piezoelectric stage, albeit at slower scanning rates.

The objective lens focuses the incoming beams down to a submicrometer-sized spot. The performance of the objective is crucial for generating the strongest possible CARS signals. Next to the nonlinear dependence of CARS on the total power density, which necessitates a clean and tight focal volume, the multicolor character of the CARS technique requires achromatic focusing. Well-corrected, high numerical aperture (NA) objectives generally fulfill these requirements. The resolving power of the CARS microscope is approximately similar to that of a two-photon fluorescence microscope, with a typical resolution of ~0.3 μm in the lateral direction and ~0.8 μm in the axial dimension for a 1.2 NA lens and near-infrared excitation.

Signal is detected either in the forward or backward direction. A wide-aperture condenser can be used for capturing the highly directional forward-propagating signal, which is spectrally filtered and focused onto a photodetector. Note that in the case of beam scanning, the signal in the forward direction is non-descanned, and the light is conveniently intercepted by a wide-aperture detector such as a photomultiplier. For highest signals in the epi-direction, the signal is detected in the non-descanned mode. Recall that because the CARS signal originates from the focal volume only, confocal detection is not required.

17.4 CARS MICROSCOPY TECHNIQUES FOR NONRESONANT BACKGROUND SUPPRESSION

The resonant CARS signal is inherently accompanied by a nonresonant background signal. For detection of weak vibrational signals, the electronic nonresonant background can be the limiting factor in CARS microscopy. The nonresonant background not only introduces an offset but also interferes with the resonant contribution, which can complicate image analysis. In this section we briefly discuss several techniques for suppressing the nonresonant background.

17.4.1 Epi-detection

Every object contributes to the nonresonant background. Both the resonant particle and the aqueous surrounding induce an electronic background signal. The aqueous environment produces a ubiquitous background that can be stronger than the resonant response from a small object in focus. Because epi-CARS (E-CARS) is based on a size-selective mechanism, the nonresonant signal from the aqueous surrounding is suppressed while the signal from small objects is retained. Excellent contrast from live cells is obtained with this relatively simple method of background suppression (Fig. 17.3) (Cheng et al., 2001b; Volkmer et al., 2001). Note that the nonresonant signal from the object itself is not reduced in E-CARS. The size-selective contrast is particularly relevant in comparatively transparent samples such as cell monolayers. In highly scattering media such as tissues, the E-CARS signal is dominated by backscattering of forward-propagating contributions, with no direct suppression of the nonresonant background.

17.4.2 Polarization-Sensitive Detection

This method is based on the different polarization properties of the electronic and resonant portions of the third-order polarization (Ahkmanov et al., 1977; Brakel & Schneider, 1988; Oudar et al., 1979). In polarization-sensitive CARS (P-CARS), an analyzer in front of the detector blocks the nonresonant signal, whereas a portion of the differently polarized resonant signal leaks through the analyzer. More on the principles of background suppression through polarization-sensitive detection can be found in Section 7.2 of Chapter 7.

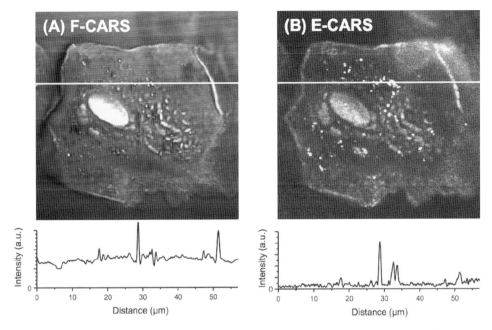

Figure 17.3. Forward-detected (F-CARS) and Epi-detected (E-CARS) images of an epithelial cell with the Raman shift tuned to the C-H stretching vibration at $2,850\,\mathrm{cm}^{-1}$. The intensity profiles of the two white lines are shown in the lower panels. Note the offset in the F-CARS image due to the nonresonant background of the solvent. The background-free E-CARS image shows better contrast for small features.

Because P-CARS detects only the resonant part of the signal, the peak positions in the CARS spectrum coincide with those in the Raman spectrum. Figure 17.4A shows the P-CARS spectrum of N-methylacetamide, a model compound containing the characteristic amide I vibration at $1,652\,\mathrm{cm}^{-1}$, which is a signature for peptides and proteins. The background-free P-CARS image of an unstained epithelial cell based on this signature is shown in Figure 17.4B (Cheng et al., 2001a). Tuning away from the Amide I band to $1,745\,\mathrm{cm}^{-1}$ results in a faint contrast (see Fig. 17.4C), proving that the image contrast was due to proteins, the distribution of which is heterogeneous in the cell. P-CARS images sometimes require longer acquisition times because besides suppression of the nonresonant background, the vibrationally resonant signal is reduced as well.

17.4.3 Time-Resolved CARS Detection

The vibrationally resonant signal can be separated from the nonresonant electronic contribution by use of pulse-sequenced detection with femtosecond pulse excitation (Kamga & Sceats, 1980; Laubereau & Kaiser, 1978). In time-resolved CARS detection, a signal-generating probe pulse is time delayed with respect to a temporally overlapped pump/Stokes pulse pair. Because of the instantaneous dephasing time of the nonresonant electronic levels, the nonresonant CARS signal exists only when the pump/Stokes pulse pair overlaps with the probe pulse. On the other hand, the vibrationally resonant CARS

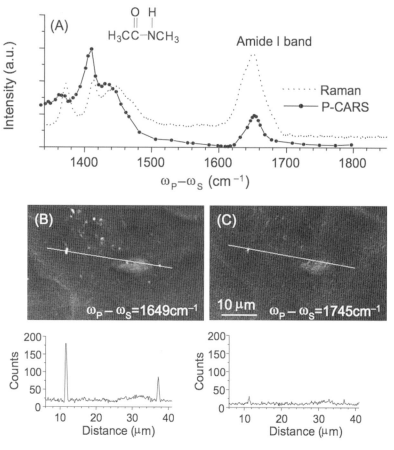

Figure 17.4. (A) P-CARS and spontaneous Raman spectra of pure N-methylacetamide liquid recorded at room temperature. (B, C) Polarization CARS images of an unstained epithelial cell with $\omega_p - \omega_s$ tuned to 1,650 cm^{-1} and 1,745 cm^{-1}, respectively. Each image was acquired by raster-scanning the sample with an acquisition time of 8 minutes. The pump and Stokes power were 1.8 and 1.0 mW, respectively, at a repetition rate of 400 kHz.

signal decays with the finite dephasing time of the vibrational mode. The dephasing time is related to the spectral width of the corresponding Raman band and is typically several hundred femtoseconds for a mode in the condensed phase (Fickenscher et al., 1992). The nonresonant background in the CARS images can thus be eliminated by introducing a suitable delay between the femtosecond pump/Stokes and the probe pulses. Time-resolved CARS imaging has been demonstrated by Volkmer and coworkers with a three-color excitation scheme (Volkmer, 2005; Volkmer et al., 2001). The disadvantages of this approach are the photodamage induced by the fs pulses and the complication of adding a third laser beam of a different color.

17.4.4 Phase Control

With femtosecond lasers, another way to reduce the electronic contribution is phase shaping of the pulses. Phase shaping of femtosecond pump and Stokes pulses suppresses the nonresonant signal by introducing a phase-mismatched coherent addition of non-resonant spectral components, whereas the resonant contributions still add up in phase (Oron et al., 2002). This method has been applied to CARS spectroscopy and microscopy with single femtosecond pulses (Dudovich et al., 2002). Next to phase-only pulse shaping, this method can be extended to controlling the polarization of the spectral components as well. In this fashion, Oron and coworkers have obtained background-free CARS signals by using a single broadband laser (Oron et al., 2003).

For CARS imaging with ps pulses, phase control can be achieved through hetero-dyning the signal with a reference beam at the anti-Stokes wavelength. Heterodyne CARS interferometry can be used to suppress the background (Marowsky & Luepke, 1990; Yacoby & Fitzgibbon, 1980) and to isolate the imaginary part of the nonlinear susceptibility $\left(\mathrm{Im}\{\chi_r^{(3)}\}\right)$. Evans and colleagues have implemented this concept for extracting the real and imaginary parts of the CARS spectra in an optical microscope (Evans et al., 2004). Another version of the CARS interferometer has been proposed for detecting background-free spectroscopic images in optical coherence tomography (Marks & Boppart, 2004; Vinegoni et al., 2004).

17.5 CARS MICROSPECTROSCOPY

In CARS imaging the signal intensity of a single vibrational band is measured from many points in the sample to create an image with chemical contrast. In keeping the focal spot fixed in space, spectral and temporally resolved spectroscopic infor-mation can be obtained from microscopic volumes. In this section two forms of CARS microspectroscopy are illuminated: multiplex-CARS and CARS correlation spectroscopy.

17.5.1 Multiplex CARS

In multiplex-CARS spectroscopy, a portion of the CARS spectrum is measured in a single shot (Akhmanov et al., 1974). To achieve this, a broadband light source (typically $>100\,\mathrm{cm}^{-1}$) is used as the Stokes beam for providing the required spectral bandwidth. The pump beam is a narrow-band laser beam that defines the spectral resolution. The pump and Stokes beam now drive multiple oscillators with different Raman frequencies, which in turn imprint their signatures in the broadband anti-Stokes signal. By spectrally resolving the signal, a multiple-band CARS spectrum can be obtained. The multiplex-CARS energy diagram is depicted in Figure 17.5A.

Otto and associates applied this spectroscopy method in a microscopic configuration for high-sensitivity vibrational measurements on porphyrins (Otto et al., 2001). The technique was later refined for spatially resolved CARS microspectroscopy (Cheng et al., 2002c; Muller & Schins, 2002). Multiplex-CARS has been used to map out differences in the thermodynamic phase of mixtures of multilamellar vesicles, by using a combination of a broadband fs and a narrow-band ps Ti:sapphire laser (Cheng et al., 2002c; Muller & Schins, 2002). Using a similar approach, Wurpel and coworkers recorded 300-cm^{-1}-wide multiplex-CARS spectra of single lipid mono- and bilayers (see Fig. 17.5B) (Wurpel et al., 2004).

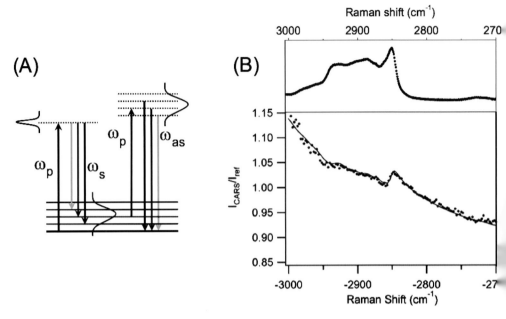

Figure 17.5. (A) Energy diagram for multiplex-CARS. A narrow-band pump and a broadband Stokes address a multifold of Raman modes that give rise to a broadband anti-Stokes signal. (B) Multiplex CARS spectrum of a single bilayer of DOPC (1,2-dioleoyl-*sn*-glycero-3-phosphocholine) on a glass–water interface. The total exposure time was 0.5 seconds. The slanted background is the tail of the resonant contribution from water. The upper panel shows the spontaneous Raman spectrum of pure DOPC for comparison. (Reprinted in part with permission from Wurpel et al., J. Phys. Chem B, Vol 108, page 3401, 2004. Copyright 2004 American Chemical Society.)

Even wider spectra can be obtained by spectrally broadening the Stokes pulses with optical fibers. Kee and Cicerone used a tapered fiber to generate Stokes light that spans more than $2,500\,\text{cm}^{-1}$, allowing single-shot recordings of CARS spectra that cover most of the molecular vibrations of interest (Kee & Cicerone, 2004).

17.5.2 CARS Correlation Spectroscopy

Although the scanning CARS microscope is capable of resolving subsecond dynamics in cells, much faster dynamical events are generally out of reach. By fixing the location of the focal spot and recording the temporal fluctuation of the signal with a high acquisition rate, rapid dynamics such as diffusion of small particles can be probed. This method is based on correlation spectroscopy and is quite similar to fluorescence correlation spectroscopy (FCS) (Elson & Magde, 1974; Schwille, 2001). Unlike in FCS, however, CARS correlation spectroscopy (CARS-CS) probes dynamics of particles through their vibrational signatures, which avoids the use of fluorescent labels.

The autocorrelation of the temporal fluctuations reflects the distribution of residence times of the particle of interest, and thus depends on the diffusion properties of the objects. Diffusion coefficients of polystyrene beads were successfully determined using the CARS-CS technique (Cheng et al., 2002b; Hellerer et al., 2002). In addition,

CARS-CS proved useful for measuring the particle concentration and viscosity of the surrounding medium.

Although the autocorrelation function (ACF) measured in CARS-CS shows much similarity to the FCS autocorrelation curve, there are some marked differences as well. The coherent addition of waves in CARS-CS gives rise to higher-order correlation terms not seen in FCS. These higher-order contributions are negligibly small when the signal is detected in the forward direction, and the ACF has the same form as in FCS. In the backward direction, however, the higher-order correlations can no longer be neglected. The ACF can be written as (Cheng et al., 2002b):

$$G_{E-CARS} = \frac{2\sqrt{2}}{\langle N \rangle} f(2\tau) + \exp\left(-\frac{2k^2 r_0^2 \tau/\tau_0}{1 + r_0^2 \tau/(z_o^2 \tau_0)}\right) [f(\tau)]^2 \qquad (17.8)$$

where

$$f(t) = \left(1 + \frac{t}{\tau_D}\right)^{-1} \left(1 + \frac{r_0^2 t}{z_0^2 \tau_D}\right)^{-1/2} \qquad (17.9)$$

where τ_D is the particle diffusion time, and r_o and z_0 are the lateral and axial widths of the Gaussian focal volume, respectively.

The first term on the right-hand side of equation (17.8) is identical to the ACF measured in FCS, while the second is unique to CARS-CS. The first term vanishes for high particle concentrations. On the other hand, the latter term is independent of the average number of particles $\langle N \rangle$ in focus, and dominates the ACF if $\langle N \rangle \gg 1$. The difference between CARS-CS and FCS is illustrated in Figure 17.6. The remarkable sensitivity of E-CARS-CS to high particle concentrations has no analog in FCS, and opens doors to probing particle diffusion in crowded environments

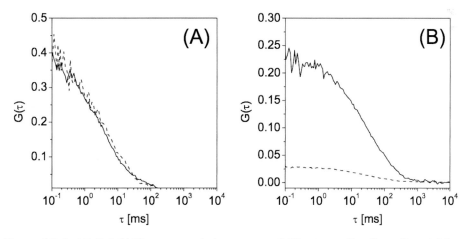

Figure 17.6. (A) Epi-CARS autocorrelation curves of 50-nm small unilamellar vesicles (SUV) of DOPC in water at different concentrations: ~4 vesicles in focus (solid line) and ~40 vesicles in focus (dashed lines). (B) Identical measurements with labeled 50-nm SUVs using fluorescence correlation spectroscopy. Note the difference in concentration dependence of the CARS-CS and FCS autocorrelation curves.

such as in cells. In addition, the ability to measure higher-order correlations makes CARS-CS a useful technique for monitoring cluster and aggregate dynamics (Palmer & Thompson, 1987).

17.6 CARS IMAGING OF BIOLOGICAL SAMPLES

The recent technological advances of CARS microscopy have significantly improved the technique's capabilities for highly sensitive imaging of biological specimens. Exciting applications are beginning to emerge. In this section we highlight several biological applications of coherent Raman imaging.

17.6.1 In Vitro Systems

CARS is very sensitive to lipids. The acyl chains are abundant in CH_2 groups, which exhibit a strong Raman active symmetric stretch vibration at $2,845$ cm^{-1}. When tuned to the symmetric stretch, single phospholipid bilayers can be visualized (Potma & Xie, 2003). The ability to image single bilayers is very useful for the study of lipid phase segregation in membranes. Phase-segregated domains, also known as lipid rafts, form if lipids of different thermodynamic phases are mixed. Lipid rafts are thought

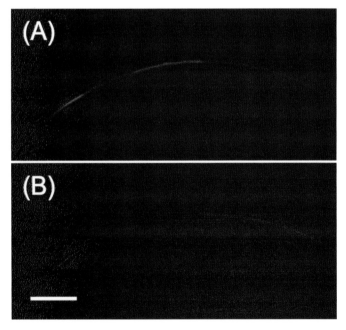

Figure 17.7. (A) Part of a giant unilamellar vesicle composed of an equimolar ratio of the phospholipids DOPC and DSPC, showing a clear phase separation of the lipids. To enhance contrast, the deuterated DSPC was used and the image was taken at the C-D stretch vibrational band at 2.090 cm^{-1}, giving rise to bright signals from the DSPC-enriched domains. (B) Same vesicle imaged at $2,140$ cm^{-1}. Note that the resonant signal has now disappeared.

to play crucial roles in such processes as protein sorting and signaling on cell membranes. With CARS, micrometer-sized lipid domains can be directly detected without the introduction of fluorescent membrane labels (Li et al., 2005; Potma & Xie, 2005). In Figure 17.7, a clear phase separation is observed between the phospholipids DOPC (1,2-dioleoyl-sn-glycero-3-phosphocholine) and deuterated DSPC (1,2-distearoyl-sn-glycero-3-phosphocholine), underlining the potential for CARS microscopy to detect lipid rafts in cells noninvasively. Details of the chain order in single lipid membranes in different thermodynamic states can be determined from multiplex CARS spectra, as shown by Wurpel and colleagues (Wurpel et al., 2004).

17.6.2 Cellular Imaging

The great advantage of CARS imaging over spontaneous Raman microscopy is its capability to capture rapid dynamics in unstained cells with chemical selectivity. Nan and associates used CARS microscopy to study the growth and transport of lipid droplets in live cells (Nan et al., 2003b). Lipid droplets contain neutral lipid in the form of triglycerides and are thought to play important roles in various cellular processes. Instead of using conventional staining of the droplets, which requires fixing of the cells, CARS microscopy allowed continuous monitoring of the droplets' dynamics. In this way, an intermediate stage in the growth process was discovered that had gone unnoticed in fluorescence studies (Fig. 17.8).

Cheng and colleagues followed apoptosis induced by L-asparaginase in unstained NIH3T3 fibroblasts with their CARS microscope (Cheng et al., 2002a). By tuning to the CH_2 lipid vibration, they were able to identify different stages in the apoptotic process, such as the compaction of cytoplasmic organelles and the rounding of the cells. Potma and associates used the resonant CARS signal from the O-H stretching vibration of water to visualize intracellular hydrodynamics (Potma et al., 2001). By

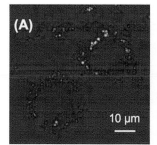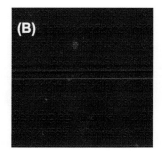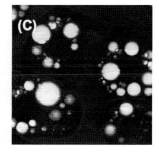

Figure 17.8. P-CARS images of 3T3-L1 preadipocytes differentiating into adipocyte cells at different stages, recorded at the C-H stretching vibration of $2,845\,cm^{-1}$. Differentiation is induced with insulin, isobutylmethylxanthine and dexamethasone 2 days after confluence (A). The dots seen in (A) are lipid droplets. Note the clearance of lipid droplets from the cytoplasm 48 hours after initiation (B). After that, cells accumulate lipid droplets again as differentiation goes on. In fully differentiated cells (C), the cytoplasm are full of large droplets (Nan et al., 2003a). Pump beam power was 15 mW and Stokes beam power was 7.5 mW at a repetition rate of 80 MHz. Image acquisition time was 2.7 s.

rapidly line scanning the laser beams, the intracellular water diffusion coefficient and the membrane permeability were determined from spatiotemporal CARS traces.

To increase the molecular specificity in CARS microscopy, isotopic substitution can be used. The Raman shift of the aliphatic and the aromatic C-D vibration bands lies in the region of 2,100 to 2,300 cm^{-1}, isolated from the Raman bands of endogenous molecules. In an early work using picosecond dye lasers, Duncan and colleagues showed that CARS microscopy could distinguish deuterated liposomes from non-deuterated ones (Duncan, 1984). The deuteration method has been applied to living cells by Holtom and coworkers, who demonstrated selective mapping of deuterated lipid vesicles in a macrophage cell (Holtom et al., 2001).

17.6.3 Tissue Imaging

In combining chemical selectivity with noninvasive imaging, CARS microscopy is particularly apt for biomedical applications. Optimization of the laser radiation, in terms of wavelength, power, and stability, is key to moving into the biomedical direction. With the aid of near-infrared ps-solid-state semiconductor sources, strong CARS signals can be generated far below the photodamaging threshold, allowing acquisition times as fast as video-rate (Evans et al., 2005).

CARS microscopy has been successfully used for imaging of nonstained axonal myelin in spinal tissues in vitro (Wang et al., 2005). In this study, both the forward and the backward CARS signals from the tissue slab were detected. Imaging of even thicker tissue requires signal detection in the epi-direction because of the tissue's limited transmission properties. The scattering properties of the tissue have an effect on the contrast observed in E-CARS, as scattering events redirect a portion of the forward-propagating photons back into the epi-direction. The detected signal is thus dominated by backscattered contributions of forward-generated signal, yielding epi-signals much stronger than those observed from relatively transparent samples. Furthermore, the contrast is unaffected by the size-selective filtering property of E-CARS, which allows a rather straightforward interpretation of the images. Preliminary studies have already demonstrated that lipid distributions can be visualized in skin tissue of live animals, underlining the vast potential for CARS biomedical imaging (Fig. 17.9) (Evans et al., 2005).

17.7 OTHER APPLICATIONS OF CARS MICROSCOPY

17.7.1 CARS and Material Sciences

Applications of CARS microscopy are not limited to biological systems. The high imaging sensitivity of CARS may prove useful for the rapid characterization of solid-state materials as well. It was shown, for instance, that CARS enables the microscopic analysis of polystyrene, poly(methyl methacrylate), and poly(ethylene terephthalate) polymer blends (Kee & Cicerone, 2004). In another study, CARS microscopy was employed for high-resolution imaging of photoresists (Potma et al., 2004). The stretching vibration of the tert-butoxyl carbonyl group of the polystyrene-derived photoresist at 880 cm^{-1} was used to chemically map lithographic line patterns with a line spacing of 200 nm (Fig. 17.10).

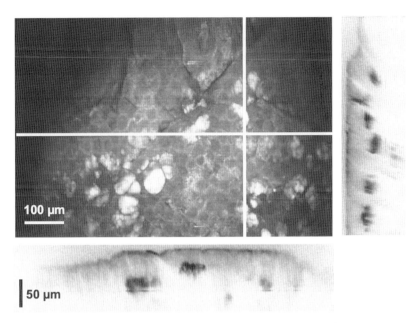

Figure 17.9. CARS skin tissue imaging of the mouse ear in vivo. The image is a two-dimensional representation of a stack of 60 depth-resolved xy images separated by 2 μm. The panels on the side and the bottom show the xy and xz cross-sections, respectively, taken at the white lines in the image. The lasers are tuned to the 2,845 cm^{-1} lipid vibration. Lipid-containing adipocytes and sebaceous glands are clearly resolved.

Figure 17.10. Chemical imaging with E-CARS of a patterned polymer film with a line spacing of 400 nm. The lasers are tuned to the tert-butoxyl carbonyl group, indicative of the chemical pattern imprinted in the polymer photoresist. The nonresonant background has been subtracted.

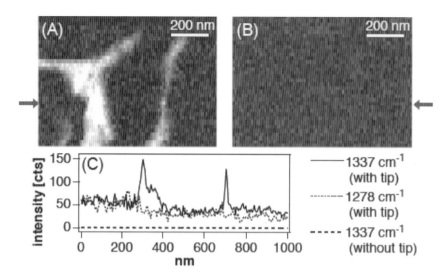

Figure 17.11. Tip-enhanced near-field images of DNA bundles on a coverslip. (A) CARS image on-resonant with the ring breathing mode of adenine at 1,337 cm^{-1}. (B) Similar image taken at an off-resonant Raman shift of 1,278 cm^{-1}. (C) Cross-section of the images taken at the arrows. (Reprinted figure with permission from Ichimura et al., Phys. Rev. Lett. Vol 92, page 220801–3 (2004). Copyright 2004 by the American Physical Society.)

17.7.2 Near-Field CARS Microscopy

The spatial resolution of the CARS microscope is restricted by the diffraction-limited focal volumes of the pump and Stokes beams. Using high-NA objectives and near-infrared radiation, lateral resolutions down to ~0.25 µm can be reached. If a higher spatial confinement is desired, coherent CARS excitation can be combined with near-field optical methods. Schaller and associates used a small-aperture optical fiber to detect the CARS radiation from areas as small as 128 nm (Schaller et al., 2002). Their CARS near-field scanning optical microscope (NSOM) proved successful in producing chemical-selective maps of lipid structures in cells. Even higher resolutions were achieved by Ichimura and coworkers by using the field enhancement effect at sharp metallic tips (Ichimura et al., 2004). Strong CARS emission was observed from the vicinity of the metal probe, bringing the resolution down to ~15 nm. The tip-enhanced CARS microscope was utilized to visualize DNA bundles on glass coverslips (Fig. 17.11).

17.8 PROSPECTS AND CONCLUSIONS

In this chapter we have given a brief overview of the experimental basics of CARS microscopy. In a relative short time, CARS microscopy has emerged from a proof-of-principle demonstration to a mature nonlinear imaging technique with tangible applications in biology and material sciences. Technological advancements, in particular the continuous evolution in laser light sources, will continue to push the envelope and improve the sensitivity and versatility of the CARS imaging technique. The potential of CARS microscopy to contribute solutions to biomedical problems will surely accelerate this development. The examples listed in this chapter are the only the beginnings of the exciting applications of this rapidly growing field.

REFERENCES

Ahkmanov, S.A., A.F. Bunkin, S.G. Ivanov, and N.I. Koroteev. 1977. Coherent ellipsometry of Raman scattered light. *JETP Lett.* 25:416–420.

Akhmanov, S.A., N.I. Koroteev, and A.I. Kholodnykh. 1974. Excitation of the coherent phonons of Eg-type in calcite by means of the active spectroscopy method. *J. Raman Spectrosc.* 2:239.

Booth, M.J., and S.W. Hell. 1998. Continuous wave excitation two-photon fluorescence microscopy exemplified with the 647-nm ArKr laser line. *J. Microsc.* 190:298–304.

Brakel, R., and F.W. Schneider. 1988. Polarization CARS spectroscopy. *In* Advances in Nonlinear Spectroscopy. Clark RJH, Hester RE, editors. John Wiley & Sons Ltd., New York. 149.

Campagnola, P.J., A.C. Millard, M. Terasaki, P.E. Hoppe, C.J. Malone, and W.A. Mohler. 2002. Three-dimensional high-resolution second harmonic generation imaging of endogenous structural proteins in biological tissues. *Biophys. J.* 82: 493–508.

Cheng, J.X., L.D. Book, and X.S. Xie. 2001a. Polarization coherent anti-Stokes Raman scattering microscopy. *Opt. Lett.* 26:1341–1343.

Cheng, J.X., Y.K. Jia, G. Zheng, and X.S. Xie. 2002a. Laser-scanning coherent anti-Stokes Raman scattering microscopy and applications to cell biology. *Biophys. J.* 83:502–509.

Cheng, J.X., S. Pautot, D.A. Weitz, and X.S. Xie. 2003. Ordering of water molecules between phopholipid bilayers visualized by coherent anti-Stokes Raman scattering microscopy. *Proc. Natl. Acad. Sci. USA* 100:9826.

Cheng, J.X., E.O. Potma, and X.S. Xie. 2002b. Coherent anti-Stokes Raman scattering correlation spectroscopy: Probing dynamical processes with chemical selectivity. *J. Phys. Chem. A* 106:8561–8568.

Cheng, J.X., A. Volkmer, L.D. Book, and X.S. Xie. 2001b. An epi-detected coherent anti-Stokes Raman scattering (E-CARS) microscope with high spectral resolution and high sensitivity. *J. Phys. Chem. B* 105:1277–1280.

Cheng, J.X., A. Volkmer, L.D. Book, and X.S. Xie. 2002c. Multiplex coherent anti-Stokes Raman scattering microspectroscopy and study of lipid vesicles. *J. Phys. Chem. B* 106:8493–8498.

Cheng, J.X., A. Volkmer, and X.S. Xie. 2002d. Theoretical and experimental characterization of coherent anti-Stokes Raman scattering microscopy. *J. Opt. Soc. Am. B* 19:1363–1375.

Cheng, J.X., and X.S. Xie. 2004. Coherent anti-Stokes Raman scattering microscopy: instrumentation, theory and applications. *J. Phys. Chem. B* 108:827–840.

Dudovich, N., D. Oron, and Y. Silberberg. 2002. Single-pulse coherently controlled nonlinear Raman spectroscopy and microscopy. *Nature* 418:512–514.

Duncan, M., J. Reintjes, and T.J. Manuccia. 1982. Scanning coherent anti-Stokes Raman microscope. *Opt. Lett.* 7:350–352.

Duncan, M.D. 1984. Molecular discrimination and contrast enhancement using a scanning coherent anti-Stokes Raman microscope. *Opt. Comm.* 50(5):307–312.

Elson, E.L., and D. Magde. 1974. Fluorescence correlation spectroscopy I. conceptual basis and theory. *Biopolymers* 13:1–27.

Evans, C.L., E.O. Potma, M. Puoris'haag, D. Côté, C. Lin, and X.S. Xie. 2005. Chemical imaging of tissue in vivo with video-rate coherent anti-Stokes Raman scattering (CARS) microscopy. *Proc. Natl. Acad. Sci. USA* 102:16807–16812.

Evans, C.L., E.O. Potma, and X.S. Xie. 2004. Coherent anti-Stokes Raman scattering spectral interferometry: determination of the real and imaginary components of nonlinear suscepstibility for vibrational microscopy. *Opt. Lett.* 29:2923–2925.

Fickenscher, M., M.G. Purucker, and A. Laubereau. 1992. Resonant vibrational dephasing invetsigated with high-precision femtosecond CARS. *Chem. Phys. Lett.* 191:182–188.

Freund, I., and M. Deutsch. 1986. Second-harmonic microscopy of biological tissue. *Opt. Lett.* 11:94–96.

Hashimoto, M., T. Araki, and S. Kawata. 2000. Molecular vibration imaging in the finger-print region by use of coherent anti-Stokes Raman scattering microscopy with a collinear configuration. *Opt. Lett.* 25:1768–1770.

Hellerer, T., A.M.K. Enejder, and A. Zumbusch. 2004. Spectral focusing: High spectral resolution spectroscopy with broad-bandwidth laser pulses. *Appl. Phys. Lett.* 85:25–27.

Hellerer, T., A. Schiller, G. Jung, and A. Zumbusch. 2002. Coherent anti-Stokes Raman scattering (CARS) correlation spectroscopy. *ChemPhysChem* 7:630–633.

Holtom, G.R., B.D. Thrall, B.-Y. Chin, H.S. Wiley, and S.D. Colson. 2001. Achieving molecular selectivity in imaging using multiphoton Raman spectroscopy techniques. *Traffic* 2:781–788.

Hopt, A., and E. Neher. 2001. Highly nonlinear photodamage in two-photon fluorescence microscopy. *Biophys. J.* 80:2029–2036.

Ichimura, T., N. Hayazawa, M. Hashimoto, Y. Inouye, and S. Kawata. 2004. Tip-enhanced coherent anti-Stokes Raman scattering for vibrational nanoimaging. *Phys. Rev. Lett.* 92:220801–220804.

Jones, D.J., E.O. Potma, J.X. Cheng, B. Burfeindt, Y. Pang, J. Ye, and X.S. Xie. 2002. Synchro-nization of two passively mode-locked, picosecond lasres within 20 fs for coherent anti-Stokes Raman scattering microscopy. *Rev. Sci. Instrum.* 73:2843–2848.

Kajzar, F., and J. Messier. 1985. Third-harmonic generation in liquids. *Phys. Rev. A* 32:2352–2363.

Kamga, F.M., and M.G. Sceats. 1980. Pulse-sequenced coherent anti-Stokes Raman scattering spectroscopy: a method for the suppression of the nonresonant background. *Opt. Lett.* 5:126–127.

Kano, H., and H. Hamaguchi. 2005. Vibrationally resonant imaging of single living cell by supercontinuum-based multiplex coherent anti-Stokes Raman scattering microspectroscopy. *Opt. Express* 13:1322–1327.

Kee, T.W., and M.T. Cicerone. 2004. Simple approach to one-laser, broadband coherent anti-Stokes Raman scattering microscopy. *Opt. Lett.* 29:2701–2703.

Killi, A., J. Dorring, U. Morgner, M.J. Lederer, J. Frei, and D. Kopf. 2005. High-speed electro-optical cavity dumping of mode-locked laser oscillators. *Opt. Express* 13:1916–1922.

Knutsen, K.P., J.C. Johnson, A.E. Miller, P.B. Petersen, and R.J. Saykally. 2004. High spectral resolution multiplex CARS spectroscopy using chirped pulses. *Chem. Phys. Lett.* 387:436–441.

Konig, K., T.W. Becker, P. Fischer, I. Riemann, and K.J. Halbhuber. 1999. Pulse-length depend-ence of cellular response to intense near-infrared laser pulses in multiphoton microscopes. *Opt. Lett.* 24:113–115.

Laubereau, A., and W. Kaiser. 1978. Vibrational dynamics of liquids and solids investigated by picosecond light pulses. *Rev. Mod. Phys.* 50:607–665.

Li, L., H. Wang, and J.X. Cheng. 2005. Quantitative coherent-anti-Stokes Raman scattering imaging of lipid distribution of co-exsiting domains. *Biophys. J.* 89:3480–3490.

Marks, D.L., and S.A. Boppart. 2004. Nonlinear interferometric vibrational imaging. *Phys. Rev. Lett.* 92(12):123905,123901–123904.

Marowsky, G., and G. Luepke. 1990. CARS-background suppression by phase-controlled nonlinear interferometry. *Appl. Phys. B* 51:49–50.

Muller, M., and J.M. Schins. 2002. Imaging the thermodynamic state of lipid membranes with multiplex CARS microscopy. *J. Phys. Chem B* 106:3715–3723.

Nan, X., J.X. Cheng, and X.S. Xie. 2003a. Vibrational imaging of lipid droplets in live fibroblast cells with coherent anti-Stokes Raman scattering microscopy. *J. Lipid Res.* 44:2202–2208.

Nan, X.L., J.X. Cheng, and X.S. Xie. 2003b. Vibrational imaging of lipid droplets in live cells using a coherent anti-Stokes Raman microscope. *J. Lipid Res., in press.*

Oron, D., N. Dudovich, and Y. Silberberg. 2002. Single-pulse phase-contrast nonlinear Raman spectroscopy. *Phys. Rev. Lett.* 89:273001–273004.

Oron, D., N. Dudovich, and Y. Silberberg. 2003. Femtosecond Phase-and-polarization control for background-free coherent anti-Stokes Raman spectroscopy. *Phys. Rev. Lett.* 90:213902,213901–213904.

Otto, C., A. Voroshilov, S.G. Kruglik, and J. Greve. 2001. Vibrational bands of luminescent zinc(II)-octaethylporphyrin using a polarization sensitive 'microscopic' multiplex CARS technique. *J. Raman Spectrosc.* 32:495–501.

Oudar, J.L., R.W. Smith, and Y.R. Shen. 1979. Polarization-sensitive coherent anti-Stokes Raman spectroscopy. *Appl. Phys. Lett.* 34:758–760.

Palmer, A.G., and N.L. Thompson. 1987. Molecular aggregation characterization by high order autocorrelation in fluorescence correlation spectroscopy. *Biophys. J.* 52:257–270.

Paulsen, H.N., K.M. Hilligsoe, J. Thogerson, S.R. Keiding, and J.J. Larsen. 2003. Coherent anti-Stokes Raman scattering microscopy with a photonic crystal based light source. *Opt. Lett.* 28:1123–1125.

Potma, E.O., W.P.d. Boeij, P.J.M.v. Haastert, and D.A. Wiersma. 2001. Real-time visualization of intracellular hydrodynamics in single living cells. *Proc. Natl. Acad. Sci. USA* 98:1577–1582.

Potma, E.O., W.P.d. Boeij, and D.A. Wiersma. 2000. Nonlinear coherent four-wave mixing in optical microscopy. *J. Opt. Soc. Am. B* 25:1678–1684.

Potma, E.O., C. Evans, X.S. Xie, J.R. Jones, and J. Ye. 2003. Picosecond-pulse amplification with an external passive optical cavity. *Opt. Lett.* 28:1835–1837.

Potma, E.O., D.J. Jones, J.X. Cheng, X.S. Xie, and J. Ye. 2002. High-sensitivity coherent anti-Stokes Raman scattering microscopy with two tightly synchronized picosecond lasers. *Opt. Lett.* 27:1168–1170.

Potma, E.O., and X.S. Xie. 2003. Detection of single lipid bilayers in coherent anti-Stokes Raman scattering (CARS) microscopy. *J. Raman. Spectrosc.* 34:642–650.

Potma, E.O., and X.S. Xie. 2005. Direct visualization of lipid phase segregation in single lipid bilayers with coherent anti-Stokes Raman scattering microscopy. *ChemPhysChem* 6: 77–79.

Potma, E.O., X.S. Xie, L. Muntean, J. Preusser, D. JOnes, J. Ye, S.R. Leone, W.D. Hinsberg, and W. Schade. 2004. Chemical imaging of photoresists with coherent anti-Stokes Raman scattering (CARS) microscopy. *J. Phys. Chem. B* 108:1296–1301.

Schaller, R.D., J. Ziegelbauer, L.F. Lee, L.H. Haber, and R.J. Saykally. 2002. Chemically selective imaging of subcellular structure in human hepatocytes with coherent anti-Stokes Raman scattering (CARS) near field scanning optical microscopy (NSOM). *J. Phys. Chem. B* 106:8489–8492.

Schonle, A., and S.W. Hell. 1998. Heating by absorption in the focus of an objective lens. *Opt. Lett.* 23:325–327.

Schwille, P. 2001. Fluorescence correlation spectroscopy and its potential for intracellular applications. *Cell Biochem. Biophys.* 34:383–408.

Vinegoni, C., J.S. Bredfeldt, D.L. Marks, and S.A. Boppart. 2004. Nonlinear optical contrast enhancement for optical coherence tomography. *Opt. Express* 12(2):331–341.

Volkmer, A. 2005. Vibrational imaging and microspectroscopies based on coherent anti-Stokes Raman scattering microscopy. *J. Phys. D* 38:R59–81.

Volkmer, A., J.X. Cheng, and X.S. Xie. 2001. Vibrational imaging with high sensitivity via epi-detected coherent anti-Stokes Raman scattering microscopy. *Phys. Rev. Lett.* 87:023901–023904.

Wang, H., Y. Fu, P. Zickmund, R. Shi, and J.X. Cheng. 2005. Coherent anti-Stokes Raman scattering imaging of axonal myelin in live spinal tissues. *Biophys. J.* 89:581–591.

Williams, R.M., W.R. Zipfel, and W.W. Webb. 2005. Interpreting second-harmonic generation imaes of collagen I fibers. *Biophys. J.* 88:1377–1386.

Wurpel, G.W.H., J.M. Schins, and M. Muller. 2004. Direct measurement of chain order in single phospholipid mono- and bilayers with multiplex CARS. *J. Phys. Chem. B* 108:3400–3403.

Yacoby, Y., and R. Fitzgibbon. 1980. Coherent cancellation of background in four-wave mixing spectroscopy. *J. Appl. Phys* 51:3072–3077.

Zumbusch, A., G.R. Holtom, and X.S. Xie. 1999. Three-dimensional vibrational imaging by coherent anti-Stokes Raman scattering. *Phys. Rev. Lett.* 82:4142–4145.

18

High-Speed Imaging Using Multiphoton Excitation Microscopy

Ki H. Kim, Karsten Bahlmann,
Timothy Ragan, Daekeun Kim,
and Peter T. C. So

18.1 INTRODUCTION

Multiphoton microscopy (Denk et al., 1990; Helmchen & Denk, 2002; So et al., 2000; Zipfel et al., 2003) produces three-dimensionally resolved images based on nonlinear optical interactions localized at the focus of a microscope objective. Multiphoton microscopy in the fluorescence mode is now the method of choice for in vivo deep tissue microscopic imaging because of its subcellular resolution, minimal phototoxicity, and excellent tissue penetration depth. For many tissues, penetration depths from a few hundred microns up to a millimeter have been reported (Helmchen & Denk, 2005). Multiphoton microscopy has become an invaluable tool in biomedical studies such as neuronal plasticity (Grutzendler et al., 2002; Lee et al., 2005; Lendvai et al., 2000), angiogenesis in solid tumors (Padera et al., 2002), and transdermal drug delivery (Yu et al., 2001).

A practical limitation of multiphoton microscopy is its frame rate, which typically lies in the range of 0.1 ~ 2 frames per second (fps) with an imaging area of several hundred micrometers on a side. While this imaging speed is sufficient in some applications, two classes of problems demand a higher frame rate. First, high-speed multiphoton microscopy is needed in the study of kinetic processes on the millisecond time scale in 3D biological specimens. For example, high-speed 3D imaging can map the 3D propagation of a calcium wave and the associated physical contraction wave through a myocyte (Fig. 18.1), the rolling of leukocytes within the blood vessel in a solid tumor (Padera et al., 2002), and neurotransmitter release at synaptic terminals (Roorda et al., 2004). Second, since the image stack of the conventional multiphoton microscope is limited to about several hundred microns on a side, the investigation volume is only $1\text{-}10 \times 10^{-3}$ mm^3. While this volume is sufficient for cellular imaging, many tissues have physiologically relevant structures that are significantly larger. For example, a neuron, with its extensive dendritic tree, can span a volume over 1 mm^3 (Fig. 18.2), and the distribution of many dermal structures such as hair follicles and

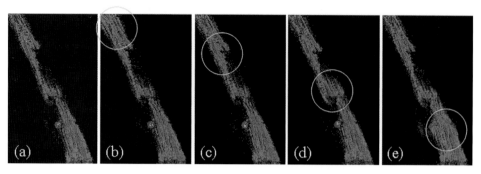

Figure 18.1. A contraction wave (indicated by white circles) propagates through a myocyte captured using the high-speed multiphoton microscope based on polygonal scanning. The images were acquired using a 40 × Zeiss Fluar 1.3 NA oil-immersion objective with dimension of 120 μm × 120 μm. The successive frames are about 400 ms apart in time.

sebaceous glands cannot be quantified in an image smaller than a few square millimeters (Fig. 18.3). Furthermore, many organs have hierarchical structures spanning length scales from submicrons to several millimeters. A recent study has shown that cardiac structures of a whole mouse heart, approximately 1 cm^3 in volume, can be imaged with micron-level resolution (Fig. 18.4). On the millimeter scale, cardiac structures such as ventricles and heart valves can be visualized. The characteristic spiral distribution of heart muscle fibers and the associated blood capillaries can be visualized on the submillimeter scale. On the subcellular micrometer scale, the distribution and the morphology of myocyte nuclei can be studied (Ragan et al., 2004). Equally important, traditional 3D microscopes with their limited frame rate realistically can study only a few hundred cells. Traditional microscopy cannot hope to provide comparable statistical accuracy and precision of quantitative assays such as flow cytometry. High-speed imaging can circumvent this difficulty by improving the number of cells that can be efficiently sampled in a tissue specimen. This approach opens the possibility of extending image cytometry into 3D.

A seminal study that demonstrates the utility of high-speed multiphoton imaging is in the area of transdermal drug delivery (Yu et al., 2002). High-speed microscopy allows imaging of a sufficiently large tissue area to provide statistically meaningful transport coefficients for hydrophilic and hydrophobic chemicals across the stratum corneum.

While high-speed multiphoton microscopy can be a powerful tool for a variety of biomedical investigations, there are significant instrumentation challenges in achieving this goal. Research in this area focuses on two directions: understanding the fundamental limitations of high-speed deep tissue multiphoton imaging and the practical implementations of high-speed multiphoton microscopes. In this chapter, we will first present the fundamental limits in the imaging speed of multiphoton microscopy. We will then discuss the two main types of high-speed multiphoton microscopes that have been realized. Since the most important applications for nonlinear optical microscopes are in the tissue area, the effects of tissue optics on high-speed microscope design will be integrated in our discussion.

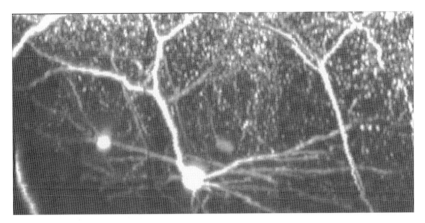

Figure 18.2. A 3D reconstructed image of a GFP-expressing neuron driven by Thy-1 promoter captured by a conventional multiphoton microscope. Images were acquired in vivo through a surgically implanted window on the mouse skull using a Zeiss Achroplan 40 × IR, 0.8 NA, long working distance, water-immersion objective. The dimension of the image is approximately 250 μm × 600 μm.

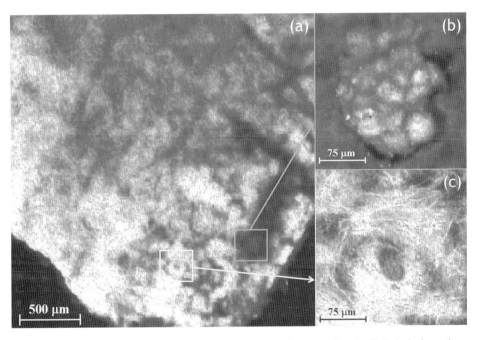

Figure 18.3. A 3D-resolved slice of ex vivo human skin at a depth slightly below the epidermal-dermal junction. Individual images were acquired using a 40 × Zeiss Fluar 1.3 NA oil-immersion objective. (a) A 2.5-mm × 2.5-mm image montage. (b) An expanded view of the cell cluster at the base of a hair follicle. (c) An expanded view of the cells forming a sebaceous gland and the collagen fiber surrounding it.

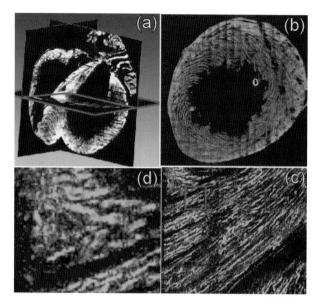

Figure 18.4. High-speed multiphoton imaging of a whole mouse heart using a polygonal scanner based system imaged using a 40 × Zeiss Fluar 1.3 NA oil-immersion objective. The mouse heart was fixed with 4% paraformaldehyde and embedded in paraffin. The vasculature and the nuclei were stained via intravital labeling before the mouse was sacrificed using the nuclear stain DAPI and a lectin-Alexa 488 linker, which stains the endothelial walls of the blood vessels. (a) X-, Y-, Z- orthogonal slices of the 3D image volume showing millimeter-scale structures of the heart, including ventricles and heart valves. (b) A cross-section of the mouse heart with a diameter of approximately 1 cm. The characteristic spiral morphology of muscle fibers is seen. (c) An expanded view of the region depicted by the red rectangle in (b). Individual capillary structure in the heart muscle is shown in green. (d) An expanded view of the region depicted by the red rectangle in (c). The size, shape, and distribution of nuclei in the heart can be clearly visualized by the blue Hoechst fluorescent labeling.

18.2 FUNDAMENTAL LIMITS OF HIGH-SPEED MULTIPHOTON MICROSCOPY

Two factors limit the frame rate of nonlinear optical microscopy. The low optical signal resulting from the nonlinear excitation process and the need for raster scanning in 3D are the primary reasons. When imaging into thick turbid specimens, other factors impose additional difficulty, such as the decrease of excitation efficiency due to spherical aberration and photon scattering, and the decrease of signal strength and image contrast due to emission photon scattering.

18.2.1 The Frame Rate Limit Imposed by the Excitation, Scanning, and Detection Subsystems

Although a high-speed multiphoton microscope has been developed based on the coherent anti-Stokes Raman scattering (CARS) process (Evans et al., 2005), the most common implementation is based on fluorescence contrast. In this chapter, we focus on fluorescence-based systems and demonstrate that the concentration of fluorophores and their finite lifetimes impose an ultimate limit on the imaging speed. Similar signal strength consideration is also the limiting factor for other modalities of high-speed multiphoton microscopy, such as CARS, but for different physical reasons.

The two-photon excitation probability of fluorophores is a quadratic function of the average excitation laser power. Fluorescence signal cannot be increased arbitrarily by increasing laser power without resulting in excitation saturation (So et al., 2000). Since typical fluorophores have a lifetime of a few nanoseconds, which is much longer than the pulse duration time of excitation laser (a few hundred femtoseconds), each fluorophore can be excited only once per laser pulse. At higher excitation probability, the excitation efficiency of fluorophores is no longer a quadratic function of excitation power due to the trapping of the fluorophores in the excited state and the depletion of fluorophores in the ground state during the femtosecond-length laser pulse. This is a process called excitation saturation. As a result, the deviation from quadratic behavior results in the broadening of the microscope point spread function (PSF) and the loss of resolution (Nagy et al., 2005a, 2005b). Numerical simulation shows that when the ground state is depleted by about 10%, the PSF already broadens significantly on the same order.

The most common laser source used in multiphoton microscopy is the titanium-sapphire laser with a typical wavelength (λ) of 800 nm, pulse width (τ_p) of 200 fs, and pulse repetition rate (f_p) of 80 MHz. The numerical aperture (NA) of the microscope objective is approximately 1. Many commonly used fluorophores have a two-photon absorption coefficient (δ_a) on the order of tens of GM, where 1 GM is 10^{-50} cm$^4 \times$ s/photon. In order to avoid excitation saturation, the probability of fluorophore excitation per laser pulse (Pr_{pulse}) must be less than 0.1 ($\text{Pr}_{\text{pulse}} < 0.1$). Pr_{pulse} is a function of these parameters:

$$\text{Pr}_{\text{pulse}} = \frac{\delta_a P_0^2}{\tau_p f_p^2} \left(\frac{NA^2}{2\hbar c \lambda} \right)^2$$

where c is the speed of light and $\hbar$ is Planck's constant.

With these conditions, the input power, P_o, where Pr_{pulse} becomes close to the saturation limit, is approximately 10 mW. Importantly, the condition that Pr_{pulse} is less than 0.1 implies that the maximum photon emission rate per fluorophore is at most 8 MHz.

In addition to excitation saturation, the instantaneous power used in nonlinear microscopes can also induce photodamage in biological specimens (reviewed in Chapter 14). The conditions for photodamage are highly specimen-dependent. Nonetheless, photodamage may place a lower limit on the acceptable incident power than the excitation saturation of fluorophores. This effect may further reduce the maximum excitation and photon emission rates.

The 3D resolution of multiphoton microscopes is contributed by the localization of fluorescence excitation at a diffraction-limited focal volume, approximately 0.1 femtoliter in size with a high NA objective. A 3D image stack is formed by raster scanning this excitation focus in the specimen and measuring fluorescence intensity at each scan location. Lateral scanning is often achieved by galvanometric mirror scanners with a line rate of approximately 500 Hz maximum, resulting in a 2-Hz frame rate for an image with 256 × 256 pixels. The depth dimension is often scanned using a piezoelectric objective positioner that has a bandwidth of only about 100 Hz due to the large mass of typical objectives. Examining the bandwidth of these scanners, it is clear that the image frame rate is typically limited by the galvanometric scanners used for lateral scanning.

The emission light generated from the specimen is collected by an optical detector synchronized with the scanning of excitation focus, allowing the reconstruction of an image stack. The emission light signal is often quite low and high-sensitivity photodetectors such as photomultiplier tubes (PMTs) are used. PMTs have very good sensitivity, characterized by the detector quantum efficiency (QE). The QE of PMTs can be estimated to be about 0.3. Accounting for the microscope objective collection solid angle (0.3 for 1.2 NA) and the transmission efficiency of detection optics (0.7), the total collection efficiency of a microscope for emission photon (ε_{col}) is approximately 0.06 for PMTs. Given the constraint of excitation saturation, the detection photon rate for an ideal (100% quantum efficiency) fluorophore is 480 kHz. One can estimate the detected photon rate given specimen fluorophore concentration and the excitation volume, which is about 0.1 fl for a high NA objective. The detected emission photon rate can be estimated to be 2.9, 29, and 290 MHz for fluorophore concentrations of 100 nM, 1 μM, and 10 μM respectively. For these photon detection rates, a shot noise limited system must have a minimum pixel dwell time of 34, 3.4, and 0.34 μs respectively to produce an image with a signal-to-noise ratio (SNR) of 10. For an image consisting of 256×256 pixels, the frame rate that can be achieved is 0.44 fps, 4.4 fps, and 44 fps. For detailed equations used in these calculations, see Table 18.1. Note that this frame rate cannot be improved with faster scanners because this rate is dictated only by the photophysics, the fluorophore concentration in the specimen, and the efficiency of the microscope detection system.

From our discussion, it is clear that one method to improve imaging speed is to increase the number of fluorophores in the excitation volume. The number of fluorophores can be increased by increasing either the fluorophore concentration or the excitation volume. Unfortunately, the fluorophore concentration is often limited by biochemical factors, such as specimen viability and probe partitioning, that are often difficult to control experimentally. Further, for the measurement of metabolites or biochemical signals, such as calcium, the probe concentration must be held sufficiently low such that the fluorescent probes do not buffer the underlying biochemical reaction and do not affect the underlying cellular signaling process to be studied.

Another way to increase fluorophore number is to increase the excitation volume. The most efficient way to increase the excitation volume is to decrease the excitation NA of the objective without affecting the detection NA. It is clear that decreasing the detection NA is not desirable as it decreases the photon collection solid angle and the detection efficiency of the overall system. Decreasing the excitation NA will increase the excitation volume, enclosing more fluorophores. However, decreasing the excitation NA with the same average input power does not increase the total photon emission rate from the excitation volume. This is a result of the fact that excitation volume increases as the fourth power of NA, while excitation efficiency per fluorophore decreases also as the fourth power of NA. However, since the excitation efficiency is decreased, average input power can be increased further until the photon emission rate per fluorophore is again at the excitation saturation limit of 8 MHz. Therefore, by decreasing the excitation NA, the detected fluorescence signal will increase with the excitation volume proportional to the inverse fourth power of the NA by appropriately adjusting input power. The use of a lower excitation NA, of course, has the disadvantage of reducing the image resolution. Therefore, this approach should be considered when imaging speed is important while lower image resolution can be tolerated. The most efficient way to decrease the excitation NA and to maintain the detection NA is to underfill the back aperture of a high NA microscope objective.

Another consideration in maximizing the imaging speed of a microscope system is the excitation polarization. In many situations, the orientation of fluorophore distribution in a biological specimen is random; the use of circularly polarized light is most efficient for excitation in this case. The use of linear polarized excitation will excite molecules that are aligned with the electric field efficiently. However, it will excite the fluorophores at the other orientation with a penalty factor equal to the fourth power of the cosine of the angle between the excitation electric field and the molecular excitation dipole.

18.2.2 The Frame Rate Limit Imposed by Optically Thick Specimens

One of the most important features of multiphoton microscopy is its excellent penetration into biological specimens. It is crucial that this important feature is not compromised in a high-speed multiphoton system.

The factors limiting the imaging depth of multiphoton microscopy are well studied. Optical spherical aberration due to refractive index mismatch between the microscope objective and the tissue is one major factor (Booth et al., 1998; Booth & Wilson, 2001; de Grauw et al., 1999). The other major factors are photon scattering and absorption by tissue constituents (Beaurepaire et al., 2002; Blanca & Saloma, 1998; Centonze & White, 1998; Dong et al., 2003; Dunn et al., 2000; Gan & Gu, 2000; Oheim et al., 2001; Ying et al., 1999). In multiphoton microscopy, photon scattering is almost always the dominant factor compared with photon absorption for biological specimens. Previous simulations and experimental studies have shown that both spherical aberration and excitation photon scattering primarily result in reducing the effective average power in the excitation focus without significantly broadening the excitation PSF for typical imaging depth. Spherical aberration can be significantly minimized by matching the refractive indices of the type of microscope objective and the tissue specimen. Scattering can also be minimized in some cases by index matching between different tissue components using optical clearing agents (Choi et al., 2005; Miller et al., 2005; Plotnikov et al., 2006). However, with the inherent tissue heterogeneity, these two effects can never be completely avoided, and typical excitation efficiency decreases exponentially with increasing imaging depth. The power delivered to the focal volume at depth z is:

$$I(z) = I_0 e^{-\frac{z}{l^s_{ex}}}$$

where I_0 is the incident power on the specimen surface and l^s_{ex} is the mean free path length of scattering at excitation wavelength in the tissue.

Since the excitation PSF is not greatly affected by tissue scattering, the reduction of excitation efficiency can be partially compensated by increasing excitation laser power, provided that a sufficiently powerful laser is available and the specimen suffers no deleterious effects such as one-photon absorption and the subsequent thermal damage (Masters et al., 2004). For deep tissue imaging, the scattering of the emission photons is also an important factor: it reduces the amount of light collected by the objective, the intermediate optical system, and the detector. The conventional multiphoton microscope optimizes the collection of these scattered photons by using a high NA objective, large intermediate optical components, and a photodetector with a large active area. The use of a large active area detector such as a PMT enables collection of a good fraction of scattered photons, in addition to most of unscattered ballistic photons. Avalanche photodiodes (APDs) are another class of high-sensitivity detectors. While APDs have higher QE (0.8) than PMTs (0.2–0.4 in blue-green), especially in the red wavelength range, the larger active area of PMTs, on the order of 50 mm^2, makes them preferred choices over APDs (with an active area <1 mm^2) for deep tissue imaging. Since the spatial information of the conventional multiphoton microscope is obtained by temporally correlating the detector signal intensity with the known raster scan position of the scanner, the collection of these scattered photons does not degrade the image resolution. Therefore, the conventional multiphoton microscope is relatively immune to the effect of emission photon scattering and is effective for turbid tissue imaging. However, it should be noted that the effect of emission photon scattering is also an exponential function of imaging depth and cannot be compensated by further increasing the laser power beyond fluorophore saturation.

18.3 IMPLEMENTATIONS OF HIGH-SPEED MULTIPHOTON MICROSCOPES

From our previous discussion, the major factors that limit the multiphoton microscope imaging speed are (1) the finite fluorophore lifetime and concentration that results in limited signal strength, (2) refractive index mismatch, scattering, and absorption in the specimen that result in excitation power and signal loss, and (3) 3D image formation requires raster scanning, but the bandwidth of scanners is limited. Note that limitation (1), and partially (2), originate from fundamental photophysical or biological constraints and cannot be improved by better instrumentation. For limitation (2), it is crucial that the design of the high-speed multiphoton microscope does not compromise the collection efficiency of scattered emission photons. Today, most high-speed designs focus on overcoming limitation (3), and these are the focus of this section. The review of the high-speed multiphoton microscopy literature reveals two classes of approaches. The first approach circumvents the bandwidth limit of galvanometric scanners by replacing them with higher-speed scanners. The second approach parallelizes the raster scanning process by simultaneously acquiring data over multiple regions in the specimen.

18.3.1 Frame Rate Improvement Based on Higher-Speed Scanners

The replacement of galvanometric scanners with higher-speed scanners is of course the most straightforward approach to improve the frame rate. The higher-speed scanners that have been used are polygonal mirror scanners (Kim et al., 1999) (Fig. 18.5) and resonant mirror scanners (Fan et al., 1999). Depending on the desired final speed, the galvanometric scanners in one or both scan directions may be replaced. The polygonal

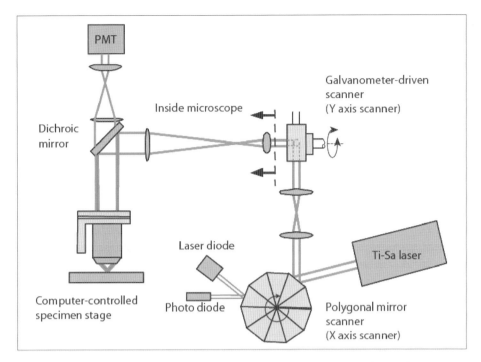

Figure 18.5. A high-speed multiphoton microscope based on a polygonal scanner. Laser incident upon the polygonal scanner is related to a galvanometric scanner via two relay lenses. The polygonal scanning provides fast scanning along the x-axis in the sample plane. The galvanometric scanner allows scanning along the slower, y-axis. The beam is directed into the microscope, filling the back-aperture of the microscope objective via a dichroic mirror and a beam expander. The fluorescence signal passes through the dichroic mirror and is detected by a photomultiplier tube (PMT).

mirror scanner is a lightweight metal cylinder rotating at high speed cushioned by an air bearing. The cylinder has mirror facets machined around its perimeter. The rotation of the cylinder sweeps the facets and generates line scans along one axis in the sample plane. It achieves a higher scanning speed than the galvanometric scanner since the polygon rotates at a constant speed rather than moving back and forth as the galvanometric scanner, which requires repeated start/stop. The resonant mirror scanner is essentially also a galvanometric system but is driven specifically at the resonance frequency. A normal galvanometric scanner allows precision point-to-point positioning by feedback control. Stable position control system requires operation at a bandwidth significantly lower than its inherent mechanical resonance frequency, typically by a factor of three. The resonance scanner system achieves higher speed by being driven at its resonance frequency. At resonance, the deflection angle of the scanner can no longer be controlled and has sinusoidal trajectory. The amplitude of this sinusoidal motion can be varied by changing the drive voltage. Both polygonal scanners and resonance scanners offer a significantly higher line rate than galvanometric scanners, up to tens of kHz. The tradeoffs in using these scanners is the loss of the precise point-to-point positioning ability that is possible with galvanometric scanners. Furthermore, the 3D reconstruction of an image stack in a typical multiphoton microscope requires synchronizing the signal acquisition circuitry of the detection system with the scanner position. However, since the phase and the frequency of the scanning trajectory for the polygonal scanner and the resonance scanner cannot be precisely controlled, the synchronization is more difficult. The instantaneous position of these scanners must be independently measured using an encoder, allowing the other components of the microscope to coordinate with the motion of scanners. The sensing circuitry is often implemented optically so that the motion of these high-speed systems is not hindered due to mechanical contact (Fan et al., 1999; Kim et al., 1999).

Finally, polygonal scanners and resonance scanners are quite equitable. These systems have comparable speeds. The polygonal scanners have the disadvantage that when the laser beam is "clipped" by the corners between its facets, the beam position is no longer well regulated. Therefore, the beam must be gated off when the corner of the polygonal rotates into the beam, resulting a period of dead time. The ratio of dead time versus active time in a polygonal system is a function of the ratio of the facet dimension versus the input laser beam diameter. The resonance scanners, on the other hand, do not have equal dwell time at different diffraction angles due to the sinusoidal nature of their motion. The dwell time in the middle of the scan is significantly shorter than at the turnaround points. This dwell time variation results in uneven SNR across the image.

Another method to achieve high-speed scanning is to use electro-optical deflectors, such as acousto-optical deflectors (AODs), instead of mechanical scanners. In an AOD, a traveling acoustic wave is generated in a crystalline material induced by piezo-actuators. This traveling wave produces a periodic refractive index variation in the crystal, forming a grating. The period of this grating determines the deflection angle of the laser beam based on Bragg phase matching.

While these electro-optical scanners can achieve point-to-point positioning with a settling time of about 10 μs, their use in multiphoton microscopy has three difficulties. First, the diffraction efficiency of AODs is often much lower than a mirror-based system and is only about 50%. A higher-powered laser is sometimes needed to use AOD-based systems for deep tissue imaging. Second, AODs introduce significant dispersion of the femtosecond laser pulses propagating through the optical crystal, which broadens the pulse temporally and reduces excitation efficiency even further. Finally, AODs also introduce significant spherical aberration in the excitation beam, further reducing both excitation efficiency and image resolution. Today, high-speed multiphoton microscopes utilizing AODs have been implemented with fairly complex laser pulse width and shape compensation subsystems (Iyer et al., 2003; Reddy & Saggau, 2005; Zeng et al., 2006). The advantages of this technique, such as rapid and accurate beam positioning in the sample, are applied to study neuronal communication by rapidly measuring optical signals at selected points of the neuronal dendrites.

With the increased scanning speed, the pixel dwell time, t_{pix}, at each location of the sample plane is reduced proportionally and the signal is reduced as well. *In general, the SNR of an optimized, ideal single focus multiphoton microscope is proportional to its pixel residence time.* Therefore, increasing imaging speed by a factor of 10 inevitably reduces SNR by a factor of about 3. As discussed previously, higher input power can be used to compensate for the signal reduction but is ultimately limited by excitation saturation and specimen photodamage threshold. In the case that equivalent SNR can be achieved, it should be noted that high-speed multiphoton microscopes based on faster scanners have similar tissue penetration depth as conventional multiphoton microscopes because both systems can use photodetectors with large active areas.

Finally, high-speed multiphoton microscopes are in general "less ideal" and have slightly lower sensitivity than conventional multiphoton microscopes because they often cannot be operated in single photon counting mode. In single photon counting mode, individual current bursts from incident single photons at the detector output are detected using an amplifier-discriminator circuit. Single photon counting mode can reject most of the dark current noise of the detector, providing excellent SNR. While single photon counting has the best SNR, its use in high-speed imaging is limited by the bandwidth of amplifier-discriminator electronics (bandwidth of about 100 MHz). Given that the photon arrival statistics are Poisson distributed, the practical average photon detection rate using a single photon counting approach is less than a few MHz. Single photon counting is further confounded by the use of pulsed lasers in nonlinear microscopy instead of continuous wave lasers. Due to the pulse structure of the laser and the lifetime of the fluorophore, most of the emission photons are generated as a bunch within a few nanoseconds after the arrival of the laser pulse at the specimen and cannot be distinguished by the amplifier-discriminator circuitry with 100-MHz bandwidth. As discussed previously, a photon detection rate as high as tens to hundreds of MHz is needed to generate images with a good SNR in high-speed multiphoton microscopes due to their shorter pixel dwell time. While the single photon counting approach has lower noise, its speed is often inadequate to handle the photon rate needed for high-speed imaging. Instead, the photocurrent from the PMT can be detected in the analog mode, where the integrated photocurrent amplitude is measured using a transimpedance amplifier circuit. The low-pass filter component in the transimpedance amplifier circuit is often designed with a time constant comparable to the pixel dwell time so that current fluctuations from the photon pulses are smoothed out. The transimpedance amplifier circuit linearly converts the integrated photocurrent signal level to a voltage amplitude. The voltage amplitude can then be quantified using an analog-to-digital converter. Since the analog detection mode measures the integrated photocurrent from the incoming emission photons, it can remain linear up to a very high photon rate. However, at a low photon rate, its SNR characteristics is not as good as single photon counting systems because some noise sources, such as detector dark current, cannot be rejected.

It is impossible to provide an exact SNR comparison between photon counting mode and analog mode since it is dependent on the exact choice of detector and the design of the signal conditioning circuitry. Nonetheless, a useful rule of thumb is that single photon counting has an SNR advantage of about 3 to 5 times when measuring tens or hundreds of photons at a 100-μs pixel dwell time. When the photon number reaches thousands, the two techniques are virtually equivalent because photon shot noise dominates in both cases. For a more in-depth discussion of detectors and their signal conditioning circuitry, please see Chapter 12.

18.3.2 Frame Rate Improvement Based on Parallelized Excitation and Detection

High-speed multiphoton microscopy based on faster scanners overcomes the speed limitation imposed by conventional, slower galvanometric scanners. However, the price for a higher frame rate with faster scanners is a shorter pixel dwell time. The practical speed these systems can achieve is often limited by image SNR but not by scanner speed. Due to the saturation and sample damaging thresholds, discussed above, an alternative to improve raster scanning bandwidth is to excite and detect signals from multiple spatial locations in parallel. The advantage of this method is that the imaging speed is enhanced by the degree of parallelization, but the pixel dwell time can in principle remain the same as in the conventional multiphoton microscope.

18.3.2.1 Implementation Methods of Parallelized High-Speed Multiphoton Microscopy

A number of methods have been developed based on the concept of parallelization. In fact, the first high-speed multiphoton microscope used line scanning (Guild & Webb, 1995) based on a concept well known in confocal microscopy (Wilson, 1990). Instead of focusing the laser light to a diffraction-limited spot at the focal plane of the objective, excitation light can be focused into a line (Fig. 18.6). By scanning along a direction perpendicular to this line in the focal plane, the lateral scanning can be quickly accomplished. In order to resolve the fluorescence signal originating from the different points along the line focus, a detector with spatial resolution, such as a charge coupled device (CCD) camera, is used. In general, if a conventional multiphoton microscope has a frame rate of N fps, a line-scan system should achieve a frame rate of N^2 fps.

While the line-scanning approach is quite attractive, there are a number of limitations that have rendered this system less desirable for multiphoton microscopy. A major

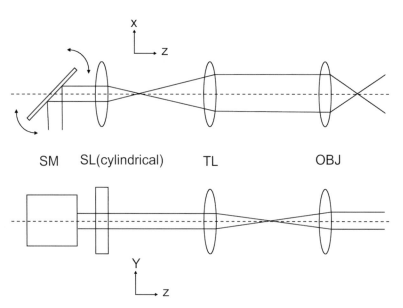

Figure 18.6. Simplified schematic of line-scan high-speed multiphoton microscope. Only the excitation beam path is shown for clarity. The top view is the ray trace on the x-z plane and the bottom view is the ray trace on the y-z plane; z is the direction of the optical axis. The parts in the diagram are SM, scanning mirror; SL, scan lens; TL, tube lens, OBJ, objective. The SL is a cylindrical lens and has curvature only on the x-z plane. The SL is a flat optical component on the y-z plane. Only one SM is needed, and its axis of rotation is parallel to the y-axis.

disadvantage of line scanning is the reduced spatial resolution. While line scanning has minimal effects on lateral resolution, axial resolution is significantly degraded in this approach (Wilson, 1990). Second, the line focus is typically several hundred times longer than the lateral dimension of the PSF in a point focus system. Therefore, the optical power to achieve equivalent excitation for the whole line is also at least several hundred times greater. For a conventional multiphoton microscope, often tens of mW of power is required at each focus for optimal excitation without excitation saturation. For deep tissue imaging, this power requirement may reach several hundred mW due to tissue aberration and scattering. Further accounting for power loss through the excitation beam path of the microscope, it is clear that the total amount of power for optimal excitation in the line focus mode is often beyond the power available from a typical femtosecond laser source. Third, while the line-scan approach can in principle improve the frame rate a few hundred times, the relatively slow read-out rate of spatially resolved detectors limits the practical speed. Fourth, the penetration depth of this multiphoton system in turbid specimens is significantly lower than in the conventional multiphoton microscope. This reduction in penetration depth is due to contrast loss when spatial resolved detectors are used in the presence of emission photon scattering and will be discussed in depth in the following section.

Extending the concept of parallel illumination beyond line focus, one may also consider simultaneously illuminating a whole plane. The depth discrimination of the multiphoton microscope results from the quadratic dependence of the fluorescence signal on the excitation photon density. The fluorescence signal decreases with the excitation photon density away from the focal plane for either point or line focus. In a wide-field microscope, the photon density is constant as a function of distance from the focal plane. Therefore, no depth discrimination is expected for a multiphoton microscope operating in wide-field illumination mode. This lack of depth discrimination has discouraged the implementation of high-speed multiphoton microscope using plane (wide-field) illumination until recently. A new method of generating a depth-resolved multiphoton excitation with wide field geometry has been devised (Oron et al., 2005; Tal et al., 2005). This idea is based on the time–bandwidth product of light; specifically, the product of the width of a light pulse and its spectral bandwidth is a constant. Therefore, a light pulse with broader spectral bandwidth has a shorter pulse width. Femtosecond laser light dispersed by a diffraction grating is focused by a lens at the back aperture of a microscope objective (Fig. 18.7). Since the different spectral component of the light is diffracted by the grating to different angles, these different spectral components also focus at different positions spatially at the back focal plane. The different spectral components are re-collimated by the objective. The collimated beams from each spectral component emerge from the objective again at different angles and do not overlap except at the focal plane. Since the different spectral components travel through different beam paths, the effective pulse width is broad almost everywhere. Only at the focal plane are these spectral components recombined and they produce a short femtosecond pulse, allowing efficient multiphoton excitation.

This is a creative concept where the depth resolution is achieved by temporal focusing instead of spatial focusing. Since depth discrimination is provided by temporal focusing, the whole specimen plane can be excited in parallel while maintaining depth discrimination. This method requires a very short pulse to produce adequate depth resolution. The depth resolution of 1.5 μm full width half maximum (FWHM) and 4.5 μm

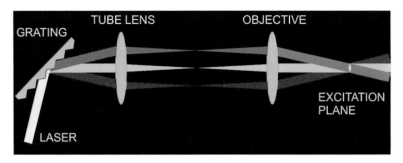

Figure 18.7. Simplified schematic of wide-field multiphoton microscope based on temporal focusing. Only the excitation beam path is shown for clarity. The detection light path is not shown, but it is identical to any wide-field microscope that images the excitation plane (EP) onto a CCD camera. After the grating, the different spectral components in the laser light travel along different optical paths until they overlap again at the EP.

FWHM is obtained with the excitation light of 10 fs pulse width, line illumination, and wide-field illumination respectively, compared with the 100 fs commonly used in conventional multiphoton microscopes. The main advantage of this method is the partial or complete elimination of lateral scanning. One disadvantage of this method is that the spatial resolution is degraded. The resolution of plane illumination based on temporal focusing is equivalent to that of line focus. Most importantly, there is inadequate laser power to efficiently excite a whole plane from typical femtosecond laser sources. This method does not provide faster imaging except for specimens labeled with probes that have a very high multiphoton cross-section, such as quantum dots.

Today, the most promising parallel excitation approach is the multifocal multiphoton microscope (MMM) (Bewersdorf et al., 1998; Buist et al., 1998). Instead of focusing light to a continuous line or a whole plane, this method creates multiple discrete foci in the specimen plane (Fig. 18.8). The multiple excitation foci can be generated by different methods such as using a microlens array. Microlens arrays are microfabricated miniature lenses of less than 1 mm in size placed in a specific pattern on a common substrate. The array generates multiple foci at its focal plane when illuminated with a collimated beam. The light from the individual foci (called beamlets) of the array are relayed via the intermediate optical elements to an objective lens. Individual beamlets are properly expanded so that they fill the back aperture of the objective at different angles. The objective generates multiple foci of excitation light in the sample plane. These multiple foci can be arranged in a helical pattern on a rotating substrate, and their trajectory covers the specimen plane. Alternatively, the foci can be arranged in a square lattice and their trajectories are scanned by an x-y mirror scanner covering the specimen plane. The fluorescence generated from the multiple excitation foci in the sample is collected by the objective. The emission light is transmitted by a dichroic mirror and is relayed to a spatially resolved detector at the image plane, collecting emission light from multiple foci simultaneously. The 2D imager accumulates signal from the specimen during the time of single frame scanning. In addition to intensity imaging, emission spectrum-resolved and fluorescence lifetime-resolved imaging (Leveque-Fort et al., 2004; Straub & Hell, 1998) have also been implemented in the MMM configuration.

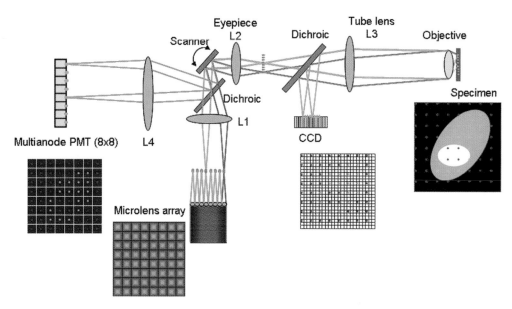

Figure 18.8. A multifocal multiphoton microscope. The excitation beam illuminates a square microlens array. The microlens array splits the excitation beam into 8 × 8 multiple beams (beamlets). In this figure, only two beamlets are ray-traced. A specimen is scanned with an 8 × 8 array of excitation foci. The emission beamlets are collected by CCD or MAPMT. For MAPMT detection, the dichroic mirror close to the CCD is removed. The CCD integrates signal during the scanning of a whole frame and reads out at the end of scan. MAPMT has only the same number of channels as the excitation beamlets. The small number of channels allows the MAPMT to be rapidly read out at each scan position of the scanner.

The multiple foci generated by a microlens array have different intensity levels, since the expanded excitation laser beam has a Gaussian intensity distribution. The foci generated by the center of excitation beam have a higher intensity than the ones generated by the edge of the excitation beam. For an n-photon process, the signal difference between a focus at the center and one on the edge is very large and is equal to the nth power of the ratio of the incident powers at each focus. To circumvent this problem, several other methods have been developed to generate multiple foci. One method is to place 50/50 beam splitters between two mirrors at specific angles. Multiple beamlets are formed as the light passes through the beam splitter multiple times (Fittinghoff et al., 2000; Nielsen et al., 2000). The advantage is that each beamlet has similar power and shape, and thus the illumination is uniform. Its disadvantage is that its configuration is more complex than using a lenslet array. The other method is to use a specially designed diffractive optical element (DOE) to generate multiple beamlets. This method has been demonstrated to generate 4 × 4 excitation foci with less than 10% intensity variation (Sacconi et al., 2003).

This multiple foci scanning approach is a compromise between the single focus method and the more parallel approaches such as line or plane scanning. By adjusting the number of foci formed, one can ensure that each laser focus has the optimal power for efficient excitation, even in thick specimens. Given that the maximum output power of available femtosecond lasers is about 2 to 4 W, the optimal number of foci for deep tissue imaging is approximately 20 to 100, providing 20 to 200 mW for each focus. Further, since the foci are well separated, the resolution of this system is not greatly different from that of a single focus system. The resolution penalty due to interference between adjacent foci can be further minimized by temporal multiplexing, where the beamlets forming different foci arrive at the specimen at different times and hence cannot interfere (Egner & Hell, 2000).

18.3.2.2 The Limitations Imposed by Using Spatially Resolved Detectors

While high-speed multiphoton microscopy based on parallelization circumvents the SNR limitation of single focus high-speed multiphoton microscopes, parallelization requires the use of spatially resolved detectors to separate the signal contribution from the different spatial locations in the specimen. Until recently, all the high-speed multiphoton microscopes based on the parallelized acquisition concept used CCD cameras as detectors. While this type of microscopes works very well for imaging thin specimens where photon scattering can be ignored, the penetration depth of this type of microscope is found to be significantly shallower in highly scattering specimens than conventional single focus multiphoton microscopes.

In turbid specimens, both the excitation and emission signal levels decrease due to photon scattering with tissue constituents. Due to the nature of the multiphoton excitation process, only unscattered light contributes to the excitation PSF, leaving its FWHM unchanged even at high penetration depth (Dong et al., 2003). On the other hand, it is the scattering of the emission photon that poses the greatest challenge.

In a conventional single focus multiphoton microscope, the spatial location of collected emission light in the image is determined by one of the excitation foci in the sample plane, because it is collected synchronously with the scanning of the excitation focus. All the emission photons from that specimen location are collected as long as they reach the detector, regardless of whether they are scattered or not. Therefore, for a detector with a sufficiently large active area, the image resolution is determined only by the excitation PSF and is independent of the emission PSF, which is affected by scattering. This advantage is further aided by the fact that the scattering cross-section decreases as a power law with the longer wavelength. Since the excitation wavelength is in the near-infrared while the emission wavelength is in the visible, the effect of scattering is much less for the excitation PSF. These factors all contribute to the deep penetration of single focus multiphoton microscopy.

On the other hand, for the parallelized microscopes, the spatial registration of an incident photon is given only by the pixel location where the 2D, spatially resolved imager intercepts it. Because the typical CCD camera pixel maps to approximately a 1-μm^2 region in the sample plane, a scattered emission photon deflected from its original path will not arrive at the correct pixel. Therefore, parallelized multiphoton microscopes are very sensitive to emission photon scattering, which is not the case for the conventional single focus systems. We have measured the distribution of these scattered emission photons and found that this distribution is very broad; the distribution of scattered emission photons has a FWHM of 25 μm at an imaging depth

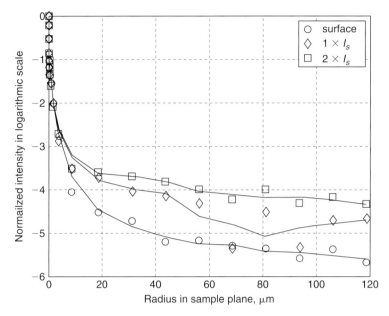

Figure 18.9. The point spread function of the scattered emission photons at the surface and depths equal to $1\times$ and $2\times$ the scattering mean free path at the emission wavelength (l_{em}^s). The FWHM of the scattered emission photon distribution is a function of (l_{em}^s) and is about 25 μm at $2 \times$ (l_{em}^s).

equal to twice the scattering mean free path length of the emission photon ($2 \times l^s$) (Fig. 18.9). The scattering mean free path is the characteristic length by which the photons are scattered. In typical tissues, the mean free path length at the blue-green range emission wavelength (l_{em}^s) is approximately 60 μm. The scattering of the emission photon results in a more rapid decay of signal strength as a function of depth and a concurrent increase of image background noise. After accounting for the loss of excitation photons as a function of depth, the effect of photon scattering on emission signal decay can be expressed as an exponential function of a dimensionless ratio of imaging depth (z) and the scattering mean free path of the emission photons (l_{em}^s). In a typical conventional multiphoton microscope, the signal decay due to emission photon scattering has a coefficient of 0.2: $S(z) \approx \exp(-0.2z/l_{em}^s)$. If all the emission photons can be collected, the coefficient would approach zero. On the other hand, if only unscattered photons are collected, the coefficient will approach one. A coefficient of 0.2 indicates that the microscope is fairly well optimized and recovers a good fraction of scattered photons. On the other hand, using the same microscope with a spatially resolved detector, a CCD camera, the coefficient approaches 1: $S(z) \approx \exp(-1 \cdot z/l_{em}^s)$, indicating that only unscattered photons are detected. Since the parallelized multiphoton microscope can detect only unscattered photons, it emission signal decays exponentially, with a characteristic length five times shorter than a typical conventional multiphoton microscope, resulting in a significantly shallower penetration depth (Fig. 18.10).

This limitation in imaging depth of a CCD-based parallelized multiphoton microscope can be overcome in another MMM configuration: a multi-anode PMT (MAPMT)

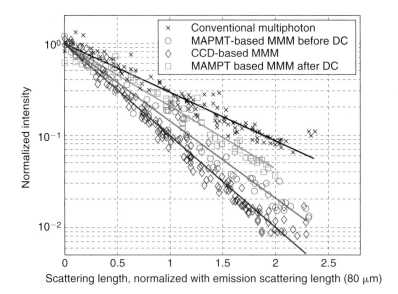

Figure 18.10. The decay of fluorescence signal as a function of depth normalized by scattering mean free path length. The measured signal decays are shown for the conventional multiphoton microscope, a CCD-based MMM, a MAPMT-based MMM, and a MAPMT-based MMM with linear deconvolution.

instead of a CCD camera can be used as photodetector for simultaneous signal collection from the multiple foci (see Fig. 18.8). The MAPMT has a segmented anode allowing spatially resolved signal detection over multiple spatial channels. A typical MAPMT has its channels arranged in square formats, such as 2×2, 4×4, or 8×8. An array of equally spaced excitation foci, typically with the same number as the MAPMT channels, is generated in the sample plane. An x-y scanner raster scans this array of foci to cover the specimen plane. The emission light generated from each excitation focus is collected by the objective and de-scanned by the intermediate optical components. The fluorescence signals from each foci are transmitted to the corresponding channel of the MAPMT. Since the emission light is de-scanned, the emission light at the detector is stationary irrespective of the scanner motion. The signal collected in each MAPMT channel is assigned to a specific pixel in the final image both as a function of MAPMT channel number and the known x-y scanner position.

As an example, for an imaging area of $360 \ \mu m \times 360 \ \mu m$ illuminated by 8×8 excitation foci, each channel in an 8×8 MAPMT collects emission light from a $45 \ \mu m \times 45\text{-}\mu m$ area in the sample plane. For a 320×320 pixel image, each focus further scans 40×40 positions within its $45 \ \mu m \times 45\text{-}\mu m$ territory.

In terms of imaging depth, this approach has two major advantages. First, a CCD pixel covers only a $1 \ \mu m \times 1\text{-}\mu m$ area in the sample plane, but the area of a MAPMT channel is more than 1,000 times larger and can collect scattered emission photons much more efficiently in this specific implementation. Specifically, the MAPMT channel should be large relative to the scattered emission photon distribution. It should be noted that the scattered photon distribution is a function of the specimen scattering coefficient and imaging depth. Therefore, the intermediate optics of

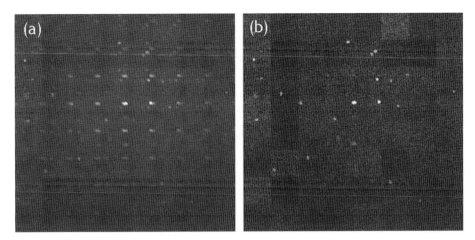

Figure 18.11. (a) A MAPMT-based MMM image (S_{acq}) of latex particles in turbid medium. Ghost images of the latex particles are seen. (b) Restored image (S_{est}) based on a linear deconvolution: $[S_{est}] = [C_{est}]^{-1}[S_{acq}]$. The convolution kernel (C_{est}) is experimentally measured at each plane, as described in text.

the microscope should be adjusted such that the MAPMT channel coverage of the specimen plane is commensurate with the measured or the estimated maximum scattered photon distribution for a given experiment. Second, while the larger coverage of the MAPMT channel allows more efficient collection of scattered emission photon, some scattered emission photons will invariably arrive at an incorrect channel. The crosstalk between the channels due to emission photon scattering can be characterized a priori by exciting at a single focus and measure the crosstalk at its neighbors. From the measurements, a convolution kernel characterizing the scattering of emission light into neighboring channels can be created. A simple linear deconvolution can effectively restore the scattered photons to the correct channel (Fig. 18.11). Furthermore, these scattered photons arriving in the neighboring channels are "encoded" by the scanner position in the image. Therefore, an image formed in the correct channel is duplicated in the images formed at the neighboring channels as "ghost" images due to these scattered photons. This additional information allows the development of other deconvolution algorithms using a maximum likelihood approach based on reducing the "ghost" images between adjacent channels. It has been demonstrated that MAPMT-based MMM can have almost equivalent imaging depth as the conventional multiphoton microscope (see Fig. 18.10). A comparison between standard CCD-based and MAPMT-based MMMs for neuronal imaging as a function of depth is presented in Figure 18.12.

18.4 CONCLUSION

Today, many implementations of high-speed multiphoton microscope have been realized and their ability to solve some unique classes of biomedical problems has been demonstrated. The most promising areas for high-speed multiphoton microscope

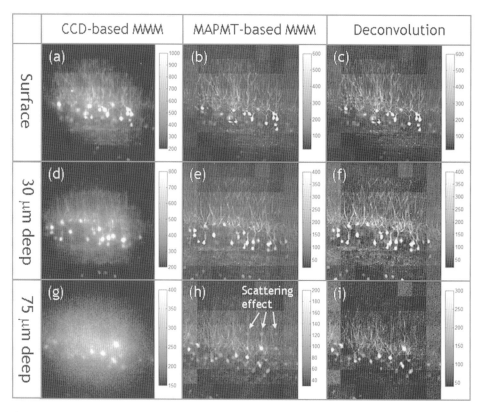

Figure 18.12. Neurons expressing GFP driven by Thy-1 promoter in a mouse brain slice are imaged at various depths with a CCD-based MMM, a MAPMT-based MMM, and a MAPMT-based MMM after deconvolution. Significant improvement in imaging depth performance is seen for MAPMT-based systems. The images were acquired using an Olympus XLUMPLFL20XW 20× water immersion with 0.95 NA. Image dimension is 160 μm × 160 μm.

are in the study of dynamical processes in 3D and in the imaging of larger tissue structures.

Although the design of high-speed multiphoton microscopes varies, most instrumental differences, such as scanner choice or focus spot geometry, are in fact minor details. There are only two fundamental factors in high-speed multiphoton microscope design: (1) the number of foci, n_{foc}, and (2) the ratio of detector channel size, d_{pix} and the size of scattered emission photon distribution, d_{sca}. Summarizing the information provided in this chapter, all high-speed multiphoton microscope designs fall within one of the following three design categories, and their ultimate performance can be fully predicted. Table 18.1 summarizes this information.

Type I microscopes include the conventional slow-speed multiphoton microscope and single focus high-speed microscope systems using fast scanners. Since these microscopes all belong to the same category, their ultimate performance is identical as long as fast enough scanners are used to achieve the desired minimum pixel residence time, t_{pix}. Type III microscopes include line scanning systems, plane scanning systems, and MMMs that use CCDs as detectors. The resolutions of these systems demand d_{pix} of the CCD to be small that is inherently incompatible with maximizing the collection of scattered photons. The speed gain of these systems by parallelization is countered by the loss of the scattered emission photons. This trade-off is characterized by the ratio:

$$\frac{n_{pix}}{\exp(z/l_{em}^s)}.$$

Since n is typically in the range of 10 to 50 constrained by the maximum laser output power, the advantage of parallelization is completely lost when imaging at a depth beyond 3 to 5 scattering length. Finally, type II microscopes represent MMM systems using pixel array with d_{pix} sufficiently large to collect most of the scattered photons, such as multi-anode PMTs. In this case, there is only a speed advantage in parallelizing and no penalty for deep imaging as long as the relationship $d_{pix} \gg d_{sca}$ holds.

In most of our discussion, we have considered imaging based on multiphoton fluorescence mode. Actually, a similar analysis also holds for other modalities. For example, second-harmonic generation is an instantaneous process and there is no excitation saturation as in fluorescence. However, there is still a maximum power that the specimen can tolerate without photodamage; the optimal excitation power, I_{ex}, is just below this damage threshold.

We have assumed using excitation, I_{ex}, that is equal to the maximum power level without saturation or damage. However, is this the optimal choice or is it better to split this power up to form more foci? The effect can be seen by considering the pixel residence time, t_{pix}, for two-photon fluorescence or second-harmonic generation:

$$t_{pix} \propto \frac{1}{n\left(\frac{I_{max}}{n}\right)^2} \propto \frac{n}{I_{max}^2}$$

where I_{max} is the maximum available power from the laser.

This equation implies that the shortest pixel residence time and the highest possible speed imaging occur when there is no parallelization and the maximum laser power from the laser is applied on the specimen. This is a consequence of the nonlinearity of the excitation process. Of course, this is never the realistic experimental condition; saturation and damage concerns always keep $I_{ex} \ll I_{max}$. Therefore, the optimal design always uses the maximum I_{ex} at each focus and splits the excess power from the laser for parallelization.

Finally, since most of the important applications of nonlinear optical microscopy are clearly in the tissue area, tissue optics must be taken into account in the design of a high-speed nonlinear optical microscope. In fact, the factors associated with tissue optics have exponential dependence, while the factors associated with parallelization and the nonlinear optical processes have linear or power law dependence. These tissue optics factors actually dominate the design considerations of nonlinear optical microscopy.

Table 18.1. A comparison of three implementations of high-speed multiphoton fluorescence microscopes[1]

Microscope type	Type I	Type II	Type III
Number of foci	$n_f = 1$	$n_f > 1$	$n_f > 1$
Detector size vs. scattered light distribution size	$d_{pix} \gg d_{sca}$	$d_{pix} \gg d_{sca}$	$d_{pix} \gg d_{sca}$
Resolution, $(z_{max} < 2\, l_{ex}^s)$ [2]	Diffraction limited	Diffraction limited	Diffraction limited
Required excitation at depth z, $I_0(z)$ [3]	$l_{ex}\exp\left(\dfrac{z}{l_{ex}^s}\right)$	$l_{ex}\exp\left(\dfrac{z}{l_{ex}^s}\right)$	$l_{ex}\exp\left(\dfrac{z}{l_{ex}^s}\right)$
Fluorescence emission at surface, $F_{em}(z)$ [4]	$F_0\exp\left(-\dfrac{bz}{l_{em}^s}\right) \approx F_0$	$F_0\exp\left(-\dfrac{bz}{l_{em}^s}\right) \approx F_0$	$F_0\exp\left(-\dfrac{b'z}{l_{em}^s}\right) \approx$ $F_0\exp\left(-\dfrac{z}{l_{em}^s}\right)$
Required excitation with z_{max}, $I_0(z_{max})$	$l_{ex}\exp\left(\dfrac{z_{max}}{l_{ex}^s}\right)$	$l_{ex}\exp\left(\dfrac{z_{max}}{l_{ex}^s}\right)$	$l_{ex}\exp\left(\dfrac{z_{max}}{l_{ex}^s}\right)$
Number of foci, $n_f = \dfrac{I_{max}}{I_0(z_{max})}$	1	$\dfrac{I_{max}}{l_{ex}}\exp\left(-\dfrac{z_{max}}{l_{ex}^s}\right)$	$\dfrac{I_{max}}{l_{ex}}\exp\left(-\dfrac{z_{max}}{l_{ex}^s}\right)$
Pixel residence time, $t_{pix} = \dfrac{SNR^2}{n_f \cdot F_{em}(z)\eta_m}$ [5]	$\dfrac{SNR^2}{F_0\eta_m}$	$\dfrac{SNR^2}{n \cdot F_0\eta_m}$	$\dfrac{SNR^2}{n \cdot F_0\eta_m}\exp\left(\dfrac{z}{l_{em}^s}\right)$
Frame rate	$\dfrac{F_0\eta_m}{SNR^2 N^2}$	$\dfrac{n_f \cdot F_0\eta_m}{SNR^2 N^2}$	$\dfrac{n_f \cdot F_0\eta_m}{SNR^2 N^2}\exp\left(-\dfrac{z}{l_{em}^s}\right)$

[1]The following variables are defined: the number of foci, n_f; detector channel size, d_{pix}; the size of scattered emission photon distribution, d_{sca}; the number of pixels in the image, $N \times N$; the desired signal to noise ratio of the image, SNR; the imaging depth, z; the maximum desired imaging depth for a given experiment, z_{max}; the maximum power delivered by the laser, I_{max}; the required incident power at the surface of the specimen for optimal imaging at depth, z, $I_0(z)$; the excitation power at the focal volume, I_{ex}; the fluorescence signal generated at the focal volume, F_0; the fluorescence signal transmitted through the tissue from depth z, $F_{em}(z)$; the specimen scattering coefficients for the excitation and emission wavelengths of the fluorophore, l_{ex}^s and l_{em}^s; photon pairs absorbed per laser pulse per molecule, n_a; the two-photon cross section of the fluorophore, δ; the fluorophore concentration, $[C]$; the fluorophore quantum efficiency, η_f; the laser repetition rate, f; the laser pulse width, τ; the excitation wavelength, λ_{ex}; the excitation volume, v; the numerical aperture of the microscopy objective, NA; the microscope collection efficiency, η_m; the pixel residence time, t_{pix}; the speed of light, c; the Planck's constant, $\hbar$; and the Avogadro's number, N_a.

Table 18.1. (*Notes continued*).

[2] Experiments have shown that the excitation PSF is not affected by scattering up to two scattering length. Some, but not all, implementations of high-speed multiphoton microscopy in these three categories can achieve diffraction limited resolution. The resolution of temporal focusing systems further depends on laser spectral bandwidth.

[3] In two-photon fluorescence mode, the optimal excitation power per focus at any depth is limited by excitation saturation (ignoring specimen photodamage) and is therefore a constant, I_{ex}. I_{ex} can be found by solving the two-photon excitation efficiency equation:

$$n_a = \frac{I_{ex}^2 \delta}{\tau f^2} \left(\frac{NA^2}{2\hbar c \lambda_{ex}} \right)^2 \approx 0.1$$

n_a is typically chosen to be about 0.1 but depends on the level of point spread function broadening that can be tolerated resulting from excitation saturation. The optimal experimental arrangement is to vary the input power on the surface of the specimen. Therefore, the excitation power, $I_0(z)$, is a function of depth.

[4] Since the optimal excitation power per focus, I_{ex} is independent of depth, the fluorescence signal generated at the focal volume is also constant:

$$F_0 = n_a f [C] v N_a \eta_f$$

The coefficient b and b' accounts for the collection efficiency of a detector channel. For example, each CCD pixel is an effective channel. When the detector channel is sufficiently large to collect most of the scattered photons ($d_{pix} \gg d_{sca}$), c approaches zero. When the detector channel is small and only collects the un-scattered ballistic photons ($d_{pix} \ll d_{sca}$), c' approaches one.

[5] Assumed the system is shot noise limited. In this case, SNR equals to the square root of the number of detected photons.

REFERENCES

Beaurepaire, E. and J. Mertz (2002). Epifluorescence collection in two-photon microscopy, Appl. Opt. 41(25): 5376–5382.

Bewersdorf, J., R. Pick and S. W. Hell (1998). Multifocal multiphoton microscopy, Opt. Lett. 23(9): 655–657.

Blanca, C. M. and C. Saloma (1998). Monte Carlo analysis of two-photon fluorescence imaging through a scattering medium, App. Opt. 37(34): 8092–8102.

Booth, M. J., M. A. A. Neil and T. Wilson (1998). Aberration correction for confocal imaging in refractive-index-mismatched media, J. Microsc 192(2): 90–98.

Booth, M. J. and T. Wilson (2001). Refractive-index-mismatch induced aberrations in single-photon and two-photon microscopy and the use of aberration correction, J. Biomed. Opt. 6(3): 266–272.

Buist, A. H., M. Muller, J. Squier and G. J. Brakenhoff (1998). Real time two-photon absorption microscopy using multi point excitation, J. Micros.-Oxf. 192: 217–226.

Centonze, V. E. and J. G. White (1998). Multiphoton excitation provides optical sections from deeper within scattering specimens than confocal imaging, Biophys. J. 75(4): 2015–2024.

Choi, B., L. Tsu, E. Chen, T. S. Ishak, S. M. Iskandar, S. Chess and J. S. Nelson (2005). Determination of chemical agent optical clearing potential using in vitro human skin, Lasers Surg Med 36(2): 72–75.

de Grauw, C. J., J. M. Vroom, H. T. M. van der Voort and H. C. Gerritsen (1999). Imaging properties in two-photon excitation microscopy and effects of refractive-index mismatch in thick specimens, Appl. Opt. 38(28): 5995–6003.

Denk, W., J. H. Strickler and W. W. Webb (1990). 2-photon laser scanning fluorescence microscopy, Science 248(4951): 73–76.

Dong, C. Y., K. Koenig and P. So (2003). Characterizing point spread functions of two-photon fluorescence microscopy in turbid medium, J Biomed Opt 8(3): 450–459.

Dong, C. Y., K. Koenig and P. T. C. So (2003). Characterizing point spread functions of two-photon fluorescence microscopy in turbid medium, J. Biomed. Opt. 8(3): 450–459.

Dunn, A. K., V. P. Wallace, M. Coleno, M. W. Berns and B. J. Tromberg (2000). Influence of optical properties on two-photon fluorescence imaging in turbid samples, App. Opt. 39(7): 1194–1201.

Egner, A. and S. W. Hell (2000). Time multiplexing and parallelization in multifocal multiphoton microscopy, J Opt Soc Am A Opt Image Sci Vis 17(7): 1192–201.

Evans, C. L., E. O. Potma, M. Puoris'haag, D. Cote, C. P. Lin and X. S. Xie (2005). Chemical imaging of tissue in vivo with video-rate coherent anti-Stokes Raman scattering microscopy, Proc Natl Acad Sci U S A 102(46): 16807–16812.

Fan, G. Y., H. Fujisaki, A. Miyawaki, R.-K. Tsay, R. Y. Tsien and M. H. Ellisman (1999). Video-Rate Scanning Two-Photon Excitation Fluorescence Microscopy and Ratio Imaging with Cameleons, Biophy. J. 78: 2412–2420.

Fittinghoff, D. N., P. W. Wiseman and J. A. Squier (2000). Widefield multiphoton and temporally decorrelated multifocal multiphoton microscopy, Opt. Exp. 7(8): 273–279.

Gan, X. and M. Gu (2000). Spatial distribution of single-photon and two-photon fluorescence light in scattering media: Monte Carlo simulation, App. Opt. 39(10): 1575–1579.

Grutzendler, J., N. Kasthuri and W.-B. Gan (2002). Long-term dendritic spine stability in the adult cortex, Nature 420(19): 812–816.

Guild, J. B. and W. W. Webb (1995). Line scanning microscopy with two-photon fluorescence excitation, Biophys J 68: 290a.

Helmchen, F. and W. Denk (2002). New developments in multiphoton microscopy, Current Opinion in Neurobiology(12): 593–601.

Helmchen, F. and W. Denk (2005). Deep tissue two-photon microscopy, Nat Methods 2(12): 932–940.

Iyer, V., B. E. Losavio and P. Saggau (2003). Compensation of spatial and temporal dispersion for acousto-optic multiphoton laser-scanning microscopy, J Biomed Opt 8(3): 460–471.

Kim, K. H., C. Buehler and P. T. C. So (1999). High-speed two-photon scanning microscope, App. Opt. 38(28): 6004–6009.

Lee, W. C., H. Huang, G. Feng, J. R. Sanes, E. N. Brown, P. T. C. So and E. Nedivi (2005). Dynamic Remodeling of Dendritic Arbors in GABAergic Interneurons of Adult Visual Cortex, PLoS Biol. 4 (2): e29.

Lendvai, B., E. A. Stern, B. Chen and K. Svoboda (2000). Experience-dependent plasticity of dendritic spines in the developing rat barrel cortex in vivo, Nature 404: 876–881.

Leveque-Fort, S., M. P. Fontaine-Aupart, G. Roger and P. Georges (2004). Fluorescence-lifetime imaging with a multifocal two-photon microscope, Opt. Lett. 29(24): 2884–2886.

Masters, B. R., P. T. So, C. Buehler, N. Barry, J. D. Sutin, W. W. Mantulin and E. Gratton (2004). Mitigating thermal mechanical damage potential during two-photon dermal imaging, J Biomed Opt 9(6): 1265–1270.

Miller, C. E., R. P. Thompson, M. R. Bigelow, G. Gittinger, T. C. Trusk and D. Sedmera (2005). Confocal imaging of the embryonic heart: how deep? Microsc Microanal 11(3): 216–223.

Nagy, A., J. Wu and K. M. Berland (2005a). Characterizing observation volumes and the role of excitation saturation in one-photon fluorescence fluctuation spectroscopy, J Biomed Opt 10(4): 44015.

Nagy, A., J. Wu and K. M. Berland (2005b). Observation volumes and {gamma}-factors in two-photon fluorescence fluctuation spectroscopy, Biophys J 89(3): 2077–2090.

Nielsen, T., M. Fricke, D. Hellweg and P. Andresen (2000). High efficiency beam splitter for multifocal multiphoton microscopy, J. Microsc 201: 368–376.

Oheim, M., E. Beaurepaire, E. Chaigneau, Jerome Mertz and S. Charpak (2001). Two-photon microscopy in brain tissue: parameters influencing the imaging depth, J. Neuro. Method. 111: 29–37.

Oron, D., E. Tal and Y. Silberberg (2005). Scanningless depth-resolved microscopy, Opt. Exp. 13(5): 1468–1476.

Padera, T. P., B. R. Stoll, P. T. C. So and R. K. Jain (2002). Conventional and High-Speed Intravital Multiphoton Laser Scanning Microscopy of Microvasculature, Lymphatics, and Leukocyte.Endothelial Interactions, Molecular Imaging 1: 9–15.

Plotnikov, S., V. Juneja, A. B. Isaacson, W. A. Mohler and P. J. Campagnola (2006). Optical clearing for improved contrast in second harmonic generation imaging of skeletal muscle, Biophys J 90(1): 328–339.

Ragan, T., K. H. Kim, K. Bahlmann and P. T. So (2004). Two-photon tissue cytometry, Methods Cell Biol 75: 23–39.

Reddy, G. D. and P. Saggau (2005). Fast three-dimensional laser scanning scheme using acousto-optic deflectors, J Biomed Opt 10(6): 064038.

Roorda, R. D., T. M. Hohl, R. Toledo-Crow and G. Miesenbock (2004). Video-rate non-linear microscopy of neuronal membrane dynamics with genetically encoded probes, J Neurophysiol 92(1): 609–621.

Sacconi, L., E. Froner, R. Antolini, M. R. Taghizadeh, A. Choudhury and F. S. Pavone (2003). Multiphoton multifocal microscopy exploiting a diffractive optical element, Opt. Lett. 28(20): 1918–1920.

So, P. T. C., C. Y. Dong, B. R. Masters and K. M. Berland (2000). Two-photon excitation fluorescence microscopy, Annu Rev Biomed Eng 2: 399–429.

Straub, M. and S. W. Hell (1998). Fluorescence lifetime three-dimensional microscopy with picosecond precision using a multifocal multiphoton microscope, Appl. Phys. Lett. 73(13): 1769–1771.

Tal, E., D. Oron and Y. Silberberg (2005). Improved depth resolution in video-rate line-scanning multiphoton microscopy using temporal focusing, Opt. Lett. 30(13): 1686–1688.

Wilson, T. (1990). Confocal Microscopy. London, Academic Press.

Ying, J., F. Liu and R. R. Alfano (1999). Spatial distribution of two-photon-excited fluorescence in scattering media, App. Opt. 38(1): 224–229.

Yu, B., C.-Y. Dong, P. T. C. So, D. Blankschtein and R. Langer (2001). In Vitro Visualization and Quantification of Oleic Acid Induced Changes in Transdermal Transport Using Two-Photon Fluorescence Microscopy, J. Invest. Dermatol. 117: 16–25.

Yu, B., K. H. Kim, P. T. C. So, D. Blankschtein and R. Langer (2002). Topographic Heterogeneity in Transdermal Transport Revealed by High-Speed Two-Photon Microscopy: Determination of Representative Skin Sample Sizes, J. Invest. Dermatol. 118: 1085–1088.

Zeng, S., X. Lv, C. Zhan, W. R. Chen, W. Xiong, S. L. Jacques and Q. Luo (2006). Simultaneous compensation for spatial and temporal dispersion of acousto-optical deflectors for two-dimensional scanning with a single prism, Opt Lett 31(8): 1091–1093.

Zipfel, W. R., R. M. Williams and W. W. Webb (2003). Nonlinear magic: multiphoton microscopy in the biosciences, Nat. Biotechnol 21(11): 1369–1137.

Nonlinear Multispectral Optical Imaging Microscopy

Concepts, Instrumentation, and Applications

Sebastian Wachsmann-Hogiu
and Daniel L. Farkas

19.1 INTRODUCTION

Nonlinear microscopy has made great advances during the past few years, as reviewed throughout this book. Simultaneously, other new approaches to microscopy have proven useful for advancing its main goals (better resolution, detection, quantitation, and applications), perhaps none more so than *spectral* microscopy. This chapter examines the confluence of these two areas, reviewing the underlying physics, the instrumentation needed for achieving nonlinear multispectral imaging with microscopic resolution, and some of the interesting applications made possible by these advances in technology.

19.2 LIGHT–MATTER INTERACTIONS

When light interacts with matter, the incident electric and magnetic fields associated with it perturb the atoms and charges in the medium. The magnitude of this interaction is proportional to the electric permittivity $\varepsilon(\omega)$ (the response of the electric charges within the material to the externally applied electric field) and magnetic permeability $\mu(\omega)$ (the extent to which an applied magnetic field is enhanced in the material) of the material. These quantities are related to the medium's complex refractive index $N(\omega)$ as follows:

$$N(\omega) = n(\omega) + i\kappa(\omega) = \sqrt{\varepsilon(\omega)\mu(\omega)} = \sqrt{[\varepsilon_1(\omega) + i\varepsilon_2(\omega)][\mu_1(\omega) + i\mu_2(\omega)]}$$
$$(19.1)$$

where ω is the frequency of light, $n(\omega)$ is the real refractive index (indicating the phase velocity of light), $\kappa(\omega)$ is the extinction coefficient (indicating the absorption of light when propagating through the medium), and $\varepsilon_{1,2}(\omega)$ and $\mu_{1,2}(\omega)$ are the real and imaginary parts of the relative permittivity and permeability, respectively.

Since the response of biological materials to the oscillating magnetic field at frequencies in the visible range ($4–10 \times 10^{14}$Hz) is very small compared to the response to the oscillating electric field at the same frequency, the permittivity of the material is the physical quantity that is most relevant for this chapter.

The response of a bulk medium to incident electromagnetic radiation is formulated in terms of induced macroscopic polarization. In the electric dipole approximation, the induced polarization may be written as a Taylor expansion on the strength of the applied electric field $E(r, t)$:

$$P(E) \approx \chi^{(1)}E + \chi^{(2)}EE + \chi^{(3)}EEE + \cdots \qquad (19.2)$$

where $\chi^{(1)}$ is the linear susceptibility (second-rank tensor describing optical linear processes like linear absorption or refraction), related to the refractive index N by $N^2 = 1 + \chi^{(1)}$. The quantity $\chi^{(2)}$ is the second-order nonlinear susceptibility (third-rank tensor describing three-wave interactions such as second-harmonic generation [SHG], sum frequency generation, and difference frequency generation), and $\chi^{(3)}$ is the third-order nonlinear susceptibilities (fourth-rank tensor describing the four-wave interactions such as third-harmonic generation [THG], coherent anti-Stokes Raman scattering [CARS], or two-photon absorption [TPA]).

The susceptibilities $\chi^{(n)}$ are bulk quantities and are related to the microscopic quantities α (polarizability), β (first-order hyperpolarizability), and γ (second-order hyperpolarizability). Although higher-order terms in the expansion (19.2) could play a role at extremely high incident intensities, for the purpose of this chapter their contributions can be neglected.

Optical spectroscopy studies the interaction of electromagnetic waves with atoms and molecules. For reasons discussed earlier, only the electric field will act on atoms or molecules by distorting or polarizing them. The induced polarization will follow the driving oscillating external driving field, and will build up for frequencies matching a transition in the atom or molecule. In other words, the induced polarization will have in-phase (far from any resonance) and out-of-phase (in resonance) components, which will determine the type of interaction. Several interactions can be distinguished: (i) refraction (phase shift), (ii) Rayleigh scattering (light is radiated in all directions by static, non-uniformly distributed dipoles), (iii) Brillouin scattering (scattering on dynamic density fluctuations associated with propagating vibrations in solids), (iv) Raman scattering (vibrational and rotational, corresponding to coupling of the electric field to molecular vibrations or rotations, which produces a change in the molecular polarizability α, (v) absorption (resonant interaction between the electric field and electrons from atomic/molecular orbitals, described by the imaginary part of $\chi^{(1)}$ for one-photon absorption and imaginary part of $\chi^{(3)}$ for two-photon absorption), and (vi) fluorescence and phosphorescence (radiative emission of light), (vii) Mie scattering (closed-surface, charge-resonance scattering; in the case of visible light the objects are micron-sized).

The main problem in imaging can be formulated in general by using the continuous-to-discrete model for the projection of a continuous object f (viewed over a region of interest S to a discrete detector space g_m (Barrett & Meyers, 2004):

$$g_m = \int_S f(r)h_m(r)dr \qquad (19.3)$$

where m is the composite index specifying the angular dependency of the system, and $h_m(r)$ describes the sensitivity of the detector to light coming from an object point whose vector is r.

The smallest separable feature X_{min} (spatial resolution) is determined by the Rayleigh criterion for the far-field diffraction limit (Abbe's diffraction limit theorem: $X_{min} = \frac{0.61*\lambda}{NA}$, where NA is the numerical aperture of the collecting lens and λ the wavelength of light), and the Nyquist sampling theorem ($X_{min} = \frac{2D}{M}$, where D is the pixel size of the detector and M the lens magnification).

19.3 MULTISPECTRAL IMAGING

Multispectral imaging is an emerging field where the advantages of optical spectroscopy as an analytical tool are combined with the power of object visualization imparted by imaging. It has proven to be a very useful tool in various areas, including remote sensing and biomedical applications. While powerful nonlinear imaging methods like two-photon absorption-excited fluorescence, second-harmonic and CARS imaging are emerging for biomedical applications, multispectral imaging clearly has the potential to add new dimensions to these methods. A general schematic of multispectral linear/nonlinear optical imaging is presented in Figure 19.1.

A multispectral image should contain spectral information on each pixel of the image. A third dimension (spectral dimension) is added to the two-dimensional spatial image, generating a three-dimensional data cube. In most applications, each dimension will require up to 10^3 pixels, the whole data cube thus containing up to 10^9 pixels; this makes recording, storing, and data analysis a technical and computational challenge. In the following, the signals for each of the pixels will be referred to as "channels."

The most efficient way of generating a multispectral image (with the highest signal-to-noise ratio [SNR]) is by using a detection architecture with a detector element at each channel of the spectral cube. Even though there are currently efforts to develop Red-Green-Blue (RGB) stacked CCD chips, physical limitations related to the absorption on the CCD chips makes this idea hardly usable for multispectral imaging, and the two-dimensional CCD chip architecture is still the best for imaging. Consequently, every single element of the detector has to measure multiple channels in a temporal sequence of measurements. This process of multiplexing consists, in general, of two steps: an encoding step and a transform step. An object (represented by a vector Ψ) is encoded by an encoding matrix W, which is specified by the experiment. The measured values υ correspond to the matrix equation:

$$\upsilon = W\Psi \tag{19.4}$$

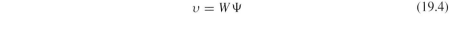
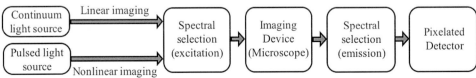

Figure 19.1. Block diagram of linear and nonlinear multispectral imaging.

If the inverse of the encoding matrix exists (W^{-1}), Ψ can be recovered by applying W^{-1} to υ:

$$\Psi = W^{-1}\upsilon \qquad (19.5)$$

The simplest encoding matrix is the identity matrix, which encodes for spatial scanning experiments like single point scanning. Other encoding matrices are possible, like the Hadamard and Fourier matrix for encoding wavelength. Even though different multiplexing methods have many common features (such as an advantage under detector-noise limited measurements and a disadvantage during photon shot-noise limited measurements), each technique is subject to a unique set of errors and artifacts.

After Ψ has been recovered, the sequence of spatial-spectral data will be represented as a spectral cube, which will contain spatial information on two dimensions (usually x, y) and spectral information on the third dimension (λ). Spectral image analysis is then performed in three steps: image segmentation, feature extraction, and classification of the spectral content based on defined libraries.

The time required for the acquisition of a full spectral cube varies from method to method, as does the recorded image size and spectral resolution. In the following, we will review some of the most common techniques used for multispectral imaging. Based on the way the spectral cube is acquired in time, they can be classified as (i) spatial scanning techniques (all the wavelengths are detected simultaneously and scanning is performed in the spatial domain), (ii) wavelength scanning techniques (a full image is detected for a narrow spectral range and scanning is performed in the wavelength domain), and (iii) non-scanning techniques (spatial and spectral data are collected in a snapshot on a single CCD detector).

19.4 SPATIAL SCANNING TECHNIQUES (SPECTRAL MULTIPLEXING)

The schemes of the commonly used spatial scanning techniques are shown in Figure 19.2. The simplest technique is the point scanning (*point scanning spectrometers*), which is widely used in confocal microscopes. The illumination of the object is made with light focused on single point with the size limited by the diffraction limit theorem. The light collected from this point is passed through a spectrometer and the wavelength λ is detected on a linear array detector. The angular dispersion/diffraction of the prism/grating is $D \equiv d\theta/d\lambda$ (where θ is the deflection angle for the wavelength λ), and the chromatic resolving power of the grating spectrometer is $R \equiv \lambda/(\Delta\lambda)_{min} = mN$, with $(\Delta\lambda)_{min}$ being the least resolvable wavelength difference (the limit of resolution, which can be identified with the Rayleigh criterion of resolution), m the diffraction order, and N the number of grating grooves. The x, y spatial information is obtained by sequential scanning along these directions.

An improvement of this technique is the line-scanning device. In this implementation, the object is wide-field illuminated, and scanning of either the specimen (*pushbroom imaging spectrometer*) or the slit (*moving slit spectrometer*) is performed to obtain the second spatial dimension. A 2D detector is required for these arrangements.

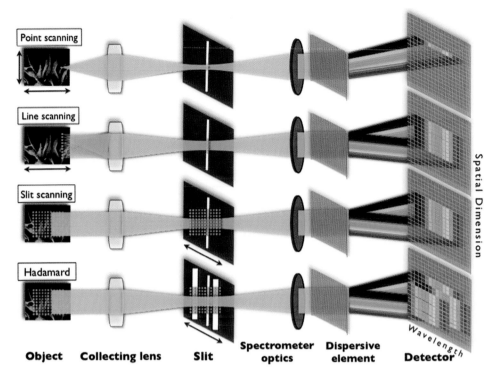

Figure 19.2. Schematics of the spatial-scanning multispectral techniques presented in this chapter. From top to bottom: point scanning, line scanning, slit scanning, and Hadamard techniques. A spectral data cube is generated that has two spatial dimensions (x and y) perpendicular on a spectral dimension λ.

19.4.1 Multiple Slit Spectrometers: Hadamard Transform Spectrometers

A conventional dispersive imaging spectrometer consists of a single entrance slit, a dispersive element (like a grating or prism), and a multichannel detector. However, increased light throughput and a higher SNR can be realized by replacing the single entrance slit with multiple slits, or in general with a mask. The first multi-slit spectrographs were reported by Golay (Golay, 1949). A more general implementation is by using special masks, called Hadamard masks, which are made up of a pattern of either transmissive and reflective or transmissive and opaque slits of various widths constructed from a Hadamard matrix. The spectral information is extracted by performing a Hadamard transform.

The Hadamard transform is an example of a generalized class of Fourier transforms and consists of a projection onto a set of square waves called Walsh functions (in comparison, the Fourier transform consists of a projection onto a set of orthogonal sinusoidal waveforms). The Hadamard transform is traditionally used in many signal processing and data compression algorithms and more recently in spectral imaging (Goelman, 1994; Hanley et al., 1998, 1999, 2000; Macgregor & Young, 1997; Tang et al., 2002; Treado & Morris, 1988, 1990; Wuttig, 2005). The encoding matrix W from (19.4) is either the Hadamard matrix $\mathbf{H}$ (consisting of either 1 or −1 elements) for transmissive and reflective masks, or a related Sylvester matrix $\mathbf{S}$ (consisting of 1 or 0 elements) for transmissive and opaque masks (Hanley, 2001). A fast Hadamard transform can be applied to reduce the number of computations from $\approx n(n-1)$ to $\approx n \log_2 n$, where n is the number of recorded channels (Harwit & Sloane, 1979).

For multispectral imaging, the most common use of Hadamard methods is by modulating a set of parallel spectrograph entrance slits by a spatial light modulator according to an $\mathbf{S}$ matrix encoding (Hanley et al., 2002). In this implementation, each column j of the $\mathbf{S}$ matrix encodes for a slit (which is opened if the matrix element is 1 and closed for matrix elements equal to 0), and the rows i of the $\mathbf{S}$ matrix encode for the pattern of opening and closing the slits in a specific experiment, creating a "moving" slit. For every i row of the matrix, an image $\sigma(x, i, \lambda)$ is recorded containing information about the x spatial position and wavelength λ. Different rows encode for the second spatial direction y, which may be recovered by an inverse Hadamard transform of $\sigma(x, i, \lambda)$ along i to obtain $\Psi(x, y, \lambda)$ from (19.5). This allows the detection of a series of spectroscopic images. The improvement in SNR is related to the expectation value for statistical error at multiplexing (Mende et al., 1993):

$$
E = \frac{2}{n+1} \left(\sum_{j=1}^{n} \bar{e}_j^2 \right)^{1/2}
\tag{19.6}
$$

where $\bar{e}_j^2$ is the mean-square signal fluctuation in the j-th detector element, and n is the degree of multiplexing (corresponds to the number of rows in $\mathbf{S}$, which is the number of measured values).

Compared to scanning, the gain in SNR due to multiplexing can be expressed as (Harwit & Sloane, 1979):

$$
Q_{multiplex} = \frac{SNR_{multiplex}}{SNR_{scan}} = \frac{(n+1)}{2\sqrt{n}}
\tag{19.7}
$$

This is equivalent to saying that the detector background noise is reduced due to multiplexing by a factor of $2\sqrt{n}/(n+1)$.

While Hadamard multiplexing has a major advantage in measurements where the noise is detector-limited, it may create masking effects and artifacts that need to be accounted for (Hanley, 2001).

19.5 WAVELENGTH SCANNING TECHNIQUES (SPATIAL MULTIPLEXING)

Diagrams of the commonly used wavelength scanning techniques are presented in Figure 19.3. The most basic approach of wavelength scanning multispectral imaging is the use of a set of narrow bandpass filters mounted on a spinning disk (*filter wheel*). As the disk rotates in synchrony with the frame rate of the CCD detector, different filters perpendicularly cross the optical path, leading to wavelength selection. The spectral resolution is determined by the bandwidth and number of the filters used. For example, a spectral resolution of 5 nm over a spectral range of 200 nm requires 40 filters. The main advantage of this technique is its simplicity and stability. Disadvantages are low spectral resolution, low versatility, and registration problems.

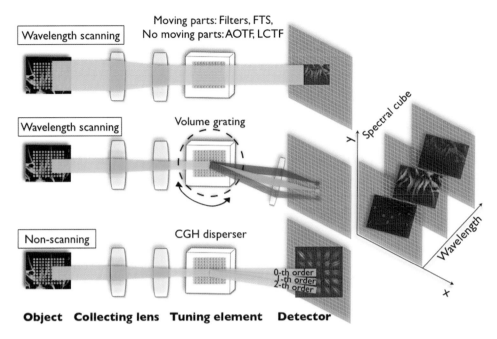

Figure 19.3. Schematics of the wavelength-scanning multispectral techniques presented in this chapter. From top to bottom: with moving parts—filter wheel, Fourier transform spectroscopy; with no moving parts—LCTF, AOTF, volume holographic gratings, and nonscanning techniques using a computer-generated hologram (CGH) are schematically represented. For all these methods, a spectral data cube is generated that has two spatial dimensions (x and y) perpendicular on a spectral dimension λ.

Other, more advanced wavelength scanning techniques (like Fourier-transform based, acousto-optical and liquid-crystal tunable filters, volumetric holographic gratings) will be discussed in detail below.

19.5.1 Fourier-Transform Imaging Spectrometers (FTIS)

In mathematics, the discrete Fourier transform (DFT) is a Fourier transform employed in signal processing and related fields to analyze the frequencies contained in a sampled signal, solve partial differential equations, and perform other operations such as convolutions. The DFT can be computed efficiently in practice using a fast Fourier transform (FFT) algorithm.

A Fourier-based spectrometer uses an interferometer for wavelength selection (Bell, 1972; Chamberlain, 1979). In the implementation by Applied Spectral Imaging (Carlsbad, CA), the interferometer is placed on the emission side of the microscope (Garini et al., 1999). The beam coming from the collecting objective is first collimated and then divided into two coherent beams. The beams are then recombined and produce an interference pattern, which depends on the wavelengths of light and the optical path difference (OPD) introduced between the two beams. A two-dimensional detector measures the intensity of this interference pattern as a function of OPD (noted D), generating an interferogram $I(D)$, which has discrete values. The subsequent FFT of the interferogram recovers the spectrum:

$$I(\lambda) = \sum_D I(D) \exp^{\frac{-i2\pi D}{\lambda}} \qquad (19.8)$$

A Michelson interferometer can be used, but a far more stable design has been found in a Sagnac type of interferometer, where the two beams travel the same path but in opposite directions (Malik et al., 1996a, 1996b). In this way, a method traditionally applied in infrared can be used in visible, where smaller wavelengths imply higher sensitivity to vibrations. By rotating the beam or the whole interferometer at very small angles, an OPD is created (within the beam splitter), which depends on the incidence angle of light on the beam splitter. This rotation causes the interference fringes to travel across the detector onto the image plane. In practice, the measurement is performed by recording successive frames on an $N x N$ elements CCD detector in synchrony with the steps of the motor used to rotate the collimated beam. To obtain meaningful spectral/spatial resolutions, each interference fringe (which represents an OPD equal to one wavelength) must be sampled by at least two frames in time and by at least two pixels in space (Nyquist theorem). Since the central fringe ($OPD = 0$) needs to travel across the image plane a total distance corresponding to $2N$ (in order to record the whole image), the $OPD_{max} = \lambda N / 2$ and the number of necessary frames is $N_f = 2N$. In this case, the chromatic resolving power of the instrument is $R = OPD_{max} / \lambda = N / 2$. On the other hand, the spectral resolution, being proportional to the maximum OPD obtained, is determined by the maximum rotational angle that does not produce vignetting. Typically, total angular amplitudes of approximately 1 to 5 degrees with a step size on the order of 60 μ rad can be achieved with this instrument, which corresponds to a spectral resolution between 4 and 8 nm. The spectral resolution decreases for increasing wavelengths. Wavelength calibration is performed by measuring a series of monochromatic sources. The time required to record a whole data cube increases linearly with the spectral resolution; it takes approximately 60 s to generate a spectral cube at 400 nm with 10-nm resolution (Garini et al., 1996). In addition, computational requirements constrain the maximum image size to approximately 500×500 pixels and long acquisition times.

Recently, a variant of FTIS that sends the excitation light through the same interferometer has been demonstrated (Heintzmann et al., 2004). This implementation, called double-pass Fourier-transform imaging spectroscopy, spectrally modulates both the excitation and emission, allowing us to obtain an excitation as well as an emission spectrum of a fluorescent sample with a single sweep of the interferometer. In addition, optical sectioning can be achieved at excitation wavelengths due to pattern excitation through the Sagnac interferometer.

Another approach, based on a scanning birefringent interferometer (Harvey & Fletcher-Holmes, 2004), addresses the problem of sensitivity to vibrations of the previously described methods. This interferometer consists of two Wollatson prisms of equal and opposite splitting angles located close to the exit pupil of the microscope. The light coming from the sample is first linearly polarized at 45 degrees to the optic axes of the prisms, and then split by the first prism into two orthogonally components of equal amplitude. The second prism refracts the two components so that they propagate parallel, and a lens focuses them in the same spot on the detector, allowing them to interfere. Since the two components are orthogonally polarized, an OPD is introduced, which can be modulated by the translation of the second Wollatson prism, according to $\Delta_{OPD} = 2bh \tan \theta$ (Harvey & Fletcher-Holmes, 2004). Here, b is the birefringence of the Wollatson prism, h is the offset of the centers of the two prisms, and θ is the prism wedge angle. While the main advantages are the improved robustness and the large spectral range (can be operated at wavelengths between 200 nm and 14 µ m), this approach has lower optical efficiency compared to the Michelson or Sagnac interferometers (maximum 50% in polarized light and 25% in unpolarized light).

The FTIS approach to spectral imaging offers high sensitivity and throughput but requires high stability and suffers from phase errors, aliasing, and incomplete modulation. Data analysis methods have evolved to include pre- and post-transform procedures to reduce artifacts, including apodizing, phase correction, zero filling, spectral averaging, and frequency-selective filtering.

19.5.2 Liquid-Crystal Tunable Filter (LCTF) Imagers

These filters select the wavelengths they transmit based on the same diffraction principles as fixed-wavelength interference filters, but with electronically controlled tuning. They have the advantage of no moving parts and random wavelength access within their spectral range. A typical LCTF is a multistage Lyot polarization interference filter, with an added liquid-crystal waveplate in each stage providing an electronically controllable variable retardance; thus, $m + 1$ polarizers are separated by m layers of liquid crystals (m is typically 3 to 10), sandwiched between birefringent crystals. The OPD in birefringent crystals is dependent upon crystal thickness d and the refractive index difference between the ordinary n_o and extraordinary n_e light rays produced at the wavelength λ of incident illumination and can be expressed as the retardation R:

$$R = \frac{2\pi d \, (n_e - n_o)}{\lambda} = \frac{2\pi d \, \Delta n}{\lambda} \tag{19.9}$$

Transmission of light through the crystal T is dependent upon the phase delay, created by the difference in propagation speed between the extraordinary and ordinary rays; this phase delay also depends on the wavelength of light λ, thus yielding the transmittance of the m-th stage of the cascaded Lyot filter (Morris et al., 1994):

$$T_m = \cos^2 \left[\frac{R_m(\lambda)}{2} \right] \tag{19.10}$$

Crystals in a typical Lyot filter are often selected for a binary sequence of retardation so that transmission is maximal at the wavelength determined by the thickest crystal retarder. Other stages in the filter serve to block the transmission of unwanted wavelengths (Spring, 2000).

All this yields versatile and relatively fast (approximately tens of milliseconds) wavelength selection. However, some of the disadvantages are limited spectral range, temperature sensitivity, polarization sensitivity, and relatively poor transmission and contrast at high spectral resolution.

19.5.3 Acousto-optic Tunable Filter (AOTF) Imagers

AOTFs provide electronically controllable, solid-state wavelength tunability of light from the ultraviolet to the near-infrared, with random-access, bandpass variability, and high throughput. Their functioning depends on light–sound (photon–phonon) interactions in a crystal with special properties (usually TeO_2), and their tuning speed, based on the underlying physics, is very high, being limited only by the speed of sound propagation in the crystal; this yields typical wavelength switching times in the tens of microseconds. These features have led to the use of AOTFs in a wide variety of spectroscopic applications. Interest in AOTFs for multispectral imaging has, however, been more recent. After remote sensing-type imaging applications for ground-based and planetary targets were demonstrated in the 1970s and 1980s, several groups have reported breadboard AOTF imaging demonstrations (Cui et al., 1993; Gao & Lin, 1993; Suhre et al., 1992), and they have been used in biologically relevant experiments such as fluorescence (Kurtz et al., 1987; Morris et al., 1994) and Raman (Treado et al., 1992) microscopy of biological samples.

Relatively poor image quality was, for a long time, a recurring difficulty with imaging AOTFs. Most of the work mentioned above reported typical spatial resolutions no better than 8 to 10 μm, and the best resolution achieved with an AOTF by the mid-1990s (Schaeberle et al., 1994) was around 2.0 μm. These results, while sufficient for many applications, were completely inadequate for high-quality light microscopy, where a spatial resolution under 0.5 μm is routinely required. As a result, although there was great potential for AOTFs in high-resolution imaging, including biological and medical, their use in such applications had to await the achievement of a higher spatial resolution (and enhanced light throughput, given the special requirements of cellular bioimaging).

Mechanisms of image blur in AOTFs have been qualitatively described, but detailed investigations of the effect and attempts to correct it have not been reported until 1996, when we presented a quantitative, experimental demonstration of AOTF image blur causes. Using these results, we made use of digital image processing techniques to computationally remove image degradation, and obtained images in which features down to a diffraction-limited 0.35 μm in size could be clearly resolved (Wachman et al., 1996). In order to preserve the speed advantage of AOTF-based imaging, an optical method was later developed (Wachman et al., 1997) that yielded similarly high AOTF spatial resolution, without a need for image processing (see also Wachman et al., 1998). The following describes the approach taken and the underlying physics.

The presence of an acoustic wave inside a medium creates a periodic modulation of its index of refraction via the elasto-optic effect. This modulation acts as a three-dimensional sinusoidal phase grating for light incident on the crystal, leading to diffraction of certain wavelengths at an angle from the incident beam direction. In an AOTF, this diffracted light is used as the filter output, and the grating frequency is electronically controlled by varying the RF frequency applied to a piezoelectric transducer bonded to one of the faces of the crystal. This results in an all-electronic, no-moving-parts, broadly tunable spectral filter with wavelength switching speeds determined by the acoustic transit time in the crystal (typically <50 μs). When driven with multiple closely spaced RF frequencies, the AOTF also provides electronically variable bandwidth control.

The interaction of acoustic and optical plane waves in an anisotropic medium may be described by a pair of coupled differential equations (Chang, 1976). The most useful solutions to these equations occur when the phase-matching conditions are satisfied:

$$k_d = k_i + k_g \tag{19.11}$$

where $k_d = 2\pi n_d/\lambda$ (for the diffracted beam), $k_i = 2\pi n_i/\lambda$ (for the incident beam), and $k_g = 2\pi f/v$ (for the grating generated by the acoustic wave in the crystal), with f the acoustic frequency, v the acoustic speed in the crystal, λ the optical wavelength, and $n_{i,d}$ the crystal indices of refraction for the incident and diffracted beams, respectively.

For the case of an o-polarized incident wave (this is the preferred polarization for imaging in a TeO$_2$ AOTF) equation (19.11) may be written:

$$\begin{aligned} n_e(\lambda, \theta_d)\cos(\theta_d) - n_o(\lambda)\cos(\theta_i) + f\lambda/v\sin(\alpha) = 0 \\ n_e(\lambda, \theta_d)\sin(\theta_d) - n_o(\lambda)\sin(\theta_i) - f\lambda/v\cos(\alpha) = 0 \end{aligned} \tag{19.12}$$

where θ_i and θ_d are the angles between the incident and diffracted beam wave-vectors and the optic axis, n_e and n_o are the extraordinary and ordinary indices of refraction, and α is the angle between the acoustic wave-vector, k_g, and the acoustic axis.

These equations determine the spectral tuning characteristics of AO tunable filters.

For AOTF *imaging* applications, image fidelity must also be considered. With the use of an AOTF as an imaging filter for o-polarized white light, every point on the object gives rise to a bundle of multichromatic parallel rays incident on the crystal at a single angle θ_i. Ideally, we would like the diffracted portion of this bundle to exit the crystal as a bundle of monochromatic parallel rays described by a unique θ_d. In this case, each point on the object plane will map to a single point on the image plane. In practice, however, it is found that the diffracted ray bundle consists of rays leaving the crystal over a range of different output angles. As a result, each object plane point maps to a distribution of image plane points, leading to a blurred image even for fixed frequency operation.

A second image degrading effect, image shift, occurs when the radiofrequency f is varied. The phase-matching equations (19.12) dictate that changes in f result in changes in both the wavelength, λ, and the diffracted angle, θ_d, for fixed θ_i and α. This leads to a shift in image position for different wavelengths. Appropriate cut of the crystal exit face, however, eliminates this almost entirely (Wachman et al., 1997; Yano & Watanabe, 1976).

It has been noted that since the transducer attached to the AOTF is of finite length, the acoustic field it produces may be described as a superposition of plane waves at various acoustic angles, α. For white light illumination at a given operating frequency f and incident beam direction, θ_i, each such frequency will produce a diffracted output at a distinct angle θ_d and wavelength λ. The spread in acoustic angle resulting from the finite transducer length consequently gives a diffracted output containing a range of angles and wavelengths, even for fixed incident beam direction, resulting in a filtered image that is blurred. AOTF image blur is, therefore, primarily attributable to acoustic beam divergence in the crystal.

For negligible incident light depletion (an approximation valid for AOTF efficiencies up to about 70%), the relationship between the acoustic angle intensity spectrum and the diffracted output intensity spectra may be derived explicitly for plane waves from the AO interaction equations (Wachman et al., 1996). For fixed f and θ_i we have:

$$I_{out}(\lambda, \theta_d) = C^2 I_{inc}(\lambda) I_\alpha(\alpha) \delta[k_d(\theta_d, \lambda) - k_i(\theta_i, \lambda) - k_g(\alpha, f)] \qquad (19.13)$$

In this equation, I_{out} is the diffracted intensity; C is a constant; I_{inc} is the wavelength spectrum of the incident light; I_α is the acoustic angle intensity spectrum, proportional to the squared magnitude of the Fourier transform of the transducer profile in the direction of light propagation; and the delta function, δ, expresses the phase-matching requirement. In the case of white light illumination (I_{inc} is constant), this equation shows that the diffracted intensity is directly proportional to the acoustic angle spectrum:

$$I_{out} \sim I_\alpha(\alpha) \qquad (19.14)$$

Since, as discussed above, both λ and θ_d are functions of α via the phase-matching equations (19.12), I_{out} can be expressed either in terms of wavelength or in terms of output angle. When expressed in terms of wavelength, $I_{out}^\lambda(\lambda)$, it may be identified as the bandpass profile of the filter; when expressed in terms of angle, $I_{out}^{\theta_d}(\theta_d)$, it may be interpreted as the image blur profile. This equation shows that both of these are determined by the Fourier transform of the transducer structure.

The effect of the acoustic angle distribution on the spectral bandpass of an AOTF is well known (Chang, 1995). Its effect on AOTF angular output, expressed by $I_{out}^{\theta_d}(\theta_d)$, has been considered in detail only in the context of AO deflectors (Pieper & Poon, 1990); however, the ramifications for imaging have not been thoroughly investigated. Indeed, for the majority of imaging AOTFs operating with a spatial resolution of 10 μm or above, the effect of the acoustic angle distribution on image quality is insignificant. However, for an image resolution of 1 μm or less, as needed in microscopy, this effect becomes highly significant. For a given operating frequency, the resulting AOTF output is well described by the diffracted intensity distribution I_{out} of equation (19.14). Changing the RF transducer profile by disconnecting one or another of the transducer ports should consequently result in substantially different AOTF images. This is illustrated below, closely following our first description of the phenomenon in 1996 (Wachman et al., 1996).

Darkfield images of a single 0.12-μm bead taken through the AOTF with a 40× objective are shown in Figure 19.4 (top) for one, two, and six transducer slices connected (transducer lengths of 0.33 μm, 0.66 μm, and 1.98 μm, respectively). The center bright spot in each picture represents the primary AOTF image of the bead; the narrower this spot, the better the AOTF resolution. The increase in resolution with longer transducer length expected from equation (19.14) is clearly evident; the center spot in the bottom picture (six slices) corresponds to a resolution of 1 μm. The secondary spots on either side of the center arise from the side-lobes in the transducer Fourier transform, and are reduced in intensity (relative to the main peak) by one to two orders of magnitude. The graphs at the bottom of Figure 19.4 present quantification of data

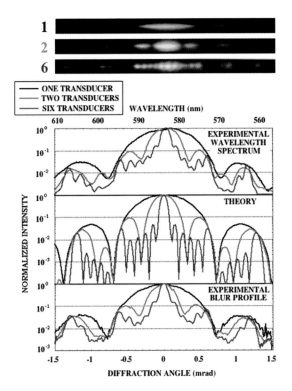

Figure 19.4. Transducer structure determines image blur in the AOTF. *Top three images*: Photographs of dark-field images of 0.12- μm opaque beads taken through the AOTF at a frequency of 74 MHz are shown using one (top), two (center), and six (bottom) transducer slices; these correspond to transducer lengths of 0.33 μm, 0.66 μm, and 1.98 μm, respectively. The differences in central peak width and sideband structure between the various transducer configurations are readily apparent. *Bottom three graphs*: These graphs quantify the images from the top of the figure and compare the results to theory. Intensity-versus-wavelength data (top) and intensity-versus-output angle data (bottom) are shown for the three top images. Theoretical curves for each of the transducer configurations, obtained by taking the squared magnitude of the Fourier transform in the direction of light propagation, are shown in the center panel.

and comparison with theory. The graphs display the dependence of intensity on AOTF output angle (proportional to distance from the central spot) for each of the images of Figure 19.4 (top three), by plotting intensity profile data of images obtained with a CCD, and are, in effect, measurements of I_{out}. The top graph displays *spectral* data taken simultaneously. These are equivalent to measurements of the spectral content of each of the images of Figure 19.4 (top), and correspond to measurements of I_{out}. According to equation (19.14), both sets of data should be proportional to the acoustic angle profiles, I_α, for each of the transducer configurations shown. The center graph in Figure 19.4 shows these theoretical results, representing the squared magnitude of the Fourier transform of each transducer profile (taking into account the slice separation of ~ 0.5 mm). The theoretical curves for these three configurations have pronounced differences in center peak width and sideband structure, reflecting the differences between their Fourier transforms. These features are also clearly evident in both the top and bottom sets of experimental data. Indeed, the detailed correspondence between measured and calculated results is a remarkable confirmation of equation (19.14). Thus, the results in Figure 19.4 quantitatively demonstrate the effect of transducer structure on the spectral and angular output characteristics of an AO imaging filter.

The diffracted intensity distributions shown result in two types of image degradation. The center peak width leads to decreased image resolution, and the sideband structure leads to decreased image contrast. With the quantitative results shown here, however, these effects could be compensated for by using either digital image processing techniques or optical corrections.

The three major limitations that have restricted the widespread use of AOTFs for imaging spectroscopy were relatively low out-of-band extinction, decreased throughput, and poor imaging quality. As described above, we have made advances on the first two of these and have significantly improved image quality, based on a quantitative, mechanistic understanding of the sources of light loss and blur, to the point that diffraction-limited microscopic imaging (with a spatial resolution of about 0.35 μm) with AOTFs is made possible (Wachman et al., 1996, 1997, 1998), enabling interesting new applications (see Section 19.8 below).

For advanced spectral imaging applications involving AOTFs, it is interesting to contemplate a microscopy workstation, similar in concept to the one we described previously (Shonat et al., 1997; Wachman et al., 1997), where both excitation and emission wavelength selection is achieved by AOTFs but fast lasers typical for nonlinear applications are used as excitation sources. This imparts great flexibility to the experimental capabilities, allows for true multispectral imaging, and could be used in a number of new nonlinear imaging experiments. One possible implementation is illustrated in Figure 19.1, with a femtosecond laser-induced white light continuum being further filtered by an AOTF used as a spectral selection device for excitation and/or emission (Wachsmann-Hogiu et al., in preparation). This is by no means the only possible choice: Saito and colleagues (2001) have proposed and demonstrated a Ti-Sapphire laser with electronic tuning via an intracavity AOTF.

19.5.4 Volume Holographic Spectral Imaging

Volume holograms are a new type of wavelength selective elements for spectral imaging consisting of a 3D periodic phase or absorption perturbation throughout the entire volume of the element. This 3D pattern gives high diffraction efficiencies (close to 100% at a single wavelength), high angle and spectral selectivity, and the ability of multiplexing many holograms in the same volume (Liu et al., 2004). Typically, a volume hologram is created by recording the interference pattern of two mutually coherent light beams.

In the first Born's approximation, the diffraction efficiency η of a transmission volume hologram is proportional to the amount of light diffracted, which is determined by the hologram thickness L and the Bragg mismatch factor $\Delta k = k_i + k_g - k_d$ (vectorial sum/difference of the incident (k_i), grating (k_g) and diffracted (k_d) wave vectors).

$$\eta \approx \operatorname{sin} c^2 \frac{L\Delta k}{2\pi} \qquad (19.15)$$

$\Delta k \neq 0$ can be caused by angular and wavelength detuning from the Bragg matching condition $\Delta k = 0$. For angular detuning $\Delta \theta$, wavelength λ, and a Bragg-matched incident angle θ in the medium, the efficiency becomes for:

$$\eta \approx \sin c^2 \left(\frac{2L \sin \theta}{\lambda} \Delta \theta \right) \tag{19.16}$$

and the angle Bragg selectivity is defined as the angular spacing between the central peak and the first null of the sinc function:

$$\Delta \theta_B = \frac{\lambda}{2L \sin \theta} \tag{19.17}$$

Similarly, it can be shown (Liu et al., 2004) that for a wavelength Bragg mismatch $\Delta \lambda$, the wavelength Bragg selectivity (the limit of resolution) can be derived as:

$$\Delta \lambda_B = \frac{\lambda^2 \cos(\theta)}{2L \sin^2(\theta)} \tag{19.18}$$

This gives, for a 45-degree incident angle, a 3-mm-thick hologram, and 550-nm wavelength, a theoretically predicted selectivity of approximately 0.03 nm. In practice, the wavelength is scanned by the rotation of the volume grating, and the achievable spectral resolution is determined by the scanning step of the rotation stage. For example, a scanning step of 0.01 degrees corresponds to 0.125-nm spectral resolution at 550 nm. Geometry considerations require that for a rotation of the hologram with an angle σ, the imaging lens and the CCD camera rotate with 2σ. The spectral coverage ranges from 400 nm to 800 nm, which is limited by the mechanical constraint on the incident angle on the hologram.

The spatial resolution is different along x and y directions. The x resolution (which is degraded by diffraction) is given by the angle Bragg selectivity of the hologram multiplied by the focal length of the collimating lens (Sinha et al., 2004). For a 10-mm focal length, the resolution along the x axis is approximately 3 μm. Since there is no angle Bragg selectivity along the y axis, the spatial resolution in this direction is diffraction-limited by the imaging optics. Although the scanning speed, as determined by the rotation stage, is relatively high (approximately 1 nm/1 ms at 550 nm), the acquisition speed is limited by the low throughput. This is, in turn, determined by the narrow bandwidth and polarization sensitivity of the hologram (Liu et al., 2002).

An intriguing feature of volume holograms is that they have the potential of recording single-shot spectral images without postcapture computation. This goal could in principle be realized by multiplexing several tens of holograms in the same material, whereas each hologram diffracts light in a narrow spectral band to a distinct direction so that image slices at different wavelengths can be projected onto different areas of the same detector.

19.6 NONSCANNING TECHNIQUES

The methods presented above are based on spatial/wavelength scanning across the respective dimension. Even though fast scanners exist, limitations imposed by the low number of photons detected in many applications make scanning spectral imaging time-costly. Applications such as remote sensing from fast-moving airplanes, monitoring simultaneous or sequential changes in cellular morphology, ion and metabolite concentrations, hormone secretion, muscle contraction, or gene expression require the use of a spectral imaging system that collects high-resolution spectral data simultaneously for all spatial locations across the cellular field (Lynch et al., 2000).

19.6.1 Computed Tomography Imaging Spectrometer (CTIS)

A computed tomography imaging spectrometer (CTIS) has been designed for this purpose (Ford et al., 2001a, 2001b; Volin et al., 1998, 2001). In its most common design, light collected from an object with an objective is directed to a 2D computer-generated hologram (CGH) placed in collimation space. This allows angular chromatic dispersion to occur, which is re-imaged in the form of a tomographic dispersion pattern on a conventional CCD detector. The dispersed pattern, which is similar to a sinogram used in medical tomographic imaging, shows the panchromatic 0-th order (the real-time image of the object) in the center, surrounded by higher orders (-1, -2, $+1$, $+2$ in Fig. 19.2) The higher orders provide broadband continuous dispersion patterns from which the spectral content can be extracted. The number of diffraction orders gives the terminology generally used for these images. For example, a so-called 5×5 image contains the 0-th, $+/-1$, and $+/-2$ diffraction orders. After digital recording, extensive image processing algorithms are performed to spatial-spectral de-multiplexing. Iterative, linear/nonlinear, maximum likelihood techniques are used along with the calibrated system transfer function to obtain the spectral data cube. A common method used to reconstruct the 3D (x, y, λ) object cube from the 2D image recorded with the detector is the multiplicative algebraic reconstruction technique (MART), which is an iterative progression from the kth estimated object cube f^k to the $(k + 1)^{st}$. This can be described by the following linear algebra equation (Lent, 1976):

$$f^{k+1} = f^k \frac{H^T g}{H^T H f^k} \tag{19.19}$$

where f and g are vectors for the 2D image and 3D object cube, respectively, and H^T is the transposed matrix of the system matrix H, which can be obtained experimentally by recording calibration images. $H^T g$ and $H^T H f^k$ are the back-projections of the collected raw image and the current image estimate, respectively. The initial estimate of the object cube, f^0, corresponds spatially to the zero-th order image.

Due to the limited size of the detector and the high number of diffracting orders measured (typically tens, which should have minimum overlap), the number of spatial resolution elements within the detector is rather small, approximately 200×200 pixels (Ford et al., 2001a, 2001b). The spatial resolution is limited by the objective lens used to collect the image, and the accumulation time ranges between 50 ms (for bright samples) and 2 s (for dim samples).

19.7 SPECTRAL IMAGE ANALYSIS

The main challenge in multispectral imaging is the ability to separate the spectral contributions from multiple chromophores that overlap spectrally and spatially. Once a spectral data cube has been produced by one of the methods described earlier, the image analysis will need to reveal spectral information about each object and classify the objects based on their spectral signatures. This problem can be addressed by applying mathematical algorithms on a spectral data cube, which need to account for the following aspects: (i) comparison of pixels along the wavelength direction, (ii) spatial correlation of influences for neighboring pixels, (iii) pixel classification (finding classes), and (iv) object classification (assignment of objects to specific classes).

A simple way to classify pixels to spectral classes can be realized by (i) building a spectral library that contains the spectra of the features of interest, (ii) measuring the Euclidian distance $d_{x,y,i}$ between the measured spectrum $I_{x,y}(\lambda)$ and each of the reference spectra $I_r(\lambda)$ (r represents the classes), for each pixel (x, y), over the whole spectral range in the experiment:

$$d_{x,y,i} = \left(\int_\lambda \left(I_{x,y}(\lambda) - I_r(\lambda) \right)^2 d\lambda \right)^{1/2} \tag{19.20}$$

(iii) selecting the reference spectrum $I_r(\lambda)$ for which $d_{x,y,i}$ is the smallest (using a least-square criterion; a smaller distance means a higher degree of similarity) and assigning the pixel to class r, and (iv) creating a pseudocolor display for each class (Garini et al., 1996). More advanced distance measurement methods, like Mahalanobis (Mark & Tunnell, 1985), take into account the variance among spectral classes. A spectral library can be created simply by selecting spectra of obvious features in the image, or more accurately by using statistical analysis methods like principal component analysis or clustering methods.

The true advantage of spectral imaging methods is their ability to discriminate spectral contributions from spectrally overlapping, co-localized dyes. If the spectra of the dyes are known, the unmixing of different, "pure" spectral components can be realized by linear combination algorithms, assuming that the spectra combine linearly. However, there are situations, such as in the case of resonance energy transfer, where quantifying individual fluorophores by linear unmixing methods is not adequate, and other unmixing algorithms need to be developed.

19.8 APPLICATIONS OF MULTIPHOTON MULTISPECTRAL IMAGING IN BIOLOGY AND MEDICINE

Multispectral imaging offers the unique capability to create spectral databases that enable demarcation of features and evaluation of objects based on their spectral content. It is, therefore, a powerful tool in all fields pertaining to biomedical imaging and research, where objective content analysis and molecular imaging are indispensable for classification and evaluation. There are numerous reports in the literature regarding applications of multispectral imaging using linear optical methods. They include multicolor spectral karyotyping of human chromosomes (Ried et al., 1997; Schrock et al.,

1996), spectral pathology and diagnosis (Levenson et al., 1997), functional imaging of brain oxygenation (Shonat et al., 1997; Wachman et al., 1997), and evaluation of melanoma in vivo (Kirkwood et al., 1999). Band sequential, tunable filters-based systems are well suited to fluorescence imaging and have been applied in multi-wavelength fluorescence in situ hybridization (FISH), for submillisecond imaging of ionic transients in neuromuscular junctions (Wachman et al., 2004), and have recently shown promise in small animal imaging applications using quantum dots (Gao et al., 2002). The latter could be particularly significant for nonlinear laser excitation applications, as quantum dots could be efficiently excited by most lasers used in two-photon microscopy.

As newly (re)discovered nonlinear optical spectroscopic processes such as two-photon excitation (Denk et al., 1990), SHG and higher harmonics generation (Campagnola et al., 2001), and Raman spectroscopy (Zumbusch et al., 1999) are proven increasingly useful in imaging microscopy in general and biomedical applications in particular (Campagnola & Loew, 2003; Masters et al., 1997, 2004; Masters & So, 1999, 2001; Piston et al., 1994; So et al., 1999, 2000; Xu et al., 1996), there is a need to combine/develop imaging techniques that take advantage of the benefits of nonlinear optical spectroscopy. So far, apart from CARS (which is discussed in another chapter), there are only scarce reports of nonlinear multispectral imaging. They include two-photon lambda stacks using LCTFs for discriminating between multiple green fluorescent protein variants and dyes (Lansford et al., 2001), investigation of intratissue and cellular two-photon excited autofluorescence using FTIS (Ulrich et al., 2004), and measuring fluorescence resonance energy transfer (Thaler et al., 2005). The most convincing technological implementation of a multicolor multiphoton imaging microscope is relatively recent (Buehler et al., 2005) and uses a high-throughput, fully parallel, single photon counting scheme along with a high-speed fiber-optic link interface to the data acquisition computer. This allows real-time 16 spectral channel wavelength stacks of 256×256 pixel images with 12-bit dynamic range to be transferred at 30 frames per second. Applications to human skin imaging were also described. Our group developed a multimodal nonlinear approach in which spectral detection is complemented by fluorescence lifetime detection (Fig. 19.5).

Imaging of proteins can provide information regarding the interactions between them. At the diffraction limit of spatial resolution obtainable with optical microscopes, spatial overlap observed in experiments suggests either co-localization or resonance energy transfer. Distinguishing between specific intermolecular relationships is possible only by performing quantitative measurements/analysis. Fluorescence resonance energy transfer (FRET), algorithms for spectral FRET (sFRET) (Thaler et al., 2005) and fluorescence lifetime analysis (Pelet et al., 2004) have been developed to quantify interacting fluorophores. Two-photon fluorescence excitation of green fluorescent protein derivatives tagging interacting proteins was used to investigate the RET. While both methods produce great results, sFRET appears to have an advantage due to faster acquisition times (producing lower phototoxicity and photobleaching). In addition, fluorescence lifetime imaging microscopy (FLIM) monitors FRET efficiencies by measuring changes in the fluorescence lifetime of the donor alone, while sFRET has the ability to measure changes in both the donor and acceptor. Often FLIM requires multiexponential curve fitting for distinguishing components undergoing FRET from those that are not, and this may be problematic in samples containing a distribution of donor–acceptor distances. In these cases, sFRET yields the correct donor and acceptor

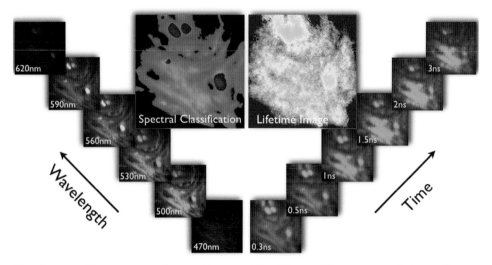

Two-Photon Fluorescence Spectral Image   Two-Photon Fluorescence Lifetime Image

Figure 19.5. Nonlinear multispectral (left) and lifetime (right) images of two-photon excited fluorescence. Fluorescence of breast tissue (H&E-stained) was measured by wide-field two-photon excitation (100 fs, 800 nm, 200 mW, 80 Mhz rep. rate, 60-μm-diameter field of view) and spectral/lifetime detection. Spectral selection at 8-nm bandwidth and 5-nm step size between 460 and 650 nm was realized by using a 2-cm imaging AOTF coupled to a Hamamatsu ORCA-ER CCD camera. Images recorded at 470, 500, 530, 560, 590, and 620 nm are shown, together with the spectral classification performed using linear discriminant analysis. The objects classified as (pseudocolor) red are red blood cells. The fluorescence lifetime image was recorded at 530 nm (20-nm bandwidth) with a 300-ps time gate and 300-ps step size at delay times ranging from 0.5 to 8 ns after the excitation pulse. A high-speed time-gated optical image intensifier (LaVision) was used for this experiment. Images recorded 0.3, 0.5, 1, 1.5, 2, and 3 ns after the excitation pulse are shown, together with the fluorescence lifetime image obtained by single exponential fit of the decay curves. The red blood cells classified as red in the spectral image have a lower fluorescence lifetime (approximately 1.2 ns, shown green in the pseudocolored image) compared with the tissue (approximately 1.8 ns, colored yellow-red).

concentrations and RET efficiencies (Thaler et al., 2005). However, since linear unmixing methods for discriminating spectral components produce erroneous results, RET needs to be accounted for in these algorithms.

19.9 PROSPECTS AND CONCLUSIONS

This brief overview of the experimental basics of multispectral imaging and its application to nonlinear imaging is aimed at summarizing the technologies that enable such imaging, their relative advantages and disadvantages, and the avenues towards extending the range and improving the quality of biological applications. As new, versatile light sources and spectral selection methods are developed and faster, low-noise detectors become available, spectral imaging approaches are expected to become invaluable for a large array of microscopy applications, including biomedical imaging.

It is important to point out that the tissue penetration of nonlinear imaging and the potential specificity of spectral imaging may provide our best approach to reaching the desirable goal of dynamically imaging single cells within the body.

REFERENCES

Barrett, H. H., and K.J. Meyers. 2004. *Foundations of Image Science*. New Jersey: Wiley and Sons.

Bell, R. J. 1972. *Introductory Fourier Transform Spectroscopy*. London: Academic Press.

Buehler, C., K.H. Kim, U. Greuter, N. Schlumpf, and P.T. C. So. 2005. Single-Photon Counting Multicolor Multiphoton Fluorescence Microscope, *Journal of Fluorescence,* 15, 41–51.

Campagnola, P. J., H. A. Clark, J. P. Wuskell, L. M. Loew, and A. Lewis. 2001. Second harmonic generation properties of fluorescent polymer encapsulated gold nanoparticles studied by high resolution non-linear optical microscopy. *Biophysical Journal* 80(1): 162A-162A.

Campagnola, P. J., and L. M. Loew. 2003. Second-harmonic imaging microscopy for visualizing biomolecular arrays in cells, tissues and organisms. *Nature Biotechnology* 21(11): 1356–1360.

Chamberlain, J. E. 1979. *The principles of Interferometric Spectroscopy*. New York: Wiley.

Chang I. C. 1976. Acousto-optic devices and applications. *IEEE Trans. Sonics Ultrason.* SU-23: 2–22.

Chang I. C. 1995. Acousto-optic devices and applications. In *Handbook of Optics,* 2nd ed., vol. II, ed. M. Bass, 12.12. New York: McGraw Hill.

Cui Y., D. Cui, and J. Tang. 1993. Study on the characteristics of an imaging spectrum system by means of an acousto-optic tunable filter. *Optical Engineering* 32, 2899–2902.

Denk, W., J. H. Strickler, and W. W. Webb. 1990. 2-Photon Laser Scanning Fluorescence Microscopy. *Science* 248(4951): 73–76.

Dixon, R. W. 1967. Acoustic diffraction of light in anisotropic media. *IEEE J. Quant. Elect.* QE-3: 85–93.

Farkas D. L., C. Du, G. W. Fisher, C. Lau, W. Niu, E. S. Wachman, and R. M. Levenson. 1998. Noninvasive image acquisition and advanced processing in optical bioimaging. *Comput Med. Imaging Graphics.* 22(2):89–102.

Ford, B. K., M. R. Descour, and R. M. Lynch. 2001a. Large-image-format computed tomography imaging spectrometer for fluorescence microscopy. *Optics Express* 9(9): 444–453.

Ford, B. K., C. E. Volin, S. M. Murphy, R. M. Lynch, and M. R. Descour. 2001b. Computed tomography-based spectral imaging for fluorescence microscopy. *Biophysical Journal* 80(2): 986–993.

Gao G., and Z. Lin. 1993. Acousto-optic supermultispectral imaging. *Applied Optics* 32, 3081–3086.

Gao X., W. C. Chan, and S. Nie. 2002. Quantum-dot nanocrystals for ultrasensitive biological labeling and multicolor optical encoding. *Journal of Biomedical Optics* 7(4):532–527.

Garini, Y., A. Gil, I. Bar-Am, D. Cabib, and N. Katzir. 1999. Signal to noise analysis of multiple color fluorescence imaging microscopy. *Cytometry.* 35(3): 214–226.

Garini, Y., M. Macville, S. du Manoir, R. A. Buckwald, M. Lavi, N. Katzir, D. Wine, I. Bar-Am, E. Schrock, D. Cabib, and T. Riedz. 1996. Spectral karyotyping. *Bioimaging* 4: 65–72.

Goelman, G. 1994. Fast Hadamard Spectroscopic Imaging Techniques. *Journal of Magnetic Resonance Series B* 104(3): 212–218.

Golay, M. J. E. 1949. Multislit spectrometry. *J. Opt. Soc. Am.* (39): 437–444.

Hanley, Q. S. 2001. Masking, photobleaching, and spreading effects in Hadamard transform imaging and spectroscopy systems. *Applied Spectroscopy* 55(3): 318–330.

Hanley, Q. S., D. J. Arndt-Jovin, and T. M. Jovin. 2002. Spectrally resolved fluorescence lifetime imaging microscopy. *Applied Spectroscopy* 56(2): 155–166.

Hanley, Q. S., P. J. Verveer, D. J. Arndt-Jovin, and T. M. Jovin. 2000. Three-dimensional spectral imaging by Hadamard transform spectroscopy in a programmable array microscope. *Journal of Microscopy-Oxford* 197: 5–14.

Hanley, Q. S., P. J. Verveer, and T. M. Jovin. 1998. Optical sectioning fluorescence spectroscopy in a programmable array microscope. *Applied Spectroscopy* 52(6): 783–789.

Hanley, Q. S., P. J. Verveer, and T. M. Jovin. 1999. Spectral imaging in a programmable array microscope by hadamard transform fluorescence spectroscopy. *Applied Spectroscopy* 53(1): 1–10.

Harvey, A. R., and D. W. Fletcher-Holmes. 2004. Birefringent Fourier-transform imaging spectrometer. *Optics Express* 12(22): 5368–5374.

Harwit, M., and N. J. A. Sloane. 1979. *Hadamard Transform Optics*. New York: Academic Press.

Heintzmann, R., K. A. Lidke, and T. M. Jovin. 2004. Double-pass Fourier transform imaging spectroscopy. *Optics Express* 12(5): 753–763.

Holmes T. 1988. Maximum-likelihood image restoration adapted for noncoherent optical imaging, *J. Opt. Soc. Am. A* 7: 666–673.

Hoyt C. 1996. Liquid crystal tunable filters clear the way for imaging multiprobe fluorescence. *Biophotonics International* (July/August): 49–51.

Kirkwood, J. M., D. L. Farkas, A. Chakraborty, K. F. Dyer, D. J. Tweardy, J. L. Abernethy, H. D. Edington, S. S. Donnelly, and D. Becker. 1999. Systemic interferon-alpha (IFN-alpha) treatment leads to Stat3 inactivation in melanoma precursor lesions. *Molecular Medicine* 5(1): 11–20.

Kurtz I., R. Dwelle, and P. Katzka. 1987. Rapid scanning fluorescence spectroscopy using an acousto-optic tunable filter. *Rev. Sci. Instrum.* 58: 1996–2003.

Lansford, R., G. Bearman, and S. E. Fraser. 2001. Resolution of multiple green fluorescent protein color variants and dyes using two-photon microscopy and imaging spectroscopy. *Journal of Biomedical Optics* 6(3): 311–318.

Lent, A. 1976. A convergent algorithm for maximum entropy image restoration. In *Image Analysis and Evaluation,* ed. Rodney Shaw, 249–257. SPIE Proceeding.

Levenson, R. M., D. L. Farkas, and S. E. Shackney. 1997. Multiparameter spectral imaging for molecular cancer staging. *American Journal of Pathology* 151(5): ST15-ST15.

Liu, W. H., G. Barbastathis, and D. Psaltis. 2004. Volume holographic hyperspectral imaging. *Applied Optics* 43(18): 3581–3599.

Liu, W. H., D. Psaltis, and G. Barbastathis. 2002. Real-time spectral imaging in three spatial dimensions. *Optics Letters* 27(10): 854–856.

Lynch, R. M., K. D. Nullmeyer, B. K. Ford, L. S. Tompkins, V. L. Sutherland, and M. R. Descour 2000. Multiparametric analysis of cellular and subcellular functions by spectral imaging. In *Molecular Imaging Reporters, Dyes, Markers and Instrumentation*, ed. Burnhop and K. Licha, 79–87. Proc. SPIE 3924.

Macgregor, A. E., and R. I. Young. 1997. Hadamard transforms of images by use of inexpensive liquid-crystal spatial light modulators. *Applied Optics* 36(8): 1726–1729.

Malik, Z., D. Cabib, R. A. Buckwald, A. Talmi, Y. Garini, and S. G. Lipson. 1996a. Fourier transform multipixel spectroscopy for quantitative cytology. *Journal of Microscopy* 182: 133–140.

Malik, Z., M. Dishi, and Y. Garini. 1996b. Fourier transform multipixel spectroscopy and spectral imaging of protoporphyrin in single melanoma cells. *Photochem Photobiol.* 63(5): 608–14.

Mark, H. L., and D. Tunnell. 1985. Qualitative near-infrared reflectance analysis using Mahalanobis distances. *Anal. Chem.* (57): 1449–85.

Masters, B. R., and P. T. C. So. 1999. Multi-photon excitation microscopy and confocal microscopy imaging of in vivo human skin: A comparison. *Microscopy and Microanalysis* 5(4): 282–289.

Masters, B. R., and P. T. C. So. 2001. Confocal microscopy and multi-photon excitation microscopy of human skin in vivo. *Optics Express* 8(1): 2–10.

Masters, B. R., P. T. C. So, C. Buehler, N. Barry, J. D. Sutin, W. W. Mantulin, and E. Gratton 2004. Mitigating thermal mechanical damage potential during two-photon dermal imaging. *Journal of Biomedical Optics* 9(6): 1265–1270.

Masters, B. R., P. T. C. So, and E. Gratton. 1997. Multiphoton excitation fluorescence microscopy and spectroscopy of in vivo human skin. *Biophysical Journal* 72(6): 2405–2412.

Mende, S. B., E. S. Claflin, R. L. Rairden, and G. R. Swenson. 1993. Hadamard Spectroscopy with a 2-Dimensional Detecting Array. *Applied Optics* 32(34): 7095–7105.

Morris H. R., C. C. Hoyt, and P. J. Treado. 1994. Imaging spectrometers for fluorescence and Raman microscopy: acousto-optic and liquid crystal tunable filters. *Applied Spectroscopy* 48, 857–866.

Pelet, S., M. J. R. Previte, L. H. Laiho, and P. T. C. So. 2004. A fast global fitting algorithm for fluorescence lifetime imaging microscopy based on image segmentation. *Biophysical Journal* 87(4): 2807–2817.

Pieper, R. J., and T. Poon. 1990. System characterization of apodized acousto-optic Bragg cells. *J Opt. Soc. Am. A.* 7, 1751–1758.

Piston, D. W., M. S. Kirby, H. P. Cheng, W. J. Lederer, and W. W. Webb. 1994. 2-Photon-Excitation Fluorescence Imaging of 3-Dimensional Calcium-Ion Activity. *Applied Optics* 33(4): 662–669.

Ried, T., M. Macville, T. Veldman, H. PadillaNash, A. Roschke, Y. Ning, S. duManoir, I. BarAm, Y. Garini, M. Liyanage, and E. Schrock. 1997. Spectral karyotyping. *Cytogenetics and Cell Genetics* 77(1–2): L4–L4.

Saito, N., S. Wada, and H. Tashiro. 2001. Dual wavelengths oscillation in an electronically tuned Ti-sapphire laser. *J. Opt. Soc. Am. B* 18(9): 1288–1296.

Schaeberle, M. D., J. F. Turner II, and P. J. Treado. 1994. Multiplexed acousto-optic tunable filter spectral imaging microscopy, *Proc. Soc. Photo-Opt. Instrum. Eng.* 2173: 11–20.

Schrock, E., S. duManoir, T. Veldman, B. Schoell, J. Wienberg, M. A. FergusonSmith, Y. Ning, D. H. Ledbetter, I. BarAm, D. Soenksen, Y. Garini, and T. Ried. 1996. Multicolor spectral karyotyping of human chromosomes. *Science* 273(5274): 494–497.

Shafer-Peltier, K. E., A. S. Haka, J. T. Motz, M. Fitzmaurice, R. R. Dasari, and M. S. Feld 2002. Model-based biological Raman spectral imaging. *J. Cell. Biochem. Suppl.* 39: 125–37.

Shonat, R. D., E. S. Wachman, W. H. Niu, A. P. Koretsky, and D. L. Farkas. 1997. Near-simultaneous hemoglobin saturation and oxygen tension maps in mouse brain using an AOTF microscope. *Biophysical Journal* 73(3): 1223–1231.

Sinha, A., G. Barbastathis, W. H. Liu, and D. Psaltis. 2004. Imaging using volume holograms. *Optical Engineering* 43(9): 1959–1972.

So, P. T. C., C. Buehler, K. H. Kim, C. Y. Dong, and B. R. Masters. 1999. Two photon imaging of skin structures. *Scanning* 21(2): 135–135.

So, P. T. C., C. Y. Dong, B. R. Masters, and K. M. Berland. 2000. Two-photon excitation fluorescence microscopy. *Annual Review of Biomedical Engineering* 2: 399–429.

Spring, K. R. 2000. The use of liquid-crystal tunable filters for fluorescence microscopy. In *Imaging Neurons: A Laboratory Manual,* ed. R. Yuste, F. Lanni, and A. Konnerth, 3.1–3.9. New York: Cold Spring Harbor Laboratory Press, Cold Spring Harbor.

Suhre, D. R., M. Gottlieb, L. H. Taylor, and N. T. Melamed. 1992. Spatial resolution of imaging noncollinear acousto-optic tunable filters. *Optical Engineering* 31: 2118–2121.

Tang, H. W., G. Q. Chen, J. S. Zhou, and Q. S. Wu. 2002. Hadamard transform fluorescence image microscopy using one-dimensional movable mask. *Analytica Chimica Acta* 468(1): 27–34.

Thaler, C., S. V. Koushik, P. S. Blank, and S. S. Vogel. 2005. Quantitative multiphoton spectral imaging and its use for measuring resonance energy transfer. *Biophysical Journal* 89(4): 2736–2749.

Treado, P. J., I. W. Levin, and E. N. Lewis. 1992. High-fidelity Raman imaging spectrometry: a rapid method using an acousto-optic tunable filter. *Applied Spectroscopy* 46: 1211–1216.

Treado, P. J., and M. D. Morris. 1988. Hadamard-Transform Raman Imaging. *Applied Spectroscopy* 42(5): 897–901.

Treado, P. J., and M. D. Morris. 1990. Hadamard-Transform Spectroscopy and Imaging. *Spectrochimica Acta Reviews* 13(5): 355–375.

Ulrich, V., P. Fischer, I. Riemann, and K. Konig. 2004. Compact multiphoton/single photon laser scanning microscope for spectral imaging and fluorescence lifetime imaging. *Scanning* 26(5): 217–225.

Volin, C. E., B. K. Ford, M. R. Descour, J. P. Garcia, D. W. Wilson, P. D. Maker, and G. H. Bearman. 1998. High-speed spectral imager for imaging transient fluorescence phenomena. *Applied Optics* 37(34): 8112–8119.

Volin, C. E., J. P. Garcia, E. L. Dereniak, M. R. Descour, T. Hamilton, and R. McMillan. 2001. Midwave-infrared snapshot imaging spectrometer. *Applied Optics* 40(25): 4501–4506.

Wachman, E. S., W. Niu, and D. L. Farkas. 1996. Imaging acousto-optic tunable filter with 0.35-micrometer spatial resolution. *Applied Optics* 35:5220–5226.

Wachman, E. S., W. H. Niu, and D. L. Farkas. 1997. AOTF microscope for imaging with increased speed and spectral versatility. *Biophysical Journal* 73(3): 1215–1222.

Wachman E.S., Farkas D.L., Niu, W. 1998. Submicron Imaging System Having an Acousto-Optic Tunable Filter, *U.S. Patents* 5,796,512 and 5,841,577 (and corresponding inter-national ones)

Wuttig, A. 2005. Optimal transformations for optical multiplex measurements in the presence of photon noise. *Applied Optics* 44(14): 2710–2719.

Xu, C., W. Zipfel, J. B. Shear, R. M. Williams, and W. W. Webb. 1996. Multiphoton fluorescence excitation: New spectral windows for biological nonlinear microscopy. *Proc. Natl. Acad. Sc.* 93(20): 10763–10768.

Yano T., and A. Watanabe. 1976. Acousto-optic TeO$_2$ tunable filter using far-off-axis anisotropic Bragg diffraction. *Applied Optics* 15, 2250–2258.

Zhou R., E. H. Hammond, and D. L. Parker. 1997. A multiple wavelength algorithm in color image analysis and its applications in stain decomposition in microscopy images. *Medical Physics* 23(12):1977–1986.

Zumbusch, A., G. R. Holtom, and X. S. Xie. 1999. Three-dimensional vibrational imaging by coherent anti-Stokes Raman scattering. *Physical Review Letters* 82(20): 4142–4145.

20

Multiphoton Polarization Microscopy

Yen Sun, Wen Lo, Jiunn Wen Su,
Sung-Jan Lin, Shiou-Hwa Jee,
and Chen-Yuan Dong

20.1 INTRODUCTION

Polarization methods have long been used in spectroscopic studies of molecular phenomena. In the standard approach, linearly polarized light is used to illuminate the molecular species under study. Sample luminescence detected at polarization directions parallel $I_\parallel$ and perpendicular $I_\perp$ to the excitation polarization can then be detected and analyzed to obtain orientation information on the specimen. The traditional approach is to analyze the polarization p or anisotropy r as defined by:

$$p = \frac{I_\parallel - I_\perp}{I_\parallel + I_\perp} \ \text{ and } \ r = \frac{I_\parallel - I_\perp}{I_\parallel + I_\perp}. \tag{20.1}$$

By analyzing the polarization or anisotropy using steady-state or time-resolved methods, orientation information or dynamics of molecular systems can be investigated (Lakowicz, 1999).

The natural extension of polarization spectroscopy is to implement polarization-resolved analysis in conjunction with optical microscopy. In this approach, spatially specific orientation information with submicron resolution can be obtained from biological specimens. A number of experimental approaches have been implemented using polarization microscopy techniques such as confocal polarization microscopy and modulated polarization microscopy (Axelrod, 1989; Massoumian et al., 2003; Wilson & Juskaitis, 1995). With the implementation of different forms of polarization microscopy, many studies in the biological and medical sciences have also been conducted. For example, collagen imaging, corneal microscopy, and the dynamics of cytoskeleton are a few examples in which polarization-resolved microscopy has been readily applied (AsiyoVogel et al., 1997; Gitirana & Trindade, 2000; Kuhn et al., 2001). In addition, polarization microscopy has been used to measure the orientation of single chromophores and quantum dots (Chung et al., 2003; Empedocles et al., 1999). Time-resolved polarization microscopy has also been used to probe dynamic phenomena inside cells such as the cytoplasmic viscosity (Fushimi & Verkman, 1991).

Although traditional polarization imaging techniques have been successful in addressing a number of significant problems, they also possess limitations. For instance, if one-photon excitation (1PE) is used for polarization microscopy, issues such as specimen photodamage and the achievable imaging depths can hinder the extension of polarization imaging to in vivo studies. With the advent of multiphoton microscopy, it is natural to combined multiphoton imaging with polarization-resolved techniques to benefit from the advantages of multiphoton excitation.

The advantages of multiphoton excitation (MPE) are well established. Using a short pulsed laser, the nonlinear optical effects such as fluorescence excitation and harmonic generation (HG) can be efficiently generated only near the focal volume of a microscopic objective where the incident photon flux is high. The point-like excitation volume generated from such processes results in confocal-like sectioning effects without the use of confocal pinholes. In addition, the point-like excitation volume also limits the specimen photodamage to the focal volume. Furthermore, the near-infrared photons used in multiphoton microscopy are absorbed and scattered less by biological tissues than the UV or visible sources used in 1PE microscopy. Therefore, specimen depth penetration is improved without the need for histological procedures. Finally, the wide spectral separation between the excitation and emission wavelengths allows complete detection of the sample luminescence. As a result, greater detection sensitivity is achieved (Callis, 1997a; Denk et al., 1990; So et al., 2000). In additional to the typical advantages of multiphoton imaging, polarization microscopy using MPE has an additional feature in the possibility of enhancing image contrast.

In order to demonstrate the difference in polarization properties between 1PE and MPE processes, one needs to examine the equations describing molecular transitions between the initial state $|i\rangle$ and the final state $|f\rangle$. In the case of 1PE, the transition probability $P_{1\gamma}$ is given by:

$$P_{1\gamma} \propto \left| \langle f | \vec{E} \cdot \vec{r} | i \rangle^2 \right| \tag{20.2}$$

while the two-photon excitation (TPE) result is:

$$P_{2\gamma} \propto \left| \sum_m \frac{\langle f | \vec{E} \cdot \vec{r} | m \rangle \langle m | \vec{E} \cdot \vec{r} | i \rangle}{\varepsilon_{mi} - \varepsilon_\gamma} \right|^2 \tag{20.3}$$

where $\vec{E}$ and $\vec{r}$ are the external electric field and position operator, respectively. For the 2PE case, ε_{mi} is the energy difference between the initial and intermediate state $|m\rangle$ and ε_γ is the photon energy.

Equations (20.2) and (20.3) are fundamentally different in that for 1PE, one electric dipole transition (EDT) is responsible to bring the molecule from the ground state $|i\rangle$ to the excited state $|f\rangle$, whereas in TPE, two EDTs, mediated by different intermediate states, are required to complete the excitation process. As is well known, the interaction of an electric dipole with an electric field $\vec{E}$ is governed by a probability of $\cos^2 \theta$, where θ is the angle between the external electric field and the transition moment. It is in the photoselection involving additional intermediate states that MPE produces different effects than 1PE. Whereas in the 1PE case a straightforward $\cos^2 \theta$ distribution adequately describes the angular distribution of excitation probability, the MPE results can become hopelessly complicated. For example, in 2PE, the lowest-order MPE effect, the electric dipole couplings of each of the intermediate states to the ground and excited states are different, rendering a general anisotropy analysis impossible. While theoretical analyses using the tensor approach have been performed, a molecule-specific analysis is needed for different cases (Andrews & Webb, 1990; Callis, 1993; Chen & Van der Meer, 1993; McClain, 1971, 1972, 1973; Wan & Johnson, 1994).

However, if one were to assume that all the excitation transition dipoles are oriented to the external electric field at the same angle θ, then insights can be gained into the effects of MPE on molecular anisotropy. Specifically, instead of the 1PE photoselection distribution of $\cos^2 \theta$, $\cos^4 \theta$ would be the angular excitation probability for 2PE. In general, the photoselection angular probability distribution is given by:

$$P_n(\theta) = \cos^{2n} \theta \tag{20.4}$$

where n is a positive integer corresponding to the order of the excitation process.

An immediate consequence of the multiphoton photoselection process is the change in initial anisotropies. As illustrated in Figure 20.1, if a molecule is excited by linearly polarized light along the z-axis, then the parallel ($\parallel$) and perpendicular ($\perp$) components of the fluorescence are given by $I_{\parallel} = I_0 \cos^2 \theta$ and $I_{\perp} = I_0 \sin^2 \theta \sin^2 \phi$, respectively. To compute $I_{\parallel}$ and $I_{\perp}$, one needs to take into account the photoselection process; the results are:

$$I_{\parallel}(\theta) = I_0 \int_0^{\pi/2} P_n \cos^2 \theta \sin \theta d\theta \tag{20.5}$$

$$I_{\perp}(\theta) = \frac{I_0}{2} \int_0^{\pi/2} P_n(\theta) \sin^2 \theta \sin \theta d\theta \tag{20.6}$$

where the factor of one half for $I_{\perp}(\theta)$ comes from the angular integration over ϕ coordinate and equations (20.4), (20.5), and (20.6) can be used to calculate r_0, the initial anisotropy.

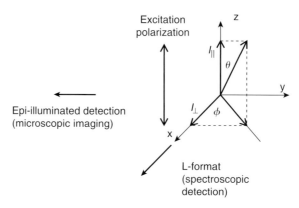

Figure 20.1. Excitation and emission detection geometry under fluorescence spectroscopy and microscopy (excitation and emission transition moments are assumed to be parallel). $I_{\parallel}$ and $I_{\perp}$ are the fluorescence emissions parallel and perpendicular to the excitation direction, respectively.

Taking into account an angle β, which accounts for the difference in the excitation and emission transition moments, one ends up with:

$$r_0(n) = \left(\frac{2n}{2n+3}\right)\left(\frac{3}{2}\cos^2\beta - \frac{1}{2}\right) \tag{20.7}$$

Therefore, the theoretical maximum initial anisotropy values can be calculated for the case when β is zero and the result is $r_0(n) = 2n/(2n+3)$.

In the 1980s, spectroscopic studies were performed to test the photoselection properties of 2PE (Scott et al., 1983). Later, in the 1990s, more anisotropic measurements were made on molecules such as 1,6-diphenylhexatriene, N-acetyl-L-tryptophanamide, and 2,5-diphenyloxazole (Lakowicz et al., 1992a, 1992b, 1992c). As multiphoton spectroscopy continued to develop, comparison of 1PE and 2PE results on tyrosine and tryptophan (Kierdaszuk et al., 1995) were made and 1PE and 2PE anisotropic studies on N-acetyl-L-tryptophanamide (NATyrA) were performed (Lakowicz et al., 1995). Later on, three-photon excitation (3PE) studies were extended to 2-5-diphenyloxazole (Gryczynski et al., 1995) and NATyrA (Gryczynski et al., 1999), and four-photon excitation (4PE) studies were performed on 2,2'-dimethyl-p-terphenyl (Gryczynski et al., 2002).

An examination of the multiphoton anisotropy studies revealed several important trends. First, many of these studies include time-resolved measurements, and the fluorescence lifetime and rotational correlation times remain practically the same under 1PE or MPE. Secondly, measurements of initial anisotropy r_0 show that in many cases, the initial anisotropies are higher for MPE than 1PE. For example, for 1,6-diphenylhexatriene, the 2PE r_0 is 0.528 at 40°C, which is higher than the theoretical 1PE limit of 0.4 (Lakowicz et al., 1992a). An even higher r_0 of 0.61 was obtained for 2-5-diphenyloxazole under 3PE (Gryczynski et al., 1995) and a 4PE initial anisotropy value of 0.70 was measured for 2,2'-dimethyl-p-terphenyl (Gryczynski et al., 2002). Compared to the maximum theoretical values of $2/5 = 0.4$ (1PE), $4/7 = 0.57$ (2PE), $2/3 = 0.67$ (3PE), and $8/11 = 0.73$ (4PE), these experimentally measured initial anisotropies remain within theoretical limits and are consistently higher than the 1PE maximum of 0.4.

However, for NATyrA, a different trend is observed. The respective 1PE and 2PE r_0 values for this molecule are 0.3 and around 0, while measurements show that the 3PE excitation anisotropy spectrum resembles that of the 1PE result (Gryczynski et al., 1999; Lakowicz et al., 1995). In this case, it is clear that $P_n(\theta) = \cos^{2n}\theta$, the angular photoselection result described by equation (20.4), is no longer valid. On the contrary, the detailed interaction with the intermediate states as described by equation (20.3) is needed in order to understand the photophysical phenomena underlying the r_0 results.

In addition to MPE anisotropy, polarization effects are also important in HG microscopy. In general, the polarization P_i of a material responds to the external electric field by the following equation:

$$P_i = \chi_{ij}E_j + \chi_{ijk}E_jE_k + \chi_{ijkl}E_jE_kE_l + \cdots \tag{20.8}$$

where χ_{ij}, χ_{ijk}, and χ_{ijkl} are the first, second, and third susceptibility tensors and E is the corresponding electric field in the tensor product operation (Campagnola & Lowe, 2003).

Since the strength of the polarization and the corresponding HG signal depend on the coupling of the χ to the external electric field, orientation properties of biomolecules can be studied by investigating the relationship between the electric field direction and the observed HG signal. In this regard, researchers have applied polarization effects in second-harmonic generation (SHG) and third-harmonic generation (THG) microscopy to study the properties of biomolecules such as collagen (Chu et al., 2004; Oron et al., 2003; Stoller et al., 2002; Williams et al., 2005). The SHG polarization effect can even be coupled to optical coherence methods to map collagen tomography (Applegate et al., 2004).

As earlier MPE anisotropy studies focused on molecular spectroscopy, it was not until recently that researchers started to implement MPE anisotropy into microscopic imaging systems (Diaspro et al., 2001; Sun et al., 2003). As multiphoton microscopes have become equipped with polarization imaging capabilities, interesting applications have begun to emerge. Studies on the changes to the stratum corneum from oleic acid and myosin rotation in skeletal muscle demonstrated the effectiveness of MPE anisotropy imaging in addressing biologically relevant questions (Borejdo et al., 2004; Sun et al., 2004).

If one further assumes that the emission dipole is aligned to the excitation moment, polarization imaging using TPE can introduce contrast in polarization imaging superior to that of 1PE microscopy. This effect is illustrated in Figure 20.2, where the angular dependence of the excitation efficiency is plotted with the excitation polarization aligned along the horizontal direction (Callis, 1997b; Johnson & Wan, 1997). Due to the $\cos^4 \theta$ dependence of 2PE efficiency, the polarization and anisotropy as defined by equation (20.1) are sharper along the excitation axis than that for 1PE. As discussed earlier, the limiting anisotropy for 2PE and 1PE is 0.4 and 0.57, respectively. If a 3PE scheme is employed, the limiting anisotropy becomes even higher, at 0.66. The broader range of polarization or anisotropy values for higher-order excitation means that polarization microscopy under MPE can generate images with sharper contrast than that achieved using 1PE microscopy.

The combination of the advantages of MPE imaging and the potentially enhanced contrast in polarization leads one to draw the conclusion that multiphoton polarization

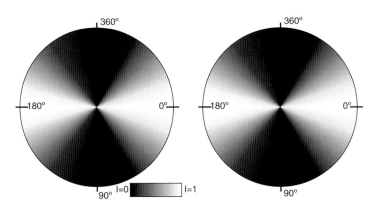

Figure 20.2. Angular dependence of photoselection between one-photon ($\cos^2 \theta$) and two-photon ($\cos^4 \theta$) excitation processes.

microscopy should be the preferred technique for polarization-resolved imaging in biological specimens.

20.2 SETTING UP A MULTIPHOTON POLARIZATION MICROSCOPE

A multiphoton polarization microscope can be easily set up by modifying an existing multiphoton imaging system (Diaspro et al., 2001; Sun et al., 2003, 2004). The major considerations would be the installation of properly placed polarization-resolved optical components and the appropriate calibration to rid the acquired images of instrument-induced polarization artifacts.

In our set-up previously described, a titanium-sapphire laser (Tsunami, Spectra Physics, Mountain View, CA, USA) pumped by a diode-pumped solid state (DPSS) source (Millennia X, Spectra Physics) is used as the excitation source. The near-infrared output of the laser is guided by an x-y mirror scanning system (Model 6220, Cambridge Technology, Cambridge, MA, USA) into a modified commercial upright microscope (E800, Nikon, Japan). The input port of the microscope was modified to accommodate a beam expander; the beam expansion is necessary to ensure overfilling of the objective's back aperture, thus achieving the needed tight focusing for optimal imaging. After beam expansion, the laser source is reflected by a dichroic mirror into the focusing objective. The multiphoton luminescence generated at the focal volume is collected by the objective in an epi-illuminated or backscattering geometry. After filtering by the dichroic and additional filters, the specimen luminescence is collected by the optical detector. In our arrangement, single photon counting photomultiplier tubes (PMTs) (R7400P, Hamamatsu, Japan) were used as detectors. If 3D scanning is desired, a sample positioning stage (H101, Prior Scientific, UK) can be used to translate the specimen after each beam scan. The experimental apparatus described here is illustrated in Figure 20.3.

In setting up a multiphoton polarization microscope, the most important issue is the proper calibration of the excitation polarization at the specimen plane and the polarization calibration of the emission detection channels. In our arrangement, the polarization and power of the titanium-sapphire source is controlled by a combination of a half-wave plate and a high-power polarizer outside of the microscope system. If a circularized polarized excitation is desired, a quarter-wave plate is used in conjunction with the half-wave plate and the polarizer. While the excitation polarization can be readily controlled outside of the microscope, proper calibration inside the microscope becomes more involved. For image acquisition, an important task is to correlate the polarization set outside of the microscope to the orientation of the microscopic images. This can be performed by placing a highly reflecting mirror on the sample side of the focusing objective (such as the S-Fluor 40x, NA 1.4, Nikon). With the excitation polarization direction fixed, the emission sheet polarizer placed in front of the PMT can be rotated until a maximum reflection is registered. That would correspond to a detection channel parallel to the excitation axis. The perpendicular direction is one in which the emission polarizer is rotated until a minimum signal is measured. Since the dichroic mirror will allow a small portion of the reflected light into the PMT, an emission polarizer placed in front of the detector can then be rotated for polarization determination.

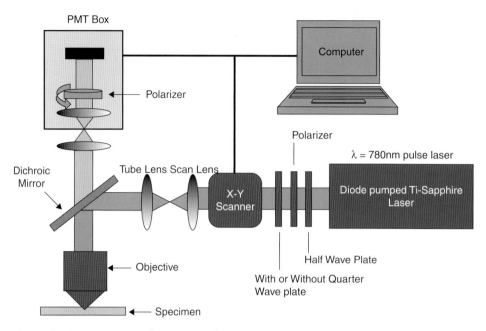

Figure 20.3. Experimental apparatus for multiphoton polarization microscopy.

In addition, a contributor to polarization-related artifacts is the dichroic mirror used for reflection of the excitation beam and transmission of sample luminescence. The reflectivity and therefore the transmission are functions of both the wavelength and polarization. In general, the width of the transmission bandwidth is the largest for p-polarized light, followed by randomly polarized and s-polarized electromagnetic waves. Therefore, to minimize unwanted polarization-induced artifacts, care must be taken to select the dichroic mirror with the minimal polarization dependence transmission coefficients in the wavelength region of interest.

Nonetheless, proper calibration of the excitation and emission polarization effects must be performed. While the excitation efficiency can be calibrated by measuring the laser reflection from the dichroic mirror, the calibration of collection efficiency can be determined by using a solution of small fluorescent molecules. Since the rotational correlation times of these molecules are fast (Buehler et al., 2000), one would expect equal steady-state luminescence to be detected regardless of the orientation of the emission polarizer. Deviations from this expectation would indicate the existence of polarization-dependent artifacts, and the difference in measurement results can be used for a quick system calibration.

A more thorough approach to calibrate for polarization-induced emission detection efficiencies is to determine the imaging system's G factor. In spectroscopic studies, the G factor can be determined simply by rotating the excitation polarization to the perpendicular or horizontal axis to generate completely randomly oriented transition moments. In the commonly used L-format of emission detection, the emission polarization can then be rotated between the parallel ($\parallel$) and perpendicular ($\perp$) directions to determine the system's G factor, defined as the ratio of detection efficiency between the vertical and horizontal axes (Lakowicz, 1999). However, in an epi-illuminated microscopic imaging system, a different approach is needed. By varying the excitation polarization between vertical (V) and horizontal (H), the measured intensity I_{VV}, I_{VH}, I_{HH}, I_{HV} given by:

$$I_{VV} = k_V S_V I_{\parallel} \quad I_{VH} = k_V S_V I_{\perp} \qquad (20.9)$$

$$I_{HH} = k_H S_H I_{\parallel} \quad I_{HV} = k_H S_V I_{\perp} \qquad (20.10)$$

can be used to determine the G factor defined by $G \equiv S_V/S_H$. In equations (20.9) and (20.10, S_V and S_H denote detection efficiencies of the vertical and horizontal axes, respectively. k_V and k_H represent other polarization-independent factors that influence the intensity measurements in the vertical and horizontal directions, respectively. Finally, algebraic manipulation using equations (20.9) and 20.(10) and the G factor definition can be used to determine a relationship between I_{VV}, I_{VH}, I_{HH}, I_{HV}, and G:

$$G = \sqrt{\left(\frac{I_{VV}}{I_{HH}}\right)\left(\frac{I_{HV}}{I_{VH}}\right)}. \qquad (20.11)$$

Once the G factor is determined, it can be used in conjunction with equations (20.9) and (20.10) to determine the calibrated intensity ratios between $I_{\parallel}$ and $I_{\perp}$, and that would allow the determinations of anisotropies (Diaspro et al., 2001).

20.3 MULTIPHOTON POLARIZATION IMAGING OF SKIN STRATUM CORNEUM AND DERMAL FIBERS

20.3.1 Skin Preparation

To illustrate multiphoton polarization microscopy, the skin stratum corneum and dermal fibers may be used as specimens. In the first example, an excised skin specimen labeled with the membrane probe 6-dodecanoyl-2-dimethylaminonaphthalene (Laurdan, D-250, Molecular Probes, Eugene, OR) was used. In labeling the skin, a stock solution of around 2 mM Laurdan was prepared. The final labeling solution was made by diluting 10 μL of the Laurdan/DMSO stock in 1 mL of PBS buffer (pH 7.4). Excised human skin was then soaked in the treatment solution for 11 hours. After labeling, the skin sample was rinsed with PBS buffer and mounted on a microscope slide for viewing.

To further illustrate the ability of multiphoton polarization imaging in resolving changes in stratum corneum, we also present images from another specimen in which the skin was immersed in the labeling solution composed of 10 μL of the Laurdan stock, 53 μL of oleic acid, and 1 mL of PBS. In this specimen, the oleic acid was used as a chemical enhancer; it is frequently used in transdermal drug delivery studies (Yu et al., 2001, 2002). In this case, the incubation time for the skin specimens (with and without oleic acid treatment) was 10.5 hours. The final example we demonstrate is dermal fiber imaging from unlabeled skin specimens.

20.3.2 Polarization-Resolved Images of the Stratum Corneum and Dermal Fibers

The polarization-resolved images obtained from the Laurdan-treated stratum corneum and dermal fibers are shown in Figure 20.4. The orientations of the excitation polarization and the detecting polarization are respectively designated as solid and dashed arrows. In our case, both the excitation and emission polarizations are varied between the horizontal and vertical directions. Therefore, unlike traditional spectroscopic methods in which the excitation polarization is fixed, the excitation polarization in our approach is given two degrees of freedom. This approach is necessary due to the sample heterogeneity inherent in microscopic imaging of biological specimens. For example, in the case of Laurdan, since its transition dipole is perpendicular to the membrane (Bagatolli & Gratton, 2000), a fixed excitation polarization would be less efficient in exciting the Laurdan molecules aligned perpendicular to the excitation polarization. To analyze the polarization properties of Laurdan molecules imbedded within the stratum corneum, six regions of the lipid layers of the stratum corneum images were

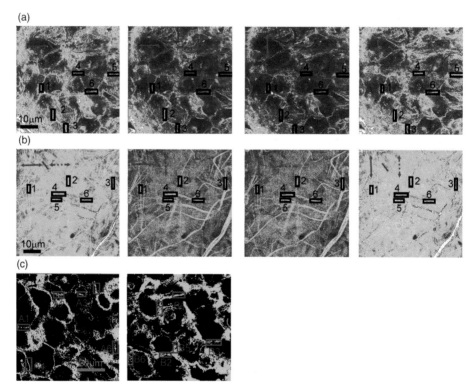

Figure 20.4. (a) Multiphoton polarization images of skin treated with Laurdan molecules (solid arrow: excitation direction; dash: emission orientation). (b) Multiphoton polarization images of skin dermal fibers under different excitation and emission polarization conditions (solid arrow: excitation direction; dash: emission orientation). (c) Without (left) and with (right) oleic acid treatment imaged by multiphoton polarization microscopy of Laurdan molecules. The boxed areas are the regions used for polarization determination.

selected: regions 1, 2, and 3 have vertical orientations, while regions 4, 5, and 6 are oriented horizontally. The results of the polarization analysis of the six regions are shown calculated according to equation (20.1) and shown in Figure 20.5a. As the table shows, for the membrane region oriented perpendicular to the excitation polarization, a higher p value was registered. Specifically, the average p values of the vertical and horizontal lipid layers were 0.27 ± 0.06 and 0.21 ± 0.06 for horizontal excitation. On the other hand, polarizations obtained for vertical excitation were 0.35 ± 0.08 for horizontal membrane orientation and 0.30 ± 0.07 for vertical lipid layer. Our results support the fact that the transition dipole moments of Laurdan are preferentially oriented perpendicular to the membrane surface.

In the case of dermal fibers within unlabeled skin, we altered the excitation and emission polarizations as for Laurdan-labeled stratum corneum; the images are shown in Figure 20.5b. Six regions, three of which are oriented horizontally and three oriented vertically, were selected for polarization analysis. Our results show that the fiber luminescence is stronger when the emission polarizer is oriented parallel to the excitation polarization.

In addition to varying the excitation polarization directions, one can also use a circularly polarized laser source for specimen excitation. This method has the advantage of equal excitation of fluorescent molecules oriented along any axis. However, the parallel and perpendicular emission directions are arbitrarily defined. We choose $I_{\parallel}$ and $I_{\perp}$ to be horizontally and vertically aligned with respect to the images. To illustrate this approach, the polarization-resolved images of the skin specimens without (left) and with (right) oleic acid treatment are shown in Figure 20.5c. For further analysis, the corresponding histograms of the p values are shown in Figure 20.5c. For p calculations, we also selected three regions to calculate the average p values for each membrane orientation; the results are also shown in Figure 20.5c. Our analysis shows that the average p values without and with oleic acid treatment were respectively -0.01 ± 0.18 and -0.02 ± 0.16. On the other hand, Figure 20.5c indicates that the average p values for the horizontal and vertical orientations were -0.08 ± 0.08 and 0.09 ± 0.11, respectively. Since $I_{\parallel}$ and $I_{\perp}$ are defined to be aligned horizontally and vertically in the image plane, the measured p values are consistent with the observation that without oleic acid treatment, the transition moments of Laurdan are preferentially aligned normal to the lipid layers. On the other hand, we found that the p values for the skin specimen treated with oleic acid were -0.01 ± 0.08 and -0.01 ± 0.08, regardless of the lipid orientations. Therefore, our results indicate that the addition of oleic acid tends to randomize lipid molecules packing in the skin.

20.4 CONCLUSION

By the proper selection and calibration of optical components, polarization-resolved imaging can be readily achieved within a multiphoton microscope. Specifically, excitation and emission polarization effects can be calibrated using a reflecting mirror. In addition, the emission polarization artifacts can be calibrated using a solution of small fluorescent molecules. The added advantages of the more stringent photoselection properties of MPE can enable polarization microscopy to be achieved at higher contrast than 1PE polarization microscopy. Using skin as a specimen, multiphoton

(a)

Stratum Corneum (Laurdan)

Excitation polarization	Membrane orientation	Area	P	P Average
↔	↕	1	0.31±0.12	
		2	0.25±0.13	0.27±0.06
		3	0.27±0.08	
	↔	4	0.21±0.12	
		5	0.20±0.10	0.21±0.06
		6	0.22±0.11	
↕	↕	1	0.33±0.16	
		2	0.29±0.10	0.30±0.07
		3	0.30±0.11	
	↔	4	0.36±0.10	
		5	0.35±0.17	0.35±0.08
		6	0.33±0.11	

(b)

Dermal Fibers

Excitation polarization	Fiber orientation	Area	P	P Average
↔	↕	1	0.29±0.10	
		2	0.24±0.16	0.27±0.08
		3	0.28±0.12	
	↔	4	0.27±0.12	
		5	0.27±0.13	0.27±0.07
		6	0.28±0.13	
↕	↕	1	0.37±0.11	
		2	0.35±0.10	0.35±0.06
		3	0.34±0.10	
	↔	4	0.34±0.10	
		5	0.32±0.11	0.33±0.06
		6	0.33±0.11	

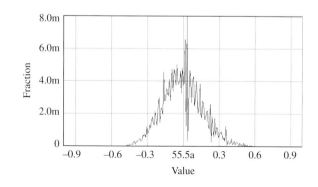

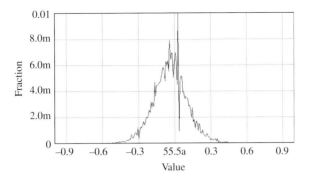

Without Oleic Acid				With Oleic Acid			
membrane orientation	area	Paverage	P total average	membrane orientation	area	Paverage	P total average
↔	1	−0.11	−0.08±0.08	↔	1	−0.02	−0.01±0.06
	2	−0.08			2	−0.01	
	3	−0.03			3	−0.01	
↕	4	−0.13	0.09±0.11	↕	4	−0.02	−0.01±0.08
	5	−0.07			5	−0.01	
	6	0.1			6	−0.01	

Figure 20.5. (a) Detailed analysis of the multiphoton polarization properties of Laurdan molecules within the stratum corneum lipid layers under different excitation and emission conditions. (b) Analysis of multiphoton polarization properties of skin dermal fibers under the given excitation and emission orientation conditions. (c) Histogram and polarization values of Laurdan molecules within the stratum corneum without and with oleic acid treatment (circularly polarized excitation).

polarization microscopy reveals that Laurdan molecules preferentially align normal to the lipid layers within the stratum corneum and that emissions from dermal fibers are strongly correlated to the fiber directions. The addition of oleic acid tends to disrupt the skin lipid molecule alignments and results in a reduction of polarization values. These examples illustrate the power of multiphoton polarization microscopy in obtaining heterogeneous structural information within biological specimens. As multiphoton microscopy continues to develop, one can envision the polarization imaging modality to be combined with other contrast-enhancement techniques in achieving a multidimensional platform for the investigation of biological and biomedical phenomena. One possibility is the combination with generalized polarization (GP) microscopy. The definition of GP

$$GP = \frac{I_{440nm} - I_{490nm}}{I_{440nm} + I_{490nm}}. \tag{20.12}$$

is specific to the relative change in fluorescence emission of Laurdan molecules between 440 and 490 nm and is sensitive to their chemical environment. Specifically, Laurdan molecules tend to red-shift in a more polar environment, and GP analysis can be used to probe changes in the chemical environment of the Laurdan molecules (Bagatolli & Gratton, 2000; Parasassi et al., 1991). While the definition of GP deviates from the traditional definition of polarization or anisotropy, it will be the development of multidimensional microscopy that will add additional benefits to life science research in the future.

ACKNOWLEDGMENTS

We acknowledge the support of National Science Council, Taiwan (NSC 92-2112-M-002-018 and NSC 92-3112-B-002-048), for the completion of this project.

REFERENCES

Andrews, D.L., Webb, B.S. 2-photon photoselection—an irreducible tensor analysis. *Journal of the Chemical Society-Faraday Transactions* 1990; 86(18):3051–3065.

Applegate, B.E., Yang, C.H., Rollins A.M., Izatt, J.A. Polarization-resolved second-harmonic-generation optical coherence tomography in collagen. *Optics Letters* 2004; 29(19): 2252–2254.

AsiyoVogel, M.N., Brinkmann, R., Notbohm, H. et al. Histologic analysis of thermal effects of laser thermokeratoplasty and corneal ablation using Sirius-red polarization microscopy. *J Cataract Refr Surg* 1997; 23:515–526.

Axelrod, D. Fluorescence Polarization Microscopy. *Method Cell Biol* 1989; 30:333–352.

Bagatolli, L.A., Gratton, E. Two photon fluorescence microscopy of coexisting lipid domains in giant unilamellar vesicles of binary phospholipids mixtures. *Biophysical J.* 2000; 78:290–305.

Borejdo, J., A. Shepard, I. Akopova, W. Grudzinski, and J. Malicka. Rotation of the lever arm of myosin in contracting skeletal muscle fiber measured by two-photon anisotropy. *Biophysical Journal* 2004; 87(6):3912–3921.

Buehler, Ch., Dong, C.Y., So, P.T.C., French, T., Gratton, E. Time-resolved polarization imaging by pump-probe (stimulated emission) fluorescence microscopy, *Biophys. J.* 2000; 79: 536–549.

Callis, P.R. On the Theory of 2-Photon Induced Fluorescence Anisotropy with Application to Indoles. *Journal of Chemical Physics* 1993; 99(1):27–37.

Callis, P.R. Two-photon-induced fluorescence. *Annual Review of Physical Chemistry* 1997a; 48:271–297.

Callis, P.R. 1997b The theory of two-photon induced fluorescence anisotropy, in Topics in Fluorescence Spectroscopy, Volume 5: Non-linear and Two-Photon-Induced Fluorescence, J. R. Lakowicz (ed.), Plenum Press, New York, pp. 1–42.

Campagnola, P.J., Loew, L.M. Second-harmonic imaging microscopy for visualizing biomolecular arrays in cells, tissues and organisms. *Nature Biotechnology* 2003; 21:1356–1360.

Chen, S.Y., Van der Meer, B.W. Theory of 2-photon induced fluorescence anisotropy decay in membranes. *Biophysical Journal*; 1993 64(5):1567–1575.

Chu, S.W., Chen, S.Y., Chern, G.W., Tsai, T.H., Chen, Y.C., Lin, B.L., Sun, C.K. Studies of $x((2))/x((3))$ tensors in submicron-scaled bio-tissues by polarization harmonics optical microscopy. *Biophysical Journal* 2004; 86(6):3914–3922.

Chung, I.H., Shimizu, K.T., Bawendi MG. Room temperature measurements of the 3D orientation of single CdSe quantum dots using polarization microscopy. *P Natl Acad Sci USA* 2003; 100:405–408.

Denk, W, Strickler JH, Webb WW. 2-photon laser scanning fluorescence microscopy. *Science* 1990; 248(4951): 73–76.

Diaspro, A., Chirico, G., Federici, F., Cannone, F., Beretta, S., Robello, M. Two-photon microscopy and spectroscopy based on a compact confocal scanning head. *Journal of Biomedical Optics* 2001; 6(3):300–310.

Empedocles, S.A, Neuhauser, R., Bawendi, M.G. Three-dimensional orientation measurements of symmetric single chromophores using polarization microscopy. *Nature* 1999; 399: 126–130.

Fushimi, K., Verkman, A.S. Low viscosity in the aqueous domain of cell cytoplasm measured by picosecond polarization microfluorimetry. *J. Cell Biol.* 1991; 112: 719–725.

Gitirana, L.D., Trindade, A.V. Direct blue staining plus polarization microscopy: An alternative dye for polarization method for collagen detection in tissue sections. *J Histotechnol* 2000; 23: 347–349.

Gryczynski, I., Malak, H., Lakowicz, J.R. 3-photon induced fluorescence of 2,5-diphenyloxazole with a femtosecond ti-sapphire laser. *Chemical Physics Letters* 1995; 245(1):30–35.

Gryczynski, I., Malak, H., Lakowicz, J.R. Three-photon excitation of N-acetyl-L-tyrosinamide. *Biophysical Chemistry* 1999; 79(1):25–32.

Gryczynski, I., Piszczek, G., Gryczynski, Z., Lakowicz, J. R. Four-photon excitation of 2,2′-dimethyl-p-terphenyl. *Journal of Physical Chemistry A* 2002; 106(5):754–759.

Johnson, C.K and Wan, C. 1997 Anisotropy decays induced by two-photon excitation, in Topics in Fluorescence Spectroscopy, Volume 5: Non-linear and Two-Photon-Induced Fluorescence, J. R. Lakowicz (ed.), Plenum Press, New York, pp. 43–85.

Kierdaszuk, B., Gryczynski, I., Modrakwojcik, A., Bzowska, A., Shugar, D., Lakowicz, J.R. Fluorescence of tyrosine and tryptophan in proteins using one-photon and 2-photon excitation. *Photochemistry and Photobiology* 1995; 61(4):319–324.

Kuhn, J.R., Wu, Z.R., Poenie, M. Modulated polarization microscopy: A promising new approach to visualizing cytoskeletal dynamics in living cells. *Biophys J* 2001; 80:972–985.

Lakowicz, J.R. Principles of Fluorescence Spectroscopy. 1999. New York: Kluwer Academic/Plenum Publishers.

Lakowicz, J.R., Gryczynski, I., Danielsen, E. Anomalous differential polarized phase angles for 2-photon excitation with isotropic depolarizing rotations. *Chemical Physics Letters* 1992a; 191(1–2):47–53.

Lakowicz, J.R., Gryczynski, I., Danielsen, E., Frisoli, J. Anisotropy spectra of indole and N-acetyl-L-tryptophanamide observed for 2-photon excitation of fluorescence. *Chemical Physics Letters* 1992b; 194(4–6):282–287.

Lakowicz, J.R., Gryczynski, I., Gryczynski, Z., Danielsen, E., Wirth, M.J. Time-resolved fluorescence intensity and anisotropy decays of 2,5-diphenyloxazole by 2-photon excitation and frequency-domain fluorometry. *Journal of Physical Chemistry* 1992c; 96(7):3000–3006.

Lakowicz, J.R., Kierdaszuk, B., Callis, P., Malak, H., Gryczynski, I. Fluorescence Anisotropy of Tyrosine Using One-and 2-Photon Excitation. *Biophysical Chemistry* 1995; 56(3): 263–271.

Massoumian, F., Juskaitis, R., Neil MAA, Wilson T. Quantitative polarized light microscopy. *J Microsc-Oxford* 2003; 209:13–22.

McClain, W.M. Excited state symmetry assignment through polarized two-photon absorption studies of fluids. *Journal of Chemical Physics* 1971; 55(6):2789–2796.

McClain, W.M. Polarization dependence of three-photon phenomena for randomly oriented molecules. *Journal of Chemical Physics* 1972; 57(6):2264–2272.

McClain, W.M. Polarization of two-photon excited fluorescence. *Journal of Chemical Physics* 1973; 58(1):324–326.

Oron, D., Tal, E., Silberberg, Y. Depth-resolved multiphoton polarization microscopy by third-harmonic generation. *Optics Letters* 2003; 28(23):2315–2317.

Parasassi, T., De Stasio, G., Ravagnan, G., Rusch, R.M., Gratton, E. Quantitation of lipid phases in phospholipids vescicles by the generalized polarization of Laurdan fluorescence. *Biophys. J.* 1991;60:179–189.

Scott, T.W., Haber, K.S., Albrecht, A.C. 2-photon photoselection in rigid solutions—a study of the B2U-a1G transition in benzene. *Journal of Chemical Physics* 1983; 78(1): 150–157.

So, P.T.C., Dong, C.Y., Masters, B.R., Berland, K.M. Two-photon excitation fluorescence microscopy. *Annual Review of Biomedical Engineering* 2000; 2:399–429.

Stoller, P., Reiser, K.M., Celliers, P.M., Rubenchik, A.M.. Polarization-modulated second harmonic generation in collagen. *Biophysical Journal* 2002; 82(6):3330–3342.

Sun, Y., Lo, W., Lin, S.J., Jee, S.H., Dong, C.Y. Multiphoton plarization and generalized polar-ization (GP) microscopy reveals oleic acid induced structural changes in intercellular lipid layers of the skin. *Optics Letters* 2004; 29(17): 2013–2015.

Sun, Y., Su, J.W., Lo, W., Lin, S.J., Jee, S.H., Dong, C.Y. Multiphoton polarization imaging of the stratum corneum and the dermis in ex-vivo human skin. *Optics Express* 2003; 11(25):3377–3384.

Wan, C.Z., Johnson, C.K. Time-resolved anisotropic 2-photon spectroscopy. *Chemical Physics* 1994; 179(3):513–531.

Williams, R.M., Zipfel, W.R. Webb, W.W. Interpreting second-harmonic generation images of collagen I fibrils. *Biophysical Journal* 2005; 88(2):1377–1386.

Wilson, T., Juskaitis, R. On the extinction coefficient in confocal polarization microscopy. *J Microsc-Oxford* 1995; 179:238–240.

Yu, B., Dong, C.Y., So, P.T.C., Blankschtein, D., Langer, R. In vitro visualization and quantifi-cation of oleic acid induced changes in transdermal transport using two-photon fluorescence microscopy. *Journal of Investigative Dermatology* 2001; 117(1):16–25.

Yu, B., Kim, K.H., So, P.T.C., Blankschtein, D., Langer, R. Topographic heterogeneity in transdermal transport revealed by high-speed two-photon microscopy: Determination of rep-resentative skin sample sizes. *Journal of Investigative Dermatology* 2002; 118(6):1085–1088.

21

Lifetime-Resolved Imaging in Nonlinear Microscopy

Wolfgang Becker and
Axel Bergmann

21.1 INTRODUCTION

Since their broad introduction in the early 1990s, confocal [139] and multiphoton laser scanning microscopes [57,136] have initiated a breakthrough in biomedical fluorescence imaging [206]. The high image quality obtained in these instruments mainly results from the fact that of out-of-focus light is strongly suppressed or, in the case of two-photon excitation, not even excited. As a result, images of high contrast are obtained, and 3D imaging becomes feasible. Moreover, the scanning technique makes detection in several wavelength channels and multispectral detection [60] relatively easy. More features, such as excitation wavelength scanning, polarization imaging, and second-harmonic imaging, have been added in the recent years. These multidimensional features make laser scanning microscopes an almost ideal choice for steady-state fluorescence imaging of biological samples [90,107,152,153].

However, the fluorescence of organic molecules is not only characterized by the emission spectrum, but it has also a characteristic lifetime. Including the fluorescence lifetime in the imaging process provides a direct approach to all effects involving energy transfer between different fluorophores and between fluorophores and their local environment. Typical examples are the probing of the local environment parameters of a fluorophore via lifetime changes, probing distances on the nanometer scale by fluorescence resonance energy transfer (FRET), and separation of fractions of the same fluorophore in different binding states to proteins, lipids, or DNA. Fluorescence lifetime imaging (FLIM) becomes particularly attractive in combination with multiphoton excitation. Multiphoton microscopes do not only provide the required pulsed excitation source, but they also avoid crosstalk of the lifetimes in different depths of thick tissue.

21.2 THE LASER SCANNING MICROSCOPE

The term "laser scanning microscope" is used for a number of very different instruments. Scanning can be accomplished by galvano-driven mirrors in the beam path, by piezo-driven mirrors, by a Nipkow disc, or by a piezo-driven sample stage. This chapter refers to microscopes with fast beam scanning by galvano-driven driven mirrors. The optical principle of these microscopes is, strongly simplified, shown in Figure 21.1.

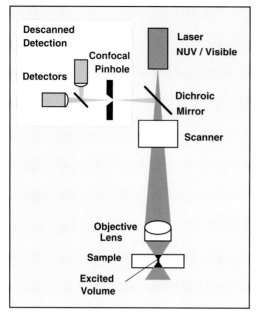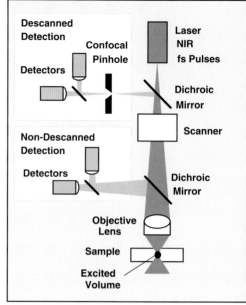

Figure 21.1. Optical principle of a laser scanning microscope. Left: One-photon excitation. Right: Two-photon excitation.

Laser scanning microscopes can be classified by the way they excite the fluorescence in the sample, and by the way they detect the fluorescence. One-photon microscopes use a near UV or visible CW laser to excite the sample. Two-photon (or multiphoton) microscopes use a femtosecond laser of high repetition rate. The fluorescence light can be detected by feeding it back through the scanner and through a confocal pinhole. The principle is termed "confocal" or "descanned" detection. A second way of detection is diverting the fluorescence directly behind the microscope objective. The principle is termed "direct" or "non-descanned" detection.

21.2.1 One-Photon Excitation with Confocal Detection

Figure 21.1, left, shows the principle of a laser scanning microscope with one-photon excitation. The laser is fed into the optical path via a dichroic mirror. It passes the optical scanner and is focused into the sample by the microscope objective. The focused laser excites fluorescence inside a double cone throughout the complete depth of the sample. The fluorescence light is collected by the objective lens.

The fluorescence light is fed back through the scanner so that the motion of the beam is cancelled or "descanned." The fluorescence light is separated from the excitation light by the dichroic mirror. The now stationary beam of fluorescence light is fed through a pinhole in the conjugate focus of the objective lens. In the plane of the pinhole, light from outside the focal plane in the sample is defocused. Out-of-focus light is therefore substantially suppressed by the pinhole. The principle is called confocal detection, and the microscope a confocal microscope.

The light passing the pinhole is usually split into several wavelength intervals and detected by several photomultiplier tubes (PMTs). X-Y imaging is achieved by scanning the laser beam; optical sectioning ("Z stack" recording) is performed by moving the sample up and down.

21.2.2 Two-Photon Excitation with Direct Detection

With a titanium-sapphire laser or another high-repetition-rate femtosecond laser, the sample can be excited by simultaneous two-photon absorption [57,73,107,136]. The excitation wavelength is twice the absorption wavelength of the molecules to be excited. Because two photons of the excitation light must be absorbed simultaneously, the excitation efficiency increases with the square of the excitation power density. Due to the high power density in the focus of a high numerical aperture (NA) microscope objective and the short pulse width of a titanium-sapphire laser, two-photon excitation works with remarkable efficiency. Excitation is obtained essentially in the volume of the diffraction pattern around the geometric focus of the objective lens. Consequently, depth resolution is an inherent feature of two-photon excitation, even if no pinhole is used. Since the scattering and the absorption at the wavelength of the two-photon excitation are small, the laser beam penetrates through relatively thick tissue. The loss on the way through the tissue can easily be compensated by increasing the laser power. The increased power does not cause much photodamage because the power density outside the focus is small.

The fluorescence photons have a shorter wavelength than the excitation photons. The scattering coefficient at the fluorescence wavelength is higher. Fluorescence photons from deep tissue layers therefore emerge from a relatively large area of the sample. To make matters worse, the surface is out of the focus of the objective lens. Therefore, the fluorescence from deep tissue layers cannot be efficiently focused into a pinhole.

The preferred detection technique for two-photon imaging is therefore direct (or non-descanned) detection. Direct detection splits off the fluorescence light immediately behind the microscope lens and directs it to a large-area detector. Consequently, an acceptable light collection efficiency is obtained even for deep layers of highly scattering samples. Two-photon imaging with non-descanned detection can therefore be used to image tissue layers several 100 µm (in extreme cases 1 mm) deep [47,61,107,137,178,182,183,184,185,192].

The absence of a pinhole in a two-photon microscope with non-descanned detection makes the optical path relatively easy to align. Two-photon microscopes can be built by upgrading a one-photon system or by attaching an optical scanner to a conventional microscope [58,59].

The downside of the large light-collection area of non-descanned detection is that the systems are very sensitive to daylight. A high background count level can, however, severely impair the accuracy of fluorescence lifetime detection. For thin samples, such as single cells, two-photon excitation is therefore used also with descanned detection. The pinhole is usually opened wide and is mainly used to suppress daylight leaking into the objective lens.

21.2.3 Two-Photon Excitation with Wide-Field Detection

The detection principles described above use single-point detectors to record the light from the scanned spot of the sample. However, laser scanning microscopes are

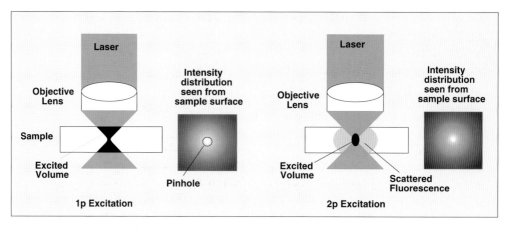

Figure 21.2. Intensity distribution around the laser focus, seen from the surface of a thick sample. Left: One-photon excitation. Fluorescence comes from the complete excitation light cone. The confocal microscope obtains a sharp image by detecting through a pinhole. Right: Two-photon excitation. Fluorescence is excited only in the focal plane. Nevertheless, scattering in a thick sample blurs the image seen from the sample surface. The two-photon microscope obtains a sharp image by assigning all photons to the pixel of the current scan position.

occasionally used also in combination with camera techniques. The differences between point detection and camera detection are illustrated in Figure 21.2.

In the one-photon microscope the laser excites fluorescence in a double cone that extends through the entire depth of the sample. A point detector peering through a pinhole in a conjugate image plane detects mainly the photons from the beam waist. An image built up by scanning is therefore virtually free of out-of-focus blur. The situation for a camera is different. The camera records the light from the complete excitation cone. Therefore, the image in the focal plane is blurred by out-of-focus light. In other words, the image is the same as in a conventional microscope.

The situation for two-photon imaging is shown in Figure 21.2, right. Two-photon excitation excites only a small spot around the geometric focus of the microscope lens. For a thin sample the camera and the point detector deliver the same image. However, two-photon imaging was originally developed for imaging sample layers deep within biological tissue. Light from deep tissue layers is strongly scattered on the way out of the sample. With a point detector, the photons can be collected from a large area of the sample surface and assigned to the current beam position. The sensitivity, the resolution, and the contrast are therefore not noticeably impaired by scattering. The situation is quite different for the camera. The camera records the image as it appears from the top of the sample. A sharp image is obtained only from the ballistic (unscattered) fluorescence photons. The effective point spread function is a sharp peak of the ballistic photons surrounded by a wide halo of scattered photons. The scattering halo can be as wide as 200 μm and then mainly causes a loss in contrast [101]. That means that the scattered photons not only are lost for the image, but they actually decrease the signal-to-noise ratio (SNR).

In the case of FLIM the situation is even worse because the scattering partially mixes the lifetimes of the pixels within the scattering radius. Nevertheless, gated and modulated cameras are used in two-photon laser scanning microscopes. They are mainly applied to multibeam scanning systems. Directing the light from the individual scanning spots to individual detectors is optically difficult, so a camera is currently the only feasible solution to multibeam scanning.

21.3 PHYSICAL BACKGROUND OF FLUORESCENCE LIFETIME IMAGING

The considerations above show that laser scanning fluorescence microscopy is an excellent technique to obtain spatial and spectral information from biological samples. FLIM does not only add a parameter to separate the signals of different fluorophores but also provides a direct approach to all processes involving energy transfer between different fluorophores and between fluorophores and their local environment. The following paragraph gives a brief summary of the practically relevant effects governing the decay of fluorescence, and their potential application. More detailed introductions into fluorescence kinetics are given in [28,147,122,161].

21.3.1 The Fluorescence Lifetime as a Separation Parameter

The practically most relevant molecular states and relaxation processes of fluorescent molecules are shown in Figure 21.3. The ground state is S0, the first excited state S1. By absorption of a photon of the energy S1–S0, the molecule transits into the S1 state.

A molecule can also be excited by absorbing two photons simultaneously. The sum of the energy of the photons must be at least the energy of the S1 state. Simultaneous two-photon excitation requires a high photon flux. Because two photons are required to excite one molecule, the excitation efficiency increases with the square of the photon flux. Efficient two-photon excitation requires a pulsed laser and focusing into a diffraction-limited spot. Due to the nonlinearity of two-photon absorption, the excitation is almost entirely confined to the central part of the diffraction pattern.

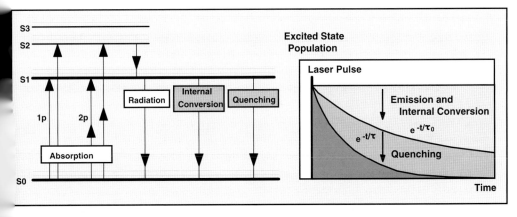

Figure 21.3. Fluorescence excitation and radiative and non-radiative decay paths.

Higher excited states, S2, S3, do exist but decay at an extremely rapid rate into the S1 state. Moreover, the electronic states of the molecules are broadened by vibration. Therefore, a molecule can be excited by almost any energy higher than the gap between S0 and S1.

Without interaction with its environment, the molecule can return from the S1 state by emitting a photon, or by internal conversion of the absorbed energy internally into heat. The probability that one of these effects occurs is independent of the time after the excitation. The fluorescence decay function measured at a large number of similar molecules is therefore single-exponential.

The lifetime the molecule had in absence of any radiationless decay processes is the "natural fluorescence lifetime," τ_n. For molecules in solution the natural lifetime is a constant for a given molecule and a given refraction index of the solvent. Because the absorbed energy can also be dissipated by internal conversion, the effective fluorescence lifetime, τ_0, is shorter than the natural lifetime, τ_n. The "fluorescence quantum efficiency" (i.e., the ratio of the number of emitted and absorbed photons) reflects the ratio of the radiative and total decay rate.

Using the fluorescence lifetime to separate the signals of different fluorophores may appear somewhat artificial at first glance. A wide variety of fluorophores are available, and normally a sample can be stained with fluorophores of distinctly different fluorescence spectra. Separating the signals by the emission wavelength is certainly easier than by using the lifetimes [89]. Nevertheless, FLIM has proved to be useful for imaging histological samples [71].

The fluorescence lifetime is particularly important when it comes to autofluorescence imaging of biological tissue. Usually a large number of endogenous fluorophores are present in tissue, most of which have poorly defined fluorescence spectra [167]. FLIM then becomes a powerful imaging tool [111,135,199].

The lifetime as a separation parameter has also been used to track the progress of photoconversion of dyes for photodynamic therapy [169] and the internalization and aggregation of dyes in cells [100], and to verify laser-based transfection of cells [195].

21.3.2 The Fluorescence Lifetime as an Indicator of the Local Environment

An excited molecule can also dissipate the absorbed energy by interaction with another molecule. The process opens an additional return path to the ground state (see Fig. 21.3). The fluorescence lifetime, τ, becomes shorter than the normally observed fluorescence lifetime, τ_0. The fluorescence intensity decreases by the same ratio as the lifetime. The effect is called "fluorescence quenching."

The quenching intensity depends linearly on the concentration of the quencher. Typical quenchers are oxygen, halogens, heavy metal ions, and a large number of organic molecules [7,122]. Many fluorescent molecules have a protonated and a deprotonated form, isomers, or can form complexes with other molecules. The fluorescence spectra of these species can be virtually identical, but the fluorescence lifetimes may be different. It is not always clear whether these effects are related to fluorescence quenching. In practice, it is important only that for almost all dyes, the fluorescence lifetime depends more or less on the concentration of ions, the oxygen concentration, the pH value, or, in biological samples, the binding to proteins, DNA, or lipids. The lifetime can therefore be used to probe the local environment of dye molecules on the molecular scale, independently of the variable and usually unknown concentration of the fluorescing molecules. Typical examples are the mapping of cell parameters [4,66,80,88,121,131,170] and probing protein or DNA structures by the environment-dependent lifetime of dyes [104,201].

Endogenous fluorophores often exist in different conformational or binding states and therefore deliver multi-exponential decay profiles. The decay components and their intensity factors depend on the environment. An example is NADH, whose lifetime increases from 400 to 600 ps to 1.4 to 2.4 ns upon binding to proteins [120,151].

21.3.3 Fluorescence Resonance Energy Transfer

A particularly efficient energy transfer process is FRET. FRET is an interaction of two fluorophore molecules with the emission band of one dye overlapping the absorption band of the other. In this case the energy from the first dye, the donor, is transferred immediately to the second one, the acceptor (Fig. 21.4, left). The energy transfer itself does not involve any light emission and absorption. Förster resonance energy transfer and resonance energy transfer (RET) are synonyms of the same effect. FRET results in an extremely efficient quenching of the donor fluorescence and, consequently, decrease of the donor lifetime (see Fig. 21.4, right). The energy transfer rate from the donor to the acceptor decreases with the sixth power of the distance. Therefore, it is noticeable only at distances shorter than 10 nm [122].

FRET has become an important tool in cell biology. It is used to verify whether labeled proteins are physically linked. By measuring the FRET efficiency, distances on the nm scale can be determined.

The obvious difficulty of steady-state FRET measurements in cells is that the concentrations of the donor and acceptor are variable and unknown. Moreover, the emission band on the donor extends into the emission band of the acceptor, and the absorption band of the acceptor extends into the absorption band of the donor. A further complication is that usually only a fraction of the donor molecules are linked with an acceptor molecule. These effects are hard to distinguish in steady-state FRET measurements.

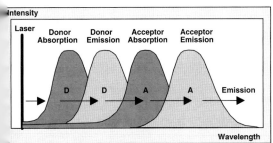
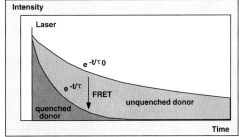

Figure 21.4. Fluorescence resonance energy transfer (FRET). The energy absorbed in the absorption band of the donor is directly transferred into the acceptor molecules and emitted via the emission band of the acceptor (left). The energy transfer results in quenching of the donor emission and a decrease of the donor lifetime (right).

Nevertheless, a number of FRET techniques based on steady-state imaging have been developed [69,85,138,154,204]. The techniques need several subsequent measurements, including images of cells containing only the donor and the acceptor, or are destructive and therefore not applicable to living cells.

FLIM-based FRET techniques have the benefit that the results are obtained from a single lifetime image of the donor [5,13,19,32,33,34,51,65,87,197]. They do not need calibration by different cells, and they are nondestructive. Moreover, FLIM is able to resolve the interacting and noninteracting donor fractions (see the section on FRET Experiments in Live Cells).

21.3.4 Fluorescence Depolarization

If the fluorescence of a sample is excited by linearly polarized light, the fluorescence is partially polarized. The fluorescence anisotropy, r, is defined as:

$$r(t) = \frac{I_p(t) - I_s(t)}{I(t)}$$

where $I_p =$ fluorescence intensity parallel to excitation, $I_s =$ fluorescence intensity perpendicular to excitation, and $I =$ total intensity.

The fluorescence anisotropy decays with the rotational relaxation time, τ_{rot}. The relaxation time is an indicator of the size of the dissolved molecules, dye–solvent interactions, aggregation states, and the binding state to proteins [54,122]. Typical rotational relaxation times of free fluorophores in water are in the range from 50 to 500 ps.

To separate the rotational relaxation from the fluorescence decay, it is essential that the correct total intensity $I(t)$ is taken for the denominator of $r(t)$. It is normally assumed that both the excitation and emission light cones have negligible NA and that the excitation is polarized perpendicularly to the plane defined by both optical axis. The total intensity is then $I(t) = I_p(t) + 2I_s(t)$. The factor of 2 results from the geometrical fact that light polarized longitudinally to the optical axis of observation is not detected [122]. The situation for a microscope lens of large NA is different. Both focusing and detection under high NA result in the conversion of transversal E vectors into longitudinal ones, and vice versa [8,162,179,180,209]. The total intensity is therefore $I(t) = I(t) + kI_p(t)$, with $k = 1$ to 2.

21.3.5 Fluorescence Correlation

Fluorescence correlation spectroscopy (FCS) is based on the excitation of a small number of molecules in a femtoliter volume, and correlating the intensity fluctuations of the fluorescence intensity. The technique dates back to a work of Magde, Elson and Webb published in 1972 [133]. FCS delivers information about the diffusion time constants of the fluorophore molecules and about the intersystem crossing rate and the triplet lifetime. In biological systems information about the binding state of the fluorophores and the mobility of dye-labeled proteins is obtained. If several detectors are used in different wavelength intervals, the autocorrelation of the signals of the individual fluorophores, or the cross-correlation between the signals of different fluorophores, can be obtained [176]. The theory and applications of FCS are described in [29,68,164,165,166,176,177,193].

A closely related technique is the photon counting histogram (PCH). The technique captures the amplitude distribution of the intensity and is therefore also called fluorescence intensity distribution analysis (FIDA). The PCH of an optical signal is obtained by recording the photons within subsequent time intervals and building up the distribution of the frequency of the counts versus the count number.

A characterization of the PCH was given in 1990 by Qian and Elson [160], and the theoretical background is described in [49,50,97,99,146,148,149,160]. The PCH delivers the average number of molecules in the focus and their "molecular brightness." Molecules of different brightness can be distinguished by fitting a model containing the relative brightness and the concentration ratio of the molecules to the measured PCH. The technique can be extended for two-dimensional histograms of the intensity recorded by two detectors in different wavelength intervals or under different polarization. 2D-FIDA delivers a substantially improved resolution of different fluorescent species [98,99]. Further improvement is achieved by using pulsed excitation and adding the fluorescence lifetime as an additional dimension of the histogram. The technique is termed fluorescence intensity and lifetime distribution analysis (FILDA) [99,148].

21.4 REQUIREMENTS FOR FLUORESCENCE LIFETIME IMAGING IN SCANNING MICROSCOPES

21.4.1 Lifetime Resolution

The lifetimes of the commonly used fluorophores are typically in the range from 1 ns to 5 ns. The lifetime components of autofluorescence are between 100 ps and 5 ns [111,135,199]. Probably low-intensity components exist with lifetimes down to the 10-ps scale [123]. The lifetime of interacting donor molecules in FRET experiments can be as short as 100 ps. Aggregates can have lifetimes down to 50 ps [100]. Aggregation of dyes to metallic nanoparticles results both in quenching and an increase of the radiative decay rate. The lifetimes can be as short as 50 ps [76,77,78,134]. Anisotropy decay times in water are in the 100-ps range but can vary over several orders of magnitude with the binding state to proteins.

Except for relatively trivial cases, the fluorescence decay functions found in cells are multi-exponential. Different fluorophores may be present in the same pixel, or a single fluorophore may exist in a bound and an unbound state. Many endogenous fluorophores exist in different conformations with different lifetimes. In almost all FRET experiments a mixture of interacting and non-interacting donor molecules is present. The decay functions in anisotropy measurements contain both the fluorescence and the anisotropy decay. The resulting double-exponential decay functions have to be recorded correctly even though anisotropy analysis itself does not necessarily require double-exponential analysis.

A lifetime imaging technique for biological applications must therefore be able to resolve multi-exponential decay functions with components down to less than 50 ps. It will be shown later that multi-exponential decay analysis actually adds more than a single dimension to laser scanning microscopy.

21.4.2 Suppression of Out-of-Focus Light

One of the most relevant features of the scanning technique is its suppression of out-of-focus light and sectioning capability. Scanning in combination with confocal detection and, more efficiently, two-photon excitation also reduces lateral crosstalk. These features become particularly important in combination with FLIM. Mixing the decay functions of different pixels or focal planes must be avoided to obtain clean lifetime results. Lateral and vertical crosstalk is avoided by point-detection FLIM techniques, which are therefore the first choice for the laser scanning microscope.

21.4.3 Scan Rates

Commercial laser scanning microscopes scan the sample with pixel dwell times down to a few hundred ns. There are two reasons for the high scanning rate. The first one is that a high frame rate is required to record fast image sequences. Of course, single frames recorded at pixel dwell times this short deliver a poor SNR. However, image correlation techniques are able to recover transient effects even from the sequence of extremely noisy images. Although generally possible, image correlation techniques have not been used in conjunction with FLIM yet. Therefore, the second benefit of high scan rates is more important: at the high excitation power density used in scanning, a considerable fraction of the fluorophore is accumulated in the triplet state. Molecules in the triplet state do not fluoresce and are lost for the build-up of the image. Typical triplet lifetimes are in the range from 10 to 100 μs. Fast scanning therefore reduces the fraction of the fluorophore molecules in the triplet state at the current scan position.

21.4.4 Efficiency

An ideal optical recording technique would record all detected photons with the same weight, an infinitely short instrument response function, negligible signal background, a time-channel width much shorter than the signal width, and a number of time channels that covers the complete duration of the signal. Under these conditions the lifetime, τ, of a single-exponential fluorescence decay can be determined with a standard deviation of:

$$\sigma_\tau = \frac{\tau}{\sqrt{N}}$$

with N = number of recorded photons [82,105].

The relation results from the simple fact that both the mean arrival time and the variance of the arrival times of the photons in a single-exponential decay is τ. The SNR of the lifetime is:

$$SNR = \frac{\tau}{\sigma_\tau} = \sqrt{N}$$

That means that a single-exponential fluorescence lifetime can ideally be derived from a given number of photons per pixel with the same accuracy as the intensity. A lifetime accuracy of 10% can be obtained from only 100 photons. The pitfall is, however, that an image with 10% intensity noise looks very pleasing, while a lifetime accuracy of 10% is not sufficient for the majority of FLIM applications. Moreover, the required number of photons, N, increases dramatically for multi-exponential lifetime analysis [105]. The resolution of double-exponential decay profiles requires at least 1,000 photons per pixel, depending on the ratio of the lifetimes and intensity factors. Therefore, the required number of photons in FLIM is normally larger than in steady-state imaging. Obtaining a large number of photons from the sample means either long exposure or high excitation power. Therefore, photobleaching [27,150] and photodamage [92,106,109] become problems in precision FLIM experiments.

It is therefore important that a lifetime detection technique comes as close as possible to the ideal SNR for a given number of detected photons. The efficiency of a lifetime technique is often characterized by the "figure of merit," F [9,105,156]. The figure of merit compares the SNR of an ideal recording device to the SNR of the technique under consideration:

$$F = \frac{SNR_{ideal}}{SNR_{real}}$$

The loss of SNR in a real technique can also be expressed by the counting efficiency. The counting efficiency, E, is the ratio of the number of photons ideally needed and the number needed by the considered technique:

$$E = \frac{1}{F^2}$$

It should be noted that the practically achieved values of F and E also depend on the numerical stability of the lifetime analysis algorithm. Moreover, F was originally defined for a single-detector device and single-exponential decay. The definition of F is therefore not directly applicable to multi-wavelength time-correlated single-photon counting (TCSPC) and multi-exponential decay analysis.

21.4.5 Multidetector Capability

FLIM on samples with several fluorophores often requires the fluorescence to be recorded in different wavelength channels. Wavelength resolution is particularly useful in autofluorescence experiments. Also, FRET experiments benefit from simultaneously recording the donor and acceptor fluorescence, and fluorescence anisotropy measurements require the fluorescence to be recorded under an angle of 0 and 90 degrees from the excitation. Photobleaching usually precludes recording these signals consecutively. Multidetector capability is therefore another important feature of a FLIM technique.

21.5 FREQUENCY-DOMAIN TECHNIQUES

FLIM techniques are usually classified into time-domain techniques and frequency-domain techniques. Time-domain techniques record the intensity of the signal as a function of time, while frequency-domain techniques measure the (electrical) amplitude and phase as a function of frequency. The time domain and the frequency domain

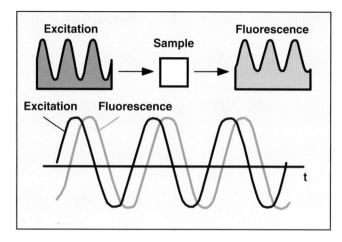

Figure 21.5. Modulation technique. The phase and the modulation degree of the fluorescence signal are used to determine the lifetime.

are connected via the Fourier transform. It is therefore commonly believed that time-domain techniques and frequency-domain techniques are generally equivalent. This assumption is, however, not correct, as a simple example shows. There are a number of time-domain techniques that resolve the waveform of an optical signal directly into a large number of time channels. There is, however, no reasonable technique to resolve the phase and amplitude spectrum of a signal simultaneously into a large number of frequency channels. The differences become even larger if the particular technical solutions or instruments are considered. The time resolution, the capability to resolve complex decay functions, and the efficiency depend on so many technical details that a general comparison of the techniques is almost impossible.

An overview about frequency-domain techniques is given in [48,202]. Frequency-domain techniques use the phase shift and the modulation degree of the fluorescence compared to the modulated excitation. The principle is shown in Figure 21.5.

The excitation light is modulated with a frequency, ω. Compared to the excitation, the fluorescence emitted by the sample has a phase shift and a reduced modulation degree. Both depend on the fluorescence lifetime and on the modulation frequency:

$$\tan \varphi_f = \omega \tau_f$$

$$\frac{M_f}{M_{ex}} = \frac{1}{\sqrt{1 + \omega^2 \tau_f^2}}$$

where ω = angular frequency of modulation, M_{ex} = modulation degree of excitation, M_f = modulation degree of fluorescence, φ_f = phase lag of fluorescence, and τ_f = fluorescence lifetime.

Both the phase and the modulation degree can be used to determine the fluorescence lifetime. The different dependence on τ_f can be used to decide whether the fluorescence decay is single-exponential. In practice, phase measurements deliver a better τ accuracy than measurements of the modulation degree. Therefore, normally the phase is used for lifetime measurements. The optimal frequency depends on the lifetime and is in the range of:

$$\omega = \frac{1}{\tau_f} \text{ or } f = \frac{1}{2\pi \tau_f}$$

Since fluorescence lifetimes are of the order of nanoseconds or picoseconds, the modulation should actually be of the order of 1 GHz. However, it turns out that lifetimes can be obtained with little loss in accuracy at much lower frequencies [46,156]. Therefore, in practice frequencies between 50 MHz and several 100 MHz are used. To resolve the components of multi-exponential decay functions, phase measurements at different frequencies are necessary.

21.5.1 Modulating the Light Source

Early modulation fluorometers modulated light from a xenon or mercury lamp by an acousto-optical modulator (AOM). For lifetime imaging in one-photon scanning microscopes, AOMs are used to modulate continuous argon and krypton lasers. Another convenient way to obtain modulated light involves electrically modulated LEDs or laser diodes. LEDs can easily be modulated at 50 to 100 MHz, and laser diodes with more than 1 GHz.

Multiphoton microscopes use a high-repetition-rate femtosecond laser for excitation. The repetition rate is normally between 78 and 90 MHz. The train of femtosecond pulses contains all harmonics of the repetition frequency. A frequency-domain system can, in principle, be synchronized to any of the harmonics. The frequency is limited by the bandwidth of the mixers, the detector, or the modulation bandwidth of the detector. By using a multichannel plate PMT with external mixers, modulation frequencies up to 10 GHz have been achieved [118].

21.5.2 Homodyne Detection

Phase-sensitive detection requires the detector signal to be multiplied with the modulation signal, $\sin \omega t$. Consider a detector signal with an amplitude, a, a phase shift, φ_f, at the modulation frequency, ω. The detector signal is multiplied by a reference signal of the same frequency, ω, and the reference phase, φ_r. The result of the multiplication is:

$$a \sin(\omega t + \varphi_f) \cdot \sin(\omega t + \varphi_r) = 0.5a \cos(\varphi_f + \varphi_r) - 0.5a \cos(2\omega t + \varphi_f + \varphi_r)$$

After filtering out the high-frequency component at 2ω the ω-independent term, $a \cos(\varphi_f + \varphi_r)$, remains. The amplitude, a, and the phase of the fluorescence, φ_f, can be obtained from two measurements at different reference phase, φ_r. This can be achieved by several mixers driven by phase-shifted reference signals or by consecutive measurements. The general principle of a point-detector frequency-domain system with homodyne detection is shown in Figure 21.6.

A high-frequency generator delivers the modulation frequency, f_{mod}. The laser intensity is modulated by an AOM driven at the frequency $f_{mod}/2$. The fluorescence light from the sample is detected by a PMT or a photodiode. The AC component of the detector signal is amplified and fed into two mixers that mix (i.e., multiply) the signal with

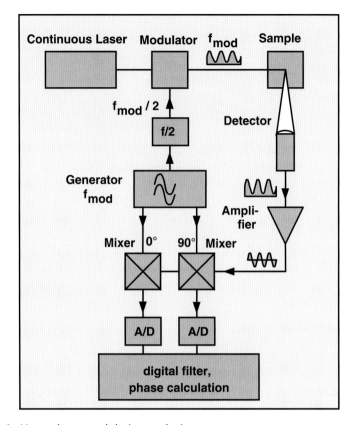

Figure 21.6. Homodyne modulation technique.

the modulation frequency at 0 and 90 degrees of phase shift. The outputs of the mixers deliver DC components that represent the 0- and 90-degree components of the amplified detector signal. After lowpass filtering the amplitude and the phase are derived from the output signals of the mixers. The operation is usually performed by digitizing the signals and calculating the phase and the amplitude from the mixer signals. The principle is analogous to a dual-phase lock-in amplifier and is therefore often called lock-in detection [44,45].

The technique can be used to record the fluorescence of several chromophores simultaneously in several parallel mixer systems. Several lasers that are modulated at different frequencies provide different excitation wavelengths. The emission of the sample is split into several wavelength ranges and detected by separate detectors. The detector signals are mixed with the modulation frequencies of the individual lasers in several parallel groups of mixers [44,45].

21.5.3 Heterodyne Detection

Heterodyne detection mixes the detection and the modulation signal with a frequency different from the modulation frequency [37,39,181,182]. The result is that the phase angle and the modulation degree are transferred to the difference frequency signals. The general principle is shown in Figure 21.7.

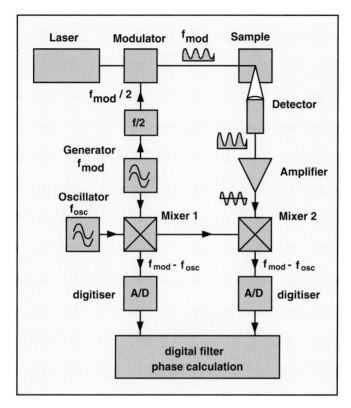

Figure 21.7. Heterodyne modulation technique.

The detector signal and the modulation frequency, f_{mod}, are mixed with an oscillator frequency, f_{osc}. The result are two signals at the difference frequency, $f_{mod} - f_{osc}$. The phase shift between the two signals is the same as between the detector signal and the modulation signal. The oscillator frequency is normally chosen to obtain a difference frequency in the kHz range. The corresponding signals are directly digitized. The consecutive results of the analog-to-digital conversion are filtered and used for phase measurement. The advantage of direct digitizing is that efficient digital filtering algorithms can be applied and the phase can be determined via fast Fourier transform [181,182]. The heterodyne technique can relatively easily be combined with two-photon excitation by a high-repetition-rate pulsed laser. The input signal of mixer 1 is then derived from the pulse train delivered by the laser.

21.5.4 Gain-Modulated Detectors

The mixers used in the signal channel of homodyne or heterodyne systems can be replaced with gain-modulated detectors [182]. The principle of a system described in [39] is shown in Figure 21.8. The Ti:Sa laser generates femtosecond pulses at 80-MHz repetition rate. The pulses are fed through the optical scanner of the microscope. The scanned laser beam is focused into the sample. The fluorescence is excited via two-photon excitation. The fluorescence is detected by PMT 1 through a direct (non-descanned) detection path.

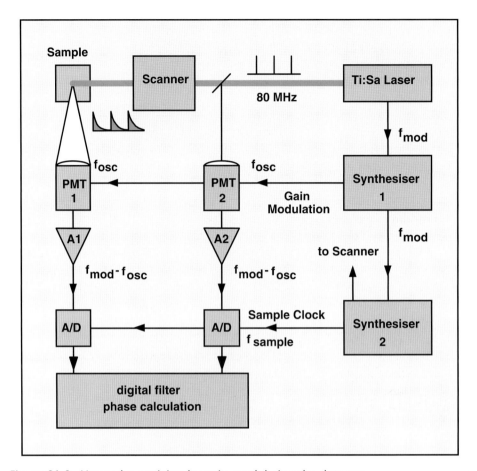

Figure 21.8. Heterodyne mixing by gain-modulating the detector.

A part of the laser light is diverted and fed to a reference photomultiplier, PMT 2. The 80-MHz pulse repetition frequency of the laser drives two frequency synthesizers. The first one generates the oscillator frequency, f_{osc}, which differs from the pulse repetition frequency by a few kHz. The oscillator frequency modulates the gain of both PMTs. Gain modulation of the PMT is achieved by applying the oscillator signal between the two early dynodes of the PMT.

The mixing products (i.e., the frequencies f_{mod}, f_{osc}, $f_{mod} + f_{osc}$, and $f_{mod} - f_{osc}$) appear at the outputs of the PMTs. The difference frequency, often called cross-correlation frequency, is separated from the high-frequency mixing products and the DC level by AC coupled preamplifiers, A1 and A2, of appropriate bandwidth. Thus, A1 delivers the down-mixed 80-MHz component of the fluorescence signal. A2 delivers a phase reference signal. The phase difference between the signals is the same as the phase difference of the fluorescence and the laser at the fundamental frequency of 80 MHz. Both signals are digitized, and the phase and the amplitude are obtained by digital filtering.

The second frequency synthesizer generates the sampling clock for the analog-to-digital converters. The sample rate is chosen substantially higher than $f_{mod} - f_{osc}$ to avoid aliasing of higher harmonics of the laser frequency. The second synthesizer also controls the scanner. The fixed relation between the pixel clock of the scanner and the sample clock avoids Moiré effects by aliasing of the two clocks.

The setup has a number of remarkable features. The most obvious benefit is that the instrument can be used for two-photon excitation. Moreover, the effective modulation frequency can easily be changed by tuning the synthesizers to higher harmonics of the laser frequency. Less obvious, but similarly important, is that the phase noise of Synthesizer 1 has little effect on the results. Phase variations have the same influence on the down-mixed fluorescence and reference signal and therefore cancel. Phase drifts in the modulation of the PMTs (i.e., by variations in the amplitude of the f_{osc} signal) are closely the same in both PMTs, and cancel. Even intensity changes of the laser can be eliminated from the result. Moreover, the modulation frequency is not limited by the bandwidth of the gain system of the PMT. Therefore, relatively high modulation frequencies can be used.

A problem of many frequency-domain systems is that the measured phase depends on the intensity of the signal. The effect, called amplitude-phase crosstalk, results from pickup of the modulation frequency by the detector signal lines and from nonlinearity in the mixers. By modulating the PMT gain in an early dynode stage (i.e., before an electrical signal is generated), these electrical effects are almost entirely avoided. The only chance of crosstalk is between the A/D converter channels.

In principle, several detectors can be operated at the same frequency but in different wavelength intervals. The system complexity increases only moderately with the number of detectors. Multidetector frequency domain systems are routinely used in optical tomography, but surprisingly they have found little application in fluorescence lifetime systems yet.

21.5.5 Modulated Image Intensifiers

Image intensifiers are vacuum tubes containing a photocathode, a multiplication system for the photoelectrons, and a two-dimensional image detection system. First-generation systems used an electron-optical imaging system that accelerated the photoelectrons to an energy of some keV and sent them to a fluorescent screen. The image from the screen was detected by a traditional camera or later with a CCD camera. First-generation devices had a relatively low gain and strong image distortions.

Second-generation image intensifiers use multichannel plates for electron multiplication (Fig. 21.9). One plate gives a typical multiplication factor of 1,000, so that a gain of 10^6 can be achieved by two plates in series. The CCD chip can also be placed inside the tube to detect the electrons directly. These electron-bombarded CCDs (EB CCDs) give higher gain than a CCD behind a fluorescent screen.

Modulation of an image intensifier is achieved by applying an RF voltage between the photocathode and the multichannel plate [84,119,181,186]. A problem with this approach can be the low conductivity of the photocathode. The time constant formed by the gate-cathode capacitance and the cathode resistance limits the modulation frequency and introduces lateral phase variation across the image. The problem can be relaxed by a conductive grid on the photocathode. Another possibility is to insert a grid between the photocathode and the microchannel plate and to apply the modulation voltage between the grid and the microchannel plate. Also, a grid outside the tube can be used. The grid then forms a capacitive bypass of the photocathode resistance.

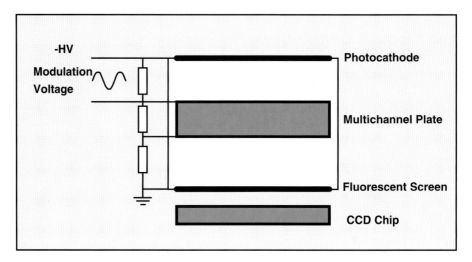

Figure 21.9. Intensified CCD camera.

Modulating the voltage across the channel plate has also been suggested [181]. The drawback of this method is the high RF amplitude required, the heating of the channel plate due to dielectric loss, and the low degree of modulation.

As for the single-point detectors, both homodyne and heterodyne detection can be used in image intensifiers. For homodyne detection the excitation light and the image intensifier are modulated at the same frequency, f_{mod}. A variable phase shifter is used to change the phase relation between the modulation of the light source and the modulation of the image intensifier. For sine wave modulation, phase of the fluorescence signal is obtained from three images acquired at different phase shifts referred to the excitation. For non-sinusoidal modulation, either a sequence of images for different phase or integral images for different profiles of a phase sweep are recorded [200].

Instead of the phase shifter, the heterodyne technique can be used. In this case the image intensifier is modulated with a frequency, f_{osc}, which differs slightly from the modulation frequency of the laser, f_{mod}. This causes the relative phase of the detector modulation to change continuously with the difference (cross-correlation) frequency, $f_{osc} - f_{mod}$. A sequence of images of different phase is read out over the period of the difference frequency [181]. The period must be considerably longer than the readout time of the camera and the luminescence lifetime of the fluorescent screen of the intensifier tube. The typical difference frequency is of the order of 10 Hz [181]. For application in a scanning microscope, the scanning must be synchronized with the difference frequency. The heterodyne technique requires a good frequency synthesizer that delivers the two modulation frequencies and the difference of both with high frequency and phase stability. As described for the point-detection technique, a high-repetition-rate pulsed laser can be used instead of a modulated laser.

An interesting development is the directly modulated CCD chip described in [140]. Results were described for 100- to 500-kHz modulation frequency. The modulation frequency could be increased to about 10 MHz. This is not enough to obtain a good efficiency for lifetimes of the order of 1 ns. If the modulation frequency could be increased by factor of 10 in the future, the chip may outperform the image intensifiers.

21.5.6 Effect of Technical Details on the System Performance

The practically achieved lifetime resolution, efficiency, and scan rate of frequency-domain instruments depend strongly on the details of the electronics and the operating conditions.

The resolution of the modulation technique mainly depends on the phase stability of the detectors and mixers. If a sufficiently large photon flux is available, lifetimes down to less than 10 ps can be measured. In FLIM applications the lifetime accuracy is usually limited by the shot noise of the photons and therefore is directly related to the detection efficiency.

Results of Monte Carlo simulations of the figure of merit, F, for single-exponential decay functions are given in [46,156]. For a detector with a separate mixer, sine-wave-modulated excitation, sine-wave mixing, and 100% modulation of the excitation, F was found to be 3.7. If the modulation is less than 100%, a fraction of unmodulated light is detected, which contributes to the noise but not to the phase signal. Therefore the SNR drops dramatically for less than 75% modulation. Interestingly, in the same setup 100% square-wave modulation of the excitation gives an F = 1.2 and Dirac pulse excitation even F = 1.1. This is a strong argument to use pulsed lasers (i.e., Ti:Sa or diode lasers) rather than modulating CW lasers.

For a gain-modulated detector with internal gain modulation, the efficiency is lower than for a detector with a separate mixer. For sine-wave modulation in the detector and Dirac excitation, F values around 4 were found. The reason of the low efficiency is inefficient mixing. Ideal mixing means to *multiply* the detected signal with the oscillator signal (i.e., to invert the polarity of the signal for the negative half-period of the oscillator signal). However, detector gain modulation at best modulates the gain between 100% and 0%. Therefore, a large fraction of the input signal is not only lost but also contributes with its shot noise to the noise of the result. It must also be expected that the details of the detector modulation have an influence on the efficiency. If the detector is modulated by applying the modulation voltage between the photocathode and the first dynode or the multichannel plate, the result is actually an efficiency modulation. The RF field then modulates the fraction of photoelectrons reaching the amplification system rather than the effective gain for the individual photons. The result is loss of photons and, consequently, loss of SNR.

Another frequently encountered problem in frequency-domain systems is amplitude–phase crosstalk [1,48,143]. Amplitude–phase crosstalk is caused by nonlinearity in the detectors, amplifiers, and mixers and by spurious pickup of the modulation frequency in the detector channel. The result is a dependence of the detected lifetime on the intensity. For intensities typical for live cell imaging, a system described in [39] had a phase error of 3 degrees at 80 MHz over an intensity interval of 1:20. If the effect was left uncompensated, it resulted in a lifetime error of 100 ps. For comparison, TCSPC systems achieve a systematic lifetime error of about 2 ps over an intensity interval of 1:1,000.

The efficiency of the techniques described above degrades at low light intensities. At the intensities obtained in laser scanning microscopy, the detector delivers a train of short pulses rather than a continuous signal (see Fig. 21.12). The pulses represent the detection of the randomly arriving photons. If this pulse train is fed to the signal processing circuitry described above, a useful phase signal is produced for only a short time after the detection of a photon. As a rule of thumb, the efficiency must be expected to drop dramatically at photon rates lower than the filter time constants in the phase measurement circuitry.

The strength of the modulation technique is that it can be used up to extremely high photon rates. The gain of the PMT or the image intensifier can be varied over several orders of magnitude to adapt the system to different intensities. Image intensifiers in particular therefore obtain short acquisition times. For point-detector scanning systems, the shortest applicable pixel dwell time is limited by the settling time of the filters in the phase and amplitude measurement block. Fast acquisition therefore requires short filter time constants, but this impairs the efficiency at low photon rates. Normally the shortest useful pixel dwell times are 20 to 100 µs.

The connection between the shortest useful pixel dwell time and the minimum useful count rate leads to an interesting conclusion. Obviously, a problem occurs if the average number of photons per pixel of a single frame becomes less than one. The only conceivable solution is then to measure the phase (or the time in the modulation period) of individual photons, and to accumulate the individual phase measurements (or photon times) and photon numbers over several frames of the scan. This approach leads directly to advanced TCSPC techniques (see the section on multidimensional TCSPC).

21.6 ANALOG TIME-DOMAIN SYSTEMS

To record a signal in the time domain, the intensity must be measured in a large number of subsequent time channels. Resolving a multi-exponential decay function requires a time channel width of the order of the shortest decay component (i.e., a sample rate of 10 to 50 GHz). However, direct digitizing of a signal with a rate this high is extremely difficult. Nevertheless, lifetime systems based on digital oscilloscopes are used in combination with slow-scan systems and low repetition-rate lasers [135]. Their resolution and capability to resolve double-exponential decay functions is limited to nanosecond decay times. The only practicable solutions are currently streak cameras, which obtain time resolution by fast deflection of the photoelectrons.

Another way to achieve high effective sample rates are sequential sampling techniques. The signal amplitude is sampled within a narrow time-gate, and the gate is sequentially shifted over the signal. The technique is used in sampling oscilloscopes and Boxcar integrators and achieves effective sample rates in the 100-GHz range. The sequential gate scan makes these instruments too slow for lifetime imaging. Sequential sampling can, however, be used in gated image intensifiers.

21.6.1 Gated Image Intensifiers

Gated image intensifiers use the same principle—often even the same intensifier tubes—as the modulated image intensifiers. Instead of a modulation signal, a gate pulse is applied between the photocathode and the microchannel plate. If the voltage of the channel plate input side is positive with respect to the cathode, the photoelectrons are transmitted. For negative voltages the electrons are driven back to the cathode. Although gating of an image intensifier looks straightforward at first glance it is anything but simple, particularly if sub-ns resolution is to be achieved. Even if a sufficiently short gating pulse can be generated electronically, the electrical field between the photocathode and the grid must follow the pulse at the same speed. Because the conductivity of the photocathode is relatively low, the time constant formed by the gate-cathode capacitance and the cathode resistance limits the switching speed. Manufacturers counteract the effect by using high-conductivity photocathodes or by adding a grid on the photocathode, behind the photocathode, or on the entrance window of the tube.

Standard gated image intensifier devices have a minimum gate width of the order of a few ns. A device with 5-ns gate width has been used to determine single-exponential decay constants down to a few ns by deconvolution [114,171,172]. The shortest gate width currently obtained with gated image intensifiers is 50 ps for repetition rates in the kHz range and 200 ps at a repetition rate of 80 MHz [157,187].

For lifetime imaging in a laser scanning microscope, the photocathode of the image intensifier is placed in a conjugate image plane of the sample. The optical scanning process then writes a fluorescence image of the scanned sample plane on the photocathode. As for all camera techniques, some loss of contrast by scattering of the fluorescence from deep sample layers must be expected (see Fig. 21.2). To obtain the shape of the fluorescence decay, a series of frames is recorded by scanning the gate delay on a frame-by-frame basis (Fig. 21.10, left). The frame period must be longer than the luminescence lifetime of the fluorescent screen of the intensifier tube.

A serious drawback of the gated image intensifier is the low counting efficiency. Due to the gating process, most of the photons collected from the sample are gated off. The counting efficiency is about the ratio of the gate width to the fluorescence lifetime. For a lifetime of 3.5 ns and a gate width of 200 ps, the efficiency is only 5.7%. The F value is 4.18. The low efficiency must be compensated by a longer acquisition time, which gives correspondingly more photobleaching in the sample.

The counting efficiency can be improved by using a large gate width and measuring the fluorescence with only two gate delays (see Fig. 21.10, right). Single-exponential

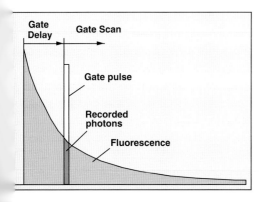
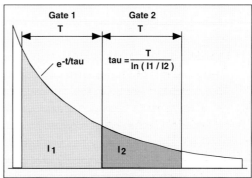

Figure 21.10. Left: Recovery of the full decay profile by scanning a narrow gate. Right: Measurement of a single-exponential approximation by recording images in two gate intervals of different delay.

decay constants can be derived from the intensities in only two time windows, analogous to multi-gate photon counting [70]. Since the measurements for the two gates have to be performed one after another, the counting efficiency of such a measurement is close to 0.5 (i.e., by a factor of two less than for multi-gate photon counting).

The consecutive recording of the frames can result in distortion of the decay profiles by photobleaching. The influence of photobleaching can be roughly determined by recording an additional image in the first time gate at the end of the measurement. The total amount of photobleaching is then determined from the two images, and a correction is applied to the data. It should be noted, however, that this procedure corrects only artifacts due to the change of the intensity during acquisition, not the change of the lifetime distribution in the sample.

The gated image intensifiers clearly have their strength in wide-field microscopy [55,62,63,171,191]. In scanning applications they are inferior to single-point detection techniques. However, gated image intensifiers are currently the only practicable FLIM solution for two-photon multibeam scanning systems [187]. Whether the increased intensity obtained in a multibeam system compensates for the poor efficiency of the gated image intensifier has not been explored yet.

21.6.2 Directly Gated CCD Chips

A directly gated camera based on a gated CCD chip was used in [141]. The principle is analogous to the directly modulated CCD. The off-on switching time is of the order of several nanoseconds, and the delay differences over the chip area are about 1 ns. The maximum repetition rate was 500 kHz. Although the quantum efficiency of a CCD can be up to 80% in the near-infrared range, the effective efficiency is reduced by the gating principle and the readout noise The currently achieved parameters are not sufficient for accurate fluorescence lifetime imaging. If the speed of the gating can be improved, the directly gated CCD may become a rugged, simple, and inexpensive wide-field FLIM detector.

21.6.3 Streak Camera Systems

A streak camera is based on an electron tube with a photocathode at the input, an electron-optical imaging and deflection system, and a fluorophore screen at the output (Fig. 21.11). Without deflection, photoelectrons from a linear pattern at the input are imaged into a linear pattern at the output. Time resolution is obtained by applying a fast voltage ramp to the deflection system. The output pattern is recorded by a CCD camera chip.

Modern streak cameras contain a multichannel plate for electron multiplication and reach a sensitivity down to the single-photon level. The deflection voltage can be either a linear ramp or a sinewave. Deflection by a ramp has the drawback of limited time resolution or limited repetition rate. Sinewave deflection reaches picosecond resolution and high repetition rates but is nonlinear and exploits only a part of the signal period.

A FLIM system based on a streak camera is described in [115,116]. The streak camera is combined with a two-photon laser scanning microscope by projecting the

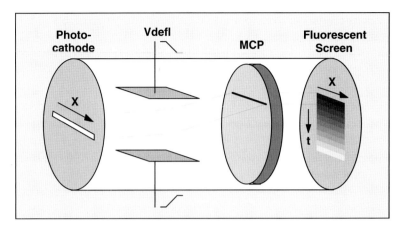

Figure 21.11. Principle of a streak camera.

individual lines of the scan on the input of the camera. This requires a special scanner that descans the image only in the Y-direction. The time-resolved intensity pattern of the whole line is accumulated in the CCD chip and read out during the beam flyback. The system has a temporal resolution of 50 ps. The fluorescence decay is recorded into a sufficiently large number of time bins so that multi-exponential decay functions can be resolved. For high MCP gain the counting efficiency can be expected close to one. Compared to photon counting techniques, the efficiency must be expected to be slightly lower because the result is not free of gain noise. Due to limitations of the trigger and deflection electronics, the instrument described works at a laser repetition rate of only 1 MHz. It is not clear whether the low repetition rate and the correspondingly high peak power cause saturation problems or increased photodamage by three-photon effects. Because the fluorescence light must be descanned and concentrated on the input slit of the camera, the optical efficiency must be expected to degrade for deep tissue layers. The general problem of camera systems (i.e., loss of contrast by scattering of fluorescence light from deep tissue images) is also present in the streak camera.

21.7 PHOTON COUNTING TECHNIQUES

Photon counting techniques consider the detector signal as a random sequence of pulses corresponding to the detection of the individual photons. The output signal of a PMT at a detection rate of 10^7 photons per second is shown in Figure 21.12, left. The pulse shape of the single-photon pulses is shown at right.

For the R5600-1 PMT, the individual pulses have a duration of less than 2 ns. Unless the photon rate is higher than 10^8 s^{-1} the signal is indeed a random sequence of photon pulses. The light intensity is represented by the density of the pulses, not by the amplitude. Obviously, the intensity of the signal is obtained best by counting the pulses in subsequent time channels.

Photon counting has a number of advantages over analog recording techniques. Figure 21.12 shows that the single photon pulses have a considerable amplitude jitter.

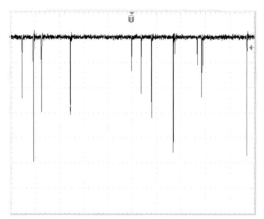 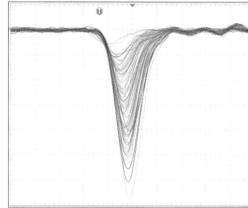

Figure 21.12. Output signal of a PMT at a photon detection rate of $10^7 \, s^{-1}$ (left, 100 ns/div), and single-photon pulses (right, 1 ns/div). R5600-1 PMT at $-900 \, V$, signal line terminated with 50 Ω.

The amplitude jitter, or gain noise, is a result of the random amplification process in the PMT. It is present in all high-gain detectors. In analog techniques the gain noise contributes to the noise of the measurement. Photon counting techniques are free of gain noise. Also, electronic noise does not contribute to the result as long as its amplitude is smaller than that of the photon pulses. Photon counting techniques therefore obtain a shot-noise-limited SNR down to the background count rate of the detector.

Another unique feature of photon counting results from the fact that the arrival time of a photon pulse can be determined with high precision. The bandwidth of a photon counting experiment is limited only by the transit time spread of the pulses in the detector, not by the width of the photon pulses. The transit-time distribution is usually an order of magnitude narrower than the width of the single-photon pulses. For the same detector, photon counting therefore obtains a significantly higher time resolution than any analog recording technique.

21.7.1 Multi-Gate Photon Counting

Multi-gate photon counting was primarily developed to overcome the count rate limitations of early TCSPC devices. The technique counts the photon pulses of a high-speed detector directly in several parallel gated counters. The gates are controlled via separate gate delays and by separate gate pulse generators (Fig. 21.13). If the measured decay curve is completely covered by consecutive gate intervals, all detected photons are counted. The figure of merit and the counting efficiency are then close to one. The counters can be made very fast. In principle, the count rates are only limited by the detector. With the commonly used PMTs, peak count rates around 100 MHz and average count rates of several 10 MHz are achieved.

The multi-gate technique has become one of the standard techniques of FLIM in laser scanning microscopes [64,80,81,170,201]. The practical implementation of the technique is described in [43,190].

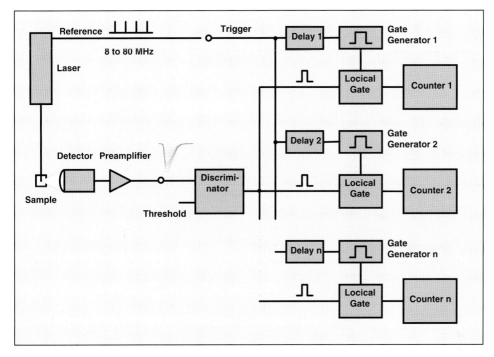

Figure 21.13. Gated photon counting within several parallel time gates.

Limitations of the multi-gate technique result from the relatively long gate duration (practically >500 ps) and from the limited number of gate and counter channels (2–8). From the point of view of signal theory, the signal waveform is heavily "undersampled." Undersampled signals cannot be reconstructed from the sample values without presumptions about the signal shape. Fortunately, fluorescence decay curves are either single exponentials or a weighted sum of a few exponentials. Ideally, the lifetime of a single exponential decay can be calculated from the photons collected in only two time windows (Fig. 21.14). The lifetime components of multi-exponential decay functions can be determined if the number of gates is increased [43,82,190].

In practice the detector signal, or the pulse density versus time, is the convolution of the fluorescence decay function with the instrument response function (IRF). The IRF is the convolution of the excitation pulse shape, the transit time spread of the detector, the pulse dispersion in the optical system, and the on-off transition time of the gating circuit. For fluorescence lifetimes shorter than 1 ns, the effective waveform cannot be considered a sum of exponentials (see Fig. 21.14). To calculate the fluorescence lifetime from the intensity in two time windows, these must be placed after the IRF where the recorded signal is exponential. This, however, discards the photons in the most intense part of the signal. Although lifetimes down to 70 ps have been measured, the efficiency decreases rapidly for lifetimes below 500 ps [82]. It is not clear whether double-exponential decay functions of FRET systems with a fast component of 100 to 300 ps can be reliably resolved. The conclusion is that multi-gate photon counting is a highly efficient, fast recording technique of moderate time resolution. The benefits of the technique can be exploited especially for extremely bright samples with fluorescence decay times in the ns range.

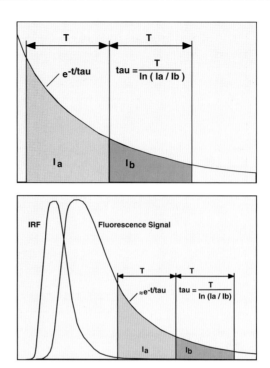

Figure 21.14. Above: Calculation of fluorescence lifetime from intensities in two time intervals, ideal IRF. Below: Calculation of fluorescence lifetime from intensities in two time intervals, real IRF.

21.7.2 Multidimensional Time-Correlated Single Photon Counting

21.7.2.1 Principle of Time-Correlated Single Photon Counting

The TCSPC technique makes use of the fact that for low-level, high-repetition-rate signals, the light intensity is so low that the probability of detecting one photon in one signal period is far less than one. Therefore, the detection of several photons in one signal period can be neglected. It is then sufficient to record the photons, measure their time in the signal period, and build up a histogram of the photon times [11,56,102,103,125,126,127,144,147,175]. The principle of classic time-correlated photon counting is shown in Figure 21.15.

The detector signal is a train of randomly distributed pulses corresponding to the detection of the individual photons. There are many signal periods without photons; other signal periods contain a single photon. Periods with more than one photon are very rare. When a photon is detected, the time of the corresponding detector pulse in the signal period is measured. The events are collected in a memory by adding a "1" in a memory location with an address proportional to the detection time. After many photons in the memory, the distribution of the detection times (i.e., the waveform of the optical pulse) builds up.

Although this principle looks complicated at first glance, TCSPC records light signals with an amazingly high time resolution and a near-ideal efficiency. As mentioned above,

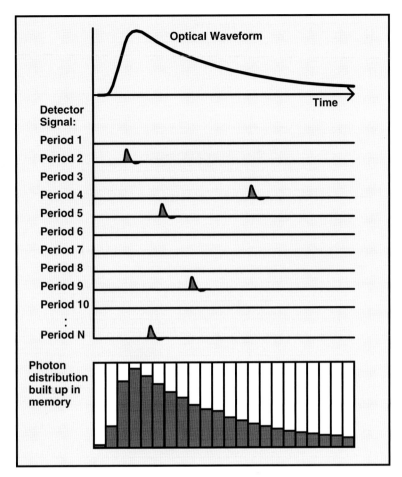

Figure 21.15. General principle of time-correlated single photon counting. The arrival times of the individual photon pulses are measured, and the photon distribution over the arrival time is built up.

the time resolution is limited only by the transit time spread of the detector. With multichannel plate PMTs, a width of the instrument response function shorter than 30 ps is achieved. The drawback of classic TCSPC devices was the limited speed of the nuclear instrumentation modules (NIM) used for signal processing. The slow acquisition can be considered a feature of the early instruments. A more severe drawback is, however, that the principle shown in Figure 21.15 is intrinsically one-dimensional: it delivers only the intensity versus time. Its application to laser scanning systems therefore requires recording a full fluorescence decay curve in one pixel, reading out the data, and then proceeding to the next pixel. Such systems have indeed been used for FLIM [40,41] but were restricted to slow scanning and low count rates.

A new generation of TCSPC devices abandoned the NIM technique entirely and integrated all building blocks an a single printed circuit board. The electronic system was optimized as a whole, resulting in time-shared operation of time-to-amplitude converter (TAC), analog-to-digital converter (ADC), and memory access. Together with new time-to-digital conversion principles, the count rate of TCSPC was increased by two orders of magnitude. Moreover, advanced TCSPC devices use a multidimensional histogramming process. They record the photon density as a function not only of the time in the signal period, but also of other parameters, such as the wavelength, spatial coordinates, location within a scanning area, time from the start of the experiment, or other externally measured variables [24].

21.7.2.2 Imaging by Multidimensional TCSPC

Figure 21.16 shows how multidimensional TCSPC is used for fluorescence lifetime imaging in laser scanning microscopes. At the input of the detection system are a number of PMTs, typically detecting the fluorescence signal in different wavelength intervals. As mentioned above, TCSPC is based on the presumption that the detection of more than one photon per laser pulse period in a single detector is unlikely. Under this condition, the detection of several photons in different detectors is unlikely as well. The single photon pulses of all detectors can therefore be combined into a common timing pulse line and sent through a single time-measurement channel. To identify the origin of the photon pulses, a "router" delivers a digital expression of the number of the PMT in which the photon was detected. The number of the detector is stored in the "channel" register of the TCSPC device.

The times of the photons in the laser pulse period are measured in the time-measurement channel of the TCSPC device. A constant-fraction discriminator removes the amplitude jitter from the pulses. The time is converted into a data word via a TAC and a fast ADC. The principle of the time-measurement channel is identical to classic TCSPC. However, new conversion principles have increased the maximum count rate by two orders of magnitude [12,15,24].

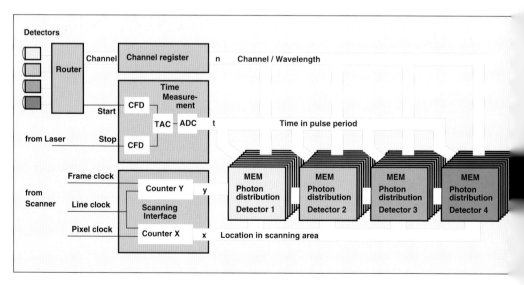

Figure 21.16. Principle of lifetime imaging by multidimensional TCSPC.
CFD = constant fraction discriminator, TAC = time-to-amplitude converter, ADC = analog-to-digital converter, MEM = memory.

The third building block of the TCSPC device is the scanning interface. The scanning interface receives the scan clock pulses of the scanner in the microscope. For each photon, it delivers the location of the laser beam in the scanning area, x and y.

The channel number, n, the time in the laser period, t, and the coordinates of the laser spot, x and y, are used to address a histogram memory in which the detection events are accumulated. Thus, in the memory the distribution of the photon density over X, Y, t, and n builds up. As described above for classic TCSPC, the width of the temporal instrument response function is determined mainly by the transit time spread of the detectors. With MCP PMTs an IRF width of less than 30 ps (full width at half-maximum) is obtained [18,19]. With an IRF this short, lifetimes down to 50 ps can be measured with almost ideal efficiency. Lifetimes down to 10 ps can certainly be measured but are observed only for fluorophores of low quantum efficiency. Their detection is therefore restricted to a few special cases [123].

The data acquisition can be run at any scanning speed of the microscope. Under typical conditions, the pixel rate is higher than the photon count rate. This makes the recording process more or less random. The acquisition process is controlled by the scanner. Therefore, no changes in the microscope hardware or software are required. The regular zoom and image rotation functions of the microscope can be used in the normal way. As many frame scans as necessary to obtain an appropriate SNR can be accumulated.

It should be pointed out that the recording process does not use any time gating, wavelength scanning, or detector multiplexing. Under reasonable operating conditions all detected photons contribute to the result, and a maximum SNR for a given fluorescence intensity and acquisition time is obtained. For all detector channels the counting efficiency, E, and the figure of merit, F, are close to one.

In practice the signal processing of a recorded photon causes a "dead time" during which the TCSPC electronics cannot process another photon. As long as the photon detection rate is small compared to the reciprocal dead time of the TCSPC module, the counting efficiency remains close to one. For a detector count rate of the reciprocal dead time the efficiency is 0.5, and the figure of merit 1.4, which is still better than for most other FLIM techniques. Currently the fastest TCSPC devices have a dead time of 100 ns and can be reasonably used up to 10-MHz detector count rate.

Multidimensional TCSPC has become a standard FLIM technique of laser scanning microscopy. TCSPC FLIM has been used for the Biorad MRC 600 [71], MRC 1024 [2,3,145] and Radiance 2000 [5,6,38,51], the Zeiss LSM 410 [13,14,169] and LSM 510 [17,18,20,33,65], the Olympus FV 300 and FV 1000, the Nikon PCM 2000, and a number of specialized or homemade two-photon scanning systems [83,95,111,198]. TCSPC FLIM is a standard option of the Leica SP2 and Zeiss LSM 510 microscopes.

The multidetector technique can be extended to 16 or 32 channels by using a multi-anode PMT with the associated routing electronics. Multispectral TCSPC FLIM based on these detectors is described in [16,31].

21.7.2.3 Sequential Recording

The scanning interface used in TCSPC scanning applications is actually a special configuration of a more general building block, called a sequencer. The sequencer is used to switch through a sequence of memory addresses by a predefined pattern. In case of scanning, the sequencer is controlled by the clock pulses of the scanner and configured to generate the x and y coordinates. The sequencer can, however, also be used to deliver a time-controlled sequence of memory addresses. Starting with an external trigger pulse, the TCSPC device then records a fast sequence of small images or single-point measurements [21,24]. The sequencing technique is also termed "double kinetic mode" or "time lapse recording." It can be combined with a memory-swapping architecture to record virtually unlimited sequences of waveforms [12]. Sequential recording has not found much application in microscopy yet. It is, however, routinely used for functional brain imaging by diffuse optical tomography techniques [128,129,130].

21.7.2.4 TCSPC Time-Tag Recording

A variation of the TCSPC technique does not build up photon distributions but stores information about each individual photon. The mode is called "time tag," "time stamp," "list," or "FIFO" mode. For each photon, the time in the signal period, the "channel" word, and the time from the start of the experiment, or "macro time," is stored in a first-in-first-out (FIFO) buffer [24]. During the measurement, the FIFO is continuously read, and the photon data are stored in the main memory or on the hard disc of a computer. The system architecture in the time-tag mode is shown in Figure 21.17. It contains the channel register, the time-measurement block, a "macro-time" clock, and the FIFO buffer for a large number of photons. The structure has some similarity to the multidimensional TCSPC described in the paragraphs above. In fact, many advanced TCSPC modules have both the histogram and the time-tag mode implemented, and the configuration can be changed by a software command [10,24]. The sequencer then turns into the macro-time clock, and the memory into the FIFO buffer.

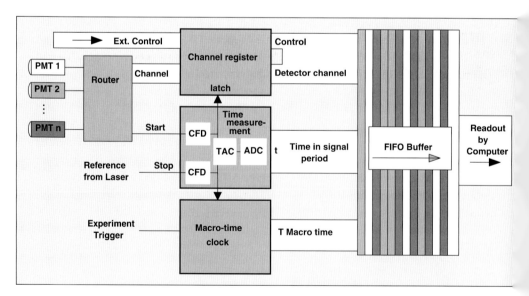

Figure 21.17. Architecture of a TCSPC module in the FIFO mode. For each individual photon, the detector channel number, the time in the signal period, and the "macro time" from the start of the experiment are stored. CFD = constant fraction discriminator, TAC = time-to-amplitude converter, ADC = analog-to-digital converter.

When a photon is detected the "micro-time" in the signal period is measured by the time-measurement block. Simultaneously, the information at the inputs of the "channel" register is latched. It contains the detector channel number for the current photon, and often a number of additional bits from external experiment control devices. The "macro-time" clock delivers the time of the photon from the start of the experiment. All these numbers are written into the FIFO.

The output of the FIFO is continuously read by the computer. Consequently, the time-tag mode delivers a continuous, and virtually unlimited, stream of photon data. It is, of course, imperative that the computer reads the photon data at a rate higher than the average photon count rate. However, modern operation systems are multitask systems, and it is unlikely that the computer reads the FIFO continuously. Moreover, in typical applications bursts of photons appear on a background of a relatively moderate count rate. Therefore, the FIFO has to be large enough to buffer the photon data for a sufficient time. In practice FIFO sizes of 64,000 to 8 million photons are used.

In principle, many multidimensional recording problems can also be solved in the time-tag mode. Synchronization with the experiment can be accomplished via the experiment trigger, the macro-time clock, and additional experiment control bits read into the "channel" register. The drawback of the time-tag mode is the large amount of data that has to be transferred into the computer and processed or stored. A single photon typically consumes four or six bytes. A single FLIM recording may deliver 10^8 to 10^9 photons, resulting in several gigabytes of data. At high count rates the bus transfer rate into the computer may be still sufficient, but the computer may be unable to process the data on-line or to write them to the hard disc. The transfer rate problem is even more severe for systems containing several TCSPC channels in one computer. Nevertheless, the time-tag mode is sometimes used for imaging and for standard FLIM experiments. A slow-scan system based on time-tag TCSPC and a piezo stage is described in [36]. TCSPC in the time-tag mode in conjunction with fast scanning was used in [39,83]. This is not objectionable as long as possible count rate limitations by the bus transfer rate, and the enormous file sizes and possible synchronization problems are taken into regard.

The time-tag mode of TCSPC is a powerful technique of fluorescence correlation spectroscopy (FCS) [29,74,133,142,164], fluorescence intensity distribution analysis (FIDA), and time-resolved single molecule spectroscopy by burst-integrated fluorescence lifetime (BIFL) detection [67,158,159]. Examples are given in the applications part of this chapter.

The time-tag mode in conjunction with multidetector capability and MHz counting capability was introduced in 1996 with the SPC-431 and SPC-432 modules of Becker & Hickl. Despite its large potential in single molecule spectroscopy, the mode did not attract much attention until sufficiently fast computers with large memories and hard disks became available.

21.8 APPLICATIONS OF FLIM IN BIOLOGY

21.8.1 Discrimination of DNA and RNA by the Lifetime of SYTO13

The probe SYTO13 binds specifically to DNA and RNA. Although there is no spectral contrast for binding to DNA or RNA, the lifetimes in both cases have been found to be different [201]. The images shown below were recorded in a lifetime microscope described in [43,190]. The instrument uses two-photon excitation by a titanium-sapphire laser of 82-MHz repetition rate. The wavelength for the experiments described below was 800 nm, the pulse width 82 fs. The fluorescence light is detected by an R1894 PMT. The single-photon pulses are recorded in four time channels by multi-gate photon counting [43,80,81,82,190]. For each pixel of the scan, a single-exponential fit is applied to the photon numbers in the four time channels.

Figure 21.18 shows a live, healthy Chinese hamster ovary (CHO) cell stained with SYTO13. Figure 21.18A is an intensity image and shows the distribution of SYTO13 in the cell. Figure 21.18B is the corresponding lifetime image.

Lifetimes of 3.4 ± 0.2 ns and 4.1 ± 0.2 ns are found for the nucleus and small spots in the cytoplasm. The nucleoli have a lifetime of 3.8 ± 0.2 ns. The 3.4-ns component is attributed to SYTO13 bound to DNA, the 4.1-ns component to SYTO13 bound to RNA. The nucleoli contain both DNA and RNA, which explains the lifetime of 3.8 ns.

Figure 21.19 shows cells in a early (A, B) and a late (C, D) phase of apoptosis. A and C are the intensity images. The lifetime images (B, D) show clusters of SYTO13 with a lifetime typical for binding to DNA (3.4 ns). The lifetime in the nucleus shortens with the progress of apoptosis. Simultaneously, the lifetime in the clusters in the cytoplasm shortens.

21.8.2 Two-Photon Imaging of the pH in Skin

The surface of mammalian skin is acidic, with a pH between 4.5 and 6. Within the stratum corneum the pH increases; it reaches neutrality at the stratum spinosum, about 10 μm below the surface. Microscopic imaging of the pH in the stratum corneum helps us to understand the barrier function of the stratum corneum and its effect on skin diseases and drug delivery.

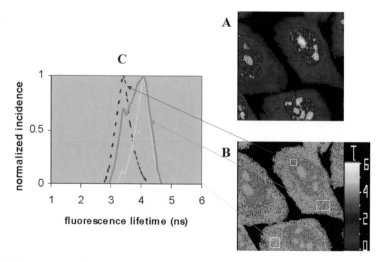

Figure 21.18. CHO cell stained with SYTO13. (A) Intensity image. (B) Lifetime image. (C) Distribution of the lifetime in the pixels of selected regions. (Images after [201], courtesy of Marc van Zandvoort, University Maastricht)

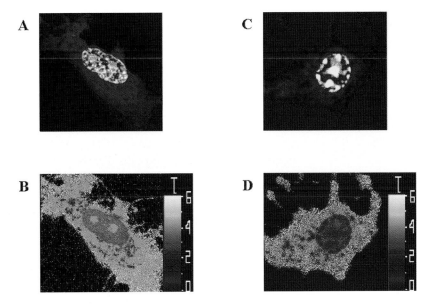

Figure 21.19. Images of CHO cells in early (A, B) and later (C, D) stage of apoptosis. (A, C) Intensity images. (B, D) Lifetime images. (Images after [201], courtesy of Marc van Zandvoort, University Maastricht)

Microscopic pH imaging can be achieved by staining the skin with a pH-sensitive fluorescent probe. These probes usually have a protonated and a de-protonated form. There is an equilibrium between both forms that depends on the pH of the local environment. If both forms have different fluorescence lifetimes, the average lifetime is a direct indicator of the pH [121,122,131]. A representative of the pH-sensitive dyes is 2′,7′-bis-(2-carboxyethyl)-5-(and-6)-carboxyfluorescein [89] (BCECF, Fig. 21.20, left). In aqueous solution the lifetimes of the protonated form and the deprotonated form are 2.75 ns and 3.90 ns, respectively [88]. In the pH range from 4.5 to 8.5 both forms exist, and the fluorescence decay function is a mixture of both decay components. The dependence of the average lifetime on the pH is shown in Figure 21.20, right.

The technique of FLIM-based pH imaging by BCECF is described in [88]. Intensity and lifetime images were obtained by two-photon excitation in a laser scanning microscope. The fluorescence light was collected by direct (non-descanned) detection. The typical photon detection rate was $2 \cdot 10^6 s^{-1}$. The frequency-domain technique shown in Figure 21.8 was used for lifetime imaging. To obtain quantitative pH values, the dependence of the natural lifetime on the refractive index was taken into account. Figure 21.21 shows images of mouse skin stained with BCECF. The upper row is from the surface of the skin, the lower row 5.1 μm deep. Intensity images are shown at the left, lifetime images in the middle, pH images at the right.

21.8.3 Two-Photon Autofluorescence FLIM of Tissue

A lifetime image of the autofluorescence of human skin is shown in Figure 21.22. The image was recorded by multidimensional TCSPC (SPC-830 module, Becker & Hickl)

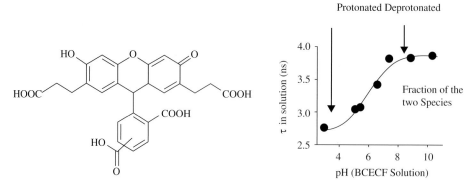

Figure 21.20. Left: Structural formula of BCECF. Right: Dependence of the average lifetime on the pH. (Courtesy of E. Gratton, University of Illinois, Urbana Champaign)

in a Zeiss LSM 410 laser scanning microscope. A Ti:Sapphire laser (MaiTai, Spectra Physics) was used for two-photon excitation of the sample.

Figure 21.22 shows an intensity image calculated from the photons of all time channels of the pixels, and a lifetime image showing the intensity as brightness and the single-exponential approximation of the lifetime as color. The lifetime distribution over the pixels is shown in the upper right corner. The lower part of the panel shows the fluorescence decay curve in a selected pixel. The fluorescence decay curve is clearly multi-exponential. A double-exponential fit delivers 79% of 188 ps and 21% of 2.2 ns. Approximating the decay functions by a single-exponential decay certainly discards useful information. Figure 21.23 shows images of the decay time of the fast component, t1, the slow component, t2, the ratio of both decay times, and the ratio of the corresponding intensity coefficients, a1/a2. The parameter range indicated in the images is displayed by a color scale from blue to red. The upper row shows the stratum corneum, 5 μm deep, the lower row the stratum spinosum, 50 mm deep.

It is worthwhile to note that t1, t2, and a1/a2 are independent intrinsic parameters of the fluorescence decay. A FLIM technique that resolves these parameters therefore adds three new dimensions to multidimensional microscopy. Figure 21.23 indicates that t1, t2, and a1/a2 show considerable variations throughout the image. It can therefore be expected that these parameters indeed contain additional information compared with single-exponential FLIM. However, the biological interpretation of the variations remains the subject of investigation. More applications of TCSPC FLIM to autofluorescence of tissue are described in [110,111,112,163].

21.8.4 FRET Experiments in Live Cells

A general problem of FRET experiments in cells is that by far not all donor molecules interact with an acceptor molecule. There are two reasons why a donor molecule may not interact. The first is that the dipoles of the donor and acceptor molecules are more or less randomly oriented. The corresponding heterogeneity of the FRET efficiency can be taken into account in distance calculations [122]. A more severe problem is that there is an unknown fraction of donor molecules that is not linked to an acceptor molecule. Surprisingly, this "underlabeling" problem [122] is rarely

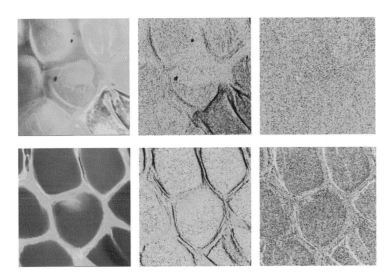

Figure 21.21. Two-photon images of mouse skin stained with BCECF. Left to right: Intensity images, lifetime images, and pH images. Lifetime range: blue to red = 2 to 5 ns. pH range: blue to red = 4 to 8. Upper row 0 μm deep, lower row 5.1 μm deep. Image size 40 μm × 40 μm. (Courtesy of E. Gratton, University of Illinois, Urbana Champaign)

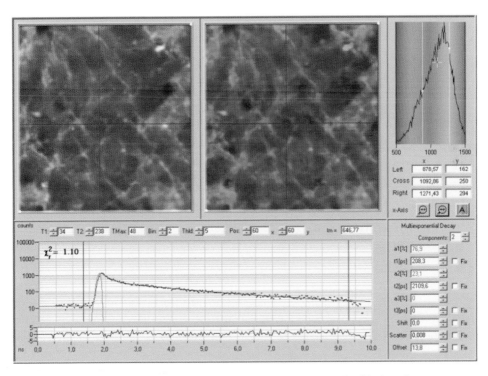

Figure 21.22. Autofluorescence of human skin. Intensity image, lifetime image, lifetime distribution, and fluorescence decay curve in selected pixel. (Courtesy of K. König and I. Riemann, Fraunhofer Institute for Biomedical Engineering, St. Ingbert, Germany)

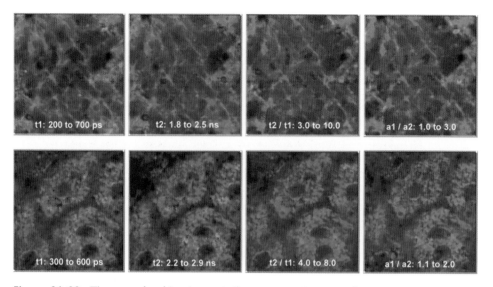

Figure 21.23. Time-resolved in vivo autofluorescence images of stratum corneum (upper row, 5 μm deep), and stratum spinosum (lower row, 50 mm deep). Left to right: fast lifetime component (t1), slow lifetime component (t2), ratio of lifetime components (t2/t1), and ratio of intensity coefficients (a1/a2). Indicated parameter range corresponds to color range blue to red. (Data courtesy of Karsten König and Iris Riemann, Fraunhofer Institute for Biomedical Engineering, St. Ingbert, Germany)

mentioned in the FRET literature. The effect on the donor decay function is shown in Figure 21.24.

The decay curves have a component from non-interacting and interacting donor molecules. The resultant decay function can be approximated by a double-exponential model. The decay analysis delivers the lifetimes of the quenched and unquenched donor molecules, τ_{fret} and τ_0, and the intensity coefficients, a_1 and a_2. From these parameters the FRET efficiency, E, the ratio of the distance and the Förster radius, r/r_0, and the ratio of the number of interacting and non-interacting donor molecules, N_{fret}/N_0, can be obtained:

$$E = 1 - \frac{\tau_{fret}}{\tau_0} \qquad (r/r_0)^6 = \frac{\tau_{fret}}{(\tau_0 - \tau_{fret})} \qquad \frac{N_{fret}}{N_0} = \frac{a_1}{a_2}$$

The predicted double-exponential decay behavior is indeed found in TCSPC FRET [5,13,18,19,32,65] and streak camera experiments [34]. The finding has implications to distance calculations based on single-exponential FLIM-FRET [70,91,154,202], and possibly even on steady-state FRET techniques. Obviously, the distance between the donor and acceptor molecules has to be calculated from τ_{fret}, not from the average or "apparent" lifetime.

Figure 21.25 shows FLIM results for a live human embryonic kidney (HEK) cell expressing two interacting proteins labeled with CFP and YFP. The images were recorded by dual-detector multidimensional TCSPC. The microscope was a Zeiss LSM 510 NLO two-photon laser scanning microscope in the Axiovert 200 version. An excitation wavelength of 860 nm was used. The non-descanned fluorescence signal

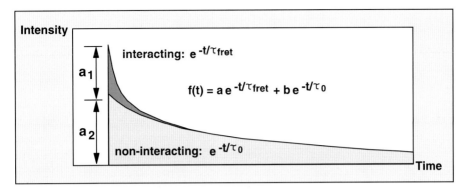

Figure 21.24. Composition of the decay curves found in FRET systems. The decay curve has a component from interacting and non-interacting donor molecules. The resulting decay function can be approximated by a double-exponential model with the lifetimes τ_{fret} and τ_0 and the intensity coefficients a_1 and a_2.

from the sample was fed out of the rear port of the Axiovert. A dual detector assembly with a dichroic beamsplitter and two Hamamatsu R3809U MCPs was attached to this port [18]. BG39 laser blocking filters and bandpass filters were inserted directly in front of the detectors. For all measurements shown below, a 510-nm dichroic mirror and bandpass filters with 480 ± 15 nm and 535 ± 13 nm transmission wavelength were used. The filters were selected to detect the fluorescence of the CFP and the YFP, respectively.

The images shown in Figure 21.25, left and right, are the lifetime images recorded simultaneously in the CFP (donor) and YFP (acceptor) channel. The color represents the average (amplitude-weighted) lifetime obtained from a double-exponential fit in the individual pixels. The average donor lifetime, tm, at the cursor position is 1.49 ns. It varies from about 1.4 ns in the region of strong FRET to about 1.9 ns in regions of weak FRET.

The YFP intensity is highest in the regions where the CFP lifetime is shortest. This is a strong indication that indeed FRET occurs between CFP and YFP. The decay in the acceptor channel is a mixture of the FRET-excited acceptor fluorescence and about 50% bleedthrough of the donor fluorescence. Because the lifetime of the YFP is longer, regions of strong FRET show an increased average lifetime. In principle, the acceptor decay could be used to derive the ratio of the directly and FRET-excited acceptor fractions, and the energy transfer rate. Whether the acceptor fluorescence can be used to improve the reliability of FRET experiments is an open question [93].

Figure 21.26 shows the donor decay function in the selected spot. The fluorescence decay is indeed double-exponential, with a fast lifetime component, t_1, of 0.59 ns and a slow component, t_2, of about 2.4 ns. Assuming that t_1 is the lifetime of the interacting donor, τ_{fret}, and t_2 the lifetime of the non-interacting donor, τ_0, a FRET efficiency of the quenched donor fraction $E = 0.754$, and a distance ratio of $(r/r_0)^6 = 0.32$ are obtained. Using the average lifetime, tm, instead of t_1 leads to $E = 0.38$, and $(r/r_0)^6 = 1.64$.

Figure 21.27 shows an image of the intensity coefficients of the fast and slow lifetime component, a_1/a_2, and an image of the ratio of the lifetime components, t_2/t_1. As shown in Figure 21.24, a_1/a_2 represents the ratio of the numbers of interacting and

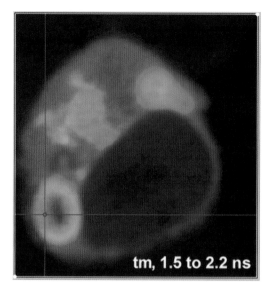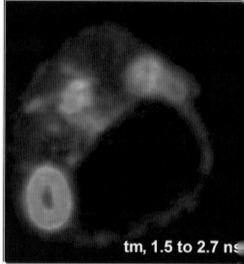

Figure 21.25. HEK cell expressing two interacting proteins labeled with CFP and YFP. Left: Lifetime image of the CFP channel. Right: Lifetime image of the YFP channel. The color represents the average (amplitude-weighted) lifetime of a double-exponential fit. The indicated lifetime range corresponds to a color scale from blue to red.

non-interacting donor molecules, N_{fret}/N_0. The ratio of the lifetimes is related to the FRET efficiency, E_{fret}. Figure 21.27 shows that the change in a_1/a_2 is considerably larger than the change in t_2/t_1 (please note the different color scales). Consequently, a large part of the variation in the average lifetime (see Fig. 21.25, left) is caused by a variable fraction of interacting donor molecules, not by a change in the distance.

A general characterization of FRET by two-photon excited TCSPC-FLIM is given in [51,52]. Applications to protein interaction related to Alzheimer's disease are described in [5,6,25,26]. Conformational changes induced by anti-inflammatory drugs are found in [132]. Interactions between the PCK and NKκB signaling pathways have been investigated in [145]. FRET between GFP and RFP and FRET cascades from GFP via Cy3 into Cy5 are demonstrated in [2]. The agglutination of red blood cells by monoclonal antibodies was studied using FRET between Alexa 488 and DiI [168]. Interactions of the neuronal PDZ protein PSD-95 with the potassium channels and SHP-1-target interaction were studied in [32,33]. A detailed description of a TCSPC-FLIM-FRET system is given in [65]. The system is used for FRET between ECPF-EYFP and FM1-43-FM4-64 in cultured neurons. FRET between ECFP and EYFP in plant cells was demonstrated in [38]. FRET measurements in plant cells are difficult because of the strong autofluorescence of the plant tissue. It could be shown that two-photon excitation can be used to keep the autofluorescence signal at a tolerable level.

21.8.5 High-Speed TCSPC FLIM

The TCSPC used for the experiments described above works up to a detector count rate of $10^7 \, s^{-1}$. At this rate 50% of the photons are lost in the dead time, resulting in a recorded rate of $5 \cdot 10^6 \, s^{-1}$. A count rate this high is rarely obtained

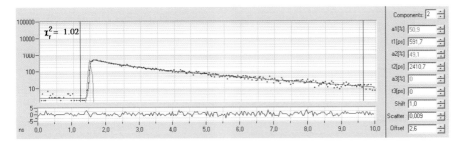

Figure 21.26. Donor fluorescence decay at the cursor position (see Figure 21.25, left). The decay is double-exponential, with 51% of 0.59 ns and 49% of 2.41 ns.

from biological specimen. The count rate for pH imaging by BCECF in skin tissue was $2 \cdot 10^6$ s^{-1}[88]. CFP-YFP FRET in *Caenorhabditis elegans* was recorded at $<10^5$ s^{-1}[39]. The CFP-YFP FRET images shown above were recorded at $50 \cdot 10^3$ s^{-1}. Two-photon autofluorescence of skin (see Fig. 21.21) delivered a rate of $60 \cdot 10^3$ s^{-1}. In most cases, higher count rates could easily be obtained by increasing the laser power. However, increasing the laser power caused noticeable photobleaching and lifetime changes.

Nevertheless, higher count rates may be obtained from samples with high fluorophore concentration, such as histological samples. If count rates in excess of 10^7 s^{-1} are available and short acquisition time is an issue, a multimodule TCSPC system can be used. The light from the sample is split into several detection channels fed to separate PMTs. Each PMT is connected to one channel of a multimodule TCSPC system as described in [20,22]. The TCSPC channels of this system have 100-ns dead time. The system can be used at a total detector count rate of $40 \cdot 10^6$ s^{-1}, or a recorded count rate of $20 \cdot 10^6$ s^{-1}. A typical result is shown in Figure 21.28.

The sample was a 16-μm cryostat section of a mouse kidney (Molecular Probes, F-24630) stained with Alexa Fluor 488 WGA and Alexa Fluor 568 phalloidin. The excitation wavelength was 860 nm. The laser power was about 400 mW at the input of the microscope, and a $100 \times$ NA $= 1.3$ lens was used. Figure 21.28, left, shows a lifetime image of the combined photon data of all four channels recorded within 10 seconds. A double-exponential Levenberg-Marquardt fit was applied to the data. The color of the image represents a single-exponential approximation of the lifetime obtained by weighting both lifetime components with their relative intensities. Figure 21.28, right, shows an image obtained by using the ratio of the two intensity coefficients as color. Images of this ratio show the concentration ratio of the fluorophores emitting the two lifetime components.

Figure 21.28 shows that TCSPC can be used to obtain high-quality double-exponential lifetime images in 10 seconds or less. Single-exponential images can certainly be obtained in less than 1 second. However, to obtain high count rates, a high excitation power has to be used. Figure 21.29 shows a sequence of recordings obtained from the same specimen as shown in Figure 21.28. The upper row shows a series of intensity images, which were calculated from subsequent measurements by summing up the fluorescence signals of the four detectors and of all time channels in each pixel. The acquisition time of each image was 20 seconds. The initial total count rate was $14 \cdot 10^6$ s^{-1}. The lower row shows the distribution of the mean lifetime over the image.

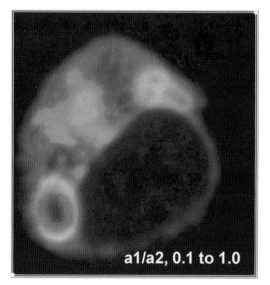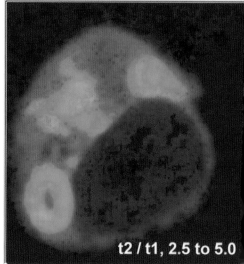

a1/a2, 0.1 to 1.0

t2 / t1, 2.5 to 5.0

Figure 21.27. Ratio of intensity coefficients, a1/a2 (left), and ratio of slow and fast lifetime component, t2/t1 (right). Blue to red corresponds to the indicated parameter range.

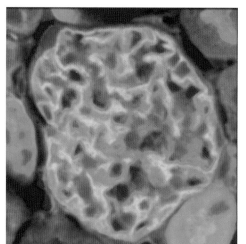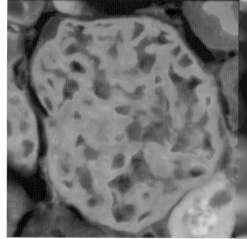

Figure 21.28. Mouse kidney sample stained with Alexa Fluor 488 wheat germ agglutinin and Alexa Fluor 568 phalloidin, recorded by four detectors connected to separate TCSPC channels. Left: Lifetime image; the color represents the amplitude-weighted mean lifetime, blue to red = 0.7 to 1.7 ns. Right: Color represents the intensity coefficients of the fast and slow lifetime component obtained by double-exponential lifetime analysis.

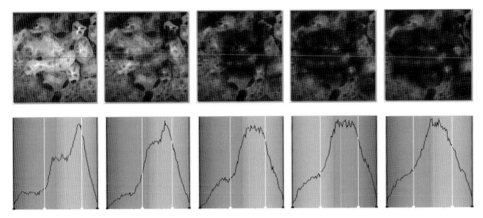

Figure 21.29. Sequence of images recorded for 20 seconds each. The initial count rate was $14 \cdot 10^6$ s−1. Upper row: Intensity images obtained from the recordings in all time channels. Lower row: Distribution of the mean lifetime in the images (from 0.5 to 2.0 ns, normalized on maximum). Photobleaching and thermal effects cause a progressive loss in intensity and a variation in the mean lifetime.

The high excitation power causes photobleaching of the sample. Since not all fluorophores are equally susceptible to photobleaching, the distribution of the mean lifetime changes during the exposure. Usually the longer lifetimes bleach more rapidly, so that the mean lifetime becomes shorter. The results show that photobleaching at high excitation power can bias lifetime measurements considerably.

A potential application of multimodule systems are high-speed multibeam scanning systems [30,42,187]. FLIM systems with 4, 8, or even 16 beams and the same number of parallel TCSPC channels appear feasible. The application of TCSPC to multibeam scanning would require the fluorescence signals from the individual beams to be directed to separate PMTs or separate channels of a multi-anode PMT. If this optical problem was solved, lifetime images could be recorded with unprecedented speed and resolution.

21.8.6 Fluorescence Correlation and Intensity Distribution Analysis

Fluorescence correlation (FCS) and intensity distribution analysis (FIDA) techniques are based on the fluctuations of the fluorescence of a small number of molecules diffusing through a femtoliter volume. Therefore, FCS and FIDA are actually single-point techniques. They are, however, related to scanning microscopy in that the small focal volume is obtained in the focus of a confocal or two-photon microscope. The instruments can, therefore, be used for steady-state imaging, FLIM, and FCS. The combination of these techniques in one instrument makes it possible to obtain the corresponding results from the same spot of the same sample, and under similar conditions. It is, however, imperative that the beam can be parked at a defined position in the sample, and that the position remains stable within a diffraction-limited spot.

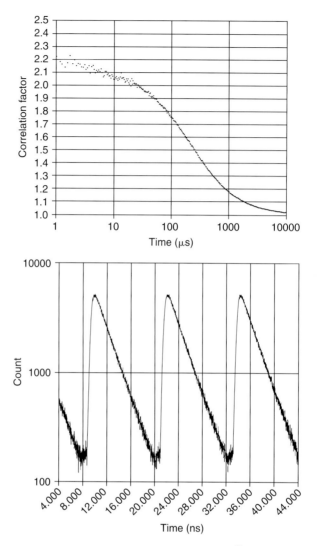

Figure 21.30. FCS curve and fluorescence decay of a 10^{-9} molar GFP solution. Both curves were obtained from a single TCSPC measurement in the time-tag mode. Average count rate $6,000\,s^{-1}$. (From [23], data courtesy of Petra Schwille and Zdenek Petrasek, Institute of Biophysics, Technical University Dresden)

FCS and FIDA results can then be calculated from time-tag data delivered by multidimensional TCSPC [10,35,203,208]. With a high-repetition-rate pulsed laser, both techniques can be combined with lifetime detection. A result for a 10^{-9} molar GFP solution is shown in Figure 21.30. A Ti:Sapphire laser was used for two-photon excitation. The fluorescence was detected by an SPCM-AQR detector (Perkin Elmer). The photon data were recorded by an SPC-830 TCSPC module (Becker & Hickl).

The photon counting histogram obtained from the same data is shown in Figure 21.31. The curves were calculated for sampling time intervals of 100 μs, 1 ms, and 10 ms. The figure shows a possible problem of the PCH in biological systems. On average, several photons per sampling-time interval must be recorded to obtain a good PCH. On the other hand, the sampling-time interval should be of the order of the diffusion time,

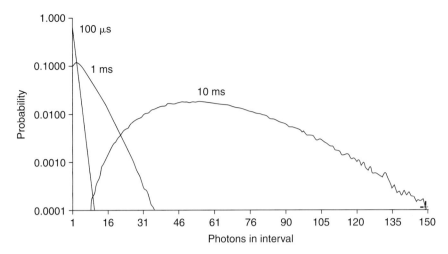

Figure 21.31. Photon counting histograms calculated from the time-tag data of Figure 21.30. Sampling-time interval 0.1 ms, 1 ms, and 10 ms.

or shorter. This requires a relatively high count rate, which is difficult to obtain from living samples.

FCS measurements in cells are more difficult than in solution. Especially in transfected cells, the fluorophore concentration cannot be accurately controlled. It is usually much higher than required for FCS. The number of molecules in the focus can easily be on the order of 100, resulting in an extremely small amplitude of the correlation function. Moreover, there is usually motion in living cells that shows up in the FCS curves at a time scale above 100 ms. Nevertheless, useful FCS data can be obtained. Figure 21.32 shows results recorded by an SPC-830 TCSPC module in a two-photon laser scanning microscope [23]. The FLIM image is shown at left. After the FLIM recording a single-point FIFO measurement was performed in the marked spot. The acquisition time of this measurement was 130 s, at an average count rate of $15 \cdot 10^3 \, \text{s}^{-1}$. A fluorescence decay curve and an FCS curve in a selected spot are shown in the middle and at right. Please note the small amplitude of the FCS curve. The maximum of the correlation coefficient is only 1.016, compared to 2.2 for the GFP solution (see Fig. 21.30).

21.8.7 Spectroscopy of Single Molecules

Multidetector TCSPC can be used to obtain several spectroscopic parameters simultaneously from a single molecule [117,159,194]. The optical setup is described in [159]. A frequency-doubled mode-locked Nd-YAG laser delivers pulses of 532-nm wavelength, 150-fs pulse width, and 64-MHz repetition rate. The laser is coupled into the beam path of a microscope and focused into the sample. The sample is mounted on a piezo-driven sample stage. The fluorescence light from the sample is collected by the microscope lens and separated from the excitation light by a dichroic mirror and a notch filter. The light is split into its 0- and 90-degree components. The 0-degree component is further split into a short-wavelength and a long-wavelength part. The signals are

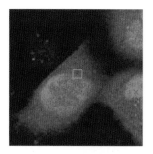

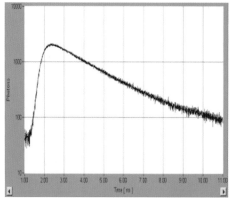 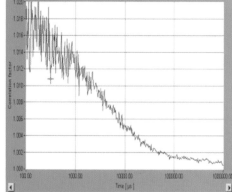

Figure 21.32. FLIM image (top), decay curve (left), and FCS curve (right) recorded in a selected spot (1). Home-made laser scanning microscope, two-photon excitation, SPCM-AQR detector, SPC-830 TCSPC module. Data courtesy of Zdenek Petrasek and Petra Schwille, Biotec TU Dresden.)

detected by single-photon avalanche photodiode (APD) modules and recorded by a single SPC-431 TCSPC module via a router. The TCSPC module records the photons in the FIFO mode.

Dioctadecyltetramethylindocarbocyanine (DiI(C18)) molecules were embedded in polymethylmethacrylate (PMMA). An image of the sample was obtained by scanning and assigning the photons to the individual pixels by their macro time. Based on this image, appropriate molecules are selected for further investigation. These molecules are then brought into the focus, and time-tag data are acquired. From the macro times of the recorded photons, traces of the emission intensity of a single molecule are build up. The traces show bright periods when the molecule cycles between the ground state and the excited singlet state, and dark periods when the molecule is in the triplet state.

Figure 21.33, left, shows a histogram of the number of photons during the bright periods. It represents a distribution of the number of S0-S1 cycles the molecule performs before it crosses into the triplet state. With a reasonable estimate of the detection probability a single-exponential fit of the curve delivers the intersystem crossing rate.

The distribution of the dark periods (see Fig. 21.33, middle) reflects the triplet decay and directly delivers the triplet lifetime. A single-exponential fit delivered a triplet lifetime of $380 \pm 30 \mu$s.

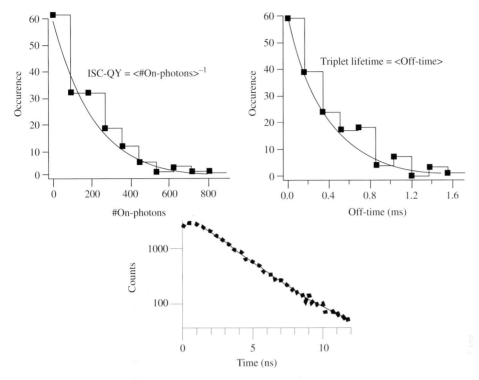

Figure 21.33. Fluorescence dynamics of a single DiI(C18) molecule embedded in a PMMA matrix. Left: Distribution of the number of photons (#On-photons) in the bright periods. With an assumed detection efficiency of 10%, an intersystem crossing rate of $4.9 \cdot 10^{-4}$ is obtained. Middle: Distribution of the duration of the dark periods (Off-time). The triplet lifetime is $380 \pm 30\mu s$. Right: Distribution of the arrival times of the photons over the time in the laser period. The fluorescence decay time is 2.33 ± 0.03 ns. (From [159])

The fluorescence decay function is obtained by building up a histogram of the photon number over the time in the laser period (the micro time) (see Fig. 21.33, right). From the decay data shown a fluorescence lifetime of 2.33 ± 0.03 ns was obtained.

The anisotropy decay can be calculated from the micro times of the photons detected under 0 and 90 degrees of polarization. Figure 21.34 shows that there is some rotational relaxation despite he solid matrix in which the molecules are embedded. A fit delivers a final anisotropy of 0.683 ± 0.003, a cone of wobbling of 12.4 ± 0.3 degrees, and a rotational correlation time of 2.7 ± 0.4 ns.

The instrument described in [159] was used to monitor conformational changes in the citrate carrier CitS and is described in [96]. A version of the instrument uses an annular aperture stop in the beam path of the microscope. With this stop, different Airy disks for different orientation of the dipoles of the molecules are obtained [179,180]. By comparing the observed intensity patterns of the molecules with calculated diffraction patterns, the 3D orientation of the molecules is derived. In [113] the lifetime of DiI molecules in a 20-nm polymer film at a glass surface is investigated by this technique. The lifetime changed from 4.7 ± 0.7 ns to 2.11 ± 0.1 ns, depending on the orientation of the molecules to the surface.

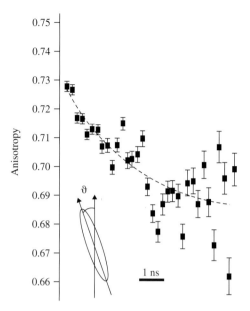

Figure 21.34. Anisotropy decay of a DiI molecule in a PMMA matrix. The molecule has a limited rotational degree of freedom. A fit delivers a final anisotropy of 0.683 ± 0.003, a cone of wobbling of 12.4 ± 0.3 degrees, and a rotational correlation time of 2.7 ± 0.4 ns [159 Prommer 2004].

21.9 CONCLUDING REMARKS

Often FLIM is considered only a way to better separate the fluorescence signals of different fluorophores in biological specimens. It is certainly correct that recording the fluorescence decay functions adds one or several new dimensions to multidimensional microscopy. However, the true potential of FLIM is much larger. The fluorescence lifetime of a fluorophore molecule depends on its local environment on the molecular scale. It can therefore be used to probe local parameters, such as pH, ion concentrations, oxygen concentration, and the binding to proteins and lipids, and to investigate protein–protein interaction. Because the fluorescence lifetime is widely independent of the concentration, these effects can be investigated independently of the variable and usually unknown fluorophore concentration.

FLIM becomes particularly interesting in combination with laser scanning microscopes. The scanning principle can be combined with confocal detection and multiphoton excitation. Thus, out-of-focus light is suppressed, or, in case of multiphoton excitation, not excited. The fluorescence decay functions are therefore obtained from a defined depth in the sample, and mixing of the lifetimes of different focal planes is avoided. With multiphoton excitation, fluorescence from layers several 100 μm deep can be recorded without causing much photodamage outside the focal plane. FLIM can therefore be applied to living tissue.

Moreover, the small detection or excitation volume of a laser scanning microscope contains a limited number of fluorophore molecules. Thus, intensity fluctuations by diffusion effects, conformational changes, and intersystem crossing become apparent and can be used to obtain additional information about the molecular environment of the fluorophore molecules.

A number of lifetime techniques have been developed and applied in combination with laser scanning microscopes. Optically, point-detection and wide-field detection techniques can be distinguished. In terms of signal processing, lifetime techniques can be classified into time-domain and frequency-domain techniques. Moreover, from the electronic point of view the techniques can be classified into analog techniques and digital (photon counting) techniques. Consequently, a wide variety of techniques exist and can be used in combination with laser scanning.

Point-detection techniques make full use of the optical sectioning capability of a scanning microscope, both for confocal and two-photon microscopes. Typical representatives are single-detector modulation techniques and photon counting techniques. Wide-field techniques are based on gated or modulated image intensifiers. They cannot reasonably be used for confocal systems. Image intensifiers work for multiphoton systems but are not free of blurring by scattering in the sample.

Time-domain techniques directly record the fluorescence decay functions in the pixels of the image. The techniques are based on either scanning a time gate over the fluorescence decay or on counting photons into a number of time channels. Frequency-domain techniques measure the degree of modulation and the phase of the fluorescence signals at one or several modulation frequencies or at the harmonics of a high-repetition-rate pulsed laser signal.

Analog techniques consider the detector signal a continuous waveform, photon counting techniques a random sequence of single-photon pulses. Consequently, analog techniques work best at a high light intensity, while photon counting has its strength in detecting faint signals.

In the typical FLIM applications the fluorophores are located in highly specific parts of the cells, and the photostability of the samples is limited. Moreover, the fluorescence comes only from a small focal volume. The low effective fluorophore concentration, the limited excitation power, and the small focal volume result in low fluorescence intensity. Thus, for the majority of laser-scanning FLIM applications the photon detection rates are well within range of photon counting.

Of all lifetime techniques photon counting delivers the best lifetime accuracy for a given number of photons emitted by the sample. Especially multidimensional TCSPC has a near-ideal efficiency, a superior time resolution, multiwavelength capability, and the capability to resolve multi-exponential decay functions. TCSPC is currently the only FLIM technique that reliably resolves the interacting and non-interacting donor fraction in FRET experiments and the lifetime components of tissue autofluorescence. Moreover, TCSPC can be used to apply both FLIM and single-molecule spectroscopy to biological specimens. It is very likely that the TCSPC technique will result in a unification of FLIM and single-molecule techniques.

Declaration of Financial Interests

Wolfgang Becker is the president of Becker & Hickl, GmbH, Berlin, Germany.

REFERENCES

[1] Alford, K., Y. Wickramasinghe. Phase-amplitude crosstalk in intensity modulated near infrared spectroscopy. Rev. Sci. Instrum. 71 (2000): 2191–2195.

[2] Ameer-Beg, S.M., N. Edme, M. Peter, P.R. Barber, T. Ng, B.Vojnovic. Imaging Protein-Protein Interactions by Multiphoton FLIM. Proc. SPIE 5139 (2003): 180–189.

[3] Ameer-Beg, S.M., P.R. Barber, R. Locke, R.J. Hodgkiss, B. Vojnovic, G.M. Tozer, J. Wilson. Application of multiphoton steady state and lifetime imaging to mapping of tumor vascular architecture in vivo. Proc. SPIE 4620 (2002): 85–95.

[4] Anderssen, R.M., K. Carlsson, A. Liljeborg, H. Brismar. Characterization of probe binding and comparison of its influence on fluorescence lifetime of two ph-sensitivy benzo[c]xantene dyes using intensity-modulated multiple-wavelength scanning technique. Analytical Biochemistry 283 (2000): 104–110.

[5] Bacskai, B.J., J. Skoch, G.A. Hickey, R. Allen, B.T. Hyman. Fluorescence resonance energy transfer determinations using multiphoton fluorescence lifetime imaging microscopy to characterize amyloid-beta plaques. J. Biomed. Opt. 8,3 (2003): 368–375.

[6] Bacskai, B.J, J. Skoch, G.A. Hickey, O. Berezovska, B.T. Hyman. Multiphoton imaging in mouse models of Alzheimer's disease. Proc. SPIE 5323 (2004): 71–76.

[7] Baeyens, W.R.G., D. de Keukeleire, K. Korkidis. *Luminescence techniques in chemical and biochemical analysis.* New York: M. Dekker, 1991.

[8] Bahlmann, K., S.W. Hell. Deplorization by high aperture focusing. Appl. Phys. Lett. 77 (2000): 612–614.

[9] Ballew, R.M., J.N. Demas. An error analysis of the rapid lifetime determination method for the evaluation of single exponential decays. Anal. Chem. 61 (1989): 30–33.

[10] Becker & Hickl GmbH. The bh TCSPC handbook. Online. Available: http://www.becker-hickl.com

[11] Becker, W., H. Stiel, E. Klose. Flexible Instrument for time-correlated single photon counting, Rev. Sci. Instrum. 62/12 (1999): 2991–2996.

[12] Becker, W., H. Hickl, C. Zander, K.H. Drexhage, M. Sauer, S. Siebert, J. Wolfrum. Time-resolved detection and identification of single analyte molecules in microcapillaries by time-correlated single photon counting. Rev. Sci. Instrum. 70 (1999): 1835–1841.

[13] Becker, W., K. Benndorff, A. Bergmann, C. Biskup, K. König, U. Tirlapur, T. Zimmer. FRET Measurements by TCSPC Laser Scanning Microscopy. Proc. SPIE 4431 (2001): 94–98.

[14] Becker, W., A. Bergmann, K. König, U. Tirlapur. Picosecond fluorescence lifetime microscopy by TCSPC imaging. Proc. SPIE 4262 (2001): 414–419.

[15] Becker, W., A. Bergmann, H. Wabnitz, D. Grosenick, A. Liebert. High count rate multichannel TCSPC for optical tomography. Proc. SPIE 4431 (2001): 249–245.

[16] Becker, W., A. Bergmann, C. Biskup, T. Zimmer, N. Klöcker, K. Benndorf. Multi-wavelength TCSPC lifetime imaging. Proc. SPIE 4620 (2002): 79–84.

[17] Becker, W., A. Bergmann, G. Weiss. Lifetime Imaging with the Zeiss LSM510. Proc. SPIE 4620 (2002): 30–35.

[18] Becker, W., A. Bergmann, C. Biskup, L. Kelbauskas, T. Zimmer, N. Klöcker, K. Benndorf. High resolution TCSPC lifetime imaging. Proc. SPIE 4963 (2003): 175–184.

[19] Becker, W., A. Bergmann, M.A. Hink, K. König, K. Benndorf, C. Biskup. Fluorescence lifetime imaging by time-correlated single photon counting. Micr. Res. Techn. 63 (2004): 58–66.

[20] Becker, W., A. Bergmann, G. Biscotti, K. König, I. Riemann, L. Kelbauskas, C. Biskup. High-Speed FLIM Data Acquisition by Time-Correlated Single Photon Counting. Proc. SPIE 5323 (2004): 27–35.

[21] Becker, W., A. Bergmann, G. Biscotti, A. Rück. Advanced time-correlated single photon counting technique for spectroscopy and imaging in biomedical systems. Proc. SPIE 5340 (2004): 104–112.

[22] Becker, W., A. Bergmann, G. Biscotti. Fluorescence lifetime imaging by multi-detector TCSPC. In *OSA Biomedical Optics Topical Meetings* on CD ROM (The Optical Sciety of America, Washington, DC) (2004): WD1.

[23] Becker, W., A. Bergmann, E. Haustein, Z. Petrasek, P. Schwille, C. Biskup, T. Anhut, I. Riemann, K. Koenig, Fluorescence lifetime images and correlation spectra obtained by multi-dimensional TCSPC, Proc. SPIE 5700 (2005): 144–152.

[24] Becker, W., Advanced time-correlated single-photon counting techniques. Springer, Berlin, Heidelberg, New York, (2005).

[25] Bereszovska, O., P. Ramdya, J. Skoch, M.S. Wolfe, B.J. Bacskai, B.T. Hyman. Amyloid precursor protein associates with a nicastrin-dependent docking site on the presenilin 1-γ-secretase complex in cells demonstrated by fluorescence lifetime imaging. J. Neurosc. 23/11 (2003): 4560–4566.

[26] Bereszovska, O., B.J. Bacskai, B.T. Hyman. Monitoring Proteins in Intact Cells, Science of Aging Knowledge Environment. SAGE KE (2003): 14.

[27] Bernas, T., M. Zarebski, R.R. Cook, J.W. Dobrucki. Minimizing photobleaching during confocal microscopy of fluorescent probes bound to chromatin: role of anoxia and photon flux. J. Microsc. 215/1 (2004): 281–296.

[28] Berland, K.M. "Basics of Fluorescence." In *Methods in Cellular Imaging*, ed. A. Periasamy, 5–19, Oxford University Press, 2001.

[29] Berland, K.M., P.T.C. So, E. Gratton. Two-photon fluorescence correlation spectroscopy: Method and application to the intracellular environment. Biophys. J. 68 (1995): 694–701.

[30] Bewersdorf, J, R. Pick, S.W. Hell. Multifocal multiphoton microscopy. Opt. Lett. 23/9 (1998): 655–657.

[31] Bird, D.K., K.W. Eliceiri, C-H. Fan, J.G. White. Simultaneous two-photon spectral and lifetime fluorescence microscopy. Appl. Opt. 43/27 (2004): 5173–5182.

[32] Biskup, C., A. Böhmer, R. Pusch, L. Kelbauskas, A. Gorshkov, I. Majoul, J. Lindenau, K. Benndorf, F-D. Böhmer. Visualization of SHP-1-target interaction, J. Cell Sci. 117 (2004): 5155–5178.

[33] Biskup, C., L. Kelbauskas, T. Zimmer, K. Benndorf, A. Bergmann, W. Becker, J.P. Ruppersberg, C. Stockklausner, N. Klöcker. Interaction of PSD-95 with potassium channels visualized by fluorescence lifetime-based resonance energy transfer imaging. J. Biomed. Opt. 9/4 (2004): 735–759.

[34] Biskup, C., T. Zimmer, K. Benndorf. FRET between cardiac Na^+ channel subunits measured with a confocal microscope and a streak camera, Nature Biotechnology, Vol. 22/2 (2004): 220–224.

[35] Böhmer, M., M. Wahl, H-J. Rahn, R. Erdmann, J. Enderlein. Time-resolved fluorescence correlation spectroscopy. Chem. Phys. Lett. 353 (2002): 439–445.

[36] Böhmer, M., F. Pampaloni, M. Wahl, H-J. Rahn, R. Erdmann. Time-resolved confocal scanning device for ultrasensitive fluorescence detection, Rev. Sci. Instrum. 72/11 (2001): 4145–4152.

[37] Booth, M.J., T. Wilson. Low-cost, frequency-domain, fluorescence lifetime confocal microscopy. J of Microscopy, 214 (2004): 36–42.

[38] Borst, J. W., M.A. Hink, A. van Hoek, A.J.W.G. Visser. Multiphoton spectroscopy in living plant cells. Proc. SPIE 4963 (2003): 231–238.

[39] Breusegem, S.Y. "In vivo investigation of protein interactions in C. Elegans by Förster resonance energy transfer microscopy." Ph.D.diss., University of Illinois at Urbana-Champaign, 2002.

[40] Brismar, H., B. Ulfhake. Fluorescence lifetime imaging measurements in confocal microscopy of neurons labeled with multiple fluorophores. Nature Biotech. 15 (1997): 373–377.

[41] Bugiel, I., K. König, H. Wabnitz. Investigations of cells by fluorescence laser scanning microscopy with subnanosecond time resolution. Lasers in the Life Sciences 3/1 (1989): 47–53.

[42] Buist, A.H., M. Müller, J. Squier, G.J. Brakenhoff. Real-time two-photon absorption microscopy using multi-point excitation. J. Microsc. 192/2 (1998): 217–226.

[43] Buurman, E.P., R. Sanders, A. Draaijer, H.C. Gerritsen, J.J.F. van Veen, P.M. Houpt, Y.K. Levine. Fluorescence lifetime imaging using a confocal laser scanning microscope. Scanning 14 (1992): 155–159.

[44] Carlsson, K., A. Liljeborg. Confocal fluorescence microscopy using spectral and lifetime information to simultaneously record four fluorophores with high channel separation. J. Microsc. 185 (1997): 37–46.

[45] Carlsson, K., Liljeborg, A., Simultaneous confocal lifetime imaging of multiple fluorophores using the intensity-modulated multiple-wavelength scanning (IMS) technique. J. Microsc. 191/2 (1998): 119–127.

[46] Carlsson, K., J.P. Philip. Theoretical investigation of the signal-to-noise ratio for different fluorescence lifetime imaging techniques. Proc. SPIE, 4622 (2002): 70–78.

[47] Centonze, V.E., J.G. White. Multiphoton excitation provides optical sections from deeper within scattering specimens than confocal imaging. Biophys. J. 75 (1998): 2015–2024.

[48] Chance, B., M. Cope, E. Gratton, N. Ramanujam, B. Tromberg. Phase measurement of light absorption and scatter in human tissue. Rev. Sci. Instrum. 69/10 (1998): 3457–3481.

[49] Chen, Y., J.D. Müller, P.T.C. So, E. Gratton. The Photon Counting Histogram in Fluorescence Fluctuation Spectroscopy. Biophys. J. 77 (1999): 553–567.

[50] Chen, Y., J.D. Müller, Q.Q. Ruan, E. Gratton. Molecular brightness characterization of EGFP in vivo by fluorescence fluctuation spectroscopy. Biophys. J. 82 (2002): 133–144.

[51] Chen, Y., A. Periasamy. Characterization of Two-photon Excitation Fluorescence Lifetime Imaging Microscopy for Protein Localization. Microsc Res. Tech. 63 (2004): 72–80.

[52] Chen, Y., A. Periasamy. Two-photon FIM-FRET microscopy for protein localization. Proc. SPIE 5323 (2004): 431–439.

[53] Cianci, G.C., K. Berland: Saturation in two-photon microscopy. Proc. SPIE 5323 (2004): 128–135.

[54] Clayton, A.H.A., Q.A. Hanley, D.J. Arndt-Jovin, V. Subramaniam, T.M. Jovin. Dynamic Fluorescence Anisotropy Imaging Microscopy in the Frequency Domain (FLIM). Biophys. J. 83 (2002): 1631–1649.

[55] Cole, M.J., J. Siegel, S.E.D. Webb, R. Jones, R. Dowling, M.J. Dayel. D. Parsons-Karavassilis, P.M. French, M.J. Lever, L.O. Sucharov, M.A. Neil, R. Juskaitas, T. Wilson. Time-domain whole-field lifetime imaging with optical sectioning. J. Microsc. 203/3 (2001): 246–257.

[56] Cova, S., M. Bertolaccini, C. Bussolati. The Measurement of Luminescence Waveforms by Single-Photon Techniques. Phys. Stat. Sol. 18 (1973): 11–61.

[57] Denk, W., J.H. Strickler, W.W.W. Webb. Two-photon laser scanning fluorescence microscopy. Science 248 (1990): 73–76.

[58] Diaspro, A. "Building a Two-Photon Microscope Using a Laser Scanning Confocal Architecture." In *Methods in Cellular Imaging,* ed. Ammasi Periasamy, 162–179. Oxford: University Press, 2001.

[59] Diaspro, A, M. Corosu, P. Ramoino, M. Robello. Adapting a Compact Confocal Microscope System to a Two-Photon Excitation Fluorescence Imaging Architecture, Micr. Res. Tech: 47 (1999): 196–205.

[60] Dickinson, M.E., C.W. Waters, G. Bearman, R. Wolleschensky, S. Tille, S.E. Fraser. Sensitive imaging of spectrally overlapping fluorochromes using the LSM 510 META BIOS, Proc. SPIE 4620 (2002): 123–136.

[61] Dong, C.-Y., K. König, T.P.C. So. Characterizing point spread functions of two-photon fluorescecne microscopy in turbid medium. J. Biomed. Opt. 8 (2003): 450–459.

[62] Dowling, K., S.C.W. Hyde, J.C. Dainty, P.M.W. French, J.D. Hares. 2-D fluorescence liefetime imaging using a time-gated image intensifier. Elsevier: Optics Communications 135 (1997): 27–31.

[63] Dowling, K., M.J. Dayel, M.J. Lever, P.M.W. French, J.D. Hares, A.K.L Dymoke-Bradshaw. Fluorescence lifetime imaging with picosecond resolution for biomedical applications. Opt. Lett. 23/10 (1998): 810–812.

[64] Draaijer A, R. Sanders, H.C. Gerritsen. "Fluorescence lifetime imaging, a new tool in confocal microscopy." In *Handbook of Biological Microscopy*, ed. J.B. Pawley, 491–504. New York: Plenum Press, 1995.

[65] Duncan, R.R., A. Bergmann, M.A. Cousin, D.K. Apps, M.J. Shipston. Multi-dimensional time-correlated single-photon counting (TCSPC) fluorescence lifetime imaging microscopy (FLIM) to detect FRET in cells. J. Microsc. 215/1 (2004): 1–12.

[66] Eftink, M.R. "Fluorescence quenching: Theory and application". In *Topics in fluorescence spectroscopy 2*, ed. J.R. Lakowicz, 53–127. New York: Plenum Press, 1991.

[67] Eggerling, C., S. Berger, L. Brand, J.R. Fries, J. Schaffer, A. Volkmer, C.A. Seidel. Data registration and selective single-molecule analysis using multi-parameter fluorescence detection. J. Biotechnol. 86 (2001): 163.

[68] Eid, J.S., J.D. Müller, E. Gratton. Data acquisition card for fluctuation correlation spec-troscopy allowing full access to the detected photon sequence. Rev. Sci. Instrum. 71/2 (2000): 361–368.

[69] Elangovan, M., H. Wallrabe, R.N. Day, M. Barroso, A. Periasamy. Characterisation of one- and two-photon fluorescence resonance energy transfer microscopy. Methods 29 (2003): 58–73.

[70] Elangovan, M., R.N. Day, A. Periasamy. Nanosecond fluorescence resonance energy transfer-fluorescence lifetime imaging microscopy to localize the protein interactions in a single cell. J. Microsc. 205/1 (2002): 3–14.

[71] Eliceiri, K.W., C.H. Fan, G.E. Lyons, J.G. White. Analysis of histology specimens using lifetime multiphoton microscopy. J. Biomed. Opt. 8/3 (2003): 376–380.

[72] Emiliani, V., D. Sanvito, M. Tramier, T. Piolot, Z. Petrasek, K. Kemnitz, C.Durieux, M.Coppey-Moisan. Low-intensity two-dimensional imaging of fluorescence lifetimes in living cells. Appl. Phys. Lett. 83/12 (2003): 2471–2473.

[73] Esposito, A., F. Federici, C. Usai, F. Cannone, G. Chirico, M. Collini, A. Diaspro. Notes on Theory and Experimental Conditions Behind Two-Photon Excitation Microscopy. Microscopy Research and Techniques 62 (2004): 12–17.

[74] S. Felekyan, R. Kühnemuth, V. Kudryavtsev, C. Sandhagen, W. Becker, C.A.M. Seidel, Full correlation from picoseconds to seconds by time-resolved and time-correlated single photon detection, Rev. Sci. Instrum. 76 (2005):083104.

[75] Gautier, I., M. Tramier, C. Durieux, J. Coppey, R.B. Pansu, J-C. Nicolas, K. Kemnitz, M. Copey-Moisan. HoMo-FRET microscopy in living cells to measure monomer-dimer transition of GFp-tagged proteins. Biophys. J. 80 (2001): 3000–3008.

[76] Geddes C.D., H. Cao, I. Gryczynski, J. Fang, J.R. Lakowicz. Metal-enhanced fluorescence (MEF) due to silver colloids on a planar surface: Potential applications of indocyanine green to in vivo imaging, J. Phys. Chem. A 107 (2003): 3443–3449.

[77] Geddes C.D., H. Cao, I. Gryczynski, J.R. Lakowicz. Enhanced photostability of ICG in close proximity to gold colloids. Stectrochimica Acta A 59 (2003): 2611–2617.

[78] Geddes, C.D., A. Parfenov, J.R. Lakowicz. Photodeposition of silver can result in metal-enhanced fluorescence. Appl. Spectroscopy 57/5 (2003): 526–531.

[79] Georgakoudi I., M.S. Feld, M.G. Müller. "Intrinsic fluorescence spectroscopy of biological tissue." In *Handbook of Biomedicl Fluorescence,* ed. M.-A. Mycek, B.W. Pogue, 109–142. New York Basel: Marcel Dekker Inc., 2003.

[80] Gerritsen, H.C., R. Sanders, A. Draaijer, Y.K. Levine. Fluorescence lifetime imaging of oxygen in cells. J. Fluoresc. 7 (1997): 11–16.

[81] Gerritsen, H.C., K. de Grauw. "One- and Two-Photon confocal Fluorescence Lifetime Imaging and Its Application." In *Methods in Cellular Imaging,* ed. A. Periasamy, 309–323. Oxford: University Press, 2001.

[82] Gerritsen, H.C., M.A.H. Asselbergs, A.V. Agronskaia, W.G.J.H.M. van Sark. Fluorescence lifetime imaging in scanning microscopes: acquisition speed, photon economy and lifetime resolution. J. Microsc. 206/3 (2002): 218–224.

[83] Gratton, E., S. Breusegem, J. Sutin, Q. Ruan, N. Barry. Fluorescence lifetime imaging for the two-photon microscope: Time-domain and frequency domain methods. J. Biomed. Opt. 8/3 (2003): 381–390.

[84] Gratton, G., B. Feddersen, M. Vande Ven. Parallel acquisition of fluorescence decay using array detectors. Proc. SPIE 1204 (1990): 21–25.

[85] Gu, Y, W.L. Di, D.P. Kelsell, D. Zicha. Quantitative fluorescecne resonance energy transfer (FRET) measurement with acceptor photobleaching and spectral unmixing. J. Microsc. 215/2 (2004): 162–173.

[86] Habenicht, A., J. Hjelm, E. Mukhtar, F. Bergström, L.B-A. Johansson. Two-photon excitation and time-resolved fluorescence: I. The proper response function for analysing single-photon counting experiments. Chem. Phys. Lett. 345 (2002): 367–375.

[87] Haj, I., P. Verveer, A. Squire, B.G. Neel, P.I.H. Bastiaens. Imaging sites of receptor dephosphorylation by PTP1B on the surface of the endoplasmic reticulum. Science 295 (2002): 1708–1710.

[88] Hanson, K.M., M.J. Behne, N.P. Barry, T.M. Mauro, E. Gratton. Two-photon fluorescence imaging of the skin stratum corneum pH gradient. Biophys. J. 83 (2002): 1682–1690.

[89] Haugland, R.P. "Handbook of fluorescent probes and research chemicals." *Molecular Probes Inc.,* 1999.

[90] Herman, B., *Fluorescence Microscopy.* 2nd. edn., New York: Springer, 1998.

[91] Herman, B., G. Gordon, N. Mahajan, V. Centonze. "Measurement of fluorescence resonance energy transfer in the optical microscope." In *Methods in Cellular Imaging,* ed. Ammasi Periasamy, 257–272. Oxford: University Press, 2001.

[92] Hopt, A., E. Neher. Highly nonlinear Photodamage in two-photon fluorescence microscopy. Biophys. J. 80 (2002): 2029–2036.

[93] Hughes, M.K.Y., S. Ameer-Beg, M. Peter, T. Ng. Use of acceptor fluorescence for determining FRET lifetimes. Proc. SPIE 5139 (2003): 88–96.

[94] Jakobs, S., V. Subramaniam, A. Schönle, T. Jovin, S.W. Hell. EGFP and DsRed expressing cultures of *escherichia coli* imaged by confocal, two-photon and fluorescence lifetime microscopy. FEBS Letters 479 (2000): 131–135.

[95] Kaneko, H., I. Putzier, S. Frings, U. B. Kaupp, Th. Gensch, Chloride Accumulation in Mammalian Olfactory Sensory Neurons, J. Neurosci., 24 (2004): 7931–7938.

[96] Kästner, C.N., M. Prummer, B. Sick, A. Renn, U.P. Wild, P. Dimroth. The citrate carrier CitS probed by single molecule fluorescence spectroscopy. Biophys. J. 84 (2003): 1651–1659.

[97] Kask, P., K. Palo, D. Ullmann, K. Gall.Fluorescence-intensity distribution analysis and its application in bimolecular detection technology. PNAS 96 (1999): 13756–13761.

[98] Kask, P., K. Palo, N. Fay, L. Brand, Ü. Mets, D. Ullmann, J. Jungmann, J. Pschorr, K. Gall. Two-dimensional fluorescence intensity distribution analysis: theory and applications. Biophys. J. 78 (2000): 1703–1713.

[99] Kask, P., C. Eggeling, K. Palo, Ü. Mets, M. Cole, K. Gall. "Fluorescence intensity distribution analysis(FIDA) and related fluorescence fluctuation techniques: theory and practice." In *Fluorescence spectroscopy imaging and probes*, ed. R. Kraayenhof, A.J.W.G. Visser, H.C. Gerritsen, 153–181. New York: Springer-Verlag Berlin Heidelberg, 2000.

[100] Kelbauskas, L., W. Dietel. Internalization of aggregated photosensitizers by tumor cells: Subcellular time-resolved fluorescence spectroscopy on derivates of pyropheophorbide-a ethers and chlorin e6 under femtosecond one- and two-photon excitation. Photochem. Photobiol. 76/6 (2002): 686–694.

[101] Kim, K.H., C. Bühler, K. Bahlmann, P.T.C. So. Signal degradation in multiple scattering medium: implication for single focus vs. multifoci two-photon microscopy. Proc. SPIE 5323 (2004): 273–278.

[102] Kinoshita, S., T. Kushida. Subnanosecond fluorescence-lifetime measuring system using single photon counting method with mode-locked laser excitation. Rev. Sci. Instrum. 52/4 (1981): 572–575.

[103] Kinoshita, S., T. Kushida. High-performance, time-correlated single-photon counting apparatus using a side-on type photomultiplier. Rev. Sci. Instrum. 53/4 (1982): 469–472.

[104] Knemeyer, J.-P., N. Marmé, M. Sauer. Probes for detection of specific DNA sequences at the single-molecule level. Anal. Chem. 72 (2002): 3717–3724.

[105] Köllner, M., J. Wolfrum. How many photons are necessary for fluorescence-lifetime measurements? Phys. Chem. Lett. 200/1,2 (1992): 199–204.

[106] König, K., P.T.C. So, W.W. Mantulin, B.J. Tromberg, E. Gratton. Two-Photon excited lifetime imaging of autofluorescence in cells during UVA and NIR photostress. J. Microsc. 183/3 (1996): 197–204.

[107] König, K. Multiphoton microscopy in life sciences. J. Microsc. 200/1 (2000): 83–104

[108] König, K. Laser tweezers and multiphoton microscopes on life science. Histochem. Cell Biol. 114 (2000): 79–92.

[109] König, K. "Cellular Response to Laser Radiation in Fluorescence Microscopes." In *Methods in Cellular Imaging*, ed. Ammasi Periasamy, 236–254. Oxford: University Press, 2001.

[110] König, K., U. Wollina, I. Riemann, C. Peuckert, K-J. Halbhuber, H. Konrad, P. Fischer, V. Fuenfstueck, T.W. Fischer, P. Elsner. Optical tomography of human skin with subcellular resolution and picosecond time resolution using intense near infrared femtosecond laser pulses. Proc. SPIE 4620 (2002): 191–201.

[111] König K, I. Riemann. High-resolution multiphoton tomography of human skin with subcellular spatial resolution and picosecond time resolution. J. Biom. Opt. 8/3 (2003): 432–439.

[112] König, K., I. Riemann, G. Ehrlich, V. Ulrich, P. Fischer. Multiphoton FLIM and spectral imaging of cells and tissue. Proc. SPIE 5323 (2004): 240–251.

[113] Kreiter, M., M. Prummer, B. Hecht, U.P. Wild. Orientation dependence of fluorescence lifetimes near an interface. J. Chem. Phys. 117 (2002): 9430–9433.

[114] Kress, M., T. Meier, T.A.A. El-Tayeb, R. Kemkemer, R. Steiner, A. Rück. Short-pulsed diode lasers as an excitation source for time-resolved fluorescence applications and confocal laser scanning microscopy in PDT. Proc. SPIE 4431 (2001): 108–113.

[115] Krishnan, R.V., A. Masuda, V.E. Centonze, B. Herman. Quantitative imaging of protein-protein interactions by multiphoton fluorescence lifetime imaging microscopy using a streak camera. J. Biomed. Opt. 8/3 (2003): 362–267.

[116] Krishnan, R.V., H. Saitoh, H. Terada, V.E. Centonze, B. Herman. Development of a multiphoton fluorescence lifetime imaging microscopy (FLIM) system using a streak camera. Rev. Sci. Instrum. 74 (2003): 2714–2721.

[117] Kühnemuth, R., C.A.M. Seidel. Principles of single molecule multiparameter fluorescence spectroscopy. Single Molecules 2 (2001): 251–254.

[118] Laczko G, I, Gryczinski, W. Wiczk, H. Malak, J.R. Lakowicz. A 10-GHz frequency-domain fluorometer. Rev. Sci. Instrum. 61 (1990): 2331–2337.

[119] Lakowicz, J.R., K. Berndt. Lifetime-selective fluorescence lifetime imaging using an rf phase-sensitive camera. Rev. Sci. Instrum. 62/7 (1991): 1727–1734.

[120] Lakowicz, J.R., H. Szmacinski, K. Nowaczyk, M.L. Johnson. Fluorescence lifetime imaging of free and protein-bound NADH. PNAS 89 (1992): 1271–1275.

[121] Lakowicz, J.R., H. Szmacinski. Fluorescence-lifetime based sensing of pH, Ca^{2+}·and glucose. Sens. Actuator Chem. 11 (1993): 133–134.

[122] Lakowicz, J.R. *Principles of Fluorescence Spectroscopy*. New York: Plenum Press, 1999.

[123] Lakowicz, J.R., B. Shen, Z. Gryczynski, S. D'Auria, I. Gryczinski. Intrinsic fluorescence from DNA can be enhanced by metallic particles. Biochemical and Biophysical research Communications 286 (2001): 875–879.

[124] Lemasters, J.J., T. Qian, D.R. Trollinger, B.J. Muller-Borer, S.P. Elmore, W.E. Cascio. "Laser Scanning confocal Microscopy Applied to Living Cells and Tissues." In *Methods in Cellular Imaging* ed. Ammasi Periasamy, 66–87. Oxford: University Press, 2001.

[125] Leskovar, B., C.C. Lo. Photon counting system for subnanosecond fluorescence lifetime measurements. Rev. Sci. Instrum. 47/9 (1976): 1113–1121.

[126] Leskovar, B. Nanosecond Fluorescence Spectroscopy. IEEE Transactions on Nuclear Science. NS-32/3 (1985): 1232–1241.

[127] Lewis, C., W.R. Ware. The Measurement of Short-Lived Fluorescence Decay Using the Single Photon Counting Method. Rev. Sci. Instrum. 44/2 (1973): 107–114.

[128] Liebert, A., H. Wabnitz, J. Steinbrink, H. Obrig, M. Möller, R. Macdonald, H. Rinneberg. Intra- and extracerebral changes of hemoglobin concentrations by analysis of moments of distributions of times of flight of photons. SPIE 5138 (2003): 126–130.

[129] Liebert, A., H. Wabnitz, M. Möller, A. Walter, R. Macdonald, H. Rinneberg, H. Obrig, J. Steinbrink. "Time-resolved diffuse NIR-reflectance topography of the adult head during motor stimulation." In *OSA Biomedical Optics Topical Meetings* on CD ROM (The Optical Sciety of America, Washington, DC) (2004): WF34.

[130] Liebert, A., H. Wabnitz, J. Steinbrink, H. Obrig, M. Möller, R. Macdonald, A. Villringer, H. Rinneberg. Time-resolved multidistance near-infrared spectroscopy at the human head: Intra- and extracerebral absorption changes from moments of distribution of times of flight of photons. Appl. Opt. 43 (2004): 3037–3047.

[131] Lin, H.-J., H. Szmacinski, J.R. Lakowicz. Lifetime-based pH sensors: indicators for acidic environments. Anal. Biochem. 269 (1999): 162–167.

[132] Lleo, A., O. Berezovska, L. Herl, S. Raju, A. Deng, B.J. Bacskai, M.P. Frosch, M. Irizarry, B.T. Hyman. Nonsteroidal anti-inflammatory drugs lower Aβ42 and change presenilin 1 conformation. Nature Medicine 10/10 (2004): 1065–1066.

[133] Magde, D., E. Elson, W.W.W. Webb. Thermodynamic fluctuations ina reacting system — measurement by fluorescence correlation spectroscopy. Phys. Rev. Lett. 29/11 (1972): 705–708.

[134] Malicka, J., I. Gryczynski, C.D. Geddes, J.R. Lakowicz. Metal-enhanced emission from indocyanine green: a new approach to in vivo imaging. J. Biomed. Opt. 8/3 (2003): 472–478.

[135] Marcu L, W.S. Grundfest, M.C. Fishbein. "Time-resolved laser-induced fluorescence spectrocopy for staging atherosclecotic lesions." In *Handbook of Biomedical Fluorescence,* ed. M-A. Mycek and B.W. Pogue, 397–430. New York Basel: Marcel Dekker Inc., 2003.

[136] Masters, B.R., P.T.C. So. Antecedents of Two-Photon Excitation Laser Scanning Microscopy. Microscopy Research and Techniques 63 (2004): 3–11.

[137] Masters, B.R., P.T.C. So, E. Gratton. Multiphoton excitation fluorescence microscopy and spectroscopy of in vivo human skin. Biophys. J. 72 (1997): 2405–2412.

[138] Mills, J.D., J.R. Stone, D.G. Rubin, D.E. Melon, D.O. Okonkwo, A. Periasamy, G.A. Helm. Illuminating protein interactions in tissue using confocal and two-photon excitation fluorescent resonance energy transfer microscopy. J. Biomed. Opt. 8/3 (2003): 347–356.

[139] Minsky, M. Memoir on inventing the confocal microscope. Scanning 10 (1988): 128–138.

[140] Mitchell, A.C., J.E. Wall, J.G. Murray, C.G. Morgan. Direct modulation of the effective sensitivity of a CCD detector: A new approach to time-resolved fluorescence imaging. J. Microsc. 206/3 (2002): 225–232.

[141] Mitchell, A.C., J.E. Wall, J.G. Murray, C.G. Morgan. Measurement of nanosecond time-resolved fluorescence with a directly gated interline CCD camera. J. Microsc. 206/3 (2002): 233–238.

[142] Moerner, W.E., D.P. Fromm D.P. Methods of single-molecule spectroscopy and microscopy, Rev. Sci. Instrum. 74 (2003): 3597–3619.

[143] Morgan, S.P., K.Y. Yong. Elimination of amplitude-phase crosstalk in frequency domain near-infrared spectroscopy. Rev. Sci. Instrum. 72/4 (2001): 1984 1987.

[144] Morton, G.A. Photon Couting. Appl. Opt. 7 (1968): 1–10.

[145] Morton, P.E., T.C. Ng, S.A. Roberts, B. Vojnovic, S.M. Ameer-Beg. Time resolved multi-photon imaging of the interaction between the PKC and NFkB signalling pathways. Proc. SPIE 5139 (2003): 216–222.

[146] Müller, J.D., Y.E. Chen, Gratton. Resolving Heterogeneity on the Single Molecular Level with the Photon-Counting Histogram. Biophys. J. 78 (2000): 474–586.

[147] O'Connor, D.V., D. Phillips. *Time Correlated Single Photon Counting*. London: Academic Press, 1984.

[148] Palo, K., L. Brand, C. Eggeling, S. Jäger, P. Kask, K. Gall. Fluorescence Intensity and Lifetime Distribution Analysis: Toward Higher Accuracy in Fluorescence Fluctuation Spectroscopy. Biophys. J. 83 (2002): 605–617.

[149] Palo, K., Ü. Mets, S. Jäger, P. Kask, K. Gall. Fluorescence intensity multiple distribution analysis: concurrent determination of diffusion times and molecular brightness. Biophys. J. 79 (2000): 2858–2866.

[150] Patterson, G.H., D.W. Piston. Photobleaching in two-photon excitation microscopy. Biophys. J. 78 (2000): 2159–2162.

[151] Paul, R.J., H. Schneckenburger. Oxygen Concentration and the Oxidation-Reduction State of Yeast: Determination of Free/Bound NADH and Flavins by Time-Resolved Spectroscopy. Springer Naturwissenschaften 83 (1996): 32–35.

[152] Pawley, J. *Handbook of biological confocal microscopy*. New York: Plenum, 1995.

[153] Periasamy, A. *Methods in Cellular Imaging*. Oxford New York: Oxford University Press, 2001.

[154] Periasamy, A., R.N. Day. "Visualizing protein interactions in living cells using digitizded GFP imaging and FRET microscopy." In *Methods in Cell Biology, ed. K.F.* Sullivan and S.A. Kay, 293–314. Academic Press, 58/1999.

[155] Perkin Elmer Optoelectronics. Single photon counting module. SPCM-AQR series.

[156] Philip, J.P., K. Carlsson. Theoretical investigation of the signal-to-noise ratio in fluorescence lifetime imaging, J. Opt. Soc. Am. A 20 (2003): 368–379.

[157] Pico Star Camera. http://www.lavision.com.

[158] Prummer, M., C. Hübner, B. Sick, B. Hecht, A. Renn, U.P. Wild. Single-Molecule Identification by Spectrally and Time-Resolved Fluorescence Detection. Anal. Chem. 72 (2000): 433–447.

[159] Prummer, M., B. Sick, A. Renn, U.P. Wild. Multiparameter microscopy and spectroscopy for single-molecule analysis. Anal. Chem. 76 (2004): 1633–1640.

[160] Qian, H., E.L. Elson. Distribution of molecular aggregation by analysis of fluctuation moments. PNAS 87 (1990): 5479–5483.

[161] Redmont, R.W. "Introduction to fluorescence and photophysics." In *Handbook of Biomedical Fluorescence,* ed. M-A. Mycek and B.W. Pogue, 1–27. New York, Basel: Marcel Dekker Inc., 2003.

[162] Richards, B., E. Wolf. Electromagnetic diffraction in optical systems II. Structure of the image field in an aplanatic system. Proc. Roy. Soc. A 253 (19059): 358–379.

[163] Riemann, I., P. Fischer, M. Kaatz, T.W. Fischer, P. Elsner, E. Dimitrov, A. Reif, K. König. Optical tomography of pigmented human skin biopies. Proc. SPIE 5312 (2004): 24–34.

[164] Rigler, R., E.S. Elson. *Fluorescence Correlation Spectroscopy.* Berlin Heidelberg: Springer, 2001.

[165] Rigler, R., Ü. Mets, J. Widengren, P. Kask. Fluorescence correlation spectroscopy with high count rate and low background: analysis of translational diffusion. European Biophysics Journal 22 (1993): 169–175.

[166] Rigler, R., J. Widengreen. Utrasensitive detection of single molecules by fluorescence correlation spectroscopy. Bioscience 3 (1990): 180–183.

[167] Richards-Kortum, R., R. Drezek, K. Sokolov, I. Pavlova, M. Follen. "Survey of endogenous biological fluorophores." In *Handbook of Biomedical Fluorescence,* ed. M-A. Mycek and B.W. Pogue, 237–264. New York Basel: Marcel Dekker Inc., 2003.

[168] Riquelme, B., D. Dumas, J. Valverde, R. Rasia, J.F. Stoltz. Analysis of the 3D structure of agglutinated erythrocyte using CellScan and Confocal microscopy. Characterisation by FLIM-FRET. Proc. SPIE 5139 (2003): 190–198.

[169] A. Rück, F. Dolp, C. Happ, R. Steiner, M. Beil. Time-resolved microspectrofluorometry and fluorescence lifetime imaging using ps pulsed laser diodes in laser scanning microscopes. Proc. SPIE 5139 (2003): 166–172.

[170] Sanders, R., A. Draaijer, H.C. Gerritsen, P.M. Houpt, Y.K. Levine. Quantitative pH Imaging in Cells Using Confocal Fluorescence Lifetime Imaging Microscopy. Analytical Biochemistry 227/2 (1995): 302–308.

[171] Schneckenburger, H., M.H. Gschwend, R. Sailer, H-P. Mock, W.S.L. Strauss. Time-gated fluorescence microscopy in cellular and molecular biology. Cellular and Molecular Biology 44/5 (1998): 795–805.

[172] Schneckenburger, H., M. Wagner, M. Kretzschmar, W.S.L. Strauss, R. Sailer. Fluorescence lifetime imaging (FLIM) of membrane markers in living cells. Biomedical Optics SPIE 5139 (2003): 1605–7422.

[173] Schönle, A., M. Glatz, S.W. Hell. Four-dimensional multiphoton microscopy with time-correlated single photon counting. Appl. Optics 39/34 (2000): 6306–6311.

[174] Schrader, M., S.W. Hell. Three-dimensional super-resolution with a 4pi-confocal microscope using image restoration. J. Appl. Phys. 84/8 (1998): 4034–4042.

[175] Schuyler, R., I. Isenberg. A Monophoton Fluorometer with Energy Discrimination. Rev. Sci. Instrum. 42/6 (1971): 813–817.

[176] P Schwille, F.J. Meyer-Almes, R Rigler. Dual-color fluorescence cross-correlation spectroscopy for multicomponent diffusional analysis in solution. Biophys. J. 72 (1997): 1878–1886.

[177] Schwille, P., U. Haupts, S. Maiti, W.W.W. Webb. Molecular Dynamics in Living Cells Observed by Fluorescence Correlation Spectroscopy with One- and Two-Photon Excitation. Biophys. J. 77 (1999): 2251–2265.

[178] Sherman, L., J.Y. Ye, O. Alberts, T.B. Norris. Adaptive correction of depth-induced aberrations in multiphoton scanning microscopy usind a deformable mirror. J. Microsc. 206/1 (2001): 65–71

[179] Sick, B., B. Hecht. Orientation of single molecules by annular Illumination. Phys. Rev. Lett. 85 (2000): 4482–4485.

[180] Sick, B., B. Hecht, U.P. Wild, L. Novotny. Probing confined fields with single molecules and vice versa. J. Microsc. 202/2 (2001): 365–373.

[181] So, P.T.C., T. French, E. Gratton. A frequency domain microscope using a fast-scan CCD camera. Proc. SPIE 2137 (1994): 83–92.

[182] So, P.T.C., T. French, W.M. Yu, K.M. Berland, C.Y. Dong, E. Gratton. Time-resolved fluorescence microscopy using two-photon excitation. Bioimaging 3 (1995): 49–63.

[183] So. P.T.C., H. Kim, I.E. Kochevar. Two-photon deep tissue ex vivo imaging of mouse dermal and subcutaneous structures. Opt. Expr. 3/9 (1998): 339–350.

[184] So, P.T.C., K.H. Kim, C. Buehler, B.R. Masters, L. Hsu, C.Y. Dong. "Basic Principles of Multiphoton Excitation Microscopy." In *Methods in Cellular Imaging,* ed. Ammasi Periasamy, 147–161. Oxford: University Press, 2001.

[185] So, P.T.C., K.H. Kim, L. Hsu, P. Kaplan, T. Hacewicz, C.Y. Dong, U. Greuter, N. Schlumpf, C. Buehler. "Twi-photon microscopy of tissues." In *Handbook of Biomedical Fluorescence*, ed. M-A. Mycek and B.W. Pogue, 181–208. New York, Basel: Marcel Dekker Inc., 2003.

[186] Squire, A., P.J. Verveer, P.I.H. Bastiaens. Multiple frequency fluorescence lifetime imaging microscopy. J. Microsc. 197/2 (2000): 136–149.

[187] Straub, M., S.W. Hell. Fluorescence lifetime three-dimensional microscopy with picosecond precision using a multifocal multiphoton microscope. Appl. Phys. Lett. 73/13 (1998): 1769–1771.

[188] Suhling K, D.M. Davis, Z. Petrasek, J. Siegel, D. Phillips. The influence of the refractive index on EGFP fluorescence lifetimes in mixtures of water and glycerol. Proc. SPIE 4259 (2001): 91–101.

[189] Suhling, K., J. Siegel, D. Phillips, P.M.W. French, S. Lévêque-Fort, S.E.D. Webb, D.M. Davis. Imaging the Environment of Green Fluorescent Protein. Biophys. J. 83 (2002): 3589–3595.

[190] Syrtsma, J., J.M. Vroom, C.J. de Grauw, H.C. Gerritsen. Time-gated fluorescence lifetime imaging and microvolume spectroscopy using two-photon excitation. J. Microsc. 191/1 (1009): 39–51.

[191] Tadrous, P.J., J. Siegel, P.M.W. French, S. Shousha, E-N. Lalani, G.W.H. Stamp. Fluorescence lifetime imaging of unstained tissues: early results in human breast cancer. Journal of Pathology 199 (2003): 309–317.

[192] Theer, P., M.T. Hasan, W. Denk. Multi-photon imaging using a Ti:sapphire regenerative amplifier. Biomedical Optics SPIE 5139 (2003): 1–6.

[193] Thomson, N.L. "Fluorescence Correlation Spectroscopy." In *Topics in Fluorescence Spectroscopy*, ed. J.R. Lakowicz, 1: 337. New York: Plenum Press, 1991.

[194] Tinnefeld, P., V. Buschmann, D-P. Herten, K.T. Han, M. Sauer. Confocal fluorescence lifetime imaging microscopy (FLIM) at the single molecule level. Single Mol. 1 (2000): 215–223.

[195] Tirlapur, U.K., K. König. Targeted transfection by femtosecond laser. Nature 418 (2002): 290–291.

[196] Tramier, M., K. Kemnitz, C. Durieux, J. Coppey, P. Denjean, B. Pansu, M. Coppey-Moisan. Restrained torsional dynamics of nuclear DNA in living proliferative mammalian cells. Biophys. J. 78 (2000): 2614–2627.

[197] Tramier, M., I. Gautier, T. Piolot, S. Ravalet, K. Kemnitz, J. Coppey, C. Durieux, V. Mignotte, M. Coppey-Moisan. Picosecond-hetero-FRET microscopy to probe protein-protein interactions in live cells. Biophys. J. 83 (2002): 3570–3577.

[198] Ulrich, V., P. Fischer, I. Riemann, K. König, Compact multiphoton / single photon laser scanning microscope for spectral imaging and fluorescence lifetime imaging, Scanning 26 (2004): 217–225.

[199] Urayama, P., M.-A. Mycek. "Fluorescence Lifetime Imaging Microscopy of Endogenous Biological Fluorescence." In *Handbook of Biomedical Fluorescence*, ed M-A. Mycek and B.W. Pogue, 211–236. New York, Basel: Marcel Dekker Inc., 2003.

[200] Van Munster, E.B., T.W.J. Gadella. ΦFLIM: a new method to avoid aliasing in frequency-domain fluorescence lifetime imaging microscopy. J. Microsc. 213/1 (2003): 29–38.

[201] Van Zandvoort, M.A.M.J., C.J. de Grauw, H.C. Gerritsen, J.L.V. Broers, M.G.A. Egbrink, F.C.S. Ramaekers, D.W. Slaaf. Discrimination of DNA and RNA in cells by a vital fluorescent probe: Lifetime imaging of SYTO13 in healthy and apoptotic cells. Cytometry 47 (2002): 226–232.

[202] Verveer, P.J., A. Squire, P.I.H. Bastiaens. "Frequency-Domain Fluorescence Lifetime Imaging Microscopy: A Window on the Biochemical Landscape of the Cell." In *Methods in Cellular Imaging* ed. Ammasi Periasamy, 273–294. Oxford: University Press, 2001.

[203] Wahl, M., I. Gregor, M. Patting, J. Enderlein. Fast calculation of fluorescence correlation data with asynchronous time-correlated single-photon counting. Opt. Expr. 11/26 (2003): 3583–3691.

[204] Wallrabe, H., M. Stanley, A. Periasamy, M. Barroso. One- and two-photon fluorescence resonance energy transfer microscopy to establish a clustered distribution of receptor-ligand complexes in endocytic membranes. Journal of Biomedical Optics 8/3 (2003): 339–346.

[205] Widengreen, J., R. Rigler. Mechanisms of photobleaching investigated by fluorescence correlation spectroscopy. Biomiaging 4 (1996): 146–159.

[206] White, J.G., W.B. Amos, M. Fordham. An evaluation of confocal versus conventional imaging of biological structures by fluorescence light microscopy. J Cell Biol. 105 (1987): 41–48.

[207] Wohland, T., R. Rigler, H. Vogel. The standard deviation in fluorescence correlation spectropscopy. Biophys. J. 80 (2001): 2987–2999.

[208] Yang, H., X.S. Xie. Probing single-molecule dynamics photon by photon. J. Chem. Phys. 117/24 (2002): 10965–10979.

[209] Youngworth, K.S., T.G. Brown. Focusing of high numerical aperture cylindrical-vector beams. Optics Express 7 (2000): 77–87.

22

Förster Resonance Energy Transfer (FRET)

Barry R. Masters and Peter T. C. So

22.1 INTRODUCTION

In this chapter we describe the theoretical and experimental basis for Förster resonance energy transfer (FRET). Since the concepts of resonance and dipole–dipole interaction are integral to the formulation of both the Perrin and the Förster theories of resonance energy transfer, their qualitative and quantitative aspects are introduced in terms of classical physics and quantum mechanics.

In these theoretical formulations of resonance energy transfer the energy is nonradiatively transferred between a donor molecule and an acceptor molecule in resonance. The donor is excited by light and the excitation energy is transferred to the acceptor molecule; simultaneously the donor molecule returns to the ground state. There are no collisions involved in this energy transfer mechanism and the transfer occurs over several molecular diameters. The trivial case of photon emission of one atom followed by absorption of that photon by another atom is excluded in our discussion. The mechanism of this resonance energy transfer involves very weak dipole–dipole interactions between the donor and the acceptor molecules. Note that when the donor is in the excited state nonradiative energy transfer competes with all the other deactivation mechanisms, including fluorescence, internal conversion, and intersystem crossing.

Since the concept of resonance in both classical and quantum physics is inextricably connected to the theories of resonance energy transfer, we include a qualitative and a quantitative review of this ubiquitous phenomenon.

We begin with the historical aspects so that the reader can see the progression from the early studies of sensitized fluorescence by metals in the gaseous state to studies of energy transfer by dye molecules in solution. Concomitant with the advances in experimental studies were several attempts to derive a theoretical model for the energy transfer. Initial models were based on classical physics, and these were superseded by model based on quantum theory.

While the emphasis of this chapter is on Förster resonance energy transfer, it is important to understand that there are a variety of energy transfer mechanisms between pairs of atoms, molecules, and molecular aggregates. They may be delineated by the strength of the molecular coupling (strong, weak, and very weak) and the mechanism of the interaction (dipole–dipole or quadrapole interaction). For example, the Dexter theory (1953) of energy transfer is appropriate for crystals and molecular aggregates such as chloroplasts.

We strongly recommend that the reader study the original papers and review articles on FRET that are listed in the references. FRET is one type of radiation energy transfer mechanism, and it is highly instructive for the reader to study the original papers that compare and contrast the various mechanisms. The differences among these energy transfer mechanisms were described by several authors: a theory of sensitized luminescence in solids (Dexter, 1953); transfer mechanisms of electronic excitation (Förster, 1959, 1960); delocalized excitation and excitation transfer (Förster, 1965); classical aspects of energy transfer in molecular systems (Kuhn, 1970); energy transfer, charge transfer, and proton transfer in molecular composite systems (Kasha, 1991); generalized Förster-Dexter theory of photoinduced intramolecular energy transfer (Lin et al., 1993); and a very good review of long-range resonance energy transfer in molecular systems (Scholes, 2003).

22.2 FRET THEORY AND EXPERIMENTAL STUDIES

Collisions of particles on the atomic and subatomic scales are associated with many advances in atomic, nuclear, and particle physics. The Rutherford scattering experiments on atomic structure, the Compton scattering experiments, and the experimental verification of elementary particles all involve scattering or collision experiments.

Similarly, collision experiments were important in the development of energy transfer mechanisms. In 1913 Franck and Hertz performed energy transfer studies in which they showed the existence of energy transfer between a colliding high-speed electron and a slower moving atom. They defined "collisions of the first kind" as those in which there was a conversion of translational energy into electronic (internal) energy. In 1921 Klein and Roseland posited the reverse process between atoms in excited states and electrons that resulted in the atoms returning to the ground state and the electrons acquiring kinetic energy. The conversion of electronic energy into kinetic energy as the result of collisions is defined as "collisions of the second kind." Franck expanded the definition to include collisions between atoms or molecules. Franck and Jordan summarized all of their experimental findings on collision processes and electronic excitation up to the year 1926 in their book *Anregung von Quantensprügen durch Stösse* [*Excitation from quantum jumps through collisions*] (Franck & Jordan, 1926).

Another recommended review of collision processes with excited atoms is the book *Resonance radiation and excited atoms* (Mitchell & Zemansky, 1934).

22.2.1 Studies of Sensitized Fluorescence

In 1922 Cario published his experimental observations on what he named "sensitized fluorescence" (Cario, 1922). It is interesting and instructive to recall his experiment. He introduced mercury vapor and thallium vapor into a quartz tube at a defined pressure and temperature. The mean-free path between collisions and the kinetic energy of the atoms was fixed at a given temperature. It was possible to alter the kinetic energy of the atoms by varying the temperature of the quartz tube, which was contained in an electric furnace. The quartz tube was illuminated with the emission line at 2,536 Å from a water-cooled, quartz-mercury lamp. A spectroscope was used to analyze the light emitted from the quartz tube. The emission from the tube was shown to be from the thallium atoms. Cario posited the following mechanism: the mercury atom absorbed the 2,536 Å radiation, which induced a transition from the lower ground state to the upper excited state. Upon collision with a thallium atom the excited mercury atom transferred its energy to the thallium atoms, exciting the thallium atoms to the excited state, which then emitted their characteristic radiation and returned to the ground state. In the absence of the mercury atoms in the quartz tube, the thallium atoms were not excited by the 2,536 Å radiation. The theoretical requirement for this energy transfer to occur between the two different atoms is that the time between collisions is the same order of magnitude as the mean lifetime of the excited state of the mercury atom. The excitation of the mercury atom is due to the absorption of the radiation. The fact that the thallium atom acquired a large velocity from collision with the excited state of the mercury atom is easily verified by measuring the Doppler shift on the emission line of the thallium atom. The high velocity following the collision would result in a line broadening due to the Doppler shift.

This type of mechanism was named "sensitized fluorescence" since the mercury atoms sensitized the thallium atoms to the 2,536 Å light from the mercury lamp. Similar results were obtained if silver atoms replaced the thallium atoms in the quartz tube.

The following year Cario and Franck expanded their investigation into this phenomenon (Cario & Franck, 1923). In their publication they detail the experimental apparatus that they used to investigate sensitized fluorescence, and more importantly they present experimental evidence that the observed energy transfer occurs simultaneously and not in a sequential mechanism involving two separate steps. Their conclusion that the energy transfer occurs simultaneously is based on the measurement of the intensities of the thallium emission lines and the measurement of Doppler broadening of emission line widths.

In order to achieve a high probability of energy transfer between two atoms or between two molecules, it is necessary that the electronic transition energy be very similar. The concept of resonance energy transfer incorporates this energy requirement.

22.3 PHYSICS BACKGROUND FOR UNDERSTANDING FRET

22.3.1 Resonance and Energy Transfer Between Two Weakly Coupled Harmonic Oscillators

The phenomenon of resonance is ubiquitous in physics and is easily understood by studying mechanical resonance in coupled pendulums. Another example of mechanical resonance is when two tuning forks of similar frequency are attached to a common base. The mechanical energy of the first vibrating tuning fork will be transferred through the base to the second tuning fork, and in absence of damping the energy will oscillate between the two tuning forks.

We are well aware of resonance in atomic and molecular spectroscopy. As the exciting radiation approaches the frequency for a transition from the ground state to the excited state, the atom or molecule will absorb a photon. Resonance is central to the theories of both Jean Perrin and Francis Perrin, and it is the central concept in Förster's theory of resonance energy transfer (FRET).

We begin with the following qualitative introduction to resonance that is based on Fermi's review article of Dirac's radiation theory (Fermi, 1932). Fermi presents a simple model for resonance: that of a pendulum and an oscillating string located near the pendulum. The pendulum could be an analogy for an atom acting as an oscillating dipole, and the oscillating string could be the radiation field. When the string and the pendulum are not connected, both oscillate independently and the total energy is the sum of the energy of the pendulum and the string. There is no interaction term.

Fermi represents the coupling term by a very thin elastic thread that connects the oscillating pendulum and the oscillating string. In the mathematical theories of both the Perrins and of Förster, they related resonance energy transfer to this coupling term, which is given by a dipole–dipole interaction.

At time $t = 0$, the string is vibrating and the pendulum is stationary. This corresponds to the conditions in which the donor atom is in the excited state and the acceptor atom is in the ground state. When the thin elastic thread is connected to the oscillating pendulum, the vibrating string energy can be transferred from the vibrating string to the pendulum. If the period of oscillation of the pendulum and the period of vibration of the string are very different, then the amplitude of oscillation of the pendulum remains very small. However, if the period of the vibration of the string is equal to the period of oscillation of the pendulum, then the amplitude of the oscillation of the pendulum will increase by a large amount, limited by damping forces such as friction. That is the condition of resonance: the frequency of the driving force is equal to the natural or "resonant" frequency of the pendulum.

Now we discuss resonance in coupled harmonic oscillators in a more quantitative manner. First we introduce some definitions that are useful to define harmonic motion. A periodic motion repeats itself over and over in a finite region of space and therefore is also bounded. The time required to return to the original position is the period of the motion, T. The frequency of the motion is the number of complete cycles per second is given as $1/T$. The angular displacement in a single period is 2π radians, and the angular frequency is $\omega = 2\pi/T$. The periodic motion of a particle can be graphed in the xy plane as a periodic function of product of angular frequency and time (ωt). A is the radius of the circular orbit, and $x = A\cos(\omega t)$ and $y = A\sin(\omega t)$. Motion that can be described by these two equations is called harmonic.

We now further develop the concept of resonance by introducing the case of two identical coupled harmonic oscillators and the concept of normal modes, and normal coordinates of a complex oscillating system (Fano & Fano, 1959; Marion, 1970). We can always describe the motion of any complex coupled oscillatory system in terms of normal coordinates. Normal coordinates have a special property: each normal coordinate oscillates with a defined frequency. Although the rectangular (x, y, z) coordinates of a complex oscillatory system are coupled, the normal coordinates have no coupling among them. We can set the initial condition of the coupled system so that only one normal coordinate varies with time, and we call that motion a normal mode of the system; it is always possible to independently excite any one of the normal modes. For a coupled oscillatory system with n degrees of freedom there are n normal modes of vibration. The total motion of the coupled oscillatory system can be described as the linear superposition of all of the normal modes.

For a single pendulum of mass M and length L, a small oscillation can be considered as a simple harmonic motion. The natural or resonant angular frequency $\omega_0 = \sqrt{g/L}$, where g is the gravitational constant. The period of the natural oscillation is:

$$T = \frac{2\pi}{\omega_0} = 2\pi\sqrt{\frac{L}{g}}. \tag{22.1}$$

Now we give the physical example of two coupled harmonic oscillators: two coupled pendulums. There are two pendulums, with a natural or resonant frequency of oscillation $v_0 = \omega_0/2\pi$, which are coupled by a very light spring. At first one pendulum is oscillating and the second pendulum is stationary. Due to resonance, the mechanical energy of the first pendulum is transferred to the second pendulum. The second pendulum will increase in the amplitude of its oscillation until all the mechanical energy of the first pendulum is transferred to the second pendulum. At that point the first pendulum is stationary and the second pendulum is oscillating at maximum amplitude. Now the process reverses and the second pendulum drives the first pendulum into oscillation. The second pendulum will become stationary and the first pendulum will oscillate with maximum amplitude. This oscillatory transfer of energy between the two pendulums is the phenomenon of "beats." In the absence of energy loss due to damping or friction the "resonance" transfer of mechanical energy will continue indefinitely. For the case of two pendulums that have slightly different resonant frequencies, the mechanical energy will still oscillate between the two pendulums; however, not all of the energy will be transferred from one pendulum to the other.

The motion of the system consisting of the two coupled pendulums can be mathematically expressed as the sum of two normal modes, which will be described in detail in the next section. The two equations describe the amplitude of the displacements of the pendulums versus time; their phase difference is 180 degrees. Therefore, when one pendulum is oscillating at maximum amplitude, the other pendulum is stationary. Then, at a later time, the amplitudes of the two pendulums are reversed. Thus, the energy is cyclically transferred between the two pendulums.

The two identical coupled pendulums or oscillators have two "normal modes" of vibration, one mode (first mode) in which the two pendulums oscillate in phase and with equal amplitude, and one mode (second mode) in which the two pendulums oscillate with equal amplitude but with opposite phase. In the first mode, both pendulums oscillate in phase, and the weak spring that connects them is never stretched. The two modes have different frequencies in general equal to v_1 and v_2.

The mathematical description of the composite system of the two coupled pendulums involves a linear combination of the oscillations of the two normal modes.

For the two coupled oscillators, the displacements for each pendulum from equilibrium positions are α and β; these displacements vary with time, and the amplitude of the oscillation of each pendulum is given by A.

We previously stated that it is always possible to define a set of coordinates (normal coordinates) that have a simple harmonic time dependence and correspond to the various normal modes of the complex oscillatory system. The two normal modes are described as follows, in which the subscripts 1 and 2 refer to the first and the second normal modes.

The first normal mode of the motion is:

$$\alpha_1 = A\cos 2\pi v_1 t \quad \text{and} \quad \beta_1 = A\cos 2\pi v_1 t. \tag{22.2}$$

The second normal mode of the motion is:

$$\alpha_2 = A \cos 2\pi \nu_2 t \quad \text{and} \quad \beta_2 = -A \cos 2\pi \nu_2 t. \tag{22.3}$$

We define two new variables, which are the sum of the displacements and the difference of the displacements.

$$\xi = \alpha + \beta \quad \text{and} \quad \eta = \alpha - \beta \tag{22.4}$$

The first normal mode is now given as:

$$\xi_1 = \alpha_1 + \beta_1 = 2A \cos 2\pi \nu_1 t, \quad \text{and} \quad \eta_1 = \alpha_1 - \beta_1 = 0. \tag{22.5}$$

The second normal mode is now given as:

$$\xi_2 = \alpha_2 + \beta_2 = 0, \quad \text{and} \quad \eta_2 = \alpha_2 - \beta_2 = 2A \cos 2\pi \nu_2 t. \tag{22.6}$$

In this representation the motion of the two normal modes is decoupled.

Now we consider the case with the following initial conditions: pendulum 2, its motion characterized by β, is at its equilibrium position and pendulum 1, its motion characterized by α, is initially displaced from its equilibrium position with amplitude A. This initial condition can be expressed as a combination of the two normal modes in the following manner:

$$\xi_3 = \xi_1 + \xi_2 = A \cos 2\pi \nu_1 t \tag{22.7}$$

$$\eta_3 = \eta_1 + \eta_2 = A \cos 2\pi \nu_2 t \tag{22.8}$$

$$\xi_3 = \alpha_3 + \beta_3 \tag{22.9}$$

$$\eta_3 = \alpha_3 - \beta_3 \tag{22.10}$$

Now we solve for α_3 and β_3

$$\alpha_3 = \frac{A}{2} (\cos 2\pi \nu_1 t + \cos 2\pi \nu_2 t) \tag{22.11}$$

$$\beta_3 = \frac{A}{2} (\cos 2\pi \nu_1 t - \cos 2\pi \nu_2 t) \tag{22.12}$$

We define new frequency variables in terms of the frequencies ν_1 and ν_2, associated with normal mode 1 and normal mode 2.

$$\nu_0 = \frac{\nu_1 + \nu_2}{2} \quad \text{and} \tag{22.13}$$

$$\delta = \frac{\nu_1 - \nu_2}{2} \tag{22.14}$$

Then we solve for ν_1 and ν_2, which are the frequencies of the two normal modes.

$$\nu_1 = \nu_0 + \delta \tag{22.15}$$

$$\nu_2 = \nu_0 - \delta \tag{22.16}$$

Finally, we arrive at the final result, which yields the periodic motion for the two pendulums in terms of their displacements.

$$\alpha_3 = A \cos(2\pi \nu_0 t) \cos(2\pi \delta t) \qquad (22.17)$$

$$\beta_3 = A \sin(2\pi \nu_0 t) \sin(2\pi \delta t) \qquad (22.18)$$

The first term in these expressions represents the carrier frequency and the second term represents the beat frequency. It is the beat frequency that determines the rate of cyclical energy transfer between the two coupled pendulums, and the time for the energy to be transferred from one pendulum to the second pendulum is one period of the beat frequency. Beats occur between two oscillations when the frequencies are very close together. They occur if the driving frequency is close to, but not exactly equal to, the natural oscillation frequency of the system. They are the slow variation of amplitude at half the normal mode frequency difference. Figure 22.1 shows the plot of the amplitude of the motion of the two pendulums as a function of time. The two

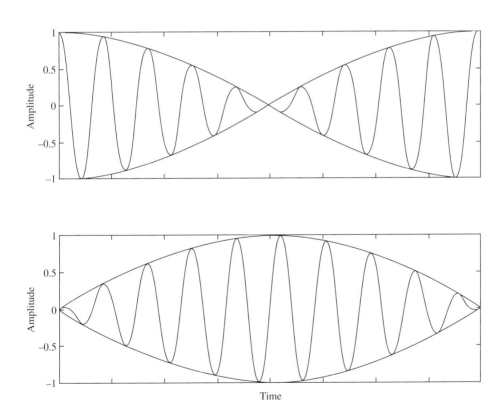

Figure 22.1. Diagrams that illustrate the formation of "beats" from the superposition of two normal modes of two weakly coupled harmonic oscillators. Each normal mode is out of phase with the other normal mode by 180 degrees (upper and lower diagrams). In each diagram the amplitude of the oscillation, normalized to 1, is plotted versus time. The lines show the envelope of the motion and correspond to the "beat" frequency. The high-frequency component oscillates with the slowly varying component. Energy is transferred back and forth between the two oscillators.

motions are 180 degrees out of phase due to the phase difference of the cosine and the sine functions.

Note that the two pendulums can exchange energy, but there is no exchange of energy between the two normal modes, which are independent. There is complete energy exchange between the two pendulums if the masses are identical and the ratio of the sum and the difference frequencies is an integer.

The motion of the combined system of two coupled pendulums is the combination of both normal modes with equal amplitudes. The coordinates are now uncoupled and are therefore independent. This is an example of the superposition of the normal modes, sometimes called the superposition principle, in which both normal frequencies are in the equations.

A mechanical analogy for the case of a donor and an acceptor molecules involved in FRET is when the energies of the two molecules differ slightly, the condition of near-resonance. For example, the energies for the first excited state differ slightly for the two molecules. In that case the first pendulum will drive the second pendulum into oscillation, but not all of the energy will be transferred. The equation of motion for the system of two nearly equal pendulums is still a linear combination of the two normal modes, but in this case there are slightly different coefficients. Furthermore, these equations yield the period of the resonance τ, which is the time required for the coordinate of one oscillator to change from its maximum amplitude to its minimum amplitude and back to its maximum amplitude during the resonance transfer of energy between the two oscillators.

There is an analogy in quantum mechanics. For near-resonance between two atoms with slightly differing energies that are weakly coupled, their combined stationary state can be represented by a linear combination of the stationary states of the individual atoms. This is correct for any pair of identical weakly coupled systems. It is also correct for any pair of weakly coupled states that have nearly equal energies.

As the driving frequency approaches the natural or "resonant" frequency of the mechanical system, the amplitude or the energy of the oscillation increases; however, it does not increase towards infinity due to the damping term in the differential equation. The quantitative description of the damped harmonic oscillator was described in Chapter 5, in which the interaction of an atom and the radiation field to explain light absorption of atoms was discussed; therefore, the quantitative details are not repeated here. The damped harmonic oscillator often appears in both classical physics and quantum mechanics (i.e., in the dispersion theory of light, in the two-photon absorption theory of Maria Göppert-Mayer, in Planck's radiation theory, and in Dirac's radiation theory). In many theories of atom–radiation interaction the atom is considered to be a charged oscillating dipole, and the radiation field is considered to be uncoupled harmonic oscillators in a cavity.

In mechanical resonance, the amount of energy transfer between two oscillators is a continuous variable. In quantum mechanical systems, the energy is treated as discrete quanta that cannot be divided into smaller units. In resonance energy transfer of weakly coupled molecules, the probability of being in the excited state is "transferred" from the donor molecule to the acceptor molecule. The energy is either in the donor excited state or in the acceptor excited state, and time-dependent perturbation theory is used to solve the time-dependent Schrödinger equation for the probability of this energy transfer.

22.3.2 Quantum Mechanical Resonance*

It was Heisenberg who first noticed that there was a quantum mechanical analogy to the known classical treatments of systems in resonance (Heisenberg, 1926, 1927). The quantum mechanical analysis of the harmonic oscillator was given in a 1926 paper by Schrödinger.

In quantum mechanical systems, the resonance phenomenon (Pauling & Wilson, 1935) occurs between a pair of stationary states that have nearly equal energies; this is a situation of weak coupling. The excited state of each oscillator in the coupled system can be represented at a given time as the linear supposition of the individual stationary states for each separate oscillator. This linear superposition of these two states is not a stationary state and evolves in time. That is the quantum analog of the superposition of two normal modes in a mechanical system of coupled oscillators For a weakly coupled system of two nearly identical oscillators at resonance, the probability for the first oscillator to be in the excited state will vary periodically with time, and a similar situation will occur for the second oscillator. However, the probability of being in an excited state will be out of phase for the two oscillators; thus, first one oscillator will have the maximum probability to be in the excited state, and then the other oscillator will have the maximum probability to be in the excited state. The quantum mechanical probability amplitude may be thought of in the corresponding classical case of the mechanical amplitude of the two coupled pendulums. The resonance frequency is the frequency at which the quantum mechanical oscillators repeatedly exchange the states of large probability amplitude.

One physical system that exhibits quantum resonance is the resonant transfer of excitation energy between similar atoms. It is observed that two similar atoms in proximity can interact in such a manner that the energy of the excited state of first atom is transferred to the second atom, thereby inducing the transition from the ground state to the excited state. This process is shown as: $A^* + A \rightarrow A + A^*$, in which the asterisk denotes the excited state of the molecule. There is a nonradiative transfer of energy between the two atoms.

The atoms can be modeled as two coupled quantum mechanical oscillators. Without coupling, $\Psi_1(\vec{r}_1, \vec{r}_2)$ is the wave function for the state in which the first atom is in the excited state and $\Psi_2(\vec{r}_1, \vec{r}_2)$ is the wave function for the state in which the second atom is in the excited state. These two states have identical energy. In the presence of coupling interaction, first-order degenerate perturbation theory is used to determine the eigenfunctions and the eigenvalues for the combined system. The excited state of the combined system has two new energy levels with different energies. The wave functions for these two energy levels are the superposition of $\Psi_1(\vec{r}_1, \vec{r}_2)$ and $\Psi_2(\vec{r}_1, \vec{r}_2)$:

$$\Psi_a = \frac{1}{\sqrt{2}}[\Psi_1 + \Psi_2] \text{ and} \tag{22.19}$$

$$\Psi_b = \frac{1}{\sqrt{2}}[\Psi_1 - \Psi_2]. \tag{22.20}$$

This resonance frequency for the exchange is given by the separation of the energy levels of the two nonstationary states formed by the superposition of unperturbed states divided by Planck's constant h. In our general discussion, we have not considered the exact nature of the interaction term that is needed to calculate the actual eigenvalues and eigenfunction and, hence, the rate of energy transfer.

22.3.3 Dipole–Dipole Interactions*

Since the Perrin and the Förster (1948) theories of resonance energy transfer involve dipole–dipole interactions, a review of the electromagnetic theory is appropriate.

First we present some material on an oscillating electric dipole. Consider a dipole on the z axis located on the origin. It consists two charges, $+q$ and $-q$, with a separation d. A point P is located at a distance $\vec{r}$ from the origin and at an angle θ from the z axis.

The electric dipole is oscillating and its moment can be described as:

$$\vec{p}(t) = \vec{p}_0 \cos(\omega t). \tag{22.21}$$

The maximum value of the electric dipole moment is $\vec{p}_0 = q\vec{d}$.

Since an accelerated electric charge radiates energy, we wish to express this in a mathematical form. The energy that is radiated by an oscillating electric dipole is determined by the Poynting vector $\vec{S}$.

$$\vec{S} = \frac{1}{\mu_0} \left(\vec{E} \times \vec{B} \right), \tag{22.22}$$

where μ_0 is the permeability of free space, $\vec{E}$ is the electric field strength, and $\vec{B}$ is the magnetic field strength, and they are connected by the vector cross-product $\times$.

The intensity of the radiation from the oscillating electric dipole is a time average over one complete cycle of the oscillation:

$$\left\langle \vec{S} \right\rangle_{timeaverage} = \left(\frac{\mu_0 \vec{p}_0^2 \omega^4}{32\pi^2 c} \right) \frac{\sin^2 \theta}{\vec{r}^2} \hat{r}, \tag{22.23}$$

where $\hat{r}$ is a unit vector of magnitude one in the direction r.

For this oscillating electric dipole the radiation is in the form of spherical waves. However, for large distances these waves can be correctly approximated by plane waves. Note that there is no radiation along the z axis ($\theta = 0$).

The total power P radiated is calculated by integrating the time-averaged Poynting vector over a sphere of radius $\vec{r}$:

$$\langle P \rangle_{total} = \frac{\mu_0 \vec{p}_0^2 \omega^4}{12\pi c}. \tag{22.24}$$

The concept of resonance is central to the development of FRET, as is the assumption of dipole–dipole interaction between the donor and the acceptor in the theories of the Perrins (father and son) and also that of Förster. Therefore it is crucial to explain exactly what is a dipole–dipole interaction.

The concept of an electric dipole is an extremely useful physical approximation. What we seek is the energy of interaction between two electric dipoles (Brau, 2004). Assume that there are two electric dipoles, $\vec{p}_1$ and $\vec{p}_2$. The first electric dipole is on the origin of the coordinate system and the second electric dipole is located at a distance $\vec{r}$ from the origin. This calculation involves two steps: the first step is to calculate the potential at point $\vec{r}$ due to the electric dipole at the origin, and the second step is to calculate the electric field from that potential. Finally, the energy of the interaction of the two dipoles is given as the scalar or dot product of the electric field and the electric dipole $\vec{p}_2$.

The potential at point $\vec{r}$ due to the electric dipole located at the origin is:

$$\Phi(r_2) = \frac{1}{4\pi\varepsilon_0} \frac{\vec{p}_1 \cdot \vec{r}}{\vec{r}^3}. \tag{22.25}$$

The electric field that results from this potential is:

$$\vec{E}(\vec{r}) = -\nabla\Phi = \frac{1}{4\pi\varepsilon_0} \left(3\frac{\vec{p}_1 \cdot \vec{r}}{\vec{r}^5}\vec{r} - \frac{\vec{p}_1}{\vec{r}^3} \right). \tag{22.26}$$

The energy of the interaction between these two dipoles is:

$$W = -\vec{E} \cdot \vec{p}_2 = \frac{1}{4\pi\varepsilon_0} \left[\frac{\vec{p}_1 \cdot \vec{p}_2}{\vec{r}^3} - 3\frac{\left(\vec{p}_1 \cdot \vec{r}\right)\left(\vec{p}_2 \cdot \vec{r}\right)}{\vec{r}^5} \right]. \tag{22.27}$$

There are several important points to note. First, the electric field from a pair of equal and opposite charges that are separated by a distance d has the identical electric field distribution as an electric dipole if the separation of the two charges is much less than the distance $\vec{R}$ from the charges. Second, if $\vec{R} >> d$, then the electric field distribution from the two charges can be identically replaced with the electric field distribution of an electric dipole. The same result is true for any distribution charges for which the sum of all the charges is neutral.

The potential of an electric dipole decreases as $1/\vec{r}^2$. The electric field strength $\vec{E}$ of the dipole decreases as $1/\vec{r}^3$. That means the electric field strength varies inversely as the cube of the distance from the electric dipole. If a point $P(x, y, z)$ is at a distance $\vec{R}$ away from an arbitrary grouping of electric charges that are electrically neutral as a whole, the potential is a dipole potential that decreases as $1/\vec{R}^2$.

The spatial distributions for both a static electric dipole and an oscillating electric dipole $\vec{p} = q \cdot \vec{d} = \vec{p}_0 \sin \omega t$ are identical.

The energy of interaction between two electric dipoles is a function of both the distance of separation between the electric dipoles $\vec{R}$ and the relative orientation of the two electric dipoles $\vec{p}_1$ and $\vec{p}_2$. For simple cases in which the separation distance $\vec{R}$ is a fixed value, and the relative orientation between the two electric dipoles is fixed, then a simple calculation yields a unique energy for the interaction. However, in applications to FRET between donor and acceptor molecules, there may be a distribution of both the distances $\vec{R}$ and the relative orientations of the electric dipoles of the acceptor and the donor molecules. For those cases it is necessary to average over all possible distributions of both the separation distance and the relative orientation in order to calculate the correct interaction energy.

If one assumes that the orientation of the acceptor molecule is random, the average relative orientation of the two electric dipoles can be derived by the Boltzmann distribution at thermal equilibrium (Atkins, 1990).

The energy of interaction of two freely rotating electric dipoles at a fixed separation r is:

$$\langle W \rangle = \frac{p_1 p_2}{4\pi \varepsilon_0 r^3} \langle \kappa \cdot f \rangle, \tag{22.28}$$

where f is a weighting factor equal to the probability of finding a specific orientation θ between the electric dipoles. The symbol $\langle \rangle$ denotes an average of the quantity within the brackets.

$$f \propto e^{-W/kT}, \tag{22.29}$$

where E is the potential energy of interactions of the two electric dipoles at a given orientation, k is the Boltzmann constant, and T is the absolute temperature.

To evaluate the Boltzmann factor, we assume $W \prec\prec kT$. The interaction energy at a given κ is:

$$W \approx \frac{p_1 p_2 \kappa}{4\pi \varepsilon_0 r^3}. \tag{22.30}$$

The next step is to expand f in a power series:

$$f \propto 1 - \frac{W}{kT} + \cdots \tag{22.31}$$

The higher-order terms are dropped.

$$\langle W \rangle \propto \frac{p_1 p_2}{4\pi \varepsilon_0 r^3} \left(\langle \kappa \rangle - \frac{p_1 p_2 \langle \kappa^2 \rangle}{4\pi \varepsilon_0 kTr^3} + \cdots \right) \tag{22.32}$$

$$\langle \kappa \rangle = 0, \quad \text{but} \quad \langle \kappa^2 \rangle \neq 0 \quad \text{since} \quad \kappa^2 \succ 0.$$

$$\langle W \rangle \propto -\frac{p_1^2 p_2^2 \langle \kappa^2 \rangle}{(4\pi \varepsilon_0)^2 kTr^6}. \tag{22.33}$$

The orientation factor κ^2 ranges from 0 to 4; therefore, $\langle \kappa^2 \rangle \approx 1$. It can be shown that for free rotations of both electric dipoles the quantity

$$\langle \kappa^2 \rangle = \frac{2}{3}. \tag{22.34}$$

The details of this derivation and the assumptions required are fully described in a comprehensive review of the role of the orientation factor in FRET theory (Clegg, 1996). When the Förster theory of resonance energy transfer is presented, the role of these equations will become apparent.

22.3.4 Fermi's Golden Rule*

The Förster theory of resonance energy transfer incorporates Fermi's Golden Rule. Despite the name, it is actually based on the work of Dirac, who developed it in 1927 as part of his time-dependent perturbation theory. Fermi used it in his lectures on quantum mechanics at the University of Chicago and called it "the golden rule," which later became associated with his name (Fermi, 1995).

Electronic transitions are calculated in quantum mechanics with Schrödinger's time-dependent perturbation theory. Perturbation theory yields approximate solutions for these quantities in the presence of an extremely small additional term added to the Hamiltonian, which is due to a perturbation or very small interaction imposed on the system (i.e., an electric field from an adjacent atom or incident light).

$$\left(\hat{H}_0 + \hat{H}'\right)\Psi = i\hbar\frac{\partial\Psi}{\partial t}. \tag{22.35}$$

where the Hamiltonian or energy operator is:

$$\hat{H} = \hat{H}_0 + \hat{H}'. \tag{22.36}$$

$\hat{H}_0$ is the Hamiltonian in the absence of the perturbation, and the term $\hat{H}'$ is a time-dependent oscillating perturbation (i.e., an oscillating electric dipole or a radiation field).

The Schrödinger equation without any perturbation has eigenfunctions φ_n and corresponding eigenvalues or energies E_n. The eigenfunctions are linear combinations of the stationary states that have eigenvalues E_n.

$$\Psi = \sum_n c_n(t)\varphi_n e^{-iE_n t/\hbar}. \tag{22.37}$$

The probability that the system originally in the initial state i will be in a final state f after a time t is: $|c_n(t)|^2$.

The transition is from a single initial ground state to a manifold of states in the final state. For the condition of exact resonance,

$$\Delta E = E_f - E_i = \hbar\omega. \tag{22.38}$$

$$|c_n(t)|^2 = 4\left|\frac{W_{fi}}{\hbar}\right|^2 \frac{\sin^2\frac{1}{2}(\omega_{fi} - \omega)t}{(\omega_{fi} - \omega)^2} \tag{22.39}$$

where

$$\omega_{fi} = \frac{E_f - E_i}{\hbar}. \tag{22.40}$$

The interaction term W_{fi} that induces the transition is defined as follows. The electric dipole moment operator is:

$$\vec{\mu} = -e\vec{r}, \tag{22.41}$$

where the charge on the electron is e, and $\vec{r}$ is a displacement.

The electric dipole transition moment is:

$$\mu_{fi} = \int \Psi_f^* \vec{\mu} \Psi_i d^3 \vec{r}. \tag{22.42}$$

The interaction term W_{fi} is given by the electric dipole transition moment $\vec{\mu}_{fi}$ and the electric field strength $\vec{E}$. The probability of a transition $P(t)$ from an initial ground state to a manifold or band of higher states is:

$$P(t) = \frac{2\pi \left| \vec{\mu}_{fi} \cdot \vec{E} \right|^2 t}{\hbar} \rho \left(\omega_{fi} = \omega \right) \tag{22.43}$$

where $\rho(\omega)$ is the density of states.

The density of final states is determined at the frequency ω of the incident radiation. The rate of transition R_{fi} is:

$$R_{fi} = \frac{dP(t)}{dt} = \frac{2\pi \left| \vec{\mu}_{fi} \cdot \vec{E} \right|^2}{\hbar} \rho \left(\omega_{fi} = \omega \right). \tag{22.44}$$

This equation is called Fermi's Golden Rule. The transition rate is given by the product of three factors:

$$\frac{2\pi}{\hbar} \times \text{ the square of the interaction } \times \text{ density of final states.}$$

For a more in-depth discussion, see chapter 5.

22.3.5 Born-Oppenheimer Approximation, the Franck-Condon Principle, and the Franck-Condon Integrals*

The Born-Oppenheimer approximation will be discussed and then the Franck-Condon integral will be described, since both are critical to the Förster mechanism of vibrational-relaxation resonance energy transfer.

In quantum mechanical terms, the electronic transition occurs between two wave functions that have the greatest overlap integral of all the vibronic states of the upper electronic state. A vibronic transition is defined as an electronic transition that derives its intensity from the vibration of a molecule.

The Born-Oppenheimer approximation, which is often termed the adiabatic approximation, is a consequence of the fact that in electronic transitions the masses of the nuclei are much greater then the masses of the electrons, and therefore electronic transitions occur in a much shorter time scale than rearrangements of the nuclei. That is why we diagram electronic transitions as vertical transitions; the nuclei remain fixed during the time of the electronic transition.

The Franck-Condon principle is a related concept. Since the masses of the nuclei are much greater than the masses of the electrons, electronic transitions occur before the nuclei can change their positions. A molecule undergoes a transition from the ground state to an upper electronic state; this transition results in an upper vibronic state that most closely resembles the vibronic ground state of the lower electronic state.

Therefore, the Franck-Condon principle is consistent with the experimental finding that the most intense transition from a ground state to an excited state is from the ground vibrational state to the vibrational state situated vertically above it. It is important to note that electronic transitions in molecules are broad and consist of many possible transitions; however, the Franck-Condon principle is related only to the most intense transition.

Now that we have presented a qualitative description of the Born-Oppenheimer approximation, we present the quantum mechanical formulation. In the approximation the total wave function for a molecule can be considered to be made up of a product of two factors at a fixed position of the nuclei. One factor depends on the positions of the nuclei $\vec{R} = (\vec{R}_i \ldots \vec{R}_m)$ and another factor depends on the positions of the electrons $\vec{r} = (\vec{r}_i \ldots \vec{r}_n)$.

$$\Psi\left(\vec{r}_1 \ldots \vec{r}_n, \vec{R}_1 \ldots \vec{R}_m\right) = \Psi_{el}\left(\vec{r}_1 \ldots \vec{r}_n, \vec{R}^{fixed}\right) \cdot \Psi_{vib}\left(\vec{r}^{fixed}, \vec{R}_1 \ldots \vec{R}_m\right).$$
(22.45)

The wave function can be expressed as a product of the electronic wave function Ψ_{el} and the vibronic wave function Ψ_{vib}. The electronic part of the wave function disregards the motion of the nuclei, which are considered to be at fixed positions. Equivalently, the vibronic part disregards the motion of the electrons.

$$\Psi = \Psi_{el}\left(\vec{r}\right) \Psi_{vib}\left(\vec{R}\right)$$
(22.46)

The electronic transition dipole moment between the initial state i and the final state f is:

$$\vec{\mu}_{fi} = -e\left\langle \Psi_f^* |\vec{r}| \Psi_i \right\rangle.$$
(22.47)

The asterisk denotes the complex conjugate.

Consider an electronic transition between a state at electronic level E and vibronic level v to a state at electronic level E' and vibronic level v', where the primed quantities represent the excited state. The transition moment can be expressed as:

$$\vec{\mu} = -e \int \int \Psi_{el'}^*\left(\vec{r}\right) \Psi_{vib'}^*\left(\vec{R}\right) r\Psi_{el}\left(\vec{r}\right) \Psi_{vib}\left(\vec{R}\right) d^3\vec{r} d^3\vec{R}.$$
(22.48)

$$\vec{\mu} = -e \int \Psi_{el'}^*\left(\vec{r}\right) \vec{r} \Psi_{el}\left(\vec{r}\right) d^3\vec{r} \cdot \int \Psi_{vib'}^*\left(\vec{R}\right) \Psi_{vib}\left(\vec{R}\right) d^3\vec{R}.$$
(22.49)

The second factor is:

$$S_{v,v'} = \int \Psi_{vib'}^*\left(\vec{R}\right) \Psi_{vib}\left(\vec{R}\right) d^3\vec{R}.$$
(22.50)

$S_{v,v'}$ is the overlap integral between the initial vibrational wave function and the final vibrational state wave function. The magnitude of the magnitude integral is a measure of the similarity between the vibrational wave function of the initial and final states. The intensity of a transition from vibrational level v of the ground electronic state to vibrational level v' of the excited electronic state is proportional to the square of the transition dipole moment. Therefore, the intensity is proportional to $S_{v,v'}^2$.

The Franck-Condon factor or integral for the transition between $E, v \rightarrow E', v'$ is $S_{v,v'}^2$. The greater the Franck-Condon factor, the greater the absorption intensity of the transition.

22.4 THEORIES AND EXPERIMENTS OF RESONANCE ENERGY TRANSFER UP TO AND INCLUDING THE WORK OF FÖRSTER

22.4.1 The Perrins' Theories and the Kallmann and London Theory of Resonance Energy Transfer

The pioneering contributions of the Perrins, father and son Jean (1870–1942) and Francis Perrin (1901–1992), to molecular luminescence provides an interesting historical insight (Berberan-Santos, 2001). The first observations of sensitized fluorescence in solution were made by J. Perrin and C. R. Choucroun in 1929. The mechanism of resonance transfer or transfer by inductive resonance is responsible for sensitized fluorescence in atoms or molecules. The phenomenon of concentration depolarization of fluorescence in solution is brought about by resonance transfer between like molecules in a series of repetitive steps.

Jean Perrin received the 1926 Nobel Prize in Physics for his elegant experiments on Brownian motion that demonstrated the existence of atoms and molecules. In 1918 Jean Perrin published a mechanism for resonance energy transfer based on classical physics; this was extended over the subsequent years (J. Perrin, 1918, 1925, 1927). His theory included the concept that nonradiative energy transfer (without the emission and reabsorption of a photon) based on electrodynamic interactive processes could take place between a molecule in the excited state and nearby molecules. Jean Perrin's theory was based on the resonance between two atoms, which are represented by charged oscillators. Their interaction is dominated by the classical dipole–dipole interaction. In the pre-quantum theory, energy could be transferred between two charged oscillators in continuous amounts of energy (J. Perrin, 1925).

Jean Perrin explained excitation transfer by resonance in terms of classical physics. The molecule is represented as a single electron that vibrates at frequency v_0. The oscillator is damped by the emission of electromagnetic radiation. The oscillator can interact with other atoms. The interaction strength decreases as the inverse third power of their separation distance, R, similar to the dipole–dipole interaction. If a second molecule is near the first one and the frequencies of both oscillators are nearly equal, then the energy of the first molecule will be transferred to the second molecule after some time before emission occurs from the second molecule. The Perrin calculation under conditions of exact resonance results in a critical transfer distance of $\lambda_0/2\pi$, where λ_0 is the wavelength of the oscillators. This model results in a transfer distance of about 1,000 Å.

The Perrin theory does not consider the role of molecular vibrations. These vibrations serve to broaden the sharp electronic transitions that occur at a single frequency into a broad frequency region. Perrin's theory also ignored the fact that there is an internal conversion between vibrational levels that occurs as a result of thermal relaxation. This process results in the Stokes' shift between the absorption and the emission spectra.

Upon excitation the molecule is in a higher vibrational level of the first electronic state; in the process of thermal relaxation the molecule relaxes to the lowest vibronic level of the same electronic state. This occurs by interaction with the vibrations of the solvent molecules.

For two polyatomic molecules in solution, a donor molecule and an acceptor molecule, the absorption spectrum of the acceptor molecule must overlap the fluorescence spectrum of the donor molecule. This is considered a weak interaction because this interaction does not affect the spectral properties of the donor or the acceptor molecules.

Jean Perrin's "transfert d'activation" or nonradiative excitation transfer was studied by the decrease of polarization of dye molecules in solution as a function of increasing concentration. As the concentration increased, the average distance between the dye molecules decreased and the enhanced probability of nonradiative excitation transfer decreased the measured polarization of the emitted luminescence.

In 1929, Kallmann and London presented a quantum mechanical theory of resonance energy transfer between atoms in the gas phase (Kallemann & London, 1929). This theory included the dipole–dipole interaction term and the parameter R_0 (the distance between the donor and the acceptor molecule at which energy transfer and deactivation by radiation of the donor excited state are equally probable). Kallmann was at the Kaiser Wilhelm Institute (which later became the Max Planck Institute) for Physical Chemistry and Electrochemistry, and London was at the Institute for Theoretical Physics at the University of Berlin. It is important since it was the first quantum mechanical theory of energy transfer and was incorporated into the theory of F. Perrin for the case of solutions, and was subsequently extended by Förster.

Their paper was written only 3 years after Schrödinger published his four classic papers on wave mechanics in 1926. These papers are discussed in the previous chapter on fluorescence and quantum theory. Kallmann and London used the quantum mechanical equations that were first published by Schrödinger in his article on energy exchange by quantum mechanics (Schrödinger, 1927). Schrödinger demonstrated that for two identical atoms with very narrow energy states, the mathematical solution is that of an oscillating system.

Kallmann and London organized their paper into two parts, a general part and a specialized part, and included an appendix on the spectral line broadening due to collisions. The general part of their paper included a discussion of quantum mechanical resonance, the quantum mechanical theory of energy exchange between two atoms, the elementary processes of energy excitation, the kinetic theory of energy excitation, and the limiting case of exact resonance. The specialized parts include discussions of excitation through collisions for specific mixtures of atoms, depolarization of resonance radiations, and quantum mechanical excitation and chemical reaction kinetics. Clearly, they covered a broad range of theoretical topics.

For nonradiative energy transfer, Kallmann and London assumed the condition of approximate resonance between the energy levels of the interacting atoms in the gas phase (i.e., no vibronic relaxation) and used second-order perturbation theory to calculate the energy of interaction. Kallmann and London, building on the quantum theories of both Dirac and Schrödinger, gave the quantum mechanical interaction term in terms of classical Coulomb interaction $W_{\alpha\beta}$ for the dipole interaction of two oscillators.

The quantum mechanical expression is:

$$W_{\alpha\beta} \propto \frac{\left(\vec{\mu}_\alpha \cdot \vec{\mu}_\beta\right)}{R^3}, \tag{22.51}$$

where $\vec{\mu}$ is the transition dipole moment for the respective atoms α and β, and R is the separation between the atoms involved in the energy transfer.

The use of second-order time-dependent perturbation theory with this interaction energy gives the R^{-6} dependence and the transition matrix expressions. They calculated that the probability of energy transfer is proportional to:

$$\frac{1}{1 + \frac{R^6}{R_0^6}}, \tag{22.52}$$

where R is the separation between the two atoms involved in the energy transfer, and R_0 is the previously defined constant.

When they make the assumption of two atoms not in perfect resonance, the value of R_0 fits the experimental results. However, when they assume exact resonance between the two atoms, which is what J. Perrin assumed in his energy transfer theory, the value of R_0 is too large to be comparable to J. Perrin's results.

Jean Perrin's son, Francis Perrin, made several important contributions to the field of molecular luminescence: the active sphere model for quenching of fluorescence, the relation between the quantum yield and the lifetime of an excited state, the theory of fluorescence polarization, including the Perrin equation, which permitted the fluorescence lifetimes of dyes in solution to be measured, and the first qualitative theory of the depolarization of fluorescence by resonance energy transfer.

In 1932 Francis Perrin published a quantum mechanical theory of energy transfer between molecules in close proximity, first in a gas phase and then in modified form for molecules in solution (F. Perrin, 1932). In this work he extended the original theory of resonance energy transfer, based on classical physics, that was developed by his father. The new theory of Francis Perrin also built on the previous theory of Kallmann and London (Kallemann & London, 1929).

Francis Perrin used the same quantum mechanical treatment for the molecular interaction as Kallmann and London did, but he extended their work by considering the molecules in solution, taking account of vibronic interactions. Francis Perrin made the conclusion that resonance energy transfer could occur over distances of one fifth of the wavelength of the emitted light. He also discussed the role of spectral overlap between the emission spectrum of the donor molecule and the absorption spectrum of the acceptor molecule.

It is instructive to further describe some of the details of the 1932 paper by Francis Perrin, which is based on exact resonance between two identical molecules. He assumed the atom has only a ground state and an excited state. The theory applied quantum mechanics (which he termed "wave mechanics" after de Broglie) to the following problem. One atom is initially in the excited state (donor) and the second atom is in the ground state at a distance from the first atom. The problem is to calculate the frequency or probability for the excitation energy of the first atom to be nonradiatively transferred to the second atom (acceptor) as a function of the interaction energy between the two atoms, which depends on the distance of their separation. He extended the treatment for energy transfer between two fixed and separated atoms to the case of molecules in solution, which includes the perturbations of solvent molecules.

Francis Perrin included the condition of quantized transfer of energy between two atoms and gave the name "d'induction quantique" to this phenomenon. First Perrin considers the case of two atoms, α and β, with the same atomic number that are separated by a great distance. A great distance is defined as the distance in which the energy of interaction between the two atoms is negligible (i.e., there is no interaction between the two atoms). He further ignored any vibronic relaxation processes. He writes the time-independent Schrödinger equation for each atom. He then shows that for the combined system of atoms α and β the solution to the wave equation, the eigenfunctions and the eigenvalues can be expressed as linear combinations of the solution for each atom. The energies of excitation, "d'activation," for each atom are equal, E_α equals E_β, and the total energy of the combined system is just the sum of the energies of each individual atom.

Second, Perrin considers the case in which the two identical atoms are brought together to a distance R. This distance is defined as that in which the potential energy of interaction is not negligible but is sufficiently great so that their electron clouds cannot overlap. Perrin assumes that the two atoms, separated by the distance R, interact through a dipole–dipole interaction (similar to the theories of Kallmann and London, and Jean Perrin). The interaction term between the two atoms is based on electrostatic interactions of the oscillating charge densities on each atom. He then proposes that the interaction between an excited atom and an atom in the ground state is based on a coupling in which the interaction is inversely dependent on the third power of the distance between the two atoms. Perrin simplifies the perturbation theory for the degenerate case by invoking symmetry. In the absence of overlap of the electron distribution of the two atoms, the time-independent Schrödinger equation yields a condition in which the energy from the excited state of one atom is transferred to the second atom.

Third, Perrin describes the transfer of excitation or "transfert d'activation." To describe this situation, Perrin use the time-dependent Schrödinger equation for two identical atoms with the initial condition that one is in the excited state and one is in the ground state. The eigenfunction for the total system is a linear combination of the wave function for the individual systems, each multiplied by a time-dependent constant, $c_\alpha(t)$ and $c_\beta(t)$. The time evolution of the system from the initial state, in which atom α is excited and atom β is in the ground state, to the final state, in which atom α is in the ground state and atom β is in the excited state, is given by the time dependence or time evolution of the time-dependent constants $c_\alpha(t)$ and $c_\beta(t)$. The results of the calculation show that the excitation will first be transferred from atom α to atom β, and then the excitation will be transferred from atom β to atom α. This oscillation of excitation energy will continue to occur and is similar to the resonance between two coupled mechanical oscillators of the same mass, which represents a mechanical analog.

Fourth, Perrin compared the lifetime for the resonance transfer of excitation energy between two identical atoms with the lifetime for spontaneous emission. He then defines a special separation distance R between atom α and atom β as R_0. The distance R_0 is defined as that distance in which the time for resonance transfer is equal to the mean lifetime for the luminescence (fluorescence). At that distance, the probability of de-excitation by energy transfer is equal to the probability of de-excitation by fluorescence.

Therefore, at the separation of R_0, one half of the energy will be emitted as radiation from the dipole oscillator, and one half of the excitation energy will be transferred nonradiatively from the donor dipole to the acceptor dipole. This value is approximately 0.22 of the wavelength of the excitation. For green excitation light, R_0 will be approximately 1,000 Å.

Finally, Perrin considers the case of resonance energy transfer between molecules in solution as well as the case that was originally analyzed by Kallemann and London, that of colliding atoms in the gas phase. In this case, vibronic relaxation becomes possible. Perrin considers the case of a very dilute solution of fluorescent molecules in which the very viscous solvent will make displacement of the molecules over short times negligible. For example, the effect of increasing concentration on the decrease of polarization in a solution of the molecule fluorescein in glycerin was investigated. These experiments demonstrated that resonance energy transfer could occur over 80 Å. By considering the effect of molecules in a solvent, Perrin calculated that the value of R_0, (the distance at which resonance energy transfer and emission can occur with equal probabilities) is in the range of 150 to 250 Å. He comments that internal vibration of the two molecules and the vibrational interaction between the molecules and the solvent could alter the absorption and the emission bands of the donor and acceptor molecules. These interactions can reduce the magnitude of the electric moments associated with the electronic transitions. For molecules in solution at room temperature, F. Perrin noted (as did Kallmann and London) that there is a broadening of the absorption and emission bands and the atomic line spectra are observed as broad bands. We now know that within the broad bands there is only a very narrow range of frequencies for which the exact condition of resonance is valid. While isolated atoms show a narrow line width (with Doppler broadening) for the absorption and the emission lines, molecules in solution show a range of wavelengths that represent the absorption and the emission bands. There is a spectral broadening due to the interaction of the molecules with the environment. This process is called "broadband dephasing."

To summarize, J. Perrin worked out the first theory of excitation transfer between identical molecules in solution in 1925–1927 to explain the concentration dependence of depolarization of fluorescent molecules in solution using classical mechanics. J. Perrin assumed the condition of exact resonance between two identical molecules: that the solvent molecules acted only as a dielectric. These assumptions resulted in a R^{-3} dependence on the rate of energy transfer. In 1927, Kallmann and London first published a quantum theory, based on dipole–dipole interactions, and a R^{-6} dependence for energy transfer between atoms in the gas phase but has a characteristic interaction distance that is too large. J. Perrin's son, F. Perrin, in 1932–1933 expanded the theory of Kallmann and London by considering energy transfer for molecules in solution with vibronic interactions. F. Perrin's theory, of course, also resulted in a R^{-6} dependence on the rate of energy transfer. but he also deduced a characteristic interaction distance that is much closer to experiment. Later he introduced the effects of collisional broadening of spectral lines to modify his theory.

22.4.2 Förster Resonance Energy Transfer

It was the seminal work of Förster, who worked in Göttingen, Germany, that provided a brilliant phenomenological theory for resonance energy transfer. The first version was based on semiclassical treatment of intramolecular resonance, and the second version was based on quantum mechanical theory. FRET is an extension of the earlier ideas of J. Perrin, F. Perrin, Kallmann and London, and Vavilov (1943).

Förster made his theory accessible to a wide range of scientists by grouping the complex physical terms of his theory into a few experimentally accessible parameters such as the overlap integral, the lifetime of the donor molecule, the index of refraction of the medium, and the orientation parameter for the donor–acceptor molecular pair. The Förster theory yields from appropriate experimental measurements the transfer efficiency and also the distance between the donor and acceptor molecules at which the energy transfer occurs between 50% of the molecular pairs. The transfer efficiency E is defined as the number of quanta transferred from the donor to the acceptor divided by the number of quanta that are absorbed by the donor.

Note that the donor molecule in the excited state can become deactivated by several mechanisms; FRET is one mechanism, and fluorescence and internal conversion are other mechanisms. The high sensitivity of FRET to molecular distance derives from the theoretical prediction that the efficiency of energy transfer shows a R^{-6} dependence on intramolecular distance between the donor molecule and acceptor molecule. In spectroscopic studies these parameters can be calculated on a per-pixel basis and correlate with cellular morphology.

While the original experimental techniques were developed for dye molecules in solution using spectroscopic measurement, they rapidly developed into microscopic techniques that are applicable to cells, and that is the reason for its great utility by cell biologists.

22.4.2.1 Förster's Classical Theory of Resonance Energy Transfer (1946)

Förster published two versions of the semiclassical theory of resonance energy transfer (Förster, 1946, 1951). His slightly modified classical theory, developed in his 1951 book *Fluoreszenz Organischer Verbindungen*, is worth reading. In his 1946 paper he did not include the orientation factor κ and the refractive index factor n, since it was an approximate theory to elucidate the basic physical concepts. The authors of this chapter have added these constants to the equations of Förster's 1946 paper so that the equations are consistent with their modern forms, where orientation and refractive index are accounted for.

We strongly recommend that the reader consult the original papers of Förster to gain insight into his thinking. For those who cannot read German, we suggest Robert Clegg's chapter on fluorescence resonance energy transfer, in which he summarized the important concepts and assumptions of the theory (Clegg, 1996).

Förster was aware of the studies of Emerson and Arnold on photosynthesis, which probably stimulated his interest in the topic of resonance energy transfer. The introduction to his 1946 paper, "Energy Migration and Fluorescence," written in German, will provide the reader with a comprehensive background of all aspects of the field. He also was also aware of the earlier works of the Perrins. One key experimental finding was the study of the depolarization of fluorescence as a function of increasing concentration of fluorescent molecules in solution. For polarization, as the concentration of the fluorescent molecules was increased, the polarization remained constant until a critical concentration was reach, at which point polarization sharply decreased. The decrease in polarization occurred at a much lower concentration than that for fluorescence intensity self-quenching. This decrease in fluorescence polarization at a lower concentration suggested the possibility of another physical process: energy transfer.

In his (1946) classical theory of resonance energy transfer, Förster assumed, as did Perrin, that there are two identical molecules that can be considered as oscillating dipoles. Since they are identical oscillators they have the same amplitude. The oscillating electric dipole moment is $\vec{\mu}(t)$ and $\vec{\mu}_0$ is the maximum value of the dipole moment. The frequency of oscillation is ω.

$$\vec{\mu}(t) = \vec{\mu}_0 \cos(\omega t). \tag{22.53}$$

An oscillating dipole will radiate energy and therefore lose energy with time. This loss of energy can be described as:

$$E = E_0 e^{-k_{rad} t}, \tag{22.54}$$

where the initial energy is E_0 and the rate of decay is k_{rad}.

The inverse of the rate of decay for the oscillating dipole to decrease in energy is a decay time τ_{rad},

$$\tau_{rad} = \frac{1}{k_{rad}}. \tag{22.55}$$

The characteristic time of radiation from the oscillating electric dipole τ_{rad} is:

$$\tau_{rad} = \frac{3\hbar c^3}{\vec{\mu}^2 \omega^3}. \tag{22.56}$$

In his 1946 paper, Förster uses the following expression, which is not exactly equal to the correct expression given above:

$$\tau_{rad} \approx \frac{\hbar c^3}{\vec{\mu}^2 \omega^3}. \tag{22.57}$$

The oscillating electric dipole from the donor molecule D creates an electric field. This electric field interacts with the acceptor molecule A, and the energy of interaction E_{int} is:

$$E_{int} = -\vec{\mu}_A \cdot \vec{E}_D. \tag{22.58}$$

Förster assumed that the energy transfer from the donor molecule to the acceptor molecule occurs during one cycle of the oscillating electric dipole. The time of the energy transfer is:

$$\tau_{transfer} = \frac{\hbar n^2 R^3}{\kappa \vec{\mu}^2}. \tag{22.59}$$

The energy of the interaction (dipole–dipole interaction) is also expressed as:

$$E_{int} = \frac{\kappa \vec{\mu}^2}{n^2 R^3} = \frac{\hbar}{\tau_{transfer}} \quad \text{or} \tag{22.60}$$

$$\tau_{transfer} = \frac{\hbar}{E_{int}}. \tag{22.61}$$

where κ is an orientation factor that depends on the relative orientation of the electric dipoles of the acceptor and the donor molecules. The refractive index is given by n. The separation between the centers of the two oscillating electric dipoles is R.

In his 1946 paper Förster wrote that the expressions for interaction energy and the energy transfer time constant differ from full equations (22.71) and (22.72) by the factors that he omitted:

$$E_{int} \approx \frac{\vec{\mu}^2}{R^3} \tag{22.62}$$

$$\tau_{transfer} \approx \frac{\hbar R^3}{\vec{\mu}^2} \tag{22.63}$$

Förster then calculated the distance R_0 for which the time for the transfer of the energy between the two identical oscillators $\tau_{transfer}$ is equal to the time of radiation of the oscillating electric dipole τ_{rad}.

$$\tau_{transfer} = \frac{\hbar n^2 R^3}{\kappa \vec{\mu}^2} \quad \text{and} \tag{22.64}$$

$$\tau_{rad} = \frac{3 \hbar c^3}{\omega^3 \vec{\mu}^2} \tag{22.65}$$

are equal to each other.

When R, the separation distance between the donor and the acceptor $R = R_0$, the time for the radiation from the oscillator is equal to the time for excitation transfer:

$$\tau_{transfer} = \tau_{rad} \equiv \tau_0 \tag{22.66}$$

$$\tau_0 = \frac{\hbar R_0^3}{\vec{\mu}^2} \frac{n^2}{k} \tag{22.67}$$

and,

$$R_0^3 = \frac{3 \kappa c^3}{n^2 \omega^3} = \frac{3 \kappa}{n^2} \lambda^3. \tag{22.68}$$

That is, the distance of separation between the two molecules for which the probability for the excited donor molecule to become de-excited by energy transfer to the acceptor molecule among all the de-excitation process is one half.

This theoretical value of R_0 is too large as compared with experimental measurements. Förster pointed out that this discrepancy is due to the broadening of the spectral lines in solution at room temperature. When this is included into the theory, the energy transfer distance R_0 conforms to the experimental measurements.

If Förster were to proceed with his theory as shown in the following paragraphs and equations, he would derive an equation that is similar to the equation that was earlier derived by the Perrins. It predicts a mean distance of energy transfer that is too long.

He did not follow the path taken earlier by the Perrins with their assumption of exact resonance between two identical atoms or molecules. Instead, Förster realized that the condition of exact resonance is not valid and results in an excessively large value for R_0. The spectral lines of dye molecules in solution are broadened due to interaction with the solvent. The false assumption of exact resonance is valid for only a small fraction of the time required for energy transfer. There is a low probability that the condition of exact resonance is valid through the energy transfer process. Förster arrived at this conclusion based on an assumption that typically the energy of interaction for very weak coupling between the donor and the acceptor molecules is much smaller than their vibrational energies. Therefore, he had to modify his results by factors that correct for the presence of a range of excited states (broad band).

Förster approximated the absorption and emission spectral lines as uniform distributions (Fig. 22.2). He defined Ω as the spectral width of the absorption curve and the spectral width of the fluorescence curve since the oscillators are identical. Due to the Stokes' shift in frequency, there is a region of spectral overlap, defined as Ω'. Förster

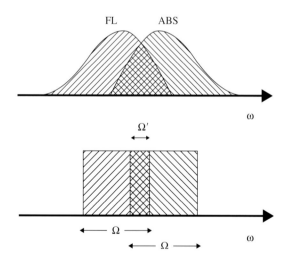

Figure 22.2. Fluorescence and absorption spectra. Upper: Intensity versus frequency for emission and absorption spectra. Note that the mirror symmetry of the emission and the absorption bands are normalized. The region of spectral overlap is indicated by the cross-hatching. Lower: An approximate schematic of the upper spectra in which the emission and absorption bands are represented by rectangles. The bandwidth of the emission and the absorption spectra are given by Ω. The width of the spectral overlap is given by Ω'.

calculated the probability to be in resonance as the probability for the two oscillators to overlap multiplied by the probability that the two oscillators are within the interaction energy frequency. The probability of spectral overlap is: Ω'/Ω.

He then calculated the probability, w, to be in exact resonance, and simultaneously for both oscillators to have the same energy within the interaction energy as:

$$w = \frac{\Omega'}{\Omega} \cdot \frac{E_{\text{int}}}{\hbar\Omega} = \frac{\Omega'}{\Omega} \cdot \frac{\frac{\kappa\vec{\mu}^2}{n^2 R^3}}{\hbar \cdot \Omega}. \tag{22.69}$$

The first term of this equation represents the overlap integral between the absorption curve and fluorescence curve. The second term represents the probability that the frequency ω is within the frequency range $E_{\text{int}}/\hbar$.

Förster calculated the rate of energy transfer, k_T, and then multiplied that rate (for narrow line spectra of exact resonance) by the factor w, which corrected the expression to include the effects of a broadband spectra:

$$\tau_{transfer} = \frac{\hbar n^2 R^3}{\kappa\vec{\mu}^2} \quad \text{and} \quad w = \frac{\Omega'}{\Omega} \cdot \frac{\frac{\kappa\vec{\mu}^2}{n^2 R^3}}{\hbar \cdot \Omega}. \tag{22.70}$$

$$k_T = \frac{w}{\tau_{transfer}} = \frac{\kappa^2\vec{\mu}^2}{\hbar n^4 R^3} \cdot \frac{\vec{\mu}^2\Omega'}{\hbar R^3\Omega^2} = \frac{\kappa^2\vec{\mu}^4\Omega'}{\hbar^2 n^4 R^6\Omega^2}. \tag{22.71}$$

He then recalculated the time, τ_0, for which the radiation from the oscillator is equal to the time for excitation transfer:

$$\tau_0 = \frac{\hbar^2 n^4 R_0^6\Omega^2}{\kappa^2\vec{\mu}^4\Omega'}. \tag{22.72}$$

The rate of energy transfer is:

$$k_T = \frac{\vec{\mu}^4\Omega'\kappa^2}{\hbar^2 R^6\Omega^2 n^4} \tag{22.73}$$

Förster included these corrective probability factors, which accounted for the probability that there is spectral overlap of the two oscillators and also, at the same time, the two oscillators are in exact resonance. Thus he derived the rate of resonance energy transfer with a R^{-6} separation dependence between the donor and the acceptor molecules.

This formation does not allow the dipole moment, $\vec{\mu}$, to be explicitly evaluated; therefore, the energy transfer rate cannot be calculated. However, the Förster distance, R_0, can be evaluated from using equations (22.65), (22.71), and (22.72) after eliminating $\vec{\mu}$ dependence:

$$R_0 = \frac{\lambda}{2\pi}\left(\frac{3\kappa}{n^2}\right)^{1/3}\left(\frac{1}{\tau_0}\frac{\Omega'}{\Omega^2}\right)^{1/6}. \tag{22.74}$$

The Förster distance contains only experimentally measurable parameters that can be evaluated. Förster's theory of resonance energy transfer, modified for broadband spectra, yielded the correct experimental value for the term $\vec{R}_0$.

22.4.2.2 Förster Quantum Theory of Resonance Energy
              Transfer (1948)*

We will begin by analyzing and summarizing Förster's seminal 1948 German paper
Zwischenmolekulare Energiewanderung und Fluoreszenz (Förster, 1948). For those
who do not read German, an English translation of this paper exists (Knox, 1993). The
Knox translation has extensive notes; the translator corrected some analytical errors
in the original German text and uses vector notation for the gradient and Laplacian
operators. Then we continue to develop Förster's theory based on his 1982 book on
fluorescence of organic compounds (Förster, 1982). In his 1982 book he presents a
second classical derivation of his theory; previously we discussed the first classical
derivation of his theory published in 1946. And finally we discuss Förster's later ideas
on delocalized excitation and excitation transfer (Förster, 1965).

   In his abstract Förster states the scope of his 1948 paper on intramolecular energy
migration (transfer) and fluorescence. He acknowledges that his present theory of
energy transfer between similar molecules in solution is an extension of the earlier
theories of J. and F. Perrin.

   He states that the measurement of three quantities of the molecule—the absorption
and emission spectrum and the emission lifetime—can be used to calculate the critical
molecular separation below which energy transfer occurs during the time the molecule
is in the excited state. These three quantities are accessible to the spectroscopist, and
thus the critical separation for energy transfer can be calculated. Förster discusses the
application of his theory of resonance energy transfer to molecular crystals and the
photosynthetic apparatus in plants.

   The paper begins with a review of the literature. An example of his erudition and
high standards for appropriate scholarly citation is the fact that this paper refers to
the prior works of others in Germany, France, and the Soviet Union. Förster separates
energy transfer in crystals with electronic conductivity (i.e., inorganic crystal phosphors
and silver halide crystals in photographic emulsions) from cases in which there is no
electron conduction or migration but which exhibit energy migration. An example of
the latter case is anthracene crystals doped with traces of naphthalene. He then describes
the effect of concentration depolarization of fluorescence in solution, and concludes
that this effect of a decrease with polarization with increasing concentration cannot
be explained by the trivial process of sequential emission and reabsorption of photons
by fluorescein. In 1924 Gaviola and Pringsheim first observed this effect. Finally, he
presents the earlier work of J. Perrin on the classical theory of energy transfer between
an excited molecule and its neighbors. F. Perrin is cited for providing a quantum theory
of energy transfer between molecules. Since the theory of F. Perrin and the experimental
studies of concentration polarization showed a discrepancy, Förster postulated that the
existence of the Stokes' shift, the difference in frequencies between absorption and
emission, which was not incorporated in the Perrin model may be the source of the
error. Förster also cites several previous publications of Vavilov in the years of 1942
and 1944 on the subject of resonance energy migration.

   Now we proceed to outline some of the key points of Förster's treatment of resonance
energy transfer. First he presents the general ideas of his model: a solution consisting
of dye molecules dissolved in a solvent that does not absorb in the same spectral region
as the dye molecules. Initially the molecule is excited and decays by both radiative

processes such as fluorescence and nonradiative processes. The quantum yield and the average lifetime of the excited state reflect the fraction of the deactivation process in the radiative pathway and in the nonradiative pathway.

Next he considers a solution composed of a number of similarly absorbing molecules. He considers the possibility that in addition to the processes previously described, there is nonradiative energy transfer between pairs of molecules. He defines the rate of transfer as the number of transfers between pairs of molecules per unit time. In his model, the process of energy transfer is limited to simultaneous processes, and not the trivial case of emission and reabsorption in which the energy transfer would be due to a series of successive processes. Förster postulated that the rate of energy transfer in his model will decrease rapidly with increasing distance of separation between the two molecules, and therefore the process of energy transfer will occur only during the lifetime of the excited state if the second molecule is present with a critical distance from the excited molecule.

Next Förster provides the quantum mechanical theory for the mechanism of energy transfer. It is important to point out that the quantum mechanical phenomenon of resonance that we have previously described was well developed and widely published prior to this 1948 publication by Förster.

In terms of quantum mechanics the initial problem is to calculate the probability for the following process involving two molecules: the first state of the system of two molecules has the molecule k in the excited state and the molecule l in the ground state, and the second state of the system has the molecule k in the ground state and the molecule l in the excited state. It is assumed that the stationary molecular vibrational states of the two molecules in the first state and in the second state can be described by the corresponding quantum numbers and the corresponding energies.

Förster assumes, as did Perrin previously, that the energy transfer is mediated by the interaction between the electronic systems of the two molecules, and that energy transfer is much slower that nuclear motion. The latter assumption permits the eigenfunctions of the molecules to be expressed in terms of the electronic coordinates, and the nuclear coordinates are neglected in the treatment. These eigenfunctions are used to describe the stationary molecular vibrational states. For the energy transfer process, the initial state corresponds to the donor being in the excited state and the acceptor being in the ground state. The final state corresponds to donor being in the ground state and the acceptor being in the excited state. Förster invokes the 1927 quantum mechanical theory of Dirac to describe the transition between these states.

Dirac's theory gives the transition probability, defined in terms of the total number of transitions per unit time for each molecule in the ground state. The transition probability is the sum of the square of the matrix transition element for the interaction (assumed to be purely electronic) over all possible electronic–vibronic transitions. Förster states that the conservation of energy is valid and constrains the possible transitions. This restriction was not present in the theoretical treatment of other investigators.

In Förster's 1948 paper there are the following definitions and assumptions: the two molecules involved in the resonance energy transfer are k and l where a prime (k') denotes the excited state and no prime denotes the ground state (k). The term F_{kl} is the probability for the transition from a state in which molecule k is excited and molecule l is in the ground state into a different state in which molecule k is in the ground state and molecule l is in the excited state. The stationary molecular vibrational states of the two molecules in these initial and final states are described by the quantum numbers v'_k, v_l and v_k, v'_l. The corresponding energies are w'_k, w_l and w_k, w'_l. The frequencies for the occurrence of individual values of the molecular vibration energy are given by two distribution functions, $g(w_l)$ and $g'(w'_k)$, which are normalized to 1 on an energy scale, a molecule in the ground and the excited state. W_0 is the energy between the lowest vibrational levels of the ground and the first excited electronic states.

Förster assumes thermal equilibrium and the molecule interacts with the solvent and undergoes vibrational relaxation. Therefore, the molecule rapidly reaches the lowest vibrational level of a given electronic state after a transition. With this assumption, the energy given up by molecule k during the transfer process is:

$$W_0 + w'_k - w_k. \tag{22.75}$$

The energy taken up by molecule l is:

$$W_0 + w'_l - w_l. \tag{22.76}$$

ΔW is defined as:

$$\Delta W = w_k + w'_l - w'_k - w_l. \tag{22.77}$$

Due to conservation of energy for this process, the total energy change

$$\Delta W = 0. \tag{22.78}$$

With these assumptions, the rate of energy transfer expressed as the number of energy transfers per unit of time is now expressed in terms of the products of the two normalized distribution functions (for the two molecules k and l) and the square of the magnitude of the matrix element for the interaction. The probability of the transition under the condition of conservation of energy is:

$$F_{kl} = \frac{2\pi}{\hbar} \int_{W=0}^{\infty} \int_{w_l=0}^{\infty} \int_{w'_k=0}^{\infty} g'\left(w'_k\right) \cdot g(w_l)$$

$$\times \left| u_{kl}\left(w'_k, w_l; W_0 - W + w'_k, W - W_0 + w_l\right) \right|^2 dw'_k \, dw_l \, dW \tag{22.79}$$

Förster then develops the interaction term, u_{kl}. He assumes that the interaction term between the two molecules is expressed as the interaction energy, $u(\vec{r}_k, \vec{r}_l)$, given by the Coulomb interaction of the moving electronic charges, for the case in which the molecular separation is large as compared to the intermolecular atomic separation; then:

$$u\left(\vec{r}_k, \vec{r}_l\right) = -\frac{e^2}{n^2 \left|\vec{r}_k - \vec{r}_l\right|} \tag{22.80}$$

in which $u(\vec{r}_k, \vec{r}_l)$, the interaction energy, is a function of the spatial coordinates of the electrons of molecule k and molecule l, the electronic charge is e, the refractive index of the solvent is n, and the dielectric constant is given as n^2.

Förster states a critical condition. Förster states that molecules k and l, which could be large organic dye molecules, must satisfy the following separation condition: when the separation distance between molecules k and l is much larger than the molecular dimensions of the molecules, then the energy of interaction reduces to the interaction energy of two electrical dipoles.

$$\mu_{kl}\left(w'_k, w_\ell; w_k, w'_l\right) = \frac{1}{n^2 \vec{R}_{kl}^{\,5}} \left\{ \vec{R}_{kl}^{\,2} \left[\vec{M}_k^*\left(w_k, w'_k\right) \cdot \vec{M}_l\left(w_l, w'_l\right) \right] \right.$$

$$\left. - 3 \left[M_k^*\left(w_k, w'_k\right) \cdot \vec{R}_{kl} \right] \left[\vec{M}_l\left(w_l, w'_l\right) \cdot \vec{R}_{kl} \right] \right\} \quad (22.81)$$

The asterisk represents the complex conjugate of the quantity. The symbol $\vec{R}_{kl}$ is the vector between the molecular centers of gravity of the two molecules and R_{kl} is its absolute value.

The matrix element of the transition moment is defined by Förster as:

$$\vec{M}_k\left(w_k, w'_k\right) = -e \int \phi_k\left(w_k, \vec{r}_k\right)^* \vec{r}_k \phi'_k\left(w'_k, \vec{r}_k\right) d\vec{r}_k \quad (22.82)$$

The corresponding molecular center of gravity is selected as the origin of the electron position vectors $\vec{r}_k$. The first variable of $\vec{M}_k$ denotes the nuclear vibrational energy in the electronic ground state, and the second variable denotes the nuclear vibrational energy in the electronic excited state.

For cases in which the energy transfer occurs more slowly than rotational Brownian motion, the rate of energy transfer must be averaged over all orientations of both molecules. For that case, Förster presents the average rate of energy transfer (number of transfers per unit time), which is expressed in terms of the minus-sixth power of the separation distance between molecules k and l, the refractive index of the solution to the minus-fourth power and products of integrals that involve the square of the transition moment (based on dipole–dipole interaction energy).

Again, following Dirac, the transition probability is given by the square of the matrix element of the interaction energy. Förster averages this quantity over all possible orientations for both molecules involved in the energy transfer:

$$\langle W \rangle^2 = \frac{2}{3n^4 R_{kl}^6} \vec{M}_k^{\,2} \vec{M}_l^{\,2}. \quad (22.83)$$

where $\vec{M}$ is the absolute value of the transition moment, and the index k and l can be dropped since the molecules are identical.

The quantum mechanical theory of Förster used Fermi's Golden Rule to give the R^{-6} dependence and the square of the transition matrix elements in the expression for the rate of energy transfer. Förster had to assume the existence of two incoherent oscillators in order to apply Fermi's Golden Rule for linear absorption.

If Förster stopped at that point, the theory of resonance energy transfer would be of theoretical interest only. The great advance was in the next section of his paper, in which he makes the crossover from the theoretical equations of Dirac's quantum mechanics to the experimentally accessible quantities such as absorption and emission spectra and excited state lifetimes—all quantities that could be measured in the laboratory. It was this reformulation of the quantum mechanical theory into a phenomenological approach that was the first major step to place this technique in the hands of spectroscopists, and later biophysicsts and biologists.

Förster states that the absorption spectrum of a molecule can be measured more easily than the fluorescence spectrum. He postulates that the transition moment is an approximately symmetric function of the vibrational energy in the ground and in the excited states. Additionally, he invokes Levschin's law of "mirror correspondence," which was published in 1931. The Levschin "mirror law of fluorescence" states that there is a approximate mirror symmetry in the intensity patterns of both the absorption and fluorescence spectra. Förster makes use of the "mirror law of fluorescence" and the fact that the overlap of the absorption and fluorescence spectra occurs only in a narrow region. Förster then defined R_0 as the critical molecular separation, below which there is energy transfer during the excited state lifetime.

Instead of presenting the original equations as Förster wrote them in his 1948 paper on the quantum mechanics of resonance energy transfer, we present them with the modern variables (Lakowicz, 1999; Valeur, 2002). Today, Förster resonance energy transfer theory permits the efficiency of energy transfer to be calculated when the following experimental quantities are measured or estimated: the overlap integral, the quantum yield of the donor molecule, the index of refraction of the solution, the angle between the dipole moments of the donor and the acceptor, and a series of universal constants.

Assume a single donor and a single acceptor separated by a distance r. The rate of resonance energy transfer k_T, which is the probability to transfer a quantum from the donor molecule to the acceptor molecule per unit time,

$$k_T\left(\vec{r}\right) = \frac{Q_D \kappa^2}{\tau_D r^6}\left(\frac{9000\,(\ln 10)}{128\pi^5 N n^4}\right)\int_0^\infty F_D\left(\lambda\right)\varepsilon_A\left(\lambda\right)\lambda^4 d\lambda \qquad (22.84)$$

where the quantum yield of the donor molecule in the absence of the acceptor molecule is Q_D, the index of refraction of the medium is n, Avogadro's number is N, and the lifetime of the donor molecule in the absence of the acceptor molecule is τ_D. The terms in the integral are as follows: $F_D(\lambda)$ is the corrected fluorescence intensity of the donor molecule in the wavelength range λ to $\lambda + \Delta\lambda$, under the condition that the total intensity is normalized to unity; $\varepsilon_A(\lambda)$ is the extinction coefficient of the acceptor molecule at the wavelength λ; and the term κ^2 is the orientation factor and describes the relative orientation of the transition dipole moments of the donor molecule and the acceptor molecule, the refractive index, n, and the inverse sixth power of the separation distance, r^{-6}, between the donor and the acceptor molecules.

An important consideration in resonance energy transfer is the correct value to use for the orientation factor κ^2. Typically this factor is given a value based on experimental conditions that may or may not be valid, and therefore an error is introduced in the calculated values of R_0 and the efficiency of resonance energy transfer E. Several authors have written on this problem and also provided calculations of κ^2 for various experimental conditions and distributions (Clegg, 1996; Lakowicz, 1999; Valeur, 2002).

The degree of spectral overlap between the emission of the donor molecule and the absorption of the acceptor molecule is expressed as the spectral overlap integral $J(\lambda)$.

$$J(\lambda) = \frac{\int_0^\infty F_D(\lambda)\varepsilon_A(\lambda)\lambda^4 d\lambda}{\int_0^\infty F_D(\lambda)d\lambda}. \qquad (22.85)$$

The Förster distance R_0 is that distance for which the rate of resonance energy transfer $k_T(r)$ is equal to the rate of excitation decay of the donor molecule in the absence of the acceptor molecule τ_D^{-1}. At this distance R_0 half of the total number of donor molecules undergo de-excitation from the excited state by resonance energy transfer, and one half of the donor molecules undergo de-excitation by radiative and nonradiative processes.

$$R_0^6 = \frac{9000(\ln 10)\kappa^2 Q_D}{128\pi^5 Nn^4} \int_0^\infty F_D(\lambda)\varepsilon_A(\lambda)\lambda^4 d\lambda \Big/ \int_0^\infty F_D(\lambda)d\lambda. \qquad (22.86)$$

The rate of resonance energy transfer can also be expressed as:

$$k_T(r) = \frac{1}{\tau_D}\left(\frac{R_0}{r}\right)^6, \qquad (22.87)$$

where all the terms have been previously defined.

Another important term is the efficiency of resonance energy transfer E:

$$E = \frac{k_T}{\tau_D^{-1} + k_T}. \qquad (22.88)$$

The efficiency of resonance energy transfer is the fraction of photons that are absorbed by the donor molecule and are transferred to the acceptor molecule.

$$E = \frac{R_0^6}{R_0^6 + r^6}. \qquad (22.89)$$

In addition to the Förster's distance, Förster's quantum mechanical treatment of FRET (1948) leads to the expressions for energy transfer rate, k_T, based on experimentally measured parameters. This expression is not derivable from his classical treatment nor from the treatments of his predecessors.

The last two sections of Förster's 1948 paper deal with energy transfer in solution, and a discussion of concentration depolarization. In this section Förster takes the rate of resonance energy transfer to be dependent on the inverse sixth power of the separation distance between molecules in the interaction term and calculates the transfer for an assumed statistical distribution of dye molecules in solution. It is important to point out that in solution there is a distribution of separation distances, and the calculation proceeds by averaging over this distribution. Using another approach Förster makes the assumption that the dye molecules are positioned on a three-dimensional lattice: simple cubic, face-centered cubic, body-centered cubic, and the diamond lattice. This model leads to the transfer of energy from molecule to molecule to be described by a differential equation similar to that used for heat conduction and diffusion.

Then he describes concentration depolarization. Förster's analysis assumes that the interaction energy can be replaced by its average over all orientations of the relative dipoles at a fixed distance. Therefore, he states that his theory is correct only for low-viscosity solutions of dye molecules in which the Brownian rotational diffusion of the molecules is much more rapid than the process of energy transfer. Förster states that an exact theory would have to incorporate the relative orientations of the molecular transition moments. Using an approximate scalar theory Förster describes concentration depolarization.

At very high concentration of dye molecules in solution, the molecular separation is of the size of the molecules. Now the interaction can no longer be correctly described by dipole–dipole interactions, and Förster's concentration depolarization theory of resonance energy transfer is no longer valid. Now the previous assumption of thermal equilibrium of vibrational motion may no longer be valid. Instead of the very weak dipole–dipole interactions of the Förster theory, when energy transfer is mediated by strong interactions, a new and different theoretical model is required. Such strong interaction models of energy transfer are based on quantum mechanical treatment of the entire assembly of molecules and were formulated by Frenkel, Peierls, and Franck and Teller. These models are very different from the very weak coupling model of Förster resonance energy transfer involving dipole–dipole interactions of a pair of molecules.

At the end of this historic paper Förster notes that chlorophyll in the photosynthetic apparatus and biological molecules in general may be ordered into some crystal lattice or similar array, and therefore other theoretical models, differing from the Förster theoretical model, such as the Frenkel exciton model, may be appropriate to describe energy transfer in these biological cases.

We now add some additional supplementary historical bibliographical notes from a review on exciton bands in molecular lamellar systems (Kasha, 1959). An exciton can be described as a wave packet that moves through an assembly of molecules and arises from a superposition of exciton states. Note that the stationary exciton states involve excitation of all of the molecules considered as locally excited. In other words, in exciton states the excitation is delocalized.

Frenkel in 1931 was the first to develop a theory of excitons in atomic lattices. Davydov in 1948 first applied exciton theory to molecular crystals (Davydov, 1962). In 1958 McRay and Kasha applied exciton theory to pigments aggregated in various geometries. In the 1948 Förster paper just discussed, a quantum mechanical theory was developed for resonance energy transfer between pairs of randomly oriented molecules.

In summary, the Förster resonance energy transfer theory is based on dipole–dipole interaction between two molecules with very weak coupling. Weak coupling of the interaction is defined in the follow way. The vibrational splitting of the energy levels of the donor is much greater than the energy of the dipole–dipole interaction. The Förster theory of resonance energy transfer provides the correct R^{-6} dependence on the rate of energy transfer and the correct value for R_0. Its wide utility is due to the fact that it can be used to measure molecular separations on the nanometer scale and that the measurement of a few spectroscopic parameters can be used to calculate molecular distances.

22.4.3 The Work of Oppenheimer and Arnold on Energy Transfer (1950)

Oppenheimer's postwar publications consisted of a paper on electrodynamics, three papers on meson physics, and one paper on biophysics. Oppenheimer published a paper in which he derived an equation for the rate of energy transfer from phycocyanin to chlorophyll (Arnold & Oppenheimer, 1950). His derivation is based on the energy transfer between a random distribution of phycocyanin molecules that are uniformly situated around a sphere that contains a chlorophyll molecule.

Much earlier, in 1941, Oppenheimer had developed a theory of internal conversion that is applicable to the process of radioactivity. In that process, a nonradiative mechanism occurs in which a nuclear transition results in the atom with the ejection of an electron. It was the genius of Oppenheimer to realize that except for scale, the processes of internal conversion of gamma rays and nonradiative energy transfer in photosynthesis are analogous. In both cases the rate varies as R^{-6}. The Arnold and Oppenheimer 1950 publication differed from the Förster theory on resonance energy transfer in that only Förster derived the important quantity R_0 in terms of phenomenological parameters.

22.4.4 Experimental Validation of R^{-6} Dependence of the Rate of Resonance Energy Transfer

Once the theoretical R dependence of the rate of resonance energy transfer was experimentally validated in 1965 by Latt and coworkers, and later in 1967 by Stryer and Haugland (Stryer & Haugland, 1967), FRET could be used by biophysicists and biologists to investigate molecular interactions at the scale of Angstroms. Initially FRET techniques were applied in solution and later in cells.

It is instructive to discuss these papers since they are often cited as experimental support of FRET. Latt and coworkers (1965) realized that in order to test the theoretical R^{-6} dependence of the rate of resonance energy transfer, the donor and acceptor molecules must have a fixed separation on a molecule. This would be possible only if the molecule was rigid and planar. At that time, they calculated the separation of the donor and the acceptor from large-scale plastic models of the molecule. Their solution to the problem was to attach the donor and the acceptor to bissteriods, which are nearly symmetrical planar molecules with a large and fixed separation between potential attachment sites. They used two different donor–acceptor pairs of molecules. For one pair of donor and acceptor molecules the measured separation agreed with the theoretical value. For the second pair of donor and acceptor molecules the measured separation was low compared to the theoretical value. The authors proposed that the low value could be due to two sources: it was most likely due to nonrandom orientations that tend to align the interacting dipoles and increase the transfer, but it could also be due to additional contributions to the energy transfer from an exciton process.

Several years later Stryer and Haugland attached a donor and an acceptor molecule to oligomers of poly-L-proline in order to test the R^{-6} dependence proposed by Förster. First the authors used ultraviolet optical rotatory dispersion (ORD) to determine that oligomers of poly-L-proline with $n = 5$–12 form type II trans helical conformation in ethanol. This permitted the authors to calculate the donor–acceptor chromophore separation distances based on known coordinates for the poly-L-proline II trans helix. They assumed that the orientations of the donor and the acceptor molecules are randomized during the excited-state lifetime. They concluded that the experimentally observed distance dependence of the transfer efficiency agrees with the Förster R^{-6} dependence, and stated that FRET can be used as a "spectroscopic ruler."

22.5 CLASSICAL AND QUANTUM MECHANICAL DESCRIPTION OF ENERGY TRANSFER AFTER FÖRSTER

A concise and modern source for the quantum mechanical formulation of Förster resonance energy transfer theory is his 1964 paper. Another good source is Valeur's fluorescence textbook (Valeur, 2002).

After the seminal work of Förster, other investigators have advanced alternative theories to explain energy transfers in a variety of physical systems. Depending on the system, the distance between donor and acceptor can range from a fraction of a nanometer to tens of nanometers. For these different distances, the form of the interaction term will vary with different strength and physical origin. For example, at longer distances, dipole–dipole (Förster) interaction predominates and exchange mechanisms are neglected. However, at shorter distances, quadruple–quadruple interaction and exchange mechanisms may dominate. In some system, several mechanisms can occur simultaneously. In particular, we present the classical theory of H. Kuhn and the quantum mechanical theory of D. L. Dexter.

In FRET the overlap of the emission spectrum of the donor molecule and the absorption spectrum of the acceptor molecule is a requirement. In molecular terms, that is a consequence of having the energy levels for electronic transitions in the donor molecule and the acceptor molecule being almost identical.

FRET is based on allowed singlet–singlet transitions of the donor and acceptor molecules due to long-range dipole–dipole Coulombic interactions. Long-range energy transfer can occur up to about 100 Å. In this type of resonance energy transfer the donor molecule is excited and then is found in the ground state; simultaneously the acceptor molecule is promoted to the excited state. The acceptor molecule then emits a photon.

In addition to the Coulombic interactions there is another mechanism of energy transfer that is based on orbital overlap between the donor molecule and the acceptor molecule. This intermolecular orbital overlap results in an exchange term in the quantum theory and corresponds to an exchange of two electrons between the donor molecule and the acceptor molecules. Exchange processes can occur in both the (Förster) singlet–singlet energy transfer and in the (Dexter) triplet–triplet energy transfer.

The Coulombic interaction is the major interaction term for allowed transitions between the donor and the acceptor molecules. This is valid for both long distances of the order of 100 Å and for short distances on the order of less than 10 Å.

For the case of forbidden electronic transitions, for example triplet-to-singlet transitions in the excited state of the donor molecule, and singlet-to-triplet electronic transitions in the ground state of the acceptor molecule, the exchange mechanism occurs at distances less than about 10 Å. The requirement of orbital overlap is the reason for this distance restriction.

Energy transfer mechanisms will now be expressed in terms of quantum theory. The system of a donor molecule and an acceptor molecule in an initial state in which the donor molecule is in the excited state following the absorption of a photon and the acceptor molecule is in the ground state, and a final state in which the donor molecule is in the ground state and the acceptor molecule is in the excited state is:

$$\Psi_i = \frac{1}{\sqrt{2}} \left(\Phi_{D^*}(1)\Phi_A(2) - \Phi_{D^*}(2)\Phi_A(1) \right) \tag{22.90}$$

$$\Psi_f = \frac{1}{\sqrt{2}} \left(\Phi_D(1)\Phi_{A^*}(2) - \Phi_D(2)\Phi_{A^*}(1) \right). \tag{22.91}$$

The numbers 1 and 2 refer to the two electrons that are involved in the electronic transition, and the asterisk denotes the excited state.

The total Hamiltonian (energy) operator for the combined system of the donor molecule and the acceptor molecule is:

$$\hat{H} = \hat{H}_D + \hat{H}_A + \hat{V}, \tag{22.92}$$

where $\hat{H}_D$ is the Hamiltonian operator for the donor molecule, $\hat{H}_A$ is the Hamiltonian operator for the acceptor molecule, and $\hat{V}$ is the contribution to the total Hamiltonian from the perturbation.

The matrix element that describes the interaction between the initial and the final states of the system is:

$$U = \left\langle \Psi_i | \hat{V} | \Psi_f \right\rangle. \tag{22.93}$$

The matrix element that describes the interaction can be expressed as the sum of two terms:

$$U = \left\langle \Phi_{D^*}(1)\Phi_A(2) | \hat{V} | \Phi_D(1)\Phi_{A^*}(2) \right\rangle - \left\langle \Phi_{D^*}(1)\Phi_A(2) | \hat{V} | \Phi_D(2)\Phi_{A^*}(1) \right\rangle. \tag{22.94}$$

The first term is the Coulombic term. In a simultaneous process the excited donor molecule is de-excited to the ground state and the acceptor molecule is forced to the excited state. The second term is the exchange term. This is a quantum mechanical process, without classical analog, in which there is an exchange of two electrons between the donor and acceptor molecules. It involves both the exchange of spin and the coordinates of the two electrons.

In FRET the second term is neglected. The first term, which is the Coulombic term, is approximated by the dipole–dipole interaction between the transition dipole moments of the donor molecule $\vec{\mu}_D$ and the acceptor molecule $\vec{\mu}_A$. This approximation is accurate only when the separation distance between the donor and acceptor molecules is much larger than the size of the molecules. The relation to classical theory is that the oscillator strength for a transition is given by the square of the transition dipole moments.

For the dipole–dipole approximation, the interaction U_{d-d} is:

$$U_{d-d} = \frac{\vec{\mu}_D \cdot \vec{\mu}_A}{|r|^3} - 3\frac{\left(\vec{\mu}_A \cdot \vec{r}\right)\left(\vec{\mu}_D \cdot \vec{r}\right)}{|r|^5}, \tag{22.95}$$

where r is the separation between the donor and the acceptor molecules.

The exchange term $U_{exchange}$ occurs over very close separations, less than 10 Å, since it requires the overlap of the electron clouds of the two molecules.

$$U_{exchange} = \left\langle \Phi_{D^*}(1)\Phi_A(2) \left| \frac{e^2}{\vec{r}_{12}} \right| \Phi_D(2)\Phi_{A^*}(1) \right\rangle \tag{22.96}$$

The separation between the two electrons is $\vec{r}_{12}$.

Finally, the rate of energy transfer is proportional to the square of the interaction term multiplied by the density of the interacting initial and final states. This density is expressed by the Franck-Condon factors. In experimental terms, the overlap integral between the emission spectrum of the donor molecule and the absorption spectrum of the acceptor molecule is equivalent to the density of the interacting initial and final states.

22.5.1 Hans Kuhn's Theory of Energy Transfer in Molecular Systems

Hans Kuhn and his coworkers at the Physikalisch-Chemisches Institute der Universitat Marburg-Lahn developed a simple quantitative model for energy transfer that does not depend on the assumption of resonance (Bücher et al., 1967; Kuhn, 1970; Kuhn & Försterling, 1999). The Kuhn derivation is instructive since it demonstrates another approach to energy transfer.

The 1999 textbook *Principles of Physical Chemistry, understanding molecules, molecular assemblies, supramolecular machines* describes a classical derivation of energy that is not based on the assumption of resonance. Kuhn states there is an excited molecule (donor, **D**) and an energy acceptor molecule, **A**, at a distance $\vec{r}$, where the transition moments are in the direction $\vec{r}$. In the classical approach the donor is a classical harmonic oscillator: frequency ν_0, charge Q, mass m, amplitude ξ_0. The average energy is:

$$\bar{E} = \frac{1}{2}k_f\xi_0^2, \tag{22.97}$$

where k_f is the spring constant. The electric field acting on the acceptor A is:

$$F = F_0 \cos 2\pi \nu_0 t, \quad \text{where} \tag{22.98}$$

$$F_0 = \frac{2Q\xi_0}{4\pi \varepsilon_0 r^3}, \tag{22.99}$$

where ε_0 is the permittivity of free space, and $Q\xi_0$ is the dipole moment, $|\vec{\mu}|$, of the harmonic oscillator.

Kuhn assumes the dipole approximation is valid (i.e., the wavelength of the light is much larger than the dimension of the donor, **D**). The mean power $\langle dE/dt \rangle$ absorbed by the acceptor **A** is proportional to the molar extinction coefficient of **A** at the wavelength of fluorescence of the donor and proportional the square of the field F_0 of the donor at the position of the acceptor.

$$\left\langle \frac{dE}{dt} \right\rangle_{energy\ trnasfer} = g\varepsilon_A F_0^2, \tag{22.100}$$

where

$$g = 3 \cdot \frac{2.303 c_0 \varepsilon_0}{2N_A}. \tag{22.101}$$

The quantity N_A is the Avogadro number, and c_0 is the speed of light in free space. The factor 3 follows because the transition moment of the acceptor molecule is parallel to the field F. However, the molar absorption coefficient ε_A of the acceptor molecule refers to a statistical spatial distribution of the transition moments of the absorbing molecules.

$$\left\langle \frac{dE}{dt} \right\rangle_{energy\ transfer} = g\varepsilon_A \left(\frac{2Q\xi_0}{4\pi \varepsilon_0 \vec{r}^3} \right)^2 = \frac{4Q^{2\xi^2}}{16\pi^2\varepsilon_0^2} \cdot \frac{1}{\vec{r}^6} \tag{22.102}$$

The average power emitted by the donor molecule (a Hertzian oscillator) with the frequency $\omega = 2\pi v_0$ is:

$$\left\langle \frac{dE}{dt} \right\rangle_{fluorescence\ of\ D} = \frac{1}{4\pi\varepsilon_0} \cdot \frac{Q^2\omega^4}{3c_0^3}\xi_0^2 = \frac{1}{4\pi\varepsilon_0} \cdot \frac{16\pi^4 Q^2 v_0^4}{3c_0^3}\xi_0^2, \qquad (22.103)$$

then,

$$\left\langle \frac{dE}{dt} \right\rangle_{fluorescence\ of\ D} = \frac{4Q^2\pi^3 v_0^4}{3\varepsilon_0 c_0^3}\xi_0^2 \qquad (22.104)$$

Kuhn defines the distance $\vec{r} = \vec{r}_0$ for which the emission is quenched to half of the value without the acceptor A,

$$\left\langle \frac{dE}{dt} \right\rangle_{energy\ trnasfer} = \left\langle \frac{dE}{dt} \right\rangle_{fluorescence\ of\ D} \qquad (22.105)$$

The final equation is valid for the following conditions: the quantum yield of fluorescence in the absence of the acceptor, ϕ, n is the refractive index of the medium, and κ is the previously defined orientation factor:

$$r_0 = \left(\frac{9\varepsilon_A\ 2.303 c_0^4}{128\pi^5 N_A v_0^4} \cdot \frac{\kappa^2\phi}{n^4} \right) \qquad (22.106)$$

The probability P for the emission of light is:

$$P = \frac{1}{1 + \left(\frac{r}{r_0}\right)^6} \qquad (22.107)$$

Kuhn's classical derivation yields the correct R^{-6} dependence for the rate of energy transfer. However, the equation for r_0 does not contain the overlap integral; it only contains the extinction coefficient for the acceptor. Since we know that the Förster derived equation for R_0 contains the overlap integral of the donor and the acceptor molecules and that the experimental studies of FRET show a dependence on the overlap integral, Förster's equation and the equation derived by Hans Kuhn are different.

22.5.2 Dexter Theory of Sensitized Luminescence

In 1953 Dexter published his theory of sensitized luminescence in solids (Dexter, 1953). While the Förster theory of resonance energy transfer is typically limited to allowed electric dipole transitions, Dexter extends the theory to include forbidden electronic transitions, which are important in inorganic solids. Dexter made a theory of energy transfer for transitions that are forbidden either by symmetry or by spin intercombinations. What stimulated interest in sensitized luminescence from inorganic solids is that resonant energy transfer can occur between an allowed electronic transition in a sensitizer molecule **S** and a forbidden electric transition in an activator molecule **A**.

Dexter delineates several mechanisms of energy transfer: those depending on the overlapping of the electric dipole fields of the sensitizer and the activator molecules, those depending on the dipole field of the sensitizer and the quadrapole field of the activator, and those that depend on exchange effects. Exchange effects are significant for magnetic dipole and higher-order multipole electronic transitions. Exchange effects are quantum mechanical processes that involve the interaction of the electronic charge clouds around both the sensitizer and the acceptor molecules. The quantum mechanical phenomenon of exchange, which does not have a classical analog, corresponds to an energy transfer mechanism that is associated with the exchange of two electrons between the donor molecule and the acceptor molecule. The Hamiltonian for the exchange process includes the electron spin wave functions, which are not included in the Förster theory of energy transfer.

For triplet–triplet energy transfer, exchange interaction is critical. In the Dexter theory of energy transfer only the exchange interaction is involved in the triplet–triplet energy transfer. Dexter theory assumes a continuum of initial and final states for the donor molecule and the acceptor molecules.

22.6 SUMMARY AND COMMENTS ON FÖRSTER'S THEORY OF RESONANCE ENERGY TRANSFER

1. Resonance energy transfer was first demonstrated by experiments in the gas phase by Frank and Cario. Soon after Heisenberg developed his quantum mechanical theory, he published two papers (1926, 1927) on the quantum theory of resonance between harmonic oscillators. Another quantum theory of resonance was published by Kallmann and London in 1929. J. Perrin made early observations of resonance energy transfer in solution, and his son F. Perrin developed a quantum theory of resonance energy transfer in 1936.

2. Förster's resonance energy transfer (1946) is a nonradiative process in which energy is transferred from the donor molecule to the acceptor molecule. The donor molecule is initially in the excited state. Förster assumed that the nonradiative energy transfer is much slower than the vibrational relaxation between higher vibrational modes of the excited state due to interaction with solvent molecules. Simultaneous with the return of the donor molecule to the ground state, the acceptor molecule is promoted from the ground state to the excited state. The rate of dipole–dipole interaction-induced energy transfer decreases as R^{-6}. The rate constant for the resonance energy transfer can be expressed as:

$$k_{D-A} = \frac{1}{\tau_D} \left(\frac{R_0}{R} \right)^6. \tag{22.108}$$

    where τ_D is the mean lifetime of the excited donor molecule and R_0 is the critical transfer distance for which the excitation transfer and spontaneous deactivations of the donor molecule occur with equal probability.

3. Very weak coupling between the donor and the acceptor molecules induces the nonradiative energy transfer. The interaction between the two molecules does not alter the absorption and the emission spectra of either molecule as compared to the spectrum of each molecule in the absence of the other molecule. If there are interactions between the donor molecules and the acceptor molecules

that alter the emission or absorption line shapes or the dipole strengths, then the Förster cannot be used.

4. The Förster theory was built on the previous theory developed by F. Perrin. The Perrin theory assumed the following: two identical oscillators are involved; initially only one oscillator is excited; the interaction is mediated by dipole–dipole coupling. Förster's theory correctly accounted for the fact that there is a manifold or distribution of vibronic states in the first electronic excited state and therefore the Perrin condition of exact resonance and single-frequency line shapes had to be modified. Förster modified the Perrin theory by calculating the probability of having the exact resonance condition for a manifold of energy states of the donor and the acceptor molecules. Because of the broadening of the spectral lines into spectral bands, the time for exact resonance is less than the time of energy transfer.

5. Dipole–dipole coupling is the mechanism of the interaction between the donor molecule and the acceptor molecule. The oscillating dipole moment of the donor molecule results in an electric field. This electric field interacts with the oscillating dipole moment of the acceptor molecule. This mechanism occurs in the near-field between the two dipoles. This dipole–dipole coupling induces the energy transfer. The dipole–dipole approximation is valid only when the separation distance between the donor and the acceptor molecules is much larger that the molecular dimensions.

6. The rate of energy transfer in FRET depends directly on the magnitudes of the transition dipole moments of the donor molecule and the acceptor molecule. Förster's quantum mechanical derivation of resonance energy transfer is based on Fermi's Golden Rule, and the Coulombic coupling between the donor molecules and the acceptor molecules is formulated in terms of time-dependent second-order perturbation theory.

7. Förster's resonance energy transfer involves singlet–singlet transitions between the donor and the acceptor molecules. The acceptor molecule must have allowed electronic transitions. This restriction is relaxed for the donor molecule. If the donor molecule has a forbidden but very long lifetime, then it can still result in appreciable rates of FRET.

8. The spectroscopic overlap integral, defined as the spectral overlap of the emission spectrum of the donor molecule and the absorption spectrum of the acceptor molecule, is a measure of the degree coupling between the donor and the acceptor molecules.

9. The dipole–dipole coupling between the donor and the acceptor molecules is mediated by the dielectric constant. In Förster's resonance energy transfer theory it is introduced as the refractive index to the minus-fourth power, n^{-4}. While the value of the refractive index is usually taken as the value for the solution, this may not be correct for all conditions.

10. The distance between the donor and the acceptor molecules is given as R. If the molecule exists in a variety of conformations, then there may be a statistical distribution for R.

11. In Förster's resonance energy transfer the rate of energy transfer is given by κ^2, in which the term κ is related to the relative orientations of the transition dipole moments of the donor molecule and the acceptor molecule. In many cases it is assumed that the donor or the acceptor molecules have free rotation; for that case the average value of κ^2 is equal to 2/3. However, that assumption may not be valid in many experimental cases.

12. In the experimental measurement of the overlap integral there is the severe problem of "spectral bleed-through" by the filters. This error may be mitigated by computer algorithms; however, it may be preferable to eliminate the problem by switching to a FRET technique that is not intensity-based. There are many varieties of FRET that are based on lifetime measurement combined with photobleaching. Each of the techniques should be carefully evaluated with respect to sensitivity.

13. The genius of Förster's theory is that he converted his classical theory and his quantum mechanical theory into a phenomenological approach that was accessible to spectroscopists. The experimentally defined "overlap integral," which is normalized to unit area on an energy scale, incorporated such terms as the transition moments of the donor and the acceptor molecules and the Franck-Condon integrals. The FRET technique become very popular and is used by chemists, biophysicists, and later biologists. Since the development of Förster's theory, there have been many new developments in nonradiative energy transfer. These developments have occurred both in theoretical models and in experimental techniques. The popularity of FRET techniques in the life sciences in part comes from the fact that they enable experimental studies of molecular interactions on the molecular scale. Today, with microscopic FRET techniques, molecular interaction can be studied within living cells.

ACKNOWLEDGMENTS

Barry R. Masters dedicates this chapter to his teachers: Michael Kasha, Dexter Easton, Hans Kuhn, David Mauzerall, and Britton Chance.

REFERENCES

Arnold, W., Oppenheimer, J. R. 1950. Internal conversion in the photosynthetic mechanism of blue-green algae. *The Journal of General Physiology*, 33: 423–435.

Atkins, P.W. 1990. *Physical Chemistry, fourth edition*. New York: W. H. Freeman and Company. pp. 657–658.

Bastiaens, P. I., Jovin, T. M. 1996. Microspectroscopic imaging tracks the intracellular processing of a signal transduction protein: fluorescent-labeled protein kinase C beta I. *Proc. Natl. Acad. Sci. U.S. A.* 93: 8407–8412.

Berberan-Santos, M. N. 2001. Pioneering contributions of Jean and Francis Perrin to molecular luminescence. In: *New Trends in Fluorescence Spectroscopy, applications to chemical and life sciences*, eds. B. Valeur, J-C. Brochon, Berlin: Springer Verlag, pp. 7–34.

Brau, C. A. 2004. *Modern Problems in Classical Electrodynamics*, New York: Oxford University Press.

Bücher, H., Kuhn, H. Mann, B., Möbius, D., von Szentpály, L., Tillmann, P. 1967. Transfer of energy and electrons in assemblies of monolayers. *Photographic Science and Engineering*, 11(4): 233–241.

Cario, G. 1922. Über Entstehung wahrer Lichtabsorption un scheinbare Koppelung von Quantensprügen. *Z. Physik*, 10: 185–199.

Cario, G., Franck, J. 1923. Über sensibilisierte Fluoreszenz von Gasen, *Z. Physik* 17: 202–212.

Clegg, R.M. 1996. Fluorescence resonance energy transfer, In: *Fluorescence Imaging Spectroscopy and Microscopy*, eds. X. F. Wang, B. Herman, New York: Wiley-Interscience, pp. 179–252.

Davydov, A. S. 1962. *Theory of Molecular Excitons*, M. Kasha, M. Oppenheimer, Jr. translators, New York: McGraw-Hill.

Dexter, D.L. 1953. A theory of sensitized luminescence in solids. *The Journal of Chemical Physics*, 21(5): 836–850.

Fano, U., Fano, L. 1959. *Basic physics of atoms and molecules*, New York: John Wiley & Sons, appendix X, pp. 398–402.

Fermi, E. 1932. Quantum theory of radiation. *Rev. Mod. Phys.* 4: 87–132.

Fermi, E. 1995. *Notes on Quantum Mechanics, a course given by Enrico Fermi at the University of Chicago, second edition*, pp. 100–103, Chicago: The University of Chicago Press.

Feynman, R.P., Leighton, R. B., Sands, M. 2006. *The Feynman Lectures on Physics, the definitive edition*, volume I, San Francisco: Addison Wesley, pp. 23-1–23-9.

Förster, T. 1946. Energiewanderung und Fluoreszenz. *Die Naturwissenschaften*, 33: 166–175.

Förster, T. 1948. Zwischenmolekulare Energiewanderung und Fluoreszenz. *Ann. Phys.* 2: 55–75.

Förster, T. 1959. Transfer mechanisms of electronic excitation. *Discuss. Faraday Soc.* 27: 7–17.

Förster, T. 1960. Transfer Mechanisms of electronic excitation energy. *Rad. Res. Supplement* 2: 326–339.

Förster, Th. 1965. Delocalized excitation and excitation transfer. In: *Modern Quantum Chemistry, Part III, Action of Light and Organic Crystals*, ed. O. Sinanoglu, 93–137. New York: Academic Press.

Förster, Th. 1982. *Fluoreszenz Organischer Verbindungen*, Göttingen: Vandenhoeck & Ruprecht.

Franck, J., Jordan, P. 1926. *Anregung von Quantensprügen durch Stösse*, Berlin: Verlag von Julius Springer.

Gottfried, K., Yan, T-M. 2004. *Quantum Mechanics, Fundamentals, second edition*. New York: Springer Verlag.

Heisenberg, W. 1926. Mehrköperproblem und Resonanz in der Quantenmechanik. *Zeitschrift für Physik*, 3: 411–426.

Heisenberg, W. 1927. Mehrköperprobleme und Resonanz in der Quantenmechanik II. *Zeitschrift für Physik*, 41: 239–267.

Jares-Erijman, E.A., Jovin, T.M. 2003. FRET imaging, *Nature Biotechnlogy*, 21(11): 1387–1395.

Jovin, T.M. and D.J. Arndt-Jovin. 1989a. FRET microscopy: digital imaging of fluorescence resonance energy transfer. Application in cell biology. In: *Cell Structure and Function by Microspectrofluorimetry*, eds. E. Kohen, J.S. Ploem, and J. G. Hirschberg, Orlando: Academic Press, pp. 99–117.

Jovin, T.M. and D.J. Arndt-Jovin. 1989b. Luminescence digital imaging microscopy. *Annu. Rev. Biophys. Biophys. Chem.* 18: 271–308.

Jurgens, L., D. Arndt-Jovin, I. Pecht, and T. M. Jovin. 1996. Proximity relationships between the type I receptor for Fce. (FceRI) and the mast cell function-associated antigen (MAFA) studied by donor photobleaching fluorescence resonance energy transfer microscopy. *Eur. J. Immunol.* 26: 84–91.

Kallmann, H., London, F. 1929. Über quantenmechnische Energieübertragung zwischen atomaren Systemen. Ein Beitrag zum Problem der anomalgrossen Wirkungsquerschnitte. *Z. Physik. Chem.* (Frankfurs) B2: 207–243

Kasha, M. 1959. Relation between exciton bands and conduction bands in molecular lamellar systems. In: *Biophysical Science-A Study Program*, ed. J. L. Oncley, New York: John Wiley & Sons, 162–169.

Kasha, M. 1991. Energy transfer, charge transfer, and proton transfer in molecular composite systems. In: *Physical and Chemical Mechanisms in Radiation Biology*, eds. W. A. Glass, M. N. Varma, New York: Plenum Press, pp. 231–255.

Knox, R. S., 1993. Intermolecular Energy Migration and Fluorescence, in: *Biological Physic*, eds. Mielczarek, E., Greenbaum, E., Knox, R. S. 148–160, New York: Am. Inst. Phys. (Robert S. Knox's translation into English of Förster, Th., 1948. Zwischenmolekulare Energiewanderung und Fluoreszenz, *Ann. Phys.* 2: 55–75).

Kuhn, H. 1970. Classical aspects of energy transfer in molecular systems. *J. Chem. Phys.* 53(1): 101–108.

Kuhn, H., Försterling, H-D., 1999. *Principles of Physical Chemistry, understanding molecules, molecular assemblies, supramolecular machines*, 862–879. New York: John Wiley and Sons.

Lakowicz, J. R. 1999. *Principles of Fluorescence Spectroscopy, second edition*, 367–394. New York: Springer Verlag, chapter 13, Energy Transfer.

Latt, S. A., Cheung, H. T., Blout, E. R. 1965. Energy transfer: a system with relatively fixed donor-acceptor separation. *J. Am. Chem. Soc.* 87: 995–1003.

Lin, S. H., Xiao, W. Z., Dietz, W. 1993. Generalized Förster-Dexter theory of photoinduced intramolecular energy transfer. *Physical Review E*, 47(5): 3698–3706.

Marion, J. B. 1970. *Classical dynamics of particles and systems, second edition*. New York: Academic Press.

Mitchell, A. C. G., Zemansky, M. W. 1934. *Resonance radiation and excited states*, Cambridge: Cambridge University Press.

Pauling, L., Wilson, Jr., E. B. 1935. *Introduction to quantum mechanics with applications to chemistry*, 314–325. New York: McGraw-Hill, reprinted by Dover Publication in 1985.

Perrin, F. 1931. Fluorescence durée élémentaire d'émission lumineuse. In: *La haute précision des mesures de longueur au laboratoire et dans l'industrie*, ed., M. A. Pérard, Paris: Libraire Scientifique Hermann et C^{fe}.

Perrin, F. 1932. Théorie quantique des transferts d'activation entre molécules de même espèce. Cas des solutions fluorescentes. [Quantum theory of the transfer of activation between molecules of the same kind in fluorescence solutions]. *Ann. Chim. Phys. (Paris)*, 10th series, 17: 283–314.

Perrin, J. 1925. Light and chemical reactions. 2me *Cons. Chim. Inst. Intern, Chim. Solvay*, Paris: Gauthier-Villars, p. 322.

Perrin, J. 1927. Fluorescence et induction molécular par résonance. *C. R. Hebd. Seances Acad. Sci.*, 184: 1097–1100.

Scholes, G.D. 2003. Long-range resonance energy transfer in molecular systems. *Annu. Rev. Phys. Chem.* 54: 57–87.

Schrödinger, E. 1927. Energieaustausch nach Wellenmechanik, *Ann. Physik*, 83: 956–968.

Stryer, L., and Haugland, R.P. 1967. Energy transfer: a spectroscopic ruler, *Proceedings of the National Academy of Sciences, USA*, 58(2): 719–726.

Valeur, B. 2002. *Molecular Fluorescence, principles and applications*. Weinheim: Wiley-VCH Verlag.

Vavilov, S. I. 1943. Theory of the influence of concentration on the fluorescence of solutions. J. Phys. U.S.S.R. 7: 141.

Weber, G., Teale, F. J. W., 1959. Electronic energy transfer in haem proteins. *Disc. Faraday Soc.* 27: 134–141.

23

Two-Photon Förster Resonance Energy Transfer (FRET) Microscopy

Ye Chen, Horst Wallrabe, and
Ammasi Periasamy

23.1 INTRODUCTION

Recent technological advances in fluorophores and detectors have made it feasible to apply Förster (fluorescence) energy transfer (FRET) in biomedical research (Clegg, 1996; Day, 1998; Kraynov et al., 2000; Ludwig et al., 1992; Periasamy, 2001; Periasamy & Day, 2005; Sekar & Periasamy, 2003; Wallrabe & Periasamy, 2005). Light microscopy FRET can provide spatial and temporal distribution of protein associations in single living cells, compared to other methods described in the literature (Guo et al., 1995). There are problems involved in using light microscopy methods to localize protein molecules, such as spectral bleedthrough or cross-talk. In principle, the bleedthrough signal is the same for one-photon (1p) or two-photon (2p) FRET microscopy; choosing the right FRET pairs and conducting qualitative and quantitative FRET data processing have been described extensively in the literature (Wallrabe & Periasamy, 2005). In this chapter we will address these issues and correction methodologies, in addition to the theory of FRET. To illustrate these methodologies, we will use transiently transfected GHFT1-5 live cells expressing cyan fluorescent protein (CFP) and yellow fluorescent protein (YFP) fused to the transcription factor CCAAT/enhancer binding protein alpha (C/EBPα). We will also explain the problems associated with various techniques, including wide-field, confocal, and multiphoton FRET microscopy.

23.2 FRET THEORY

In the late 1940s Förster proposed the theory of FRET, which described how energy is transferred directly from a fluorophore in the excited state (the donor, D) to a non-identical acceptor (A) fluorophore (Förster, 1965; Lakowicz, 1999; Stryer, 1978). The transfer of excited state energy occurs without the production of heat and does not require that a collision occur between donor and acceptor.

FRET is a quantum mechanical effect that occurs when the energy from the donor (D) is transferred directly to an acceptor (A) (Förster, 1965; Lakowicz, 1999). A high efficiency of energy transfer depends on four conditions: (i) when the emission spectrum of the donor has significant overlaps with the absorption spectrum of

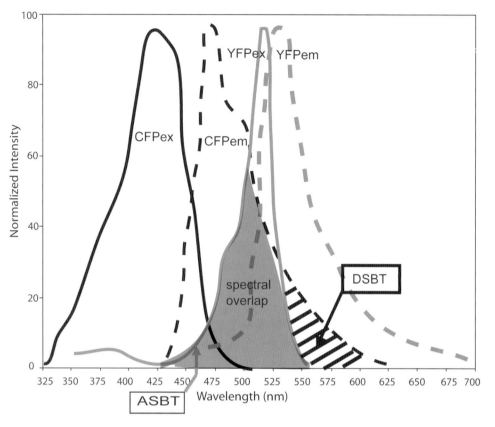

Figure 23.1. Absorption and emission spectra of CFP and YFP. The cyan emission (shaded by lines) in the YFP emission channel represents the donor spectral bleedthrough (DSBT) in the FRET channel, and the ASBT arrow shows the donor excitation wavelength exciting acceptor molecules to cause acceptor spectral bleedthrough (ASBT) in the FRET channel. The spectral overlap integral (J) is shaded in gray. CFPex 457 nm, CFPem 485/30 nm, YFPex 514 nm, YFPem 540/40 nm. (Adapted from Wallrabe & Periasamy, 2005)

the acceptor (Fig. 23.1); (ii) when the two fluorophores are within $\sim$1 to $\sim$10 nm of each other; (iii) when the donor emission dipole moment, the acceptor absorption dipole moment, and their separation vectors are in a favorable mutual orientation, and (iv) when the emission of the donor has a reasonably high or equal quantum yield as the acceptor.

FRET is a measure of the efficiency of the energy transfer (E), which is the fraction of absorbed photons that are transferred without radiation to the acceptor, calculated as (Förster, 1965; Lakowicz, 1999):

$$E = \frac{k_T}{(k_T + k_{Dr} + k_{Dnr})} \qquad (23.1)$$

where k_T is the rate of energy transfer and k_{Dr} and k_{Dnr} are the radiative and nonradiative deactivation of the excited donor state without transfer.

Qualitatively, FRET provides proof of proximity within the 1- to 10-nm range. Quantitatively, there are a number of applications, including its utility as a "spectroscopic ruler" based on Förster's basic rate equation for a donor and acceptor pair at a distance r from each other, given by (Förster, 1965; Stryer, 1978):

$$k_T = (k_{Dr} + k_{Dnr}) \left(\frac{R_0}{r} \right)^6 \qquad (23.2)$$

where R_0 is the Förster distance. The Förster distance is the distance between the donor and the acceptor at which half the excitation energy of the donor is transferred to the acceptor while the other half is dissipated by all other processes, including light emission.

Therefore, R_0 can also be defined as the donor–acceptor distance at which the transfer efficiency is 50%. The efficiency drops if the distance between D and A molecules changes from the Förster distance. Energy transfer efficiency E can also be calculated using equation (23.3).

$$E = \frac{R_0^6}{(R_0^6 + r^6)} \qquad (23.3)$$

Förster's equation relates the energy transfer rate (Förster, 1965) with the distance between D and A fluorophores and their spectroscopic properties. Two events occur when incident light at the donor absorption wavelength excites a protein labeled with donor and acceptor fluorophores: (i) the donor molecules nonradiatively transfer the energy to the acceptor and (ii) the donor's fluorescence quenches or decreases (and so will the donor lifetime decrease) and the acceptor channel fluorescence increases. There are several ways to calculate energy transfer efficiency (E) and distance (r) for measuring FRET using various light microscopy techniques, including acceptor photobleaching (Bastiaens et al., 1996; Day et al., 2001; Kenworthy, 2005), the use of correction algorithms (Chen et al., 2005; Elangovan et al., 2003; Gordon et al., 1998), and the lifetime method (Elangovan et al., 2002; Gadella et al., 1993).

23.3 FRET MICROSCOPY

Different microscopy techniques allow us to visualize protein molecules under live conditions and to investigate their 2-dimensional or 3-dimensional distribution (Periasamy & Day, 1999). Two-photon excitation microscopy uses infrared laser light with variable excitation wavelengths and has the ability to excite almost any selected FRET pair in addition to discriminating the out-of-focus information. The use of infrared laser light excitation instead of ultraviolet laser light also reduces the phototoxicity in live cells. It has been cited in the literature (Patterson & Piston, 2000) that fluorophore molecules do bleach with 2p excitation under various conditions, but in general photobleaching is considerably less in 2p than in 1p excitation for thick specimens (Denk et al., 1995). The advantages and disadvantages of various FRET techniques are outlined in Table 23.1.

Table 23.1. Comparison of wide-field, confocal and multiphoton FRET microscopy

	Wide-field FRET	Confocal FRET	Multiphoton FRET
System	Requires excitation, emission, and neutral-density filter wheels, a single dichroic mirror to excite and to detect the D and A molecules.	Respective filter sets and dichroic mirrors come with the system.	Respective emission filter sets and dichroic mirrors come with the system.
Light source	Arc lamp. Any FRET pair combination can be used.	Visible laser light source. Selected FRET pair can be used.*	Infrared pulsed light source (700–1,000 nm). Any FRET pair can be used.
Detector	CCD camera with high quantum efficiency (70–90%).	Photomultiplier tubes with about 8–20% quantum efficiency.	Photomultiplier tubes with about 8–20% quantum efficiency.
Advantages	Simple to use and available in many laboratories. Any FRET pair combination can be used.	Simple to use. Laser beam scanning, protein localizations at different optical sections. Good system for FRET microscopy.	It's easy to use with computerized tunable laser, protein localizations at different optical sections, deep tissue protein imaging.
Disadvantages	Optical sectioning is possible but requires dedicated digital deconvolution software. Requires spectral bleedthrough correction.	Only selected FRET pairs can be used due to the availability of limited excitation wavelengths. Requires spectral bleedthrough correction.	Manual tuning is difficult to maintain for inexperienced laser users. There is potential for appreciable excitation of the acceptor by the donor wavelength in GFP-derived FRET pairs and a potential for back-spectral bleed through. Once these limitations have been removed by the choice of a suitable FRET pair (e.g. Alexa 488/Alexa 555), multiphoton FRET is a powerful alternative, particularly for thicker tissue specimens. A major application for multiphoton excitation is found in lifetime FRET (FLIM/FRET).

* A useful alternative to the laser-based confocal microscope is the arc lamp-based spinning disk confocal system, which allows variable excitation wavelengths by using filters.

Table 23.2. FRET fluorophore pairs for FRET and FLIM (Adapted from Wallrabe and Periasamy 2005)

Donor	Acceptor	Used in cited references
BFP	GFP	Daelemans et al., 2004; Periasamy & Day, 1998, 1999
BFP	YFP	Day et al., 2003
CFP	YFP	Bacskai et al., 2003; Becker et al., 2004; Biskup et al., 2004; Chan et al., 2004; Chan et al., 2001; Chen et al., 2003; Chen & Periasamy, 2004, 2005; Elangoven et al., 2002, 2003; Krishnan et al., 2003; Larson et al., 2003; Nashmi et al., 2003; Oliveria et al., 2003; Ottolia et al., 2004; Rizzo et al., 2004; Sekar & Periasamy, 2003
YFP	YFP	Sharma et al., 2004
GFP	Rhod-2	Periasamy, 2001
FITC	Rhod-2	Periasamy, 2001
Cy3	Cy5, Cy5.5	Hohng et al., 2004
Alexa488	Alexa555	Elangovan et al., 2003; Mills et al., 2003;
FITC	Alexa546 Cy3	Periasamy, 2001; Wallrabe et al., 2003
Alexa350	Alexa488 Alexa594	Watrob et al., 2003
Fluorescein	Cy5	Panyi et al., 2003
Fluorescein	Rhodamine	Hiller et al., 2003
	Texas Red	Klostermeier et al., 2004
Rhodamine	NBD	Lear et al., 2004
DCIA	NBD	Fernandes et al., 2004
IAEDANS	DABCYL	Mueller et al., 2004
Trp	dansyl	Gustiananda et al., 2004

Basically, a FRET signal is generated by exciting the double-labeled (donor and acceptor) protein(s) of interest within the cell with the donor excitation wavelength and collecting both donor and acceptor emission. When all the conditions for FRET are satisfied, the donor signal will be quenched due to the transfer of energy and the acceptor signal will be sensitized (increased). The acceptor emission channel contains the FRET signal, which is usually contaminated by spectral bleedthrough, which should be corrected as explained in the following sections.

The burgeoning number of publications (Berney & Danuser, 2003; Periasamy & Day, 2005) on the use of FRET is a testament to the utility of this microscopy technique becoming increasingly the starting point to investigate cellular structure and function in four dimensions and to proceed from there to more quantitative analyses. Moreover, the advent of green fluorescent protein (GFP)and its derivatives (Day et al., 2006; Shaner et al., 2004; Wang et al., 2004) made implementing the FRET methodology more attractive for various biological applications, as listed in the literature (Table 23.2).

23.4 DIFFERENT MICROSCOPIC TECHNIQUES OF MEASURING
       FRET SIGNALS

23.4.1 Photobleaching Method (pbFRET)

Even though pbFRET is destructive, it is a convenient method to demonstrate the occurrence of FRET if proper controls are used. Photobleaching methods, in which either the
donor or acceptor molecules are selectively photodestructed, provide a sensitive means
for establishing and quantitating energy transfer efficiencies in cells (Bastiaens et al.,
1996; Day et al., 2001; Jurgens et al., 1996; Kenworthy, 2005). The photobleaching
time of a fluorophore is inversely proportional to the excited state lifetime; any process
such as energy transfer that shortens the lifetime will decrease the photobleaching rate.
   In the acceptor photobleaching scenario, destruction of the acceptor leads to a corresponding increase in the donor emission quantum yield (i.e., intensity). In the acceptor
bleaching method, the difference between quenched donor and unquenched donor
is used to calculate the energy transfer efficiency (Kenworthy et al., 2000; Wouters
et al., 1998).

$$E = 1 - (I_{DA}/I_D) \qquad (23.4)$$

where I_{DA} is the intensity of the quenched donor in the presence of an acceptor and I_D
is the intensity of the unquenched donor after bleaching the acceptor.
   This approach has the advantage of using a double-labeled specimen only, and no
bleedthrough is involved. However, it is generally not suitable for live cell imaging,
since the exposure to extended laser energy is believed to cause detrimental effects,
apart from cellular processes proceeding during the period of bleaching with potential
spatial rearrangements of the proteins of interest. During the bleaching time, the cells
themselves may also move and cause image shifting between images collected before
and after bleaching. It is then necessary to perform image registration before estimating
E% using equation (23.4). Figure 23.2 demonstrates before and after photobleaching
and shows where images have shifted. Because of image registration issues in the
acceptor photobleaching method, it is not recommended for quantitative FRET analysis.
However, it is undeniably a simple method for qualitative FRET analysis, with the
mentioned caveats in live cell imaging.
   Multiphoton FRET microscopy can also be used for pbFRET in addition to wide-
field and confocal microscopy. It is advisable to bleach the acceptor molecule using
confocal lasers or an arc lamp for GFP-derived FRET pairs (e.g., CFP and YFP), as the
2p acceptor excitation wavelength may bleach the donor molecules (see Table 23.1).

23.4.2 Algorithm Method

FRET signals are detected by exciting double-labeled (donor and acceptor) specimens with the donor excitation wavelength. When FRET occurs, the donor emission
is decreased and the acceptor emission is increased. Unfortunately, because of the
spectral overlap, there is a cross-talk in the acceptor emission channels. We call this
spectral bleedthrough (SBT). The details of the corrections and the relevant biological

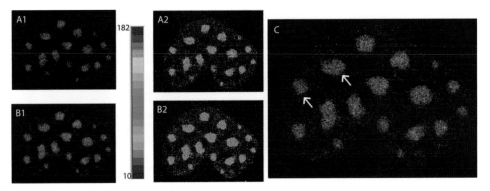

Figure 23.2. Acceptor photobleaching method to demonstrate the occurrence of FRET. Dimerization of CFP-YFP-C/EBPΔ244 protein molecules in mouse pituitary GHFT1-5 cell nucleus. Even though it is a good technique to demonstrate the occurrence of FRET, there is a possibility of image shift (C, red color), which generates errors in the E% calculation. This requires an image registration algorithm to align the pixels before and after bleaching. A1, A2: donor images before acceptor photobleaching (quenched donor). B1, B2: after acceptor photobleaching. C: overlay of A1 and B1; A2 and B2 are pseudocolored images of A1 and B1. The increase in signals after photobleaching the acceptor molecules clearly represented in B2 with increase in gray-level intensity as indicated in the color bar.

Table 23.3. Data acquisition and the respective symbols

One-Letter Symbol	Fluorophores	Excitation	Emission
a	Donor	D_{ex}	D_{em}
b	Donor	D_{ex}	A_{em}
c	Acceptor	D_{ex}	A_{em}
d	Acceptor	A_{ex}	A_{em}
e	Donor + Acceptor	D_{ex}	D_{em}
f	Donor + Acceptor	D_{ex}	A_{em}
g	Donor + Acceptor	A_{ex}	A_{em}

D_{ex}, D_{em}: Donor excitation, Donor emission; A_{ex}, A_{em}: Acceptor excitation, Acceptor emission

applications have been fully discussed in the literature (Chen et al., 2005; Elangovan et al., 2003; Mills et al., 2003; Wallrabe et al., 2003).

Briefly, the main SBTs are (i) the emission of donor fluorescence overlapping with the acceptor emission channel (DSBT) and (ii) acceptor fluorescence signal detected in the acceptor emission channel due to donor excitation wavelength exciting acceptor fluorescence (ASBT). To remove these two SBTs, seven images are acquired, as shown in Table 23.3: three double-labeled images (donor excitation/donor and acceptor channel (e & f), acceptor excitation/acceptor channel (g)); two single-labeled donor images (donor excitation/donor and acceptor channels (a & b)); and two single-labeled acceptor images (donor excitation/acceptor channel (c) and acceptor excitation/acceptor channel (d)), all with appropriate filters for PFRET data analysis as described in the literature (Chen et al., 2005; Elangovan et al., 2003; Fig. 23.3). This approach works on the

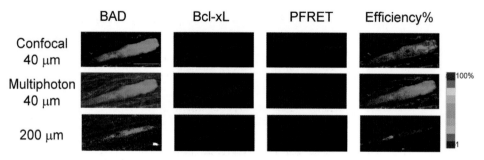

Figure 23.3. Multiphoton FRET microscopy/deep tissue FRET imaging. Using serial sections of rat brain tissue, we simultaneously labeled injured tissue with donor only (BAD/Alexa488), with acceptor only (Bcl-xL/Alexa555), and with both donor and acceptor. The specimens were then examined for axons demonstrating vacuolization or formation of retraction bulbs, morphological characteristics of axonal injury (Mills et al., 2003). Then we used confocal and multiphoton microscopy 6 hours after injury to demonstrate energy transfer consistent with BAD-Bcl-xL heterodimerization. The same tissue was used for C-FRET and 2p-FRET imaging using Biorad Radiance2100 confocal/multiphoton microscopy. The confocal PFRET image was obtained at 40-microns depth and was unable to produce a decent signal beyond that depth. On the other hand, we obtained two-photon (or multiphoton) PFRET images up to 200 microns depth. As shown in the figure, the signal appears to be less when we moved from 40 to 200 microns. This can be attributed to the concentration of the fluorophore deep inside the tissue and also to the visible FRET signal lost in the tissue before reaching the detector. Bar = 10 μm; color bar represents percentage of energy transfer efficiency; Lens 20× MIMM NA 0.75; confocal ExD 488, EmD 528/30, ExA 543, EmA 590/70; two-photon ExD 790, ExA 730. (Adapted from Chen et al., 2003)

assumption that the double-labeled cells and single-labeled donor and acceptor cells, imaged under the same conditions, exhibit the same SBT characteristics. The hurdle to be overcome is the fact that there are three different cells (D, A, and D+A), where individual pixel locations cannot be compared. What can be compared, however, are pixels with matching fluorescence levels. The algorithm follows fluorescence levels pixel by pixel to establish the level of SBT in the single-labeled cells, and then applies these values as a correction factor to the appropriate matching pixels of the double-labeled cell with the same fluorescence levels. Equation (23.5) through (23.10) show how to calculate the two main bleedthrough DSBT and ASBT.

$$rd_{(j)} = \frac{\sum_{i=1}^{i=m} b_i / a_i}{m} \qquad (23.5)$$

$$DSB_{(j)} = \sum_{p=1}^{p=n} (e_p * rd_{(j)}) \qquad (23.6)$$

$$DSBT = \sum_{j=1}^{j=k} DSBT_{(j)} \qquad (23.7)$$

where j is the jth range of intensity, $rd_{(j)}$ is the donor bleedthrough ratio for the jth intensity range, m is the number of pixels in 'a' for the jth range, $a_{(i)}$ is the intensity of pixels i, $DSBT_{(j)}$ is the donor bleedthrough factor for the range j, n is the number of pixels in 'e' for the jth range, $e_{(p)}$ is the intensity of pixels (p), k is the number of range, and DSBT is the total donor bleedthrough.

$$ra_{(j)} = \frac{\sum_{j=1}^{i=m} (c_i/d_i)}{m} \tag{23.8}$$

$$ASBT_{(j)} = \sum_{p=1}^{p=n} (g_p * ra_{(j)}) \tag{23.9}$$

$$ASBT = \sum_{j=1}^{j=k} ASBT_{(j)} \tag{23.10}$$

where j is the jth range of intensity, $ra_{(j)}$ is the acceptor bleedthrough ratio for the jth intensity range, m is the number of pixels in 'd' for the jth range, $d_{(i)}$ is the intensity of pixels i, $ASBT_{(j)}$ is the acceptor bleedthrough factor for the range j, n is the number of pixels in 'g' for the jth range, $g_{(p)}$ is the intensity of pixels (p), k is the number of range, and ASBT is the total acceptor bleedthrough.

Then, in PFRET, the contamination-free FRET signal is (Chen et al., 2005; Elangovan et al., 2003):

$$PFRET = uFRET - DSBT - ASBT \tag{23.11}$$

where uFRET is uncorrected FRET (signal in the FRET/acceptor channel) (see Figs. 23.1 and 23.3).

Conventionally, energy transfer efficiency (E) is calculated by ratioing the donor image in the presence (I_{DA}) and absence (I_D) of acceptor (Periasamy & Day, 1999). To execute this calculation, the acceptor in the double-labeled specimen either has to be bleached (equation (23.4)) or the donor fluorescence averages of two different cells (single and double label) with most likely different FRET dynamics are used in the efficiency calculation. When using the algorithm as described, the I_D image is indirectly obtained by using the PFRET image (Elangovan et al., 2003). The sensitized emission in the acceptor channel is due to the quenching of the donor or energy transferred signal from the donor molecule in the presence of acceptor. Therefore, I_D is obtained by adding the PFRET to the intensity of the donor in the presence of acceptor. This I_D is from the same cell used to obtain the I_{DA}. So, the new energy transfer efficiency (E_n) is shown in equation (23.12):

$$E_n = 1 - \left[\frac{I_{DA}}{I_{DA} + PFRET} \right] \tag{23.12}$$

In addition, we have to correct the variation of the detector spectral sensitivity of donor and acceptor channels and the donor (Q_D) and acceptor (Q_A) quantum yields with PFRET signal. Equation (23.12) is improved to:

$$E_n = 1 - \left\{ \frac{I_{DA}}{I_{DA} + PFRET * \Psi_{DD}/\Psi_{AA} * Q_D/Q_A} \right\} \tag{23.13}$$

where

$$\left(\frac{\Psi_{DD}}{\Psi_{AA}}\right) = \left[\frac{\text{PMT gain of donor channel}}{\text{PMT gain of acceptor channel}} \times \frac{\text{spectral sensitivity of donor channel}}{\text{spectral sensitivity of acceptor channel}}\right]$$

To estimate the distance r between donor and acceptor, deduced from equation (23.3), r can be calculated as below:

$$r = R_0 \left(\frac{1}{E} - 1\right)^{1/6} \tag{23.14}$$

It has been reported in the literature (Gordon et al., 1998; Raheenen et al., 2004) that FRET back-bleedthrough to the donor channel is also possible due to the overlap spectrum. In our experience we do not see any SBT of this kind for 1p excitation. This could be attributed the fact that the FRET occurs at the peak absorption and peak emission (Chen et al., 2005; Herman et al., 2001). We have investigated the issues in our system and found that with the appropriate excitation intensity the back-bleedthrough signals are negligible. On the other hand, we observed back-bleedthrough signals in 2p-FRET microscopy because the donor 2p excitation wavelength excites the acceptor molecule and vice versa (Chen et al., 2005). This is particularly the case when various GFP vectors are used as FRET pairs. It is advisable to choose a non-GFP–derived FRET pair for the 2p-FRET microscopy system (see Table 23.1 and Fig. 23.3).

There are other methods available to correct SBT in the literature (Gordon et al., 1998; Tron et al., 1984; Raheenen et al., 2004; Xia & Liu, 2001). The basic idea is the same as we described above. The main difference is the way to calculate the SBT ratio ra and rd. In their methods, ra and rd are independent of different fluorescence intensity levels (Chen & Periasamy, 2005a). Accordingly, equations (23.5) through (23.10) can be simplified as:

$$DSBT = e^* \frac{b}{a} \tag{23.15}$$

$$ASBT = g^* \frac{c}{d} \tag{23.16}$$

To some extent, especially in wide-field FRET microscopy, it is true that the ratio ra and rd is constant and is independent of fluorescence intensity level. Equations (23.15) and (23.16) can appropriately be applied to calculate DSBT and ASBT. However, in our experience in 2p- and confocal FRET microscopy, the ratio between bleedthrough and fluorescence intensity is not constant or stable (Fig. 23.4). We compared the DSBT ratio for different FRET microscopy systems as shown in Figure 23.4. The data in Panel A were plotted using equation (23.5) and b/a in equation (23.15) for Panel B. There is no difference in the DSBT plot in Figure 23.4 (I), and we conclude that the wide-field FRET microscopy system produces a constant bleedthrough ratio. On the other hand, in Figure 23.4 (II) (2p-FRET microscopy), even though the ratio is almost constant, the standard deviation is almost close to the mean value in the lower intensity level (Panel A). In Panel B the standard deviation is higher and may introduce errors in the SBT correction compared to Panel A. Figure 23.4 (III) shows similar results for confocal FRET microscopy as Figure 23.4 (II), except that the SBT ratio depends on the change in fluorescence intensity level.

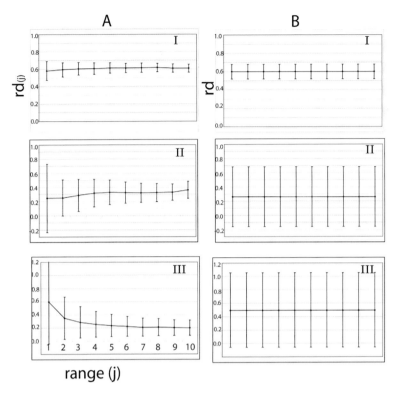

range (j)

Figure 23.4. Comparison of donor spectral bleedthrough ratio using intensity-based and constant ratio method. Panel A: Intensity range-based method (Chen et al., 2005). Panel B: Constant ratio method (Chen & Periasamy 2005a; Gordon et al., 1998). I, wide-field FRET microscopy; II, two-photon FRET microscopy; III, laser scanning confocal microscopy. Dimerization of CFP-YFP-C/EBPΔ244 protein molecules in mouse pituitary GHFT1-5 cell nucleus was used to demonstrate the differences between these two algorithms.

It is important to note that the simplified correction method (equations 23.15 and 23.16) is quite suitable for wide-field FRET microscopy, but for 2p and confocal systems this method may cause over- or under-correction of SBT and may produce inconsistent results. This may be attributed to the spectral sensitivity of the detectors used in the microscopy systems. On the other hand, the error will be much reduced by using equations (23.5) to (23.10) for confocal and 2p-FRET imaging.

23.4.3 Two-Photon FRET-FLIM Microscopy

Wide-field, confocal, and 2p microscopy techniques use intensity measurements to reveal fluorophore concentration and protein distributions in cells. Recent advances in camera sensitivities and resolutions have improved the ability of these techniques to detect dynamic cellular events. Unfortunately, even with these improvements in technology, currently available techniques do not have the temporal resolution to fully characterize the organization and dynamics of complex cellular structures. In contrast, time-resolved fluorescence microscopy (or lifetime) allows the measurement of dynamic events at very high temporal resolution. Fluorescence lifetime imaging

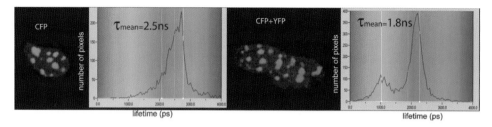

Figure 23.5. Two-photon FLIM-FRET microscopy. The donor lifetime images were acquired in the absence (CFP) and the presence of acceptor (CFP+YFP). The natural lifetime of the donor (2.5 ns) was reduced to 1.8 ns (mean value) due to FRET. Lifetime measurements are the accurate values of the distance distribution of the dimerization of C/EBPΔ244 protein molecules in mouse pituitary GHFT1-5 cell nucleus (Day et al., 2003). Becker & Hickl lifetime board (SPC-730; www.becker-hickl.de) was integrated with Biorad Radiance2100 Confocal/multiphoton microscopy system. Ex 820 nm; Em 480/40 nm; Lens 60× W NA 1.2.

(FLIM) was developed to image and study the environmental behavior of living specimens and to study the dynamic behavior from single cells to single molecules (Dowling et al., 1998; Gadella et al., 1993; Gerritsen et al., 2002; Hanley et al., 2002; Redford & Clegg, 2005; Straub & Hell, 1998).

The combination of FRET and lifetime (FRET-FLIM) provides high spatial (nanometer) and temporal (nanoseconds) resolution (Backsai et al., 2003; Chen & Periasamy, 2005b; Elangovan et al., 2002; Krishnan et al., 2003). The presence of acceptor molecules within the local environment of the donor that permit energy transfer will influence the fluorescence lifetime of the donor. By measuring the donor lifetime in the presence and the absence of an acceptor, one can more accurately estimate the distance between the donor- and acceptor-labeled proteins. While 1p-FRET produces "apparent" E% (i.e., efficiency calculated on the basis of all donors (FRET and non-FRET)), the double-label lifetime data in 2p-FLIM-FRET usually exhibits two different lifetimes (FRET and non-FRET), allowing a more precise estimate of distance based on the comparison with the single-label non-FRET donor lifetime (Fig. 23.5). The former (1p-FRET) may be sufficiently accurate for many situations, but the latter (FRET-FLIM) may be vital for establishing comparative distances of several proteins from a protein of interest.

The energy transfer efficiency (E) is calculated using the following equations (Elangovan et al., 2002):

$$E = 1 - \frac{\tau_{DA}}{\tau_D} \tag{23.17}$$

$$k_T = \left(\frac{1}{\tau_D}\right)\left(\frac{R_0}{r}\right)^6 \tag{23.18}$$

$$r = R_0 \left\{\frac{1}{E} - 1\right\}^{1/6} \tag{23.19}$$

$$R_0 = 0.211 \left(\kappa^2 n^{-4} Q_D J \lambda\right)^{1/6} \tag{23.20}$$

where τ_{DA} is the lifetime of donor in the presence of acceptor, which can be measured using double-labeled specimen with donor excitation and donor emission filter settings. τ_D is the lifetime of donor in the absence of acceptor, which can be measured using single-labeled donor specimen. R_0 is the Förster distance; Q_D is the quantum yield of the donor; n is the refractive index; κ^2 is a factor describing the relative dipole orientation; and Jλ) expresses the degree of spectral overlap between the donor emission and the acceptor absorption (see Fig. 23.1).

Suitable fluorophores have been developed for FRET and FLIM (see Table 23.2). It has been reported that some of the GFP variants have two component lifetimes when fused to the proteins, and this may be an issue in FRET-FLIM imaging (Chen & Periasamy, 2005b; Rizzo et al., 2004). Since the donor molecule lifetime is also tracked to compare with the energy transfer process, it is important to verify whether the donor molecule in the absence of acceptor has single exponential decays.

23.5 CONCLUSION

In this chapter we have described the differences and difficulties that can occur in the use of the wide-field, confocal, and multiphoton FRET microcopy system to localize and quantitate protein–protein interactions in living specimens. Two-photon FRET microscopy is a better technique for deep tissue FRET imaging (Chen et al., 2003; Mills et al., 2003) and for the reasons explained, it is important to choose appropriate FRET pairs for specific microscopy systems. For quantitative analyses of intensity-based FRET imaging it is important to use an algorithm-based SBT correction methodology. As outlined, 2p excitation greatly supports FRET-FLIM approaches. With 2p-FRET-FLIM any FRET pair can be used since the spectral bleedthrough (forward or backward) is not an issue, except one has to be mindful that the donor excitation wavelength may also excite the acceptor. Acceptor photobleaching methodology is simple and suitable for the qualitative verification of the occurrence of FRET with limitations in live cell imaging. In conclusion, we can say that a wide range of FRET imaging systems are available to the biomedical researcher, each with its advantages and disadvantages. However, there is usually one approach that will best fit the individual circumstances.

ACKNOWLEDGMENTS

The author wishes to acknowledge Prof. Richard Day for providing the cells for the experiment and funding provided by the National Center for Research Resources (NCRR-NIH, RR021202) and the University of Virginia.

REFERENCES

Bacskai, B. J., J. Skoch, G. A. Hickey, R. Allen and B. T. Hyman. Fluorescence resonance energy transfer determinations using multiphoton fluorescence lifetime imaging microscopy to characterize amyloid-beta plaques. J. Biomed. Opt. 8:368–375, 2003.

Bastiaens, P. I., I. V. Majoul, P. J. Verveer, H. D. Soling and T. M. Jovin. Imaging the intracellular trafficking and state of the AB5 quaternary structure of cholera toxin. EMBO J. 15:4246–4253, 1996.

Becker, W., A. Bergmann, M.A. Hink, K. Konig, K. Benndorf and C. Biskup. Fluorescence lifetime imaging by time-correlated single-photon counting. Microsc. Res. Tech. 63:58–66, 2004.

Berney, C. and G. Danuser. FRET or no FRET: a quantitative comparison. Biophys. J. 84:3992–4010, 2003.

Biskup, C., L. Kelbauskas, T. Zimmer, K. Benndorf, A. Bergmann, W. Becker, J.P. Ruppersberg, C. Stockklausner and N. Klocker. Interaction of PSD-95 with potassium channels visualized by fluorescence lifetime-based resonance energy transfer imaging. J Biomed. Opt. 9:753–759, 2004.

Chan, F.K., R.M. Siegel, D. Zacharias, R. Swofford, K.L. Holmes, R.Y. Tsien and M.J. Lenardo. Fluorescence resonance energy transfer analysis of cell surface receptor interactions and signaling using spectral variants of the green fluorescent protein. Cytometry. 44:361–368, 2001.

Chen, Y., J.D. Mills and A. Periasamy. Protein localization in living cells and tissues using FRET and FLIM. Differentiation. 71:528–41, 2003.

Chen, Y. and A. Periasamy. Characterization of two-photon excitation fluorescence lifetime imaging microscopy for protein localization. Microsc. Res. Tech. 63:72–80, 2004.

Chen, Y., M. Elangovan and A. Periasamy. FRET data analysis: The algorithm. In: Molecular Imaging: FRET Microscopy and Spectroscopy, edited by. A. Periasamy and R. N. Day. New York: Academic-Elsevier Press, 126–145, 2005.

Chen, Y. and A. Periasamy. Intensity Range Based Quantitative FRET Data Analysis to Localize Protein Molecules in Live Cell Nuclei. J. Fluoresc. 2006. 16:95–104.

Chen, Y. and A. Periasamy. Time-correlated single photon counting fluorescence lifetime imaging FRET microscopy for protein localization. In: Molecular Imaging: FRET Microscopy and Spectroscopy, edited by A. Periasamy and R. N. Day. New York: Academic-Elsevier Press, 239–259, 2005.

Clegg, R. M. Fluorescence resonance energy transfer. In: Fluorescence Imaging Spectroscopy and Microscopy, edited by X. F. Wang and B. Herman. New York: John Wiley & Sons, 179–251, 1996.

Day, R. N. Visualization of Pit-1 transcription factor interactions in the living cell nucleus by fluorescence resonance energy transfer microscopy. Mol. Endocrinol. 12:1410–1419, 1998.

Day, R. N., A. Periasamy and F. Schaufele. Fluorescence resonance energy transfer microscopy of localized protein interactions in the living cell nucleus. Methods 25:4–18, 2001.

Day, R.N., T.C. Voss, J.F. Enwright, 3rd, C.F. Booker, A. Periasamy and F. Schaufele. Imaging the localized protein interactions between Pit-1 and the CCAAT/enhancer binding protein alpha in the living pituitary cell nucleus. Mol. Endocrinol. 17:333–345, 2003.

Denk, W., D. W. Piston and W. W. Webb. Two-photon molecular excitation in laser-scanning microscopy. In: Handbook of Biological Confocal Microscopy, edited by J. B. Pawley. New York: Plenum Press, 445–458, 1995.

Dowling, K., M. J. Dayel, M. J. Lever, P. M. W. French, J. D. Hares and A. K. L. Dymoke-Bradshaw. Fluorescence lifetime imaging with picosecond resolution for biomedical applications. Opt. Lett. 23:810–812, 1998.

Elangovan, M., R. N. Day and A. Periasamy. Nanosecond fluorescence resonance energy transfer-fluorescence lifetime imaging microscopy to localize the protein interactions in a single living cell. J. Microsc. 205:3–14, 2002.

Elangovan, M., H. Wallrabe, Y. Chen, R. N. Day, M. Barroso and A. Periasamy. Characterization of one- and two-photon excitation fluorescence resonance energy transfer microscopy. Methods 29:58–73, 2003.

Fernandes, F., L.M. Loura, R. Koehorst, R.B. Spruijt, M.A. Hemminga, A. Fedorov, and M. Prieto. Quantification of Protein-Lipid Selectivity using FRET: Application to the M13 Major Coat Protein. Biophys. J. 87:344–352, 2004.

Förster, T. Delocalized excitation and excitation transfer. In: Modern Quantum Chemistry Part III: Action of Light and Organic Crystals, edited by O. Sinanoglu. New York: Academic Press, 93–137, 1965.

Gadella, T. W. J., T. M. Jovin and R. M. Clegg. Fluorescence Lifetime Imaging Microscopy (FLIM)—Spatial-Resolution of Microstructures on the Nanosecond Time-Scale. Biophys. Chem. 48:221–239, 1993.

Gerritsen, H. C., M. A. Asselbergs, A. V. Agronskaia and W. G. Van Sark. Fluorescence lifetime imaging in scanning microscopes: acquisition speed, photon economy and lifetime resolution. J. Microsc. 206:218–224, 2002.

Gordon, G. W., G. Berry, X. H. Liang, B. Levine and B. Herman. Quantitative fluorescence resonance energy transfer measurements using fluorescence microscopy. Biophys. J. 74:2702–2713, 1998.

Guo, C., S. K. Dower, D. Holowka and B. Baird. Fluorescence resonance energy transfer reveals interleukin (IL)-1-dependent aggregation of IL-1 type I receptors that correlates with receptor activation. J. Biol. Chem. 270:27562–27568, 1995.

Gustiananda, M., J.R. Liggins, P.L. Cummins and J.E. Gready. Conformation of prion protein repeat peptides probed by FRET measurements and molecular dynamics simulations. Biophys. J. 86:2467–2483, 2004.

Hanley, Q. S., D. J. Arndt-Jovin and T. M. Jovin. Spectrally resolved fluorescence lifetime imaging microscopy. Appl. Spectrosc. 56:155–166, 2002.

Herman, B., G. Gordon, N. Mahajan and V. E. Centonze. Measurement of Fluorescence Resonance Energy Transfer in the Optical Microscope. In: Methods in Cellular Imaging, edited by A. Periasamy. New York: Oxford University Press, 257–272, 2001.

Hiller, D.A., J.M. Fogg, A.M. Martin, J.M. Beechem, N.O. Reich and J.J. Perona. Simultaneous DNA binding and bending by EcoRV endonuclease observed by real-time fluorescence. Biochemistry. 42:14375–14385, 2003.

Hohng, S., C. Joo and T. Ha. Single-Molecule Three-Color FRET. Biophys. J. 87:1328–1337, 2004.

Jurgens, L., D. Arndt-Jovin, I. Pecht and T. M. Jovin. Proximity relationships between the type I receptor for Fc epsilon (Fc epsilon RI) and the mast cell function-associated antigen (MAFA) studied by donor photobleaching fluorescence resonance energy transfer microscopy. Eur. J. Immunol. 26:84–91, 1996.

Kenworthy, A. K. Photobleaching FRET microscopy. In: Molecular Imaging: FRET Microscopy and Spectroscopy, edited by A. Periasamy and R. N. Day. New York: Oxford University Press, 146–164, 2005.

Kenworthy, A. K., N. Petranova and M. Edidin. High-resolution FRET microscopy of cholera toxin B-subunit and GPI-anchored proteins in cell plasma membranes. Mol. Biol. Cell 11:1645–1655, 2000.

Klostermeier, D., P. Sears, C.H. Wong, D.P. Millar and J.R. Williamson. A three-fluorophore FRET assay for high-throughput screening of small-molecule inhibitors of ribosome assembly. Nucleic Acids Res. 32:2707–2715, 2004.

Kraynov, V. S., C. Chamberlain, G. M. Bokoch, M. A. Schwartz, S. Slabaugh and K. M. Hahn. Localized Rac activation dynamics visualized in living cells. Science 290:333–337, 2000.

Krishnan, R. V., A. Masuda, V. E. Centonze and B. Herman. Quantitative imaging of protein-protein interactions by multiphoton fluorescence lifetime imaging microscopy using a streak camera. J. Biomed. Opt. 8:362–367, 2003.

Lakowicz, J. R. Principles of Fluorescence Spectroscopy. 2nd ed., Plenum Press, New York. 1999.

Larson, D.R., Y.M. Ma, V.M. Vogt, and W.W. Webb, Direct measurement of Gag-Gag interaction during retrovirus assembly with FRET and fluorescence correlation spectroscopy. J Cell Biol, 2003. 162(7): p. 1233–1244.

Lear, J.D., A. Stouffer, H. Gratkowski, V. Nanda, and W.F. DeGrado. Association of a model transmembrane peptide containing Gly in a heptad sequence motif. Biophys. J. 87:3421–3429, 2004.

Ludwig, M., N. F. Hensel and R. J. Hartzman. Calibration of a resonance energy transfer imaging system. Biophys. J. 61:845–857, 1992.

Mills, J. D., J. R. Stone, D. G. Rubin, D. E. Melon, D. O. Okonkwo, A. Periasamy and G. A. Helm. Illuminating protein interactions in tissue using confocal and two-photon excitation fluorescent resonance energy transfer microscopy. J. Biomed. Opt. 8:347–356, 2003.

Mueller, B., C.B. Karim, I.V. Negrashov, H. Kutchai and D.D. Thomas. Direct detection of phospholamban and sarcoplasmic reticulum Ca-ATPase interaction in membranes using fluorescence resonance energy transfer. Biochemistry. 43:8754–8765, 2004.

Nashmi, R., M.E. Dickinson, S. McKinney, M. Jareb, C. Labarca, S.E. Fraser and H.A. Lester. Assembly of alpha4beta2 nicotinic acetylcholine receptors assessed with functional fluorescently labeled subunits: effects of localization, trafficking, and nicotine-induced upregulation in clonal mammalian cells and in cultured midbrain neurons. J Neurosci. 23:11554–11567.

Panyi, G., M. Bagdany, A. Bodnar, G. Vamosi, G. Szentesi, A. Jenei, L. Matyus, S. Varga, T.A. Waldmann, R. Gaspar and S. Damjanovich. Colocalization and nonrandom distribution of Kv1.3 potassium channels and CD3 molecules in the plasma membrane of human T lymphocytes. Proc. Natl. Acad. Sci. U.S.A. 100: 2592–2597, 2003.

Patterson, G. H. and D. W. Piston. Photobleaching in two-photon excitation microscopy. Biophy. J. 78:2159–2162, 2000.

Periasamy, A. Methods in Cellular Imaging. New York, Oxford University Press. 2001.

Periasamy, A. and R.N. Day. FRET imaging of Pit-1 protein interactions in living cells. J. Biomed. Opt. 3:154–160, 1998.

Periasamy, A. and R. N. Day. Visualizing protein interactions in living cells using digitized GFP imaging and FRET microscopy. Methods Cell Biol 58:293–314, 1999.

Periasamy, A. and R. N. Day. Molecular Imaging: FRET Microscopy and Spectroscopy. New York, Academic-Elsevier Press. 2005.

Raheenen, J.-v., M. Langeslag and K. Jalink. Correcting Confocal Acquisition to optimize imaging of fluorescence resonance energy transfer by sensitized emission. Biophys. J. 86:2517–2529, 2004.

Redford, G. and R. M. Clegg. Real-Time Fluorescence Lifetime Imaging and FRET using Fast Gated Image Intensifiers. In: Molecular Imaging: FRET Microscopy and Spectroscopy. Eds. A. Periasamy and R. N. Day, Academic-Elsevier Press, NY, 2005.

Rizzo, M. A., G. H. Springer, Granada, B. and D. W. Piston. An improved cyan fluorescent protein variant useful for FRET. Nature Biotech. 22:445–449, 2004.

Sekar, R. B. and A. Periasamy. Fluorescence resonance energy transfer (FRET) microscopy imaging of live cell protein localizations. J. Cell Biol. 160:629–633, 2003.

Shaner, N.C., Campbell, R.E., Steinbach, P.A., Giepmans, B.N., Palmer, A.E., and Tsien, R.Y., 2004, Improved monomeric red, orange and yellow fluorescent proteins derived from Discosoma sp. red fluorescent protein. Nat. Biotechnol. 22:1567–1572.

Sharma, P., R. Varma, R.C. Sarasij, Ira, K. Gousset, G. Krishnamoorthy, M. Rao and S. Mayor. Nanoscale organization of multiple GPI-anchored proteins in living cell membranes. Cell. 116:577–589, 2004.

Straub, M. and S. W. Hell. Fluorescence lifetime three-dimensional microscopy with picosecond precision using a multifocal multiphoton microscope. Appl. Phy. Lett. 73:1769–1771, 1998.

Stryer, L. Fluorescence energy transfer as a spectroscopic ruler. Annu. Rev. Biochem. 47:819–846, 1978.

Tron, L., J. Szollosi, S. Damjanovich, S. Helliwell, D. Arndt-Jovin and T. Jovin. 1984. Flow cytometric measurement of fluorescence resonance energy transfer on cell surfaces. Quantitative evaluation of the transfer efficiency on a cell-by-cell basis. Biophys. J. 45:939–946.

Wallrabe, H., M. Elangovan, A. Burchard, A. Periasamy and M. Barroso. Confocal FRET microscopy to measure clustering of ligand-receptor complexes in endocytic membranes. Biophys J 85:559–571, 2003.

Wallrabe, H. and A. Periasamy. Imaging protein molecules using FRET and FLIM microscopy. Curr. Opin. Biotechnol. 16:19–27, 2005.

Wang, L., Jackson, W.C., Steinbach, P.A., and Tsien, R.Y., 2004, Evolution of new nonantibody proteins via iterative somatic hypermutation. Proc. Natl. Acad. Sci. U.S.A. 101:16745–16749.

Watrob, H. M., C. P. Pan and M. D. Barkley. Two-step FRET as a structural tool. J Am Chem Soc 125:7336–7343, 2003.

Wouters, F. S., P. I. Bastiaens, K. W. Wirtz and T. M. Jovin. FRET microscopy demonstrates molecular association of non-specific lipid transfer protein (nsL-TP) with fatty acid oxidation enzymes in peroxisomes. Embo J 17:7179–7189, 1998.

Xia, Z. and Y. Liu. 2001. Reliable and global measurement of fluorescence resonance energy transfer using fluorescence microscopes. Biophys. J. 81:2395–2402.

24

Diffraction Unlimited Far-Field Fluorescence Microscopy

Andreas Schönle, Jan Keller,
Benjamin Harke, and Stefan W. Hell

24.1 INTRODUCTION

For centuries, light microscopy has been an important tool in many fields of science. With the quality of available optical components becoming better over time, it has delivered ever-sharper images of small structures. However, in 1873, Ernst Abbe discovered a fundamental limitation known as the diffraction barrier (Abbe, 1873): Far-field light microscopes would never be able to resolve features smaller than approximately half the wavelength of the light used. However, many modern imaging applications obviously necessitate a higher resolution. Since the diffraction barrier had been accepted as an unalterable fact, the use of focused visible light was not considered an option in these situations. Consequently, alternative methods were developed to deliver images with much higher resolution. Two prominent examples are electron microscopy, which focused (matter) waves of much shorter wavelengths, and atomic force microscopy, which gave up focusing altogether. Nevertheless, far-field light microscopy has remained an irreplaceable tool, especially in the life sciences, due to a number of exclusive advantages: it has the ability to image samples in three dimensions with minimal negative impact on the (living) specimen. Unlike some other methods, it is not restricted to surfaces and does not require any special sample preparation. Finally, it can be combined with fluorescence to image features of interest with high specificity. Mapped with a confocal or multiphoton excitation microscope (Denk et al., 1990; Sheppard & Kompfner, 1978; Wilson & Sheppard, 1984), fluorescence emission readily measures protein 3D distributions, or those of other fluorescently labeled molecules within the complex inside of biological specimens. The development of a microscope that retains all these advantages while not being limited by diffraction would therefore almost certainly lead to exciting new discoveries and become a crucial tool for scientists not only in biology. However, due to the fundamental nature of Abbe's law, all attempts to break this barrier seemed futile.

The problem of attaining resolution truly beyond the diffraction limit in a far-field optical microscope was tackled by several scientists in the 20th century (Lukosz, 1966; Toraldo di Francia, 1952) but never solved. It was realized only in the mid-1990s that the solution lies in not only using the dye as a mere contrast agent but also using its interaction with light for resolution improvement (Hell, 1994). By exploiting the spectroscopic properties of fluorescent dyes, methods could be devised (Hell, 1997;

Hell & Kroug, 1995; Hell & Wichmann, 1994) that broke Abbe's diffraction barrier eventually, and in consequence have the potential to provide molecular resolution with regular lenses and visible light (Hell, 2003; Hell et al., 2003; Westphal & Hell, 2005). Here we will give an introduction to the underlying principle and review some of the recent experimental results that successfully applied this technique to real-world imaging problems.

24.2 ABBE'S DIFFRACTION LIMIT

Abbe's discovery was that the minimum size, Δr, of the intensity maximum at the focal spot in the focal plane of a lens is given by

$$\Delta r \cong \frac{\lambda}{2NA} \tag{24.1}$$

where λ is the vacuum wavelength and $NA = n \sin \alpha$ is called the numerical aperture; n denotes the refractive index of the medium and α the aperture angle of the lens (Fig. 24.1).

This limitation stems from the diffraction of light. It is obvious that for a beam-scanning microscope, Δr is at the same time an upper bound for the resolution limit: two objects closer than this distance cannot be distinguished by the sweeping focal spot.

In fact, a generalized form of equation (24.1) holds for every far-field imaging method using focusing optics in its excitation and detection path, regardless of the exact optical arrangement and illumination mode. In this article we will concentrate on incoherent linear imaging—that is, on imaging systems that are linear with regard to the intensity of the object and the image. In other words, the image of a set of objects can be written as the sum of the images of each object. In addition, we assume that imaging is approximately space-invariant—that is, a point object creates the same image, regardless of where it is located in the sample space. Imaging modes to which these assumptions apply largely simplify image interpretation and any kind of image deconvolution and restoration. Methods not fulfilling the linearity condition include all coherent imaging modes where the phase of the excitation light is retained in the detected light. This is true for all methods that are based on reflected or scattered light, irrespective of whether the actual scattering is a single or multiphoton (Barad et al., 1997; Gannaway, 1978; Hellwarth & Christensen, 1974; Zumbusch et al., 1999) induced process.

With these assumptions made, all parameters of image formation, including the achievable resolution, are fully determined by the image of a point object, the so-called (intensity) point spread function (PSF) or effective focal spot. The image is then given by the convolution of the object with the PSF:

$$I = h \otimes G \tag{24.2}$$

where I, h, and G denote the image, the PSF, and the object, respectively, with the object consisting of, for example, a distribution of fluorophores in space.

Such a convolution is nothing else but a "smearing out" of the object with the focal spot, therefore losing information about small features. As mentioned above, in most cases the full-width-half-maximum (FWHM) of the PSF is a good approximation of

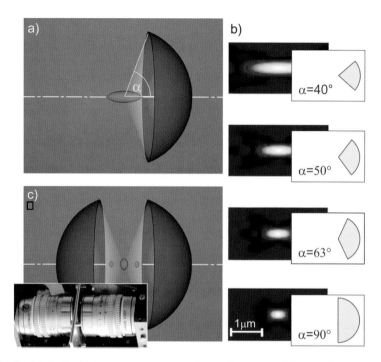

Figure 24.1. (a) A single lens creates a wavefront that resembles a spherical cap. Due to diffraction, the main focal intensity maximum, shown as an elongated blob centered at the focal point, has a finite size and is elongated along the optic axis. (b) Its extent and elongation sensitively depends on the semi-aperture angle, following Abbe's equation. The larger the aperture angle α, the smaller and the more isotropic the focal spot becomes. Most prominent is the axial compression of the spot, which, in first approximation, scales inversely with $\sin^2 \alpha$. For the largest commonly available aperture angle ($\alpha \sim 70$ degrees), it is about half a wavelength wide (xy) and ~ 3 times longer along the optic axis (z). The focal spot expected for a $\alpha = 90$ degrees is displayed at the bottom. Note, however, that aperture angles of $\alpha \sim 80$ degrees are technically not feasible. (c) The largest technically possible aperture and thus the highest realistically possible "classical" resolution are achieved by using two opposing lenses and thus creating two coherent counter-propagating wavefronts. As a result the central focal spot (i.e., the main diffraction maximum) becomes more spherical. The two side maxima and the elongation along the lateral direction are a result of the missing parts of the wavefront.

the system's resolution, but for a more detailed understanding and accurate assessment of the microscope's resolving power, a more detailed analysis will be needed. Nevertheless, it is obvious that minimizing the PSF's extent is tantamount to maximizing the resolution. Therefore, equation (24.1) suggests two possibilities to improve the resolution of imaging systems. Without changes in the principal method, the spot size can be decreased only by using shorter wavelengths and/or larger aperture angles (Abbe, 1873; Born & Wolf, 1993); see Figure 24.1 for an illustration of the PSF's dependence on the aperture angle. The lens half-aperture is technically limited to ~ 70 degrees, and therefore the largest feasible aperture, as implemented in a 4Pi microscope (Hell & Stelzer, 1992b), is given by

$$4\pi(1 - \cos\alpha) \cong 0.65 \cdot 4\pi$$

or 65% of the full solid angle. Apertures that are even larger may facilitate image interpretation but cannot push the resolution limits much further. On the other hand, the wavelength λ cannot be reduced below 350 nm because shorter wavelengths are not compatible with live cell imaging. In fact, λ < 450 nm is rather difficult to handle in optical systems (e.g., because of dispersion). For typical wavelengths in the visible range, confocal microscopes optimally resolve 180 nm in the focal plane (x,y) and merely 500 to 800 nm along the optic axis (z) (Pawley, 1995). The recently realized 4Pi microscopy of type C, when combined with image restoration, can unambiguously resolve approximately ∼100 nm axially and ∼120 nm laterally (Gugel et al., 2004). The practical improvement of lateral resolution is the result of the better applicability of restoration algorithms due to better axial discrimination.

We will now see why this is already close to the best performance theoretically possible, and therefore why all attempts to improve the resolution beyond this limit by merely changing the optical setup proved ineffectual due to the diffraction limit. This can be best understood in the frequency domain. Here, the resolving power is described by the optical transfer function (OTF) giving the strength with which the spatial frequencies are transferred from the object to the image. The OTF is given by the Fourier transform of the PSF (Goodman, 1968; Wilson 7 Sheppard, 1984), and equation (24.2) can then be transformed using the convolution theorem

$$\hat{I} = \hat{h} \cdot \hat{G} = o \cdot \hat{G} \qquad (24.3)$$

where the hat signifies a 3D Fourier transform and o is the OTF.

This illustrates the significance of the OTF. On the right-hand side the multiplication will lead to the complete loss of all object frequencies where $o(k)$ is zero, and object features smaller than the corresponding spatial extent will disappear. Hence, the ultimate resolution limit is given by the highest frequency where the OTF is non-zero (i.e., the extent of the support of the OTF). We will now see that for fundamental reasons this support cannot extend beyond a frequency determined by the wavelengths of exciting and emitted light, which is nothing more than Abbe's diffraction limit.

Let us assume excitation by a (quasi) monochromatic wave passing through the setup with wave vector $k = 2\pi n/\lambda$. The field directly after the lens is then given by a spherical wavefront with its cutoff angle given by the half-aperture angle of the lens, as seen in Figure 24.1a. Assuming that the focal length of the lens is much larger than the extent of the focal region we are interested in and the wavelength, the field around the geometrical focus can be approximated by a superposition of plane waves. They interfere constructively in the focal spot and their wave vectors have a modulus k and angles $\theta < \alpha$ with respect to the optic axis. This so-called Debye approximation is commonly used to calculate the focal field (Richards & Wolf, 1959). The Fourier transform of the electric field, $C(k)$, is therefore given by a spherical cap with radius k. Following the electric dipole approximation, the excitation of an arbitrarily oriented dye molecule is proportional to the absolute square of the electric field, corresponding to an autocorrelation of the field in frequency space. This signifies that frequencies of up to $2k$ are contained in the excitation OTF. When imaging is employed in the detection path as well, a spatially varying detection probability for the emitted light is also introduced. Since we assumed incoherent imaging (i.e., we excluded any phase dependence), the effective PSF is given by the multiplication of the excitation PSF with its normalized detection counterpart (i.e., the detection PSF) (Wilson & Sheppard, 1984). This is equivalent to a convolution in frequency space. The achievable cutoff is therefore at $2k_{ex} + 2k_{det}$, where k_{ex} and k_{det} are the frequencies corresponding to the excitation and detection wavelengths in the medium, respectively. As an example, if fluorescence is excited at 488 nm and detected at around 530 nm, the cutoff is given by $(2n/488 \text{ nm} + 2n/530 \text{ nm})^{-1} \cong 127 \text{ nm}/n$, corresponding to approximately 80 nm for oil immersion and 96 nm for water immersion. Therefore, all far-field light microscopes relying solely on improvements of the instrument have an ultimate resolution limit of about 100 nm in all directions, even when all conceivable spatial frequencies are transmitted and restored in the image.

Figure 24.2 illustrates these considerations and relates them to the performance of available optical setups. The half-aperture angle theoretically can reach up to 90 degrees, but note that such an angle would make sense only if the light field was a scalar. With increasing aperture angle the transverse nature of electromagnetic waves comes more and more into play; for aperture angles >80 degrees, the vectorial effects of focusing are so predominant that attaining such high angles is usually not attractive. Besides, the adverse effect of spherical aberrations induced by refractive index mismatches becomes exacerbated with larger angles (Egner & Hell, 1999; Hell et al., 1993). Nevertheless, let us assume that light was a scalar and that both excitation and detection are performed with a 90-degree lens. In this case $C_{ex}(k)$ and $C_{det}(k)$ would be complete spherical surfaces of radius k_{ex} and k_{det}. Their autocorrelation results in excitation and detection OTFs with support regions given by spheres of radius $2k_{ex}$ and $2k_{det}$. Calculations (Schönle & Hell, 2002) or geometrical considerations (Sheppard et al., 1993) reveal that within its support, the OTF is proportional to $1/k$. Convolution of these two spheres gives the effective OTF that features the maximum support region achievable in far-field microscopy using focusing optics for excitation and detection: it is non-zero on a sphere of radius $2k_{ex} + 2k_{det}$. However, even here, large frequencies close to the edge of the support region are strongly suppressed. The PSF of the system is also readily calculated using Huygens' principle and the spherical symmetry of the problem:

$$h_{eff} = h_{ex} \cdot h_{det} = \left[\frac{\sin(k_{ex}r)}{k_{ex}r} \right]^2 \left[\frac{\sin(k_{det}r)}{k_{det}r} \right]^2 \tag{24.4}$$

For the practically relevant example of a confocal microscope (where the fluorescence is imaged on a nearly point-like detector), the cutoff angle for the cap $C(k)$ is given by the semi-aperture angle of the lens. The modulus of $C(k)$ usually decreases with larger angle, an effect called apodization. As the excitation and detection are achieved through the same lens, the excitation and detection OTFs differ just by a scaling factor given by the ratio of the respective wavelengths. Figure 24.2 shows that the support region of the confocal OTF covers only a fraction of the theoretical maximum: it is squeezed along the z-direction, indicating a degraded axial resolution. Even within its support region, the OTF drops off toward the edges. The latter can be addressed by modifying the excitation or detection wavefronts in a specific way. For example, high frequencies can be boosted compared to lower ones while leaving the support unchanged. Several methods of this kind were proposed. They were mostly based on the introduction of aperture wavefront modifications (Toraldo di Francia, 1952) such

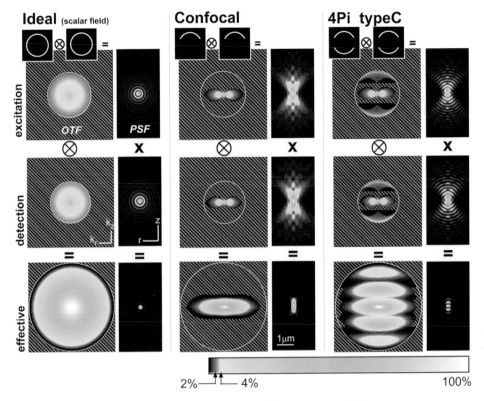

Figure 24.2. PSFs and corresponding OTFs and their supports in frequency space. The panels displaying the PSFs are $2.5 \times 5\,\mu$ m with the geometrical focus in the center. The OTFs are displayed up to a wave-vector $2\pi/80$ nm in positive and negative x- and z-directions. The columns refer to ideal scalar focusing with a full solid angle as aperture (left), a confocal microscope with a numerical aperture of 1.4 (oil) (middle), and the microscope with the largest aperture currently technically feasible, the 4Pi type C microscope (right). The insets at the top indicate the formation of the excitation and detection OTFs for the respective types of microscope. For the OTFs, a white circle indicates the maximum achievable region of support, while the diagonal hatch pattern indicates regions of no support for the focusing mode investigated. Excitation and detection wavelengths of 488 nm and 530 nm respectively were assumed for all calculations. The color look-up table was adjusted to emphasize small values of the PSF and OTF. Please note that the choice of scale for the excitation and detection OTFs (first and second row) is somewhat arbitrary because they feature a singularity at the origin. However, the more important effective OTFs are finite everywhere and the value at the origin is the volume integral of the PSF.

as annular apertures (Hegedus & Sarafis, 1986; Wilson & Sheppard, 1984). The use of patterned illumination (Gustafsson, 2000; Heintzmann & Cremer, 1998) also falls into this category. Since the support is unchanged, these methods do not break the diffraction barrier. Thus, it is apparent that confocal and patterned illumination microscopy improve the resolution over conventional epifluorescence imaging (at most) by a factor of two. As shown in Figure 24.2, this factor stems from the convolution of the excitation with the detection OTF—that is, from the fact that the fluorescence light is *imaged* by the lens on a detector. (Note that in the case of patterned illumination the detector recording the fluorescence image is a camera, but not a point as in a confocal microscope; this does not change the OTF support that is maximally possible, of course.)

A significant improvement of the axial resolution over the confocal arrangement is achieved by using opposing lenses coherently for both excitation and detection—that is, by 4Pi microscopy (Egner et al., 2002; Hell, 1990; Hell & Stelzer, 1992a; Hell et al., 1997). A 4Pi (type C) microscope fills in most of the missing wavefront (see Fig. 24.1c) and effectively doubles the aperture. The region of support is therefore much closer to the theoretical limit than in the confocal case (Egner & Hell, 2005). However, the missing parts still result in lateral stretching and the appearance of side-lobes (a modulation along the z-axis) of the PSF. Also, the effective OTF now features depressions. The depth of these depressions is the critical parameter that determines whether the larger support region results in a practical resolution increase (Nagorni & Hell, 2001).

At this point, it is important to note again that confocal, 4Pi, and I^5M microscopy (Gustafsson et al., 1999) as well as all techniques involving the creation of excitation patterns and reconstruction of images from several exposures, are limited in their OTF support and thus in their ability to resolve features smaller than $[2n/\lambda_{ex} + 2n/\lambda_{det}]^{-1}$. We here refer to this limit as the classical resolution limit. Of all far-field microscopes currently in use, the (multiphoton) 4Pi type C microscope has the largest possible aperture. It therefore exploits almost the full potential of diffraction-limited imaging. But again, none of these methods break the diffraction barrier, because breaking implies that there is no "hard" limit. Breaking the diffraction barrier actually means that the method has the potential of producing an OTF with an infinitely large bandwidth.

24.3 BREAKING THE DIFFRACTION BARRIER

If changing the optical setup cannot improve the resolution any further, there is only one other part of the imaging process in a far-field focusing microscope that could potentially be modified to solve the problem: the interaction of light with the sample, in our case with the fluorescent dye. The fact that nonlinear interactions of light with the dye molecules lead to spatial frequencies beyond those that result from the use of the illumination wavelength is known from the analysis of multiphoton excitation microscopy. For example, using multiphoton (m-photon) excitation, the probability of producing a fluorescence photon depends on the mth power of the illumination intensity (Bloembergen, 1965; Göppert-Mayer, 1931; Sheppard & Kompfner, 1978). In the frequency domain, this corresponds to an m-fold convolution of the excitation OTF with itself, extending the support region to m times higher frequencies, mk_{ex}, At a first glance it might therefore seem as though multiphoton excitation increases the microscope's resolving power, but it can be readily understood why this is not the case. If m-photon excitation is used to excite the same dye, the excitation energy has to be split between the photons (Denk et al., 1990). Consequently, the photon energy is m times lower and thus the wavelength of the excitation light is m times larger. This of course results in an m times larger focal intensity distribution. While it is true that the nonlinear dependence of the excitation probability on this intensity distribution results in an excitation PSF narrower than the latter, it will always remain wider than its one-photon counterpart. Following the same arguments as in the previous section, it is quickly seen that the theoretical cutoff is given by $2m(k_{ex}/m) + 2k_{det} = 2k_{ex} + 2k_{det}$ and thus remains the same as in the one-photon case. Here, the multiplication by m stems from the m correlations or convolutions in frequency space and the division from the longer excitation wavelength. Obviously, this equally holds for excitation with several photons of different energies, such as two-color two-photon excitation (Lakowicz et al., 1996). In practice, multiphoton excitation usually even results in decreasing resolution because higher frequencies within the support are damped (Schrader et al., 1997). In addition, multiphoton excitation, especially for $m > 3$, requires very high intensities (Hell et al., 1996; Xu et al., 1996), leading to complicated setups and damage to the sample. Bleaching is largely restricted to the focal plane, but there it can be much stronger than for one-photon excitation. For some dyes, however, the excitation wavelength can be chosen to be shorter than $m\lambda_{ex}$, and when imaging deep into scattering samples such as brain slices, multiphoton imaging can feature a superior signal-to-noise ratio (SNR). In these cases, multiphoton excitation is a valuable method for reasons unrelated to the classical resolution issue.

After recognizing that methods that resulted in photon energy subdivision would prevent resolution increase, alternatives were considered as well. One such alternative is the use of multiple photons, where the detection of a fluorescence photon occurs only after the consecutive (not simultaneous) absorption of multiple individual photons. Because each individual photon transition is still linear by itself, photon energy subdivision does not occur in such a scheme and high intensities are avoided. Indeed, such concepts could be pinpointed theoretically (Hänninen et al, 1996; Schönle et al., 1999; Schönle & Hell, 1999). Unfortunately, their realization is hampered by the requirement for very specific dyes or excitation conditions.

In fact, a much simpler approach already led to a breakthrough in super-resolving microscopy and, consequently, to the first images with resolution that is truly beyond the diffraction limit: <u>St</u>imulated <u>E</u>mission <u>D</u>epletion microscopy (STED) has produced images with a spatial resolution of down to $\lambda/25$ using focused light and regular lenses (Westphal et al., 2003b). Moreover, this method is just one possible realization of an entire class of imaging methods based on a common concept termed RESOLFT (Hell, 2003; Hell et al., 2004). The common ground in RESOLFT imaging is that the required optical nonlinearity is provided by a spatially modulated *reversible* <u>s</u>aturable <u>o</u>ptically <u>l</u>inear (<u>f</u>luorescence) <u>t</u>ransition between two molecular states. We shall see that since there is no physical limit to the degree of saturation, there is no longer a theoretical limit to the resolution. Therefore, concepts based on the RESOLFT principle, such as STED or ground state depletion microscopy (Hell & Kroug, 1995; Hell & Wichmann, 1994), truly break the diffraction barrier. Due to their principle of operation, they can achieve nanoscale resolution in *all* directions (Heintzmann et al., 2002; Hell, 1997; Hell et al., 2003). Therefore, and because they are based on the use of visible light, there is no fundamental reason why these methods should not eventually bring nanometer resolution to live cell imaging. This opens up the prospect of bridging the resolution gap between electron and current light microscopy.

24.3.1 The Principle of RESOLFT Microscopy

The principle for breaking Abbe's barrier with RESOLFT is readily understood. Let us assume that the sample consists of a density distribution of small entities with at least two states. While we will call them molecules in the remainder of this section, these entities could be fluorescent dye molecules, quantum dots, or any other construct that meets the conditions outlined below. A focal intensity distribution featuring a zero intensity point (or at least a strong gradient in space) effects a *saturated depletion* of one of the states everywhere except at the location of the zero point. This state should be essential to the signal generation so that the signal can be read out exclusively from regions where the state is not depleted (the immediate vicinity of the zero point). The recorded signal measures the density of the molecules in these regions and the resolution is therefore determined by their extent. After depletion is turned off, the state should be repopulated—that is, the saturated transition must be reversible. Because the linear transition depleting the state is saturated, vast nonlinearities are introduced, which in turn introduces new frequencies and thus breaks the diffraction barrier. Importantly, any saturable transition between two states where the molecule can be returned to its initial state is a potential candidate for breaking the diffraction barrier (Dyba & Hell, 2002; Hell, 1997; Hell et al., 2003).

Figure 24.3 illustrates this concept in more detail. The molecular states involved are labeled A and B. In fluorophores, typical examples for these states are the ground and first excited electronic states, conformational or isomeric states. While the transition A→B is induced or at least boosted by light, the transition B→A can be spontaneous but may also be induced by light, heat, or any other mechanism. In addition, the molecule should be capable of producing a detectable signal indicating its state. In fluorescent microscopy, this means that, for example, the dye can fluoresce only (or much more intensively) in state A or changes its emission wavelength between states A and B. Let us assume that some mechanism transfers all or a certain fraction of the molecules inside a region of interest such as the diffraction-limited excitation volume into state A. Our goal is to transform this into a diffraction-*un*limited distribution of molecules in state A. We are then able to associate the signal measured with this smaller region of the sample and thus break the diffraction limit. The crucial step herein is to illuminate the sample with light that depletes state A (i.e., drives the transition from A to B). The intensity distribution $I(r)$ of the illuminating light is of course diffraction-limited but designed in such a way that it features one or several positions with zero intensity. Consequently, after a certain period of illumination, the probability $p_A(r)$ of finding molecules in state A depends strongly on $I(r)$. While in the zero points it will still be equal to unity because no transitions to B are induced, it will drop off where $I(r)$ increases. We define the saturation intensity I_{sat} as the intensity value at which 50% of the molecules are transferred to state B; $p_A(r)$ thus drops to 50% where $I(r) = I_{sat}$. If we now chose the maximum intensity I_{max} to be equal to the saturation intensity, this will be the case only in the maxima of $I(r)$ (see Fig. 24.3a). If we increase I_{max} further and further beyond I_{sat} (see Fig. 24.3b–d), $I(r) = I_{sat}$ will hold for smaller and smaller values of r. However, this intersection marks the FWHM of the distribution $p_A(r)$, and therefore we are able to arbitrarily squeeze this distribution through this process. For any given setup, the achieved spatial width of state A molecules merely depends on the value by which the maximum intensity exceeds the saturation intensity. Of course, the exact form of the distribution $p_A(r)$ depends on the depleting mechanism and the kinetics involved. In Figure 24.3 we assumed an exponential dependence (see lower left) on the illuminating intensity, as it occurs in pulse-based cyclic RESOLFT schemes described in more detail below.

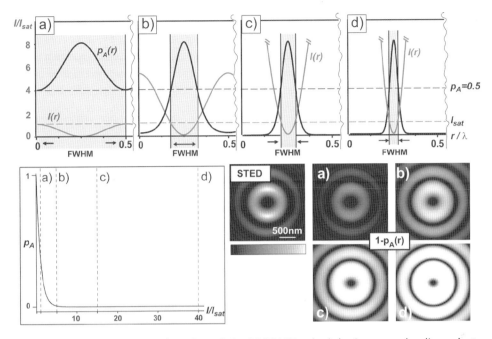

Figure 24.3. Schematic explanation of the RESOLFT principle. In general, a linear but reversible optical transition from state A to state B is driven by an irradiating, spatially modulated intensity I(r). Ideally the modulation is perfect (i.e., the minima feature [virtually] zero intensity) and thus all molecules remain in state A at these positions. The probability of finding initially excited molecules in A after illumination, $p_A(r)$, depends on the irradiation intensity I. For pulse-based implementations it is given by equation (24.7), and this typical saturation behavior is plotted on the lower left. The vertical lines indicate the values of the intensity maxima of I(r) in panels (a–d). Examination of panels (a–d) reveals how saturation leads to diffraction-unlimited resolution. (a) At $I_{max} = I_{sat}$, the modulation of the light is approximately replicated in $p_A(r)$ and the width of its maximum will be limited by diffraction, because I(r) is diffraction-limited and the relationship between I and p_A is nearly linear. (b) With increasing maximum intensity $I_{max} = 5I_{sat}$ the intersections of I(r) with the horizontal line marking the saturation intensity move closer to the intensity minima. Therefore, the points where $p_A(r)$ drops to 0.5 move closer together, reducing the FWHM. (c, d) Further increasing the maximum intensity to $I_{max} = 15I_{sat}$ and $I_{max} = 40I_{sat}$ leads to a further reduction of the FWHM. This process is only limited by the maximum available power, possible photodamage, and the quality of the intensity zeros. There are a number of possible mechanisms how the diffraction-unlimited spatial probability distributions created here can be translated into imaging with arbitrary resolution. On the lower right, 2D resolution improvement is illustrated. A possible intensity distribution I(r) of the x-component in the focal plane of a microscope is shown in the panel marked STED. We assumed a wavelength of 700 nm. Panels (a–d) plot the probability of depleting state A, $1 - p_A(r)$. Note the decreasing size of the "hole" in the center, which represents the region from which fluorescence is still possible.

It is straightforward to translate such a diffraction-unlimited distribution of the probability of molecules in state A into diffraction-unlimited resolution. Let us assume that the signal stems exclusively from molecules in state A. We can now repeat the cycle described above while scanning these "zero(s)" through the sample. The fluorescence signal automatically stems just from the immediate vicinity of the positions where I(r) is zero. If there are many zeros, the signal (fluorescence light) can be recorded simultaneously from the many points. Although the fluorescence is collected with a regular lens, Abbe's diffraction limit does not apply, because we know the exact location of the maxima of $p_A(r)$ in the sample and can assign the fluorescence to each of them. The resolution is then solely determined by the FWHM of these maxima; the lens used for fluorescence collection acts merely as a condenser.

At the same time this shows that scanning is mandatory in a far-field optical system with a broken diffraction barrier, because—and Abbe was perfectly correct in this regard—the objective lens cannot transmit the higher spatial frequencies of the object through the lens. Importantly, this does not imply that a RESOLFT microscope needs to be a single spot scanning microscope. By contrast, multiple spots are possible as well, as long as the nodal points of I(r) are further apart than Abbe's resolution limit. When using wide-field detection on a camera, each nodal point will produce a diffraction-limited, blurred spot probably extending over several pixels. However, resolution is retained because each of these spots can be directly assigned to a nodal point of which the position in the sample is exactly known. More complex schemes could involve nodal lines or other shapes that would have to be combined in appropriate ways to raster the sample and assign fluorescence to diffraction-unlimited regions in the sample. The term "wide field" is somewhat inappropriate in this context, because the element of scanning (with spots, lines, or regions of zero-depletion intensity) is still a crucial part of the concept.

We shall see below that complete depletion of A (or complete darkness of B) is *not* required. The important requirement is that the non-nodal region features a constant, notably lower probability to emit fluorescence, so that it can be distinguished from its sharp counterpart. Even if the signal stems from state B and consequently $1 - p_A(r)$ is read out, the concept can deliver superresolution (Heintzmann et al., 2002) but the images must be inverted, and therefore we expect this version to be heavily challenged by SNR issues: the "bright" light from the non-nodal regions, while not carrying any information about the object, contributes with a substantial amount of photon shot noise.

24.3.2 Basic Theory of RESOLFT Microscopy

Next we derive an estimate for the achievable resolution at a given intensity I_{max} (Hell, 2003, 2004). We assume a cyclic RESOLFT scheme with a cycle consisting of:

1. Transferring the molecules into state A
2. Illuminating them with light, effecting the saturated depletion of state A except at the zero points
3. Reading out the signal, now exclusively from the zero points and their immediate proximity
4. Translating the zero intensity point(s) further (i.e., scanning)

We use a simple kinetic model that explains all crucial phenomena observed in such RESOLFT implementations. Let us denote the spontaneous rates of A→B and B→A with k_{AB}^s and k_{BA}^s, respectively, and allow for light-driven rates proportional to the depleting intensity: $\sigma_{AB}I$ and $\sigma_{BA}I$. The time evolution of the normalized populations of the two states n_A and n_B is then given by:

$$\frac{dn_A}{dt} = -k_{AB}n_A + k_{BA}n_B = -\frac{dn_B}{dt} \tag{24.5}$$

with $k_{AB} = k_{AB}^s + \sigma_{AB}I$ and $k_{BA} = k_{BA}^s + \sigma_{BA}I$.

If a molecule that is initially in state A is illuminated by a light pulse of duration τ during phase (2), the probability that it remains in A is given by:

$$N_A = (1 - N_{eq})\exp(-k_{tot}\tau) + N_{eq} \tag{24.6}$$

where the equilibrium population and the relaxation rate are given by $N_{eq} = k_{BA}/k_{tot}$ and $k_{tot} = k_{BA} + k_{AB}$, respectively.

To derive an upper limit for the resolution increase we assume optimal conditions. The transition into state A is induced by light of a different wavelength and can be completely suppressed during the depletion and read-out process. In addition, we assume that no signal is generated during the depletion process and that the detected signal from a molecule (in state A) depends solely on the excitation probability multiplied by the probability of it remaining in state A. The latter is given by:

$$p_A = \exp\left(\frac{-\ln 2I}{I_{sat}}\right) \tag{24.7}$$

The saturation intensity is then given by:

$$I_{sat} = \frac{\ln 2}{\sigma\tau} \tag{24.8}$$

A complete RESOLFT cycle starts with all the molecules in their nonsignaling state B. Our aim is to produce the narrowest possible spatial probability distribution of molecules being in state A around the origin. The first step is therefore the creation of a diffraction-limited distribution using the alternate wavelength and far-field focusing. For the sake of simplicity we approximate the focal intensity distribution by a Gaussian, and after illumination we have:

$$N_A^{(1)}(r) = \exp\left(\frac{-r^2}{a^2}\right) \tag{24.9}$$

where a is the $1/e$ radius. It is related to the focal spot's FWHM by $\Delta r = 2a\sqrt{\ln 2}$.

The focal distribution of the depleting light is also diffraction-limited and approximated by a Gaussian:

$$I(r) = I_{max}\left[1 - \exp\left(\frac{-r^2}{a^2}\right)\right] \tag{24.10}$$

Using equation (24.7) we can calculate N_A after the depletion pulse:

$$N_A^{(2)}(r) = N_A^{(1)}(r)p_A(r) = \exp\left(\frac{-r^2}{a^2}\right)\exp\left(-\xi\left[1 - \exp\left(\frac{-r^2}{a^2}\right)\right]\right) \tag{24.11}$$

where we defined the saturation factor $\xi = \ln 2I/I_{sat}$. This definition was chosen to match the context of pulse-based RESOLFT schemes and therefore differs slightly from earlier ones (Hell, 2003; Hell & Schönle, 2006).

The FWHM of N_A depends only on the saturation factor and is obtained by setting the right-hand side of this equation equal to 0.5 and solving for r:

$$-\ln 2 = \frac{-r^2}{a^2} - \xi\left[1 - \exp\left(\frac{-r^2}{a^2}\right)\right] \tag{24.12}$$

At large saturation factors we assume $r << \Delta r$, and with a linear approximation of the exponential in equation (24.12) we obtain:

$$\Delta r(\xi) = \frac{\Delta r}{\sqrt{1 + \xi}} = \frac{\lambda}{2NA\sqrt{1 + \xi}} \tag{24.13}$$

This equation is the essence of RESOLFT. For a vanishing saturation factor $\xi = 0$, it assumes Ernst Abbe's diffraction limited form, whereas for $\xi \to \infty$, the spot becomes infinitely small.

We conclude this section by considering some of the adverse effects encountered in practical RESOLFT microscopy. Commonly, the state A cannot be completely emptied by even very intense illumination (e.g., because there is an excitation by the same beam and $\sigma_{BA} \neq 0$ in equations (24.5) and (24.6)). In addition, state B may also contribute to the signal but m times weaker than state A. Both can be considered by including a constant offset in equation (24.7):

$$p_A = \delta + (1 - \delta)\exp\left(\frac{-\ln 2I}{I_{sat}}\right) \tag{24.14}$$

with $\delta = 1/m + \sigma_{BA}/\sigma_{AB}$. This would result in the image consisting of a superresolved image plus a (weak) conventional image. The frequency content of the image is unchanged. Therefore, if δ is sufficiently small so as not to swamp the image with noise, the conventional contribution can be subtracted (Hell, 1997; Hell & Kroug, 1995). A non-negligible spontaneous decay of the molecule into state B that takes place during the depletion step and produces a detectable signal (e.g., if it is the same fluorescent transition used as a signal later) has a similar effect. A quick integration of equation (24.6) over the depletion pulse yields equation (24.14) with $\delta = k^s_{BA}/k_{tot}$. Finally, a non-negligible spontaneous rate k^s_{AB} introduces an additional steady-state component, as discussed in equation (24.16) below. The essence is that the concept will work as long as the depletion rate is sufficiently large as compared to the spontaneous rates in the molecule. In other words, the spontaneous rates dictate the intensities needed for a certain resolution improvement, a conclusion that will prove important in the choice of suitable marker candidates in RESOLFT microscopy.

Another common problem is imperfections of the intensity zeros. Aberrations can cause the intensity distribution to have insufficient "depth" (i.e., residual intensity at the zero points). Equation (24.10) would then become:

$$I(r) = I_{max}\left[1 - (1 - \gamma)\exp\left(\frac{-r^2}{a^2}\right)\right] \tag{24.15}$$

which turns out to have a much more serious impact on performance: the maximum signal in the intensity minima drops by a factor $\exp(-\xi\gamma)$ as a result. However, following the same calculation as above, the resolution of the system remains virtually unchanged. Alternative realizations of RESOLFT use continuous wave illumination (Hofmann et al., 2005) and the exponential dependence in equation (24.7) is replaced by:

$$p_A = \frac{1}{(1 + I/I_{sat})} \qquad (24.16)$$

with the saturation intensity now given by $I_{sat} = k_{BA}/(k_{AB} + k_{BA})$, and a similar analysis as above essentially reproduces equation (24.13) (Hell & Schönle, 2006). For this scheme, imperfections in the zero points are however more severe because they lead to a hard resolution limit, depending on the quality of the zero points.

While RESOLFT is far more intuitively explained in the way presented above, it is also helpful to take a look at the frequency space in order to relate these findings to the concepts and results presented in the first part of this chapter. If the detection PSF is neglected, the effective PSF of the system is given by $h(r) = N_A^{(2)}(r)$. For an arbitrary form of the depleting intensity distribution, setting $I(r) = I_{max} f(r)$ and using equation (24.7), we can develop the effective PSF in a Taylor series:

$$h(r) = \exp(-\xi) \exp\left(\frac{-r^2}{a^2}\right)\left(1 + \xi(1-f) + \frac{\xi^2}{2!}(1-f)^2 + \frac{\xi^3}{3!}(1-f)^3 + \cdots\right)$$
$$(24.17)$$

With $g(r) = 1 - f(r)$ we obtain the OTF after Fourier transformation:

$$o(k) = \widehat{g} + \xi\widehat{g}\otimes\widehat{g} + \frac{\xi^2}{2!}\widehat{g}\otimes\widehat{g}\otimes\widehat{g} + \frac{\xi^3}{3!}\widehat{g}\otimes\widehat{g}\otimes\widehat{g}\otimes\widehat{g} + \cdots \qquad (24.18)$$

At low intensities, ξ is so small that only the first term is relevant and the convolution does not extend the support at all. But the larger the maximum intensity, the more important higher orders of the Taylor series will become. These involve multiple autoconvolutions of the function g, extending the support further and further. This shows that while the form of the intensity distribution is a central parameter of the practical realization, conceptually the diffraction barrier is broken due to the nonlinearities introduced by saturating the single-photon transition. Again, we would like to note that the origin of these nonlinearities fundamentally differs from nonlinear interactions connected with m-photon excitation, m-th harmonics generation, coherent anti-Stokes Raman scattering (Shen, 1984; Sheppard & Kompfner, 1978), and so forth. In the latter cases the nonlinear signal stems from the simultaneous action of *more than one photon* at the sample, which would work only at high focal intensities.

To conclude this section let us calculate the OTF for our simple model above. Fourier transforming the effective PSF is straightforward. Starting from equation (24.11) and using the same linear approximation as in equation (24.12) we obtain:

$$o(k) \propto \exp\left(\frac{-a^2 k^2}{1+\xi}\right) \qquad (24.19)$$

which essentially means that the attenuation of the modulus of the OTF at large frequencies is antiproportional to the square root of the saturation factor. This is of course equivalent to the conclusion from equation (24.13).

24.3.3 Superresolution with STED Microscopy

STED microscopy was the first implementation of the RESOLFT concept. The two states A and B are realized as the first excited and electronic ground state of a dye molecule, and the mechanism driving molecules into the ground state is stimulated emission. As of today, STED microscopy is realized in (partially confocalized) spot-scanning systems due to a number of technical advantages. However, it has been conceptually clear from the outset that nonconfocalized operation is viable as well (Hell & Wichmann, 1994). The principle idea, a schematic setup, and an exemplary measurement of the resolution increase are shown in Figure 24.3. The dye is excited by a pulse of a first laser beam within the diffraction-limited excitation volume. Before the dye fluoresces, a second laser pulse with its wavelength at the red edge of the emission spectrum illuminates regions outside the geometric focus. Stimulated emission takes place, and if the intensity of this second pulse is increased beyond the saturation level, fluorescence is allowed only from confined regions where the intensity of the second beam is zero. With typical saturation intensities ranging from 1 to 100 MW/cm^2, saturation factors of up to 120 have been reported (Klar et al., 2000, 2001). This potentially yields a 10-fold resolution improvement over the diffraction barrier, but imperfections in the doughnut have usually limited the improvement to 5- to 7-fold in the initial experiments (Klar et al., 2001).

Figure 24.4c shows imaging of a single JA 26 dye molecule spin-coated on a glass slide (Westphal & Hell, 2005). In this example a resolution potential of down to 16 nm was demonstrated with STED wavelengths of $\lambda = 750$ to 800 nm. Besides the unprecedented lateral resolution increase, this also proves that STED is single molecule-sensitive. In fact, individual molecules have been switched on and off by STED upon command in other experiments (Kastrup & Hell, 2004; Westphal et al., 2003a).

In all STED applications, the focal intensity distribution of the depleting beam is designed by manipulating its wavefront before it enters the back-aperture of the objective lens. For the image shown in Figure 24.4c the STED beam was optimized for resolution increase along the lateral coordinate perpendicular to the polarization of the depleting light. This is achieved by using a one-dimensional phase plate (explained further down in Fig. 24.7b), which leads to an intensity distribution with two strong peaks at either side of the excitation maximum (Keller et al., in preparation). Of course, in many applications a more isotropic resolution gain is preferred over a maximal resolution along a single axis. In addition, for the one-dimensional phase plate, the resolution gain obviously depends on the orientation of the transition dipole of the molecule with respect to the light field. Saturation will be more easily achieved for dipoles oriented along the direction of the focal field.

Almost isotropic lateral resolution is attained by combining two one-dimensional phase plates that are oriented perpendicularly to each other. Figure 24.5 displays such an arrangement, yielding two perpendicularly polarized, complementary STED intensity distributions. Combined with circularly polarized excitation, such a setting yields almost uniform resolution in the focal plane (Fig. 24.6). Here, a series of xy images

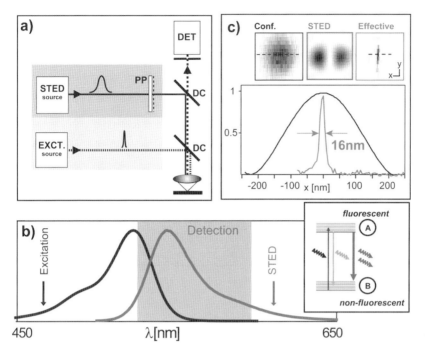

Figure 24.4. Stimulated emission depletion (STED), the first implementation of the RESOLFT principle. (a) Schematic of a point-scanning STED microscope. Excitation and depletion beams are combined using appropriate dichroic mirrors (DC). The excitation beam forms a diffraction-limited excitation spot in the sample, while the depletion beam is manipulated using a phase-plate (PP) or any other device to modify the wavefront in such a way that it forms an intensity distribution with a nodal point in the excitation maximum. The upper row in (c) shows the measured reference (confocal) PSF, the STED-PSF (see Fig. 24.6) with a line-shaped central minimum, and the resulting effective PSF measured by scanning a single dye molecule through the focal region. The graph compares a gray-value plot of the effective PSFs in the confocal and the STED mode, illustrating the fundamental sharpening of the focal spot in the x-direction down to 16 nm (i.e., a small fraction of the wavelength employed). (b) The dye molecules are excited into the S_1 (state A) by an excitation laser pulse while the fluorescence is detected over most of the emission spectrum. Molecules can be quenched back into the ground state S_0 (state B) using stimulated emission before they fluoresce. Stimulated emission is generated by irradiating the molecules with a light pulse at the red edge of the emission spectrum, shortly after the excitation pulse.

acquired with different STED-beam powers demonstrates how resolution increases for increasing saturation factors (Schönle et al., in preparation). However, a more careful analysis shows that even this combination of two incoherent beams causes the resolution to depend on the orientation of the transition dipole; hence spikes result along the x and y direction of the OTF when imaging randomly oriented molecules.

Generally speaking, with the principle method established, the question arises as to the optimal STED PSF for a given application. Therefore, a rigorous analysis of the critical parameters and an assessment of optimal wavefront modifications are in order. Such an analysis was recently conducted (Keller et al., 2007), and some results of this study are displayed in Figure 24.7. The maximum intensity in the back-aperture of the objective lens was assumed to be finite. An optimization algorithm was used to

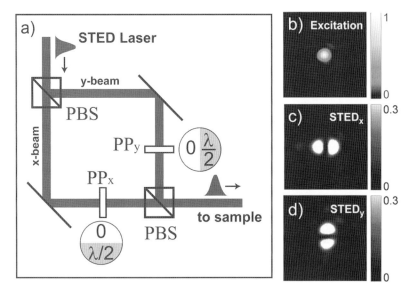

Figure 24.5. (a) Schematic of the combination of two 1D-mode phase-plates (see Fig. 24.7) for quasi-isotropic lateral resolution. This setup replaces the phase-plate in Figure 24.4a. The STED beam is split by a polarizing beam-splitter (PBS) into two beams of perpendicular polarization and equal intensity. Both pass identical phase-plates, which are oriented according to the polarization and recombined using another PBS. When focused into the sample, this results in two 1D-mode STED-PSFs, one of them, for example, x-polarized (c) and one y-polarized (d). The excitation light is circularly polarized and forms an isotropic diffraction-limited spot (b). The effective PSF of such a setup will depend on the orientation of the molecule because x-oriented molecules are not quenched on the y-axis, and vice versa.

maximize a figure of merit (FoM); the phase and the amplitude in the back-aperture of the lens were allowed to vary freely while a strict intensity zero in the focal spot was enforced. The FoM was chosen to reflect the tasks of increasing the resolution along a single or two or three spatial dimensions. This is achieved by creating large intensities at a short distance r_{FoM} from the geometrical focal point. It turned out that provided it is chosen around the anticipated final resolution, the choice of r_{FoM} only marginally influences the outcome of the optimization. In all situations the optimal manipulation involved only phase changes and no attenuation of the beam.

For linearly polarized light, the phase plates intuitively chosen in earlier applications (Klar et al., 2001) were confirmed as being optimal for the purpose. However, linearly polarized STED beams always result in orientation-dependent depletions, even if they are incoherently combined. Therefore, a circularly polarized STED beam is advisable. In this case, the algorithm yields a "ring-shaped phase step of π" for a 3D resolution gain. Similarly, it converges to a helical 0 to 2π phase ramp for an optimal focal plane resolution (2D) improvement. Unlike the ring-shaped phase step of π, the latter does not improve the axial resolution over a confocal or conventional microscope but delivers a superb lateral resolution. For saturation factors $\sim$50, the most isotropic 3D resolution is gained by using the ring step of π yielding a FWHM of $\sim$120 nm along all directions. The concomitant focal volume reduction is by a factor of 7, as measured in FWHM volume. The smallest focal volume can be achieved by incoherently combining (c)

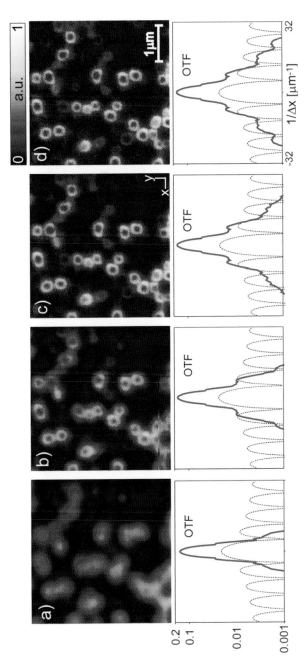

Figure 24.6. Images of a wetted Al_2O_3 matrix featuring z-oriented holes (Whatman plc, Brentford, UK) with a spin cast of a dyed (JA 26) polymethyl methacrylate solution. The rings formed in this way are ~250 nm in diameter and are barely resolved in confocal mode. The STED intensity was chosen corresponding to (b) 84, (c) 187, and (d) 290 MW/cm^2 in the unaberrated STED focus. This corresponds approximately to intensities of I_{sat}, $2I_{sat}$, and $3I_{sat}$ respectively in the lobe maxima of the two aberrated STED PSFs. The resolution increase is immediately visible to the eye. The smaller effective spot size also results in an extended OTF, as seen in the graphs of the second row, which show a profile along the x-direction. Please note the logarithmic scales. The Fourier transform of the image (red) is given by the product of OTF and the Fourier transform of the object. Because we image copies of (almost) the same object, this effect can be directly observed in the graphs. The Fourier transform of a ring with a diameter of 275 nm and a width of 50 nm (shown as a dashed line) exhibits the same modulation as the images' Fourier transform. The OTF exhibits an extension of its support for increasing STED powers as expected.

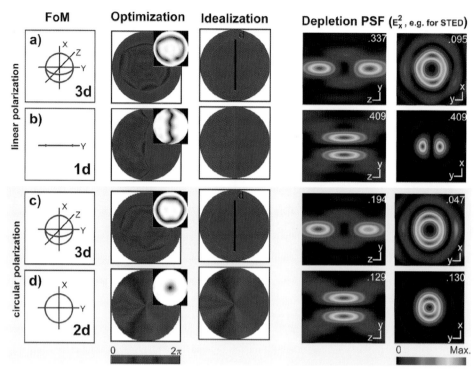

Figure 24.7. Optimal manipulation of the STED beam's wavefront to create the most effective STED PSFs assuming a limited amplitude of the incoming wavefront at the back-aperture. An optimization algorithm varies the phase and amplitude of the focused wavefront while optimizing a figure of merit (FoM). Depending on whether resolution should be increased in 1, 2, or 3 dimensions, different FoMs are used. For linearly (x-) polarized light, the minimum amplitude of the field's x-component on (a) a sphere of radius r_{FoM} (3D mode) and (b) two points on the y-axis at $y = \pm r_{FoM}$ (1D mode) were maximized. Idealizing the results confirm that the "rings" and "linear valley" phase-plates chosen in many applications are indeed the optimal ones. The intensity of the corresponding STED-PSF's x-component for a numerical aperture of 1.2 is shown on the right; the numbers give the normalized maximum intensity. The latter is defined as the relative intensity as compared to the focal maximum of a perfectly focused wavefront with the same amplitude. For linear polarization the 2D mode is not shown. For circular polarization, the 1D and 2D modes result in the same phase-plate, representing the helical phase ramp shown in (d). The 3D mode shown in (c) delivers the same "rings" as in (a).

and (d), with the available power being split 30/70 between the shapes. This results in a 16-times-reduced focal volume; the resulting PSF is significantly elongated along the optic axis.

Importantly, STED has been successfully applied to the imaging of biological samples. Subdiffraction images with 3-fold enhanced axial and doubled lateral resolution have been obtained with membrane-labeled bacteria and live budding yeast cells (Klar et al., 2000). More recently, STED microscopy was applied to the recycling of synaptic vesicle resident proteins in neuronal cells (Willig et al., 2006). Using the optimized phase change of Figure 24.7d and circularly polarized light, the resolution in the focal plane could be uniformly improved by more than a factor of 3. By resolving for the first

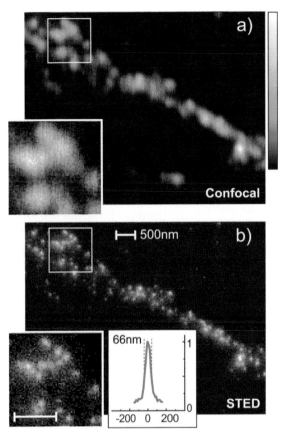

Figure 24.8. STED microscopy of synaptic vesicles in individual boutons of primary cultured hippocampal neurons from rats; the vesicle membrane resident protein synaptotagmin I has been labeled with a primary and secondary antibody. Unlike in the confocal image (a), the STED image (b) discloses individual patches pertinent to individual vesicles. A typical gray-value plot is shown in the inset. The profile of a vesicle exhibits a FWHM of approximately 66 nm. Because the vesicle itself is about 40 nm in diameter, this indicates a lateral resolution of <60 nm. All scale bars in the figure are 500 nm long. (Adapted from Willig et al., 2006)

time protein agglomeration at a resolution <70 nm (Fig. 24.8), STED microscopy has thus solved an important neurobiological problem in a unique way. By direct inspection and quantitative analysis of superresolved images, it could be concluded that the synaptic vesicle resident protein synaptotagmin I is not dispersed in the presynaptic membrane following vesicular fusion.

It is to be expected that the ultrasmall detection volumes created by STED will also be useful in a number of sensitive bioanalytical techniques. Fluorescence correlation spectroscopy (FCS) (Magde et al., 1972) relies on small focal volumes to detect rare molecular species or interactions in concentrated solutions (Eigen & Rigler, 1994; Elson & Rigler, 2001). While volume reduction can be obtained by nanofabricated structures (Levene et al., 2003), STED may prove instrumental to attaining ultrasmall spherical volumes at the nanoscale inside samples that do not allow mechanical confinement. The latter fact is particularly important in order to avoid an alteration of the measured fluctuations by the nanofabricated surface walls. In fact, the viability of STED-FCS has recently been shown in experiments (Kastrup et al., 2005). In a particular implementation, STED-FCS has witnessed a reduction of the focal volume by a factor of five along the optic axis along with the concomitant reduction of the axial diffusion time (Fig. 24.9). These experiments showed that for particular dye/wavelength combinations, the evaluation of the STED-FCS data might be complicated by background from the outer wings of the fluorescence spot, where STED does not completely suppress the signal (see Fig. 24.9b). More sophisticated data analysis methods will have to be developed. Still, being the only method known to date that noninvasively "squeezes" a fluorescence volume to the zeptoliter scale, STED may well be a viable pathway to improving the sensitivity of fluorescence-based bioanalytical techniques (Laurence & Weiss, 2003; Weiss, 2000).

The results discussed in this section, including the application of STED to cell biology (Willig et al., 2006), indicate that the basic physical obstacles of this concept have been overcome. In many of these experiments, the samples were mounted in an aqueous buffer (Dyba & Hell, 2002; Dyba et al., 2003). The required intensities are by two to three orders of magnitude lower than those used in multiphoton microscopy (Denk et al., 1990). This clearly shows the potential of STED to obtain subdiffraction resolution in live-cell imaging. Another important step toward far-reaching applicability of STED microscopy was the demonstration of the suitability of laser diodes both for excitation and for depletion (Westphal et al., 2003a).

Although the most fundamental hurdles toward attaining a 3D resolution of the order of a few tens of nanometers have already been taken, several issues remain to be addressed. Due to the considerably smaller detection volumes, the signal per pixel is reduced and the number of pixels to be recorded increases. Therefore, it will be important to incorporate STED into fast, parallelized scanning systems. Also, at high saturation levels, the photo-stability of the marker may become an issue. For several current markers, this was considerably improved by stretching the depleting pulse to >300 ps (Dyba & Hell, 2003), but it might not be easily possible to attain saturation factors $\xi > 200$ (i.e., spatial resolutions <15 nm).

24.3.4 RESOLFT Techniques

Importantly, the RESOLFT concept is not restricted to the process of stimulated emission (i.e., to STED) but can draw on any reversible (linear) transition that is driven by light. In retrospect, stimulated emission just happened to be a popular and obvious transition in common fluorescent dyes taking place at laser wavelengths available. In a way, STED appeared to be the most promising candidate, next to the depletion of the dye's ground state by transiently shelving it in its dark triplet state (Hell & Kroug, 1995). In fact, an even simpler method seems to be the depletion of the ground state by the saturation of the first excited state. Unlike STED, it can be accomplished without an additional wavelength. This version of employing a reversible saturable transition has indeed been proposed (Heintzmann et al., 2002), but it has to deal with a number of serious issues hampering its real-world application. For example, the fluorescence emission maps the spatially extended, "majority population" in state B (the excited state), meaning that the superresolved images (represented by state A) are negative images hidden under a bright signal from state B. In addition, the dye is pumped to an

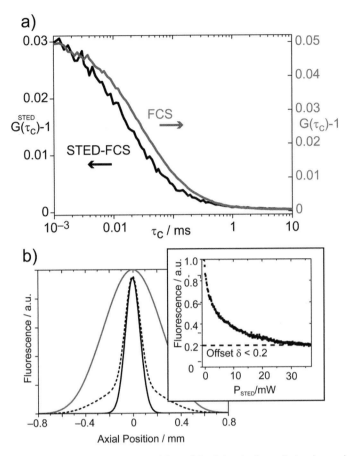

Figure 24.9. STED-FCS curves measured in a 50-nM solution of single molecules of the oxazine dye MR121. The STED PSF was created using circularly polarized light in the 3D mode (i.e., a phase-plate as shown in Figure 24.7c). The STED power was approximately 22 mW, which should result in a nearly isotropic effective PSF. (a) FCS curves with and without the STED beam switched on. Fitting the data with a model for 3D diffusion results in lateral and axial diffusion times of ~25 μs and ~400 μs respectively for the confocal mode. When STED is switched on, the axial diffusion time is reduced to ~20 μs, suggesting the creation of a spherical spot. Unexpectedly, the amplitude of the FCS curve decreases for STED. This indicates a larger number of molecules in the (5-fold-smaller) focal volume. (b) A closer look reveals that in this experiment the STED-PSF cannot completely quench fluorescence even for very large STED powers but leaves a fluorescence offset δ (inset). This results in an axial profile of the effective PSF that features "wings" that are weakly correlated with the signal from the central sharp maximum that is responsible for the short average diffusion time. Such uncorrelated background reduces the amplitude of the FCS curve and can account for the surprising effect.

excited state rather than into the ground state as in STED. Therefore, a large number of dye molecules undergo excitation–emission cycles repeatedly in order to image the comparatively small zero-intensity regions. This should lead to stronger photobleaching as well as to a significant triplet-state buildup. Nevertheless, due to the simplicity of raw data acquisition, making negative images may prove to be an attractive method for imaging very bright and photostable samples.

The intensities required for large saturation factors may limit the saturation of the transition between the S_0 and S_1 states of a molecule. A promising path to circumvent this problem is based on a fact already mentioned in Section 24.3.2. The saturation factor depends on the ratio of the light-induced rates to be saturated and the competing spontaneous rates. If the applicable intensity is limited for any reason, further resolution increase can be achieved only by finding molecular processes where high saturation levels are attained at lower intensities. In other words, we want to find transitions where the competing spontaneous rates are much slower than typical fluorescence lifetimes.

In fact, one such transition exists in every fluorescent dye: intersystem crossing into the triplet state. This phenomenon is exploited in a concept called ground state depletion (GSD). In this version of the RESOLFT concept, the ground state (now state A) is depleted by targeting an excited state (B) with a comparatively long lifetime (Hell, 1997; Hell & Kroug, 1995), such as the metastable triplet state T_1. In many fluorophores T_1 can be reached through the S_1 with a quantum efficiency of 1% to 10% (Lakowicz, 1983). The relaxation of the T_1 is 10^3 to 10^5 times slower than that of the S_1, because it is a spin-forbidden transition. In a steady-state scheme, saturation is defined by equation (24.16), and we therefore expect saturation intensities of $I_{sat} = 0.1$ to 100 kW/cm^2. The signal to be measured (from the intensity zero) is the fluorescence of the molecules that remained in the singlet system; this measurement can be accomplished through a synchronized further excitation (Hell & Kroug, 1995). For many fluorophores, this approach is not straightforward, because T_1 is involved in the process of photobleaching. However, there are potential alternatives like the metastable states of rare earth metal ions that are fed through chelates.

Even slower rates are found in compounds with two (semi-)stable states (Dyba & Hell, 2002; Hell et al., 2003). If the rate k_{BA} and the spontaneous rate k_{AB} almost vanish, large saturation factors are attained at extremely low intensities. The lowest useful intensity is then determined by the slowest acceptable imaging speed, which is ultimately determined by the switching *rate*. Several such compounds have been reported in the literature. One example is a photoswitchable coupled molecular system, based on a photochromic diarylethene derivative and a fluorophore (Irie et al., 2002). Using the kinetic parameters reported for this construct, equations (24.16) and (24.13) predict that focusing of less than $100\,\mu$W of deep-blue "switch-off light" to an area of 10^{-8} cm^2 for $50\,\mu$s should yield better than 5 nm spatial resolution. On the other hand, fluorescent proteins have many advantages over synthetic dyes, especially in live-cell imaging. Many of them feature dark states with light-driven transitions (Hell, 1997; Hell et al., 2003). If the spontaneous lifetimes of these states are longer than 10 ns, such proteins may permit much larger saturation factors than those used in STED microscopy today. The most attractive solution, however, is fluorescent proteins that can be switched on and off at different wavelengths (Hell et al., 2003)—for example, asFP595 (Lukyanov et al., 2000) and dronpa (Ando et al., 2004). Insertion of the published data into equation (24.16) predicted saturated depletion of the fluorescence state with intensities of less than a few W/cm^2, and therefore RESOLFT was recently implemented using asFP595. Intensities of approximately 30 W/cm^2 were shown to break the diffraction barrier (Hofmann et al., 2005). In this initial implementation, residual intensity in the minima hindered the realization of larger saturation factors and the resolution increase was limited to a factor of $\sim$1.7. Other possible challenges of such very-low-intensity depletion microscopes may be switching fatigue (Irie et al., 2002) and overlapping action spectra

(Lukyanov et al., 2000). Nevertheless, the prospect of attaining nanoscale resolution with regular lenses and focused light is an incentive to surmount these challenges by strategic fluorophore modification, and this or similar types of fluorescent proteins are a good starting point for these efforts (Andresen et al., 2005).

24.3.5 Conclusion

All far-field microscopes relying solely on focusing for spatial resolution are fundamentally limited by diffraction. Confocal, two-photon and more recently patterned illumination, 4Pi-microscopy, and related techniques have pushed their resolving power close to its theoretical maximum in the 100-nm range. However, none of these improvements can bring about resolution beyond this limit. In order to break this barrier first formulated by Ernst Abbe almost one and a half centuries ago, the spectroscopy of the contrast agent can be exploited in a specific way. The RESOLFT concept is based on this idea. Relying on a reversible saturation of the transition between two states of a marker RESOLFT provides diffraction unlimited resolution with visible light and with regular lenses. The resolution is limited by the attainable level of saturation of the (linear) optical transition in the marker molecule. Therefore, the rather "hard" theoretical resolution barrier is replaced by a set of operational parameters that depend on practical conditions such as available laser power, cross-sections, and the stability of the dye and the sample.

STED microscopy is the most advanced implementation of the RESOLFT principle to date. It has been well characterized by its application to single fluorescent molecules and has been applied to imaging fixed and live, albeit simple, biological specimens. More recently STED was a key tool to solve a longstanding problem in neurobiology (Willig et al., 2006). This demonstrates both the maturity of the method and the fact that with its superior resolving power, it allows the application of life imaging to a whole class of new problems and could thus lead to many important new discoveries.

Resolution improvements by a factor of up to 8 have already been demonstrated, and the resulting fluorescent volumes are the smallest that have so far been created with focused light. The resolutions achieved in these experiments were limited by the power of the lasers used, suboptimal intensity distributions in the sample, and dye instabilities. Future research on spectroscopy conditions and on practical aspects will surely bring about substantial improvements.

Nevertheless, STED could ultimately turn out to be limited to resolutions of 15 nanometers in routine applications. The reason is that due to the fast spontaneous rates competing with the saturation, very high laser intensities would be necessary to push beyond this limit, ruling out samples and markers that are damaged by such intense radiation. Fortunately, the intensities required for obtaining a high saturation level can be fundamentally lowered in other implementations of the RESOLFT concept. This is particularly true when utilizing optically bistable markers such as photoswitchable dyes and photochromic fluorescent proteins. Both are very promising candidates for providing high levels of saturation at ultralow intensities of light, and dedicated synthesis or protein engineering might eventually uncover a whole new range of suitable markers. Lower intensities also have the practical advantage of enabling massive parallelization of RESOLFT without the need for extremely powerful lasers. This could prove essential to achieving acquisition speeds needed for 3D life-cell imaging.

Ultimately, the only fundamental limit to the resolution of a RESOLFT microscope is the extent of the marker. We therefore believe that the results reported here could eventually be considered the first steps in the development of far-field "nanoscopes" operating with regular lenses and visible light.

ACKNOWLEDGMENTS

The authors thank all members of the Department of NanoBiophotonics for contributions to the work reviewed herein and valuable discussions. A fair part of the results described in this chapter have been adopted from original work with significant contributions by K. Willig, V. Westphal, L. Kastrup, and C. Eggeling. We thank J. Jethwa for critical reading of the manuscript.

REFERENCES

Abbe E. 1873. Beiträge zur Theorie des Mikroskops und der mikroskopischen Wahrnehmung. Arch. f. Mikr. Anat. 9:413–420.

Ando R, Mizuno H, Miyawaki A. 2004. Regulated Fast Nucleocytoplasmic Shuttling Observed by Reversible Protein Highlighting. Science 306(5700):1370–1373.

Andresen M, Wahl MC, Stiel AC, Gräter F, Schäfer LV, Trowitzsch S, Weber G, Eggeling C, Grubmüller H, Hell SW and others. 2005. Structure and mechanism of the reversible photoswitch of a fluorescent protein. Proc. Natl. Acad. Sci. USA 102(37):13070–13074.

Barad Y, Eisenberg H, Horowitz M, Silberberg Y. 1997. Nonlinear scanning laser microscopy by third harmonic generation. Appl. Phys. Lett. 70:922–924.

Bloembergen N. 1965. Nonlinear optics. New York/Amsterdam: Benjamin.

Born M, Wolf E. 1993. Principles of Optics. Oxford: Pergamon Press.

Denk W, Strickler JH, Webb WW. 1990. Two-photon laser scanning fluorescence microscopy. Science 248:73–76.

Dyba M, Hell SW. 2002. Focal spots of size l/23 open up far-field fluorescence microscopy at 33 nm axial resolution. Phys. Rev. Lett. 88:163901.

Dyba M, Hell SW. 2003. Photostability of a fluorescent marker under pulsed excited-state depletion through stimulated emission. Appl. Optics 42(25):5123–5129.

Dyba M, Jakobs S, Hell SW. 2003. Immunofluorescence stimulated emission depletion microscopy. Nature Biotechnol. 21(11):1303–1304.

Egner A, Hell SW. 1999. Equivalence of the Huygens-Fresnel and Debye approach for the calculation of high aperture point-spread-functions in the presence of refractive index mismatch. J. Microsc. 193:244–249.

Egner A, Hell SW. 2005. Fluorescence microscopy with super-resolved optical sections [Review]. Trends in Cell Biology 15(4):207–215.

Egner A, Jakobs S, Hell SW. 2002. Fast 100-nm resolution 3D-microscope reveals structural plasticity of mitochondria in live yeast. Proc. Natl. Acad. Sci. USA 99:3370–3375.

Eigen M, Rigler R. 1994. Sorting single molecules: applications to diagnostics and evolutionary biotechnology. Proc. Natl. Acad. Sci. USA 91:5740–5747.

Elson EL, Rigler R, editors. 2001. Fluorescence correlation spectroscopy. Theory and applications. Berlin: Springer.

Gannaway JN. 1978. Second-harmonic imaging in the scanning optical microscope. Opt. Quant. Electron. 10:435–439.

Goodman JW. 1968. Introduction to Fourier optics: McGraw Hill.

Göppert-Mayer M. 1931. Über Elementarakte mit zwei Quantensprüngen. Ann. Phys. (Leipzig) 9:273–295.

Gugel H, Bewersdorf J, Jakobs S, Engelhardt J, Storz R, Hell SW. 2004. Cooperative 4Pi excitation and detection yields 7-fold sharper optical sections in live cell microscopy. Biophys. J. 87:4146–4152.

Gustafsson MGL. 2000. Surpassing the lateral resolution limit by a factor of two using structured illumination microscopy. J. Microsc. 198(2):82–87.

Gustafsson MGL, Agard DA, Sedat JW. 1999. I5M: 3D widefield light microscopy with better than 100 nm axial resolution. J. Microsc. 195:10–16.

Hänninen PE, Lehtelä L, Hell SW. 1996. Two- and multiphoton excitation of conjugate dyes with continuous wave lasers. Opt. Commun. 130:29–33.

Hegedus ZS, Sarafis V. 1986. Superresolving filters in confocally scanned imaging systems. J. Opt. Soc. Am. A 3(11):1892–1896.

Heintzmann R, Cremer C. 1998. Laterally modulated excitation microscopy: improvement of resolution by using a diffraction grating. SPIE Proc. 3568:185–195.

Heintzmann R, Jovin TM, Cremer C. 2002. Saturated patterned excitation microscopy - A concept for optical resolution improvement. J. Opt. Soc. Am. A: Optics and Image Science, and Vision 19(8):1599–1609.

Hell S, Stelzer EHK. 1992a. Properties of a 4Pi-confocal fluorescence microscope. J. Opt. Soc. Am. A 9:2159–2166.

Hell SW, S. W. Hell, assignee. 1990 18. 12. 90. Double-scanning confocal microscope. European Patent patent 0491289.

Hell SW. 1994. Improvement of lateral resolution in far-field light microscopy using two-photon excitation with offset beams. Opt. Commun. 106:19–22.

Hell SW. 1997. Increasing the resolution of far-field fluorescence light microscopy by point-spread-function engineering. In: Lakowicz JR, editor. Topics in Fluorescence Spectroscopy. New York: Plenum Press. p 361–422.

Hell SW. 2003. Toward fluorescence nanoscopy. Nature Biotechnol. 21(11):1347–1355.

Hell SW. 2004. Strategy for far-field optical imaging and writing without diffraction limit. Phys. Lett. A 326(1–2):140–145.

Hell SW, Bahlmann K, Schrader M, Soini A, Malak H, Gryczynski I, Lakowicz JR. 1996. Three-photon excitation in fluorescence microscopy. J. Biomed. Opt. 1(1):71–73.

Hell SW, Dyba M, Jakobs S. 2004. Concepts for nanoscale resolution in fluorescence microscopy. Curr. Opin. Neurobio. 14(5):599–609.

Hell SW, Jakobs S, Kastrup L. 2003. Imaging and writing at the nanoscale with focused visible light through saturable optical transitions. Appl. Phys. A 77:859–860.

Hell SW, Kroug M. 1995. Ground-state depletion fluorescence microscopy, a concept for breaking the diffraction resolution limit. Appl. Phys. B 60:495–497.

Hell SW, Reiner G, Cremer C, Stelzer EHK. 1993. Aberrations in confocal fluorescence microscopy induced by mismatches in refractive index. J. Microsc. 169:391–405.

Hell SW, Schönle A. 2006. Far-Field Optical Microscopy with Diffraction-Unlimited Resolution. in press.

Hell SW, Schrader M, van der Voort HTM. 1997. Far-field fluorescence microscopy with three-dimensional resolution in the 100 nm range. J. Microsc. 185(1):1–5.

Hell SW, Stelzer EHK. 1992b. Fundamental improvement of resolution with a 4Pi-confocal fluorescence microscope using two-photon excitation. Opt. Commun. 93:277–282.

Hell SW, Wichmann J. 1994. Breaking the diffraction resolution limit by stimulated emission: stimulated emission depletion microscopy. Opt. Lett. 19(11):780–782.

Hellwarth R, Christensen P. 1974. Nonlinear optical microscopic examination of structures in polycrystalline ZnSe. Opt. Commun. 12:318–322.

Hofmann M, Eggeling C, Jakobs S, Hell SW. 2005. Breaking the diffraction barrier in fluorescence microscopy at low light intensities by using reversibly photoswitchable proteins. Proc. Natl. Acad. Sci. USA 102(49):17565–17569.

Irie M, Fukaminato T, Sasaki T, Tamai N, Kawai T. 2002. A digital fluorescent molecular photoswitch. Nature 420(6917):759–760.

Kastrup L, Blom H, Eggeling C, Hell SW. 2005. Fluorescence Fluctuation Spectroscopy in Subdiffraction Focal Volumes. Phys. Rev. Lett. 94:178104.

Kastrup L, Hell SW. 2004. Absolute optical cross section of individual fluorescent molecules. Angew. Chem. Int. Ed. 43:6646–6649.

Keller J, Schönle A, Hell SW. Efficient fluorescence inhibition patterns for RESOLFT microscopy. Opt. Exp. 15: 3361–3371.

Klar TA, Engel E, Hell SW. 2001. Breaking Abbe's diffraction resolution limit in fluorescence microscopy with stimulated emission depletion beams of various shapes. Phys. Rev. E 64:066613, 1–9.

Klar TA, Jakobs S, Dyba M, Egner A, Hell SW. 2000. Fluorescence microscopy with diffraction resolution limit broken by stimulated emission. Proc. Natl. Acad. Sci. USA 97:8206–8210.

Lakowicz JR. 1983. Principles of Fluorescence Spectroscopy. New York: Plenum Press.

Lakowicz JR, Gryczynski I, Malak H, Gryczynski Z. 1996. Two-color two-photon excitation of fluorescence. Photochem. Photobiol. 64:632–635.

Laurence TA, Weiss S. 2003. How to detect weak pairs. Science 299(5607):667–668.

Levene MJ, Korlach J, Turner SW, Foquet M, Craighead HG, Webb WW. 2003. Zero-Mode Waveguides for Single-Molecule Analysis at High Concentrations. Science 299:682–686.

Lukosz W. 1966. Optical systems with resolving powers exceeding the classical limit. J. Opt. Soc. Am. 56:1463–1472.

Lukyanov KA, Fradkov AF, Gurskaya NG, Matz MV, Labas YA, Savitsky AP, Markelov ML, Zaraisky AG, Zhao X, Fang Y and others. 2000. Natural animal coloration can be determined by a nonfluorescent green fluorescent protein homolog. J. Biol. Chem. 275(34): 25879–25882.

Magde D, Elson EL, Webb WW. 1972. Thermodynamic fluctuations in a reacting system - measurement by fluorescence correlation spectroscopy. Phys. Rev. Lett. 29(11):705–708.

Nagorni M, Hell SW. 2001. Coherent use of opposing lenses for axial resolution increase in fluorescence microscopy. I. Comparative study of concepts. J. Opt. Soc. Am. A 18(1):36–48.

Pawley J. 1995. Handbook of biological confocal microscopy. Pawley J, editor. New York: Plenum Press.

Richards B, Wolf E. 1959. Electromagnetic diffraction in optical systems II. Structure of the image field in an aplanatic system. Proc. R. Soc. Lond. A 253:358–379.

Schönle A, Hänninen PE, Hell SW. 1999. Nonlinear fluorescence through intermolecular energy transfer and resolution increase in fluorescence microscopy. Ann. Phys. (Leipzig) 8(2): 115–133.

Schönle A, Harke B, Westphal V, Hell SW. *In preparation*.

Schönle A, Hell SW. 1999. Far-field fluorescence microscopy with repetetive excitation. Eur. Phys. J. D 6:283–290.

Schönle A, Hell SW. 2002. Calculation of vectorial three-dimensional transfer functions in large-angle focusing systems. J. Opt. Soc. Am. A 19(10):2121–2126.

Schrader M, Bahlmann K, Hell SW. 1997. Three-photon excitation microscopy: theory, experiment and applications. Optik 104(3):116–124.

Shen YR. 1984. The principles of nonlinear optics: John Wiley & Sons.

Sheppard CJR, Gu M, Kawata Y, Kawata S. 1993. Three-dimensional transfer functions for high-aperture systems. J. Opt. Soc. Am. A 11(2):593–596.

Sheppard CJR, Kompfner R. 1978. Resonant scanning optical microscope. Appl. Optics 17:2879–2882.

Toraldo di Francia G. 1952. Supergain antennas and optical resolving power. Nuovo Cimento Suppl. 9:426–435.

Weiss S. 2000. Shattering the diffraction limit of light: A revolution in fluorescence microscopy? Proc. Natl. Acad. Sci. USA 97(16):8747–8749.

Westphal V, Blanca CM, Dyba M, Kastrup L, Hell SW. 2003a. Laser-diode-stimulated emission depletion microscopy. Appl. Phys. Lett. 82(18):3125–3127.

Westphal V, Hell SW. 2005. Nanoscale Resolution in the Focal Plane of an Optical Microscope. Phys. Rev. Lett. 94:143903.

Westphal V, Kastrup L, Hell SW. 2003b. Lateral resolution of 28 nm ($\lambda/25$) in far-field fluorescence microscopy. Appl. Phys. B 77(4):377–380.

Willig KI, Rizzoli SO, Westphal V, Jahn R, Hell SW. 2006. STED-microscopy reveals that synaptotagmin remains clustered after synaptic vesicle exocytosis. Nature 440(7086): 935–939.

Wilson T, Sheppard CJR. 1984. Theory and practice of scanning optical microscopy. New York: Academic Press.

Xu C, Zipfel W, Shear JB, Williams RM, Webb WW. 1996. Multiphoton fluorescence excitation: New spectral windows for biological nonlinear microscopy. Proc. Natl. Acad. Sci. USA 93:10763–10768.

Zumbusch A, Holtom GR, Xie XS. 1999. Three-dimensional vibrational imaging by coherent anti-stokes Raman scattering. Phys. Rev. Lett. 82(20):4142–4145.

Two-Photon Fluorescence Correlation Spectroscopy

Suzette Pabit and Keith Berland

25.1 INTRODUCTION

With advances in high-sensitivity fluorescence microscopy, fluorescence correlation spectroscopy (FCS) and related fluctuation spectroscopy methods have become important and practical research tools that allow researchers to investigate the chemical and physical properties of biomolecules in a variety of complex environments (Bacia & Schwille, 2003; Chen et al., 2001; Hess et al., 2002; Krichevsky & Bonnet, 2002; Muller et al., 2003; Rigler & Elson, 2001; Schwille, 2001; Thompson et al., 2002; Webb, 2001). Specifically, fluctuation spectroscopy measurements can provide quantitative information about mobility, concentration, interactions, chemical kinetics, and physical dynamics of biomolecules both in vitro and within living cells and tissues, and the various FCS methods are being increasingly applied in many new biomedical research areas. This chapter provides an introduction to FCS, focusing primarily on the basic theoretical and experimental considerations involved in FCS measurements and analysis. Several different classes of FCS applications are also introduced to provide an overview of some key FCS capabilities.

25.2 BACKGROUND AND THEORY

The basic approach of fluorescence fluctuation spectroscopy experiments, conceptualized in Figure 25.1, is to observe random spontaneous fluctuations in fluorescence intensity measured from a minute optically defined observation volume (Elson & Magde, 1973; Magde et al., 1972; Thompson, 1991). Fluctuations can arise from multiple sources such as molecular diffusion in and out of the observation volume or any physical or chemical dynamics that modify the photophysical properties of the fluorescing molecules in the volume. While individual fluctuation events are random occurrences, on average the amplitude and time scale of the observed fluctuation dynamics reflect the underlying physical and chemical dynamics of the molecules under observation. Thus, with appropriate statistical analysis tools to analyze the observed fluctuations, one can recover significant information about the mobility, dynamics, and interactions of biomolecules.

In order for fluctuations to be experimentally resolved, one must observe relatively small numbers of molecules at a time, which necessitates the use of only minute measurement volumes. The observation volume is optically defined in three dimensions

(a)

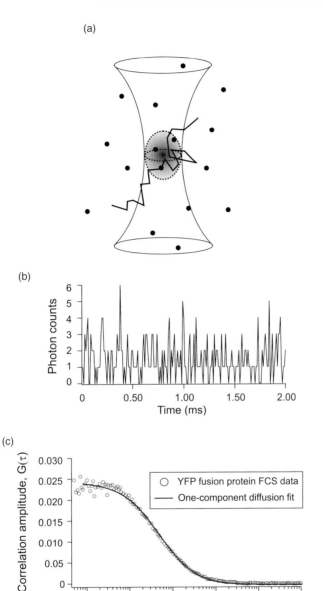

(b)

(c)

Figure 25.1. Conceptual representation of an FCS experiment. (a) An observation volume is optically defined using two-photon excited fluorescence. The cone represents the profile of the focused laser, and the shaded region represents the observation volume from which the fluorescence signal is detected (the peak laser intensity is at the center of the shaded region). The black dots represent molecules as they randomly diffuse about the sample. The jagged line represents the random-walk path of a single molecule. (b) The fluorescence signal will fluctuate as molecules diffuse into and out of the volume and may also fluctuate due to chemical dynamics. (c) Correlation functions are computed from the fluctuating fluorescence signal, and curve fitting to specific physical models is used to determine system parameters such as diffusion coefficients and molecular concentrations. The correlation curve shown is for the yellow fluorescent protein (YFP) fused with a ~60-kilodalton protein with diffusion coefficient of D ~ 0.04 μm^2/ms.

using either two-photon or confocal microscopy. This chapter will focus exclusively on two-photon FCS, in which two-photon excitation optically defines a subfemtoliter sample region from which the majority of the fluorescence is detected.

25.2.1 Two-Photon Fluorescence Signals, Observation Volumes, and Gamma Factors

The basic theory of two-photon excited fluorescence is well known (Denk et al., 1990; Xu & Webb, 1997). The instantaneous rate for absorption of photon pairs follows the familiar intensity squared dependence and is given by:

$$W(\mathbf{r}, t) = \frac{\sigma_2 I^2(t) S^2(\mathbf{r})}{2},$$

where σ_2 is the two-photon absorption cross-section, $I(t)$ is the peak laser flux (in photons/cm^2/sec) of the focused excitation source, and $S(\mathbf{r})$ is a dimensionless distribution function representing the spatial profile or point spread function (PSF) of the laser excitation. For most fluctuation spectroscopy applications the relevant fluctuation time scales are much longer than the laser pulse width and pulse repetition rate, and one can therefore conveniently work with the average excitation rate

$$\langle W(\mathbf{r}) \rangle = f_p \int W(\mathbf{r}, t)\, dt$$

where f_p is the laser pulse repetition rate and the angular brackets represent the time average quantity (Xu & Webb, 1997).

The measured fluorescence signal from a unit volume is directly proportional to the average molecular excitation rate multiplied by the local molecular concentration, $C(\mathbf{r}, t)$. One can therefore express the total measured fluorescence signal $F(t)$ as:

$$F(t) = \kappa \int \langle W(\mathbf{r}) \rangle C(\mathbf{r}, t) d\mathbf{r}. \tag{25.1}$$

The factor κ accounts for the fluorescence quantum yield and the detection efficiency of the instrumentation. We note that the total measured fluorescence signal is finite for two-photon excitation, even in the absence of confocal detection pinholes, due to the quadratic dependence of the molecular excitation rate on the local excitation power. However, it is important to recognize that the fluorescence signal is measured from an "open" sample region with no physical boundaries.

In the context of fluctuation spectroscopy, it is useful to define a quantity referred to as the "observation volume" that serves as an estimate of the size of the sample region from which most of the measured fluorescence signal arises. This is typically accomplished by dividing the total fluorescence signal of equation (25.1) by the fluorescence per unit volume generated by molecules located at the center of the focused laser beam. Specifically, the measurement volume is defined as:

$$V_{psf} = \frac{1}{\langle W(0) \rangle} \int \langle W(\mathbf{r}) \rangle\, d\mathbf{r} = \int \langle \hat{W}(\mathbf{r}) \rangle d\mathbf{r} \tag{25.2}$$

where $\left\langle \hat{W}(\mathbf{r}) \right\rangle = \langle W(\mathbf{r}) \rangle / \langle W(0) \rangle$ is the normalized fluorescence excitation probability that defines the profile of the observation volume.

Having defined the volume, V_{psf}, it is semantically very convenient to describe molecules as being located either inside or outside of the volume. For lack of better terminology we will routinely use such descriptions in this chapter. However, it is important to remember that V_{psf} is a *normalized probability rather than a container size* and that for the open sample volume defined by two-photon excitation it is not generally possible to rigorously define whether specific molecules reside inside or outside of the volume. Moreover, the actual size of the sample region that contributes significantly to the measured fluorescence signal will be larger than V_{psf}.

Due to the probabilistically defined volume, V_{psf} is not by itself fully sufficient to characterize the observation volume. Higher-order moments of the distribution function $\left\langle \hat{W}(\mathbf{r}) \right\rangle$ that defines the volume are also important for modeling fluctuation spectroscopy measurements. In the context of FCS the additional required parameter is referred to as the gamma factor (Elson & Magde, 1974; Mertz et al., 1995; Thompson, 1991). This parameter is a measure of the uniformity of the fluorescence signal from molecules located at various locations within the volume and the effective steepness of the boundary defining the volume, and is defined as:

$$\gamma = \frac{\int \left\langle \hat{W}(\mathbf{r}) \right\rangle^2 d\mathbf{r}}{\int \left\langle \hat{W}(\mathbf{r}) \right\rangle d\mathbf{r}} \tag{25.3}$$

For the optically defined volumes characteristic of two-photon or confocal microscopy, the gamma factor always has a value less than one. We note that some authors prefer to incorporate the gamma factor into their definition of the volume, defining an effective detection volume as $V_{eff} = V_{psf}/\gamma$ (Mertz et al., 1995; Schwille et al., 1999; Webb, 2001). To avoid confusion in reading the literature, one should be careful to note which definition of the volume is being used. In this manuscript we use volume to refer to V_{psf} as defined in equation (25.2).

Using the volume notation, the average measured fluorescence signal of equation (25.1) can be rewritten conveniently as $\langle F \rangle = \psi \langle C \rangle V_{psf}$. Here we have introduced the molecular brightness parameter, $\psi = \kappa \langle W(0) \rangle$, which depends explicitly on the molecular properties, the excitation conditions (i.e., laser power, pulse width, and beam waist), and the measurement instrumentation. The molecular brightness specifies the average number of fluorescence photons per molecule per second measured from molecules located in the center and at the focal plane of the excitation laser. The molecular brightness is a very useful parameter for investigating molecular interactions as well as for predicting the signal-to-noise ratios of fluctuation spectroscopy measurements, with higher brightnesses leading to better signal statistics (Chen et al., 2002; Koppel, 1974; Saffarian & Elson, 2003). The expression for the total fluorescence is sometimes further simplified as

$$\langle F \rangle = \psi N$$

where N is calculated by multiplying the concentration by the volume and is referred to as the number of molecules within the volume.

However, since V_{psf} does not represent a physical volume, the statement that there are specifically N molecules within the volume, each contributing ψ photons per second, is not conceptually accurate, although it does yield a numerically correct value for the total measured fluorescence signal.

25.2.2 Fluorescence Correlation Spectroscopy

In fluctuation spectroscopy measured fluorescence fluctuations serve as reporters of local concentration fluctuations within the observation volume. As noted above, concentration fluctuations can originate from various sources such as diffusion, and chemical, conformational, or photophysical dynamics that alter the fluorescent properties of the sample molecules. Since the laser excitation conditions are effectively constant on the relevant fluctuation time scales, all the fluctuation dynamics are contained within the concentration term in equation (25.1). Specifically, the fluorescence fluctuations are typically written as

$$\delta F(t) = \psi \int \delta C(\mathbf{r}, t) \hat{W}(\mathbf{r}) d\mathbf{r}$$

where $\delta C(\mathbf{r}, t) = C(\mathbf{r}, t) - \langle C \rangle$ represents the local concentration fluctuations and $\langle C \rangle$ is the average concentration.

Individual fluctuations are random events; therefore, statistical analysis is required to recover useful information from fluctuation measurements. For brevity we will here discuss only autocorrelation and cross-correlation analysis, which are the most mature and widely used fluctuation spectroscopy analysis tools. We note, however, that the broad capabilities of fluctuation spectroscopy continue to advance rapidly, and other approaches to analyzing the fluctuation data can have important advantages for some applications. We briefly discuss some of these alternative analysis procedures at the end of this chapter.

In FCS, the statistical analysis is performed by computing correlation functions. The fluorescence autocorrelation function is generally defined as:

$$G(\tau) = \frac{\langle \delta F(\mathbf{r}, t) \delta F(\mathbf{r}', t + \tau) \rangle}{\langle F \rangle^2} \qquad (25.4)$$

The correlation function mathematically compares the signal with itself on various time scales, τ. Theoretical forms for the correlation function are readily derived and can be used for curve fitting of measured FCS data to recover the desired experimental parameters. While the exact mathematical form of the correlation function depends on the profile of the observation volume as well as the underlying physical dynamics, it is generally possible to represent the normalized correlation function for a single molecular species as:

$$G(\tau) = \frac{\gamma}{\langle C \rangle V_{psf}} A(\tau), \qquad (25.5)$$

where $G(0) = \gamma / \langle C \rangle V_{psf}$ represents the amplitude of the correlation function and $A(\tau)$ represents the temporal relaxation profile ($A(0) = 1$).

The correlation amplitude is thus inversely proportional to the sample concentration. Equation (25.5) can be further generalized to account for both multiple molecular species and multiple detection channels, resulting in a correlation function that has a similar form although weighted by the molecular brightness of the individual species, written:

$$G_{mn}(\tau) = \frac{\gamma}{V} \frac{\sum_i \psi_{i,m} \psi_{i,n} \langle C_i \rangle A_i(\tau)}{\left(\sum_i \psi_{i,m} \langle C_i \rangle\right)\left(\sum_i \psi_{i,n} \langle C_i \rangle\right)} \tag{25.6}$$

where the indices m and n represent detector channels and the index i represents the various molecular species. The quantity $\psi_{i,m}$ therefore represents the molecular brightness of molecular species i in detector channel m.

Equation (25.6) is quite general, accurately describing the signal for both auto- and cross-correlation measurements with multiple molecular species. For autocorrelation measurements the indices m and n refer to the same detector channel, whereas for cross-correlation they refer to two unique detector channels.

25.2.2.1 Analysis of Brownian Diffusion

Diffusion of molecules into and out of the observation volume is the most ubiquitous source of fluorescence fluctuations and thus the best starting point for considering specific mathematical expressions for the correlation function. The correlation function is derived by solving the diffusion equation and pairing that solution with an appropriate expression for the spatial profiles of the observation volume (Berland et al., 1995; Krichevsky & Bonnet, 2002; Thompson, 1991). While wave optics computations are required to generate physically precise representations of the focused laser profile, most fluctuation spectroscopy researchers make use of the much simpler three-dimensional Gaussian (3DG) function, $S_{3DG}(\mathbf{r})$, to describe the laser profile. The true volume profiles are significantly different from the 3DG distribution, yet in most circumstances correlation functions derived from more accurate profiles are not experimentally distinguishable from the 3DG-based correlation functions (Muller et al., 2003). Therefore, with appropriate volume calibration the 3DG-based profiles are sufficient for most applications. In cylindrical coordinates, the 3DG profile is defined as:

$$S_{3DG}(\rho, z) = e^{-2\rho^2/\omega_0^2} e^{-2z^2/z_0^2}$$

with $1/e^2$ radial and axial beam waists ω_0 and z_0 respectively (Qian & Elson, 1990; Rigler et al., 1993).

For two-photon excitation with diffusion coefficient D, the corresponding temporal dependence of the correlation function is found to be $A(\tau) = 1/(1 + 8D\tau/\omega_0^2)(1 + 8D\tau/z_0^2)^{1/2}$, which yields:

$$G(\tau) = \frac{\gamma_{3DG}}{\langle C \rangle V_{3DG}} \frac{1}{(1 + 8D\tau/\omega_0^2)(1 + 8D\tau/z_0^2)^{1/2}} \tag{25.7}$$

with $V_{3DG} = \frac{\pi^{3/2}}{8}\omega_0^2 z_0$ and $\gamma_{3DG} = 1/2\sqrt{2}$ (Berland & Shen, 2003; Schwille et al., 1999). This solution is also sometimes written in terms of a diffusion time defined as

$\tau_D = \omega_0^2/8D$, which is related to the average time a molecule will reside within the observation volume before diffusing out. We note that the diffusion time for two-photon excitation has half the value of the equivalent expression for one-photon excitation. Correlation functions have also been derived that account for transport mechanisms that are not purely diffusive such as flow and anomalous diffusion and may be used in FCS analysis when appropriate (Feder et al., 1996; Kohler et al., 2000; Magde & Elson, 1978; Thompson, 1991).

25.2.2.2 Chemical and Conformational Dynamics

One of the more unusual capabilities of fluctuation spectroscopy is to measure chemical kinetics without perturbing the system away from equilibrium. This is accomplished by measuring the relaxation rate of spontaneous local microscopic fluctuations, which match the relaxation kinetics of bulk sample perturbations. In order for kinetics events to be visible, the reaction investigated must result in some change to the fluorescent properties of the molecules, such as chemical quenching. For example, molecules that undergo chemical transitions between fluorescence and nonfluorescent (i.e., quenched) states will introduce additional blinking fluctuations to the measured fluorescence signal as they pass through the observation volume. The dynamics of the blinking will reflect the kinetics of the underlying transitions between the bright and dark molecular states. In a similar manner, conformational dynamics that result in fluorescence quenching can also be measured using FCS. In general, an exact representation of such dynamics requires a solution of a reaction diffusion equation (Elson & Magde, 1974; Else & Webb, 1975; Krichevsky & Bonnet, 2002; Magde & Elson, 1978; Malvezzi-Campeggi et al., 2001; Thompson, 1991). However, provided the diffusion coefficient is not appreciably altered by the chemical reaction or conformational dynamics, then the correlation function is altered only by a multiplicative term, $G(\tau) = G(\tau)_D \cdot X(\tau)$ (Palmer & Thompson, 1987). Here $G_D(\tau)$ is the correlation function for normal diffusive motion from equation (25.7) and $X(\tau)$ accounts for the chemical or physical kinetic process. For transitions between dark and bright states, this factor can be written explicitly as:

$$X_{\text{kinetics}}(\tau) = \frac{1 - A + A\exp(-\tau/\tau_R)}{1 - A} \tag{25.8}$$

where $\tau_R = 1/(k_d + k_b)$ is the relaxation time of transitions between dark (d) and bright (b) states, k_d is rate of conversion from bright to dark, and k_b is the reverse rate (Malvezzi-Canpeggi et al., 2002; Widengren et al., 1995). The quantity A, not to be confused with $A(\tau)$ from equations (25.5) and (25.6), is the average fraction of molecules in the dark state $A = k_d/(k_d + k_b)$.

   To be observable by FCS, the time scale of the kinetics should typically be on the order of or faster than translational diffusion through the observation volume.

25.2.2.3 Photophysical Dynamics in FCS

Photophysical dynamics such as photobleaching, triplet state crossing, and excitation saturation can each significantly affect FCS measurements. Photobleaching and triplet crossing appear in FCS measurements because each results in an apparent loss of the molecule from the observation volume before it actually leaves the volume. If not properly accounted for in FCS analysis, this can look like faster diffusion processes. Since each effect also reduces the average effective population of fluorescent molecules in the volume, bleaching and triplet crossing can also increase the amplitude of measured FCS curves. The best available model for bleaching or triplet state crossing also results in a multiplicative factor that is mathematically equivalent to equation (25.8), with A again representing the dark fraction (e.g., bleached molecules or triplet state population) (Nagy et al., 2005; Widengren et al., 1995; Widengren & Rigler, 1996). This model has been fairly successful in fitting experimental data, although there are conditions for which it may not be sufficiently accurate (Swift et al., 2004).

Excitation saturation can also play a dramatic role in FCS measurements by altering the size and shape of the observation volume (Cianci et al., 2004; Nagy et al., 2005). This can occur even with relatively low excitation power. One must therefore use caution to properly account for such effects or risk incorrect interpretation of measured data. It is particularly important to pay attention to saturation if one makes measurements with different fluorophores simultaneously, since saturation can result in fluorophore-specific alterations in the instrument calibration. Saturation also causes the instrument calibration to be dependent on excitation power. Significant progress has recently been made in quantitatively describing for these effects, with new physical models accurately accounting for the scaling of the volume calibration with differing excitation conditions (Berland & Shen, 2003; Nagy et al., 2005).

25.3 INSTRUMENTATION AND ANALYSIS

The basics of two-photon microscopy and instrumentation are discussed elsewhere in the volume. The key consideration in performing FCS measurements is that to achieve reasonable signal-to-noise ratios in measured data, one needs to optimize the sensitivity of the instrument and minimize background signals. Even though FCS may look at several molecules at a time, it is really a single molecule measurement technology in that the instrument must have the sensitivity to detect single molecules. Thus, in addition to the standard ultrafast near-infrared lasers used for two-photon microscopy, one must use a combination of high numerical aperture (NA) lenses, high transmission optics, and high sensitivity detectors. The instrument must also contain some mechanism to block stray light from reaching the detectors to keep the background levels minimized. A schematic representation of a typical FCS setup in the authors' lab is shown in Figure 25.2.

25.3.1 Instrumentation

25.3.1.1 Excitation Sources

Commercial mode-locked Ti:sapphire lasers are the standard excitation source for two-photon FCS. The ~80-MHz repetition rates ensure that the excitation is effectively continuous on the time scale of measured fluctuations. It is also important that the laser operate with stable power and pointing, since any significant fluctuations in the excitation could otherwise mask the fluorescence fluctuations one wants to measure. We control the power at the sample by rotating a broadband half-wave plate in front

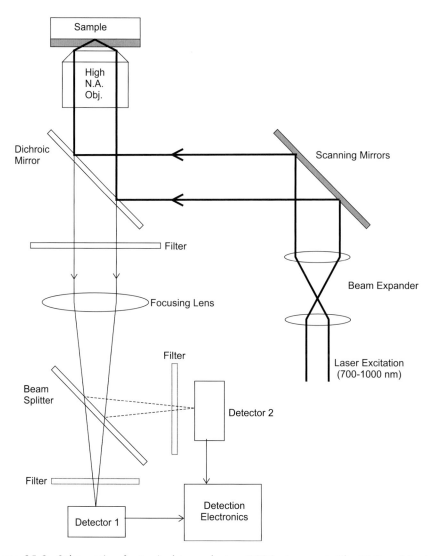

Figure 25.2. Schematic of a typical two-photon FCS instrument. The Ti:Sapphire laser is ported to a microscope with a high NA objective lens in an epi-fluorescence geometry. The collected fluorescence is filtered and delivered to high-sensitivity avalanche photodiode detectors. The pulsed photon counting signals are then delivered to computer data acquisition boards either to record the full photon arrival history or to compute the correlation functions. Details of the components are described in the text and in Berland, 2004.

of a cube polarizer. This ensures pointing stability for varied excitation powers. One generally uses average laser powers below 10 mW at the sample and often just a few milliwatts or less, particularly for live cell work.

25.3.1.2 Microscope

The most important part of the optical configuration is a high NA objective lens. For dual-color applications, one should also use a lens that is very highly corrected for chromatic aberrations. The high NA lens ensures both a small observation volume, which maximizes the correlation amplitude for a given concentration, and high collection efficiency of the fluorescence emission. Best focusing is achieved when the back-aperture of the lens is overfilled, which may require some beam expansion, depending on the particular lens. Our system uses $4\times$ beam expansion before the microscope. In working with immersion lenses one must also take care to avoid bubbles in the immersion fluid, which can alter the focal dimensions and influence the instrument calibration. One should also be aware that the depth of focus into the sample can influence the focused laser profile due to optical aberrations. When possible, it is therefore useful to calibrate the instrument and make measurements at relatively similar sample depths.

Since one needs to achieve single molecule sensitivity, non-descanned detectors are required. One should also select filters and dichroic mirrors with high transmission and strong rejection of scattered laser light. If the microscope contains scanning mirrors for imaging, it is essential that they can be stably parked in a single position for FCS measurements, a condition not attainable on all commercial systems. Use of scanning optics is particularly convenient for live cell applications of FCS, since they allow use of two-photon imaging to identify particular cellular regions of interest in which to park the laser for FCS measurements.

25.3.1.3 Detectors

Photon counting detectors with high quantum efficiency are needed for fluctuation measurements. Currently, the best available option is single photon counting avalanche photodiodes (APDs). They have high sensitivity through most of the visible spectrum, although their blue sensitivity is quite limited. They have low dark count rates and relatively fast response times and also supply a convenient output pulse train that is compatible with most commercial correlator hardware. The major drawback of APDs is that they can have significant afterpulsing that distorts the short time scale (microseconds) part of measured correlation curves. The afterpulsing characteristics can vary significantly between individual APD units and must be accounted for (Enderlein & Gregor, 2005; Hobel & Ricka, 1994; Zhao et al., 2003).

25.3.1.4 Acquisition Electronics

A variety of correlator platforms are commercially available. In recent years, it has become more common to record the full photon arrival history for a complete experiment rather than simply correlation curves. The full photon history allows one to analyze data not only by analyzing correlation functions but also with a variety of powerful distribution analysis tools, which are briefly described at the end of this chapter (Chen et al., 1999; Kask et al., 1999; Muller, 2004).

25.3.1.5 Samples

For fluctuations to be resolved, experiments must be configured such that relatively few molecules are within the observation volume at any given time. For the subfemtoliter volumes characteristic of two-photon excitation, this condition is achieved with sample concentrations in the range from subnanomolar up to a few micromolar. Significantly higher fluorophore concentrations result in prohibitively low correlation amplitudes. For live cell FCS measurements with overexpressed fluorescent fusion proteins, this generally means selecting cells with relatively low expression levels. When working in vitro with such low sample concentration, one must be aware of the possibility of sample losses to the container walls, which may significantly alter the actual sample concentration. This can be minimized to some extent by pretreating surfaces with blocking buffers containing molecules such as casein or bovine serum albumin prior to loading the samples. Dilution experiments can also be helpful for evaluating the extent of sample loss when working with new samples (Muller et al., 2003).

FCS measurements are exquisitely sensitive, and one must therefore also ensure that samples are very clean and free of dust and fluorescent contaminants. Since the contribution of different species to the overall correlation amplitude varies with the square of their molecular brightness, even a few bright contaminant particles can significantly distort the measured correlation data.

25.3.2 Instrument Calibration

25.3.2.1 Calibrating the Beam Waists

Equation (25.7), which describes the temporal relaxation of FCS curves, contains only the ratio of the diffusion coefficient and the squared beam waist—that is, the diffusion time $\tau_D = \omega_0^2/8D$. To uniquely measure a diffusion coefficient, one must therefore measure the beam waist by some other means, which is the purpose of calibrating the instrument. This is achieved by measuring FCS curves for a monodisperse sample with a known diffusion coefficient. Fitting of the calibration FCS curves with the known diffusion coefficient held fixed then allows recovery of the radial and axial beam waists ω_0 and z_0. For subsequent measurements, these quantities are then held fixed while the diffusion coefficient and other relevant system parameters are allowed to vary.

Calibration is generally simple and effective, although one must recognize that many variables can influence the calibration. For example, the beam waists are dependent on excitation wavelength. Also, any saturation or photobleaching that occurs during the calibration measurement will distort the recovered values of the beam waists. The degree of saturation and bleaching will vary for different fluorophores, and perhaps also with the diffusion coefficient of the molecule the fluorophore is attached to. Therefore, if possible the system should be calibrated with the same fluorescent molecule and the same laser power that will be used in subsequent experiments and in a similar environment. When this is not practical, recently introduced models for the scaling of the volume calibration with excitation conditions can be of use in properly calibrating the instrument (Nagy et al., 2005).

25.3.2.2 Molecular Brightness

In addition to calibrating the beam waists, one can also calibrate the molecular brightness ψ of a particular fluorophore. For monodisperse samples, the molecular brightness can be determined from the autocorrelation amplitude $G(0)$ and the average fluorescence intensity $\langle F \rangle$, and is computed as $\psi = \langle F \rangle \cdot G(0)/\gamma_{3DG}$. As noted above, the molecular brightness depends on both intrinsic molecular properties (two-photon cross-section and fluorescence quantum yield) and instrument-dependent parameters (incident light flux and detection efficiency). It specifies the average number of photons detected per second from a molecule in the observation volume, typically reported in counts per second per molecule. Using the same instrument at the same excitation wavelength and the same excitation power, a particular fluorophore should yield the same molecular brightness parameter regardless of concentration. We find that ensuring that the measured brightness matches previously recorded values is a particularly useful procedure to verify proper instrument alignment and calibration. For a carefully maintained system, the molecular brightness varies only minimally from day to day. Determining the molecular brightness of different fluorophores on a particular instrument is also a valuable first step for using FCS to investigate molecular interactions.

25.3.3 Data Analysis

When properly acquired and analyzed, FCS data can be extremely useful for measuring the physical dynamics and interactions of various experimental systems. On the other hand, the mathematical properties of correlation functions are such that measured correlation functions more often than not look fine even when the underlying data used to compute FCS curves are corrupted or not physically meaningful. Accurate information recovery from FCS measurements thus requires detailed knowledge about which types of observed phenomena can be accurately interpreted and resolved by FCS, as well as correct selection of physical models and implementation of curve fitting procedures.

25.3.3.1 Data Rejection

FCS theory is based on the measurement of equilibrium fluctuations, and the theory assumes that measurements are performed for long times relative to the intrinsic sample fluctuation time scales such that they measure large numbers of fluctuation events. This assumption is valid over a wide range of experimental conditions for which FCS analysis works well. However, it is also not uncommon in some systems to observe large short-duration spikes or dips in the fluorescence signal. For example, if protein samples contain a small number of large bright aggregates that pass through the volume only rarely, this will lead to occasional bright bursts of fluorescence that will dramatically alter the measured correlation curves. A variety of phenomena can also cause rare but significant bursts or dips in fluorescence intensity measured from within living cells, such as large nonfluorescent vesicles passing through the volume that temporarily exclude the fluorescent species of interest. Fluorescent spikes and other rare events will generally significantly alter the profile and time scale of FCS curves, yet such rare events do not satisfy the fundamental assumption that one is measuring spontaneous *equilibrium* fluctuations, or the assumption that one has measured very large numbers of fluctuation events. In other words, FCS curves corrupted by large but rare fluctuation events should generally not be analyzed, as the fundamental assumptions of the FCS theory are not met. As a rule of thumb, it is generally safe to conclude that data should be rejected when large spikes are observed during data acquisition that suddenly and significantly alter the measured correlation curves. A thorough discussion of error analysis and bias in FCS measurements was recently presented by Saffarian and colleagues (Saffarian & Elson, 2003). This problem has seen only limited treatment in the literature but is likely the largest source of errors in FCS applications. As such, there is certainly room for additional theoretical progress in treating this topic.

25.3.3.2 Fitting Models

Table 25.1 contains a collection of several of the most common fitting functions used in FCS analysis. Parameters recovered from FCS analysis can be meaningful only if the fitting model accurately reflects the underlying fluctuation dynamics. Often the dynamic processes are known, and curve fitting is then a straightforward process. Caution is warranted in fitting FCS data when the appropriate physical model is not known a priori, since in many cases FCS data can be reasonably well fit with multiple physical models. Experiments may need to be specifically designed in order to help discriminate between different fitting models. For example, chemical dynamics should generally not depend on excitation power, but bleaching or triplet dynamics should. Similarly, chemical dynamics may be concentration dependent, whereas photophysical processes generally are not. New models for photophysical processes in FCS measurements provide an important capability to acquire FCS data under varied excitation conditions, facilitating experiments designed to discriminate between various physical models (Nagy et al., 2005).

25.3.3.3 Curve Fitting

Information recovery from FCS measurements requires curve fitting of measured correlation data to an appropriate physical model for the underlying fluctuations. Nonlinear least-squares algorithms such as Levenberg-Marquardt are typically employed, and the goodness of fit is evaluated using the reduced chi-squared value, given by:

$$\chi_v^2 = \frac{\sum_i \left[(y(x_i) - y_i)^2 / \sigma_i^2 \right]}{v - p} \tag{25.9}$$

where the fitting function values are $y(x_i)$, y_i are values of the experimental data points, and σ_i is the standard deviation of the experimental point i. The number of significant data points and free parameters are given by v and p respectively.

Appropriate determination of experimental errors can play an important role in the accuracy of fitting results. A number of authors have recently discussed appropriate methods for error estimation in FCS (Saffarian & Elson, 2003; Wohland et al., 2001). The simplest approach, which is often used with good success, is to measure several FCS curves for short times and compute the average and standard deviation from this collection of data sets (Wohland et al., 2001).

Curve fitting of FCS data is complicated to a degree by the relatively simple shapes of measured correlation curves. Accurate and stable curve fitting analysis generally requires procedures to limit the total number of free fitting parameters for fitting any given FCS data set. This can sometimes be achieved with independent measurements of certain system parameters prior to full analysis of data sets. Global analysis is another particularly powerful approach for analysis of multiple FCS curves in which some parameter is systematically varied, and global fitting effectively limits the number of independent free parameters, as well.

Table 25.1. Fitting Models for Fluorescence Correlation Spectroscopy

Experimental system	Autocorrelation Function	Parameters	Ref.
One-component diffusion	$G_D(\tau) = \dfrac{\gamma_{3DG}}{CV_{3DG}} \dfrac{1}{(1 + 8D\tau/\omega_0^2)(1 + 8D\tau/z_0^2)^{1/2}} = \dfrac{\gamma_{3DG} A(\tau)}{CV_{3DG}}$	C, D	Schwille, Haupts et al., 1999; Thompson, 1991
Two-component diffusion	$G(\tau) = \dfrac{\gamma_{3DG}}{V_{3DG}} \dfrac{\psi_1^2 C_1 A_1(\tau) + \psi_2^2 C_2 A_2(\tau)}{(\psi_1 C_1 + \psi_2 C_2)^2}$	C_1, C_2, D_1, D_2	Thompson, 1991
Diffusion with triplet state kinetics or chemical kinetics	$G(\tau) = G_D(\tau) \dfrac{1 - A + A \exp(-\tau/\tau_R)}{1 - A}$	C, D, A, τ_R	Widengren, Mets et al., 1995
Diffusion with flow, velocity v	$G(\tau) = G_D(\tau) \exp\left(-2\left(\dfrac{\tau \cdot v}{\omega_0}\right)^2\right)$	C, D, v	Kohler, Schwille et al., 2000; Magde & Elson, 1978
Anomalous diffusion	$G_D(\tau) = \dfrac{\gamma_{3DG}}{CV_{3DG}} (1 + (\tau/\tau_D)^\alpha)^{-1} \left(1 + \dfrac{\omega_0^2}{z_0^2}(\tau/\tau_D)^\alpha\right)^{-1/2}$	C, τ_D, α	(Banks and Fradin, Biophysical J)

   Curve fitting of autocorrelation data can also be complicated by detector afterpulsing effects that systematically corrupt the short time scale correlation data. When the short time data are of interest, one must either use cross-correlation measurements to eliminate the afterpulsing effects or apply correction algorithms to correct for the afterpulsing (Enderlein & Grego, 2005; Hobel & Ricka, 1994; Zhao et al., 2003). When the short time data are not of interest, it is also possible to exclude the distorted short time points from the curve fitting analysis.

25.4 APPLICATIONS OF TWO-PHOTON FCS MEASUREMENTS

With the routine achievement of single molecule sensitivity in fluorescence microscopy, fluctuation spectroscopy techniques have become widely used in a great variety of research applications, and a comprehensive review of FCS applications is beyond the scope of this chapter (Haustein & Schwille, 2003, 2004; Hess et al., 2002; Maiti et al., 1997; Rigler & Elson, 2001; Thompson et al., 2002). We here provide an overview of some of the main classes of FCS experiments to highlight the capabilities of these methods in three major areas: (i) measurements of concentration and molecular mobility in living cells; (ii) measurements of molecular interactions, both in vitro and in living cells; and (iii) measurements of chemical kinetics and conformational dynamics.

25.4.1 Measuring Molecular Mobility and Concentration

25.4.1.1 Mobility

Measurements of molecular concentration and diffusion coefficients (or other mobility-related parameters) are perhaps the most straightforward use of FCS. Molecular mobility can be measured on time scales ranging from microseconds to seconds, for sample concentrations ranging from subnanomolar to micromolar. Such measurement capabilities can be of interest in a wide range of experimental systems, although the possibility of performing FCS measurements within living cells is perhaps among the most exciting directions of this type of research (Berland et al., 1995; Schwille et al., 1999). Two-photon FCS is particularly well suited for applications in living systems due to reduced photobleaching and phototoxicity and provides a powerful tool for addressing questions about cellular function at a molecular level. By investigating the shape of measured correlation curves through curve fitting analysis, one can also determine whether molecular mobility is purely diffusive or exhibits other mobility modes, such as multicomponent diffusion, anomalous diffusion, and active transport or flow. This provides useful characterization of local microenvironments and may also reveal how particular molecules are localized within or trafficked between various regions of the cell. Figure 25.3 shows a series of FCS curves representing different mobility regimes: (i) diffusion motion on time scales characteristic of small and large molecules in solution, in cells, and within membranes; (ii) anomalous diffusion, which can be distinguished by the shape of the curve; and (iii) flow or active transport, which has a steeper relaxation profile than purely diffusive motion.

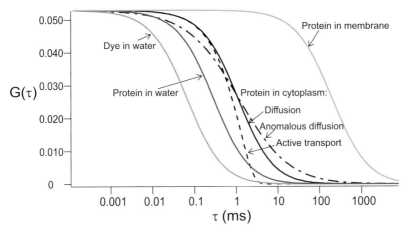

Figure 25.3. Several different mobility regimes in FCS data, including purely diffusive motion, anomalous diffusion, and active transport. The characteristic shape of the correlation curve is fixed with diffusion curves simply shifting in time for different values of the diffusion coefficient. Pure diffusion curves are shown for diffusion coefficients of 0.3, 0.07, 0.018, and 0.0001 $\mu m^2/ms$, representing respectively some characteristic values of D for dye molecules in water, the green fluorescent protein (GFP) in water, GFP in the cytoplasm of a mammalian cell, and protein diffusion in cellular membranes. Also shown for reference are curves for active transport with 2-msec time constant (- - -), and anomalous diffusion with an anomalous exponent of $\gamma = 0.7$ ($- \cdot - \cdot -$). The differences in temporal relaxation profiles for these different physical processes are readily apparent and serve as the basis for discriminating between differing mobility modes in FCS applications. All computed curves were normalized to the same amplitude to emphasize the differences in temporal relaxation profiles. In actual measurements, the amplitude information is also useful for recovering molecular concentrations.

    While still in the early phases, FCS measurements are beginning to make important contributions to our understanding of molecular motion within living cells. An example of novel findings from early live cell FCS measurements is the observation of anomalous diffusion, which can be due to environmental heterogeneities, transient nonspecific interactions with other cellular structures, local confinement (Haustein & Schwille, 2003) or cytoplasmic crowding (Weiss et al., 2004). Other examples include distinguishing between diffusive and actively directed motions of proteins in live cells (Gennerich & Schild, 2000; Kohler et al., 2000) and characterizing the partitioning of protein isoforms between those diffusing freely in the cytoplasm and those partially immobilized on the plasma membrane (Ruan et al., 2002). We note that most live cell applications typically make use of genetically encoded fluorescent fusion proteins (Tsien, 1998), providing the capability to probe the dynamics and concentration of specific molecules without needing to introduce exogenous fluorophores into live cells.

25.4.1.2 Concentration

One advantage of using FCS as an analytical tool is its ability to measure the concentration of fluorescent species, a useful first step in any analytical investigation (Larson et al., 2003). Once the effective volume has been determined by calibrating the beam waists, the average sample concentration $\langle C \rangle$ can be easily calculated from the autocorrelation amplitude $G(0)$. When using the 3DG model to define the observation volume, the average concentration is computed as $\langle C \rangle = \gamma_{3DG}/G(0) \cdot V_{3DG}$. It is difficult to directly measure the $G(0)$ value due to shot noise and detector afterpulsing. Therefore, the best estimate for $G(0)$ is usually recovered by extrapolation of curve fitting results for the measured $G\tau$ to zero time (Berland et al., 1996; Icenogle & Elson, 1983; Palmer & Thompson, 1989).

It should be noted that a number of experimental realities can distort FCS concentration measurements. One source of problems is stray light or scattered laser light reaching the detector, as well as the dark count levels on the detector. These background signal sources do not distort the temporal profile of measured correlation curves, but they can lead to distortions in the correlation amplitude and correspondingly the recovered sample concentration. Provided that one can independently measure these background levels and that they remain constant, one can make a simple correction to determine the actual correlation amplitude. The corrected correlation amplitude is given by:

$$G(0)_{corrected} = G(0)_{measured} \cdot \left(\frac{\langle F \rangle + \langle B \rangle}{\langle F \rangle} \right)^2 \tag{25.10}$$

where the quantity $\langle F \rangle + \langle B \rangle$ is the actual measured signal level including fluorescence and background contributions, and $\langle B \rangle$ is the independently measured background signal level (Schwille et al., 1999; Thompson, 1991).

Background due to ambient lighting is easy to measure by blocking the laser excitation. Background signals due to scattered laser light can be measured by illuminating the sample with the laser out of mode-lock, which effectively eliminates the fluorescence signal but will not alter the scattered light signal.

Concentration measurements in living specimens are sometimes further complicated by two other problems. First, fluorescent background signals that are not due to the fluorophore of interest (e.g., cellular autofluorescence) can be more difficult or impossible to correct for, particularly since such signals may not be constant in time and may exhibit their own diffusive motion. Second, the equations that allow determination of molecular concentrations from the correlation curves contain the implicit assumption that all molecules are mobile and a substantial population of immobile or stationary molecules can significantly distort the recovered concentration. Caution is thus warranted when using FCS to measure concentrations within live cells and tissues. By comparing autofluorescence levels with the average signal level from the fluorophore of interest, one can determine whether autofluorescence will be a major concern or not.

25.4.2 Measuring Molecular Interactions with FCS

Measuring the interactions between specific biomolecules (e.g., protein–protein interactions, protein–ligand and protein–DNA binding) is an extremely important capability for biomedical research, and FCS provides powerful tools to quantify molecular associations both in vitro and in vivo. The two main classes of FCS assays for quantifying molecular interactions with FCS are diffusion-based assays and molecular brightness assays. We describe each of these methods below.

25.4.2.1 Diffusion-Based Assays

Diffusion-based FCS assays are much like fluorescence anisotropy measurements in which the binding of a fluorescent molecule experiences a change in its diffusive behavior upon interaction with a binding partner, with FCS monitoring translational diffusion as opposed to the rotational motion probed by anisotropy (Lakowicz, 1999). These methods are very effective provided there is an appreciable change in the diffusion coefficient of the fluorescently labeled molecule upon binding to its partner. Clear resolution of interactions with this assay requires changes in the diffusion coefficient on the order of a factor of two (Meseth et al., 1999). Since diffusion coefficients are not particularly sensitive to small changes in molecular weight, this approach is typically constrained to systems in which the fluorescent molecule interacts with a much larger molecule, resulting in minimum molecular weight increases from 5 to 10 times upon binding.

Most experimental systems probably contain mixtures of bound and unbound fractions of interacting partners, and the relative fractions will vary as molecular concentrations are titrated. Therefore, full quantitative characterization of binding interactions requires fitting FCS curves for multicomponent diffusion as described in equation (25.6). There are typically at least four parameters for any given FCS measurement (the diffusion coefficient and concentration for both the bound and unbound species), which are too many fitting parameters to get reliable results when fitting individual FCS curves. One effective solution to this problem is to determine the bound and unbound diffusion coefficients in independent measurements, and then use curve fitting with the known diffusion coefficients to recover the concentrations of the bound and unbound fractions. An alternative approach that can also work well is to determine the diffusion coefficient for the unbound molecule independently, and then apply global analysis to fit a series of measured FCS curves (e.g., from a titration experiment) using the diffusion coefficient of the bound species as a global fitting parameter. Using either of these curve fitting strategies together with a series of FCS measurements performed while titrating molecular concentrations also allows for the determination of binding energies, as shown in Figure 25.4, which shows the shift in the FCS curve due to faster diffusion in going from bound to unbound states.

25.4.2.2 Molecular Brightness Assays

While diffusion-based assays are convenient and effective for measuring interactions between molecules with largely different molecular weights, one is very often interested in measuring interactions between molecules of similar or the same molecular weight. Moreover, diffusion-based methods are not particularly useful for characterizing molecular interactions within living cells, since one cannot confidently distinguish interactions from local environmental conditions (e.g., higher viscosity) using diffusion measurements alone. The key advantage of molecular brightness assays is that they do not rely on diffusion measurements explicitly, and thus are independent of molecular weights and viscosity. The basic concept of the molecular brightness assay is quite simple: the single molecule sensitivity of fluctuation spectroscopy methods is exploited to measure the average fluorescence signal from each individually diffusing molecule. A molecular complex with n fluorescent subunits will have a

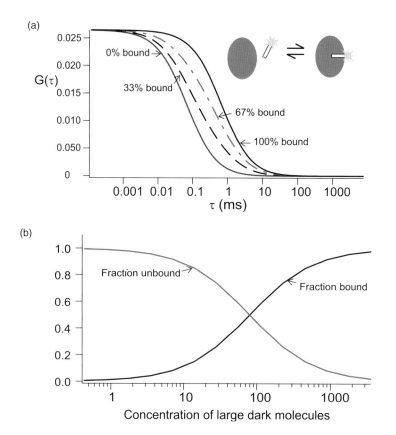

Figure 25.4. Binding assays based on diffusion detect the decrease in the diffusion coefficient upon binding of a small fluorescent molecule to a much larger target molecule. (a) The correlation curves shift towards the right with decreasing diffusion coefficients. Mixtures of bound and unbound species result in two-component diffusion, which has a slightly different relaxation profile from the single-component curves. Computed curves shown are for diffusion coefficients of 0.3 and 0.03 $\mu m^2/ms$ for the free and bound states respectively. Curves for this somewhat dramatic 10× decrease in diffusion coefficient are plotted to allow clear visual resolution of the four curves for fully bound, 67% bound, 33% bound, and fully unbound. Real experiments would generally exhibit significantly smaller increases in measured diffusion coefficients. Changes in D of at least a factor of two are typically required for clear resolution of multiple diffusing species. (b) When a series of diffusion curves are fit for the bound and free fractions as the concentration of the nonfluorescent molecule is titrated, one can then use these values to plot the binding curves and recover dissociation constants and therefore binding energies.

molecular brightness of n times the brightness of the individual species alone. Thus, for example, homodimers of a fluorescent protein would have a molecular brightness equivalent to twice the brightness of the monomers. This principle is demonstrated in Figure 25.5, which shows the ability to distinguish EGFP monomers from EGFP dimers using FCS.

When investigating molecular interactions, samples are typically not monodisperse, but rather contain mixtures of bound and unbound species. In using FCS to measure the molecular brightness as described above, one does not recover the brightnesses

(a)

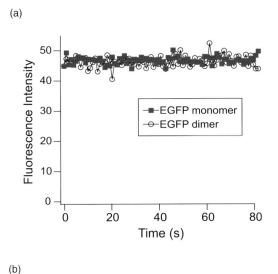

(b)

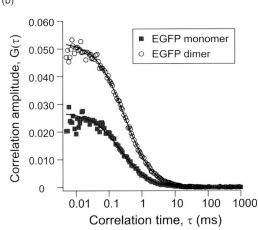

Figure 25.5. Molecular brightness assays can be used to measure molecular interactions. This is demonstrated here using a model system with the enhanced green fluorescent protein (EGFP). Two samples were prepared consisting of either EGFP monomers or EGFP dimers (clone established using a linker peptide), and prepared such that the average fluorescence signal from each sample was equivalent (a). The total protein concentration for each sample is thus the same, although the molar concentration of the monomer sample is twice that of the dimeric sample. This is immediately apparent from the FCS measurements of these two samples (b), which show that the $G(0)$ for the dimeric sample is twice as large as that for the monomeric sample, corresponding to twice the molecular brightness, determined by multiplying the average fluorescence signal by the $G(0)$ value. This interaction could not be clearly resolved using diffusion analysis.

of the individual species, but rather an average brightness for the mixture, which is a weighted average of the component brightnesses. In some cases this is sufficient to determine the sample composition if one knows something about the possible subunit structures. If not, additional analysis techniques such as photon counting histograms may be required to fully resolve the sample composition.

25.4.2.3 Cross-Correlation Spectroscopy

Fluorescence cross-correlation spectroscopy is conceptually similar to standard FCS, but the fluorescence signal is resolved into two distinct detector channels. The detector channels may be set up identically or with unique spectral properties. There are two main motivations for using cross-correlation spectroscopy. The first is to investigate fast dynamics, for which the short time autocorrelation values can be corrupted by detector afterpulsing. Detector afterpulsing is not a problem in cross-correlation measurements since the afterpulses from the two independent detectors are uncorrelated. For this style of cross-correlation measurement, two identical detection channels are used with a 50-50 beam splitter to send half the fluorescence signal to each detector. This approach can offer an important advantage in specialized applications, although it does reduce measured signal-to-noise ratios since the molecular brightness will be half its normal value in each detector channel.

Much more common in the context of biomedical research is the use of dual-color cross-correlation spectroscopy, which is used to detect the interactions between molecular species labeled with spectrally distinct fluorophores (Berland, 2004; Eigen & Rigler, 1994; Schwille et al., 1997). The idea of the cross-correlation measurement is that only molecular complexes that contain both color subunits can contribute to the cross-correlation signal, and the amplitude of the cross-correlation thus serves as a direct reporter of the dual-color species concentration. Figure 25.6 conceptualizes a dual-color cross-correlation experiment. There are several important considerations in

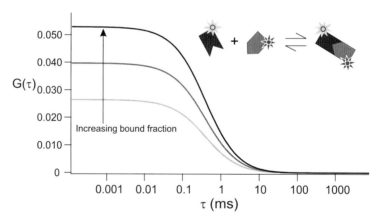

Figure 25.6. In dual-color cross-correlation spectroscopy, interacting molecular species are labeled with spectrally distinct fluorescent probes that are resolved in two separate detector channels. In contrast to the detection of interactions using diffusion, here both molecular species are fluorescently labeled. The cross-correlation signal between the two channels is then computed. Fluctuations in the two channels can be correlated only if the two different color species are physically interacting. Dual-color measurements partially resemble the molecular brightness assays in that it is the amplitude of the cross-correlation curves that represents the degree of molecular interactions, with the cross-correlation amplitude increasing with increasing molecular association between the dual-color species. For cases where it may be of interest, the diffusion coefficient of the dual-color complex can also be recovered from the temporal relaxation of the cross-correlation curves.

designing dual-color experiments, including the need to ensure simultaneous excitation of each probe with similar efficiency and the capability to spectrally resolve the two fluorescence signals with minimal cross-talk between the detector channels (Berland, 2004; Heinze et al., 2000). Two-photon excitation can be particularly advantageous for dual-color measurements since it is possible to find spectrally distinct fluorophores that are easily excited with a single two-photon excitation wavelength (Berland, 2004; Kim et al., 2005; Schwille et al., 1997). Dual-color FCS measurements are particularly promising for measurements of mobility and interactions within living cells, although these measurements are still challenging due to the spectral overlap of genetically encoded fluorescent proteins. Recent progress has been made using one color fluorescent protein together with a second dye-labeled molecule introduced into the cell using cell loading techniques (Kim et al., 2005).

25.4.3 Conformational Dynamics and Binding Kinetics

FCS was introduced to investigate the binding kinetics of ethidium bromide to DNA (Magde et al., 1972). The ability to measure chemical kinetics and conformational dynamics in samples that are not perturbed away from their equilibrium states remains one of the more powerful and unusual capabilities of FCS measurements. The key idea is that any chemical or conformational reactions that result in modified fluorescence properties of the component molecules (e.g., changes in quantum yield) will introduce additional fluorescence fluctuations that occur as the molecules traverse the observation volume. These fluctuations are distinct from the fluctuations due to diffusion in and out of the volume, and their time scales will reflect the transitions between different states, as illustrated in Figure 25.7. Given an appropriate model for the underlying dynamics, the reaction rates are easily recovered from FCS analysis. Such measurements have been particularly useful for investigating the photophysical properties of fluorescent proteins and dyes (Doose et al., 2005; Haupts et al., 1998; Heikal et al., 2001; Hess et al., 2004; Malvezzi-Campeggi et al., 2001; Schenk et al., 2004; Schwille et al., 2000; Widengren et al., 1995; Widengren & Schwille, 2000). With appropriately prepared fluorescence labeling conditions, such that conformational changes result in fluorescence quenching or FRET, FCS has also proven to be a powerful method to investigate the conformational dynamics of proteins and nucleic acids (Bonnet et al., 1998; Chattopadhyay et al., 2002, 2005). One constraint on the use of such methods is that the time scale of the reaction is limited by the diffusion time of the molecule through the observation volume, typically on the order of a millisecond. Faster reaction rates are easily resolved, but intermediate-rate kinetics are often more difficult to access using FCS.

25.5 OTHER FLUORESCENCE FLUCTUATION SPECTROSCOPY TOOLS

FCS is the most mature of the fluctuation spectroscopy methods, but many alternative analysis procedures have recently been introduced that can provide important advantages in certain applications and thus provide an important complement to FCS measurements. These include such methods as photon counting histograms (PCH) and

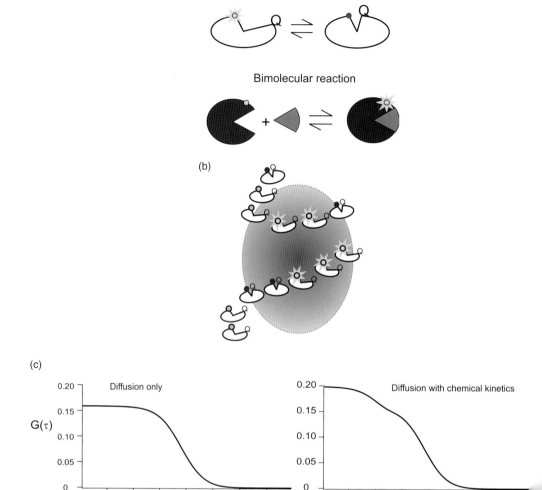

Figure 25.7. Chemical kinetics and conformational dynamics can be measured using FCS, provided that the reactions involved result in a change to the fluorescent properties of the component molecules. (a) illustrates two such scenarios. First, a molecule is labeled with a fluorophore and quencher (Q) such that the fluorescence is quenched via a FRET interaction for one conformation and unquenched for the second conformation. Second, a chemical reaction or interaction can also modify the fluorescence yield of a molecule. In either case, the interconversion rates of the molecule between bright and dark states will reflect the chemical or conformational dynamics of the molecule. As the molecule traverses the observation volume via diffusive motion (b), the fluorescence signal will flicker due to conversion between the bright and dark states. Analysis of these additional fluctuation time scales allows recovery of the kinetic parameters associated with the forward and backward reaction rates. (c) A comparison of autocorrelation curves for purely diffusive motion and for diffusion with chemical kinetics. The chemical/conformational dynamics were computed for a kinetic relaxation time of 10 μs. The diffusion time in each case is 670 μs. Banks, Daniel S. and Cecile Fradin (2005). "Anomalous Diffusion of Proteins Due to Molecular Crowding." Biophysical Journal, 89, 2960–2971.

related distribution analysis methods, cumulant analysis, and various forms of image correlation spectroscopy (Chen et al., 1999; Digman, Brown et al., 2005; Hebert et al., 2005; Heinze et al., 2002; Kask et al., 1999; Laurence et al., 2004; Muller, 2004; Petersen et al., 1993; Winkler et al., 1999; Wiseman & Petersen, 1999). Each of these methods are similar to FCS in that fluorescence fluctuations are measured from well-defined observation volumes and the fluctuation are recorded either as a function of time or position in space. The difference thus lies mainly in the approach to analyzing the fluctuation data. A detailed discussion of these methods is beyond the scope of this work, but we briefly summarize some of the general advantages of alternate approaches and refer the reader to the literature for additional details.

25.5.1 Distribution Analysis

Photon counting histograms (PCH) and related distribution analysis methods focus on the amplitude distributions in fluorescence fluctuations (Chen et al., 1999; Kask et al., 1999). The key advantage of these methods is the possibility of resolving populations of molecular species with differing molecular brightnesses but similar diffusion times. A similar approach is taken in the newly introduced cumulant analysis (Muller, 2004). Each of these methods has proven quite effective for characterizing molecular interactions both in vivo and in vitro (Chen, Muller et al., 2000, 2002; Chen, Tekmen et al., 2005; Chen, Wei et al., 2003, 2005; Muller, Chen et al., 2000). These methods have also recently been extended to make use of dual-color approaches (Chen, Tekmen et al., 2005).

25.5.2 Image Correlation Spectroscopy

In many cellular systems, particularly for membrane-bound molecules, diffusion is too slow for efficient FCS measurements. Image correlation spectroscopy (ICS), an alternative approach based on FCS ideas but monitoring spatial fluctuations rather than temporal fluctuations, has also proven to be very useful for investigating molecular dynamics and interactions in living cells (Broek et al., 1999; Petersen et al., 1993; Rocheleau & Petersen, 2000; Wiseman, Capani et al., 2002; Wiseman & Petersen, 1999; Wiseman, Squier et al., 2000). Although significantly slower than for standard FCS applications, ICS can also achieve temporal resolution appropriate for membrane-bound molecules by correlating successively acquired image frames, providing spatially resolved dynamic information (Hebert et al., 2005; Srivastava & Petersen, 1998; Wiseman, Brown et al., 2000; Wiseman, Squier et al., 2000). Recently, FCS and scanning approaches have begun to converge with the introduction of methods that exploit the hidden temporal information encoded within raster-scanned images that arises due to the time delays between acquiring data at different pixels (Digman, Brown et al., 2005; Digman, Sengupta et al., 2005).

25.6 SUMMARY

FCS and related fluctuation spectroscopy techniques are versatile tools for investigating molecular dynamics and interactions in a variety of experimental systems, both in vitro and in vivo. Two-photon FCS couples single molecule sensitivity with inherent depth discrimination and minimal biological invasiveness, providing an important and practical research tool for biomedical investigations within living cells. We have introduced the basic theory behind these measurements as well as some of the key applications of these methods with the goal of making the FCS literature more accessible to the readers. We have also briefly introduced some alternative fluctuation spectroscopy analysis tools, as these methods provide important complements to FCS analysis procedures. Many exciting applications of fluctuation spectroscopy as well as many important practical considerations are beyond the scope of this chapter. Readers planning to implement these methods are encouraged to consult the technical literature for additional information and experimental protocols.

ACKNOWLEDGMENT

The authors are partially supported by the National Institutes of Health and Emory University.

REFERENCES

Bacia, K. and P. Schwille (2003). "A dynamic view of cellular processes by in vivo fluorescence auto- and cross-correlation spectroscopy." Methods 29(1): 74–85.

Banks, Daniel S. and Cecile Fradin (2005). "Anomalous Diffusion of Proteins Due to Molecular Crowding." Biophysical Journal, 89, 2960–2971.

Berland, K. and G. Q. Shen (2003). "Excitation saturation in two-photon fluorescence correlation spectroscopy." Applied Optics 42(27): 5566–5576.

Berland, K. M. (2004). "Detection of specific DNA sequences using dual-color two-photon fluorescence correlation spectroscopy." Journal Of Biotechnology 108(2): 127–136.

Berland, K. M., P. T. C. So, et al. (1996). "Scanning two-photon fluctuation correlation spectroscopy: Particle counting measurements for detection of molecular aggregation." Biophysical Journal 71(1): 410–420.

Berland, K. M., P. T. C. So, et al. (1995). "2-Photon Fluorescence Correlation Spectroscopy— Method And Application To The Intracellular Environment." Biophysical Journal 68(2): 694–701.

Bonnet, G., O. Krichevsky, et al. (1998). "Kinetics of conformational fluctuations in DNA hairpin-loops." Proceedings Of The National Academy Of Sciences Of The United States Of America 95(15): 8602–8606.

Broek, W. V., Z. P. Huang, et al. (1999). "High-order autocorrelation with imaging fluorescence correlation spectroscopy: Application to IgE on supported planar membranes." Journal Of Fluorescence 9(4): 313–324.

Chattopadhyay, K., E. L. Elson, et al. (2005). "The kinetics of conformational fluctuations in an unfolded protein measured by fluorescence methods." Proceedings Of The National Academy Of Sciences Of The United States Of America 102(7): 2385–2389.

Chattopadhyay, K., S. Saffarian, et al. (2002). "Measurement of microsecond dynamic motion in the intestinal fatty acid binding protein by using fluorescence correlation spectroscopy." Proceedings Of The National Academy Of Sciences Of The United States Of America 99(22): 14171–14176.

Chen, Y., J. D. Muller, et al. (2001). Two-photon Fluorescence Fluctuation Spectroscopy. New Trends in Fluorescence Spectroscopy: Applications to Chemical and Life Sciences. B. Valeur and J.-C. Brochon. Berlin, Springer-Verlag.

Chen, Y., J. D. Muller, et al. (2002). "Molecular brightness characterization of EGFP in vivo by fluorescence fluctuation spectroscopy." Biophysical Journal 82(1): 133–144.

Chen, Y., J. D. Muller, et al. (1999). "The photon counting histogram in fluorescence fluctuation spectroscopy." Biophysical Journal 77(1): 553–567.

Chen, Y., J. D. Muller, et al. (2000). "Probing ligand protein binding equilibria with fluorescence fluctuation spectroscopy." Biophysical Journal 79(2): 1074–1084.

Chen, Y., M. Tekmen, et al. (2005). "Dual-color photon-counting histogram." Biophysical Journal 88(3): 2177–2192.

Chen, Y., L.-N. Wei, et al. (2005). "Unraveling Protein-Protein Interactions in Living Cells with Fluorescence Fluctuation Brightness Analysis." Biophysical Journal 88: 4366–4377.

Chen, Y., L. N. Wei, et al. (2003). "Probing protein oligomerization in living cells with fluorescence fluctuation spectroscopy." Proceedings Of The National Academy Of Sciences Of The United States Of America 100(26): 15492–15497.

Cianci, G. C., J. Wu, et al. (2004). "Saturation Modified Point Spread Functions in Two-Photon Microscopy." Microscopy Research and Technique 64: 135–141.

Denk, W., J. H. Strickler, et al. (1990). "2-Photon Laser Scanning Fluorescence Microscopy." Science 248(4951): 73–76.

Digman, M. A., C. M. Brown, et al. (2005). "Measuring Fast Dynamics in Solutions and Cells with a Laser Scanning Microscope." Biophysical Journal 89: 1317–1327.

Digman, M. A., P. Sengupta, et al. (2005). "Fluctuation correlation spectroscopy with a laser-scanning microscope: Exploiting the hidden time structure." Biophysical Journal 88(5): L33-L36.

Doose, S., J. M. Tsay, et al. (2005). "Comparison of photophysical and colloidal properties of biocompatible semiconductor nanocrystals using fluorescence correlation spectroscopy." Analytical Chemistry 77(7): 2235–2242.

Eigen, M. and R. Rigler (1994). "Sorting Single Molecules—Application To Diagnostics And Evolutionary Biotechnology." Proceedings Of The National Academy Of Sciences Of The United States Of America 91(13): 5740–5747.

Elson, E. L. and D. Magde (1974). "Fluorescence Correlation Spectroscopy.1. Conceptual Basis And Theory." Biopolymers 13(1): 1–27.

Elson, E. L. and W. W. Webb (1975). "Concentration Correlation Spectroscopy - New Biophysical Probe Based On Occupation Number Fluctuations." Annual Review Of Biophysics And Bioengineering 4: 311–334.

Enderlein, J. and I. Gregor (2005). "Using fluorescence lifetime for discriminating detector afterpulsing in fluorescence-correlation spectroscopy." Review Of Scientific Instruments 76(3).

Feder, T. J., I. BrustMascher, et al. (1996). "Constrained diffusion or immobile fraction on cell surfaces: A new interpretation." Biophysical Journal 70(6): 2767–2773.

Gennerich, A. and D. Schild (2000). "Fluorescence correlation spectroscopy in small cytosolic compartments depends critically on the diffusion model used." Biophysical Journal 79(6): 3294–3306.

Haupts, U., S. Maiti, et al. (1998). "Dynamics of fluorescence fluctuations in green fluorescent protein observed by fluorescence correlation spectroscopy." Proceedings Of The National Academy Of Sciences Of The United States Of America 95(23): 13573–13578.

Haustein, E. and P. Schwille (2003). "Ultrasensitive investigations of biological systems by fluorescence correlation spectroscopy." Methods 29(2): 153–166.

Haustein, E. and P. Schwille (2004). "Single-molecule spectroscopic methods." Current Opinion In Structural Biology 14(5): 531–540.

Hebert, B., S. Costantino, et al. (2005). "Spatiotemporal image correlation Spectroscopy (STICS) theory, verification, and application to protein velocity mapping in living CHO cells." Biophysical Journal 88(5): 3601–3614.

Heikal, A. A., S. T. Hess, et al. (2001). "Multiphoton molecular spectroscopy and excited-state dynamics of enhanced green fluorescent protein (EGFP): acid-base specificity." Chemical Physics 274(1): 37–55.

Heinze, K. G., A. Koltermann, et al. (2000). "Simultaneous two-photon excitation of distinct labels for dual-color fluorescence crosscorrelation analysis." Proceedings Of The National Academy Of Sciences Of The United States Of America 97(19): 10377–10382.

Heinze, K. G., M. Rarbach, et al. (2002). "Two-photon fluorescence coincidence analysis: Rapid measurements of enzyme kinetics." Biophysical Journal 83(3): 1671–1681.

Hess, S. T., A. A. Heikal, et al. (2004). "Fluorescence photoconversion kinetics in novel green fluorescent protein pH sensors (pHluorins)." Journal Of Physical Chemistry B 108(28): 10138–10148.

Hess, S. T., S. H. Huang, et al. (2002). "Biological and chemical applications of fluorescence correlation spectroscopy: A review." Biochemistry 41(3): 697–705.

Hobel, M. and J. Ricka (1994). "Dead-Time And Afterpulsing Correction In Multiphoton Timing With Nonideal Detectors." Review Of Scientific Instruments 65(7): 2326–2336.

Icenogle, R. D. and E. L. Elson (1983). "Fluorescence Correlation Spectroscopy And Photo-bleaching Recovery Of Multiple Binding Reactions.1. Theory And FCS Measurements." Biopolymers 22(8): 1919–1948.

Kask, P., K. Palo, et al. (1999). "Fluorescence-intensity distribution analysis and its application in biomolecular detection technology." Proceedings Of The National Academy Of Sciences Of The United States Of America 96(24): 13756–13761.

Kim, S. A., K. G. Heinze, et al. (2005). "Two-Photon Cross-Correlation Analysis of Intracellular Reactions with Variable Stoichiometry." Biophysical Journal 88: 4319–4336.

Kohler, R. H., P. Schwille, et al. (2000). "Active protein transport through plastid tubules: velocity quantified by fluorescence correlation spectroscopy." Journal Of Cell Science 113: 3921–3930.

Koppel, D. E. (1974). "Statistical Accuracy In Fluorescence Correlation Spectroscopy." Physical Review A 10(6): 1938–1945.

Krichevsky, O. and G. Bonnet (2002). "Fluorescence correlation spectroscopy: the technique and its applications." Reports on Progress in Physics 65: 251–297.

Lakowicz, J. R. (1999). Principles of Fluorescence Spectroscopy. New York, Kluwer Academic/Plenum Publishers.

Larson, D. R., W. R. Zipfel, et al. (2003). "Water-soluble quantum dots for multiphoton fluorescence imaging in vivo." Science 300(5624): 1434–1436.

Laurence, T. A., A. N. Kapanidis, et al. (2004). "Photon arrival-time interval distribution (PAID): A novel tool for analyzing molecular interactions." Journal Of Physical Chemistry B 108(9): 3051–3067.

Magde, D. and E. L. Elson (1978). "Fluorescence Correlation Spectroscopy.3. Uniform Translation And Laminar-Flow." Biopolymers 17(2): 361–376.

Magde, D., W. W. Webb, et al. (1972). "Thermodynamic Fluctuations In A Reacting System - Measurement By Fluorescence Correlation Spectroscopy." Physical Review Letters 29(11): 705–708.

Maiti, S., U. Haupts, et al. (1997). "Fluorescence correlation spectroscopy: Diagnostics for sparse molecules." Proceedings Of The National Academy Of Sciences Of The United States Of America 94(22): 11753–11757.

Malvezzi-Campeggi, F., M. Jahnz, et al. (2001). "Light-induced flickering of DsRed provides evidence for distinct and interconvertible fluorescent states." Biophysical Journal 81(3): 1776–1785.

Mertz, J., C. Xu, et al. (1995). "Single-molecule detection by two-photon-excited fluorescence." Opt. Lett. 20(24): 2532–4.

Meseth, U., T. Wohland, et al. (1999). "Resolution of fluorescence correlation measurements." Biophysical Journal 76(3): 1619–1631.

Muller, J. D. (2004). "Cumulant analysis in fluorescence fluctuation spectroscopy." Biophysical Journal 86(6): 3981–3992.

Muller, J. D., Y. Chen, et al. (2000). "Resolving heterogeneity on the single molecular level with the photon-counting histogram." Biophysical Journal 78(1): 474–486.

Muller, J. D., Y. Chen, et al. (2003). Fluorescence correlation spectroscopy. Methods In Enzymology. 361: 69–92.

Nagy, A., J. Wu, et al. (2005). "Observation Volumes and gamma-Factors in Two-Photon Fluorescence Fluctuation Spectroscopy." Biophysical Journal 89: 2077–2090.

Palmer, A. G. and N. L. Thompson (1987). "Theory Of Sample Translation In Fluorescence Correlation Spectroscopy." Biophysical Journal 51(2): 339–343.

Palmer, A. G. and N. L. Thompson (1989). "Fluorescence Correlation Spectroscopy For Detecting Submicroscopic Clusters Of Fluorescent Molecules In Membranes." Chemistry And Physics Of Lipids 50(3–4): 253–270.

Petersen, N. O., P. L. Hoddelius, et al. (1993). "Quantitation Of Membrane-Receptor Distributions By Image Correlation Spectroscopy - Concept And Application." Biophysical Journal 65(3): 1135–1146.

Qian, H. and E. L. Elson (1990). "Analysis of confocal laser-microscope optics for 3-D fluorescence correlation spectroscopy." Applied Optics 30(10): 1185–95.

Rigler, R. and E. Elson, Eds. (2001). Fluorescence Correlation Spectroscopy Theory and Applications. Springer Series in Chemical Physics. New York, Springer.

Rigler, R., U. Mets, et al. (1993). "Fluorescence Correlation Spectroscopy With High Count Rate And Low-Background - Analysis Of Translational Diffusion." European Biophysics Journal With Biophysics Letters 22(3): 169–175.

Rocheleau, J. V. and N. O. Petersen (2000). "Sendai virus binds to a dispersed population of NBD-GD1a." Bioscience Reports 20(3): 139–155.

Ruan, Q. Q., Y. Chen, et al. (2002). "Cellular Characterization of adenylate kinase and its isoform: Two-photon excitation fluorescence imaging and Fluorescence correlation spectroscopy." Biophysical Journal 83: 3177–3187.

Saffarian, S. and E. Elson (2003). "Statistical analysis of Fluorescence Correlation Spectroscopy: The Standard Deviation and Bias." Biophys. J 84: 2030–2042.

Schenk, A., S. Ivanchenko, et al. (2004). "Photodynamics of red fluorescent proteins studied by fluorescence correlation spectroscopy." Biophysical Journal 86(1): 384–394.

Schwille, P. (2001). "Fluorescence Correlation Spectroscopy and Its Potential for Intracellular Applications." Cell Biochemistry and Biophysics 34: 383–408.

Schwille, P., U. Haupts, et al. (1999). "Molecular dynamics in living cells observed by fluorescence correlation spectroscopy with one- and two-photon excitation." Biophysical Journal 77(4): 2251–2265.

Schwille, P., J. Korlach, et al. (1999). "Fluorescence correlation spectroscopy with single-molecule sensitivity on cell and model membranes." Cytometry 36(3): 176–182.

Schwille, P., S. Kummer, et al. (2000). "Fluorescence correlation spectroscopy reveals fast optical excitation-driven intramolecular dynamics of yellow fluorescent proteins." Proceedings Of The National Academy Of Sciences Of The United States Of America 97(1): 151–156.

Schwille, P., F. J. MeyerAlmes, et al. (1997). "Dual-color fluorescence cross-correlation spectroscopy for multicomponent diffusional analysis in solution." Biophysical Journal 72(4): 1878–1886.

Srivastava, M. and N. O. Petersen (1998). "Diffusion of transferrin receptor clusters." Biophysical Chemistry 75(3): 201–211.

Swift, K. M., S. A. Beretta, et al. (2004). "Triplet state effects in the application of FCS to binding equilibria." Biophysical Journal 86(1): 159A–159A.

Thompson, N. L. (1991). Fluorescence correlation spectroscopy. Topics in Fluorescence Spectroscopy. J. R. Lakowicz. New York, Plenum: 337–378.

Thompson, N. L., A. M. Lieto, et al. (2002). "Recent advances in fluorescence correlation spectroscopy." Current Opinion In Structural Biology 12(5): 634–641.

Tsien, R. Y. (1998). "The green fluorescent protein." Annu. Rev. Biochem. 67: 509–544.

Webb, W. W. (2001). "Fluorescence Correlation Spectroscopy: Inception, Biophysical Experimentations, and Prospectus." Applied Optics 40(24): 3969–3983.

Widengren, J., U. Mets, et al. (1995). "Fluorescence Correlation Spectroscopy Of Triplet-States In Solution - A Theoretical And Experimental-Study." Journal Of Physical Chemistry 99(36): 13368–13379.

Widengren, J. and R. Rigler (1996). "Mechanisms of photobleaching investigated by fluorescence correlation spectroscopy." Bioimaging 4(3): 149–157.

Widengren, J. and P. Schwille (2000). "Characterization of photoinduced isomerization and back-isomerization of the cyanine dye Cy5 by fluorescence correlation spectroscopy." Journal of Physical Chemistry A 104(27): 6416–6428.

Winkler, T., U. Kettling, et al. (1999). "Confocal fluorescence coincidence analysis: An approach to ultra high-throughput screening." Proceedings Of The National Academy Of Sciences Of The United States Of America 96(4): 1375–1378.

Wiseman, P. W., C. M. Brown, et al. (2004). "Spatial mapping of integrin interactions and dynamics during cell migration by Image Correlation Microscopy." Journal Of Cell Science 117(23): 5521–5534.

Wiseman, P. W., F. Capani, et al. (2002). "Counting dendritic spines in brain tissue slices by image correlation spectroscopy analysis." Journal Of Microscopy-Oxford 205: 177–186.

Wiseman, P. W. and N. O. Petersen (1999). "Image correlation spectroscopy. II. Optimization for ultrasensitive detection of preexisting platelet-derived growth factor-beta receptor oligomers on intact cells." Biophysical Journal 76(2): 963–977.

Wiseman, P. W., J. A. Squier, et al. (2000). "Two-photon image correlation spectroscopy and image cross-correlation spectroscopy." Journal Of Microscopy-Oxford 200: 14–25.

Wohland, T., R. Rigler, et al. (2001). "The standard deviation in fluorescence correlation spectroscopy." Biophysical Journal 80(6): 2987–2999.

Xu, C. and W. W. Webb (1997). Multiphoton excitation of molecular fluorophores and nonlinear laser microscopy. Topics in Fluorescence Spectroscopy. J. Lakowicz, Plenum. 5: 471–540.

Zhao, M., L. Jin, et al. (2003). "Afterpulsing and its correction in fluorescence correlation spectroscopy experiments." Applied Optics 42(19): 4031–4036.

26

Photobleaching and Recovery with Nonlinear Microscopy

Edward Brown, Ania Majewska,
and Rakesh K. Jain

26.1 INTRODUCTION

Fluorescence recovery after photobleaching (FRAP), or fluorescence photobleaching recovery (FPR), describes a family of related techniques that measure transport properties of fluorescently labeled molecules. This is done by monitoring the evolution of a fluorescence signal after a spatially localized population of fluorophores is bleached by light. The classical FRAP experiment with one-photon excitation uses a focused laser beam to bleach a disk-shaped region of fluorescently labeled molecules in a thin sample such as a cell membrane or a lamellipodium (Fig. 26.1) (Axelrod et al., 1976). The same laser beam, greatly attenuated, generates fluorescence signal from that region as unbleached fluorophores diffuse in. The recovery in fluorescence signal is recorded by a photomultiplier tube (PMT) or similar detector, producing a fluorescence-versus-time curve (Fig. 26.2). Simple analytical formulas can often fit the fluorescence recovery curve, with the recovery time revealing the two-dimensional diffusion coefficient of the fluorescent molecule, and the extent of fluorescence recovery revealing the fraction of fluorophores that are mobile. More complex analysis with multiple diffusion coefficients can be used if the dynamics of the system warrant (Feder et al., 1996; Periasamy & Verkman, 1998). The three-dimensional hourglass profile of a focused laser beam (Born & Wolf, 1980) renders analytical solutions difficult unless the sample is relatively thin along the optical axis of the laser beam. Therefore, conventional one-photon FRAP is often limited to analysis of diffusion in thin two-dimensional systems such as cell membranes, where the intersection of the membrane with the laser beam produces a simple disk of photobleaching. In spite of this limitation, conventional one-photon FRAP has provided extensive insight into myriad cellular processes involving lateral membrane diffusion (Brown et al., 2005; Reits & Neefjes, 2001).

One-photon FRAP can be accurately extended to thick, three-dimensional samples as fluorescence recovery after photobleaching with spatial Fourier analysis (SFA-FRAP), which involves imaging the bleached distribution using an epifluorescence microscope and a CCD camera (Fig. 26.3). After the photobleaching laser pulse, the CCD camera and microscope generates a series of images of the bleached region, each of which is a two-dimensional representation of the complex three-dimensional bleached distribution, generated by looking "down the barrel" of the bleached hourglass. The spatial

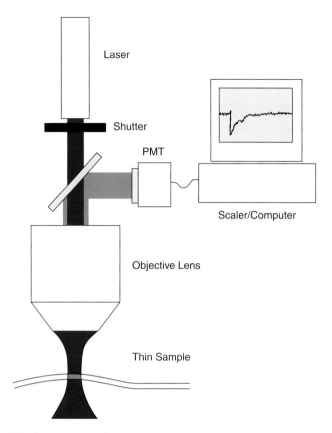

Figure 26.1. FRAP instrumentation.

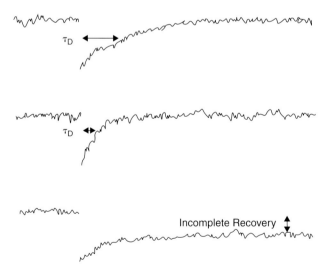

Figure 26.2. Cartoon of typical FRAP curves. The characteristic recovery time D reveals the diffusion coefficient of the fluorescent molecule, with rapidly diffusing molecules producing shorter recovery times (second curve). An incomplete fluorescence recovery after a single FRAP experiment reveals the presence of a subpopulation of immobile fluorophores (i.e., the "immobile fraction") (third curve).

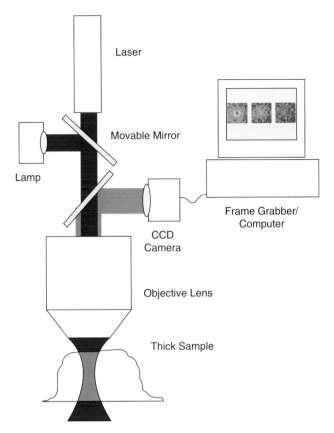

Figure 26.3. SFA-FRAP instrumentation.

Fourier transform of each image is calculated, and the temporal evolution of the various spatial frequency components directly yields the diffusion coefficient of the fluorescent molecule in the region of interest (Berk et al., 1993; Tsay & Jacobson, 1991). Neither the bleaching nor the imaging is axially confined; therefore, the resultant diffusion coefficient is an average over the visible depth of the sample and is not three-dimensionally resolved. SFA-FRAP has been used to measure diffusion and convection in 50-micron-thick normal and tumor tissue growing in the rabbit ear chamber (Chary & Jain, 1989). It has also been used to study the characteristic pore size of the tumor extracellular matrix (Pluen et al., 2001), the effects of matrix modification (Brown et al., 2003), as well as the diffusive properties of tumor biopsies (Brown et al., 2004) and hydrogels (Kosto et al., 2004). More recently, it has demonstrated that artificial cartilage integrates and matures better when cultured adjacent to tissues with high diffusivities (Tognana et al., 2005).

   In summary, to determine diffusion coefficients with high spatial resolution, single-photon FRAP is limited to thin samples. Measurement of diffusion coefficients with low spatial resolution in thick samples is possible with SFA-FRAP. However, neither of these techniques produces diffusion coefficients with high three-dimensional resolution, nor can they penetrate beyond the relatively shallow $\sim$100-μm imaging depth of conventional epifluorescence imaging.

26.2 MULTIPHOTON FRAP

Multiphoton FRAP (MPFRAP) (Brown et al., 1999) is performed on a modified multiphoton laser-scanning microscope (Fig. 26.4) and uses the intrinsic spatial confinement of multiphoton excitation (Denk et al., 1990) to measure diffusion coefficients with three-dimensional resolution in thick samples. A high-intensity bleaching flash from a mode-locked laser bleaches out a fraction of fluorophores at the laser focus within a three-dimensionally confined volume of less than a femtoliter (10^{-15} L) via multiphoton excitation. The same laser beam, greatly attenuated, generates fluorescence from the focal volume as unbleached fluorophores diffuse in, and the recovery in fluorescence signal is recorded by a PMT, producing a fluorescence-versus-time curve (Fig. 26.5). Simple analytical formulas can fit the fluorescence recovery curve, generating the three-dimensional diffusion coefficient of the fluorescent molecule, or multiple coefficients, or anomalous subdiffusion parameters if the dynamics of the system warrant (Brown et al., 1999). The intrinsic three-dimensional confinement of multiphoton excitation allows the use of detectors without confocal apertures, thereby enabling the collection of highly scattered fluorescence light. This, combined with the relatively long excitation wavelength used, gives the multiphoton

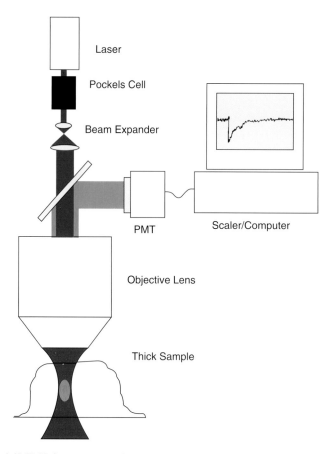

Figure 26.4. MPFRAP instrumentation.

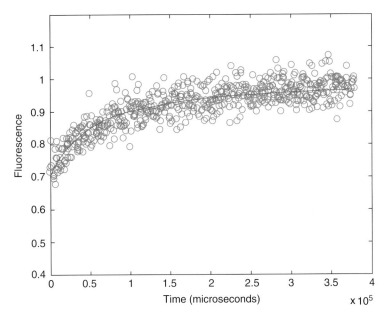

Figure 26.5. MPFRAP recovery curve. MPFRAP was performed on 2M MW FITC-Dextran infused into an MU89 murine melanoma growing in the dorsal skinfold chamber of a SCID mouse (Jain, Brown et al., 2003). The solid line is the least chi-squared fit, producing a diffusion coefficient at 22C of 4.7×10^{-9} cm^{2}s^{-1}.

laser-scanning microscope (MPLSM) (and hence MPFRAP) a relatively high depth penetration, with imaging depths as high as 0.5 mm being reported (Brown et al., 2001; Kleinfeld et al., 1998).

26.2.1 Detailed MPFRAP Instrumentation

An MPFRAP apparatus (Brown et al., 1999) consists of a laser, a beam expander, a Pockels cell, a dichroic mirror, an objective lens, a PMT, and a data recording system such as a scaler that counts the electronic pulses produced by the PMT each time it detects a photon. All of these components, except the Pockels cell and scaler, are typically contained in an MPLSM (Majewska, Yiu et al., 2000). The beam from the laser source, typically a mode-locked Ti:Sapphire, first passes through the Pockels cell, then the beam expander, dichroic mirror, and objective lens, and finally into the sample.

The beam expander serves to overfill the objective lens (Zipfel et al., 2003), ensuring that the cross-sectional area of the laser beam is as large or larger than the back-aperture of the objective lens. This produces a uniformly illuminated back-aperture and thereby results in the highest-resolution focus at the sample (Born & Wolf, 1980). The intrinsic spatial confinement of multiphoton excitation produces a three-dimensionally restricted bleach volume whose size depends on the numerical aperture (NA) of the objective lens used and wavelength of excitation light, and is typically less than a micron radially and 1 micron axially (Zipfel et al., 2003). This extremely small bleached volume dissipates rapidly (hundreds of microseconds for smaller fluorescently labeled molecules such as FITC-BSA). Consequently, MPFRAP requires a beam modulation system with response times as fast as one microsecond, as is offered by a Pockels cell. This device operates by passing the laser beam through a crystal and applying a voltage across the crystal. This serves to rotate the plane of polarization of incoming laser light to an extent dictated by the applied voltage. The laser light exits the crystal and passes through a fixed polarizer, thereby converting a rotation in plane of polarization to a variation in intensity.

26.2.2 Procedure

An MPFRAP experiment consists of the following steps:

1. The Pockels cell is set to a low transmission state producing the "monitoring beam," which generates fluorescence from the sample that is collected by the objective lens and detected by the PMT. The scaler monitors the output of the PMT for a short duration (tens of microseconds to milliseconds), recording the pre-bleach signal.
2. The PMT is then gated (i.e., the dynode voltage is set to zero) to avoid flash artifacts, and the transmissivity of the Pockels cell is switched to a high level, which photobleaches a fraction of the fluorophores at the sample. The Pockels cell returns back to low transmissivity (after a total bleach time that depends upon the sample dynamics; see the section on limitations below), and the PMT is then ungated.
3. The fluorescence generated by the low-intensity laser beam is then recorded continuously. As unbleached fluorophores diffuse into the focal volume, the fluorescence signal recovers back to equilibrium levels.

26.2.3 Constraints

A number of steps must be taken to measure accurate diffusion coefficients with MPFRAP:

1. The power of the monitoring beam must be kept low enough to avoid significant photobleaching. Photobleaching by the monitoring beam can be easily quantified by performing an MPFRAP experiment with no bleach pulse. If the resultant highest safe monitoring power is too low to allow a sufficient signal-to-noise ratio, the monitoring beam can be cycled between zero and a higher power that causes limited photobleaching, thereby periodically recording fluorescence at higher signal rates while limiting the total photobleaching by the monitoring beam. Generation of MPFRAP curves in the regime where significant photobleaching occurs during monitoring is discussed extensively in the article by Waharte and coworkers (Waharte et al., 2005).
2. The duration of the bleaching flash must be significantly shorter than the subsequent diffusive recovery time, or underestimation of the diffusion recovery time can occur. Generation of MPFRAP curves where significant diffusion occurs during the bleach pulse is also discussed extensively in the article by Waharte and coworkers (Waharte et al., 2005).

3. If the fluorescence curve does not recover fully during the measurement period, overestimation of the immobile fraction and underestimation of the diffusion recovery time can result.

4. In MPFRAP the fluorescence excitation rate is assumed to scale as $\sim \langle I^b \rangle$ where I is the intensity of the bleach beam, b is the number of photons absorbed in a bleaching event (e.g., 2 for two-photon excitation), and $<>$ denotes a time average. This rate has an upper limit because fluorescent molecules have excited state lifetimes of $\tau_L \sim 10$ ns and hence cannot be excited at a rate faster than $1/\tau_L$. Furthermore, the typical pulsed lasers used for MPFRAP have a duty cycle of $\tau_D = 12.5$ ns. Consequently, when the excitation rate of fluorophores during the bleaching pulse approaches a significant fraction of $1/\tau_L$ or $1/\tau_D$, the rate of excitation will drop off from $\sim \langle I^b \rangle$, a phenomenon known as excitation saturation. MPFRAP curves generated in this saturation regime can produce erroneously low diffusion coefficients

   To avoid excitation saturation, a series of MPFRAP curves can be generated at increasing bleach powers. When the bleach depth parameter (see the section on analysis below) measurably deviates from $\langle I^b \rangle$ dependence ($\sim 10\%$ or more, for a good rule of thumb), the bleaching is subject to excitation saturation and the diffusion coefficients will be erroneously low.

5. In order to derive a diffusion coefficient from a measured recovery time, the geometry of the bleached region must be known. In a MPLSM, the excitation probability distribution is well represented by a 3D Gaussian function. The e^{-2} widths of this Gaussian are typically measured by imaging a subresolution (~ 10 nm or less) fluorescent bead in x, y, and z.

26.2.4 Theory

In the absence of excitation saturation, the local rate of two-photon excitation is proportional to the square of the local laser intensity. For an overfilled objective lens this produces a three-dimensional Gaussian excitation profile given by:

$$\left\langle I^2(r,z) \right\rangle = \left\langle I^2(0,0) \right\rangle \exp \left[-\frac{4r^2}{w_r^2} - \frac{4z^2}{w_z^2} \right] \qquad (26.1)$$

where $\langle I^2(0,0) \rangle$ is the square of the excitation intensity at the center of the focal volume, w_r and w_z are the radial and axial characteristic beam radii, and $<>$ represents a time average.

Consequently, the concentration distribution of unbleached fluorophores immediately after the photobleaching pulse is given by:

$$c(r,z,t=0) = c_0 e^{\left[-q_b \delta \langle I_{bl}^2(r,z) \rangle \Delta t/2 \right]} \qquad (26.2)$$

where c_0 is the original equilibrium concentration of dye molecules, δ is the two-photon absorption cross-section ($cm^4 s^{-1}$), q_b is the quantum efficiency of photobleaching (i.e., the fraction of excited dye molecules that photobleach), Δt is the duration of the bleaching flash, and the factor of ½ takes into account the fact that two excitation photons disappear to make one absorption event.

For free diffusion this concentration distribution then evolves over time in a manner described by the diffusion equation:

$$\frac{\partial c(\vec{r},t)}{\partial t} = -D\nabla c(\vec{r},t) \tag{26.3}$$

where ∇ is the gradient operator and the constant of proportionality D is the diffusion coefficient $(cm^2 s^{-1})$. In its integral form in cylindrical coordinates this is given by:

$$c(r,z,t) = \int G(r,z,r',z';t)\, c(r',z';0) 2\pi\, r'dr'dz' \tag{26.4}$$

where G(r,z,r',z';t) is the Green's function of the diffusion equation in cylindrical coordinates (i.e., the response of the diffusion equation (26.3) to a delta function concentration distribution at r',z', and at t = 0). This is given by:

$$G(r,z,r',z';t) = \frac{I_0(rr'/2Dt)}{(4\pi Dt)^{3/2}} e^{-(r^2+r'^2+(z-z')^2)/4Dt} \tag{26.5}$$

where I_0 is the modified Bessel function of the first kind. Equations (26.2) and (26.5) can be substituted into equation (26.4). Using the series expansion of the exponential function in equation (26.5) ($e^{-ax} = \sum_{n=0}^{\infty} \frac{(-ax)^n}{n!}$) yields the concentration of unbleached fluorophores as a function of position and time after the photobleaching pulse:

$$c(r,z,t) = c_0 \sum_{n=0}^{\infty} \frac{(-\beta)^n \exp\left(-\frac{4nz^2}{w_z^2}\frac{1}{1+(16nDt/w_z^2)} - \frac{4nr^2}{w_r^2}\frac{1}{1+(16nDt/w_r^2)}\right)}{n!\left(1+\left(16nDt/w_z^2\right)\right)^{1/2}\left(1+\left(16nDt/w_r^2\right)\right)} \tag{26.6}$$

where $\beta = q_b \delta \langle I_{bl}^2(0,0)\rangle \Delta t$ is defined as the bleach depth parameter and is equal to the peak effective rate of photobleaching at the center of the focal volume. Experimentally, this evolving concentration distribution is monitored by illuminating it with the same laser beam, greatly attenuated. The fluorescence generated is then given by:

$$F(t) = \frac{q_f \delta}{2}\int \left\langle I_m^2(r,z)\right\rangle c(r,z,t) 2\pi\, rdrdz \tag{26.7}$$

where $\langle I_m^2(r,z)\rangle$ is the time average of the square of the local monitoring intensity and q_f is the fluorescence action cross-section. Once again the factor of ½ accounts for the use of two excitation photons to generate one excitation event.

Equations (26.1) and (26.6) can be substituted into equation (26.7), yielding an expression for the temporal evolution of the fluorescence signal (Brown et al., 1999):

$$F(t) = F_\infty \sum_{n=0}^{\infty} \frac{(-\beta)^n}{n!}\frac{1}{(1+n+2nt/\tau_D)}\frac{1}{\sqrt{1+n+2nt/(R\tau_D)}} \tag{26.8}$$

where F_∞ is the fluorescence signal at t = ∞, and τ_D is the three-dimensional diffusion recovery time. The diffusion coefficient is given by $D = w_r^2/8\tau_D$ and R is the square of the ratio of the axial e^{-2} half-width to the radial e^{-2} half-width.

Due to the limited signal levels caused by the restrictions on bleaching and monitoring powers as described in the section on limitations above, MPFRAP experiments are typically performed with a series of several flashes at a location of interest, and the resultant curves are averaged together to increase the signal-to-noise ratio. Therefore, this technique is insensitive to immobile fluorophores, as they are bleached out during the repetitive flashes and do not contribute to recovery kinetics. If an MPFRAP curve is performed with a single flash and monitor sequence, then the fraction of immobile fluorophores in the sample is given by $(F_0 - F_\infty)/F_\infty$, where F_0 is the pre-bleach fluorescent signal (Brown et al., 1999).

The aforementioned analysis is given for the case of free diffusion described by equation (26.3), where the diffusion coefficient is constant in space and time. In complex biological systems, this is often not the case. The anomalous subdiffusion model describes the case in which the fluorophores are diffusing in a landscape of potential energy traps (i.e., locations that are "sticky"). As time passes, the diffusing dye molecules are more likely to encounter these "sticky" sites, causing the effective diffusion coefficient to decrease with time (Periasamy & Verkman, 1998). Therefore, unlike normal free diffusion, in which the mean square displacement increases linearly with time:

$$\left\langle r^2 \right\rangle = 6Dt \tag{26.9}$$

in anomalous subdiffusion the mean square displacement of a diffusing particle increases with a power law (Bouchard & Georges, 1990):

$$\left\langle r^2 \right\rangle = 6\Gamma t^\alpha \tag{26.10}$$

where $\alpha < 1$.

Mathematically, the MPFRAP model (equation (26.8)) can be modified to account for anomalous subdiffusion simply by replacing terms of the form Dt with terms of the form $(Dt)^\alpha$ Feder et al., 1996; Periasamy & Verkman, 1998).

26.2.5 Discussion of Diffusion Measurements

A summary of various applications of MPFRAP is provided in Table 26.1. In the original MPFRAP paper, the technique was used to reveal anomalous subdiffusion within the cytoplasm of living cells (Brown et al., 1999). Recently, MPFRAP has been used to study diffusion in free solution, where the resultant diffusion coefficient was used to deduce the tortuosity that nerve growth factor encounters within the striatum of rat brain (Stroh et al., 2003). MPFRAP has also been used in artificial matrices constructed by multiphoton excitation of a light-excitable cross-linker, where it revealed that these matrices hindered diffusion by orders of magnitude relative to free solution, with a hindrance that could be controlled by the cross-linking illumination dose, and that certain dyes could exhibit anomalous subdiffusion (see below) within the matrices (Basu et al., 2004, 2005). MPFRAP has been performed on patterned lipid bilayers, revealing that molecules within these useful two-dimensional constructs retain their motility (Orth et al., 2003). MPFRAP data analyzed using two diffusing components revealed that ablation of the liver fatty acid binding protein (L-FABP)

Table 26.1. Applications of MPFRAP

Use Of MPFRAP	Reference
The diffusion of calcein in the cytoplasm of RBL cells exhibits anomalous subdiffusion.	Brown, Wu et al., 1999
Nerve growth factor experiences significant tortuosity as it diffuses through rat brain striatum.	Stroh, Zipfel et al., 2003
Gels manufactured by multiphoton-induced crosslinking exhibit diffusive hindrance orders of magnitude higher than free diffusion, and this hindrance can be adjusted by adjusting the crosslinking dose.	Basu, Rodionov et al., 2005
Certain fluorescent dyes exhibit anomalous subdiffusion within gels manufactured by multiphoton-induced crosslinking.	Basu, Wolgemuth et al., 2004
Fluorescent probes within specially constructed patterned lipid bilayers retain their lateral diffusive motility.	Orth, Kameoka et al., 2003
Ablation of the liver fatty acid binding protein (L-FABP) reduces cytoplasmic diffusion of NBD-stearic acid by 2-fold while leaving membrane diffusion unchanged.	Atshaves, McIntosh et al., 2004
Myo1a dynamics depend upon its ATPase activity and the integrity of its motor domain.	Waharte, Brown et al., 2005
Ezrin exists in one cytoplasmic and three different membrane-bound populations.	Coscoy, Waharte et al., 2002
Spines are unlikely to compartmentalize electrical signaling.	Svoboda, Tank et al., 1996
Diffusional equilibration between spine and dendrite is dependent on spine morphology.	Majewska, Brown et al., 2000
Rapid changes in spine morphology change the diffusion of GFP between spine and dendrite.	Majewska, Tashiro et al., 2000
MPFRAP can be used to measure calcium signaling in spines in order to extrapolate to the behavior of calcium in the absence of exogenous dye.	Hasan, Friedrich et al., 2004; Majewska, Brown et al., 2000; Pologruto, Yasuda et al., 2004; Sobczyk, Scheuss et al., 2005
Calbindin is relatively immobile in dendritic spines.	Schmidt, Schwaller et al., 2005
Parvalibumin diffuses freely between the spine and dendrite.	Schmidt, Brown et al., 2003
GFP diffusion between plant plastids is possible.	Kohler, Cao et al., 1997

reduced cytoplasmic diffusion of NBD-stearic acid by twofold while leaving membrane diffusion unchanged (Atshaves et al., 2004). MPFRAP has also been applied to quasi–one-dimensional intestinal microvilli to determine that the protein ezrin exists in three different membrane-bound states (Coscoy et al., 2002), and more recently to discover that Myo1a dynamics depend upon its ATPase activity and the integrity of its motor domain (Waharte et al., 2005).

26.3 MPFRAP FOR COMPARTMENTALIZATION ANALYSIS

MPFRAP can also be used to measure the diffusional coupling between two connected compartments. A characteristic time scale for the diffusional coupling can be obtained by bleaching one compartment and monitoring the recovery in fluorescence as it is refilled with fluorophore from the unbleached compartment. This time scale can then be used to determine parameters such as the resistivity or the characteristic pore size of the separating barrier.

26.3.1 Instrumentation

Due to slower recovery times, compartmentalization analysis can often be performed in line-scan mode on an MPLSM. In line-scan mode, the excitation beam is scanned repeatedly along a single line that intersects an object of interest, and an x-versus-t image of the fluorescence is generated using the MPLSM software, without the need for a separate scaler.

26.3.2 Procedure

A MPFRAP compartmentalization analysis is typically performed as follows:

1. After imaging the structure of interest (for example, a dendritic spine and its parent dendrite), one of the compartments (typically the smaller one) is chosen for bleaching and monitoring. The Pockels cell is set to a low "monitoring beam" to measure the pre-bleach fluorescence, and a line scan of that compartment is begun.
2. If possible, the PMT is then gated (i.e., the dynode voltage is set to zero) to avoid flash artifacts, and the Pockels cell produces the "bleach beam," which photobleaches a fraction of the fluorophores in the compartment. The modulator returns back to the low state (after a total bleach time that depends upon the sample dynamics, as described above), and the PMT is subsequently ungated.
3. The compartment is then monitored via the continuing line scan, and the fluorescence signal recovers back to equilibrium levels as unbleached fluorophores diffuse from the unbleached compartment.

26.3.3 Constraints

1. Bleaching during the monitoring phase, bleach duration, and total acquisition time must be evaluated as in points 1 to 3 in Section 26.2.3 on MPFRAP limitations.

2. MPFRAP compartmentalization analysis typically assumes that the communicating compartments are well mixed—that is, that diffusional equilibrium within the compartments is much faster than between compartments. This assumption is usually based upon the known diffusion coefficient of the fluorophore, which can often be determined with conventional free-diffusion MPFRAP as described above.

3. When the communicating compartments are well mixed, the initial spatial distribution of bleached molecules is irrelevant because the bleached compartment undergoes diffusive mixing before significant communication with other compartments can occur. Consequently, neither excitation saturation nor the precise geometry of the focal spot is a significant concern.

26.3.4 Mathematical Analysis

The fluorescence recovery curve in a well-mixed, photobleached compartment diffusionally coupled to a larger well-mixed compartment is described by (Majewska et al., 2000; Svoboda et al., 1996):

$$F(t) = F(\infty) - \Delta F_0 e^{-t/\tau} \qquad (26.11)$$

where $F(\infty)$ is the fluorescence at $t = \infty$, ΔF_0 is the change in fluorescence following the bleach pulse, and τ is the time scale of diffusion between the two compartments. The recovery between compartments provides insight into the characteristic resistivity of the coupling pathway, the number of coupling pathways, the diffusion coefficient of the tracer, and so forth, depending upon the geometry of the system.

26.3.5 Discussion of Compartmentalization Analysis

MPFRAP within cellular compartments has been extensively used in neurobiology to determine the coupling of dendritic spines to their parent dendrite. Dendritic spines are small synaptic compartments with a peculiar mushroom-like shape whereby a thick stalk connects the spherical spine head to the dendritic shaft. Because of their small size, spines have traditionally been very difficult to study in live tissue. Recently MPFRAP has been used to probe the diffusional coupling between spines and their parent dendrites. In most cases MPFRAP has been used to determine the diffusion equilibration time of exogenous indicators between the two compartments (Majewska, Brown et al., 2000; Majewska, Tashiro et al., 2000; Pologruto et al., 2004; Sobczyk et al., 2005; Svoboda et al., 1996). Initial studies showed that after bleaching in the spine head, diffusion of fluorescent dextrans from the dendrite was fast enough to make spines unlikely to compartmentalize electrical signaling (Svoboda et al., 1996). Subsequent studies showed that diffusional equilibration between spine and dendrite was dependent on spine morphology, with longer spine necks acting as a greater barrier to intercompartmental diffusion (Majewska, Brown et al., 2000). Interestingly, rapid changes in spine morphology changed the diffusion time of GFP between spine and dendrite (Majewska, Tashiro et al., 2000). MPFRAP has been used to characterize the behavior of indicators used to measure calcium signaling in spines in order to extrapolate to the behavior of calcium in the absence of exogenous dye (Hasan et al., 2004; Majewska, Brown et al.,

2000; Pologruto et al., 2004; Sobczyk et al., 2005). Additionally, MPFRAP has been used to characterize the diffusion of endogenous calcium buffers between the compartments, showing that calbindin is relatively immobile (Schmidt et al., 2005), while parvalbumin diffuses freely between the spine and dendrite (Schmidt et al., 2003).

Outside of the field of neuroscience, compartment MPFRAP has been used to determine the diffusional coupling of plant plastids (Kohler et al., 1997). Although plant plastids have been thought to be isolated, independent units, MPFRAP showed that GFP diffusion between plant plastids was possible. This suggests that an interplastid communication system exists and might be responsible for coordinating the function of plastid compartments.

26.4 OVERVIEW

26.4.1 Advantages

Overall, MPFRAP allows the three-dimensionally resolved measurement of diffusion coefficients with significant depth penetration in thick, intact tissues. Prior techniques are limited to thin samples (FRAP) or can investigate thicker samples but lack three-dimensional resolution (SFA-FRAP). However, high spatial resolution is not always an advantage. Lower-resolution techniques such as SFA-FRAP offer an intrinsic spatial averaging over the myriad microdomains and heterogeneities of biological tissue. This is in contrast to MPFRAP, where each individual experiment may probe a specific microdomain and hence display a high experiment-to-experiment variability. This represents the intrinsic variability in the tissue due to its heterogeneous structure of cells, blood vessels, collagen, and so forth. The specific needs of the experiment will dictate which technique is most useful.

26.4.2 Pitfalls

As discussed above, care must be exerted to avoid bleaching during monitoring, diffusion during the bleach pulse, and excitation saturation. These concerns are especially true in scattering samples where signals are low and the inclination to increase laser powers and photobleaching doses is high. As with all light-based techniques, the potential for photodamage to alter the measured parameter is a major consideration. Repeated experiments, performed at the same location, that yield identical results are a good indicator that photodamage is not altering the measured property.

26.4.3 Future Directions

While the improvement in depth penetration of MPLSM over traditional epifluorescence and confocal microscopy is considerable, in an absolute sense a depth penetration of 0.5 mm is still a serious limitation. Fortunately a great deal of progress has been made in the creation of endoscopic devices that allow microscopy of many interior surfaces of laboratory animals without surgical intervention. A side-viewing microfabricated laser-scanning microscope has been demonstrated (Dickensheets & Kino, 1996), as have forward-viewing GRIN lens-based microscopes (Kunkel et al., 1997; Levene et al., 2004). These and other endoscopic advances promise to bring MPFRAP deeper into living tissue and hence enhance its capabilities.

ACKNOWLEDGEMENTS

This work was supported by the NIH (P01CA80134 and R24CA85140 to R.K.J.), the Department of Defense (an Era of Hope Scholar Award to E.B.), the Whitaker Foundation (a Biomedical Engineering Grant to E.B.), and the Burroughs Wellcome Fund (a Career Award in the Biomedical Sciences to A.M.). This chapter is based on the previous review: Brown, Majewska, and Jain. Single and Multiphoton Fluorescence Recovery After Photobleaching. *In:* Yuste and Konnerth, eds. *Imaging in Neuroscience and Development. A Laboratory Manual.* Cold Spring Harbor, NY: Cold Spring Harbor Laboratory Press; (2005) 429–438.

REFERENCES

Atshaves, B. P., A. M. McIntosh, et al. (2004). "Liver fatty acid-binding protein gene ablation inhibits branched-chain fatty acid metabolism in cultured primary hepatocytes." J Biol Chem 279(30): 30954–65.

Axelrod, D., D. E. Koppel, et al. (1976). "Mobility measurement by analysis of fluorescence photobleaching recovery kinetics." Biophys J 16(9): 1055–1069.

Basu, S., V. Rodionov, et al. (2005). "Multiphoton-excited microfabrication in live cells via Rose Bengal cross-linking of cytoplasmic proteins." Opt Lett 30(2): 159–161.

Basu, S., C. W. Wolgemuth, et al. (2004). "Measurement of normal and anomalous diffusion of dyes within protein structures fabricated via multiphoton excited cross-linking." Biomacromolecules 5(6): 2347–2357.

Berk, D. A., F. Yuan, et al. (1993). "Fluorescence photobleaching with spatial Fourier analysis: measurement of diffusion in light-scattering media." Biophys J 65(6): 2428–2436.

Born, M. and E. Wolf (1980). Principles of Optics. New York, Pergamon Press.

Bouchard, J. and A. Georges (1990). "Anomalous diffusion in disordered media; statistical mechanisms, models and physical applications." Phys. Rep. 195: 127–293.

Brown, E., R. Campbell, et al. (2001). "In vivo measurement of gene expression, angiogenesis, and physiological function in tumors using multiphoton laser scanning microscopy." Nat Med 7(7): 864–868.

Brown, E., A. Majewska, et al. (2005). Single and Multiphoton Fluorescence Recovery after Photobleaching. Imaging in Neuroscience and Development, A Laboratory Manual. R. Y. a. A. Konnerth. Cold Spring Harbor, Cold Spring Harbor Laboratory: 429–438.

Brown, E., T. D. McKee, et al. (2003). "Dynamic imaging of collagen and its modulation in tumors in vivo using second harmonic generation." Nat Med 9(6): 796–801.

Brown, E. B., Y. Boucher, et al. (2004). "Measurement of macromolecular diffusion coefficients in human tumors." Microvasc Res 67(3): 231–236.

Brown, E. B., E. S. Wu, et al. (1999). "Measurement of molecular diffusion in solution by multiphoton fluorescence photobleaching recovery." Biophys J 77(5): 2837–2849.

Chary, S. R. and R. K. Jain (1989). "Direct measurement of interstitial convection and diffusion of albumin in normal and neoplastic tissues by fluorescence photobleaching." Proc Natl Acad Sci U S A 86(14): 5385–5389.

Coscoy, S., F. Waharte, et al. (2002). "Molecular analysis of microscopic ezrin dynamics by two-photon FRAP." Proc Natl Acad Sci U S A 99(20): 12813–12818.

Denk, W., J. H. Strickler, et al. (1990). "Two-photon laser scanning fluorescence microscopy." Science 248(4951): 73–76.

Dickensheets, D. and G. Kino (1996). "Micromachined scanning confocal optical microscope." Opt Lett 21(10): 764–766.

Feder, T. J., I. Brust-Mascher, et al. (1996). "Constrained diffusion or immobile fraction on cell surfaces: a new interpretation." Biophys J 70(6): 2767–2773.

Hasan, M. T., R. W. Friedrich, et al. (2004). "Functional fluorescent Ca2+ indicator proteins in transgenic mice under TET control." PLoS Biol 2(6): e163.

Jain, R. K., E. Brown, et al. (2003). Intravital Microscopy of Normal and Diseased Tissue in the Mouse. Live Cell Imaging: A Laboratory Manual. G. R. Spector D. Cold Spring Harbor, Cold Spring Harbor Press.

Kleinfeld, D., P. P. Mitra, et al. (1998). "Fluctuations and stimulus-induced changes in blood flow observed in individual capillaries in layers 2 through 4 of rat neocortex." Proc Natl Acad Sci U S A 95(26): 15741–15746.

Kohler, R. H., J. Cao, et al. (1997). "Exchange of protein molecules through connections between higher plant plastids." Science 276(5321): 2039–2042.

Kosto, K. B., S. Panuganti, et al. (2004). "Equilibrium partitioning of Ficoll in composite hydrogels." J Colloid Interface Sci 277(2): 404–409.

Kunkel, E. J., U. Jung, et al. (1997). "TNF-alpha induces selectin-mediated leukocyte rolling in mouse cremaster muscle arterioles." Am J Physiol 272(3 Pt 2): H1391–400.

Levene, M. J., D. A. Dombeck, et al. (2004). "In vivo multiphoton microscopy of deep brain tissue." J Neurophysiol 91(4): 1908–1912.

Majewska, A., E. Brown, et al. (2000). "Mechanisms of calcium decay kinetics in hippocampal spines: role of spine calcium pumps and calcium diffusion through the spine neck in biochemical compartmentalization." J Neurosci 20(5): 1722–1734.

Majewska, A., A. Tashiro, et al. (2000). "Regulation of spine calcium dynamics by rapid spine motility." J Neurosci 20(22): 8262–8268.

Majewska, A., G. Yiu, et al. (2000). "A custom-made two-photon microscope and deconvolution system." Pflugers Arch 441(2–3): 398–408.

Orth, R. N., J. Kameoka, et al. (2003). "Creating biological membranes on the micron scale: forming patterned lipid bilayers using a polymer lift-off technique." Biophys J 85(5): 3066–3073.

Periasamy, N. and A. S. Verkman (1998). "Analysis of fluorophore diffusion by continuous distributions of diffusion coefficients: application to photobleaching measurements of multicomponent and anomalous diffusion." Biophys J 75(1): 557–567.

Pluen, A., Y. Boucher, et al. (2001). "Role of tumor-host interactions in interstitial diffusion of macromolecules: cranial vs. subcutaneous tumors." Proc Natl Acad Sci U S A 98(8): 4628–4633.

Pologruto, T. A., R. Yasuda, et al. (2004). "Monitoring neural activity and [Ca2+] with genetically encoded Ca2+ indicators." J Neurosci 24(43): 9572–9579.

Reits, E. A. and J. J. Neefjes (2001). "From fixed to FRAP: measuring protein mobility and activity in living cells." Nat Cell Biol 3(6): E145–E147.

Schmidt, H., E. B. Brown, et al. (2003). "Diffusional mobility of parvalbumin in spiny dendrites of cerebellar Purkinje neurons quantified by fluorescence recovery after photobleaching." Biophys J 84(4): 2599–2608.

Schmidt, H., B. Schwaller, et al. (2005). "Calbindin D28k targets myo-inositol monophosphatase in spines and dendrites of cerebellar Purkinje neurons." Proc Natl Acad Sci U S A 102(16): 5850–5855.

Sobczyk, A., V. Scheuss, et al. (2005). "NMDA receptor subunit-dependent [Ca2+] signaling in individual hippocampal dendritic spines." J Neurosci 25(26): 6037–6046.

Stroh, M., W. R. Zipfel, et al. (2003). "Diffusion of nerve growth factor in rat striatum as determined by multiphoton microscopy." Biophys J 85(1): 581–588.

Svoboda, K., D. W. Tank, et al. (1996). "Direct measurement of coupling between dendritic spines and shafts." Science 272(5262): 716–719.

Tognana, E., F. Chen, et al. (2005). "Adjacent tissues (cartilage, bone) affect the functional integration of engineered calf cartilage in vitro." Osteoarthritis Cartilage 13(2): 129–138.

Tsay, T. T. and K. A. Jacobson (1991). "Spatial Fourier analysis of video photobleaching measurements. Principles and optimization." Biophys J 60(2): 360–368.

Waharte, F., C. M. Brown, et al. (2005). "A two-photon FRAP analysis of the cytoskeleton dynamics in the microvilli of intestinal cells." Biophys J 88(2): 1467–1478.

Zipfel, W. R., R. M. Williams, et al. (2003). "Live tissue intrinsic emission microscopy using multiphoton-excited native fluorescence and second harmonic generation." Proc Natl Acad Sci U S A 100(12): 7075–7080.

27

Femtosecond Laser Nanoprocessing

Karsten König

27.1 INTRODUCTION

Multiphoton microscopy based on the application of tight-focused near-infrared (NIR) femtosecond laser beams has been considered a valuable tool for vital cell imaging (Denk et al., 1990). Typically, 80-MHz/90-MHz mode-locked titanium sapphire lasers with about 1 W mean output power have been employed as a laser source to realize two-photon fluorescence imaging and second-harmonic generation (SHG) microscopy (König et al., 2000; Mertz, 2008). The laser beam has to be attenuated to provide <10-mW laser power and <130-pJ pulse energy, respectively, at the sample. However, when increasing the mean power at the sample to the range of 50 mW to 250 mW and 0.5 nJ to 3 nJ pulse energy, respectively, destructive effects occur (König, 2001; König et al., 1997, 1999; Oehring et al., 2000; Tirlapur et al., 2001). At these power and pulse energy levels, the transient laser intensity reaches the TW/cm^2 range, which is sufficient to induce multiphoton ionization and plasma formation. When working near the threshold for optical breakdown, material ablation can be realized within the central part of the diffraction-limited subfemtoliter multiphoton interaction volume without any collateral effects (König et al., 1999, 2001, 2002). In fact, nanodissection and hole drilling in human chromosomes with ablation zones as low as sub-70 nm have been realized without out-of-focus effects using a laser wavelength more than 10 times larger ($\lambda = 800$ nm) (König et al. 2001). Multiphoton femtosecond laser microscopy offers the possibility of breaking the "Abbe barrier" of diffraction-limited nanoprocessing and of enabling laser nanobiotechnology and laser nanomedicine.

The possible diffraction-limited minimum beam spot size is inversely proportional to the laser wavelength. So far, most of laser-induced ablation, cutting, and drilling effects in the submicron range have been realized with ultraviolet (UV) radiation. The ArF excimer laser at 193 nm plays the most important role for material processing (e.g., in the semiconductor industry to process sub-200-nm features into silicon), as well as in the refractive corneal LASIK surgery for photoablation of stromal tissue with high submicron accuracy (Basting, 2001; Krueger et al., 1995). The 193-nm light penetration depth is on the order of 250 nm and the collateral damage zone is less than 100 nm (Pettit & Ediger, 1996). The KrF and XeCl excimer lasers, with 248-nm and 308-nm emission wavelengths, are not useful for medical procedures such as LASIK due to the large damage zone of several tens of microns, respectively (Basting, 2001). Photoablation is based on UV-induced photochemical decomposition due to the fact that the high photon energy exceeds the dissociation energy of molecular bonds (e.g., C—C and C—O: 3.6 eV). The application of 193-nm photons causes bond-breaking and kinetic as well as thermal energy to the dissociation products (Srinivasan, 1986; Trokel et al., 1983). Users of UV radiation face the problem of out-of-focus absorption, scattering, mutagenic risk, and low micrometer-sized light penetration depth (König, 2001). Intratissue UV surgery without destroying the surface is impossible. In fact, in order to preserve the outermost epithelium layer in LASIK surgery, mechanical cutting devices called microkeratomes have to be employed prior to excimer laser treatment for stromal ablation.

In contrast, the use of NIR lasers, with their high light penetration depth on the order of several millimeters, allows precise intratissue surgery due to the lack of efficient absorbers in the 700- to 1,200-nm spectral range. Hemoglobin, melanin, water, and proteins are the major intratissue NIR absorbers. At 1,064 nm, the absorption coefficient of water is on the order of 0.1 cm^{-1}, whereas the values increase up to 12,000 cm^{-1} and $2,740 \text{ cm}^{-1}$ at the absorption peaks of 2.94μm and 6.1μm, respectively. Proteins absorb mainly around 6 to 8μm. At 6.1μm, collagen absorbs twice as well as water (Vogel & Venugopalan, 2003). The optical behavior of melanin is determined by scattering. The extinction coefficient drops monotonically with increasing VIS/NIR wavelength and is at 800 nm roughly 10% of the value at 400 nm (Kollias & Bager, 1985). The major absorption band ("Soret band") of hemoglobin is around 400 nm, with further absorption bands in the 500- to 600-nm spectral range. A minor peak of hemoglobin is around 760 nm, whereas oxidized hemoglobin absorbs also in the broad NIR range of 800 to 1,000 nm. Most of the cells and nonpigmented, vascular-free tissue areas can be considered as nearly transparent within the 700- to 1,000-nm spectral region, which corresponds to the tuning range of titanium:sapphire lasers. This means that NIR radiation can be absorbed by these transparent objects only via a multiphoton process, which occurs typically within the subfemtoliter focal volume of high laser intensity.

NIR femtosecond laser pulses of high microjoule (μJ) pulse energy of amplified laser systems from IntraLase Inc. (Irvine, CA) have been clinically employed as microsurgery tools to replace the microkeratome and to generate "optical flaps" (Ratkay-Traub et al., 2003). The mechanism is based on photodisruptive effects such as the formation and dynamics of large cavitation bubbles of several tens of micrometers as well as destructive shock waves (Heisterkamp et al., 2002, 2003). However, an additional excimer laser is still used for stromal ablation. Several patients complained about visual problems and bleeding after the "femtosecond-LASIK" procedure (Principe et al., 2004). According to studies by Le Harzic and coworkers (2005a, 2005b), about 20% of the incident NIR femtosecond laser photons are not involved in the ablation procedure and transmit towards the retina. When using μJ laser pulses of amplified laser systems in combination with low numerical aperture (NA) focusing optics, destructive self-focusing effects, white light generation, and collateral photodisruptive damage must be taken into account.

This review focuses on the application of low-energy NIR femtosecond laser pulses in the sub-10-nJ range from nonamplified laser systems, which are typically used in two-photon microscopy. Working with NA > 1 objectives, precise sub-200-nm nanoprocessing can be performed without destructive collateral effects. Applications in the field of nanostructuring of biomaterials, molecular pathology, genomics, proteomics, gene therapy, intracellular surgery, and refractive eye surgery are presented. These examples demonstrate clearly that femtosecond laser scanning microscopes can be used not only as high-resolution 3D imaging devices but also as nanomanipulation and nanosurgery tools.

27.2 PRINCIPLE OF FEMTOSECOND LASER ABLATION

The young woman and later Nobel Prize winner Maria Göppert-Mayer predicted in her Ph.D. thesis under the supervision of Max Born in Göttingen in 1931 the major principle of two-photon effects ("Zwei-Quanten-Sprüngen"). She hypothesized that an atom can simultaneously absorb two photons (nonresonant two-photon absorption) with a certain low probability (Göppert-Mayer, 1931). With the availability of laser sources, Kaiser and Garret (1961) finally realized the first two-photon effect in certain crystalline structures and proved her hypothesis of this two-quantum event. Following the work of Sheppard and Kompfner (1978), Denk and associates (1990) built up the first two-photon microscope and imaged fluorescent probes in living cells using a femtosecond dye laser. Two-photon microscopes have become a valuable tool in modern research institutions. Long-term embryo imaging with NIR femtosecond lasers (Squirrel et al., 1999) and two-photon multiphoton tomography of patients with dermatological disorders based on the clinical femtosecond laser system DermaInspect (JenLab GmbH, Jena) have been realized (König & Riemann, 2003). However, biosafe work with femtosecond laser scanning microscopes requires operation within a certain laser intensity window in dependence on laser wavelength (Koester et al., 1999; König et al., 1997, 1999; Oehring et al., 2000; Tirlapur et al., 2001). König and associates (1995) reported on two-photon intracellular destructive effects induced by highly focused NIR continuous wave (CW) laser beams ("laser tweezers") at 100-mW power. When working with NIR titanium:sapphire femtosecond lasers, the researchers realized the first precise nanoablation of metaphase chromosomes inside living cells (König, 2000; König et al., 1999, 2001).

Because most of the mode-locked Ti:sapphire lasers provide mean powers on the order of 1 W, the transient laser intensity at the target can be easily increased up to TW/cm^2 (10^{12} W/cm^2), where molecules can be dissociated and ionized. Multiphoton ionization (Fig. 27.1) within the focal volume occurs with a probability proportional to I^n if the sum of the energies of the n photons involved in the nonlinear absorption process exceeds the binding energy E^* of the electron. The first quasi-free/free electrons produced by ultrashort laser pulses act as seed electrons. They gain kinetic energy through photon absorption ("inverse Bremsstrahlung") and can create further free electrons through collisions. This leads to cascade (avalanche) ionization. Optical breakdown occurs at a critical free electron density of 10^{18} to 10^{20} cm^{-3} and is accompanied by the formation of a luminescent plasma (Barnes & Rieckhoff, 1968; Bloembergen, 1974; Kennedy et al., 1997; Vogel, 2003; Vogel et al., 1999). The formation of quasi-free electrons in water with an energy gap of about 6.5 eV can be realized with a five-photon NIR absorption process (Grand et al., 1979; Williams et al., 1984). Interestingly, the threshold for plasma formation within intracellular regions requires at least one order less laser intensity. This is likely due to an initial two-photon and three-photon absorption process in biological targets as well as to different

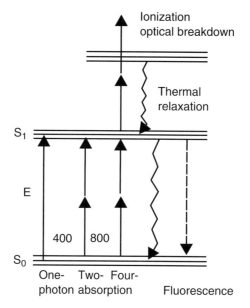

Figure 27.1. Multiphoton absorption processes. Optical breakdown occurs at a typical free electron density of 10^{20} cm^{-3} and is determined by seed electrons provided by multiphoton ionization at TW/cm^2 NIR laser intensities in the case of femtosecond lasers. By fine-tuning the laser power, ablation without energy in excess can be realized. In the case of nanosecond pulses, a sharp optical breakdown threshold exists and ablation is accompanied by destructive photomechanical collateral effects. (Scheme modified from Vogel et al., 2002)

ionization potentials. Major targets are mitochondria (Oehring et al., 2000). Plasma-like broadband luminescence with complex subnanosecond decay time and a spectral maximum around 500 nm accompanied by destructive effects have been detected within living cells at 0.9 to 1.6 TW/cm^2 laser intensity (König et al., 1997).

The formation of plasma and its use for ultraprecise ablation were theoretically described by Vogel and Venugopalan (2003) with the assumption that the optical breakdown occurs in water. Figure 27.1 (modified from Vogel et al., 2003) shows the calculated temporal evolution of the free-electron density during nanosecond laser radiation at 1,064 nm versus femtosecond laser radiation at 532 nm. Optical breakdown of water was assumed to occur at a critical free electron density of 10^{20} cm^{-3}. In the case of nanosecond pulses, the ionization avalanche process proceeds very rapidly if a first seed electron at a high NIR light intensity is provided. This results in a sharp optical breakdown threshold. In contrast, the free electron density induced by femtosecond laser pulses is determined by seed electrons provided by multiphoton ionization and characterized by a slow increase with laser intensity. By fine-tuning the laser power threshold, more precise ablation without destructive mechanical energy can be realized. If the laser power exceeds the threshold, the excess energy is transformed into photodisruptive side effects.

When working near the threshold, the required high TW/cm^2 intensity is achieved within the central part of the submicron laser spot only. This enables cutting and drilling effects with a precision below Abbe's diffraction limit. In contrast to nanosecond laser ablation, destructive mechanical photodisruptive side effects, mainly due to the formation and the collapse of cavitation bubbles as well as shock wave generation, are significantly reduced in the case of femtosecond laser ablation.

In our studies, we often used 640,000 laser pulses of 1-nJ pulse energy for nanodissection and hole drilling, which corresponds to a time period of 1/125 sec. When considering a pulse width of 270 fs at the target, the real exposure time was 173 ns and the applied total energy 640 μJ. Note that only a part of the incident photons will be absorbed.

27.3 A FEMTOSECOND LASER NANOPROCESSING MICROSCOPE

In principle, nearly every two-photon microscope can be used to realize nanoprocessing with femtosecond lasers. However, most of the commercial two-photon systems are based on confocal microscopes with an additional feature for the launch of femtosecond NIR laser beams.

We used the system FemtoCut (JenLab GmbH, www.jenlab.de) for most of our recent studies (Fig. 27.2). The system provides the features of nonlinear imaging and nanoprocessing. It is based on a conventional low-cost fluorescence microscope such as an Axiovert or Axioskop from Zeiss equipped with high NA objectives. A special compact scanning module is attached either to the side microscope port or to the back port. The beam-scanning module consists of beam deflection mirrors, scan optics including an 1:6 beam expander, a motorized beam attenuator, a trigger diode, a beam shutter, and a fast x,y galvoscanner (GSI Lumonics, USA). A piezo-driven z-tuning system with an accuracy of about 40 nm and a working distance of 330 μm is employed. The microscopes contain high-sensitivity photon detectors with short rise time in combination with special beam splitters and shortpass filters for imaging. Using scanning

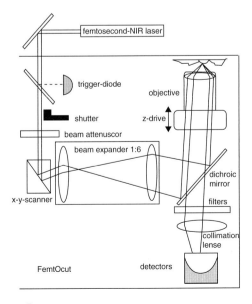

Figure 27.2. The FemtoCut system.

synchronization, the photomultiplier tube (PMT) signal can be correlated to the position of the scanner (x,y) and the objective (z). This enables the formation of 3D images. Single-photon sensitivity can be achieved. The software allows operation in different scan modes: (i) single point illumination (spot scan) to enable drilling nanoeffects, (ii) line scans to realize cutting nanoeffects, and (iii) region-of-interest (ROI) scanning to perform material (tissue) ablation.

A variety of turnkey femtosecond lasers have been attached to the system FemtOcut: (i) the 80-MHz MaiTai (Spectra Physics, USA), tunable from 750 nm to 850 nm, (ii) the 80-MHz Ti:sapphire laser Vitesse at 800 nm, (iii) the tunable 90-MHz Chameleon (730–960 nm, Coherent, Santa Clara, CA). In addition, (iv) a novel laser system based on the use of Ytterbium-doped glass as an active medium and semiconductor saturable absorber mirrors (SESAM) with pulse energies up to 400 nJ with a typical repetition rate up to 1 MHz and a laser wavelength of 1040 nm has been employed (Riemann et al., 2005).

A typical pulse width at the target of 200 to 300 fs was determined with an autocorrelator in combination with a special nonlinear diode detector that was placed on the specimen stage and due to the measurement of the beam, which was transmitted through the whole system including the 40×, NA 1.3 objective (Ulrich et al. 2004).

Cell surgery and targeted laser transfection was performed on living cells growing within the miniaturized cell chamber MiniCeM-grid (MobiTec, Göttingen) with a silicon ring and two round 0.17-μ m-thick glass windows (see Fig. 27.2). One of these windows had an etched grid for cell tracing. The laser exposure was performed through the glass window. The cells can be monitored for up to 5 days within these chambers. The medium can be changed by injection needles perforating the silicon gasket. Chinese hamster (*Cricetulus griseus*) ovary epithelial cells (CHO-K1), potoroo (*Potorous tridactylis*) kidney epithelial cells (PTK2), the human malignant melanoma cell line SK-Mel-28 cells, NG108-15 neuroblastoma (mouse) × glioma (rat) hybridoma cells, rat epithelium cells, and human dental pulpa stem cells (DPSC) have been used.

27.4 NANOPROCESSING OF BIOMATERIALS

Biomaterials with micro- and nanostructured surfaces play an important role in tissue engineering, drug testing, and stem cell handling. It was shown that functional nanostructured surfaces and nanofibers may influence the differentiation of stem cells (Anderson et al., 2004; Flaim et al., 2005; Silva et al., 2004). The control of the differentiation procedure is the crucial step for the clinical use of stem cells.

Most of the micro- and nanostructures for biomedicine, in particular polymers, have been engineered using conventional UV lamps and excimer lasers. Here we demonstrate that a variety of materials, including metal (gold) films, silicon, photoresists, copolymers, and polymers, can be processed with NIR sub-nJ and nJ laser pulses (0.3 nJ to 3 nJ) and μs beam dwell times per pixel. Using the mode of single line scanning, cutting effects with sub-200-nm full-width half-maximum (FWHM) could be realized at 800-nm laser wavelength. The cutting profile was measured with atomic force microscopes and electron microscopes.

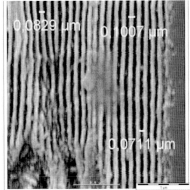

Figure 27.3. Nanomachined silicon wafer with 2D sub-80 nm-ripple nanostructures.

Interestingly, during the laser processing of silicon waver at atmospheric pressure, we found two types of nanostructures (Le Harzic et al., 2005b). The first was a superficial layer containing nonhomogeneously distributed laser-induced nanocones representing noncoherent structures. After removing this layer by etching with ammonium fluoride, a second highly coherent "ripple" structure was detected. These symmetric features possessed a wavelength (distance) of 50 to 80 nm. which is one order below "Abbe's diffraction limit." Note that normally deep UV sources are required to realize sub-100-nm features on wafers.

We fabricated a multiphoton-induced nanostructured "chess-pattern" 2D array with sub-100-nm ripples using ROI scanning (Fig. 27.3). These arrays are under investigation to try to influence adhesion, cell migration, reproduction behavior, and differentiation of stem cells.

27.5 NANODISSECTION OF CHROMOSOMES AND DNA MOLECULES

Interestingly, laser-fabricated sub-100-nm features can also be realized in natural polymers, such as DNA. In particular, multiphoton effects near the intensity threshold were used for highly accurate nanoprocessing of human metaphase chromosomes and elongated single DNA molecules without photothermal, photomechanical, or photochemical damage to the surrounding. Human metaphase chromosomes were prepared from peripheral blood by standard methods.

As depicted in Figure 27.4, clear nanocuts and nanoholes in chromosomes and stretched DNA molecules could be created with 800-nm radiation and the mode of line scanning and single-point illumination. It was possible to cut incisions in the chromosome number 1 with a minimum FWHM cut size of 80 nm (König et al., 2001). Material of only $0.003 \mu m^3$ was removed, corresponding to 1/400 of the chromosome volume. The cut removes about 65,000 base pairs when assuming homogenous basepair distribution along the 263-Mbp chromosome.

The micrometer-sized chromosomes are condensed by a factor of 10^4 in order to contain the several-centimeter-long DNA molecule. When using noncondensed fiber DNA, an 80-nm laser cut knocks out less than 10 base pairs (König, 2005).

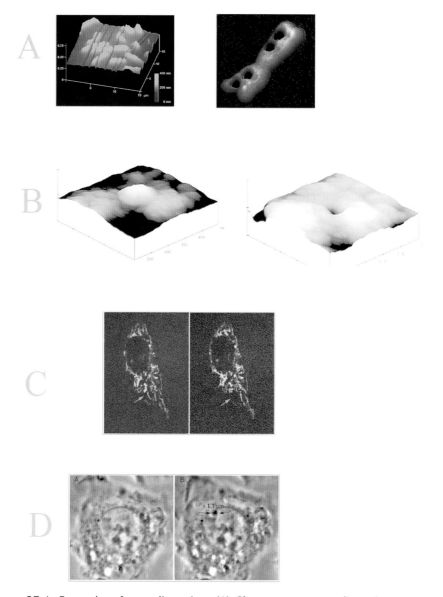

Figure 27.4. Examples of nanodissection. (A) Chromosome nanodissection. (B) Optical knockout of specific centromeric genomic regions using 70-nm metallic nanoparticles. (C) Destruction of a single intracellular mitochondrion without collateral damage. (D) Laser perforation of the nuclear membrane within a living cell.

27.6 LASER PROCESSING WITH NANOPARTICLES AND OPTICAL KNOCKOUT OF GENES

Nanoparticles, such as polymer nanobeads, gold particles, and fluorescent quantum dots, have been employed to probe biological targets and to detect (image) them (e.g., Park et al., 2002). In addition, nanoparticle-assisted laser inactivation based on local heating effects has been used for protein destruction and cancer therapy (El-Sayed et al., 2005). Conventional procedures include also the chromophore-assisted laser inactivation (CALI) method, where special light-activated absorbers linked to the target induce destructive effects by photochemical reactions, such as the formation of reactive singlet oxygen (Jay & Sakurai, 1999). In principle, the submicron-sized dermal and ocular melanosomes and the tattoo pigments can also be considered as nanoparticles, which can be selectively thermally destroyed for hair and tattoo removal and the destruction of malfunctioned retinal pigment epithelium (RPE) with pulsed laser beams (Brinkmann et al., 2000).

The absorption behavior of nanoparticles depends on the particle morphology and its size. For example, gold spheres with typical diameters of 50 to 100 nm have an absorption maximum in the green spectral range. When using ellipsoidal particles, the absorption maximum is red-shifted according to the ratio of the axis.

The desired therapeutic effect is based on selective thermal heating ("selective photothermolysis") proposed by Anderson and Parrish in 1983 (Anderson & Parrish, 1983). So far, this is based on strong UV/IR one-photon absorption and laser pulses with pulse widths τ_{pulse} shorter than the characteristic thermal diffusion time τ_d of the heated target volume according to the formula $\tau_{pulse} \leq \tau_d = 1/\kappa\mu_{\alpha}$, with the thermal diffusivity κ and typical τ_d values of ns for single melanosomes and μs for capillaries. Under this condition, the temperature increase is almost confined to the absorbing volume. When using ultrashort laser pulses shorter than the thermal diffusion time, such as femtosecond laser pulses, temperature gradients inside the nanotarget can be created that may result in nano-/microexplosions.

Interestingly, certain nanoparticles, such as silver spheres, possess absorption bands in the violet spectral range. We used femtosecond laser pulses at 800 nm to realize selective nonlinear absorption by silver nanoparticles and used this novel femtosecond laser/nanoparticle interaction to knock out genomic regions of interest.

For that purpose, centromeric DNA regions of interest have been labeled by in situ hybridization (ISH) with metallic nanoparticles. By designing the right size and composition, we fabricated about 70-nm-sized gold/silver particles with an absorption maximum of 400 nm. We were able to deposit energy selectively within the particle at laser intensities below the threshold intensity for destruction of nonlabeled DNA and performed intrachromosomal drilling effects. Interestingly, nanoholes with diameters in the range of the particle size have been realized (Garwe et al., 2005). Using this novel technology we were able to create for the first time sub-70-nm holes in human chromosomes and to knock out specific genomic regions of interest without collateral damage (see Fig. 27.4). These NIR laser-produced features represent a dimension of only 1/12 of the laser wavelength!

27.7 CONTAMINATION-FREE CELL SELECTION FOR MOLECULAR PATHOLOGY

NIR MHz femtosecond laser pulses have been used to isolate single cells or cell clusters of interest from chemically fixed sections, cryosections, and living cell monolayers placed on thin 1-μm polymer/plastic films. The samples were visualized by real-time CCD camera imaging, two-photon fluorescence imaging, or reflection/transmission imaging. A target of interest such as a single cell with a morphology different from the

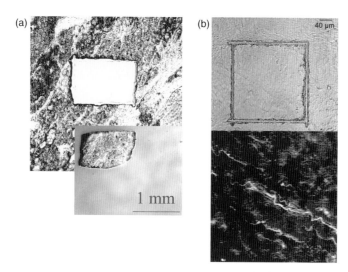

Figure 27.5. Cell isolation from tissue biopsies (left) and intratissue marking by laser microdissection (right).

surrounding cells was determined according to these images and isolated by laser cutting the foil and extracellular matrix elements. The applied laser power determined the cut size and depth. In part, the focal plane was varied in typical steps of 2 μm. The laser-isolated piece of foil with the target was taken by needles, by tweezers, or by gravitation-driven fall into the collection tubes. The isolated biological sample can be used to analyze DNA, RNA, and proteins or to culture particular living cells without cells of no interest. Figure 27.5, left, demonstrates the isolation of cell clusters from a biopsy.

A second method to isolate targets of interest from "contamination" is the laser removal of surrounding areas of no interest by ablation. This method can also be employed for standard histological sections placed on conventional glass slides, for cell containers, tissue-engineered surfaces, and so forth, without any foil.

It should be mentioned that low-cost laser systems based on nitrogen lasers at 337 nm are very successful on the market for cell isolation (Westphal et al., 2002). Femtosecond lasers may provide advantages over UV systems when biosafety and high precision are required.

Femtosecond laser pulses can be also employed to mark specific tissue areas. An example is seen in Figure 27.5, right, where fiber-like autofluorescent structures of interest from a cardiac tissue biopsy were marked with four line cuts and stained with fuchsin-eosin and antibodies to determine if these structures were elastic fibers (König et al., 2005a).

27.8 INTRACELLULAR SURGERY AND OPTICAL NANOINJECTION

Intracellular surgery was performed by positioning the multiphoton interaction volume inside the cell. Optical knocking out of single organelles such as an intracellular mitochondrion and nanodissection of particular chromosomes have been realized in living cells (see Fig. 27.4). Also, laser dissection of 0.3- to 0.5-μm cell–cell connections has been performed.

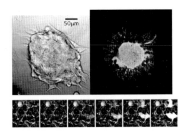

Figure 27.6. Optical nanoinjection of chemical compounds into single melanoma cells as well as specific intratissue cells of a melanoma spheroid by transient laser perforation of the cellular membrane.

Optical nanoinjection based on femtosecond lasers at the single-point illumination mode was realized by poking tiny holes into the cellular membrane and in part in the nuclear membrane. As a result, a typical cavitation bubble of only 1- to 5-μm diameter was observed, which was self-repaired within 5 s (see Fig. 27.4).

We used the nanoprocessing tool to load membrane-impermeable dyes and nanoparticles into living cells and deep-tissue cells such as 13.3-nm-sized 500-kDa FITC dextran macromolecules (10 μM in medium, 22) in NG108-15 melanoma cells introduced by laser puncture (Fig. 27.6). The cells remained alive after this laser procedure. The macromolecules accumulated in the cytoplasm of laser-perforated cells at a concentration of about 0.7 μM as determined by two-photon fluorescence intensity measurements. This corresponds to about 40 molecules in an 0.1-fl excitation volume (Stracke et al., 2005).

Loading of membrane-impermeable compounds into intratissue cells can be performed if the chemical agents can diffuse into the spheroids (Robinson et al., 2003) and tissues. Otherwise the drilling of small channels from the surface into the bulk material can be performed. Nanosurgery of intracelluar structures and the loading of the membrane-permeable protein-binding dye cascade blue acetyl azide within melanoma cell spheroids are depicted in Figure 27.6. The highest intracellular dye concentration was found 40 minutes after laser nanoinjection.

Loading in plant tissue with the dye propidium iodide by nanodissection of cell walls was performed in *Elodea densa*. In addition, optical destruction of single deep-tissue chloroplasts without collateral side effects has been demonstrated (Tirlapur & König 1999, 2002a, 2002b).

Watanabe and colleagues reported on femtosecond laser disruption for precise material removal and modification of intracellular organelles such as mitochondria (Watanabe et al. 2005). Maxwell and coworkers (2005) described nanoprocessing of intracellular structures of fluorophore-labelled cytoskeletal filaments and nuclei and the cutting of dendrites in living *C. elegans* without affecting the neighboring neurons.

27.9 TARGETED LASER TRANSFECTION

The laser-induced transient opening in the cellular membrane can also be used to introduce foreign DNA and to realize targeted transfection of particular cells of interest.

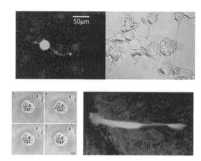

Figure 27.7. Targeted transfection of fibroblasts and neuroblastoma/glioma hybrid cells after laser spot illumination using DNA plasmids for the biosynthesis of green fluorescent proteins. Note the formation of a tiny transient 2- to 4-μm bubble after the laser spot illumination of the membrane.

Most of the current techniques such as electroporation, chemical membrane permeabilization, and viral carrier-mediated transfer do not allow the transfection of individual cells. Targeted transfection can be performed with mechanical microdelivery systems (microneedles).

Early laser studies on optoporation were conducted with UV lasers (Greulich & Weber, 1992; Tao et al., 1987) but faced problems with phototoxicity, low transfection efficiency, and poor beam quality. Also, photodynamic opening of the cellular membrane with the disadvantage of additional photosensitizing probes (Schneckenburger et al., 2002) and photoacoustic opening by focusing intense laser pulses into the cellular microenvironment (Shirahata et al., 2001) have been realized.

We developed a transfection technology based on femtosecond laser optoporation (Tirlapur & König, 2002a, 2002b) based on a single transient submicron hole in the cellular membrane and the diffusion-controlled transfer of DNA plasmids, such as the 4.7-kilobases vector pEEGFP-N1, into the cell. An additional laser optoporation of the nuclear membrane was not required due to the incorporation of the foreign DNA from the cytoplasm into the cellular DNA during cell division. Successful targeted transfection without any cell damage occurred when the laser-exposed cell started to synthesize the green fluorescent protein (GFP) within 1 to 2 days after the laser procedure. Transfection was realized in a variety of cells, including fibroblast cells, hybrid cells, and human dental pulp stem cells (Fig. 27.7).

27.10 INTRATISSUE NANOSURGERY

NIR laser pulses provide the possibility to deposit energy inside transparent materials including tissues Optical knocking out of deep-tissue cells and deep-tissue organelles can be performed. It is also possible to dissect single elastic fibers and collagen bundles in the tissue without destroying any superficial cell layers.

The first nanosurgery studies have been performed in a vital *Drosophila* larva. Two intratissue cuts with 1-nJ pulses and 80-μs exposure time were performed to dissect internal structures (Fig. 27.8).

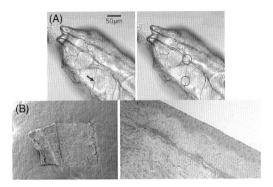

Figure 27.8. Intratissue nanosurgery. (A) Nanodissection of tissue structures inside a vital Drosophila larva. (B) Flap generation and intraocular tissue removal with 80-MHz femtosecond nanojoule laser pulses.

An interesting application of femtosecond laser nanoprocessing is intraembryonal surgery, as demonstrated on live gastrulating Drosophila embryos by Suppatto and colleagues (2005). They combined two-photon fluorescence microscopy, THG microscopy, and multiphoton ablation microscopy to perform and to control intravital nano-/microdissections that alter the embryo structural integrity without perturbing the cell movements in adjacent cells.

Yanik and associates (2004) reported on functional regeneration after femtosecond laser axotomy in the larval-stage worm *C. elegans*. They cut single circumferential 0.3-μm-thick axons of D-type motor neurons at their midbody positions inside the worm with 100% efficiency and monitored nerve regeneration and development.

27.11 OCULAR REFRACTIVE SURGERY WITH NANOJOULE FEMTOSECOND LASER PULSES

The most interesting current clinical laser application is refractive surgery, where extracellular matrix (ECM) components of the stroma have to be removed with submicron precision. Most exciting is the approach to realize intraocular nanosurgery without destruction of the epithelium layer. In order to find the intratissue target, two-photon tomography with pJ laser pulses can be performed with submicron spatial resolution. Epithelium, Bowman's membrane, and stroma of porcine and rabbit eyes can be clearly imaged if excited with NIR femtosecond laser pulses. By positioning the laser focus in the intratissue area of interest, cutting and ablation can be realized at an increased laser intensity of 1 to10 nJ. Clear intraocular cuts without significant damage zones with a size of 0.6 μm were demonstrated. Also, flaps could be produced by scanning a squared bottom deep in the stroma to remove the hydrophilic ECM protein collagen and by preparing the walls with x,z and y,z scans (König, 2003; König et al., 2002; see Fig. 27.8).

Studies on living New Zealand albino rabbits show a damage-free epithelium layer and fast healing effects after intratissue femtosecond laser procedures. It is very likely that the tiny microbubbles with a diameter of 5 μm or less that occur during the ablation process with sub-10-nJ pulses have been "recycled" by the tissue.

27.12 DISCUSSION

Multiphoton-ionization with nanojoule NIR femtosecond laser pulses opens the possibility of performing precise intramolecular, intracellular, and intratissue nanoprocessing without damage to surface layers and adjacent cells, in contrast to UV and NIR laser pulses for conventional microprocessing and surface ablation, due to selective photothermal effects, thermomechanical photodisruption, or photoablation.

The reported laser-induced sub-80-nm nanostructures in silicon, chromosomes, and DNA molecules present the finest NIR laser-produced "sub-Abbe" features.

Nanojoule femtosecond lasers can be used in molecular pathology, genomics, and proteomics to isolate living cells or tissue compartments of interest from histological sections.

Interestingly, as an alternative to expensive high-energy microjoule laser pulses of kHz laser amplifiers in refractive surgery, less-expensive nonamplified femtosecond laser systems providing nanojoule laser pulses at high MHz repetition rates can be used in ocular surgery. Low-energy pulses in combination with high NA focusing optics have also the advantage of reduced collateral damage and the absence of uncontrolled beam path due to self-focusing effects. In the future, a compromise between laser ablation time, NA, and laser pulse energy has to be found for optimal treatment of ocular tissues.

Potential applications of future laser nanomedicine and laser nanobiotechnology with femtosecond lasers include highly precise intracellular and intratissue surgery (König et al., 2005b), tissue engineering, stem cell research, gene therapy, and the production of DNA vaccines.

ACKNOWLEDGMENTS

The authors wish to thank Dr. Katja Schencke-Layland (Children's Hospital Los Angeles), Dr. Wolfgang Fritzsche, Dr. Andrea Csaki (IPHT Jena), Dr. Frank Garwe, Dr. Sven Martin (JenLab GmbH), Dr. Bagui Wang (FSU Jena), Herbert Schuck and Daniel Sauer from the Fraunhofer IBMT for their scientific contributions and the DFG (grant KO 1361/10-4) as well as the BMBF (PhoNaChi project 13N8467) for financial support.

Declaration of Financial Interests

Karsten König is the co-founder of the spin-off company JenLab GmbH.

REFERENCES

Anderson, Danile G, Levenberg, Shulamit, Langer, Robert. "Nanoliter-scale synthesis of arrayed biomaterials and application to human embryonic stem cells". Nature Biotechnology 22, 863–866, 2004.

Anderson Richard Rox, Parrish John A. "Selective photothermolysis: precise microsurgery by selective absorption of pulses radiation". Science 220:524–527, 1983.

Barnes Peter A., Riechhoff KE. "Laser-induced underwater sparks". Appl. Phys. Lett. 13: 282–288, 1968.

Basting, Dirk. "Excimer laser technology: laser sources, optics, systems and applications". Göttingen: Lambda Physics AG, 2001.

Bloembergen N. "Laser-induced electric breakdown in solids". IEEE QE 10:375–386, 1974.

Brinkmann R, Hüttmann G, Rögener J, Roider J, Birngruber R, Lin CP. Origin of retinal pigment epithelium cell damage by pulsed laser irradiance in the nanosecond to microsecond time regime. Lasers Surg Med 27:451–464, 2000.

Denk, Winfried, Strickler, James H., Webb, Watt W. "Two-photon laser scanning fluorescence microscope". Science 248:73–76, 1990.

El-Sayed IH, Huang X, El-Sayed MA. "Selective laser photo-thermal therapy of epithelial carcinoma using anti-EGFR antibody conjugated gold nanoparticles. Cancer Lett. 27, 2005.

Flaim, Christopher J, Chien, Shu, Bhatia, Sangeeta N. "An extracellular matrix microarray for probing cellular differentiation". Nature Methods 2, 119–125, 2005.

Garwe, Frank; Czaki, Andrea; Maubach, Gunter; Steinbrück, Andrea; Weise, Anja; König, Karsten; Fritzsche, Wolfgang. "Laser pulse energy conversion on sequence-specifically bound metal nanoparticles and its application for DNA manipulation. Nanoprocessing with nanojoule near infrared laser pulses". Med. Laser Appl. 20:201–206, 2005.

Göppert-Mayer, Maria. "Über Elementarakte mit zwei Quantensprüngen". Ann Phys 9:273–294, 1931.

Grand Denise, Bernas A, Amouyal E. "Photoionization of aqueous indole, conducting band edge and energy gap in liquid water". Chem. Phys. 44:73–79, 1979.

Greulich, Karl-Otto, Weber G. "The light microscope on its way from an analytical to preparative tool". J. Microsc. 167:127–151, 1992.

Heisterkamp Alexander, Ripken T, Thanongsak Mamom, Drommer Wolfgang, Welling Herbert, Ertmer Wolfgang, Lubatschowski Holger. "Nonlinear side effects of fs-pulses inside corneal tissue during photodisruption". Appl. Phys. B 74, 1–7, 2002.

Heisterkamp Alexander, Thanongsak Mamom, Kermani Omid, Drommer Wolfgang, Welling Herbert, Ertmer Wolfgang, Lubatschowski Holger. "Intrastromal refractive surgery with ultrashort laser pulses–in vivo study on rabbit eyes". Graefes Archives Clin Experiment Ophthal 241, 511–517, 2003.

Jay DG, Sakurai T. Chromophore-assisted laser inactivation (CALI) to elucidate cellular mechanisms of cancer. Biochim Biophys Acta 29, 1424(2–3):M39–48, 1999.

Kaiser Wolfgang, Garret CGB. "Two-photon excitation in CaF2:Eu2+". Phys Rev Lett 7:229–231, 1961.

Kennedy Paul K, Hammer DX, Rockwell Benjamin A. "Laser-induced breakdown in aqueous media". Progr. Quantum. Electron. 21:155–248, 1997.

Koester Helmut J, Baur Dagmar, Uhl Rainer, Hell Stefan W. Ca^{2+} fluorescence imaging with pico- and femtosecond two-photon excitation: signal and photodamage. Biophys J 77: 2226–2236, 1999.

Kollias Nikiforos, Bager A. "Spectroscopic characteristics of human melanin in vivo". J Invest Derm 85:38–42, 1985.

König Karsten, Liang Hong, Berns Michael W, Tromberg Bruce J. "Cell damage by near-IR microbeams". Nature 377:20–21, 1995.

König, Karsten, So, Peter; Mantulin William W., Gratton, Enrico. "Cellular response to near-infrared femtosecond laser pulses in two-photon microscopes". Opt. Lett. 22:135–136, 1997.

König, Karsten, Becker Thomas W., Riemann, Iris; Fischer, Peter; Halbhuber, Karl-Jürgen "Pulse-length dependence of cellular response to intense near-infrared laser pulses in multiphoton microscopes". Opt. Lett. 24:113–115, 1999.

König Karsten, Riemann Iris, Fischer Peter, Halbhuber Karl-Jürgen "Intracellular nanosurgery with near infrared femtosecond laser pulses". Cell Mol Biol 45:195–201, 1999.

König Karsten "Multiphoton microscopy in life sciences". J Microsc 200:83–104, 2000.

König Karsten, Riemann Iris, Fritzsche Wolfgang "Nanodissection of human chromosomes with near infrared femtosecond laser pulses". Opt Lett 26:819–821, 2001.

König Karsten "Cellular response to laser radiation in fluorescence microscopes". In: A. Perisami (editor), University Press, Oxford, Cellular Imaging: 236–251, 2001.

König Karsten, Krauss Oliver, Riemann Iris "Intratissue surgery with 80 MHz nanojoule femtosecond laser pulses in the near infrared". Opt Express 10:171–176, 2002.

König Karsten, Riemann Iris "High-resolution multiphoton tomography of human skin with subcellular spatial resolution and picosecond time resolution". J Biomed Opt 8:432–439, 2003.

König Karsten "High-resolution multiphoton imaging and nanosurgery of the cornea using femtosecond laser pulses". F. Fankhauser, S. Kwasniewska (eds.). Lasers in Ophthalmology. Kugler Publ., The Hague 79–89, 2003.

König Karsten "Multiphoton multicolor fluorescence in situ hybridization and nanoprocessing of chromosomes with near-infrared femtosecond laser pulses". Hemmerich P, Dieckmann S (Eds): Visions of the Cell Nucleus. American Scientific Publishers, Stevenson Ranch, California, USA, 2005.

König Karsten, Schenke-Layland Katja, Riemann Iris, Stock Ulrich A. Multiphoton autofluorescence imaging of intratissue elastic fibers. Biomaterials 26:495–500, 2005a.

König, Karsten. Riemann, Iris, Stracke, Frank, LeHarzic, Ronan "Nanoprocessing with nanojoule near infrared laser pulses". Med. Laser Appl. 20:169–184, 2005b.

Krueger Ronald R., Binder Perry S., McDonnell PJ. The effects of excimer laser photoablation on the cornea. In: Salz JJ (editor). Corneal laser surgery. pp. 11–17, 1995.

Le Harzic, Ronan, Bückle, Rainer, Wüllner, Christian, Donitzky, Christof, König, Karsten "Laser safety aspects of refractive eye surgery with femtosecond laser pulses", Med. Laser Appl. 20:233–238, 2005a.

Le Harzic, Ronan; Schuck, Herbert; Sauer, Daniel, Anhut, Tiemo, Velten, Thomas, König, Karsten "Sub-100nm nanostructering of silicon by ultrashort laser pulses". Optics Express. 13:6651–6656, 2005b.

Maxwell, Ina, Chung, Samuel, Mazur, Eric "Nanoprocessing of subcellular targets using femtosecond laser pulses. Nanoprocessing with nanojoule near infrared laser pulses". Med. Laser Appl. 20:193–200, 2005.

Mertz, Jerome "Second and third harmonic microscopy". In: Handbook of Biological Nonlinear Optical Microscopy. Eds. Barry R. Masters and Peter T.C. So, Oxford University Press, 2008

Oehring Hartmut, Riemann Iris, Fischer Peter, Halbhuber Karl-Jürgen, König Karsten. Ultrastructure and reproduction behaviour of single CHO-K1 cells exposed to near infrared femtosecond laser pulses. Scanning 22:263–270, 2000.

Park SJ, Taton TA, Mirkin CA. "Array-based electrical detection of DNA with nanoparticle probes." Science 295:1503–1506, 2002.

Pettit GH, Ediger MN. Corneal tissue absorption of ultraviolet radiation at 193 nm and 213 nm. Applied Optics 35:3386–3391, 1996.

Principe AH, Lin DY, Small KW, Aldave AJ. "Molecular hemorrhage after laser in situ keratomileusis (LASIK) with femtosecond laser flap creation". J. Refract. Surg. 19:94–103, 2004.

Ratkay-Traub I, Feruncz IE, Juhasz T, Kurtz R, Krüger RR. First clinical results with the femtosecond neodynium-glass laser in refractive surgery. J Refract Surg 19:94–103, 2003.

Riemann Iris, Killi Andreas, Anhut Tiemo, Le Harzic Ronan, Morgner Uwe, König Karsten. "Imaging and nanosurgery of biological specimen with a new diode pumped femtosecond laser at a wavelength of 1040 nm", Beyer E, Dausinger F, Ostendorf A, Otto A (eds.).

Procced. 3rd international WLT-conference on lasers in manufacturing, Munich, June 2005. AT-Fachverlag GmbH, Stuttgart, pp. 781–784, 2005.

Robinson Elisabeth E., Zazzali Kathleen M., Corbett Siobhan A., Foty Ramsey A. Foty. Alpha5beta1 integrin mediates strong tissue cohesion. J Cell Sci, 116:377–386, 2003.

Schneckenburger Herbert, Hendinger Anita, Sailer Reinhard, Strauss Wolfgang S. L., Schmitt Michael. "Laser-assisted optoporation of single cells". J Biomed Opt 7:410–416, 2002.

Sheppard Colin, Kompfner R. "Resonant scanning optical microscope". Appl Opt 17: 2879–2882, 1978.

Shirahata Y, Ohkohchi N, Itagak H, Satumo S. New technique for gene transfection using laser irradiation. J Invest Med 49:184–190, 2001.

Silva, Gabriel A., Czeisler, Catherine, Niece, Krisat L., Beniash, E., Harrington, Daniel A., Kessler, John A., Stupp, Samuel I. Selective differentiation of neural progenitor cells by high-epitope density nanofibers. Science 303, 1352–1355, 2004.

Squirrel Jayne M., Wokosin David L., White John G., Barister Barry D. "Long-term two-photon fluorescence imaging of mammalian embryos without comprimising viability". Nature Biotechnol 17:763–767, 1999.

Srinivasan Radhika. "Ablation of polymers and biological tissue by ultraviolet lasers". Science 234:559–565, 1986.

Stracke, Frank, Riemann, Iris, König, Karsten. Optical nanoinjection of macromolecules into vital cells. J Photochem Photobiol 81:136–142, 2005.

Supatto, Willy; Debarre, Delphine; Farge, Emmanuel; Beaurepaire, Emmanuel. "Femtosecond pulse-induced microprocessing of live Drosophila embryos. Nanoprocessing with nanojoule near infrared laser pulses". Med. Laser Appl. 20:207–216, 2005.

Tao Wen, Wilkinson Joyce, Stambridge Eric J., Berns Michael W. "Directed gene transfer into human cultured cells facilitated by laser micropuncture of the cell membranes". PNAS 84:4180–4184, 1987.

Tirlapur Uday K, König Karsten. "Near infrared femtosecond laser pulses as a novel non-invasive means for dye-permeation and 3D imaging of localized dye-coupling in the Arabidopsis root meristem". Plant J 20:363–370, 1999.

Tirlapur, Uday K, König, Karsten, Peuckert, Christiane, Krieg, Reimar, Halbhuber, Karl-Jürgen. Femtosecond near-infrared laser pulses elicit generation of reactive oxygen species in mammalian cells leading to apoptosis-like death. Experimental Cell Research 263:88–97, 2001.

Tirlapur Uday K, König Karsten. "Femtosecond near-infrared laser pulses as a versatile non-invasive tool for intra-tissue nanoprocessing in plants without compromising viability". Plant J 31:365–374, 2002a.

Tirlapur, Uday K., König, Karsten. "Targeted transfection by femtosecond laser". Nature 418:290–291, 2002b.

Trokel Stephen L., Srinivasan Radhika, Braren B. "Excimer laser surgery of the cornea". Am J Ophthalmol 96:710–715, 1983.

Ulrich Volker, Fischer Peter, Riemann Iris, König Karsten. Compact multiphoton/single photon laser scanning microscope for spectral imaging and fluorescence lifetime imaging. Scanning 26:217–225, 2004.

Vogel Alfred et al. "Energy balance of optical breakdown in water at nanosecond to femtosecond time scales". Appl. Phys. B 68:271–280, 1999.

Vogel Alfred, Venugopalan Vasan. "Mechanisms of pulsed ablation of biological tissues". Chem Rev 103:577–644, 2003.

Watanabe, Wataru; Matsungata, Sachihiro, Shimada, Tomoko. "Femtosecond laser disruption of mitochondria in living cells. Nanoprocessing with nanojoule near infrared laser pulses". Med. Laser Appl. 20:185–191, 2005.

Westphal Götz, Burgemeister R, Friedemann G, Wellmann A, Wernert N, Wollscheid V, Becker B, Vogt T, Knüchel R, Stolz W, Schütze K. "Noncontact Laser Catapulting: A basic procedure for functional genomics and proteomics". Methods Enzymol. 356:80–99, 2002.

Williams Frances, Varama SP, Hillenius S. Liquid water as a lone-pair amorphous semiconductor. JOSA B 1:67–72, 1984.

Yanik Mehmet Fatih, Cinar Hulusi, Cinar Hediye Nese, Chisholm Andrew D., Jin Yishi, Ben-Yakar A. "Nanosurgery: functional regeneration after laser axotomy". Nature 432:822, 2004.

Part IV

Biomedical Applications of Nonlinear Optical Microscopy

Current and Emerging Applications of Multiphoton Microscopy

Bruce J. Tromberg

The eminent physicist Freeman Dyson, who has written widely about the role of technology in science and society, observed, "There is a great satisfaction in building good tools for other people to use." The explosion of multiphoton microscopy (MPM) applications, comprehensively presented in this section, should provide immense satisfaction for the developers of MPM. Dyson also observed, "The technologies with the greatest impact on human life are simple", and MPM is certainly no exception. However, in the world of photonics, "simple" is a relative concept. While the original MPM idea remains simple and elegant, early MPM systems based on unstable picosecond mode-locked lasers, home-built data acquisition systems, and custom-built microscopes have been transformed from "complex" to "simple" by cleverly engineered commercial components. This movement towards technological simplicity has allowed scientists to focus their attentions on important new applications and discoveries. In fact, applications of MPM have grown, much like the physics of the technique, in a nonlinear manner since its introduction by Watt Webb and colleagues in 1990. In 1998 there were only 60 scientific publications with the keywords "multiphoton microscopy (MPM)" in the text. In 2005 there were approximately 1,000 and over 5,400 with "two-photon fluorescence." Perhaps even more dramatic is fact that only a handful of papers mentioned "second-harmonic generation (optical) imaging" prior to the year 2000, while in 2005 there were nearly 1,500 (Google Scholar, 2006).

MPM is an enabling technology that allows the visualization and measurement of biological processes uniquely, often with unprecedented performance. In order to appreciate how MPM has achieved this status, and to continue to match MPM with important applications, it is necessary to understand basic principles of non-linear optics, as well as the emerging field of tissue optics. MPM utilizes the high instantaneous excitation intensity of focused, ultrashort pulsed near-infrared laser light (e.g., femtosecond, or 10^{-15} s pulses) to electronically select the molecular origin of image-forming signals. These electronic processes include "two-photon fluorescence" (TPF) as well as "second-harmonic generation" (SHG), a signal that is derived nearly exclusively from fibrillar collagen in biological tissues. MPM is most commonly employed using TPF from a variety of exogenous fluorescent probes (e.g., organic dyes, fluorescent proteins, and nanoparticles) in order to provide contrast for imaging studies that span from single cells to thick intact tissues. TPF autofluorescence, primarily from NAD(P)H and flavoproteins in cells as well as the structural proteins elastin and collagen, can also provide important endogenous information without the use of exogenous probes. Both SHG and TPF signals are "blue-shifted" from the source, and SHG appears at exactly half the excitation wavelength. The combination of SHG and TPF allows for highly specific visualization of cell–extracellular matrix interactions with submicron resolution up to depths of approximately 0.5 mm in tissue (Zoumi et al., 2002). Because two-photon molecular absorption spectra are typically broad, multiple probes can be excited simultaneously, and MPM investigations based on spectral detection and fluorescence resonance energy transfer (FRET) are steadily expanding.

CURRENT MPM APPLICATIONS

The applications presented in this section are in six broad areas that have been uniquely advanced by MPM technology. In Chapter 33, Barry R. Masters describes the use of MPM to monitor intrinsic NAD(P)H and flavoprotein signals for studying cellular metabolism. TPF excitation allows substantially deeper penetration into tissues compared to the single-photon UV excitation used in conventional wide-field and confocal microscopes. Studies in cornea, pancreatic islets, neural tissues, and skin highlight the ability of intrinsic signal TPF to monitor both oxidative and glycolytic energy metabolism with high spatial and temporal resolution. Single- and multi-wavelength excitation strategies are used to optimize sensitivity to changes in cellular metabolic state, although one of the unique advantages of MPM is the ability to excite NAD(P)H and FAD^+ simultaneously with a single near-infrared wavelength. Additional information regarding NAD(P)H and flavoprotein binding, aggregation, and compartmentalization dynamics can be determined from fluorescence spectra and lifetime imaging.

In Chapter 32, Barry R. Masters and Peter T. C. So extend metabolic imaging concepts to detailed studies of skin. They describe the role of MPM in imaging both morphology and biochemistry, although the highly scattering keratinized epidermis limits signals to <200 μm in depth. TPF from cellular NAD(P)H and flavoproteins, and SHG/TPF from the extracellular matrix proteins collagen and elastin, respectively, constitute the primary MPM intrinsic signal components. The combination of MPM with exogenous fluorescent probes is a particularly powerful approach for quantifying transdermal transport properties, and TPF of exogenous "targeted" probes that can be topically applied is expected to have a significant impact on assessing skin pathologies.

In Chapter 28, David Kleinfeld and colleagues highlight seven seminal applications of TPF in mammalian neurobiology. MPM allows imaging hundreds of micrometers deep inside the brain of live animals, which, with the use of appropriate fluorescent probes, allows for the recording of both electrical and hemodynamic activity of entire neural networks. Important applications range from intracellular imaging of calcium dynamics and long-term neurodegeneration studies, to understanding the coupling between blood flow and neuronal activity.

In Chapter 29, Rakesh Jain and colleagues have focused on developing chronic in vivo models for dynamic imaging of tumor-associated blood and lymphatic vessels, vascular permeability, and tumor/host cell interactions in vessels and stroma. An additional important MPM capability in tumor biology is the intrinsic SHG signal derived from fibrillar collagen, a major constituent of the extracellular matrix that influences tumor cell metastasis and drug transport.

Developmental biology applications, described in Chapter 31 by Mary Dickinson and Irina Larina, span from imaging early cleavage-stage embryos to following development in the nervous, visual, and cardiovascular systems. Particularly important factors in these studies are related to imaging tissues longitudinally without damage for extended periods, and MPM has dramatically extended the safe, non-perturbative period for imaging embryogenesis far beyond that of confocal microscopy. Thus, the combination of MPM and proper selection of fluorescent reporters, particularly fluorescent protein-expressing animals, is essential to this field.

In Chapter 30, Ian Parker and Michael Cahalan review recent advances in immunology that allow visualization of lymphocyte mobility, chemotaxis, and antigen recognition in isolated lymphoid organs and in vivo lymph nodes in anesthetized mice. MPM allows both direct visualization and quantitation of the roles of lymphocytes (T and B) and dendritic cells in a lymph node at the beginning of an immune response. No other imaging approach is able to characterize motion, morphology, and function with comparable spatial and temporal resolution. Critical technical challenges involve visualization and tracking of cell positions in three spatial dimensions with sufficient contrast and resolution to quantify cell motility, proliferation, signaling, and cell–cell interactions.

A common MPM enabling feature in each application area is the ability to image single cells and cell populations with high resolution in the context of thick, 3D tissues (both perfused tissues and live animals). An additional important characteristic is the potential to use MPM for long-term, nondestructive, longitudinal sampling. Although Kleinfeld and Jain estimate that $\sim 1,000\ \mu m$ is the MPM depth limit, most studies probe tissues at less than half that depth. Parker and Cahalan point out that the primary factors behind the explosive growth of MPM in immunology are the "order-of-magnitude improvement in depth (vs. confocal microscopy) . . . and the relatively innocuous nature of the excitation light that permits long-term imaging with minimal photodamage." Because of signal attenuation, there is a tradeoff between imaging depth and speed, the latter being highly desirable in the case of capturing fast events, such as action potentials in neural tissue, blood flow in tumors and developing organs, and the homing of circulating T cells into lymph nodes.

MPM AND TISSUE OPTICS

Improving imaging depth and speed, enhancing contrast, and expanding field of view without loss of resolution are critical technical challenges for advancing MPM in virtually every application. The depth limits of MPM are a strong function of the intensity of light scattering in tissue [Dunn et al., 2000]. Light penetration is greatest in the red/NIR (600–1,000 nm) spectral region due to the fact that hemoglobin absorption diminishes by up to $\sim$100-fold between 600 and 700 nm and light scattering intensity drops off with increasing wavelength. In transparent media such as the eye, nonlinear imaging methods are primarily limited by the working distance of microscope objectives and tissue power exposure limits. In highly turbid materials such as skin, light scattering severely limits practical imaging depths to <200 μm for unstained tissues. The fundamental origin of light scattering in tissue is the microscopic discontinuity of refractive index both within and between cells. Photons encountering variations in refractive index will change direction (i.e., "scatter"). While the average tissue refractive index is approximately 1.39 to 1.41, this value can range from extremes of extracellular fluid (1.35) to melanin (1.7). Each scattering event results in the loss of coherent or ballistic excitation intensity in the focal plane and a commensurate quadratic diminution of MPM signal (Dunn et al., 2000).

Light scattering in tissue is defined by the scattering length, l_{sc}, the average distance between scattering events (approximately 20–50 μm in the NIR), and its reciprocal, the tissue scattering coefficient, μ_s. The reduced scattering coefficient, μ_s', and its reciprocal, the transport scattering length, l^*, account for the angular dependence of light scattering in tissue and is determined by $\mu_s' = \mu_s(1-g)$ where g = $\langle\cos\theta\rangle$ and θ is the average scattering angle. For most tissues, scattering is highly forward-directed and g ~0.8 to 0.95 with typical NIR values of μ_s' on the order of 1 mm^{-1}. Importantly, tissue scattering is dependent on wavelength, and $\mu_s'(\lambda)$ approximates Mie scattering in the NIR, such that $\mu_s' = a\lambda^{-b}$ where "a" and "b" parameters (pre-factor and scatter power) are proportional to scatter density and size, respectively. Because this relationship is valid only for reduced scattering properties, it is not, strictly speaking, a direct indicator of MPM penetration depth. However, the wavelength dependence of μ_s' is proportional to $\mu_s(\lambda)$ and can be readily measured in vivo for various tissues using noninvasive, broadband photon migration technologies (Bevilacqua et al., 2000; Cerussi et al., 2002). Thus, the relative benefit of longer-wavelength MPM sources can be estimated from multi-wavelength photon migration measurements of tissue optical properties.

For example, Figure IV.1 shows the wavelength dependence of μ_s' we have measured in vivo in human breast tissue, human breast tumor (adenocarcinoma), rat brain, human skin, and rabbit skeletal muscle using broadband diffuse optical spectroscopy (DOS). The transport scattering lengths, l^*, at 800 nm (a typical MPM excitation wavelength) are 0.56, 1.50, 1.75, 1.78, and 1.92 mm for skin, brain, muscle, breast, and tumor, respectively. These data clearly show the significant wavelength and tissue dependence of light scattering. In addition, they explain the relative difficulty of acquiring in vivo MPM images at depth in highly scattering skin structures. However, by moving to 1,000-nm excitation, with all other factors remaining equal, significant improvements in imaging depth of approximately 52% for skin, 11% for brain, 29% for muscle, and 40% for tumor can theoretically be achieved. Less than 10% improvement is expected in breast tissue because of the relatively flat wavelength dependence of scattering.

EMERGING MPM APPLICATIONS

What are the ideal excitation wavelengths for in vivo MPM applications, and are they accessible with current technologies? Consider the rapidly growing field of photonic crystal fibers and fiber lasers. Fiber lasers operating at 1.0, 1.3, and 1.5 μm have been reported in academic laboratories and are just now appearing commercially

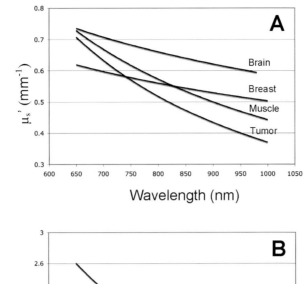

Figure IV.1 Wavelength dependence of the reduced scattering coefficient for various tissues.

(e.g., PolarOnyx Corporation, Sunnyvale, CA; Del Mar Photonics, San Diego, CA). Although much of the driving force behind source development in this spectral region is derived from the Optical Coherence Tomography community, MPM applications stand to benefit substantially from this effort.

For the MPM imaging applications in tumors described by Rakesh Jain and colleagues, a 1,300-nm source would penetrate up to ∼100% deeper than "conventional" 800-nm excitation (assuming animal tumor models can be developed with optical properties similar to humans). TPF from a near-infrared fluorophore, such as Cy 5.5 or Alexafluor, is detectable from much greater depths in tissue than probes such as fluorescein and GFP since emission in the ∼700-nm "tissue optical window" region will be detected. This confers an additional advantage since 1/e light penetration depths in tissue improve by ∼10 to 50 fold for NIR versus green light. If typical images are acquired from 400- to 500-μm depths using 800-nm excitation, these factors combined could lead to routine MPM imaging from *mm* depths using 1,300-nm excitation. Shifting to longer-wavelength excitation sources can also provide greater flexibility in terms of imaging speed. The 2-fold increase in l^* at 1,300 nm would double the excitation intensity (compared to 800 nm) at an equivalent depth, resulting in 4-fold greater fluorescence intensity. Since a NIR fluorophore would be used, the attenuation of the emission light from the point of excitation in the tissue back to the detector would be negligible. Thus, overall improvements in S/N of 8- to 10-fold or greater could easily be realized (at equivalent depths), reducing the dwell time per pixel and overall frame rate proportionally. Similar improvements are expected for skin (150%), brain (28%), and muscle (77%) on the basis of increased l^* at 1,300 nm (vs. 800 nm).

Alternative strategies for improving imaging speed and depth that are expected to have practical impact on MPM applications are suggested by the development of turnkey $\sim$ 10-fs sources and chirped mirrors with significant dispersion compensation (e.g., 300 fs^2/bounce at 800 nm), making sub-20-fs excitation at the sample a reality (e.g., Femtolasers, Vienna, Austria). We have recently demonstrated that as the source pulse duration at the sample is reduced from 400 fs to below 20 fs, the intensities of TPF and SHG increase close to the theoretical τ_p^{-1} limit [Tang et al., 2006]. This significant increase in MPM signal intensity has important practical implications. With the high excitation efficiency of sub-20-fs laser pulses, one can potentially interrogate deeper tissue structures. This was confirmed by comparing sub-20-fs to 120-fs pulses in ex vivo skin specimens where maximum imaging depths were $\sim$ 160% greater for the shorter (sub-20-fs) pulses [Tang et al., 2006]. Similarly, at a fixed depth, the increased excitation efficiency for sub-20-fs pulses can increase S/N, allowing for proportionally reduced signal integration times and faster data acquisition.

When using $\sim$10-fs sources to improve MPM efficiency, the impact of the intrinsically broad source spectral bandwidth should also be considered. In TPF, this can reduce the efficiency of two-photon absorption if the laser spectrum is broader than the absorption window of the fluorophore. For SHG, no effect is expected since the phase-matching condition is not critical for SHG interactions over the microscopy length scale (i.e., focal depths <20 μm). It is important to note that, if made bandwidth-limited, $\sim$ 20-fs pulses would have $\sim$ 34-nm spectral width (assuming sech^2 pulse shape), which fits in the absorption window of most fluorophores (Tang et al., 2006). Thus, depending on the application, a pulse width and spectral region can likely be selected that allows one to optimize MPM performance in terms of imaging depth, speed, and contrast.

With rapid commercial developments in fiber lasers, photonic crystal fibers, MEMS mirrors, table-top broadband sub-20-fs pulse sources, and spectrometers, new technologic components are poised to revolutionize MPM. In addition to optimizing TPF and SHG signals and reducing costs, it is expected that these tools will drive the growth of MPM micro-endoscope technologies. These advances will facilitate the "translation" of MPM technologies to the clinic, where endogenous TPF and SHG signals from skin, cartilage (Yeh et al., 2005), ophthalmic (Yeh et al., 2002), and cardiovascular (Zoumi et al., 2004) tissues have been reported, particularly in the case of imaging wounds, degenerative diseases, and wound repair. Fiber-based systems could be routinely used for remote "optical biopsy" measurements in body cavities, lumens (e.g., vascular, aerodigestive), and interstices (e.g., joints, tumors). MPM fiber micro-endoscopes with MEMS scanning mirrors have been demonstrated by Min Gu (Fu et al., 2006) and Mark Schnitzer (Piyawattanametha et al., 2006) and colleagues, and in vivo images have been acquired from deep brain hippocampal vessels in mice (Flusberg et al., 2005).

In summary, MPM has rapidly emerged as *the* standard approach for high-resolution imaging of cellular, tissue, and vascular structures in thick, intact tissues. This section highlights important, fundamental problems in dermatology, neuroscience, tumor biology, developmental biology, and immunology that are ideally suited to MPM solutions. Future applications of MPM are expected to depend on advances in optical technologies, such as laser sources, photonic crystal fibers, dispersion compensation optics, and scanners that facilitate greater imaging depth and speed, and probe/labeling technologies that enhance contrast and sensitivity. As the availability of state-of-the-art components expands, long-term interrogation of animal models, ex vivo tissue preparations, and engineered tissues at 1- to 2-mm depths will become routine. With increasing understanding of endogenous TPF and SHG signals, and the appearance of clinically approved molecular imaging agents, MPM technologies will be translated to the clinic, providing physicians with new strategies for disease detection, surgical guidance, and assessment of therapeutic response in individual patients.

ACKNOWLEDGMENTS

This work was supported by the Laser Microbeam and Medical Program (LAMMP) through the NIH (P41-RR01192), the Air Force Office of Scientific Research (AFOSR MFEL program), and the Beckman Foundation. Special thanks to my colleagues Shuo Tang for helpful discussions and Albert Cerussi, Sheng-Hao Tseng, Anthony Durkin, Jangwoen Lee, and David Cuccia for contributing tissue optical property measurements to this review.

REFERENCES

Bevilacqua, F, Berger, AJ, Cerussi, AE, Jakubowski, D, Tromberg, BJ *Broadband Absorption Spectroscopy in Turbid Media by Combined Frequency-Domain and Steady-State Methods.* Applied Optics, 39, 6498–6507, 2000.

Cerussi, AE, Tromberg, BJ "Photon Migration Spectroscopy," in Biomedical Optics Handbook, Tuan Vo-Dinh, Ed., CRC Press (Boca Raton, FL), 2002.

Dunn, AK, Wallace, VP, Coleno, M, Berns, MW, Tromberg, BJ *Influence of Optical Properties on Two-Photon Fluorescence Imaging in Turbid Samples*, Applied Optics, 39, 1–8, 2000.

Flusberg BA, Jung JC, Cocker ED, Anderson EP, Schnitzer MJ., *In vivo brain imaging using a portable 3.9 gram two-photon fluorescence microendoscope.* Opt Lett. 30. 2272, 2005.

Fu, L., Jain, A., Xie, H., Cranfield, C., Gu, M., *Nonlinear optical endoscopy based on a double-clad photonic crystal fiber and a MEMS mirror,* Optics Express 14, 1027, 2006.

Piyawattanametha, W., Barretto, R.P.J., Ko, T.H., Flusberg, B.A., Cocker, E.D., Ra, H., Lee, D., Solgaard, O., Schnitzer, M.J., *Fast-scanning two-photon fluorescence imagig based on a microelectromechanical systems two-dimensional scanning mirror,* Opt. Let. 31, 2018, 2006.

Tang, S., Krasieva, T.B., Chen, Z., Tempea, G., Tromberg, B.J., *Effect of pulse duration on two-photon excited fluorescence and second harmonic generation in nonlinear optical microscopy,* J. Biomed. Opt. 11, 02051, 2006.

Vogel, A., Venugopalan, V., *Mechanisms of pulsed laser ablation of biological tissues*, Chem Rev. 103 577–644, 2003.

Yeh, A, Nassif, N, Zoumi, A, Tromberg, B *Selective Corneal Imaging Using Combined Second Harmonic Generation And Two-Photon Excited Fluorescence,* Optics Letters, 27, 2082–2084, 2002.

Yeh, AT, Hammer-Wilson, M, Van Sickle, D, Benton, H, Zoumi, A, Tromberg, BJ, Peavy, G *Nonlinear Optical Microscopy of Articular Cartilage*, Osteoarthritis and Cartilage, 13 (4) 345–352, 2005.

Zoumi, A, Yeh, A, Tromberg, BJ *Imaging Cells And Extracellular Matrix In Vivo Using Second-Harmonic Generation And Two-Photon Excited Fluorescence*, PNAS, 99, 11014–11019, 2002.

Zoumi, A, Lu, X, Kassab, GS, Tromberg, BJ *Imaging Coronary Artery Microstructure Using Second-Harmonic and Two-Photon Fluorescence Microscopy*, Biophys. J. 87, 2778–2786, 2004.

28

Pioneering Applications of Two-Photon Microscopy to Mammalian Neurophysiology

Seven Case Studies

Q.-T. Nguyen, G. O. Clay, N. Nishimura,
C. B. Schaffer, L. F. Schroeder, P. S. Tsai, and
D. Kleinfeld

It is commonly assumed, although insufficiently acknowledged, that major advances in neuroscience are spurred by methodological innovations. Novel techniques may appear quite daunting at first, but their successful application not only places them among mainstream methods, but also stimulates further developments in new directions. Such is the case for two-photon laser scanning microscopy (TPLSM) (Denk et al., 1990), which was introduced in neuroscience in the early 1990s (Denk et al., 1994). Since that time, the impact of TPLSM in neurobiology has been nothing short of remarkable. Several excellent reviews on TPLSM have already covered the general aspects of this technique (Denk et al., 1995; Denk & Svoboda, 1997; Stutzmann and Parker, 2005; Zipfel et al., 2003), details of the instrumentation (Tsai et al., 2002), and recent advances (Brecht et al., 2004). The object of this chapter is to showcase seminal applications of TPLSM in neuroscience, with a particular emphasis on mammalian neurobiology. The seven case studies presented here not only expound the broad range of applications of TPLSM but also show that TPLSM has been used to resolve significant issues in mammalian neurobiology that were previously unanswered due to lack of an appropriate technique.

28.1 QUALITATIVE BACKGROUND

The inherent qualities of TPLSM that have allowed groundbreaking advances in neuroscience derive entirely from using a high-repetition-rate mode-locked laser, typically a femtosecond pulsed Ti:Sapphire oscillator. This device generates trains of intense short light pulses ($\leq 100\,\text{fs}$) at rates ranging from a few kilohertz to one gigahertz. Although the average power of a femtosecond laser in a TPLSM is in the order of a few hundred milliwatts or less, the density of incident photons during each pulse is such that fluorophores that are normally excited by photons with wavelength λ (energy hc/λ, where h is Planck's constant and c is the speed of light) can be excited by the near-simultaneous arrival of two photons with double the wavelength and half the energy. The probability that the nonlinear, two-photon absorption process occurs will increase as the square of the intensity of the incident light.

Two-photon laser scanning microscopes provide images with a resolution markedly better than that of wide-field fluorescence microscopes in scattering samples; a comparison between nonlinear and confocal imaging techniques with the same preparation is given by Kang and colleagues (Kang et al., 2005). Two-photon laser scanning microscopes take advantage of the square dependence of two-photon absorption with excitation intensity by tightly focusing the laser beam with a high numerical aperture (NA) objective. This creates a volume of less than 1 μm^3 in which two-photon excitation takes place. The resulting natural optical sectioning effect allows micrometer to submicrometer resolution in three dimensions. A volume is probed by sweeping the focal point of the beam. Within the focal plane, this is achieved by systematically varying the incident angle of the laser beam in the back-aperture of the objective with galvanometric mirrors (Tan et al., 1999), resonant fiber optics (Helmchen et al., 2001), or acousto-optical deflectors (Bullen et al., 1997; Roorda et al., 2004). Between planes, this is achieved by changing the height of the objective relative to the sample.

The confinement of excitation in TPLSM is particularly advantageous for neurobiological applications. First, the small excitation volume alleviates fluorophore bleaching or phototoxic damages outside the focal plane. This feature is critical for long-term time-lapse imaging of cellular morphology or observation of neuronal activity in subcellular compartments. Second, the small excitation volume greatly reduces out-of-focus absorption due to scattering of incident photons, which is advantageous when imaging deep inside a specimen. Incidentally, the ability to optically section solely with the incident beam simplifies the design of two-photon microscopes in comparison with conventional laser confocal scanning systems.

Of considerable importance for in vivo neurobiological imaging applications is the use of laser wavelengths that fall in the near-infrared range. Photons in these wavelengths are better able to penetrate neural tissue, which is heavily scattering, than photons with shorter wavelengths. Consequently, excitation light that is generated by common femtosecond lasers can reach several hundred of microns under the surface of the brain while still being able to excite fluorescent molecules that are normally stimulated with shorter wavelengths. This was particularly apparent in the case of astrocytes imaged in a brain slice (Kang et al., 2005), where TPLSM provided excellent resolution of cells as deep as 150 μm inside the specimen, while imaging with a laser confocal microscope was limited to the immediate surface of the neural tissue and was further hampered by phototoxicity.

28.2 IN VITRO SUBCELLULAR APPLICATIONS OF TPLSM

TPLSM was first employed to measure calcium dynamics in subcellular compartments of single neurons with unprecedented spatial resolution (Kaiser et al., 2004; Noguchi et al., 2005; Svoboda et al., 1996; Yuste & Denk, 1995). These experimental results bear on theories that associate activity-dependent, long-lasting cognitive processes such as learning and memory with possible physiological and anatomical changes in brain cells. It has been proposed that in certain categories of neurons, subcellular

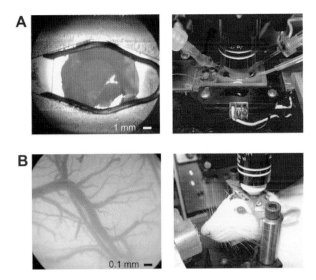

Figure 28.1. In vitro and in vivo mammalian preparations used in neurobiological experiments involving TPLSM. (A) Brain slice preparation. Left picture: 400-μm thick slice from the forebrain of an adult rat. The slice was placed in a recording chamber and held in place by an anchor with nylon threads straddling the slice. Right picture: Optical/electrophysiological recording setup with an extracellular stimulating electrode on the left and an intracellular recording electrode on the right. (B) Anesthetized rodent preparation. Left picture: Bright-field image of the vasculature at the surface of the cerebral cortex of a rat viewed through the imaging craniotomy. Bottom picture: In vivo recording setup. Notice the optical window placed above the craniotomy and under the microscope objective. The nose of the rat is facing a nozzle providing a constant flow of anesthetic.

appendages present on neuronal extensions, called dendritic spines, play a central role in the integration of input signals coming from other neurons. Further, many theories posit that dendritic spines are essential in memory formation and storage (Martin et al., 2000). The geometry and size of dendritic spines could, in principle, allow local retention of calcium, an ion known to be critical for many long-term intracellular events. However, basic physiological properties of dendritic spines could not be measured prior to the advent of TPLSM since the volume of dendritic spines is in the order of one cubic micrometer or less. Seminal experiments on dendritic spines using in vitro brain slice preparations (Fig. 28.1A) were among the first applications of TPLSM in neurobiology (Svoboda et al., 1996; Yuste & Denk, 1995). Calcium regulation in subcellular compartments is still a much-studied topic with TPLSM (Kaiser et al., 2004), sometimes in combination with two-photon activation of caged neurotransmitters (Noguchi et al., 2005).

28.3 IN VIVO CELLULAR APPLICATIONS OF TPLSM

It was quickly realized that the ability to use TPLSM to image deep inside neural tissue would allow investigators to link neuronal activity and morphology at higher levels of organization in the brain. It has long been proposed that the connectivity of neural circuits is not fixed but varies according to the pattern of its inputs. TPLSM was used to validate this hypothesis in vitro (Engert & Bonhoeffer, 1999; Maletic-Savatic et al., 1999). Experimental evidence of activity-dependent remodeling of neuronal wiring in response to sensory manipulation was provided by TPLSM imaging in the cerebral cortex of developing rats (Grutzendler et al., 2002; Lendvai et al., 2000; Trachtenberg et al., 2002).

The ability to image deep inside the brain of live, anesthetized animals has opened the way to record the electrical activity of whole neural networks using TPLSM (see Fig. 28.1B), providing that neurons in these networks are loaded with an appropriate optical reporter. Calcium-sensitive fluorescent dyes (Grynkiewicz et al., 1985) are currently the indicators of choice because their fluorescence varies robustly with changes in the concentration of intracellular calcium that, in turn, is strongly correlated with neuronal activity. Furthermore, many calcium-sensitive fluorescent dyes exist in a cell-permeant form, which facilitates their entry into neurons. However, a recurrent problem in in vivo experiments is the ability to label potentially thousands of cells of interest with intracellular dyes in a volume that corresponds to the extent of the network of interest. A method that involves intracerebral perfusion of dyes is described by Stosiek and colleagues (Stosiek et al., 2003). Using this approach, Ohki and coworkers managed to accurately map domains of common neuronal responses involved in vision in the adult rat cerebral cortex (Ohki et al., 2005). Other labeling approaches include the use of modified, nonpathogenic viruses to deliver a genetically engineered fluorescent dye (Lendvai et al., 2000) and the development of transgenic mice that express a fluorescent protein in specific neurons (Margrie et al., 2003). In the latter case, labeled cells correspond to neurons with a unique phenotype. These cells can be unequivocally localized and targeted for intracellular microelectrode recording under visual control provided by TPLSM (Margrie et al., 2003).

28.4 TPLSM FOR HEMODYNAMIC AND NEUROPATHOLOGIC ASSESSMENTS

The coupling between blood flow and neuronal activity is, at present, only partially understood. TPLSM provides a tool to aid studies in this area. Fluorescent dye injected into the bloodstream acts as a contrast agent, allowing TPLSM to visualize the motion of red blood cells. This technique provides a straightforward means to measure cerebral blood flow in vivo (Chaigneau et al., 2003; Kleinfeld, 2002; Kleinfeld et al., 1998). A related application of TPLSM is to study neurovascular diseases in animal models. For example, TPLSM in combination with neuropathological markers has been used to monitor the progression of Alzheimer's disease in a minimally invasive fashion before and after immunological treatment (Backsai et al., 2001, 2002). The proven ability of TPLSM to monitor the effectiveness of potential cures for a major neurodegenerative disease in vivo highlights the potency and versatility of this technique.

28.4.1 Case Studies of the Application of TPLSM to Mammalian Neurophysiology

We consider seven case studies highlighting novel findings in mammalian neurophysiology that relied on TPLSM. These serve to illustrate the power of the technique, as well as the importance of correlating TPLSM-based observations with measurements performed with other techniques such as intracellular electrophysiology. Our examples range from subcellular dynamics to population responses.

28.4.1.1 Calcium Dynamics in Dendritic Spines

Two-photon imaging is a particularly powerful tool to study neural activity at the scale of individual dendrites and dendritic spines in vitro. These experiments are performed in brain slice preparations (see Fig. 28.1A). Brain slices are obtained by cutting specific regions of isolated brains, such as the hippocampus, into sections with a thickness usually ranging from 100 to 500 μm. When brain slices are bathed in the appropriate saline, neurons remain viable and their connections can be functional for several hours. An important advantage of optical experiments done in brain slices is the absence of biological motion artifacts, such as respiration or heartbeat. The thinness of brain slices is also advantageous in many respects, while the relative transparency of slices and their flatness facilitates optical imaging. Pharmacological compounds such as ion channel blockers have better access to their target than in the whole brain. Finally, intracellular recordings in slices are much easier to perform and more stable than those in vivo.

The size of spines is on the order of what can be resolved with diffraction-limited optical imaging. Thus, dynamic studies of spine morphology and physiology require the ability to image with micrometer-scale resolution into a highly scattering sample. This requirement is met by TPLSM used in conjunction with fluorescent probes that are sensitive to the intracellular environment. Yuste and Denk (Yuste & Denk, 1995) injected single neurons in brain slices with the calcium-sensitive dye Calcium Green and monitored fluorescence changes in individual spines and the contiguous dendritic shafts in response to electrical stimulation. TPLSM enabled them to resolve individual spines and the shaft between the spines. Stimulation of the cell soma elicited calcium changes in individual spines via opening of local calcium channels (Figs. 28.2A and 28.2B), definitively establishing that the dendritic spine serves as a basic unit of computation in the mammalian nervous system. A key result of this study was that the confluence of presynaptic and postsynaptic spiking led to a nonlinear enhancement in the concentration of intracellular calcium within the postsynaptic spine.

Kaiser and colleagues (Kaiser et al., 2004) extended the approach of Yuste and Denk (1995) by injecting two connected neurons with calcium-sensitive fluorescent dyes with different emission wavelengths. They subsequently identified a synaptic connection between the axon of one neuron and one dendrite on the other neuron based on cellular morphology. Since the calcium dyes labeling either side of the synapse were spectrally distinct, the authors were able to independently identify the calcium dynamics on the presynaptic and postsynaptic side of the neural connection (see Figs. 28.2C to 28.2E).

28.4.1.2 Structure–Function Relationship of Dendritic Spines

Fundamental characterization of dendritic spine properties can be carried further by combining imaging with other optical processes that utilize the spatial confinement

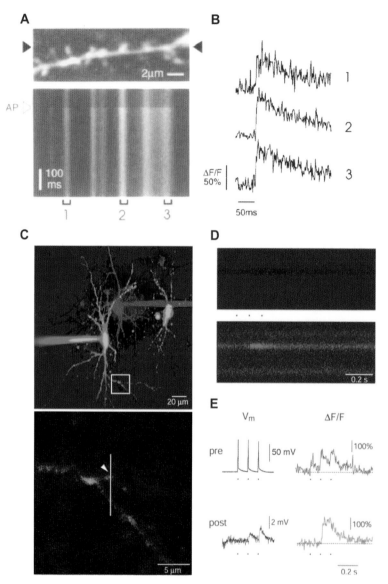

Figure 28.2. Subcellular calcium dynamics in neurons imaged by TPLSM. (A) Image of dendritic segment of a neuron labeled with the calcium-sensitive fluorescent dye Calcium Green (top) and repetitive line-scans between black arrows (bottom) during an action potential elicited by a 3-ms, 50-mV pulse applied to the soma (white arrow in line-scan data). Increases in fluorescence, associated with increases in calcium ion concentration, were observed in all dendrites. (B) Fluorescence changes in three dendritic spines as a function of time. In contrast to these data, the calcium ion transients induced by stimulation that was not sufficient to generate an action potential resulted in measurable calcium ion concentrations only in a subset of dendritic spines. (C) Two-photon fluorescence image (top) of a pyramidal neuron (red, labeled with Rhodamine-2) and a bitufted interneuron (green, labeled with Oregon Green Bapta-1). For both dyes, the fluorescence increased with increases in calcium ion concentration. Image of a synapse (bottom) between the pyramidal cell and the interneuron, from the region in panel A, indicated with a white box. The arrowhead indicates the location of the synapse, and the white line indicates the region where the line-scan data shown in panel D were taken.

of two-photon interactions. Photobleaching and photoactivation processes that result from two-photon absorption can be used to induce and observe dynamics with a spatial resolution sufficient to study subcellular structures. In work by Svoboda and associates (Svoboda et al., 1996), dendrites of neurons in rat hippocampal slices were loaded with fluorescein-dextran via an intracellular micropipette. TPLSM enabled visualization of individual spines in neurons inside the optically scattering brain slice (see Fig. 28.1A). The authors achieved the necessary time resolution to image diffusion through the use of a line-scan pattern, in which the focus of the femtosecond laser beam was repeatedly scanned along a single line through the spine or dendrite of interest (Figs. 28.3A to 28.3C). The time resolution was determined by the speed of the scanning mirrors, which was ~2 ms/line in these experiments. During line-scan imaging, fluorescein in either the dendrite or synapse was photobleached or photoreleased by a high laser power during a single line scan. The subsequent recovery of fluorescence or decrease of fluorescence was used as a measure of the diffusion between the unbleached and bleached areas. The use of different combinations of photobleaching and imaging in the spine and its dendritic shaft enabled the authors to demonstrate that chemical compartmentalization takes place in spines, and to estimate the electrical resistance of the spine neck.

The role of specific components involved in synaptic signaling, such as NMDA (N-methyl-d-aspartate) receptors, can also be investigated using TPLSM. Presynaptic neurons transform their electrical activity into the release of packets of neurotransmitter molecules, which will target cells across the synapse. NMDA receptors are protein complexes located on the postsynaptic side of a synapse that transduce pulses of the neurotransmitter glutamate into excitatory electrical potentials in the postsynaptic cell. NMDA receptors are thought to be critical for neuronal plasticity because of their additional dependence of the postsynaptic voltage and their large permeability to calcium, which leads to an increase in intracellular calcium in the postsynaptic cell when NMDA receptors are activated. Localized glutamate release can be mimicked in vitro by using caged glutamate, an inert derivative of glutamate that can release ("uncage") glutamate when it is optically activated by absorption of a UV photon or, equivalently, by two photons in the visible range.

Noguchi and colleagues (Noguchi et al., 2005) combined two-photon fluorescence imaging of a calcium indicator with two-photon photon-uncaging of glutamate to understand the role of calcium signaling during NMDA receptor activation in hippocampal

---

Figure 28.2 (*Continued*). (D) Line-scans taken during the stimulation of three action potentials in the presynaptic, pyramidal neuron (top) and the postsynaptic interneuron (bottom). Three action potentials were evoked in the pyramidal cell, giving rise to a calcium ion increase in the presynaptic cell (top image). (E) Electrical recordings of action potentials in the cell soma (left) and changes in fluorescence measured in a single synapse induced by calcium dynamics (right) in the presynaptic (top) and postsynaptic (bottom) cell. Three action potentials were induced in the presynaptic cell, giving rise to three transient increases in calcium ion concentration. In the postsynaptic cell, two excitatory postsynaptic potentials were measured, but only one led to a measured increase in calcium ion concentration (perhaps indicating the presence another synaptic contact between these two cells). (A and B adapted from Yuste & Denk, 1995; C-E adapted from Kaiser et al., 2004)

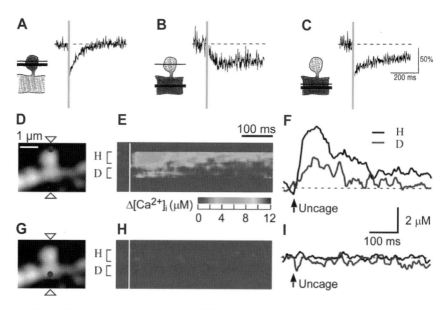

Figure 28.3. Two-photon imaging of diffusion dynamics in dendritic spines. (A–C) Photobleaching and diffusional recovery of fluorescence. Schematics indicate the position of fluorescence measurement (thin line) and photobleaching (thick lines) through spines and associate dendritic shafts. Graphs show the time course of fluorescence before and after photobleaching, indicated by gray bar. (D–I) Measurement of calcium concentration in response to stimulation by photo-uncaging glutamate. Red dots indicate position of uncaging of glutamate and arrowheads indicate position of line-scan for fluorescence signal on stacked images of spines filled with Alexa Fluor 594 (D, G). (E–H) Changes in calcium concentration derived from line-scan images of fluorescence of calcium-sensitive Oregon Green-BAPTA-5N, with bar indicating time of glutamate uncaging. H and D show regions average for F and I. (A–C adapted from Svoboda et al., 1996, D–H from Noguchi et al., 2005)

slices. Two-photon uncaging of glutamate allowed the precise excitation of NMDA receptors on a single spine head, unlike previous attempts that used electrical stimulation and failed to confine electrical excitation to individual synapses. The combination of optical activation of single synapses with an optical measurement of intracellular calcium enabled them to measure calcium changes that were induced by photochemical stimulation of a single synapse and the attached dendritic shaft. Calcium concentration was estimated with the calcium-sensitive fluorescent dye Oregon Green, and individual neurons were filled with a mixture of both calcium-sensitive and calcium-insensitive fluorescent dye. Both probes were excited with 830-nm femtosecond laser light, but their emission spectra were far enough apart to determine changes in calcium concentration via the ratio of emitted light. A second femtosecond laser at 720-nm wavelength was used to photo-uncage glutamate. Electrical current generated by the influx of ion inside the cell was also measured by conventional microelectrode intracellular recording (see Figs. 28.3D to 28.3I). Combined with pharmacological blockage of various NMDA receptors, these experiments determined that excitation of a single synapse does lead to calcium increases in the dendritic shaft, and suggested that the extent of the calcium movement between synapse and dendrite depended on the shape of the spine head and neck (see Figs. 28.3E to 28.3H).

28.4.1.3 Activity-Dependent Plasticity of the Neuronal Architectonics

Persistent changes in neuronal circuitry in response to varying inputs may occur by at least three different mechanisms: dendritic and axonal extension and pruning (Martin et al., 2000), synapse formation and elimination (Ramón y Cajal, 1893), and potentiation or depression of existing synapses (Ziv & Smith, 1996). Testing these hypotheses ideally requires that one observe a single neuron over periods of time that range from seconds to months. Early attempts to image morphological changes in peripheral neurons used wide-field microscopy (Purves & Hadley, 1985). These efforts were hampered by the poor z-axis spatial resolution inherent to conventional fluorescence imaging. In contrast, TPLSM offers the ability to image individual neurons over several months, with sufficient resolution to track changes in neuronal morphology as small as dendritic spines.

The cerebral cortex of mice includes a region called the vibrissa (whisker) somatosensory cortex, where sensory neurons are organized into functional maps that correspond to the grid-like organization of the large vibrissae on the face of the animal (Woolsey & van der Loos, 1970). Trachtenberg and coworkers (Trachtenberg et al., 2002) were able to image individual green fluorescent protein (GFP)-expressing neurons in this region over many days. They found that axons and dendrites were remarkably stable over several months. However, the morphological changes in spines varied according to three time scales: transient (<1 day), semi-stable (2 to 7 days), and stable (>8 days) (Figs. 28.4A to 28.4E). Interestingly, the evolution of spine morphology is not the same in all regions of cerebral cortex. Grutzendler and colleagues (Grutzendler et al., 2002) performed a similar study in the visual cortex of adult mouse on the same layer 2/3 dendritic arbors using TPLSM and found that 98% of the spines were stable over days and 82% were stable over months (see Fig. 28.4F).

Trachenberg and coworkers (Trachenberg et al., 2002) further assessed spine turnover in response to sensory stimulation of the vibrissae. The investigators trimmed every other vibrissa from the mystacial pad of the mouse, such that each trimmed vibrissa was surrounded by untrimmed vibrissae. This chessboard deprivation is known to produce a robust rearrangement of the cortical neuronal map of the sensory responses to individual vibrissae. The results from this study showed an increase in the turnover of spines a few days after trimming, suggesting a role for spine formation/elimination in rewiring cortical circuits (see Fig. 28.4G).

28.4.1.4 Calcium Imaging of Populations of Single Cells

While calcium imaging of individually loaded cells has yielded important insights into the calcium dynamics and neural activity in single cells, the bulk loading of cell populations with calcium indicator dyes has been, until recently, problematic for in vivo studies. The ability to load tens to hundreds of cells in a local population in vivo was demonstrated by Stosiek and associates (Stosiek et al., 2003) using a pressure-ejection method they called multicell bolus loading. In brief, a micropipette filled with dye-containing solution was placed deep in rodent cortex and was used to pressure-eject ~400 fl of dye into the extracellular space (Fig. 28.5A). This resulted in a ~300-μm-diameter spherical region of stained cells (see Fig. 28.5B). Using this

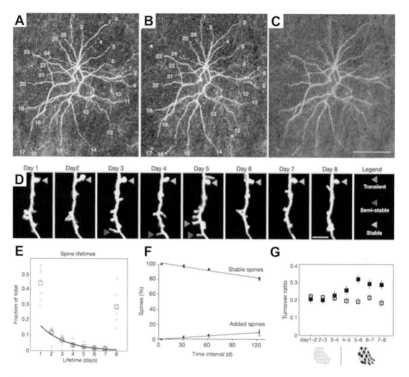

Figure 28.4. Long-term in vivo imaging of neuronal structure. (A–C) Adult mouse dendritic branches were stable over weeks. (A) Dorsal view of an apical tuft at imaging day 16; 26 branch tips are labeled for reference. (B) Imaging day 32. (C) Overlay (day 16, red; day 32, green). Scale bar, 100 μm. (D) Images of a dendritic segment acquired over eight sequential days. Spines appeared and disappeared with broadly distributed lifetimes. Examples of transient, semi-stable, and stable spines (with lifetimes of 1 day, 2 to 7 days, and 8 days, respectively) are indicated with blue, red, and yellow arrowheads, respectively. Scale bar, 5μm. (E) Spine lifetimes in adult mouse barrel cortex. Lifetimes are defined as the number of sequential days (from a total of eight) over which a spine existed. Individual neurons (gray diamonds) and the average (black squares) are shown. The fraction of spines with lifetimes of 2 to 7 days was fitted with a single time constant (thick black line). The fractions of spines with lifetimes of less than 1 day (transient spines) and greater than 8 days (stable spines) were significantly greater than predicted from the exponential fit, and therefore constitute distinct kinetic populations. (F) Spine lifetimes in adult mouse visual cortex. Percentage of spines that remained stable or were added as a function of imaging interval. Data are presented as mean ±SD. Scale bars, 1 mm. (G) Sensory experience modulated spine turnover ratio (the fraction of spines that turn over between successive imaging sessions). Chessboard deprivation of whiskers occurred immediately after imaging day 4. Turnover ratio increased after deprivation within (solid squares) but not outside (open squares) the barrel cortex. Error bars show SEM. (A–E and G adapted from Trachtenberg et al., 2002, F from Grutzendler et al., 2002)

method, cells could be loaded with a variety of membrane-permeant calcium indicators, although at lower indicator concentrations compared to cells individually loaded by intracellular injections (i.e., $<100\,\mu$M compared to a few millimolar, respectively). Nonetheless, spontaneous calcium transients could be recorded optically with TPLSM from a population of cells (see Fig. 28.5C).

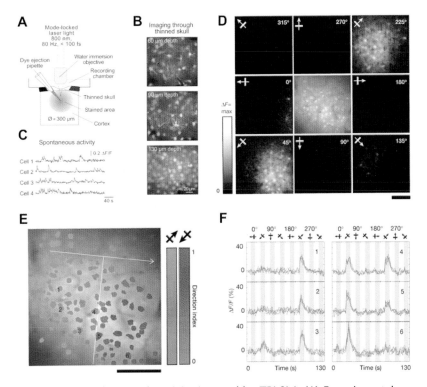

Figure 28.5. Neuronal network activity imaged by TPLSM. (A) Experimental arrangement for pressure-ejection loading of a neuronal population with a calcium indicator in vivo. (B) Images taken through a thinned skull of a P13 mouse at increasing depth after pressure-ejection loading with calcium indicator Fura-PE3 AM. (C) Spontaneous Ca^{2+} transients recorded in a different experiment through a thinned skull in individual neurons (P5 mouse) located 70 μm below the cortical surface, after loading with calcium indicator Calcium Green-1 AM. Scale bar = 20 μm. (D) Direction discontinuity in cat visual cortex visualized by population loading of cells with calcium indicator in vivo, and visual stimulation with drifting square wave gratings at different orientations. Single-condition maps (ΔF) imaged 180 μm below the pia mater layer are shown in the outer panels. The central panel shows an anatomical image reconstructed by averaging over all frames during the visual stimulation protocol. Cells were activated almost exclusively by stimuli of one orientation moving in either direction (45 degrees and 225 degrees). To the non-preferred stimuli, such as 90 degrees, the calcium responses were so small and the noise was so low that the single-condition maps are almost indistinguishable from zero. Scale bar = 100 μm. (E) Cell-based direction map; 100% of cells had significant responses. Color specifies preferred direction (green, 225 degrees; red, 45 degrees). The cells responding to both 45 degrees and 225 degrees are displayed as gray, according to their direction index (see color scale on right). The vertical white line below the arrow indicates approximate position of the direction discontinuity. Scale bar = 100 μm. (F) Single-trial time courses of six cells, numbered 1 to 6 as in part E. Five trials (out of ten) are superimposed. (A–C adapted from Stosiek et al., 2003, D–F adapted from Ohki et al., 2005)

The ability to quickly and simultaneously load an entire population of cells with a functional fluorescent indicator opens the door to optical evaluation of network dynamics and functional architecture. Previously, such studies were largely limited to the use of multisite electrodes, voltage-sensitive dyes, or intrinsic optical imaging of the hemodynamic response to neuronal activity. In the case of multisite electrode recordings, the recording sites are tens to hundreds of microns apart, and hence provide only sparse sampling of a cell population. Conversely, voltage-sensitive dyes and intrinsic optical imagining have insufficient spatial resolution to distinguish the activation of individual cells in vivo.

The use of calcium indicator imaging to elucidate functional architecture was demonstrated by Ohki and associates (Ohki et al., 2005) in cat visual cortex. A population of cells was loaded with the AM form of a calcium indicator using multicell bolus loading (see Fig. 28.5D). The AM form is permeant through the cell membrane and then is trapped inside the cell (Grynkiewicz et al., 1985). The response of individual cells to visual stimuli at different orientations was characterized. It was found that sharp boundaries existed between cell populations with different tunings to oriented bars in the visual field (see Fig. 28.5E). Importantly, single-cell resolution of neuronal activity over a population of cells in vivo, as demonstrated by the response curves in Figure 28.5F, enabled investigations of the local heterogeneity in cell populations and the precision and sharpness of cortical maps.

28.4.1.5 Targeted Intracellular Recording In Vivo

The state of a neuron is represented by many dynamic variables, including intracellular calcium (see above) and other chemical species. However, the state variable closely tied to neuronal input/output relations is the transmembrane voltage. The current lack of suitable optical indicators of membrane voltage with rapid kinetics and sufficient signal-to-noise ratio precludes the use of TPLSM to replace conventional, direct measurement of membrane potential using intracellular microelectrodes. However, TPLSM is particularly useful as a tool to target specific neurons for intracellular recording in the brain of mammals in vivo, a technique that was not possible until recently.

Traditionally, a microelectrode is inserted in the brain, literately "blindly," until it contacts and then penetrates a cell. During the recording session, the neuron is filled with a dye through the micropipette. Classification of the cell phenotype is done postmortem using the position and morphology of the cell, which are determined using standard, albeit laborious, histological techniques. As demonstrated by Margrie and coworkers (Margrie et al., 2003), in vivo intracellular recordings could be drastically improved by combining transgenic mice that expressed an intrinsic fluorescent indicator in a specific population of neurons and TPLSM, which allowed precise visualization of these cells in the z-plane.

Margrie and coworkers (Margrie et al., 2003) specifically targeted cortical inhibitory interneurons. These are small cells that do not have a stereotypical pattern of electrical activity that would make them identifiable using "blind" intracellular or extracellular recording techniques. The animals belonged to a strain of mice that was genetically altered so that their cortical inhibitory interneurons selectively expressed GFP, an exogenous protein. Mice were prepared with a cranial window covered with a recording chamber filled with agarose to dampen cardiovascular pulsations of the cortex. The chamber had an edge open on one side to allow insertion of an electrode at an oblique angle. TPLSM was used to visualize a targeted interneuron and to guide the intracellular recording microelectrode. To facilitate its localization, the electrode was filled with a fluorescent dye that was also excited by the laser but emitted in a different color than that of the labeled cells (Fig. 28.6A). Proof that the electrode had penetrated and thus

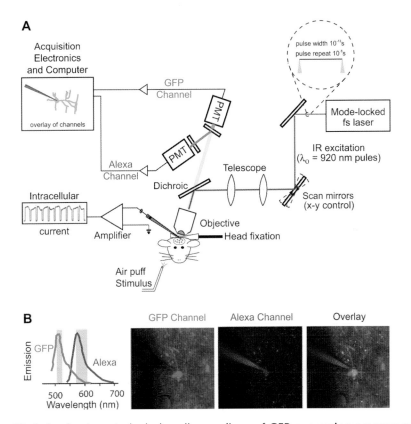

Figure 28.6. In vivo targeted whole-cell recordings of GFP-expressing neurons using TPLSM guidance. (A) The two-photon excitation beam simultaneously stimulated GFP-labeled neurons and the fluorophore Alexa 594 in the patch-clamp pipette. Emission signals from these two sources were spectrally separated by a dichroic mirror and detected separately by two photomultipliers (PMTs). To facilitate the visual guidance of the electrode during its descent towards the targeted cell, the computer generated a composite picture by overlaying the two fluorescence signals using different colors for the cell and the electrode. During final approach of the micropipette electrode on the cell, the resistance of the electrode was used to detect contact between pipette and cell membrane. An example trace shows the change in electrode resistance upon contact of the pipette with the targeted neuron. Since electrode resistance depends on the pressure exerted by the pipette on the cell, modulation of electrode resistance occurred at the heartbeat frequency. (B) TPLSM image of a GFP-expressing interneuron in the superficial layers of vibrissa sensory cortex of mouse. GFP channel: the green channel showing the soma and dendrites of the labeled neuron. Alexa channel: the micropipette that contained the soluble dye Alexa 594. Also shown are the overlay of these two channels and the emission spectra of the two fluorophores. (Adapted from Margrie et al., 2003)

recorded the targeted cell was provided by injecting dye present inside the electrode into the cell; penetrated cells were thus double-labeled (see Fig. 28.6B). At this point, intracellular responses to sensory stimulation were recorded.

The application of TPLSM to guide intracellular electrodes in specific classes of mammalian neurons in vivo will enable future experiments to reach a level of sophistication that was once achieved exclusively in invertebrate preparations and opens the possibility for many future variations of this method. One can anticipate alternative ways to label cells, for instance with retrogradely transported dextran dyes injected into a target zone. Also, this technique could be adapted for intrinsic indicators of intracellular ion concentration (Ohki et al., 2005; Stosiek et al., 2003) to achieve combined intracellular and functional imaging from targeted neurons in vivo.

28.4.1.6 Measurement of Vascular Hemodynamics

Noninvasive blood flow-based imaging techniques are critical to unravel neurovascular coupling (i.e., the relationship between neuronal activity and blood flow). The exchange of nutrients, metabolites, and heat between neurons, astrocytes, and the bloodstream occurs at the level of individual capillaries, vessels 5 to 8 μm in diameter in which red blood cells (RBCs) move in single file. TPLSM, used in combination with electrophysiological recordings, provides the necessary spatial and temporal resolutions to study neurovascular coupling at the level of individual capillaries in vivo. Cortical blood flow can be visualized in anesthetized animals by injecting a fluorescent dye in the bloodstream and by subsequently imaging blood vessels through either a thinned skull or a window-capped craniotomy (Kleinfeld et al., 1998) (Fig. 28.7A). Red blood cells appear as dark spots against the fluorescent plasma (see Fig. 28.7B). Line scans along the length of a capillary of interest can reveal RBC orientation, velocity, and linear density (see Figs. 28.7B and 28.7C). Capillary blood flow in cortex has been visualized in this manner at depths up to 600 μm below the surface of the cerebral cortex. In rat, this allows access to the capillaries of layer 4 neurons, which receive tactile sensory inputs from the large mystacial vibrissae on the face of the rat. This procedure has also been applied in rat olfactory bulb (Chaigneau et al., 2003), a region that contains neurons that respond to odorants. In order to correlate neuronal and hemodynamic activity, electrophysiology is used to make a functional map of neuronal activity in response to a stimulus. The stimulus-induced neuronal activation map can then be compared to blood flow changes made in response to the same stimulus protocol.

Several important conclusions have been drawn from TPLSM studies on neuronal blood flow. First, the natural variability in capillary blood flow was significant and included stalls (see Fig. 28.7D) and even sustained flow reversals. These fluctuations in RBC flow constituted a physiological noise floor that placed limits on the ultimate sensitivity of blood flow-based imaging techniques (Chaigneau et al., 2003). In particular, stimulus-induced changes in cortical RBC speed were found to be on the order of trial-to-trial variability. Nonetheless, significant changes in blood flow around excited neuronal populations were observed to occur both in rat somatosensory cortex and olfactory bulb within 1 to 2 seconds of the application of relevant stimulation (whisker movement and release of odorant, respectively) (see Figs. 28.7E to 28.7H). These

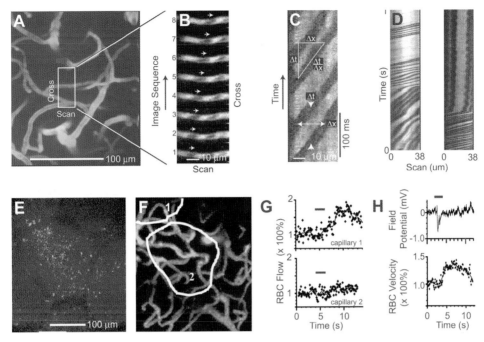

Figure 28.7. TPLSM studies of neurovascular coupling in rat cerebral cortex and olfactory bulb. (A) Horizontal section of capillaries in cortex of a rat constructed from a set of 100 planar TPLSM scans acquired every 1 m at depths of 310 to 410 μm below the pia mater layer. (B) Successive line-scan images of a small vessel acquired every 16 ms. Unstained RBCs showed up as dark shadows against the fluorescently labeled blood plasma. The change in position of an RBC is indicated by the series of arrows. (C) Derivation of instantaneous RBC velocity from a stack of line-scan images made from a 34-μm-long segment of capillary. (D) Examples of irregular flow in 1-second-long line-scan data taken through a straight section of capillary 240 μm below the cortical surface. Notice the change in speed in the first image and the complete stall in the second. (E) Odor-specific populations of a cluster of olfactory bulb neurons called glomerulus were labeled with Oregon Green dextran and imaged with TPLSM. (F) The vasculature of the same region of the olfactory bulb was imaged with TPLSM after injecting the blood stream with fluorescein dextran. The outlines of two adjacent glomeruli were superposed and two capillaries are indicated. (G) One of the two capillaries indicated in G showed an increase in blood flow when an almond odorant was applied to the bulb. The presentation of the stimulus (odorant) is indicated by a black bar. Note that the capillaries are interconnected and separated by ~200 μm. (H) Field potential and blood flow rate measurements in a single glomerulus show that vascular responses occurred 1 to 2 seconds after the neuronal response. (Figure adapted from Kleinfeld et al., 1998, and Chaigneau et al., 2003)

changes in capillary blood flow were well localized to brain regions associated with specific neuronal populations. In the olfactory bulb, differential hemodynamic activity could be distinguished at capillary separation distances of 100 to 200 μm (Chaigneau et al., 2003), though there were no obvious features in the vascular architecture that would support such selective activity. Finally, these studies suggested that the spatial specificity of stimulus-induced rearrangements of blood flow was likely to be based on a decrease in the impedance of a network of capillaries.

28.4.1.7 Long-Term Imaging of Neurodegeneration

The accumulation of amyloid-β plaques in the brain is a hallmark of Alzheimer's disease. Unfortunately, in vivo detection of these plaques remains difficult. Currently, clinical diagnosis relies principally on neurological examinations, with only post-mortem confirmation of the underlying pathology (Bacskai et al., 2001). For the development of therapeutic strategies, techniques that allow changes in the neuropathology to be observed over time are critical. Bacskai and associates (Bacskai et al., 2001) recently demonstrated the use of TPLSM to image amyloid-β in the brains of transgenic mice. These mice have been genetically engineered to express a mutant human amyloid-β precursor protein that leads to the formation of plaques similar to those observed in humans with Alzheimer's disease. These plaques were fluorescently labeled by topically applying Thioflavine S or fluorescein-labeled anti-amyloid-β antibodies through a small cranial window over the brain. The vasculature was further labeled by intravenous injection of Texas Red-dextran. Two-photon imaging revealed amyloid-β plaques as well as amyloid accumulation around cerebral vessels (i.e., amyloid angiopathy) (Figs. 28.8A through 28.8C). This technique was then used to test the efficacy of immunotherapy for clearing existing amyloid-β plaques. Animals were imaged before and several days after treatment with an antibody against amyloid-β, administered topically to the brain. The fluorescent vascular images were used as a reference to ensure the same brain areas were imaged before and after treatment. Previous histological studies had observed that immunotherapy reduced plaque accumulation in transgenic mice but could not determine whether old plaques were cleared, new ones prevented, or both. In the work of Bacskai and coworkers (Bacskai et al., 2001), however, the same regions of the brain could be repeatedly imaged, which allowed the effect of the therapy to be observed over time on a plaque-by-plaque basis. Within a few days after treatment, large reductions in the number of plaques were observed while amyloid angiopathy remained unaffected (see Figs. 28.8D and 28.8E). With an appropriate animal preparation that allows repeated imaging over the course of days, weeks, or months, and with easily administered fluorescent markers, TPLSM is a powerful and promising tool for monitoring the progression of neural disease in animal models and will become an important part of testing of future preclinical therapeutic strategies.

28.5 CONCLUSION

Since its introduction in neuroscience about a decade ago, TPLSM has been the object of numerous methodological improvements that have broadened the use of this technique. The next scientific challenges are likely to push the limits of TPLSM towards rapid frame rates in order to image fast neuronal events such as individual action potentials. However, progress in this direction will depend ultimately on the availability of new functional indicators. In TPLSM, the frame rate is determined by the pixel dwell time, which is governed by the minimal acceptable signal-to-noise ratio (Tan et al., 1999). Cells of interest will have to be brightly labeled if they are to be imaged faster and deeper inside the brain. Promising dyes include genetically engineered fluorescent voltage sensors (Siegel & Isacoff, 1997), inducible calcium sensor proteins (Hasan et al., 2004), and second-harmonic activated voltage reporting dyes (Dombeck et al.,

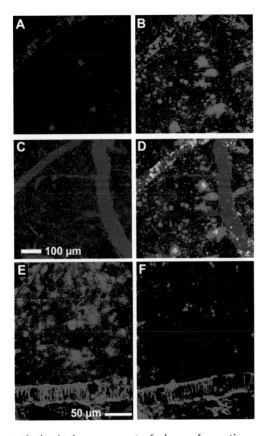

Figure 28.8. Neuropathological assessment of plaque formation and clearance in an animal model of Alzheimer's disease. Projections of two-photon fluorescence image stacks of cerebral amyloid-beta pathology and clearance through immunotherapy in a live, 20-month-old transgenic mouse. (A) Imaging of Thioflavine-S showing amyloid plaques as well as amyloid angiopathy. (B) Visualization of fluorescein-labeled anti-amyloid-beta antibody, showing diffuse amyloid-beta deposits as well as plaques. Both the Thioflavine-S and anti-amyloid-beta antibody were applied to the brain surface for 20 minutes to achieve this labeling. (C) Fluorescence angiography obtained by injecting Texas-Red dextran into the tail vein of the mouse. (D) Merged representation of (A), (B), and (C). (E, F) Images of fluorescently labeled anti-amyloid-antibody reactivity taken before (E) and 3 days after (F) immunotherapy. Amyloid-beta plaques were dramatically reduced, while the amyloid angiopathy was largely unaffected. (Figure adapted from Backsai et al., 2001).

2004). A concomitant issue is how to deliver dyes into neurons of a living animal. Recent techniques that have been evaluated for this task include ballistic delivery of micrometer-size metal particles coated with dye (Kettunen et al., 2002) or electrophoresis (Bonnot et al., 2005). Improvement in the pulse pattern as well as peak intensity of femtosecond lasers will be undoubtedly helpful (Kawano et al., 2003). However, two-photon microscopy is, in practice, limited to ~1,000 μm down in cerebral cortex. To image at this depth, laser power has to be increased so much to counteract scattering that out-of-focus fluorescence resulting from excitation of the surface of the brain above the focal point starts to dominate (Theer et al., 2003).

Despite its tremendous potential, widespread adoption of TPLSM in neurobiology has been limited by its cost, of which a great deal is that of the femtosecond laser, and the fact that only a small handful of companies currently market a complete TPLSM. This situation has forced many investigators to build their own two-photon systems (Tsai et al., 2002). We can hope that in the near future TPLSM will become more readily available and more affordable. This will undoubtedly have beneficial consequences for the development of new contrast agents specifically designed for neurobiological applications with two-photon excitation.

ACKNOWLEDGMENTS

Work in the Kleinfeld laboratory that involves the use of nonlinear microscopy has been supported by grants from the David and Lucille Packard Foundation (99-8326), the NIH (EB003832, MH72570, and NS41096), and the NSF (DBI-0455027). In addition, G.O.C. was supported in part by a NSF/IGERT grant, N.N. was supported in part by a NSF predoctoral training grant, C.B.S. was supported in part by the LJIS training grant from the Burroughs-Wellcome Trust, L.F.S. was supported in part by a NIH/MST grant, and P.S.T. was supported in part by a NIH/NIMH training grant. D.K. takes this opportunity to thank Winfried Denk and Jeff A. Squier to introducing him to nonlinear microscopy and optics.

REFERENCES

Bacskai, B.J., S.T. Kajdasz, R.H. Christie, C. Carter, D. Games, P. Seubert, D. Schenk, and B.T. Hyman. 2001. Imaging of amyloid-beta deposits in brains of living mice permits direct observation of clearance of plaques with immunotherapy. Nature Med. 7: 69–372.

Bacskai, B.J., W.E. Klunk, C.A. Mathis, and B.T. Hyman. 2002. Imaging amyloid-beta deposits in vivo. J. Cereb. Blood Flow & Metab. 22:1035–1041.

Bonnot, A., G.Z. Mentis, J. Skoch, and M.J. O'Donovan. 2005. Electroporation loading of calcium-sensitive dyes into the CNS. J. Neurophysiol. 93:1793–808.

Brecht, M., M.S. Fee, O. Garaschuk, F. Helmchen, T.W. Margrie, K. Svoboda, and P. Osten. 2004. Novel approaches to monitor and manipulate single neurons in vivo. J. Neurosci. 24:9223–9227.

Bullen, A., S.S. Patel, and P. Saggau. 1997. High-speed, random-access fluorescence microscopy: I. High-resolution optical recording with voltage-sensitive dyes and ion indicators. Biophys. J. 73:477–491.

Chaigneau, E., M. Oheim, E. Audinat, and S. Charpak. 2003. Two-photon imaging of capillary blood flow in olfactory bulb glomeruli. Proc. Natl. Acad. Sci. USA. 100:13081–13086.

Denk, W., J.H. Strickler, and W.W. Webb. 1990. Two-photon laser scanning microscopy. Science. 248:73–76.

Denk, W., K.R. Delaney, A. Gelperin, D. Kleinfeld, B.W. Strowbridge, D.W. Tank and R. Yuste. 1994. Anatomical and functional imaging of neurons using 2-photon laser scanning microscopy. J. Neurosci. Methods. 54:151–162.

Denk, W., D.W. Piston, and W.W. Webb. 1995. Two-photon molecular excitation in laser-scanning microscopy. In Handbook of Biological Confocal Microscopy. J.W. Pawley, editor. Plenum Press, New York. 445–458.

Denk, W. and K. Svoboda. 1997. Photon upmanship: Why multiphoton imaging is more than a gimmick. Neuron. 18:351–357.

Dombeck, D.A., M. Blanchard-Desce, and W.W. Webb. 2004. Optical recording of action potentials with second-harmonic generation microscopy. J. Neurosci. 24:999–1003.

Engert, F. and T. Bonhoeffer. 1999. Dendritic spine changes associated with hippocampal long-term synaptic plasticity. Nature. 399:66–70.

Grutzendler. J., N. Kasthuri, and W.-B. Gan. 2002. Long-term dendritic spine stability in the adult cortex. Nature. 420:812–816.

Grynkiewicz, G., M. Poenie, and R.Y. Tsien. 1985. A new generation of Ca2+ indicators with greatly improved fluorescence properties. J. Biol. Chem. 260:3440–3450.

Hasan, M.T., R.W. Friedrich, T. Euler, M.E. Larkum, G. Giese, M. Both, J. Duebel, J. Waters, H. Bujard, O. Griesbeck, R.Y. Tsien, T. Nagai, A. Miyawaki, and W. Denk. 2004. Functional fluorescent Ca2+ indicator proteins in transgenic mice under TET control. PLoS Biol. 2: e163.

Helmchen, F., M.S. Fee, D.W. Tank, and W. Denk. 2001. A miniature head-mounted two-photon microscope: High-resolution brain imaging in freely moving animals. Neuron. 31: 903–912.

Kaiser, K.M.M., J. Lübke, Y. Zilberter, and Sakmann, B. 2004. Postsynaptic calcium influx at single synaptic contacts between pyramidal neurons and bitufted interneurons in layer 2/3 of rat neocortex is enhanced by backpropagating action potentials. J. Neurosci. 24:1319–1329.

Kang, J., G. Arcuino, M. Nedergaard. 2005. Calcium imaging of identified astrocytes in hippocampal slices. In Imaging in Neuroscience and Development, R. Yuste, and A. Konnerth, editors. Cold Spring Harbor Laboratory Press, Cold Spring Harbor. 289–297.

Kawano, H., Y. Nabekawa, A. Suda, Y. Oishi, H. Mizuno, A. Miyawaki, and K. Midorikawa. 2003. Attenuation of photobleaching in two-photon excitation fluorescence from green fluorescent protein with shaped excitation pulses. Biochem. Biophys. Res. Comm. 311:592–596.

Kettunen, P., J. Demas, C. Lohmann, N. Kasthuri, Y. Gong, R.O. Wong, and W.-B. Gan. 2002. Imaging calcium dynamics in the nervous system by means of ballistic delivery of indicators. J. Neurosci. Methods 119:37–43.

Kleinfeld, D., P.P. Mitra, F. Helmchen, and W. Denk. 1998. Fluctuations and stimulus-induced changes in blood flow observed in individual capillaries in layers 2 through 4 of rat neocortex. Proc. Natl. Acad. Sci. USA. 95:15741–15746.

Kleinfeld, D. 2002. Cortical blood flow through individual capillaries in rat vibrissa S1 cortex: Stimulus induced changes in flow are comparable to the underlying fluctuations in flow. In Brain Activation and Cerebral Blood Flow Control, Excerpta Medica, International Congress Series 1235. M. Tomita, editor. Elsevier Science.

Lendvai, B., E.A. Stern, B. Chen, and K. Svoboda. 2000. Experience-dependent plasticity of dendritic spines in the developing rat barrel cortex in vivo. Nature. 404:876–881.

Maletic-Savatic, M., R. Malinow, and K. Svoboda. 1999. Rapid dendritic morphogenesis in CA1 hippocampal dendrites induced by synaptic activity. Science. 283:1923–1927.

Margrie, T.W., A.H. Meyer, A. Caputi, H. Monyer, M.T. Hasan, A.T. Schaefer, W. Denk, and M. Brecht. 2003. Targeted whole-cell recordings in the mammalian brain in vivo. Neuron. 39:911–918.

Martin, S.J., P.D. Grimwood, and R.G. Morris. 2000. Synaptic plasticity and memory: an evaluation of the hypothesis. Annu. Rev. Neurosci. 23:649–711.

Noguchi, J., M. Matsuzaki, G.C.R. Ellis-Davies, and H. Kasai. 2005. Spine-neck geometry determines NMDA receptor-dependent Ca2+ signaling in dendrites. Neuron. 46:609–622.

Ohki, K., S. Chung, Y.H. Ch'ng, P. Kara, and C.R. Reid. 2005. Functional imaging with cellular resolution reveals precise micro-architecture in visual cortex. Nature. 433:597–603.

Purves, D., and R.D. Hadley. 1985. Changes in the dendritic branching of adult mammalian neurones revealed by repeated imaging in situ. Nature. 315:404–406.

Ramón y Cajal, S. 1893. Neue Darstellung vom histologischen Bau des Centralnervensystems. Arch. Anat. Physiol. Anat. Abt. Suppl. 319:428.

Roorda, R.D., T.M. Hohl, R. Toledo-Crow, and G. Miesenbock. 2004. Video-rate nonlinear microscopy of neuronal membrane dynamics with genetically encoded probes. J. Neurophysiol. 92:609–621.

Siegel, M.S., and E.Y. Isacoff. 1997. A genetically encoded probe of membrane voltage. Neuron. 19:735–741.

Stosiek, C., O. Garashuk, K. Holthoff, and A. Konnerth. 2003. In vivo two-photon calcium imaging of neuronal networks. Proc. Natl. Acad. Sci. USA. 100:7319–7324.

Stutzmann, G.E., and I. Parker. 2005. Dynamic multiphoton imaging: A live view from cells to systems. Physiology (Bethesda). 20:15–21.

Svoboda, K., D.W. Tank, and W. Denk. 1996. Direct measurement of coupling between dendritic spines and shafts. Science. 272:716–719.

Tan, Y.P., I. Llano, A. Hopt, F. Wurriehausen, and E. Neher. 1999. Fast scanning and efficient photodetection in a simple two-photon microscope. J. Neurosci. Methods 92: 123–135.

Theer, P., M.T. Hasan, and W. Denk. 2003. Two-photon imaging to a depth of 1000 microm in living brains by use of a Ti:Al2O3 regenerative amplifier. Opt. Lett. 15:1022–1024.

Trachtenberg, J.T., B.E. Chen, G.W. Knott, G. Feng, J.R. Sanes, E. Welker, and K. Svoboda. 2002. Long-term in vivo imaging of experience-dependent synaptic plasticity in adult cortex. Nature. 420:788–794.

Tsai, P. S., N. Nishimura, E. J. Yoder, E. M. Dolnick, G. A. White, and D. Kleinfeld. 2002. Principles, design, and construction of a two photon laser scanning microscope for in vitro and in vivo brain imaging. In In Vivo Optical Imaging of Brain Function. R.D. Frostig, editor. CRC Press, Boca Raton. 113–171.

Woolsey, T.A., and H. van der Loos. 1970. The structural organization of layer IV in the somatosensory region (SI) of mouse cerebral cortex: the description of a cortical field composed of discrete cytoarchitectonic units. Brain Res. 17:205–242.

Yuste, R., and W. Denk. 1995. Dendritic spines as basic functional units of neuronal integration. Nature. 375:682–684.

Zipfel, W.R., R.M. Williams, and W.W. Webb. 2003. Nonlinear magic: multiphoton microscopy in the biosciences. Nat. Biotechnol. 21:1369–1377.

Ziv, N. E., and S.J. Smith. 1996. Evidence for a role of dendritic filopodia in synaptogenesis and spine formation. Neuron. 17:91–102.

29

Applications of Nonlinear Intravital Microscopy in Tumor Biology

Rakesh K. Jain, Michael F. Booth,
Timothy P. Padera, Lance L. Munn,
Dai Fukumura, and Edward Brown

29.1 INTRODUCTION

The past few decades have produced a wealth of information about the molecular underpinnings of cancer. Various genes associated with oncogenesis and tumor angiogenesis have been identified, and this has led to the development of a wide variety of therapeutic agents. As we enter the post-genomic era, the grand challenges remaining in tumor biology are (i) to discover the relationships between the expression of novel genes and their function in an intact organism and (ii) to deliver newly discovered therapeutics to their targets in vivo in optimal quantities (Jain et al., 2002).

Gene expression, physiological function, and drug delivery are typically measured with destructive techniques (which provide very limited insight into dynamics) or techniques with poor spatial resolution, on the scale of millimeters (which precludes the study of biology at the cellular and subcellular levels, on the scale of 1 to 10 micrometers). The ideal solution to this dilemma would be a variant of light microscopy that could reach any location in a living, intact tumor. However, in vivo light microscopy is exceptionally difficult due to limited light penetration through the body. Optical access to the entire tumor may be obtained by imaging the tissue after excision and/or sectioning (i.e., ex vivo), but this removes the tissue from its blood supply, disturbs the ongoing physiological processes within the tissue, and loses dynamic information. To gain optical access while preserving the tissue's blood supply and metabolic microenvironment, tumors can be grown in chronic animal window models such as the dorsal skinfold chamber (Fig. 29.1) and cranial window (Fig. 29.2), or they can be carefully exteriorized for acute observation (Fig. 29.3) (Jain et al., 2002; Leunig et al., 1992; Monsky et al., 2002; Yuan et al., 1994). Unfortunately, even in these preparations, both epifluorescence and confocal microscopy are limited to the outer 50 to 100 microns of the optically accessible tumor surface due to light scattering and absorption within the tumor tissue itself (Jain et al., 2002).

Multiphoton laser-scanning microscopy (MPLSM) has recently emerged as a useful technique for high-resolution, nondestructive, chronic imaging within living tumors (Fig. 29.4). MPLSM has markedly superior depth penetration in tumors (Brown et al., 2001) and in combination with chronic window models allows repetitive imaging at depths of several hundred microns within living tumor tissue while retaining submicron

Figure 29.1. Dorsal skinfold chamber.

Figure 29.2. Cranial window.

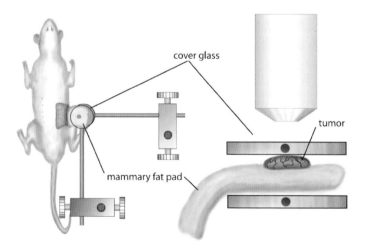

Figure 29.3. Acute mammary fat pad preparation.

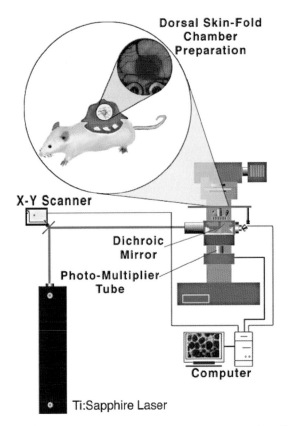

Figure 29.4. A multiphoton laser-scanning microscope. (Reproduced from Brown et al., 2001)

spatial resolution. As summarized in Table 29.1, MPLSM has been used to dynamically image cells within tumors, to localize and quantify drug carriers, to image and quantify blood vessels, and to quantify both red blood cell velocity and tumor vessel permeability (Brown et al., 2001). It has also been used to image and quantify tumor-associated lymphatic vessels (Padera et al., 2002a) and to quantify diffusive transport in the tumor extracellular matrix (Alexandrakis et al., 2004). The contrast mechanism provided by second-harmonic generation has proven extremely useful for the imaging and characterization of the tumor extracellular matrix (Ahmed et al., 2002), as it is altered with enzymes and hormones (Brown et al., 2003). In this chapter we briefly review the biology of solid tumors, describe conventional techniques of wide-field microscopy and the advantages afforded by MPLSM, survey some of the in vivo applications of MPLSM in tumor biology and provide a detailed description of the methodologies involved, discuss some of the limitations of MPLSM, and provide a perspective on the future development of this technology.

29.2 AN OVERVIEW OF TUMOR BIOLOGY

A solid tumor is an organ comprising cancer cells and non-neoplastic host stromal cells embedded in an extracellular matrix and nourished by a vascular network

Table 29.1. Examples of MPLSM techniques developed in authors' laboratory

Technique	References
Blood vessel parameters	
3D microangiography in tumors using fluorescent tracers and quantum dots	Brown et al., 2001; Padera et al., 2002b; Stroh et al., 2005
3D vascular volume fractions	Brown et al., 2001
Automated vessel tracing and change analysis	Abdul-Karim et al., 2003; Tyrrell et al., 2005
Blood flow of individual tumor microvessels	Brown et al., 2001
Leukocyte–endothelial interactions using high-speed MPLSM	Brown et al., 2001; Padera et al., 2002b
Microvascular permeability of individual vessels in tumors	Brown et al., 2001
Nanoparticle delivery through tumor vessels	Campbell et al., 2002; Stroh et al., 2005
Regulation of angiogenesis and vessel maturation	Gardavtsev et al., 2004; Kashiwagi et al., 2005
Normalization of tumor vasculature by anti-angiogenic agents	Tong et al., 2004; Winkler et al., 2004
Creation of long-lasting tissue-engineered blood vessels	Koike et al., 2004
Endothelial–perivascular cell interaction	Kashiwagi et al., 2005
Lymph vessel parameters	
3D microlymphangiography	Isaka et al., 2004; Ny et al., 2005; Padera et al., 2002a, 2002b
Regulation of lymphangiogenesis and metastasis	Isaka et al., 2004; Padera et al., 2002a
Visualization of lymphatic valves and cells in lymphatic vessels	Isaka et al., 2004
Detection of metastasizing tumor cells in lymph nodes	Hoshida et al., 2006
Extravascular parameters	
Imaging of collagen fibers in tumors using second-harmonic generation	Brown et al., 2003
Two-photon fluorescence correlation microscopy for interstitial transport measurements in tumors	Alexandrakis et al., 2004
Tissue nitric oxide distribution in normal and tumor tissues	Kashiwagi et al., 2005
Gene expression using fluorescent reporter gene	Brown et al., 2002; Duda et al., 2004; Huang et al., 2005
Identification of specific population of cells using fluorescent reporter transgenic mice	Brown et al., 2001; Duda et al., 2004
Cell trafficking using fluorescent gene transfection or Qd-TAT conjugation	Brown et al., 2001; Fukumura et al., 2003; Stroh et al., 2005

(Fukumura, 2005; Jain, 2002). Vascular function is an important determinant of the microenvironment inside the tumor. There are two circulation systems in the body, blood vessels and lymph vessels, both of which are important for tumor growth, invasion, and metastasis, and both of which affect the treatment of tumors. Blood vessels supply nutrients to tumor cells and remove waste from tumor tissue. Therefore, angiogenesis (the formation of new blood vessels) is crucial for the growth of solid tumors beyond $1 \sim 2$ mm in diameter (Folkman, 2000). Furthermore, blood vessels are one route by which tumor cells metastasize to distant organs. On the other hand, blood vessels are the route by which therapeutic molecules can be efficiently delivered to tumors, and this is a prerequisite for successful treatment; thus, angiogenesis and microvascular function in tumors must be explored if the delivery of bloodborne antitumor agents is to be improved.

Tumor vessels are chaotic in both morphology and function (Fig. 29.5) (Jain et al., 2002). Normal vessels have a well-organized architecture, whereas tumor vessels have a tortuous shape, irregular surface, enlarged diameter, and heterogeneous spatial distribution. Normal microvessels differentiate into arterioles, capillaries, and venules, each of which has a characteristic morphology and function. Instead of these differentiated functional units, tumors contain a mesh-like network of immature vessels. In normal vessels, perivascular cells such as pericytes or vascular smooth muscle cells regulate vascular tone, but in most tumor vessels the perivascular cells are either missing or abnormally structured, which leads to poor regulation of vascular tension in tumors (Jain, 2003; Morikawa et al., 2002). As a result of these structural abnormalities, blood flow in the average tumor vessel is very slow and often static and may even change direction over time (Jain, 1988). Furthermore, some tumor vessels lack oxygen, despite the fact that blood perfusion is well maintained (Helmlinger et al., 1997). The abnormal, inhomogenous blood flow creates a physiological barrier to the delivery of therapeutic agents to tumors (Jain, 1998) and can lead to the hypoxia and acidosis that are often seen in tumors (Helmlinger et al., 1997). Such a severe metabolic microenvironment

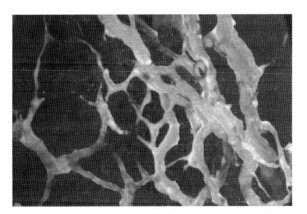

Figure 29.5. Angiography. The vasculature of an LS174T colon adenocarcinoma implanted in the dorsal skinfold chamber of a SCID mouse was highlighted with intravenous injection of 2M MW FITC-Dextran. A series of 15 images were taken at 5-micron depth intervals and merged via maximum intensity projection. The image is 550 microns across.

reduces the efficacy of antitumor therapies and, furthermore, makes tumor cells more aggressive (Harris, 2002).

Tumor vessels also have high vascular permeability and generally low leukocyte adhesion (Jain et al., 2002). Leaked plasma proteins, such as fibronectin, form a provisional matrix that is optimal for tumor angiogenesis, and poor leukocyte–endothelial interaction may restrict the immune surveillance of tumors.

The lymphatic system is also abnormal in solid tumors (Fig. 29.6). Due to the mechanical stress produced by proliferating tumor cells, there are no functional lymphatic vessels inside solid tumors (Padera et al., 2002a, 2004). The high permeability of intratumor blood vessels and the impairment of lymphatic drainage cause significant elevation of interstitial fluid pressure and oncotic pressure in solid tumors (Jain, 1994, 2005). As a result, the pressure gradient between blood vessel and tumor tissue is lost; this constitutes an additional physiological barrier to the delivery of therapeutic agents to tumors (Tong et al., 2004). On the other hand, in peripheral regions of the tumor, lymph-angiogenesis, lymphatic hypertrophy, and lymphatic dilatation are often found (Alitalo & Carmeliet, 2002; Isaka et al., 2004; Padera et al., 2002a). Dysfunction of the lymphatic valves allows retrograde flow in these lymphatic vessels (Isaka et al., 2004). Tumor cells can invade these peripheral lymphatic vessels and form metastases within

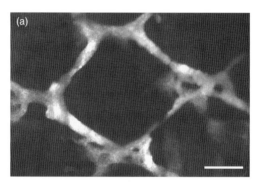

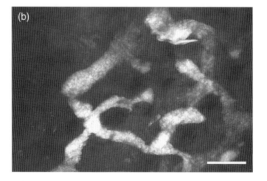

Figure 29.6. Lymphangiography. (A) Initial lymphatics of a non-tumor-bearing mouse tail display a distinctive hexagonal pattern. Image is a single MPSLM slice with a scale bar of 100 μm. (B) Functional lymphatics at the edge of a tumor grown in the mouse tail have a distorted hexagonal architecture. The tumor alters the lymphatic morphology, even causing the lymphatics inside the tumor to become nonfunctional. Image is a maximum intensity projection with a scale bar of 100 μm.

the lymphatic system. Altogether, abnormal blood and lymphatic vessels contribute to the unique microenvironment of tumors, promote tumor growth and metastasis, and hinder the efficacy of various antitumor therapies.

Tumors contain not only transformed neoplastic cells but also stromal cells, such as vascular cells, fibroblasts, and inflammatory cells. It is now well recognized that genetic mutations in cancer cells cannot fully explain the biology of tumors (Hanahan & Weinberg, 2000). Stromal cells have been shown to influence many steps of tumor progression, such as tumor cell proliferation, invasion, angiogenesis, metastasis, and even malignant transformation (Brown et al., 2001; Elenbaas & Weinberg, 2001; Fukumura et al., 1998; Li et al., 2003; Liotta & Kohn, 2001; Orimo et al., 2005; Pollard, 2004; Ruiter et al., 2001; Tlsty, 2001). Crosstalk between the heterotypic cell populations via soluble factors, interaction with the extracellular matrix, and direct cell-to-cell contact plays an important role in the induction, selection, and expansion of the neoplastic cells. Intravital microscopy, together with cellular and molecular approaches, has successfully revealed these interactions. For an example, to study the contribution of stromal cells to angiogenic gene expression in tumors, we have engineered transgenic mice bearing a green fluorescent protein (GFP) construct driven by the vascular endothelial growth factor (VEGF) promoter (Fukumura et al., 1998). We found strong VEGF promoter activation in the stromal cells, as indicated by increased GFP fluorescence. MPLSM observation revealed a close association between the VEGF-expressing stromal cells and tumor blood vessels, indicating that stromal cells produce angiogenic factors and assist with the regulation of tumor vessel formation and function (Fig. 29.7) (Brown et al., 2001). Furthermore, we have found significant differences in angiogenesis, microcirculatory function, and the expression of pro- and anti-angiogenic factors when the same tumor cells are grown in different host organs (Fukumura et al., 1997b; Gohongi et al., 1999; Hobbs et al., 1998; Kashiwagi et al., 2005; Monsky et al., 2002). These variations are probably due to differences in the stromal cell population and local microenvironment of different organs. Tumor–stromal interaction influences the biology of both the tumor cells and the stromal cells; in particular, it influences the expression of positive and negative regulators of angiogenesis. This interaction depends on tumor type and location, may vary during the course of treatment, and can influence the efficiency of various treatments. A better understanding of host–tumor interactions, and of the formation and function of blood and lymphatic vessels during the genesis, growth, and treatment of tumors, will help us to develop improved treatment strategies.

29.3 INTRAVITAL IMAGING WITH CONVENTIONAL WIDE-FIELD MICROSCOPY

Wide-field microscopes, in combination with window models, have been used for intravital microscopy of normal and diseased tissue for over 80 years. In 1924, Sandison reported a method for intravital microscopy using the rabbit ear chamber model (Sandison, 1924). Ide and colleagues were the first to observe tumor vasculature in this model (Ide et al., 1939). Since then, various animal models and intravital microscopy techniques have been developed, and these have provided dramatic insight into tumor biology (reviewed in Jain et al., 2002, 2002). Wide-field intravital microscopy has been employed to measure tumor angiogenesis (Ide et al., 1939; Leunig et al., 1992), blood

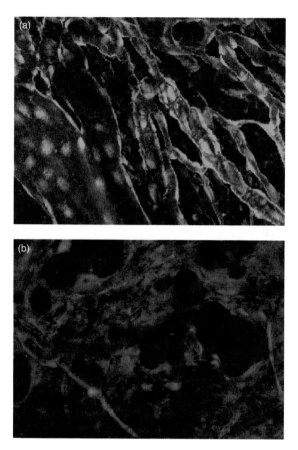

Figure 29.7. Imaging tumors grown in GFP reporter mice. (A) Endothelial cells within a TG1-1 murine mammary adenocarcinoma, implanted in the dorsal skinfold chamber of a *TIE2ᵖ-GFP* mouse. The cells of the host animal express GFP under control of the endothelial-specific TIE2 promoter (green). Vessels were highlighted with intravenous injection of TMR-Dextran (red). Image is 550 microns across and is a maximum intensity projection of 15 images taken at 5-micron depth interval. (Image courtesy of Dr. Dan Duda) (B) VEGF-expressing host cells within an MCaIV murine mammary adenocarcinoma, implanted in the dorsal skinfold chamber of a *VEGFᵖ-GFP* mouse. The cells of the host animal express GFP under control of the VEGF promoter (green). Vessels were highlighted with intravenous injection of TMR-Dextran (blue) while collagen in the extracellular matrix was imaged via second-harmonic generation (red). Image is 550 microns across.

flow (Endrich et al., 1979; Leunig et al., 1992), vascular permeability (Gerlowski & Jain, 1986; Yuan et al., 1993), pore cutoff size (Hobbs et al., 1998; Yuan et al., 1995), and leukocyte–endothelial interaction (Fukumura et al., 1995; Sasaki et al., 1991). Wide-field lymphangiography has been used to highlight peritumor lymphatic structures and to measure flow velocity in the superficial lymphatic network with and without the regulation of deep lymphatics (Berk et al., 1996; Hagendoorn et al., 2004; Leu et al., 1994; Swartz et al., 1996). Wide-field techniques have been used to measure extravascular parameters: fluorescence photobleaching with spatial Fourier analysis to measure interstitial diffusion and binding (Berk et al., 1993, 1997), phosphorescence quenching to determine tissue oxygen level (Helmlinger et al., 1997; Torres Filho et al., 1994), fluorescence ratio imaging to measure tissue pH (Dellian et al., 1996; Helmlinger et al., 1997), and fluorescent reporter imaging to locate gene expression (Fukumura et al., 1998). Wide-field microscopes are still in use worldwide for the study of tissue structure and function inside living tumors.

However, although wide-field microscopy is well established, powerful, user-friendly, and relatively inexpensive, it has significant limitations for the study of living tumors. The biggest of these are its surface-weighted imaging capability and its relatively high signal-to-noise ratio, caused by the scattering and absorption of light outside the focal plane and by tissue autofluorescence. Solid tumors are three-dimensional structures, and many of their biological features differ from the periphery to the inside. Typically, the tumor periphery has higher vessel density than the center (Fukumura et al., 1997b). Although hyperplastic lymphatic vessels are often found at the margins of solid tumors, there are no functional lymphatics inside the tumors (Padera et al., 2002a) due (at least in part) to vascular collapse induced by elevated solid stress (Padera et al., 2004). At the tumor–host interface, activated fibroblasts form a sheet of VEGF-expressing cells (Fukumura et al., 1998), while inside the tumor the VEGF-expressing stromal cells behave like perivascular cells (Brown et al., 2001). Thus, a reliance on two-dimensional, surface-weighted imaging techniques may lead one to miss important biological information.

29.4 MULTIPHOTON LASER-SCANNING MICROSCOPY

MPLSM, discussed throughout this book, has superior depth penetration, sensitivity, and signal-to-noise ratio and produces less photobleaching and photodamage. This allows MPLSM to overcome many of the limitations of conventional microscopy. In combination with sophisticated animal models, genetic and/or synthetic fluorescent tracers, and computer-assisted data analysis, MPLSM is opening new doors in the study of tumor biology and is facilitating the development of new strategies for tumor diagnosis and treatment.

A typical instrument used for MPLSM of living specimens is shown in Figure 29.4. The key component is the light source: a Ti:Sapphire laser producing a train of output pulses, each with a length of several hundred femtoseconds (10^{-15} seconds) and a tunable wavelength in the infrared range (700 to 900 nanometers or longer). The laser light is scanned by a computer-controlled X-Y scanner (at typical scan rates of 1 or more seconds per frame), and the scanned beam is directed by a dichroic mirror into the back-aperture of a microscope objective (which must be transparent to infrared as well as visible light). Multiphoton absorption and fluorescence occurs only at the focal point of the objective lens, where the intensity of infrared photons is sufficiently high. Fluorescence at visible wavelengths (generally shorter than 750 nm) is collected by the same objective and routed back through the dichroic to an array of filters and dichroics that route the signal light to one or more photomultiplier tubes (PMTs), each of which detects a different wavelength channel. A typical system can detect two or three wavelengths simultaneously. The signals from the PMTs are gathered, digitized, and displayed by a computer system. To acquire data in a three-dimensional region within the sample, either the objective lens or the sample stage is moved vertically, using a computer-controlled stepper motor, during the interval between one scanned image and the next.

29.5 SURGICAL PREPARATIONS

While the imaging depth of the MPLSM is a vast improvement over conventional light microscopy techniques, it is nonetheless extremely difficult to image through the skin or overlying tissues and still retain any significant depth penetration within an underlying tumor. Therefore, most in vivo tumor biology work with the MPLSM relies either on chronic window preparations, in which the tumor is grown under a surgically implanted window, or on acute exteriorization, in which the tumor is grown in a mouse without a window, and then temporarily surgically exteriorized for imaging purposes. We will briefly summarize the methodology of three such preparations: the chronic dorsal skinfold chamber and cranial window, and the acute mammary fat pad preparation. These three preparations have been used for the vast majority of MPLSM experiments in tumor biology.

29.5.1 Dorsal Skinfold Chamber (Chronic)

Dorsal skinfold chambers are implanted in mice using the following procedure (Leunig et al., 1992) (see Fig. 29.1):

1. The mouse is anesthetized (90 mg/kg body weight of ketamine and 9 mg/kg body weight of xylazine), hair is removed from the back of the animal, and two identical titanium frames are used to sandwich a pinch of skin, which is gently stretched from the back of the mouse and held in place with sutures.
2. One layer of skin is excised in a 15-mm-diameter circle, and the remaining layer is covered with a glass coverslip laid in a depression built into one of the frames, fastened with a retaining ring.
3. Animals are allowed to recover from surgery for 48 hours.
4. To implant tumor cells, the animals are positioned in a transparent plastic tube (ID = 25 mm) with the dorsal skinfold chamber extending from a slot running down one side of the tube.
5. The coverslip of the chamber is gently removed with a spanner wrench, and 2 μL of a suspension of tumor cells ($\sim 2 \times 10^5$ cells), or a $\sim$1-mm piece of tumor tissue is implanted at the center of the dorsal chamber. The chamber is sealed with a new coverslip.
6. To image the chamber, the mouse is anesthetized and laid on its side on a heated stage and the chamber is gently clamped to the microscope stage with a custom-built metal frame.

29.5.2 Cranial Window (Chronic)

Cranial windows are implanted in mice using the following procedure (Yuan et al., 1994) (see Fig. 29.2):

1. The mouse is anesthetized and its head is fastened within a stereotactic apparatus.
2. A longitudinal incision of the skin is made between the occiput and the forehead. A circular region of skin is removed from the top of the skull, and the periosteum underneath is scraped off to the temporal crests.

3. A 6-mm circle is drawn over the frontal and parietal regions of the skull using a marker. Using a high-speed drill with a 0.5-mm bur tip, a groove is made along the edges of the drawn circle. This groove is made deeper by careful repetitive sweeps along the circle until the bone flap becomes loose.
4. The bone flap is gently separated from the dura mater underneath and removed. Surgical Gelfoam is placed on the cut edge and the dura mater is continuously rinsed with saline.
5. A small incision is made close to the sagittal sinus. Using iris microscissors, the dura and arachnoid membranes are removed from the surface of both hemispheres without damaging the sagittal sinus.
6. A 1-mm piece of the tumor tissue is placed in the center of the window, which is sealed with a 7-mm circular coverglass glued to the bone with cyanoacrylate glue.
7. To image the cranial window, the animal is anesthetized and laid on a simple stereotactic apparatus.

29.5.3 Mammary Fat Pad Preparation (Acute)

The mouse mammary fat pad serves as an orthotopic site (natural organ site of growth for its tumor type) for breast cancer. Tumors growing in the mammary fat pad are prepared as follows (Monsky et al., 2002) (see Fig. 29.3):

1. The mouse is anesthetized, and using a syringe, a suspension of breast carcinoma cells is injected into the mammary fat pad just inferior to the nipple of female mice.
2. After 4 to 6 weeks, tumors grow to ~6 mm in diameter. A incision is made along the midline of the mouse through skin and fascia, and a flap is gently elevated, avoiding the disruption of vasculature.
3. The flap is then glued, skin side down, to a ~1-cm-high pedestal and a glass coverslip is placed over the tumor using a metal ring support.
4. Imaging occurs as long as anesthesia lasts, and the preparation is kept moist with saline.

29.6 IMAGING TUMOR-ASSOCIATED BLOOD AND LYMPHATIC VESSELS: BACKGROUND

One of the most popular applications of MPLSM in tumor biology continues to be the in vivo imaging of tumor-associated blood and lymphatic vessels (see Figs. 29.5 and 29.6). Brown and colleagues highlighted the vasculature of a tumor growing in the dorsal skinfold chamber of a mouse via intravenous injection of 2M MW FITC-Dextran and imaged the tumor vessels with MPLSM (Brown et al., 2001). The low phototoxicity of MPLSM (Squirrell et al., 1999), combined with the chronic dorsal skinfold chamber, allowed repetitive imaging of three-dimensional stacks of the tumor vasculature over several days and weeks. MPLSM imaging of the tumor vasculature has been used to reveal angiogenic properties of a candidate tumor-suppressor protein (Garkavtsev et al., 2004) as well as vascular normalization by vascular endothelial growth factor (VEGF) receptor blockade (Tong et al., 2004). MPLSM was extended to the study of the lymphatic system by Padera and coworkers using an injection of FITC-Dextran into the peritumor interstitium to highlight peritumor lymphatics and reveal an absence of functional intratumoral lymphatics (Padera et al., 2002a). MPLSM lymphangiography has more recently been used to understand the abnormal function of peritumoral lymphatics induced by VEGF-C (Isaka et al., 2004) and the effects of compressive forces on the morphology and function of tumor-associated blood and lymphatic vessels (Padera et al., 2004).

29.7 IMAGING TUMOR-ASSOCIATED BLOOD AND LYMPHATIC VESSELS: APPLICATION

To image the blood vessels of a tumor, the vasculature must be filled with a large-molecular-weight fluorescent indicator dye. Intravenous injection of dye occurs as follows:

1. The mouse is anesthetized.
2. A 30-gauge needle is attached to a 1-mL syringe filled with sterile saline.
3. The needle is inserted into one end of a 6-inch length of P.E. 10 tubing.
4. Using forceps, a 30-gauge needle is broken off its plastic base, and the blunt (broken) end is inserted into the open end of the P.E. 10 tubing.
5. Saline is pushed through the tubing, removing all the air.
6. Using forceps, the 30-gauge needle is inserted into a visible vein in the tail of the mouse. The veins are exceptionally shallow. When the needle has entered the vein, blood will backfill an inch or so of the tubing.
7. Approximately 0.01 mL of saline is pushed into the mouse, thereby ensuring that the needle is not clogged.
8. The syringe of saline with its needle is gently removed from the tubing, and a new syringe and needle, filled with 10 mg/mL of 2M MW FITC-Dextran in saline, is inserted. Avoid introducing air bubbles into the tube.
9. The FITC-Dextran solution is gently injected until it has completely filled the tube. The tube will now be visibly yellow along its entire length.
10. Approximately 0.2 mL of FITC-Dextran is injected into the mouse, and then the needle is removed from the tail.

This will ensure very brightly labeled vasculature visible for several hundred microns into the tumor. Note that only functional vessels are labeled: vessels with no blood flow will not be labeled. Depending upon the relative permeability of the tumor vessels (which are considerably more permeable than most healthy vessels), imaging will be crisp and clean for up to an hour before significant leakage of dye into the interstitium occurs.

To image functional lymphatic vessels, a fluorescently labeled high-molecular-weight tracer must be introduced into the interstitium of the tissue of interest. This procedure, called lymphangiography, has been adapted from clinical medicine for use in the mouse. For intravital imaging, this requires the tissue be in an accessible location, such as in the ear, tail, or leg skin of the mouse. The dorsal skin chamber has also been used to image functional lymphatic vessels. The procedure is as follows:

1. Using forceps, a 30-gauge needle is broken off its plastic base, and the blunt (broken) end is inserted into the one end of a ∼50-cm length of P.E. 10 tubing.
2. A 30-gauge needle is inserted into the other end of the P.E. 10 tubing and connected to a three-way Luer-lok valve.
3. A 10-cc syringe filled with water is attached to the three-way Luer-lok valve.
4. The remaining port of the three-way Luer-lok valve is connected, using appropriate materials, to a standing column of water, which will provide a steady input hydrostatic pressure for lymphangiography. All air bubbles should be eliminated from the system using the 10-cc syringe, and flow should be steady. The pressure applied should range from 10 to 60 cm of water.
5. To perform lymphangiography, a small air bubble is created in the very tip of the P.E. 10 tubing in order to prevent mixing of the fluorescent molecule and the water in the system. 2 million MW fluorescent dextran (2.5 mg/mL) is then drawn up into the tubing (∼10–15 μL). Large-molecular-weight molecules are transported preferentially through lymphatics compared to blood vessels, a property that makes large molecules ideal for lymphangiography.
6. The mouse is anesthetized (Leunig et al., 1992) and secured to a heated stage, and its tail is taped to the stage using double-sided tape.
7. The needle at the end of the P.E. 10 tubing is then placed in the interstitium of the tissue of interest. Common tissues are the tail, ear, and dorsal skinfold chamber. These injections can also be made in the edge of a tumor in order to highlight the functional peritumor and tumor margin lymphatic vessels. After the large molecules enter the interstitial space they will be taken up by functional lymphatic vessels, which will be visible using the microscopy techniques described in this text.

An alternative to the constant-pressure lymphangiography described above is to connect a syringe directly to a 30-gauge needle (in lieu of the three-way Luer-lok valve) and perform the injections manually. Great care must be taken to avoid applying too much pressure during manual injections, as this may destroy the tissue and give unreliable results. Manual pressure injection cannot be used to measure lymphatic function, but only to identify functional lymphatic vessels.

29.8 IMAGING CELLS WITHIN THE LIVING TUMOR: BACKGROUND

Since its first application to the in vivo imaging of tumors in 2001 (Brown et al., 2001), one of the primary applications of the MPLSM in tumor biology has been the dynamic imaging of tumor and host cells (see Fig. 29.7). Initially, MPLSM was used to investigate the spatial distribution of VEGF-expressing fluorescent host cells within a tumor. This preparation was created when a nonfluorescent tumor cell line was grown in a dorsal skinfold chamber implanted on a transgenic mouse that expresses GFP under control of the VEGF promoter. In these VEGF-GFP mice, the high depth penetration and three-dimensional resolution of MPLSM revealed that VEGF-expressing cells from the host animal reside within the tumor and form three-dimensional sleeve-like structures around tumor blood vessels (Brown et al., 2001). In the same work, the dynamic behavior of host immune cells within slow-flowing tumor vessels was quantified with time-lapse imaging using intravenous injection of rhodamine 6G to selectively label nucleated cells within the bloodstream. This allowed the quantification of key parameters describing the host/tumor immune response (Fukumura et al., 1997a) such as the total leukocyte flux, rolling fraction, and adhering density. This analysis was extended to high-flow vessels using high-speed MPLSM based upon a spinning multifaceted primary scan mirror (Padera et al., 2002b). The creation of transgenic VEGF-GFP mice combined with in vivo MPLSM has also allowed the investigation of the fate of nonendothelial stromal cells after tumor transplantation (Duda et al., 2004). This work revealed that the ability of a tumor xenograft to recruit a new vasculature depends upon the presence of stromal cells within the xenograft that derive from its previous host.

In addition to imaging stromal cells within tumors, MPLSM has been used to image the tumor cells themselves. In addition to studying the spatial distribution of GFP-labeled tumor cells with relation to the tumor vasculature in a teratoma model (Brown et al., 2001), MPLSM of GFP-labeled gliosarcoma cells has been used in conjunction with near-infrared fluorescence imaging to localize the proteolysis of self-quenched fluorescent probes (Bogdanov et al., 2002). MPLSM imaging of mammary tumor cells in an acute mammary fat pad preparation has also been used in a series of publications dissecting the biochemical machinery of tumor cell metastasis (Ahmed et al., 2002; Wang et al., 2002, 2004; Wyckoff et al., 2004).

29.9 IMAGING CELLS WITHIN THE LIVING TUMOR: APPLICATION

With the exception of the quantification of leukocyte/endothelial interactions (LEI), the aforementioned applications do not need any detailed expertise beyond the creation of the surgical or window preparation, which has already been discussed, and the generation of the transgenic mice, which is beyond the scope of this work. Therefore we will restrict our discussion to the methodology of LEI analysis, which is as follows:

1. The tail vein of an anesthetized mouse is cannulated with a saline-filled syringe by performing items 1 through 7 described previously for tumor angiography.
2. The saline-filled syringe is gently removed from the P.E. 10 tube and replaced with one filled with 0.1% rhodamine-6G in saline. Avoid introducing air bubbles into the tube.
3. The rhodamine-6G solution is gently injected until it has completely filled the tube.
4. Approximately 0.2 mL of rhodamine-6G is injected into the mouse; the needle is then removed from the tail.
5. A series of 10 to 15 high-temporal-resolution (2 or more frames per second) images are generated of a given vessel, covering a total duration of 30 seconds. The numbers of rolling and adhering leukocytes are counted, as well as the total flux of cells.

29.10 TUMOR VESSEL PERMEABILITY: BACKGROUND

Elevation in blood vessel permeability is a key hallmark of angiogenesis. As a result of elevated levels of VEGF and other pro-angiogenic factors, tumor blood vessels are exceptionally leaky, with characteristic pore sizes ranging from 100 nm to 2 microns (Hobbs et al., 1998). The effective permeability of tumor vessels has been measured with conventional epifluorescence microscopy by the intravenous injection of small-molecular-weight fluorescence tracers, followed by the monitoring of extravascular fluorescence as it increases due to extravasation (Yuan et al., 1994). This analysis produces an effective permeability averaged over the vessels in a field of view of the epifluorescence microscope. In 2001 Brown and associates extended this technique to MPLSM, and the intrinsic three-dimensional resolution of MPLSM allowed the effective permeability of individual tumor vessel segments to be quantified (Brown et al., 2001).

29.11 TUMOR VESSEL PERMEABILITY: APPLICATION

The effective vascular permeability of individual vessel segments can be measured by MPLSM using the following procedure (Brown et al., 2001):

1. A low-molecular-weight tracer (e.g., TMR-labeled bovine serum albumin [TMR-BSA]) is injected intravenously following the procedure described above for tumor angiography.
2. A single optical section containing the vessel segment of interest is recorded periodically for up to 20 minutes using MPLSM.
3. Using image-processing software, the fluorescence intensity, $F(r)$, along a line perpendicular to the vessel is calculated, and the effective permeability, P, of the vessel segment adjacent to that line is given by:

$$P = \frac{\partial}{\partial t} \frac{\int_{r=R}^{\infty} F(r) r \, dr}{(F_v - F_i)R}$$

   where R is the radius of the vessel segment, F_v is the fluorescent signal from the cell-free layer within the vessel, and F_i is the fluorescence signal immediately outside the vessel.
4. To minimize errors that can result from a finite integration length and due to fluorescence contribution from material leaking out of nearby vessels, the value of P is calculated from the t = 0 limit of the fluorescence-versus-time curve.

29.12 SECOND-HARMONIC GENERATION IN TUMORS: BACKGROUND

Several organic molecules produce a strong intrinsic signal known as second-harmonic generation (SHG), which can be imaged with a typical MPLSM setup (Campagnola et al., 2002). SHG is the direct combination of two excitation photons into one photon of exactly half the excitation wavelength. Due to the coherent nature of SHG, the total signal does not necessarily scale linearly with the number of scatterers, and its emission pattern is anisotropic (Moreaux et al., 2000). Fortuitously, fibrillar collagen

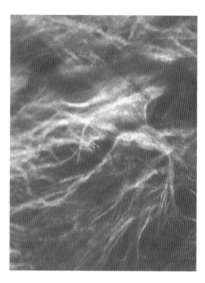

Figure 29.8. Imaging extracellular matrix. Collagen fibers in the extracellular matrix of an LS174T colon adenocarcinoma growing in the dorsal skinfold chamber of a SCID mouse were imaged with second-harmonic generation. Image is 275 microns across.

type I produces strong SHG, which can be imaged in tumors in vivo with reasonable excitation intensities (Fig. 29.8). Collagen type I is one of the key constituents of the tumor extracellular matrix and influences tumor cell metastasis and drug transport in tumors (Netti et al., 2000). Ahmed and coworkers studied tumor cell metastasis in vivo in relation to fibers imaged with SHG (Ahmed et al., 2002), while Brown and associates determined that the SHG signal in tumors was due to fibrillar collagen type I and that it scaled linearly with collagen content in the concentration range present in tumors, and used this knowledge to quantify in vivo the matrix modification produced during 2 weeks of systemic treatment with the hormone relaxin (Brown et al., 2003).

29.13 SECOND-HARMONIC GENERATION IN TUMORS: APPLICATION

The tumor collagen matrix can be imaged without extrinsic staining using SHG as follows:

1. The fluorophores in the system are evaluated, and an appropriate excitation wavelength is chosen that adequately excites those fluorophores while generating sufficient SHG in an unoccupied part of the visible spectrum. For example, to image fibrillar collagen simultaneously with TMR-dextran-labeled vasculature and GFP-labeled cells, ~860-nm excitation is appropriate, producing SHG at ~430 nm.
2. An emission filter and dichroic are chosen to transmit at exactly half the excitation wavelength and separate this emission wavelength from the emission of the fluorophores in the system. In the example cited above, 860-nm excitation requires a 430DF30-nm emission filter (2HG) along with 525DF100 (GFP) and 610DF75 (TMR) emission filters and 460LP and 575LP dichroics.

3. The dynamics of acute collagen modification can be quantified by generating repetitive images of a collagen matrix after addition of a collagen-degrading reagent: for example, one image every minute after superfusion of 100 micro-liters of 1% bacterial collagenase to a tumor growing in a dorsal skinfold chamber, or one image of a tumor every 2 days after implantation of an osmotic pump expressing the hormone relaxin (Brown et al., 2003). Image stacks are generated under identical conditions (e.g., excitation power, thickness of stack) to facilitate comparison from time point to time point.

29.14 LIMITATIONS OF MULTIPHOTON LASER-SCANNING MICROSCOPY

While MPLSM has produced important advances in tumor biology, it retains significant limitations. While the depth penetration of MPLSM ($\sim$1 mm) is a significant achieve-ment, it is not sufficient to penetrate most organs, let alone the entire body. The 3D imaging capability and high spatial resolution of MPLSM come at the expense of time resolution: it is difficult to raise the scanning speed without sacrificing signal-to-noise and/or compromising the 3D imaging. An ironic drawback of MPLSM is that the very short depth of focus produced by the physics of multiphoton absorption is not always desirable: because fluorophores are excited only within a thin focal plane, it can be dif-ficult to navigate through tissues using MPLSM, or to locate small, sparse fluorescent features. Another current limitation is the difficulty of exciting certain fluorescent probes using MPLSM, particularly those probes with a multiphoton excitation cross-section that is highest at infrared wavelengths of 1,000 nm or longer, beyond the tuning range of Ti:Sapphire lasers. Fortunately, the development of longer-wavelength, higher-power laser sources and other technological improvements is continuing at a rapid pace, and this should serve to mitigate some of these difficulties in the near future.

29.15 FUTURE DIRECTIONS

New MPLSM designs that improve the optical depth penetration are always welcome, as are design changes that make it possible to access more organs in mice and other animals, including humans. For example, a fiberoptic endoscope that employs MPLSM could theoretically allow access to the upper and lower digestive organs, airways, and (with minimally invasive surgery) many internal organs. Unfortunately, the design of fiberoptic multiphoton endoscopes is intrinsically difficult because of the physics of glass fibers: nonlinear interactions between high-intensity short-pulsed light and the glass create a broadening of the pulses that is difficult if not impossible to compensate for. However, multiphoton microscopes incorporating inflexible but thin GRIN lenses may be applicable to some endoscopic experiments (Jung & Schnitzer, 2003; Levene et al., 2004), and the use of microstructured optical fibers may eventually permit the design of long, flexible MPLSM endoscopes (Gobel et al., 2004).

Today, MPLSM is primarily aimed at the visualization of exogenous or genetically engineered fluorescent molecules. The ability to directly visualize unlabeled endo-geous molecules (e.g., collagen fibers, by SHG) or clinically relevant molecules (e.g., chemotherapeutic agents, some of which may have specific fluorescent signatures) will expand the usefulness of MPLSM, especially in clinical settings. MPLSM designs that permit better gathering and analysis of spectroscopic data, and that can distinguish fluorescent particles based on details of their emission spectrum, would assist in such studies.

Much of the effort in tumor-imaging research is concerned with the design of more relevant optically accessible tumor models. For example, although many studies have made use of transplanted tumors in animal models, and these tumors are known to be similar to spontaneous tumors in many ways, recent studies have shown that spontaneous tumors have a different growth rate, and respond differently to antiangiogenic agents, than do tumors that are serially transplanted from the original spontaneous tumors (Izumi et al., 2003). In the future, engineered mice with tissue-specific and/or inducible promoters that can generate spontaneous tumors in optically accessible regions will allow us to remedy this shortcoming by studying spontaneous tumors directly (Hagendoorn et al., 2006). Current work is also limited by the limited availability of fluorescent reporters (e.g., *VEGF*^{p}*-GFP* mice, *TIE2*^{p}*-GFP* mice). However, this is a rapidly advancing area of research, and new reporter mice and tumor cell lines appear weekly. In combination with novel imaging and surgical methods, these advances will allow us to continuously improve our understanding of tumor biology.

ACKNOWLEDGMENTS

This work was supported by a Program Project Grant (P01CA80134) and Bioengineering Research Partnership Grant (R24 CA85140) to R.K.J., D.F., and L.M., an R01 (HL64240) to L.M., and an R01 (CA96915) to D.F., a Department of Defense Era of Hope Scholar Award to E.B., and a Whitaker Foundation Biomedical Engineering Grant to E.B.

This work was partially based on the previous related reviews:

1. Jain RK, Brown EB, Munn LL, Fukumura D. Intravital microscopy of normal and diseased tissues in mice. In: Goldman RD, Spector DL Editors. Live Cell Imaging: A Laboratory Manual, Cold Spring Harbor Laboratory Press, Cold Spring Harbor, NY, 2004: Chapter 24, 435–466.
2. Fukumura D. Role of microenvironment on gene expression, angiogenesis and microvascular functions in tumors. In: Meadows GG. editor. Integration/Interaction of Oncologic Growth, Vol. 15. In: Kaiser HE. series editor. Cancer Growth And Progression series, Springer Science + Business Media B.V., Dordrecht, the Netherlands, 2005: Chapter 2, 23–36.

REFERENCES

Abdul-Karim, M. A., K. Al-Kofahi, et al. (2003). "Automated tracing and change analysis of angiogenic vasculature from in vivo multiphoton confocal image time series." Microvascular Research 66(2): 113–125.

Ahmed, F., J. Wyckoff, et al. (2002). "GFP expression in the mammary gland for imaging of mammary tumor cells in transgenic mice." Cancer Res 62(24): 7166–7169.

Alexandrakis, G., E. B. Brown, et al. (2004). "Two-photon fluorescence correlation microscopy reveals the two-phase nature of transport in tumors." Nat Med 10(2): 203–207.

Alitalo, K. and P. Carmeliet (2002). "Molecular mechanisms of lymphangiogenesis in health and disease." Cancer Cell 1: 219–227.

Berk, D. A., M. A. Swartz, et al. (1996). "Transport in lymphatic capillaries. II. Microscopic velocity measurement with fluorescence photobleaching." American Journal of Physiology 270(1 Pt 2): H330–H337.

Berk, D. A., F. Yuan, et al. (1993). "Fluorescence photobleaching with spatial Fourier analysis: measurement of diffusion in light-scattering media." Biophysical Journal 65(6): 2428–36.

Berk, D. A., F. Yuan, et al. (1997). "Direct in vivo measurement of targeted binding in a human tumor xenograft." Proceedings of the National Academy of Sciences of the USA 94: 1785–1790.

Bogdanov, A. A., Jr., C. P. Lin, et al. (2002). "Cellular activation of the self-quenched fluorescent reporter probe in tumor microenvironment." Neoplasia 4(3): 228–236.

Brown, E., R. Campbell, et al. (2001). "In vivo measurement of gene expression, angiogenesis, and physiological function in tumors using multiphoton laser scanning microscopy." Nat Med 7(7): 864–868.

Brown, E., T. D. McKee, et al. (2003). "Dynamic imaging of collagen and its modulation in tumors in vivo using second harmonic generation." Nat Med 9(6): 796–801.

Campagnola, P. J., A. C. Millard, et al. (2002). "Three-dimensional high-resolution second-harmonic generation imaging of endogenous structural proteins in biological tissues." Biophys J 82(1 Pt 1): 493–508.

Campbell, R. B., D. Fukumura, et al. (2002). "Cationic charge determines the distribution of liposomes between the vascular and extravascular compartments of tumors." Cancer Res 62(23): 6831–6836.

Dellian, M., G. Helmlinger, et al. (1996). "Fluorescence ratio imaging of interstitial pH in solid tumours: effect of." British Journal of Cancer 74(8): 1206–1215.

Duda, D. G., D. Fukumura, et al. (2004). "Differential transplantability of tumor-associated stromal cells." Cancer Res 64(17): 5920–5924.

Elenbaas, B. and R. A. Weinberg (2001). "Heterotypic signaling beteen epithelial tumor cells and fibroblasts in carcinoma formation." Experimental Cell Research 264: 169–184.

Endrich, B., M. Intaglietta, et al. (1979). "Hemodynamic characteristics in microcirculatory blood channels during early tumor growth." Cancer Res 39: 17–23.

Folkman, J. (2000). Tumor angiogenesis. Cancer Medicine, 5th Edition. J. F. Holand, E. I. Frei, R. C. J. Bast et al. Ontario, B.C. Decker Inc.: 132–152.

Fukumura, D. (2005). Role of microenvironment on gene expression, angiogenesis and microvascular functions in tumors. Integration/Interaction of Oncologic Growth. G. G. Meadows. Dordrecht, Springer Science + Business Media B.V. 15: 23–36.

Fukumura, D., Y. Chen, et al. (1998). "VEGF expression in tumors is regulated by oxygen gradient rather than absolute level of hypoxia itself." Proceedings of American Association of Cancer Research 39: 271.

Fukumura, D., H. A. Salehi, et al. (1995). "Tumor necrosis factor a-induced leukocyte adhesion in normal and tumor vessels: Effect of tumor type, transplantation site, and host strain." Cancer Research 55: 4824–4829.

Fukumura, D., A. Ushiyama, et al. (2003). "Paracrine regulation of angiogenesis and adipocyte differentiation during in vivo adipogenesis.[erratum appears in Circ Res. 2004 Jan 9;94(1):e16]." Circulation Research 93(9): e88–97.

Fukumura, D., F. Yuan, et al. (1997a). "Role of nitric oxide in tumor microcirculation. Blood flow, vascular permeability, and leukocyte-endothelial interactions." Am J Pathol 150(2): 713–725.

Fukumura, D., F. Yuan, et al. (1997b). "Effect of host microenvironment on the microcirculation of human colon adenocarcinoma." American Journal of Pathology 151: 679–688.

Garkavtsev, I., S. V. Kozin, et al. (2004). "The candidate tumour suppressor protein ING4 regulates brain tumour growth and angiogenesis." Nature 428(6980): 328–332.

Gerlowski, L. and R. K. Jain (1986). "Microvascular permeability of normal and neoplastic tissues." Microvasc Res 31: 288–305.

Gobel, W., A. Nimmerjahn, et al. (2004). "Distortion-free delivery of nanojoule femtosecond pulses from a Ti:sapphire laser through a hollow-core photonic crystal fiber." Opt Lett 29(11): 1285–1287.

Gohongi, T., D. Fukumura, et al. (1999). "Tumor-host interactions in the gallbladder suppress distal angiogenesis and tumor growth: Involvement of transforming growth factor b1." Nature Medicine 5(10): 1203–1208.

Hagendoorn, J., T. P. Padera, et al. (2004). "Endothelial Nitric Oxide Synthase Regulates Microlymphatic Flow via Collecting Lymphatics." Circ Res 95(2): 204–209.

Hagendoorn, J., R. Tong, et al. (2006). "Onset of abnormal blood and lymphatic vessel function and interstitial hypertension in early stages of carcinogenesis." Cancer Res 66(7): 3360–3364.

Hanahan, D. and R. A. Weinberg (2000). "The hallmarks of cancer." Cell 100: 57–70.

Harris, A. L. (2002). "Hypoxia – A key regulatory factor in tumor growth." Nat Rev Cancer 2: 38–47.

Helmlinger, G., F. Yuan, et al. (1997). "Interstitial pH and pO_2 gradients in solid tumors *in vivo*: high-resolution measurements reveal a lack of correlation." Nature Medicine 3(2): 177–182.

Hobbs, S. K., W. L. Monsky, et al. (1998). "Regulation of transport pathways in tumor vessels: role of tumor type and microenvironment." Proc Natl Acad Sci U S A 95(8): 4607–4612.

Hoshida T, Isaka N, Hagendoorn J, di Tomaso E, Chen YL, Pytowski B, Fukumura D, Padera TP, Jain RK. (2006). Imaging steps of lymphatic metastasis reveals that VEGF-C increases metastasis by increasing delivery of cancer cells to lymph nodes: Therapeutic implications. Cancer Research, 66: 8065–8075.

Huang, P., T. D. McKee, et al. (2005). "Green fluorescent protein (GFP)-expressing tumor model derived from a spontaneous osteosarcoma in a vascular endothelial growth factor (VEGF)-GFP transgenic mouse." Comparative Medicine 55(3): 236–243.

Ide, A. G., N. H. Baker, et al. (1939). "Vascularization of the Brown-Pearce rabbit epithelioma transplant as seen in the transparent ear chamber." Am J Roentgenol 42: 891–899.

Isaka, N., T. P. Padera, et al. (2004). "Peritumor lymphatics induced by vascular endothelial growth factor-C exhibit abnormal function." Cancer Res 64(13): 4400–4404.

Izumi, Y., E. di Tomaso, et al. (2003). "Responses to antiangiogenesis treatment of spontaneous autochthonous tumors and their isografts." Cancer Res 63(4): 747–751.

Jain, R. K. (1988). "Determinants of tumor blood flow: A review." Cancer Res. 48: 2641–2658.

Jain, R. K. (1994). "Barriers to drug delivery in solid tumors." Scientific American 271: 58–65.

Jain, R. K. (1998). "The next frontier of molecular medicine: Delivery of therapeutics." Nature Medicine 4(6): 655–657.

Jain, R. K. (2002). "Angiogenesis and lymphangiogenesis in tumors: Insights from intravital microscopy." Cold Spring Harbor Symposia on Quantitative Biology 67: 239–248.

Jain, R. K. (2003). "Molecular regulation of vessel maturation." Nature Medicine 9(6): 685–693.

Jain, R. K. (2005). "Normalization of tumor vasculature: An emerging concept in antiangiogenic therapy." Science 307: 58–62.

Jain, R. K., L. L. Munn, et al. (2001). Transparent window models and intravital microscopy. Tumor models in cancer research. B. A. Teicher. Totowa, Humana Press Inc: 647–671.

Jain, R. K., L. L. Munn, et al. (2002). "Dissecting tumour pathophysiology using intravital microscopy." Nat Rev Cancer 2(4): 266–276.

Jung, J. C. and M. J. Schnitzer (2003). "Multiphoton endoscopy." Opt Lett 28(11): 902–904.

Kashiwagi, S., Y. Izumi, et al. (2005). "NO mediates mural cell recruitment and vessel morphogenesis in murine melanomas and tissue-engineered blood vessels." J. Clin. Invest. 115(7): 1816–1827.

Koike, N., D. Fukumura, et al. (2004). "Tissue engineering: creation of long-lasting blood vessels." Nature 428(6979): 138–139.

Leu, A. J., D. A. Berk, et al. (1994). "Flow velocity in the superficial lymphatic network of the mouse tail." American Journal of Physiology 267(4 Pt 2): H1507–13.

Leunig, M., F. Yuan, et al. (1992). "Angiogenesis, microvascular architecture, microhemodynamics, and interstitial fluid pressure during early growth of human adenocarcinoma LS174T in SCID mice." Cancer Res 52(23): 6553–6560.

Levene, M. J., D. A. Dombeck, et al. (2004). "In vivo multiphoton microscopy of deep brain tissue." J Neurophysiol 91(4): 1908–1912.

Li, G., K. Satyamoorthy, et al. (2003). "Function and regulation of melanoma–stromal fibroblast interactions: when seeds meet soil." Oncogene 22: 3162–3171.

Liotta, L. A. and E. C. Kohn (2001). "The microenvironment of the tumour–host interface." Nature 411: 375–379.

Monsky, W. L., C. Mouta Carreira, et al. (2002). "Role of host microenvironment in angiogenesis and microvascular functions in human breast cancer xenografts: mammary fat pad versus cranial tumors." Clin Cancer Res 8(4): 1008–1013.

Moreaux, L., O. Sandre, et al. (2000). "Membrane imaging by second-harmonic generation microscopy." J. Opt. Soc. Am. B 17(10): 1685–1694.

Morikawa, S., P. Baluk, et al. (2002). "Abnormalities in pericytes on blood vessels and endothelial sprouts in tumors." American Journal of Pathology 160(3): 985–1000.

Netti, P. A., D. A. Berk, et al. (2000). "Role of extracellular matrix assembly in interstitial transport in solid tumors." Cancer Res 60(9): 2497–503.

Ny, A., M. Koch, et al. (2005). "A genetic Xenopus laevis tadpole model to study lymphangiogenesis." Nature Medicine 11(9): 998–1004.

Orimo, A., P. B. Gupta, et al. (2005). "Stromal Fibroblasts Present in Invasive Human Breast Carcinomas Promote Tumor Growth and Angiogenesis through Elevated SDF-1/CXCL12 Secretion." Cell 121(3): 335.

Padera, T. P., A. Kadambi, et al. (2002a). "Lymphatic metastasis in the absence of functional intratumor lymphatics." Science 296(5574): 1883–1886.

Padera, T. P., B. Stoll, et al. (2002b). "Conventional and High-Speed Intravital Multiphoton Laser Scanning Microscopy of Microvasculature, Lymphatics, and Leukocyte-Endothelial Interactions." Molecular Imaging 1: 9–15.

Padera, T. P., B. R. Stoll, et al. (2004). "Pathology: cancer cells compress intratumour vessels." Nature 427(6976): 695.

Pollard, J. W. (2004). "Tumour-educated macrophages promote tumour progression and metastasis." Nat Rev Cancer 4: 71–78.

Ruiter, D. J., J. H. van Krieken, et al. (2001). "Tumour metastasis: is tissue an issue?" Lancet Oncology 2(2): 109–112.

Sandison, J. C. (1924). "A new method for the microscopic study of living growing tissues by the introduction of a transparent chamber in the rabbit's ear." Anat Rec 28: 281–287.

Sasaki, A., R. J. Melder, et al. (1991). "Preferential localization of human A-LAK cells in tumor microcirculation: A novel mechanism of adoptive immunotherapy." J Nat Cancer Inst 83: 433–437.

Squirrell, J. M., D. L. Wokosin, et al. (1999). "Long-term two-photon fluorescence imaging of mammalian embryos without compromising viability." Nat Biotechnol 17(8): 763–767.

Stroh, M., J. P. Zimmer, et al. (2005). "Quantum dots spectrally distinguish multiple species within the tumor milieu in vivo." Nature Medicine 11(6): 678–682.

Swartz, M. A., D. A. Berk, et al. (1996). "Transport in lymphatic capillaries. I. Macroscopic measurements using residence time distribution theory." American Journal of Physiology 270(1 Pt 2): H324–H329.

Tlsty, T. D. (2001). "Stromal cells can contribute oncogenic signals." Cancer Biology 11: 97–104.

Tong, R. T., Y. Boucher, et al. (2004). "Vascular normalization by vascular endothelial growth factor receptor 2 blockade induces a pressure gradient across the vasculature and improves drug penetration in tumors." Cancer Res 64(11): 3731–3736.

Torres Filho, I. P., M. Leunig, et al. (1994). "Noninvasive measurement of microvascular and interstitial oxygen profiles in a human tumor in SCID mice." Proceedings of the National Academy of Sciences of the USA 91(6): 2081–2085.

Tyrrell, J. A., V. Mahadevan, et al. (2005). "A 2-D/3-D model-based method to quantify the complexity of microvasculature imaged by in vivo multiphoton microscopy." Microvascular Research 70: 165–178.

Wang, W., S. Goswami, et al. (2004). "Identification and testing of a gene expression signature of invasive carcinoma cells within primary mammary tumors." Cancer Res 64(23): 8585–8594.

Wang, W., J. B. Wyckoff, et al. (2002). "Single cell behavior in metastatic primary mammary tumors correlated with gene expression patterns revealed by molecular profiling." Cancer Res 62(21): 6278–6288.

Winkler, F., S. V. Kozin, et al. (2004). "Kinetics of vascular normalization by VEGFR2 blockade governs brain tumor response to radiation: role of oxygenation, angiopoietin-1, and matrix metalloproteinases." Cancer Cell 6(6): 553–563.

Wyckoff, J., W. Wang, et al. (2004). "A paracrine loop between tumor cells and macrophages is required for tumor cell migration in mammary tumors." Cancer Res 64(19): 7022–7029.

Yuan, F., M. Dellian, et al. (1995). "Vascular permeability in a human tumor xenograft: Molecular size dependence and cutoff size." Cancer Res 55(September 1): 3752–3756.

Yuan, F., M. Leunig, et al. (1993). "Microvascular permeability of albumin, vascular surface area, and vascular volume measured in human adenocarcinoma LS174T using dorsal chamber in SCID mice." Microvascular Research 45: 269–289.

Yuan, F., H. A. Salehi, et al. (1994). "Vascular permeability and microcirculation of gliomas and mammary carcinomas transplanted in rat and mouse cranial windows." Cancer Res 54(17): 4564–4568.

30

Immunology Based on Nonlinear Optical Microscopy

Ian Parker and Michael D. Cahalan

30.1 INTRODUCTION

The immune system is probably the most disseminated and mobile system in our bodies. Its component parts (largely individual cells) are in continuous circulation, endlessly surveying peripheral tissues for possible infection, and returning to mount defenses. A key part of the immune response involves cell–cell interactions: for example, the presentation of antigens by dendritic cells (DCs) to T cells within lymph nodes. Until recently, studies of these interactions have been constrained by methodological limitations, and two distinct lines of investigation have dominated the field: (i) in vivo experiments that examine the behavior of populations of cells in living animals during an immune response and (ii) in vitro experiments utilizing individual cells in artificial environments. However, the immune response is the sum of many complex and dynamic individual cellular behaviors that are shaped by a host of environmental factors. Whereas in vivo experiments maintain this natural environment, they cannot resolve the behaviors of individual cells. This approach can be termed "snapshot immunology," as it has previously been limited to histologically fixed preparations or to analysis of cells taken from an animal at a particular time. Conversely, in vitro experiments provide information at subcellular and molecular levels and with real-time single-cell monitoring, but they cannot adequately recapitulate the complexity and dynamics of intact tissue environments.

Despite the wealth of knowledge of the molecular and physiological aspects of the immune system, we know surprisingly little of how cells coordinate the initiation or suppression of an immune response. A pressing need has existed for techniques that permit real-time observation of single cells and molecules within intact tissues. Recent applications of multiphoton imaging technology now go a long way toward making this possible. The purpose of this chapter is to provide a guide to the application of nonlinear biophotonic techniques to the field of immunology. In addition, we briefly summarize some key observations and generalizations that have emerged and are influencing the way immunologists think about the immune response.

30.1.1 A Brief Primer on Immunology for Non-Biologists

The immune system protects us from certain death caused by infectious diseases (see Note 1). Divided into a primitive "innate" immune response and a more advanced "adaptive" immune response, the mammalian immune response is coordinated by a variety of cell types that are derived from bone marrow. Innate immunity includes the ability of neutrophils and macrophages to ingest and destroy bacteria and to produce inflammatory reactions. Adaptive immunity is mediated by lymphocytes and allows the organism to adjust the response according to the unpredictable encounter with myriad foreign antigens. Such antigens might include protein fragments on bacteria, viruses, or other pathogens. The ability of T and B lymphocytes to recognize countless numbers of antigens is achieved by lymphocytes' unique capacity to reorganize the DNA sequences for specific antigen receptors through a process of somatic cell recombination. During development in the thymus, enzymes called recombinases perform genetic surgery, through which gene sequences are rearranged to generate novel T-cell receptor (TCR) proteins with the ability to recognize a different antigen. Thymocytes with broadly diverse T-cell receptors are first generated and then pared down by a process of positive and negative selection to derive the mature T-cell repertoire that consists of billions of distinct T cells, each bearing a different receptor that can recognize a particular antigen. This strategy of "preparedness" permits recognition of a wide range of potential antigens that might never be encountered during the lifetime of an organism. But the immune response is always ready. If a foreign protein is encountered, it is first gobbled up by "professional" antigen-presenting cells, such as DCs, partially digested down to a peptide antigen fragments that become bound to self-major histocompatibility complex (MHC) proteins on the surface of the DC, and then transported into a lymphoid organ by the DC.

How is the response to antigen coordinated within the body? As mentioned above, cells of the immune system can migrate between the blood, the secondary lymphoid organs (including lymph nodes and spleen), and other tissues of the body. Antigen recognition occurs inside a lymphoid organ by direct cell contact between an antigen-specific T cell and an antigen-bearing DC. For example, an immune response to vaccination or infection occurs first by cell-to-cell contacts between DCs that bring antigen into the lymph node and T cells, and then by contact between helper T cells and B cells. In the contact zone between the cells, an "immunological synapse" is formed as receptors come together with adapter molecules, cytoskeletal elements, and other proteins. This initial recognition phase is followed by an intracellular signalling cascade that results in new gene expression, secretion of cytokines, and cell proliferation to expand the clone of cells that recognize the particular antigen.

Of course, the range of possible responses is much more sophisticated than just the clonal expansion of antigen-specific T cells. One subset of T cells, characterized by expression of a surface membrane protein called CD8, can become killers that seek and destroy virus-infected cells to clear a viral infection. Such cells can also recognize foreign cells in the body, leading to rejection of improperly matched organ transplants. Another subset that expresses a different membrane protein, CD4, becomes helper T cells. These are required for B lymphocytes to generate an antibody response. Once fully activated by antigen and by contact with a helper T cell, B cells terminally differentiate to form plasma cells that are factories for the production of a specific antibody against the antigen that initiated the response. Specificity is guaranteed by the requirement for a specific TCR to be in contact with an antigen-specific B cell that has taken up the same antigen. In B cells (but not T cells) antigen receptor genes can be expressed as surface membrane antigen receptors or as a secreted form (the antibody), and the immunoglobulin genes undergo further mutations to expand the diversity of antibodies that are produced. As well as providing help for B-cell differentiation, activated T cells migrate out of the lymph node to become effector T cells that invade tissue sites of inflammation, where they recognize antigen, and produce further responses by secretion of lymphokines.

In summary, coordination of the response to foreign antigens involves migration of immune system cells through the body, recognition of antigen by direct cell-to-cell contacts, and secretion of proteins that result in a variety of subsequent effector responses.

30.2 SEEING INSIDE LYMPHOID ORGANS

Ideally, of course, it would be advantageous and most relevant to investigate the cellular interactions, signaling cascades, and effector function of the immune system entirely within an in vivo context. Although the field is yet far from this goal, recent advances in microscopy; new probes, including a variety of indicator dyes and fluorescent proteins expressed in transgenic mice; and new experimental preparations have created exciting opportunities for the study of lymphocyte motility, chemotaxis, and antigen recognition in the physiological context of the native tissue environment. In particular, two-photon laser microscopy provides a near-ideal technique to visualize the cellular dynamics of lymphocytes and DCs deep within lymphoid organs and other tissues (1–4). This method has now been applied to image cells of the immune system within isolated organ and tissue preparations, including lymph node (1,5–13), thymus (14–16), spleen (17), intestinal tissue (18), and brain slices (19). In addition, for the first time immune cells have been seen using intravital preparations of lymph nodes in living, anesthetized mice (5,9).

We concentrate here on the lymph node, which provides the location for encounters between T cells and DCs that are required for homeostatic maintenance of T cells, for encounters between T cells and B cells that lead to antibody production, for the establishment of tolerance to self-antigen, and for the generation of the protective immune response. It has long been appreciated that the lymph node is a highly structured organ designed to enhance cellular interactions, but visualizing cellular motility and encounters under physiological conditions presents several challenges. Chief among these, specific cell types must be appropriately labeled and imaged noninjuriously in three dimensions over extended time periods deep (hundreds of μm) within an optically scattering tissue.

30.3 LINEAR VERSUS NONLINEAR IMAGING MODALITIES

Conventional linear (i.e., single-photon) fluorescence imaging techniques that can be used to provide multi-dimensional (x, y, z, time, wavelength) information from thick tissues or organs include wide-field microscopy in conjunction with deconvolution processing and confocal microscopy. In both cases, short wavelength light (usually UV, blue or green) is used to excite fluorophore in labeled cells, and the resulting emission at a slightly longer wavelength is used to form the image. This results in several disadvantages that severely restrict the applicability of linear fluorescence imaging for live cell studies. A general problem is that the short (high-energy) excitation wavelengths exacerbate phototoxicity and cause rapid bleaching. More specifically, light scattering limits the depth to which satisfactory images can be obtained within biological tissues. The effects of light scattering can be mitigated either by re-assigning the positions of emitted photons by deconvolution, or by blocking the detection of scattered photons by confocal microscopy. However, neither approach is a panacea, and the practicable depth penetration is only a few tens of μm. Deconvolution strategies appear not to have been utilized for imaging leukocytes in living tissues, and while confocal microscopy has been employed to image T cells and DCs in superficial regions of excised lymph nodes (20), the cells in that study showed little motility, possibly as a result of laser-induced photodamage. Confocal microscopy was used recently to image highly motile NKT cells at the surface of liver sinusoids (21).

The application of nonlinear multiphoton excitation techniques substantially overcomes many of these problems and has enabled a revolution in immunological studies. In this regard, immunologists are relatively "late adopters," having been preceded, among others, by neuroscientists and developmental biologists. Nonetheless, following initial publications in 2002 (1,2,14), there has been an explosive growth in the numbers of papers describing multiphoton immuno-imaging. The primary factors driving this are the order-of-magnitude improvement in depth to which good images can be obtained in living tissue, and the relatively innocuous nature of the excitation light that permits long-term imaging with minimal photodamage.

The technology of multiphoton excitation is extensively described elsewhere in this volume (see Chapter 5). The essence is that a fluorophore is excited by the near-simultaneous absorption of energy from two photons, each of which contributes half of the energy required to induce fluorescence. Because of this quadratic dependence on excitation intensity, fluorescence is constrained to the focal spot formed by the microscope objective, thereby providing an inherent "optical sectioning" effect (Fig. 30.1A). However, in order to achieve practicable fluorescence signals, the photon density in the focal spot must be incredibly high, yet without damaging the specimen. This is achieved by use of femtosecond or picosecond pulsed lasers, which concentrate their output into brief bursts with enormous instantaneous power. Typically, each pulse lasts ~100 fs, with intervening gaps of ~100 ns between pulses. So, the temporal compression is by a factor of one hundred thousand, and the two-photon evoked fluorescence is correspondingly enhanced by the same factor as compared to a continuous laser of equivalent average power (see Fig. 30.1B).

Multiphoton microscopy thus provides images that represent an optical section within a tissue. These are closely analogous to confocal sections, but the means by which the sectioning effect is achieved are very different. In confocal microscopy fluorescence excitation occurs throughout the depth of the tissue and out-of-focus fluorescence is rejected by the pinhole; in multiphoton microscopy there essentially is no out-of-focus signal. However, the principal features of multiphoton excitation that

(A) **Spatial compression of photons by objective lens**

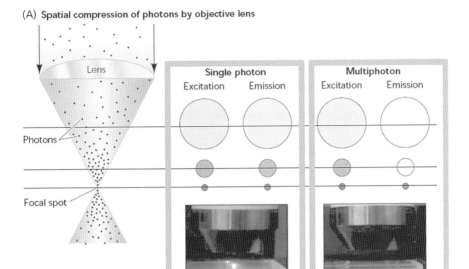

(B) **Temporal compression of photons during femtosecond pulses**

Continuous laser

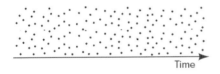

Femtosecond-pulsed laser

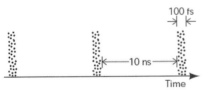

Figure 30.1. Principles and imaging modalities of multiphoton microscopy. (A) Differences between single-photon and multiphoton laser excitation. In both cases, the spatial compression of photons by an objective lens is greatest in the focal spot, and the photon density rapidly decreases away from this region. However, with conventional single-photon excitation, fluorophore excitation occurs throughout the beam path, resulting in fluorescence emission above and below the plane of focus. Multiphoton excitation restricts emission to the focal spot owing to the quadratic relationship between excitation intensity and fluorescence emission. Photographs illustrate fluorescence in a cuvette of fluoresceine resulting from focused laser beams using single-photon excitation at 488 nm (*left*) and two-photon excitation with 780-nm, femtosecond pulses (*right*). (B) Temporal compression of photons into 100-fs packets achieves the high instantaneous power needed for multiphoton excitation, with an average power not much greater than a continuous laser beam as would be used for confocal imaging. (Used with permission from (4))

make it attractive for biological imaging in general, and for immuno-imaging in particular, accrue secondarily from the use of excitation wavelengths that are roughly twice as long (half the energy) as would be used for conventional linear excitation by wide-field or confocal microscopy. Ranked subjectively in order of importance, these are:

1. Light scattering by particles (primarily organelles) in biological tissue decreases with increasing wavelength (22), and absorption by hemoglobin and other proteins is minimized at long wavelengths. Thus, the infrared wavelengths (700–1,200 nm) used for two-photon imaging allows at least a five-fold deeper tissue penetration than confocal imaging (1,5,20,21,23). This is of crucial importance for imaging into intact, complex tissues such as the lymph node, where cells and cell interactions may be localized within structures hundreds of micrometers below the surface.
2. Infrared wavelengths cause negligible photodamage or phototoxicity. Long-term imaging is thereby facilitated because damage processes (by multiphoton excitation) are largely confined to those cells lying at the focal plane, and cells above and below experience only the innocuous infrared light. For the same reason, photobleaching of fluorescent probes is minimized and, unlike confocal microscopy, occurs only in those cells actually being imaged.
3. The two-photon excitation spectra of most fluorophores are broader than for one-photon excitation (24), so a single excitation wavelength can be used to efficiently and simultaneously excite multiple probes with distinct emission wavelengths.

30.4 INSTRUMENTATION AND BIOLOGICAL PREPARATIONS FOR IMMUNO-IMAGING

30.4.1 Lasers

The availability of lasers that produce brief (ca. 1 ps or shorter) pulses at infrared wavelengths has been the key factor driving the development of two-photon microscopy, and progressive improvements in tuning flexibility and "user friendliness" have increased their applications by biologists. Two-photon immuno-imaging studies to date have utilized mode-locked Ti:sapphire lasers, which have advantages of high output power (approaching 2W) combined with the ability to tune over a range of about 700 to 1,000 nm. The latest generation feature a briefcase-sized laser head, run from a regular wall outlet, and are operated entirely by computer control, with no mechanical adjustments and no requirement for periodic alignment or maintenance. Their only significant disadvantages are high cost (ca. $150,000) and inability to tune to wavelengths >1 μm. Various designs of diode-pumped solid-state (DPSS) femtosecond lasers offer a less expensive alternative at long wavelengths that would be applicable for imaging fluorescent proteins or for three-photon imaging, but the inability of DPSS lasers to tune over a broad range restricts their versatility.

Average excitation intensities at the specimen must typically be restricted to a few tens of mW in order to minimize photodamage, so lasers with outputs >1 W may seem like overkill. However, it is wise to purchase the most powerful laser that can be afforded. Firstly, the maximum output is specified at the peak of the tuning range, and the power falls precipitously toward the edges of the curve. Secondly, scattering losses require higher excitation powers when imaging deep into tissues, and we often find that the imaging depth is limited by lack of excitation power rather than by degradation of image resolution. Because Ti:sapphire lasers operate properly only when producing near their specified output, it is necessary to attenuate the beam so as to regulate excitation at the specimen. This can readily be accomplished using a continuously variable reflective neutral density filter, but it is difficult to manually adjust the intensity to counteract the effect of increasing tissue depth while capturing deep image stacks. A more elegant approach would be to use a Pockels cell as a computer-controlled attenuator and shutter to regulate excitation intensity in parallel with command signals to the microscope focus motor.

30.4.2 Image Scanning

Lymphocytes migrate along apparently random paths in lymphoid tissue with mean velocities around 12 μm min^{-1}. In order to follow a given cell over several minutes it is, therefore, necessary to acquire image sequences (z-stacks) throughout a volume sufficiently large that the cell has only a small chance of escaping, at a rate fast enough to unambiguously identify a given cell from one time point to the next. In practice, this necessarily involves some compromise. A fundamental limit is set by the time required to image a single plane (x-y image). Commercially available microscopes typically require 0.5 to 1 s to scan an image with good (512 $\times$ 480-pixel) resolution, and even though we employ a custom resonant scanning system operating at 30 frames s^{-1} (25), it is usually necessary to average 10 to 15 consecutive frames to improve the signal-to-noise ratio. The overall acquisition speed is then set by the axial (z) depth of the imaged volume and the z spacing between image planes. A z spacing of about 2.5 μm (appreciably less than the $\sim$10-μm diameter of a T cell) is sufficient to locate cell positions without danger of a cell being lost between planes, giving a time resolution of around 20 s per z-stack for an axial depth of 50 μm at 0.5 s per plane. The image acquisition time is long compared to the time required to physically refocus the microscope between z-steps, and either stepper-motor focus drives or piezoelectric devices are suitable.

In contrast to conventional scanning systems that construct an image by raster scanning a single laser spot, multifocal two-photon microscopy simultaneously scans multiple spots to achieve a corresponding improvement in frame rate (26). However, as discussed below, it is then necessary to use an imaging detector, so that scattered light degrades the image rather than contributing to the signal.

30.4.3 Objective Lenses

Microscope objectives for biological two-photon imaging must have good infrared transmission and correction and possess a high numerical aperture (NA) both to form a smaller diffraction-limited excitation spot and to collect as much emitted fluorescence as possible. Moreover, it is convenient to use "dipping" water immersion objectives so that the specimen (excised tissue or in vivo preparation) can be superfused with warmed, oxygenated solution. We have had good success with the Olympus 20$\times$, 0.95 NA dipping objective, which combines high NA, wide field of view, and long (2 mm) working distance, and a new generation of Olympus water dipping objectives offer even higher NA at higher magnifications.

Remaining problems are that even with two-photon microscopy, imaging is restricted to at best about 400 μm into lymphoid organs, and access to internal organs for in vivo imaging is hampered by the physical size and working distance of microscope objectives. A promising way to circumvent both these problems is to extend the imaging range by means of small probes based on coherent fiberoptic bundles (27) or gradient-index lenses (28) that may, for example, be inserted into the abdominal cavity, or even inserted into the matrix of a lymph node through a small incision in the capsule.

30.4.4 Detectors

In a single-spot scanned two-photon microscope all emitted fluorescence, even that which is highly scattered, provides useful signal since it originates at a known point. Optimal detection from lymphoid tissue is thus achieved by using large-area non-descanned detectors, ideally placed close to the back-aperture of the objective to maximize collection on noncollimated (highly scattered) photons. These criteria effectively limit the choice of detectors to photomultiplier tubes (PMTs), which, although having lower quantum efficiency than semiconductor devices (ca. 0.4 vs. 0.9 at visible wavelengths), are available with wide active areas and low dark counts. Multiple PMTs are typically employed together with appropriate dichroic mirrors and emission filters to monitor simultaneously the fluorescence from different probes with differing emission spectra: for example, to visualize interactions between green-labeled dendritic cells and red T cells (6,7), or for Ca^{2+} measurement using the ratiometric probe indo-1 (15). An alternative would be to use an imaging detector with high quantum efficiency (e.g., a cooled, back-illuminated CCD camera) to integrate an image during each raster scan, but this is likely to be unsatisfactory for deep imaging into tissues as scattered light will degrade the image quality.

30.4.5 A Multiphoton Microscope System for Immuno-imaging

Figure 30.2 shows a multiphoton microscope system constructed specifically for immuno-imaging. This was developed from a previously published design (23), incorporating several evolutionary improvements. The system is based around an Olympus BX 71 frame, which can be extended to provide extra clearance around the specimen for use in live-animal imaging experiments. The scan head and detector head are constructed as separate modules (see Fig. 30.2A,B) that plug onto the microscope frame using standard Olympus dovetail mounts.

The scanner (see Fig. 30.2C,D) is based on a resonant mirror (CRS; GSI Lumonics, Bedford, MA) that oscillates at about 8 KHz, deflecting the laser beam in the x axis at a rate of 64 μs per scan line. A second, orthogonal galvanometer-driven mirror scans the beam in the y axis at 30 Hz, providing a final image resolution of about 500×450 pixels at a frame rate of 30 s^{-1}. The scanned laser beam enters the microscope through a scan lens and tube lens and is directed to the objective via a fully reflecting mirror that can be rotated out of position when desired for direct viewing of the specimen by transmitted light. Emitted fluorescence captured through the objective is directed into

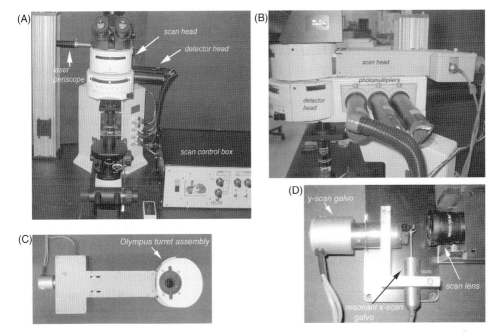

Figure 30.2. A video-rate, multiphoton microscope system for immuno-imaging. (A) Overview of the system, constructed around an upright Olympus BX 71 microscope. The scanner and detector head are constructed as two modules, based on Olympus reflected-light illuminators and turret heads, and plug onto the microscope frame using standard Olympus dovetail mounts. A periscope mirror (at left) brings the beam from a femtosecond laser into the scan head. Control of the scan mirrors and detector sensitivities is achieved manually via the electronics control box (lower right). (B) Side view of the scan and detector heads. The photomultiplier detectors are located as close as possible to the back-aperture of the objective lens, for increased collection of noncollimated (scattered) fluorescence emission. A primary dichroic in a rotating turret reflects visible wavelengths toward the detector unit while allowing transmission of IR excitation light from the scanner unit mounted above. Within the detector housing are three further dichroic/filter cubes, mounted on a dovetail slide for easy interchange, that direct light to four photomultipliers. (C) Scan head module removed from the microscope. (D) Interior view of the scan head. The beam from the femtosecond laser is directed first by the y-scan galvanometer mirror driven by a 30-Hz sawtooth waveform, and then by a resonant mirror operating at about 8.9 KHz. A 10× microscope ocular serves as the scan lens.

the detector head by a visible-reflecting dichroic mirror in a rotating turret. Four photomultiplier detectors are located as close as possible to the back-aperture of the objective lens for increased collection of noncollimated (scattered) fluorescence emission. The primary dichroic reflects visible wavelengths toward the detector unit while allowing transmission of infrared excitation light from the scanner unit mounted above. Within the detector housing are three further dichroic/filter cubes, mounted on a dovetail slide for easy interchange, that direct light to four photomultipliers. Photomultiplier signals and synchronization pulses from the scanner unit are connected to a Raven acquisition board (BitFlow Inc., Woburn, MA) that permits simultaneous acquisition from four channels at pixel clock rates up to 20 MHz. Custom software routines (Video Savant 4.0: IO Industries, London, ON, Canada) interlace alternate left–right and right–left line scans, correct for image distortion arising from the sinusoidal resonant x-scan, and display and store to disc in real-time single-color and composite images. The software can be configured to control external devices such as shutters and microscope focus, allowing automated collection of x, y, z, t multidimensional data sets with up to four independent emission wavelength channels. For tracking the relatively slow (tens of μm per minute) motility of lymphocytes within tissue, several video frames are averaged to provide improved image quality for each z plane, but the ability to capture images at 30 frames per second enhances the versatility of the system for applications such as visualizing the homing of circulating T cells (5).

30.4.6 Multiphoton Imaging of Lymph Nodes Ex Vivo and In Vivo

Multiphoton microscopy has been used to visualize cellular dynamics and interactions in intact lymphoid organs after surgical removal from the animal (1,5–11), in tissue slices (12–14,17), and within organs in the intact animal (5,9). For explant imaging, the tissue is generally maintained by superfusion with a physiological solution that is warmed close to body temperature and is bubbled with a 95% O_2/5% CO_2 mix to ensure adequate oxygenation. Figure 30.3A,B illustrates the system we developed for in vivo imaging of the mouse inguinal lymph node. This maintains intact circulation, allowing, for example, the visualization of T cells as they home into the node from blood vessels (see Fig. 30.3C).

30.5 IMAGE PROCESSING AND ANALYSIS

30.5.1 Data Storage

The sheer size and complexity of multidimensional (x, y, z, t, intensity, wavelength) data sets generated by imaging cells in three spatial dimensions over time create problems for both of data storage and analysis. Typically, we acquire images at a rate of one plane (500 × 450 pixels) per 300 to 500 ms, focusing through a z-stack of about 40 planes at a separation of 1 to 3 μm every 20 to 30 s. If captured as 24-bit color images (allowing simultaneous acquisition of three emission wavelength channels), this translates to a data throughput of about 5 GB per hour, or as much as 50 GB from a good day's experiment. Internal hard drives thus fill up quickly. One solution is to use multiple external USB2 drives, but a preferable approach employs a network-attached hard drive array. In the latter case, data can conveniently be accessed from multiple locations, and data security can be ensured by configuring the drives as a redundant RAID 1 or 5 array so that failure of a single drive does not result in loss.

30.5.2 Data Processing and Display

Visualization of cell positions in three spatial dimensions over time raises the particular problem of how to represent depth within the imaging volume on a two-dimensional monitor screen. One approach is to display "top" and "side" views of the imaging volume created by forming maximum-intensity projections along orthogonal axes.

However, this representation becomes confusing at higher densities of cells, as it is increasingly difficult to locate a given cell in both views. We thus devised a color-coded depth representation in which the axial (z) depth of cells within an image stack is encoded by a pseudocolor scheme, with cells at increasingly superficial depths shown as increasingly "warm" colors (5). This works well for experiments involving a single cell type (where color is not required as a means of distinguishing different cell populations). Axial velocities can readily be measured from the times required for a cell to traverse an axial distance corresponding to a defined color transition, and it is easy to discriminate cases where cells contact one another from instances where cells may appear to touch but in fact cross paths with a wide separation in the z axis.

An alternative approach involves the use of programs such as Imaris (Bitplane AG, Zurich) and Volocity (Improvision, Warwick, UK), which can display an interactive, three-dimensional reconstruction of the image volume in real time (e.g., Fig. 30.3C). The image can thus be rotated to provide a view from any desired arbitrary perspective, and by means of colored or liquid crystal spectacles it is even possible to create a stereoscopic depth effect.

30.5.3 Cell Tracking and Analysis

In principle, cell tracking can be automated using programs such as Imaris and Volocity. However, our experience is that the image quality may not be sufficiently good for the programs to consistently identify labeled cells against the background, and it is then necessary to resort to manual tracking. In cases where cell movement is expected to be isotropic (i.e., similar in x, y, and z axes) it may be sufficient to track cells in only two dimensions. This can readily be achieved by forming maximum-intensity projections along a desired axis (i.e., "top" or "side" views) and manually tracking cells over time using routines such as the "track points" function in MetaMorph (Fig. 30.4A,B). For three-dimensional tracking cells can be traced manually using Imaris or Volocity, or by means of the freely available PIC-Viewer software (Strathclyde Imaging: http://spider.science.strath.ac.uk/PhysPharm/showPage.php?deptID=2&pageName= software_imaging).

Measurements of centroid coordinates of cells tracked at defined intervals can readily be processed to yield information on "instantaneous" velocities (see Fig. 30.4C,D), cumulative distance traveled, and displacement from starting point. Instantaneous lymphocyte velocities are typically in the range of several to a few tens of μm min^{-1}, with T cells showing characteristically higher mean velocities than B cells (see Fig. 30.4G,H). Moreover, the distribution of velocities exhibited by individual cells varies between types of lymphocytes. B cells tend to move at a roughly constant speed, whereas T cells progress by a series of lunges and pauses. Distribution plots of velocities derived from cell population measurements hint at this difference (i.e., the spread of velocities is wider for T cells than B cells) but do not discriminate whether the variation arises between cells, or within individual cells over time (1). Instead, the differing motility characteristics are more readily apparent in representative plots of instantaneous velocity as a function of time for individual cells (see Fig. 30.4C,D) and can be quantified by performing a power spectrum analysis of the instantaneous velocity fluctuations. In the case of T cells, power spectra reveal a marked peak at a period of about 2 min, corresponding to their cyclical lunge–pause behavior (5). These variations in motility

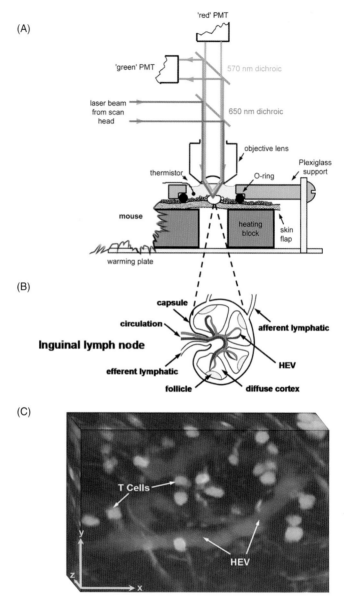

Figure 30.3. Imaging system for in vivo two-photon microscopy of the inguinal node of an anesthetized mouse. (A) The multiphoton microscope and intravital imaging chamber. The system illustrates two-color imaging, employing dual photomultipliers (PMT) with a dichroic mirror (560 nm) separating the emission wavelengths of CFSE (green) and CMTMR (orange). The path of the infrared femtosecond laser beam is indicated in red. The inguinal lymph node is exposed on a skin flap to which a rubber O-ring is glued to form a watertight imaging chamber filled with PBS. A Plexiglas support holds the O-ring to reduce movement artifacts caused by respiration, and imaging is achieved through a water-dipping objective lens. The chamber is warmed by a heating block, and the temperature of the fluid is maintained at 35° to 36°C. (B) The structure of a lymph node, showing the relative positions of B-cell follicles, T-cell areas in the diffuse cortex, high endothelial blood vessels (HEVs), overlying capsule, and lymphatic and circulatory connections. (C) Intravital imaging of vessels and T cells in a living lymph node.

are accompanied by changes in cell shape: T cells round up when paused and elongate while lunging. Thus, measurements of a simple "shape index" (ratio of the length of the long axis of the cell to its short axis) show cyclical variations that correlate well with the velocity fluctuations (see Fig. 30.4E). In contrast, B cells generally remain more nearly spherical, with a shape index close to 1 (see Fig. 30.4H).

The movements of lymphocytes (in two dimensions, $x - y$ or $x - z$) can be conveniently depicted as a "flower plot" representation in which the initial positions of several cells in the imaging field are all normalized to coordinates of 0,0. The arrangement of cell tracks radiating out from the origin then graphically indicates the motility pattern (Fig. 30.5A,B). If cells were to show consistent motion without turns, this would be apparent as a series of radial lines; directed motion (e.g., bulk flow or motion directed by a uniform chemokine gradient) would give rise to tracks biased in a particular direction; and Brownian-type motion would give rise to random-walk patterns. For T cells in the node parenchyma (see Fig. 30.5A) and for B cells in follicles (see Fig. 30.5B) the latter, random-walk patterns are, in fact, observed (1,5). For a random walk, the mean displacement from the origin increases as a square root function of time (see Fig. 30.5C), yielding a straight-line relationship when mean displacement is plotted against square root of time (see Fig. 30.5D). Moreover, the slope of this line then provides a "motility coefficient" M (calculated as $M = x^2/4t$ for two-dimensional motion or $M = x^2/6t$ for motion in three dimensions, where $x =$ mean distance from origin at time t). This is analogous to the diffusion coefficient of a molecule or particle undergoing random Brownian motion and gives a measure of the amount of "territory" that would be randomly explored by a cell in any given time (1). The motility coefficient of T cells ("roaming range") is typically two to three times greater than that of B cells (see Fig. 30.5D). However, the motility coefficient reflects both instantaneous cell velocity as well as the frequency with which cells make random turns. An alternative measure, which is independent of velocity, can be derived by plotting paired measurements of the mean absolute displacement of cells from their origins at different time points against the cumulative path lengths traversed by the cells at those times.

Surprisingly, despite the many studies of lymphocyte chemotaxis in vitro and our understanding of chemokine requirements for lymphocyte homing, there are yet few observations of directional migration of lymphocytes from ex vivo or in vivo microscopy of lymph nodes. One example, described further in Section 30.7.2, is the directed migration of antigen-engaged B cells to the boundary between the B-cell follicle and the T-cell zone (8). A simple test for directed motion is to plot net displacement of cells from their starting point versus cumulative path length on double-logarithmic coordinates. In the case of linear motion, a slope of 1 is expected, whereas the net displacement increases as the square root of path length (slope of 0.5 on double-log axes) for random motion. Thus, antigen-engaged cells showed a higher slope than did simultaneously imaged naïve cells (8).

Figure 30.3 (Continued).
The image is a 3D reconstruction (scale bars, 30 μm in all axes) showing CFSE-labeled naïve T cells (green) in the vicinity of a HEV (red), stained by intravenous injection of tetramethyl-rhodamine dextran. (Reproduced with permission from (5))

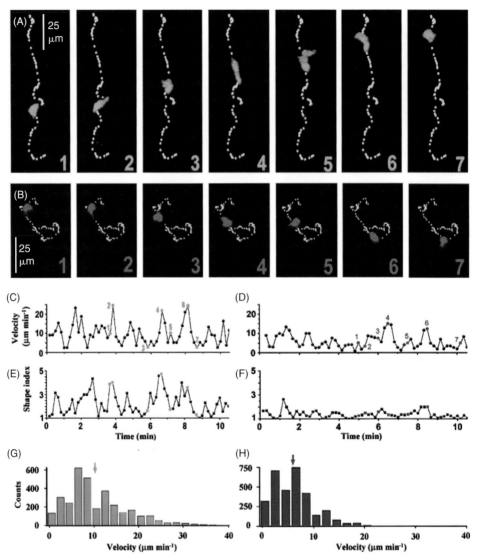

Figure 30.4. Tracking the morphology and motility of naïve T and B cells in inguinal and cervical lymph nodes. Image volumes of 200 μm × 200 μm × 50 μm (*xyz*) were acquired at 10-s intervals and then collapsed as maximum-intensity projections along the *z* axis for display and measurement. The imaging volume contained 25 to 100 labeled cells, in addition to a much larger unlabeled population from the recipient mouse. All motility experiments were performed at 36°C. Lymphocytes remained highly motile for >8 hours within the superfused nodes and could be continuously imaged for up to 1 hour without impairment. Quantitative measurements of T- and B-cell velocities, shapes, and trajectories were made by tracking the movements of individual cells in the *xy* plane (parallel to the overlying capsule) through stacks of 3D time-lapse images. (A, B) Image sequences illustrate typical T- and B-cell motility patterns, respectively. Yellow dots mark the positions of each cell at successive 10-s intervals. (C) Changes in the instantaneous cell velocity of the T cell shown in (A). (D) Changes in the instantaneous cell velocity of the B cell shown in (B). (E, F) Corresponding measurements of the cell shape index (long axis divided by short axis) of the T and B cells. (G, H) Histograms showing, respectively, the distributions of T- and B-cell velocities, with mean velocities indicated by the arrows. (Reproduced with permission from (1))

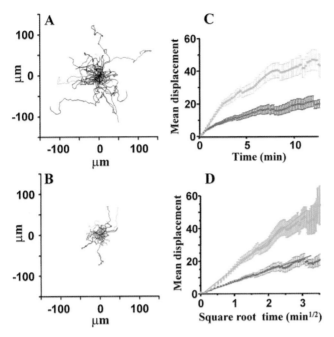

Figure 30.5. T and B cells move within the lymph node by a random walk. (A) Superimposed tracks of 43 T cells in the diffuse cortex, normalized to their starting coordinates. (B) Corresponding tracks of 41 B cells in a primary follicle. (C) Mean displacement plot for T (green) and B (red) cells from (A) and (B). (D) The data from (C), replotted as a function of the square root of time. Points fall on straight lines, consistent with random movement, and the slope of each line represents the motility coefficient M, given by $M = x^2/4t$, where $x =$ mean distance from origin at time t. (Reproduced with permission from (1))

30.6 FLUORESCENT PROBES AND CELL LABELING

Except for a few special cases in which cells are directly visible by bright-field microscopy or confocal microscopy near the surface (the eye, organ surfaces, thin tissues), cells of the immune system must be imaged where they carry out their function: within lymphoid organs. For tissue imaging, it is necessary to make specific cell types visible by fluorescent labeling. Several approaches have been employed for visualizing living cells and structures within lymphoid organs (Table 30.1).

30.6.1 Adoptive Transfer of In Vitro-Labeled Cells

Several studies have relied on cell-tracker fluorescent dyes to label cells. Usually, cells are labeled in vitro and then injected into the animal and subsequently imaged in the tissue environment. Cell trackers are usually loaded into cells using a cell-permeant acetoxymethyl (or acetate) ester. Within the cell, ubiquitous cytoplasmic esterases react with the cell-permeant molecule to generate a fluorescent product that is nonpermeant and thereby concentrated within the cell. The product then reacts with intracellular proteins, forming covalent bonds with a variety of constituents. In the case of CFSE, the parent compound is nonfluorescent; a succinimidyl ester moiety reacts

Table 30.1. Techniques of cell labeling for immuno-imaging

Labeling Method	Cell Type	References	Advantages	Limitations
Cell tracker in vitro (CFSE, CMTMR)	T cell	1, 20	Bright, stable, multi-color, can be used to monitor proliferation	Dye diluted, too dim to see after multiple rounds of proliferation
	Thymocyte	14		
	B cell	1		
	Dendritic Cell	8, 9, 20		
Cell tracker in vivo (CFSE)	Dendritic Cell	6	Labels migratory DCs that carry antigen, no in vitro step	Does not label tissue-resident DCs
Genetically encoded fluorescent protein (GFP, YFP)	Dendritic Cell	11	Promoter-driven labeling of specific cell types, no dilution	Requires generating and characterizing a new mouse strain
	NKT cell	21		
Antibody injection	Dendritic Cell	10	Applicable to a wide array of readily available antibodies	Unbound antibody remains as background; injection may cause tissue damage
Quantum dots	Not yet used, except marking blood vessels	33, 38	Very bright and photostable, multi-colors	Must be loaded into cell or conjugated to surface
Ca^{2+} indicator dye	Thymocyte	15	Provides readout of second messenger	Dye does not persist in cells
	T cell	18		

with amino groups (lysines and arginines) and forms a green fluorescent covalently linked conjugate. In most cases, the labeling is completely benign. Indeed, several rounds of division can be tracked by progressive dilution of the dye. This approach has been employed by immunologists to track the number of times that lymphocytes are able to divide in vitro or in vivo, by taking advantage of the fact that the original dye content is split equally into the daughter cells each time the cell divides. Starting from a uniformly labeled population of resting lymphocytes, flow cytometry reveals populations of cells that have divided, each division yielding cells with half the fluorescence intensity. Other cell trackers such as the red dye CMTMR, also loaded by a cell-permeant ester, react with glutathione and other thiols within cells via a chloromethyl group.

Labeled cells derived from a donor animal are then introduced into the recipient animal by injection into the bloodstream or beneath the skin. In such an adoptive transfer experiment, cells that have been purified and separately labeled green or red can be combined and injected together into the tail vein of a mouse. For example, T cells can be labeled green and B cells red, or vice versa, and then injected together. The fluorescently labeled T or B lymphocytes then home from the bloodstream into appropriate tissue locations within secondary lymphoid organs, and each can then be tracked in a single experiment. The percentage of fluorescent cells within the lymphoid organ can be adjusted by varying the number of adoptively transferred cells and typically is <1% of the total cells present, permitting the behavior of individual cells to be tracked readily. This approach provides a powerful internal control and can be applied to lymphocytes from transgenic mice with differing antigen specificity or to varying subsets of cells.

Cell-tracker dyes have also been used to label DCs in vitro, with subsequent visualization in lymph nodes following subcutaneous injection and migration. The in vitro-derived DCs can be exposed in vitro to a specific antigen to initiate an immune response involving antigen-specific T cells, usually transferred from a TCR transgenic mouse into the recipient. This approach has enabled T-cell/DC interactions to be tracked using a defined antigen dose to initiate the immune response.

Cell-tracker dyes can also be employed for in vivo labeling of cells, as we have done with DCs. By including CFSE in an adjuvant formulation (alum, cytokines, antigen, dye) that was injected beneath the skin, we succeeded in labeling a subset of DCs that migrated into the lymph node, optionally bearing an antigen (ovalbumin) specific for a naïve T cell (from the ovalbumin-specific TCR transgenic mouse DO11.10).

30.6.2 Genetically Encoded Fluorescent Proteins

Depending upon the experimental goal, cell-tracker dyes are especially useful because they can be used to track not only the cell but also the number of cell divisions that have occurred, because the dye concentration is halved every time the cell divides. However, this fact also limits the use of such dyes because after too many rounds of cell division, the dye is diluted to the point where it becomes too dim to detect. Up to 7 to 10 rounds of division can be detected, either by flow cytometry of cells recovered from lymphoid organs or by direct tissue imaging. Thus, an alternative is needed for longer-term imaging of immune response effector cells that divide many times before acquiring a particular function (e.g., antibody-secreting plasma cells, or cytotoxic effector memory T cells) or to provide labeling of resident cells that have not been subjected to an in vitro step during labeling. For this purpose, proteins related to green fluorescent protein (GFP) are now starting to be used as genetically encoded markers in transgenic mice. Fluorescent proteins include GFP and other variants with differing fluorescence emission (YFP, CFP), the unrelated red fluorescent protein DS-Red, and customized indicator proteins such as the chameleons and pericams that change their fluorescence characteristics in response to a change in calcium concentration. The expression of fluorescent proteins can be directed to occur only within certain cells or during specific states of activation by driving expression with specific promoter elements upstream of the coding region. In recent work, a transgenic mouse was developed in which DCs express YFP driven by the promoter for the specific surface molecule CD11c (11).

30.6.3 Quantum Dots

Quantum dots are nanocrystalline spheres that exhibit very bright, photostable fluorescence with narrow (10-nm) emission spectra that depend solely upon the size of the particles. Upon excitation in a two-photon microscope, a quantum dot emits fluorescence that is 100 to 1,000 times brighter than any conventional dye, yet it is no larger than many protein molecules. Recent advances have made these particles soluble in an aqueous environment present in a living cell (29–36). Containing a semiconductor CdSe core from 3 to 6 nanometers in diameter (with emission maximum at 525 nm and 655 nm, respectively), these particles can be conjugated with different organic molecules such as streptavidin or antibodies and are available commercially (Fischer, Quantum Dot™ Corporation). The narrow emission spectrum causes very little interference when probes of different emission wavelengths (colors) are used. Also, the bright emission will enable single-particle detection or imaging deep into tissue. These particles retain fluorescence for long periods of time. Recent application of quantum dots to biological investigations include single-particle tracking of receptor dynamics, in vivo whole-animal (low-resolution) imaging following injection of quantum dots into the venous circulation, and immunofluorescent labeling of breast cancer markers (33,34,37). Thus far, immuno-imaging of quantum dots has been restricted to use as a marker of the vasculature (38), but we anticipate that this will become a valuable tool for specific labeling, taking advantage of the ability to conjugate quantum dots to specific antibodies for molecular labeling and visualization. In addition, individual quantum dots are visible as single particles by electron microscopy, and it is possible to image living cells and examine the same cells at the ultrastructural level following fixation. We anticipate that quantum dots will provide a flexible platform for in vivo imaging.

30.6.4 Probes of Cell Signaling and Activation

Thus far, most work has focused on tracking cells as they migrate and interact within lymphoid organs. However, two recent studies have succeeded in using calcium indicators in T cells (15,18). T cells of the intestinal mucosa are constantly exposed to antigens in food to which they become tolerized, a form of hyporesponsiveness that dampens or prevents an immune response to commonly encountered antigens. Human tissue biopsies were treated with Indo-1 AM ester, imaged, and shown to exhibit Ca^{2+} signals to treatment with thapsigargin or removal of external $[Ca^{2+}]_i$(18). The study validates the feasibility of imaging Ca^{2+} signals in human T cells within a biopsied tissue. $[Ca^{2+}]_i$ signals have also been imaged in the thymic tissue environment by allowing dye-loaded thymocytes to crawl within a thymus slice. Positive selection of T cells within the thymus guarantees that thymocytes will survive to maturity if they receive signals from direct contact with stromal cells that are interspersed within the thymus. Other thymocytes with TCRs that do not recognize stromal cells die. During positive selection, it was shown that Ca^{2+} signaling induced by contact with stromal cells cause thymocytes to stop moving, allowing prolonged interaction with stromal cells (15). $[Ca^{2+}]_i$ control of cell motility, first demonstrated in vitro (39), thus appears to play an important role in the maturation of T cells. In addition to monitoring early signaling events following TCR engagement, we anticipate that transgenic mouse strains with reporter gene expression will soon be employed to indicate cells that are producing cytokines as one outcome of an immune response.

30.6.5 Landmarks for Navigation in the Tissue Environment

Lymph nodes and other lymphoid organs are not "bags of cells"; rather, they are complex organs that contain blood vessels, afferent and efferent lymphatic vessels, structural elements, and a variety of cell types that include both resident and transient populations. During an imaging experiment, it is very useful to have landmarks to know where one is imaging. In addition to labeling the cells with cell-tracker dyes or genetically encoded fluorescent proteins as described above, it is possible to stain cells within lymphoid organs by perfusing labeled antibodies or lectins through the blood, or by injection of these into the isolated organ prior to imaging. Antigens can also be tracked directly using fluorescently conjugated proteins that are antigenic for T cells from transgenic animals (7). Blood or lymphatic vessels can be visualized during intravital imaging by injection of fluorescently tagged high-molecular-weight dextrans (e.g., Fig. 30.3C), or quantum dots into the vascular space to mark vessels such as high-endothelial venules that are sites of homing into the lymph node, or subcutaneously to mark afferent lymphatics that drain the surrounding tissue. The capsule and reticular fibers can be readily labeled by soaking the lymph node in medium containing a reactive dye, such as CMTMR (1). Collagen fibers can be visualized even without staining by second-harmonic generation using an excitation wavelength of 900 nm (12,23). Generally, it is useful to track multiple colors simultaneously and to use at least one of these to image a tissue landmark.

30.6.6 Transgenic Animal Models

The adoptive transfer procedure for studying an immune response enables cells to be mixed and matched to highlight key events during the immune response. T cells that are specific for a particular antigen are normally very rare, but transgenic TCR animals have been created in which all T cells possess a particular TCR directed against a specific antigen, providing a supply of naïve T cells that have never encountered antigen and making it possible, after specific labeling, to focus attention on a specific immune response. Similarly, B-cell transgenic animals are also available. Furthermore, specific gene knockouts are providing an alternative to pharmacological approaches to test for functional roles of specific molecules, although thus far these have not been investigated by imaging methods. An additional use of transgenic technology is to tag specific cell types or even specific molecules within cells with fluorescent proteins that can be directly imaged.

30.7 CELL BEHAVIOR

Two-photon microscopy has now made it possible to image cells of the immune system in real time within lymphoid organs. Here we highlight a few of the observations and

Table 30.2. Summary of immunological topics studied by multiphoton imaging

Key Findings	Location	References
Basal motility of T and B lymphocytes, thymocytes	Lymph node, RTOC	1, 14
Autonomous motility of naïve T cells, homing events	Lymph node (intravital)	5
Directional migration (chemotaxis)	Lymph node, thymus lobe	12, 43
Motility arrest during antigen activation	Lymph node, RTOC	1, 14, 15
Frequency of contacts between DCs and T cells	Lymph node	6, 14
Resident DC networks	Lymph node	11
Stages of CD4$^+$ T-cell activation: T-cell/DC interactions	Lymph node	7
Stages of CD8$^+$ T-cell activation T-cell/DC interactions	Lymph node (intravital)	9
Stages of CD4$^+$ T-cell activation: T-cell/B-cell interactions	Lymph node	12
Cell proliferation	Lymph node	1, 7, 12
$[Ca^{2+}]_i$ imaging	Thymus slice	15
Tolerance and priming responses compared	Lymph node	10, 13, 38
Cytolytic effector function	Brain slice	19

novel findings on T lymphocytes, B lymphocytes, and DCs in the context of their roles in the immune response. Several key results are summarized in Table 30.2.

30.7.1 T Lymphocytes

T lymphocytes are the cells of the body that detect the presence of foreign antigens. Specialized during differentiation to express a unique TCR corresponding to a particular antigen bound to an antigen-presenting cell, CD4$^+$ helper T cells respond to a specific antigen challenge in two ways: by secreting cytokines to coordinate the host response to antigen and by proliferating to expand the pool of antigen-specific T cells. CD8$^+$ cytotoxic T cells recognize foreign or virus-infected cells in the body and respond by secreting molecules that rapidly kill the invading or infected cells. Initially, it came as a great surprise to many immunologists that T lymphocytes are highly motile in vivo

(1,5). We discovered that T cells move rapidly in an amoeboid manner in their native environment of the lymph node. The cells crawl over each other and also make contact with fibrous elements in the lymph node, pausing every few minutes before continuing their journey. When actively moving, T cells are elongated and move with velocities that exceed 20 μm/min, averaging 10 to 12 μm/min (see Fig. 30.4A,G). For a cell that has a diameter of 7 μm when rounded up, the motion of T cells is very rapid indeed. Tracking the cells in x, y, and z revealed that they meander through the T-cell zone of the lymph node in a random manner, without indication of directed migration (see Fig. 30.5A). The net displacement from any given position was shown to be proportional to the square root of time, like diffusion, and this enabled a "motility coefficient" to be defined (see Fig. 30.5C,D). Both resting cells from a wild-type mouse and naïve $CD4^+$ T cells from a transgenic mouse that had never encountered antigen migrated equally well. The robust motility of T cells suggested that they might be scanning the environment for the presence of antigen. $CD8^+$ T cells move just as vigorously in the lymph node (9). Following activation by antigen T cells are also highly motile, pausing only to divide and then immediately regaining motility following cell division.

30.7.2 B Lymphocytes

B lymphocytes are cells that when activated by antigen can differentiate to become plasma cells that secrete copious amounts of antibody molecules. Like T cells, B cells are antigen-specific and only produce antibody directed against the foreign antigen that activated particular clones of B cells. B cells in the lymph node occupy a spatially separate compartment called the follicle (Fig. 30.6A). When imaged together in the absence of antigen, T and B cells appear to respect the invisible boundary at the follicle edge that separates them, each cell type migrating within its own area. Normally, B cells migrate within the follicle at a speed slower than that of T cells and on more circuitous paths (see Figs. 30.4B, 30.5B). In the absence of antigen, these paths appear random, without any particular orientation. Yet antigen-activated B cells require contact with antigen-specific T cells to differentiate fully to produce antibodies. How they come into contact has been visualized in a recent study (8). Upon encountering soluble antigen, B cells near the follicle edge were seen to proceed along relatively straight paths directly toward the follicle edge, where they then continued to migrate but without deviating far from the T-cell zone (see Fig. 30.6B). A quantitative analysis was made by tracking cells through randomly selected boxes and counting the numbers of lymphocytes that crossed the sides of the square facing the follicle/T-cell boundary versus those that crossed the opposite sides (see Fig. 30.6C). For boxes located close (<140 μm) to the boundary, a preponderance of antigen-engaged cells crossed the sides facing the boundary, but at greater distances this directionality disappeared (8).

The directional migration of antigen-engaged B cells requires upregulation of the chemokine receptor CCR7 and the presence of the chemokine CCL21, a ligand for CCR7 that is present with a concentration gradient that tapers from the T zone, extending into the follicle. Although chemotaxis has been suspected for some time to direct the traffic of lymphocytes in secondary lymphoid organs, this is the first instance in which directed motility has been observed (12). B-cell migration to the follicle edge enables antigen-specific B cells to encounter antigen-specific T cells, an interaction that drives B cells to differentiate and secrete antibodies.

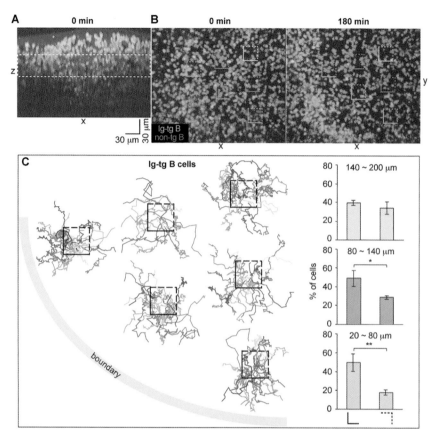

Figure 30.6. Antigen-engaged B cells show directed migration toward the boundary of the B-cell follicle. (A) An x-z projection (side view) within a lymph node showing the location of a B-cell follicle containing adoptively transferred B cells (green and red). The collagen-rich lymph node capsule is visualized by second-harmonic emission (blue). The dashed white rectangle shows the region from which the cell tracks in (C) were measured. (B) Images (x-y projections) captured at different times showing antigen-engaged B cells (green) clustering at the follicle–T zone boundary, whereas naive B cells (red) remain in the follicle. (C) Tracks of antigen-engaged B cells originating within arbitrarily placed boxes in the follicle. Note that B cells near the follicular boundary tend to move toward the boundary, whereas cells deeper in the follicle display apparently random tracks. The histograms plot the percentage of cells that moved across the sides of the square (solid lines) that face the boundary (left histograms) and the percentage of cells that moved across the opposite sides (dashed lines) of the squares (right histograms). Cells originating close (20–80 μm) from the follicular boundary showed more directed motion than those further (140–200 μm) from the boundary. (Reproduced from (42))

30.7.3 Dendritic Cells

DCs initiate the adaptive immune response by capturing antigen in the periphery, for example at a site of immunization, injury, or infection, and then transporting the antigen into lymph nodes and presenting antigenic peptides along with co-stimulatory interactions to specific T cells. Effective antigen responses require not only the antigen but also a "danger signal" that results in DCs becoming activated at a peripheral location.

An adjuvant, such as complete Freund's adjuvant or alum, provides the danger signal by activating a set of receptors called toll-like receptors (TLRs) present on DCs and other antigen-presenting cells. Engagement of TLRs shifts the behavior of DCs from a mode that specializes in antigen uptake by endocytosis to one that favors trafficking to the lymph node and antigen presentation to specific T cells. Once DCs become activated locally, the antigen-bearing dendritic cells can migrate via lymphatic vessels into nearby lymph nodes.

There are several populations of dendritic cells within a lymph node, some located in the T-cell zone, some in the follicle, some resident and some newly arrived carrying antigen from the periphery. Four different labeling techniques (see Table 30.1) have permitted visualization of DCs in the T-cell zone. We employed an in vivo approach to label the pool of DCs that traffic from the skin into the lymph node by including CFSE in the alum adjuvant injected subcutaneously. At the adjuvant site, cells could take up CFSE and become labeled in vivo. Because inflammatory cytokines were present to promote DC maturation, optionally with specific antigen, the labeled DCs could then migrate via afferent lymphatic vessels into nearby draining lympssh nodes. These DCs tended to cluster near high-endothelial venules, outside B-cell follicles, ideally situated for interactions with T cells. A limitation of the in vivo labeling method—the flip side of its strength in being able to visualize those DCs from the periphery—is that the network of resident DCs are not visible. Two recent studies have imaged resident DCs (10,11), and these have also observed dynamic ruffling and extensions of DC processes. Viewed by time-lapse within the lymph node, living in vivo labeled DCs were observed for the first time to extend and retract long processes that greatly expanded the swept volume within which migrating T cells could come into contact. The DCs themselves usually migrated slowly, but motile T cells interacted with them at a very high rate and apparently by random collisions rather than by chemotaxis.

30.8 GENERAL INSIGHTS FROM IMMUNO-IMAGING

Apart from the sheer pleasure in being able to see living cells and their responses to antigen inside lymphoid organs, several insights have been gained by direct immuno-imaging that were otherwise unsuspected or merely surmised from examination of static pictures of fixed tissue (snapshot immunology). Seeing the activity within lymphoid organs has changed the way many immunologists think about how an adaptive immune response is initiated and regulated.

30.8.1 Stochastic T-Cell Repertoire Scanning

Lymphoid organs are organized to promote cell–cell interactions that guarantee continuous in vivo surveillance of an unimaginably wide range of potential foreign antigens. To scan the potentially limitless number of possible antigens, a diversity of TCRs, each unique to a given naïve T cell, is generated by gene rearrangement during development. This process can lead, at least theoretically, to 10^{18} different receptor types (40). Within any given lymph node, there are perhaps 10^6 to 10^7 T lymphocytes, nominally each with differing antigen specificity. Thus, to find a particular antigen that might be present within a particular lymphoid organ, it is also critical that T and B lymphocytes must come and go, entering lymphoid organs from the blood at sites of homing in high-endothelial venules, and leaving via efferent lymphatic vessels ultimately to return to the blood via the thoracic duct in order to take up residence elsewhere in the body. The problem is, how are antigens detected within the expanse of a lymph node if T cells specific for that particular antigen are so rare? Visualization of T cells together with DCs in the absence of antigen led to the concept of stochastic repertoire scanning. We have proposed that stochastic repertoire scanning, mediated by robust T-cell motility in combination with DC probing, allows antigen detection within the lymph node. By tracking DCs together with T cells, we counted the number of contacts per DC and their duration. With adequate spatial resolution of fine dendritic processes and tracking in 3D, we discovered that a single DC is able to contact more than 5,000 T cells per hour, an order of magnitude larger than another study that used CD8 cells and in vitro-derived DCs (8). In the absence of antigen, such contacts lasted an average of 3 to 4 minutes. This means that every 3 to 4 minutes, about 250 newly arrived T cells come into contact with a DC, allowing a very large number of cells to be scanned for the presence of antigen on a fairly small number of antigen-bearing DCs. If antigen is present, then longer interactions that lead to T-cell activation are promoted. The random walk by T cells must occupy a substantial energy cost; the benefit is that through the random encounters between T cells and DCs, our immune system is continuously providing for detection of potentially harmful antigens.

30.8.2 Stages of T-Cell Activation

Over the past decade, a new concept of antigen recognition has emerged: the immunological synapse. In vitro studies have identified the molecular players that initiate the intracellular signaling cascade that leads to activation of T lymphocytes. Subcellular co-localization of key signaling molecules showed that TCRs, integrins, CD4 (or CD8), kinases, phosphatases, and cytoskeletal proteins cluster in a signaling complex at the zone of contact with an antigen-presenting cell. The surface proteins interact with MHC and other proteins on the antigen-presenting cell, forming what is now termed the immunological synapse. In liquid culture systems in vitro, T cells and antigen-presenting cells make contact and remain together for hours (39,41–43). A very different picture of the in vivo cell–cell interactions has begun to emerge from two studies that imaged the interaction between T cells and DCs in the lymph node (7,9). The presence of antigen, brought to the draining lymph node by DCs from the adjuvant injection site under the skin, triggers a sequence of behavioral changes in T cells. Instead of immediately forming stable interactions as seen in vitro, antigen activation takes place in distinct behavioral stages for both CD4$^+$ and CD8$^+$ T cells in the lymph node. Figure 30.7 illustrates these stages for CD4$^+$ T cells. Initially, the T cells dance in the vicinity of a DC, making contact intermittently and with several partners (7). Later, after a few hours, these contacts stabilize, and clusters are formed consisting of about a dozen T cells stably attached to DCs. Still later, during the activation sequence, T cells dissociate from DCs and begin to swarm locally. We have even been able to watch the individual antigen-activated T cells divide in vivo. Over the next few days, T cells divide five to eight times and then exit the lymph node to take up residence in distant lymphoid organs as central memory cells. A similar sequence of events was observed with CD8$^+$ cells, the initial phase of interaction being, if anything, more subtle than for CD4$^+$ T cells (9).

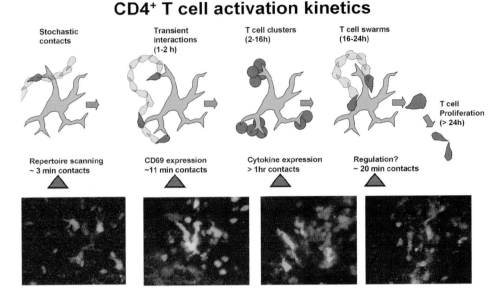

Figure 30.7. Interactions between antigen-bearing DC (green) and CD4$^+$ T cells (red), illustrating the successive kinetic stages involved in antigen recognition and T-cell activation. (Adapted and reproduced with permission from (6))

30.8.3 Comparison of T-Cell/DC and T-Cell/B-Cell Interactions

Two types of cell–cell interactions essential to the immune response have been characterized in lymph node thus far. As described above, the interaction of T cells with DCs passes through distinct behavioral stages from transient interactions to more stable clusters that later dissociate (see Fig. 30.7). Antigen-presenting DCs can form stable contacts with 10 or more antigen-specific T cells during the stable cluster stage. In contrast, the interaction between antigen-specific T and B cells, as recently visualized for the first time (12), is generally monogamous. Initially separated into distinct zones within the lymph node, B cells that encounter antigen begin to migrate in a directional fashion toward the T-cell zone, the first evidence that chemotaxis can guide the motion of cells inside the lymph node (see Fig. 30.6). Then, T and B cells form conjugate pairs. Interestingly, and in distinct contrast to T-cell/DC clusters, these T-cell/B-cell conjugate pairs are highly motile (Fig. 30.8). In every instance, B cells lead the way and drag firmly attached T cells along behind them. The very different choreographies of T cells with DCs and with B cells are another fascinating aspect of immune response dynamics.

30.9 FUTURE DIRECTIONS

No other imaging modality (e.g., positron emission tomography [PET], magnetic resonance imaging [MRI], bioluminescence imaging) can approach the spatial (subcellular to molecular) and temporal (milliseconds) resolution afforded by optical imaging. Moreover, this technique is not limited to mere visualization but also permits quantitative analysis and testing of specific mechanisms. These initial studies, however, have merely scratched the surface of what is possible. We have succeeded in imaging T

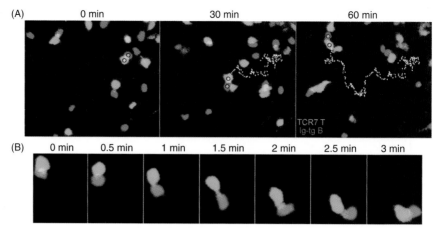

Figure 30.8. Interactions between antigen-engaged B Cells and helper T cells. (A) Time-lapse images of Ig-tg B cells interacting with TCR7 CD4$^+$ T cells after immunization, showing T cells moving along behind B cells. The tracks follow a B-cell–T-cell conjugate pair that remained bound to each other for more than 1 h. (B) Time-lapse images showing the dynamics of a B-cell–T-cell conjugate at higher magnification. (Reproduced from (42))

lymphocytes interacting with DCs and with B lymphocytes in a lymph node at the beginning of an immune response. The immune response, however, also involves additional cell movements and responses that take place in other locations and over a longer time scale. Moreover, we have thus far been able to visualize only motion and morphology. The field is now poised to use functional imaging to investigate the molecular events that underlie lymphocyte homing, motility, and antigen recognition deep within tissues in the living animal. We also look forward to examining the interactions between T cells and target cells during killing (effector function). Finally, immuno-imaging will no doubt be extended to image and understand the action of immunosuppressive drugs, infections (viral, bacterial, parasite), tumor immunotherapy, cell repopulation following bone marrow or stem cell transplant, autoimmune diseases, and modeling the immune response for vaccine development.

ACKNOWLEDGMENTS

The authors wish to acknowledge funding support from the National Institutes of Health (GM-41514 and GM-48071) and very helpful discussions with Mark Miller and Grace Stutzmann.

NOTE

The field of immunology is vast and complex. A classic text, *Immunobiology: The Immune System in Health and Disease*, edited by Janeway, Travers, Walport, and Shlomchik from Garland Science, is now in its sixth edition.

REFERENCES

[1] Miller MJ, Wei SH, Parker I, Cahalan MD. 2002. Two-photon imaging of lymphocyte motility and antigen response in intact lymph node. *Science* 296: 1869–1873.

[2] Cahalan MD, Parker I, Wei SH, Miller MJ. 2002. Two-photon tissue imaging: seeing the immune system in a fresh light. *Nat Rev Immunol* 2: 872–880.

[3] Cahalan MD, Parker I, Wei SH, Miller MJ. 2003. Real-time imaging of lymphocytes in vivo. *Curr Opin Immunol* 15: 372–377.

[4] Stutzmann GE, Parker I. 2005. Dynamic multiphoton imaging: a live view from cells to systems. *Physiology (Bethesda)* 20: 15–21.

[5] Miller MJ, Wei SH, Cahalan MD, Parker I. 2003. Autonomous T cell trafficking examined in vivo with intravital two-photon microscopy. *Proc Natl Acad Sci U S A* 100: 2604–2609.

[6] Miller MJ, Hejazi AS, Wei SH, Cahalan MD, Parker I. 2004. T cell repertoire scanning is promoted by dynamic dendritic cell behavior and random T cell motility in the lymph node. *Proc Natl Acad Sci U S A* 101: 998–1003.

[7] Miller MJ, Safrina O, Parker I, Cahalan MD. 2004. Imaging the single cell dynamics of CD4+ T cell activation by dendritic cells in lymph nodes. *J Exp Med* 200: 847–856.

[8] Bousso P, Robey E. 2003. Dynamics of CD8+ T cell priming by dendritic cells in intact lymph nodes. *Nat Immunol* 4: 579–585.

[9] Mempel TR, Henrickson SE, Von Andrian UH. 2004. T-cell priming by dendritic cells in lymph nodes occurs in three distinct phases. *Nature* 427: 154–159.

[10] Hugues S, Fetler L, Bonifaz L, Helft J, Amblard F, Amigorena S. 2004. Distinct T cell dynamics in lymph nodes during the induction of tolerance and immunity. *Nat Immunol* 5: 1235–1242.

[11] Lindquist RL, Shakhar G, Dudziak D, Wardemann H, Eisenreich T, et al. 2004. Visualizing dendritic cell networks in vivo. *Nat Immunol* 5: 1243–1250.

[12] Okada T, Miller MJ, Parker I, Krummel MF, Neighbors M, et al. 2005. Antigen-engaged B cells undergo chemotaxis toward the T zone and form motile conjugates with helper T cells. *PLoS Biol* 3: 1062–1069.

[13] Zinselmeyer BH, Dempster J, Gurney AM, Wokosin D, Miller M, et al. 2005. In situ characterization of CD4+ T cell behavior in mucosal and systemic lymphoid tissues during the induction of oral priming and tolerance. *J Exp Med* 201: 1815–1823.

[14] Bousso P, Bhakta NR, Lewis RS, Robey E. 2002. Dynamics of thymocyte-stromal cell interactions visualized by two-photon microscopy. *Science* 296: 1876–1880.

[15] Bhakta NR, Oh DY, Lewis RS. 2005. Calcium oscillations regulate thymocyte motility during positive selection in the three-dimensional thymic environment. *Nat Immunol* 6: 143–151.

[16] Robey EA, Bousso P. 2003. Visualizing thymocyte motility using 2-photon microscopy. *Immunol Rev* 195: 51–57.

[17] Wei SH, Miller MJ, Cahalan MD, Parker I. 2002. Two-photon imaging in intact lymphoid tissue. *Adv Exp Med Biol* 512: 203–208.

[18] Tutsch E, Griesemer D, Schwarz A, Stallmach A, Hoth M. 2004. Two-photon analysis of calcium signals in T lymphocytes of intact lamina propria from human intestine. *Eur J Immunol* 34: 3477–3484.

[19] Nitsch R, Pohl EE, Smorodchenko A, Infante-Duarte C, Aktas O, Zipp F. 2004. Direct impact of T cells on neurons revealed by two-photon microscopy in living brain tissue. *J Neurosci* 24: 2458–2464.

[20] Stoll S, Delon J, Brotz TM, Germain RN. 2002. Dynamic imaging of T cell-dendritic cell interactions in lymph nodes. *Science* 296: 1873–1876.

[21] Geissmann F, Cameron TO, Sidobre S, Manlongat N, Kronenberg M, et al. 2005. Intravascular immune surveillance by CXCR6+ NKT cells patrolling liver sinusoids. *PLoS Biol* 3: e113.

[22] Johnsen S, Widder EA. 1999. The physical basis of transparency in biological tissue: ultrastructure and the minimization of light scattering. *J Theor Biol* 199: 181–198.

[23] Mempel TR, Scimone ML, Mora JR, von Andrian UH. 2004. In vivo imaging of leukocyte trafficking in blood vessels and tissues. *Curr Opin Immunol* 16: 406–417.

[24] Dickinson ME, Simbuerger E, Zimmermann B, Waters CW, Fraser SE. 2003. Multiphoton excitation spectra in biological samples. *J Biomed Opt* 8: 329–338.

[25] Nguyen QT, Callamaras N, Hsieh C, Parker I. 2001. Construction of a two-photon microscope for video-rate Ca(2+) imaging. *Cell Calcium* 30: 383–393.

[26] Leveque-Fort S, Fontaine-Aupart MP, Roger G, Georges P. 2004. Fluorescence-lifetime imaging with a multifocal two-photon microscope. *Opt Lett* 29: 2884–2886.

[27] Gobel W, Kerr JN, Nimmerjahn A, Helmchen F. 2004. Miniaturized two-photon microscope based on a flexible coherent fiber bundle and a gradient-index lens objective. *Opt Lett* 29: 2521–2523.

[28] Levene MJ, Dombeck DA, Kasischke KA, Molloy RP, Webb WW. 2004. In vivo multiphoton microscopy of deep brain tissue. *J Neurophysiol* 91: 1908–1912.

[29] Dubertret B, Skourides P, Norris DJ, Noireaux V, Brivanlou AH, Libchaber A. 2002. In vivo imaging of quantum dots encapsulated in phospholipid micelles. *Science* 298: 1759–1762.

[30] Jovin TM. 2003. Quantum dots finally come of age. *Nat Biotechnol* 21: 32–33.

[31] Seydel C. 2003. Quantum dots get wet. *Science* 300: 80–81.

[32] Jaiswal JK, Mattoussi H, Mauro JM, Simon SM. 2003. Long-term multiple color imaging of live cells using quantum dot bioconjugates. *Nat Biotechnol* 21: 47–51.

[33] Larson DR, Zipfel WR, Williams RM, Clark SW, Bruchez MP, et al. 2003. Water-soluble quantum dots for multiphoton fluorescence imaging in vivo. *Science* 300: 1434–1436.

[34] Wu X, Liu H, Liu J, Haley KN, Treadway JA, et al. 2003. Immunofluorescent labeling of cancer marker Her2 and other cellular targets with semiconductor quantum dots. *Nat Biotechnol* 21: 41–46.

[35] Lidke DS, Nagy P, Heintzmann R, Arndt-Jovin DJ, Post JN, et al. 2004. Quantum dot ligands provide new insights into erbB/HER receptor-mediated signal transduction. *Nat Biotechnol* 22: 198–203.

[36] Michalet X, Pinaud FF, Bentolila LA, Tsay JM, Doose S, et al. 2005. Quantum dots for live cells, in vivo imaging, and diagnostics. *Science* 307: 538–544.

[37] Dahan M, Levi S, Luccardini C, Rostaing P, Riveau B, Triller A. 2003. Diffusion dynamics of glycine receptors revealed by single-quantum dot tracking. *Science* 302: 442–445.

[38] Shakhar G, Lindquist RL, Skokos D, Dudziak D, Huang JH, et al. 2005. Stable T cell-dendritic cell interactions precede the development of both tolerance and immunity in vivo. *Nat Immunol* 6: 707–714.

[39] Negulescu PA, Krasieva TB, Khan A, Kerschbaum HH, Cahalan MD. 1996. Polarity of T cell shape, motility, and sensitivity to antigen. *Immunity* 4: 421–430.

[40] Janeway CA, Travers, P., Walport, M., Shlomchik, M.J. 2004. Immunobiology: the immune system in health and disease, 6th edition.

[41] Bromley SK, Burack WR, Johnson KG, Somersalo K, Sims TN, et al. 2001. The immunological synapse. *Annu Rev Immunol* 19: 375–396.

[42] Monks CR, Freiberg BA, Kupfer H, Sciaky N, Kupfer A. 1998. Three-dimensional segregation of supramolecular activation clusters in T cells. *Nature* 395: 82–86.

[43] Witt CM, Raychaudhuri S, Schaefer B, Chakraborty AK, Robey EA. 2005. Directed migration of positively selected thymocytes visualized in real time. *PLoS Biol* 3: 1062–1069.

31

Multiphoton Imaging in Animal Development

Irina V. Larina and Mary E. Dickinson

31.1 INTRODUCTION

Vital time-lapse microscopy is a powerful and exciting way to study development. Many protocols for live cell imaging are now available for a large variety of developmental systems, and the advent of multicolor fluorescent proteins and turnkey instrumentation has attracted many scientists to this active field. Multiphoton microscopy is particularly well suited for developmental studies and has been used to answer many interesting questions about the mechanisms of invertebrate and vertebrate ontogeny. In this chapter we have highlighted a selection of recent studies where multiphoton microscopy has been used to study animal development. We describe animal models, pointing out their advantages and limitations for microscopic imaging, and review major achievements of multiphoton imaging in studying development of the different physiological systems.

31.2 ADVANTAGES OF MULTIPHOTON IMAGING FOR DEVELOPING SYSTEMS

The dynamic analysis of a developing organism requires that the microscopist be a "silent bystander" in order to observe normal cellular processes and behaviors. Thus, image resolution and acquisition speed must be balanced with the amount of energy applied to the sample, to preserve vitality. As has been described in other chapters in this volume (see Chapter 5) multiphoton imaging relies on the quasi-simultaneous absorption of two or more near-infrared (NIR) photons by a fluorophore, producing visible-range fluorescence. The use of NIR excitation is more compatible with living processes as it is absorbed less by cells, reducing the chance of cellular damage. Combined with deep penetration and the inherent optical sectioning capability afforded by nonlinear absorption, this makes this method a powerful approach for vital, deep tissue imaging. The advantages of this technique over traditional confocal microscopy are perhaps best described by Squirrell and colleagues (Squirrell et al., 1999), who monitored the development of hamster embryos under confocal and two-photon illumination. They observed healthy embryo development using two-photon microscopy for over 24 hours, while confocal microscopy inhibited normal development after 8 hours, and embryos imaged with the two-photon method could even be transplanted back into recipient host females to produce viable offspring. Multiphoton imaging has developed into a significant area of imaging and has been used to study animal development in a variety of different species. Here we discuss some of these findings.

31.3 IMAGING ANIMAL DEVELOPMENT

The success of any live imaging study hinges on three important parameters: access, contrast, and vitality. Access to the cells of interest depends on the ability to resolve the cells or process without causing damage. Light is a versatile medium, but it is not without limitations of penetration and resolution. That being said, optical imaging provides the best resolution available for live specimens, and the transparency and small size of many types of embryos allows for sufficient light penetration to cells of interest. Contrast in fluorescence imaging experiments is a very significant parameter. Cells and molecules can be labeled in a variety of ways that are compatible with normal development, and cleverly designed fluorochromes can provide access to otherwise invisible processes. In addition, the choice of the fluorochrome and the method of introduction often has as much to do with the scientific question as it does the specimen that is used. For instance, systems with well-established genetic methods can easily make use of genetically encoded markers, whereas other systems may rely on local application or injection of dyes. Last, the methods used for imaging and labeling cells must be compatible with normal cellular processes; otherwise, the lessons learned can be misleading. Table 31.1 provides a brief summary of some of the most popular developmental systems, indicating advantages and disadvantages for live cell microscopic imaging.

31.4 MULTIPHOTON IMAGING IN EARLY-CLEAVAGE-STAGE EMBRYOS

The early cleavage of cells in the fertilized zygote requires precise cell-cycle control. It is known that the early mammalian embryo undergoes cleavage without interphase growth and the initial volume is divided by cell division. Aiken and colleagues (Aiken et al., 2004) have used two-photon excitation and confocal microscopy to study pre-implantation development of mouse zygotes. To determine when individual cells begin to grow during early development, changes in nuclear and cytoplasmic volumes were studied. Nuclei in live embryos were stained with Hoechst 33258, the cytoplasm was marked with Calcein AM, and cell membranes were labeled with the styryl dye FM4-64 to provide easily discernible boundaries between these structures. Microscopic analysis revealed that the total cytoplasmic volume of the conceptus remains constant during pre-implantation development, while the nucleo-cytoplasmic ratio increases exponentially. However, both the volume and nucleo-cytoplasmic ratio of individual blastomeres depended on their position: the internal subpopulation of blastomeres was smaller and had a larger nucleo-cytoplasmic ratio than cells of outer subpopulation. This could be significant, since later in development, cells of the inner cell mass give rise to the embryo proper and can give rise to pluripotent stem cells, whereas more peripheral cells give rise to extra-embryonic cell types.

Table 31.1. Developmental systems utilized for live cell microscopic imaging

Animal models	Advantages	Limitations
C. elegans	• Optically transparent • Short regeneration time • Genome sequence is completed	• Limited number of cells, primitive body organization
Drosophila	• Very well-established genetics • Well-characterized mutants • Embryos develop externally • Relatively short life cycle	• Simple body plan
Zebrafish	• Vertebrate • Development is rapid • Embryos develop externally and are transparent • Eggs are relatively large and can be easily microinjected • Many mutants have been isolated through screens	• Limited strategies to mutate specific genes
Xenopus laevis	• Vertebrate • Embryos are very large and easy to handle • Embryos develop externally • Eggs can be easily microinjected	• Embryos are very turbid and scattering • Lack of genetic data • Long reproduction and maturation cycle
Chicken	• Vertebrate • Embryos develop externally • Embryos can remain intact in their natural environment during imaging • Genome is sequenced	• Limited genetic manipulations are possible • Lengthy generation time • Antibody availability is limited
Mouse	• Mammals • Very well-established genetic system • Well-characterized mutants are available • Reliable genome manipulation methods are available • Protocols for embryo culture are available for some stages	• Embryos develop in utero and can be cultured for limited time only and only for certain stages

In the early Drosophila embryo, nuclear cleavage gives rise to a syncytium, where many nuclei are contained within a single membrane. Individual cells are formed when nuclei migrate to the periphery of the ellipsoid-shaped embryo and membranes extend between these nuclei to create an epithelium of individual cells with a single nucleus. Shortly thereafter, morphogenetic movements of the entire epithelium ensue as the embryo gastrulates. Gastrulation requires considerable motion, and it is not known how the forces to produce this motion are established. A group of researchers in France combined two-photon excited fluorescence and third-harmonic generation imaging to study morphogenetic movements in Drosophila embryos (Debarre et al., 2004; Supatto et al., 2005). They used two-photon microscopy for imaging of embryos expressing nuclear or actin-associated green fluorescent protein (GFP), while third-harmonic generation microscopy allowed for imaging of unstained embryos. In their experiments, the same femtosecond NIR laser source was used for pulse-induced ablation of tissue as for microscopic analysis of morphogenetic movements resulting from these ablations. Highly localized ablations were produced without damaging the membrane surrounding the embryo by focusing the laser in a particular region of an embryo. Mechanical behavior of embryonic tissues was analyzed and compared in native and ablation-affected embryos. It was observed that ablation of the dorsal side of the embryo can inhibit lateral cell migration. They also studied mechanosensitivity of *twist* gene expression and demonstrated that photoablation inhibits expression of *twist*, demonstrating that developmental gene expression can be mechanically induced. These studies show that mechanical force is important for establishing the early Drosophila body plan, and those changes in force or motion can affect developmental gene expression.

31.5 MULTIPHOTON IMAGING OF THE DEVELOPING NERVOUS SYSTEM

Nearly every animal has some type of nervous or sensory system. For different species, these can range from simple indicators of environmental change to complex systems for cognitive thought and articulated motion. In higher vertebrates, for instance, the development of the nervous system is highly dynamic. It requires the interconnection of cells over long distances and phases of development that can last beyond embryonic stages, into larval or postnatal stages, sometimes even into adulthood. Models of many human neural defects and diseases can be found in different animal models, and imaging can be a productive way to study etiology. Multiphoton imaging has become a popular method for developmental neuroscientists, who often peer into regions with dense assemblies of neurons, glia, and axons or perform studies in cultured neurons that are very sensitive to environmental conditions.

Fragile X syndrome is a mental retardation disorder caused by a single mutation in the FMR1 gene that results in the absence of the fragile X mental retardation protein (FMRP). It has been reported that this syndrome is associated with abnormalities in the development of dendritic spines, but little is known about the neuropathology of this disorder, and the requirement for FMRP during maturation and pruning of dendritic spines has not been demonstrated in the intact brain. Nimchinsky and colleagues (Nimchinsky et al., 2001) characterized development of dendritic spines in FMR1 knockout mice with two-photon laser scanning microscopy. The aim of this work was to study involvement of FMRP in synaptogenesis in the intact brain. Neurons were labeled with enhanced GFP (EGFP) by injection of viral vector into the barrel cortex. Two-photon imaging was performed in fixed tissue sections and 3D reconstructions of the somatosensory cortex of mutant and wild-type mice were compared at different developmental stages. Analysis revealed differences in the spine length between the genotypes, demonstrating that FMRP is involved in the normal process of dendritic spine growth.

During early development, mammalian dendrites produce long and thin protrusions called filopodia. Dendritic filopodia were first described more than 70 years ago, but the function and some aspects of behavior of dendritic filopodia remain unknown. Portera-Calliau and colleagues (Portera-Cailliau et al., 2003) studied dynamic behavior of dendritic protrusions during early postnatal development in mouse neocortex. Time-lapse two-photon imaging was performed on acute slices of postnatal mouse brain using a patch pipette to deliver Alexa 488 intracellularly to layer 5 pyrimidal neurons. The lengths and density of dendritic protrusions were measured and dynamic changes in dendrite morphology were assessed using time-lapse analysis at different stages of postnatal development. Two-photon imaging revealed different types of motility of dendritic filopodia. It was observed that density, motility, and average length of filopodia are greater in dendritic growth cones than in dendritic shafts. Furthermore, blocking synaptic transmission led to an increase in density and length of dendritic filopodia in shafts but not in growth cones. Blocking of ionotropic glutamate receptors affected the density and turnover of shaft filopodia but did not influence growth cone filopodia. These studies established two populations of filopodia with different mechanisms of regulation.

Studies using a similar approach but performed in brain slices from the primary visual cortex showed that the motility of dendritic protrusions and spine density are reduced by both sensory and spontaneous activity deprivation (Konur & Yuste, 2004). Maximal spine and filopodial motility was observed during early postnatal stages (P11–P13) when synaptogenesis is highest, and after P15 there was sharp decrease in motility that was accompanied by disappearance of highly motile filopodia, an increase in the density of protrusions, and the formation of mature spines. These studies suggest that visual activity is an important factor controlling filopodial activity during synaptogenesis.

Axonal filopodia are highly dynamic structures, but it is not clear what role neuronal activity plays in regulating filopodial motility during development. To address this question, Tashiro and associates (Tashiro et al., 2003) studied dynamics of mossy fiber filopodia in postnatal hippocampus. Time-lapse two-photon imaging of mossy fibers transfected with EGFP was performed on cultured slices of mouse hippocampus. Imaging of mossy terminals with filopodial extensions revealed that the predominant type of filopodial motion is elongation and retraction with the changes in length as rapid as 5.4 μm/min. The motility is actin-based since it is blocked by actin polymerization inhibitor cytochalasin D. It was observed that filopodial motility decreases with development, possibly because filopodia are stabilized after the formation of synaptic contacts, and the developmental reduction of filopodial motility is dependent on kainite receptors. The regulation of filopodial motility by kainite receptor is bidirectional: low concentrations of kainate induce increases in filopodial motility, while higher concentrations inhibit dynamics. As a result of this study, a two-step model of synaptogenesis was proposed: axonal filopodia are highly dynamic to allow effective searches for postsynaptic targets, and once the target is found, the filopodia are stabilized to maintain the contacts.

Dendritic arbors are formed by a series of events that include extension, branching, and the formation of synapses, as well as retraction and elimination of branches and synapses. Little is known about the cell-signaling events that lead to the developmental regulation or dendritic arbor formation. Williams and Truman (Williams & Truman, 2004) performed in vivo time-lapse multiphoton imaging to study dendritic elaboration of sensory neurons in live intact Drosophila. The authors reconstructed metamorphic reorganization of GFP-expressing dendritic arborizing (DA) neurons during metamorphosis. They observed repositioning of the cell body and outgrowth of the adult arbor of persistent larval DA sensory neurons. Treatment with the insect juvenile hormone involved in regulation of metamorphosis had varying effects on dendritic outgrowth: early treatments maintained the retraction program and inhibited extension programs, but later treatments did not affect extension programs while retraction programs were prolonged. These studies suggest that the control of retraction and extension programs may be directly regulated by the hormonal changes necessary for metamorphosis into the adult form.

There are data suggesting that neuron axon growth is coordinated with neuromuscular synaptogenesis. To test this hypothesis in vivo, Javaherian and Cline (Javaherian & Cline, 2005) used time-lapse two-photon imaging to follow the growth of neuron axons in vivo in *Xenopus laevis* for 3 days to investigate the role of Candidate Plasticity Gene 15 (CPG15) in both axonal arbor growth and synapse maturation. GFP or yellow fluorescent protein (YFP) labeling of motor neuron axons was combined with labeling of presynaptic vesicle clusters by CFP-tagged synaptophysin and postsynaptic acetylcholine receptors with Texas Red-tagged α-bungarotoxin (TR-α BTX). It was observed that new axon branches emerge from sites of presynaptic vesicle clusters that are mainly opposed to acetylcholine receptors, suggesting that regulation of neuromuscular synaptogenesis is coordinated with regulation of motor neuron axon arbor development. Transfection of neurons with CPG15 showed that CPG15 expression promotes the development of motor neuron axon terminal arbor by enhancing neuromuscular synaptogenesis and the addition of new axon branches.

Real-time two-photon microscopy has also been used to observe differences in axon growth and guidance among different types of neurons in the developing brain of postnatal mice for 3 weeks in vivo (Portera-Cailliau et al., 2005). Figure 31.1 shows two-photon imaging of thalamocortical (A) and Cajal-Retzius (B) axonal dynamics in the developing mouse neocortex. Membrane-bound GFP expressed under the Thy-1 allowed the visualization of thalamocortical (TC) neurons and Cajal-Retzius (CR) interneurons. For long-term imaging of the developing brain, the skull was removed over the somatosensory and visual cortices, leaving the dura intact, and was replaced with a round coverglass, which served as a window for imaging during the entire course of the experiment. Observation of axons in superficial layers of cortex revealed that TC and CR axons undergo different structural dynamics. At early stages of outgrowth, TC axons grew rapidly in straight paths and did not produce distinct growth cones, but branching increased significantly as development proceeded. In contrast, at earlier stages CR axons grew slowly and had large growth cones, but at later stages they continued to grow without branching. These studies suggested that dynamic patterns of axon growth depend more on the neuronal cell type and less on the cellular environment,

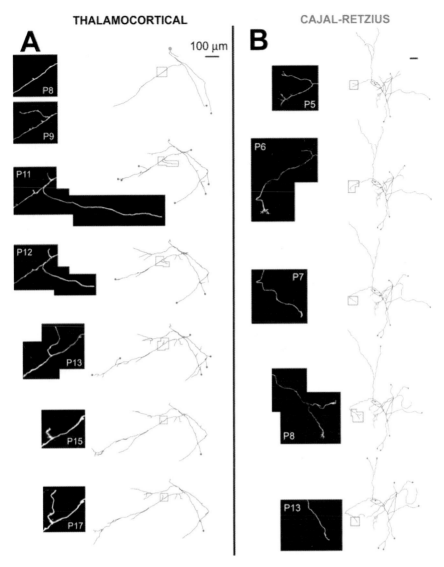

Figure 31.1. Two-photon imaging of thalamocortical (A) and Cajal-Retzius (B) axonal dynamics in the developing mouse neocortex. (Adapted from Portera-Cailliau, Weimer, et al., 2005)

and demonstrated a great potential of multiphoton microscopy to study the development of mammalian brain in vivo.

In another series of intravital experiments, the dynamics of dendritic spines at later developmental stages was studied (Zuo et al., 2005). Using two-photon microscopy, the cerebral cortex of living mice were imaged repeatedly at different ages. To access the cerebral cortex without disrupting neuronal structures, the skull over the cortex was thinned. These studies utilized transgenic mice expressing YFP in a subset of cortical neurons that allowed the researchers to track the formation and elimination of dendritic spines over time. It was observed that changes in spine morphology occur throughout life. However, dendritic spines become progressively stabilized from young adolescence to adulthood and the majority of adult spines are maintained over nearly the entire lifespan of an animal.

Intracellular calcium controls a variety of cellular processes, such as excitability, contraction, and secretion, and is widely used as an indicator of neuronal activity. Since changes in intracellular calcium concentration can be visualized using fluorescent dyes, two-photon imaging has contributed substantially to the study of calcium dynamics and neuronal network activity deep within biological tissues (Helmchen & Waters, 2002; Nguyen et al., 2001; Svoboda et al., 1997). In vivo calcium imaging in zebrafish larva was performed by Brustein and colleagues (Brustein et al., 2003). In their experiments, bolus injection of membrane-permeable calcium indicator into the spinal cord resulted in rapid staining of the entire spinal cord. Two-photon imaging allowed for the monitoring of calcium signaling simultaneously from a large population of fluorescently labeled neurons with single-cell resolution. Iontophoretic applications of glutamate, GABA, or glycine resulted in calcium transients, highlighting developmental changes in neuronal networks during development.

Two-photon microscopy has also been used in conjunction with fluorescent protein-based calcium reporters to study the development of functional spatial maps in the zebrafish olfactory bulb (Li et al., 2005). Exposure to different odorants resulted in the activation of distinct areas of the olfactory bulb. Two-photon microscopy made it possible to acquire detailed 3D reconstructions of activity patterns with high resolution. It was observed that the functional topographic map of the olfactory bulb is established very early, but the number of odor-responsive units increased during development, while the overall spatial organization of olfactory bulb activity was maintained invariable.

31.6 MULTIPHOTON IMAGING OF RETINAL DEVELOPMENT

In the retina, two functionally distinct types of retinal ganglion cells are organized as separate networks in a mosaic fashion. Lohman and Wong (Lohmann & Wong, 2001) used multiphoton microscopy to characterize the formation of dendritic contacts between ganglion cells during retinal development in ferrets. Imaging of retinal ganglion cells stained with Sulfurhodamine 101 or Oregon Green was performed in extracted retinas. Multiphoton microscopy revealed dendritic patterns and formation of stratification at different stages of development. The authors observed that retinal ganglion cells of the same sign display a variety of dendritic contacts, while ganglion cell pairs of different signs do not form dendritic contacts. These observations indicate that dendro-dendritic contacts are formed between retinal ganglion cells of the same type and may lead to intracellular signaling events that regulate the pattern of ganglion cells within a mosaic during retinal development.

Multiphoton microscopy has also been used to monitor cell divisions during neurogenesis in the zebrafish retina in vivo (Das et al., 2003). Bodipy-FL-C_5-Ceramide staining provided an outline of all cells in the retinal neuroepithelium, H2B-GFP plasmid was used to label cell nuclei, and retinal ganglion cells were labeled by membrane-tagged GFP driven by zebrafish Brn3c promoter. These researchers were able to reconstruct cell divisions and analyze the orientation of the division plane in different areas of the retina during neurogenesis. Two-photon microscopy allowed for observation of neurogenesis over a long period of time without the significant photodamage that usually results from confocal microscopy.

31.7 MULTIPHOTON IMAGING OF DEVELOPING CARDIOVASCULAR SYSTEM

Zebrafish have proven to be a robust model for studying angiogenesis, and many studies have employed multiphoton microscopy for long-term analysis and deep tissue imaging. For instance, using GFP driven by the *fli1* promoter, angiogenic network formation in wild-type zebrafish was compared to that of a mutant *mib*[ta52] with impaired vascular development (Lawson & Weinstein, 2002). This mutant is characterized by severe cranial hemorrhaging and defects in segmental vessel patterning due to disruption of arterial-venous identity. Time-lapse multiphoton imaging of EGFP-labeled embryonic blood vessels showed that the pattern of the vascular growth is similar to the development of nervous system: they observed filopodial activity and pathfinding behavior of the vessels that was analogous to that of neuronal growth cones. More recent studies indicate that vascular and neuronal development is regulated by an overlapping set of signals that may account for the similarities in axonal outgrowth patterns and mechanisms. Semaphorins, for instance, have repulsive activity that inhibits axonal growth, shaping the route of the growth cone. Using multiphoton microscopy, Torres-Vazquez and colleagues (Torres-Vazquez et al., 2004) demonstrated that blood vessel pathfinding is regulated by a similar mechanism that requires semaphorin-plexin signaling. Figure 31.2. shows multiphoton imaging of developing vessels in zebrafish. Loss of PlexinD1 function results in intersegmental vessel patterning defects. Similar mispatterning was observed when semaphorins' expression in the developing somites was disrupted, leading to inhibited growth of intersegmental vessels.

The same transgenic GFP marker line was also used to visualize formation of the angiogenic network in the developing vertebrate trunk and to investigate effect of blood circulation on this process in vivo (Isogai et al., 2003). A combination of confocal and multiphoton time-lapse multiphoton laser scanning microscopy revealed that first, the primary artery-derived angiogenic network forms by emergence, elongation, and interconnection of primary sprouts. Next, the secondary vein-derived network develops by emergence and elongation of secondary sprouts that interact with the primary network. To study the role of blood circulation in early vascular development, patterning of the vascular network was imaged in wild-type and *silent heart* mutant embryos. Comparison of the mutant and the wild-type embryos suggested that blood flow is not required for the formation of the primary as well as secondary angiogenic networks since these processes proceeded normally in the mutant embryos without circulation, although circulatory flow plays an important role during the interconnection of the primary and secondary networks.

In another study, *fli1*-GFP fish were used to screen for cardiovascular defects using confocal and multiphoton microscopy identifying a mutation in the phospholipase C gamma-1 (plcg1) gene (Lawson et al., 2003). *plcg1*[y10] mutants exhibit arterial defects but have normal veins. Time-lapse laser-scanning microscopy revealed that primary artery-derived segmental vessels fail to form in the mutant, while secondary vein-derived vessels sprout from the posterior cardinal vein, similar to wild-type embryos.

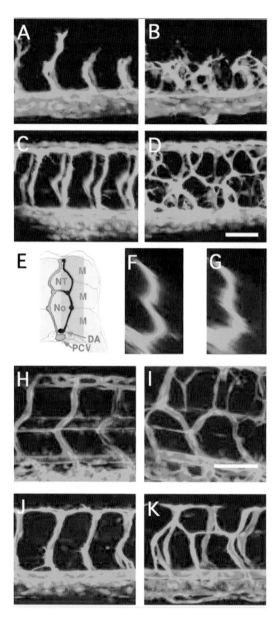

Figure 31.2. Multiphoton imaging of developing vessels in zebrafish. Control (A, C, and F) and PlexinD1 (B, D, and G) at approximately 24 hpf (A and B), 48 hpf (C, D, F and G), and 3 dpf (H and I). Control (J) and morphants with impaired Semaphorins (K) at 48 hpf. The shading in the diagram in (E) corresponds to the region of the trunk imaged in (F) and (G). Abbreviations in (E) and NT = neural tube; No = notochord; M = myotomes/somites; DA = dorsal aorta; PCV = posterior cardinal vein. (Adapted from Torres-Vazques, Gitler, et al., 2004)

This work shows that time-lapse multiphoton imaging can be a powerful tool for the early screening of cardiovascular phenotypes and the identification of signaling pathways controlling angiogenesis.

Tools for studying dynamic aspects of cardiovascular development in the mouse embryo are emerging. Recently, the movement of GFP-labeled blood cells has been imaged in early mouse embryos, allowing for the measurement of flow-derived forces using confocal microscopy. Endothelial specific markers for the in vivo characterization of angiogenesis in the embryo have also been recently reported, allowing for time-lapse imaging of vessel development and patterning. These markers add to the already available Tie-2:GFP mice (Sato mice), and studies are ongoing using multiphoton microscopy to understand angiogenesis in living embryos.

31.8 CONCLUSION

Multiphoton imaging has become a popular method for understanding dynamic events during embryogenesis. As more cell type-specific promoters are defined and more fluorescent protein-expressing animals are generated, more studies become possible. There is a long history of understanding embryonic development using microscopy, and until recently much of this work has been limited to small, externally developing transparent embryos. Now, with the advent of numerous multicolor markers for fluorescence microscopy and the development with more and more culture protocols for embryos and extracted tissues, more of what we are learning about cellular interactions in even the most complex systems is coming from direct observation. As we begin to tie these studies together with molecular aspects of cell signaling, markers that can act as real-time reporters of cellular activity are becoming more popular, enabling developmental biologists to decode the cryptic communications between cells.

REFERENCES

Aiken, C. E. M., P. P. L. Swoboda, et al. (2004). "The direct measurement of embryogenic volume and nucleo-cytoplasmic ratio during mouse pre-implantation development." Reproduction 128(5): 527–535.

Brustein, E., N. Marandi, et al. (2003). " "In vivo" monitoring of neuronal network activity in zebrafish by two-photon Ca2+ imaging." Pflugers Archiv European Journal of Physiology 446(6): 766.

Das, T., B. Payer, et al. (2003). "In Vivo Time-Lapse Imaging of Cell Divisions during Neurogenesis in the Developing Zebrafish Retina." Neuron 37(4): 597.

Debarre, D., W. Supatto, et al. (2004). "Velocimetric third-harmonic generation microscopy: micrometer-scale quantification of morphogenetic movements in unstained embryos." Optics Letters 29(24): 2881–2883.

Helmchen, F. and J. Waters (2002). "Ca2+ imaging in the mammalian brain in vivo." European Journal of Pharmacology 447(2–3): 119.

Isogai, S., N. D. Lawson, et al. (2003). "Angiogenic network formation in the developing vertebrate trunk." Development 130(21): 5281–5290.

Javaherian, A. and H. T. Cline (2005). "Coordinated Motor Neuron Axon Growth and Neuromuscular Synaptogenesis Are Promoted by CPG15 In Vivo." Neuron 45(4): 505.

Konur, S. and R. Yuste (2004). "Developmental regulation of spine and filopodial motility in primary visual cortex: Reduced effects of activity and sensory deprivation." Journal of Neurobiology 59(2): 236–246.

Lawson, N. D., J. W. Mugford, et al. (2003). "phospholipase C gamma-1 is required downstream of vascular endothelial growth factor during arterial development." Genes Dev. 17(11): 1346–1351.

Lawson, N. D. and B. M. Weinstein (2002). "In Vivo Imaging of Embryonic Vascular Development Using Transgenic Zebrafish." Developmental Biology 248(2): 307.

Li, J., J. A. Mack, et al. (2005). "Early Development of Functional Spatial Maps in the Zebrafish Olfactory Bulb." J. Neurosci. 25(24): 5784–5795.

Lohmann, C. and R. O. L. Wong (2001). "Cell-type specific dendritic contacts between retinal ganglion cells during development." Journal of Neurobiology 48(2): 150–162.

Nguyen, Q. T., N. Callamaras, et al. (2001). "Construction of a two-photon microscope for video-rate Ca2+ imaging." Cell Calcium 30(6): 383.

Nimchinsky, E. A., A. M. Oberlander, et al. (2001). "Abnormal Development of Dendritic Spines in FMR1 Knock-Out Mice." J. Neurosci. 21(14): 5139–5146.

Portera-Cailliau, C., D. T. Pan, et al. (2003). "Activity-Regulated Dynamic Behavior of Early Dendritic Protrusions: Evidence for Different Types of Dendritic Filopodia." J. Neurosci. 23(18): 7129–7142.

Portera-Cailliau, C., R. M. Weimer, et al. (2005). "Diverse Modes of Axon Elaboration in the Developing Neocortex." PLoS Biology 3(8): 1473–1487.

Squirrell, J. M., D. L. Wokosin, et al. (1999). "Long-term two-photon fluorescence imaging of mammalian embryos without compromising viability." Nat Biotech 17(8): 763.

Supatto, W., D. Debarre, et al. (2005). "In vivo modulation of morphogenetic movements in Drosophila embryos with femtosecond laser pulses." PNAS 102(4): 1047–1052.

Svoboda, K., W. Denk, et al. (1997). "In vivo dendritic calcium dynamics in neocortical pyramidal neurons." Nature 385: 161–165.

Tashiro, A., A. Dunaevsky, et al. (2003). "Bidirectional Regulation of Hippocampal Mossy Fiber Filopodial Motility by Kainate Receptors: A Two-Step Model of Synaptogenesis." Neuron 38(5): 773.

Torres-Vazquez, J., A. D. Gitler, et al. (2004). "Semaphorin-Plexin Signaling Guides Patterning of the Developing Vasculature." Developmental Cell 7(1): 117.

Williams, D. W. and J. W. Truman (2004). "Mechanisms of Dendritic Elaboration of Sensory Neurons in Drosophila: Insights from In Vivo Time Lapse." J. Neurosci. 24(7): 1541–1550.

Zuo, Y., A. Lin, et al. (2005). "Development of Long-Term Dendritic Spine Stability in Diverse Regions of Cerebral Cortex." Neuron 46(2): 181

32

Nonlinear Microscopy Applied to Dermatology

Barry R. Masters and Peter T. C. So

32.1 INTRODUCTION

The application of nonlinear optical microscopy for tissue imaging has an early start in dermatology and is the subject of an increasing number of studies that are based on nonlinear optical microscopy. The term "dermatology" (Greek *derma* [skin] + Greek *logos* [study]) was created in the late 1700s (Holubar & Schmidt, 1992). The early adaptation of nonlinear microscopy in dermal studies originates from two factors. First, unlike many internal organs, the accessibility of skin allows nonlinear microscopic imaging in the format of standard laboratory microscopes and does not require surgery or the use of endoscopic technology. Second, skin is a highly scattering tissue with large layer-by-layer variation in refractive index. Therefore, while there is good progress in noninvasive skin imaging using linear optical technology, the superior performance of nonlinear microscopy in highly scattering materials enables more rapid progress to be made in this field.

For biomedical applications, the ideal case is to study in vivo human skin using minimally invasive optical imaging and spectroscopy techniques. In practice, this is not always possible, as in vivo imaging techniques do not provide resolution and biochemical information content comparable to that of traditional histology. Furthermore, some measurements, such as the determination of tissue genetic expression profiles, inevitably require specimen excision. When optical imaging can be obtained together with histological skin preparations from the same body region, very good comparison can still be made (Rajadhyaksha, González, et al., 1999). Alternatively, a better understanding of skin physiology can also be obtained by using two different in vivo optical techniques, including nonlinear optical microscopy methods, to study the same region of human in vivo skin (Masters & So, 1999, 2001). In this chapter, we are discussing human skin unless otherwise stated.

This chapter reviews the applications of linear and nonlinear microscopy to the field of dermatology. An overview of skin anatomy and physiology is provided. The majority of this chapter focuses on optical biopsy of skin tumors, one of the most important applications of nonlinear optical imaging in dermatology. Within this subject area, we will provide an overview of skin tumor physiology, procedures in clinical diagnosis, and current applications of linear and nonlinear optical imaging in this field. We will subsequently provide an overview of other biomedical applications of nonlinear microscopy in dermatology: skin regeneration, transdermal drug delivery, skin photodamage, and skin photoaging and protection.

32.2 ANATOMY AND PHYSIOLOGY

Ex vivo structures of skin are well known through extensive histological studies (Burkitt et al., 1993; Stenn, 1988). The fine structure of skin is further derived from electron microscopy studies. In vivo studies based on fluorescence microscopy are being supplemented by three-dimensional techniques such as in vivo confocal microscopy. More recently, nonlinear microscopic techniques, such as multiphoton excitation microscopy and spectroscopy, and second-harmonic generation microscopy have become the methods of choice to elucidate in vivo skin structure and function.

The skin is the largest organ of the body and forms a continuous external surface of the body (Bolognia et al., 2003). It is a self-renewing interface between the environment and the body. The skin provides a protective environment to the body, a fact that Rudolph Virchow cited more than 100 years ago. Specific structures in the skin provide the organism with the following functions: barrier function, injury repair, photoprotection, thermoregulation, cutaneous (Latin *cutis* [skin]) circulation, cutaneous sensation, and immunologic protection.

Skin color is a function of the degree of oxygenation of the blood in the cutaneous circulation, the thickness of the cornified layer, non-melanin pigmentation, and the activity of the melanin-producing cells. Skin color is also a function of light scattering from the skin. Subcutaneous veins have a deep blue appearance, which is mainly due to the higher scattering of blue light. The actual color of the venous blood from the subcutaneous vessel is less blue than the appearance of the vein. Racial variations in skin color depend on differences in the amount, type, and distribution of melanin.

The skin contains an extensive vascular supply, which plays a key role in its thermoregulation. The cutaneous blood supply is regulated by the thermoregulatory requirements. Temperature regulation is controlled by the hypothalamus, which regulates both the blood supply and the sweat glands, which produce up to 10 liters of sweat each day. The skin has dual thermoregulatory functions. The primary regulation is evaporative cooling after eccrine sweating and vasodilation and vasoconstriction; in addition, the skin acts as an insulator. The skin is also a major sensory surface and contains a rich nerve supply and many sensory receptors that respond to stimuli for touch, pressure, vibration, hot and cold, and pain.

Typically images of skin provide a picture of uniform morphology (Greek *morpho* [form, shape, structure] + Greek *logos* [study]); however, there are important regional variations. The thickness, color, surface appearance, and flexibility of the skin and the density of its appendages vary from region to region. The skin of the scalp, the eyelids, the trunk, the genitalia, the palms, and the soles of the feet show modified characteristics.

To introduce skin morphology, we discuss the stratified components of the epidermis, the dermis, and the subcutaneous tissue (Fig. 32.1). We will further examine the functional structures in the skin, including the adipose tissue, arteries and veins, nerves, and Pacinian corpuscles and other sensory organs within the subcutaneous tissues. There are special adnexa (e.g., hair follicles and shafts, sweat glands, and sebaceous glands);

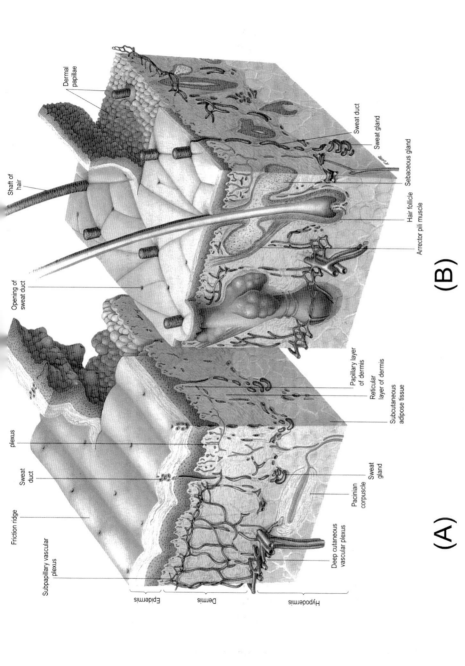

Figure 32.1. Gross structure of the skin that compares the structure of thick, hairless skin found in the plantar and palmar regions (A), and the structure of thin, hairy skin (B). The structures in the epidermal, dermal, and subdermal regions are shown. In each of the three-dimensional drawings the epidermis has been peeled back to permit the visualization of the interdigitation between the dermal and the epidermal papillae. Note that in (A) the interdigitations of the dermal and the epidermal papillae follow the friction ridges, which contain the outlets of the sweat ducts arranged in a linear array along the friction ridges. In (B) the surface is covered with triangles and the interdigitations are of less amplitude as compared to thick, hairless skin. (Adapted from S. Strandring, ed., *Gray's Anatomy, the anatomical basis of clinical practice, 39th edition,* Elsevier, Churchill Livingstone, p. 158)

the ducts of sweat glands traverse the skin and exit on the skin surface. Finally, skin stem cell and epithelial layer replacement processes are discussed.

32.2.1 Stratified Skin Structure: Epidermis and Dermis

32.2.1.1 Epidermis

The epidermis (Greek *epidermis* [the outer skin], from *epi* [on] + *derma* [skin]) is the superficial epithelial portion of the skin and consists of a keratinized, stratified squamous epithelium (Fig. 32.2). The thickness of the epidermis ranges from 0.07 to 1.4 mm. A rete ridge structure is observed at the epidermal–dermal junction that includes the dermal papilla and its capillary loop (see Fig. 32.1). The papillary dermis contains dermal extensions that project into the epidermis. Each dermal papilla contains a single capillary loop, which provides for the nutritional requirements of the epidermis. The epidermis is usually divided into distinct layers (stratified epithelia) that are parallel to the skin surface. These layers are, from the innermost layer to the superficial layer, the basal layer, composed of a single layer of cells (stratum basale); the spinous or prickle cell layer, composed of two to five cell layers (stratum spinosum); the granular layer, composed of one to three cell layers (stratum granulosum); a clear layer (stratum lucidum); and the surface layer (stratum corneum). The stratum lucidum is present only in very thick skin and appears as a homogeneous layer between the stratum granulosum and the superficial cornified layer. The cells in the stratum corneum have lost their

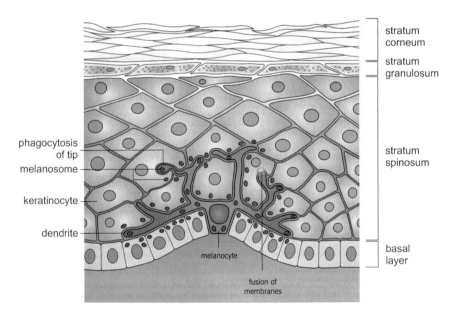

Figure 32.2. Cellular structure of normal human skin. The figure shows a melanocyte located in the basal layer of the epidermis. The melanocyte forms melanosomes, which are transferred from the dendrites of the melanosome into the adjacent keratinocytes of the epidermis. The cellular organization of the epidermis consists of stratum corneum, stratum granulosum, stratum spinosum, and the basal epithelial cell layer, which is located adjacent to the dermis. (Adapted from J. L. Bolognia et al., 2003, *Dermatology*, volume 1, Mosby, p. 937)

nuclei; they are transformed into non-nucleated horny plates or squama (Latin, scales) that are mainly composed of the fibrous protein keratin.

Epidermis contains three major populations of cells: keratinocytes (the majority of the cells), melanocytes, and Langerhans cells. In the epidermis the density of keratinocytes near the skin surface is about 50,000 cells/mm^2. The melanocytes have long dendritic processes that are used to transport melanosomes to other keratinocytes (see Fig. 32.2). Melanocytes are located above the nuclei of basal epithelial cells. The Langerhans cells play a major role in the immunological functions of the epidermis. There are also Merkel cells residing in the basal cell layer, which contain neurosecretory granules that are associated with free nerve endings in thick skin. With white-light microscopy, histological staining techniques are used to increase the contrast of these cells to make them visible. The keratinocytes synthesize keratin (Greek *kerat*−[horn] + in) protein and many regulatory molecules called cytokines. Keratin is a group of proteins that forms the intermediate filaments in epithelial cells. The epidermal keratinocytes (squamous epithelial cells) are derived from epidermal stem cells. They require about 30 days to differentiate and migrate through the 50 μm of epidermis to reach the surface layer of the stratum corneum. However, the normal epithelial turnover time can range from 30 to 75 days, depending on the skin thickness. It should be noted that keratinocytes take on different names, such as basal cells and corneocytes, as they migrate and differentiate through the different epidermal strata. Normally the rate of production of basal epithelial cells in the basal layer is equal to the loss of cells from the cornified layer. With abrasion the thickness of the cornified layer increases. The chemical mediators and biochemical mechanisms that regulate the relative rates of cell division and cell differentiation of basal epithelial cells represent an active area of research. Understanding of this regulatory function can have an important impact on several skin diseases and recurrent ulcers in lower extremities of the elderly (Limat et al., 1996).

The basal epithelial cells and the basement membrane form the dermal–epidermal junction, which separates the epidermis from the dermis. Basal cells (keratinocytes in the basal layer) are derived from stem cells in the hair follicles and populate the basal layer. Cells in the stem cell region show a very slow rate of cell division. When these stem cells divide, they form transit amplifying cells. The basal cells are attached to the basal lamina with hemidesmosomes. The bright junction, as observed with electron microscopy, between the epidermis and the dermis is the lamina lucida, which is bound by the basal lamina. Desmosomes connect adjacent basal epithelial cells together as well as connecting keratinocytes. The basal epithelial cells are columnar to cuboidal in shape, with large nuclei and prominent nucleoli. The long axes of the basal epithelial cell nuclei are oriented perpendicular to the surface of the skin.

Only in the layer of basal epithelial cells are dividing keratinocytes observed. After a number of cell divisions, these transit amplifying cells become committed to differentiation. These basal epithelial cells then differentiate and migrate towards the surface of the skin. These cells become flatter and larger as they migrate towards the skin surface, where they are shed. The prickle cell (keratinocytes in the spinosum) stratum is composed of several layers of closely packed cells that interdigitate with each other and are connected by desmosomes. Four layers of flattened keratinocyte cells form the granular layer. The nuclei and cellular organelles are beginning to degenerate in the granular layer. The clear layer occurs only in thick palmar (refers to the palm of

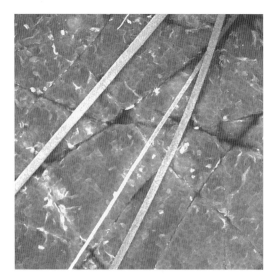

Figure 32.3. In vivo confocal microscopic autofluorescence image of stratum corneum of the volar surface of the human forearm. Microscope objective: 10×, NA 0.3. Horizontal field, 1,000 μm. Diameter of thick hair, 40 μm. Squames have a diameter of 30 to 40 μm. The confocal images show the interlacing network of creases and the triangular areas between them. Many polygonal, scale-like keratocytes (squames) of the superficial cornified layer are visible. (Adapted from Masters, Bioimages, 1996)

the hand) or plantar (relating to the sole of the foot) skin. The epithelial cells adjacent to the stratum corneum located in the stratum granulosum have flattened nuclei, with their long axis parallel to the surface of the skin. At the surface of the skin, the dead differentiated keratinocytes (corneocytes) form the stratum corneum, presenting a barrier layer that is critical to prevent fluid loss and resist microbiological invasion and chemical penetration.

The cornified stratum corneum consists of closely packed layers of flattened polyhedral squames (Fig. 32.3). The area of the squames ranges from 800 to 1,100 μm^2. In thick skin the squames can be 50 layers thick, but in thin skin they are only 3 layers thick. The squames are sometimes organized into columns. In some species these columns are very distinct; for example, the mouse ear epidermis shows a prominent columnar patterning of the packed layers of flattened polyhedral squames. In tissues where the mitotic rate is high, such as the palmar-plantar epidermis, the columns are less evident. In human skin the stacking of the keratinocytes into columns is less defined compared to that in rodents.

In addition to keratinocytes, Langerhans cells are found in the epidermis. Langerhans cells are dendritic cells that were first identified in 1868 by Paul Langerhans. Working as a medical student, he stained a section of skin with gold chloride and made the first observation of dendritic cells that later were named after him. These cells are usually located within the stratum spinosum. Langerhans cells are derived from bone marrow. Nerve fibers are closely associated with Langerhans cells, and their adjacent nerves can release neuropeptides that can modulate the functions of the Langerhans cells. While the Langerhans cells are the major source of immunologic cells in the epidermis, the keratinocytes also participate in the skin immune responses, such as in contact hypersensitivity.

The third major cell type in the epidermis is the melanocyte. Melanocytes (melano- + Greek *kytos* [cell]) are dendritic cells that contain melanosomes and are the site of melanin biosynthesis. Melanocytes are derived from the neural crest. What determines the color of skin is the amount and the type of pigment production in the melanocytes. Skin color is not a function of the density of the melanocytes, but their type of activity. Dark skin contains larger melanosomes that contain more melanin (Greek *melan-* [black]). Melanocytes synthesize two types of melanin: brown-black eumelanin and yellow-red phaeomelanin. Melanocytes are located in the basal cell layer of the epidermis, as well as in the basal layer of the hair matrix (base region of the hair follicle). In normal skin about one cell in ten in the basal epithelium is a melanocyte. However, the density of melanocytes can vary from about 900 melanocytes per mm^2 on the back to about 1,500 melanocytes per mm^2 in the genital region. While the cell body of the melanocyte, the part containing the cell nucleus, is similar in size to the basal cells, the long dendritic processes that extend outward and upward from the main nuclear region can extend over a length of five to ten basal cells. The nuclei of melanocytes are significantly larger than the nuclei of basal epithelial cells.

Melanocytes play a protective role in skin. Repeated exposure to ultraviolet light results in an increased number of melanocytes as well as an increase in the number of melanosomes that are transferred to keratinocytes. The melanin accumulates above the keratinocyte nuclei in the direction of the skin surface. It is thought that this protects the dividing nucleus of the basal epithelial cells from sunlight.

32.2.1.2 Dermis

Below the epidermal junction is the dermis (Greek *derma* [skin]), a dense collagenous tissue that contains a vascular system consisting of blood and lymphatic vessels, nerves and nerve endings, and many sensory receptors; it consists of a mucopolysaccharide matrix that contains collagen and elastin fibers (see Fig. 32.1). The fibroblasts synthesize the collagen, elastin, and ground substance. The superficial layer of the dermis interdigitates with the epidermis. Near the epidermal–dermal junction the collagen is tightly packed, but in deeper regions of the dermis the collagen is more loosely arranged. The thickness of the dermis varies from 0.5 mm on the eyelids to over 5 mm over the back.

The layers of the dermis of increasing depth include the papillary layer, the reticular layer, and the hypodermis or superficial fascia. The papillary layer is located adjacent to and just below the epidermis and contains sensory nerves and blood vessels. It provides mechanical anchorage to the epidermis and metabolic support and supplies sensory nerve endings and blood vessels. The superficial surface of the dermis forms many papillae or rete ridges, which interdigitate with the rete pegs of the base of the epidermis. This interdigitation forms the dermal–epidermal junction. In thick skin such as on the fingertips and the soles of the foot, they are large and arranged into parallel lines that follow the patterns of ridges and grooves on the skin surface. Dermal papillae contain vascular loops and specialized nerve endings and are most developed in the hand and the foot. The reticular layer, which makes up the main bulk of the dermis, is contiguous with the deep aspect of the papillary layer. It contains bundles of thicker collagen fibers compared to the papillary layer. The hypodermis, also known as the superficial fascia, is a layer of loose connective tissue that merges with the deep aspect of the dermis. Often there is adipose tissue present.

32.2.2 Glands, Vascular System, Hair and Nails

Within the skin over many regions of the body (not on the palms or the soles of the feet) are pilosebaceous units. They consist of the hair and its follicle, an associated arrector pili muscle, a sebaceous gland, and sometimes an apocrine gland (a sweat gland associated with the hair follicle). The apocrine glands produce thick secretions and open into the hair follicles. The hair follicle is of great importance; it is the only permanently regenerating organ in the mammalian body. The hair follicle is a reservoir for both epithelial and melanocyte stem cells.

There are also eccrine (sweat) glands unconnected to the hair follicles. Eccrine sweat glands consist of coiled tubes that open directly onto the skin surface. The eccrine sweat glands secrete a dilute salt solution and thus provide the major source of evaporative heat loss.

The vascular system provides nutrition and the nerve fibers provide sensation and sensing. The dermis also contains the following cells: macrophages, white blood cells, and dermal dendritic cells, which contribute to the immunological function of the skin.

Hair and nails are derived from the epidermis. The skin contains hair follicles, which generate the hair shafts that grow out from the skin. Vellus hair, a very fine hair, occurs over many parts of the body. Terminal hair occurs on the head, face, arms, chest, back, and legs.

32.2.3 Stem Cells and Epithelia

The human body contains several epithelial tissues that are renewable through the presence of stem cells. The epithelium of the gastrointestinal tract is renewed by stem cells that reside in the crypts. In the skin, the epidermis is renewed by stem cells in the hair follicles. Cell proliferation occurs in the basal epithelial layer of skin. In the cornea, the corneal epithelium is renewed by stem cells that reside in the limbus of the cornea.

The skin stem cells are located in the bulge region of the hair follicles. The bulge region of the hair follicle is in the outer root sheath near the attachment of the arrector pili muscle. It is this muscle that erects the hair shaft. These stem cells can be identified as label-retaining keratinocytes. Therefore, hair follicles can be clinically used as a source of stem cells for the epidermis (Limat et al., 1996).

There is a homology between cell division and differentiation in both the cornea and the skin. An example of the study of cell lineage and differentiation of corneal epithelial cells in adult rats illustrates how these processes can be studied in a stratified epithelium (Beebe & Masters, 1996). First, cells synthesizing DNA incorporated 5-bromo-deoxyuridine (BrdU) into their DNA. Corneas were fixed and permeabilized, and the BrdU-labeled nuclei were stained with a monoclonal antibody to BrdU and a rhodamine-labeled secondary antibody. Using three-dimensional confocal microscopy, the full-thickness cornea in whole mounts was studied at different times after BrdU labeling. Cell division of BrdU-labeled cells produced pairs of daughter cells in the basal cell layer. When daughter cells migrated out of the basal cell layer and began the process of terminal differentiation, they did so as pairs of cells. The synchronous differentiation of daughter cells was observed at pairs of labeled nuclei (cells) in the full thickness of the corneal epithelium from 2 to 14 days after BrdU labeling.

32.3 SKIN CANCER AND ITS DIAGNOSIS

One of the promising applications of nonlinear optical microscope is in the diagnosis and treatment of skin cancer based on noninvasive optical biopsy (bio- + Greek *opsis* [vision]).

32.3.1 Tumor Biology

A neoplasm (neo- + Greek *plasma* [thing formed]) is an abnormal tissue that grows by cellular proliferation, shows a partial or complete lack of structural organization and functions compared to normal tissue, and usually forms a distinct mass of tissue that may be either benign (benign tumor) or malignant (cancer). Neoplasms of the skin include a wide variety of pathological conditions (Bolognia et al., 2003).

We limit our discussion to the following: actinic keratosis, basal cell carcinoma (BCC), squamous cell carcinoma (SCC), benign epidermal tumors, atypical melanocytic nevus (AMN), and melanoma (melano- + Greek *-oma* [tumor]). Non-melanoma skin cancer includes both BCC and SCC. Carcinoma (Greek *karkinoma*, from *karkinos* [cancer] + *-oma* [tumor]) is a term covering all malignant neoplasms that are derived from epithelial cells.

Most common forms of cancer occur in epithelial cells, and that is the focus of this section. Normal epithelial cell function involves a regulated balance of cell proliferation and cell differentiation that maintains and renews epithelial tissue. The cell cycle is regulated by a series of checkpoints and regulatory mechanisms. The signal for a cell to transform from a quiescent condition to a proliferative condition is controlled by extracellular signals. Loss of regulation of oncogenes results in increased cell proliferation. Tumor suppressor genes inhibit the cell cycle. It is the equilibrium between cell proliferation and cell death that results in normal homeostasis.

In order for a tumor to expand its mass, it is necessary to provide the growing tumor with adequate amounts of oxygen and nutrients. Therefore, the formation of new blood vessels (angiogenesis) is a requirement. Angiogenesis is stimulated by growth factors such as fibroblast growth factor (FGF) and vascular endothelial growth factor (VEGF), which acts to stimulate endothelial proliferation.

From the point of view of pathology, cancer in epithelial tissues is associated with a disruption of tissue organization. Epithelial cancers may escape from their primary location and move to other locations with the growth of secondary tumors; this is the process of forming metastases.

These neoplasms progress in different manners and present a range of risks and outcomes. It is critical to correctly diagnose life-threatening neoplasms of the skin from benign epidermal tumors such as seborrheic keratosis and epidermal nevus. Seborrheic keratoses (SKs) are benign skin lesions that are very common in individuals over

30 years old. The number of SKs increases with age. They occur only on hair-bearing skin: the face, neck, trunk, upper back, and extremities. They are single or multiple brown papules or plaques and can be confused with melanoma; therefore, it may be necessary to perform a biopsy and a histopathologic examination to arrive at a correct diagnosis.

BCC and SCC are common forms of non-melanoma skin cancer. BCCs are locally invasive, with a slow growth rate, and rarely metastasize. They grow in skin that contains hair follicles. The most common malignancy of the eyelid is BCC. Upon microscopic examination of a biopsy specimen, BCC appears as discontinuous buds of cancer cells. Histological examination indicates that BCC is composed of aggregates of neoplastic cells that form lobules, islands, or nests. Abundant melanin pigment is seen within the tumor cells. SCC originates from epidermal keratinocytes in sun-exposed skin and develops in a set of progressive stages that include actinic keratosis, invasive cancer, and eventually metastasis. It is thought that ultraviolet light induces mutations in keratinocytes that results in SCC. Lesions that are precursors to SCC occur on the face, the hairless scalp, the arms, and the dorsal aspects of the hands. At least 95% of cancers in the head and neck are SCCs. They are an aggressive epithelial malignancy and are usually found in the oral cavity. Some cancers tend to invade a particular organ or tissue; for example, SCC is the most common type of carcinoma of the esophagus. Once the tumor crosses the epidermal basement membrane, it can invade fat, bone, muscle, and cartilage and also metastasize. Tumor progression can lead to metastasis of SCC, with new tumors forming in regional lymph nodes and internal organs such as the lungs.

BCC and SCC are two types of tumors that can be treated with Mohs surgery. This micrographic surgical technique provides for the examination of 100% of the tumor margin, is tissue-sparing, and leads to the least complicated tissue reconstruction and thus is used for the removal of high-risk cutaneous neoplasms. Mohs surgery is controversial for the surgical treatment of melanoma since some clinicians think that the diagnosis of melanoma is unreliable in frozen sections.

There are a large number of benign melanocytic neoplasms. Hyperpigmentation of the skin can occur by sun-induced melanogenesis and an increased transport of melanosomes from melanocytes to keratinocytes. This can result in freckles and/or tanning of the skin. There are also numerous pigmented lesions of the skin that are termed melanocytic nevi or more commonly moles. A nevus (pl. nevi) (Latin *nae-vus* [mole, birthmark]) is a circumscribed malformation of the skin, especially if colored by hyperpigmentation or increased vascularity. It can be a benign localized overgrowth of melanin-forming cells of the skin present at birth or appearing early in life.

In addition to these common lesions or moles, there are lesions that are labeled atypical melanocytic nevus (AMN). Synonyms include atypical mole, atypical nevus, dysplastic melanocytic nevus (an older, outdated term), and the mole of familial atypi-cal mole and melanoma syndrome. The risk related to melanoma is related to the total number of nevi and their size. These nevi have morphologic changes such as asym-metry, irregular borders, and distinct color variation. In patients with large numbers of atypical melanocytic nevi, some of the nevi will develop into melanoma. There is no complete agreement about the histopathologic criteria for the diagnosis of atypical melanocytic nevi; however, there are some general correlates. Malignant melanoma can be distinguished from AMN by the following: larger size, increased asymmetry, irregular margins with notching, and striking variations in color and pigmentation. Unfortunately, there are still lesions that can be diagnosed only by histopathological examination.

To distinguish melanocytic nevi from melanoma, the role of the histopathologist is to identify atypical cells from observation of fixed, stained biopsy specimens (Kumar et al., 2005). Factors that are usually associated with atypical cells include hyperchromatism (hyper + Greek *chroma* [color]; i.e., increased pigmentation), pleomorphism (pleo- + Greek *morphe* [form]); i.e., occurring in more than one morphologic form), prominent nucleoli, increased nuclear-to-cytoplasmic ratio, and abnormal mitosis. The diagnosis of malignancy depends on many factors: the morphology of the lesion, the type of cells in the specimen, and the overall clinical situation. While atypical cells are often found in malignancies, atypical cells can also be found in benign skin disorders.

Melanoma is a malignant tumor arising from melanocytes. Melanoma is a significant public health problem and represents a very common type of cancer in young adults. Early detection is extremely important, and surgical treatment results in a high frequency of cures (>90%) for most patients. Exposure to sunlight is a prime factor for the induction of cutaneous malignant melanoma.

The most common type of melanoma is superficial spreading melanoma (SSM), which begins as a brown to black macule with color variations and irregular notched borders. Following a slow horizontal growth phase that is limited to the epidermis, the disease progresses with a rapid vertically oriented formation of a papular nodule. The second most common type of melanoma is nodular melanoma, which is frequently observed on the trunk, head, and neck of patients in their sixth decade of life.

Malignant melanoma is clinically characterized by the following features. The key clinical sign is a change in size, color, or shape in a pigmented lesion. In sharp contrast to benign nevi, melanomas are characterized by wide variations in pigmentation; therefore, the lesions can appear to be black, brown, red, dark blue, and gray. In addition, their borders are irregular and often notched. Malignant melanomas may also show a region of spontaneous regression, which results in a lesion with less pigmented regions.

In epidemiological studies of skin disease we often see the terms "incidence" and "prevalence," which refer to disease frequencies. Prevalence is the amount of disease that is present in a given population. Incidence is defined as the number of new cases of a given disease during a given time period in a given population. The incidence rate of a disease is defined as the number of healthy persons who become diseased during a specified time period, divided by the number of person-years in the population.

BCC and SCC are the most common human cancers. In Caucasians, BCC is the most common skin cancer. The non-melanoma skin cancer incidence in the United States is 232 per 100,000 (lightly pigmented skin color) and 3.4 per 100,000 (darkly pigmented skin color). The corresponding ratio of BCC to SCC is 4:1 and 1:1. The incidence of melanoma in the United States is about 13 per 100,000 population. The incidence rate in Australia is the highest worldwide: 35 per 100,000 individuals in 2000.

32.3.2 Diagnosis: a Primer on the Process of Disease Determination

32.3.2.1 Process of Clinical Diagnosis, Biopsy, and Laboratory Studies

It is an image (visual, photographic, or digital) of the skin that is the basis for diagnosing a skin disease (Kumar et al., 2005). This could be a visual image of the skin surface, a microscopic image of the skin surface together with the underlying tissue, or a microscopic image of excised tissue that has been fixed, sectioned, and stained. While the unaided eye can observe and detect the skin over the whole body, it is unable to resolve microscopic structures. The trained dermatologist is able to differentiate normal variations in skin morphology in diverse populations of different pigmentation, gender, and age from atypical skin lesions that require biopsy for further examination. Characteristic features of skin lesions include size, color, shape, thickness, and pigment distribution. Often it is time-dependent changes in skin morphology that provide diagnostic clues to pathology. Is the appearance of the skin lesion changing between observations? These changes could affect the size, thickness, or pigment distribution.

If a lesion is suspicious, a biopsy may be performed. The fixed tissue specimen is examined by a pathologist, who provides a diagnosis. A biopsy is the surgical removal of a sample of the tissue, with the associated risk for infection, scarring, and a wound healing process. The microscopic observation of a fixed, stained biopsy specimen provides the trained pathologist with a full-section, transverse view of a tissue specimen. The distribution and morphology of cells, organelles, pigmentation, and nuclei provide a wealth of information that is not routinely accessible to the dermatologist using the naked eye or a dermatoscope to examine the skin surface. For example, a biopsy specimen may generate the following diagnoses: benign, various degrees of atypical morphology, or melanoma. While the goal of the diagnosis is to determine with certainty the disease state, there is some degree of subjectivity in the diagnosis, and several different pathologists may not reach the same diagnosis.

The goal of any diagnostic technique is to minimize the number of false-negative and false-positive results. False negatives are when a pathological specimen is diagnosed as normal; false positives are when a normal specimen is diagnosed as pathological. An ideal diagnostic technique will have high sensitivity (the ability to detect a small indicator of pathology) and high selectivity (the ability to clearly separate normal tissue from pathological tissue).

The entire field of dermatologic diagnosis is based on the visual examination of the skin and the expert correlation of the clinical appearance with histological appearance (Kumar et al., 2005). Excisional in toto biopsy, which removes a specimen that contains the full thickness of the skin together with subcutaneous tissue, is used for the histological examination of an atypical pigmented lesion, as it removed the tumor and surrounding tissue should melanoma be confirmed. Biopsy is usually indicated when the diagnosis is not definitive. For the cases of atypical pigmented lesions, biopsy is indicated since the histopathological examination is more accurate than visual observation for the diagnosis of melanoma. Note that melanoma can become life-threatening!

32.4 ROLE OF OPTICS IN DERMATOLOGY: DIAGNOSIS

Optical techniques may provide additional information to perform a diagnosis. It is legitimate to ask: Why seek to develop optical diagnostic techniques when the gold standard of excisional biopsy, fixing, staining and microscopic histopathologic examination is available? One feature of optical diagnostic techniques is their noninvasive nature. A strong driving force in the development of optical diagnostic techniques is that their implementation may reduce the number of required excisional biopsies or, in the best case, provide an alternative to a biopsy. However, it should be pointed out that a false diagnosis can result in significant patient harm or even death. Nonetheless, there are also many cases for which multiple biopsies are undesirable for the patient. For example, patients with atypical melanocytic nevi may have 50 to 100 nevi, and many of them will require biopsy examination over a period of years. For such patients, total body photography of the entire skin is recommended. In other cases, such as biopsy of the eyelid, there are tissue-sparing considerations. Ideally it would be desirable to use optical techniques to replace excisional biopsies. While work is progressing in this important area of optical biopsy and spectroscopy, the current techniques suffer from several problems: reduced resolution in the plane of the skin and in the transverse plane, reduced depth of penetration, and reduced contrast compared to histopathologic examination of a biopsy specimen. Another possibility is that noninvasive optical methods will not replace excisional biopsies but may serve to reduce their number. Alternatively, optical methods, especially nonlinear microscopy, may serve to increase the information that can be obtained in histopathological examination of skin biopsy specimens. This potential advantage may derive from the improved resolution and new contrast mechanisms of new forms of nonlinear microscopy compared to conventional fluorescence and epi-illumination microscopy.

32.4.1 Optical Properties of Skin

There are few tissue or organ systems in which the clinician can readily see the disease state and pathology. The ophthalmologist, with the help of a slit-lamp, can readily observe the eye. The dermatologist, with either the naked eye or a dermatoscope, can observe the skin of the whole body.

An understanding of the optics of human skin and its regional variations is important for an understanding of light–tissue interactions in dermatology. The optical properties of human skin are discussed in a variety of papers (Anderson & Parrish, 1981; Troy & Thennadil, 2001).

The optical properties of melanin have been studied (Sardar et al., 2001). Melanin has been studied by stepwise two-photon femtosecond laser excitation (Hoffmann et al., 2001; Teuchner et al., 1999, 2000, 2003). Stepwise two-photon excitation differs from two-photon excitation. Stepwise two-photon excitation requires a real intermediate energy level (i.e., a one-photon absorption at about half the energy of the transition from the ground state to the excited fluorescent state). In two-photon excitation spectroscopy there is a virtual intermediate state. In these studies of stepwise two-photon excitation, the absorption spectra, emission spectra, and fluorescence decay behavior were investigated.

32.4.2 Optical Imaging of Skin

Many optical imaging techniques have been developed for the imaging and disease diagnosis of skin with increasing resolution, depth of penetration, and information content. However, with these improvements come limitations: these new techniques are also often more difficult to use clinically and more expensive. Importantly, with the increased resolution and information content, the visualization and interpretation of these data are no longer trivial. With these new optical imaging modalities, a new knowledge base must be built to correlate these new image data with diagnosis based on traditional histopathology.

32.4.2.1 Dermatoscopy

Unaided clinical observation is useful for the diagnosis of most melanocytic lesions. However, there are subsets of lesions that are difficult to diagnosis. For those cases, dermatoscopy, or microscopy of the surface of the skin, is helpful and is widely accepted clinically. Dermatoscopy (Greek *derma* [skin] + Greek *skopeo* [to view]) refers to the inspection of the skin with the aid of a lens or by epi-illumination microscopy. Goldman performed some of the early microscopic investigations of skin (Goldman, 1951; Goldman & Younker, 1947).

Modern dermatoscopes use epi-illumination and sometimes crossed polarizers (illumination) and analyzers (image) to reduce surface reflections and increase the image contrast (Jacques et al., 2002). Sometimes an index matching fluid is placed on the skin that renders the stratum corneum transparent: the fluids act as a clearing agent to increase the visibility of the deeper layers of the skin. Dermatoscopes provide a magnification range from about 6- to 100-fold. With an index matching fluid or clearing fluid in place, dermatoscopes can help to image structures within the epidermis, the dermo-epidermal border, and the superficial dermis.

The most important application of the dermatoscope is to differentiate early stages of melanoma from benign lesions. The goal is to correctly distinguish melanocytic skin tumors from non-melanocytic pigmented skin lesions. Examples of the latter class are seborrheic keratosis and heavily pigmented BCC. With the dermatoscope the clinician can observe the pigmented network in melanocytic tumors, which corresponds to pigmented rete ridges in histopathological sections.

32.4.2.2 High-Resolution Three-Dimensional Optical Imaging and Spectroscopy of Skin: Linear Optics

The gold standard for the diagnosis of skin cancer is based on expert microscopic evaluation of a biopsy section. Higher-resolution microscopy of a biopsy specimen can be an important tool to help assess lineage. For example, it is generally accepted that cells that are differentiated from sebaceous tissue show a coarsely vacuolated cytoplasm and scalloped nuclei.

Optical low-coherence tomography is a technique that has potential for skin imaging (Yeh et al., 2004), with resolution on the order of about 10 microns. This is an interference technique in which the contrast of the image is due to light scattering at interfaces between regions of different refractive index.

For submicron-level resolution, confocal reflected light microscopes based on the Nipkow disk and an incoherent light source can obtain optical sections of the skin from the surface to deeper layers (Bertrand, 1994; Bertrand & Corcuff, 1994; Corcuff et al., 1993; Corcuff & Lévêque, 1993; New et al., 1991). Other investigators developed laser-scanning confocal microscopes to image in vivo human skin by reflected light (Rajadhyaksha et al., 1995; Rajadhyaksha, Anderson, & Webb, 1999; Rajadhyaksha, González, et al., 1999; Rajadhyaksha et al., 2001). The depth of the deepest images is a function of the optical system; the degree of pigmentation, which will affect light scatter and absorption; and the vascular system.

Confocal microscopy of in vivo human skin can acquire a single two-dimensional image or stacks of sequential two-dimensional images that can be converted into three-dimensional images. Confocal microscopy can acquire images in the fluorescent mode using intrinsic autofluorescence, with extrinsic fluorescent probes, or with genetically expressed fluorescent proteins. Alternatively, reflected-light confocal microscopy can be used to image in vivo human skin. Figures 32.4 and 32.5 are examples of in vivo confocal microscopy of human skin based on autofluorescence. Figure 32.6 is an example of the three-dimensional confocal reflected microscopy of in vivo human skin (Masters, Gonnord, & Corcuff, 1996). Laser scanning confocal microscopy can also be used to obtain three-dimensional images of human skin (Masters, 1996; Masters, Aziz, et al., 1997).

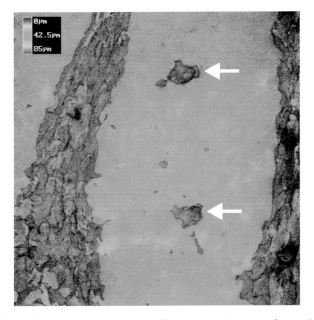

Figure 32.4. 3D confocal microscopic autofluorescence image of anterior region of the index finder. The excitation light was 488 nm, the fluorescence emission was isolated with a 515-nm long-pass filter. The microscope objective was 20×, NA 0.5. The horizontal field width is 500 μm. This is a pseudocolored depth-coded image. The red color indicates the surface of the stratum corneum; dark blue indicates an apparent depth located 85 μm below the surface. The openings of two sweat glands that are located on a friction ridge are shown (arrows).

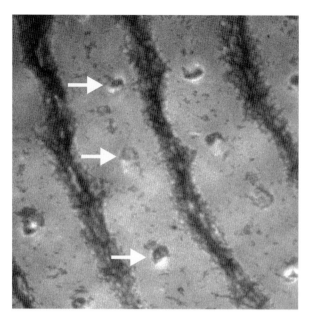

Figure 32.5. A confocal fluorescence light image of the hairless skin of the volar surface of a human digit, showing friction ridges along which lines of sweat ducts open as pores (arrows). The excitation light was 365 nm and the fluorescence emission was isolated with a 515-nm long-pass filter. (Adapted from Masters, Bioimages, 1996)

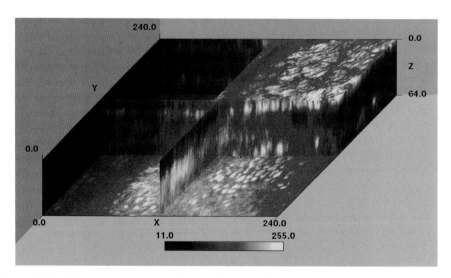

Figure 32.6. 3D reconstruction of in vivo human skin. Reflected-light confocal microscopy of in vivo human skin from the volar surface of the human arm. The color bar indicates the intensity of the reflected light. The reconstruction is $240 \times 240 \times 64 \,\mu m$. The 3D image shows the surface of the skin; the thickness of the stratum corneum; the dark round nuclei of the stratum spinosum; the bright oval nuclei of the basal cells; and the papillary dermis. (Adapted from Masters et al. J. Microscopy, 185: 333, 1997, figure 4)

Spectral imaging is another potentially useful tool to aid in the study of human melanoma and its precursors (Farkas & Becker, 2001). Images obtained with a variety of wavelengths can be analyzed and recombined to yield images with higher information content compared to those obtained with a single wavelength.

Spectroscopic oblique-incidence reflectometry is another technique that has been applied to skin cancer detection (Garcia-Uribe et al., 2004).

32.4.2.3 High-Resolution Three-Dimensional Optical Imaging and Spectroscopy of Skin: Nonlinear Optics

Recently, imaging techniques based on nonlinear techniques have shown particular promise in skin imaging (Alfano & Masters, 2004). Nonlinear optical techniques such as multiphoton fluorescence excitation microscopy, second-harmonic generation microscopy, and coherent anti-Stokes Raman microscopy all have shown potential as diagnostic tools for dermatology.

The application of multiphoton excitation microscopy and spectroscopy to in vivo human skin was a milestone in minimally invasive skin imaging (Masters, So, & Gratton, 1997, 1998a, 1998b). Multiphoton excitation microscopy at 730 nm and at 960 nm was used to image in vivo human skin autofluorescence from the superficial layers to a depth of about 200 μm. At specific depths in the skin the emission spectra and the fluorescence lifetime were measured. Cell borders and cell nuclei were observed in the skin. With 730-nm excitation, the primary source of skin autofluorescence was the reduced pyridine nucleotides, NAD(P)H. This study demonstrated the use of multiphoton excitation microscopy for functional imaging (two-photon redox imaging) of the metabolic states of in vivo human skin.

In order to compare confocal microscopy and multiphoton excitation microscopy of in vivo human skin, the two techniques were compared on the same region of in vivo human skin (Masters, So, Gratton, 1998a, 1998b; Masters & So, 1999, 2001; Masters, So, Kim, et al., 1999; Masters et al., 2004). Tandem scanning confocal microscopy using a white-light source was compared with multiphoton excitation microscopy, which used an 80-MHz pulse train of femtosecond laser pulses at 780-nm wavelength (Fig. 32.7). Below the dermal–epidermal border the multiphoton excitation microscope provides images with higher contrast compared to the tandem scanning confocal microscope. The use of multiphoton excitation microscope has the advantage of showing the corneocytes in the stratum corneum and also the collagen/elastin fibers in the dermis of in vivo human skin. The authors note that with the two-photon excitation microscope, thermal damage can result from one-photon absorption in the diffraction-limited volume. In general, when biochemical information is not required, the authors state that reflected-light confocal microscopy is preferred for morphological studies of in vivo human skin on the basis of cost, simplicity of instrumentation, and ease of use.

Another group also investigated the use of two-photon fluorescence and confocal video microscopy of in vivo human skin (Hendriks & Lucassen, 2001). Their instrument consisted of one microscope objective and separate light sources, detectors, scan units, and control electrons. It has the important advantage of simultaneously imaging the identical region of in vivo human skin with two different imaging modalities. Their results are similar to that of the previously described studies (Masters, So, Gratton,

Confocal
Microscopy

Multiphoton
Microscopy

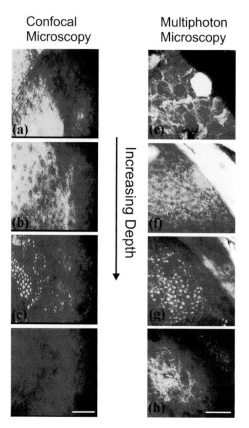

Figure 32.7. Comparison of reflected-light confocal microscopy (a–d) and multiphoton excitation microscopy (e–f) of the same region of in vivo human skin. The images were acquired from the volar surface of the human forearm. Images were acquired at the following regions: 10 μm below the surface of the stratum corneum (a, e), cells of the stratum spinosum (b, f), cells of the basal epithelial layer (c, g), and within the dermis (d, h). Scale bars are 50 μm. Note the similarity of the images at equivalent depths within the skin that are acquired with both confocal microscopy and with multiphoton excitation microscopy. Imaging below the dermal–epidermal junction demonstrates the enhanced depth penetration of multiphoton excitation microscopy compared to confocal microscopy. (Adapted from Masters and So, Microsc. Microanal. 5: 286, 1999, Figure 3)

1998a, 1998b; Masters & So, 1999, 2001; Masters, So, Kim, et al., 1999). With the use of two imaging modalities, the images contain complementary information, which can be advantageous.

While in vivo human skin was studied by a few groups using nonlinear microscopy and spectroscopy (Masters, So, & Gratton, 1997), many groups performed nonlinear microscopy on samples of ex vivo human skin (Dong et al., 2002). Sometimes these skin specimens are obtained from a skin bank, and the site of origin of the skin, as well as the age and gender of the skin donor, is unknown. Another confounding unknown is the duration over which the donor skin specimen has been stored, and what alterations of the optical and morphological properties of the ex vivo skin occurred prior to the experiments.

Alternatively, some investigators performed their studies on ex vivo human skin obtained from biopsies. Two-photon three-dimensional mapping of tissue endogenous fluorescence is a technique based on fluorescence excitation spectra (Hsu et al., 2001). A group at the University of Jena developed a microscopy system for imaging ex vivo human skin obtained from biopsies, based on multiphoton excitation microscopy of skin autofluorescence and second-harmonic generation imaging (König & Riemann, 2003). The nonlinear-induced autofluorescence is due to the naturally occurring fluorophores. The second-harmonic generation was used to image collagen structures.

In order to understand the signal observed in nonlinear microscopy, spectral resolved microscopy is performed to examine ex vivo human skin and was used to characterize the endogenous fluorescence and second-harmonic generating species based on their emission spectra (Laiho et al., 2005). They identified the following species that are autofluorescent: tryptophan, NAD(P)H, melanin, and elastin. The second-harmonic generation signal is from collagen.

Both second-harmonic and third-harmonic generation microscopies are active research frontiers for skin imaging (Chu et al., 2002; Gauderon et al., 2001; Mohler et al., 2003; Sun, 2005). Similar to multiphoton fluorescence, they provide three-dimensional resolved imaging with deep penetration in tissues. Their advantages over fluorescence method are that no extrinsic probes are needed and there is no tissue damage if an excitation wavelength is chosen out of any tissue absorption band. The disadvantages of these techniques are that the second-harmonic signal is generated only in biological crystalline structures that lack centrosymmetry, and in the skin only collagen matrix has this property to be imaged; and third-harmonic signal is generated at any region where there is an index of refraction mismatch, providing an image that is similar to that generated by confocal reflected light microscopy and that does not have biochemical information. Nonetheless, collagen, as the major constituent in the dermis, can serve as a major target for cancer diagnosis using second-harmonic generation. It is known that many cancer types upregulate matrix metalloproteinases (MMPs), resulting in a degradation of matrix component and facilitating cellular invasion into the dermis. Recent studies combining second-harmonic generation and polarization resolved imaging have allowed a better understanding of how the molecular conformation of collagen affects SHG (Chu et al., 2004). This information may be potentially used in diagnostics by distinguishing different collagen types in tissues and their structural states on the nanoscale.

In addition to fluorescence and second-harmonic contrast, coherent anti-Stokes Raman scattering microscopy (CARS) is another technique with potential for diagnostic skin imaging. CARS has the great advantage of chemical specificity, and it does not require extrinsic probes. CARS microscopy can be tuned to a specific vibrational band of specific molecules and can be used to map out the chemical nature of cellular components. CARS imaging of skin structures has also been performed at video rate (Evans et al., 2005). For this study, CARS signal is tuned to the vibration band corresponding to lipid-rich structures in the skin, providing complementary information to fluorescence and second-harmonic generation imaging.

While two-photon excitation fluorescence microscopy provides an important method for in vivo high-resolution optical imaging of human skin morphology and biochemistry and metabolism, there is the possibility of thermal mechanical damage. The major mechanism is the formation of cavitation due to melanin absorption at the epidermal–dermal border, which results in thermal–mechanical damage to the skin. An important technique for mitigating thermal–mechanical damage during two-photon excitation microscopy of in vivo human skin was recently developed (Masters et al., 2004). The authors describe a new technique to mitigate thermal–mechanical damage based on the use of a laser pulse picker that reduces the laser repetition rate by selecting a fraction of pulses from the laser pulse train. Since the laser pulse picker decreases the laser average power while it maintains the laser pulse peak power, the damage is minimized while the two-photon fluorescence excitation efficiency is maximized. The study of thermal–mechanical damage is relatively easy since this damage process involves gross morphological changes on the cytological level. It should be noted that other optical damage mechanisms, such as ones related to the formation of reactive oxygen species, may also be induced during optical biopsy of skin, although there is no definitive study today.

32.5 OTHER NONLINEAR MICROSCOPY APPLICATIONS IN DERMATOLOGY

Skin cancer diagnosis is an important application area where nonlinear optical microscopy is making an impact on dermatology. New applications of nonlinear optical techniques in other areas of dermatology are being investigated today. Some of the most promising directions are in regenerative medicine, studying transdermal drug delivery process, and studying photoaging of skin.

32.5.1 Regenerative Medicine

Regenerative medicine is one of the most important new areas in bioengineering and medicine. While many diseases can be cured by pharmaceutical treatment or localized corrective surgery, there are many cases where systematic failure of the whole organ can occur due to either the aggressive nature of the disease or acute trauma. In these cases, the only treatment is often the replacement of the whole organ, but this process is often severely limited by the availability of donors. Therefore, regenerative medicine focuses on the creation of artificial tissue constructs that can either directly replace the diseased organ or can induce a proper "wound healing" response where the body is induced to regenerate the diseased organ (Yannas, 2000, 2005).

One of the earliest applications of regenerative medicine is in the area of skin regeneration (Heimbach et al., 1988; Yannas et al., 1982). Regeneration matrix material based on collagen was developed to treat full-thickness skin wounds. Full-thickness wounds normally heal by contraction without synthesis of new skin. With the regeneration matrix, new skin can be synthesized. One of the early applications of skin regeneration matrix is in the treatment of patients with severe burns. The use of this new technology significantly reduces mortality and morbidity rates. Furthermore, motion constriction in patients is minimized by reducing scarring.

While regenerative medicine in skin is routinely used in clinics today to facilitate wound healing, current research focuses on obtaining better understanding of the wound healing mechanisms in skin and optimizing these regeneration matrices for different applications. In vivo multiphoton imaging has been performed in a guinea pig model to monitor the wound closure process without regeneration matrix implantation over a period of 1 month (Navarro et al., 2004). In vivo multiphoton images were compared with ex vivo histopathology, and the authors found that multiphoton imaging can provide comparable information to that provided by histopathology to study wound healing without biopsy. A number of studies have been performed to study cell–matrix interaction in engineered tissue models (Agarwal et al., 1999, 2000; Torkian et al., 2004; Yeh et al., 2004). The matrix component of the model can be visualized by second-harmonic generation, while the cellular components can be imaged using multiphoton fluorescence based on live cell-tracking fluorescent probes. The imaging of these tissue models using nonlinear optical microscopy has also been applied in the study of photodynamic therapy mechanisms (Chiu et al., 2005).

Multiphoton excitation microscopy has also been used to investigate sulfur mustard lesions in skin (Werrlein et al., 2003, 2004, 2005).

32.5.2 Transdermal Drug Delivery

Transdermal delivery of therapeutics is an alternative to conventional methods such as oral and subcutaneous administration. Transdermal drug delivery is convenient and painless and can avoid first-pass drug metabolism (Finnin & Morgan, 1999). The primary barrier to transdermal delivery is the stratum corneum, which comprises closely packed corneocytes and their lipid multilamellae that form an "oily" environment. Therefore, stratum corneum allows the transport of only small hydrophobic drugs such as scopolamine, clonidine, nitroglycerine, estradiol, fentanyl, and nicotine. Many other drugs, which are either large or hydrophilic, cannot be delivered well through the skin. The use of a chemical enhancer that can reversibly lower the resistance has been studied to enhance transdermal drug delivery (Peck et al., 1998). While transdermal drug delivery is routinely used today, the pharmacokinetics and mechanisms of drug transport through the stratum corneum in the presence and absence of chemical enhancers are not known.

Multiphoton imaging has been shown to be a promising method to study transdermal drug delivery because of its high spatial resolution, depth discrimination, and long penetration length (Grewal et al., 2000). By simultaneously performing multiphoton imaging probe distribution as a function of skin depth and bulk chemical measurement of the total transport rate through ex vivo skin specimens, the microscale transport properties of the permeant in skin, such as the vehicle-to-skin partition coefficient, the skin diffusion coefficient, and the skin barrier diffusion length, can be measured as a function of probe hydrophobicity and enhancer presence (Yu et al., 2001, 2002). Equally important, the transport pathway differences for hydrophobic and hydrophilic permeants in the stratum corneum have been quantified by three-dimensional imaging of stratum corneum distribution of permeants within and outside corneocytes (Yu et al., 2003).

In addition to quantifying transdermal transport kinetics and pathways, multiphoton fluorescence microscopy was vital in understanding the acidification of the stratum corneum. A steep gradient is present in the stratum corneum and stratum granulosum boundary, with the acidity of stratum corneum in the range of pH 4.5 to 6 and the rest of the epidermis at neutral pH. The presence of this acidity gradient is found to be important for the stratum corneum's barrier function. The combination of multiphoton microscopy and fluorescence lifetime-resolved imaging allows three-dimensional resolved quantification of pH distribution in the stratum corneum using pH probe, 2',7'-Bis(carboxyethyl)-5,6-carboxyfluorescein (Hanson et al., 2002). Importantly, NHE1, a Na^+/H^+ antiporter of the keratinocytes, has been shown to be responsible for maintaining this acidity gradient (Behne et al., 2002).

32.5.3 Skin Photodamage, Photoaging, and Protection

The effect of sunlight on skin physiology is not completely understood. While sunlight is necessary, overexposure to sunlight has been shown to be the primary risk factor in the development of melanoma. Further, even in the absence of malignant transformation, overexposure to sunlight is also responsible for photoaging, a process characterized by skin wrinkling and a loss of elasticity. On the molecular level, acute exposure to ultraviolet light causes a complex cascade of changes in skin biochemistry, including an inflammation-like process and the production of reactive oxygen species. Chronic ultraviolet exposure at low doses does not result in inflammation but nonetheless results in long-term changes in skin physiology. In either case, the most prominent changes on the molecular level due to photoaging is elastosis, a long-term transformation of elastin in the skin from a more fibrous to a more globular form (Gonzalez et al., 1999; Ibbotson et al., 1999; Kochevar et al., 1994; Kollias et al., 1998; Wong et al., 2001).

Nonlinear microscopy may find uses in the quantification of the photoaging process. Elastin cross-links are fluorescent, and coarse elastin fibers can be seen in the dermis. Collagen fibers in the skin can be visualized using second-harmonic generation microscopy. Since the elastin luminescence is a fluorescent process, it has a finite lifetime, and since the collagen luminescence is a second-harmonic generation process, it is instantaneous and has no lifetime. Quantitative measurement of the distribution of these two extracellular matrix components can be accomplished based on global analysis of time-resolved images of the dermis (Pelet et al., 2004).

In addition to the quantification of elastin–collagen distribution in the dermis, nonlinear microscopy has also been applied to quantify the reactive oxygen species formation in the skin under ultraviolet irradiation (Hanson & Clegg, 2002). The effect of sunprotective agents in the modification of reactive oxygen species generation has also been studied (Hanson & Clegg, 2003).

32.6 CONCLUSIONS

Linear and nonlinear optical techniques can have an impact on important aspects of dermatology. Nonlinear microscopy has been applied to basic science studies such as cell proliferation and cell differentiation in skin. These techniques have also found many uses in applied studies, including the study of transdermal penetration of drugs and pathogens and the development of tissue regeneration substrates.

Besides scientific studies, the application of nonlinear microscopy for clinical diagnosis of skin cancer is particularly promising. When the dermatologist observes an atypical skin lesion, the safe and prudent procedure is to perform a biopsy. A pathologist is required to examine the fixed, stained specimen and determine if it represents a normal or a pathological state. For difficult determinations, a second pathologist may be required to give a judgment. The goal of an optical biopsy, one based on nonlinear microscopy and spectroscopy, is to make a determination of normalcy or pathology in vivo. Since the consequences of a false determination are extremely serious (i.e., overlooking a malignant tumor), the specificity of the optical biopsy must be very high. That sets a very high standard for the development of an optical biopsy technique based on nonlinear microscopy and spectroscopy. Nevertheless, nonlinear microscopy and spectroscopy has the potential to develop into an emerging clinical technique in the field of dermatology. The advantages of nonlinear microscopies are that the use of infrared light is less damaging than ultraviolet light when used to excite the autofluorescence of in vivo skin, photobleaching is largely limited to the focal volume, and the infrared light has a deeper penetration then ultraviolet light. Multiphoton excitation microscopy can be used to image several intrinsic probes in human skin: NAD(P)H, flavoproteins, and elastin. Since the raster-scanning process in most nonlinear optical microscopic and spectroscopic techniques is rather slow, typically only a small region, such as a suspect lesion, can be examined. Improving the imaging speed and the field of view of these nonlinear optical techniques presents a challenge to the instruments of the future.

REFERENCES

Agarwal, A., V. P. Wallace, B. J. Tromberg and S. C. George. 1999. Collagen remodeling by human lung fibroblasts in 3-D gels. *American Journal of Respiratory and Critical Care Medicine* 159(3): A199–A199.

Agarwal, A., M. Coleno, V. P. Wallace, B. J. Tromberg and S. C. George. 2000. Two-photon scanning microscopy of epithelial cell-modulated collagen density in 3-D gels. *Faseb Journal* 14(4): A445–A445

Alfano, R. R., Masters, B. R. eds. 2004. *Biomedical Optical Biopsy. Vol. 2. Classic Articles in Optics and Photonics on CD-ROM Series*, Washington, DC: Optical Society of America.

Anderson, R. R., Parrish, J. A. 1981. The optics of human skin. *J. Invest. Dermatol.* 77: 13–19.

Anderson, R. R., Webb, R. H., Rajadhyaksha, M. United States Patent, 5,880,880, March 9, 1999. Three-dimensional scanning confocal laser microscope.

Beebe, D. C., Masters, B. R. 1996. Cell lineage and the differentiation of corneal epithelial cells. *Invest Ophthalmol Vis Sci.* 37: 1815–1825.

Behne, M. J., J. W. Meyer, K. M. Hanson, N. P. Barry, S. Murata, D. Crumrine, R. W. Clegg, E. Gratton, W. M. Holleran, P. M. Elias and T. M. Mauro. 2002. NHE1 regulates the stratum corneum permeability barrier homeostasis. Microenvironment acidification assessed with fluorescence lifetime imaging. *J Biol Chem* 277(49): 47399–47406.

Bertrand, C. 1994. *Développement d'une nouvelle méthode d'imagerie cutanée in vivo par microscopie confocale tandem*. Thése de Doctorat de L'Universite de Saint-Etienne.

Bertrand, C., Corcuff, P. 1994. In vivo spatial-temporal visualization of the human skin by real-time confocal microscopy. *Scanning* 16:150–154.

Bolognia, J.L., Jorizzo, J. L., Rapini, R. P., 2003. *Dermatology*, London: Mosby.

Burkitt, H. G., Young, B., Heath, J. W. 1993. *Wheater's Functional Histology, A text and colour atlas. Third edition*. London: Churchill Livingstone.

Chiu, L. L., C. H. Sun, A. T. Yeh, B. Torkian, A. Karamzadeh, B. Tromberg and B. J. Wong. 2005. Photodynamic therapy on keloid fibroblasts in tissue-engineered keratinocyte-fibroblast co-culture. *Lasers Surg Med* 37(3): 231–264.

Chu, S. W., I. H. Chen, T. M. Liu, C. K. Sun, S. P. Lee, B. L. Lin, P. C. Cheng, M. X. Kuo, D. J. Lin and H. L. Liu. 2002. Nonlinear bio-photonic crystal effects revealed with multimodal nonlinear microscopy. *J Microsc* 208(Pt 3): 190–200.

Chu, S. W., S. Y. Chen, G. W. Chern, T. H. Tsai, Y. C. Chen, B. L. Lin and C. K. Sun. 2004. Studies of chi(2)/chi(3) tensors in submicron-scaled bio-tissues by polarization harmonics optical microscopy. *Biophys J* 86(6): 3914–3922.

Corcuff, P., Bertrand, C., Lévêque, J.-L. 1993. Morphometry of human epidermis in vivo by real-time confocal microscopy. *Arch. Dermatol. Res.* 285, 475–481.

Corcuff, P., Lévêque, J.-L. 1993. In vivo vision of the human skin with the tandem scanning microscope. *Dermatology* 186: 50–54.

Corcuff, P., Gonnord, G., Pierard, G. E., Lévêque, J.L. 1996 In vivo confocal microscopy of human skin: a new design for cosmetology and dermatology. *Scanning* 18(3): 351–355.

Dong, C-Y, Yu, B., Hsu, L., So, P. T. C. 2002. Characterization of two-photon point-spread-function in skin imaging applications. in: *Multiphoton Microscopy in the Biomedical Sciences II*, Ammasi Periasamy, Peter T. C. So, Eds, Proceedings of SPIE Vol. 4620:1–8.

Ehlers, A., Riemann, I., Anhut, T., Kobow, J., König, K. 2005. Multiphoton tomography of epidermis and dermis. in: *Multiphoton Microscopy in the Biomedical Sciences V*, edited by Ammasi Periasamy, Peter T. C. So, eds. Proceedings of SPIE Vol. 5700, 197–204.

Evans, C. L., E. O. Potma, M. Puoris'haag, D. Cote, C. P. Lin and X. S. Xie. 2005. Chemical imaging of tissue in vivo with video-rate coherent anti-Stokes Raman scattering microscopy. *Proc Natl Acad Sci U S A* 102(46): 16807–16812.

Farkas, D. L., Becker, D. 2001. Applications of spectral imaging: detection and analysis of human melanoma and its precursors. *Pigment Cell Res.* 14: 2–8.

Finnin, B. C. and T. M. Morgan. 1999. Transdermal penetration enhancers: applications, limitations, and potential. *J Pharm Sci* 88(10): 955–958.

Garcia-Uribe, A., Kehtarnavaz, N., Marquez, G., Prieto, V., Duvic, M., Wang, L. V. 2004. Skin cancer detection by spectroscopic oblique-incidence reflectometry: classification and physiological origins. *Appl. Opt.* 43(13): 2643–2650.

Gauderon, R., P. B. Lukins and C. J. Sheppard. 2001. Simultaneous multichannel nonlinear imaging: combined two-photon excited fluorescence and second-harmonic generation microscopy. *Micron* 32(7): 685–689.

Goldman, L., Younker, W. 1947. Studies in microscopy of surface of skin. Preliminary report of techniques. *J. Invest. Dermatol.* 9: 11–16.

Goldman, L. 1951. Some investigative studies of pigmented nevi with cutaneous microscopy. *J. Invest. Dermatol.* 16: 407–427.

Gonzalez, S., M. Moran and I. E. Kochevar. 1999. Chronic photodamage in skin of mast cell-deficient mice. *Photochem Photobiol* 70(2): 248–253.

Grewal, B. S., A. Naik, W. J. Irwin, G. Gooris, C. J. de Grauw, H. G. Gerritsen and J. A. Bouwstra. 2000. Transdermal macromolecular delivery: real-time visualization of iontophoretic and chemically enhanced transport using two-photon excitation microscopy. *Pharm Res* 17(7): 788–795.

Hanson, K. M., M. J. Behne, N. P. Barry, T. M. Mauro, E. Gratton and R. M. Clegg. 2002. Two-photon fluorescence lifetime imaging of the skin stratum corneum pH gradient. *Biophys J* 83(3): 1682–1690.

Hanson, K. M. and R. M. Clegg. 2002. Observation and quantification of ultraviolet-induced reactive oxygen species in ex vivo human skin. *Photochem Photobiol* 76(1): 57–63.

Hanson, K. M., R. M. Clegg. 2003. Bioconvertible vitamin antioxidants improve sunscreen photoprotection against UV-induced reactive oxygen species. *J Cosmet Sci* 54(6): 589–598.

Heimbach, D., A. Luterman, J. Burke, A. Cram, D. Herndon, J. Hunt, M. Jordan, W. McManus, L. Solem, G. Warden and et al. 1988. Artificial dermis for major burns. A multi-center randomized clinical trial. *Ann Surg* 208(3): 313–320.

Hendriks, R., Lucassen, G. 2001. Two-photon fluorescence and confocal video microscopy of in vivo human skin. in: *Multiphoton Microscopy in the Biomedical Sciences*, Ammasi Periasamy, Peter T. C. So, eds., Proceedings of SPIE. 4262: 287–293.

Hoffmann, K., Stücker, M., Altmeyer, P., Teuchner, K., Leupold, D. 2001. Selective femtosecond pulse-excitation of melanin fluorescence in tissue. *J. Invest. Dermatol.* 116: 629–630.

Holubar, K., Schmidt, C. 1992. "Dermatology:" the name of the game and its development. *J. Invest. Dermatol.* 98, 403–404.

Hsu, L., Hancewicz, T. M., Kaplan, P. D., Berland, K. M., So, P. T. C. 2001. Two-photon 3-D mapping of tissue endogeneous fluorescence species based on fluorescence excitation spectra. in: *Multiphoton Microscopy in the Biomedical Sciences*, Ammasi Periasamy, Peter T. C. So, eds., Proceedings of SPIE 4262.

Ibbotson, S. H., Moran M., NNash. J. F., and Kochevar I. E. 1999. The effects of radicals compared with UVB as initiating species for the induction of chronic cutaneous photodamage. *J Invest Dermatol* 112(6): 933–938.

Jacques, S. L., Ramella-Roman, J. C., Lee, K. 2002. Imaging skin pathology with polarized light. *J. Biomed. Opt.* 7(3): 329–340.

Kim, D., Kim, K. H., Yazdanfar, S., So, P. T. C. 2005. Optical biopsy in high-speed hand-held miniaturized multifocal multiphoton microscopy. in: *Multiphoton Microscopy in the Biomedical Sciences V*, Ammasi Periasamy, Peter T. C. So, eds. Proceedings of SPIE 5700: 14–22.

Kim, K. H. , So, P. T. C., Kochevar, I. E. , Masters, B R. , Gratton, E. 1998. Two-photon fluorescence and confocal reflected light imaging of thick tissue structures. in: *Optical Investigations of Cells In Vitro and In Vivo*, Daniel L. Farkas, Robert C Leif, Bruce J. Tromberg, eds., Proceedings of SPIE. 3260: 46–57.

Kochevar, I. E., Moran, M., and Granstein, R. D. 1994. Experimental photoaging in C3H/HeN, C3H/HeJ, and Balb/c mice: comparison of changes in extracellular matrix components and mast cell numbers. *J Invest Dermatol* 103(6): 797–800.

Kollias, N., R. Gillies, M. Moran, Kocheva, I. E., Anderson, R. R. 1998. Endogenous skin fluorescence includes bands that may serve as quantitative markers of aging and photoaging. *Journal of Investigative Dermatology* 111: 776–780.

König, K, Wollina, U., Riemann, I., Peuckert, C., Halbhuber, K-J., Konrad, H., Fischer, P., Fünfstück, V., Fischer, T. W. , Elsner, P. 2002. Optical tomography of human skin with subcellular spatial and picosecond time resolution using intense near infrared femtosecond laser pulses, in: *Multiphoton Microscopy in the Biomedical Sciences II*, Ammasi Periasamy, Peter T C. So, eds., Proceedings of SPIE, 4620: 191–196.

König, K, Riemann, I. High-resolution multiphoton tomography of human skin with subcellular spatial resolution and picosecond time resolution. 2003. *J. Biomed. Opt.* 8(3): 432–439.

Kumar, V., Abbas, A. K., Fausto, N., 2005. *Robbins and Cotran Pathologic Basis of Disease*, Philadelphia: Elsevier.

Laiho, L., H., Pelet, S., Hancewicz, T. M., Kaplan, P. D., So, P. T. C. 2005. Two-photon 3-D mapping of ex vivo human skin endogenous fluorescence spies based on fluorescence emission spectra. *J. Biomed. Opt.* 10(2): 024016–1–10.

Limat, A., Mauri, D., Hunziker, T. 1996. Successful treatment of chronic leg ulcers with epidermal equivalents generated from cultured autologous outer root sheath cells. *J Invest Dermatol* 107: 128–135.

Masters, B.R. 1996. Three-dimensional confocal microscopy of human skin in vivo: autofluorescence of normal skin. *Bioimages* 4(1): 1–7.

Masters, B. R., Aziz, D. J., Gmitro, A. F., Kerr, J. H., O'Grady, T. C., Goldman, L. 1997. Rapid observation of unfixed, unstained human skin biopsy specimens with confocal microscopy and visualization. *J. Biomed. Opt.* 2(4): 437–445. This paper provides a noninvasive optical technique for the rapid evaluation of biopsy specimens without the tissue preparation of fixing, sectioning and staining) that is required with conventional histological examination.

Masters, B. R., Gonnord, G., Corcuff, P. 1996. Three-dimensional microscopic biopsy of in vivo human skin: a new technique based on a flexible confocal microscope. *J. Microscopy*, 185(3): 329–338.

Masters, B. R., So, P. T. C. 1999. Multi-photon excitation microscopy and confocal microscopy imaging of in vivo human skin: a comparison. *Microsc. Microanal.* 5: 282–289.

Masters, B. R., So, P. T. C., 2001. Confocal microscopy and multi-photon excitation microscopy of human skin in vivo. *Optics Express*, 8(1): 2–10.

Masters, B. R., So, P. T. C., Buehler, C., Barry, N., Sutin, J. D., Mantulin, W. W., Gratton, E. 2004. Mitigating thermal mechanical damage potential during two-photon dermal imaging. *J. Biomed. Opt.* 9(6): 1265–1270.

Masters, B. R., So, P. T. C., Gratton, E. 1997. Multiphoton excitation fluorescence microscopy and spectroscopy of in vivo human skin. *Biophys. J.* 72: 2405–2412.

Masters, B. R., So, P. T. C., Gratton, E. 1998a. Multiphoton excitation microscopy of in vivo human skin. *Ann. N. Y. Acad. Sci.* 838: 58–67.

Masters, B. R., So, P. T. C., Gratton, E. 1998b. Optical biopsy of in vivo human skin:multi-photon excitation microscopy. *Lasers Med Sci* 13: 196–203.

Masters, B. R., So, P. T. C., Kim, K. H., Buehler, C., Gratton, E. 1999. Multiphoton excitation microscopy, confocal microscopy, and spectroscopy of living cells and tissues: functional metabolic imaging of human skin in vivo, in *Methods in Enzymology, Confocal Microscopy*, vol. 307, P. M. Conn, ed., New York: Academic Press, Chapter 29, pp. 513–536.

Mohler, W., Millard, A. C., Campagnola, P. J. 2003. Second harmonic generation imaging of endogenous structural proteins. *Methods* 29(1): 97–109.

Navarro, F.A., So, P.T. Nirmalan R., Kropf, N, Sakaguchi F. Park, C. S., Lee, H. B and Orgill, D. P. 2004. Two-photon confocal microscopy: a nondestructive method for studying wound healing. *Plast Reconstr Surg* 114(1): 121–128.

New, K. C., Petroll, W. M., Boyde, A., Martin, L., Corcuff, P., Lévêque, J. L., Lemp, M. A., Cavanagh, H. D., Jester, J. V. 1991. In vivo imaging of human teeth and skin using real-time confocal microscopy, Scanning, 13: 369–372.

Peck, K. D, Hsu, J., Li, S. K, Ghane, A. H., Higuchi, W. I. 1998. Flux enhancement effects of ionic surfactants upon passive and electroosmotic transdermal transport. *J Pharm Sci* 87(9): 1161–1169.

Pelet, S., Previte, M. J., Laiho, L. H, So, P. T. 2004. A fast global fitting algorithm for fluorescence lifetime imaging microscopy based on image segmentation. *Biophys J* 87(4): 2807–2817.

Rajadhyaksha, M., Grossman, M., Esterowitz, D., Webb, R.H., and R. Rox Anderson. 1995. *In vivo* confocal scanning laser microscopy of human skin: melanin provides strong contrast. J. Invest. Dermatol. 104: 946–952.

Rajadhyaksha, M., Anderson, R. R., Webb, R.H. 1999. Video-rate confocal scanning laser microscope for imaging human tissue in vivo. *Appl. Opt.* 38(10): 2105–2115.

Rajadhyaksha, M., González, S., Zavislan, J. M., Anderson, R. R., Webb, R.H. 1999. *In vivo* confocal scanning laser microscopy of human skin II: advances in instrumentation and comparison with histology. *J. Invest.Dermatol.* 113: 293–303.

Rajadhyaksha, M., Menaker, G., Flotte, T., Dwyer, P. J., González, S. 2001. Confocal examination of nonmelanoma cancers in thick excisions to petentially guide Mohs micrographic surgery without frozen histopathology. *J. Invest. Dermatol.* 117: 1137–1143.

Sardar, D. K., Mayo, M. L., Glickman, R. D. 2001. Optical characterization of melanin. *J. Biomed. Opt.* 6(4): 404–411.

So, P. T. C., Dong, C. Y., Masters, B. R., Berland, K. M. 2000. Two-photon excitation fluorescence microscopy. *Annu. Rev. Biomed. Eng*, 399–429.

So, P. T. C., Kim, H., Kochevar, I. E. 1998. Two-photon deep tissue ex vivo imaging of mouse dermal and subcutaneous structures, *Optics Express*, 3(9): 339–350.

Stenn, K. S., 1988. *The skin, in: Cell and Tissue Biology, A textbook of histology*, Leon Weiss, ed., sixth edition, Baltimore: Urban & Schwarzenberg.

Sun, Y., Su, J-W, Lo, W., Lin, S-J.,Jee, S-H., Dong, C-Y, 2003. Multiphoton polarization imaging of the stratum corneum and the dermis in ex-vivo human skin. *Optics Express*, 11(25): 3377–3384.

Sun, C. K. 2005. Higher harmonic generation microscopy. *Adv Biochem Eng Biotechnol* 95: 17–56.

Teuchner, K., Freyer, W., Leupold, D., Volkmer, A., Birch, D. J. , Altmeyer, P., Stücker, M., Hoffmann, K. 1999. Femtosecond two-photon excited fluorescence of melanin. *Photochem. Photobiol.* 70(2): 146–151.

Teuchner, K., Ehlert, J., Freyer, W., Leupold, D., Altmeyer, P., Stücker, M., Hoffmann, K. 2000. Fluorescence studies of melanin by stepwise two-photon femtosecond laser excitation. *Journal of Fluorescence*, 10(3): 275–281.

Teuchner, K., Mueller, S., Freyer, W., Leupold, D., Altmeyer, P., Stücker, M., Hoffmann, K. 2003. Femtosecond two-photon excited fluorescence of melanin. *Multiphoton Absorption and Nonlinear Transmission Processes: Materials, Theory and Applications*, Proceedings of SPIE Vol. 4797: 211–219.

Torkian, B. A., A. T. Yeh, R. Engel, C. H. Sun, B. J. Tromberg and B. J. Wong. 2004. Modeling aberrant wound healing using tissue-engineered skin constructs and multiphoton microscopy. *Arch Facial Plast Surg* 6(3): 180–187.

Troy, T. L., Thennadil, S. N. Optical properties of human skin in the near infrared wavelength range of 1000 to 2200 nm. J. Biomed. Opt. 6(2): 167–176, (2001).

Tung, C-K., Chiu, T-K, Lo, W., Wnag, P-H., Lin S-J, Jee, S-H, Dong, C. Y. 2003. Effects of index mismatch induced spherical aberration on two-photon imaging in skin and tissue-like constructs. *Multiphoton Microscopy in the Biomedical Sciences III*, Ammasi Periasamy, Peter T. C. So, eds., Proceedings of SPIE. 4963: 95–104.

Werrlein, R. J., Madren-Whalley, J. S. 2003. Multiphoton microscopy: an optical approach to understanding and resolving sulfur mustard lesions. *J. Biomed. Opt.* 8(3): 396–409.

Werrlein, R. J., Braue, C. R. 2004. Sulfur mustard disrupts human a3b1 (alpha sub 3, beta sub 1)-integrin receptors in concert with a6b4 (alpha sub 6 beta sub 4) integrin receptors and collapse of the keratin K5/K14 cytoskeleton. in: *Multiphoton Microscopy in the Biomedical Sciences IV*, Ammasi Periasamy, Peter T. C. So, eds., Proceedings of SPIE. 5323: 297–305.

Werrlein, R J., Braue, C R., Dillman, J. F. 2005. Multiphoton imaging of the disruptive nature of sulfur mustard lesions. in: *Multiphoton Microscopy in the Biomedical Sciences V*, Ammasi Periasamy, Peter T. C. So,eds. , Proceedings of SPIE. 5700: 240–248.

White, W. M., Rajadhyaksha, M., González, S., Fabian, R. L., Anderson, R R. 1999. Non-invasive imaging of human oral mucosa in vivo by confocal reflectance microscopy. *The Laryngoscope*, 109: 1709–1717.

Wong, W. R., S. Kossodo and I. E. Kochevar. 2001. Influence of cytokines on matrix metalloproteinases produced by fibroblasts cultured in monolayer and collagen gels. *J Formos Med Assoc* 100(6): 377–382.

Yannas, I. V. 2000. Synthesis of organs: in vitro or in vivo? *Proc Natl Acad Sci U S A* 97(17): 9354–9356.

Yannas, I. V. 2005. Facts and theories of induced organ regeneration. *Adv Biochem Eng Biotechnol* 93: 1–38.

Yannas, I. V., Burke, J. F., Orgill, D. P, Skrabut, E. M. 1982. Wound tissue can utilize a polymeric template to synthesize a functional extension of skin. *Science* 215(4529): 174–176.

Yeh, A. T., Kao B., Jung W. G., Chen, Z, Nelso, J. S., Tromber, B. J. 2004. Imaging wound healing using optical coherence tomography and multiphoton microscopy in an in vitro skin-equivalent tissue model. *J Biomed Opt* 9(2): 248–253.

Yu, B., Dong, C. Y., So, P. T. C., Blankschtein D., Langer R.. 2001. In vitro visualization and quantification of oleic acid induced changes in transdermal transport using two-photon fluorescence microscopy. *J Invest Dermatol* 117(1): 16–25.

Yu, B., Hean Kim, K., So, P. T., Blankschtein, D., Langer, R. 2002. Topographic heterogeneity in transdermal transport revealed by high-speed two-photon microscopy: determination of representative skin sample sizes. *J Invest Dermatol* 118(6): 1085–1088.

Yu, B., Kim K. H., So, P. T., Blankschtein, D., Langer, R. 2003. Visualization of oleic acid-induced transdermal diffusion pathways using two-photon fluorescence microscopy. *J Invest Dermatol* 120(3): 448–455.

33

Cellular Metabolism Monitored by NAD(P)H Imaging with Two-Photon Excitation Microscopy

Barry R. Masters

33.1 INTRODUCTION

The history, techniques, limitations, and applications of the optical method to monitor oxidative metabolism based on autofluorescence of cells are the subjects of this chapter. In particular, microscopic techniques based on one- and two-photon excitation are compared. The intent is to provide a critical account of the methodology and its limitations, and to describe some of the important applications that are based on two-photon excitation microscopy.

The concept of optical monitoring of redox "states" or reduction–oxidation "states" of a chemical reaction rests on the physical chemical principle of an electrochemical cell and the relation of the net Gibbs free energy to the electromotive force (EMF) that drives the net chemical reactions in each half of the electrochemical cell. The equilibrium constant of a set of coupled chemical reactions is determined by the Gibbs free energy of each chemical reaction. In electrochemistry, the ratio of the concentrations, or more correctly the activities (concentration of a chemical species multiplied by its activity coefficient), determines the EMF, which can be measured with a voltmeter.

In principle the ratio of the activities could alternatively be measured by an optical technique (i.e., absorption or fluorescence measurements). An optical technique that can measure the concentration or activity ratio of a coupled chemical reaction that has an equilibrium value dependent on the ratio of an oxidized and a reduced molecule requires that the two forms of the molecules have different optical properties (i.e., different absorption coefficients) or different emission properties, such as quantum yield or fluorescent lifetimes. We review absorption studies and then microspectrofluorometry as an introduction to the "optical method" (Chance, 1991). The term "optical method" refers to Chance's use of optical techniques to monitor chemical concentrations within cells or biological fluids. In the 1930s Warburg coined the term "optical method." Prior to that discussion we include a brief review of oxidative metabolism.

33.1.1 Oxidative Metabolism

First we give some nomenclature of some biomolecules. Nicotinamide adenine dinucleotide exists in the oxidized form as NAD^+ and in the reduced form as NADH. Similarly, nicotinamide adenine dinucleotide phosphate exists in the oxidized form as

$NADP^+$ and in the reduced form as NADPH. The coenzyme flavin adenine dinucleotide exists in the oxidized form as FAD and in the reduced form as $FADH_2$.

In mitochondria the production of ATP is optimized by transferring electrons from NADH and $FADH_2$ via a series of electron carriers: the electron transport chain or the respiratory chain. At each electron transfer step a small amount of the free energy in NADH or $FADH_2$ is released. In the process of oxidative phosphorylation, this free energy is stored as a proton concentration gradient and as an electric potential difference across the inner mitochondrial membrane. In the mitochondria, the synthesis of ATP from ADP and P_i (inorganic phosphate) is driven by these gradients, in which the movement of protons back across the inner membrane is coupled to mitochondrial ATP production.

Now we explore some historical developments in the use of optical methods to monitor cellular oxidative metabolism.

33.1.2 Absorption Studies

Historically, the first use of optical properties to identify and quantify biological molecules in biological fluids (i.e., blood and urine) and cells was based on the absorption of specific wavelengths of light (MacMunn, 1880, 1914). These early studies were made in the transmission mode of optical microscopes. Later researchers began to characterize the fluorescence properties of biological tissues, and as specific biological molecules were isolated and purified their absorption and emission (spectral and lifetime) properties were characterized.

In an important early example, the ultraviolet absorption bands of the mitochondrial cytochromes were investigated by Keilin, who initiated this research in the early 1920s (Keilin, 1966). David Keilin used a low-dispersion prism spectroscope fitted to a Carl Zeiss microscope. Keilin studied slices of plant and animal tissues as well as suspensions of bacterial and yeast. Keilin observed the oxidation and reduction of cytochromes within living tissues and cells based on changes in the absorption spectrum of visible light.

In the early 1930s Otto Warburg constructed an absorption spectrophotometer that worked in the wavelength range from the ultraviolet to the visible. Warburg investigated substrate-level phosphorylation by measuring the change in absorption at 334 nm that occurred from the reduction of pyridine nucleotides. That work resulted in the next generation of tissue spectrometers developed in Sweden (T. Caspersson, 1950, 1954) and in the United States (Chance & B. Thorell, 1959; Chance, 1951). In 1934, Hugo Theorell had a duplicate of Warburg's spectrophotometer constructed in his laboratory in Stockholm.

Caspersson studied the ultraviolet absorption bands of DNA in cell nuclei in both normal and cancer cells. The advantage of microspectrographic methods is that cellular and intracellular organelles can be studied in the living state, in real time, with noninvasive optical techniques. Today variations of the optical method are important in tissue diagnostics (Alfano & Masters, 2004; Richards-Kortum & Sevick-Muraca, 1996).

Another important example of optical studies of oxidative metabolism in tissues is Glenn Millikan's tissue spectrometer, developed from 1937 to 1941. He was able to measure hemoglobin and myoglobin deoxygenation in the stimulated or ischemic cat soleus muscle (Millikan, 1942). Glenn Millikan showed the way for Britton Chance and other investigators to use optical measurements to monitor tissue oxygenation. An early important study demonstrated the strong absorption band of NADH in yeast cells that could be altered by oxidizing and reducing substrates, and the concentration changes could be detected by absorption spectrophotometry (Chance & Williams, 1955).

The pulse oximeter, first developed by Glenn Millikan at the Johnson Research Foundation, University of Pennsylvania, in 1942, is based on the absorption spectra of blood and is used to determine the ratio of oxygenated and deoxygenated hemoglobin, thus providing an optical indicator of tissue oxygenation. Today the pulse oximeter is a common device in the hospital.

33.1.3 Fluorescence Studies

The origins of intracellular microspectrofluorometry began with the important work of Otto Warburg, who discovered NAD(P)H fluorescence. In the 1930s Warburg coined the term "optical method." Between 1938 and 1943 Warburg studied the NAD(P)H fluorescence of living cells, in which the ultraviolet excitation resulted in fluorescence with an emission peak at 460 nm. In 1951 L. N. M. Duysens in Utrecht investigated the fluorescence emission from yeast cells for a Ph.D. thesis.

Drs. Bo Thorell and Britton Chance, and Herrick Baltscheffsky discovered that the cell autofluorescence was due to NAPH and NADPH, written as NAD(P)H, which are in the mitochondrial matrix space (Chance & Baltscheffsky, 1958). In 1959, Chance and Thorell were able to measure the fluorescence from a single mitochondrion within a single cell. The NAD(P)H fluorescence intensity occurs in two cellular compartments, the mitochondrial and the cytosolic; this duality confounds the interpretation of fluorescence studies. However, in some tissues (e.g., cardiac myocytes), the NAD(P)H fluorescence is predominantly from the mitochondrial space. A wide variety of instrumentation for the study of cellular metabolism in living cells and tissues is the subject of a review by Masters and Chance (Masters & Chance, 1999).

It was the characterization of the fluorescence properties of the reduced and the oxidized pyridine nucleotide, NAD(P), and the flavoproteins, FAD, both key components of oxidative metabolism, that permitted the use of fluorescence signals to optically monitor the concentration of NAD(P)H in biological cells and tissues (Chance, 1976, 1991; Chance et al., 1979). As Chance explained in his seminal paper that described experiments from isolated mitochondria at temperatures below $-80°C$, it is possible to perform oxidation–reduction ratio studies in freeze-trapped samples in two and three dimensions (Chance et al., 1979). Chance and his coworkers assumed that the flavoproteins and the NAD(P)H signals are confined to the mitochondrial matrix space and that they represent the major fluorochromes that contribute to the fluorescence. Chance then stated the second set of assumptions: the ratio of the two fluorescence intensities from the flavoproteins and the NAD(P)H is near oxidation–reduction equilibrium, and that this ratio, suitably normalized, approximated the oxidation–reduction ratio of the oxidized flavoproteins/NAD(P)H. Chance states that the goal of the ratio measurements is to develop a more precise indicator of the oxidation–reduction state than is possible with the measurements of the NAD(P)H signal alone. In addition, the authors assume that the ratio of the two fluorescent intensities would permit a measure of the oxidation–reduction state that is independent of the factors that would affect the measurement of flavoprotein or NAD(P)H fluorescence alone. While these assumptions are all reasonable, there is a paucity of experimental evidence, if any at all, that validates these assumptions. Therefore, the field is confounded by the difficulty in interpreting measurements of NAD(P)H fluorescence from cells and tissues.

The intellectual jump from the measurements of NAD(P)H fluorescence in cells and tissues to the determination of "redox states" derives from the previous discussion on the EMF and Gibbs free energy of electrochemical cells. In this chapter we will re-examine the validity of both the measurements and their interpretation as well as the limitations of the methodology.

It will be shown that optical "redox" imaging of cells, tissues, and organs is based on the intrinsic fluorescent probes of cellular metabolism (Eng et al., 1989; Evans et al., 2005). Cellular metabolism can be noninvasively monitored through the "optical method" based on the fluorescence intensity of intrinsic probes (Chance, 1991). The intrinsic fluorescence probes that report on cellular metabolism are the reduced pyridine nucleotides and the oxidized flavoproteins.

Although the "optical method" facilitates the noninvasive monitoring of cell metabolism, the methodology and the interpretation of the measurements is confounded by a variety of difficulties. The most problematical are the lack of linearity between fluorescence intensity measurements and the concentration of the fluorophore, the duality of compartments of the fluorophores, and the inability to accurately determine the concentration or activity ratios of the reduced and the oxidized forms of the fluorophores NAD(P)H and its oxidized form NAD^+, or the reduced and oxidized forms of the flavoproteins. While the cellular fluorescence of the reduced form NAD(P)H is readily measured, the extremely low fluorescence from the oxidized form, NAD^+, is difficult to measure, and thus their ratio is difficult to determine in cells and in vivo tissues. While NADH and NADPH are fluorescent, the oxidized forms, NAD^+ and NAD^+P, are not fluorescent; therefore, changes in the ratio of the oxidized to the reduced forms result in changes in the intensity of the autofluorescence signal. The absorption and emission spectra of NADH and NADPH are similar, and both forms are denoted as NAD(P)H. The fluorescence from the reduced form NAD(P)H is much greater than from the oxidized form $NAD(P)^+$. NAD(P)H has a low quantum yield and a small absorption cross-section and its absorption is in the ultraviolet region of the spectrum.

Oxidative metabolism is used by biochemists to describe the intracellular oxidation of substrates that are coupled with the production of ATP and oxidized coenzymes (NAD and FAD). The autofluorescence from NAD(P)H can also be used to investigate cell and tissue glucose metabolism. During glucose metabolism NAD(P)H can increase in both the cell cytoplasmic and the mitochondrial compartments. Glucose metabolism can occur by way of glycolysis or in the citric acid cycle; in both pathways NAD(P)H levels will increase and that increase will result in an increase in the intensity of the autofluorescence that is related in a complex fashion to the concentrations in both cell compartments. A highly recommended critical review of the field is available (Masters & Chance, 1999).

In the next section we review a recent paper that compares the two-photon spectroscopy of NAD(P)H and flavoprotein in solution and in cells. Then we will discuss and compare single-photon and two-photon excitation of NAD(P)H in a variety of tissues together with microanalytical techniques in an attempt to validate the "optical method."

33.2 TWO-PHOTON FLUORESCENCE SPECTROSCOPY AND MICROSCOPY OF NAD(P)H AND FLAVOPROTEIN

33.2.1 Introduction

The spectral characterization of NAD(P)H and flavoproteins (FPs) measured with live cells is extremely important for the correct interpretation of two-photon excitation microscopy. Flavins and flavin adenine dinucleotides (FADs) and mononucleotides (FMNs) are cofactors for a large group of enzymes involved in oxidation–reduction reactions. The oxidized flavoproteins and the reduced NAD(P)H are fluorescent, and therefore ratiometric fluorometry can be used to monitor cell and tissue metabolism with minimal interference from absorption of other chromophores, light scattering in tissues, and variation in mitochondrial density and flavoprotein concentrations. Single-photon excitation required two wavelengths to excite each of the chromophores. The advantage of two-photon excitation is that a single wavelength can be used to excite both NAD(P)H and flavoproteins (Xu & Webb, 1997). Flavoprotein fluorescence has several advantages compared to NAD(P)H fluorescence. It can be excited over longer excitation wavelengths, it is more resistant to photobleaching, and it is located in the mitochondrial space.

33.2.2 Techniques

Huang and associates have studied both in solution and in isolated adult dog cadiomyocytes the two-photon excitation spectra of NAD(P)H, FAD, and lipoamide dehydrogenase (LipDH) over the wavelength range of 720 to 1,000 nm (Huang et al., 2002). The two-photon excitation spectra of these compounds were measured using the techniques previously described (Xu & Webb, 1997). Multiphoton excitation laser-scanning microscopy was performed with a multiphoton microscope previously described (Williams, Webb 2000).

33.2.3 Summary of Key Findings

1. The autofluorescence from cells is mainly due to contributions from intra-cellular NAD(P)H, riboflavin, flavin co-enzymes, and flavoproteins that are located in the mitochondria (unvalidated assumption by Huang et al.). Co-enzymes fluoresce in the reduced form, NAD(P)H, and do not fluoresce in the oxidized form, NAD(P). The flavins fluoresce in the oxidized form, FAD, and do not fluoresce in the reduced form, $FADH_2$. NAD(P)H is produced by several enzymes and is used in several metabolic processes: in the respiratory chain and in biosynthesis as well as detoxification reactions.
2. The use of multiphoton excitation microscopy permits both NAD(P)H and flavoproteins to be simultaneously excited with one near-infrared wavelength.
3. Confounding factors such as inner-filter effects, poor depth of penetration of exciting light into highly scattering tissues, and light-scattering effects that are associated with single photon redox fluorometry can be minimized.
4. Multiphoton excitation microscopy improves the optical sectioning capability compared to single-photon excitation.

5. Single ratiometric fluorescence techniques cannot differentiate the mitochondrial NAD(P)H fluorescence from the cytosolic NAD(P)H signal. With the submicron resolution of multiphoton excitation microscopy, the mitochondrial NAD(P)H fluorescence can be isolated spatially.
6. With two-photon excitation of both the NAD(P)H and flavoproteins, it is difficult to separate the overlap of the signals.
7. The authors state, but do not demonstrate, that two-photon excitation ratio imaging of NAD(P)H and flavoproteins may be improved by the use of lifetime images, since lifetimes are not subject to the multitude of confounding problems that make intensity measurement difficult to interpret.

33.2.4 Limitations of Technique and Interpretation

The steady state and the time-resolved emission spectroscopy of reduced nicotinamides, reduced β-nicotinamide adenine dinucleotide (NADH), and reduced β-nicotinamide mononucleotide (NAMH) using one- and two-photon excitation are critical for the interpretation of tissue autofluorescence (Kierdaszuk et al., 1996). NADH has a single-photon absorption maximum near 350 nm and a broad emission centered near 450 nm. The two-photon excitation of NADH occurs in the wavelength region of 570 to 740 nm. Although NADH has a low two-photon cross-section, there is an advantage of single-photon ultraviolet excitation: the long wavelengths used for two-photon excitation result in reduced background autofluorescence in biological samples.

The fluorescence intensity and the quantum yield of NADH increase with binding to dehydrogenases (Velick, 1958). The fluorescence lifetime of NADH in solution in the unbound state is 0.4 ns; in the bound state with proteins the lifetime increases to 1 to 5 ns (Gafni & Brand, 1976; Lakowicz et al., 1992).

The key questions for both one-photon and two-photon excitation of NAD(P)H are: Is the fluorescence signal proportional to the concentration of NAD(P)H, and does the total measured intensity derive from different pools of NAD(P)H?

33.3 LINEAR OPTICAL METHODS TO NONINVASIVELY MONITOR CELL METABOLISM

33.3.1 Applications in Ophthalmology

The absolute measurement of NAD(P)H concentrations in both the cytoplasmic and the mitochondrial compartment of cells confounds the interpretation of the alterations of the fluorescence intensity of NAD(P)H following external biochemical alterations of cell metabolism. Nevertheless, it is possible to carefully design biological experiments that can be calibrated and thus provide real-time, noninvasive biological information.

Autofluorescence from frozen corneas demonstrated evidence of mitochondrial fluorescence and differing metabolic conditions in epithelium and endothelium (Chance & Lieberman, 1978; Nissen et al., 1980). The next advance was to demonstrate that in the living ex vivo rabbit cornea (Masters et al., 1981; Masters, Falk & Chance, 1982; Masters, Chance & Fischbarg, 1982; Masters, 1984a, 1984b, 1985, 1993; Masters & Chance, 1999).

We now describe a series of papers that investigate the single-photon excitation of NAD(P)H in ocular tissues: a paper that illustrates how to use external calibration in order to calibrate the "optical method," a paper that shows the limits of confocal microscopy with single-photon ultraviolet excitation of ex vivo corneal tissues, and finally two papers that compare the "optical method" or single-photon excitation fluorescence to monitor NAD(P)H concentrations with microanalytical techniques based on enzyme cycling methods that provide the amounts of various enzyme cofactors such as NAD^+, NADH, $NADP^+$, and NADPH.

An example of this paradigm is the in vivo study of the effects of contact lenses on the oxygen concentration and epithelial oxidative metabolism of a living rabbit (Masters, 1988). The key message is that appropriate experimental design, instrumentation, and analysis can provide useful biological information in studies that employ NAD(P)H fluorescence as an indicator of cellular oxidative metabolism. The purpose of the study on rabbits is to demonstrate that the oxygen concentration at the corneal epithelial surface under various contact lenses can be noninvasively monitored in vivo, and the resulting degree of hypoxia can be determined and correlated with the oxygen transmission properties of the contact lens.

33.3.2 Techniques

An optically sectioning, fluorometer microscope was constructed with a compact nitrogen laser as the excitation source of light pulses at 337 nm. A fiberoptic coupled the laser to the microscope. The microscope depth resolution was 6 μm with the 100×, 1.2 NA water immersion objective. The scattered light and the NAD(P)H fluorescence from the cornea were measured as functions of depth in the cornea.

The effect of various oxygen concentrations on the corneal epithelium was first calibrated with a series of oxygen mixtures. After equilibration with hydrated air the NAD(P)H fluorescence was measured, and then a hydrated calibration gas mixture with a known percentage of oxygen was equilibrated with the cornea and the change in NAD(P)H fluorescence was measured. A series of contact lenses were made with calibrated values of oxygen transmissibility and thickness. The lenses were placed on the cornea and the baseline and final equilibrium values of the NAD(P)H of the corneal epithelium were measured.

From the calibration curves, and the knowledge of the oxygen transmissibility and contact lens thickness, the following quantities were calculated: oxygen flux (μL/cm^2/min) measured with 21% oxygen gradient across the contact lens, percent oxygen at the epithelial surface, and percent hypoxia of the corneal epithelium.

33.3.3 Summary of Key Findings with One-Photon Excitation

1. The oxygen concentration under the various contact lenses was very similar to the values obtained with a platinum oxygen electrode placed under the contact lens.
2. The human cornea is different from the rabbit cornea in its oxygen consumption rate and in the blink frequency rate. The higher blink frequency of the human cornea results in a higher aeration of the corneal epithelium.

3. The results are dependent on the tightness of the fit of the contact lens. In the present study all of the various contact lenses differed only in material composition, size, and shape, and the fit was identical for all contact lenses.

4. The results of the study indicate that even the most permeable contact lens resulted in an epithelial hypoxia of 7.7%. The steady-state oxygen concentration under this contact lens was 7.2%, which is similar to that under the closed lid corneal epithelium.

5. This study show how optical measurements of NAD(P)H fluorescence intensity can be used in a calibrated manner to determine cell and tissue oxygenation in vivo.

Now we present a study illustrating the limits of single-photon confocal imaging of ex vivo cornea based on fluorescence of NAD(P)H as well as reflection imaging at the same ultraviolet wavelength. This example of single-photon, ultraviolet, three-dimensional confocal fluorescence microscopy of the in vitro rabbit cornea is presented as a benchmark for redox metabolic imaging (Masters, Kriete, & Kukulies, 1993).

33.4 IMAGING NAD(P)H WITH SINGLE-PHOTON CONFOCAL MICROSCOPY

33.4.1 Techniques

A Zeiss laser scanning confocal microscope was fitted with two argon-ion lasers that provided wavelengths in the regions of 364, 488, and 514 nm. A Zeiss water objective of 25×, with a numerical aperture of 0.8 corrected for the UV, was used to image optical sections through the full thickness of the ex vivo rabbit cornea. A confocal microscope was used in both the reflected and the fluorescent modes to image in situ epithelial and endothelial cells. An excitation wavelength of 364 nm and emission at 400 to 500 nm were used to image the fluorescence from the reduced pyridine nucleotides. First the 364-nm light was used to image in the reflected-light mode the endothelial and epithelial cell layers, and then the microscope was switched to the fluorescent mode and images were acquired with emission light in the 400- to 500-nm range. These wavelengths correspond to the excitation and the emission of the reduced pyridine nucleotides NAD(P)H. Reflected-light images and XZ-scans of the anterior cornea were obtained with 488-nm light. In order to suppress the central bright reflected-light spot on each image, which was due to reflection of stray light from the collection lens, the collection lens in front of the photomultiplier was slightly tilted.

33.4.2 Summary of Key Findings with One-Photon Excitation

1. To demonstrate the optical quality of the laser-scanning microscope with UV light, images were acquired of the in vitro cornea that were perpendicular to the corneal surface. With 488-nm reflected light (backscattered light) we observed in the full thickness of the stroma small highly reflecting fibers about 1μ m in diameter within the stromal fibroblast cells.

2. The superficial epithelial and the endothelial cells were imaged with 364-nm light in both the reflected light mode and the fluorescent light mode using the emission in the region of 400 to 500 nm. For both epithelial cells and the

endothelial cells the fluorescent mode showed bright regions from the reduced pyridine nucleotides, and dark regions due to the nuclei.

3. This study demonstrates for the first time the feasibility of UV confocal fluorescence microscopy for imaging the intrinsic reduced pyridine nucleotides in cells of the corneal epithelium and endothelium in an in vitro rabbit cornea. The imaging of corneal cells in both reflected light and with 364-nm excitation provides an optical technique to investigate the heterogeneity of cellular respiration in the corneal layers.

33.4.3 Validation and Limitations of the Technique

Invasive and destructive enzyme cytochemical techniques have been applied for metabolic mapping in living cells (Broonacker & Van Noorden, 2001). In order to validate in situ corneal redox fluorometry, the redox state and phosphorylation potential of free trapped rabbit corneal epithelial and endothelium were studied using quantitative histochemical methods (Masters, Riley, Fischbarg & Chance, 1983; Masters, Ghosh, Wilson & Matschinsky, 1989). The results were compared with noninvasive measurements using an optically sectioning fluorometer microscope designed and constructed by Masters (Masters, 1988).

We now present a paper that compares the "optical method" of fluorescent monitoring of NAD(P)H levels with microanalytical techniques to directly measure the amounts of the pyridine nucleotides in ocular tissues.

Cyanide is a reversible respiratory inhibitor that binds to the ferric form of cytochrome oxidase in the mitochondrial respiratory chain and prevents the generation of ATP by oxidative phosphorylation. Since NADH is continuously synthesized and there is no functioning pathway to oxidize it in the presence of cyanide, the concentration of NADH increases. A decrease in the availability of oxygen causes a similar result. A second oxidative pathway in the cornea is the hexose–monophosphate shunt, which converts glucose to pentose and leads to the reduction of $NADP^+$ to NADPH. NADPH is required for glutathione-coupled redox reactions, providing protection against oxidative damage for the corneal cells.

33.4.4 Techniques

Fresh enucleated rabbit eyes were frozen in freon-12, cooled by liquid nitrogen, or exposed for 1 hour in 1 mM NaCN to block oxidation and then freeze-trapped. Corneas were sectioned and freeze-dried and samples of individual layers were dissected, weighed, and analyzed for NADH, NAD^+, NADPH, $NADP^+$, ATP, ADP, and P_i. The microanalytical techniques used sample weights between 0.05 and 0.1 μg. The concentrations of pyridine and adenine nucleotides are expressed in units of millimoles per kilogram of dry weight of cornea. The fluorescence of the corneal epithelium and endothelium in the normoxic state and in the anoxic state induced with cyanide was measured with an optically sectioning microscope designed and constructed in the laboratory (Masters, 1988).

33.4.5 Summary of Key Findings

1. The aerobic epithelium showed a ratio for NAD^+/NADPH of 1.85 ± 0.08; in anoxia this ratio was 0.68 ± 40.2. The aerobic epithelium and endothelium

contain two redox couples, $NAD^+/NADH$ (I) and $NADP^+/NADPH$ (II); the former is predominantly oxidized and the latter is predominantly reduced.

2. Cyanide-induced anoxia caused the fluorescence from the corneal epithelium to be increased by $+153\%$ compared to the normoxic condition. There was no significant change in the fluorescence intensity from the stromal region. The lack of any change in the stromal region could be due to the extremely low signal from the NADH in the stromal keratocytes compared to the background signal.

3. Comparison of microchemical analysis with optical measurement of intensity from the corneal epithelium is still unresolved. Anoxia resulted in the epithelial NADH concentration increasing by $+16\%$; however, the fluorescence intensity increased by 158%, indicating a large disparity between the two techniques.

4. The fluorescence intensity is proportional to the quantum yield. The quantum yield of the reduced pyridine nucleotides is significantly increased when it is bound to dehydrogenase in the mitochondria, and this resulted in an increased fluorescence intensity (Avi-Dor et al., 1962; Salmon et al., 1982).

5. The discrepancy between the microanalytical results and the fluorometric results may have the following explanation. The fluorometric measurements of NADPH measure the increase of the mitochondrial pool of bound (high quantum yield) NADH. The microanalytical method measures the total increase of NADH. While the analytical method measured the total increase of NADH ($+16\%$), the fraction of the bound NADH underwent a large increase. The microanalytical method cannot discriminate between the bound and the free NADH; however, the optical fluorometric method is extremely sensitive to the amount of bound NADH.

6. In an earlier study of the effects of histotoxic anoxia on the pyridine nucleotides of rabbit cornea, similar results were found (Masters, Riley, Fischbarg & Chance, 1983). The pyridine nucleotides from the epithelium and the endothelium of the rabbit cornea were measured with an enzyme cycling assay. Sodium azide (10 mM) was applied to include histotoxic anoxia, which caused the epithelial $NAD^+/NADH$ ratio to decrease from 2.56 to 1.08. With anoxia the epithelial NADH increased from $1,080 \pm 88$ ng/cm^3 to $1,722 \pm 79$ ng/cm^3. Under anoxic conditions the NAD^+ decreased from $2,766 \pm 157$ ng/cm^3 (normoxic condition) to $1,863 \pm 84$ ng/cm^3 (anoxic condition).

7. In two independent studies of NAD(P)H levels during normoxic and hypoxic conditions, the analytical determinations did not correlate well with the measurement of fluorescent intensity changes of the chromophore. The significance of this disparity between the "optical method" and the microanalytical method points out a severe limitation of the "optical method."

We now complete our discussion of single-photon excitation of pyridine nucleotides with UV excitation, and in the next part we will discuss two-photon excitation of NAD(P)H in a variety of cells and tissues: ex vivo rabbit cornea, pancreatic islets, astrocytes and neurons in brain tissue slices, and in vivo human skin.

33.5 NONLINEAR OPTICAL METHODS TO NONINVASIVELY MONITOR CELL METABOLISM

In early 1994 I initiated a series of experimental studies that I had previously performed with single-photon confocal microscopes and spectrometers in the laboratory of Watt Webb in order to determine the efficacy of using two-photon excitation microscopy to measure the NAD(P)H concentration in freshly excised rabbit corneas. In these studies I worked during several visits to the Webb laboratory with David Piston over a period of 1 year. I shall now describe and summarize the techniques and the results and conclusions (Piston et al., 1995).

33.5.1 Imaging NAD(P)H with Two-Photon Excitation Microscopy

The posited advantages of two-photon excitation microscopy include: (i) photobleaching predominately occurs in the focal volume, (ii) increased penetration depth due to the use of near-infrared light as compared to UV excitation in the single-photon technique, and (iii) decreased cell and tissue damage with near-infrared light compared to UV excitation in the single-photon technique. The experimental goals included using two-photo excitation microscopy to image the basal epithelial cells, which are approximately 40 μm below the surface of the ex vivo rabbit cornea, and to determine if the bulk of the fluorescence signal was from the reduced pyridine adenine nucleotide NAD(P)H.

The instrumentation has been described (Piston, Masters, and Webb 1995). The two-photon excitation microscope was build around a Zeiss Universal microscope with a 50×, 1.0 NA water immersion objective lens (Leitz W 50/1.0) with a working distance of 1.7 mm. The laser produced 150-fs pulses of 705-nm light at a repetition rate of 76 MHz. The average excitation power at the sample was 10 mW for epithelial imaging and 15 mW for endothelial imaging. The calculated full-width half-maximum (FWHM) axial resolution of the two-photon excitation microscopy measurements with the 50 × 1.0 NA objective was 1.2 μm, and the lateral FWHM resolution was 0.4 μm. These calculations were verified to be within 10% by XZ imaging 0.2 – μm fluorescent beads over the entire field of view.

The following is a summary of key findings of this study:

1. Autofluorescence images with submicrometer transverse resolution were obtained throughout the full 400–μm thickness of the ex vivo rabbit cornea (Fig. 33.1).
2. Autofluorescence images of the basal epithelial cells showed high-contrast images with unprecedented detail of the cytoplasmic invaginations of the nucleus (Fig. 33.2).
3. The fluorescence signal was shown to be confined to the cytoplasmic region and not the nucleus. The optical sections through the basal cells contained dark regions in the center of each cell. These dark regions were shown to be basal cell nuclei by staining with Hoechst 33342, a stain with a high affinity for DNA.
4. The autofluorescence was confirmed to be from NAD(P)H, as was previously demonstrated using single-photon fluorescence. The addition of cyanide

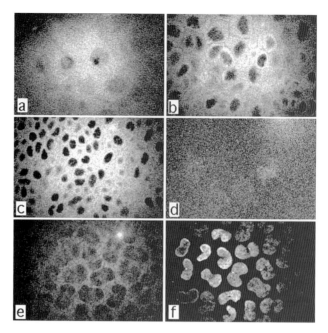

Figure 33.1. Typical two-photon excitation microscopy images from the ex vivo rabbit cornea. Each image is 150 × 100 μm. (a) Superficial epithelial cells at the surface of the cornea. (b) Wing cells located below the superficial epithelial cells. (c) Basal epithelial cells about 50 μm below the surface of the cornea. (d) Corneal stroma. (e) Endothelial cells at the posterior surface of the cornea located about 400 μm from the anterior corneal surface. (f) The corneal endothelial cells stained with Hoechst-33342 that stains the nuclei of the endothelial cells. Images (a) to (e) are based on two-photon excitation of intrinsic NADP(H) in the full thickness of the ex vivo rabbit cornea. (Adapted from Piston, Masters & Webb, Journal of Microscopy, 178: 20–27, 1994)

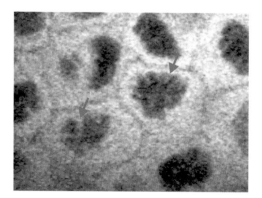

Figure 33.2. Typical two-photon excitation microscopy images from the ex vivo rabbit cornea. Each image is 150 × 100 μm. The focal plane is located in the center of the basal epithelial cell layer. The bright regions correspond to fluorescence from the intrinsic NAD(P)H, and the dark oval regions correspond to the nuclei of the basal epithelial cells. The invaginations of the cytoplasm into the basal cell nuclei are clearly visible and indicated by the arrows. (Adapted from Piston, Masters & Webb, Journal of Microscopy, 178: 20–27, 1994).

reversibly increased the fluorescence from the cells in the corneal epithelium by a factor of two.

5. With the addition of cyanide (10 μM in bicarbonate Ringer's solution) to the cornea, the individual keratocytes within the corneal stroma became visible. These very thin cells with long processes were not visible in the absence of cyanide treatment. This is the first instance of keratocyte metabolic state imaging in the ex vivo cornea.

6. There was no observable photobleaching in these experiments. Even after 100 3-second scans of the basal epithelial cells there was no measurable reduction in the NAD(P)H autofluorescence nor any visible changes in cellular morphology. Over a period of several hours of nearly continuous NAD(P)H imaging of a single ex vivo eye, no photobleaching artifacts were observed.

7. Two-photon excitation microscopy was used to obtain 3-D imaging of the entire thickness of the ex vivo rabbit cornea at near diffraction-limited resolution.

8. The increase of the intensity of the cellular autofluorescence induced by cyanide exposure (anoxia) as measured with two-photon excitation microscopy is similar to the previous studies based on single-photon cellular autofluorescence.

33.5.2 Two-Photon Excitation Microscopy of NAD(P)H in Pancreatic Islets

Two-photon excitation microscopy can provide a real-time analysis of glucose metabolism in cells and tissues based on the measurements of fluorescence intensity from NAD(P)H. As compared to studies with single-photon UV confocal microscopy, two-photon excitation microscopy with near-infrared light should greatly reduce the total photobleaching and photodamage to living cells and tissues (Masters, Kriete & Kukulies, 1993). One interesting application area is to use TPEM to investigate glucose metabolism within beta cells of intact pancreatic islets. Islets of Langerhans are cells in the pancreas that secrete insulin. They are named after Paul Langerhans (1847–1888), a German pathologist-anatomist who discovered these cells during his studies for his doctorate at the Berlin Pathological Institute in 1869.

I present a brief summary of the mechanisms of blood glucose regulation. Both insulin and glucagon are produced in the islets of Langerhans within the pancreas. Insulin is synthesized by the beta cells in the islets. Glucagon is synthesized by the alpha cells in the islets. These peptide hormones have opposite roles in blood glucose regulation: insulin acts to reduce the level of glucose, glucagon acts to increase the level of glucose. The level of blood glucose is regulated by the adjustment of the relative levels of insulin and glucagon in the blood. Insulin secretion from pancreatic beta cells is strongly coupled to glucose metabolism.

Many of these studies were carried out in the laboratory of David Piston at Vanderbilt University. Unfortunately, these studies are confounded by the fact that the NAD(P)H autofluorescence signal is located in both the cytoplasm and the mitochondria. A second confounding issue is that the studies were carried out in excised islets. It would be useful to compare the results from studies of excised intact islets with similar studies that on in vivo islets using intravital microscopy.

In a series of studies two-photon excitation microscopy was used to monitor NAD(P)H levels by measurement of fluorescent intensity changes for both intact pancreatic islets and dispersed beta cells (Bennett et al, 1996; Patterson et al, 2000; Piston & Knobel, 1999a, 1999b; Piston et al., 1999; Rocheleau et al, 2002, 2004).

Two-photon excitation microscopy used a laser that produced 150-fs pulses of 710-nm light. The microscope objective was a Zeiss 40× Plan Neofluar 1.3 NA objective. The excitation power at the specimen was 3 mW and the maximum photobleaching was about 1%. The fluorescence from NAD(P)H was collected with a filter that passed light in the region 380 to 550 nm.

The following is a summary of key findings from ex vivo rat islets:

1. Concentrations of NAD(P)H cannot be absolutely determined from measurements of autofluorescence intensity since, besides NAD(P)H, other components such as collagen, elastin, and FAD contribute to the fluorescence. The beta-cell autofluorescence is reduced to only 50% when glucose is inhibited; this indicated that another fluorescence component not related to glucose metabolism is present in the cells.

2. At the two-photon excitation wavelength of 710 nm, FAD does not contribute to the autofluorescence. Therefore, glucose-stimulated changes in autofluorescence can be attributed to NAD(P)H.

3. The observed autofluorescence was largely due to NAD(P)H because its responses to cyanide and mannoheptulose exposure were consistent with results from other techniques, such as quantitative histochemistry.

4. Beta-cell heterogeneity may be less important in the intact islet than has been predicted from studies of dispersed beta cells, and this supports the role of glucokinase as the rate-limiting enzyme in the beta-cell glucose response (Bennett et al., 1996).

5. Glucokinase is essential for glucose-stimulated metabolic responses in beta cells within intact mice islets, and intercellular coupling within the islet is of minor or no importance in glucose-stimulated metabolic responses (Piston et al., 1999).

6. Two-photon excitation microscopy of mouse pancreatic islet beta cells indicated that the large mitochondrial change in glucose-stimulated NAD(P)H is the major component in the total autofluorescence signal, but there is also a smaller and more rapid cytoplasmic component (Patterson et al., 2000).

7. Based on two-photon excitation microscopy of mouse pancreatic islet beta cells in which NAD(P)H autofluorescence intensity is measured, a model is proposed for glucose-stimulated insulin secretion that includes allosteric inhibition of tricarboxylic acid cycle enzymes and a pH dependence of mitochondrial pyruvate transport (Rocheleau et al., 2002).

33.5.3 Two-Photon Excitation Spectroscopy of NAD(P)H in Processes of Astrocytes and Neurons in Brain Tissue Slices

Multiphoton excitation microscopy permits subcellular metabolic imaging based on the intrinsic fluorescence of NADH as an indictor of both oxidative and glycolytic energy metabolism. These studies are performed on the hippocampal brain slice preparation. The dominant fluorophores in this specimen are the protein-bound pyridine nucleotides, NADH, and nicotinamide adenine dinucleotide phosphate, NADPH.

Key findings of this study are as follows:

1. Two-photon fluorescence imaging of nicotinamide adenine dinucleotide (NADH) provides the sensitivity and spatial three-dimensional resolution to resolve metabolic signatures in processes of astrocytes and neurons deep in highly scattering brain tissue slices. This functional imaging reveals spatiotemporal partitioning of glycolytic and oxidative metabolism between astrocytes and neurons during focal neural activity that establishes a unifying hypothesis for neurometabolic coupling in which early oxidative metabolism in neurons is eventually sustained by late activation of the astrocyte–neuron lactate shuttle. The model integrates existing views of brain energy metabolism and is in accord with known macroscopic physiological changes in vivo (Kasischke et al., 2004).
2. Early neuronal oxidative metabolism follows increased neuronal activity.
3. Astrocytic glycolysis is activated only after the depletion of substrates for oxidative metabolism.

33.5.4 Multiphoton Excitation Fluorescence Microscopy and Spectroscopy of NAD(P)H from in Vivo Human Skin

Britton Chance investigated the NADH spectrofluorometry of rat skin using single-photon excitation (Pappajohn et al., 1972). Thirty years later this work was extended using two-photon excitation microscopy (Masters, So & Gratton, 1997).

In our investigations of multiphoton excitation fluorescence microscopy and spectroscopy of in vivo human skin, we have measured the emission spectra and the fluorescence lifetime with 730-nm excitation and characterized the fluorescent species in the skin at specific positions (Masters, So & Gratton, 1997, 1998, 1999). Over a period of years I and Peter So and coworkers investigated two-photon excitation microscopy of in vivo human skin (Dong et al., 2000; Kim et al., 1998; Masters, 1996a, 1996b, 1996c, 1997; Masters, So, Mauntulin & Gratton, 1998a, 1998b, 1998c; Masters et al., 1999, 2000).

The two-photon microscope was previously described (So et al., 1995, 1996). Emission spectra were obtained by use of a scanning monochromometer. Emission spectra were taken at two depths: the surface (0–50 μm) and below the epidermis (100–150 μm). The fluorescence lifetime is determined using frequency domain spectroscopy techniques (So et al., 1995).

The following is a summary of key findings:

1. The spectroscopic data (emission spectra and lifetime measurements) suggest that the reduced pyridine nucleotides, NAD(P)H, are the primary source of the skin autofluorescence at 730-nm excitation.
2. With 960-nm excitation, a two-photon fluorescence emission at 520 nm indicates the presence of a variable, position-dependent intensity component of

flavoprotein. A second fluorescence emission component, which starts at 425 nm, is observed with 960-nm excitation.

3. The mean fluorescence lifetime for both excitation wavelengths is between 0.5 ns and 3 ns, which is consistent with NAD(P)H contributing to the 730-nm excitation. The fluorescence lifetime measured at 960 nm is consistent with the flavoprotein contribution.

33.6 CHAPTER SUMMARY AND FUTURE DIRECTIONS

The advantages of two-photon excitation microscopy applied to the intensity measurements of cellular NAD(P)H and flavoproteins include minimal photodamage of cells and tissues, deeper depth of penetration compared to single-photon UV excitation, and photobleaching that is confined to the focal volume.

The use of two-photon excitation microscopy of cellular NAD(P)H does not eliminate the major confounding issues that occur when monitoring intensity of fluorescence: inner filter effects, change in quantum yields and fluorescence with binding of NAD(P)H to enzymes, shifts in emission peak of NAD(P)H with binding to enzymes, and fluorescence from both the cytosolic and the mitochondrial compartments (Rothstein et al., 2005; Schneckenburger & König, 1992).

Within these serious limitations, investigators have designed experiments that yielded significant biological information from a variety of cells and tissues. The use of lifetime imaging has the potential to mitigate some of the problems that plague intensity measurements (Ramanujan et al., 2005; So et al., 1995, 1998a, 1998b, 1998c). However, these techniques will not overcome the limitations cited in this chapter.

In summary, the use of the optical method to monitor oxidative metabolism in cells and tissues provides real-time, noninvasive alterations of NAD(P)H fluorescence; however, without an independent calibration for each tissue and cell type and an understanding of all the confounding conditions, the technique is only a qualitative indicator of oxidative metabolism In addition, it should be noted that biochemical techniques are very different from optical techniques to monitor metabolism; each set of techniques has its limitations, and all these techniques require correct calibration for a specific tissue or organ system.

REFERENCES

Alfano, R. A., Masters, B. R. Eds., 2004. *Biomedical Optical Biopsy Vol. 2. Classic Articles in Optics and Photonics on CD-ROM Series*, Washington, D.C.: Optical Society of America.

Avi-Dor, Y., Olson, J. M., Doherty, M. D., Kaplan, N. O. 1962. Fluorescence of pyridine nucleotides in mitochondria. *J. Biol. Chem.* 237: 2377–2383.

Bennett, B. D., Jetton, T. L., Ying, G., Magnuson, M. A., Piston, D. W. 1996. Quantitative subcellular imaging of glucose metabolism within intact pancreatic islets. *The Journal of Biological Chemistry*, 271(7): 3647–3651.

Boonacker, E., Van Noorden, C. J. F., 2001. Enzyme cytochemical techniques for metabolic mapping in living cells, with special reference to proteolysis. *The Journal of Histochemisty & Cytochemistry*, 49(12): 1473–1486.

Chance, B. 1976. Pyridine nucleotide as an indicator of the oxygen requirements for energy-linked functions of mitochondria. *Circulation Research, Supplement I*, 38(5): I-31 to I-38.

Chance, B. 1991. Optical Method. *Annu. Rev. Biophys. Biophys. Chem.* 20, 1–28. D. M. Engelman, (ed.). Palo Alto, CA. Annual Reviews.

Chance, B., Baltscheffsky, H. 1958. Respiratory enzymes in oxidative phosphorylation. VII. Binding of intramitochondial reduced pyridine nucleotide. *The Journal of Biological Chemistry*, 233(3): 736–739.

Chance, B., Lieberman, M. 1978. Intrinsic fluorescence emission from the cornea at low temperatures: evidence of mitochondrial signals and their differing redox states in epithelial and endothelial sides. *Exp. Eye Res.* 26: 111–117 (1978).

Chance, B., Schoener, B., Oshino, R., Itshak, F., Nakase, Y. 1979. Oxidation-reduction ratio studies of mitochondria in freeze-trapped samples. NADH and flavoprotein fluorescence signals. *The Journal of Biological Chemistry*, 254 (11): 4764–4771.

Chance, B., Thorell, B. 1959. Localization and kinetics of reduced pyridine nucleotide in living cells by microfluorometry. *J. Biol.Chem.* 234: 3044–3050.

Chance, B., Williams, G. R. 1955. Respiratory enzymes in oxidative phosphorylation. III. *J. Biol Chem.* 217(1): 409–427.

Dong, C-Y., Kim, K-H., Buehler, C., Hsu, L., Kim, H., So, P. T. C., Masters, B. R., Gratton, E., Kochevar, I.E., 2000. Probing Deep-Tissue Structures by Two-Photon Fluorescence Microscopy, in: *Emerging Tools for Single Cell Analysis*, G. Durack, J. P. Robinson, eds. New York, Wiley-Liss, Inc. pp. 221–237.

Eng, J., Lynch, R. M., Balaban, R. S. 1989. Nicotinamide adenine dinucleotide fluorescence spectroscopy and imaging of isolated cardiac myocytes. *Biophysical J.* 55: 621–630.

Evans N. D., Gnudi L., Rolinski O. J, Birch D. J., Pickup J. C. 2005. Glucose-dependent changes in NAD(P)H-related fluorescence lifetime of adipocytes and fibroblasts in vitro: potential for non-invasive glucose sensing in diabetes mellitus. *J Photochem Photobiol B*. 80(2): 122–129.

Gafni, A., Brand, L. 1976. Fluorescence decay studies of reduced nicotinamide adenine dinucleotide in solution and bound to liver alcohol dehydrogenase. *Biochem.* 15: 3165–3171.

Galeotti, T., van Rossum, G. D., Mayer, D. H., Chance, B., 1970. On the fluorescence of NAD(P)H in whole-cell preparations of tumors and normal tissues. *Eur. J. Biochem.* 17: 485–496.

Huang, S., Heikal, A. A., Webb, W. W. 2002. Two-photon fluorescence spectroscopy and microscopy of NAD(P)H and flavoprotein. *Biophysical Journal*, 82: 2811–2825.

Kasischke K. A., Vishwasrao H. D., Fisher P. J., Zipfel W. R., Webb W. W. 2004. Neural activity triggers neuronal oxidative metabolism followed by astrocytic glycolysis. *Science*, 305(5680): 99–103.

Keilin D. 1966. *The History of Cell respiration and Cytochrome*. London, Cambridge University Press.

Kierdaszuk, B., Malak, H., Gryczynski, I., Callis, P., Lakowicz, J. 1996. Fluorescence of reduced nicotinamides using one- and two-photon excitation. *Biophysical Journal*, 62: 1–13.

Kim, K. H., So, P. T. C., Kochevar, I. E., Masters, B. R., Gratton, E. 1998, Two-Photon Fluorescence and Confocal Reflected Light Imaging of Thick Tissue Structures, in: *Optical Investigations of Cells In Vitro and In Vivo*, Daniel L. Farkas, Robert C. Leif, Bruce J. Tromberg, eds., Proceedings of SPIE, Vol. 3260, 46–57.

Lakowicz, J. R., Szmacinski, H., Nowaczyk, K., Johnson, M. L. 1992. Fluorescence lifetime imaging of free and protein-bound NADH. *Proc. Nat. Acad. Sci.*, 89: 1271–1275.

MacMunn, C. A. 1880. *The Spectroscope in Medicine*. London, Churchill.

MacMunn, C. A. 1914. *Spectrum Analysis Applied to Biology and Medicine*. London, Longmans, Green & Co.

Masters, B. R. 1984a. Noninvasive corneal redox fluorometry, in: *Current Topics in Eye Research*, eds. J. A. Zadunaisky, H Davson, 4, 140–200. New York: Academic Press.

Masters, B. R., 1984b. Noninvasive redox fluorometry: how light can be used to monitor alterations of corneal mitochondrial function. *Current Eye Research* 3: 23–26.

Masters, B. R., 1985. A noninvasive optical method to measure oxygen tension at the corneal epithelium. *Current Eye Research* 4(6): 725–727.

Masters, B. R. 1988. Effects of contact lenses on the oxygen concentration and epithelial mitochondrial redox state of rabbit cornea measured noninvasively with an optically sectioning redox fluorometer microscope. In *The Cornea: Transactions of the World Congress of the Cornea III*. ed. Cavanagh, H. D., New York: Raven Press, Ltd, 281–286.

Masters, B. R., 1993. Specimen preparation and chamber for confocal microscopy of the eye, *Scanning Microscopy*, 7(2): 645–651.

Masters, B. R. 1996a. Three-Dimensional Optical Functional Imaging of Tissue With Two-Photon Excitation Laser Scanning Microscopy, in: *Analytical Use of Fluorescent Probes in Oncology*, Ed. E. Kohen, J.G. Hirschberg, New York: Plenum Press, Chapter 20, pp. 205–211.

Masters, B. R. 1996b. Optical metabolic imaging of ocular tissue with two-photon excitation laser scanning microscopy, *Biomedical Optical Spectroscopy and Diagnostics*, Technical Digest, 3:157–161, Washington, D.C.: Optical Society of America.

Masters, B. R. 1996c. Optical biopsy of ocular tissue with two-photon excitation laser scanning microscopy, *Biomedical Optical Spectroscopy ('96), Trends in Optics and Photonics*, Vol. 3, Washington, D.C.: Optical Society of America, pp. 157–161.

Masters, B. R. 1997. Three-dimensional optical functional imaging of tissue with two-photon excitation laser scanning microscopy, *Applications of Optical Engineering to the Study of Cellular Pathology*, vol. 1, E. Kohen, J. G. Hirschberg, eds. Trivandrun, India: Research Signpost, pp. 173–182.

Masters, B. R. and Chance, B. 1999. Redox Confocal Imaging: Intrinsic Fluorescent Probes of Cellular Metabolism, In: *Fluorescent and Luminescent Probes, Second Edition*, Ed. W. T. Mason, London, U.K.: Academic Press, chapter 28, pp. 361–374.

Masters, B. R., Chance, B. and Fischbarg, J., 1981. Corneal redox fluorometry: a noninvasive probe of corneal redox states and function. *Trends in Biochemical Sciences*, 6: 282–284.

Masters, B. R., Falk, S. and Chance, B., 1982. In vivo flavoprotein measurements of rabbit corneal normoxic-anoxic transitions. *Current Eye Research*, 1: 623–627.

Masters, B. R., Chance, B., Fischbarg, J. 1982. Noninvasive fluorometric study of rabbit corneal redox states and function. In: Noninvasive probes of tissue metabolism, ed. J. S. Cohen, New York: John Wiley & sons, Inc, chapter 4, 79–118.

Masters, B. R., Ghosh, A. K., Wilson, J., Matschinsky, F. M., 1989. Pyridine nucleotides and phosphorylation potential of rabbit corneal epithelium and endothelium. *Invest Ophthalmol Vis Sci*, 30: 861–868.

Masters, B. R., Kriete, A., Kukulies, J. 1993. Ultraviolet confocal fluorescence microscopy of the in vitro cornea: redox metabolic imaging. *Applied Optics*, 32(4): 592–596.

Masters, B. R., Riley, M. V., Fischbarg, J. and Chance, B., 1983. Pyridine nucleotides of rabbit cornea with histotoxic anoxia: chemical analysis, noninvasive fluorometry and physiological correlates. *Experimental Eye Research*, 36: 1–9.

Masters, B. R., So, P. T. C., Gratton, E. 1997. Instrument Development of In Vivo Human Deep Tissue Multi-Photon Fluorescence Microscopy and Spectroscopy, *Cell Vision*, 4(2): 130–131.

Masters, B. R., So, P. T. C., Gratton, E. 1998. Multiphoton Excitation Microscopy of In Vivo Human Skin: Functional and Morphological Optical Biopsy Based on Three-Dimensional Imaging, Lifetime Measurements and Fluorescence Spectroscopy. *Symposium on Advances in Optical Biopsy and Optical Mammography*, Annals of the New York Academy of Sciences, 770: 58–67.

Masters, B. R., So, P. T. C., Mantulin, W., Gratton, E. 1998a, Tissue Microscopy and Spectroscopy: A Two-Photon Approach, *Biomedical Optical Spectroscopy and Diagnostics*, Technical Digest, Washington, D.C.: Optical Society of America, pp. 31–33.

Masters, B. R., So, P. T. C., Mantulin, W., Gratton, E. 1998b, Tissue Microscopy and Spectroscopy: A Two-Photon Approach, OSA Trends in Optics and Photonics, Vol. 22, Eva

M. Sevick-Muraca, Joseph A. Izatt, and Marwood N. Ediger, eds., *Biomedical Optical Spectroscopy and Diagnostics*, Washington, DC, Optical Society of America, pp. 43–45.

Masters, B. R., So, P. T. C., Mantulin, W., Gratton, E. 1998c, Tissue Microscopy and Spectroscopy: A Two-Photon Approach, OSA Trends in Optics and Photonics, Vol. 21, *Advances in Optical Imaging and Photon Migration*, J. G. Fujimoto and M. S. Patterson, eds. Washington, D.C.: Optical Society of America, pp. 417–419.

Masters, B. R., So, P. T. C., Gratton, E. 1999. Multi-photon excitation microscopy and spectroscopy of cells, tissues, and human skin in vivo. In: *Fluorescent and Luminescent Probes, Second Edition*, Ed. W. T. Mason, London, U.K.: Academic Press, chapter 31, pp. 414–432.

Masters, B. R., So, P. T. C., Kim, K., Buehler, C., Gratton, E. 1999. Multiphoton Excitation Microscopy, Confocal Microscopy, and Spectroscopy of Living Cells and Tissues; Functional Metabolic Imaging of Human Skin in vivo, in: *Methods in Enzymology, vol. 307, Confocal Microscopy*, Ed. P. Michael Conn, New York: Academic Press, pp. 513–536.

Masters, B. R., So, P. T. C., 2000. Multiphoton Excitation Microscopy of Human Skin in Vivo: Early Development of an Optical Biopsy. in *Saratov Fall Meeting '99: Optical Technologies in Biophyscis and Medicine*, V. V. Tuchin, D. A. Zimnyakov, A. B. Pravdin, eds. Proceedings of SPIE, 4001, 156–164.

Masters, B. R., So, P. T. C., Gratton, E. 1997. Multiphoton excitation fluorescence microscopy and spectroscopy of in vivo human skin.*Biophysical J*. 72: 2405–2412.

Millikan, G. A. 1942. The oximeter, an instrument for measuring continuously the oxygen saturation in arterial blood in man. *Rev. Sci. Instrum.* 13: 434–444.

Nissen, P., Lieberman, M., Fischbarg, J., Chance, B. 1980. Altered redox states in corenal epithelium and endothelium: NADH fluorescence in rat and rabbit ocular tissue. *Exp. Eye Res*. 30: 691–697.

Pappajohn, D. J., Penneys, R., Chance, B. 1972. NADH spectrofluorometry of rat skin. *Journal of Applied Physiology*, 33(5): 684–687.

Patterson, G. H., Knobel, S. M., Arkhammar, P., Thastrup, O., Piston, D. W. 2000. Separation of the glucose-stimulated cytoplasmic and mitochondrial NAD(P)H responses in pancreatic islet β cells, *PNAS*, 97(10): 5203–5207.

Piston, D. W., Knobel, S. M. 1999a. Real-time Analysis of Glucose Metabolism by Microscopy. *Trends Endocrinol Metab*. 10(10): 413–417.

Piston D. W., Knobel S. M. 1999b. Quantitative imaging of metabolism by two-photon excitation microscopy. *Methods Enzymol*. 307: 351–368.

Piston, D. W., Knobel, S. M., Postic, C., Shelton, K. D., Magnunson, M. A. 1999. Adenovirus-mediated knockout of a conditional glucokinase gene in isolated pancreatic islets reveals an essential role for proximal metabolic coupling events in glucose-stimulated insulin secretion. *The Journal of Biological Chemistry,* 274(2): 1000–1004.

Piston, D. W., Masters, B. R., Webb, W. W., 1995. Three-dimensionally resolved NAD(P)H cellular metabolic redox of the in situ cornea with two-photon excitation laser scanning microscopy, *J. Microscopy*, 178: 20–27.

Ramanujan, V. K., Zhang, J-H, Biener, E., Herman, B. 2005. Multiphoton fluorescence lifetime contrast in deep tissue imaging: prospects in redox imaging and disease diagnosis. *Journal of Biomedical Optics*, 10(5): 051407.

Richards-Kortum, R., Sevick-Muraca, E. 1996. Quantitative optical spectroscopy for tissue diagnosis. *Annu Rev Phys Chem* 47: 555–606.

Rocheleau, J. V., Head, W. S., Nicholson, W. E., Powers, A. C., Piston, D. W. 2002. Pancreatic islet β-cells transiently metabolize pyruvate. *The Journal of Biological Chemistry*, 277(34): 30914–30920.

Rocheleau, J. V., Head, W. S., Piston, D. W. 2004. Quantitative NAD(P)H/Flavoprotein autofluorescence imaging reveals metabolic mechanisms of pancreatic islet pyruvate response. *J. Biol. Chem*. 279(30): 31780–31787.

Rothstein, E. C., Carroll, S., Combs C. A., Jobsis, P. D., Balaban, R. S. 2005. Skeletal muscle NAD(P)H two-photon fluorescence microscopy in vivo: topology and optical inner filters. *Biophys*J. 88(3): 2165–2176.

Salmon, J.-M., Kohen, E., Viallet, P., Hirschberg, J. G., Wouters, A. W., Kohen, C., Thorell, B. 1982. Microspectrofluorometric approach to the study of free/bound NAD(P)H as a metabolic indicator in various cell types. *Photochem. Photobiol.* 36: 585–593.

Schneckenburger, H., König, K. 1992. Fluorescence decay kinetics and imaging of NADP(H) and flavins as metabolic indicators. *Optical Engineering*, 31(7): 1147–1451.

So, P. T. C., French, W. M., Yu, K. M., Berland, C., Dong, C. Y., Gratton, E. 1996. Two-photon fluorescence microscopy: time-resolved and intensity imaging. In: *Fluorescence Imaging and Mic*roscopy. X. F. Wang and B. Herman, eds. Chemical Analysis Series, Vol. 137, New York: John Wiley and Sons, 351–374.

So, P. T. C., French, W. M., Yu, K. M., Berland, C., Dong, C. Y., Gratton, E. 1995. Time-resolved fluorescence microscopy using two-photon excitation. *Bioimaging*. 3: 49–63.

So, P. T. C., Masters, B. R., Gratton, E., Kochevar, I. E. 1998a, Two-Photon Optical Biopsy of Thick Tissues, in: *Biomedical Optical Spectroscopy and Diagnostics, Technical Digest*, Washington, D.C.: Optical Society of America, pp. 28–30.

So, P. T. C., Masters, B. R., Gratton, E., Kochevar, I. E. 1998b, Two-Photon Optical Biopsy of Thick Tissues, *OSA Trends in Optics and Photonics, Vol. 22, Biomedical Optical Spectroscopy and Diagnostics*, Eva M. Sevick-Muraca, Joseph A. Izatt, and Marwood N. Ediger, eds., Washington, D.C.: Optical Society of America, pp. 40–42.

So, P. T. C., Masters, B. R., Gratton, E., Kochevar, I. E. 1998c, Tissue Microscopy and Spectroscopy: A Two-Photon Approach, *OSA Trends in Optics and Photonics, Vol. 21, Advances in Optical Imaging and Photon Migration*, J. G. Fujimoto and M. S. Patterson, eds. Washington, D.C.: Optical Society of America, pp. 420–422.

Velick, S. F. 1958. Fluorescence spectra and polarization of glyceraldehyde-3-phosphate and lactic dehydrogenase coenzyme complexes. *J. Biol. Chem.* 233: 1455–1467.

Williams, R. M., Webb, W. W. 2000. Single granule pH cycling in antigen-induced mast cell secretion. J. Cell Sci. 113: 3839–3850.

Xu, C., Webb, W. W. 1997. Multiphoton excitation of molecular fluorophores and nonlinear laser microscopy. In: *Topics in fluorescence Spectroscopy*. Vol. 5. J. R. Lakowicz, ed., New York, Plenum Press, 471–540.

Index